VOGEL's QUANTITATIVE CHEMICAL ANALYSIS

VOGEL's

TEXTBOOK OF
QUANTITATIVE CHEMICAL ANALYSIS

FIFTH EDITION
Revised by the following members of
The School of Chemistry,
Thames Polytechnic, London

G H Jeffery, B Sc, Ph D, C Chem, F R S C
Former Principal Lecturer and Deputy Head of the School of Chemistry

J Bassett, M Sc, C Chem, F R S C
Former Senior Lecturer in Inorganic Chemistry

J Mendham, M Sc, C Chem, M R S C
Principal Lecturer in Analytical Chemistry

R C Denney, B Sc, Ph D, C Chem, F R S C, M B I M
Principal Lecturer in Organic Chemistry

Longman
Scientific &
Technical

Copublished in the United States with
John Wiley & Sons, Inc., New York

Longman Scientific & Technical
Longman Group UK Limited
Longman House, Burnt Mill, Harlow
Essex CM20 2JE, England
and Associated Companies throughout the world

Copublished in the United States with
John Wiley and Sons Inc, 605 Third Avenue, New York NY 10158

First published in 1939
New impressions 1941, 1942, 1943, 1944, 1945, 1946, 1947, 1948
Second edition 1951
New impressions 1953, 1955, 1957, 1958, 1959, 1960
Third edition 1961 (published under the title *A Text-book of Quantitative Inorganic Analysis Including Elementary Instrumental Analysis*)
New impressions 1962, 1964, 1968, 1969, 1971, 1974, 1975
Fourth edition 1978
New impressions 1979, 1981, 1983, 1985, 1986, 1987
Fifth edition 1989

British Library Cataloguing in Publication Data

Vogel, Arthur Israel
    Vogel's textbook of quantitative chemical
    analysis. – 5th ed.
    1. Quantitative analysis
    I. Title   II. Jeffery, G. H.
    545

ISBN 0-582-44693-7

Library of Congress Cataloguing in Publication Data

Vogel, Arthur Israel
    [Textbook of quantitative chemical analysis]
    Vogel's textbook of quantitative chemical analysis. – 5th ed./
    revised by ... G. H. Jeffery ... [et al.]
        p.   cm.
    Rev. ed. of: Vogel's textbook of quantitative inorganic analysis.
    4th ed. 1978.
    Includes bibliographies and index.
    ISBN 0-470-21517-8
    1. Chemistry, Analytic–Quantitative.   2. Chemistry, Inorganic.
    I. Jeffery, G. H. 1909–   .   II. Vogel, Arthur Israel. Vogel's
    textbook of quantitative inorganic analysis.   III. Title.
    QD101.2.V63   1989
    545–dc20   89-12296
                                                      CIP

Set in 10/11pt Lasercomp Times New Roman

Printed in Great Britain
by Bath Press, Avon

CONTENTS

CHAPTER 3 COMMON APPARATUS AND BASIC TECHNIQUES   71

CHAPTER 8 COLUMN AND THIN–LAYER LIQUID CHROMATOGRAPHY   216

CHAPTER 9 GAS CHROMATOGRAPHY   235

PART D TITRIMETRY AND GRAVIMETRY 255

CHAPTER 10 TITRIMETRIC ANALYSIS 257

THEORETICAL CONSIDERATIONS 257

NEUTRALISATION TITRATIONS 262

CHAPTER 11 GRAVIMETRY   417

QUANTITATIVE SEPARATIONS BASED UPON PRECIPITATION METHODS   433

PRACTICAL GRAVIMETRIC ANALYSIS   446

CATIONS   446

CHAPTER 13 CONDUCTIMETRY 519

CHAPTER 14 COULOMETRY 529

COULOMETRY AT CONSTANT CURRENT: COULOMETRIC TITRATIONS 534

PART F SPECTROANALYTICAL METHODS   643

CHAPTER 17 COLORIMETRY AND SPECTROPHOTOMETRY   645

EXPERIMENTAL COLORIMETRIC DETERMINATIONS   672

CATIONS   678

ANIONS   699

PREFACE TO FIRST EDITION

In writing this book, the author had as his primary object the provision of a complete up-to-date text-book of quantitative inorganic analysis, both theory and practice, at a moderate price to meet the requirements of University and College students of all grades. It is believed that the material contained therein is sufficiently comprehensive to cover the syllabuses of all examinations in which quantitative inorganic analysis plays a part. The elementary student has been provided for, and those sections devoted to his needs have been treated in considerable detail. The volume should therefore be of value to the student throughout the whole of his career. The book will be suitable *inter alia* for students preparing for the various Intermediate B.Sc. and Higher School Certificate Examinations, the Ordinary and Higher National Certificates in Chemistry, the Honours and Special B.Sc. of the Universities, the Associateship of the Institute of Chemistry, and other examinations of equivalent standard. It is hoped, also, that the wide range of subjects discussed within its covers will result in the volume having a special appeal to practising analytical chemists and to all those workers in industry and research who have occasion to utilise methods of inorganic quantitative analysis.

The kind reception accorded to the author's *Text Book of Qualitative Chemical Analysis* by teachers and reviewers seems to indicate that the general arrangement of that book has met with approval. The companion volume on *Quantitative Inorganic Analysis* follows essentially similar lines. Chapter I is devoted to the theoretical basis of quantitative inorganic analysis, Chapter II to the experimental technique of quantitative analysis, Chapter III to volumetric analysis, Chapter IV to gravimetric analysis (including electro-analysis), Chapter V to colorimetric analysis, and Chapter VI to gas analysis; a comprehensive Appendix has been added, which contains much useful matter for the practising analytical chemist. The experimental side is based essentially upon the writer's experience with large classes of students of various grades. Most of the determinations have been tested out in the laboratory in collaboration with the author's colleagues and senior students, and in some cases this has resulted in slight modifications of the details given by the original authors. Particular emphasis has been laid upon recent developments in experimental technique. Frequently the source of certain apparatus or chemicals has been given in the text; this is not intended to convey the impression that these materials cannot be obtained from other sources, but merely to indicate that the author's own experience is confined to the particular products mentioned.

The ground covered by the book can best be judged by perusal of the Table of Contents. An attempt has been made to strike a balance between the classical and modern procedures, and to present the subject of analytical chemistry as it is today. The theoretical aspect has been stressed throughout, and numerous cross-references are given to Chapter I (the theoretical basis of quantitative inorganic analysis).

No references to the original literature are given in the text. This is because the introduction of such references would have considerably increased the size and therefore the price of the book. However, a discussion on the literature of analytical chemistry is given in the Appendix. With the aid of the various volumes mentioned therein — which should be available in all libraries of analytical chemistry — and the Collective Indexes of *Chemical Abstracts* or of *British Chemical Abstracts*, little difficulty will, in general, be experienced in finding the original sources of most of the determinations described in the book.

In the preparation of this volume, the author has utilised pertinent material wherever it was to be found. While it is impossible to acknowledge every source individually, mention must, however, be made of Hillebrand and Lundell's *Applied Inorganic Analysis* (1929) and of Mitchell and Ward's *Modern Methods in Quantitative Chemical Analysis* (1932). In conclusion, the writer wishes to express his thanks: to Dr. G. H. Jeffery, A.I.C., for reading the galley proofs and making numerous helpful suggestions; to Mr. A. S. Nickelson, B.Sc., for reading some of the galley proofs; to his laboratory steward, Mr. F. Mathie, for preparing a number of the diagrams, including most of those in Chapter VI, and for his assistance in other ways; to Messrs. A. Gallenkamp and Co., Ltd., of London, E.C.2, and to Messrs. Fisher Scientific Co, of Pittsburgh, Pa., for providing a number of diagrams and blocks;* and to Mr. F. W. Clifford, F.L.A., Librarian to the Chemical Society, and his able assistants for their help in the task of searching the extensive literature.

Any suggestions for improving the book will be gratefully received by the author.

ARTHUR I. VOGEL
Woolwich Polytechnic, London, SE18
June, 1939

---

* Acknowledgment to other firms and individuals is made in the body of the text.

PREFACE TO THE FIFTH EDITION

We consider ourselves most fortunate to have had the opportunity to continue the collaboration we enjoyed over the previous Fourth Edition of Arthur I. Vogel's *Textbook of Quantitative Inorganic Analysis* and to prepare this Fifth Edition.

It will not have gone unnoticed by readers familiar with earlier editions that the title has now been altered to *Vogel's Textbook of 'Quantitative Chemical Analysis'*. This has been done because the growth and development of analytical chemistry has now totally blurred the boundaries which rather artificially existed between inorganic and organic chemistry. As a result we have made a deliberate policy to incorporate a number of useful organic analytical applications and experiments in the new text. It says much for the foresight of Dr Vogel that he clearly anticipated this development as in the Third Edition he incorporated organic fluorescence and an introductory chapter on infrared spectroscopy, and we have built upon this basis. As a result this volume is a far more substantial revision than that which was given to the Fourth Edition and we believe that it will be of value to an even wider readership in both academic and industrial circles.

One change that will be evident to many chemists is the deletion of normalities and equivalents from the body of the text. This has been done because current teaching and all our external contacts indicated that there was little long-term benefit in retaining them. However, there are many older readers who still employ this system and because of this we have retained a detailed explanation of normalities and equivalents as an Appendix.

The other changes we have made are almost too numerous to list separately in a Preface. As far as possible all subject areas have been up-dated and numerous references given to research papers and other textbooks to enable the reader to study particular topics more extensively if required.

Part A, dealing with the Fundamentals of Quantitative Chemical Analysis, has been extended to incorporate sections of basic theory which were originally spread around the body of the text. This has enabled a more logical development of theoretical concepts to be possible. Part B, concerned with errors, statistics, and sampling, has been extensively rewritten to cover modern approaches to sampling as well as the attendant difficulties in obtaining representative samples from bulk materials. The statistics has been restructured to provide a logical, stepwise approach to a subject which many people find difficult.

The very extensive changes that have taken place over recent years and the broad application to organic separations necessitated a major revision of Part C covering solvent extraction and chromatographic procedures. These particular chapters now incorporate a number of separations of organic materials, most of which can be fairly readily carried out even in modestly equipped laboratories.

The traditional areas of 'wet' chemistry came under very close scrutiny and it was felt that whilst the overall size of Part D could be justifiably reduced, the chapter on titrimetry required modification to include a section on titrations in non-aqueous solvents as these are of particular application to organic materials. It was also felt that environmentally important titrations such as those for dissolved oxygen and chemical oxygen demand should be introduced for the first time. By way of contrast to this we considered that gravimetry has greatly diminished in application and justified a substantial reduction in volume. This in no way undermines its importance in terms of teaching laboratory skills, but the original multitude of precipitations has been substantially pruned and experimental details abbreviated.

Electroanalytical methods is another area which has changed substantially in recent years and this has been reflected in the treatment given to Part E. Apart from a revision of the theory and the circuit diagrams, modifications have been made to the experiments and the chapters have been reorganised in a more logical sequence. Because of the obvious overlap in theory and application, amperometry has now been incorporated into the chapter on voltammetry. Even more substantial changes have been made to the spectroanalytical methods in Part F, in which all chapters have received a major revision, especially to include more organic applications where possible. Details of Fourier transform techniques and derivative spectroscopy are included for the first time, along with a general up-date on instrument design. The growing importance of quantitative infrared spectrophotometry has well justified the re-introduction of a chapter dealing more extensively with this topic. Similarly the extensive and rapid growth of procedures and applications in atomic absorption spectroscopy has necessitated another major revision in this area.

A full revision has been made to the appendices and some of those used in the Fourth Edition have now been incorporated into the main text where appropriate. At the same time other tables have been extended to include more organic compounds and additional appendices include correlation tables for infrared, absorption characteristics for ultraviolet/visible, and additional statistical tables, along with the essential up-dated atomic weights.

In carrying out this revision we owe a great debt to the many companies and individuals who have so willingly helped us not only in giving permission to reproduce their tables and diagrams but who have often gone to considerable trouble to provide us with current information and special photographs and illustrations. We have also paid special attention to the many ideas, suggestions and corrections made by readers who took the trouble to write to us when the Fourth Edition was published. Most of these were constructive and useful, especially the one from Papua–New Guinea pointing out to us the difficulty of producing a 'flesh-coloured precipitate'! We have done our best to avoid such misleading errors on this occasion. Nevertheless we will be pleased to learn of any errors which may have inadvertently crept into the text and/or suggestions for further improvement. We greatly hope that this edition will continue to maintain the very high standards for quantitative analysis which Dr Arthur I. Vogel helped to establish with the First Edition some 50 years ago.

Finally, we wish to express our especial thanks to our friends and colleagues who have so willingly helped us with data, sources of material and discussion throughout the revision of this book. We are grateful to Thames Polytechnic for its continued support and enthusiasm in our work and particularly to the School of Chemistry in which we have all spent many years. Needless to say, we owe a great debt to our wives who have once again encouraged, assisted and tolerated us over the many months spent discussing, writing, revising, checking and proof-reading this edition. We hope that along with us they will feel that the final result more than justifies the efforts that have been put into it and that we have produced a book which will continue to be of substantial value in the teaching and application of analytical chemistry.

G. H. JEFFERY, J. BASSETT, J. MENDHAM, R. C. DENNEY
Thames Polytechnic, Woolwich, London, SE18 6PF, England
August, 1988

A NOTE ON UNITS AND REAGENT PURITY

SI units have been used throughout this book, but as the 'litre' (L) has been accepted as a special name for the cubic decimetre (dm^3) although this is not strictly speaking an SI term we have felt that it is appropriate to employ it throughout this book. Similarly we have chosen to use millilitres (mL) instead of cubic centimetres (cm^3).

Concentrations of solutions are usually expressed in terms of moles per litre: a molar solution (M) has one mole of solute per L.

It should also be emphasised that unless otherwise stated all reagents employed in the analytical procedures should be of appropriate 'analytical grade' or 'spectroscopic grade' materials. Similarly, where solutions are prepared in water this automatically means 'distilled' or 'deionised' water from which all but very minor impurities will have been removed.

PART A
FUNDAMENTALS OF
QUANTITATIVE CHEMICAL
ANALYSIS

CHAPTER 1
INTRODUCTION

1.1 CHEMICAL ANALYSIS

'The resolution of a chemical compound into its proximate or ultimate parts; the determination of its elements or of the foreign substances it may contain': thus reads a dictionary definition.

This definition outlines in very broad terms the scope of analytical chemistry. When a completely unknown sample is presented to an analyst, the first requirement is usually to ascertain what substances are present in it. This fundamental problem may sometimes be encountered in the modified form of deciding what impurities are present in a given sample, or perhaps of confirming that certain specified impurities are absent. The solution of such problems lies within the province of **qualitative analysis** and is outside the scope of the present volume.

Having ascertained the nature of the constituents of a given sample, the analyst is then frequently called upon to determine how much of each component, or of specified components, is present. Such determinations lie within the realm of **quantitative analysis**, and to supply the required information a variety of techniques is available.

1.2 APPLICATIONS OF CHEMICAL ANALYSIS

In a modern industrialised society the analytical chemist has a very important role to play. Thus most manufacturing industries rely upon both qualitative and quantitative chemical analysis to ensure that the raw materials used meet certain specifications, and also to check the quality of the final product. The examination of raw materials is carried out to ensure that there are no unusual substances present which might be deleterious to the manufacturing process or appear as a harmful impurity in the final product. Further, since the value of the raw material may be governed by the amount of the required ingredient which it contains, a quantitative analysis is performed to establish the proportion of the essential component: this procedure is often referred to as **assaying**. The final manufactured product is subject to **quality control** to ensure that its essential components are present within a pre-determined range of composition, whilst impurities do not exceed certain specified limits. The semiconductor industry is an example of an industry whose very existence is dependent upon very accurate determination of substances present in extremely minute quantities.

The development of new products (which may be mixtures rather than pure materials, as for example a polymer composition, or a metallic alloy) also

3

requires the services of the analytical chemist. It will be necessary to ascertain the composition of the mixture which shows the optimum characteristics for the purpose for which the material is being developed.

Many industrial processes give rise to pollutants which can present a health problem. Quantitative analysis of air, water, and in some cases soil samples, must be carried out to determine the level of pollution, and also to establish safe limits for pollutants.

In hospitals, chemical analysis is widely used to assist in the diagnosis of illness and in monitoring the condition of patients. In farming, the nature and level of fertiliser application is based upon information obtained by analysis of the soil to determine its content of the essential plant nutrients, nitrogen, phosphorus and potassium, and of the trace elements which are necessary for healthy plant growth.

Geological surveys require the services of analytical chemists to determine the composition of the numerous rock and soil samples collected in the field. A particular instance of such an exercise is the qualitative and quantitative examination of the samples of 'moon rock' brought back to Earth in 1969 by the first American astronauts to land on the moon.

Much legislation enacted by governments relating to such matters as pollution of the atmosphere and of rivers, the monitoring of foodstuffs, the control of substances hazardous to health, the misuse of drugs, and many others are dependent upon the work of analytical chemists for implementation.

When copper(II) sulphate is dissolved in distilled water, the copper is present in solution almost entirely as the hydrated copper ion $[Cu(H_2O)_6]^{2+}$. If, however, a natural water (spring water or river water) is substituted for the distilled water, then some of the copper ions will interact with various substances present in the natural water. These substances may include acids derived from vegetation (such as humic acids and fulvic acid), colloidal materials such as clay particles, carbonate ions (CO_3^{2-}) and hydrogencarbonate ions (HCO_3^-) derived from atmospheric carbon dioxide, and various other cations and anions leached from the rocks with which the water has been in contact. The copper ions which become adsorbed on colloidal particles, or those which form an organic complex with (for example) fulvic acid, will no longer show the usual behaviour of hydrated copper(II) ions and thus their biological and geological effects are modified. For the investigation of such problems in natural waters, it is therefore necessary for the analyst to devise procedures whereby the various copper-containing species in the solution can be identified, and the distribution of the copper among them determined. Such procedures are referred to as 'speciation'.

1.3 SAMPLING

The results obtained for the proportion of a certain constituent in a given sample may form the basis of assessing the value of a large consignment of the commodity from which the sample was drawn. In such cases it is absolutely essential to be certain that the sample used for analysis is truly representative of the whole. When dealing with a homogeneous liquid, sampling presents few problems, but if the material under consideration is a solid mixture, then it is necessary to combine a number of portions to ensure that a representative sample is finally selected for analysis. The analyst must therefore be acquainted with the normal standard sampling procedures employed for different types of materials.

1.4 TYPES OF ANALYSIS

With an appropriate sample available, attention must be given to the question of the most suitable technique or techniques to be employed for the required determinations. One of the major decisions to be made by an analyst is the choice of the most effective procedure for a given analysis, and in order to arrive at the correct decision, not only must he be familiar with the practical details of the various techniques and of the theoretical principles upon which they are based, he must also be conversant with the conditions under which each method is reliable, aware of possible interferences which may arise, and capable of devising means of circumventing such problems. He will also be concerned with questions regarding the accuracy and the precision to be expected from given methods and, in addition, he must not overlook such factors as time and costing. The most accurate method for a certain determination may prove to be lengthy or to involve the use of expensive reagents, and in the interests of economy it may be necessary to choose a method which, although somewhat less exact, yields results of sufficient accuracy in a reasonable time.

Important factors which must be taken into account when selecting an appropriate method of analysis include (*a*) the nature of the information which is sought, (*b*) the size of sample available and the proportion of the constituent to be determined, and (*c*) the purpose for which the analytical data are required.

The nature of the information sought may involve requirement for very detailed data, or alternatively, results of a general character may suffice. With respect to the information which is furnished, different types of chemical analysis may be classified as follows:

1. *proximate analysis*, in which the amount of each element in a sample is determined with no concern as to the actual compounds present;
2. *partial analysis*, which deals with the determination of selected constituents in the sample;
3. *trace constituent analysis*, a specialised instance of partial analysis in which we are concerned with the determination of specified components present in very minute quantity;
4. *complete analysis*, when the proportion of each component of the sample is determined.

On the basis of sample size, analytical methods are often classified as:

1. *macro*, the analysis of quantities of 0.1 g or more;
2. *meso* (*semimicro*), dealing with quantities ranging from 10^{-2} g to 10^{-1} g;
3. *micro*, for quantities in the range 10^{-3} g to 10^{-2} g;
4. *submicro*, for samples in the range 10^{-4} g to 10^{-3} g;
5. *ultramicro*, for quantities below 10^{-4} g.

The term 'semimicro' given as an alternative name for classification (2) is not very apt, referring as it does to samples larger than micro.

A **major constituent** is one accounting for 1–100 per cent of the sample under investigation; a **minor constituent** is one present in the range 0.01–1 per cent; a **trace constituent** is one present at a concentration of less than 0.01 per cent.

With the development of increasingly sophisticated analytical techniques it has become possible to determine substances present in quantities much lower than the 0.01 per cent upper level set for trace constituents. It is therefore necessary to make further subdivisions: **trace** corresponds to 10^2–10^4 μg per gram, or 10^2–10^4 parts per million (ppm), **microtrace** to 10^2–10^{-1} pg per gram, (10^{-4}–10^{-7} ppm), **nanotrace** to 10^2–10^{-1} fm per gram (10^{-7}–10^{-10} ppm).

When the sample weight is small (0.1–1.0 mg), the determination of a trace component at the 0.01 per cent level may be referred to as **subtrace analysis**. If the trace component is at the microtrace level, the analysis is termed **sub-microtrace**. With a still smaller sample (not larger than 0.1 mg) the determination of a component at the trace level is referred to as **ultratrace analysis**, whilst with a component at the microtrace level, the analysis is referred to as **ultra-microtrace**.

The purpose for which the analytical data are required may perhaps be related to process control and quality control. In such circumstances the objective is checking that raw materials and finished products conform to specification, and it may also be concerned with monitoring various stages in a manufacturing process. For this kind of determination methods must be employed which are quick and which can be readily adapted for routine work: in this area instrumental methods have an important role to play, and in certain cases may lend themselves to automation. On the other hand, the problem may be one which requires detailed consideration and which may be regarded as being more in the nature of a research topic.

1.5 USE OF LITERATURE

Faced with a research-style problem, the analyst will frequently be dealing with a situation which is outside his normal experience and it will be necessary to seek guidance from published data. This will involve consultation of multi-volume reference works such as Kolthoff and Elving, *Treatise on Analytical Chemistry*; Wilson and Wilson, *Comprehensive Analytical Chemistry*; Fresenius and Jander, *Handbuch der analytischen Chemie*; of a compendium of methods such as Meites, *Handbook of Analytical Chemistry*; or of specialised monographs dealing with particular techniques or types of material. Details of recognised procedures for the analysis of many materials are published by various official bodies, as for example the American Society for Testing Materials (ASTM), the British Standards Institution and the Commission of European Communities. It may be necessary to seek more up-to-date information than that available in the books which have been consulted and this will necessitate making use of review publications (e.g. *Annual Reports of the Chemical Society*; reviews in *The Analyst* and *Analytical Chemistry*), and of abstracts (e.g. *Analytical Abstracts*; *Chemical Abstracts*), and referring to journals devoted to analytical chemistry and to specific techniques: see Section 1.7.*

Such a literature survey may lead to the compilation of a list of possible procedures and the ultimate selection must then be made in the light of the criteria previously enunciated, and with special consideration being given to questions of possible interferences and to the equipment available.

---

* Selected Bibliographies and References are given at the end of each part of the book; for Part A, see Sections 3.38 and 3.39.

1.6 COMMON TECHNIQUES

The main techniques employed in quantitative analysis are based upon (*a*) the quantitative performance of suitable chemical reactions and either measuring the amount of reagent needed to complete the reaction, or ascertaining the amount of reaction product obtained; (*b*) appropriate electrical measurements (e.g. potentiometry); (*c*) the measurement of certain optical properties (e.g. absorption spectra). In some cases, a combination of optical or electrical measurements and quantitative chemical reaction (e.g. amperometric titration) may be used.

The quantitative execution of chemical reactions is the basis of the traditional or 'classical' methods of chemical analysis: gravimetry, titrimetry and volumetry. In **gravimetric analysis** the substance being determined is converted into an insoluble precipitate which is collected and weighed, or in the special case of **electrogravimetry** electrolysis is carried out and the material deposited on one of the electrodes is weighed.

In **titrimetric analysis** (often termed volumetric analysis in certain books), the substance to be determined is allowed to react with an appropriate reagent added as a standard solution, and the volume of solution needed for complete reaction is determined. The common types of reaction which are used in titrimetry are (*a*) neutralisation (acid–base) reactions; (*b*) complex-forming reactions; (*c*) precipitation reactions; (*d*) oxidation–reduction reactions.

Volumetry is concerned with measuring the volume of gas evolved or absorbed in a chemical reaction.

Electrical methods of analysis (apart from electrogravimetry referred to above) involve the measurement of current, voltage or resistance in relation to the concentration of a certain species in solution. Techniques which can be included under this general heading are (i) **voltammetry** (measurement of current at a micro-electrode at a specified voltage); (ii) **coulometry** (measurement of current and time needed to complete an electrochemical reaction or to generate sufficient material to react completely with a specified reagent); (iii) **potentiometry** (measurement of the potential of an electrode in equilibrium with an ion to be determined); (iv) **conductimetry** (measurement of the electrical conductivity of a solution).

Optical methods of analysis are dependent either upon (i) measurement of the amount of radiant energy of a particular wavelength absorbed by the sample, or (ii) the emission of radiant energy and measurement of the amount of energy of a particular wavelength emitted. Absorption methods are usually classified according to the wavelength involved as (*a*) **visible spectrophotometry** (colorimetry), (*b*) **ultraviolet spectrophotometry**, and (*c*) **infrared spectrophotometry**.

Atomic absorption spectroscopy involves atomising the specimen, often by spraying a solution of the sample into a flame, and then studying the absorption of radiation from an electric lamp producing the spectrum of the element to be determined.

Although not strictly absorption methods in the sense in which the term is usually employed, **turbidimetric and nephelometric methods** which involve measuring the amount of light stopped or scattered by a suspension should also be mentioned at this point.

Emission methods involve subjecting the sample to heat or electrical treatment so that atoms are raised to excited states causing them to emit energy: it is the intensity of this emitted energy which is measured. The common excitation techniques are:

(*a*) *emission spectroscopy*, where the sample is subjected to an electric arc or spark plasma and the light emitted (which may extend into the ultraviolet region) is examined;

(*b*) *flame photometry*, in which a solution of the sample is injected into a flame;

(*c*) *fluorimetry*, in which a suitable substance in solution (commonly a metal–fluorescent reagent complex) is excited by irradiation with visible or ultraviolet radiation.

Chromatography is a separation process employed for the separation of mixtures of substances. It is widely used for the identification of the components of mixtures, but as explained in Chapters 8 and 9, it is often possible to use the procedure to make quantitative determinations, particularly when using Gas Chromatography (GC) and High Performance Liquid Chromatography (HPLC).

1.7 INSTRUMENTAL METHODS

The methods dependent upon measurement of an electrical property, and those based upon determination of the extent to which radiation is absorbed or upon assessment of the intensity of emitted radiation, all require the use of a suitable instrument, e.g. polarograph, spectrophotometer, etc., and in consequence such methods are referred to as 'instrumental methods'. Instrumental methods are usually much faster than purely chemical procedures, they are normally applicable at concentrations far too small to be amenable to determination by classical methods, and they find wide application in industry. In most cases a microcomputer can be interfaced to the instrument so that absorption curves, polarograms, titration curves, etc., can be plotted automatically, and in fact, by the incorporation of appropriate servo-mechanisms, the whole analytical process may, in suitable cases, be completely automated.

Despite the advantages possessed by instrumental methods in many directions, their widespread adoption has not rendered the purely chemical or 'classical' methods obsolete; the situation is influenced by three main factors.

1. The apparatus required for classical procedures is cheap and readily available in all laboratories, but many instruments are expensive and their use will only be justified if numerous samples have to be analysed, or when dealing with the determination of substances present in minute quantities (trace, subtrace or ultratrace analysis).

2. With instrumental methods it is necessary to carry out a calibration operation using a sample of material of known composition as reference substance.

3. Whilst an instrumental method is ideally suited to the performance of a large number of routine determinations, for an occasional, non-routine, analysis it is often simpler to use a classical method than to go to the trouble of preparing requisite standards and carrying out the calibration of an instrument.

Clearly, instrumental and classical methods must be regarded as supplementing each other.

1.8 OTHER TECHNIQUES

In addition to the main general methods of analysis outlined above there are also certain specialised techniques which are applied in special circumstances. Among these are X-ray methods, methods based upon the measurement of radioactivity, mass spectrometry, the so-called kinetic methods, and thermal methods.

X-ray methods. When high-speed electrons collide with a solid target (which can be the material under investigation), X-rays are produced. These are often referred to as **primary X-rays,** and arise because the electron beam may displace an electron from the inner electron shells of an atom in the target, and the electron lost is then replaced by one from an outer shell; in this process energy is emitted as X-rays. In the resultant X-ray emission it is possible to identify certain emission peaks which are characteristic of elements contained in the target. The wavelengths of the peaks can be related to the atomic number of the elements producing them, and thus provide a means of identifying elements present in the target sample. Further, under controlled conditions, the intensity of the peaks can be used to determine the amounts of the various elements present. This is the basis of **electron probe microanalysis**, in which a small target area of the sample is pinpointed for examination. This has important applications in metallurgical research, in the examination of geological samples, and in determining whether biological materials contain metallic elements.

When a beam of primary X-rays of short wavelength strikes a solid target, by a similar mechanism to that described above, the target material will emit X-rays at wavelengths characteristic of the atoms involved: the resultant emission is termed **secondary or fluorescence radiation**. The sample area can be large, and quantitative results obtained by examining the peak heights of the fluorescence radiation can be taken as indicative of sample composition. **X-ray fluorescence analysis** is a rapid process which finds application in metallurgical laboratories, in the processing of metallic ores, and in the cement industry.

Crystalline material will diffract a beam of X-rays, and X-ray powder diffractometry can be used to identify components of mixtures. These X-ray procedures are examples of **non-destructive methods of analysis**.

Radioactivity. Methods based on the measurement of radioactivity belong to the realm of radiochemistry and may involve measurement of the intensity of the radiation from a naturally radioactive material; measurement of induced radioactivity arising from exposure of the sample under investigation to a neutron source (**activation analysis**); or the application of what is known as the **isotope dilution technique**.

Typical applications of such methods are the determination of trace elements in (a) the investigation of pollution problems; (b) the examination of geological specimens; (c) quality control in the manufacture of semiconductors.

Mass spectrometry. In this technique, the material under examination is vaporised under a high vacuum and the vapour is bombarded by a high-energy electron beam. Many of the vapour molecules undergo fragmentation and produce ions of varying size. These ions can be distinguished by accelerating them in an electric field, and then deflecting them in a magnetic field where they follow paths dictated by their mass/charge ratio (m/e) to detection and recording equipment: each kind of ion gives a peak in the **mass spectrum**. Non-volatile inorganic materials can be examined by vaporising them by subjecting them to a high-voltage electric spark.

Mass spectrometry can be used for gas analysis, for the analysis of petroleum products, and in examining semiconductors for impurities. It is also a very useful tool for establishing the structure of organic compounds.

Kinetic methods. These methods of quantitative analysis are based upon the fact that the speed of a given chemical reaction may frequently be increased by the addition of a small amount of a catalyst, and within limits, the rate of the catalysed reaction will be governed by the amount of catalyst present. If a calibration curve is prepared showing variation of reaction rate with amount of catalyst used, then measurement of reaction rate will make it possible to determine how much catalyst has been added in a certain instance. This provides a sensitive method for determining sub-microgram amounts of appropriate substances.

The method can also be adapted to determine the amount of a substance in solution by adding a catalyst which will destroy it completely, and measuring the concomitant change in for example, the absorbance of the solution for visible or ultraviolet radiation. Such procedures are applied in clinical chemistry.

Optical methods. Those of particular application to *organic compounds* are:

1. Use of a **refractometer** to make measurements of the refractive index of liquids. This will often provide a means of identifying a pure compound, and can also be used (in conjunction with a calibration curve) to analyse a mixture of two liquids.
2. Measurement of the **optical rotation** of optically active compounds. Polarimetric measurements can likewise be used as a method of identifying pure substances, and can also be employed for quantitative purposes.

Thermal methods. Changes in weight, or changes in energy, recorded as a function of temperature (or of time) can provide valuable analytical data. For example, the conditions can be established under which a precipitate produced in a gravimetric determination can be safely dried. Common techniques include the recording as a function of temperature or time of (*a*) change in weight (**Thermogravimetry, TG**); (*b*) the difference in temperature between a test substance and an inert reference material (**Differential Thermal Analysis, DTA**); (*c*) the energy necessary to establish a zero temperature difference between a test substance and a reference material (**Differential Scanning Calorimetry, DSC**).

1.9 FACTORS AFFECTING THE CHOICE OF ANALYTICAL METHODS

An indication has been given in the preceding sections of a number of techniques available to the analytical chemist. The techniques have differing degrees of sophistication, of sensitivity, of selectivity, of cost and also of time requirements, and an important task for the analyst is the selection of the best procedure for a given determination. This will require careful consideration of the following criteria.

(a) The type of analysis required: elemental or molecular, routine or occasional.

(b) Problems arising from the nature of the material to be investigated, e.g. radioactive substances, corrosive substances, substances affected by water.

(c) Possible interference from components of the material other than those of interest.

(d) The concentration range which needs to be investigated.

(e) The accuracy required.

(f) The facilities available; this will refer particularly to the kinds of instrumentation which are at hand.

(g) The time required to complete the analysis; this will be particularly relevant when the analytical results are required quickly for the control of a manufacturing process. This may mean that accuracy has to be a secondary rather than a prime consideration, or it may require the use of expensive instrumentation.

(h) The number of analyses of similar type which have to be performed; in other words, does one have to deal with a limited number of determinations or with a situation requiring frequent repetitive analyses?

(i) Does the nature of the specimen, the kind of information sought, or the magnitude of the sample available indicate the use of non-destructive methods of analysis as opposed to the more commonly applied destructive methods involving dissolution of the sample (possibly in acid) prior to the application of normal analytical techniques?

Some information relevant to the choice of appropriate methods is given in condensed form in Table 1.1, which is divided into three sections: the 'classical' techniques; a selection of instrumental methods; some 'non-destructive' methods.

Table 1.1 Conspectus of some common quantitative analytical methods

Method	Speed	Relative cost	Concentration range (pC)*	Accuracy
Gravimetry	S	L	1–2	H
Titrimetry	M	L	1–4	H
Coulometry	S–M	L–M	1–4	H
Voltammetry	M	M	3–10	M
Potentiometry	M–F	L–M	1–7	M
Spectrophotometry	M–F	L–M	3–6	M
Atomic spectrometry	F	M–H	3–9	M
Emission (plasma) spectrometry	F	H	5–9	M
Chromatography (GLC; HPLC)	F	M–H	3–9	M
Neutron activation	S	H	†(a)	M
X-ray fluorescence	F	H	†(b)	H

* $pC = \log_{10}\dfrac{1}{\text{Concn}}$, where Concentration is expressed in moles per litre.

† Concentration range has little significance: detection values are (a) 10^{-5}–10^{-12} g; (b) 10^{-3}–10^{-6} g. Abbreviations: F, Fast; H, High; L, Low; M, Moderate; S, Slow.

1.10   INTERFERENCES

Whatever the method finally chosen for the required determination, it should ideally be a **specific method**; that is to say, it should be capable of measuring the amount of desired substance accurately, no matter what other substances may be present. In practice few analytical procedures attain this ideal, but many methods are **selective**; in other words, they can be used to determine any of a small group of ions in the presence of certain specified ions. In many instances the desired selectivity is achieved by carrying out the procedure under carefully controlled conditions, particularly with reference to the pH of the solution.

Frequently, however, there are substances present that prevent direct measurement of the amount of a given ion; these are referred to as **interferences**, and the selection of methods for separating the interferences from the substance to be determined are as important as the choice of the method of determination. Typical separation procedures include the following:

(a) *Selective precipitation.* The addition of appropriate reagents may convert interfering ions into precipitates which can be filtered off, careful pH control is often necessary in order to achieve a clean separation, and it must be borne in mind that precipitates tend to adsorb substances from solution and care must be taken to ensure that as little as possible of the substance to be determined is lost in this way.

(b) *Masking.* A complexing agent is added, and if the resultant complexes are sufficiently stable they will fail to react with reagents added in a subsequent operation: this may be a titrimetric procedure or a gravimetric precipitation method.

(c) *Selective oxidation (reduction).* The sample is treated with a selective oxidising or reducing agent which will react with some of the ions present: the resultant change in oxidation state will often facilitate separation. For example, to precipitate iron as hydroxide, the solution is always oxidised so that iron(III) hydroxide is precipitated: this precipitates at a lower pH than does iron(II) hydroxide and the latter could be contaminated with the hydroxides of many bivalent metals.

(d) *Solvent extraction.* When metal ions are converted into chelate compounds by treatment with suitable organic reagents, the resulting complexes are soluble in organic solvents and can thus be extracted from the aqueous solution. Many ion-association complexes containing bulky ions which are largely organic in character (e.g. the tetraphenylarsonium ion $(C_6H_5)_4As^+$) are soluble in organic solvents and can thus be utilised to extract appropriate metals ions from aqueous solution. Such treatment may be used to isolate the ion which is to be determined, or alternatively, to remove interfering substances.

(e) *Ion exchange.* Ion exchange materials are insoluble substances containing ions which are capable of replacement by ions from a solution containing electrolytes. The phosphate ion is an interference encountered in many analyses involving the determination of metals; in other than acidic solutions the phosphates of most metals are precipitated. If, however, the solution is passed through a column of an anion exchange resin in the chloride form, then phosphate ions are replaced by chloride ions. Equally, the determination of phosphates is difficult in the presence of a variety of metallic ions, but if the solution is passed through a column of a cation exchange resin in the protonated form, then the interfering cations are replaced by hydrogen ions.

(*f*) *Chromatography.* The term chromatography is applied to separation techniques in which the components of solutions travel down a column at different rates, the column being packed with a suitable finely divided solid termed the stationary phase, for which such diverse materials as cellulose powder, silica gel and alumina are employed. Having introduced the test solution to the top of the column, an appropriate solvent (the mobile phase) is allowed to flow slowly through the column. In **adsorption chromatography** the solutes are adsorbed on the column material and are then eluted by the mobile phase: the less easily adsorbed components are eluted first and the more readily adsorbed components are eluted more slowly, thus effecting separation. In **partition chromatography** the solutes are partitioned between the mobile phase and a film of liquid (commonly water) firmly absorbed on the surface of the stationary phase. A typical example is the separation of cobalt from nickel in solution in concentrated hydrochloric acid: the stationary phase is cellulose powder, the mobile phase, acetone containing hydrochloric acid; the cobalt is eluted whilst the nickel remains on the column. If compounds of adequate volatility are selected, then '**gas chromatography**' may be carried out in which the mobile phase is a current of gas, e.g. nitrogen. For liquids it is frequently possible to dispense with a column and to use the adsorbent spread as a thin layer on a glass plate (**thin layer chromatography**) and in some cases a roll or a sheet of filter paper without any added adsorbent may be used (**paper chromatography**): these techniques are especially useful for handling small amounts of material. Of particular interest in this field are the developments associated with **high performance liquid chromatography (HPLC)** and with **ion chromatography.**

1.11 DATA ACQUISITION AND TREATMENT

Once the best method of dealing with interferences has been decided upon and the most appropriate method of determination chosen, the analysis should be carried out in duplicate and preferably in triplicate. For simple classical determinations the experimental results must be recorded in the analyst's notebook. However, many modern instruments employed in instrumental methods of analysis are interfaced with computers and the analytical results may be displayed on a visual display unit, whilst a printer will provide a printout of the pertinent data which can be used as a permanent record.

A simple calculation will then convert the experimental data into the information which is sought: this will usually be the percentage of the relevant component in the analytical sample. When using computer-interfaced instruments the printout will give the required percentage value. The results thus obtained will be subject to a degree of uncertainty as is true for every physical measurement, and it is necessary to establish the magnitude of this uncertainty in order that meaningful results of the analysis can be presented.

It is, therefore, necessary to establish the **precision** of the results, by which we mean the extent to which they are reproducible. This is commonly expressed in terms of the numerical difference between a given experimental value and the mean value of all the experimental results. The **spread** or **range** in a set of results is the numerical difference between the highest and lowest results: this

13

figure is also an indication of the precision of the measurements. However, the most important measures of precision are the standard deviation and the variance: these are discussed in Chapter 4.

The difference between the most probable analytical result and the true value for the sample is termed the **systematic error** in the analysis: it indicates the **accuracy** of the analysis.

1.12 SUMMARY

Summarising, the following steps are necessary when confronted with an unfamiliar quantitative determination.

1. Sampling.
2. Literature survey and selection of possible methods of determination.
3. Consideration of interferences and procedures for their removal.

Pooling the information gathered under headings (2) and (3), a final selection will be made of the method of determination and of the procedure for eliminating interferences.

4. Dissolution of sample.
5. Removal or suppression of interferences.
6. Performance of the determination.
7. Statistical analysis of the results.

For References and Bibliography see Sections 3.38 and 3.39.

CHAPTER 2
FUNDAMENTAL THEORETICAL PRINCIPLES OF REACTIONS IN SOLUTION

Many of the reactions of qualitative and quantitative chemical analysis take place in solution; the solvent is most commonly water but other liquids may also be used. It is, therefore, necessary to have a general knowledge of the conditions which exist in solutions, and also of the factors which influence chemical reactions.

2.1 CHEMICAL EQUILIBRIUM

If a mixture of hydrogen and iodine vapour is heated to a temperature of about 450 °C in a closed vessel, the two elements combine and hydrogen iodide is formed. It is found, however, that no matter how long the duration of the experiment, some hydrogen and iodine remain uncombined. If pure hydrogen iodide is heated in a closed vessel to a temperature of about 450 °C, the substance decomposes to form hydrogen and iodine, but again, no matter how prolonged the heating, some hydrogen iodide remains unchanged. This is an example of a **reversible reaction** in the gaseous phase.

$$H_2(g) + I_2(g) \rightleftharpoons 2HI(g)$$

An example of a reversible reaction in the liquid phase is afforded by the esterification reaction between ethanol and acetic (ethanoic) acid forming ethyl acetate and water. Since, however, ethyl acetate undergoes conversion to acetic acid and ethanol when heated with water, the esterification reaction never proceeds to completion.

$$C_2H_5OH + CH_3COOH \rightleftharpoons CH_3COOC_2H_5 + H_2O$$

It is found that after the elapse of a sufficient time interval, all reversible reactions reach a state of **chemical equilibrium**. In this state the composition of the equilibrium mixture remains constant, provided that the temperature (and for some gaseous reactions, the pressure also) remains constant. Furthermore, provided that the conditions (temperature and pressure) are maintained constant, the same state of equilibrium may be obtained from either direction of a given reversible reaction. In the equilibrium state, the two opposing reactions are taking place at the same rate so that the system is in a state of **dynamic equilibrium**.

It must be emphasised that the composition of a given equilibrium system can be altered by changing the conditions under which the system is maintained and it is necessary to consider the effect of changes in (*a*) the temperature, (*b*) the pressure and (*c*) the concentration of the components. According to the **Le Chatelier–Braun Principle**: 'If a constraint is applied to a system in equilibrium, the system will adjust itself so as to nullify the effect of the constraint', and the effect of the factors referred to can be considered in the light of this statement.

(**a**) **Temperature.** The formation of ammonia from its elements is a reversible process:

$$N_2(g) + 3H_2(g) \rightleftharpoons 2NH_3(g)$$

in which the forward reaction is accompanied by the evolution of heat (energy), and is said to be an exothermic reaction; the reverse reaction absorbs heat and is said to be endothermic. If the temperature of an equilibrium mixture of nitrogen, hydrogen and ammonia is increased, then the reaction which absorbs heat will be favoured, and so ammonia is decomposed.

(**b**) **Pressure.** Referring to the hydrogen iodide equilibrium system, the stoichiometric coefficients of the molecules on each side of the equation for the reaction are equal, and there is no change in volume when reaction occurs. Therefore, if the pressure of the system is doubled, thus halving the total volume, the two sides of the equation are equally affected, and so the composition of the equilibrium mixture remains unchanged.

In the nitrogen, hydrogen, ammonia equilibrium system, there is a decrease in volume when ammonia is produced, and hence an increase in pressure will favour the formation of ammonia. Any gaseous equilibrium in which a change in volume takes place will be affected by a change in pressure. For equilibrium in the liquid phase, moderate changes in pressure have practically no effect on the volume owing to the small compressibility of liquids, and so moderate pressure changes do not affect the equilibrium.

(**c**) **Concentration of reagents.** If hydrogen is added to the equilibrium mixture resulting from the thermal decomposition of hydrogen iodide, it is found that more hydrogen iodide is present when equilibrium is restored. In accordance with the Le Chatelier–Braun Principle, the system has reacted to remove some of the added hydrogen.

2.2 THE LAW OF MASS ACTION

Guldberg and Waage (1867) clearly stated the Law of Mass Action (sometimes termed the Law of Chemical Equilibrium) in the form: 'The velocity of a chemical reaction is proportional to the product of the active masses of the reacting substances'. 'Active mass' was interpreted as concentration and expressed in moles per litre. By applying the law to homogeneous systems, that is to systems in which all the reactants are present in one phase, for example in solution, we can arrive at a mathematical expression for the condition of equilibrium in a reversible reaction.

Consider first the simple reversible reaction at constant temperature:

$$A + B \rightleftharpoons C + D$$

The rate of conversion of A and B is proportional to their concentrations, or

$$r_1 = k_1 \times [A] \times [B]$$

where k_1 is a constant known as the rate constant or rate coefficient, and the square brackets (see footnote Section 2.21) denote the concentrations (mol L^{-1}) of the substances enclosed within the brackets.

Similarly, the rate of conversion of C and D is given by:

$$r_2 = k_2 \times [C] \times [D]$$

At equilibrium, the two rates of conversion will be equal:

$$r_1 = r_2$$

or

$$k_1 \times [A] \times [B] = k_2 \times [C] \times [D]$$

or

$$\frac{[C] \times [D]}{[A] \times [B]} = \frac{k_1}{k_2} = K$$

K is the **equilibrium constant** of the reaction at the given temperature.

The expression may be generalised. For a reversible reaction represented by:

$$p_1 A_1 + p_2 A_2 + p_3 A_3 + \ldots \rightleftharpoons q_1 B_1 + q_2 B_2 + q_3 B_3 + \ldots$$

where p_1, p_2, p_3 and q_1, q_2, q_3 are the stoichiometric coefficients of the reacting species, the condition for equilibrium is given by the expression:

$$\frac{[B_1]^{q_1} \times [B_2]^{q_2} \times [B_3]^{q_3} \ldots}{[A_1]^{p_1} \times [A_2]^{p_2} \times [A_3]^{p_3} \ldots} = K$$

This result may be expressed in words: when equilibrium is reached in a reversible reaction, at constant temperature, the product of the concentrations of the resultants (the substances on the right-hand side of the equation) divided by the product of the concentrations of the reactants (the substances on the left-hand side of the equation), each concentration being raised to a power equal to the stoichiometric coefficient of the substance concerned in the equation for the reaction, is constant.

The equilibrium constant of a reaction can be related to the changes in Gibbs Free Energy (ΔG), enthalpy (ΔH) and entropy (ΔS) which occur during the reaction by the mathematical expressions:

$$\Delta G^\ominus = -RT \ln K^\ominus = -2.303 RT \log_{10} K^\ominus$$

$$\frac{d \ln K^\ominus}{dT} = \frac{\Delta H^\ominus}{RT^2}$$

$$\Delta G^\ominus = \Delta H^\ominus - T \Delta S^\ominus$$

In these expressions, the superscript symbol ($\ominus$) indicates that the quantities concerned relate to a so-called 'standard state'. For the derivation and the significance of these expressions, a textbook of physical chemistry[1] should be consulted, but briefly a reaction will be spontaneous when ΔG is negative, it will be at equilibrium when ΔG is zero, and when ΔG is positive the reverse reaction will be spontaneous. It follows that a reaction is favoured when heat is produced, i.e. it is an exothermic reaction so that the enthalpy change ΔH is negative. It is also favoured by an increase in entropy, that is when ΔS is positive. A knowledge of the values of the equilibrium constants of certain selected systems can be of great value to the analyst; for example in dealing with acid–base interactions, with solubility equilibria, with systems involving complex ions, with oxidation–reduction systems, and with many separation problems: note however that equilibrium constants do not give any indication of the *rate* of reaction. These matters are dealt with in detail in succeeding sections of this chapter, and in other pertinent chapters.

2.3 FACTORS AFFECTING CHEMICAL REACTIONS IN SOLUTION

There are three main factors whose influence on chemical reactions in solution need to be considered: (*a*) the nature of the solvent; (*b*) temperature; and (*c*) the presence of catalysts.

(a) Nature of the solvent. Reactions in aqueous solution generally proceed rapidly because they involve interaction between ions. Thus the precipitation of silver chloride from a chloride solution by the addition of silver nitrate solution can be formulated

$$Ag^+ + Cl^- = AgCl(solid)$$

Reactions between molecules in solution, for example the formation of ethyl acetate from acetic acid and ethanol, are generally comparatively slow. It is therefore convenient to classify solvents as **ionising solvents** if they tend to produce solutions in which the solute is ionised, and as **non-ionising solvents** if they give solutions in which the solute is not ionised. Common ionising solvents include water, acetic acid, hydrogen chloride, ammonia, amines, bromine trifluoride and sulphur dioxide. Of these solvents, the first four are characterised by a capability of giving rise to hydrogen ions, as for example with water:

$$2H_2O \rightleftharpoons H_3O^+ + OH^-$$

and with ammonia:

$$2NH_3 \rightleftharpoons NH_4^+ + NH_2^-$$

These four solvents can thus be termed **protogenic solvents**, whilst bromine trifluoride and sulphur dioxide which do not contain hydrogen are **non-protonic solvents**. Non-ionising solvents include hydrocarbons, ethers, esters and higher alcohols; the lower alcohols, especially methanol and ethanol, do show slight ionising properties with appropriate solutes.

(b) Temperature. Reaction rates increase rapidly with rising temperature, and in some analytical procedures it is necessary to heat the solution to ensure that the required reaction takes place with sufficient rapidity.

An example of such behaviour is the titration of acidified oxalate solutions with potassium permanganate solution. When potassium permanganate solution is added to a solution of an oxalate containing sulphuric acid at room temperature, reaction proceeds very slowly, and the solution sometimes acquires a brown tinge due to the formation of manganese(IV) oxide. If, however, the solution is heated to about 70 °C before adding any permanganate solution, then the reaction becomes virtually instantaneous, and no manganese(IV) oxide is produced.

(c) Catalysts. The rates of some reactions can be greatly increased by the presence of a catalyst. This is a substance that alters the rate of a reaction without itself undergoing any net change: it follows that a small amount of the catalyst can influence the conversion of large quantities of the reactants. If the reaction under consideration is reversible, then the catalyst affects both the forward and back reactions, and although the reaction is speeded up, the position of equilibrium is unchanged.

An example of catalytic action is provided by the titration of oxalates with potassium permanganate solution referred to above. It is found that even though the oxalate solution is heated, the first few drops of permanganate solution are only slowly decolorised, but as more permanganate solution is added the decoloration becomes instantaneous. This is because the reaction between oxalate ions and permanganate ions is catalysed by the Mn^{2+} ions formed by the reduction of permanganate ions:

$$MnO_4^- + 8H^+ + 5e^- \rightleftharpoons Mn^{2+} + 4H_2O$$

Other examples are the use of osmium(VIII) oxide (osmium tetroxide) as catalyst in the titration of solutions of arsenic(III) oxide with cerium(IV) sulphate solution, and the use of molybdate(VI) ions to catalyse the formation of iodine by the reaction of iodide ions with hydrogen peroxide. Certain reactions of various organic compounds are catalysed by several naturally occurring proteins known as enzymes.

The determination of trace quantities of many substances can be accomplished by examining the rate of a chemical reaction for which the substance to be determined acts as a catalyst. By comparing the observed rate of reaction with rates determined for the same reaction, with known quantities of the same catalyst present, the unknown concentration can be calculated. Likewise a catalyst may be used to convert a substance for which no suitable analytical reaction exists for the conditions under which the substance is present, to a product which can be determined. Alternatively, the substance to be determined may be destroyed by adding a catalyst, and the resultant change in some measured property, for example the absorption of light, enables the amount of substance present to be evaluated. Thus, uric acid in blood can be determined by measurement of the absorption of ultraviolet radiation at a wavelength of 292 nm, but the absorption is not specific. The absorption meter reading is recorded, and then the uric acid is destroyed by addition of the enzyme uricase. The absorption reading is repeated, and from the difference between the two results, the amount of uric acid present can be calculated.

2.4 ELECTROLYTIC DISSOCIATION

Aqueous solutions of many salts, of the common 'strong acids' (hydrochloric, nitric and sulphuric), and of bases such as sodium hydroxide and potassium hydroxide are good conductors of electricity, whereas pure water shows only a very poor conducting capability. The above solutes are therefore termed electrolytes. On the other hand, certain solutes, for example ethane-1,2-diol (ethylene glycol) which is used as 'antifreeze', produce solutions which show a conducting capability only little different from that of water: such solutes are referred to as non-electrolytes. Most reactions of analytical importance occurring in aqueous solution involve electrolytes, and it is necessary to consider the nature of such solutions.

Salts. The structure of numerous salts in the solid state has been investigated by means of X-rays and by other methods, and it has been shown that they are composed of charged atoms or groups of atoms held together in a crystal lattice; they are said to be ionic compounds. When these salts are dissolved in a solvent of high dielectric constant such as water, or are heated to the melting point, the crystal forces are weakened and the substances dissociate into the pre-existing charged particles or ions, so that the resultant liquids are good conductors of electricity; they are referred to as **strong electrolytes**. Some salts, however, exemplified by cyanides, thiocyanates, the halides of mercury and cadmium, and by lead acetate, give solutions which show a significant electrical conductance, but which is not as great as that shown by solutions of strong electrolytes of comparable concentration. Solutes showing this behaviour are referred to as **weak electrolytes**: they are generally covalent compounds which undergo only limited ionisation when dissolved in water:

$$BA \rightleftharpoons B^+ + A^-$$

Acids and bases. An acid may be defined as a substance which, when dissolved in water, undergoes dissociation with the formation of hydrogen ions as the only positive ions:

$$HCl \rightleftharpoons H^+ + Cl^-$$

$$HNO_3 \rightleftharpoons H^+ + NO_3^-$$

Actually the hydrogen ion H^+ (or proton) does not exist in the free state in aqueous solution; each hydrogen ion combines with one molecule of water to form the hydroxonium ion, H_3O^+. The hydroxonium ion is a hydrated proton. The above equations are therefore more accurately written:

$$HCl + H_2O \rightleftharpoons H_3O^+ + Cl^-$$

$$HNO_3 + H_2O \rightleftharpoons H_3O^+ + NO_3^-$$

The ionisation may be attributed to the great tendency of the free hydrogen ions H^+ to combine with water molecules to form hydroxonium ions. Hydrochloric and nitric acids are almost completely dissociated in aqueous solution in accordance with the above equations; this is readily demonstrated by freezing-point measurements and by other methods.

   Polyprotic acids ionise in stages. In sulphuric acid, one hydrogen atom is almost completely ionised:

$$H_2SO_4 + H_2O \rightleftharpoons H_3O^+ + HSO_4^-$$

The second hydrogen atom is only partially ionised, except in very dilute solution:

$$HSO_4^- + H_2O \rightleftharpoons H_3O^+ + SO_4^{2-}$$

Phosphoric(V) acid also ionises in stages:

$$H_3PO_4 + H_2O \rightleftharpoons H_3O^+ + H_2PO_4^-$$

$$H_2PO_4^- + H_2O \rightleftharpoons H_3O^+ + HPO_4^{2-}$$

$$HPO_4^{2-} + H_2O \rightleftharpoons H_3O^+ + PO_4^{3-}$$

The successive stages of ionisation are known as the primary, secondary, and tertiary ionisations respectively. As already mentioned, these do not take place to the same degree. The primary ionisation is always greater than the secondary, and the secondary very much greater than the tertiary.

Acids of the type of acetic acid (CH_3COOH) give an almost normal freezing-point depression in aqueous solution; the extent of dissociation is accordingly small. It is usual, therefore, to distinguish between acids which are completely or almost completely ionised in solution and those which are only slightly ionised. The former are termed **strong acids** (examples: hydrochloric, hydrobromic, hydriodic, iodic(V), nitric and perchloric [chloric(VII)] acids, primary ionisation of sulphuric acid), and the latter are called **weak acids** (examples: nitrous acid, acetic acid, carbonic acid, boric acid, phosphorous (phosphoric(III)) acid, phosphoric(V) acid, hydrocyanic acid, and hydrogen sulphide). There is, however, no sharp division between the two classes.

A base was originally defined as a substance which, when dissolved in water, undergoes dissociation with the formation of hydroxide ions OH^- as the only negative ions. Thus sodium hydroxide, potassium hydroxide, and the hydroxides of certain bivalent metals are almost completely dissociated in aqueous solution:

$$NaOH \rightarrow Na^+ + OH^-$$

$$Ba(OH)_2 \rightarrow Ba^{2+} + 2OH^-$$

These are **strong bases**. Aqueous ammonia solution, however, is a weak base. Only a small concentration of hydroxide ions is produced in aqueous solution:

$$NH_3 + H_2O \rightleftharpoons NH_4^+ + OH^-$$

General concept of acid and bases. The Brønsted–Lowry theory. The simple concept given in the preceding paragraphs suffices for many of the requirements of quantitative inorganic analysis in aqueous solution. It is, however, desirable to have some knowledge of the general theory of acids and bases proposed independently by J. N. Brønsted and by T. M. Lowry in 1923, since this is applicable to all solvents. According to this theory, an acid is a species having a tendency to lose a proton, and a base is a species having a tendency to add on a proton. This may be represented as:

Acid $\rightleftharpoons$ Proton + Conjugate base

$$A \rightleftharpoons H^+ + B \tag{a}$$

It must be emphasised that the symbol H^+ represents the proton and not the 'hydrogen ion' of variable nature existing in different solvents (OH_3^+, NH_4^+, $CH_3CO_2H_2^+$, $C_2H_5OH_2^+$, etc.); the definition is therefore independent of solvent. The above equation represents a hypothetical scheme for defining A and B and not a reaction which can actually occur. Acids need not be neutral molecules (e.g., HCl, H_2SO_4, CH_3CO_2H), but may also be anions (e.g., HSO_4^-, $H_2PO_4^-$, $HOOC \cdot COO^-$) and cations (e.g. NH_4^+, $C_6H_5NH_3^+$, $Fe(H_2O)_6^{3+}$). The same is true of bases where the three classes can be illustrated by NH_3, $C_6H_5NH_2$, H_2O; CH_3COO^-, OH^-, HPO_4^{2-}, $OC_2H_5^-$; $Fe(H_2O)_5(OH)^{2+}$.

Since the free proton cannot exist in solution in measurable concentration, reaction does not take place unless a base is added to accept the proton from the acid. By combining the equations $A_1 \rightleftharpoons B_1 + H^+$ and $B_2 + H^+ \rightleftharpoons A_2$, we obtain

$$A_1 + B_2 \rightleftharpoons A_2 + B_1 \tag{b}$$

$A_1 - B_1$ and $A_2 - B_2$ are two conjugate acid–base pairs. This is the most important expression for reactions involving acids and bases; it represents the transfer of a proton from A_1 to B_2 or from A_2 to B_1. The stronger the acid A_1 and the weaker A_2, the more complete will be the reaction (*b*). The stronger acid loses its proton more readily than the weaker; similarly, the stronger base accepts a proton more readily than does the weaker base. It is evident that the base or acid conjugate to a strong acid or a strong base is always weak, whereas the base or acid conjugate to a weak acid or weak base is always strong.

In aqueous solution a Brønsted–Lowry acid A

$$A + H_2O \rightleftharpoons H_3O^+ + B$$

is strong when the above equilibrium is virtually complete to the right so that [A] is almost zero. A strong base is one for which [B], the equilibrium concentration of base other than hydroxide ion, is almost zero.

Acids may thus be arranged in series according to their relative combining tendencies with a base, which for aqueous solutions (in which we are largely interested) is water:

$$\underset{\text{Acid}_1}{HCl} + \underset{\text{Base}_2}{H_2O} \rightleftharpoons \underset{\text{Acid}_2}{H_3O^+} + \underset{\text{Base}_1}{Cl^-}$$

This process is essentially complete for all typical 'strong' (i.e. highly ionised) acids, such as HCl, HBr, HI, HNO_3, and $HClO_4$. In contrast with the 'strong' acids, the reactions of a typical 'weak' or slightly ionised acid, such as acetic acid or propionic (propanoic) acid, proceeds only slightly to the right in the equation:

$$\underset{\text{Acid}_1}{CH_3COOH} + \underset{\text{Base}_2}{H_2O} \rightleftharpoons \underset{\text{Acid}_2}{H_3O^+} + \underset{\text{Base}_1}{CH_3COO^-}$$

The typical strong acid of the water system is the hydrated proton H_3O^+, and the role of the conjugate base is minor if it is a sufficiently weak base, e.g. Cl^-, Br^-, and ClO_4^-. The conjugate bases have strengths that vary inversely as the strengths of the respective acids. It can easily be shown that the basic ionisation constant of the conjugate base $K_{B,conj.}$ is equal to $K_w/K_{A,conj.}$, where K_w is the ionic product of water.

Scheme (*b*) includes reactions formerly described by a variety of names, such as dissociation, neutralisation, hydrolysis and buffer action (see below). One acid–base pair may involve the solvent (in water $H_3O^+ - H_2O$ or $H_2O - OH^-$), showing that ions such as H_3O^+ and OH^- are in principle only particular examples of an extended class of acids and bases though, of course, they do occupy a particularly important place in practice. It follows that the properties of an acid or base may be greatly influenced by the nature of the solvent employed.

Another definition of acids and bases is due to G. N. Lewis (1938). From the experimental point of view Lewis regarded all substances which exhibit 'typical' acid–base properties (neutralisation, replacement, effect on indicators, catalysis), irrespective of their chemical nature and mode of action, as acids or bases. He related the properties of acids to the acceptance of electron pairs, and bases as donors of electron pairs, to form covalent bonds regardless of whether protons are involved. On the experimental side Lewis' definition brings together a wide range of qualitative phenomena, e.g. solutions of BF_3, BCl_3,

$AlCl_3$, or SO_2 in an inert solvent cause colour changes in indicators similar to those produced by hydrochloric acid, and these changes are reversed by bases so that titrations can be carried out. Compounds of the type of BF_3 are usually described as **Lewis acids** or **electron acceptors**. The Lewis bases (e.g. ammonia, pyridine) are virtually identical with the Brønsted–Lowry bases. The great disadvantage of the Lewis definition of acids is that, unlike proton-transfer reactions, it is incapable of general quantitative treatment.

The implications of the theory of the complete dissociation of strong electrolytes in aqueous solution were considered by Debye, Hückel and Onsager, and they succeeded in accounting quantitatively for the increasing molecular conductivity of a strong electrolyte producing singly charged ions with decreasing concentration of the solution over the concentration range $0–0.002M$. For full details, textbooks of physical chemistry must be consulted.

It is important to realise that whilst complete dissociation occurs with strong electrolytes in aqueous solution, this does not mean that the effective concentrations of the ions are identical with their molar concentrations in any solution of the electrolyte: if this were the case the variation of the osmotic properties of the solution with dilution could not be accounted for. The variation of colligative, e.g. osmotic, properties with dilution is ascribed to changes in the activity of the ions; these are dependent upon the electrical forces between the ions. Expressions for the variations of the activity or of related quantities, applicable to dilute solutions, have also been deduced by the Debye–Hückel theory. Further consideration of the concept of activity follows in Section 2.5.

2.5 ACTIVITY AND ACTIVITY COEFFICIENT

In the deduction of the Law of Mass Action it was assumed that the effective concentrations or active masses of the components could be expressed by the stoichiometric concentrations. According to thermodynamics, this is not strictly true. The rigorous equilibrium equation for, say, a binary electrolyte:

$$AB \rightleftharpoons A^+ + B^-$$

is
$$\frac{(a_{A^+} \times a_{B^-})}{a_{AB}} = K_t$$

where a_{A^+}, a_{B^-}, and a_{AB} represent the **activities** of A^+, B^-, and AB respectively, and K_t is the **true** or **thermodynamic, dissociation constant**. The concept of activity, a thermodynamic quantity, is due to G. N. Lewis. The quantity is related to the concentration by a factor termed the activity coefficient:

Activity = Concentration × Activity coefficient

Thus at any concentration

$a_{A^+} = y_{A^+}.[A^+], a_{B^-} = y_{B^-}.[B^-],$ and $a_{AB} = y_{AB}.[AB]$

where y refers to the activity coefficients,* and the square brackets to the

---

* The symbol used is dependent upon the method of expressing the concentration of the solution. The recommendations of the IUPAC Commision on Symbols, Terminology and Units (1969) are as follows: concentration in moles per litre (molarity), activity coefficient represented by y, concentration in mols per kilogram (molality), activity coefficient represented by γ, concentration expressed as mole fraction, activity coefficient represented by f.

concentrations. Substituting in the above equation, we obtain:

$$\frac{y_{A^+} \cdot [A^+] \times y_{B^-} \cdot [B^-]}{y_{AB} \cdot [AB]} = \frac{[A^+] \cdot [B^-]}{[AB]} \times \frac{y_{A^+} \times y_{B^-}}{y_{AB}} = K_t$$

This is the rigorously correct expression for the Law of Mass Action as applied to weak electrolytes.

The activity coefficient varies with the concentration. For ions it also varies with the ionic charge, and is the same for all dilute solutions having the same **ionic strength**, the latter being a measure of the electrical field existing in the solution. The term ionic strength, designated by the symbol I, is defined as equal to one half of the sum of the products of the concentration of each ion multiplied by the square of its charge number, or $I = 0.5 \Sigma c_i z_i^2$, where c_i is the ionic concentration in moles per litre of solution and z_i is the charge number of the ion concerned. An example will make this clear. The ionic strength of 0.1 M HNO_3 solution containing $0.2M$ $Ba(NO_3)_2$ is given by:

$$0.5\{0.1 \text{ (for } H^+) + 0.1 \text{ (for } NO_3^-)$$
$$+ 0.2 \times 2^2 \text{ (for } Ba^{2+}) + 0.2 \times 2 \text{ (for } NO_3^-)\} = 0.5\{1.4\} = 0.7$$

It can be shown on the basis of the Debye–Hückel theory that for aqueous solutions at room temperature:

$$\log y_i = -\frac{0.505 z_i^2 . I^{0.5}}{1 + 3.3 \times 10^7 a . I^{0.5}}$$

where y_i is the activity coefficient of the ion, z_i is the charge number of the ion concerned, I is the ionic strength of the solution, and a is the average 'effective diameter' of all the ions in the solution. For very dilute solutions ($I^{0.5} < 0.1$) the second term of the denominator is negligible and the equation reduces to:

$$\log y_i = -0.505 z_i^2 . I^{0.5}$$

For more concentrated solutions ($I^{0.5} > 0.3$) an additional term BI is added to the equation; B is an empirical constant. For a more detailed treatment of the Debye–Hückel theory a textbook of physical chemistry should be consulted.[1]

2.6 SOLUBILITY PRODUCT

For sparingly soluble salts (i.e. those of which the solubility is less than 0.01 mol per L) it is an experimental fact that the mass action product of the concentrations of the ions is a constant at constant temperature. This product K_s is termed the 'solubility product'. For a binary electrolyte:

$$AB \rightleftharpoons A^+ + B^-$$

$$K_{s(AB)} = [A^+] \times [B^-]$$

In general, for an electrolyte $A_p B_q$, which ionises into pA^{q+} and qB^{p-} ions:

$$A_p B_q = pA^{q+} + qB^{p-}$$

$$K_{s(A_p B_q)} = [A^{q+}]^p \times [B^{p-}]^q$$

A plausible deduction of the solubility product relation is the following. When excess of a sparingly soluble electrolyte, say silver chloride, is shaken up with water, some of it passes into solution to form a saturated solution of the salt and the process appears to cease. The following equilibrium is actually present (the silver chloride is completely ionised in solution):

$$AgCl(solid) \rightleftharpoons Ag^+ + Cl^-$$

The rate of the forward reaction depends only upon the temperature, and at any given temperature:

$$r_1 = k_1$$

where k_1 is a constant. The rate of the reverse reaction is proportional to the activity of each of the reactants; hence at any given temperature:

$$r_2 = k_2 \times a_{Ag^+} \times a_{Cl^-}$$

where k_2 is another constant. At equilibrium the two rates are equal, i.e.

$$k_1 = k_2 \times a_{Ag^+} \times a_{Cl^-}$$

or $$a_{Ag^+} \times a_{Cl^-} = k_1/k_2 = K_{s(AgCl)}$$

In the very dilute solutions with which we are concerned, the activities may be taken as practically equal to the concentrations so that $[Ag^+] \times [Cl^-] = \text{const.}$

It is important to note that the solubility product relation applies with sufficient accuracy for purposes of quantitative analysis only to saturated solutions of slightly soluble electrolytes and with *small* additions of other salts. In the presence of moderate concentrations of salts, the ionic concentration, and therefore the ionic strength of the solution, will increase. This will, in general, lower the activity coefficients of both ions, and consequently the ionic concentrations (and therefore the solubility) must increase in order to maintain the solubility product constant. This effect, which is most marked when the added electrolyte does not possess an ion in common with the sparingly soluble salt, is termed the **salt effect**.

It will be clear from the above short discussion that two factors may come into play when a solution of a salt containing a common ion is added to a saturated solution of a slightly soluble salt. At moderate concentrations of the added salt, the solubility will generally decrease, but with higher concentrations of the soluble salt, when the ionic strength of the solution increases considerably and the activity coefficients of the ions decrease, the solubility may actually increase. This is one of the reasons why a very large excess of the precipitating agent is avoided in quantitative analysis.

The following examples illustrate the method of calculating solubility products from solubility data and also the reverse procedure.

Example 1. The solubility of silver chloride is 0.0015 g per L. Calculate the solubility product.

The relative molecular mass of silver chloride is 143.3. The solubility is therefore $0.0015/143.3 = 1.05 \times 10^{-5}$ mol per L. In a saturated solution, 1 mole of AgCl will give 1 mole each of Ag^+ and Cl^-. Hence $[Ag^+] = 1.05 \times 10^{-5}$ and $[Cl^-] = 1.05 \times 10^{-5}$ mol L^{-1}.

$$K_{s(AgCl)} = [Ag^+] \times [Cl^-] = (1.05 \times 10^{-5}) \times (1.05 \times 10^{-5})$$
$$= 1.1 \times 10^{-10} \text{ mol}^2 \text{ L}^{-2}$$

Example 2. Calculate the solubility product of silver chromate, given that its solubility is $2.5 \times 10^{-2}\,g\,L^{-1}$.

$$Ag_2CrO_4 \rightleftharpoons 2Ag^+ + CrO_4^{2-}$$

The relative molecular mass of Ag_2CrO_4 is 331.7; hence the solubility $=$ $2.5 \times 10^{-2}/331.7 = 7.5 \times 10^{-5}\,mol\,L^{-1}$.

Now 1 mole of Ag_2CrO_4 gives 2 moles of Ag^+ and 1 mole of CrO_4^{2-}; therefore

$$K_{s(Ag_2CrO_4)} = [Ag^+]^2 \times [CrO_4^{2-}] = (2 \times 7.5 \times 10^{-5})^2 \times (7.5 \times 10^{-5})$$
$$= 1.7 \times 10^{-12}\,mol^3\,L^{-3}$$

Example 3. The solubility product of magnesium hydroxide is $3.4 \times 10^{-11}\,mol^3\,L^{-3}$. Calculate its solubility in grams per L.

$$Mg(OH)_2 \rightleftharpoons Mg^{2+} + 2OH^-$$

$$[Mg^{2+}] \times [OH^-]^2 = 3.4 \times 10^{-11}$$

The relative molecular mass of magnesium hydroxide is 58.3. Each mole of magnesium hydroxide, when dissolved, yields 1 mole of magnesium ions and 2 moles of hydroxyl ions. If the solubility is $x\,mol\,L^{-1}$, $[Mg^{2+}] = x$ and $[OH^-] = 2x$. Substituting these values in the solubility product expression:

$$x \times (2x)^2 = 3.4 \times 10^{-11}$$

or $x = 2.0 \times 10^{-4}\,mol\,L^{-1}$

$$= 2.0 \times 10^{-4} \times 58.3$$
$$= 1.2 \times 10^{-2}\,g\,L^{-1}$$

The great importance of the solubility product concept lies in its bearing upon precipitation from solution, which is, of course, one of the important operations of quantitative analysis. The solubility product is the ultimate value which is attained by the ionic concentration product when equilibrium has been established between the solid phase of a difficultly soluble salt and the solution. If the experimental conditions are such that the ionic concentration product is different from the solubility product, then the system will attempt to adjust itself in such a manner that the ionic and solubility products are equal in value. Thus if, for a given electrolyte, the product of the concentrations of the ions in solution is arbitrarily made to exceed the solubility product, as for example by the addition of a salt with a common ion, the adjustment of the system to equilibrium results in precipitation of the solid salt, provided supersaturation conditions are excluded. If the ionic concentration product is less than the solubility product or can arbitrarily be made so, as (for example) by complex salt formation or by the formation of weak electrolytes, then a further quantity of solute can pass into solution until the solubility product is attained, or, if this is not possible, until all the solute has dissolved.

2.7 QUANTITATIVE EFFECTS OF A COMMON ION

An important application of the solubility product principle is to the calculation of the solubility of sparingly soluble salts in solutions of salts with a common ion. Thus the solubility of a salt MA in the presence of a relatively large amount of the common M^+ ions,* supplied by a second salt MB, follows from the definition of solubility products:

$$[M^+] \times [A^-] = K_{s(MA)}$$

or $\qquad [A^-] = K_{s(MA)}/[M^+]$

The solubility of the salt is represented by the $[A^-]$ which it furnishes in solution. It is clear that the addition of a common ion will *decrease* the solubility of the salt.

Example 4. Calculate the solubility of silver chloride in (a) $0.001 M$ and (b) $0.01 M$ sodium chloride solutions respectively ($K_{s(AgCl)} = 1.1 \times 10^{-10} mol^2 L^{-2}$).

In a saturated solution of silver chloride $[Cl^-] = \sqrt{1.1 \times 10^{-10}} = 1.05 \times 10^{-5} mol L^{-1}$; this may be neglected in comparison with the excess of Cl^- ions added.

For (a) $[Cl^-] = 1 \times 10^{-3}$, $[Ag^+] = 1.1 \times 10^{-10}/1 \times 10^{-3}$

$$= 1.1 \times 10^{-7} mol L^{-1}$$

For (b) $[Cl^-] = 1 \times 10^{-2}$, $[Ag^+] = 1.1 \times 10^{-10}/1 \times 10^{-2}$

$$= 1.1 \times 10^{-8} mol L^{-1}$$

Thus the solubility is decreased 100 times in $0.001 M$ sodium chloride and 1000 times in $0.01 M$ sodium chloride. Similar results are obtained for $0.001 M$ and $0.01 M$ silver nitrate solutions.

Example 5. Calculate the solubilities of silver chromate in $0.001 M$ and $0.01 M$ silver nitrate solutions, and in $0.001 M$ and $0.01 M$ potassium chromate solutions (Ag_2CrO_4: $K_s = 1.7 \times 10^{-12} mol^3 L^{-3}$, solubility in water $= 7.5 \times 10^{-5} mol L^{-1}$).

$$[Ag^+]^2 \times [CrO_4^{2-}] = 1.7 \times 10^{-12}$$

or $\qquad [CrO_4^{2-}] = 1.7 \times 10^{-12}/[Ag^+]^2$

For $0.001 M$ silver nitrate solution: $[Ag^+] = 1 \times 10^{-3}$

$\therefore [CrO_4^{2-}] = 1.7 \times 10^{-12}/1 \times 10^{-6} = 1.7 \times 10^{-6} mol L^{-1}$.

For $0.01 M$ silver nitrate solution: $[Ag^+] = 1 \times 10^{-2}$

$[CrO_4^{2-}] = 1.7 \times 10^{-12}/1 \times 10^{-4} = 1.7 \times 10^{-8} mol L^{-1}$.

The solubility product equation gives:

$$[Ag^+] = \sqrt{1.7 \times 10^{-12}/[CrO_4^{2-}]}$$

For $[CrO_4^{2-}] = 0.001$, $[Ag^+] = \sqrt{1.7 \times 10^{-12}/1 \times 10^{-3}}$

$$= 4.1 \times 10^{-5} mol L^{-1}$$

---

*This enables us to neglect the concentration of M^+ ions supplied by the sparingly soluble salt itself, and thus to simplify the calculation.

For $[CrO_4^{2-}] = 0.01$, $[Ag^+] = \sqrt{1.7 \times 10^{-12}/1 \times 10^{-2}}$

$$= 1.3 \times 10^{-5} \, mol \, L^{-1}$$

This decrease in solubility by the common ion effect is of fundamental importance in gravimetric analysis. By the addition of a suitable excess of a precipitating agent, the solubility of a precipitate is usually decreased to so small a value that the loss from solubility influences is negligible. Consider a specific case — the determination of silver as silver chloride. Here the chloride solution is added to the solution of the silver salt. If an exactly equivalent amount is added, the resultant saturated solution of silver chloride will contain 0.0015 g per L (Example 1). If 0.2 g of silver chloride is produced and the volume of the solution and washings is 500 mL, the loss, owing to solubility, will be 0.000 75 g or 0.38 per cent of the weight of the salt; the analysis would then be 0.38 per cent too low. By using an excess of the precipitant, say, to a concentration of 0.01 M, the solubility of the silver chloride is reduced to $1.5 \times 10^{-5} \, g \, L^{-1}$ (Example 4), and the loss will be $1.5 \times 10^{-5} \times 0.5 \times 100/0.2 = 0.0038$ per cent. Silver chloride is therefore very suitable for the quantitative determination of silver with high accuracy.

It should, however, be noted that as the concentration of the excess of precipitant increases, so too does the ionic strength of the solution. This leads to a decrease in activity coefficient values with the result that to maintain the value of K_s *more* of the precipitate will dissolve. In other words there is a limit to the amount of precipitant which can be safely added in excess. Also, addition of excess precipitant may sometimes result in the formation of soluble complexes causing some precipitate to dissolve.

2.8 FRACTIONAL PRECIPITATION

In the previous section the solubility product principle has been used in connection with the precipitation of one sparingly soluble salt. It is now necessary to examine the case where two slightly soluble salts may be formed. For simplicity, consider the situation which arises when a precipitating agent is added to a solution containing two anions, both of which form slightly soluble salts with the same cation, e.g. when silver nitrate solution is added to a solution containing both chloride and iodide ions. The questions which arise are: which salt will be precipitated first, and how completely will the first salt be precipitated before the second ion begins to react with the reagent?

The solubility products of silver chloride and silver iodide are respectively $1.2 \times 10^{-10} \, mol^2 \, L^{-2}$ and $1.7 \times 10^{-16} \, mol^2 \, L^{-2}$; i.e.

$$[Ag^+] \times [Cl^-] = 1.2 \times 10^{-10} \tag{1}$$

$$[Ag^+] \times [I^-] = 1.7 \times 10^{-16} \tag{2}$$

It is evident that silver iodide, being less soluble, will be precipitated first since its solubility product will be first exceeded. Silver chloride will be precipitated when the Ag^+ ion concentration is greater than

$$\frac{K_{s(AgCl)}}{[Cl^-]} = \frac{1.2 \times 10^{-10}}{[Cl^-]}$$

and then both salts will be precipitated simultaneously. When silver chloride commences to precipitate, silver ions will be in equilibrium with both salts, and equations (1) and (2) will be simultaneously satisfied, or

$$[Ag^+] = \frac{K_{s(AgI)}}{[I^-]} = \frac{K_{s(AgCl)}}{[Cl^-]} \tag{3}$$

and

$$\frac{[I^-]}{[Cl^-]} = \frac{K_{s(AgI)}}{K_{s(AgCl)}} = \frac{1.7 \times 10^{-16}}{1.2 \times 10^{10}} = 1.4 \times 10^{-6} \tag{4}$$

Hence when the concentration of the iodide ion is about one-millionth part of the chloride ion concentration, silver chloride will be precipitated. If the initial concentration of both chloride and iodide ions is $0.1\,M$, then silver chloride will be precipitated when

$$[I^-] = 0.1 \times 1.4 \times 10^{-6} = 1.4 \times 10^{-7}\,M = 1.8 \times 10^{-5}\,g\,L^{-1}$$

Thus an almost complete separation is theoretically possible. The separation is feasible in practice if the point at which the iodide precipitation is complete can be detected. This may be done: (a) by the use of an adsorption indicator (see Section 10.75(c)), or (b) by a potentiometric method with a silver electrode (see Chapter 15).

For a mixture of bromide and iodide:

$$\frac{[I^-]}{[Br^-]} = \frac{K_{s(AgI)}}{K_{s(AgBr)}} = \frac{1.7 \times 10^{-16}}{3.5 \times 10^{-13}} = \frac{1}{2.0 \times 10^3}$$

Precipitation of silver bromide will occur when the concentration of the bromide ion in the solution is 2.0×10^3 times the iodide concentration. The separation is therefore not so complete as in the case of chloride and iodide, but can nevertheless be effected with fair accuracy with the aid of adsorption indicators (Section 10.75(c)).

2.9 EFFECT OF ACIDS ON THE SOLUBILITY OF A PRECIPITATE

For sparingly soluble salts of a strong acid the effect of the addition of an acid will be similar to that of any other indifferent electrolyte but if the sparingly soluble salt MA is the salt of a weak acid HA, then acids will, in general, have a solvent effect upon it. If hydrochloric acid is added to an aqueous suspension of such a salt, the following equilibrium will be established:

$$M^+ + A^- + H^+ \rightleftharpoons HA + M^+$$

If the dissociation constant of the acid HA is very small, the anion A^- will be removed from the solution to form the undissociated acid HA. Consequently more of the salt will pass into solution to replace the anions removed in this way, and this process will continue until equilibrium is established (i.e. until $[M^+] \times [A^-]$ has become equal to the solubility product of MA) or, if sufficient hydrochloric acid is present, until the sparingly soluble salt has dissolved completely. Similar reasoning may be applied to salts of acids, such as phosphoric(V) acid ($K_1 = 7.5 \times 10^{-3}\,mol\,L^{-1}$; $K_2 = 6.2 \times 10^{-8}\,mol\,L^{-1}$; $K_3 = 5 \times 10^{-13}\,mol\,L^{-1}$), oxalic acid ($K_1 = 5.9 \times 10^{-2}\,mol\,L^{-1}$; $K_2 = 6.4 \times 10^{-5}\,mol\,L^{-1}$), and arsenic(V) acid. Thus the solubility of, say, silver phosphate(V) in dilute nitric acid is due to the removal of the PO_4^{3-} ion as

HPO_4^{2-} and/or $H_2PO_4^-$:

$$PO_4^{3-} + H^+ \rightleftharpoons HPO_4^{2-}; \quad HPO_4^{2-} + H^+ \rightleftharpoons H_2PO_4^-$$

With the salts of certain weak acids, such as carbonic, sulphurous, and nitrous acids, an additional factor contributing to the increased solubility is the actual disappearance of the acid from solution either spontaneously, or on gentle warming. An explanation is thus provided for the well-known solubility of the sparingly soluble sulphites, carbonates, oxalates, phosphates(V), arsenites(III), arsenates(V), cyanides (with the exception of silver cyanide, which is actually a salt of the strong acid $H[Ag(CN)_2]$), fluorides, acetates, and salts of other organic acids in strong acids.

The sparingly soluble sulphates (e.g. those of barium, strontium, and lead) also exhibit increased solubility in acids as a consequence of the weakness of the second-stage ionisation of sulphuric acid ($K_2 = 1.2 \times 10^{-2}$ mol L^{-1}):

$$SO_4^{2-} + H^+ \rightleftharpoons HSO_4^-$$

Since, however, K_2 is comparatively large, the solvent effect is relatively small; this is why in the quantitative separation of barium sulphate, precipitation may be carried out in slightly acid solution in order to obtain a more easily filterable precipitate and to reduce co-precipitation (Section 11.5).

The precipitation of substances within a controlled range of pH is discussed in Section 11.10.

2.10 EFFECT OF TEMPERATURE ON THE SOLUBILITY OF A PRECIPITATE

The solubility of the precipitates encountered in quantitative analysis increases with rise of temperature. With some substances the influence of temperature is small, but with others it is quite appreciable. Thus the solubility of silver chloride at 10 and 100 °C is 1.72 and 21.1 mg L^{-1} respectively, whilst that of barium sulphate at these two temperatures is 2.2 and 3.9 mg L^{-1} respectively. In many instances, the common ion effect reduces the solubility to so.small a value that the temperature effect, which is otherwise appreciable, becomes very small. Wherever possible it is advantageous to filter while the solution is hot; the rate of filtration is increased, as is also the solubility of foreign substances, thus rendering their removal from the precipitate more complete. The double phosphates of ammonium with magnesium, manganese or zinc, as well as lead sulphate and silver chloride, are usually filtered at the laboratory temperature to avoid solubility losses.

2.11 EFFECT OF THE SOLVENT ON THE SOLUBILITY OF A PRECIPITATE

The solubility of most inorganic compounds is reduced by the addition of organic solvents, such as methanol, ethanol, propan-1-ol, acetone, etc. For example, the addition of about 20 per cent by volume of ethanol renders the solubility of lead sulphate practically negligible, thus permitting quantitative separation. Similarly calcium sulphate separates quantitatively from 50 per cent ethanol. Other examples of the influence of solvents will be found in Chapter 11.

2.12 ACID–BASE EQUILIBRIA IN WATER

Consider the dissociation of a weak electrolyte, such as acetic acid, in dilute aqueous solution:

$$CH_3COOH + H_2O \rightleftharpoons H_3O^+ + CH_3COO^-$$

This will be written for simplicity in the conventional manner:

$$CH_3COOH \rightleftharpoons H^+ + CH_3COO^-$$

where H^+ represents the hydrated hydrogen ion. Applying the Law of Mass Action, we have:

$$[CH_3COO^-] \times [H^+]/[CH_3COOH] = K$$

K is the equilibrium constant at a particular temperature and is usually known as the **ionisation constant** or **dissociation constant**. If 1 mole of the electrolyte is dissolved in V litres of solution ($V = 1/c$, where c is the concentration in moles per litre), and if α is the degree of ionisation at equilibrium, then the amount of un-ionised electrolyte will be $(1 - \alpha)$ moles, and the amount of each of the ions will be α moles. The concentration of un-ionised acetic acid will therefore be $(1 - \alpha)/V$, and the concentration of each of the ions α/V. Substituting in the equilibrium equation, we obtain the expression:

$$\alpha^2/(1-\alpha)V = K \quad \text{or} \quad \alpha^2 c/(1-\alpha) = K$$

This is known as **Ostwald's Dilution Law**.

Interionic effects are, however, not negligible even for weak acids and the activity coefficient product must be introduced into the expression for the ionisation constant:

$$K = \frac{\alpha^2 c}{(1-\alpha)} \cdot \frac{y_{H^+} \cdot y_{A^-}}{y_{HA}}; \quad A^- = CH_3COO^-$$

Reference must be made to textbooks of physical chemistry (see Bibliography, Section 3.39) for details of the methods used to evaluate true dissociation constants of acids.

From the point of view of quantitative analysis, sufficiently accurate values for the ionisation constants of weak monoprotic acids may be obtained by using the classical Ostwald Dilution Law expression: the resulting 'constant' is sometimes called the 'concentration dissociation constant'.

2.13 STRENGTHS OF ACIDS AND BASES

The Brønsted–Lowry expression for acid–base equilibria (see Section 2.4)

$$A_1 + B_2 \rightleftharpoons A_2 + B_1 \tag{b}$$

leads, upon application of the Law of Mass Action, to the expression:

$$K = \frac{[A_2] [B_1]}{[A_1] [B_2]} \tag{5}$$

where the constant K depends on the temperature and the nature of the solvent. This expression is strictly valid only for extremely dilute solutions: when ions are present the electrostatic forces between them have appreciable effects on

31

the properties of their solutions, and deviations are apparent from ideal laws (which are assumed in the derivation of the Mass-Action Law by thermodynamic or kinetic methods); the deviations from the ideal laws are usually expressed in terms of activities or activity coefficients. For our purpose, the deviations due to interionic attractions and ionic activities will be regarded as small for small ionic concentrations and the equations will be regarded as holding in the same form at higher concentrations, provided that the total ionic concentration does not vary much in a given set of experiments.

To use the above expression for measuring the strength of an acid, a standard acid–base pair, say A_2–B_2, must be chosen, and it is usually convenient to refer acid–base strength to the solvent. In water the acid–base pair H_3O^+–H_2O is taken as the standard. The equilibrium defining acids is therefore:

$$A + H_2O \rightleftharpoons B + H_3O^+ \tag{c}$$

and the constant

$$K' = \frac{[B][H_3O^+]}{[A][H_2O]} \tag{6}$$

gives the strength of A, that of the ion H_3O^+ being taken as unity. Equation (c) represents what is usually described as the dissociation of the acid A in water, and the constant K' is closely related to the dissociation constant of A in water as usually defined and differing only in the inclusion of the term $[H_2O]$ in the denominator. The latter term represents the 'concentration' of water molecules in liquid water (55.5 moles per litre on the ordinary volume concentration scale). When dealing with dilute solutions, the value of $[H_2O]$ may be regarded as constant, and equation (6) may be expressed as:

$$K_a = \frac{[B][H^+]}{[A]} \tag{7}$$

by writing H^+ for H_3O^+ and remembering that the hydrated proton is meant. This equation defines the strength of the acid A. If A is an uncharged molecule (e.g. a weak organic acid), B is the anion derived from it by the loss of a proton, and (7) is the usual expression for the ionisation constant. If A is an anion such as $H_2PO_4^-$, the dissociation constant $[HPO_4^{2-}][H^+]/[H_2PO_4^-]$ is usually referred to as the second dissociation constant of phosphoric(V) acid. If A is a cation acid, for example the ammonium ion, which interacts with water as shown by the equation

$$NH_4^+ + H_2O \rightleftharpoons NH_3 + H_3O^+$$

the acid strength is given by $[NH_3][H^+]/[NH_4^+]$.

On the above basis it is, in principle, unnecessary to treat the strength of bases separately from acids, since any protolytic reaction involving an acid must also involve its conjugate base. The basic properties of ammonia and various amines in water are readily understood on the Brønsted–Lowry concept.

$$H_2O = H^+ + OH^-$$
$$\underline{NH_3 + H^+ = NH_4^+}$$
$$NH_3 + H_2O \rightleftharpoons NH_4^+ + OH^-$$

The basic dissociation constant K_b is given by:

$$K_b = \frac{[NH_4^+][OH^-]}{[NH_3]} \tag{8}$$

Since $[H^+][OH^-] = K_w$ (the ionic product of water), we have

$$K_b = K_w/K_a$$

The values of K_a and K_b for different acids and bases vary through many powers of ten. It is often convenient to use the dissociation constant exponent pK defined by

$$pK = \log_{10} 1/K = -\log_{10} K$$

The larger the pK_a value is, the weaker is the acid and the stronger the base.

For very weak or slightly ionised electrolyes, the expression $\alpha^2/(1-\alpha)V = K$ reduces to $\alpha^2 = KV$ or $\alpha = \sqrt{KV}$, since α may be neglected in comparison with unity. Hence for any two weak acids or bases at a given dilution V (in L), we have $\alpha_1 = \sqrt{K_1 V}$ and $\alpha_2 = \sqrt{K_2 V}$, or $\alpha_1/\alpha_2 = \sqrt{K_1}/\sqrt{K_2}$. Expressed in words, for any two weak or slightly dissociated electrolytes at equal dilutions, the degrees of dissociation are proportional to the square roots of their ionisation constants. Some values for the dissociation constants at $25\,°C$ for weak acids and bases are collected in Appendix 7.

2.14 DISSOCIATION OF POLYPROTIC ACIDS

When a polyprotic acid is dissolved in water, the various hydrogen atoms undergo ionisation to different extents. For a diprotic acid H_2A, the primary and secondary dissociations can be represented by the equations:

$$H_2A \rightleftharpoons H^+ + HA^-$$

$$HA^- \rightleftharpoons H^+ + A^{2-}$$

If the acid is a weak electrolyte, the Law of Mass Action may be applied, and the following expressions obtained:

$$[H^+][HA^-]/[H_2A] = K_1 \tag{9}$$

$$[H^+][A^{2-}]/[HA^-] = K_2 \tag{10}$$

K_1 and K_2 are known as the **primary** and **secondary dissociation constants** respectively. Each stage of the dissociation process has its own ionisation constant, and the magnitudes of these constants give a measure of the extent to which each ionisation has proceeded at any given concentration. The greater the value of K_1 relative to K_2, the smaller will be the secondary dissociation, and the greater must be the dilution before the latter becomes appreciable. It is therefore possible that a diprotic (or polyprotic) acid may behave, so far as dissociation is concerned, as a monoprotic acid. This is indeed characteristic of many polyprotic acids.

A triprotic acid H_3A (e.g. phosphoric(V) acid) will similarly yield three dissociation constants, K_1, K_2, and K_3, which may be derived in an analogous manner:

$$[H^+][H_2A^-]/[H_3A] = K_1 \tag{9'}$$

$$[H^+][HA^{2-}]/[H_2A^-] = K_2 \tag{10'}$$

$$[H^+][A^{3-}]/[HA^{2-}] = K_3 \tag{11}$$

Application of these theoretical considerations to situations encountered in practice may be illustrated by numerical examples.

Example 6. Calculate the concentrations of HS^- and S^{2-} in a solution of hydrogen sulphide.

A saturated aqueous solution of hydrogen sulphide at 25 °C, at atmospheric pressure, is approximately $0.1\,M$, and for H_2S the primary and secondary dissociation constants may be taken as $1.0 \times 10^{-7}\,\mathrm{mol\,L^{-1}}$ and $1 \times 10^{-14}\,\mathrm{mol\,L^{-1}}$ respectively.

In the solution the following equilibria are involved:

$$H_2S + H_2O \rightleftharpoons HS^- + H_3O^+; \quad K_1 = [H^+][HS^-]/[H_2S] \tag{d}$$

$$HS^- + H_2O \rightleftharpoons S^{2-} + H_3O^+; \quad K_2 = [H^+][S^{2-}]/[HS^-] \tag{e}$$

$$H_2O \rightleftharpoons H^+ + OH^-$$

Electroneutrality requires that the total cation concentration must equal total anion concentration and hence, taking account of charge numbers,

$$[H^+] = [HS^-] + 2[S^{2-}] + [OH^-] \tag{f}$$

but since in fact we are dealing with an acid solution, $[H^+] > 10^{-7} > [OH^-]$ and we can simplify equation (e) to read

$$[H^+] = [HS^-] + 2[S^{2-}] \tag{g}$$

The 0.1 mol H_2S is present partly as undissociated H_2S and partly as the ions HS^- and S^{2-}, and it follows that

$$[H_2S] + [HS^-] + [S^{2-}] = 0.1 \tag{h}$$

The very small value of K_2 indicates that the secondary dissociation and therefore $[S^{2-}]$ are extremely minute, and ignoring $[S^{2-}]$ in equation (g) we are left with the result

$$[H^+] \approx [HS^-]$$

Since K_1 is also small, $[H^+] \ll [H_2S]$ and so equation (h) can be reduced to

$$[H_2S] \approx 0.1$$

Using these results in equation (d) we find

$$[H^+]^2/0.1 = 1 \times 10^{-7}; \quad [H^+] = [HS^-] = 1.0 \times 10^{-4}\,\mathrm{mol\,L^{-1}}.$$

From equation (e) it then follows that

$$(1.0 \times 10^{-4})[S^{2-}]/(1.0 \times 10^{-4}) = 1 \times 10^{-14}$$

and $[S^{2-}] = 1 \times 10^{-14}\,\mathrm{mol\,L^{-1}}$.

2.15 COMMON ION EFFECT

The concentration of a particular ion in an ionic reaction can be increased by the addition of a compound which produces that ion upon dissociation. The particular ion is thus derived from the compound already in solution and also from the added reagent, hence the name 'common ion'. If the original compound is a weak electrolyte, the Law of Mass Action will be applicable. The result is that there is a higher concentration of this ion in solution than that derived from the original compound alone, and new equilibrium conditions will be produced. Examples of the calculation of the common ion effect are given below. In general, it may be stated that if the total concentration of the common ion is only slightly greater than that which the original compound alone would furnish, the effect is small; if, however, the concentration of the common ion is very much increased (e.g. by the addition of a completely dissociated salt), the effect is very great, and may be of considerable practical importance. Indeed, the common ion effect provides a valuable method for controlling the concentration of the ions furnished by a weak electrolyte.

Example 7. Calculate the sulphide ion concentration in a 0.25M hydrochloric acid solution saturated with hydrogen sulphide.

This concentration has been chosen since it is that at which the sulphides of certain heavy metals are precipitated. The total concentration of hydrogen sulphide may be assumed to be approximately the same as in aqueous solution, i.e. 0.1M; the $[H^+]$ will be equal to that of the completely dissociated HCl, i.e. 0.25M, but the $[S^{2-}]$ will be reduced below 1×10^{-14} (see *Example 6*).

Substituting in equations (*d*) and (*e*) (*Example 6*), we find:

$$[HS^-] = \frac{K_1 \times [H_2S]}{[H^+]} = \frac{1.0 \times 10^{-7} \times 0.1}{0.25} = 4.0 \times 10^{-8} \text{ mol L}^{-1}$$

$$[S^{2-}] = \frac{K_2 \times [HS^-]}{[H^+]} = \frac{(1 \times 10^{-14}) \times (4 \times 10^{-8})}{0.25} = 1.6 \times 10^{-21} \text{ mol L}^{-1}$$

Thus by changing the acidity from $1.0 \times 10^{-4} M$ (that present in saturated H_2S water) to 0.25 M, the sulphide ion concentration is reduced from 1×10^{-14} to 1.6×10^{-21}.

Example 8. What effect has the addition of 0.1 mol of anhydrous sodium acetate to 1 L of 0.1 M acetic acid upon the degree of dissociation of the acid?

The dissociation constant of acetic acid at 25 °C is 1.75×10^{-5} mol L^{-1} and the degree of ionisation α in 0.1 M solution may be computed by solving the quadratic equation:

$$\frac{[H^+] \times [CH_3COO^-]}{[CH_3COOH]} = \frac{\alpha^2 c}{(1-\alpha)} = 1.75 \times 10^{-5}$$

For our purpose it is sufficiently accurate to neglect α in $(1 - \alpha)$ since α is small:

$$\therefore \alpha = \sqrt{K/c} = \sqrt{1.75 \times 10^{-4}} = 0.0132$$

Hence in 0.1 M acetic acid,

$[H^+] = 0.00132$, $[CH_3COO^-] = 0.00132$,

and $\qquad\qquad\qquad [CH_3COOH] = 0.0987$ mol L^{-1}

The concentrations of sodium and acetate ions produced by the addition of the completely dissociated sodium acetate are:

$[Na^+] = 0.1$, and $[CH_3COO^-] = 0.1 \, mol \, L^{-1}$ respectively.

The acetate ions from the salt will tend to decrease the ionisation of the acetic acid, and consequently the acetate ion concentration derived from it. Hence we may write $[CH_3COO^-] = 0.1$ for the solution, and if α' is the new degree of ionisation, $[H^+] = \alpha'c = 0.1\alpha'$, and $[CH_3COOH] = (1 - \alpha')c = 0.1$, since α' is negligibly small.

Substituting in the mass action equation:

$$\frac{[H^+] \times [CH_3COO^-]}{[CH_3COOH]} = \frac{0.1\alpha' \times 0.1}{0.1} = 1.75 \times 10^{-5}$$

or $\quad \alpha' = 1.75 \times 10^{-4}$

$[H^+] = \alpha'c = 1.75 \times 10^{-5} \, mol \, L^{-1}$

The addition of a tenth of a mole of sodium acetate to a $0.1 \, M$ solution of acetic acid has decreased the degree of ionisation from 1.32 to 0.018 per cent, and the hydrogen ion concentration from 0.00132 to $0.000018 \, mol \, L^{-1}$.

Example 9. What effect has the addition of 0.5 mol of ammonium chloride to 1 L of $0.1 \, M$ aqueous ammonia solution upon the degree of dissociation of the base?

(Dissociation constant of NH_3 in water $= 1.8 \times 10^{-5} \, mol \, L^{-1}$)

In $0.1 \, M$ ammonia solution $\alpha = \sqrt{1.8 \times 10^{-5}/0.1} = 0.0135$. Hence $[OH^-] = 0.00135$, $[NH_4^+] = 0.00135$, and $[NH_3] = 0.0986 \, mol \, L^{-1}$. Let α' be the degree of ionisation in the presence of the added ammonium chloride. Then $[OH^-] = \alpha'c = 0.1\alpha'$, and $[NH_3] = (1 - \alpha')c = 0.1$, since α' may be taken as negligibly small. The addition of the completely ionised ammonium chloride will, of necessity, decrease the $[NH_4^+]$ derived from the base and increase $[NH_3]$, and as a first approximation $[NH_4^+] = 0.5$.

Substituting in the equation:

$$\frac{[NH_4^+] \times [OH^-]}{[NH_3]} = \frac{0.5 \times 0.1\alpha'}{0.1} = 1.8 \times 10^{-5}$$

$\alpha' = 3.6 \times 10^{-5}$ and $[OH^-] = 3.6 \times 10^{-6} \, mol \, L^{-1}$

The addition of half a mole of ammonium chloride to 1 litre of a $0.1 \, M$ solution of aqueous ammonia has decreased the degree of ionisation from 1.35 to 0.0036 per cent, and the hydroxide ion concentration from 0.00135 to $0.0000036 \, mol \, L^{-1}$.

2.16 THE IONIC PRODUCT OF WATER

Kohlrausch and Heydweiller (1894) found that the most highly purified water that can be obtained possesses a small but definite conductivity. Water must therefore be slightly ionised in accordance with the equation:

$$H_2O \rightleftharpoons H^+ + OH^-*$$

Applying the Law of Mass Action to this equation, we obtain, for any given temperature:

$$\frac{a_{H^+} \times a_{OH^-}}{a_{H_2O}} = \frac{[H^+].[OH^-]}{[H_2O]} \times \frac{y_{H^+}.y_{OH^-}}{y_{H_2O}} = \text{a constant}$$

Since water is only slightly ionised, the ionic concentrations will be small, and their activity coefficients may be regarded as unity; the activity of the un-ionised molecules may also be taken as unity. The expression thus becomes:

$$\frac{[H^+] \times [OH^-]}{[H_2O]} = \text{a constant}$$

In pure water or in dilute aqueous solutions, the concentration of the undissociated water may be considered constant. Hence:

$$[H^+] \times [OH^-] = K_w$$

where K_w is the **ionic product of water**. It must be pointed out that the assumption that the activity coefficients of the ions are unity and that the activity coefficient of water is constant applies strictly to pure water and to very dilute solutions (ionic strength < 0.01); in more concentrated solutions, i.e. in solutions of appreciable ionic strength, the activity coefficients of the ions are affected (compare Section 2.5), as is also the activity of the un-ionised water. The ionic product of water will then not be constant, but will depend upon the ionic strength of the solution. It is, however, difficult to determine the activity coefficients, except under specially selected conditions, so that in practice the ionic product K_w, although not strictly constant, is employed.

The ionic product varies with the temperature, but under ordinary experimental conditions (at about 25 °C) its value may be taken as 1×10^{-14} with concentrations expressed in mol L^{-1}. This is sensibly constant in dilute aqueous solutions. If the product of $[H^+]$ and $[OH^-]$ in aqueous solution momentarily exceeds this value, the excess ions will immediately combine to form water. Similarly, if the product of the two ionic concentrations is momentarily less than 10^{-14}, more water molecules will dissociate until the equilibrium value is attained.

The hydrogen and hydroxide ion concentrations are equal in pure water; therefore $[H^+] = [OH^-] = \sqrt{K_w} = 10^{-7}$ mol L^{-1} at about 25 °C. A solution in which the hydrogen and hydroxide ion concentrations are equal is termed an **exactly neutral solution**. If $[H^+]$ is greater than 10^{-7}, the solution is **acid**, and if less than 10^{-7}, the solution is **alkaline** (or basic). It follows that at ordinary temperatures $[OH^-]$ is greater than 10^{-7} in alkaline solution and less than this value in acid solution.

---

* Strictly speaking the hydrogen ion H^+ exists in water as the hydroxonium ion H_3O^+ (Section 2.4). The electrolytic dissociation of water should therefore be written:

$$2H_2O \rightleftharpoons H_3O^+ + OH^-$$

For the sake of simplicity, the more familiar symbol H^+ will be retained.

In all cases the reaction of the solution can be quantitatively expressed by the magnitude of the hydrogen ion (or hydroxonium ion) concentration, or, less frequently, of the hydroxide ion concentration, since the following simple relations between $[H^+]$ and $[OH^-]$ exist:

$$[H^+] = \frac{K_w}{[OH^-]}, \quad \text{and} \quad [OH^-] = \frac{K_w}{[H^+]}$$

The variation of K_w with temperature is shown in Table 2.1.

Table 2.1 Ionic product of water at various temperatures

Temp. ($^\circ$C)	$K_w \times 10^{14}$	Temp. ($^\circ$C)	$K_w \times 10^{14}$
0	0.12	35	2.09
5	0.19	40	2.92
10	0.29	45	4.02
15	0.45	50	5.47
20	0.68	55	7.30
25	1.01	60	9.61
30	1.47		

2.17 THE HYDROGEN ION EXPONENT

For many purposes, especially when dealing with small concentrations, it is cumbersome to express concentrations of hydrogen and hydroxyl ions in terms of moles per litre. A very convenient method was proposed by S. P. L. Sørensen (1909). He introduced the hydrogen ion exponent pH defined by the relationships:

$$pH = \log_{10} 1/[H^+] = -\log_{10}[H^+], \quad \text{or} \quad [H^+] = 10^{-pH}$$

The quantity pH is thus the logarithm (to the base 10) of the reciprocal of the hydrogen ion concentration, or is equal to the logarithm of the hydrogen ion concentration with negative sign. This method has the advantage that all states of acidity and alkalinity between those of solutions containing, on the one hand, $1 \, mol \, L^{-1}$ of hydrogen ions, and on the other hand, $1 \, mol \, L^{-1}$ of hydroxide ions, can be expressed by a series of positive numbers between 0 and 14. Thus a neutral solution with $[H^+] = 10^{-7}$ has a pH of 7; a solution with a hydrogen ion concentration of $1 \, mol \, L^{-1}$ has a pH of 0 ($[H^+] = 10^0$); and a solution with a hydroxide-ion concentration of $1 \, mol \, L^{-1}$ has $[H^+] = K_w/[OH^-] = 10^{-14}/10^0 = 10^{-14}$, and possesses a pH of 14. A neutral solution is therefore one in which pH = 7, an acid solution one in which pH < 7, and an alkaline solution one in which pH > 7. An alternative definition for a neutral solution, applicable to all temperatures, is one in which the hydrogen ion and hydroxide ion concentrations are equal. In an acid solution the hydrogen ion concentration exceeds the hydroxide ion concentration, whilst in an alkaline or basic solution, the hydroxide ion concentration is greater.

Example 10. (i) Find the pH of a solution in which $[H^+] = 4.0 \times 10^{-5} \, mol \, L^{-1}$.

$$pH = \log_{10} 1/[H^+] = \log 1 - \log[H^+]$$
$$= \log 1 - \log 4.0 \times 10^{-5}$$
$$= 0 - \bar{5}.602$$
$$= \underline{4.398}$$

(ii) Find the hydrogen ion concentration corresponding to pH = 5.643.

$$pH = \log_{10} 1/[H^+] = \log 1 - \log[H^+] = 5.643$$
$$\therefore \log[H^+] = -5.643$$

This must be written in the usual form containing a negative characteristic and a positive mantissa:

$$\log[H^+] = -5.643 = \bar{6}.357$$

By reference to a calculator or to tables of antilogarithms we find $[H^+] = 2.28 \times 10^{-6}$ mol L^{-1}.

(iii) Calculate the pH of a 0.01 M solution of acetic acid in which the degree of dissociation is 12.5 per cent.

The hydrogen ion concentration of the solution is 0.125×0.01

$$= 1.25 \times 10^{-3} \text{ mol L}^{-1}$$
$$pH = \log_{10} 1/[H^+] = \log 1 - \log[H^+]$$
$$= 0 - \bar{3}.097$$
$$= \underline{2.903}$$

The hydroxide ion concentration may be expressed in a similar way:

$$pOH = -\log_{10}[OH^-] = \log_{10} 1/[OH^-], \quad \text{or} \quad [OH^-] = 10^{-pOH}$$

If we write the equation:

$$[H^+] \times [OH^-] = K_w = 10^{-14}$$

in the form:

$$\log[H^+] + \log[OH^-] = \log K_w = -14$$

then $pH + pOH = pK_w = 14$

This relationship should hold for all dilute solutions at about 25 °C.

Figure 2.1 will serve as a useful mnemonic for the relation between $[H^+]$, pH, $[OH^-]$, and pOH in acid and alkaline solution.

The logarithmic or exponential form has also been found useful for expressing other small quantities which arise in quantitative analysis. These include: (i) dissociation constants (Section 2.13), (ii) other ionic concentrations, and (iii) solubility products (Section 2.6).

(i) For any acid with a dissociation constant of K_a:

$$pK_a = \log 1/K_a = -\log K_a$$

Similarly for any base with dissociation constant K_b:

$$pK_b = \log 1/K_b = -\log K_b$$

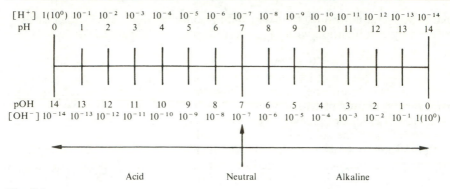

Fig. 2.1

(ii) For any ion I of concentration $[I]$:

$$pI = \log 1/[I] = -\log[I]$$

Thus, for $[Na^+] = 8 \times 10^{-5} \, mol \, L^{-1}$, $pNa = 4.1$.

(iii) For a salt with a solubility product K_s:

$$pK_s = \log 1/K_s = -\log K_s.$$

2.18 THE HYDROLYSIS OF SALTS

Salts may be divided into four main classes:

(1) those derived from strong acids and strong bases, e.g. potassium chloride;
(2) those derived from weak acids and strong bases, e.g. sodium acetate;
(3) those derived from strong acids and weak bases, e.g. ammonium chloride; and
(4) those derived from weak acids and weak bases, e.g. ammonium formate or aluminium acetate.

When any of these from classes (2) to (4) is dissolved in water, the solution, as is well known, is not always neutral in reaction. Interaction may occur with the ions of water, and the resulting solution will be neutral, acid, or alkaline according to the nature of the salt.

With an aqueous solution of a salt of class (1), neither do the anions have any tendency to combine with the hydrogen ions nor do the cations with the hydroxide ions of water, since the related acids and bases are strong electrolytes. The equilibrium between the hydrogen and hydroxide ions in water:

$$H_2O \rightleftharpoons H^+ + OH^- \qquad\qquad (i)$$

is therefore not disturbed and the solution remains neutral.

Consider, however, a salt MA derived from a weak acid HA and a strong base BOH {class (2)}. The salt is completely dissociated in aqueous solution:

$$MA \rightarrow M^+ + A^-$$

A very small concentration of hydrogen and hydroxide ions, originating from the small but finite ionisation of water, will be initially present. HA is a weak acid, i.e. it is dissociated only to a small degree; the concentration of A^- ions which can exist in equilibrium with H^+ ions is accordingly small. In order to maintain the equilibrium, the large initial concentration of A^- ions must be reduced by combination with H^+ ions to form undissociated HA:

$$H^+ + A^- \rightleftharpoons HA \qquad (j)$$

The hydrogen ions required for this reaction can be obtained only from the further dissociation of the water; this dissociation produces simultaneously an equivalent quantity of hydroxyl ions. The hydrogen ions are utilised in the formation of HA; consequently the hydroxide ion concentration of the solution will increase and the solution will react alkaline.

It is usual in writing equations involving equilibria between completely dissociated and slightly dissociated or sparingly soluble substances to employ the ions of the former and the molecules of the latter. The reaction is therefore written:

$$A^- + H_2O \rightleftharpoons OH^- + HA \qquad (k)$$

This equation can also be obtained by combining (i) and (j), since both equilibria must co-exist. This interaction between the ion (or ions) of a salt and water is called '**hydrolysis**'.

Consider now the salt of a strong acid and a weak base {class (3)}. Here the initial high concentration of cations M^+ will be reduced by combination with the hydroxide ions of water to form the little-dissociated base MOH until the equilibrium:

$$M^+ + OH^- \rightleftharpoons MOH$$

is attained. The hydrogen ion concentration of the solution will thus be increased, and the solution will react acid. The hydrolysis is here represented by:

$$M^+ + H_2O \rightleftharpoons MOH + H^+$$

For salts of class (4), in which both the acid and the base are weak, two reactions will occur simultaneously

$$M^+ + H_2O \rightleftharpoons MOH + H^+; \quad A^- + H_2O \rightleftharpoons HA + OH^-$$

The reaction of the solution will clearly depend upon the relative dissociation constants of the acid and the base. If they are equal in strength, the solution will be neutral; if $K_a > K_b$, it will be acid, and if $K_b > K_a$, it will be alkaline.

Having considered all the possible cases, we are now in a position to give a more general definition of hydrolysis. Hydrolysis is the interaction between an ion (or ions) of a salt and water with the production of (a) a weak acid or a weak base, or (b) of both a weak acid and a weak base.

The phenomenon of salt hydrolysis may be regarded as a simple application of the general Brønsted–Lowry equation

$$A_1 + B_2 \rightleftharpoons A_2 + B_1$$

Thus the equation for the hydrolysis of ammonium salts

$$NH_4^+ + H_2O \rightleftharpoons NH_3 + H_3O^+$$

is really identical with the expression used to define the strength of the ammonium ion as a Brønsted–Lowry acid (see Section 2.4) and the constant K_a for NH_4^+ is in fact what is usually termed the hydrolysis constant of an ammonium salt.

The hydrolysis of the sodium salt of a weak acid can be treated similarly. Thus for a solution of sodium acetate

$$CH_3COO^- + H_2O \rightleftharpoons CH_3COOH + OH^-$$

the hydrolysis constant is

$$[CH_3COOH][OH^-]/[CH_3COO^-] = K_h = K_w/K_a$$

where K_a is the dissociation constant of acetic (ethanoic) acid.

2.19 HYDROLYSIS CONSTANT AND DEGREE OF HYDROLYSIS

Case 1. Salt of a weak acid and a strong base. The equilibrium in a solution of salt MA may be represented by:

$$A^- + H_2O \rightleftharpoons OH^- + HA$$

Applying the Law of Mass Action, we obtain:

$$\frac{a_{OH^-} \times a_{HA}}{a_{A^-}} = \frac{[OH^-].[HA]}{[A^-]} \times \frac{y_{OH^-} \cdot y_{HA}}{y_{A^-}} = K_h \tag{12}$$

where K_h is the **hydrolysis constant**. The solution is assumed to be dilute so that the activity of the un-ionised water may be taken as constant, and the approximation that the activity coefficient of the un-ionised acid is unity and that both ions have the same activity coefficient may be introduced. Equation (12) then reduces to:

$$K_h = \frac{[OH^-] \times [HA]}{[A^-]} \tag{13}$$

This is often written in the form:

$$K_h = \frac{[Base] \times [Acid]}{[Unhydrolysed\ salt]}$$

The free strong base and the unhydrolysed salt are completely dissociated and the acid is very little dissociated.

The degree of hydrolysis is the fraction of each mole of anion A^- hydrolysed at equilibrium. Let 1 mole of salt be dissolved in V L of solution, and let x be the degree of hydrolysis. The concentrations in $mol\,L^{-1}$ are:

$$[HA] = [OH^-] = x/V; \quad [A^-] = (1-x)/V$$

Substituting these values in equation (13):

$$K_h = \frac{[OH^-] \times [HA]}{[A^-]} = \frac{x/V \times x/V}{(1-x)/V} = \frac{x^2}{(1-x)V}$$

This expression enables us to calculate the degree of hydrolysis at the dilution V; it is evident that as V increases, the degree of hydrolysis x must increase.

The two equilibria:

$$H_2O \rightleftharpoons H^+ + OH^- \quad \text{and} \quad HA \rightleftharpoons H^+ + A^-$$

must co-exist with the hydrolytic equilibrium:

$$A^- + H_2O \rightleftharpoons HA + OH^-$$

Hence the two relationships:

$$[H^+] \times [OH^-] = K_w \quad \text{and} \quad [H^+] \times [A^-]/[HA] = K_a$$

must hold in the same solution as:

$$[OH^-] \times [HA]/[A^-] = K_h$$

But $\quad \dfrac{K_w}{K_a} = \dfrac{[H^+] \times [OH^-] \times [HA]}{[H^+] \times [A^-]} = \dfrac{[OH^-] \times [HA]}{[A^-]} = K_h$

therefore $\quad K_w/K_a = K_h$

or $\quad pK_h = pK_w - pK_a$

The hydrolysis constant is thus related to the ionic product of water and the ionisation constant of the acid. Since K_a varies slightly and K_w varies considerably with temperature, K_h and consequently the degree of hydrolysis will be largely influenced by changes of temperature.

The hydrogen ion concentration of a solution of a hydrolysed salt can be readily calculated. The amounts of HA and of OH^- ions formed as a result of hydrolysis are equal; therefore, in a solution of the pure salt in water, $[HA] = [OH^-]$. If the concentration of the salt is c mol L^{-1}, then:

$$\frac{[HA] \times [OH^-]}{[A^-]} = \frac{[OH^-]^2}{c} = K_h = \frac{K_w}{K_a}$$

and $\quad [OH^-] = \sqrt{c.K_w/K_a}$

or $\quad [H^+] = \sqrt{K_w.K_a/c}, \quad \text{since} \quad [H^+] = K_w/[OH^-]$

and $\quad pH = \frac{1}{2}pK_w + \frac{1}{2}pK_a + \frac{1}{2}\log c$

To be consistent we should use $pc = -\log c$ so that the equation becomes:

$$pH = \frac{1}{2}pK_w + \frac{1}{2}pK_a - \frac{1}{2}pc \tag{14}$$

Equation (14) can be employed for the calculation of the pH of a solution of a salt of a weak acid and a strong base. Thus the pH of a solution of sodium benzoate (0.05 mol L^{-1}) is given by:

$$pH = 7.0 + 2.10 - \frac{1}{2}(1.30) = 8.45$$

(Benzoic acid: $K_a = 6.37 \times 10^{-5}$ mol L^{-1}; $pK_a = 4.20$)

Such a calculation will provide useful information as to the indicator which should be employed in the titration of a weak acid and a strong base (see Section 10.13).

Example 11. Calculate: (i) the hydrolysis constant, (ii) the degree of hydrolysis, and (iii) the hydrogen ion concentration of a solution of sodium acetate (0.01 mol L^{-1}) at the laboratory temperature.

$$K_h = \frac{K_w}{K_a} = \frac{1.0 \times 10^{-14}}{1.75 \times 10^{-5}} = 5.7 \times 10^{-10}$$

The degree of hydrolysis x is given by:

$$K_h = \frac{x^2}{(1-x)V}$$

Substituting for K_h and $V \, (= 1/c)$, we obtain:

$$5.7 \times 10^{-10} = \frac{x^2 \times 0.01}{(1-x)}$$

Solving this quadratic equation, $x = 0.000238$ or 0.0238 per cent.

If the solution were completely hydrolysed, the concentration of acetic (ethanoic) acid produced would be 0.01 mol L^{-1}. But the degree of hydrolysis is 0.0238 per cent, therefore the concentration of acetic acid is $2.38 \times 10^{-6} \text{ mol L}^{-1}$. This is also equal to the hydroxide ion concentration produced, i.e. pOH $= 5.62$.

pH $= 14.0 - 5.62 = 8.38$

The pH may also be calculated from equation (14):

pH $= \frac{1}{2}pK_w + \frac{1}{2}pK_a - \frac{1}{2}pc = 7.0 + 2.38 - \frac{1}{2}(2) = 8.38$

Case 2. Salt of a strong acid and a weak base. The hydrolytic equilibrium is represented by:

$$M^+ + H_2O \rightleftharpoons MOH + H^+$$

By applying the Law of Mass Action along the lines of Case 1, the following equations are obtained:

$$K_h = \frac{[H^+] \times [MOH]}{[M^+]} = \frac{[Acid] \times [Base]}{[Unhydrolysed \, salt]} = \frac{K_w}{K_b}$$

$$= \frac{x^2}{(1-x)V}$$

K_b is the dissociation constant of the base. Furthermore, since $[MOH]$ and $[H^+]$ are equal:

$$K_h = \frac{[H^+] \times [MOH]}{[M^+]} = \frac{[H^+]^2}{c} = \frac{K_w}{K_b}$$

$$[H^+] = \sqrt{c \cdot K_w/K_b},$$

or   pH $= \frac{1}{2}pK_w - \frac{1}{2}pK_b + \frac{1}{2}pc$ (15)

Equation (15) may be applied to the calculation of the pH of solutions of salts of strong acids and weak bases. Thus the pH of a solution of ammonium chloride (0.2 mol L^{-1}) is:

pH $= 7.0 - 2.37 + \frac{1}{2}(0.70) = 4.98$

(Ammonia in water: $K_b = 1.8 \times 10^{-5} \text{ mol L}^{-1}$; $pK_b = 4.74$)

Case 3. Salt of a weak acid and a weak base. The hydrolytic equilibrium is expressed by the equation:

$$M^+ + A^- + H_2O \rightleftharpoons MOH + HA$$

Applying the Law of Mass Action and taking the activity of un-ionised water as unity, we have:

$$K_h = \frac{a_{MOH} \times a_{HA}}{a_{M^+} \times a_{A^-}} = \frac{[MOH].[HA]}{[M^+].[A^-]} \times \frac{y_{MOH} \cdot y_{HA}}{y_{M^+} \cdot y_{A^-}}$$

By the usual approximations, i.e. by assuming that the activity coefficients of the un-ionised molecules and, less justifiably, of the ions are unity, the following approximate equation is obtained:

$$K_h = \frac{[MOH] \times [HA]}{[M^+] \times [A^-]} = \frac{[Base] \times [Acid]}{[Unhydrolysed\ salt]^2}$$

If x is the degree of hydrolysis of 1 mole of the salt dissolved in V litres of solution, then the individual concentrations are:

$$[MOH] = [HA] = x/V; \quad [M^+] = [A^-] = (1-x)/V$$

leading to the result

$$K_h = \frac{x/V \cdot x/V}{(1-x)/V \cdot (1-x)/V} = \frac{x^2}{(1-x)^2}$$

The degree of hydrolysis and consequently the pH is independent of the concentration of the solution.*

It may be readily shown that:

$$K_h = K_w/K_a \times K_b$$

or $\quad pK_h = pK_w - pK_a - pK_b$

This expression enables us to calculate the value of the degree of hydrolysis from the dissociation constants of the acid and the base.

The hydrogen ion concentration of the hydrolysed solution is calculated in the following manner:

$$[H^+] = K_a \times \frac{[HA]}{[A^-]} = K_a \times \frac{x/V}{(1-x)/V} = K_a \times \frac{x}{(1-x)}$$

But $\quad x/(1-x) = \sqrt{K_h}$

Hence $\quad [H^+] = K_a\sqrt{K_h} = \sqrt{K_w \times K_a/K_b}$

or $\quad pH = \frac{1}{2}pK_w + \frac{1}{2}pK_a - \frac{1}{2}pK_b$ \hfill (16)

If the ionisation constants of the acid and the base are equal, that is $K_a = K_b$, $pH = \frac{1}{2}pK_w = 7.0$ and the solution is neutral, although hydrolysis may be considerable. If $K_a > K_b$, $pH < 7$ and the solution is acid, but when $K_b > K_a$, $pH > 7$ and the solution reacts alkaline.

The pH of a solution of ammonium acetate is given by:

$$pH = 7.0 + 2.38 - 2.37 = 7.1$$

i.e. the solution is approximately neutral. On the other hand, for a dilute

---

* This applies only if the original assumptions as to activity coefficients are justified. In solutions of appreciable ionic strength, the activity coefficients of the ions will vary with the total ionic strength.

solution of ammonium formate:

pH $= 7.0 + 1.88 - 2.37 = 6.51$

(Formic acid: $K_a = 1.77 \times 10^{-4}\,\mathrm{mol\,L^{-1}}$; $pK_a = 3.75$)

i.e. the solution has a slightly acid reaction.

2.20 BUFFER SOLUTIONS

A solution of hydrochloric acid ($0.0001\,\mathrm{mol\,L^{-1}}$) should have a pH equal to 4, but the solution is extremely sensitive to traces of alkali from the glass of the containing vessel and to ammonia from the air. Likewise a solution of sodium hydroxide ($0.0001\,\mathrm{mol\,L^{-1}}$), which should have a pH of 10, is sensitive to traces of carbon dioxide from the atmosphere. Aqueous solutions of potassium chloride and of ammonium acetate have a pH of about 7. The addition to 1 L of these solutions of 1 mL of a solution of hydrochloric acid ($1\,\mathrm{mol\,L^{-1}}$) results in a change of pH to 3 in the former case and in very little change in the latter. The resistance of a solution to changes in hydrogen ion concentration upon the addition of small amounts of acid or alkali is termed **buffer action**; a solution which possesses such properties is known as a **buffer solution**. It is said to possess 'reserve acidity' and 'reserve alkalinity'. Buffer solutions usually consist of solutions containing a mixture of a weak acid HA and its sodium or potassium salt (A^-), or of a weak base B and its salt (BH^+). A buffer, then, is usually a mixture of an acid and its conjugate base. In order to understand buffer action, consider first the equilibrium between a weak acid and its salt. The dissociation of a weak acid is given by:

$$HA \rightleftharpoons H^+ + A^-$$

and its magnitude is controlled by the value of the dissociation constant K_a:

$$\frac{a_{H^+} \times a_{A^-}}{a_{HA}} = K_a, \quad \text{or} \quad a_{H^+} = \frac{a_{HA}}{a_{A^-}} \times K_a \tag{17}$$

The expression may be approximated by writing concentrations for activities:

$$[H^+] = \frac{[HA]}{[A^-]} \times K_a \tag{18}$$

This equilibrium applies to a mixture of an acid HA and its salt, say MA. If the concentration of the acid be c_a and that of the salt be c_s, then the concentration of the undissociated portion of the acid is ($c_a - [H^+]$). The solution is electrically neutral, hence $[A^-] = c_s + [H^+]$ (the salt is completely dissociated). Substituting these values in the equilibrium equation (18), we have:

$$[H^+] = \frac{c_a - [H^+]}{c_s + [H^+]} \times K_a \tag{19}$$

This is a quadratic equation in $[H^+]$ and may be solved in the usual manner. It can, however, be simplified by introducing the following further approximations. In a mixture of a weak acid and its salt, the dissociation of the acid is repressed by the common ion effect, and $[H^+]$ may be taken as negligibly small by comparison with c_a and c_s. Equation (19) then reduces to:

$$[H^+] = \frac{c_a}{c_s}.K_a, \quad \text{or} \quad [H^+] = \frac{[\text{Acid}]}{[\text{Salt}]} \times K_a \tag{20}$$

$$\text{or} \quad pH = pK_a + \log\frac{[\text{Salt}]}{[\text{Acid}]} \tag{21}$$

The equations can be readily expressed in a somewhat more general form when applied to a Brønsted–Lowry acid A and its conjugate base B:

$$A \rightleftharpoons H^+ + B$$

(e.g. CH_3COOH and CH_3COO^-, etc.). The expression for pH is:

$$pH = pK_a + \log\frac{[B]}{[A]}$$

where $K_a = [H^+][B]/[A]$.

Similarly for a mixture of a weak base of dissociation constant K_b and its salt with a strong acid:

$$[OH^-] = \frac{[\text{Base}]}{[\text{Salt}]} \times K_b \tag{22}$$

$$\text{or} \quad pOH = pK_b + \log\frac{[\text{Salt}]}{[\text{Base}]} \tag{23}$$

Confining attention to the case in which the concentrations of the acid and its salt are equal, i.e. of a half-neutralised acid then $pH = pK_a$. Thus the pH of a half-neutralised solution of a weak acid is equal to the negative logarithm of the dissociation constant of the acid. For acetic (ethanoic) acid, $K_a = 1.75 \times 10^{-5}$ mol L^{-1}, $pK_a = 4.76$; a half-neutralised solution of, say 0.1M acetic acid will have a pH of 4.76. If we add a small concentration of H^+ ions to such a solution, the former will combine with acetate ions to form undissociated acetic acid:

$$H^+ + CH_3COO^- \rightleftharpoons CH_3COOH$$

Similarly, if a small concentration of hydroxide ions be added, the latter will combine with the hydrogen ions arising from the dissociation of the acetic acid and form water; the equilibrium will be disturbed, and more acetic acid will dissociate to replace the hydrogen ions removed in this way. In either case, the concentration of the acetic acid and acetate ion (or salt) will not be appreciably changed. It follows from equation (21) that the pH of the solution will not be materially affected.

Example 12. Calculate the pH of the solution produced by adding 10 mL of 1M hydrochloric acid to 1 L of a solution which is 0.1M in acetic (ethanoic) acid and 0.1M in sodium acetate ($K_a = 1.75 \times 10^{-5}$ mol L^{-1}).

The pH of the acetic acid–sodium acetate buffer solution is given by the equation:

$$pH = pK_a + \log\frac{[\text{Salt}]}{[\text{Acid}]} = 4.76 + 0.0 = 4.76$$

47

The hydrogen ions from the hydrochloric acid react with acetate ions forming practically undissociated acetic acid, and neglecting the change in volume from 1000 mL to 1010 mL we can say

$CH_3COO^- = 0.1 - 0.01 = 0.09$

$CH_3COOH = 0.1 + 0.01 = 0.11$

and $pH = 4.76 + \log 0.09/0.11 = 4.76 - 0.09 = \underline{4.67}$

Thus the pH of the acetic acid–sodium acetate buffer solution is only altered by 0.09 pH unit on the addition of the hydrochloric acid. The same volume of hydrochloric acid added to 1 litre of water (pH = 7) would lead to a solution with $pH = -\log(0.01) = 2$; a change of 5 pH units. This example serves to illustrate the regulation of pH exercised by buffer solutions.

A solution containing equal concentrations of acid and its salt, or a half-neutralised solution of the acid, has the maximum 'buffer capacity'. Other mixtures also possess considerable buffer capacity, but the pH will differ slightly from that of the half-neutralised acid. Thus in a quarter-neutralised solution of acid, [Acid] = 3 [Salt]:

$pH = pK_a + \log \frac{1}{3} = pK_a + \bar{1}.52 = pK_a - 0.48$

For a three-quarter-neutralised acid, [Salt] = 3 [Acid]:

$pH = pK_a + \log 3 = pK_a + 0.48$

In general, we may state that the buffering capacity is maintained for mixtures within the range 1 acid:10 salt and 10 acid:1 salt and the approximate pH range of a weak acid buffer is:

$pH = pK_a \pm 1$

The concentration of the acid is usually of the order $0.05-0.2 \, \text{mol L}^{-1}$. Similar remarks apply to weak bases. It is clear that the greater the concentrations of acid and conjugate base in a buffer solution, the greater will be the buffer capacity. A quantitative measure of buffer capacity is given by the number of moles of strong base required to change the pH of 1 litre of the solution by 1 pH unit.

The preparation of a buffer solution of a definite pH is a simple process once the acid (or base) of appropriate dissociation constant is found: small variations in pH are obtained by variations in the ratios of the acid to the salt concentration. One example is given in Table 2.2.

Before leaving the subject of buffer solutions, it is necessary to draw attention to a possible erroneous deduction from equation (21), namely that the hydrogen-ion concentration of a buffer solution is dependent only upon the ratio of the concentrations of acid and salt and upon K_a, and not upon the actual concentrations; otherwise expressed, that the pH of such a buffer mixture should not change upon dilution with water. This is approximately although not strictly true. In deducing equation (18), concentrations have been substituted for activities, a step which is not entirely justifiable except in dilute solutions. The exact expression controlling buffer action is:

$$a_{H^+} = \frac{a_{HA}}{a_{A^-}} \times K_a = \frac{c_a \cdot y_a}{c_s \cdot y_{A^-}} \times K_a \qquad (24)$$

Table 2.2 pH of acetic acid–sodium acetate buffer mixtures

10 mL mixtures of x mL of $0.2M$ acetic acid and y mL of $0.2M$ sodium acetate

Acetic acid (x mL)	Sodium acetate (y mL)	pH
9.5	0.5	3.48
9.0	1.0	3.80
8.0	2.0	4.16
7.0	3.0	4.39
6.0	4.0	4.58
5.0	5.0	4.76
4.0	6.0	4.93
3.0	7.0	5.13
2.0	8.0	5.36
1.0	9.0	5.71
0.5	9.5	6.04

The activity coefficient y_a of the undissociated acid is approximately unity in dilute aqueous solution. Expression (24) thus becomes:

$$a_{H^+} = \frac{[\text{Acid}]}{[\text{Salt}] \times y_{A^-}} \times K_a \tag{25}$$

$$\text{or}\quad pH = pK_a + \log[\text{Salt}]/[\text{Acid}] + \log y_{A^-} \tag{26}$$

This is known as the Henderson–Hasselbalch equation.

If a buffer solution is diluted, the ionic concentrations are decreased and so, as shown in Section 2.5, the ionic activity coefficients are increased. It follows from equation (26) that the pH is increased.

Buffer mixtures are not confined to mixtures of monoprotic acids or monoacid bases and their salts. We may employ a mixture of salts of a polyprotic acid, e.g. NaH_2PO_4 and Na_2HPO_4. The salt NaH_2PO_4 is completely dissociated:

$$NaH_2PO_4 \rightleftharpoons Na^+ + H_2PO_4^-$$

The ion $H_2PO_4^-$ acts as a monoprotic acid:

$$H_2PO_4^- \rightleftharpoons H^+ + HPO_4^{2-}$$

for which K ($\equiv K_2$ for phosphoric acid) is 6.2×10^{-8} mol L^{-1}. The addition of the salt Na_2HPO_4 is analogous to the addition of, say, acetate ions to a solution of acetic acid, since the tertiary ionisation of phosphoric acid ($HPO_4^{2-} \rightleftharpoons H^+ + PO_4^{3-}$) is small ($K_3 = 5 \times 10^{-13}$ mol L^{-1}). The mixture of NaH_2PO_4 and Na_2HPO_4 is therefore an effective buffer over the range pH 7.2 ± 1.0 ($= pK \pm 1$). It will be noted that this is a mixture of a Brønsted–Lowry acid and its conjugate base.

Buffer solutions find many applications in quantitative analysis, e.g. many precipitations are quantitative only under carefully controlled conditions of pH, as are also many compleximetric titrations: numerous examples of their use will be found throughout the book.

2.21 COMPLEX IONS

The increase in solubility of a precipitate upon the addition of excess of the precipitating agent is frequently due to the formation of a complex ion. A

complex ion is formed by the union of a simple ion with either other ions of opposite charge or with neutral molecules as shown by the following examples.

When potassium cyanide solution is added to a solution of silver nitrate, a white precipitate of silver cyanide is first formed because the solubility product of silver cyanide:

$$[Ag^+] \times [CN^-] = K_{s(AgCN)} \tag{27}$$

is exceeded. The reaction is expressed:

$$CN^- + Ag^+ = AgCN$$

The precipitate dissolves upon the addition of excess of potassium cyanide, the complex ion $[Ag(CN)_2]^-$ being produced:

$$AgCN(solid) + CN^-(excess) \rightleftharpoons [Ag(CN)_2]^- {}^*$$

(or $AgCN + KCN = K[Ag(CN)_2]$ – a soluble complex salt)

This complex ion dissociates to give silver ions, since the addition of sulphide ions yields a precipitate of silver sulphide (solubility product 1.6×10^{-49} mol^3 L^{-3}), and also silver is deposited from the complex cyanide solution upon electrolysis. The complex ion thus dissociates in accordance with the equation:

$$[Ag(CN)_2]^- \rightleftharpoons Ag^+ + 2CN^-$$

Applying the Law of Mass Action, we obtain the dissociation constant of the complex ion:

$$\frac{[Ag^+] \times [CN^-]^2}{[\{Ag(CN)_2\}^-]} = K_{diss.} \tag{28}$$

which has a value of 1.0×10^{-21} mol^2 L^{-2} at the ordinary temperature. By inspection of this expression, and bearing in mind that excess of cyanide ion is present, it is evident that the silver ion concentration must be very small, so small in fact that the solubility product of silver cyanide is not exceeded.

The inverse of equation (28) gives us the stability constant or formation constant of the complex ion:

$$K = \frac{[\{Ag(CN)_2\}^-]}{[Ag^+] \times [CN^-]^2} = 10^{21} \text{ mol}^{-2} \text{ L}^2 \tag{29}$$

Consider now a somewhat different type of complex ion formation, viz. the production of a complex ion with constituents other than the common ion present in the solution. This is exemplified by the solubility of silver chloride in ammonia solution. The reaction is:

$$AgCl + 2NH_3 \rightleftharpoons [Ag(NH_3)_2]^+ + Cl^-$$

Here again, electrolysis, or treatment with hydrogen sulphide, shows that silver

---

*Square brackets are commonly used for two purposes: to denote concentrations and also to include the whole of a complex ion; for the latter purpose curly brackets (braces) are sometimes used. With careful scrutiny there should be no confusion regarding the sense in which the square brackets are used: with complexes there will be no charge signs *inside* the brackets.

ions are present in solution. The dissociation of the complex ion is represented by:

$$[Ag(NH_3)_2]^+ \rightleftharpoons Ag^+ + 2NH_3$$

and the dissociation constant is given by:

$$K_{diss.} = \frac{[Ag^+] \times [NH_3]^2}{[\{Ag(NH_3)_2\}^+]} = 6.8 \times 10^{-8} \, mol^2 \, L^{-2}$$

The stability constant $K = 1/K_{diss.} = 1.5 \times 10^7 \, mol^{-2} \, L^2$

The magnitude of the dissociation constant clearly shows that only a very small silver ion concentration is produced by the dissociation of the complex ion.

The stability of complex ions varies within very wide limits. It is quantitatively expressed by means of the **stability constant**. The more stable the complex, the greater is the stability constant, i.e. the smaller is the tendency of the complex ion to dissociate into its constituent ions. When the complex ion is very stable, e.g. the hexacyanoferrate(II) ion $[Fe(CN)_6]^{4-}$, the ordinary ionic reactions of the components are not shown.

The application of complex-ion formation in chemical separations depends upon the fact that one component may be transformed into a complex ion which no longer reacts with a given reagent, whereas another component does react. One example may be mentioned here. This is concerned with the separation of cadmium and copper. Excess of potassium cyanide solution is added to the solution containing the two salts when the complex ions $[Cd(CN)_4]^{2-}$ and $[Cu(CN)_4]^{3-}$ respectively are formed. Upon passing hydrogen sulphide into the solution containing excess of CN^- ions, a precipitate of cadmium sulphide is produced. Despite the higher solubility product of CdS ($1.4 \times 10^{-28} \, mol^2 \, L^{-2}$ as against $6.5 \times 10^{-45} \, mol^2 \, L^{-2}$ for copper(II) sulphide), the former is precipitated because the complex cyanocuprate(I) ion has a greater stability constant ($2 \times 10^{27} \, mol^{-4} \, L^4$ as compared with $7 \times 10^{10} \, mol^{-4} \, L^4$ for the cadmium compound).

2.22 COMPLEXATION

The processes of complex-ion formation referred to above can be described by the general term **complexation**. A complexation reaction with a metal ion involves the replacement of one or more of the coordinated solvent molecules by other nucleophilic groups. The groups bound to the central ion are called **ligands** and in aqueous solution the reaction can be represented by the equation:

$$M(H_2O)_n + L = M(H_2O)_{(n-1)}L + H_2O$$

Here the ligand (L) can be either a neutral molecule or a charged ion, and successive replacement of water molecules by other ligand groups can occur until the complex ML_n is formed; n is the coordination number of the metal ion and represents the maximum number of monodentate ligands that can be bound to it.

Ligands may be conveniently classified on the basis of the number of points of attachment to the metal ion. Thus simple ligands, such as halide ions or the molecules H_2O or NH_3, are **monodentate**, i.e. the ligand is bound to the metal ion at only one point by the donation of a lone pair of electrons to the metal.

When, however, the ligand molecule or ion has two atoms, each of which has a lone pair of electrons, then the molecule has two donor atoms and it may be possible to form two coordinate bonds with the same metal ion; such a ligand is said to be **bidentate** and may be exemplified by consideration of the tris(ethylenediamine)cobalt(III) complex, $[Co(en)_3]^{3+}$. In this six-coordinate octahedral complex of cobalt(III), each of the bidentate ethylenediamine* molecules is bound to the metal ion through the lone pair electrons of the two nitrogen atoms. This results in the formation of three five-membered rings, each including the metal ion; the process of ring formation is called **chelation**.

Multidentate ligands contain more than two coordinating atoms per molecule, e.g. 1,2-diaminoethanetetra-acetic acid (ethylenediaminetetra-acetic acid, EDTA),† which has two donor nitrogen atoms and four donor oxygen atoms in the molecule, can be hexadentate.

In the foregoing it has been assumed that the complex species does not contain more than one metal ion, but under appropriate conditions a binuclear complex, i.e. one containing two metal ions, or even a polynuclear complex, containing more than two metal ions may be formed. Thus interaction between Zn^{2+} and Cl^- ions may result in the formation of binuclear complexes, e.g. $[Zn_2Cl_6]^{2-}$, in addition to simple species such as $ZnCl_3^-$ and $ZnCl_4^{2-}$. The formation of bi- and poly-nuclear complexes will clearly be favoured by a high concentration of the metal ion; if the latter is present as a trace constituent of a solution, polynuclear complexes are unlikely to be formed.

2.23 STABILITY OF COMPLEXES

The thermodynamic stability of a species is a measure of the extent to which this species will be formed from other species under certain conditions, provided that the system is allowed to reach equilibrium. Consider a metal ion M in solution together with a monodentate ligand L, then the system may be described by the following stepwise equilibria, in which, for convenience, coordinated water molecules are not shown:

$M + L \rightleftharpoons ML$; $K_1 = [ML]/[M][L]$

$ML + L \rightleftharpoons ML_2$; $K_2 = [ML_2]/[ML][L]$

$ML_{(n-1)} + L \rightleftharpoons ML_n$; $K_n = [ML_n]/[ML_{(n-1)}][L]$

The equilibrium constants $K_1, K_2, \ldots, K_n$ are referred to as **stepwise stability constants**.

An alternative way of expressing the equilibria is as follows:

$M + L \rightleftharpoons ML$; $\beta_1 = [ML]/[M][L]$

$M + 2L \rightleftharpoons ML_2$; $\beta_2 = [ML_2]/[M][L]^2$

$M + nL \rightleftharpoons ML_n$; $\beta_n = [ML_n]/[M][L]^n$

The equilibrium constants $\beta_1, \beta_2, \ldots, \beta_n$ are called the **overall stability constants** and are related to the stepwise stability constants by the general expression

$\beta_n = K_1 \times K_2 \times \ldots K_n$

---

* Ethane-1,2-diamine.
† 1,2-Bis[bis(carboxymethyl)amino]ethane.

In the above equilibria it has been assumed that no insoluble products are formed nor any polynuclear species.

A knowledge of stability constant values is of considerable importance in analytical chemistry, since they provide information about the concentrations of the various complexes formed by a metal in specified equilibrium mixtures; this is invaluable in the study of complexometry, and of various analytical separation procedures such as solvent extraction, ion exchange, and chromatography.[2,3]

2.24 METAL ION BUFFERS

Consider the equation for complex formation

$$M + L \rightleftharpoons ML; \quad K = [ML]/[M][L]$$

and assume that ML is the only complex to be formed by the particular system. The equilibrium constant expression can be rearranged to give:

$$[M] = 1/K \times [ML]/[L]$$

$$\log[M] = \log 1/K + \log[ML]/[L]$$

$$pM = \log K - \log[ML]/[L]$$

This shows that the pM value of the solution is fixed by the value of K and the ratio of complex-ion concentration to that of the free ligand. If more of M is added to the solution, more complex will be formed and the value of pM will not change appreciably. Likewise, if M is removed from the solution by some reaction, some of the complex will dissociate to restore the value of pM. This recalls the behaviour of buffer solutions encountered with acids and bases (Section 2.20), and by analogy, the complex–ligand system may be termed a **metal ion buffer**.

2.25 FACTORS INFLUENCING THE STABILITY OF COMPLEXES

The stability of a complex will obviously be related to (a) the complexing ability of the metal ion involved, and (b) characteristics of the ligand, and it is important to examine these factors briefly.

(**a**) **Complexing ability of metals.** The relative complexing ability of metals is conveniently described in terms of the **Schwarzenbach classification**, which is broadly based upon the division of metals into Class A and Class B Lewis acids, i.e. electron acceptors. Class A metals are distinguished by an order of affinity (in aqueous solution) towards the halogens $F^- \gg Cl^- > Br^- > I^-$, and form their most stable complexes with the first member of each group of donor atoms in the Periodic Table (i.e. nitrogen, oxygen and fluorine). Class B metals coordinate much more readily with I^- than with F^- in aqueous solution, and form their most stable complexes with the second (or heavier) donor atom from each group (i.e. P, S, Cl). The Schwarzenbach classification defines three categories of metal ion acceptors:

1. Cations with noble gas configurations. The alkali metals, alkaline earths and aluminium belong to this group which exhibit Class A acceptor properties. Electrostatic forces predominate in complex formation, so interactions

53

between small ions of high charge are particularly strong and lead to stable complexes.

2. Cations with completely filled d sub-shells. Typical of this group are copper(I), silver(I) and gold(I) which exhibit Class B acceptor properties. These ions have high polarising power and the bonds formed in their complexes have appreciable covalent character.

3. Transition metal ions with incomplete d sub-shells. In this group both Class A and Class B tendencies can be distinguished. The elements with Class B characteristics form a roughly triangular group within the Periodic Table, with the apex at copper and the base extending from rhenium to bismuth. To the left of this group, elements in their higher oxidation states tend to exhibit Class A properties, while to the right of the group, the higher oxidation states of a given element have a greater Class B character.

The concept of '**hard**' and '**soft**' **acids and bases** is useful in characterising the behaviour of Class A and Class B acceptors. A soft base may be defined as one in which the donor atom is of high polarisability and of low electronegativity, is easily oxidised, or is associated with vacant, low-lying orbitals. These terms describe, in different ways, a base in which the donor atom electrons are not tightly held, but are easily distorted or removed. Hard bases have the opposite properties, i.e. the donor atom is of low polarisability and high electronegativity, is difficult to reduce, and is associated with vacant orbitals of high energy which are inaccessible.

On this basis, it is seen that Class A acceptors prefer to bind to hard bases, e.g. with nitrogen, oxygen and fluorine donor atoms, whilst Class B acceptors prefer to bind to the softer bases, e.g. P, As, S, Se, Cl, Br, I donor atoms. Examination of the Class A acceptors shows them to have the following distinguishing features; small size, high positive oxidation state, and the absence of outer electrons which are easily excited to higher states. These are all factors which lead to low polarisability, and such acceptors are called hard acids. Class B acceptors, however, have one or more of the following properties: low positive or zero oxidation state, large size, and several easily excited outer electrons (for metals these are the d electrons). These are all factors which lead to high polarisability, and Class B acids may be called soft acids.

A general principle may now be stated which permits correlation of the complexing ability of metals: 'Hard acids tend to associate with hard bases and soft acids with soft bases'. This statement must not, however, be regarded as exclusive, i.e. under appropriate conditions soft acids may complex with hard bases or hard acids with soft bases.

(*b*) **Characteristics of the ligand.** Among the characteristics of the ligand which are generally recognised as influencing the stability of complexes in which it is involved are (i) the basic strength of the ligand, (ii) its chelating properties (if any), and (iii) steric effects. From the point of view of the analytical applications of complexes, the chelating effect is of paramount importance and therefore merits particular attention.

The term **chelate effect** refers to the fact that a chelated complex, i.e. one formed by a bidentate or a multidenate ligand, is more stable than the *corresponding* complex with monodentate ligands: the greater the number of points of attachment of ligand to the metal ion, the greater the stability of the complex. Thus the complexes formed by the nickel(II) ion with (a) the monodentate NH_3 molecule, (b) the bidentate ethylenediamine (1,2-diamino-ethane), and (c) the hexadentate ligand 'penten' $\{(H_2N \cdot CH_2 \cdot CH_2)_2N \cdot CH_2 \cdot CH_2 \cdot N(CH_2 \cdot CH_2 \cdot NH_2)_2\}$ show an overall stability constant value for the ammonia complex of 3.1×10^8, which is increased by a factor of about 10^{10} for the complex of ligand (b), and is approximately ten times greater still for the third complex.

The most common steric effect is that of inhibition of complex formation owing to the presence of a large group either attached to, or in close proximity to, the donor atom.

A further factor which must also be taken into consideration from the point of view of the analytical applications of complexes and of complex-formation reactions is the rate of reaction: to be analytically useful it is usually required that the reaction be rapid. An important classification of complexes is based upon the rate at which they undergo substitution reactions, and leads to the two groups of **labile** and **inert** complexes. The term labile complex is applied to those cases where nucleophilic substitution is complete within the time required for mixing the reagents. Thus, for example, when excess of aqueous ammonia is added to an aqueous solution of copper(II) sulphate, the change in colour from pale to deep blue is instantaneous; the rapid replacement of water molecules by ammonia indicates that the Cu(II) ion forms kinetically labile complexes. The term inert is applied to those complexes which undergo slow substitution reactions, i.e. reactions with half-times of the order of hours or even days at room temperature. Thus the Cr(III) ion forms kinetically inert complexes, so that the replacement of water molecules coordinated to Cr(III) by other ligands is a very slow process at room temperature.

Kinetic inertness or lability is influenced by many factors, but the following general observations form a convenient guide to the behaviour of the complexes of various elements.

(i)   Main group elements usually form labile complexes.

(ii) With the exception of Cr(III) and Co(III), most first-row transition elements form labile complexes.

(iii) Second- and third-row transition elements tend to form inert complexes.

For a full discussion of the topics introduced in this section a textbook of inorganic chemistry (e.g. Ref. 4) or one dealing with complexes (e.g. Ref. 2), should be consulted.

2.26 COMPLEXONES

The formation of a single complex species rather than the stepwise production of such species will clearly simplify complexometric titrations and facilitate the detection of end points. Schwarzenbach[2] realised that the acetate ion is able to form acetato complexes of low stability with nearly all polyvalent cations, and that if this property could be reinforced by the chelate effect, then much stronger complexes would be formed by most metal cations. He found that the aminopolycarboxylic acids are excellent complexing agents: the most important of these is 1,2-diaminoethanetetra-acetic acid (ethylenediaminetetra-acetic acid). The formula (I) is preferred to (II), since it has been shown from measurements of the dissociation constants that two hydrogen atoms are probably held in the form of zwitterions. The values of pK are respectively $pK_1 = 2.0$, $pK_2 = 2.7$,

$pK_3 = 6.2$, and $pK_4 = 10.3$ at $20\,^\circ$C; these values suggest that it behaves as a dicarboxylic acid with two strongly acidic groups and that there are two ammonium protons of which the first ionises in the pH region of about 6.3 and the second at a pH of about 11.5. Various trivial names are used for *ethylenediaminetetra-acetic* acid and its sodium salts, and these include Trilon B, Complexone III, Sequestrene, Versene, and Chelaton 3; the disodium salt is most widely employed in titrimetric analysis. To avoid the constant use of the long name, the abbreviation EDTA is utilised for the disodium salt.

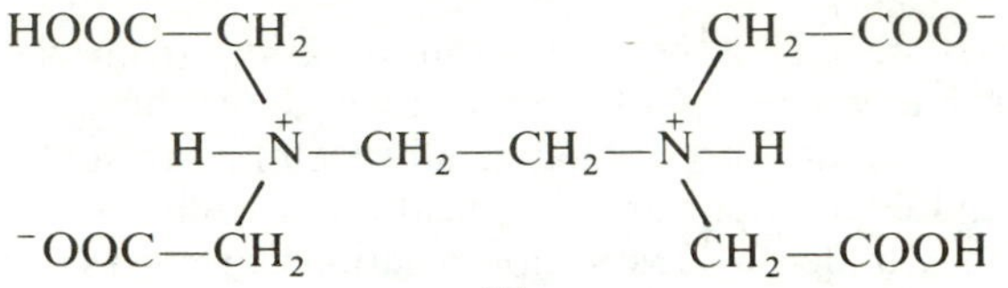

$$
\begin{array}{ccc}
\text{HOOC}-\text{CH}_2 & & \text{CH}_2-\text{COO}^- \\[2pt]
\qquad\searrow & & \nearrow \\[2pt]
\text{H}-\overset{+}{\text{N}}-\text{CH}_2-\text{CH}_2-\overset{+}{\text{N}}-\text{H} \\[2pt]
\nearrow & & \searrow \\[2pt]
{}^-\text{OOC}-\text{CH}_2 & & \text{CH}_2-\text{COOH}
\end{array}
$$

(I)

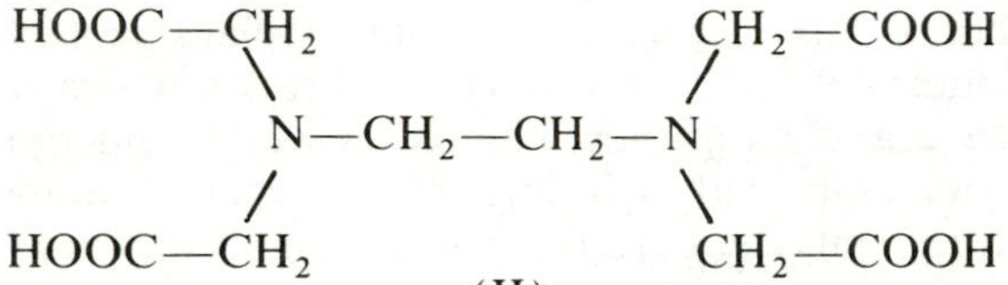

$$
\begin{array}{ccc}
\text{HOOC}-\text{CH}_2 & & \text{CH}_2-\text{COOH} \\[2pt]
\qquad\searrow & & \nearrow \\[2pt]
\text{N}-\text{CH}_2-\text{CH}_2-\text{N} \\[2pt]
\nearrow & & \searrow \\[2pt]
\text{HOOC}-\text{CH}_2 & & \text{CH}_2-\text{COOH}
\end{array}
$$

(II)

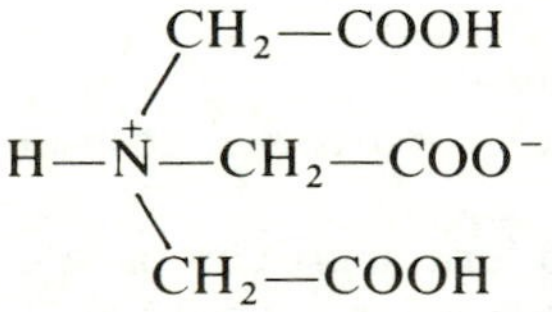

$$
\begin{array}{c}
\quad\;\text{CH}_2-\text{COOH} \\[2pt]
\nearrow \\[2pt]
\text{H}-\overset{+}{\text{N}}-\text{CH}_2-\text{COO}^- \\[2pt]
\searrow \\[2pt]
\quad\;\text{CH}_2-\text{COOH}
\end{array}
$$

(III)

$$
\begin{array}{c}
\qquad\text{CH}_2-\text{COOH} \\
\quad\text{CH}_2\;\;\text{N} \\
\text{H}_2\text{C}\quad\text{CH}\;\;\text{CH}_2-\text{COOH} \\
\text{H}_2\text{C}\quad\text{CH}\;\;\text{CH}_2-\text{COOH} \\
\quad\text{CH}_2\;\;\text{N} \\
\qquad\text{CH}_2-\text{COOH}
\end{array}
$$

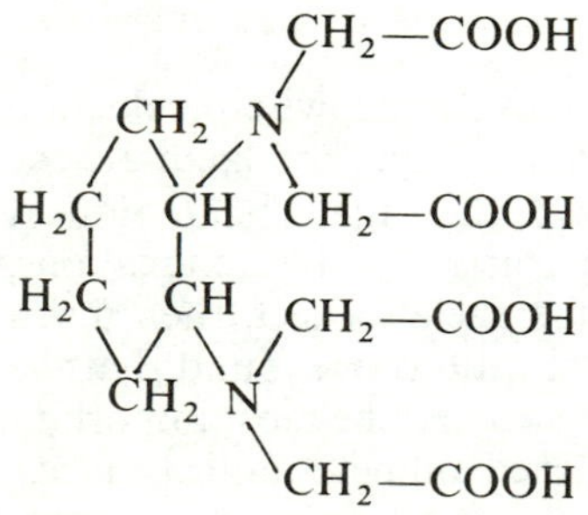

(IV)

$$CH_2COO^-$$

$$\overset{+}{H}N$$

$$(CH_2)_2 \quad CH_2COOH$$

$$O$$

$$(CH_2)_2$$

$$O$$

$$(CH_2)_2 \quad CH_2COOH$$

$$\overset{+}{H}N$$

$$CH_2COO^-$$

$$CH_2COO^-$$

$$\overset{+}{H}N$$

$$(CH_2)_2 \quad CH_2COOH$$

$$CH_2COO^-$$

$$\overset{+}{H}N$$

$$(CH_2)_2 \quad CH_2COO^-$$

$$\overset{+}{H}N$$

$$(CH_2)_2 \quad CH_2COOH$$

$$\overset{+}{H}N$$

$$CH_2COO^-$$

(V) EGTA   (VI) TTHA

Other complexing agents (complexones) which are sometimes used include (a) nitrilotriacetic acid (III) (NITA or NTA or Complexone I; this has $pK_1 = 1.9$, $pK_2 = 2.5$, and $pK_3 = 9.7$), (b) trans-1,2-diaminocyclohexane-N,N,N',N'-tetra-acetic acid (IV): this should presumably be formulated as a zwitterion structure like (I); the abbreviated name is CDTA, DCyTA, DCTA or Complexone IV, (c) 2,2'-ethylenedioxybis{ethyliminodi(acetic acid)} (V) also known as ethylene glycolbis(2-aminoethyl ether)N,N,N',N'-tetra-acetic acid (EGTA), and (d) triethylenetetramine-N,N,N',N'',N''',N'''-hexa-acetic acid (TTHA) (VI). CDTA often forms stronger metal complexes than does EDTA and thus finds applications in analysis, but the metal complexes are formed rather more slowly than with EDTA so that the end-point of the titration tends to be drawn out with the former reagent. EGTA finds analytical application mainly in the determination of calcium in a mixture of calcium and magnesium and is probably superior to EDTA in the calcium/magnesium water-hardness titration (Section 10.61) TTHA forms 1:2 complexes with many trivalent cations and with some divalent metals, and can be used for determining the components of mixtures of certain ions without the use of masking agents (see Section 10.47).

However, EDTA has the widest general application in analysis because of its powerful complexing action and commercial availability. The spatial structure of its anion, which has six donor atoms, enables it to satisfy the coordination number of six frequently encountered among the metal ions and to form strainless five-membered rings on chelation. The resulting complexes have similar structures but differ from one another in the charge they carry.

To simplify the following discussion EDTA is assigned the formula H_4Y: the disodium salt is therefore Na_2H_2Y and affords the complex-forming ion H_2Y^{2-} in aqueous solution; it reacts with all metals in a 1:1 ratio. The reactions with cations, e.g. M^{2+}, may be written as:

$$M^{2+} + H_2Y^{2-} \rightleftharpoons MY^{2-} + 2H^+ \qquad (l)$$

57

For other cations, the reactions may be expressed as:

$$M^{3+} + H_2Y^{2-} \rightleftharpoons MY^- + 2H^+ \qquad\qquad (m)$$

$$M^{4+} + H_2Y^{2-} \rightleftharpoons MY + 2H^+ \qquad\qquad (n)$$

or $\quad M^{n+} + H_2Y^{2-} \rightleftharpoons (MY)^{(n-4)+} + 2H^+ \qquad\qquad (o)$

One mole of the complex-forming H_2Y^{2-} reacts in all cases with one mole of the metal ion and in each case, also, two moles of hydrogen ion are formed. It is apparent from equation (o) that the dissociation of the complex will be governed by the pH of the solution; lowering the pH will decrease the stability of the metal–EDTA complex. The more stable the complex, the lower the pH at which an EDTA titration of the metal ion in question may be carried out. Table 2.3 indicates minimum pH values for the existence of EDTA complexes of some selected metals.

Table 2.3 Stability with respect to pH of some metal–EDTA complexes

Minimum pH at which complexes exist	Selected metals
1–3	Zr^{4+}; Hf^{4+}; Th^{4+}; Bi^{3+}; Fe^{3+}
4–6	Pb^{2+}; Cu^{2+}; Zn^{2+}; Co^{2+}; Ni^{2+}; Mn^{2+}; Fe^{2+}; Al^{3+}; Cd^{2+}; Sn^{2+}
8–10	Ca^{2+}; Sr^{2+}; Ba^{2+}; Mg^{2+}

It is thus seen that, in general, EDTA complexes with metal ions of the charge number 2 are stable in alkaline or slightly acidic solution, whilst complexes with ions of charge numbers 3 or 4 may exist in solutions of much higher acidity.

2.27 STABILITY CONSTANTS OF EDTA COMPLEXES

The stability of a complex is characterised by the stability constant (or formation constant) K:

$$M^{n+} + Y^{4-} \rightleftharpoons (MY)^{(n-4)+} \qquad\qquad (p)$$

$$K = [(MY)^{(n-4)+}]/[M^{n+}][Y^{4-}] \qquad\qquad (q)$$

Some values for the stability constants (expressed as log K) of metal–EDTA complexes are collected in Table 2.4: these apply to a medium of ionic strength $I = 0.1$ at 20 °C.

Table 2.4 Stability constants (as log K) of metal–EDTA complexes

Mg^{2+}	8.7	Zn^{2+}	16.7	La^{3+}	15.7
Ca^{2+}	10.7	Cd^{2+}	16.6	Lu^{3+}	20.0
Sr^{2+}	8.6	Hg^{2+}	21.9	Sc^{3+}	23.1
Ba^{2+}	7.8	Pb^{2+}	18.0	Ga^{3+}	20.5
Mn^{2+}	13.8	Al^{3+}	16.3	In^{3+}	24.9
Fe^{2+}	14.3	Fe^{3+}	25.1	Th^{4+}	23.2
Co^{2+}	16.3	Y^{3+}	18.2	Ag^+	7.3
Ni^{2+}	18.6	Cr^{3+}	24.0	Li^+	2.8
Cu^{2+}	18.8	Ce^{3+}	15.9	Na^+	1.7

In equation (q) only the fully ionised form of EDTA, i.e. the ion Y^{4-}, has been taken into account, but at low pH values the species HY^{3-}, H_2Y^{2-}, H_3Y^- and even undissociated H_4Y may well be present; in other words, only a part of the EDTA uncombined with metal may be present as Y^{4-}. Further, in equation (q) the metal ion M^{n+} is assumed to be uncomplexed, i.e. in aqueous solution it is simply present as the hydrated ion. If, however, the solution also contains substances other than EDTA which can complex with the metal ion, then the whole of this ion uncombined with EDTA may no longer be present as the simple hydrated ion. Thus, in practice, the stability of metal–EDTA complexes may be altered (a) by variation in pH and (b) by the presence of other complexing agents. The stability constant of the EDTA complex will then be different from the value recorded for a specified pH in pure aqueous solution; the value recorded for the new conditions is termed the **'apparent'** or **'conditional' stability constant**. It is clearly necessary to examine the effect of these two factors in some detail.

(a) pH effect. The apparent stability constant at a given pH may be calculated from the ratio K/α, where α is the ratio of the total uncombined EDTA (in all forms) to the form Y^{4-}. Thus K_H, the apparent stability constant for the metal–EDTA complex at a given pH, can be calculated from the expression

$$\log K_H = \log K - \log \alpha \tag{30}$$

The factor α can be calculated from the known dissociation constants of EDTA, and since the proportions of the various ionic species derived from EDTA will be dependent upon the pH of the solution, α will also vary with pH; a plot of $\log \alpha$ against pH shows a variation of $\log \alpha = 18$ at pH $= 1$ to $\log \alpha = 0$ at pH $= 12$: such a curve is very useful for dealing with calculations of apparent stability constants. Thus, for example, from Table 2.4, $\log K$ of the EDTA complex of the Pb^{2+} ion is 18.0 and from a graph of $\log \alpha$ against pH, it is found that at a pH of 5.0, $\log \alpha = 7$. Hence from equation (30), at a pH of 5.0 the lead–EDTA complex has an apparent stability constant given by:

$$\log K_H = 18.0 - 7.0 = 11.0$$

Carrying out a similar calculation for the EDTA complex of the Mg^{2+} ion ($\log K = 8.7$), for the same pH (5.0), it is found:

$$\log K_H(\text{Mg(II)} - \text{EDTA}) = 8.7 - 7.0 = 1.7$$

These results imply that at the specified pH the magnesium complex is appreciably dissociated, whereas the lead complex is stable, and clearly titration of an Mg(II) solution with EDTA at this pH will be unsatisfactory, but titration of the lead solution under the same conditions will be quite feasible. In practice, for a metal ion to be titrated with EDTA at a stipulated pH the value of $\log K_H$ should be greater than 8 when a metallochromic indicator is used.

As indicated by the data quoted in the previous section, the value of $\log \alpha$ is small at high pH values, and it therefore follows that the larger values of $\log K_H$ are found with increasing pH. However, by increasing the pH of the solution the tendency to form slightly soluble metallic hydroxides is enhanced owing to the reaction:

$$(\text{MY})^{(n-4)+} + n\text{OH}^- \rightleftharpoons \text{M(OH)}_n + Y^{4-}$$

The extent of hydrolysis of $(MY)^{(n-4)+}$ depends upon the characteristics of the metal ion, and is largely controlled by the solubility product of the metallic hydroxide and, of course, the stability constant of the complex. Thus iron(III) is precipitated as hydroxide ($K_{sol} = 1 \times 10^{-36}$) in basic solution, but nickel(II), for which the relevant solubility product is 6.5×10^{-18}, remains complexed. Clearly the use of excess EDTA will tend to reduce the effect of hydrolysis in basic solutions. It follows that for each metal ion there exists an optimum pH which will give rise to a maximum value for the apparent stability constant.

(*b*) **The effect of other complexing agents.** If another complexing agent (say NH_3) is also present in the solution, then in equation (*q*) [M^{n+}] will be reduced owing to complexation of the metal ions with ammonia molecules. It is convenient to indicate this reduction in effective concentration by introducing a factor β, defined as the ratio of the sum of the concentrations of all forms of the metal ion not complexed with EDTA to the concentration of the simple (hydrated) ion. The apparent stability constant of the metal–EDTA complex, taking into account the effects of both pH and the presence of other complexing agents, is then given by:

$$\log K_{HZ} = \log K - \log \alpha - \log \beta. \tag{31}$$

2.28 ELECTRODE POTENTIALS

When a metal is immersed in a solution containing its own ions, say, zinc in zinc sulphate solution, a potential difference is established between the metal and the solution. The potential difference E for an electrode reaction

$$M^{n+} + ne \rightleftharpoons M$$

is given by the expression:

$$E = E^{\ominus} + \frac{RT}{nF} \ln a_{M^{n+}} \tag{32}$$

where R is the gas constant, T is the absolute temperature, F the Faraday constant, n the charge number of the ions, $a_{M^{n+}}$ the activity of the ions in the solution, and $E^{\ominus}$ is a constant dependent upon the metal. Equation (32) can be simplified by introducing the known values of R and F, and converting natural logarithms to base 10 by multiplying by 2.3026; it then becomes:

$$E = E^{\ominus} + \frac{0.0001984T}{n} \log a_{M^{n+}}$$

For a temperature of 25 °C ($T = 298$ K):

$$E = E^{\ominus} + \frac{0.0591}{n} \log a_{M^{n+}} \tag{33}$$

For many purposes in quantitative analysis, it is sufficiently accurate to replace $a_{M^{n+}}$ by $c_{M^{n+}}$, the ion concentration (in moles per litre):

$$E = E^{\ominus} + \frac{0.0591}{n} \log c_{M^{n+}} \tag{34}$$

The latter is a form of the **Nernst equation**.

If in equation (33), $a_{M^{n+}}$ is put equal to unity, E is equal to $E^\ominus$. $E^\ominus$ is called the **standard electrode potential** of the metal; both E and $E^\ominus$ are expressed in volts.

In order to determine the potential difference between an electrode and a solution, it is necessary to have another electrode and solution of accurately known potential difference. The two electrodes can then be combined to form a voltaic cell, the e.m.f. of which can be directly measured. The e.m.f. of the cell is the difference of the electrode potentials at zero current; the value of the unknown potential can then be calculated. The primary reference electrode is the **normal** or **standard hydrogen electrode** (see also Section 15.2). This consists of a piece of platinum foil, coated electrolytically with platinum black, and immersed in a solution of hydrochloric acid containing hydrogen ions at unit activity. (This corresponds to $1.18 M$ hydrochloric acid at $25\,^\circ$C.) Hydrogen gas at a pressure of one atmosphere is passed over the platinum foil through the side tube C (Fig. 2.2) and escapes through the small holes B in the surrounding glass tube A. Because of the periodic formation of bubbles, the level of the liquid inside the tube fluctuates, and a part of the foil is alternately exposed to the solution and to hydrogen. The lower end of the foil is continuously immersed in the solution to avoid interruption of the electric current. Connection between the platinum foil and an external circuit is made with mercury in D. The platinum black has the property of adsorbing large quantities of atomic hydrogen, and it permits the change from the gaseous to the ionic form and the reverse process to occur without hindrance; it therefore behaves as though it were composed entirely of hydrogen, that is, as a hydrogen electrode. Under fixed conditions, viz. hydrogen gas at atmospheric pressure and unit activity of hydrogen ions in the solution in contact with the electrode, the hydrogen electrode possesses a definite potential. By convention, the potential of the standard hydrogen electrode is equal to zero at all temperatures. Upon connecting the standard hydrogen electrode with a metal electrode consisting of a metal in contact with a solution of its ions of unit activity and measuring the cell e.m.f. the **standard electrode potential** of the metal may be determined. The cell is usually written as

$$\text{Pt}, \text{H}_2 | \text{H}^+(a = 1) \, \| \, M^{n+}(a = 1) | M$$

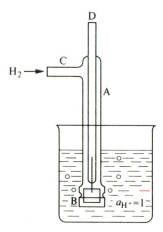

Fig. 2.2

In this scheme, a single vertical line represents a metal–electrolyte boundary at which a potential difference is taken into account: the double vertical broken lines represent a liquid junction at which the potential is to be disregarded or is considered to be eliminated by a salt bridge.

When reference is made to the electrode potential of a zinc electrode, it is the e.m.f. of the cell:

$$Pt,H_2 | H^+(a = 1) \| Zn^{2+} | Zn$$

or the e.m.f. of the half-cell $Zn^{2+} | Zn$ which is meant. The cell reaction is:

$$H_2 + Zn^{2+} \rightarrow 2H^+(a = 1) + Zn$$

and the half-cell reaction is written as:

$$Zn^{2+} + 2e \rightleftharpoons Zn$$

The electrode potential of the $Fe^{3+}, Fe^{2+} | Pt$ electrode is the e.m.f. of the cell:

$$Pt,H_2 | H^+(a = 1) \| Fe^{3+}, Fe^{2+} | Pt$$

or the e.m.f. of the half-cell $Fe^{3+}, Fe^{2+} | Pt$. The cell reaction is:

$$\tfrac{1}{2}H_2 + Fe^{3+} \rightarrow H^+(a = 1) + Fe^{2+}$$

and the half-cell reaction is written:

$$Fe^{3+} + e \rightleftharpoons Fe^{2+}$$

The convention is adopted of writing all half-cell reactions as reductions:

$$M^{n+} + ne \rightarrow M$$

e.g. $Zn^{2+} + 2e \rightarrow Zn;$ $E^{\ominus} = -0.76$ volt

When the activity of the ion M^{n+} is equal to unity (approximately true for a $1M$ solution), the electrode potential E is equal to the standard potential $E^{\ominus}$. Some important standard electrode potentials referred to the standard hydrogen electrode at $25\,^{\circ}C$ (in aqueous solution) are collected in Table 2.5.[5]

Table 2.5 Standard electrode potentials at $25\,^{\circ}C$

Electrode reaction	$E^{\ominus}$ (volts)	Electrode reaction	$E^{\ominus}$ (volts)
$Li^+ + e = Li$	-3.045	$Tl^+ + e = Tl$	-0.336
$K^+ + e = K$	-2.925	$Co^{2+} + 2e = Co$	-0.277
$Ba^{2+} + 2e = Ba$	-2.90	$Ni^{2+} + 2e = Ni$	-0.25
$Sr^{2+} + 2e = Sr$	-2.89	$Sn^{2+} + 2e = Sn$	-0.136
$Ca^{2+} + 2e = Ca$	-2.87	$Pb^{2+} + 2e = Pb$	-0.126
$Na^+ + e = Na$	-2.714	$2H^+ + 2e = H_2$	0.000
$Mg^{2+} + 2e = Mg$	-2.37	$Cu^{2+} + 2e = Cu$	$+0.337$
$Al^{3+} + 3e = Al$	-1.66	$Hg^{2+} + 2e = Hg$	$+0.789$
$Mn^{2+} + 2e = Mn$	-1.18	$Ag^+ + e = Ag$	$+0.799$
$Zn^{2+} + 2e = Zn$	-0.763	$Pd^{2+} + 2e = Pd$	$+0.987$
$Fe^{2+} + 2e = Fe$	-0.440	$Pt^{2+} + 2e = Pt$	$+1.2$
$Cd^{2+} + 2e = Cd$	-0.403	$Au^{3+} + 3e = Au$	$+1.50$

It may be noted that the standard hydrogen electrode is rather difficult to manipulate. In practice, electrode potentials on the hydrogen scale are usually determined indirectly by measuring the e.m.f. of a cell formed from the electrode in question and a convenient reference electrode whose potential with respect to the hydrogen electrode is accurately known. The reference electrodes generally used are the calomel electrode and the silver–silver chloride electrode (see Sections 15.3–4).

When metals are arranged in the order of their standard electrode potentials, the so-called **electrochemical series** of the metals is obtained. The greater the negative value of the potential, the greater is the tendency of the metal to pass into the ionic state. A metal will normally displace any other metal below it in the series from solutions of its salts. Thus magnesium, aluminium, zinc, or iron will displace copper from solutions of its salts; lead will displace copper, mercury, or silver; copper will displace silver.

The standard electrode potential is a quantitative measure of the readiness of the element to lose electrons. It is therefore a measure of the strength of the element as a reducing agent in aqueous solution; the more negative the potential of the element, the more powerful is its action as a reductant.

It must be emphasised that standard electrode potential values relate to an *equilibrium* condition between the metal electrode and the solution. Potentials determined under, or calculated for, such conditions are often referred to as 'reversible electrode potentials', and it must be remembered that the Nernst equation is only strictly applicable under such conditions.

2.29 CONCENTRATION CELLS

An electrode potential varies with the concentration of the ions in the solution. Hence two electrodes of the same metal, but immersed in solutions containing different concentrations of its ions, may form a cell. Such a cell is termed a **concentration cell**. The e.m.f. of the cell will be the algebraic difference of the two potentials, if a salt bridge be inserted to eliminate the liquid–liquid junction potential. It may be calculated as follows. At $25\,°C$:

$$E = \frac{0.0591}{n} \log c_1 + E^\ominus - \left(\frac{0.0591}{n} \log c_2 + E^\ominus \right)$$

$$= \frac{0.0591}{n} \log \frac{c_1}{c_2}, \quad \text{where } c_1 > c_2$$

As an example consider the cell:

$$\overset{-}{\text{Ag}} \left| \begin{array}{c} \text{AgNO}_3 \text{ aq.} \\ [\text{Ag}^+] = 0.00475\,M \end{array} \right\| \begin{array}{c} \text{AgNO}_3 \text{ aq.} \\ [\text{Ag}^+] = 0.043\,M \end{array} \left| \overset{+}{\text{Ag}} \right.$$

$$\overset{\leftarrow}{E_2} \qquad\qquad\qquad\qquad\qquad\qquad \overset{\rightarrow}{E_1}$$

Assuming that there is no potential difference at the liquid junction:

$$E = E_1 - E_2 = \frac{0.0591}{1} \log \frac{0.043}{0.004\,75} = 0.056 \text{ volt}$$

63

2.30 CALCULATION OF THE e.m.f. OF A VOLTAIC CELL

An interesting application of electrode potentials is to the calculation of the e.m.f. of a voltaic cell. One of the simplest of galvanic cells is the Daniell cell. It consists of a rod of zinc dipping into zinc sulphate solution and a strip of copper in copper sulphate solution; the two solutions are generally separated by placing one inside a porous pot and the other in the surrounding vessel. The cell may be represented as:

$$Zn\,|\,ZnSO_4\,aq.\,\|\,CuSO_4\,aq.\,|\,Cu$$

At the zinc electrode, zinc ions pass into solution, leaving an equivalent negative charge on the metal. Copper ions are deposited at the copper electrode, rendering it positively charged. By completing the external circuit, the current (electrons) passes from the zinc to the copper. The chemical reactions in the cell are as follows:

(*a*) zinc electrode: $Zn \rightleftharpoons Zn^{2+} + 2e$;
(*b*) copper electrode: $Cu^{2+} + 2e \rightleftharpoons Cu$.

The net chemical reaction is:

$$Zn + Cu^{2+} = Zn^{2+} + Cu$$

The potential difference at each electrode may be calculated by the formula given above, and the e.m.f. of the cell is the algebraic difference of the two potentials, the correct sign being applied to each.

As an example we may calculate the e.m.f. of the Daniell cell with molar concentrations of zinc ions and copper(II) ions:

$$E = E^{\ominus}_{(Cu)} - E^{\ominus}_{(Zn)} = +0.34 - (-0.76) = 1.10 \text{ volts}$$

The small potential difference produced at the contact between the two solutions (the so-called liquid–junction potential) is neglected.

2.31 OXIDATION–REDUCTION CELLS

Reduction is accompanied by a gain of electrons, and oxidation by a loss of electrons. In a system containing both an oxidising agent and its reduction product, there will be an equilibrium between them and electrons. If an inert electrode, such as platinum, is placed in a redox system, for example, one containing Fe(III) and Fe(II) ions, it will assume a definite potential indicative of the position of equilibrium. If the system tends to act as an oxidising agent, then $Fe^{3+} \rightarrow Fe^{2+}$ and it will take electrons from the platinum, leaving the latter positively charged; if, however, the system has reducing properties ($Fe^{2+} \rightarrow Fe^{3+}$), electrons will be given up to the metal, which will then acquire a negative charge. The magnitude of the potential will thus be a measure of the oxidising or reducing properties of the system.

To obtain comparative values of the 'strengths' of oxidising agents, it is necessary, as in the case of the electrode potentials of the metals, to measure under standard experimental conditions the potential difference between the platinum and the solution relative to a standard of reference. The primary standard is the standard or normal hydrogen electrode (Section 2.28) and its potential is taken as zero. The standard experimental conditions for the redox system are those in which the ratio of the activity of the oxidant to that of the reductant is unity. Thus for the $Fe^{3+} - Fe^{2+}$ electrode, the redox cell would be:

$$Pt, H_2 \left| H^+(a = 1) \right| \left| \begin{array}{l} Fe^{3+}(a = 1) \\ Fe^{2+}(a = 1) \end{array} \right| Pt$$

The potential measured in this way is called the **standard reduction potential**. A selection of standard reduction potentials is given in Table 2.6.

The standard potentials enable us to predict which ions will oxidise or reduce other ions at unit activity (or molar concentration). The most powerful oxidising agents are those at the upper end of the table, and the most powerful reducing agents at the lower end. Thus permanganate ion can oxidise Cl^-, Br^-, I^-, Fe^{2+} and $[Fe(CN)_6]^{4-}$; Fe^{3+} can oxidise H_3AsO_3 and I^- but not $Cr_2O_7^{2-}$ or Cl^-. It must be emphasised that for many oxidants the pH of the medium is of great importance, since they are generally used in acidic media. Thus in measuring the standard potential of the $MnO_4^- - Mn^{2+}$ system; $MnO_4^- + 8H^+ + 5e = Mn^{2+} + 4H_2O$, it is necessary to state that the hydrogen-ion activity is unity; this leads to $E^{\ominus} = +1.52$ volts. Similarly, the value of $E^{\ominus}$ for the $Cr_2O_7^{2-} - Cr^{3+}$ system is $+1.33$ volts. This means that the $MnO_4^- - Mn^{2+}$ system is a better oxidising agent than the $Cr_2O_7^{2-} - Cr^{3+}$ system. Since the standard potentials for $Cl_2 - 2Cl^-$ and $Fe^{3+} - Fe^{2+}$ systems are $+1.36$ and 0.77 volt respectively, permanganate and dichromate will oxidise Fe(II) ions but only permanganate will oxidise chloride ions; this explains why dichromate but not permanganate (except under very special conditions) can be used for the titration of Fe(II) in hydrochloric acid solution. Standard potentials do not give any information as to the speed of the reaction: in some cases a catalyst is necessary in order that the reaction may proceed with reasonable velocity.

Standard potentials are determined with full consideration of activity effects, and are really limiting values. They are rarely, if ever, observed directly in a potentiometric measurement. In practice, measured potentials determined under defined concentration conditions (formal potentials) are very useful for predicting the possibilities of redox processes. Further details are given in Section 10.90.

2.32 CALCULATION OF THE STANDARD REDUCTION POTENTIAL

A reversible oxidation–reduction system may be written in the form

Oxidant + $ne \rightleftharpoons$ Reductant

or

Ox + $ne \rightleftharpoons$ Red

(*oxidant* = substance in oxidised state, *reductant* = substance in reduced state). The electrode potential which is established when an inert or unattackable electrode is immersed in a solution containing both oxidant and reductant is given by the expression:

$$E_T = E^{\ominus} + \frac{RT}{nF} \ln \frac{a_{Ox}}{a_{Red}}$$

where E_T is the observed potential of the redox electrode at temperature T

Table 2.6 Standard reduction potentials at 25 °C

Half-reaction	$E^{\ominus}$, volts
$F_2 + 2e \rightleftharpoons 2F^-$	+2.65
$S_2O_8^{2-} + 2e \rightleftharpoons 2SO_4^{2-}$	+2.01
$Co^{3+} + e \rightleftharpoons Co^{2+}$	+1.82
$Pb^{4+} + 2e \rightleftharpoons Pb^{2+}$	+1.70
$MnO_4^- + 4H^+ + 3e \rightleftharpoons MnO_2 + 2H_2O$	+1.69
$Ce^{4+} + e \rightleftharpoons Ce^{3+}$ (nitrate medium)	+1.61
$BrO_3^- + 6H^+ + 5e \rightleftharpoons \frac{1}{2}Br_2 + 3H_2O$	+1.52
$MnO_4^- + 8H^+ + 5e \rightleftharpoons Mn^{2+} + 4H_2O$	+1.52
$Ce^{4+} + e \rightleftharpoons Ce^{3+}$ (sulphate medium)	+1.44
$Cl_2 + 2e \rightleftharpoons 2Cl^-$	+1.36
$Cr_2O_7^{2-} + 14H^+ + 6e \rightleftharpoons 2Cr^{3+} + 7H_2O$	+1.33
$Tl^{3+} + 2e \rightleftharpoons Tl^+$	+1.25
$MnO_2 + 4H^+ + 2e \rightleftharpoons Mn^{2+} + 2H_2O$	+1.23
$O_2 + 4H^+ + 4e \rightleftharpoons 2H_2O$	+1.23
$IO_3^- + 6H^+ + 5e \rightleftharpoons \frac{1}{2}I_2 + 3H_2O$	+1.20
$Br_2 + 2e \rightleftharpoons 2Br^-$	+1.07
$HNO_2 + H^+ + e \rightleftharpoons NO + H_2O$	+1.00
$NO_3^- + 4H^+ + 3e \rightleftharpoons NO + 2H_2O$	+0.96
$2Hg^{2+} + 2e \rightleftharpoons Hg_2^{2+}$	+0.92
$ClO^- + H_2O + 2e \rightleftharpoons Cl^- + 2OH^-$	+0.89
$Cu^{2+} + I^- + e \rightleftharpoons CuI$	+0.86
$Hg_2^{2+} + 2e \rightleftharpoons 2Hg$	+0.79
$Fe^{3+} + e \rightleftharpoons Fe^{2+}$	+0.77
$BrO^- + H_2O + 2e \rightleftharpoons Br^- + 2OH^-$	+0.76
$BrO_3^- + 3H_2O + 6e \rightleftharpoons Br^- + 6OH^-$	+0.61
$MnO_4^{2-} + 2H_2O + 2e \rightleftharpoons MnO_2 + 4OH^-$	+0.60
$MnO_4^- + e \rightleftharpoons MnO_4^{2-}$	+0.56
$H_3AsO_4 + 2H^+ + 2e \rightleftharpoons H_3AsO_3 + H_2O$	+0.56
$Cu^{2+} + Cl^- + e \rightleftharpoons CuCl$	+0.54
$I_2 + 2e \rightleftharpoons 2I^-$	+0.54
$IO^- + H_2O + 2e \rightleftharpoons I^- + 2OH^-$	+0.49
$[Fe(CN)_6]^{3-} + e \rightleftharpoons [Fe(CN)_6]^{4-}$	+0.36
$UO_2^{2+} + 4H^+ + 2e \rightleftharpoons U^{4+} + 2H_2O$	+0.33
$IO_3^- + 3H_2O + 6e \rightleftharpoons I^- + 6OH^-$	+0.26
$Cu^{2+} + e \rightleftharpoons Cu^+$	+0.15
$Sn^{4+} + 2e \rightleftharpoons Sn^{2+}$	+0.15
$TiO^{2+} + 2H^+ + e \rightleftharpoons Ti^{3+} + H_2O$	+0.10
$S_4O_6^{2-} + 2e \rightleftharpoons 2S_2O_3^{2-}$	+0.08
$2H^+ + 2e \rightleftharpoons H_2$	0.00
$V^{3+} + e \rightleftharpoons V^{2+}$	-0.26
$Cr^{3+} + e \rightleftharpoons Cr^{2+}$	-0.41
$Bi(OH)_3 + 3e \rightleftharpoons Bi + 3OH^-$	-0.44
$Fe(OH)_3 + e \rightleftharpoons Fe(OH)_2 + OH^-$	-0.56
$U^{4+} + e \rightleftharpoons U^{3+}$	-0.61
$AsO_4^{3-} + 3H_2O + 2e \rightleftharpoons H_2AsO_3^- + 4OH^-$	-0.67
$[Sn(OH)_6]^{2-} + 2e \rightleftharpoons [HSnO_2]^- + H_2O + 3OH^-$	-0.90
$[Zn(OH)_4]^{2-} + 2e \rightleftharpoons Zn + 4OH^-$	-1.22
$[H_2AlO_3]^- + H_2O + 3e \rightleftharpoons Al + 4OH^-$	-2.35

relative to the standard or normal hydrogen electrode taken as zero potential, $E^{\ominus}$ is the standard reduction potential,* n the number of electrons gained by

---

* $E^{\ominus}$ is the value of E_T at unit activities of the oxidant and reductant. If both activities are variable, e.g. Fe^{3+} and Fe^{2+}, $E^{\ominus}$ corresponds to an activity ratio of unity.

the oxidant in being converted into the reductant, and a_{Ox} and a_{Red} are the activities of the oxidant and reductant respectively.

Since activities are often difficult to determine directly, they may be replaced by concentrations; the error thereby introduced is usually of no great importance. The equation therefore becomes:

$$E_T = E^{\ominus} + \frac{RT}{nF} \ln \frac{c_{Ox}}{c_{Red}}$$

Substituting the known values of R and F, and changing from natural to common logarithms, at a temperature of 25 °C ($T = 298$K):

$$E_{25^{\circ}} = E^{\ominus} + \frac{0.0591}{n} \log \frac{[Ox]}{[Red]}$$

If the concentrations (or, more accurately, the activities) of the oxidant and reductant are equal, $E_{25^{\circ}} = E^{\ominus}$, i.e. the standard reduction potential. It follows from this expression that, for example, a ten-fold change in the ratio of the concentrations of the oxidant to the reductant will produce a change in the potential of the system of $0.0591/n$ volts.

2.33 EQUILIBRIUM CONSTANTS OF OXIDATION–REDUCTION REACTIONS

The general equation for the reaction at an oxidation–reduction electrode may be written:

$$p\text{A} + q\text{B} + r\text{C} \ldots + ne \rightleftharpoons s\text{X} + t\text{Y} + u\text{Z} + \ldots$$

The potential is given by:

$$E = E^{\ominus} + \frac{RT}{nF} \ln \frac{a_A^p \cdot a_B^q \cdot a_C^r \ldots}{a_X^s \cdot a_Y^t \cdot a_Z^u \ldots}$$

where a refers to activities, and n to the number of electrons involved in the oxidation–reduction reaction. This expression reduces to the following for a temperature of 25 °C (concentrations are substituted for activities to permit ease of application in practice):

$$E = E^{\ominus} + \frac{0.0591}{n} \log \frac{c_A^p \cdot c_B^q \cdot c_C^r \ldots}{c_X^s \cdot c_Y^t \cdot c_Z^u \ldots}$$

It is, of course, possible to calculate the influence of the change of concentration of certain constituents of the system by the use of the latter equation. Consider, for example, the permanganate reaction:

$$\text{MnO}_4^- + 8\text{H}^+ + 5e \rightleftharpoons \text{Mn}^{2+} + 4\text{H}_2\text{O}$$

$$E = E^{\ominus} + \frac{0.0591}{5} \log \frac{[\text{MnO}_4^-] \times [\text{H}^+]^8}{[\text{Mn}^{2+}]} \quad \text{(at 25 °C)}$$

The concentration (or activity) of the water is taken as constant, since it is assumed that the reaction takes place in dilute solution, and the concentration of the water does not change appreciably as the result of the reaction. The equation may be written in the form:

$$E = E^{\ominus} + \frac{0.0591}{5} \log \frac{[MnO_4^-]}{[Mn^{2+}]} + \frac{0.0591}{5} \log[H^+]^8$$

This enables us to calculate the effect of change in the ratio $[MnO_4^-]/[Mn^{2+}]$ at any hydrogen ion concentration, other factors being maintained constant. In this system, however, difficulties are experienced in the calculation owing to the fact that the reduction products of the permanganate ion vary at different hydrogen ion concentrations. In other cases no such difficulties arise, and the calculation may be employed with confidence. Thus in the reaction:

$$H_3AsO_4 + 2H^+ + 2e \rightleftharpoons H_3AsO_3 + H_2O$$

$$E = E^{\ominus} + \frac{0.0591}{2} \log \frac{[H_3AsO_4] \times [H^+]^2}{[H_3AsO_3]} \quad \text{(at 25 °C)}$$

or $E = E^{\ominus} + \dfrac{0.0591}{2} \log \dfrac{[H_3AsO_4]}{[H_3AsO_3]} + \dfrac{0.0591}{2} \log[H^+]^2$

It is now possible to calculate the equilibrium constants of oxidation–reduction reactions, and thus to determine whether such reactions can find application in quantitative analysis. Consider first the simple reaction:

$$Cl_2 + 2Fe^{2+} \rightleftharpoons 2Cl^- + 2Fe^{3+}$$

The equilibrium constant is given by:

$$\frac{[Cl^-]^2 \times [Fe^{3+}]^2}{[Cl_2] \times [Fe^{2+}]^2} = K$$

The reaction may be regarded as taking place in a voltaic cell, the two half-cells being a $Cl_2,2Cl^-$ system and a Fe^{3+},Fe^{2+} system. The reaction is allowed to proceed to equilibrium, and the total voltage or e.m.f. of the cell will then be zero, i.e. the potentials of the two electrodes will be equal:

$$E^{\ominus}_{Cl_2,2Cl^-} + \frac{0.0591}{2} \log \frac{[Cl_2]}{[Cl^-]^2} = E^{\ominus}_{Fe^{3+},Fe^{2+}} + \frac{0.0591}{1} \log \frac{[Fe^{3+}]}{[Fe^{2+}]}$$

Now $E^{\ominus}_{Cl_2,2Cl^-} = 1.36$ volts and $E^{\ominus}_{Fe^{3+},Fe^{2+}} = 0.75$ volt, hence

$$\log \frac{[Fe^{3+}]^2 \times [Cl^-]^2}{[Fe^{2+}]^2 \times [Cl_2]} = \frac{0.61}{0.02965} = 20.67 = \log K$$

or $K = 4.7 \times 10^{20}$

The large value of the equilibrium constant signifies that the reaction will proceed from left to right almost to completion, i.e. an iron(II) salt is almost completely oxidised by chlorine.

Consider now the more complex reaction:

$$MnO_4^- + 5Fe^{2+} + 8H^+ \rightleftharpoons Mn^{2+} + 5Fe^{3+} + 4H_2O$$

The equilibrium constant K is given by:

$$K = \frac{[Mn^{2+}] \times [Fe^{3+}]^5}{[MnO_4^-] \times [Fe^{2+}]^5 \times [H^+]^8}$$

The term $4H_2O$ is omitted, since the reaction is carried out in dilute solution, and the water concentration may be assumed constant. The hydrogen ion concentration is taken as molar. The complete reaction may be divided into two half-cell reactions corresponding to the partial equations:

$$MnO_4^- + 8H^+ + 5e \rightleftharpoons Mn^{2+} + 4H_2O \qquad (35)$$

and $\quad Fe^{2+} \rightleftharpoons Fe^{3+} + e \qquad (36)$

For (35) as an oxidation–reduction electrode, we have:

$$E = E^\ominus + \frac{0.0591}{5} \log \frac{[MnO_4^-] \times [H^+]^8}{[Mn^{2+}]}$$
$$= 1.52 + \frac{0.0591}{5} \log \frac{[MnO_4^-] \times [H^+]^8}{[Mn^{2+}]}$$

The partial equation (36) may be multiplied by 5 in order to balance (35) electrically:

$$5Fe^{2+} \rightleftharpoons 5Fe^{3+} + 5e \qquad (37)$$

For (37) as an oxidation–reduction electrode:

$$E = E^\ominus + \frac{0.0591}{5} \log \frac{[Fe^{3+}]^5}{[Fe^{2+}]^5} = 0.77 + \frac{0.0591}{5} \log \frac{[Fe^{3+}]^5}{[Fe^{2+}]^5}$$

Combining the two electrodes into a cell, the e.m.f. will be zero when equilibrium is attained, i.e.

$$1.52 + \frac{0.0591}{5} \log \frac{[MnO_4^-] \times [H^+]^8}{[Mn^{2+}]} = 0.77 + \frac{0.0591}{5} \log \frac{[Fe^{3+}]^5}{[Fe^{2+}]^5}$$

$$\text{or} \quad \log \frac{[Mn^{2+}] \times [Fe^{3+}]^5}{[MnO_4^-] \times [Fe^{2+}]^5 \times [H^+]^8} = \frac{5(1.52 - 0.77)}{0.0591} = 63.5$$

$$K = \frac{[Mn^{2+}] \times [Fe^{3+}]^5}{[MnO_4^-] \times [Fe^{3+}]^5 \times [H^+]^8} = 3 \times 10^{63}$$

This result clearly indicates that the reaction proceeds virtually to completion. It is a simple matter to calculate the residual Fe(II) concentration in any particular case. Thus consider the titration of 10 mL of a 0.1 M solution of iron(II) ions with 0.02 M potassium permanganate in the presence of hydrogen ions, concentration 1 M. Let the volume of the solution at the equivalence point be 100 mL. Then $[Fe^{3+}] = 0.01 M$, since it is known that the reaction is practically complete, $[Mn^{2+}] = \frac{1}{5} \times [Fe^{3+}] = 0.002 M$, and $[Fe^{2+}] = x$. Let the excess of permanganate solution at the end-point be one drop or 0.05 mL; its concentration will be $0.05 \times 0.1/100 = 5 \times 10^{-5} M = [MnO_4^-]$. Substituting these values in the equation:

$$K = \frac{(2 \times 10^{-3}) \times (1 \times 10^{-2})^5}{10^{-5} \times x^5 \times 1^8} = 3 \times 10^{63}$$

or $\quad x = [Fe^{2+}] = 5.8 \times 10^{-15} \text{ mol L}^{-1}$

It is clear from what has already been stated that standard reduction potentials may be employed to determine whether redox reactions are sufficiently complete for their possible use in quantitative analysis. It must be emphasised, however, that these calculations provide no information as to the speed of the reaction, upon which the application of that reaction in practice will ultimately depend. This question must form the basis of a separate experimental study, which may include the investigation of the influence of temperature, variation of pH and of the concentrations of the reactants, and the influence of catalysts. Thus, theoretically, potassium permanganate should quantitatively oxidise oxalic acid in aqueous solution. It is found, however, that the reaction is extremely slow at the ordinary temperature, but is more rapid at about 80 °C, and also increases in velocity when a little manganese(II) ion has been formed, the latter apparently acting as a catalyst.

It is of interest to consider the calculation of the equilibrium constant of the general redox reaction, viz.:

$$a \, Ox_I + b \, Red_{II} \rightleftharpoons b \, Ox_{II} + a \, Red_I$$

The complete reaction may be regarded as composed of two oxidation–reduction electrodes. $a \, Ox_I$, $a \, Red_I$ and $b \, Ox_{II}$, $b \, Red_{II}$ combined together into a cell; at equilibrium, the potentials of both electrodes are the same:

$$E_1 = E_1^\ominus + \frac{0.0591}{n} \log \frac{[Ox_I]^a}{[Red_I]^a}$$

$$E_2 = E_2^\ominus + \frac{0.0591}{n} \log \frac{[Ox_{II}]^b}{[Red_{II}]^b}$$

At equilibrium, $E_1 = E_2$, and hence:

$$E_1^\ominus + \frac{0.0591}{n} \log \frac{[Ox_I]^a}{[Red_I]^a} = E_2^\ominus + \frac{0.0591}{n} \log \frac{[Ox_{II}]^b}{[Red_{II}]^b}$$

$$\text{or} \quad \log \frac{[Ox_{II}]^b \times [Red_I]^a}{[Red_{II}]^b \times [Ox_I]^a} = \log K = \frac{n}{0.0591}(E_1^\ominus - E_2^\ominus)$$

This equation may be employed to calculate the equilibrium constant of any redox reaction, provided the two standard potentials $E_1^\ominus$ and $E_2^\ominus$ are known; from the value of K thus obtained, the feasibility of the reaction in analysis may be ascertained.

It can readily be shown that the concentrations at the equivalence point, when equivalent quantities of the two substances Ox_I and Red_{II} are allowed to react, are given by:

$$\frac{[Red_I]}{[Ox_I]} = \frac{[Ox_{II}]}{[Red_{II}]} = \sqrt[(a+b)]{K}$$

This expression enables us to calculate the exact concentration at the equivalence point in any redox reaction of the general type given above, and therefore the feasibility of a titration in quantitative analysis.

For References and Bibliography see Sections 3.38 and 3.39.

CHAPTER 3
COMMON APPARATUS AND BASIC TECHNIQUES

3.1 INTRODUCTION

In this chapter the more important basic techniques and the apparatus commonly used in analytical operations will be described. It is essential that the beginner should become familiar with these procedures, and also acquire dexterity in handling the various pieces of apparatus. The habit of clean, orderly working must also be cultivated, and observance of the following points will be helpful in this direction.

1. The bench must be kept clean and a bench-cloth must be available so that any spillages of solid or liquid chemicals (solutions) can be removed immediately.
2. All glassware must be scrupulously clean (see Section 3.8), and if it has been standing for any length of time, must be rinsed with distilled or de-ionised water before use. The outsides of vessels may be dried with a lint-free glass-cloth which is reserved exclusively for this purpose, and which is frequently laundered, but the cloth should not be used on the insides of the vessels.
3. Under no circumstances should the working surface of the bench become cluttered with apparatus. All the apparatus associated with some particular operation should be grouped together on the bench; this is most essential to avoid confusion when duplicate determinations are in progress. Apparatus for which no further immediate use is envisaged should be returned to the locker, but if it will be needed at a later stage, it may be placed at the back of the bench.
4. If a solution, precipitate, filtrate, etc., is set aside for subsequent treatment, the container must be labelled so that the contents can be readily identified, and the vessel must be suitably covered to prevent contamination of the contents by dust: in this context, bark corks are usually unsuitable; they invariably tend to shed some dust. For temporary labelling, a 'Chinagraph' pencil or a felt tip pen which will write directly on to glass is preferable to the gummed labels which are used when more permanent labelling is required.
5. Reagent bottles must never be allowed to accumulate on the bench; they must be replaced on the reagent shelves immediately after use.
6. It should be regarded as normal practice that all determinations are performed in duplicate.
7. A stiff covered notebook of A4 size must be provided for **recording experimental observations as they are made**. A double page should be devoted

to each determination, the title of which, together with the date, must be clearly indicated. One of the two pages must be reserved for the experimental observations, and the other should be used for a brief description of the procedure followed, but with a full account of any special features associated with the determination, In most cases it will be found convenient to divide the page on which the experimental observations are to be recorded into two halves by a vertical line, and then to halve the right-hand column thus created with a second vertical line. The left-hand side of the page can then be used to indicate the observations to be made, and the data for duplicate determinations can be recorded side by side in the two right-hand columns.

The record must conclude with the calculation of the result of the analyses, and in this connection the equation(s) for the principal chemical reaction(s) involved in the determination should be shown together with a clear exposition of the procedure used for calculating the result. Finally, appropriate comments should be made upon the degree of accuracy and the precision achieved.

Many modern instruments used in the analytical laboratory are interfaced with a computer and a printer provides a permanent record of the experimental data and the final result may even be given. This printout should be permanently attached to the observations page of the laboratory record book, and it should be regarded as normal practice to perform a 'rough' calculation to confirm that the printed result is of the right order.

8. **Safety procedures must be observed in the laboratory at all times**. Many chemicals encountered in analysis are poisonous and must be carefully handled. Whereas the dangerous properties of concentrated acids and of widely recognised poisons such as potassium cyanide are well known, the dangers associated with organo-chlorine solvents, benzene and many other chemicals are less apparent.

Many operations involving chemical reactions are potentially dangerous, and in such cases recommended procedures must be carefully followed and obeyed. All laboratory workers should familiarise themselves with local safety requirements (in some laboratories, the wearing of safety spectacles may be compulsory), and with the position of first-aid equipment.

For further guidance it is recommended that some study should be made of books devoted to hazards and safe practices in chemical laboratories. Some institutions and organisations issue booklets dealing with these matters and further information will be found in citations 12–17 of the Bibliography, Section 3.39.

BALANCES

3.2 THE ANALYTICAL BALANCE

One of the commonest procedures carried out by the analyst is the measurement of mass. Many chemical analyses are based upon the accurate determination of the mass of a sample, and that of a solid substance produced from it (gravimetric analysis), or upon ascertaining the volume of a carefully prepared standard solution (which contains an accurately known mass of solute) which is required to react with the sample (titrimetric analysis). For the accurate measurement of mass in such operations, an analytical balance is employed; the operation is referred to as weighing, and invariably reference is made to the weight of the object or material which is weighed.

The weight of an object is the force of attraction due to gravity which is exerted upon the object:

$$w = mg$$

where w is the weight of the object, m its mass, and g is the acceleration due to gravity. Since the attraction due to gravity varies over the earth's surface with altitude and also with latitude, the weight of the object is variable, whereas its mass is constant. It has however become the custom to employ the term 'weight' synonymously with mass, and it is in this sense that 'weight' is employed in quantitative analysis.

The analytical balance is thus one of the most important tools of the analytical chemist, and it is one which of recent years has undergone radical changes. These changes have been prompted by the desire to produce an instrument which is more robust, less dependent upon the experience of the operator, less susceptible to the environment, and above all, one which will hasten the weighing operation. In meeting these requirements, the design of the balance has been fundamentally altered, and the conventional free-swinging, equal-arm, two-pan chemical balance together with its box of weights is now an uncommon sight.

An important development was the replacement of the two-pan balance with its three knife edges by a **two-knife single-pan balance**. In this instrument one balance pan and its suspension is replaced by a counterpoise, and dial-operated ring weights are suspended from a carrier attached to the remaining pan support: see Fig. 3.1. In this system all the weights are permanently in position on the carrier when the beam is at rest, and when an object to be weighed is placed upon the balance pan, weights must be *removed* from the carrier to compensate for the weight of the object. Weighing is completed by allowing the beam to assume its rest position, and then reading the displacement of the beam on an optical scale which is calibrated to read weights below 100 mg. Weighing is thus accomplished by **substitution**; many such manually operated balances are still in service in analytical laboratories.

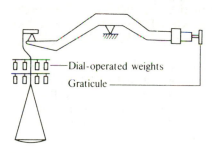

Dial-operated weights

Graticule

Fig. 3.1

The standard modern instrument however is the **electronic balance**, which provides convenience in weighing coupled with much greater freedom from mechanical failure, and greatly reduced sensitivity to vibration. The operations of selecting and removing weights, smooth release of balance beam and pan support, noting the reading of weight dials and of an optical scale, returning the beam to rest, and replacing weights which have been removed, are eliminated. With an electronic balance, operation of a single on–off control permits the operator to read the weight of an object on the balance pan immediately from a digital display: most balances of this type can be coupled to a printer which gives a printed record of the weight. The majority of balances incorporate a **tare** facility which permits the weight of a container to be balanced out, so that when material is added to the container, the weight recorded is simply that of the material used. Many balances of this type incorporate a self-testing system which indicates that the balance is functioning correctly each time it is switched on, and also include a built-in weight calibration system. Operation of the calibration control leads to display of the weight of the standard incorporated within the balance, and thus indicates whether any correction is necessary. A more satisfactory calibration procedure is to check the balance readings against a series of calibrated analytical weights.

Electronic balances operate by applying an electromagnetic restoring force to the support to which the balance pan is attached, so that when an object is added to the balance pan, the resultant displacement of the support is cancelled out. The magnitude of the restoring force is governed by the value of the current flowing in the coils of the electromagnetic compensation system, and this is proportional to the weight placed on the balance pan: a microprocessor converts the value of the current into the digital display in grams.

The balance must of course be protected from draughts and from dust, and the balance pan is situated within an enclosure provided with glass doors which can be opened to provide access to the pan. The remainder of the balance, including the electrical components, is contained in a closed compartment attached to the rear of the pan compartment.

Electronic balances are available to cover weight ranges of

1. up to about 150 g and reading to 0.1 mg (macrobalance),
2. up to about 30 g and reading to 0.01 mg (semimicro balance),
3. up to about 20 g and reading to 1 μg (microbalance),
4. up to 5 g and reading to 0.1 μg (ultramicro balance).

Thus a wide variety of analytical balances is available.

3.3 OTHER BALANCES

For many laboratory operations it is necessary to weigh objects or materials which are far heavier than the upper weight limit of a macro analytical balance, or small amounts of material for which it is not necessary to weigh to the limit of sensitivity of such a balance: this type of weighing is often referred to as a 'rough weighing'. A wide range of electronic balances is available for such purposes with characteristics such as, for example,

Maximum capacity	Reading to
350 g	0.01 g
3500 g	0.1 g
6 kg	0.1 g

With these top pan balances it is not necessary to shield the balance pan from gentle draughts, and weighings can be accomplished very rapidly and with the usual facility of the results being recorded with a printer.

3.4 WEIGHTS, REFERENCE MASSES

Although with modern balances it is not necessary to make use of a box of weights in the weighing process, as indicated in Section 3.2 a set of weights is desirable for checking the accuracy of a balance.

For scientific work the fundamental standard of mass is the international prototype kilogram, which is a mass of platinum–iridium alloy made in 1887 and deposited in the International Bureau of Weights and Measures near Paris. Authentic copies of the standard are kept by the appropriate responsible authorities* in the various countries of the world; these copies are employed for the comparison of secondary standards, which are used in the calibration of weights for scientific work. The unit of mass that is almost universally employed in laboratory work, however, is the **gram**, which may be defined as the one-thousandth part of the mass of the international prototype kilogram.

An ordinary set of analytical weights contains the following: grams, 100, 50, 30, 20, 10, 5, 3, 2, 1; milligrams, 500–100 and 50–10 in the same 5, 3, 2, 1 sequence. The weights from 1 g upwards are constructed from a non-magnetic nickel–chromium alloy (80% Ni, 20% Cr), or from austenitic stainless steel; plated brass is sometimes used but is less satisfactory. The fractional weights are made from the same alloys, or from a non-tarnishable metal such as gold or platinum. For handling the weights a pair of forceps, preferably ivory-tipped, are provided and the weights are stored in a box with suitably shaped compartments.

Analytical weights can be purchased which have been manufactured to 'Class A' standard; this is the only grade of laboratory weights officially recognised in the United Kingdom. In 'Class A' weights the following tolerances are permitted: 100 g, 0.5 mg; 50 g, 0.25 mg; 30 g, 0.15 mg; 20 g, 0.10 mg; 10 g–100 mg, 0.05 mg; 50–10 mg, 0.02 mg.

The National Bureau of Standards at Washington recognises the following classes of precision weights:

Class M   For use as reference standards, for work of the highest precision, and where a high degree of constancy over a period of time is required.

Class S   For use as working reference standards or as high-precision analytical weights.

Class S-1   Precision analytical weights for routine analytical work.

Class J   Microweight standards for microbalances.

3.5 CARE AND USE OF ANALYTICAL BALANCES

No matter what type of analytical balance is employed, due attention must be paid to the manner in which it is used. The following remarks apply particularly to electronic balances.

---

*The National Physical Laboratory (NPL) in Great Britain, the National Bureau of Standards (NBS) in USA, etc.

1. Never exceed the stated maximum load of the balance.
2. The balance must be kept clean. Remove dust from the pan and from the floor of the pan compartment with a camel hair brush.
3. Objects to be weighed should never be handled with the fingers; always use tongs or a loop of clean paper.
4. Objects to be weighed should be allowed to attain the temperature of the balance before weighing, and if the object has been heated, sufficient time must be allowed for cooling. The time required to attain the temperature of the balance varies with the size, etc., of the object, but as a rule 30–40 minutes is sufficient.
5. No chemicals or objects which might injure the balance pan should ever be placed directly on it. Substances must be weighed in suitable containers, such as small beakers, weighing bottles or crucibles, or upon watch glasses. Liquids and volatile or hygroscopic solids must be weighed in tightly closed vessels, such as stoppered weighing bottles.

   The addition of chemicals to the receptacle must be done outside the balance case. It is good practice to weigh the chosen receptacle on the analytical balance, to transfer it to a rough balance, to add approximately the required amount of the necessary chemical, and then to return the receptacle to the analytical balance for re-weighing, thus giving the exact weight of substance taken.
6. Nothing must be left on the pan when the weighing has been completed. If any substance is spilled accidentally upon the pan or upon the floor of the balance compartment, it must be removed at once.
7. Exposure of the balance to corrosive atmospheres must be avoided.

The actual weighing process will include the following steps.

1. Brush the balance pan lightly with a camel hair brush to remove any dust.
2. With the balance at rest, place the object to be weighed, which must be at or close to room temperature, on the pan, and close the pan compartment case.
3. Set the on/off control of the balance to the 'on' position and record the weight shown on the digital display: if the balance is linked to a printer, confirm that the printed result agrees with the digital display. Return the control to the 'off' position.
4. When all weighings have been completed, remove the object which has been weighed, clear up any accidental spillages, and close the pan compartment.

The above remarks apply particularly to analytical balances of the macrobalance range; microbalances and ultramicro balances must be handled with special care, particularly with regard to the temperature of objects to be weighed.

3.6 ERRORS IN WEIGHING

The chief sources of error are the following:

1. Change in the condition of the containing vessel or of the substance between successive weighings.
2. Effect of the buoyancy of the air upon the object and the weights.
3. Errors in recording the weights.

1. The first source of error is occasioned by change in weight of the containing vessel: (a) by absorption or loss of moisture, (b) by electrification of the

surface caused by rubbing, and (c) by its temperature being different from that of the balance case. These errors may be largely eliminated by wiping the vessel gently with a linen cloth, and allowing it to stand at least 30 minutes in proximity to the balance before weighing. The electrification, *which may cause a comparatively large error, particularly if both the atmosphere and the cloth are dry*, is slowly dissipated on standing; it may be removed by subjecting the vessel to the discharge from an antistatic gun. Hygroscopic, efflorescent, and volatile substances must be weighed in completely closed vessels. Substances which have been heated in an air oven or ignited in a crucible are generally allowed to cool in a desiccator containing a suitable drying agent. The time of cooling in a desiccator cannot be exactly specified, since it will depend upon the temperature and upon the size of the crucible as well as upon the material of which it is composed. Platinum vessels require a shorter time than those of porcelain, glass, or silica. It has been customary to leave platinum crucibles in the desiccator for 20–25 minutes, and crucibles of other materials for 30–35 minutes before being weighed. It is advisable to cover crucibles and other open vessels.

2. When a substance is immersed in a fluid, its true weight is diminished by the weight of the fluid which it displaces. If the object and the weights have the same density, and consequently the same volume, no error will be introduced on this account. If, however, as is usually the case, the density of the object is different from that of the weights, the volumes of air displaced by each will be different. If the substance has a lower density than the weights, as is usual in analysis, the former will displace a greater volume of air than the latter, and it will therefore weigh less in air than in a vacuum. Conversely, if a denser material (e.g. one of the precious metals) is weighed, the weight in a vacuum will be less than the apparent weight in air.

Consider the weighing of 1 litre of water, first *in vacuo*, and then in air. It is assumed that the flask containing the water is tared by an exactly similar flask, that the temperature of the air is 20 °C and the barometric pressure is 101 325 Pa (760 mm of mercury). The weight of 1 litre of water *in vacuo* under these conditions is 998.23 g. If the water is weighed in air, it will be found that 998.23 g are too heavy. We can readily calculate the difference. The weight of 1 litre of air displaced by the water is 1.20 g. Assuming the weights to have a density of 8.0, they will displace $998.23/8.0 = 124.8$ mL, or $124.8 \times 1.20/1000 = 0.15$ g of air. The net difference in weight will therefore be $1.20 - 0.15 = 1.05$ g. Hence the weight in air of 1 litre of water under the experimental conditions named is $998.23 - 1.05 = 997.18$ g, a difference of 0.1 per cent from the weight *in vacuo*.

Now consider the case of a solid, such as potassium chloride, under the above conditions. The density of potassium chloride is 1.99. If 2 g of the salt are weighed, the apparent loss in weight (= weight of air displaced) is $2 \times 0.0012/1.99 = 0.0012$ g. The apparent loss in weight for the weights is $2 \times 0.0012/8.0 = 0.000\,30$ g. Hence 2 g of potassium chloride will weigh $0.0012 - 0.000\,30 = 0.000\,90$ g less in air than *in vacuo*, a difference of 0.05 per cent.

It must be pointed out that for most analytical purposes where it is desired to express the results in the form of a percentage, the ratio of the weights in air, so far as solids are concerned, will give a result which is practically the same as that which would be given by the weights *in vacuo*. Hence no buoyancy

77

correction is necessary in these cases. However, where absolute weights are required, as in the calibration of graduated glassware, corrections for the buoyancy of the air must be made (compare Section 3.16). Although an electronic balance does not employ any weights, the above remarks apply to weights recorded by the balance because the balance scale will have been established by reference to metal (stainless steel) weights used in air.

Now consider the general case. It is evident that the weight of an object *in vacuo* is equal to the weight in air *plus* the weight of air displaced by the object *minus* the weight of air displaced by the weights. It can easily be shown that if W_v = weight *in vacuo*, W_a = apparent weight in air, d_a = density of air, d_w = density of the weights, and d_b = density of the body, then:

$$W_v = W_a + d_a \left(\frac{W_v}{d_b} - \frac{W_a}{d_w} \right)$$

The density of the air will depend upon the humidity, the temperature, and the pressure. For an average relative humidity (50 per cent) and average conditions of temperature and pressure in a laboratory, the density of the air will rarely fall outside the limits 0.0011 and 0.0013 $\mathrm{g\,mL^{-1}}$. It is therefore permissible for analytical purposes to take the weight of 1 mL of air as 0.0012 g.

Since the difference between W_v and W_a does not usually exceed 1 to 2 parts per thousand, we may write:

$$W_v = W_a + d_a \left(\frac{W_a}{d_b} - \frac{W_a}{d_w} \right)$$

$$= W_a + W_a \left\{ 0.0012 \left(\frac{1}{d_b} - \frac{1}{8.0} \right) \right\} = W_a + kW_a/1000$$

where

$$k = 1.20 \left(\frac{1}{d_b} - \frac{1}{8.0} \right)$$

If a substance of density d_b weighs W_a grams in air, then $W_a . k$ milligrams are to be added to the weight in air in order to obtain the weight *in vacuo*. The correction is positive if the substance has a density lower than 8.0, and negative if the density of the substance is greater than 8.0.

3. The correct reading of weights is best achieved by checking weights as they are added to the balance *and* as they are removed from the balance. In the case of electronic balances any digital displays should be read at least twice.

GRADUATED GLASSWARE

3.7 UNITS OF VOLUME

For scientific purposes the convenient unit to employ for measuring reasonably large volumes of liquids is the cubic decimetre (dm^3), or, for smaller volumes, the cubic centimetre (cm^3). For many years the fundamental unit employed was the *litre*, based upon the volume occupied by one kilogram of water at 4 °C (the temperature of maximum density of water): the relationship between the litre as thus defined and the cubic decimetre was established as

1 litre = 1.000 028 dm^3 or   1 millilitre = 1.000 028 cm^3.

In 1964 the *Conférence Générale des Poids et des Mésures* (CGPM) decided to accept the term **litre** as a special name for the cubic decimetre, and to discard the original definition. With this new meaning of the term litre (L), the millilitre (mL) and the cubic centimetre (cm^3) are identical.

3.8 GRADUATED APPARATUS

The most commonly used pieces of apparatus in titrimetric (volumetric) analysis are graduated flasks, burettes, and pipettes. Graduated cylinders and weight pipettes are less widely employed. Each of these will be described in turn.

Graduated apparatus for quantitative analysis is generally made to specification limits, particularly with regard to the accuracy of calibration. In the United Kingdom there are two grades of apparatus available, designated Class A and Class B by the British Standards Institution. The tolerance limits are closer for Class A apparatus, and such apparatus is intended for use in work of the highest accuracy: Class B apparatus is employed in routine work. In the United States, specifications for only one grade are available from the National Bureau of Standards at Washington, and these are equivalent to the British Class A.

Cleaning of glass apparatus. Before describing graduated apparatus in detail, reference must be made to the important fact that all such glassware must be perfectly clean and free from grease, otherwise the results will be unreliable. One test for cleanliness of glass apparatus is that on being filled with distilled water and the water withdrawn, only an unbroken film of water remains. If the water collects in drops, the vessel is dirty and must be cleaned. Various methods are available for cleaning glassware.

Many commercially available detergents are suitable for this purpose, and some manufacturers market special formulations for cleaning laboratory glassware; some of these, e.g. 'Decon 90' made by Decon Laboratories of Portslade, are claimed to be specially effective in removing contamination due to radioactive materials.

'Teepol' is a relatively mild and inexpensive detergent which may be used for cleaning glassware. The laboratory stock solution may consist of a 10 per cent solution in distilled water. For cleaning a burette, 2 mL of the stock solution diluted with 50 mL of distilled water are poured into the burette, allowed to stand for $\frac{1}{2}$ to 1 minute, the detergent run off, the burette rinsed three times with tap water, and then several times with distilled water. A 25 mL pipette may be similarly cleaned using 1 mL of the stock solution diluted with 25–30 mL of distilled water.

A method which is frequently used consists in filling the apparatus with 'chromic acid cleaning mixture' (**CARE**), a nearly saturated solution of powdered sodium dichromate or potassium dichromate in concentrated sulphuric acid, and allowing it to stand for several hours, preferably overnight; the acid is then poured off, the apparatus thoroughly rinsed with distilled water, and allowed to drain until dry. [It may be mentioned that potassium dichromate is not very soluble in concentrated sulphuric acid (about 5 g per litre), whereas sodium dichromate ($Na_2Cr_2O_7,2H_2O$) is much more soluble (about 70 g per litre); for this reason, as well as the fact that it is much cheaper, the latter is usually preferred for the preparation of 'cleaning mixture'. From time to time it is advisable to filter the sodium dichromate–sulphuric acid mixture through a little glass wool placed in the apex of a glass funnel: small particles or sludge, which are often present and may block the tips of burettes, are thus removed. A more efficient cleaning liquid is a mixture of concentrated sulphuric acid and fuming nitric acid; this may be used if the vessel is very greasy and dirty, but must be handled with extreme caution.

A very effective degreasing agent, which it is claimed is much quicker-acting than 'cleaning mixture' is obtained by dissolving 100 g of potassium hydroxide in 50 mL of water, and after cooling, making up to 1 litre with industrial methylated spirit.[6a]

3.9 TEMPERATURE STANDARD

The capacity of a glass vessel varies with the temperature, and it is therefore necessary to define the temperature at which its capacity is intended to be correct: in the UK a temperature of 20 °C has been adopted. A subsidiary standard temperature of 27 °C is accepted by the British Standards Institution, for use in tropical climates where the ambient temperature is consistently above 20 °C. The US Bureau of Standards, Washington, in compliance with the view held by some chemists that 25 °C more nearly approximates to the average laboratory temperature in the United States, will calibrate glass volumetric apparatus marked either 20 °C or 25 °C.

Taking the coefficient of cubical expansion of soda glass as about 0.000 030 and of borosilicate glass about 0.000 010 per 1 °C, Part A of Table 3.1 gives the correction to be added when the sign is $+$, or subtracted when the sign is $-$, to or from the capacity of a 1000 mL flask correct at 20 °C in order to obtain the capacity at other temperatures.

In the use of graduated glassware for measurement of the volume of liquids, the expansion of the liquid must also be taken into consideration if temperature corrections are to be made. Part B of Table 3.1 gives the corrections to be added or subtracted in order to obtain the volume occupied at 20 °C by a volume of water which at the tabulated temperature is contained in an accurate 1000 mL flask having a standard temperature of 20 °C. It will be seen that the allowance for the expansion of water is considerably greater than that for the expansion

Table 3.1 Temperature corrections for a 1 L graduated flask

Temperature (°C)	(A) Expansion of glass		(B) Expansion of water	
	Correction (mL)		Correction (mL)	
	Soda glass	Borosilicate glass	Soda glass	Borosilicate glass
5	−0.39	−0.15	+1.37	+1.61
10	−0.26	−0.10	+1.24	+1.40
15	−0.13	−0.05	+0.77	+0.84
20	0.00	0.00	0.00	0.00
25	+0.13	+0.05	−1.03	−1.11
30	+0.26	+0.10	−2.31	−2.46

of the glass. For dilute (e.g. 0.1 M) aqueous solutions, the corrections can be regarded as approximately the same as for water, but with more concentrated solutions the correction increases, and for non-aqueous solutions the corrections can be quite large.[6b]

3.10 GRADUATED FLASKS

A graduated flask (known alternatively as a volumetric flask or a measuring flask), is a flat-bottomed, pear-shaped vessel with a long narrow neck. A thin line etched around the neck indicates the volume that it holds at a certain definite temperature, usually 20 °C (both the capacity and temperature are clearly marked on the flask); the flask is then said to be graduated *to contain*. Flasks with one mark are always taken *to contain* the volume specified. A flask may also be marked *to deliver* a specified volume of liquid under certain definite conditions; these are, however, not suitable for exact work and are not widely used. Vessels intended to contain definite volumes of liquid are marked C or TC or In, while those intended to deliver definite volumes are marked D or TD or Ex.

The mark extends completely around the neck in order to avoid errors due to parallax when making the final adjustment; the lower edge of the meniscus of the liquid should be tangential to the graduation mark, and both the front and the back of the mark should be seen as a single line. The neck is made narrow so that a small change in volume will have a large effect upon the height of the meniscus: the error in adjustment of the meniscus is accordingly small.

The flasks should be fabricated in accordance with BS 5898 (1980)* and the opening should be ground to standard (interchangeable) specifications and fitted with an interchangeable glass or plastic (commonly polypropylene) stopper. They should conform to either Class A or Class B specification BS 1792 (1982); examples of permitted tolerances for Class B grade are as follows:

Flask size	5	25	100	250	1000 mL
Tolerance	0.04	0.06	0.15	0.30	0.80 mL

For Class A flasks the tolerances are approximately halved: such flasks may be purchased with a works calibration certificate, or with a British Standard Test (BST) Certificate.

Graduated flasks are available in the following capacities: 1, 2, 5, 10, 20, 50, 100, 200, 250, 500, 1000, 2000 and 5000 mL. They are employed in making up standard solutions to a given volume; they can also be used for obtaining, with the aid of pipettes, aliquot portions of a solution of the substance to be analysed.

3.11 PIPETTES

Pipettes are of two kinds: (i) those which have one mark and *deliver* a small, constant volume of liquid under certain specified conditions (transfer pipettes);

---

*Many modern British Standards are closely linked to the specifications laid down by the International Standardisation Organisation based in Geneva; in the above example the relevant reference is to ISO 384-1978.[7]

(ii) those in which the stems are graduated and are employed to deliver various small volumes at discretion (graduated or measuring pipettes).

The **transfer pipette** consists of a cylindrical bulb joined at both ends to narrower tubing: a calibration mark is etched around the upper (suction) tube, while the lower (delivery) tube is drawn out to a fine tip. The graduated or measuring pipette is usually intended for the delivery of pre-determined variable volumes of liquid: it does not find wide use in accurate work, for which a burette is generally preferred. Transfer pipettes are constructed with capacities of 1, 2, 5, 10, 20, 25, 50 and 100 mL; those of 10, 25 and 50 mL capacity are most frequently employed in macro work. They should conform to BS 1583 (1986); ISO 648-1984 and should carry a colour code ring at the suction end to identify the capacity [BS 5898 (1980)]: as a safety measure an additional bulb is often incorporated above the graduation mark. They may be fabricated from lime-soda or Pyrex glass, and some high-grade pipettes are manufactured in Corex glass (Corning Glass Works, USA). This is glass which has been subjected to an ion exchange process which strengthens the glass and also leads to a greater surface hardness, thus giving a product which is resistant to scratching and chipping. Pipettes are available to Class A and Class B specifications: for the latter grade typical tolerance values are:

Pipette capacity	5	10	25	50	100 mL
Tolerance	0.01	0.04	0.06	0.08	0.12 mL

whilst for Class A, the tolerances are approximately halved.

To use such pipettes, a suitable **pipette filler** is first attached to the upper or suction tube. These devices are obtainable in various forms, a simple version consisting of a rubber or plastic bulb fitted with glass ball valves which can be operated between finger and thumb: these control the entry and expulsion of air from the bulb and thus the flow of liquid into and out of the pipette. **Suction by mouth must never be used to fill a pipette with liquid chemicals or with a solution containing chemicals.**

The pipette is then rinsed with a little of the liquid to be transferred, and then filled with the liquid to about 1–2 cm above the graduation mark. Any adhering liquid is removed from the outside of the lower stem by wiping with a piece of filter paper, and then by careful manipulation of the filler, the liquid is allowed to run out slowly until the bottom of the meniscus just reaches the graduation mark: the pipette must be held vertically and with the graduation mark at eye-level. Any drops adhering to the tip are removed by stroking against a glass surface. The liquid is then allowed to run into the receiving vessel, the tip of the pipette touching the wall of the vessel. When the continuous discharge has ceased, the jet is held in contact with the side of the vessel for 15 seconds (**draining time**). At the end of the draining time, the tip of the pipette is removed from contact with the wall of the receptacle; the liquid remaining in the jet of the pipette must not be removed either by blowing or by other means.

A pipette will not deliver constant volumes of liquid if discharged too rapidly. The orifice must be of such size that the time of outflow is about 20 seconds for a 10 mL pipette, 30 seconds for a 25 mL pipette, and 35 seconds for a 50 mL pipette.

Graduated pipettes consist of straight, fairly narrow tubes with no central bulb, and are also constructed to a standard specification [BS 6696 (1986)];

they are likewise colour-coded in accordance with ISO 1769. Three different types are available:

Type 1   delivers a measured volume from a top zero to a selected graduation mark;

Type 2   delivers a measured volume from a selected graduation mark to the jet: i.e. the zero is at the jet;

Type 3   is calibrated to *contain* a given capacity from the jet to a selected graduation mark, and thus to *remove* a selected volume of solution.

For Type 2 pipettes the final drop of liquid remaining in the tip must be expelled, which is contrary to the usual procedure. Such pipettes are therefore distinguished by a white or sand-blasted ring near the top of the pipette.

For dealing with smaller volumes of solution, **micropipettes**, often referred to as **syringe pipettes**, are employed. These can be of a 'push-button' type, in which the syringe is operated by pressing a button on the top of the pipette: the plunger travels between two fixed stops and so a remarkably constant volume of liquid is delivered. Such pipettes are fitted with disposable plastic tips (usually of polythene or polypropylene) which are not wetted by aqueous solutions, thus helping to ensure constancy of the volume of liquid delivered. The liquid is contained entirely within the plastic tip and so, by replacing the tip, the same pipette can be employed for different solutions. Such pipettes are available to deliver volumes of 1 to 1000 μL, and the delivery is reproducible to within about 1 per cent.

Microlitre syringe pipettes are available with capacities ranging from 10 to 250 μL and with the body of the pipette calibrated. When fitted with a needle tip they are particularly useful for introducing liquids into gas chromatographs (Chapter 9).

Micrometer syringe pipettes are fitted with a micrometer head which operates the plunger of the syringe, and when fitted with a stainless steel needle tip can be used for the dropwise addition of liquid; the volume added is recorded by the micrometer.

Automatic pipettes. The Dafert pipette (Fig. 3.2) is an automatic version of a transfer pipette. One side of the two-way tap is connected to a reservoir containing the solution to be dispensed. When the tap is in the appropriate position, solution fills the pipette completely, excess solution draining away

Fig. 3.2

through the overflow chamber. The pipette now contains a definite volume of solution which is delivered to the receiver by appropriate manipulation of the tap. These pipettes, are available in a range of sizes from 5–100 mL and are useful in routine work.

Autodispensers are also useful for measuring definite volumes of solutions on a routine basis. Solution is forced out of a container by depressing a syringe plunger: the movement of the plunger and hence the volume of liquid dispensed are controlled by means of a moveable clamp. The plunger is spring-loaded so that, when released, it returns to its original position and is immediately ready for operation again.

Tilting pipettes, which are attached to a reagent bottle, are only suitable for delivering approximate volumes of solution.

3.12 BURETTES

Burettes are long cylindrical tubes of uniform bore throughout the graduated length, terminating at the lower end in a glass stopcock and a jet; in cheaper varieties, the stopcock may be replaced by a rubber pinch valve incorporating a glass sphere. A diaphragm-type plastic burette tap is marketed: this can be fitted to an ordinary burette and provides a delicate control of the outflow of liquid. The merits claimed include: (a) the tap cannot stick, because the liquid in the burette cannot come into contact with the threaded part of the tap; (b) no lubricant is generally required; (c) there is no contact between ground glass surfaces; and (d) burettes and taps can be readily replaced. Burette taps made of polytetrafluoroethylene (PTFE or Teflon) are also available; these have the great advantage that no lubricant is required.

It is sometimes advantageous to employ a burette with an extended jet which is bent twice at right angles so that the tip of the jet is displaced by some 7.5–10 cm from the body of the burette. Insertion of the tip of the burette into complicated assemblies of apparatus is thus facilitated, and there is a further advantage, that if heated solutions have to be titrated the body of the burette is kept away from the source of heat. Burettes fitted with two-way stopcocks are useful for attachment to reservoirs of stock solutions.

As with other graduated glassware, burettes are produced to both Class A and Class B specifications in accordance with the appropriate standard [BS 846 (1985); ISO 385 (1984)], and Class A burettes may be purchased with BST Certificates. All Class A and some Class B burettes have graduation marks which completely encircle the burette; this is a very important feature for the avoidance of parallax errors in reading the burette. Typical values for the tolerances permitted for Class A burettes are:

Total capacity	5	10	50	100 mL
Tolerance	0.02	0.02	0.05	0.10 mL

For Class B, these values are approximately doubled. In addition to the volume requirements, limits are also imposed on the length of the graduated part of the burette and on the drainage time.

When in use, a burette must be firmly supported on a stand, and various types of burette holders are available for this purpose. The use of an ordinary laboratory clamp is not recommended: the ideal type of holder permits the burette to be read without the need of removing it from the stand.

Lubricants for glass stopcocks. The object of lubricating the stopcock of a burette is to prevent sticking or 'freezing' and to ensure smoothness in action. The simplest lubricant is pure Vaseline, but this is rather soft, and, unless used sparingly, portions of the grease may readily become trapped at the point where the jet is joined to the barrel of the stopcock, and lead to blocking of the jet. Various products are available commercially which are better suited to the lubrication of burette stopcocks. **Silicone-containing lubricants must be avoided** since they tend to 'creep' with consequent contamination of the walls of the burette.

To lubricate the stopcock, the plug is removed from the barrel and two thin streaks of lubricant are applied to the length of the plug on lines roughly midway between the ends of the bore of the plug. Upon replacing in the barrel and turning the tap a few times, a uniform thin film of grease is distributed round the ground joint. A spring or some other form of retainer may be subsequently attached to the key to lessen the chance of it becoming dislodged when in use.

Reference is again made to the Teflon stopcocks and to the diaphragm type of burette tap which do not require lubrication.

Mode of use of a burette. If necessary, the burette is thoroughly cleaned using one of the cleaning agents described in Section 3.8 and is then well rinsed with distilled water. The plug of the stopcock is removed from the barrel, and after wiping the plug and the inside of the barrel dry, the stopcock is lubricated as described in the preceding paragraph. Using a small funnel, about 10 mL of the solution to be used are introduced into the burette, and then after removing the funnel, the burette is tilted and rotated so that the solution flows over the whole of the internal surface; the liquid is then discharged through the stopcock. After repeating the rinsing process, the burette is clamped *vertically* in the burette holder and then filled with the solution to a little above the zero mark. The funnel is removed, and the liquid discharged through the stopcock until the lowest point of the liquid meniscus is just below the zero mark; the jet is inspected to ensure that all air bubbles have been removed and that it is completely full of liquid. To read the position of the meniscus, the eye must be at the same level as the meniscus, in order to avoid errors due to parallax. In the best type of burette, the graduations are carried completely round the tube for each millilitre (mL) and half-way round for the other graduation marks: parallax is thus easily avoided. To aid the eye in reading the position of the meniscus a piece of white paper or cardboard, the lower half of which is blackened either by painting with dull black paint or by pasting a piece of dull black paper upon it, is employed. When this is placed so that the sharp dividing line is 1–2 mm below the meniscus, the bottom of the meniscus appears to be darkened and is sharply outlined against the white background; the level of the liquid can then be accurately read. A variety of 'burette readers' are available from laboratory supply houses, and a home-made device which is claimed to be particularly effective has been described by Woodward and Redman.[6c] For all ordinary purposes readings are made to 0.05 mL, but for precision work, readings should be made to 0.01–0.02 mL, using a lens to assist the estimation of the subdivisions.

To deliver liquid from a burette into a conical flask or other similar receptacle, place the fingers of the left hand behind the burette and the thumb in front, and hold the tap between the thumb and the fore and middle fingers. In this way, there is no tendency to pull the plug out of the barrel of the stopcock, and the operation is under complete control. Any drop adhering to the jet after the liquid has been discharged is removed by bringing the side of the receiving vessel into contact with the jet. During the delivery of the liquid, the flask may be gently rotated with the right hand to ensure that the added liquid is well mixed with any existing contents of the flask.

3.13 WEIGHT BURETTES

For work demanding the highest possible accuracy in transferring various quantities of liquids, weight burettes are employed. As their name implies, they are weighed before and after a transfer of liquid. A very useful form is shown diagrammatically in Fig. 3.3(a). There are two ground-glass caps of which the lower one is closed, whilst the upper one is provided with a capillary opening; the loss by evaporation is accordingly negligible. For hygroscopic liquids, a small ground-glass cap is fitted to the top of the capillary tube. The burette is roughly graduated in 5 mL intervals. The titre thus obtained is in terms of weight loss of the burette, and for this reason the titrants are prepared on a weight/weight basis rather than a weight/volume basis. The errors associated with the use of a volumetric burette, such as those of drainage, reading, and change in temperature, are obviated, and weight burettes are especially useful when dealing with non-aqueous solutions or with viscous liquids.

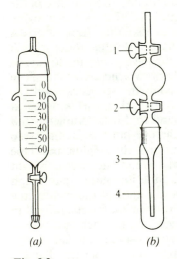

(a)   *(b)*

Fig. 3.3

An alternative form of weight burette due to Redman[6d] consists of a glass bulb, flattened on one side so that it will stand on a balance pan. Above the flattened side is the stopcock-controlled discharge jet, and a filling orifice which is closed with a glass stopper. The stopper and short neck into which it fits are pierced with holes, by alignment of which air can be admitted, thus permitting discharge of the contents of the burette through the delivery jet.

The **Lunge–Rey pipette** is shown in Fig. 3.3(*b*). There is a small central bulb (5–10 mL capacity) closed by two stopcocks 1 and 2; the pipette 3 below the stopcock has a capacity of about 2 mL, and is fitted with a ground-on test-tube 4. This pipette is of particular value for the weighing out of corrosive and fuming liquids.

3.14 PISTON BURETTES

In piston burettes, the delivery of the liquid is controlled by movement of a tightly fitting plunger within a graduated tube of uniform bore. They are particularly useful when the piston is coupled to a motor drive, and in this form serve as the basis of automatic titrators. These instruments can provide automatic plotting of titration curves, and provision is made for a variable rate of delivery as the end point is approached so that there is no danger of overshooting the end point.

3.15 GRADUATED (MEASURING) CYLINDERS

These are graduated vessels available in capacities from 2 to 2000 mL. Since the area of the surface of the liquid is much greater than in a graduated flask, the accuracy is not very high. Graduated cylinders cannot therefore be employed for work demanding even a moderate degree of accuracy. They are, however, useful where only rough measurements are required.

3.16 CALIBRATION OF VOLUMETRIC APPARATUS

For most analytical purposes, volumetric apparatus manufactured to Class A standard will prove to be satisfactory, but for work of the highest accuracy it is advisable to calibrate all apparatus for which a *recent* test certificate is unavailable. The calibration procedure involves determination of the weight of water contained in or delivered by the particular piece of apparatus. The temperature of the water is observed, and from the known density of water at that temperature, the volume of water can be calculated. Tables giving density values are usually based on weights *in vacuo* (Section 3.6), but the data given in Table 3.2 are based on weighings in air with stainless-steel weights, and these can be used to calculate the relevant volume directly from the observed weight of water. It is suggested that the data given in the table be plotted on a graph so that the volume of 1 gram of water at the exact temperature at which the calibration was performed can be ascertained. Fuller tables are given in BS 6696 (1986).

Table 3.2 Volume of 1 g of water at various temperatures

Temp. (°C)	Volume (mL)	Temp. (°C)	Volume (mL)
10.00	1.0013	22.00	1.0033
12.00	1.0015	24.00	1.0037
14.00	1.0017	26.00	1.0044
16.00	1.0021	28.00	1.0047
18.00	1.0023	30.00	1.0053
20.00	1.0027		

In all calibration operations, the apparatus to be calibrated must be carefully cleaned and allowed to stand adjacent to the balance which is to be employed, together with a supply of distilled or de-ionised water, so that they assume the temperature of the room. Flasks will also need to be dried, and this can be accomplished by rinsing twice with a little acetone and then blowing a current of air through the flask to remove the acetone.

Graduated flask. After allowing the clean dry flask to stand in the balance room for an hour it is stoppered and weighed. A small filter funnel, the stem of which has been drawn out so that it reaches below the graduation mark of the flask, is then inserted into the neck and de-ionised (distilled) water, which has also been standing in the balance room for an hour, is added slowly until the mark is reached. The funnel is then carefully removed, taking care not to wet the neck of the flask above the mark, and then, using a dropping tube, water is added dropwise until the meniscus stands on the graduation mark. The stopper is replaced, the flask re-weighed, and the temperature of the water noted. The true volume of the water filling the flask to the graduation mark can be calculated with the aid of Table 3.2.

Pipette. The pipette is filled with the distilled water which has been standing in the balance room for at least an hour, to a short distance above the mark. Water is run out until the meniscus is exactly on the mark, and the out-flow is then stopped. The drop adhering to the jet is removed by bringing the surface of some water contained in a beaker in contact with the jet, and then removing it without jerking. The pipette is then allowed to discharge into a clean, weighed stoppered flask (or a large weighing bottle) and held so that the jet of the pipette is in contact with the side of the vessel (it will be necessary to incline slightly either the pipette or the vessel). The pipette is allowed to drain for 15 seconds after the outflow has ceased, the jet still being in contact with the side of the vessel. At the end of the draining time the receiving vessel is removed from contact with the tip of the pipette, thus removing any drop adhering to the outside of the pipette and ensuring that the drop remaining in the end is always of the same size. To determine the instant at which the outflow ceases, the motion of the water surface down the delivery tube of the pipette is observed, and the delivery time is considered to be complete when the meniscus comes to rest slightly above the end of the delivery tube. The draining time of 15 seconds is counted from this moment. The receiving vessel is weighed, and the temperature of the water noted. The capacity of the pipette is then calculated with the aid of Table 3.2. At least two determinations should be made.

Burette. If it is necessary to calibrate a burette, it is essential to establish that it is satisfactory with regard to (*a*) leakage, and (*b*) delivery time, before undertaking the actual calibration process. To test for leakage, the plug is removed from the barrel of the stopcock and both parts of the stopcock are carefully cleaned of all grease; after wetting well with de-ionised water, the stopcock is reassembled. The burette is placed in the holder, filled with distilled (de-ionised) water, adjusted to the zero mark, and any drop of water adhering to the jet removed with a piece of filter paper. The burette is then allowed to stand for 20 minutes, and if the meniscus has not fallen by more than one scale division, the burette may be regarded as satisfactory as far as leakage is concerned.

To test the delivery time, again separate the components of the stopcock, dry, grease and reassemble, then fill the burette to the zero mark with distilled water, and place in the holder. Adjust the position of the burette so that the jet comes inside the neck of a conical flask standing on the base of the burette stand, but does not touch the side of the flask. Open the stopcock fully, and note the time taken for the meniscus to reach the lowest graduation mark of the burette: this should agree closely with the time marked on the burette, and in any case, must fall within the limits laid down by BS 846 (1985).

If the burette passes these two tests, the calibration may be proceeded with. Fill the burette with the distilled water which has been allowed to stand in the balance room to acquire room temperature: ideally, this should be as near to 20 °C as possible. Weigh a clean, dry stoppered flask of about 100 mL capacity, then, after adjusting the burette to the zero mark and removing any drop adhering to the jet, place the flask in position under the jet, open the stopcock fully and allow water to flow into the flask. As the meniscus approaches the desired calibration point on the burette, reduce the rate of flow until eventually it is discharging dropwise, and adjust the meniscus exactly to the required mark. Do not wait for drainage, but remove any drop adhering to the jet by touching the neck of the flask against the jet, then re-stopper and re-weigh the flask. Repeat this procedure for each graduation to be tested; for a 50 mL burette, this will usually be every 5 mL. Note the temperature of the water, and then, using Table 3.2, the volume delivered at each point is calculated from the weight of water collected. The results are most conveniently used by plotting a calibration curve for the burette.

WATER FOR LABORATORY USE

3.17 PURIFIED WATER

From the earliest days of quantitative chemical measurements it has been recognised that some form of purification is required for water which is to be employed in analytical operations, and with increasingly lower limits of detection being attained in instrumental methods of analysis, correspondingly higher standards of purity are imposed upon the water used for preparing solutions. Standards have now been laid down for water to be used in laboratories,[8] which prescribe limits for non-volatile residue, for residue remaining after ignition, for pH and for conductance. The British Standard 3978 (1987) (ISO 3696-1987) recognises three different grades of water.

(a) *Grade 3* is suitable for ordinary analytical purposes and may be prepared by single distillation of tap water, by de-ionisation, or by reverse osmosis: see below.

(b) *Grade 2* is suitable for more sensitive analytical procedures, such as atomic absorption spectroscopy and the determination of substances present in trace quantities. Water of this quality can be prepared by redistillation of Grade 3 distilled water, or by the distillation of de-ionised water, or of the product of reverse osmosis procedures.

(c) *Grade 1* water is suitable for the most stringent requirements including high-performance liquid chromatography and the determination of substances present in ultratrace amounts. Such water is obtained by subjecting

Grade 2 water to reverse osmosis or de-ionisation, followed by filtration through a membrane filter of pore size 0.2 μm to remove particulate matter. Alternatively, Grade 2 water may be redistilled in an apparatus constructed from fused silica.

The standards laid down for the three grades of water are summarised in Table 3.3.

Table 3.3 Standards for water to be used in analytical operations

Parameter	Grade of water		
	1	2	3
pH at 25 °C	*	*	5.0–7.5
Electrical conductance, mS m^{-1} at 25 °C	0.01	0.1	0.5
Oxidisable matter (equivalent to mg oxygen L^{-1})	†	0.08	0.4
Absorbance at 254 nm, 1 cm cell	0.001	0.01	‡
Residue after evaporation, mg kg^{-1}	†	1	2
SiO$_2$ content, mg L^{-1}	0.01	0.02	‡

* pH measurements in highly purified water are difficult; results are of doubtful significance.
† Not applicable.
‡ Not specified.

For many years the sole method of purification available was by distillation, and distilled water was universally employed for laboratory purposes. The modern water-still is usually made of glass and is heated electrically, and provision is made for interrupting the current in the event of failure of the cooling water, or of the boiler-feed supply; the current is also cut off when the receiver is full.

Pure water can also be obtained by allowing tap water to percolate through a mixture of ion exchange resins: a strong acid resin which will remove cations from the water and replace them by hydrogen ions, and a strong base resin (OH$^-$ form) which will remove anions. A number of units are commercially available for the production of **de-ionised water**, and the usual practice is to monitor the quality of the product by means of a conductance meter. The resins are usually supplied in an interchangeable cartridge, so that maintenance is reduced to a minimum. A mixed-bed ion exchange column fed with distilled water is capable of producing water with the very low conductance of about $2.0 \times 10^{-6} \Omega^{-1}$ cm^{-1} (2.0 μs cm^{-1}), but in spite of this very low conductance, the water may contain traces of organic impurities which can be detected by means of a spectrofluorimeter. For most purposes, however, the traces of organic material present in de-ionised water can be ignored, and it may be used in most situations where distilled water is acceptable.

An alternative method of purifying water is by **reverse osmosis**. Under normal conditions, if an aqueous solution is separated by a semi-permeable membrane from pure water, osmosis will lead to water entering the solution to dilute it. If, however, sufficient pressure is applied to the solution, i.e. a pressure in excess of its osmotic pressure, then water will flow through the membrane *from* the solution; the process of reverse osmosis is taking place. This principle has been adapted as a method of purifying tap water. The tap water, at a pressure of 3–5 atmospheres, is passed through a tube containing the semi-permeable membrane. The permeate which is collected usually still contains traces of inorganic material and is therefore not suitable for operations requiring very pure water, but it will serve for many laboratory purposes, and is very suitable for further purification by ion exchange treatment. The water produced by reverse osmosis is passed first through a bed of activated charcoal which removes organic contaminants, and is then passed through a mixed-bed ion exchange column and the resultant effluent is finally filtered through a sub-micron filter membrane to remove any last traces of colloidal organic particles.

The **high-purity water** thus produced typically has a conductance of about $0.5 \times 10^{-6}\,\Omega^{-1}\,cm^{-1}$ ($0.5\,\mu S\,cm^{-1}$) and is suitable for use under the most stringent requirements. It will meet the purity required for trace-element determinations and for operations such as ion chromatography. It must however be borne in mind that such water can readily become contaminated from the vessels in which it is stored, and also by exposure to the atmosphere. For the determination of organic compounds the water should be stored in containers made of resistant glass (e.g. Pyrex), or ideally of fused silica, whereas for inorganic determinations the water is best stored in containers made from polythene or from polypropylene.

3.18 WASH BOTTLES

A wash bottle is a flat-bottomed flask fitted up to deliver a fine stream of distilled water or other liquid for use in the transfer and washing of precipitates. A convenient size is a 500–750 mL flask of Pyrex or other resistant glass; it should be fitted up as shown in Fig. 3.4. A rubber bung is used, and the jet should deliver a fine stream of water; a suitable diameter of the orifice is 1 mm. Thick string, foam rubber, or other insulating material, held in place by copper wire, should be wrapped round the neck of the flask in order to protect the hand when hot water is used. In order to protect the mouth from scalding by the back rush of steam through the mouth-piece when the blowing is stopped, it is convenient to use a three-holed rubber stopper; a short piece of glass tubing open at both ends is inserted in the third hole. The thumb is kept over this tube whilst the water is being blown out, and is removed immediately before the mouth pressure is released. All-glass wash bottles, fitted with ground-glass joints, can be purchased. They should be used with organic solvents that attack rubber.

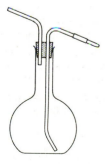

Fig. 3.4

A **polythene wash bottle** is available commercially and is inexpensive. It is fitted with a plastic cap carrying a plastic jet, and has flexible sides. The bottle can be held in the hand; application of slight pressure by squeezing gives an easily controllable jet of water. It is more or less unbreakable and is inert to many wash liquids. A polythene wash bottle should be used only for cool liquids.

Polythene wash bottles are sometimes charged with wash liquids other than water. Attention must be drawn to the fact that the components of some wash solutions may pass into the polythene and may be released into the space in the bottle when it is set aside: repeated fillings and rinsings may be required to remove the chemicals from the bottle. It is safer to label the wash bottle and to reserve it for the special wash liquid. Such wash solutions include a weakly acid solution saturated with hydrogen sulphide, dilute aqueous ammonia, saturated bromine water, and dilute nitric acid.

GENERAL APPARATUS

3.19 GLASSWARE, CERAMICS, PLASTIC WARE

In the following sections, a brief account of general laboratory apparatus relevant to quantitative analysis will be given. The commonest materials of construction of such apparatus are glass, porcelain, fused silica, and various plastics; the merits and disadvantages of these are considered below.

Glassware. In order to avoid the introduction of impurities during analysis, apparatus of resistance glass should be employed. For most purposes Pyrex glass (a borosilicate glass) is to be preferred. Resistance glass is very slightly affected by all solutions, but, in general, attack by acid solutions is less than that by pure water or by alkaline solutions; for this reason the latter should be acidified whenever possible, if they must be kept in glass for any length of time. Attention should also be given to watch, clock, and cover glasses; these should also be of resistance glass. As a rule, glassware should not be heated with a naked flame; a wire gauze should be interposed between the flame and the glass vessel.

For special purposes, Corning Vycor glass (96 per cent silica) may be used. It has great resistance to heat and equally great resistance to thermal shock, and is unusually stable to acids (except hydrofluoric acid), water, and various solutions.

The most satisfactory **beakers** for general use are those provided with a spout. The advantages of this form are: (a) convenience of pouring, (b) the spout forms a convenient place at which a stirring rod may protrude from a covered beaker, and (c) the spout forms an outlet for steam or escaping gas when the beaker is covered with an ordinary clock glass. The size of a beaker must be selected with due regard to the volume of the liquid which it is to contain. The most useful sizes are from 250 to 600 mL.

Conical (or Erlenmeyer's) flasks of 200–500 mL capacity find many applications, for example, in titrations.

Funnels should enclose an angle of 60°. The most useful sizes for quantitative analysis are those with diameters of 5.5, 7 and 9 cm. The stem should have an internal diameter of about 4 mm and should not be more than 15 cm long. For filling burettes and transferring solids to graduated flasks, a short-stem, wide-necked funnel is useful.

Porcelain apparatus. Porcelain is generally employed for operations in which hot liquids are to remain in contact with the vessel for prolonged periods. It is usually considered to be more resistant to solutions, particularly alkaline solutions, than glass, although this will depend primarily upon the quality of the glaze. Shallow porcelain basins with lips are employed for evaporations. Casseroles are lipped, flat-bottomed porcelain dishes provided with handles; they are more convenient to use than dishes.

   Porcelain crucibles are very frequently utilised for igniting precipitates and heating small quantities of solids because of their cheapness and their ability to withstand high temperatures without appreciable change. Some reactions, such as fusion with sodium carbonate or other alkaline substances, and also evaporations with hydrofluoric acid, cannot be carried out in porcelain crucibles owing to the resultant chemical attack. A slight attack of the porcelain also takes place with pyrosulphate fusions.

Fused-silica apparatus. Two varieties of silica apparatus are available commercially, the translucent and the transparent grades. The former is much cheaper and can usually be employed instead of the transparent variety. The advantages of silica ware are: (*a*) its great resistance to heat shock because of its very small coefficient of expansion, (*b*) it is not attacked by acids at a high temperature, except by hydrofluoric acid and phosphoric acid, and (*c*) it is more resistant to pyrosulphate fusions than is porcelain. The chief disadvantages of silica are: (*a*) it is attacked by alkaline solutions and particularly by fused alkalis and carbonates, (*b*) it is more brittle than ordinary glass, and (*c*) it requires a much longer time for heating and cooling than does, say, platinum apparatus. Corning Vycor apparatus (96 per cent silica glass) possesses most of the merits of fused silica and is transparent.

Plastic apparatus. Plastic materials are widely used for a variety of items of common laboratory equipment such as aspirators, beakers, bottles, Buchner

Table 3.4 Plastics used for laboratory apparatus

Material	Appearance†	Highest temperature (°C)	Chemical reagents‡ Acids		Alkalis		Attacking organic solvents§
			Weak	**Strong**	**Weak**	**Strong**	
Polythene (L.D.)	TL	80–90	R	R*	V	R	1, 2
Polythene (H.D.)	TL–O	100–110	V	R*	V	V	2
Polypropylene	T–TL	120–130	V	R*	V	V	2
TPX (Polymethylpentene)	T	170–180	V	R*	V	V	1, 2
Polystyrene	T	85	V	R*	V	V	Most
PTFE (Teflon)	O	250–300	V	V	V	V	V
Polycarbonate	T	120–130	R	A	F	A	Most
PVC [Poly(vinyl chloride)]	T–O	50–70	R	R*	R	R	2, 3, 4
Nylon	TL–O	120	R	A	R	F	V

† O = opaque; T = transparent; TL = translucent.
‡ A = attacked; F = fairly resistant; R = resistant; R* = generally resistant but attacked by oxidising mixtures; V = very resistant.
§ 1 = hydrocarbons; 2 = chlorohydrocarbons; 3 = ketones; 4 = cyclic ethers; V = very resistant.

funnels and flasks, centrifuge tubes, conical flasks, filter crucibles, filter funnels, measuring cylinders, scoops, spatulas, stoppers, tubing, weighing bottles, etc.; such products are often cheaper than their glass counterparts, and are frequently less fragile. Although inert towards many chemicals, there are some limitations on the use of plastic apparatus, not the least of which is the generally rather low maximum temperature to which it may be exposed: salient properties of the commonly used plastic materials are summarised in Table 3.4.

Attention is drawn to the extremely inert character of Teflon, which is so lacking in reactivity that it is used as the liner in pressure digestion vessels in which substances are decomposed by heating with hydrofluoric acid, or with concentrated nitric acid (see Section 3.31).

3.20 METAL APPARATUS

Crucibles and basins required for special purposes are often fabricated from various metals, amongst which platinum holds pride of place by virtue of its general resistance to chemical attack.

Platinum. Platinum is used mainly for crucibles, dishes and electrodes; it has a very high melting point (1773 °C), but the pure metal is too soft for general use, and is therefore always hardened with small quantities of rhodium, iridium, or gold. These alloys are slightly volatile at temperatures above 1100 °C, but retain most of the advantageous properties of pure platinum, such as resistance to most chemical reagents, including molten alkali carbonates and hydrofluoric acid (the exceptions are dealt with below), excellent conductivity of heat, and extremely small adsorption of water vapour. A 25 mL platinum crucible has an area of 80–100 cm² and, in consequence, the error due to volatility may be appreciable if the crucible is made of an alloy of high iridium content. The magnitude of this loss will be evident from Table 3.5, which gives the approximate loss in weight of crucibles expressed in mg/100 cm²/hour at the temperature indicated. An alloy consisting of 95 per cent platinum and 5 per cent gold is referred to as a 'non-wetting' alloy and fusion samples are readily removed from crucibles composed of this alloy; removal is assisted by keeping the crucible tilted while the melt is solidifying. Crucibles made of this alloy are used in preparing samples for investigation by X-ray fluorescence.

A recent development is the introduction of ZGS (Zirconia Grain Stabilised) platinum. This is produced by the addition of a small amount of zirconia (zirconium(IV) oxide) to molten platinum, which leads to modification of the microstructure of the solid material with increased hot strength and greater resistance to chemical attack. Whereas the recommended operating temperature for pure platinum is 1400 °C, the ZGS material can be used up to 1650 °C.

Table 3.5 Weight loss of platinum crucibles

Temp. (°C)	Pure Pt	99% Pt–1% Ir	97.5% Pt–2.5% Ir
900	0.00	0.00	0.00
1000	0.08	0.30	0.57
1200	0.81	1.2	2.5

Apparatus can also be constructed from 'TRIM' which consists of palladium coated with ZGS platinum; this permits the production of stouter apparatus with the corrosion resistance of ZGS platinum at an appreciably cheaper price.

Platinum crucibles should be supported, when heated, upon a platinum triangle. If the latter is not available, a silica triangle may be used. Nichrome and other metal triangles should be avoided; pipe-clay triangles may contain enough iron to damage the platinum. Hot platinum crucibles must always be handled with platinum-tipped crucible tongs; unprotected brass or iron tongs produce stains on the crucible. Platinum vessels must not be exposed to a luminous flame, nor may they be allowed to come into contact with the inner cone of a gas flame; this may result in the disintegration of the surface of the metal, causing it to become brittle, owing, probably, to the formation of a carbide of platinum.

It must be appreciated that at high temperatures platinum permits the flame gases to diffuse through it, and this may cause the reduction of some substances not otherwise affected. Hence if a covered crucible is heated by a gas flame there is a reducing atmosphere in the crucible: in an open crucible diffusion into the air is so rapid that this effect is not appreciable. Thus if iron(III) oxide is heated in a covered crucible, it is partly reduced to metallic iron, which alloys with the platinum; sodium sulphate is similarly partly reduced to the sulphide. It is, advisable, therefore, in the ignition of iron compounds or sulphates to place the crucible in a slanting position with free access of air.

Platinum apparatus may be used without significant loss for:

1. Fusions with (a) sodium carbonate or fusion mixture, (b) borax and lithium metaborate, (c) alkali bifluorides, and (d) alkali hydrogensulphates (slight attack in the last case above 700 °C, which is diminished by the addition of ammonium sulphate).
2. Evaporations with (a) hydrofluoric acid, (b) hydrochloric acid in the absence of oxidising agents which yield chlorine, and (c) concentrated sulphuric acid (a slight attack may occur).
3. Ignition of (a) barium sulphate and sulphates of metals which are not readily reducible, (b) the carbonates, oxalates, etc., of calcium, barium and strontium, and (c) oxides which are not readily reducible, e.g. CaO, SrO, Al_2O_3, Cr_2O_3, Mn_3O_4, TiO_2, ZrO_2, ThO_2, MoO_3, and WO_3. (BaO, or compounds which yield BaO on heating, attack platinum.)

Platinum is *attacked* under the following conditions, and such operations must not be conducted in platinum vessels:

1. Heating with the following liquids: (a) aqua regia, (b) hydrochloric acid and oxidising agents, (c) liquid mixtures which evolve bromine or iodine, and (d) concentrated phosphoric acid (slight, but appreciable, action after prolonged heating).
2. Heating with the following solids, their fusions, or vapours: (a) oxides, peroxides, hydroxides, nitrates, nitrites, sulphides, cyanides, hexacyanoferrate(III), and hexacyanoferrate(II) of the alkali and alkaline-earth metals (except oxides and hydroxides of calcium and strontium); (b) molten lead, silver, copper, zinc, bismuth, tin, or gold, or mixtures which form these metals upon reduction; (c) phosphorus, arsenic, antimony, or silicon, or mixtures which form these elements upon reduction, particularly phosphates, arsenates,

95

and silicates in the presence of reducing agents; (*d*) sulphur (slight action), selenium, and tellurium; (*e*) volatile halides (including iron(III) chloride), especially those which decompose readily; (*f*) all sulphides or mixtures containing sulphur and a carbonate or hydroxide; and (*g*) substances of unknown composition: (*h*) heating in an atmosphere containing chlorine, sulphur dioxide, or ammonia, whereby the surface is rendered porous.

Solid carbon, however produced, presents a hazard. It may be burnt off at low temperatures, with free access to air, without harm to the crucible, but it should never be ignited strongly. Precipitates in filter paper should be treated in a similar manner; strong ignition is only permissible after *all* the carbon has been removed. Ashing in the presence of carbonaceous matter should not be conducted in a platinum crucible, since metallic elements which may be present will attack the platinum under reducing conditions.

Cleaning and preservation of platinum ware.   All platinum apparatus (crucibles, dishes, etc.) should be kept clean, polished, and in proper shape. If, say, a platinum crucible becomes stained, a little sodium carbonate should be fused in the crucible, the molten solid poured out on to a dry stone or iron slab, the residual solid dissolved out with water, and the vessel then digested with concentrated hydrochloric acid: this treatment may be repeated, if necessary. If fusion with sodium carbonate is without effect, potassium hydrogensulphate may be substituted; a slight attack of the platinum will occur. Disodium tetraborate may also be used. In some cases, the use of hydrofluoric acid or potassium hydrogenfluoride may be necessary. Iron stains may be removed by heating the covered crucible with a gram or two of pure ammonium chloride and applying the full heat of a burner for 2–3 minutes.

All platinum vessels must be handled with care to prevent deformation and denting. Platinum crucibles must on no account be squeezed with the object of loosening the solidified cake after a fusion. Box-wood formers can be purchased for crucibles and dishes; these are invaluable for re-shaping dented or deformed platinum ware.

Platinum-clad stainless steel laboratory ware is available for the evaporation of solutions of corrosive chemicals. These vessels have all the corrosion-resistance properties of platinum up to about 550 °C. The main features are: (1) much lower cost than similar apparatus of platinum; (2) the overall thickness is about four times that of similar all-platinum apparatus, thus leading to greater mechanical strength; and (3) less susceptible to damage by handling with tongs, etc.

Silver apparatus.   The chief uses of silver crucibles and dishes in the laboratory are in the evaporation of alkaline solutions and for fusions with caustic alkalis; in the latter case, the silver is slightly attacked. Gold vessels (m.p. 1050 °C) are more resistant than silver to fused alkalis. Silver melts at 960 °C, and care should therefore be taken when it is heated over a bare flame.

Nickel ware.   Crucibles and dishes of nickel are employed for fusions with alkalis and with sodium peroxide (**CARE!**). In the peroxide fusion a little nickel is introduced, but this is usually not objectionable. No metal entirely withstands the action of fused sodium peroxide. Nickel oxidises in air, hence nickel apparatus cannot be used for operations involving weighing.

Iron ware. Iron crucibles may be substituted for those of nickel in sodium peroxide fusions. They are not so durable, but are much cheaper.

Stainless-steel ware. Beakers, crucibles, dishes, funnels, etc., of stainless steel are available commercially and have obvious uses in the laboratory. They will not rust, are tough, strong, and highly resistant to denting and scratching.

3.21 HEATING APPARATUS

Various methods of heating are required in the analytical laboratory ranging from gas burners, electric hot plates and ovens to muffle furnaces.

Burners. The ordinary Bunsen burner is widely employed for the attainment of moderately high temperatures. The maximum temperature is attained by adjusting the regulator so as to admit rather more air than is required to produce a non-luminous flame; too much air gives a noisy flame, which is unsuitable.

Owing to the differing combustion characteristics and calorific values of the gaseous fuels which are commonly available [natural gas, liquefied petroleum (bottled) gas], slight variations in dimensions, including jet size and aeration controls, are necessary: for maximum efficiency it is essential that, unless the burner is of the 'All Gases' type which can be adjusted, the burner should be the one intended for the available gas supply.

Hot plates. The electrically heated hot plate, preferably provided with three controls — 'Low', 'Medium' and 'High' — is of great value in the analytical laboratory. The heating elements and the internal wiring should be totally enclosed; this protects them from fumes or spilled liquids. Electric hot plates with 'stepless' controls are also marketed; these permit a much greater selection of surface temperatures to be made. A combined electric hot plate and magnetic stirrer is also available. For some purposes a steam bath may be used.

Electric ovens. The most convenient type is an electrically heated, thermostatically controlled drying oven having a temperature range from room temperature to about $250-300\,°C$; the temperature can be controlled to within $\pm 1-2\,°C$. They are used principally for drying precipitates or solids at comparatively low controlled temperatures, and have virtually superseded the steam oven.

Microwave ovens. These also find application for drying and heating operations. They are particularly useful for determining the moisture content of materials since the elimination of water takes place very rapidly on exposure to microwave radiation.

Muffle furnaces. An electrically heated furnace of muffle form should be available in every well-equipped laboratory. The maximum temperature should be about $1200\,°C$. If possible, a thermocouple and indicating pyrometer should be provided; otherwise the ammeter in the circuit should be calibrated, and a chart constructed showing ammeter and corresponding temperature readings. Gas-heated muffle furnaces are marketed; these may give temperatures up to about $1200\,°C$.

Air baths. For drying solids and precipitates at temperatures up to $250\,°C$ in which acid or other corrosive vapours are evolved, an electric oven should not be used. An air bath may be constructed from a cylindrical metal (copper, iron, or nickel) vessel with the bottom of the vessel pierced with numerous holes. A silica triangle, the legs of which are appropriately bent, is inserted inside the bath for supporting an evaporating dish, crucible, etc. The whole is heated by a Bunsen flame, which is shielded from draughts. The insulating layer of air prevents bumping by reducing the rate at which heat reaches the contents of the inner dish or crucible. An air bath of similar construction but with special heat-resistant glass sides may also be used; this possesses the obvious advantage of visibility inside the air bath.

Infrared lamps and heaters. Infrared lamps with internal reflectors are available commercially and are valuable for evaporating solutions. The lamp may be mounted immediately above the liquid to be heated: the evaporation takes place rapidly, without spattering and also without creeping. Units are obtainable which permit the application of heat to both the top and bottom of a number of crucibles, dishes, etc., at the same time; this assembly can char filter papers in the crucibles quite rapidly, and the filter paper does not catch fire.

Immersion heaters. An immersion heater consisting of a radiant heater encased in a silica sheath, is useful for the direct heating of most acids and other liquids (except hydrofluoric acid and concentrated caustic alkalis). Infrared radiation passes through the silica sheath with little absorption, so that a large proportion of heat is transferred to the liquid by radiation. The heater is almost unaffected by violent thermal shock due to the low coefficient of thermal expansion of the silica.

Heating mantles. These consist of a flexible 'knitted' fibre glass sheath which fits snugly around a flask and contains an electrical heating element which operates at black heat. The mantle may be supported in an aluminium case which stands on the bench, but for use with suspended vessels the mantle is supplied without a case. Electric power is supplied to the heating element through a control unit which may be either a continuously variable transformer or a thyristor controller, and so the operating temperature of the mantle can be smoothly adjusted.

Heating mantles are particularly designed for the heating of flasks and find wide application in distillation operations. For details of the distillation procedure and description of the apparatus employed, a textbook of practical organic chemistry should be consulted.[9]

Crucibles and beaker tongs. Apparatus such as crucibles, evaporating basins and beakers which have been heated need to be handled with suitable tongs. Crucible tongs should be made of solid nickel, nickel steel, or other rustless ferro-alloy. For handling hot platinum crucibles or dishes, platinum-tipped tongs must be used. Beaker tongs are available for handling beakers of 100–2000 mL capacity. The tongs have jaws: an adjustable screw with locknut limits the span of these jaws and enables the user to adjust them to suit the container size.

3.22 DESICCATORS AND DRY BOXES

It is usually necessary to ensure that substances which have been dried by heating (e.g. in an oven, or by ignition) are not unduly exposed to the atmosphere, otherwise they will absorb moisture more or less rapidly. In many cases, storage in the dry atmosphere of a desiccator, allied to minimum exposure to the atmosphere during subsequent operations, will be sufficient to prevent appreciable absorption of water vapour. Some substances, however, are so sensitive to atmospheric moisture that all handling must be carried out in a 'dry box'.

A **desiccator** is a covered glass container designed for the storage of objects in a dry atmosphere; it is charged with some drying agent, such as anhydrous calcium chloride (largely used in elementary work), silica gel, activated alumina, or anhydrous calcium sulphate ('Drierite'). Silica gel, alumina and calcium sulphate can be obtained which have been impregnated with a cobalt salt so that they are self-indicating: the colour changes from blue to pink when the desiccant is exhausted. The spent material can be regenerated by heating in an electric oven at $150-180\,°C$ (silica gel); $200-300\,°C$ (activated alumina) $230-250\,°C$ (Drierite); and it is therefore convenient to place these drying agents in a shallow dish which is situated at the bottom of the desiccator, and which can be easily removed for baking as required.

The action of desiccants can be considered from two points of view. The amount of moisture that remains in a closed space, containing incompletely exhausted desiccant, is related to the vapour pressure of the latter, i.e. the vapour pressure is a measure of the extent to which the desiccant can remove moisture, and therefore of its efficiency. A second factor is the weight of water that can be removed per unit weight of desiccant, i.e. the drying capacity. In general, substances that form hydrates have higher vapour pressures but also have greater drying capacities. It must be remembered that a substance cannot be dried by a desiccant of which the vapour pressure is greater than that of the substance itself.

The relative efficiencies of various drying agents will be evident from the data presented in Table 3.6. These were determined by aspirating properly conditioned air through **U**-tubes charged with the desiccants; they are applicable, strictly, to the use of these desiccants in absorption tubes, but the figures may reasonably be applied as a guide for the selection of desiccants for desiccators. It would appear from the table that a hygroscopic material such as ignited alumina should not be allowed to cool in a covered vessel over 'anhydrous' calcium chloride; anhydrous magnesium perchlorate or phosphorus pentoxide is satisfactory.

Table 3.6 Comparative efficiency of drying agents

Drying agent	Residual water (mg per L of air)	Drying agent	Residual water (mg per L of air)
$CaCl_2$ (gran. 'anhyd.' tech.)	1.5	Al_2O_3	0.005
NaOH (sticks)	0.8	$CaSO_4$	0.005
H_2SO_4 (95%)	0.3	Molecular sieve	0.004
Silica gel	0.03	H_2SO_4	0.003
KOH (sticks)	0.014	$Mg(ClO_4)_2$	0.002
		P_2O_5	0.00002

The normal (or Scheibler) desiccator is provided with a porcelain plate having apertures to support crucibles, etc.: this is supported on a constriction situated roughly halfway up the wall of the desiccator. For small desiccators, a silica triangle, with the wire ends suitably bent, may be used. The ground edge of the

99

desiccator should be lightly coated with white Vaseline or a special grease in order to make it air-tight; too much grease may permit the lid to slide.

There is however controversy regarding the effectiveness of desiccators. If the lid is briefly removed from a desiccator then it may take as long as two hours to remove the atmospheric moisture thus introduced, and to re-establish the dry atmosphere: during this period, a hygroscopic substance may actually gain in weight while in the desiccator. It is therefore advisable that any substance which is to be weighed should be kept in a vessel with as tightly fitting a lid as possible while it is in the desiccator.

The problem of the cooling of hot vessels within a desiccator is also important. A crucible which has been strongly ignited and immediately transferred to a desiccator may not have attained room temperature even after one hour. The situation can be improved by allowing the crucible to cool for a few minutes before transferring to the desiccator, and then a cooling time of 20–25 minutes is usually adequate. The inclusion in the desiccator of a metal block (e.g. aluminium), upon which the crucible may be stood, is also helpful in ensuring the attainment of temperature equilibrium.

When a hot object, such as a crucible, is placed in a desiccator, about 5–10 seconds should elapse for the air to become heated and expand before putting the cover in place. When re-opening, the cover should be slid open very gradually in order to prevent any sudden inrush of air due to the partial vacuum which exists owing to the cooling of the expanded gas content of the desiccator, and thus prevent material being blown out of the crucible.

A desiccator is frequently also employed for the thorough drying of solids for analysis and for other purposes. Its efficient operation depends upon the condition of the desiccant; the latter should therefore be renewed at frequent intervals, particularly if its drying capacity is low. For dealing with large quantities of solid a vacuum desiccator is advisable.

Convenient types of **'vacuum' desiccators** are illustrated in Fig. 3.5. Large surfaces of the solid can be exposed; the desiccator may be evacuated, and drying is thus much more rapid than in the ordinary Scheibler type. These desiccators are made of heavy glass, plastics, or even metal, and are designed to withstand reduced pressure; nevertheless, no desiccator should be evacuated unless it is surrounded by an adequate guard in the form of a stout wire cage.

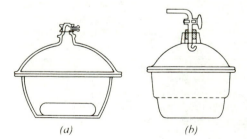

(a)   *(b)*

Fig. 3.5

For most purposes the 'vacuum' produced by an efficient water pump (20–30 mm mercury) will suffice; a guard tube containing desiccant should be inserted between the pump and the desiccator. The sample to be dried should be covered with a watch or clock glass, so that no mechanical loss ensues as a result of the removal or admission of air. Air must be admitted slowly into an exhausted desiccator: if the substance is very hygroscopic, a drying train should be attached to the stopcock. In order to maintain a satisfactory vacuum within the desiccator, the flanges on both the lid and the base must be well lubricated with Vaseline or other suitable grease. In some desiccators an elastomer ring is incorporated in a groove in the flange of the lower component of the desiccator: when the pressure is reduced, the ring is compressed by the lid of the desiccator, and an air-tight seal is produced without the need for any grease. The same desiccants are used as with an ordinary desiccator.

Dry boxes (glove boxes) which are especially intended for the manipulation of materials which are very sensitive to atmospheric moisture (or to oxygen), consist of a plastic or metal box provided with a window (of glass or clear plastic) on the upper side, and sometimes also on the side walls. A pair of rubber or plastic gloves are fitted through air-tight seals through the front side of the box, and by placing the hands and forearms into the gloves, manipulations may be carried out inside the box. One end of the box is fitted with an air-lock so that apparatus and materials can be introduced into the box without disturbing the atmosphere inside. A tray of desiccant placed inside the box will maintain a dry atmosphere, but to counter the unavoidable leakages in such a system, it is advisable to supply a slow current of dry air to the box; inlet and outlet taps are provided to control this operation. If the box is flushed out before use with an inert gas (e.g. nitrogen), and a slow stream of the gas is maintained while the box is in use, materials which are sensitive to oxygen can be safely handled. For a detailed discussion of the construction and uses of glove boxes, see Ref. 10.

3.23 STIRRING APPARATUS

Many operations involving solutions of reagents require the thorough mixing of two or more reactants, and apparatus suitable for this purpose ranges from a simple glass stirring rod to electrically operated stirrers.

Stirring rods. These are made from glass rod 3–5 mm in diameter, cut into suitable lengths. Both ends should be rounded by heating in the Bunsen or blowpipe flame. The length of the stirring rod should be suitable for the size and the shape of the vessel for which it is employed, e.g. for use with a beaker provided with a spout, it should project 3–5 cm beyond the lip when in a resting position.

A short piece of Teflon or of rubber tubing (or a rubber cap) fitted tightly over one end of a stirring rod of convenient size gives the so-called 'policeman'; it is used for detaching particles of a precipitate adhering to the side of a vessel which cannot be removed by a stream of water from a wash bottle: it should not, as a rule, be employed for stirring, nor should it be allowed to remain in a solution.

Boiling rods. Boiling liquids and liquids in which a gas, such as hydrogen sulphide, sulphur dioxide etc., has to be removed by boiling can be prevented from superheating and 'bumping' by the use of a boiling rod (Fig. 3.6). This consists of a piece of glass tubing closed at one end and sealed approximately 1 cm from the other end; the latter end is immersed in the liquid. When the rod

Fig. 3.6

is removed, the liquid in the open end must be shaken out and the rod rinsed with a jet of water from a wash bottle. This device should not be used in solutions which contain a precipitate.

Stirring may be conveniently effected with the so-called **magnetic stirrer**. A rotating field of magnetic force is employed to induce variable-speed stirring action within either closed or open vessels. The stirring is accomplished with the aid of a small cylinder of iron sealed in Pyrex glass, polythene, or Teflon, which is caused to rotate by a rotating magnet.

The usual type of glass paddle stirrer is also widely used in conjunction with an electric motor fitted with either a transformer-type, or a solid-state, speed controller. The stirrer may be either connected directly to the motor shaft or to a spindle actuated by a gear box which forms an integral part of the motor housing; by these means, wide variation in stirrer speed can be achieved.

Under some circumstances, e.g. the dissolution of a sparingly soluble solid, it may be more advantageous to make use of a **mechanical shaker**. Various models are available, ranging from 'wrist action shakers' which will accommodate small-to-moderate size flasks, to those equipped with a comparatively powerful electric motor and capable of shaking the contents of large bottles vigorously.

3.24   FILTRATION APPARATUS

The simplest apparatus used for filtration is the filter funnel fitted with a filter paper. The funnel should have an angle as close to 60° as possible, and a long stem (15 cm) to promote rapid filtration. Filter papers are made in varying grades of porosity, and one appropriate to the type of material to be filtered must be chosen (see Section 3.34).

In the majority of quantitative determinations involving the collection and weighing of a precipitate, it is convenient to be able to collect the precipitate in a crucible in which it can be weighed directly, and various forms of **filter crucible** have been devised for this purpose. Sintered glass crucibles are made of resistance glass and have a porous disc of sintered ground glass fused into the body of the crucible. The filter disc is made in varying porosities as indicated by numbers from 0 (the coarsest) to 5 (the finest); the range of pore diameter for the various grades is as follows:

Porosity	0	1	2	3	4	5
Pore diameter μm	200–250	100–120	40–50	20–30	5–10	1–2

Porosity 3 is suitable for precipitates of moderate particle size, and porosity 4 for fine precipitates such as barium sulphate. These crucibles should not be heated above about 200 °C.

Silica crucibles of similar pattern are also available, and, although expensive, have certain advantages in thermal stability.

Filter crucibles with a porous filter base are available in porcelain (porosity 4), in silica (porosities 1, 2, 3, 4), and in alumina (coarse, medium and fine porosities): these have the advantage as compared with sintered crucibles, of being capable of being heated to much higher temperatures. Nevertheless, the heating must be gradual otherwise the crucible may crack at the join between porous base and glazed side.

For filtering large quantities of material, a **Buchner funnel** is usually employed; alternatively, one of the modified funnels shown diagrammatically in Fig. 3.7 may be used. Here (a) is the ordinary porcelain Buchner funnel, (b) is the 'slit sieve' glass funnel. In both cases, one or (better) two good-quality filter papers are placed on the plate; the glass type is preferable since it is transparent and it is easy to see whether the funnel is perfectly clean. Type (c) is a Pyrex funnel with a sintered glass plate; no filter paper is required so that strongly acidic and weakly alkaline solutions can be readily filtered with this funnel. In all cases the funnel of appropriate size is fitted into a filter flask (d), and the filtration conducted under the diminished pressure provided by a filter pump or vacuum line.

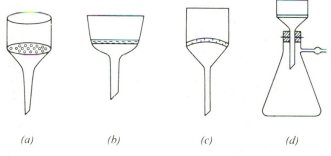

(a)   (b)   (c)   (d)

Fig. 3.7

One of the disadvantages of the porcelain Buchner funnel is that, being of one-piece construction, the filter plate cannot be removed for thorough cleaning and it is difficult to see whether the whole of the plate is clean on both sides. In a modern polythene version, the funnel is made in two sections which can be unscrewed, thus permitting inspection of both sides of the plate.

In some circumstances, separation of solid from a liquid is better achieved by use of a **centrifuge** than by filtration, and a small, electrically driven centrifuge is a useful piece of equipment for an analytical laboratory. It may be employed for removing the mother liquor from recrystallised salts, for collecting difficultly filterable precipitates, and for the washing of certain precipitates by decantation. It is particularly useful when small quantities of solids are involved; centrifuging, followed by decantation and re-centrifuging, avoids transference losses and yields the solid phase in a compact form. Another valuable application is for the separation of two immiscible phases.

3.25 WEIGHING BOTTLES

Most chemicals are weighed *by difference* by placing the material inside a stoppered weighing bottle which is then weighed. The requisite amount of substance is shaken out into a suitable vessel (beaker or flask), and the weight of substance taken is determined by re-weighing the weighing bottle. In this way, the substance dispensed receives the minimum exposure to the atmosphere during the actual weighing process: a feature of some importance if the material is hygroscopic.

The most convenient form of weighing bottle is one fitted with an external cap and made of glass, polythene or polycarbonate. A weighing bottle with an *internally* fitting stopper is not recommended; there is always the danger that small particles may lodge at the upper end of the bottle and be lost when the stopper is pressed into place.

If the substance is unaffected by exposure to the air, it may be weighed on a watch glass, or in a disposable plastic container. The weighing funnel (Fig. 3.8) is very useful, particularly when the solid is to be transferred to a flask: having weighed the solid into the scoop-shaped end which is flattened so that it will stand on the balance pan, the narrow end is inserted into the neck of the flask and the solid washed into the flask with a stream of water from a wash bottle.

Fig. 3.8

Woodward and Redman[6c] have described a specially designed weighing bottle which will accommodate a small platinum crucible: when a substance has been ignited in the crucible, the crucible is transferred to the weighing bottle and subsequently weighed in this. This device obviates the need for a desiccator.

If the substance to be weighed is a liquid, it is placed in a weighing bottle fitted with a cap carrying a dropping tube.

REAGENTS AND STANDARD SOLUTIONS

3.26 REAGENTS

The purest reagents available should be used for quantitative analysis; the analytical reagent quality is generally employed. In Great Britain 'AnalaR' chemicals from BDH Chemicals conform to the specifications given in the handbook '*AnalaR' Standards for Laboratory Chemicals*.[11] In the USA the American Chemical Society committee on Analytical Reagents has established standards for certain reagents, and manufacturers supply reagents which are labelled 'Conforms to ACS Specifications'. In addition, certain manufacturers market chemicals of high purity, and each package of these analysed chemicals has a label giving the manufacturer's limits of certain impurities.

With the increasingly lower limits of detection being achieved in various types of instrumental analysis, there is an ever growing demand for reagents of correspondingly improved specification, and some manufacturers are now offering a range of specially purified reagents such as the BDH Ltd 'Aristar' chemicals, specially purified solvents for spectroscopy (e.g. BDH Ltd 'Spectrosol') and specially prepared reagents for chromatography.

In some instances, where a reagent of the requisite purity is not available, it may be advisable to weigh out a suitable portion of the appropriate *pure* metal (e.g. the Johnson, Matthey 'Specpure' range), and to dissolve this in the appropriate acid.

It must be remembered that the label on a bottle is not an infallible guarantee of the purity of a chemical, for the following reasons:

(*a*) Some impurities may not have been tested for by the manufacturer.
(*b*) The reagent may have been contaminated after its receipt from the manufacturers either by the stopper having been left open for some time, with the consequent exposure of the contents to the laboratory atmosphere or by the accidental return of an unused portion of the reagent to the bottle.
(*c*) In the case of a solid reagent, it may not be sufficiently dry. This may be due either to insufficient drying by the manufacturers or to leakage through the stoppers during storage, or to both of these causes.

However, if the analytical reagents are purchased from a manufacturing firm of repute, the instructions given (*a*) that no bottle is to be opened for a longer time than is absolutely necessary, and (*b*) that no reagent is to be returned to the bottle after it has been removed, the likelihood of any errors arising from some of the above possible causes is considerably reduced. Liquid reagents should be poured from the bottle; a pipette should never be inserted into the reagent bottle. Particular care should be taken to avoid contamination of the stopper of the reagent bottle. When a liquid is poured from a bottle, the stopper should never be placed on the shelf or on the working bench; it may be placed upon a clean watchglass, and many chemists cultivate the habit of holding the stopper between the thumb and fingers of one hand. The stopper should be returned to the bottle immediately after the reagent has been removed, and all reagent bottles should be kept scrupulously clean, particularly round the neck or mouth of the bottle.

If there is any doubt as to the purity of the reagents used, they should be tested by standard methods for the impurities that might cause errors in the determinations. It may be mentioned that not all chemicals employed in quantitative analysis are available in the form of analytical reagents; the purest commercially available products should, if necessary, be purified by known methods: see below. The exact mode of drying, if required, will vary with the reagent; details are given for specific reagents in the text.

3.27 PURIFICATION OF SUBSTANCES

If a reagent of adequate purity for a particular determination is not available, then the purest available product must be purified: this is most commonly done by recrystallisation from water. A known weight of the solid is dissolved in a volume of water sufficient to give a saturated or nearly saturated solution at the boiling point: a beaker, conical flask or porcelain dish may be used. The hot solution is filtered through a fluted filter paper placed in a short-stemmed funnel, and the filtrate collected in a beaker: this process will remove insoluble material which is usually present. If the substance crystallises out in the funnel, it should be filtered through a hot-water funnel. The clear hot filtrate is cooled rapidly by immersion in a dish of cold water or in a mixture of ice and water, according to the solubility of the solid; the solution is constantly stirred in order to promote the formation of small crystals, which occlude less mother liquor than larger crystals. The solid is then separated from the mother liquor by filtration, using one of the Buchner-type funnels shown in Fig. 3.7 (Section 3.24). When all the liquid has been filtered, the solid is pressed down on the funnel with a wide glass stopper, sucked as dry as possible, and then washed with small portions of the original solvent to remove the adhering mother liquor. The recrystallised solid is dried upon clock glasses at or above the laboratory temperature according to the nature of the material; care must of course be taken to exclude dust. The dried solid is preserved in glass-stoppered bottles. It should be noted that unless great care is taken when the solid is removed from the funnel, there is danger of introducing fibres from the filter paper, or small particles of glass from the glass filter disc: scraping of the filter paper or of the filter disc must be avoided.

Some solids are either too soluble, or the solubility does not vary sufficiently with temperature, in a given solvent for direct crystallisation to be practicable. In many cases, the solid can be precipitated from, say, a concentrated aqueous solution by the addition of a liquid, miscible with water, in which it is less soluble. Ethanol, in which many inorganic compounds are almost insoluble, is generally used. Care must be taken that the amount of ethanol or other solvent added is not so large that the impurities are also precipitated. Potassium hydrogencarbonate and antimony potassium tartrate may be purified by this method.

Many organic compounds can be purified by recrystallisation from suitable organic solvents, and here again, precipitation by the addition of another solvent in which the required compound is insoluble, may be effective; while liquids can be purified by fractional distillation.

Sublimation. This process is employed to separate volatile substances from non-volatile impurities. Iodine, arsenic(III) oxide, ammonium chloride and a number of organic compounds can be purified in this way. The material to be purified is gently heated in a porcelain dish, and the vapour produced is condensed on a flask which is kept cool by circulating cold water inside it.

Zone refining. This is a purification technique originally developed for the refinement of certain metals, and is applicable to all substances of reasonably low melting point which are stable at the melting temperature. In a zone refining apparatus, the substance to be purified is packed into a column of glass or stainless steel, which may vary in length from 15 cm (semimicro apparatus) to 1 metre. An electric ring heater which heats a narrow band of the column is allowed to fall slowly by a motor-controlled drive, from the top to the bottom of the column. The heater is set to produce a molten zone of material at a temperature 2–3 °C above the melting point of the substance, which travels slowly down the tube with the heater. Since impurities normally lower the melting point of a substance, it follows that the impurities tend to flow down the column in step with the heater, and thus to become concentrated in the lower part of the tube. The process may be repeated a number of times (the apparatus may be programmed to reproduce automatically a given number of cycles), until the required degree of purification has been achieved.

3.28 PREPARATION AND STORAGE OF STANDARD SOLUTIONS

In any analytical laboratory it is essential to maintain stocks of solutions of various reagents: some of these will be of accurately known concentration (i.e. standard solutions) and correct storage of such solutions is imperative.

Solutions may be classified as:

1. reagent solutions which are of approximate concentration;
2. standard solutions which have a known concentration of some chemical;
3. standard reference solutions which have a known concentration of a primary standard substance (Section 10.6);
4. standard titrimetric solutions which have a known concentration (determined either by weighing or by standardisation) of a substance other than a primary standard.

The IUPAC Commission on Analytical Nomenclature refers to (3) and (4) respectively as Primary Standard Solutions and Secondary Standard Solutions.

For **reagent solutions** as defined above (i.e. 1) it is usually sufficient to weigh out approximately the amount of material required, using a watchglass or a plastic weighing container, and then to add this to the required volume of solvent which has been measured with a measuring cylinder.

To prepare a **standard solution** the following procedure is followed. A short-stemmed funnel is inserted into the neck of a graduated flask of the appropriate size. A suitable amount of the chemical is placed in a weighing bottle which is weighed, and then the required amount of substance is transferred from the weighing bottle to the funnel, taking care that no particles are lost. After the weighing bottle has been re-weighed, the substance in the funnel is washed down with a stream of the liquid. The funnel is thoroughly washed, inside and out, and then removed from the flask; the contents of the flask are dissolved, if necessary, by shaking or swirling the liquid, and then made up to the mark: for the final adjustment of volume, a dropping tube drawn out to form a very fine jet is employed.

If a watch glass is employed for weighing out the sample, the contents are transferred as completely as possible to the funnel, and then a wash bottle is used to remove the last traces of the substance from the watch glass. If the weighing scoop (Fig. 3.8; Section 3.25) is used, then of course a funnel is not needed provided that the flask is of such a size that the end of the scoop is an easy fit in the neck.

If the substance is not readily soluble in water, it is advisable to add the material from the weighing bottle or the watchglass to a beaker, followed by distilled water; the beaker and its contents are then heated gently with stirring until the solid has dissolved. After allowing the resulting concentrated solution to cool a little, it is transferred through the short-stemmed funnel to the graduated flask, the beaker is rinsed thoroughly with several portions of distilled water, adding these washings to the flask, and then finally the solution is made up to the mark: it may be necessary to allow the flask to stand for a while before making the final adjustment to the mark to ensure that the solution is at room temperature. **Under no circumstances may the graduated flask be heated.**

107

In some circumstances it may be considered preferable to prepare the standard solution by making use of one of the concentrated volumetric solutions supplied in sealed ampoules which only require dilution in a graduated flask to produce a standard solution.

Solutions which are comparatively stable and unaffected by exposure to air may be stored in 1 litre or 2.5 litre bottles; for work requiring the highest accuracy, the bottles should be Pyrex, or other resistance glass, and fitted with ground-glass stoppers: the solvent action of the solution being thus considerably reduced. It is however necessary to use a rubber bung instead of a glass stopper for alkaline solutions, and in many instances a polythene container may well replace glass vessels. It should be noted, however, that for some solutions as, for example, iodine and silver nitrate, glass containers only may be used, and in both these cases the bottle should be made of dark (brown) glass: solutions of EDTA (Section 10.49) are best stored in polythene containers.

The bottle should be clean and dry: a little of the stock solution is introduced, the bottle well rinsed with this solution, drained, the remainder of the solution poured in, and the bottle immediately stoppered. If the bottle is not dry, but has recently been thoroughly rinsed with distilled water, it may be rinsed successively with three small portions of the solution and drained well after each rinsing; this procedure is, however, less satisfactory than that employing a clean and dry vessel. Immediately after the solution has been transferred to the stock bottle, it should be labelled with: (1) the name of the solution; (2) its concentration; (3) the data of preparation; and (4) the initials of the person who prepared the solution, together with any other relevant data. Unless the bottle is completely filled, internal evaporation and condensation will cause drops of water to form on the upper part of the inside of the vessel. For this reason, the bottle must be thoroughly shaken before removing the stopper.

For expressing concentrations of reagents, the molar system is universally applicable, i.e. the number of moles of solute present in 1 L of solution. Concentrations may also be expressed in terms of normality if no ambiguity is likely to arise (see Appendix 17).

Solutions liable to be affected by access of air (e.g. alkali hydroxides which

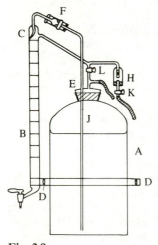

Fig. 3.9

absorb carbon dioxide; iron(II) and titanium(III) which are oxidised) may be stored in the apparatus shown diagrammatically in Fig. 3.9. A is a large storage bottle of 10–15 litres capacity. B is a 50 mL burette provided with an automatic filling device at C (the point of the drawn-out tube is adjusted to be exactly at the zero mark of the burette), D is the burette–bottle clamp, E is a two-holed ground-glass joint, F is a ground-glass tension joint, a rubber tube is connected to a hydrogen cylinder and to the **T**-joint below L, H is a Bunsen valve, and J is hydrogen. The burette is filled by closing tap K and passing hydrogen through the rubber tube attached to the **T**-piece (below tap L) with tap L closed; taps L and K are opened, and the excess of liquid allowed to siphon back.

Another apparatus for the storage of standard solutions is shown in Fig. 3.10 which is self-explanatory. The solution is contained in the storage bottle A, and the 50 mL burette is fitted into this by means of a ground-glass joint B. To fill the burette, tap C is opened and the liquid pumped into the burette by means of the small bellows E. F is a small guard tube; this is filled with soda-lime or 'Carbosorb' when caustic alkali is contained in the storage bottle. Bottles with a capacity up to 2 litres are provided with standard ground-glass joints; large bottles, up to 15 L capacity, can also be obtained. With both of these storage vessels, for strongly alkaline solutions, the ground-glass joints should be replaced by rubber bungs or rubber tubing.

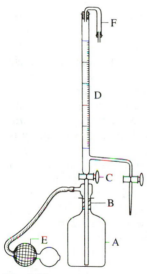

Fig. 3.10

The Dafert pipette (Fig. 3.2; Section 3.11) is a convenient apparatus for dispensing fixed volumes of a standard solution, as are also the various liquid dispensers which are available.

SOME BASIC TECHNIQUES

3.29 PREPARATION OF THE SUBSTANCE FOR ANALYSIS

Presented with a large quantity of a material to be analysed, the analyst is immediately confronted with the problem of selecting a representative sample for the analytical investigations. It may well be that the material is in such large pieces that comminution is necessary in order to produce a specimen suitable for handling in the laboratory. These important factors are considered in Chapter 5 (Sections 5.2; 5.4), and as explained therein, the material is usually dried at 105–110 °C before analysis.

3.30 WEIGHING THE SAMPLE

If necessary refer to Section 3.5 dealing with the operation of a chemical balance, and to Sections 3.25 and 3.22 which are concerned with the use of weighing bottles and desiccators respectively.

The material, prepared as above, is usually transferred to a weighing bottle which is stoppered and stored in a desiccator. Samples of appropriate size are withdrawn from the weighing bottle as required, the bottle being weighed before and after the withdrawal, so that the weight of substance is obtained by difference.

Attention is drawn to vibro-spatulas which are useful adjuncts to the weighing-out of powders. The spatula is connected to the electric mains, and the powder is placed on the blade of the spatula. When the current is switched on, the blade is caused to vibrate and to deposit solid gradually into the beaker or other container over which it is held: the intensity of the vibration may be adjusted.

3.31 SOLUTION OF THE SAMPLE

Most organic substances can be dissolved readily in a suitable organic solvent and some are directly soluble in water or can be dissolved in aqueous solutions of acids (basic materials) or of alkalis (acidic materials). Many inorganic substances can be dissolved directly in water or in dilute acids, but materials such as minerals, refractories, and alloys must usually be treated with a variety of reagents in order to discover a suitable solvent: in such cases the preliminary qualitative analysis will have revealed the best procedure to adopt. Each case must be considered on its merits; no attempt at generalisation will therefore be made. It is however of value to discuss the experimental technique of the simple process of solution of a sample in water or in acids, and also the method of treatment of insoluble substances.

For a substance which dissolves readily, the sample is weighed out into a beaker, and the beaker immediately covered with a clockglass of suitable size (its diameter should not be more than about 1 cm larger than that of the beaker) with its convex side facing downwards. The beaker should have a spout in order to provide an outlet for the escape of steam or gas. The solvent is then added by pouring it carefully down a glass rod, the lower end of which rests against the wall of the beaker; the clockglass is displaced somewhat during this process. If a gas is evolved during the addition of the solvent (e.g. acids with carbonates, metals, alloys, etc.), the beaker must be kept covered as far as possible during the addition. The reagent is then best added by means of a pipette or by means of a funnel with a bent stem inserted beneath the clockglass at the spout of the beaker; loss by spirting or as spray is thus prevented. When the evolution of gas has ceased and the substance has completely dissolved, the underside of the clockglass is well rinsed with a stream of water from a wash bottle, care being taken that the washings fall on to the side of the beaker and not directly into the solution. If warming is necessary, it is usually best to carry out the dissolution in a conical flask with a small funnel in the mouth; loss of liquid by spirting is thus prevented and the escape of gas is not hindered. When using volatile solvents, the flask should be fitted with a reflux condenser.

It may often be necessary to reduce the volume of the solution, or sometimes to evaporate completely to dryness. Wide and shallow vessels are most suitable, since a large surface is thus exposed and evaporation is thereby accelerated. Shallow beakers of resistance glass, Pyrex evaporating dishes, porcelain basins or casseroles, silica or platinum basins may be employed; the material selected will depend upon the extent of attack of the hot liquid upon it and upon the constituents being determined in the subsequent analysis. Evaporations should be carried out on the steam bath or upon a low-temperature hot plate; slow evaporation is preferable to vigorous boiling, since the latter may lead to some mechanical loss in spite of the precautions to be mentioned below. During evaporations, the vessel must be covered by a Pyrex clockglass of slightly larger diameter than the vessel, and supported either on a large all-glass triangle or upon three small U-rods of Pyrex glass hanging over the rim of the container. Needless to say, at the end of of the evaporation the sides of the vessel, the lower side of the clockglass and the triangle and glass hooks (if employed) should be rinsed with distilled water into the vessel.

For evaporation at the boiling point either a conical flask with a short Pyrex funnel in the mouth or a round-bottomed flask inclined at an angle of about 45° may be employed; in the latter the drops of liquid, etc., thrown up by the ebullition or by effervescence will be retained by striking the inside of the flask, while gas and vapour will escape freely. When organic solvents are employed the flask should be fitted with a 'swan-neck' tube and a condenser so that the solvent is recovered.

Consideration must be given to the possibility of losses occurring during the concentration procedure; for example, boric acid, halogen acids and nitric acid are lost from boiling aqueous solutions.

Substances which are insoluble (or only slightly soluble) in water can often be dissolved in an appropriate acid, but the possible loss of gaseous products must be borne in mind. The evolution of carbon dioxide, hydrogen sulphide and sulphur dioxide from carbonates, sulphides and sulphites respectively will be immediately apparent, but less obvious are losses of boron and silicon as the corresponding fluorides during evaporations with hydrofluoric acid, or loss of halogen by the treatment of halides with a strong oxidising agent such as nitric acid.

Concentrated hydrochloric acid will dissolve many metals (generally those situated above hydrogen in the electrochemical series), as well as many metallic oxides. Hot concentrated nitric acid dissolves most metals, but antimony, tin and tungsten are converted to slightly soluble acids thus providing a separation of these elements from other components of alloys. Hot concentrated sulphuric acid dissolves many substances and many organic materials are charred and then oxidised by this treatment.

A mixture of hydrochloric and nitric acids (3:1 by volume) known as *aqua regia* is a very potent solvent largely due to its oxidising character, and the addition of oxidants such as bromine or hydrogen peroxide frequently increases the solvent action of acids.

Hydrofluoric acid is mainly used for the decomposition of silicates; excess

111

hydrofluoric acid is removed by evaporation with sulphuric acid leaving a residue of metallic sulphates. Complexes of fluoride ions with many metallic cations are very stable and so the normal properties of the cation may not be exhibited. It is therefore essential to ensure complete removal of fluoride, and to achieve this, it may be necessary to repeat the evaporation with sulphuric acid two or three times. **Hydrofluoric acid must be handled with great care**; it causes serious and painful burns of the skin.

Perchloric acid attacks stainless steels and a number of iron alloys that do not dissolve in other acids. **Perchloric acid must be used with great care**; the hot concentrated acid gives explosive reactions with organic materials or easily oxidised inorganic compounds, and it is recommended that if frequent reactions and evaporations involving perchloric acid are to be performed, a fume cupboard which is free from combustible organic materials should be used. A mixture of perchloric and nitric acids is valuable as an oxidising solvent for many organic materials to produce a solution of inorganic constituents of the sample. For safety in such operations, the substance should be treated first with concentrated nitric acid, the mixture heated, and then careful additions of small quantities of perchloric acid can be made until the oxidation is complete. Even then, the mixture should not be evaporated because the nitric acid evaporates first allowing the perchloric acid to reach dangerously high concentrations. If a mixture of nitric, perchloric and sulphuric acids (3:1:1 by volume) is used, then the perchloric acid is also evaporated leaving a sulphuric acid solution of the components to be analysed. In this operation the organic part of the material under investigation is destroyed and the process is referred to as '**wet ashing**'.

Substances which are insoluble or only partially soluble in acids are brought into solution by fusion with the appropriate reagent. The most commonly used fusion reagents, or **fluxes** as they are called, are anhydrous sodium carbonate, either alone or, less frequently, mixed with potassium nitrate or sodium peroxide; potassium pyrosulphate, or sodium pyrosulphate; sodium peroxide; sodium hydroxide or potassium hydroxide. *Anhydrous* lithium metaborate has found favour as a flux, especially for materials containing silica;[12] when the resulting fused mass is dissolved in dilute acids, no separation of silica takes place as it does when a sodium carbonate melt is similarly treated. Other advantages claimed for lithium metaborate are the following.

1. No gases are evolved during the fusion or during the dissolution of the melt, and hence there is no danger of losses due to spitting.
2. Fusions with lithium metaborate are usually quicker (15 minutes will often suffice), and can be performed at a lower temperature than with other fluxes.
3. The loss of platinum from the crucible is less during a lithium metaborate fusion than with a sodium carbonate fusion.
4. Many elements can be determined directly in the acid solution of the melt without the need for tedious separations.

Naturally, the flux employed will depend upon the nature of the insoluble substance. Thus acidic materials are attacked by basic fluxes (carbonates, hydroxides, metaborates), whilst basic materials are attacked by acidic fluxes (pyroborates, pyrosulphates, and acid fluorides). In some instances an oxidising medium is useful, in which case sodium peroxide or sodium carbonate mixed with sodium peroxide or potassium nitrate may be used. The vessel in which fusion is effected must be carefully chosen; platinum crucibles are employed for sodium carbonate, lithium metaborate and potassium pyrosulphate; nickel or silver crucibles, for sodium hydroxide or potassium hydroxide; nickel, gold, silver, or iron crucibles for sodium carbonate and/or sodium peroxide; nickel crucibles for sodium carbonate and potassium nitrate (platinum is slightly attacked).

For the preparation of samples for X-ray fluorescence spectroscopy, lithium metaborate is the preferred flux because lithium does not give rise to interfering X-ray emissions. The fusion may be carried out in platinum crucibles or in crucibles made from specially prepared graphite: these graphite crucibles can also be used for the vacuum fusion of metal samples for the analysis of occluded gases.

To carry out the fusion, a layer of flux is placed at the bottom of the crucible, and then an intimate mixture of the flux and the finely divided substance added; the crucible should be not more than about half-full, and should, generally, be kept covered during the whole process. The crucible is very gradually heated at first, and the temperature slowly raised to the required temperature. The final temperature should not be higher than is actually necessary; any possible further attack of the flux upon the crucible is thus avoided. When the fusion, which usually takes 30–60 minutes, has been completed, the crucible is grasped by means of the crucible tongs and gently rotated and tilted so that the molten material distributes itself around the walls of the container and solidifies there as a thin layer. This procedure greatly facilitates the subsequent detachment and solution of the fused mass. When cold, the crucible is placed in a casserole, porcelain dish, platinum basin, or Pyrex beaker (according to the nature of the flux) and covered with water. Acid is added, if necessary, the vessel is covered with a clockglass, and the temperature is raised to 95–100 °C and maintained until solution is achieved.

Many of the substances which require fusion treatment to render them soluble will in fact dissolve in mineral acids if the digestion with acid is carried out under pressure, and consequently at higher temperatures than those normally achieved. Such drastic treatment requires a container capable of withstanding the requisite pressure, and also resistant to chemical attack: these conditions are met in **acid digestion vessels** (bombs). These comprise a stainless-steel pressure vessel (capacity 50 mL) with a screw-on lid and fitted with a Teflon liner. They may be heated to 150–180 °C and will withstand pressures of 80–90 atmospheres; under these conditions decomposition of refractory materials may be accomplished in 45 minutes. Apart from the saving in time which is achieved, and the fact that the use of expensive platinum ware is obviated, other advantages of the method are that no losses can occur during the treatment, and the resulting solution is free from the heavy loading of alkali metals which follows the usual fusion procedures. A recent modification is the construction of vessels made entirely of Teflon which can be heated in a microwave oven, with even more rapid reaction times. A full discussion of decomposition techniques is given in Ref. 13.

A decomposition procedure applicable to organic compounds containing elements such as halogens, phosphorus or sulphur, consists in combustion of the organic material in an atmosphere of oxygen; the inorganic constituents are thus converted to forms which can be determined by titrimetric or spectrophotometric procedures. The method was developed by Schöniger[14,15] and is usually referred to as the Schöniger Oxygen Flask Method. A number

113

of reviews of the procedures have been published[16,17] giving considerable details of all aspects of the subject.

In outline the procedure consists of carefully weighing about 5–10 mg of sample on to a shaped piece of paper (Fig. 3.11c) which is folded in such a way that the tail (wick) is free. This is then placed in a platinum basket or carrier suspended from the ground-glass stopper of a 500 mL or 1 litre flask. The flask, containing a few millilitres of absorbing solution (e.g. aqueous sodium hydroxide), is filled with oxygen and then sealed with the stopper with the platinum basket attached.

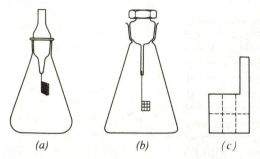

(a)   *(b)*   *(c)*

Fig. 3.11 Conventional flasks for microdeterminations: (a) air leak design; (b) stopper design; (c) filter paper for wrapping sample. Reproduced by permission from A. M. B. Macdonald, in *Advances in Analytical Chemistry and Instrumentation*, C. N. Reilly (Ed.), Vol. 4, Interscience, New York, 1965, p. 75.

The wick of the sample paper can either be ignited before the stopper is placed in the flask neck, or better still ignited by remote electrical control, or by an infrared lamp. In any case combustion is rapid and usually complete within 5–10 seconds. After standing for a few minutes until any combustion cloud has disappeared, the flask is shaken for 2–3 minutes to ensure that complete absorption has taken place. The solution can then be treated by a method appropriate to the element being determined.

Organic sulphur is converted to sulphur trioxide and sulphur dioxide by the combustion, absorbed in hydrogen peroxide, and the sulphur determined as sulphate.

The combustion products of organic halides are usually absorbed in sodium hydroxide containing some hydrogen peroxide. The resulting solutions may be analysed by a range of available procedures. For chlorides the method most commonly used is that of argentimetric potentiometric titration[18] (see Section 15.20), whilst for bromides a mercurimetric titration[19] is comparable with the argentimetric method.

Phosphorus from organophosphorus compounds, which are combusted to give mainly orthophosphate, can be absorbed by either sulphuric acid or nitric acid and readily determined spectrophotometrically either by the molybdenum blue method or as the phosphovanadomolybdate (Section 17.39).

Procedures have also been devised for the determination of metallic constituents. Thus, mercury is absorbed in nitric acid and titrated with sodium diethyldithiocarbamate, whilst zinc is absorbed in hydrochloric acid and determined by an EDTA titration (see Section 10.65).

The simplest method for decomposing an organic sample is to heat it in an open crucible until all carbonaceous matter has been oxidised leaving a residue of inorganic components, usually as oxide. The residue can then be dissolved in dilute acid giving a solution which can be analysed by appropriate procedures. This technique is referred to as **dry ashing**; it is obviously inapplicable when the inorganic component is volatile. Under these conditions the wet ashing procedure described under perchloric acid must be used. A full discussion of the destruction of organic matrices is given in Ref. 20.

3.32 PRECIPITATION

The conditions for precipitation of inorganic substances are given in Section 11.6. Precipitations are usually carried out in resistance-glass beakers, and the solution of the precipitant is added slowly (for example, by means of a pipette, burette, or tap funnel) and with efficient stirring of the suitably diluted solution. The addition must always be made without splashing; this is best achieved by allowing the solution of the reagent to flow down the side of the beaker or precipitating vessel. Only a moderate excess of the reagent is generally required; a very large excess may lead to increasing solubility (compare Section 2.6) or contamination of the precipitate. After the precipitate has settled, a few drops of the precipitant should always be added to determine whether further precipitation occurs. As a general rule, precipitates are not filtered off immediately after they have been formed; most precipitates, with the exception of those which are definitely colloidal, such as iron (III) hydroxide, require more or less digestion (Section 11.5) to complete the precipitation and make all particles of filterable size. In some cases digestion is carried out by setting the beaker aside and leaving the precipitate in contact with the mother liquor at room temperature for 12–24 hours; in others, where a higher temperature is permissible, digestion is usually effected near the boiling point of the solution. Hot plates, water baths, or even a low flame if no bumping occurs, are employed for the latter purpose; in all cases the beaker should be covered with a clockglass with the convex side turned down. If the solubility of the precipitate is appreciable, it may be necessary to allow the solution to attain room temperature before filtration.

3.33 FILTRATION

This operation is the separation of the precipitate from the mother liquor, the object being to get the precipitate and the filtering medium quantitatively free from the solution. The media employed for filtration are: (1) filter paper; (2) porous fritted plates of resistance glass, e.g. Pyrex (sintered-glass filtering crucibles), of silica (Vitreosil filtering crucibles), or of porcelain (porcelain filtering crucibles): see Section 3.24.

The choice of the filtering medium will be controlled by the nature of the precipitate (filter paper is especially suitable for gelatinous precipitates) and also by the question of cost. The limitations of the various filtering media are given in the account which follows.

3.34 FILTER PAPERS

Quantitative filter papers must have a very small ash content; this is achieved during manufacture by washing with hydrochloric and hydrofluoric acids. The sizes generally used are circles of 7.0, 9.0, 11.0, and 12.5 cm diameter, those of 9.0 and 11.0 cm being most widely employed. The ash of a 11 cm circle should not exceed 0.0001 g; if the ash exceeds this value, it should be deducted from the weight of the ignited residue. Manufacturers give values for the average ash per paper: the value may also be determined, if desired, by igniting several filter papers in a crucible. Quantitative filter paper is made of various degrees of porosity. The filter paper used must be of such texture as to retain the smallest particles of precipitate and yet permit of rapid filtration. Three textures are generally made, one for very fine precipitates, a second for the average precipitate which contains medium-sized particles, and a third for gelatinous precipitates and coarse particles. The speed of filtration is slow for the first, fast for the third, and medium for the second. 'Hardened' filter papers are made by further treatment of quantitative filter papers with acid; these have an extremely small ash, a much greater mechanical strength when wet, and are more resistant to acids and alkalis: they should be used in all quantitative work. The characteristics of the Whatman series of hardened ashless filter papers are shown in Table 3.7.

Table 3.7 'Whatman' quantitative filter papers

Filter paper	Hardened ashless		
Number	540	541	542
Speed	Medium	Fast	Slow
Particle size retention	Medium	Coarse	Fine
Ash (%)	0.008	0.008	0.008

The size of the filter paper selected for a particular operation is determined by the bulk of the precipitate, and not by the volume of the liquid to be filtered. The entire precipitate should occupy about a third of the capacity of the filter at the end of the filtration. The funnel should match the filter paper in size; the folded paper should extend to within 1–2 cm of the top of the funnel, but never closer than 1 cm.

A funnel with an angle as nearly 60° as possible should be employed; the stem should have a length of about 15 cm in order to promote rapid filtration. The filter paper must be carefully fitted into the funnel so that the upper portion beds tightly against the glass. To prepare the filter paper for use, the dry paper is usually folded exactly in half and exactly again in quarters. The folded paper is then opened so that a 60° cone is formed with three thicknesses of paper on the one side and a single thickness on the other; the paper is then adjusted to fit the funnel. The paper is placed in the funnel, moistened thoroughly with water, pressed down tightly to the sides of the funnel, and then filled with water. If the paper fits properly, the stem of the funnel will remain filled with liquid during the filtration.

To carry out a filtration, the funnel containing the properly fitted paper is placed in a funnel stand (or is supported vertically in some other way) and a clean beaker placed so that the stem of the funnel just touches the side; this will prevent splashing. The liquid to be filtered is then poured down a glass rod into the filter, directing the liquid against the side of the filter and not into the apex; the lower end of the stirring rod should be very close to, but should not quite touch, the filter paper on the side having three thicknesses of paper. The paper is never filled completely with the solution; the level of the liquid should not rise closer than to within 5–10 mm of the top of the paper. A precipitate which tends to remain in the bottom of the beaker should be removed by holding the glass rod across the beaker, tilting the beaker, and directing a jet of water from a wash bottle so that the precipitate is rinsed into the filter funnel. This procedure may also be adopted to transfer the last traces of the precipitate in the beaker to the filter. Any precipitate which adheres firmly to the side of the beaker or to the stirring rod may be removed with a rubber tipped rod or 'policeman' (Section 3.23).

Filtration by suction is rarely necessary: with gelatinous and some finely divided precipitates, the suction will draw the particles into the pores of the paper, and the speed of filtration will actually be reduced rather than increased.

3.35 CRUCIBLES WITH PERMANENT POROUS PLATES

Reference has already been made in Section 3.24 to these crucibles and to crucibles with a porous base. In use, the crucible is supported in a special holder, known as a crucible adapter, by means of a wide rubber tube (Fig. 3.12); the bottom of the crucible should be quite free from the side of the funnel and from the rubber gasket, the latter in order to be sure that the filtrate does not come into contact with the rubber. The adapter passes through a one-holed rubber bung into a large filter flask of about 750 mL capacity. The tip of the funnel must project below the side arm of the filter flask so that there is no risk that the liquid may be sucked out of the filter flask. The filter flask should be coupled with another flask of similar capacity, and the latter connected to a water filter pump; if the water in the pump should 'suck back', it will first enter the empty flask and the filtrate will not be contaminated. It is advisable also to have some sort of pressure regulator to limit the maximum pressure under which filtration is conducted. A simple method is to insert a glass tap in the second filter flask, as in the figure; alternatively, a glass **T**-piece may be introduced between the receiver and the pump, and one arm closed either by a glass tap or by a piece of heavy rubber tubing ('pressure' tubing) carrying a screw clip.

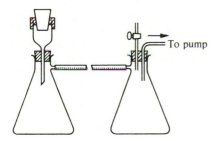

To pump

Fig. 3.12

When the apparatus is assembled, the crucible is half-filled with distilled water, then gentle suction is applied to draw the water through the crucible. When the water has passed through, suction is maintained for 1–2 minutes to remove as much water as possible from the filter plate. The crucible is then placed on a small ignition dish or saucer or upon a shallow-form Vitreosil capsule and dried to constant weight at the same temperature as that which will be subsequently used in drying the precipitate. For temperatures up to about 250 °C a thermostatically controlled electric oven should be used. For higher temperatures, the crucible may be heated in an electrically heated muffle furnace. In all cases the crucible is allowed to cool in a desiccator before weighing.

When transferring a precipitate into the crucible, the same procedure is employed as described in Section 3.34 referring to the use of filter papers: care must be taken that the liquid level in the crucible is never less than 1 cm from the top of the crucible.

Care must be taken with both sintered glass and porous base crucibles to avoid attempting the filtration of materials that may clog the filter plate. A new crucible should be washed with concentrated hydrochloric acid and then with distilled water. The crucibles are chemically inert and are resistant to all solutions which do not attack silica; they are attacked by hydrofluoric acid, fluorides, and strongly alkaline solutions.

Crucibles fitted with permanent porous plates are cleaned by shaking out as much of the solid as possible, and then dissolving out the remainder of the solid with a suitable solvent. A hot $0.1M$ solution of the tetrasodium salt of the ethylenediaminetetra-acetic acid is an excellent solvent for many of the precipitates [except metallic sulphides and hexacyanoferrates(III)] encountered in analysis. These include barium sulphate, calcium oxalate, calcium phosphate, calcium oxide, lead carbonate, lead iodate, lead oxalate, and ammonium magnesium phosphate. The crucible may either be completely immersed in the hot reagent or the latter may be drawn by suction through the crucible.

3.36 WASHING PRECIPITATES

Most precipitates are produced in the presence of one or more soluble compounds. Since the latter are frequently not volatile at the temperature at which the precipitate is ultimately dried, it is necessary to wash the precipitate to remove such material as completely as possible. The minimum volume of the washing liquid required to remove the objectionable matter should be used, since no precipitate is absolutely insoluble. Qualitative tests for the removal of the impurities should be made on small volumes of the filtered washing solution. Furthermore, it is better to wash with a number of small portions of the washing liquid, which are well drained between each washing, than with one or two large portions, or by adding fresh portions of the washing liquid while solution still remains on the filter (see Section 11.8).

The ideal washing liquid should comply as far as possible with the following conditions.

1. It should have no solvent action upon the precipitate, but dissolve foreign substances easily.
2. It should have no dispersive action on the precipitate.
3. It should form no volatile or insoluble product with the precipitate.
4. It should be easily volatile at the temperature of drying of the precipitate.
5. It should contain no substance which is likely to interfere with subsequent determinations in the filtrate.

In general, pure water should not be used unless it is certain that it will not dissolve appreciable amounts of the precipitate or peptise it. If the precipitate is appreciably soluble in water, a common ion is usually added, since any electrolyte is less soluble in a dilute solution containing one of its ions than it is in pure water (Section 2.7); as an example the washing of calcium oxalate with dilute ammonium oxalate solution may be cited. If the precipitate tends to become colloidal and pass through the filter paper (this is frequently observed with gelatinous or flocculent precipitates), a wash solution containing an electrolyte must be employed (compare Section 11.3). The nature of the electrolyte is immaterial, provided it has no action upon the precipitate during washing and is volatilised during the final heating. Ammonium salts are usually selected for this purpose: thus ammonium nitrate solution is employed for washing iron(III) hydroxide. In some cases it is possible to select a solution which will both reduce the solubility of the precipitate and prevent peptisation; for example, the use of dilute nitric acid with silver chloride. Some precipitates tend to oxidise during washing; in such instances the precipitate cannot be allowed to run dry, and a special washing solution which re-converts the oxidised compounds into the original condition must be employed, e.g. acidified hydrogen sulphide water for copper sulphide. Gelatinous precipitates, like aluminium hydroxide, require more washing than crystalline ones, such as calcium oxalate.

In most cases, particularly if the precipitate settles rapidly or is gelatinous, **washing by decantation** may be carried out. As much as possible of the liquid above the precipitate is transferred to the prepared filter (either filter paper or filter crucible), observing the usual precautions, and taking care to avoid, as far as possible, disturbing the precipitate. Twenty to fifty millilitres of a suitable wash liquid is added to the residue in the beaker, the solid stirred up and allowed to settle. If the solubility of the precipitate allows, the solution should be heated, since the rate of filtration will thus be increased. When the supernatant liquid is clear, as much as possible of the liquid is decanted through the filtering medium. This process is repeated three to five times (or as many times as is necessary) before the precipitate is transferred to the filter. The main bulk of the precipitate is first transferred by mixing with the wash solution and pouring off the suspension, the process being repeated until most of the solid has been removed from the beaker. Precipitate adhering to the sides and bottom of the beaker is then transferred to the filter with the aid of a wash bottle as described in Section 3.34, using a 'policeman' if necessary to transfer the last traces of precipitate. Finally, a wash bottle is used to wash the precipitate down to the bottom of the filter paper or to the plate of the filter crucible.

In all cases, tests for the completeness of washing must be made by collecting a small sample of the washing solution after it is estimated that most of the impurities have been removed, and applying an appropriate qualitative test. Where filtration is carried out under suction, a small test-tube is placed under the crucible adapter.

3.37 DRYING AND IGNITING PRECIPITATES

After a precipitate has been filtered and washed, it must be brought to a constant composition before it can be weighed. The further treatment will depend both upon the nature of the precipitate and upon that of the filtering medium; this treatment consists in drying or igniting the precipitate. Which of the latter two terms is employed depends upon the temperature at which the precipitate is heated. There is, however, no definite temperature below or above which the precipitate is said to be dried or ignited respectively. The meaning will be adequately conveyed for our purpose if we designate *drying* when the temperature is below 250 °C (the maximum temperature which is readily reached in the usual thermostatically controlled, electric drying-oven), and *ignition* above 250 °C up to, say 1200 °C. Precipitates that are to be dried should be collected on filter paper, or in sintered-glass or porcelain filtering crucibles. Precipitates that are to be ignited are collected on filter paper, porcelain filtering crucibles, or silica filtering crucibles. Ignition is simply effected by placing in a special ignition dish and heating with the appropriate burner; alternatively, these crucibles (and, indeed, any type of crucible) may be placed in an electrically heated muffle furnace, which is equipped with a pyrometer and a means for controlling the temperature.

Attention is directed to the information provided by thermogravimetric analysis[21,22] concerning the range of temperature to which a precipitate should be heated for a particular composition. In general, thermal gravimetric curves seem to suggest that in the past precipitates were heated for too long a period and at too high a temperature. It must, however, be borne in mind that in some cases the thermal gravimetric curve is influenced by the experimental conditions of precipitation, and even if a horizontal curve is not obtained, it is possible that a suitable weighing form may be available over a certain temperature range. Nevertheless, thermograms do provide valuable data concerning the range of temperature over which a precipitate has a constant composition under the conditions that the thermogravimetric analysis was made; these, at the very least, provide a guide for the temperature at which a precipitate should be dried and heated for quantitative work, but due regard must be paid to the general chemical properties of the weighing form.

Although precipitates which require ignition will usually be collected in porcelain or silica filtering crucibles, there may be some occasions where filter paper has been used, and it is therefore necessary to describe the method to be adopted in such cases. The exact technique will depend upon whether the precipitate may be safely ignited in contact with the filter paper or not. It must be remembered that some precipitates, such as barium sulphate, may be reduced or changed in contact with filter paper or its decomposition products.

A. Incineration of the filter paper in the presence of the precipitate. A silica crucible is first ignited to constant weight (i.e. to within 0.0002 g) at the same temperature as that to which the precipitate is ultimately heated. The well-drained filter paper and precipitate are carefully detached from the funnel; the filter paper is folded so as to enclose the precipitate completely, care being taken not to tear the paper. The packet is placed point-down in the weighed crucible, which is supported on a pipe-clay, or better, a silica triangle resting on a ring stand. The crucible is slightly inclined, and partially covered with the lid, which should rest partly on the triangle. A *very small flame* is then placed under the crucible lid; drying thus proceeds quickly and without undue risk. When the moisture has been expelled, the flame is increased slightly so as to carbonise the paper *slowly*. The paper should not be allowed to inflame, as this may cause a mechanical expulsion of fine particles of the precipitate owing to the rapid escape of the products of combustion: if, by chance, it does catch fire, the flame should be extinguished by momentarily placing the cover on the mouth of the crucible with the aid of a pair of crucible tongs. When the paper has completely carbonised and vapours are no longer evolved, the flame is moved to the back (bottom) of the crucible and the carbon slowly burned off while the flame is gradually increased.* After all the carbon has been burned away, the crucible is covered completely (if desired, the crucible may be placed in a vertical position for this purpose) and heated to the required temperature by means of a Bunsen burner. Usually it takes about 20 minutes to char the paper, and 30–60 minutes to complete the ignition

When the ignition is ended, the flame is removed and, after 1–2 minutes, the crucible and lid are placed in a desiccator containing a suitable desiccant (Section 3.22), and allowed to cool for 25–30 minutes. The crucible and lid are then weighed. The crucible and contents are then ignited at the same temperature for 10–20 minutes, allowed to cool in a desiccator as before, and weighed again. The ignition is repeated until constant weight is attained. Crucibles should always be handled with clean crucible tongs and preferably with platinum-tipped tongs.

It is important to note that 'heating to constant weight' has no real significance unless the periods of heating, cooling of the *covered* crucible, and weighing are duplicated.

B. Incineration of the filter paper apart from the precipitate. This method is employed in all those cases where the ignited substance is reduced by the burning paper; for example, barium sulphate, lead sulphate, bismuth oxide, copper oxide, etc. The funnel containing the precipitate is covered by a piece of qualitative filter paper upon which is written the formula of the precipitate and the name of the owner; the paper is made secure by crumpling its edges over the rim of the funnel so that they will engage the outer conical portion of the funnel. The funnel is placed in a drying oven maintained at 100–105 °C, for 1–2 hours or until completely dry. A sheet of glazed paper about 25 cm square (white or black, to contrast with the colour of the precipitate) is placed on the bench away from all draughts. The dried filter is removed from the funnel, and as much as possible of the precipitate is removed from the paper and allowed to drop on a clockglass resting upon the glazed paper. This is readily done by very gently rubbing the sides of the filter paper together, when the bulk of the precipitate becomes detached and drops upon the clockglass. Any small particles of the precipitate which may have fallen upon the glazed paper are brushed into the clockglass with a small camel-hair brush. The clockglass containing the precipitate is then covered with a larger clockglass or with a beaker. The filter paper is now carefully folded and placed inside a weighed porcelain or silica crucible. The crucible is placed on a triangle and the filter paper incinerated as detailed above. The crucible is allowed to cool, and the filter ash subjected to a suitable chemical treatment in order to convert any reduced or changed material into the form finally desired. The cold crucible is then placed upon the glazed paper and the main part of the precipitate carefully transferred from the clockglass to the crucible. A small camel-hair brush will assist in the transfer. Finally, the precipitate is brought to constant weight by heating to the necessary temperature as detailed under A.

---

* If the carbon on the lid is oxidised only slowly, the cover may be heated separately in a flame. It is, of course, held in clean crucible tongs.

3.38 REFERENCES FOR PART A

1. P W Atkins *Physical Chemistry*, 3rd edn, Oxford University Press, Oxford, 1987
2. (*a*) A Ringbom *Complexation in Analytical Chemistry*, Interscience Publishers, New York, 1963
   (*b*) R Pribil *Applied Complexometry*, Pergamon Press, Oxford, 1982
   (*c*) G Schwarzenbach and H Flaschka *Complexometric Titrations*, 2nd edn, Methuen and Co, London, 1969
3. (*a*) *Stability Constants of Metal Ion Complexes*, Special Publications Nos 17, 25, The Chemical Society, London, 1964
   (*b*) *Stability Constants of Metal Ion Complexes*, Part A, *Inorganic Ligands*, E. Högfeldt (Ed) 1982; Part B, *Organic Ligands*, D. Perrin (Ed) 1979; IUPAC Publication, Pergamon Press, Oxford
4. (*a*) F A Cotton and G Wilkinson *Advanced Inorganic Chemistry*, 4th edn, Interscience Publishers, London, 1980
   (*b*) A G Sharpe *Inorganic Chemistry*, 2nd edn, Longman Group Ltd, London, 1986
5. A J Bard, R Parsons and J Jordan *Standard Potentials in Aqueous Solution*, IUPAC Publication, Marcel Dekker Inc, New York, 1985
6. C Woodward and H N Redman *High Precision Titrimetry*, Society for Analytical Chemistry, London, 1973, (*a*) p. 5; (*b*) p. 14; (*c*) p. 11; (*d*) p. 10; (*e*) p. 12
7. *British Standards for graduated apparatus (with International Standardisation Organisation equivalents):*

(*a*) Graduated flasks	BS 5898 (1980); ISO 384-1978	
	BS 1792 (1982); ISO 1042-1981	
(*b*) Pipettes	BS 1583 (1986); ISO 648-1984	
(*c*) Burettes	BS 846 (1985); ISO 385-1984	
(*d*) Calibration procedures	BS 6696 (1986); ISO 4787-1985	

   British Standards Institution, London.
8. *Water for Laboratory Use*, BS 3978 (1987), British Standards Institution, London
9. B S Furniss, A J Hannaford, V Rogers, P W G Smith and A R Tatchell *Vogel's Practical Organic Chemistry*, 4th edn, Longman Group Ltd, London, 1978
10. D F Shriver *The Manipulation of Air-Sensitive Materials*, McGraw-Hill Book Co., New York, 1969
11. D J Bucknell (Ed) '*AnalaR*' *Standards for Laboratory Chemicals*, 8th edn, BDH Chemicals, Poole, 1984
12. C O Ingamells Rapid chemical analysis of silicate rocks. *Talanta*, 1964, **11**, 665
13. R Bock (Trans I Marr) *Handbook of Decomposition Methods in Chemistry*, International Book Co., Glasgow, 1979
14. W Schöniger *Mikrochim. Acta.* 1955, 123
15. W Schöniger *Mikrochim. Acta.* 1956, 869
16. A M G Macdonald In *Advances in Analytical Chemistry and Instrumentation*, Vol 4, C N Reilley (Ed) Interscience, New York, 1965
17. A M G Macdonald *Analyst*, 1961, **86**, 3
18. Analytical Methods Committee *Analyst*, 1963, **88**, 415
19. R C Denney and P A Smith *Analyst*, 1974, **99**, 166
20. T Gorsuch *The Destruction of Organic Matter*, Pergamon Press, Oxford, 1970
21. C Duval *Inorganic Thermogravimetric Analysis*, Elsevier, Amsterdam, 1963
22. *Wilson and Wilson Comprehensive Analytical Chemistry*, Vol. XII, *Thermal Analysis*, Part A, J Paulik and F Paulik (Eds), 1981; Part D, J Seslik (Ed), 1984; Elsevier Science Publishers, Amsterdam.

3.39 SELECTED BIBLIOGRAPHY FOR PART A

1. O Budevsky *Foundations of Chemical Analysis*, Ellis Horwood, Chichester, 1979

2. I M Kolthoff and P J Elving. *Treatise on Analytical Chemistry*. Part 1, Theory and Practice, Vol. 1, 2nd edn, Interscience Publishers, New York, 1978

3. H A Laitinen and W E Harris *Chemical Analysis*, 2nd edn, McGraw-Hill, New York, 1975

4. *Wilson and Wilson Comprehensive Analytical Chemistry*, Vols 1A (1959); 1B (1960), Elsevier Publishing Co., Amsterdam

5. Q Fernando and M D Ryan *Calculations in Analytical Chemistry*, Harcourt Brace Jovanovich, New York, 1982

6. F R Hartley, C Burgess and R M Alcock *Solution Equilibria*, John Wiley and Sons, Chichester, 1980

7. S Kotrly and L Sucha *Handbook of Chemical Equilibria in Analytical Chemistry*, Ellis Horwood, Chichester, 1985

8. L Meites *An Introduction to Chemical Equilibrium and Kinetics*, Pergamon Press, Oxford, 1981

9. R W Ramette *Chemical Equilibrium and Analysis*, Addison–Wesley, London, 1981

10. D R Crow *Principles and Applications of Electrochemistry*, 2nd edn, Chapman and Hall, Andover, 1979

11. J Robbins *Ions in Solution*, Clarendon Press, Oxford, 1972

12. *Guide to Safe Practices in Chemical Laboratories*, Royal Society of Chemistry, London, 1986

13. L Bretherick (Ed) *Hazards in the Chemical Laboratory*, 4th edn, Royal Society of Chemistry, London, 1986

14. L Bretherick (Ed) *Laboratory Hazards Bulletin*, Royal Society of Chemistry, London

15. N T Freeman and J Whitehead *Introduction to Safety in the Chemical Laboratory*, Academic Press, London, 1982

16. M Richardson (Ed) *Toxic Hazard Assessment of Chemicals*, Royal Society of Chemistry, London, 1986

17. G Weiss (Ed) *Hazardous Chemicals Data Book*, 2nd edn, Noyes Publications, New Jersey, 1986

18. National Bureau of Standards Handbook 44: 1979, *Specifications, Tolerances and other Technical Requirements for Weighing and Measuring as adopted by the National Conference on Weights and Measures*. NBS, Washington, DC, 1979

19. *Reagent Chemicals*, 7th edn, American Chemical Society, Washington, DC, 1986

20. J Coetzee (Ed) *Recommended Methods for Purification of Solvents and Tests for Impurities*, Pergamon Press, Oxford, 1982

21. A Weissberger (Ed) *Techniques of Chemistry*, Vol. 2, *Organic Solvents. Physical Properties and Methods of Purification*, 4th edn, J A Riddick, W B Bunger and T K Sakano. John Wiley, New York, 1986

22. P G Jeffrey and D Hutchinson *Chemical Methods of Rock Analysis*, 3rd edn, Pergamon Press, Oxford, 1981

23. P J Potts *A Handbook of Silicate Rock Analysis*, 2nd edn. Blackie and Son Ltd, Glasgow, 1987

PART B
ERRORS, STATISTICS, AND SAMPLING

CHAPTER 4
ERRORS AND STATISTICS

4.1 LIMITATIONS OF ANALYTICAL METHODS

The function of the analyst is to obtain a result as near to the true value as possible by the correct application of the analytical procedure employed. The level of confidence that the analyst may enjoy in his results will be very small unless he has knowledge of the accuracy and precision of the method used as well as being aware of the sources of error which may be introduced. Quantitative analysis is not simply a case of taking a sample, carrying out a single determination and then claiming that the value obtained is irrefutable. It also requires a sound knowledge of the chemistry involved, of the possibilities of interferences from other ions, elements and compounds as well as of the statistical distribution of values. The purpose of this chapter is to explain some of the terms employed and to outline the statistical procedures which may be applied to the analytical results.

4.2 CLASSIFICATION OF ERRORS

The errors which affect an experimental result may conveniently be divided into 'systematic' and 'random' errors.

Systematic (determinate) errors. These are errors which can be avoided, or whose magnitude can be determined. The most important of them are:

1. *Operational and personal errors.* These are due to factors for which the individual analyst is responsible and are not connected with the method or procedure: they form part of the 'personal equation' of an observer. The errors are mostly physical in nature and occur when sound analytical technique is not followed. Examples are: mechanical loss of materials in various steps of an analysis; underwashing or overwashing of precipitates; ignition of precipitates at incorrect temperatures; insufficient cooling of crucibles before weighing; allowing hygroscopic materials to absorb moisture before or during weighing; and use of reagents containing harmful impurities.

   Personal errors may arise from the constitutional inability of an individual to make certain observations accurately. Thus some persons are unable to judge colour changes sharply in visual titrations, which may result in a slight overstepping of the end point.
2. *Instrumental and reagent errors.* These arise from the faulty construction of balances, the use of uncalibrated or improperly calibrated weights, graduated

glassware, and other instruments; the attack of reagents upon glassware, porcelain, etc., resulting in the introduction of foreign materials; volatilisation of platinum at very high temperatures; and the use of reagents containing impurities.

3. *Errors of method*. These originate from incorrect sampling and from incompleteness of a reaction. In gravimetric analysis errors may arise owing to appreciable solubility of precipitates, co-precipitation, and post-precipitation, decomposition, or volatilisation of weighing forms on ignition, and precipitation of substances other than the intended ones. In titrimetric analysis errors may occur owing to failure of reactions to proceed to completion, occurrence of induced and side reactions, reaction of substances other than the constituent being determined, and a difference between the observed end point and the stoichiometric equivalence point of a reaction.

4. *Additive and proportional errors*. The absolute value of an additive error is independent of the amount of the constituent present in the determination. Examples of additive errors are loss in weight of a crucible in which a precipitate is ignited, and errors in weights. The presence of this error is revealed by taking samples of different weights.

The absolute value of a proportional error depends upon the amount of the constituent. Thus a proportional error may arise from an impurity in a standard substance, which leads to an incorrect value for the molarity of a standard solution. Other proportional errors may not vary linearly with the amount of the constituent, but will at least exhibit an increase with the amount of constituent present. One example is the ignition of aluminium oxide: at 1200 °C the aluminium oxide is anhydrous and virtually non-hygroscopic; ignition of various weights at an appreciably lower temperature will show a proportional type of error.

Random (indeterminate) errors. These errors manifest themselves by the slight variations that occur in successive measurements made by the same observer with the greatest care under as nearly identical conditions as possible. They are due to causes over which the analyst has no control, and which, in general, are so intangible that they are incapable of analysis. If a *sufficiently large number of observations* is taken it can be shown that these errors lie on a curve of the form shown in Fig. 4.1 (Section 4.9). An inspection of this error curve shows: (*a*) small errors occur more frequently than large ones; and (*b*) positive and negative errors of the same numerical magnitude are equally likely to occur.

4.3 ACCURACY

The accuracy of a determination may be defined as the concordance between it and the true or most probable value. It follows, therefore, that systematic errors cause a constant error (either too high or too low) and thus affect the accuracy of a result. For analytical methods there are two possible ways of determining the accuracy; the so-called absolute method and the comparative method.

Absolute method. A synthetic sample containing known amounts of the constituents in question is used. Known amounts of a constituent can be obtained by weighing out pure elements or compounds of known stoichiometric composition. These substances, primary standards, may be available commercially or they may be prepared by the analyst and subjected to rigorous purification by recrystallisation, etc. The substances must be of known purity. The test of the accuracy of the method under consideration is carried out by taking varying amounts of the constituent and proceeding according to specified instructions. The amount of the constituent must be varied, because the determinate errors in the procedure may be a function of the amount used. The difference between the mean of an adequate number of results and the amount of the constituent actually present, usually expressed as parts per thousand, is a measure of the accuracy of the method in the absence of foreign substances.

The constituent in question will usually have to be determined in the presence of other substances, and it will therefore be necessary to know the effect of these upon the determination. This will require testing the influence of a large number of elements, each in varying amounts — a major undertaking. The scope of such tests may be limited by considering the determination of the component in a specified range of concentration in a material whose composition is more or less fixed both with respect to the elements which may be present and their relative amounts. It is desirable, however, to study the effect of as many foreign elements as feasible. In practice, it is frequently found that separations will be required before a determination can be made in the presence of varying elements; the accuracy of the method is likely to be largely controlled by the separations involved.

Comparative method. Sometimes, as in the analysis of a mineral, it may be impossible to prepare solid synthetic samples of the desired composition. It is then necessary to resort to standard samples of the material in question (mineral, ore, alloy, etc.) in which the content of the constituent sought has been determined by one or more supposedly 'accurate' methods of analysis. This comparative method, involving secondary standards, is obviously not altogether satisfactory from the theoretical standpoint, but is nevertheless very useful in applied analysis. Standard samples can be obtained from various sources (see Section 4.5).

If several fundamentally different methods of analysis for a given constituent are available, e.g. gravimetric, titrimetric, spectrophotometric, or spectrographic, the agreement between at least two methods of essentially different character can usually be accepted as indicating the absence of an appreciable systematic error in either (a systematic error is one which can be evaluated experimentally or theoretically).

4.4 PRECISION

Precision may be defined as the concordance of a series of measurements of the same quantity. Accuracy expresses the correctness of a measurement, and precision the 'reproducibility' of a measurement (the latter definition will be modified later). Precision always accompanies accuracy, but a high degree of precision does not imply accuracy. This may be illustrated by the following example.

A substance was known to contain 49.10 ± 0.02 per cent of a constituent A. The results obtained by two analysts using the same substance and the same analytical method were as follows.

Analyst (1) % A 49.01; 49.25; 49.08; 49.14

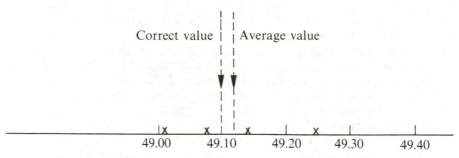

The arithmetic mean is 49.12% and the results range from 49.01% to 49.25%.

Analyst (2) % A 49.40; 49.44; 49.42; 49.42

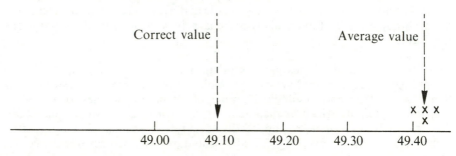

The arithmetic mean is 49.42% and the results range from 49.40% to 49.44%.

We can summarise the results of the analyses as follows.

(*a*) The values obtained by Analyst 1 are accurate (very close to the correct result), but the precision is inferior to the results given by Analyst 2. The values obtained by Analyst 2 are very precise but are not accurate.

(*b*) The results of Analyst 1 face on both sides of the mean value and could be attributed to random errors. It is apparent that there is a constant (systematic) error present in the results of Analyst 2.

Precision was previously described as the reproducibility of a measurement. However, the modern analyst makes a distinction between the terms '**reproducible**' and '**repeatable**'. On further consideration of the above example:

(*c*) If Analyst 2 had made the determinations on the same day in rapid succession, then this would be defined as 'repeatable' analysis. However, if the determinations had been made on separate days when laboratory conditions may vary, this set of results would be defined as '**reproducible**'.

Thus, there is a distinction between a within-run precision (repeatability) and a between-run precision (reproducibility).

130

4.5 MINIMISATION OF ERRORS

Systematic errors can often be materially reduced by one of the following methods.

1. *Calibration of apparatus and application of corrections.* All instruments (weights, flasks, burettes, pipettes, etc.) should be calibrated, and the appropriate corrections applied to the original measurements. In some cases where an error cannot be eliminated, it is possible to apply a correction for the effect that it produces; thus an impurity in a weighed precipitate may be determined and its weight deducted.

2. *Running a blank determination.* This consists in carrying out a separate determination, the sample being omitted, under exactly the same experimental conditions as are employed in the actual analysis of the sample. The object is to find out the effect of the impurities introduced through the reagents and vessels, or to determine the excess of standard solution necessary to establish the end-point under the conditions met with in the titration of the unknown sample. A large blank correction is undesirable, because the exact value then becomes uncertain and the precision of the analysis is reduced.

3. *Running a control determination.* This consists in carrying out a determination under as nearly as possible identical experimental conditions upon a quantity of a standard substance which contains the same weight of the constituent as is contained in the unknown sample. The weight of the constituent in the unknown can then be calculated from the relation:

$$\frac{\text{Result found for standard}}{\text{Result found for unknown}} = \frac{\text{Weight of constituent in standard}}{x}$$

where x is the weight of the constituent in the unknown.

In this connection it must be pointed out that standard samples which have been analysed by a number of skilled analysts are commercially available. These include certain primary standards (sodium oxalate, potassium hydrogenphthalate, arsenic(III) oxide, and benzoic acid) and ores, ceramic materials, irons, steels, steel-making alloys, and non-ferrous alloys.

Many of these are also available as BCS Certified Reference Materials (CRM) supplied by the Bureau of Analysed Samples Ltd, Newham Hall, Middlesborough, UK, who also supply EURONORM Certified Reference Materials (ERCM), the composition of which is specified on the basis of results obtained by a number of laboratories within the EEC. BCS Reference Materials are obtainable from the Community Bureau of Reference, Brussels, Belgium. In the USA similar reference materials are supplied by the National Bureau of Standards.

4. *Use of independent methods of analysis.* In some instances the accuracy of a result may be established by carrying out the analysis in an entirely different manner. Thus iron may first be determined gravimetrically by precipitation as iron(III) hydroxide after removing the interfering elements, followed by ignition of the precipitate to iron(III) oxide. It may then be determined titrimetrically by reduction to the iron(II) state, and titration with a standard solution of an oxidising agent, such as potassium dichromate or cerium(IV) sulphate. Another example that may be mentioned is the determination of the strength of a hydrochloric acid solution both by titration with a standard

131

solution of a strong base and by precipitation and weighing as silver chloride. If the results obtained by the two radically different methods are concordant, it is highly probable that the values are correct within small limits of error.

5. *Running parallel determinations.* These serve as a check on the result of a single determination and indicate only the precision of the analysis. The values obtained for constituents which are present in not too small an amount should not vary among themselves by more than three parts per thousand. If larger variations are shown, the determinations must be repeated until satisfactory concordance is obtained. Duplicate, and at most triplicate, determinations should suffice. It must be emphasised that good agreement between duplicate and triplicate determinations does not justify the conclusion that the result is correct; a constant error may be present. The agreement merely shows that the accidental errors, or variations of the determinate errors, are the same, or nearly the same, in the parallel determinations.

6. *Standard addition.* A known amount of the constituent being determined is added to the sample, which is then analysed for the total amount of constituent present. The difference between the analytical results for samples with and without the added constituent gives the recovery of the amount of added constituent. If the recovery is satisfactory our confidence in the accuracy of the procedure is enhanced. The method is usually applied to physico-chemical procedures such as polarography and spectrophotometry.

7. *Internal standards.* This procedure is of particular value in spectroscopic and chromatographic determinations. It involves adding a fixed amount of a reference material (the internal standard) to a series of known concentrations of the material to be measured. The ratios of the physical value (absorption or peak size) of the internal standard and the series of known concentrations are plotted against the concentration values. This should give a straight line. Any unknown concentration can then be determined by adding the same quantity of internal standard and finding where the ratio obtained falls on the concentration scale.

8. *Amplification methods.* In determinations in which a very small amount of material is to be measured this may be beyond the limits of the apparatus available. In these circumstances if the small amount of material can be reacted in such a way that every molecule produces two or more molecules of some other measurable material, the resultant amplification may then bring the quantity to be determined within the scope of the apparatus or method available.

9. *Isotopic dilution.* A known amount of the element being determined, containing a radioactive isotope, is mixed with the sample and the element is isolated in a pure form (usually as a compound), which is weighed or otherwise determined. The radioactivity of the isolated material is measured and compared with that of the added element: the weight of the element in the sample can then be calculated.

4.6 SIGNIFICANT FIGURES AND COMPUTATIONS

The term 'digit' denotes any one of the ten numerals, including the zero. A significant figure is a digit which denotes the amount of the quantity in the place in which it stands. The digit zero is a significant figure except when it is the first figure in a number. Thus in the quantities 1.2680 g and 1.0062 g the zero is significant, but in the quantity 0.0025 kg the zeros are not significant figures; they serve only to locate the decimal point and can be omitted by proper choice of units, i.e. 2.5 g. The first two numbers contain five significant figures, but 0.0025 contains only two significant figures.

Observed quantities should be recorded with one uncertain figure retained. Thus in most analyses weights are determined to the nearest tenth of a milligram, e.g. 2.1546 g. This means that the weight is less than 2.1547 g and more than 2.1545 g. A weight of 2.150 g would signify that it has been determined to the nearest milligram, and that the weight is nearer to 2.150 g than it is to either 2.151 g or 2.149 g. The digits of a number which are needed to express the precision of the measurement from which the number was derived are known as significant figures.

There are a number of rules for computations with which the student should be familiar.

1. Retain as many significant figures in a result or in any data as will give only one uncertain figure. Thus a volume which is known to be between 20.5 mL and 20.7 mL should be written as 20.6 mL, but not as 20.60 mL, since the latter would indicate that the value lies between 20.59 mL and 20.61 mL. Also, if a weight, to the nearest 0.1 mg, is 5.2600 g, it should not be written as 5.260 g or 5.26 g, since in the latter case an accuracy of a centigram is indicated and in the former a milligram.

2. In rounding off quantities to the correct number of significant figures, add one to the last figure retained if the following figure (which has been rejected) is 5 or over. Thus the average of 0.2628, 0.2623, and 0.2626 is 0.2626 (0.2625_7).

3. In addition or subtraction, there should be in each number only as many significant figures as there are in the least accurately known number. Thus the addition

$168.11 + 7.045 + 0.6832$

should be written

$168.11 + 7.05 + 0.68 = 175.84$

The sum or difference of two or more quantities cannot be more precise than the quantity having the largest uncertainty.

4. In multiplication or division, retain in each factor one more significant figure than is contained in the factor having the largest uncertainty. The percentage precision of a product or quotient cannot be greater than the percentage precision of the least precise factor entering into the calculation. Thus the multiplication

$1.26 \times 1.236 \times 0.6834 \times 24.8652$

should be carried out using the values

$1.26 \times 1.236 \times 0.683 \times 24.87$

and the result expressed to three significant figures.

4.7 THE USE OF CALCULATORS AND MICROCOMPUTERS

The advent of reasonably priced hand-held calculators has replaced the use of both logarithms and slide-rules for statistical calculations. In addition to the normal arithmetic functions, a suitable calculator for statistical work should enable the user to evaluate the mean and standard deviation (Section 4.8), linear regression and correlation coefficient (Section 4.16). The results obtained by the use of the calculator must be carefully scrutinised to ascertain the number of significant figures to be retained, and should always be checked against a 'rough' arithmetical calculation to ensure there are no gross computational errors. Microcomputers are used for processing large amounts of data. Although computer programming is outside the scope of this book it should be pointed out that standard programs now exist in BASIC, and other high-level computer languages (see Bibliography, Section 5.7).

The microcomputer may also be interfaced with most types of electronic equipment used in the laboratory. This facilitates the collection and processing of the data, which may be stored on floppy or hard discs for later use.

There is a large amount of commercial software available for performing the statistical calculations described later in this chapter, and for more advanced statistical tests beyond the scope of this text.

4.8 MEAN AND STANDARD DEVIATION

When a quantity is measured with the greatest exactness of which the instrument, method, and observer are capable, it is found that the results of successive determinations differ among themselves to a greater or lesser extent. The average value is accepted as the most probable. This may not always be the true value. In some cases the difference may be small, in others it may be large; the reliability of the result depends upon the magnitude of this difference. It is therefore of interest to enquire briefly into the factors which affect and control the trustworthiness of chemical analysis.

The absolute error of a determination is the difference between the observed or measured value and the true value of the quantity measured. It is a measure of the accuracy of the measurement.

The relative error is the absolute error divided by the true value; it is usually expressed in terms of percentage or in parts per thousand. The true or absolute value of a quantity cannot be established experimentally, so that the observed result must be compared with the most probable value. With pure substances the quantity will ultimately depend upon the relative atomic mass of the constituent elements. Determinations of the relative atomic mass have been made with the utmost care, and the accuracy obtained usually far exceeds that attained in ordinary quantitative analysis; the analyst must accordingly accept their reliability. With natural or industrial products, we must accept provisionally the results obtained by analysts of repute using carefully tested methods. If several analysts determine the same constituent in the same sample by different methods, the most probable value, which is usually the average, can be deduced from their results. In both cases, the establishment of the most probable value involves the application of statistical methods and the concept of precision.

In analytical chemistry one of the most common statistical terms employed is the **standard deviation** of a population of observations. This is also called the root mean square deviation as it is the square root of the mean of the sum of the squares of the differences between the values and the mean of those values (this is expressed mathematically below) and is of particular value in connection with the normal distribution.

If we consider a series of n observations arranged in ascending order of magnitude:

$$x_1, x_2, x_3, \ldots, x_{n-1}, x_n,$$

the arithmetic mean (often called simply the mean) is given by:

$$\bar{x} = \frac{x_1 + x_2 + x_3 \ldots + \ldots + x_{n-1} + x_n}{n}$$

The spread of the values is measured most efficiently by the standard deviations defined by:

$$s = \sqrt{\frac{(x_1 - \bar{x})^2 + (x_2 - \bar{x})^2 + \ldots (x_n - \bar{x})^2}{n-1}}$$

In this equation the denominator is $(n-1)$ rather than n when the number of values is small.

The equation may also be written as:

$$s = \sqrt{\frac{\Sigma(x - \bar{x})^2}{n-1}}$$

The square of the standard deviation is called the variance. A further measure of precision, known as the Relative Standard Deviation (R.S.D.), is given by:

$$\text{R.S.D.} = \frac{s}{\bar{x}}$$

This measure is often expressed as a percentage, known as the **coefficient of variation** (C.V.):

$$\text{C.V.} = \frac{s \times 100}{\bar{x}}$$

Example 1. Analyses of a sample of iron ore gave the following percentage values for the iron content: 7.08, 7.21, 7.12, 7.09, 7.16, 7.14, 7.07, 7.14, 7.18, 7.11. Calculate the mean, standard deviation and coefficient of variation for the values.

Results (x)	$x - \bar{x}$	$(x - \bar{x})^2$
7.08	−0.05	0.0025
7.21	0.08	0.0064
7.12	−0.01	0.0001
7.09	−0.04	0.0016
7.16	0.03	0.0009
7.14	0.01	0.0001
7.07	−0.06	0.0036
7.14	0.01	0.0001
7.18	0.05	0.0025
7.11	−0.02	0.0004

$\Sigma x = 71.30$ $\Sigma(x - \bar{x})^2 = 0.0182$
Mean $\bar{x}$ 7.13 per cent

$$s = \sqrt{\frac{0.0182}{9}}$$

$$= \sqrt{0.0020}$$

$$= \pm 0.045 \text{ per cent}$$

$$\text{C. V.} = \frac{0.045 \times 100}{7.13} = 0.63 \text{ per cent}$$

The mean of several readings ($\bar{x}$) will make a more reliable estimate of the true mean (μ) than is given by one observation. The greater the number of measurements (n), the closer will the sample average approach the true mean. The standard error of the mean s_x is given by:

$$s_x = \frac{s}{\sqrt{n}}$$

In the above example,

$$s_x = \pm \frac{0.045}{\sqrt{10}} = \pm 0.014$$

and if 100 measurements were made,

$$s_x = \pm \frac{0.045}{\sqrt{100}} = \pm 0.0045$$

Hence the *precision* of a measurement may be improved by increasing the number of measurements.

4.9 DISTRIBUTION OF RANDOM ERRORS

In the previous section (4.8) it has been shown that the spread of a series of results obtained from a given set of measurements can be ascertained from the value of the standard deviation. However, this term gives no indication as to the manner in which the results are distributed.

If a large number of replicate readings, at least 50, are taken of a continuous variable, e.g. a titrimetric end-point, the results attained will usually be distributed about the mean in a roughly symmetrical manner. The mathematical model that best satisfies such a distribution of random errors is called the Normal (or Gaussian) distribution. This is a bell-shaped curve that is symmetrical about the mean as shown in Fig. 4.1.

The curve satisfies the equation:

$$\frac{1}{\sigma\sqrt{2\pi}} e^{\frac{-(x-\mu)^2}{2\sigma}}$$

It is important to know that the Greek letters σ and μ refer to the standard deviation and mean respectively of a total population, whilst the Roman letters s and $\bar{x}$ are used for samples of populations, irrespective of the values of the population mean and the population standard deviation.

With this type of distribution about 68 per cent of all values will fall within

136

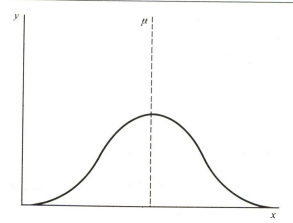

Fig. 4.1

one standard deviation on either side of the mean, 95 per cent will fall within two standard deviations, and 99.7 per cent within three standard deviations.

From the worked example (Example 1 in Section 4.8) for the analysis of an iron ore sample, the standard deviation is found to be ±0.045 per cent. If the assumption is made that the results are normally distributed, then 68 per cent (approximately seven out of ten results) will be between ±0.045 per cent and 95 per cent will be between ±0.090 per cent of the mean value. It follows that there will be a 5 per cent probability (1 in 20 chance) of a result differing from the mean by more than ±0.090 per cent, and a 1 in 40 chance of the result being 0.090 per cent *higher* than the mean.

4.10 RELIABILITY OF RESULTS

Statistical figures obtained from a set of results are of limited value by themselves. Analysis of the results can be considered in two main categories: (*a*) the reliability of the results; and (*b*) comparison of the results with the true value or with other sets of data (Section 4.12).

A most important consideration is to be able to arrive at a sensible decision as to whether certain results may be rejected. It must be stressed that values should be rejected only when a suitable statistical test has been applied or when there is an obvious chemical or instrumental reason that could justify exclusion of a result. Too frequently, however, there is a strong temptation to remove what may appear to be a 'bad' result without any sound justification. Consider the following example.

Example 2. The following values were obtained for the determination of cadmium in a sample of dust: 4.3, 4.1, 4.0, 3.2 μg g^{-1}. Should the last value, 3.2, be rejected?

The Q test may be applied to solve this problem.

$$Q = \frac{|\text{Questionable value} - \text{Nearest value}|}{\text{Largest value} - \text{Smallest value}}$$

$$Q = \frac{|3.2 - 4.0|}{4.3 - 3.2} = \frac{0.8}{1.1} = 0.727$$

137

If the calculated value of Q exceeds the critical value given in the Q table (Appendix 14), then the questionable value may be rejected.

In this example Q calculated is 0.727 and Q critical, for a sample size of four, is 0.831. Hence the result $3.2 \ \mu g \ g^{-1}$ should be retained. If, however, in the above example, three additional measurements were made, with the results:

$4.3, 4.1, 4.0, 3.2, 4.2, 3.9, 4.0 \ \mu g \ g^{-1}$

Then

$$Q = \frac{|3.2 - 3.9|}{4.3 - 3.2} = \frac{0.7}{1.1} = 0.636$$

The value of Q critical for a sample size of seven is 0.570, so rejection of the value $3.2 \ \mu g \ g^{-1}$ is justified.

It should be noted that the value Q has no regard to algebraic sign.

4.11 CONFIDENCE INTERVAL

When a small number of observations is made, the value of the standard deviation s, does not by itself give a measure of how close the sample mean $\bar{x}$ might be to the true mean. It is, however, possible to calculate a confidence interval to estimate the range within which the true mean may be found. The limits of this confidence interval, known as the confidence limits, are given by the expression:

Confidence limits of μ, for n replicate measurements, $\mu = \bar{x} \pm \dfrac{ts}{\sqrt{n}}$ (1)

where t is a parameter that depends upon the number of degrees of freedom (v) (Section 4.12) and the confidence level required. A table of the values of t at different confidence levels and degrees of freedom (v) is given in Appendix 12.

Example 3. The mean $(\bar{x})$ of four determinations of the copper content of a sample of an alloy was 8.27 per cent with a standard deviation $s = 0.17$ per cent. Calculate the 95% confidence limit for the true value.

From the t-tables, the value of t for the 95 per cent confidence level with $(n - 1)$, i.e. three degrees of freedom, is 3.18.

Hence from equation (1), the 95 per cent confidence level,

$$95\% (\text{C.L.}) \text{ for } \mu = 8.27 \pm \frac{3.18 \times 0.17}{\sqrt{4}}$$

$$= 8.27 \pm 0.27 \text{ per cent}$$

Thus, there is 95 per cent confidence that the true value of the copper content of the alloy lies in the range 8.00 to 8.54 per cent.

If the number of determinations in the above example had been 12, then the reader may wish to confirm that

$$95\% (\text{C.L.}) \text{ for } \mu = 8.27 \pm \frac{2.20 \times 0.17}{\sqrt{12}}$$

$$= 8.27 \pm 0.11 \text{ per cent}$$

Hence, on increasing the number of replicate determinations both the values of t and $s/\sqrt{n}$ decrease with the result that the confidence interval is smaller. There is, however, often a limit to the number of replicate analyses that can be sensibly performed. A method for estimating the optimum number of replicate determinations is given in Section 4.15.

4.12 COMPARISON OF RESULTS

The comparison of the values obtained from a set of results with either (a) the true value or (b) other sets of data makes it possible to determine whether the analytical procedure has been accurate and/or precise, or if it is superior to another method.

There are two common methods for comparing results: (a) Student's t-test and (b) the variance ratio test (F-test).

These methods of test require a knowledge of what is known as the number of degrees of freedom. In statistical terms this is the number of independent values necessary to determine the statistical quantity. Thus a sample of n values has n degrees of freedom, whilst the sum $\Sigma(x - \bar{x})^2$ is considered to have $n - 1$ degrees of freedom, as for any defined value of $\bar{x}$ only $n - 1$ values can be freely assigned, the nth being automatically defined from the other values.

(a) Student's t-test. This is a test[1] used for small samples; its purpose is to compare the mean from a sample with some standard value and to express some level of confidence in the significance of the comparison. It is also used to test the difference between the means of two sets of data $\bar{x}_1$ and $\bar{x}_2$.

The value of t is obtained from the equation:

$$t = \frac{(\bar{x} - \mu)\sqrt{n}}{s} \tag{2}$$

where μ is the true value.

It is then related to a set of t-tables (Appendix 12) in which the probability (P) of the t-value falling within certain limits is expressed, either as a percentage or as a function of unity, relative to the number of degrees of freedom.

Example 4. t-Test when the true mean is known.

If $\bar{x}$ the mean of the 12 determinations $= 8.37$, and μ the true value $= 7.91$, say whether or not this result is significant if the standard deviation is 0.17.

From equation (2)

$$t = \frac{(8.37 - 7.91)\sqrt{12}}{0.17} = 9.4$$

From t-tables for *eleven* degrees of freedom (one less than those used in the calculation)

for $P = 0.10$ (10 per cent)	0.05 (5 per cent)	0.01 (1 per cent)
$t = 1.80$	2.20	3.11

and as the calculated value for t is 9.4 the result is highly significant. The t-table tells us that the probability of obtaining the difference of 0.46 between the experimental and true result is less than 1 in 100. This implies that some particular bias exists in the laboratory procedure.

139

Had the calculated value for t been less than 1.80 then there would have been no significance in the results and no apparent bias in the laboratory procedure, as the tables would have indicated a probability of greater than 1 in 10 of obtaining that value. It should be pointed out that these values refer to what is known as a double-sided, or two-tailed, distribution because it concerns probabilities of values both less and greater than the mean. In some calculations an analyst may only be interested in one of these two cases, and under these conditions the t-test becomes single-tailed so that the probability from the tables is halved.

(b) **F-test.** This is used to compare the precisions of two sets of data,[2] for example, the results of two different analytical methods or the results from two different laboratories. It is calculated from the equation:

$$F = \frac{s_A^2}{s_B^2} \tag{3}$$

N.B. The larger value of s is always used as the numerator so that the value of F is always greater than unity. The value obtained for F is then checked for its significance against values in the F-table calculated from an F-distribution (Appendix 13) corresponding to the numbers of degrees of freedom for the two sets of data.

Example 5. F-test comparison of precisions.

The standard deviation from one set of 11 determinations was $s_A = 0.210$, and the standard deviation from another 13 determinations was $s_B = 0.641$. Is there any significant difference between the precision of these two sets of results?

From equation (3)

$$F = \frac{(0.641)^2}{(0.210)^2} = \frac{0.411}{0.044} = 9.4$$

for

$$P = 0.10 \quad 0.05 \quad 0.01$$

$$F = 2.28 \quad 2.91 \quad 4.71$$

The first value (2.28) corresponds to 10 per cent probability, the second value (2.91) to 5 per cent probability and the third value (4.71) to 1 per cent probability.

Under these conditions there is less than one chance in 100 that these precisions are similar. To put it another way, the difference between the two sets of data is highly significant.

Had the value of F turned out to be less than 2.28 then it would have been possible to say that there was no significant difference between the precisions, at the 10 per cent level.

4.13 COMPARISON OF THE MEANS OF TWO SAMPLES

When a new analytical method is being developed it is usual practice to compare the values of the mean and precision of the new (test) method with those of an established (reference) procedure.

The value of t when comparing two sample means $\bar{x}_1$ and $\bar{x}_2$ is given by the expression:

$$t = \frac{\bar{x}_1 - \bar{x}_2}{s_p\sqrt{1/n_1 + 1/n_2}} \tag{4}$$

where s_p the pooled standard deviation, is calculated from the two sample standard deviations s_1 and s_2, as follows:

$$s_p = \sqrt{\frac{(n_1 - 1)s_1^2 + (n_2 - 1)s_2^2}{n_1 + n_2 - 2}} \tag{5}$$

It should be stressed that there must not be a significant difference between the precisions of the methods. Hence the F-test (Section 4.12) is applied prior to using the t-test in equation (5).

Example 6. Comparison of two sets of data.
The following results were obtained in a comparison between a new method and standard method for the determination of the percentage nickel in a special steel.

	New method	Standard method
Mean	$\bar{x}_1 = 7.85$ per cent	$\bar{x}_2 = 8.03$ per cent
Standard deviation	$s_1 = \pm 0.130$ per cent	$s_2 = \pm 0.095$ per cent
Number of samples	$n_1 = 5$	$n_2 = 6$

Test at the 5 per cent probability value if the new method mean is significantly different from the standard reference mean.
The F-test must be applied to establish that there is no significant difference between the precisions of the two methods.

$$F = \frac{s_A^2}{s_B^2} = \frac{(0.130)^2}{(0.095)^2} = 1.87$$

The F-value ($P = 5$ per cent) from the tables (Appendix 13) for four and five degrees of freedom respectively for s_A and $s_B = 5.19$.
Thus, the calculated value of F (1.87) is less than the tabulated value; therefore the methods have comparable precisions (standard deviations) and so the t-test can be used with confidence.
From equation (5) the pooled standard deviation s_p is given by:

$$s_p = \sqrt{\frac{(5-1) \times 0.0169 + (6-1) \times 0.0090}{9}} = \pm 0.112$$

and from equation (4)

$$t = \frac{7.85 - 8.03}{0.112\sqrt{1/5 + 1/6}} = \frac{0.18}{0.112 \times 0.605} = 2.66$$

At the 5 per cent level, the tabulated value of t for $(n_1 + n_2 - 2)$, i.e. nine degrees of freedom, is 2.26.
Since $t_{\text{calculated}} 2.66 > t_{\text{tabulated}} 2.26$, there is a significant difference, at the specified probability, between the mean results of the two methods.

141

4.14 PAIRED *t*-TEST

Another method of validating a new procedure is to compare the results using samples of varying composition with the values obtained by an accepted method.

The manner of performing this calculation is best illustrated by an example:

Example 7. The *t*-test using samples of differing composition (the paired *t*-test).

Two different methods, A and B, were used for the analysis of five different iron compounds.

Sample	1	2	3	4	5
Method A	17.6	6.8	14.2	20.5	9.7 per cent Fe
Method B	17.9	7.1	13.8	20.3	10.2 per cent Fe

It should be apparent that in this example it would not be correct to attempt the calculation by the method described previously (Section 4.13).

In this case the differences (d) between each pair of results are calculated and $\bar{d}$, the mean of the difference, is obtained. The standard deviation s_d of the differences is then evaluated. The results are tabulated as follows.

Method A	Method B	d	$d - \bar{d}$	$(d - \bar{d})^2$
17.6	17.9	+ 0.3	0.2	0.04
6.8	7.1	+ 0.3	0.2	0.04
14.2	13.8	− 0.4	0.5	0.25
20.5	20.3	− 0.2	− 0.3	0.09
9.7	10.2	+ 0.5	0.4	0.16
		$\sum d = 0.5$		$\sum(d - \bar{d})^2 = 0.58$
		$\therefore \bar{d} = 0.1$		

$$s_d = \sqrt{\frac{0.58}{4}} = \pm 0.38$$

Then *t* is calculated from the equation

$$t = \frac{\bar{d}\sqrt{n}}{s_d} = \frac{0.10\sqrt{5}}{0.38} = 0.58_9$$

The tabulated value of *t* is 2.78 ($P = 0.05$) and since the calculated value is less than this, there is no significant difference between the methods.

4.15 THE NUMBER OF REPLICATE DETERMINATIONS

To avoid unnecessary time and expenditure, an analyst needs some guide to the number of repetitive determinations needed to obtain a suitably reliable result from the determinations performed. The larger the number the greater the reliability, but at the same time after a certain number of determinations any improvement in precision and accuracy is very small.

Although rather involved statistical methods exist for establishing the number of parallel determinations, a reasonably good assessment can be made by establishing the variation of the value for the absolute error Δ obtained for an increasing number of determinations.

$$\Delta = \frac{ts}{n}$$

The value for t is taken from the 95 per cent confidence limit column of the t-tables for $n - 1$ degrees of freedom.

The values for Δ are used to calculate the reliability interval L from the equation:

$$L = \frac{100\Delta}{z} \text{ per cent}$$

where z is the approximate percentage level of the unknown being determined. The number of replicate analyses is assessed from the magnitude of the change in L with the number of determinations.

Example 8. Ascertain the number of replicate analyses desirable (*a*) for the determination of approximately 2 per cent Cl^- in a material if the standard deviation for determinations is 0.051, (*b*) for approximately 20 per cent Cl^- if the standard deviation of determinations is 0.093.

(*a*) For 2 per cent Cl^-:

Number of determinations	$\Delta = \dfrac{ts}{n}$	$L = \dfrac{100\Delta}{z}$	Difference (per cent)
2	$12.7 \times 0.051 \times 0.71 = 0.4599$	22.99	
3	$4.3 \times 0.051 \times 0.58 = 0.1272$	6.36	16.63
4	$3.2 \times 0.051 \times 0.50 = 0.0816$	4.08	2.28
5	$2.8 \times 0.051 \times 0.45 = 0.0642$	3.21	0.87
6	$2.6 \times 0.051 \times 0.41 = 0.0544$	2.72	0.49

(*b*) For 20 per cent Cl^-:

Number of determinations	$\Delta = \dfrac{ts}{n}$	$L = \dfrac{100\Delta}{z}$	Difference (per cent)
2	$12.7 \times 0.093 \times 0.71 = 0.838$	4.19	
3	$4.3 \times 0.093 \times 0.58 = 0.232$	1.16	3.03
4	$3.2 \times 0.093 \times 0.50 = 0.148$	0.74	0.42
5	$2.8 \times 0.093 \times 0.45 = 0.117$	0.59	0.15
6	$2.6 \times 0.093 \times 0.41 = 0.099$	0.49	0.10

In (*a*) the reliability interval is greatly improved by carrying out a third analysis. This is less the case with (*b*) as the reliability interval is already narrow. In this second case no substantial improvement is gained by carrying out more than two analyses.

This subject is dealt with in more detail by Eckschlager,[4] and Shewell[5] has discussed other factors which influence the value of parallel determinations.

4.16 CORRELATION AND REGRESSION

When using instrumental methods it is often necessary to carry out a calibration procedure by using a series of samples (standards) each having a *known* concentration of the analyte to be determined. A **calibration curve** is constructed by measuring the instrumental signal for each standard and plotting this response against concentration (See Sections 17.14 and 17.21). Provided the same experimental conditions are used for the measurement of the standards and for the test (unknown) sample, the concentration of the latter may be determined from the calibration curve by graphical interpolation.

There are two statistical tests that should be applied to a calibration curve:

(*a*) to ascertain if the graph is linear, or in the form of a curve;
(*b*) to evaluate the best straight line (or curve) throughout the data points.

Correlation coefficient. In order to establish whether there is a linear relationship between two variables x_1 and y_1 the Pearson's correlation coefficient r is used.

$$r = \frac{n\Sigma x_1 y_1 - \Sigma x_1 \Sigma y_1}{\sqrt{[n\Sigma x_1^2 - (\Sigma x_1)^2][n\Sigma y_1^2 - (\Sigma y_1)^2]}} \tag{6}$$

where n is the number of data points.

The value of r must lie between $+1$ and -1: the nearer it is to $+1$, or in the case of negative correlation to -1, then the greater the probability that a definite linear relationship exists between the variables x and y. Values of r that tend towards zero indicate that x and y are not linearly related (they may be related in a non-linear fashion).

Although the correlation coefficient r would easily be calculated with the aid of a modern calculator or computer package, the following example will show how the value of r can be obtained.

Example 9. Quinine may be determined by measuring the fluorescence intensity in $1M\ H_2SO_4$ solution (Section 18.4). Standard solutions of quinine gave the following fluorescence values. Calculate the correlation coefficient r.

Concentration of quinine (x_1)	0.00	0.10	0.20	0.30	0.40 μg mL^{-1}
Fluorescence intensity (y_1)	0.00	5.20	9.90	15.30	19.10 arbitrary units

The terms in equation (6) are found from the following tabulated data.

x_1	y_1	x_1^2	y_1^2	$x_1 y_1$
0.00	0.00	0.00	0.00	0.00
0.10	5.20	0.01	27.04	0.52
0.20	9.90	0.04	98.01	1.98
0.30	15.30	0.09	234.09	4.59
0.40	19.10	0.16	364.81	7.64
$\Sigma x_1 = 1.00$	$\Sigma y_1 = 49.5$	$\Sigma x_1^2 = 0.30$	$\Sigma y_1^2 = 723.95$	$\Sigma x_1 y_1 = 14.73$

Therefore

$(\Sigma x_1)^2 = 1.000;$ $(\Sigma y_1)^2 = 2450.25;$ $n = 5$

Substituting the above values in equation (6), then

$$r = \frac{5 \times 14.73 - 1.00 \times 49.5}{\sqrt{(5 \times 0.30 - 1.000)(5 \times 723.95 - 2450.25)}} = \frac{24.15}{\sqrt{584.75}} = 0.9987$$

Hence, there is a very strong indication that a linear relation exists between fluorescence intensity and concentration (over the given range of concentration).

It must be noted, however, that a value of r close to either $+1$ or -1 does not necessarily confirm that there is a linear relationship between the variables. It is sound practice first to plot the calibration curve on graph paper and ascertain by visual inspection if the data points could be described by a straight line or whether they may fit a smooth curve.

The significance of the value of r is determined from a set of tables (Appendix 15). Consider the following example using five data $(x_1 y_1)$ points: From the table the value of r at 5 per cent significance value is 0.878. If the value of r is greater than 0.878 or less than -0.878 (if there is negative correlation), then the chance that this value could have occurred from random data points is less than 5 per cent. The conclusion can, therefore, be drawn that it is likely that x_1 and y_1 are linearly related. With the value of $r = 0.998_7$ obtained in the example given above there is confirmation of the statement that the linear relation between fluorescence intensity and concentration is highly likely.

4.17 LINEAR REGRESSION

Once a linear relationship has been shown to have a high probability by the value of the correlation coefficient (r), then the best straight line through the data points has to be estimated. This can often be done by visual inspection of the calibration graph but in many cases it is far better practice to evaluate the best straight line by linear regression (the method of least squares).

The equation of a straight line is

$$y = ax + b$$

where y, the **dependent** variable, is plotted as a result of changing x, the **independent** variable. For example, a calibration curve (Section 21.16) in atomic absorption spectroscopy would be obtained from the measured values of absorbance (y-axis) which are determined by using known concentrations of metal standards (x-axis).

To obtain the regression line 'y on x', the slope of the line (a) and the intercept on the y-axis (b) are given by the following equations.

$$a = \frac{n\Sigma x_1 y_1 - \Sigma x_1 \Sigma y_1}{n\Sigma x_1^2 - (\Sigma x_1)^2} \tag{7}$$

and $b = \bar{y} - a\bar{x}$ (8)

where $\bar{x}$ is the mean of all values of x_1 and $\bar{y}$ is the mean of all values of y_1.

Example 10. Calculate by the least squares method the equation of the best straight line for the calibration curve given in the previous example.

From Example 9 the following values have been determined.

145

$\Sigma x_1 = 1.00$; $\Sigma y_1 = 49.5$; $\Sigma x_1^2 = 0.30$; $\Sigma x_1 y_1 = 14.73$; $(\Sigma x_1)^2 = 1.000$; the number of points $(n) = 5$

and the values

$$\bar{x} = \frac{\Sigma x_1}{n} = \frac{1.00}{5} = 0.2$$

and

$$\bar{y} = \frac{\Sigma y_1}{n} = \frac{49.5}{5} = 9.9$$

By substituting the values in equations (7) and (8), then

$$a = \frac{5 \times 14.73 - 1.00 \times 49.5}{(5 \times 0.30) - (1.00)^2} = \frac{24.15}{0.5} = 48.3$$

and

$$b = 9.9 - (48.3 \times 0.2) = 0.24$$

So the equation of the straight line is

$$y = 48.3x + 0.24$$

If the fluorescence intensity of the test solution containing quinine was found to be 16.1, then an estimate of the concentration of quinine ($x\ \mu g\ mL^{-1}$) in this unknown could be

$$16.10 = 48.3x + 0.24$$

$$x = \frac{15.86}{48.30} = 0.32_8\ \mu g\ mL^{-1}$$

The determination of errors in the slope a and the intercept b of the regression line together with multiple and curvilinear regression is beyond the scope of this book but references may be found in the Bibliography, page 156.

4.18 COMPARISON OF MORE THAN TWO MEANS (ANALYSIS OF VARIANCE)

The comparison of more than two means is a situation that often arises in analytical chemistry. It may be useful, for example, to compare (a) the mean results obtained from different spectrophotometers all using the same analytical sample; (b) the performance of a number of analysts using the same titration method. In the latter example assume that three analysts, using the same solutions, each perform four replicate titrations. In this case there are two possible sources of error: (a) the random error associated with replicate measurements; and (b) the variation that may arise between the individual analysts. These variations may be calculated and their effects estimated by a statistical method known as the Analysis of Variance (ANOVA), where the **square of the standard deviation**, s^2, is termed the **variance**, V. Thus $F = \dfrac{s_1^2}{s_2^2}$

where $s_1^2 > s_2^2$, and may be written as $F = \dfrac{V_1}{V_2}$ where $V_1 > V_2$.

An Analysis of Variance calculation is best illustrated by using specific values in situation (*b*) just referred to.

Example 11. Three analysts were each asked to perform four replicate titrations using the same solutions. The results are given below.

Titre (mL)

Analyst A	Analyst B	Analyst C
22.53	22.48	22.57
22.60	22.40	22.62
22.54	22.48	22.61
22.62	22.43	22.65

To simplify the calculation it is sound practice to subtract a common number, e.g. 22.50, from each value. The sum of each column is then determined. *Note:* this will have no effect on the final values.

	Analyst A	Analyst B	Analyst C
	0.03	−0.02	0.07
	0.10	−0.10	0.12
	0.04	−0.02	0.11
	0.12	−0.07	0.15
Sum =	0.29	−0.21	0.45

The following steps have to be made in the calculation:

(*a*) *The grand total*

$$T = 0.29 - 0.21 + 0.45$$
$$= 0.53$$

(*b*) *The correction factor (C.F.)*

$$\text{C.F.} = \frac{T^2}{N} = \frac{(0.53)^2}{12} = 0.0234$$

where N is the total number of results.

(*c*) *The total sum of squares.* This is obtained by squaring each result, summing the totals of each column and then subtracting the correction factor (C.F.).

	Analyst A	Analyst B	Analyst C
	0.0009	0.0004	0.0049
	0.0100	0.0100	0.0144
	0.0016	0.0004	0.0121
	0.0144	0.0049	0.0225
Sum =	0.0269	0.0157	0.0539

147

Total sum of squares $= (0.0269 + 0.0157 + 0.0539) - \text{C.F.}$

$$= 0.0965 - 0.0234 = 0.0731$$

(d) *The between-treatment (analyst) sum of squares.* The sum of the squares of each individual column is divided by the number of results in each column, and then the correction factor is subtracted.

Between sum of squares $= \frac{1}{4}(0.29^2 + 0.21^2 + 0.45^2) - 0.0234$

$$= 0.0593$$

(e) *The within-sample sum of squares.* The between sum of squares is subtracted from the total sum of squares.

$$0.0731 - 0.0593 = 0.0138$$

(f) *The degrees of freedom (v).* These are obtained as follows:

The total number of degrees of freedom $v = N - 1 = 11$
The between-treatment degrees of freedom $v = C - 1 = 2$
The within-sample degrees of freedom $v = (N-1) - (C-1) = 9$

where C is the number of columns (in this example, the number of analysts).

(g) A table of Analysis of Variance (ANOVA table) may now be set up.

Source of variation	Sum of squares	d.f.	Mean square
'Between analysts'	0.0593	2	$0.0593/2 = 0.0297$
'Within titrations'	0.0138	9	$0.0138/9 = 0.00153$
Total	0.0731	11	

(h) The *F*-test is used to compare the two mean squares:

$$F_{2,9} = \frac{0.0297}{0.00153} = 19.41$$

From the *F*-tables (Appendix 13), the value of *F* at the 1 per cent level for the given degrees of freedom is 8.02. The calculated result (19.41) is higher than 8.02; hence there is a significant difference in the results obtained by the three analysts. Having ascertained in this example there is a significant difference between the three analysts, the next stage would be to determine whether the mean result is different from the others, or whether all the means are significantly different from each other.

The procedure adopted to answer these questions for the example given above is as follows:

(a) Calculate the titration means for each analyst. The mean titration values are $\bar{x}(A) = 22.57$ mL; $\bar{x}(B) = 22.45$ mL; and $\bar{x}(C) = 22.61$ mL.
(b) Calculate the quantity defined as the 'least significant difference', which is given by $s\sqrt{2/n}\, t_{0.05}$ where s is the square root of the Residual Mean Square, i.e. the 'within-titration' Mean Square. Hence $s = \sqrt{0.00153}$; n is the number of results in each column (in this example, 4); t is the 5 per cent value from the t-tables (Appendix 12), with the same number of degrees of

freedom as that for the Residual term, i.e. the 'within-titration' value. In this example the number of degrees of freedom is 9, so the least significant difference is given by

$$\sqrt{0.00153} \times \sqrt{2/4} \times 2.26 = 0.06 \, \text{mL}$$

If the titration means are arranged in increasing order, then $\bar{x}(B) < \bar{x}(A) < \bar{x}(C)$, and $\bar{x}(C) - \bar{x}(B)$ and $\bar{x}(A) - \bar{x}(B)$ are both greater than 0.06, whereas $\bar{x}(C) - \bar{x}(A)$ is less than 0.06. Hence there is no significant difference between analysts A and C, but the results of analyst B are significantly different from those of both A and C.

It should be noted that in this example the performance of only *one* variable, the three analysts, is investigated and thus this technique is called a one-way ANOVA. If *two* variables, e.g. the three analysts with four *different* titration methods, were to be studied, this would require the use of a two-way ANOVA. Details of suitable texts that provide a solution for this type of problem and methods for multivariate analysis are to be found in the Bibliography, page 156.

4.19 THE VALUE OF STATISTICS

Correctly used, statistics is an essential tool for the analyst. The use of statistical methods can prevent hasty judgements being made on the basis of limited information. It has only been possible in this chapter to give a brief resumé of some statistical techniques that may be applied to analytical problems. The approach, therefore, has been to use specific examples which illustrate the scope of the subject as applied to the treatment of analytical data. There is a danger that this approach may overlook some basic concepts of the subject and the reader is strongly advised to become more fully conversant with these statistical methods by obtaining a selection of the excellent texts now available.

In addition there is the rapidly developing subject of Chemometrics, which may be broadly defined as the application of mathematical and statistical methods to design and/or to optimise measurement procedures, and to provide chemical information by analysing relevant data. Space does not permit an inclusion in this book of such topics as experimental design and instrumental optimisation techniques or more sophisticated subjects as pattern recognition. There is no doubt however, that a knowledge of the scope of Chemometrics will be increasingly important for any competent analytical chemist. Details of some useful texts, both introductory and more advanced, are given in the Bibliography (Section 5.8). The reader should be aware, however, that some signal-processing techniques are included in this book, e.g. information will be found on derivative spectroscopy (Section 17.12) and Fourier transform methods (Section 19.2).

For References and Bibliography see Sections 5.7 and 5.8.

CHAPTER 5
SAMPLING

5.1 THE BASIS OF SAMPLING

The purpose of analysis is to determine the quality or composition of a material; and for the analytical results obtained to have any validity or meaning it is essential that adequate sampling procedures be adopted. Sampling is the process of extracting from a large quantity of material a small portion which is truly representative of the composition of the whole material.

Sampling methods fall into three main groups:

1. those in which all the material is examined;
2. casual sampling on an *ad hoc* basis;
3. methods in which portions of the material are selected based upon statistical probabilities.

Procedure (1) is normally impracticable, as the majority of methods employed are destructive, and in any case the amount of material to be examined is frequently excessive. Even for a sample of manageable size the analysis would be very time-consuming, it would require large quantities of reagents, and would monopolise instruments for long periods.

Sampling according to (2) is totally unscientific and can lead to decisions being taken on inadequate information. In this case, as the taking of samples is entirely casual, any true form of analytical control or supervision is impossible.

For these reasons the only reliable basis for sampling must be a mathematical one using statistical probabilities. This means that although not every item or every part of the sample is analysed, the limitations of the selection are carefully calculated and known in advance. Having calculated the degree of acceptable risk or margin of variation, the sampling plan is then chosen that will give the maximum information and control that is compatible with a rapid turnover of samples. For this reason, in the case of sampling from batches the selection of individual samples is carried out according to special random tables[6] which ensure that personal factors do not influence the choice.

5.2 SAMPLING PROCEDURE

The sampling procedure may involve a number of stages prior to the analysis of the material. The sampling stages are outlined in Fig. 5.1.

For the most part, bulk materials are non-homogeneous, e.g. minerals, sediments, and foodstuffs. They may contain particles of different composition which are not uniformly distributed within the material. In this case, a number

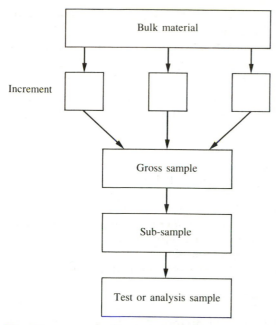

Fig. 5.1

of increments is taken in a random manner from points in the bulk material, so that each part has an equal chance of being selected. The combination of these increments then forms the gross sample. The gross sample is often too large for direct analysis and must be divided further to produce a sub-sample. The sub-sample may require treatment, for example reduction in particle size or thorough mixing, before the analytical sample can be obtained. The analytical sample should retain the same composition of the gross sample.

It must be stressed, however, that the whole object may be the analytical sample, e.g. a specimen of moon-rock. Ideally this sample would be analysed by non-destructive methods. Occasionally the bulk material may be homogeneous (some water samples) and then only one increment may be needed to determine the properties of the bulk. This increment should be of suitable size to provide samples for replicate analyses.

5.3 SAMPLING STATISTICS

The errors arising in sampling, particularly in the case of heterogeneous solids, may be the most important source of uncertainty in the subsequent analysis of the material. If we represent the standard deviation of the sampling operation (the sampling error) by s_S and the standard deviation of the analytical procedures (the analytical error) by s_A, then the overall standard deviation s_T (the total error) is given by

$$s_T = \sqrt{s_S^2 + s_A^2} \qquad (1)$$

or

$$s_T = \sqrt{V_S + V_A}$$

where V represents the appropriate variance. The separate evaluation of both V_S (the sampling variance) and V_A (the analytical variance) may be achieved by using the analysis of variance procedure (See Section 4.18). A comparison can be made of the between-sample variance — an estimate of the sampling error — and the within-sample variance — an estimate of the analytical error.

Example 1. If the sampling error is ± 3 per cent and the analytical error is ± 1 per cent, from equation (1) we can see that the total error s_T is given by

$$s_T = \sqrt{3^2 + 1^2} = \pm 3.16 \text{ per cent}$$

If, in the above example, the analytical error was ± 0.2 per cent then the total error s_T would be equal to ± 3.006 per cent. Hence the contribution of the analytical error to the total error is virtually insignificant. Youden[7] has stated that once the analytical uncertainty is reduced to one-third of the sampling uncertainty, further reduction of the former is not necessary. It is most important to realise that if the sampling error is large, then a rapid analytical method with relatively low precision may suffice.

In designing a sampling plan the following points should be considered:[8]

(*a*) the number of samples to be taken;
(*b*) the size of the sample;
(*c*) should individual samples be analysed or should a sample composed of two or more increments (composite) be prepared.

If the composition of the bulk material to be sampled is unknown, it is sensible practice to perform a preliminary investigation by collecting a number of samples and determining the analyte of interest.

The confidence limits (see Section 4.11) are given by the relationship

$$\mu = \bar{x} \pm t s_s \Big/ \sqrt{n} \tag{2}$$

where s_s is the standard deviation of individual samples, $\bar{x}$ is the mean of the analytical results and serves as an estimate of the true mean μ, and n is the number of samples taken.

Example 2. An estimate of the variability of nickel in a consignment of an ore, based on 16 determinations, was found to be $\pm 1.5\%$. How many samples should be taken to give (at the 95 per cent confidence level) a sampling error of less than 0.5 per cent nickel?

The value 0.5 per cent is in fact the difference between the sample mean $\bar{x}$ and the actual value μ. If this value is represented by E, then equation (2) may be written as

$$E = t\, s_s / \sqrt{n}$$

and, therefore,

$$n = \left(\frac{t\, s_s}{E}\right)^2$$

From the tables (Appendix 12) the value of t for $(n-1)$, 15 degrees of freedom at the 95 per cent confidence level is 2.13.

$$\therefore n = \left(\frac{2.13 \times 1.5}{0.5}\right)^2 \approx 41$$

Hence, from this test it has been shown that at least 41 samples are required if the specifications given in the above example are to be satisfied.

The other major problem concerned with sampling is that of the sample size. The size of the sample taken from a heterogeneous material is determined by the variation in particle size, and the precision needed in the results of the analysis.

The major source of error in sampling can arise from the taking of increments from the bulk material. It can be shown from random sampling theory that the accuracy of the sample is determined by its total size. Hence, the sampling variance, V, is inversely proportional to the mass of the sample. However, this statement is not true if the bulk material consists of varying particle sizes; then the number of increments taken will influence the sampling accuracy. The sampling variance, V, is inversely proportional to the number of sampling increments (n):

$$V = \frac{k}{n} \tag{3}$$

where k is a constant dependent on the size of the increment and variation within the bulk material.

5.4 SAMPLING AND PHYSICAL STATE

Many of the problems occurring during sampling arise from the physical nature of the materials to be studied.[9] Although gases and liquids can, and do, present difficulties, the greatest problems of adequate sampling undoubtedly arise with solids.

Gases. Few problems arise over homogeneity of gas mixtures where the storage vessel is not subjected to temperature or pressure variations. Difficulties may arise if precautions are not taken to clear valves, taps and connecting lines of any other gas prior to passage of the sample. Similarly care must be taken that no gaseous components will react with the sampling and analytical devices.

Liquids. In most cases general stirring or mixing is sufficient to ensure homogeneity prior to sampling. Where separate phases exist it is necessary to determine the relative volumes of each phase in order to compare correctly the composition of one phase with the other. The phases should in any case be individually sampled as it is not possible to obtain a representative sample of the combined materials even after vigorously shaking the separate phases together.

Solids. It is with solids that real difficulties over homogeneity arise. Even materials that superficially have every appearance of being homogeneous in fact may have localised concentrations of impurities and vary in composition. The procedure adopted to obtain as representative a sample as possible will depend greatly upon the type of solid. This process is of great importance since, if it is not satisfactorily done, the labour and time spent in making a careful analysis

153

of the sample may be completely wasted. If the material is more or less homogeneous, sampling is comparatively simple. If, however, the material is bulky and heterogeneous, sampling must be carried out with great care, and the method will vary somewhat with the nature of the bulk solid.

The underlying principle of the sampling of material in bulk, say of a truckload of coal or iron ore, is to select a large number of portions in a systematic manner from different parts of the bulk and then to combine them. This large sample of the total weight is crushed mechanically, if necessary, and then shovelled into a conical pile. Every shovelful must fall upon the apex of the cone and the operator must walk around the cone as he shovels; this ensures a comparatively even distribution. The top of the cone is then flattened out and divided into quarters. Opposite quarters of the pile are then removed, mixed to form a smaller conical pile, and again quartered. This process is repeated, further crushing being carried out if necessary, until a sample of suitable weight (say, 200–300 g) is obtained.

If the quantity of material is of the order of 2–3 kg or less, intermixing may be accomplished by the method known as 'tabling'. The finely divided material is spread on the centre of a large sheet of oilcloth or similar material. Each corner is pulled in succession over its diagonal partner, the lifting being reduced to a minimum; the particles are thus caused to roll over and over on themselves, and the lower portions are constantly brought to the top of the mass and thorough intermixing ensues. The sample may then be rolled to the centre of the cloth, spread out, and quartered as before. The process is repeated until a sufficiently small sample is obtained. The final sample for the laboratory, which is usually between 25 and 200 g in weight, is placed in an air-tight bottle. This method produces what is known as the 'average sample' and any analysis on it should always be compared with those of a second sample of the same material obtained by the identical routine.

Mechanical methods also exist for dividing up particulate material into suitably sized samples. Samples obtained by these means are usually representative of the bulk material within limits of less than ± 1 per cent, and are based upon the requirements established by the British Standards Institution. Sample dividers exist with capacities of up to 10 L and operate either by means of a series of rapidly rotating sample jars under the outlet of a loading funnel, or by a rotary cascade from which the samples are fed into a series of separate compartments. Sample dividers can lead to a great deal of time-saving in laboratories dealing with bulk quantities of powders or minerals.

The sampling of metals and alloys may be effected by drilling holes through a representative ingot at selected points; *all* the material from the holes is collected, mixed, and a sample of suitable size used for analysis. Turnings or scrapings from the outside are not suitable as these frequently possess superficial impurities from the castings or moulds.

In some instances in which grinding presents problems it is possible to obtain a suitable homogeneous sample by dissolving a portion of the material in an appropriate solvent.

Before analysis the representative solid sample is usually dried at 105–110 °C, or at some higher specified temperature if necessary, to constant weight. The results of the analysis are then reported on the 'dry' basis, viz. on a material dried at a specified temperature. The loss in weight on drying may be determined, and the results may be reported, if desired on the original 'moist' basis; these figures will only possess real significance if the material is not appreciably hygroscopic and no chemical changes, other than the loss of water, take place on drying.

In a course of systematic quantitative analysis, such as that with which we are chiefly concerned in the present book, the unknowns supplied for analysis are usually portions of carefully analysed samples which have been finely ground until uniform.

It should be borne in mind that although it is possible to generalise on sampling procedures, all industries have their own established methods for obtaining a record of the quantity and/or quality of their products. The sampling procedures for tobacco leaves will obviously differ from those used for bales of cotton or for coal. But although the types of samples differ considerably the actual analytical methods used later are of general application.

5.5 CRUSHING AND GRINDING

If the material is hard (e.g. a sample of rock), it is first broken into small pieces on a hard steel plate with a hardened hammer. The loss of fragments is prevented by covering the plate with a steel ring, or in some other manner. The small lumps may be broken in a 'percussion' mortar (also known as a 'diamond' mortar) (Fig. 5.2). The mortar and pestle are constructed entirely of hard tool steel. One or two small pieces are placed in the mortar, and the pestle inserted into position; the latter is struck lightly with a hammer until the pieces have been reduced to a coarse powder. The whole of the hard substance may be treated in this manner. The coarse powder is then *ground* in an agate mortar in small quantities at a time. A mortar of mullite is claimed to be superior to one of agate: mullite is a homogeneous ceramic material that is harder, more resistant to abrasion, and less porous than agate. A synthetic sapphire mortar and pestle (composed essentially of a specially prepared form of pure aluminium oxide) is marketed; it is extremely hard (comparable with tungsten carbide) and will grind materials not readily reduced in ceramic or metal mortars. Mechanical (motor-driven) mortars are available commercially.

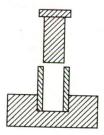

Fig. 5.2

5.6 HAZARDS IN SAMPLING

The handling of many materials is fraught with hazards[10] and this is no less so when sampling materials in preparation for chemical analysis. The sampler must always wear adequate protective clothing and if possible have detailed prior knowledge of the material being sampled. When dangers from toxicity

155

exist the necessary antidotes and treatment procedures should be available and established before sampling commences.[11] In no instances should naked flames be allowed anywhere near the sampling area.

Apart from the toxic nature of many gases, the additional hazards are those of excessive release of gas due to pressure changes, spontaneous ignition of flammable gases and sudden vaporisation of liquefied gases.

With liquids, dangers frequently arise from easily volatilised and readily flammable liquids. In all cases precautions should be greater than under normal circumstances due to the unpredictable nature and conditions of taking samples. The sampler must always be prepared for the unexpected, as can arise, for example, if a container has built up excess pressure, or if the wrong liquid has been packed. Toxic and unknown liquids should never be sucked along tubes or into pipettes by mouth.

Even the sampling of solids must not be casually undertaken, and the operator should always use a face mask as a protection until it is established that the powdered material is not hazardous.

It should be borne in mind that sampling of radioactive substances is a specialist operation at all times and should be carried out only under strictly controlled conditions within restricted areas. In almost all instances the operator must be protected against the radioactive emanations from the substance he is sampling.

Correct sampling of materials is therefore of importance in two main respects; firstly to obtain a representative portion of the material for analysis, and secondly to prevent the occurrence of accidents when sampling hazardous materials.

5.7 REFERENCES FOR PART B

1. W G Gosset 'Student', *Biometrika*, 1908, **6**, 1
2. J Mandel *The Statistical Analysis of Experimental Data*, Interscience, New York, 1964
3. C J Brookes, I G Betteley and S M Loxston *Mathematics and Statistics for Chemists*, John Wiley, New York, 1966, p. 304
4. K Eckschlager *Errors, Measurements and Results in Chemical Analysis*, Van Nostrand Reinhold, London; 1969
5. C T Shewell *Anal. Chem.*, 1959, **31**(5), 21A
6. J Murdoch and J A Barnes (1970). *Statistical Tables for Science, Engineering and Management*, 2nd edn, Macmillan, London, pp. 30–33
7. W J Youden *J. Assoc. Off. Anal. Chem.*, 1967, **50**, 1007
8. B Kratochvil and J. K. Taylor. *Anal. Chem.*, 1981, **53**, 925A
9. C R N Strouts, J H Gilfillan and H N Wilson *Sampling in Analytical Chemistry: The Working Tools*, Vol I, Oxford University Press, London, 1955, Chapter 3
10. N Irving Sax (Ed) *Dangerous Properties of Industrial Materials*, 3rd edn, Reinhold, New York, 1968
11. G D Muir (Ed) *Hazards in the Chemical Laboratory*, Royal Institute of Chemistry, London, 1971

5.8 SELECTED BIBLIOGRAPHY FOR PART B

1. J C Miller and J N Miller *Statistics for Analytical Chemistry*, 2nd edn, Wiley, Chichester, 1988
2. C Chatfield *Statistics for Technology*, 2nd edn, Chapman and Hall, London, 1984

3. O L Davies and P L Goldsmith *Statistical Methods in Research and Production*, Longman, London, 1982
4. M J Moroney *Facts from Figures*, 3rd edn, Penguin Books, Harmondsworth, 1965
5. G E P Box, W G Hunter and J S Hunter *Statistics for Experimentalists*, Wiley, New York, 1978
6. V Barnett and T Lewis *Outliers in Statistical Data*, Wiley, New York, 1978
7. R Caulcutt and R Boddy *Statistics for Analytical Chemists*, Chapman and Hall, London, 1983
8. P Sprent *Quick Statistics*, Penguin Books, Harmondsworth, 1981
9. B E Cooper *Statistics for Experimentalists*, Pergamon Press, Oxford, 1975
10. B R Kowalski (Ed) *Chemometrics*: *Theory and Application*, American Chemical Society, Washington, DC, 1977
11. D L Massart, A Dijkstra and L Kaufman *Evaluation and Optimisation of Laboratory Methods and Analytical Procedures*, Elsevier, Amsterdam, 1978
12. C R Hicks *Fundamental Concepts in the Design of Experiments*, 2nd edn, Holt, Rinehart and Winston, New York, 1973
13. K Varmuza *Pattern Recognition in Chemistry*, Springer-Verlag, Berlin, 1980
14. D L Massart, B G M Vandeginste, S N Deming, Y Michotte and L Kaufman *Chemometrics*: *A Textbook*, Elsevier, Amsterdam, 1987
15. S N Deming and S L Morgan *Experimental Design*: *A Chemometric Approach*, Elsevier, Amsterdam, 1987
16. J D Lee and T D Lee *Statistics and Computer Methods in BASIC*, Van Nostrand, London, 1983
17. D McCormick and A Roach *Measurements, Statistics and Computation*, ACOL–Wiley, Chichester, 1987
18. R G Brereton Chemometrics in analytical chemistry, a review, *Analyst*, 1987, **112**, 1635
19. C L Wilson and D W Wilson (Eds) Sampling, In *Comprehensive Analytical Chemistry*, Vol 1A, *Classical Analysis*, Elsevier, Amsterdam, 1959, Chapter II.3
20. H A Laitinen Sampling, In *Chemical Analysis: An Advanced Text and Reference*, McGraw-Hill, New York, 1960, Chapter 27
21. N V Steere (Ed) *Handbook of Laboratory Safety*, Chemical Rubber Co., Cleveland, Ohio, 1967
22. G Kateman and F W Pijpers *Quality Control in Analytical Chemistry*, Wiley, Chichester, 1981
23. R Smith and G V James *The Sampling of Bulk Materials*, Royal Society of Chemistry, London, 1981
24. B W Woodget and D Cooper *Samples and Standards*, ACOL–Wiley, Chichester, 1987
25. R Anderson *Sample Pretreatment and Separation*, ACOL–Wiley, Chichester, 1987

PART C
SEPARATIVE TECHNIQUES

CHAPTER 6
SOLVENT EXTRACTION

6.1 GENERAL DISCUSSION

Liquid–liquid extraction is a technique in which a solution (usually aqueous) is brought into contact with a second solvent (usually organic), essentially immiscible with the first, in order to bring about a transfer of one or more solutes into the second solvent. The separations that can be performed are simple, clean, rapid, and convenient. In many cases separation may be effected by shaking in a separatory funnel for a few minutes. The technique is equally applicable to trace level and large amounts of materials.

In the case of inorganic solutes we are concerned largely with samples in aqueous solution so that it is necessary to produce substances, such as neutral metal chelates and ion-association complexes, which are capable of extraction into organic solvents. For organic solutes, however, the extraction system may sometimes involve two immiscible organic solvents rather than the aqueous–organic type of extraction.

The technique of liquid–liquid extraction has, of course, been widely used to separate the components of organic systems; in particular, solvent extraction may be employed to effect a 'clean-up' and to achieve concentration of the solutes of interest prior to analysis. This is illustrated by the clean-up steps which have been used for the analysis of organochlorine pesticides, at the 0.1 μg g^{-1} level, in animal fats and dairy products. In one such procedure the solution of animal fat in hexane (25 mL) is extracted with three successive portions (10 mL) of dimethylformamide (DMF) saturated with hexane. The combined DMF extracts are washed with hexane (10 mL) saturated with DMF to remove any remaining traces of fat; after separation of the hexane it is equilibrated with further DMF (10 mL) to reduce loss of pesticide residue. The pesticide compounds are finally partitioned back into clean hexane after adding water to the combined DMF extracts. The hexane layer is used for the quantitative analysis of the extracted pesticide residues using gas chromatography with an electron-capture detector [Section 9.2(4)]. It is of interest to note that the extraction procedure involves two organic phases (i.e. no aqueous phase is involved) and that miscibility is minimised by saturating each solvent with the other. Extraction procedures for organic species, however, do not in general possess the same degree of selectivity as may be achieved for metal-containing systems, and the chief analytical application of solvent extraction is for the determination of metals as minor and trace constituents in various inorganic and organic materials.

Although solvent extraction has been used predominantly for the isolation and pre-concentration of a *single* chemical species prior to its determination, it may also be applied to the extraction of groups of metals or classes of organic compounds, prior to their determination by techniques such as atomic absorption or chromatography.

To understand the fundamental principles of extraction, the various terms used for expressing the effectiveness of a separation must first be considered. For a solute A distributed between two immiscible phases a and b, the Nernst Distribution (or Partition) Law states that, provided its molecular state is the same in both liquids and that the temperature is constant:

$$\frac{\text{Concentration of solute in solvent } a}{\text{Concentration of solute in solvent } b} = \frac{[A]_a}{[A]_b} = K_D$$

where K_D is a constant known as the **distribution (or partition) coefficient**. The law, as stated, is not thermodynamically rigorous (e.g. it takes no account of the activities of the various species, and for this reason would be expected to apply only in very dilute solutions, where the ratio of the activities approaches unity), but is a useful approximation. The law in its simple form does not apply when the distributing species undergoes dissociation or association in either phase. In the practical applications of solvent extraction we are interested primarily in the fraction of the total solute in one or other phase, quite regardless of its mode of dissociation, association, or interaction with other dissolved species. It is convenient to introduce the term distribution ratio D (or extraction coefficient E):

$$D = (C_A)_a/(C_A)_b$$

where the symbol C_A denotes the concentration of A in all its forms as determined analytically.

A problem often encountered in practice is to determine what is the most efficient method for removing a substance quantitatively from solution. It can be shown that if V mL of, say, an aqueous solution containing x_0 g of a solute be extracted n times with v-mL portions of a given solvent, then the weight of solute x_n remaining in the water layer is given by the expression:

$$x_n = x_0\left(\frac{DV}{DV + v}\right)^n$$

where D is the distribution ratio between water and the given solvent. It follows, therefore, that the best method of extraction with a given volume of extracting liquid is to employ the liquid in several portions rather than to utilise the whole quantity in a single extraction.

This may be illustrated by the following example. Suppose that 50 mL of water containing 0.1 g of iodine are shaken with 25 mL of carbon tetrachloride. The distribution coefficient of iodine between water and carbon tetrachloride at the ordinary laboratory temperature is 1/85, i.e. at equilibrium the iodine concentration in the aqueous layer is 1/85th of that in the carbon tetrachloride layer. The weight of iodine remaining in the aqueous layer after one extraction with 25 mL, and also after three extractions with 8.33 mL of the solvent, can be calculated by application of the above formula. In the first case, if x_1 g of iodine remains in the 50 mL of water, its concentration is $x_1/50$ g mL^{-1}; the concentration in the carbon tetrachloride layer will be $(0.1 - x_1)/25$ g mL^{-1}.

Hence:

$$\frac{x_1/50}{(0.1 - x_1)/25} = \frac{1}{85}, \text{ or } x_1 = 0.002\,30\,g$$

The concentration in the aqueous layer after three extractions with 8.33 mL of carbon tetrachloride is given by:

$$x_3 = 0.1\left(\frac{(1/85) \times 50}{(50/85) + 8.33}\right)^3 = 0.000\,028\,7\,g$$

The extraction may, therefore, be regarded as virtually complete.

If we confine our attention to the distribution of a solute A between water and an organic solvent, we may write the percentage extraction $E_\%$ as:

$$E_\% = \frac{100[A]_o V_o}{[A]_o V_o + [A]_w V_w} = \frac{100D}{D + (V_w/V_o)}$$

where V_o and V_w represent the volumes of the organic and aqueous phases respectively. Thus, the percentage of extraction varies with the volume ratio of the two phases and the distribution coefficient.

If the solution contains two solutes A and B it often happens that under the conditions favouring the complete extraction of A, some B is extracted as well. The effectiveness of separation increases with the magnitude of the **separation coefficient** or **factor** β, which is related to the individual distribution ratios as follows:

$$\beta = \frac{[A]_o/[B]_o}{[A]_w/[B]_w} = \frac{[A]_o/[A]_w}{[B]_o/[B]_w} = \frac{D_A}{D_B}$$

If $D_A = 10$ and $D_B = 0.1$, a single extraction will remove 90.9 per cent of A and 9.1 per cent of B (ratio 10:1); a second extraction of the same aqueous phase will bring the total amount of A extracted up to 99.2 per cent, but increases that of B to 17.4 per cent (ratio 5.7:1). More complete extraction of A thus involves an increased contamination by B. Clearly, when one of the distribution ratios is relatively large and the other very small, almost complete separation can be quickly and easily achieved. If the separation factor is large but the smaller distribution ratio is of sufficient magnitude that extraction of both components occurs, it is necessary to resort to special techniques to suppress the extraction of the unwanted component.

6.2 FACTORS FAVOURING SOLVENT EXTRACTION

It is well known that hydrated inorganic salts tend to be more soluble in water than in organic solvents such as benzene, chloroform, etc., whereas organic substances tend to be more soluble in organic solvents than in water unless they incorporate a sufficient number of hydroxyl, sulphonic, or other hydrophilic groupings. In solvent extraction analysis of metals we are concerned with methods by which the water solubility of inorganic cations may be masked by interaction with appropriate (largely organic) reagents; this will in effect remove some or all of the water molecules associated with the metal ion to which the water solubility is due.

Ionic compounds would not be expected to extract into organic solvents from aqueous solution because of the large loss in electrostatic solvation energy which would occur. The most obvious way to make an aqueous ionic species extractable is to neutralise its charge. This can be done by formation of a neutral metal chelate complex or by ion association; the larger and more hydrophobic the resulting molecular species the better will be its extraction.

In chelation complexes (sometimes called inner complexes when uncharged) the central metal ion coordinates with a polyfunctional organic base to form a stable ring compound, e.g. copper(II) 'acetylacetonate' or iron(III) 'cupferrate':

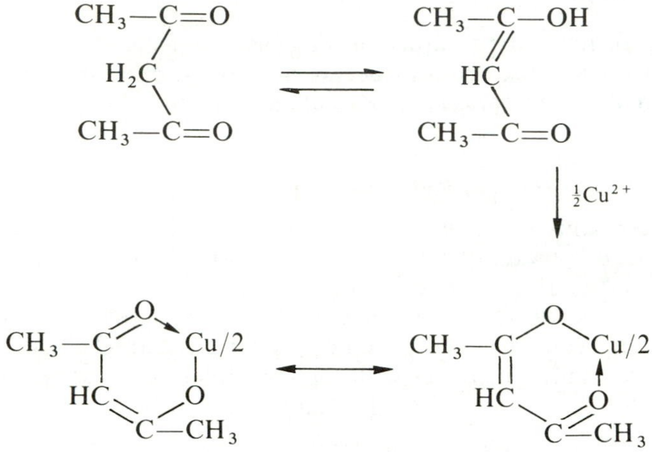

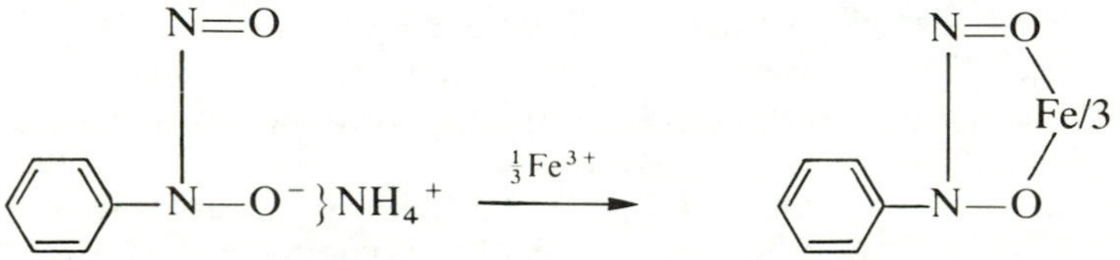

$$
\begin{array}{ccc}
CH_3-C=O & & CH_3-C-OH \\
/ & & \parallel \\
H_2C & \rightleftharpoons & HC \\
\backslash & & \backslash \\
CH_3-C=O & & CH_3-C=O
\end{array}
$$

$$\downarrow \tfrac{1}{2}Cu^{2+}$$

The factors which influence the stability of metal ion complexes have been discussed in Section 2.23, but it is appropriate to emphasise here the significance of the chelate effect and to list the features of the ligand which affect chelate formation:

1. *The basic strength of the chelating group.* The stability of the chelate complexes formed by a given metal ion generally increase with increasing basic strength of the chelating agent, as measured by the pK_a values.
2. *The nature of the donor atoms in the chelating agent.* Ligands which contain donor atoms of the soft-base type form their most stable complexes with the relatively small group of Class B metal ions (i.e. soft acids) and are thus more selective reagents. This is illustrated by the reagent diphenylthiocarbazone (dithizone) used for the solvent extraction of metal ions such as Pd^{2+}, Ag^+, Hg^{2+}, Cu^{2+}, Bi^{3+}, Pb^{2+}, and Zn^{2+}.
3. *Ring size.* Five- or six-membered conjugated chelate rings are most stable since these have minimum strain. The functional groups of the ligand must be so situated that they permit the formation of a stable ring.

4. *Resonance and steric effects.* The stability of chelate structures is enhanced by contributions of resonance structures of the chelate ring: thus copper acetylacetonate (see formula above) has greater stability than the copper chelate of salicylaldoxime. A good example of steric hindrance is given by 2,9-dimethylphenanthroline (neocuproin), which does not form a complex with iron(II) as does the unsubstituted phenanthroline; this hindrance is at a minimum in the tetrahedral grouping of the reagent molecules about a univalent tetracoordinated ion such as that of copper(I). A nearly specific reagent for copper is thus available.

The choice of a satisfactory chelating agent for a particular separation should, of course, take all the above factors into account. The critical influence of pH on the solvent extraction of metal chelates is discussed in the following section.

6.3 QUANTITATIVE TREATMENT OF SOLVENT EXTRACTION EQUILIBRIA

The solvent extraction of a neutral metal chelate complex formed from a chelating agent HR according to the equation

$$M^{n+} + nR^- \rightleftharpoons MR_n$$

may be treated quantitatively on the basis of the following assumptions: (*a*) the reagent and the metal complex exist as simple unassociated molecules in both phases; (*b*) solvation plays no significant part in the extraction process; and (*c*) the solutes are uncharged molecules and their concentrations are generally so low that the behaviour of their solutions departs little from ideality. The dissociation of the chelating agent HR in the aqueous phase is represented by the equation

$$HR \rightleftharpoons H^+ + R^-$$

The various equilibria involved in the solvent-extraction process are expressed in terms of the following thermodynamic constants:

Dissociation constant of complex, $K_c = [M^{n+}]_w [R^-]_w^n / [MR_n]_w$
Dissociation constant of reagent, $K_r = [H^+]_w [R^-]_w / [HR]_w$
Partition coefficient of complex, $p_c = [MR_n]_w / [MR_n]_o$
Partition coefficient of reagent, $p_r = [HR]_w / [HR]_o$

where the subscripts c and r refer to complex and reagent, and w and o to aqueous and organic phase respectively.

The distribution ratio, i.e. the ratio of the amount of metal extracted as complex into the organic phase to that remaining in all forms in the aqueous phase, is given by

$$D = [MR_n]_o / \{[MR_n]_w + [M^{n+}]_w\}$$

which can be shown[1] to reduce to

$$D = K[HR]_o^n / [H^+]_w^n$$

where $K = (K_r p_r)^n / K_c p_c$

If the reagent concentration remains virtually constant

$$D = K^* / [H^+]_w^n \text{ where } K^* = K[HR]_o^n$$

165

and the percentage of solute extracted, E, is given by

$$\log E - \log(100 - E) = \log D$$
$$= \log K^* + n\text{pH}$$

The distribution of the metal in a given system of the above type is a function of the pH alone. The equation represents a family of sigmoid curves when E is plotted against pH, with the position of each along the pH axis depending only on the magnitude of K^* and the slope of each uniquely depending upon n. Some theoretical extraction curves for divalent metals showing how the position of the curves depends upon the magnitude of K^* are depicted in Fig. 6.1; Fig. 6.2 illustrates how the slope depends upon n. It is evident that a ten-fold change in reagent concentration is exactly offset by a ten-fold change in hydrogen ion concentration, i.e. by a change of a single unit of pH: such a change of pH is much easier to effect in practice. If $\text{pH}_{1/2}$ is defined as the pH value at 50 per cent extraction ($E_\% = 50$) we see from the above equation that

$$\text{pH}_{1/2} = -\frac{1}{n}\log K^*$$

The difference in $\text{pH}_{1/2}$ values of two metal ions in a specific system is a measure of the ease of separation of the two ions. If the $\text{pH}_{1/2}$ values are sufficiently far

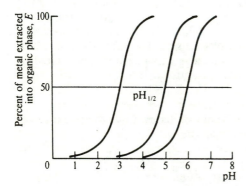

Fig. 6.1

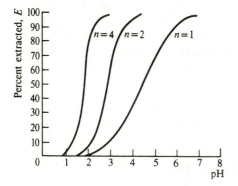

Fig. 6.2

166

apart, then excellent separation can be achieved by controlling the pH of extraction. It is often helpful to plot the extraction curves of metal chelates. If one takes as the criterion of a successful single-stage separation of two metals by pH control a 99 per cent extraction of one with a maximum of 1 per cent extraction of the other, for bivalent metals a difference of two pH units would be necessary between the two $pH_{1/2}$ values; the difference is less for tervalent metals. Some figures for the extraction of metal dithizonates in chloroform are given in Table 6.1. If the pH is controlled by a buffer solution, then those metals with $pH_{1/2}$ values in this region, together with all metals having smaller $pH_{1/2}$ values, will be extracted. The $pH_{1/2}$ values may be altered (and the selectivity of the extraction thus increased) by the use of a competitive complexing agent or of masking agents. Thus in the separation of mercury and copper by extraction with dithizone in carbon tetrachloride at pH 2, the addition of EDTA forms a water-soluble complex which completely masks the copper but does not affect the mercury extraction. Cyanides raise the $pH_{1/2}$ values of mercury, copper, zinc, and cadmium in dithizone extraction with carbon tetrachloride.

Table 6.1

Metal ion	Cu(II)	Hg(II)	Ag	Sn(II)	Co	Ni, Zn	Pb
Optimum pH of extraction	1	1–2	1–2	6–9	7–9	8	8.5–11

6.4 SYNERGISTIC EXTRACTION

The phenomenon in which two reagents, when used together, extract a metal ion with enhanced efficiency compared to their individual action is called synergism. A common form of synergistic extraction is that in which a metal ion, M^{n+}, is extracted by a mixture of an acidic chelating reagent, HR, and an uncharged basic reagent, S. The joint action of the reagents is especially pronounced in those cases where the coordination capacity of the metal ion is not fully achieved in the MR_n chelate; then the extractant S gives a mixed complex, MR_nS_x, which is extracted with much greater efficiency than the parent chelate. This concept has been usefully applied in the case of the reagent dithizone by using it in combination with bases such as pyridine and 1,10-phenanthroline. Thus, although the complex formed by manganese(II) with dithizone alone is of no analytical value because it decomposes rapidly, the red complex formed by manganese(II) with dithizone and pyridine is sufficiently stable to light and oxidation for it to be used in a sensitive photometric procedure for determining trace amounts of manganese.[2] Again the normally very slow reaction of nickel(II) with dithizone, H_2Dz, is greatly accelerated by the addition of nitrogen bases such as 1,10-phenanthroline (phen). The complex formed according to the equation:

$$Ni^{2+} + 2H_2Dz + phen \rightleftharpoons [Ni(HDz)_2(phen)] + 2H^+$$

is the basis of a very sensitive extraction–photometric method for nickel[3] (see also Section 6.17).

In addition to systems of the above type, i.e. involving adduct formation, various other types of synergistic extraction systems are recognised and have been reviewed.[4] An example is the synergistic influence of zinc in the extraction and AAS determination of trace cadmium in water.[5]

6.5 ION ASSOCIATION COMPLEXES

An alternative to the formation of neutral metal chelates for solvent extraction is that in which the species of analytical interest associates with oppositely charged ions to form a neutral extractable species.[6] Such complexes may form clusters with increasing concentration which are larger than just simple ion pairs, particularly in organic solvents of low dielectric constant. The following types of ion association complexes may be recognised.

1. Those formed from a reagent yielding a large organic ion, e.g. the tetraphenylarsonium, $(C_6H_5)_4As^+$, and tetrabutylammonium, $(n\text{-}C_4H_9)_4N^+$ ions, which form large ion aggregates or clusters with suitable oppositely charged ions, e.g. the perrhenate ion, ReO_4^-. These large and bulky ions do not have a primary hydration shell and cause disruption of the hydrogen-bonded water structure; the larger the ion the greater the amount of disruption and the greater the tendency for the ion association species to be pushed into the organic phase.

   These large ion extraction systems lack specificity since any relatively large unhydrated univalent cation will extract any such large univalent anion. On the other hand polyvalent ions, because of their greater hydration energy, are not so easily extracted and good separations are possible between MnO_4^-, ReO_4^- or TcO_4^- and CrO_4^{2-}, MoO_4^{2-} or WO_4^{2-}, for example.

2. Those involving a cationic or anionic chelate complex of a metal ion. Thus chelating agents having two uncharged donor atoms, such as 1:10-phenanthroline, form cationic chelate complexes which are large and hydrocarbon-like. Tris(phenanthroline) iron(II) perchlorate extracts fairly well into chloroform, and extraction is virtually complete using large anions such as long-chain alkyl sulphonate ions in place of ClO_4^-. The determination of anionic detergents using ferroin has been described.[7]

   Dagnall and West[8] have described the formation and extraction of a blue *ternary* complex, Ag(I)-1,10-phenanthroline-bromopyrogallol red (BPR), as the basis of a highly sensitive spectrophotometric procedure for the determination of traces of silver (Section 6.16). The reaction mechanism for the formation of the blue complex in aqueous solution was investigated by photometric and potentiometric methods and these studies led to the conclusion that the complex is an ion association system, $(Ag(phen)_2)_2BPR^{2-}$, i.e. involving a cationic chelate complex of a metal ion (Ag^+) associated with an anionic counter ion derived from the dyestuff (BPR). Ternary complexes have been reviewed by Babko.[9]

   Types (1) and (2) represent extraction systems involving coordinately unsolvated large ions and differ in this important respect from type (3).

3. Those in which solvent molecules are directly involved in formation of the ion association complex. Most of the solvents (ethers, esters, ketones and alcohols) which participate in this way contain donor oxygen atoms and the coordinating ability of the solvent is of vital significance. The coordinated solvent molecules facilitate the solvent extraction of salts such as chlorides and nitrates by contributing both to the size of the cation and the resemblance of the complex to the solvent.

A class of solvents which shows very marked solvating properties for inorganic compounds comprises the esters of phosphoric(V) (orthophosphoric) acid. The functional group in these molecules is the semipolar phosphoryl group, $\geq P^{\pm}-O^-$, which has a basic oxygen atom with good steric availability. A typical compound is tri-*n*-butyl phosphate (TBP), which has been widely used in solvent extraction on both the laboratory and industrial scale; of particular note is the use of TBP for the extraction of uranyl nitrate and its separation from fission products.

The mode of extraction in these 'oxonium' systems may be illustrated by considering the ether extraction of iron(III) from strong hydrochloric acid solution. In the aqueous phase chloride ions replace the water molecules coordinated to the Fe^{3+} ion, yielding the tetrahedral $FeCl_4^-$ ion. It is recognised that the hydrated hydronium ion, $H_3O^+(H_2O)_3$ or $H_9O_4^+$, normally pairs with the complex halo-anions, but in the presence of the organic solvent, solvent molecules enter the aqueous phase and compete with water for positions in the solvation shell of the proton. On this basis the primary species extracted into the ether (R_2O) phase is considered to be $[H_3O(R_2O)_3^+, FeCl_4^-]$ although aggregation of this species may occur in solvents of low dielectric constant.

The principle of ion-pair formation has long been used for the extraction of many metal ions, but not the alkali metals, due to the lack of complexing agents forming stable complexes with them. A significant development of recent years, however, has been the application of the so-called **'crown ethers'** which form stable complexes with a number of metal ions, particularly the alkali metal ions. These crown ethers are macrocyclic compounds containing 9–60 atoms, including 3–20 oxygen atoms, in the ring. Complexation is considered to result mainly from electrostatic ion–dipole attraction between the metal ion situated in the cavity of the ring and the oxygen atoms surrounding it. The ion-pair extraction of Na^+, K^+ and Ca^{2+} with some organic counter-ions and dicyclohexyl-18-crown-6 as complex-forming reagent has been described.[10]

6.6 EXTRACTION REAGENTS

This section provides a brief review of a number of chelating and other extraction reagents, as well as some organic solvents, with special interest as to their selective extraction properties. The handbook of Cheng *et al.* should be consulted for a more detailed account of organic analytical reagents.[11]

Acetylacetone (pentane-2,4-dione), $CH_3CO \cdot CH_2 \cdot COCH_3$. Acetylacetone is a colourless mobile liquid, b.p. 139 °C, which is sparingly soluble in water $(0.17 \, g \, mL^{-1}$ at 25 °C) and miscible with many organic solvents. It is useful both as a solution (in carbon tetrachloride, chloroform, benzene, xylene, etc.) and as the pure liquid. The compound is a β-diketone and forms well-defined chelates with over 60 metals. Many of the chelates (acetylacetonates) are soluble in organic solvents, and the solubility is of the order of grams per litre, unlike that of most analytically used chelates, so that macro- as well as micro-scale separations are possible. The selectivity can be increased by using EDTA as a masking agent. The use of acetylacetone as both solvent and extractant [e.g. for Al, Be, Ce, Co(III), Ga, In, Fe, U(VI), etc.] offers several advantages over its use in solution in carbon tetrachloride, etc.: extraction may be carried out at a lower pH than otherwise feasible because of the higher reagent concentration; and often the solubility of the chelate is greater in acetylacetone than in many organic solvents. The solvent generally used is carbon tetrachloride; the organic layer is heavier than water.

An interesting application is the separation of cobalt and nickel: neither $Co(II)$ nor $Ni(II)$ forms extractable chelates, but $Co(III)$ chelate is extractable; extraction is therefore possible following oxidation.

Thenoyltrifluoroacetone (TTA), $C_4H_3S \cdot CO \cdot CH_2 \cdot COCF_3$. This is a crystalline solid, m.p. 43 °C; it is, of course, a β-diketone, and the trifluoromethyl group increases the acidity of the enol form so that extractions at low pH values are feasible. The reactivity of TTA is similar to that of acetylacetone: it is generally used as a 0.1–0.5 M solution in benzene or toluene. The difference in extraction behaviour of hafnium and zirconium, and also among lanthanides and actinides, is especially noteworthy.

Other fluorinated derivatives of acetylacetone are trifluoroacetylacetone ($CF_3COCH_2COCH_3$) and hexafluoroacetylacetone ($CF_3COCH_2COCF_3$), which form stable volatile chelates with aluminium, beryllium, chromium(III) and a number of other metal ions. These reagents have consequently been used for the solvent extraction of such metal ions, with subsequent separation and analysis by gas chromatography [see Section 9.2(2)].

8-Hydroxyquinoline (oxine). Oxine is a versatile organic reagent and forms chelates with many metallic ions. The chelates of doubly and triply charged metal ions have the general formulae $M(C_9H_6ON)_2$ and $M(C_9H_6ON)_3$; the oxinates of the metals of higher charge may differ somewhat in composition, e.g. $Ce(C_9H_6ON)_4$; $Th(C_9H_6ON)_4 \cdot (C_9H_7ON)$; $WO_2(C_9H_6ON)_2$; $MoO_2(C_9H_6ON)_2$; $U_3O_6(C_9H_6ON)_6 \cdot (C_9H_7ON)$. Oxine is generally used as a 1 per cent (0.07 M) solution in chloroform, but concentrations as high as 10 per cent are advantageous in some cases (e.g. for strontium).

8-Hydroxyquinoline, having both a phenolic hydroxyl group and a basic nitrogen atom, is amphoteric in aqueous solution; it is completely extracted from aqueous solution by chloroform at pH < 5 and pH > 9; the distribution coefficient of the neutral compound between chloroform and water is 720 at 18 °C. The usefulness of this sensitive reagent has been extended by the use of masking agents (cyanide, EDTA, citrate, tartrate, etc.) and by control of pH.

Dimethylglyoxime. The complexes with nickel and with palladium are soluble in chloroform. The optimum pH range for extraction of the nickel complex is 4–12 in the presence of tartrate and 7–12 in the presence of citrate (solubility 35–50 μg Ni mL^{-1} at room temperature); if the amount of cobalt exceeds 5 mg some cobalt may be extracted from alkaline solution. Palladium(II) may be extracted out of ca 1M-sulphuric acid solution.

1-Nitroso-2-naphthol. The reagent forms extractable complexes (chloroform) with $Co(III)$ in an acid medium and with $Fe(II)$ in a basic medium.

Cupferron (ammonium salt of N-nitroso-N-phenylhydroxylamine). The reagent is used in cold aqueous solution (about 6 per cent). Metal cupferrates are soluble in diethyl ether and in chloroform, and so the reagent finds wide application in solvent-extraction separation schemes. Thus Fe(III), Ti, and Cu may be extracted from 1.2 M HCl solution by chloroform: numerous other elements may be extracted largely in acidic solution.

Diphenylthiocarbazone (dithizone), $C_6H_5 \cdot N{=}N \cdot CS \cdot NH \cdot NH \cdot C_6H_5$. The compound is insoluble in water and dilute mineral acids, and is readily soluble in dilute aqueous ammonia. It is used in dilute solution in chloroform or carbon tetrachloride. Dithizone is an important selective reagent for quantitative determinations of metals: colorimetric (and, of course, spectrophotometric) analyses are based upon the intense green colour of the reagent and the contrasting colours of the metal dithizonates in organic solvents. The selectivity is improved by the control of pH and the use of masking agents, such as cyanide, thiocyanate, thiosulphate, and EDTA.

The use of dithizone in combination with various organic bases for synergistic extraction has already been indicated (Section 6.4).

Sodium diethyldithiocarbamate, $\{(C_2H_5)_2N \cdot CS \cdot S\}^- Na^+$. This reagent is generally used as a 2 per cent aqueous solution; it decomposes rapidly in solutions of low pH. It is an effective extraction reagent for over 20 metals into various organic solvents, such as chloroform, carbon tetrachloride, and ethanol. The selectivity is enhanced by the control of pH and the addition of masking agents.

Ammonium pyrrolidine dithiocarbamate (APDC). The reagent is a white to pale yellow crystalline substance and is commonly supplied in bottles which contain a piece of ammonium carbonate in a muslin bag. In this form it is stable for at least one year at temperatures below 30 °C, but the finely divided material exposed to the ordinary atmosphere is much less stable.

$$M.W. = 164.28$$

The stability of the reagent in acid solution, together with its ability to complex a wide range of metals, make it a very useful general extracting reagent, especially for heavy metals. The chief applications of APDC in quantitative analysis are as follows:

(a) The separation and/or concentration (can be 100-fold or more) of heavy metals prior to their determination by atomic absorption spectrophotometry[12];
(b) The extraction and determination of metals by spectrophotometry (UV/visible) or for removing heavy metals prior to spectrophotometric determination of other elements (e.g. magnesium, calcium, aluminium).

Although APDC complexes are soluble in many organic solvents, it is found that 4-methylpent-2-one (isobutyl methyl ketone) and heptan-2-one (n-pentyl methyl ketone) are, in general, the most satisfactory for direct nebulisation into the air/acetylene flame used in atomic absorption spectroscopy.

Tri-n-butyl phosphate, $(n\text{-}C_4H_9)_3PO_4$. This solvent is useful for the extraction of metal thiocyanate complexes, of nitrates from nitric acid solution (e.g. cerium, thallium, and uranium), of chloride complexes, and of acetic acid from aqueous solution. In the analysis of steel, iron(III) may be removed as the soluble 'iron(III) thiocyanate'. The solvent is non-volatile, non-flammable, and rapid in its action.

Tri-*n*-octylphosphine oxide, $(n\text{-}C_8H_{17})_3PO$. This compound (TOPO) dissolved in cyclohexane $(0.1M)$ is an excellent extraction solvent. Thus the distribution ratio of U(VI) is of the order of 10^5 times greater for TOPO than for tri-*n*-butyl phosphate. The following elements are completely extracted from $1M$-hydrochloric acid: Cr(VI) as $H_2Cr_2O_7,2TOPO$; Zr(IV) as $ZrCl_4,2TOPO$; Ti(IV); U(VI) as $UO_2(NO_3)_2,2TOPO$; Fe(III); Mo(VI) and Sn(IV). If the hydrochloric acid concentration is increased to $7M$, Sb(III), Ga(III) and V(IV) are completely extracted.

Poly (macrocyclic) compounds. The analytical application of compounds such as crown polyethers and cryptands is based on their ability to function as ligands and form stable stoichiometric complexes with certain cations. Special importance is due to their preference for alkali metal ions which do not form complexes with many other ligands. A number of these compounds are commercially available and their properties and analytical applications have been described by Cheng *et al.*[11]

Cetyltrimethylammonium bromide (CTMB). Quaternary ammonium salts having one or more long-chain alkyl groups (e.g. CTMB and the corresponding chloride) have important applications as analytical reagents. These applications are mainly based on the ability of the quaternary ammonium ion to act (*a*) as a cationic reagent for the ion-pair extraction of metals as complex anions and (*b*) as a cationic micelle-forming reagent for photometric determination of metals.

$$\left[CH_3(CH_2)_{15}-\underset{\underset{CH_3}{|}}{\overset{\overset{CH_3}{|}}{N}}-CH_3 \right]^+ \; Br^- \qquad \text{M.W.} = 364.45$$

In the latter function, the reagent behaves as a surfactant and forms a cationic micelle at a concentration above the critical micelle concentration $(1 \times 10^{-4}M$ for CTMB). The complexation reactions occurring on the surface of the micelles differ from those in simple aqueous solution and result in the formation of a complex of higher ligand to metal ratio than in the simple aqueous system; this effect is usually accompanied by a substantial increase in molar absorptivity of the metal complex.

CTMB is commercially available as a colourless hygroscopic powder, readily soluble in water, alcohol and acetone. The aqueous solution foams strongly when shaken.

6.7 SOME PRACTICAL CONSIDERATIONS

Solvent extraction is generally employed in analysis to separate a solute (or solutes) of interest from substances which interfere in the ultimate quantitative analysis of the material; sometimes the interfering solutes are extracted selectively. Solvent extraction is also used to concentrate a species which in aqueous solution is too dilute to be analysed.

The choice of solvent for extraction is governed by the following considerations.

1. A high distribution ratio for the solute and a low distribution ratio for undesirable impurities.
2. Low solubility in the aqueous phase.
3. Sufficiently low viscosity and sufficient density difference from the aqueous phase to avoid the formation of emulsions.
4. Low toxicity and flammability.
5. Ease of recovery of solute from the solvent for subsequent analytical processing. Thus the b.p. of the solvent and the ease of stripping by chemical reagents merit attention when a choice is possible.

Sometimes mixed solvents may be used to improve the above properties. Salting-out agents may also improve extractability.

Extraction. Extraction may be accomplished in either a batch operation or a continuous operation. Batch extraction, the simplest and most widely used method, is employed where a large distribution ratio for the desired separation is readily obtainable. A small number of batch extractions readily remove the desired component completely and may be carried out in a simple separatory funnel. The two layers are shaken in the separatory funnel until equilibrium is attained, after which they are allowed to settle completely before separating. The extraction and separation should be performed at constant temperature, since the distribution ratio as well as the volumes of the solvent are influenced by temperature changes. It must be borne in mind that too violent agitation of the extraction mixture often serves no useful purpose: simple repeated inversions of the vessel suffice to give equilibrium in a relatively few inversions. If droplets of aqueous phase are entrained in the organic extract it is possible to remove them by filtering the extract through a dry filter paper: the filter paper should be washed several times with fresh organic solvent.

When the distribution ratio is low, continuous methods of extraction are used. This procedure makes use of a continuous flow of immiscible solvent through the solution; if the solvent is volatile, it is recycled by distillation and condensation and is dispersed in the aqueous phase by means of a sintered glass disc or equivalent device. Apparatus is available for effecting such continuous extractions with automatic return of the volatilised solvent (see the Bibliography, Section 9.10).

Stripping. Stripping is the removal of the extracted solute from the organic phase for further preparation for the detailed analysis. In many analytical procedures involving an extraction process, however, the concentration of the desired solute is determined directly in the organic phase.

Where other methods of analysis are to be employed, or where further separation steps are necessary, the solute must be removed from the organic phase to a more suitable medium. If the organic solvent is volatile (e.g. diethyl ether) the simplest procedure is to add a small volume of water and evaporate the solvent on a water bath; care should be taken to avoid loss of a volatile solute during the evaporation. Sometimes adjustment of the pH of the solution, change in valence state, or the use of competitive water-soluble complexing reagents may be employed to prevent loss of the solute. When the extracting solvent is non-volatile the solute is removed from the solvent by chemical means, e.g. by shaking the solvent with a volume of water containing acids or other reagents, whereby the extractable complex is decomposed. The metal ions are then quantitatively back-extracted into the aqueous phase.

Impurities present in the organic phase may sometimes be removed by backwashing. The organic extract when shaken with one or more small portions of a fresh aqueous phase containing the optimum reagent concentration and of correct pH will result in the redistribution of the impurities in favour of the aqueous phase, since their distribution ratios are low: most of the desired element will remain in the organic layer.

Completion of the analysis. The technique of solvent extraction permits the separation and often the pre-concentration of a particular element or substance (or of a group of elements or substances). Following this separation procedure, the final step of the analysis involves the quantitative determination of the species of interest by an appropriate technique.

Spectrophotometric methods may often be applied directly to the solvent extract utilising the absorption of the extracted species in the ultraviolet or visible region. A typical example is the extraction and determination of nickel as dimethylglyoximate in chloroform by measuring the absorption of the complex at 366 nm. Direct measurement of absorbance may also be made with appropriate ion association complexes, e.g. the ferroin–anionic detergent system, but improved results can sometimes be obtained by developing a chelate complex after extraction. An example is the extraction of uranyl nitrate from nitric acid into tributyl phosphate and the subsequent addition of dibenzoylmethane to the solvent to form a soluble coloured chelate.

Further techniques which may be applied directly to the solvent extract are flame spectrophotometry and atomic absorption spectrophotometry (AAS).[13] The direct use of the solvent extract in AAS may be advantageous since the presence of the organic solvent generally enhances the sensitivity of the method. However, the two main reasons for including a chemical separation in the preparation of a sample for AAS are:

(a) the concentration of the element to be determined is below the detection limit after normal preparation of the sample solution; and

(b) it is necessary to separate the species of interest from an excessive concentration of other solutes, affecting the nebuliser and burner system, or from a very strong chemical interference effect.

Solvent extraction is probably the separation technique which is most widely used in conjunction with AAS. It often allows the extraction of a number of elements in one operation and, because of the specific nature of AAS, non-selective reagents such as the thiocarbamate derivatives (e.g. APDC) may be used for the liquid–liquid extraction (see Section 6.18).

Multi-element analyses involving solvent extraction and high performance liquid chromatography (HPLC) have also been described. The extracts, containing metal–chelate complexes with sulphur-containing reagents, such as dithizone and diethyldithiocarbamate, were used directly for determination of the metals by HPLC.[14]

Automation of solvent extraction. Although automatic methods of analysis do not fall within the scope of the present text, it is appropriate to emphasise here that solvent extraction methods offer considerable scope for automation. A fully automated solvent extraction procedure, using APDC, for the determination of trace metals in water has been described.[15] Details of the use of automatic analysers are best obtained by referring to the appropriate manufacturers' manuals.

SOME APPLICATIONS

6.8 DETERMINATION OF BERYLLIUM AS THE ACETYLACETONE COMPLEX

Discussion. Beryllium forms an acetylacetone complex, which is soluble in chloroform, and yields an absorption maximum at 295 nm. The excess of acetylacetone in the chloroform solution may be removed by rapid washing with 0.1M-sodium hydroxide solution. It is advisable to treat the solution containing up to 10 μg of Be with up to 10 mL of 2 per cent EDTA solution: the latter will mask up to 1 mg of Fe, Al, Cr, Zn, Cu, Pb, Ag, Ce, and U.

Procedure.   Prepare a solution containing 10 μg of beryllium in 50 mL. (**CARE: Beryllium compounds are toxic.**) Use beryllium sulphate, $BeSO_4,4H_2O$. To 50.0 mL of this solution contained in a beaker, add dilute hydrochloric acid until the pH is 1.0, and then introduce 10.0 mL of 2 per cent EDTA solution. Adjust the pH to 7 by the addition of 0.1M sodium hydroxide solution. Add 5.0 mL of 1 per cent aqueous acetylacetone and readjust the pH to 7–8. After standing for 5 minutes, extract the colourless beryllium complex with three 10 mL portions of chloroform. Wash the chloroform extract rapidly with two 50 mL portions of 0.1M sodium hydroxide in order to remove the excess of acetylacetone. To determine the absorbance at 295 nm (in the ultraviolet region of the spectrum) it may be necessary to dilute the extract with chloroform. Measure the absorbance using 1.0 cm absorption cells against a blank.
     Repeat the determination with a solution containing 100 μg of iron(III) and of aluminium ion; the absorbance is unaffected.

6.9 DETERMINATION OF BORON USING FERROIN

Discussion. The method is based upon the complexation of boron as the bis(salicylato)borate(III) anion (A), (borodisalicylate), and the solvent extraction into chloroform of the ion-association complex formed with the ferroin.

(A)

The intensity of the colour of the extract due to ferroin is observed spectrophotometrically and may be related by calibration to the boron content of the sample.
     The method has been applied to the determination of boron in river water and sewage,[16] the chief sources of interference being copper(II) and zinc ions, and anionic detergents. The latter interfere by forming ion-association complexes with ferroin which are extracted by chloroform; this property

may, however, be utilised for the joint determination of boron and anionic detergents by the one procedure. The basis of this joint determination is that the ferroin–anionic detergent complex may be immediately extracted into chloroform, whereas the formation of the borodisalicylate anion from boric acid and salicylate requires a reaction time of one hour prior to extraction using ferroin. The absorbance of the chloroform extract obtained after zero minutes thus gives a measure of the anionic detergent concentration, whereas the absorbance of the extract after a one-hour reaction period corresponds to the amount of boron plus anionic detergent present. Interference due to copper(II) ions may be eliminated by masking with EDTA.

Reagents. *Sulphuric acid solution,* 0.05M.
Sodium hydroxide solution, 0.1M.
Sodium salicylate solution, 10 per cent w/v.
EDTA solution, 1 per cent w/v. Use the disodium salt of EDTA.
Ferroin solution, 2.5×10^{-2} M. Dissolve 0.695 g of iron(II) sulphate heptahydrate and 1.485 g of 1,10-phenanthroline hydrate in 100 mL of distilled water.
Boric acid solution, 2.5×10^{-4} M. Dissolve 61.8 mg of boric acid in 1 L of distilled water; dilute 250 mL of this solution to 1 L to give the standard boric acid solution.

Use analytical reagent grade materials whenever possible and store the solutions in polythene bottles.

Procedure. *(a) Zero-minutes reaction time.* Neutralise a measured volume of the sample containing $1–2$ mg L^{-1} of boron with sodium hydroxide or sulphuric acid (0.05M) to a pH of 5.5 (use a pH meter). Note the change in volume and hence calculate the volume correction factor to be applied to the final result. Measure 100 mL of the neutralised sample solution into a flask, add 10 mL of 10 per cent sodium salicylate solution and 17.5 mL of 0.05M sulphuric acid, and mix the solutions thoroughly. Adjust the pH of the solution to pH 6 to 7 with 0.1M sodium hydroxide and transfer the solution immediately to a separatory funnel; wash the flask with 20 mL of distilled water and add the washings to the rest of the solution. Add by pipette 1 mL of 1 per cent EDTA solution and 1 mL of $2.5 \times 10^{-2}M$ ferroin solution and again throughly mix the solution. Add 50 mL of chloroform and shake the funnel for 30 seconds to mix the phases thoroughly. Allow the layers to separate and transfer the chloroform layer to another separatory funnel. Wash the chloroform by shaking it vigorously for 30 seconds with 100 mL of water and repeat this process with a second 100 mL of water. Filter the chloroform phase through cotton-wool and measure the absorbance, A_o, against pure chloroform at 516 nm in a 1 cm cell [this 1 cm cell reading is used to calculate the boron concentration on the basis of equation (1) (see below), but if the zero-minutes reading is to be used for determination of anionic detergent concentration a 2 cm cell reading is more suitable].

(b) One-hour reaction time. Measure a second 100 mL of neutralised sample solution into a flask, add 10 mL of 10 per cent sodium salicylate solution and 17.5 mL of 0.05 M sulphuric acid solution. Mix the solutions thoroughly, allow the mixture to stand for one hour and adjust the pH of the solution to $6–7$ with 0.1 M sodium hydroxide. Now proceed as previously described, under *(a)*,

to obtain the absorbance, A_{1h}. The absorbance, A, to be used in the calculation of the boron concentration is obtained from the following equation:

$$A = A_{1h} - (A_0 - A_{(blank)0}) \tag{1}$$

$A_{(blank)0}$ is determined by repeating procedure (a), i.e. zero-minutes, using 100 mL of distilled water in place of the sample solution.

Calculate the amount of boron present by reference to a calibration graph of absorbance against boron concentration (mg L^{-1}). Multiply the result obtained by the appropriate volume correction factor arising from neutralisation of the sample.

Calibration. Take 5, 10, 25, 50, 75 and 100 mL of the standard boric acid solution ($2.5 \times 10^{-4}M$) and make each up to 100 mL with distilled water; this yields a boron concentration range up to 2.70 mg L^{-1}. Continue with each solution as described under procedure (b), i.e. one-hour reaction time, except that the initial neutralisation of the boron solution to pH 5.5 is not necessary. Construct a calibration graph of absorbance at 516 nm against boron concentration, mg L^{-1}. For maximum accuracy, the calibration should be carried out immediately prior to the analysis of samples.

6.10 DETERMINATION OF COPPER AS THE DIETHYLDITHIOCARBAMATE COMPLEX

Discussion. Sodium diethyldithiocarbamate (B) reacts with a weakly acidic or ammoniacal solution of copper(II) in low concentration to produce a brown colloidal suspension of the copper(II) diethyldithiocarbamate. The suspension may be extracted with an organic solvent (chloroform, carbon tetrachloride or butyl acetate) and the coloured extract analysed spectrophotometrically at 560 nm (butyl acetate) or 435 nm (chloroform or carbon tetrachloride).

$$(C_2H_5)_2N-\overset{\displaystyle S}{\underset{\displaystyle S^-\}Na^+}{\overset{\|}{C}}} \qquad (B)$$

Many of the heavy metals give slightly soluble products (some white, some coloured) with the reagent, most of which are soluble in the organic solvents mentioned. The selectivity of the reagent may be improved by the use of masking agents, particularly EDTA.

The reagent decomposes rapidly in solutions of low pH.

Procedure. Dissolve 0.0393 g of pure copper(II) sulphate pentahydrate in 1 L of water in a graduated flask. Pipette 10.0 mL of this solution (containing about 100 μg Cu) into a beaker, add 5.0 mL of 25 per cent aqueous citric acid solution, render slightly alkaline with dilute ammonia solution and boil off the excess of ammonia; alternatively, adjust to pH 8.5 using a pH meter. Add 15.0 mL of 4 per cent EDTA solution and cool to room temperature. Transfer to a separatory funnel, add 10 mL of 0.2 per cent aqueous sodium diethyldithio-carbamate solution, and shake for 45 seconds. A yellow–brown colour develops in the solution. Pipette 20 mL of butyl acetate (ethanoate) into the funnel and shake for 30 seconds. The organic layer acquires a yellow colour. Cool, shake for 15 seconds and allow the phases to separate. Remove the lower aqueous layer; add 20 mL of 5 per cent sulphuric acid (v/v), shake for 15 seconds, cool, and separate the organic phase. Determine the absorbance at 560 nm in 1.0 cm absorption cells against a blank. All the copper is removed in one extraction.

Repeat the experiment in the presence of 1 mg of iron(III); no interference can be detected.

6.11 DETERMINATION OF COPPER AS THE 'NEO-CUPROIN' COMPLEX

Discussion. 'Neo-cuproin' (2,9-dimethyl-1,10-phenanthroline) can, under certain conditions, behave as an almost specific reagent for copper(I). The complex is soluble in chloroform and absorbs at 457 nm. It may be applied to the determination of copper in cast iron, alloy steels, lead–tin solder, and various metals.

Procedure. To 10.0 mL of the solution containing up to 200 μg of copper in a separatory funnel, add 5.0 mL of 10 per cent hydroxylammonium chloride solution to reduce Cu(II) to Cu(I), and 10 mL of a 30 per cent sodium citrate solution to complex any other metals which may be present. Add ammonia solution until the pH is about 4 (Congo red paper), followed by 10 mL of a 0.1 per cent solution of 'neo-cuproin' in absolute ethanol. Shake for about 30 seconds with 10 mL of chloroform and allow the layers to separate. Repeat the extraction with a further 5 mL of chloroform. Measure the absorbance at 457 nm against a blank on the reagents which have been treated similarly to the sample.

6.12 DETERMINATION OF IRON AS THE 8-HYDROXYQUINOLATE

Discussion. Iron(III) (50–200 μg) can be extracted from aqueous solution with a 1 per cent solution of 8-hydroxyquinoline in chloroform by double extraction when the pH of the aqueous solution is between 2 and 10. At a pH of 2–2.5 nickel, cobalt, cerium(III), and aluminium do not interfere. Iron(III) oxinate is dark-coloured in chloroform and absorbs at 470 nm.

Procedure. Weigh out 0.0226 g of hydrated ammonium iron(III) sulphate and dissolve it in 1 L of water in a graduated flask; 50 mL of this solution contain 100 μg of iron. Place 50.0 mL of the solution in a 100 mL separatory funnel, add 10 mL of a 1 per cent oxine (analytical grade) solution in chloroform and shake for 1 minute. Separate the chloroform layer. Transfer a portion of the latter to a 1.0 cm absorption cell. Determine the absorbance at 470 nm in a spectrophotometer, using the solvent as a blank or reference. Repeat the extraction with a further 10 mL of 1 per cent oxine solution in chloroform, and measure the absorbance to confirm that all the iron was extracted.

Repeat the experiment using 50.0 mL of the iron(III) solution in the presence of 100 μg of aluminium ion and 100 μg of nickel ion at pH 2.0 (use a pH meter to adjust the acidity) and measure the absorbance. Confirm that an effective separation has been achieved.

Note. Some typical results are given below. Absorbance after first extraction 0.605; after second extraction 0.004; in presence of 100 μg Al and 100 μg Ni the absorbance obtained is 0.602.

6.13   DETERMINATION OF LEAD BY THE DITHIZONE METHOD*

Discussion. Diphenylthiocarbazone (dithizone) behaves in solution as a tautomeric mixture of (C) and (D):

(C)   (D)

It functions as a monoprotic acid ($pK_a = 4.7$) up to a pH of about 12; the acid proton is that of the thiol group in (C). 'Primary' metal dithizonates are formed according to the reaction:

$$M^{n+} + nH_2Dz \rightleftharpoons M(HDz)_n + nH^+$$

Some metals, notably copper, silver, gold, mercury, bismuth, and palladium, form a second complex (which we may term 'secondary' dithizonates) at a higher pH range or with a deficiency of the reagent:

$$2M(HDz)_n \rightleftharpoons M_2Dz_n + nH_2Dz$$

In general, the 'primary' dithizonates are of greater analytical utility than the 'secondary' dithizonates, which are less stable and less soluble in organic solvents.

Dithizone is a violet–black solid which is insoluble in water, soluble in dilute ammonia solution, and also soluble in chloroform and in carbon tetrachloride to yield green solutions. It is an excellent reagent for the determination of small (microgram) quantities of many metals, and can be made selective for certain metals by resorting to one or more of the following devices.

(a) Adjusting the pH of the solution to be extracted. Thus from acid solution ($0.1–0.5M$) silver, mercury, copper, and palladium can be separated from other metals; bismuth can be extracted from a weakly acidic medium; lead and zinc from a neutral or faintly alkaline medium; cadmium from a strongly basic solution containing citrate or tartrate.

(b) Adding a complex-forming agent or masking agent, e.g. cyanide, thiocyanate, thiosulphate, or EDTA.

It must be emphasised that dithizone is an extremely sensitive reagent and is applicable to quantities of metals of the order of micrograms. Only the purest dithizone may be used, since the reagent tends to oxidise to diphenylthiocarbadiazone, $S=C(N=NC_6H_5)_2$: the latter does not react with metals, is insoluble in ammonia solution, and dissolves in organic solvents to give yellow or brown solutions. Reagents for use in dithizone methods of analysis must be of the highest purity. De-ionised water and redistilled acids are recommended: ammonia solution should be prepared by passing ammonia gas into water. Weakly basic and neutral solutions can frequently be freed from

---

* This experiment is not recommended for elementary students or students having little experience of analytical work.

179

reacting heavy metals by extracting them with a fairly strong solution of dithizone in chloroform until a green extract is obtained. Vessels (of Pyrex) should be rinsed with dilute acid before use. Blanks must always be run.

Only one example of the use of dithizone in solvent extraction will be given in order to illustrate the general technique involved.

Procedure. Dissolve 0.0079 g of pure lead nitrate in 1 L of water in a graduated flask. To 10.0 mL of this solution (containing about 50 μg of lead) contained in a 250 mL separatory funnel, add 75 mL of ammonia–cyanide–sulphite mixture (Note 1), adjust the pH of the solution to 9.5 (pH meter) by the cautious addition of hydrochloric acid **(CARE!),*** then add 7.5 mL of a 0.005 per cent solution of dithizone in chloroform (Note 2), followed by 17.5 mL of chloroform. Shake for 1 minute, and allow the phases to separate. Determine the absorbance at 510 nm against a blank solution in a 1.0 cm absorption cell. A further extraction of the same solution gives zero absorption indicative of the complete extraction of the lead. Almost the same absorbance is obtained in the presence of 100 μg of copper ion and 100 μg of zinc ion.

Notes. (1) This solution is prepared by diluting 35 mL of concentrated ammonia solution (sp. gr. 0.88) and 3.0 mL of 10 per cent potassium cyanide solution (**caution**) to 100 mL, and then dissolving 0.15 g of sodium sulphite in the solution.

(2) One millilitre of this solution is equivalent to about 20 μg of lead. The solution should be freshly prepared using the analytical-grade reagent, ideally taken from a new or recently opened reagent bottle.

6.14 DETERMINATION OF MOLYBDENUM BY THE THIOCYANATE METHOD

Discussion. Molybdenum(VI) in acid solution when treated with tin(II) chloride [best in the presence of a little iron(II) ion] is converted largely into molybdenum(V): this forms a complex with thiocyanate ion, probably largely $Mo(SCN)_5$, which is red in colour. The latter may be extracted with solvents possessing donor oxygen atoms (3-methylbutanol is preferred). The colour depends upon the acid concentration (optimum concentration $1M$) and the concentration of the thiocyanate ion ($\ll 1$ per cent, but colour intensity is constant in the range 2–10 per cent); it is little influenced by excess of tin(II) chloride. The molybdenum complex has maximum absorption at 465 nm.

Reagents. *Standard molybdenum solution.* Dissolve 0.184 g of ammonium molybdate $(NH_4)_6[Mo_7O_{24}]4H_2O$ in 1 L of distilled water in a graduated flask: this gives a 0.01 per cent Mo solution containing 100 μg Mo mL^{-1}. Alternatively, dissolve 0.150 g of molybdenum trioxide in a few millilitres of dilute sodium hydroxide solution, dilute with water to about 100 mL, render slightly acidic with dilute hydrochloric acid, and then dilute to 1 L with water in a graduated flask: this is a 0.0100 per cent solution. It can be diluted to 0.001 per cent with 0.1M hydrochloric acid.

Ammonium iron(II) sulphate solution. Dissolve 10 g of the salt in 100 mL of very dilute sulphuric acid.

---

* **It is essential that the pH of the mixture does not fall below 9.5,** even temporarily, as there is always the possibility that HCN could be liberated.

Tin(II) chloride solution. Dissolve 10 g of tin(II) chloride dihydrate in 100 mL of $1M$ hydrochloric acid.

Potassium thiocyanate solution. Prepare a 10 per cent aqueous solution from the pure salt.

Procedure. Construct a calibration curve by placing 1.0, 2.0, 3.0, 4.0, and 5.0 mL of the 0.001 per cent Mo solution (containing 10 μg, 20 μg, 30 μg, 40 μg, and 50 μg Mo respectively) in 50 mL separatory funnels and diluting each with an equal volume of water. Add to each funnel 2.0 mL of concentrated hydrochloric acid, 1.0 mL of the ammonium iron(II) sulphate solution, and 3.0 mL of the potassium thiocyanate solution; shake gently and then introduce 3.0 mL of the tin(II) chloride solution. Add water to bring the total volume in each separatory funnel to 25 mL and mix. Pipette 10.0 mL of redistilled 3-methylbutanol into each funnel and shake individually for 30 seconds. Allow the phases to separate, and carefully run out the lower aqueous layer. Remove the glass stopper and pour the alcoholic extract through a small plug of purified glass wool in a small funnel and collect the organic extract in a 1.0 cm absorption cell. Measure the absorbance at 465 nm in a spectrophotometer against a 3-methylbutanol blank. Plot absorbance against μg of Mo. A straight line is obtained over the range 0–50 μg Mo: Beer's law is obeyed (Section 17.2).

Determine the concentration of Mo in unknown samples supplied and containing less than 50 μg Mo per 10 mL: use the calibration curve, and subject the unknown to the same treatment as the standard solutions.

The above procedure may be adapted to the determination of **molybdenum in steel**. Dissolve a 1.00 g sample of the steel (accurately weighed) in 5 mL of 1:1 hydrochloric acid and 15 mL of 70 per cent perchloric acid. Heat the solution until dense fumes are evolved and then for 6–7 minutes longer. Cool, add 20 mL of water, and warm to dissolve all salts. Dilute the resulting cooled solution to volume in a 1 L flask. Pipette 10.0 mL of the diluted solution into a 50 mL separatory funnel, add 3 mL of the tin(II) chloride solution, and continue as detailed above. Measure the absorbance of the extract at 465 nm with a spectrophotometer, and compare this value with that obtained with known amounts of molybdenum. Use the calibration curve prepared with equal amounts of iron and varying quantities of molybdenum. If preferred, a mixture of 3-methylbutanol and carbon tetrachloride, which is heavier than water, can be used as extractant.

Note. Under the above conditions of determination the following elements interfere in the amount specified when the amount of Mo is 10 μg (error greater than 3 per cent): V, 0.4 mg, yellow colour [interference prevented by washing extract with tin(II) chloride solution]; Cr(VI), 2 mg, purple colour; W(VI), 0.15 mg, yellow colour; Co, 12 mg, slight green colour; Cu, 5 mg; Pb, 10 mg; Ti(III), 30 mg (in presence of sodium fluoride).

6.15 DETERMINATION OF NICKEL AS THE DIMETHYLGLYOXIME COMPLEX

Discussion. Nickel (200–400 μg) forms the red dimethylglyoxime complex in a slightly alkaline medium; it is only slightly soluble in chloroform (35–50 μg Ni mL^{-1}). The optimum pH range of extraction of the nickel complex is 7–12 in the presence of citrate. The nickel complex absorbs at 366 nm and also at 465–470 nm.

Procedure. Weigh out 0.135 g of pure ammonium nickel sulphate $(NiSO_4,(NH_4)_2SO_4,6H_2O)$ and dissolve it in 1 L of water in a graduated flask. Transfer 10.0 mL of this solution (Ni content about 200 μg) to a beaker containing 90 mL of water, add 5.0 g of citric acid, and then dilute ammonia solution until the pH is 7.5. Cool and transfer to a separatory funnel, add 20 mL of dimethylglyoxime solution (Note 1) and, after standing for a minute or two, 12 mL of chloroform. Shake for 1 minute, allow the phases to settle out, separate the red chloroform layer, and determine the absorbance at 366 nm in a 1.0 cm absorption cell against a blank. Extract with a further 12 mL of chloroform and measure the absorbance of the extract at 366 nm; very little nickel will be found.

Repeat the experiment in the presence of 500 μg of iron(III) and 500 μg of aluminium ions; no interference will be detected, but cobalt may interfere (Note 2).

Notes. (1) The dimethylglyoxime reagent is prepared by dissolving 0.50 g of dimethylglyoxime in 250 mL of ammonia solution and diluting to 500 mL with water.

(2) Cobalt forms a brown soluble dimethylglyoxime complex which is very slightly extracted by chloroform; the amount is only significant if large amounts of Co (> 2–3 mg) are present. If Co is suspected it is best to wash the organic extract with ca 0.5M ammonia solution: enough reagent must be added to react with the Co and leave an excess for the Ni. Large amounts of cobalt may be removed by oxidising with hydrogen peroxide, complexing with ammonium thiocyanate (as a 60 per cent aqueous solution), and extracting the compound with a pentyl alcohol–diethyl ether (3:1) mixture. Copper(II) is extracted to a small extent, and is removed from the extract by shaking with 0.5M ammonia solution. Copper in considerable amounts is not extracted if it is complexed with thiosulphate at pH 6.5. Much Mn tends to inhibit the extraction of Ni; this difficulty is overcome by the addition of hydroxylammonium chloride. Iron(III) does not interfere.

6.16 DETERMINATION OF SILVER BY EXTRACTION AS ITS ION ASSOCIATION COMPLEX WITH 1,10–PHENANTHROLINE AND BROMOPYROGALLOL RED

Discussion. Silver can be extracted from a nearly neutral aqueous solution into nitrobenzene as a blue ternary ion association complex formed between silver(I) ions, 1,10-phenanthroline and bromopyrogallol red. The method is highly selective in the presence of EDTA, bromide and mercury(II) ions as masking agents and only thiosulphate appears to interfere.[8]

Reagents. *Silver nitrate solution, 10^{-4}M.* Prepare by dilution of a standard 0.1 M silver nitrate solution.
1,10-Phenanthroline solution. Dissolve 49.60 mg of analytical grade 1,10-phenanthroline in distilled water and dilute to 250 mL.
Ammonium acetate (ethanoate) solution, 20 per cent. Dissolve 20 g of the analytical grade salt in distilled water and dilute to 100 mL.
Bromopyrogallol red solution, 10^{-4}M. Dissolve 14.0 mg of bromopyrogallol red and 2.5 g of ammonium acetate in distilled water and dilute to 250 mL. This solution should be discarded after five days.
EDTA solution, 10^{-1}M. Dissolve 3.7225 g of analytical grade disodium salt in distilled water and dilute to 100 mL.
Sodium nitrate solution, 1M. Dissolve 8.5 g of analytical grade sodium nitrate in distilled water and dilute to 100 mL.

Nitrobenzene, analytical grade.
Sodium hydroxide, analytical grade pellets.

Procedure. *Calibration.* Pipette successively 1, 2, 3, 4 and 5 mL of $10^{-4}M$ silver nitrate solution, 1 mL of 20 per cent ammonium acetate solution, 5 mL of $10^{-3}M$ 1,10-phenanthroline solution, 1 mL of $10^{-1}M$ EDTA solution and 1 mL of 1 M sodium nitrate solution into five 100 mL separatory funnels. Add sufficient distilled water to give the same volume of solution in each funnel, then add 20 mL of nitrobenzene and shake by continuous inversion for one minute. Allow about 10 minutes for the layers to separate, then transfer the lower organic layers to different 100 mL separatory funnels and add to the latter 25 mL of $10^{-4}M$ bromopyrogallol red solution. Again shake by continuous inversion for one minute and allow about 30 minutes for the layers to separate. Run the lower nitrobenzene layers into clean, dry 100 mL beakers and swirl each beaker until all cloudiness disappears (Note 1). Finally transfer the solutions to 1 cm cells and measure the absorbance at 590 nm against a blank carried through the same procedure, but containing no silver. Plot a calibration curve of absorbance against silver content (μg).

1 mL of $10^{-4}M$ $AgNO_3$ = 10.788 μg of Ag

Determination. To an aliquot of the silver(I) solution containing between 10 and 50 μg of silver, add sufficient EDTA to complex all those cations present which form an EDTA complex. If gold is present ($\not> 250\,\mu$g) it is masked by adding sufficient bromide ion to form the $AuBr_4^-$ complex. Cyanide, thiocyanate or iodide ions are masked by adding sufficient mercury(II) ions to complex these anions followed by sufficient EDTA to complex any excess mercury(II). Add 1 mL of 20 per cent ammonium acetate solution, etc., and proceed as described under Calibration.

Note. (1) More rapid clarification of the nitrobenzene extract is obtained if the beakers contain about five pellets of sodium hydroxide. The latter is, however, a source of instability of the colour system and its use is, therefore, not recommended.

6.17 DETERMINATION OF NICKEL BY SYNERGISTIC EXTRACTION*

Discussion. Using dithizone and 1,10-phenanthroline (Note 1), nickel is rapidly and quantitatively extracted over a broad pH range (from 5.5 to at least 11.0) to give a highly coloured mixed ligand complex having an absorption band centred at 520 nm. The complex is sufficiently stable to permit the removal of excess dithizone by back-extraction with 0.1 M sodium hydroxide, so that a 'monocolour' method is applicable. The molar absorptivity of this complex is 4.91×10^4 mol^{-1} L cm^{-1}, which makes the method significantly more sensitive than any other method for determination of nickel.[3]

Procedure. To 10 mL of a solution (Note 2) containing from 1 to 10 μg of nickel(II) add 5 mL of a phthalate or acetate (ethanoate) buffer of pH 6.0 or, if the sample solution is acidic, use dilute ammonia to adjust the pH. To this solution now add 15 mL of a chloroform solution of dithizone ($7 \times 10^{-5}M$) and 1,10-phenanthroline ($3 \times 10^{-5}M$). Shake the phases for five minutes in a

---

* This experiment is not recommended for students having little experience of analytical work.

separatory funnel, allow them to settle and separate the aqueous and chloroform layers. Remove excess dithizone from the chloroform layer by back-extraction with 10 mL of 0.1 M sodium hydroxide: vigorous shaking for about one minute is sufficient. Again separate the chloroform layer and measure its absorbance in a 1 cm cell at 520 nm against a similarly treated blank. Construct a calibration curve (which should be a straight line through the origin) using standard nickel(II) solutions containing 2, 4, 6, 8 and 10 μg in 10 mL.

Notes. (1) The reagent solution should be freshly prepared using analytical-grade dithizone and 1,10-phenanthroline, preferably taken from new or recently opened reagent bottles.

(2) Glassware should be rinsed with dilute acid and then several times with de-ionised water. All aqueous solutions must be made up using de-ionised water.

6.18 EXTRACTION AND DETERMINATION OF LEAD, CADMIUM, AND COPPER USING AMMONIUM PYROLLIDINE DITHIOCARBAMATE

Discussion. Because of the specific nature of atomic absorption spectroscopy (AAS) as a measuring technique, non-selective reagents such as ammonium pyrollidine dithiocarbamate (APDC) may be used for the liquid–liquid extraction of metal ions. Complexes formed with APDC are soluble in a number of ketones such as methyl isobutyl ketone which is a recommended solvent for use in atomic absorption and allows a concentration factor of ten times. The experiment described illustrates the use of APDC as a general extracting reagent for heavy metal ions.

Reagents and solutions. *APDC solution.* Dissolve 1.0 g of APDC in water, dilute to 100 mL and filter (Note 1).
Standard solutions. Prepare, by appropriate dilution of standardised stock solutions of lead, cadmium and copper(II) ions, *mixed* aqueous standards of Pb^{2+}, Cd^{2+} and Cu^{2+} containing the following concentrations of each metal ion: 0.4, 0.6, 0.8 and 1.0 mg L^{-1} (Note 2).

Procedure. To 100 mL of sample solution, containing 0.5 to 1.0 mg L^{-1} of Pb^{2+}, Cd^{2+} and Cu^{2+} ions, add 10 mL of APDC solution and adjust to pH 5 with dilute acetic (ethanoic) acid or sodium hydroxide. Transfer the solution to a separatory funnel and extract the complex into 8 mL of methyl isobutyl ketone (MIBK) by vigorously shaking the phases for 30 seconds. Allow the mixture to stand for about two minutes, transfer the aqueous phase to another separatory funnel and repeat the extraction with 2 mL of MIBK. Discard the aqueous phase, which should now be colourless, combine the extracts in the first funnel, mix, and filter through a cotton-wool plug into a small dry beaker. Standard extracts should be prepared from the mixed aqueous standards using the same procedure. Aspirate the standard extracts successively into the flame, followed by the sample extract. In each case read the separate absorbances obtained using lead, cadmium and copper hollow cathode lamps (Note 3). Plot a calibration curve for each metal ion and use this to determine the concentrations of Pb^{2+}, Cd^{2+}, and Cu^{2+} in the sample solution.

When the AAS measurements have been completed, aspirate de-ionised water for several minutes to ensure thorough cleaning of the nebuliser–burner system.

Notes. (1) The solution should be freshly prepared using analytical reagent grade APDC (see Section 6.6).

(2) De-ionised water must be used in the preparation of all aqueous solutions.

(3) The detailed experimental procedure for determinations by AAS is given in Section 21.14.

For References and Bibliography see Sections 9.9 and 9.10.

CHAPTER 7
ION EXCHANGE

7.1 GENERAL DISCUSSION

The term ion exchange is generally understood to mean the exchange of ions of like sign between a solution and a solid highly insoluble body in contact with it. The solid (ion exchanger) must, of course, contain ions of its own, and for the exchange to proceed sufficiently rapidly and extensively to be of practical value, the solid must have an open, permeable molecular structure so that ions and solvent molecules can move freely in and out. Many substances, both natural (e.g. certain clay minerals) and artificial, have ion exchanging properties, but for analytical work synthetic organic ion exchangers are chiefly of interest, although some inorganic materials, e.g. zirconyl phosphate and ammonium 12-molybdophosphate, also possess useful ion exchange capabilities and have specialised applications.[17] All ion exchangers of value in analysis have several properties in common, they are almost insoluble in water and in organic solvents, and they contain active or counter-ions that will exchange reversibly with other ions in a surrounding solution without any appreciable physical change occurring in the material. The ion exchanger is of complex nature and is, in fact, polymeric. The polymer carries an electric charge that is exactly neutralised by the charges on the counter-ions. These active ions are cations in a cation exchanger and anions in an anion exchanger. Thus a **cation exchanger** consists of a polymeric anion and active cations, while an **anion exchanger** is a polymeric cation with active anions.

A widely used cation exchange resin is that obtained by the copolymerisation of styrene (A) and a small proportion of divinylbenzene (B), followed by sulphonation; it may be represented as (C):

186

The formula enables us to visualise a typical cation exchange resin. It consists of a polymeric skeleton, held together by linkings crossing from one polymer chain to the next: the ion exchange groups are carried on this skeleton. The physical properties are largely determined by the degree of cross-linking. This cannot be determined directly in the resin itself: it is often specified as the moles per cent of the cross-linking agent in the mixture polymerised. Thus 'polystyrene sulphonic acid, 5 per cent DVB' refers to a resin containing nominally 1 mole in 20 of divinylbenzene: the true degree of cross-linking probably differs somewhat from the nominal value, but the latter is nevertheless useful for grading resins. Highly cross-linked resins are generally more brittle, harder, and more impervious than the lightly cross-linked materials; the preference of a resin for one ion over another is influenced by the degree of cross-linking. The solid granules of resin swell when placed in water to give a gel structure, but the swelling is limited by the cross-linking. In the above example the divinylbenzene units 'weld' the polystyrene chains together and prevent it from swelling indefinitely and dispersing into solution. The resulting structure is a vast sponge-like network with negatively charged sulphonate ions attached firmly to the framework. These fixed negative charges are balanced by an equivalent number of cations: hydrogen ions in the hydrogen form of the resin and sodium ions in the sodium form of the resin, etc. These ions move freely within the water-filled pores and are sometimes called mobile ions; they are the ions which exchange with other ions. When a cation exchanger containing mobile ions C^+ is brought into contact with a solution containing cations B^+ the latter diffuse into the resin structure and cations C^+ diffuse out until equilibrium is attained. The solid and the solution then contain both cations C^+ and B^+ in numbers depending upon the position of equilibrium. The same mechanism operates for the exchange of anions in an anion exchanger.

Anion exchangers are likewise cross-linked, high-molecular-weight polymers. Their basic character is due to the presence of amino, substituted amino, or quaternary ammonium groups. The polymers containing quaternary ammonium groups are strong bases; those with amino or substituted amino groups possess weak basic properties. A widely used anion exchange resin is prepared by copolymerisation of styrene and a little divinylbenzene, followed by chloro-methylation (introduction of the $—CH_2Cl$ grouping, say, in the free *para* position) and interaction with a base such as trimethylamine. A hypothetical formulation of such a polystyrene anion exchange resin is given as (D).

(D)

Numerous types of both cation and anion exchange resins have been prepared, but only a few can be mentioned here. Cation exchange resins include that prepared by the copolymerisation of methacrylic acid (E) with glycol

$$CH_2=C(CH_3)-COOH$$
(E)

$$CH_2=C(CH_3)-COOCH_2$$
$$\mid$$
$$CH_2=C(CH_3)-COOCH_2$$
(F)

bismethacrylate (F) (as the cross-linking agent); this contains free —COOH groups and has weak acidic properties. Weak cation exchange resins containing free —COOH and —OH groups have also been synthesised. Anion exchange resins containing primary, secondary, or tertiary amino groups possess weakly basic properties. We may define a **cation exchange resin** as a high-molecular-weight, cross-linked polymer containing sulphonic, carboxylic, phenolic, etc., groups as an integral part of the resin and an equivalent number of cations: an **anion exchange resin** is a polymer containing amine (or quaternary ammonium) groups as integral parts of the polymer lattice and an equivalent number of anions, such as chloride, hydroxyl, or sulphate ions.

The fundamental requirements of a useful resin are:

1. the resin must be sufficiently cross-linked to have only a negligible solubility;
2. the resin must be sufficiently hydrophilic to permit diffusion of ions through the structure at a finite and usable rate;
3. the resin must contain a sufficient number of accessible ionic exchange groups and it must be chemically stable;
4. the swollen resin must be denser than water.

A new polymerisation technique yields a cross-linked ion exchange resin having a truly macroporous structure quite different from that of the conventional homogeneous gels already described. An average pore diameter of 130 nm is not unusual and the introduction of these **macroreticular** resins (e.g. the Amberlyst resins developed by the Rohm and Haas Co.) has extended the scope of the ion exchange technique. Thus, the large pore size allows the more complete removal of high-molecular-weight ions than is the case with the gel-type resins. Macroporous resins are also well suited for non-aqueous ion exchange applications.[18]

New types of ion exchange resins have also been developed to meet the specific needs of high-performance liquid chromatography (HPLC) (Chapter 8). These include pellicular resins and microparticle packings (e.g. the Aminex-type resins produced by Bio-Rad). A review of the care, use and application of the various ion exchange packings available for HPLC is given in Ref. 19.

Some of the commercially available ion exchange resins are collected in Table 7.1. These resins, produced by different manufacturers, are often interchangeable and similar types will generally behave in a similar manner. For a more comprehensive list of ion exchange resins and their properties, reference may be made to the booklet published by BDH Ltd (see the Bibliography, Section 9.10).

Finally, mention should be made to the development of silica-based ion exchange packings for HPLC. Their preparation is similar to that for the

Table 7.1 Comparable ion exchange materials

Type	Duolite International Ltd	Rohm & Haas Co., USA	Dow Chemical Co., USA	Bio-Rad Labs Ltd, Watford, UK
Strong acid cation exchangers	Duolite C225 Duolite C255 Duolite C26C*	Amberlite 120 Amberlite 200*	Dowex 50	AG50W AGMP-50*
Weak acid cation exchangers	Duolite C433 Duolite C464*	Amberlite 84 Amberlite 50		Bio-Rex 70*
Strong base anion exchangers	Duolite A113 Duolite A116 Duolite A161*	Amberlite 400 Amberlite 410 Amberlite 900*	Dowex 1 Dowex 2	AG1 AGMP-1*
Weak base anion exchangers	Duolite A303 Duolite A378*	Amberlite 45 Amberlite 68 Amberlite 93*		AG3-X4A
Chelating resins	Duolite ES466*	Amberlite 718*		Chelex 100

* Macroporous/macroreticular resins.

bonded-phase packings (Chapter 8) with the ion exchange groups being subsequently introduced into the organic backbone. The small particle size ($5-10 \, \mu$m diameter) and narrow distribution of such packings provide high column efficiencies and typical applications include high-resolution analysis of amino acids, peptides, proteins, nucleotides, etc. These silica-based packings are preferred when column efficiency is the main criterion but, when capacity is the main requirement, the resin microparticle packings should be selected. The chromatographic properties of ion exchange packings for analytical separations have been compared.[20]

7.2 ACTION OF ION EXCHANGE RESINS

Cation exchange resins* contain free cations which can be exchanged for cations in solution (soln).

$$(Res.A^-)B^+ + C^+ \, (soln) \rightleftharpoons (Res.A^-)C^+ + B^+ \, (soln)$$

If the experimental conditions are such that the equilibrium is completely displaced from left to right the cation C^+ is completely fixed on the cation exchanger. If the solution contains several cations (C^+, D^+, and E^+) the exchanger may show different affinities for them, thus making separations possible. A typical example is the displacement of sodium ions in a sulphonate resin by calcium ions:

$$2(Res.SO_3^-)Na^+ + Ca^{2+} \, (soln) \rightleftharpoons (Res.SO_3)_2^- Ca^{2+} + 2Na^+ \, (soln)$$

---

* These will be represented by $(Res.A^-)B^+$, where Res. is the basic polymer of the resin, A^- is the anion attached to the polymeric framework, B^+ is the active or mobile cation: thus a sulphonated polystyrene resin in the hydrogen form would be written as $(Res.SO_3^-)H^+$. A similar nomenclature will be employed for anion exchange resins, e.g. $(Res. NMe_3^+)Cl^-$.

The reaction is reversible; by passing a solution containing sodium ions through the product, the calcium ions may be removed from the resin and the original sodium form regenerated. Similarly, by passing a solution of a neutral salt through the hydrogen form of a sulphonic resin, an equivalent quantity of the corresponding acid is produced by the following typical reaction:

$$(\text{Res.SO}_3^-)\text{H}^+ + \text{Na}^+\text{Cl}^- \text{ (soln)} \rightleftharpoons (\text{Res.SO}_3^-)\text{Na}^+ + \text{H}^+\text{Cl}^- \text{ (soln)}$$

For the strongly acidic cation exchange resins, such as the cross-linked polystyrene sulphonic acid resins, the exchange capacity is virtually independent of the pH of the solution. For weak acid cation exchangers, such as those containing the carboxylate group, ionisation occurs to an appreciable extent only in alkaline solution, i.e. in their salt form; consequently the carboxylic resins have very little action in solutions below pH 7. These carboxylic exchangers in the hydrogen form will absorb strong bases from solution:

$$(\text{Res.COO}^-)\text{H}^+ + \text{Na}^+\text{OH}^- \text{ (soln)} \rightleftharpoons (\text{Res.COO}^-)\text{Na}^+ + \text{H}_2\text{O}$$

but will have little action upon, say, sodium chloride; hydrolysis of the salt form of the resin occurs so that the base may not be completely absorbed even if an excess of resin is present.

Strongly basic anion exchange resins, e.g. a cross-linked polystyrene containing quaternary ammonium groups, are largely ionised in both the hydroxide and the salt forms. Some of their typical reactions may be represented as:

$$2(\text{Res.NMe}_3^+)\text{Cl}^- + \text{SO}_4^{2-} \text{ (soln)} \rightleftharpoons (\text{Res.NMe}_3^+)_2\text{SO}_4^{2-} + 2\text{Cl}^- \text{ (soln)}$$

$$(\text{Res.NMe}_3^+)\text{Cl}^- + \text{OH}^- \text{ (soln)} \rightleftharpoons (\text{Res.NMe}_3^+)\text{OH}^- + \text{Cl}^- \text{ (soln)}$$

$$(\text{Res.NMe}_3^+)\text{OH}^- + \text{H}^+\text{Cl}^- \text{ (soln)} \rightleftharpoons (\text{Res.NMe}_3^+)\text{Cl}^- + \text{H}_2\text{O}$$

These resins are similar to the sulphonate cation exchange resins in their activity, and their action is largely independent of pH. Weakly basic ion exchange resins contain little of the hydroxide form in basic solution. The equilibrium of, say,

$$(\text{Res.NMe}_2) + \text{H}_2\text{O} \rightleftharpoons (\text{Res.NHMe}_2)^+\text{OH}^-$$

is mainly to the left and the resin is largely in the amine form. This may also be expressed by stating that in alkaline solutions the free base $\text{Res.NHMe}_2 \cdot \text{OH}$ is very little ionised. In acidic solutions, however, they behave like the strongly basic ion exchange resins, yielding the highly ionised salt form:

$$(\text{Res.NMe}_2) + \text{H}^+\text{Cl}^- \rightleftharpoons (\text{Res.NHMe}_2^+)\text{Cl}^-$$

They can be used in acid solution for the exchange of anions, for example:

$$(\text{Res.NHMe}_2^+)\text{Cl}^- + \text{NO}_3^- \text{ (soln)} \rightleftharpoons (\text{Res.NHMe}_2^+)\text{NO}_3^- + \text{Cl}^- \text{ (soln)}$$

Basic resins in the salt form are readily regenerated with alkali.

Ion exchange equilibria. The ion exchange process, involving the replacement of the exchangeable ions A_r of the resin by ions of like charge B_s from a solution, may be written:

$$A_r + B_s \rightleftharpoons B_r + A_s$$

The process is a reversible one and for ions of like charge the selectivity coefficient, K, is defined by:

$$K_A^B = \frac{[B]_r[A]_s}{[A]_r[B]_s}$$

where the terms in brackets represent the concentrations of ions A and B in either the resin or solution phase. The values of selectivity coefficients are obtained experimentally and provide a guide to the relative affinities of ions for a particular resin. Thus if $K_A^B > 1$ the resin shows a preference for ion B, whereas if $K_A^B < 1$ its preference is for ion A; this applies to both anion and cation exchanges.

The relative selectivities of strongly acid and strongly basic polystyrene resins, with about 8 per cent DVB, for singly charged ions are summarised in Table 7.2. It should be noted that the relative selectivities for certain ions may vary with a change in the extent of cross-linking of the resin; for example, with a 10 per cent DVB resin the relative selectivity values for Li^+ and Cs^+ ions are 1.00 and 4.15, respectively.

Table 7.2

Cation	Relative selectivity	Anion	Relative selectivity
Li^+	1.00	F^-	0.09
H^+	1.26	OH^-	0.09
Na^+	1.88	Cl^-	1.00
NH_4^+	2.22	Br^-	2.80
K^+	2.63	NO_3^-	3.80
Rb^+	2.89	I^-	8.70
Cs^+	2.91	ClO_4^-	10.0

The extent to which one ion is absorbed in preference to another is of fundamental importance: it will determine the readiness with which two or more substances, which form ions of like charge, can be separated by ion exchange and also the ease with which the ions can subsequently be removed from the resin. The factors determining the distribution of inorganic ions between an ion exchange resin and a solution include:

1. *Nature of exchanging ions.* (*a*) At low aqueous concentrations and at ordinary temperatures the extent of exchange increases with increasing charge of the exchanging ion, i.e.

    $$Na^+ < Ca^{2+} < Al^{3+} < Th^{4+}$$

    (*b*) Under similar conditions and constant charge, for singly charged ions the extent of exchange increases with decrease in size of the hydrated cation

    $$Li^+ < H^+ < Na^+ < NH_4^+ < K^+ < Rb^+ < Cs^+$$

    while for doubly charged ions the ionic size is an important factor but the incomplete dissociation of salts of such cations also plays a part

    $$Cd^{2+} < Be^{2+} < Mn^{2+} < Mg^{2+} = Zn^{2+} < Cu^{2+} = Ni^{2+}$$
    $$< Co^{2+} < Ca^{2+} < Sr^{2+} < Pb^{2+} < Ba^{2+}$$

    (*c*) With strongly basic anion exchange resins, the extent of exchange for singly charged anions varies with the size of the hydrated ion in a similar manner to that indicated for cations. In dilute solution multicharged anions are generally absorbed preferentially.

(d) When a cation in solution is being exchanged for an ion of different charge the relative affinity of the ion of higher charge increases in direct proportion to the dilution. Thus to exchange an ion of higher charge on the exchanger for one of lower charge in solution, exchange will be favoured by increasing the concentration, while if the ion of lower charge is in the exchanger and the ion of higher charge is in solution, exchange will be favoured by high dilutions.

2. *Nature of ion exchange resin.* The absorption* of ions will depend upon the nature of the functional groups in the resin. It will also depend upon the degree of cross-linking: as the degree of cross-linking is increased, resins become more selective towards ions of different sizes (the volume of the ion is assumed to include the water of hydration); the ion with the smaller hydrated volume will usually be absorbed preferentially.

Exchange of organic ions. Although similar principles apply to the exchange of organic ions, the following features must also be taken into consideration.

1. The sizes of organic ions differ to a much greater extent than is the case for inorganic ions and may exceed 100-fold or even 1000-fold the average size of inorganic ions.

2. Many organic compounds are only slightly soluble in water so that non-aqueous ion exchange has an important role in operations with organic substances.[21]

Clearly the application of macroreticular (macroporous) ion exchange resins will be often advantageous in the separation of organic species.

Ion exchange capacity. The total ion exchange capacity of a resin is dependent upon the total number of ion-active groups per unit weight of material, and the greater the number of ions, the greater will be the capacity. The **total ion exchange capacity** is usually expressed as millimoles per gram of exchanger. The capacities of the weakly acidic and weakly basic ion exchangers are functions of pH, the former reaching moderately constant values at pH above about 9 and the latter at pH below about 5. Values for the total exchange capacities, expressed as $mmol\,g^{-1}$ of dry resin, for a few typical resins are: Duolite C225 (Na$^+$ form), 4.5–5; Zerolit 226 (H$^+$ form), 9–10; Duolite A113 (Cl$^-$ form), 4.0; Amberlite IR-45, 5.0. The total exchange capacity expressed as $mmol\,mL^{-1}$ of the wet resin is about $\frac{1}{3} - \frac{1}{2}$ of the $mmol\,g^{-1}$ of the dry resin. These figures are useful in estimating very approximately the quantity of resin required in a determination: an adequate excess must be employed, since the 'break-through' capacity is often much less than the total capacity of the resin. In most cases a 100 per cent excess is satisfactory.

The exchange capacity of a cation exchange resin may be measured in the laboratory by determining the number of milligram moles of sodium ion which are absorbed by 1 g of the dry resin in the hydrogen form. Similarly, the exchange capacity of a strongly basic anion exchange resin is evaluated by measuring the amount of chloride ion taken up by 1 g of dry resin in the hydroxide form.

---

* The term absorption is used whenever ions or other solutes are taken up by an ion exchanger. It does not imply any specific types of forces responsible for this uptake.

It should be noted that large-size ions may not be absorbed by a medium cross-linked resin so that its effective capacity is seriously reduced. A resin with larger pores should be used for such ions.

Changing the ionic form: some widely used resins. It is frequently necessary to convert a resin completely from one ionic form to another. This should be done after regeneration, if this is being practised to 'clean' the resin (for example, if the 'standard' grade of ion exchanger is used). An excess of a suitable salt solution should be run through a column of the resin. Ready conversion will occur if the ion to be introduced into the resin has a higher, or only a slightly lower, affinity than that actually on the resin. When replacing an ion of lower charge number on the exchanger by one of higher charge number, the conversion is assisted by using a dilute solution of replacing salt (preferably as low as $0.01\,M$), while to substitute a more highly charged ion in the exchanger by one of lower charge number, a comparatively concentrated solution should be used (say, a $1\,M$ solution).

Strongly acidic cation exchangers are usually supplied in the hydrogen or sodium forms, and strongly basic anion exchangers in the chloride or hydroxide forms; the chloride form is preferred to the free base form, since the latter readily absorbs carbon dioxide from the atmosphere and becomes partly converted into the carbonate form. Weakly acidic cation exchangers are generally supplied in the hydrogen form, while weakly basic anion exchange resins are available in the hydroxide or chloride forms.

The resins are available in 'standard' grade, in a purer 'chromatographic' grade, and in some cases in a more highly purified 'analytical' grade: 'standard' grade materials should be subjected to a preliminary 'cleaning' by a regeneration procedure (see Section 7.8).

Strongly acidic cation exchangers (polystyrene sulphonic acid resins). These resins (Duolite C225, Amberlite 120, etc.) are usually marketed in the sodium form* and to convert them into the hydrogen form (which, it may be noted, are also available commercially) the following procedure may be used.

The cleaned standard grade resin (which may of course be replaced by one of the purer grades) is treated with $2\,M$ or with 10 per cent hydrochloric acid; one bed volume of the acid is passed through the column in 10–15 minutes. The effluent should then be strongly acid to methyl orange indicator; if it is not, further acid must be used (about three bed volumes may be required). The excess of acid is drained to almost bed level and the remaining acid washed away with distilled or de-ionised water, the volume required being about six times that of the bed. This operation occupies about 20 minutes: it is complete when the final 100 mL of effluent requires less than 1 mL of $0.2\,M$ sodium hydroxide to neutralise its acidity using methyl orange as indicator. The resin can now be employed for the exchange of its hydrogen ions for cations present in a given solution. Tests on the effluent show that its acidity, due to the exchange, rises to a maximum, which is maintained until the capacity is exhausted when the acidity of the treated solution falls. Regeneration is then necessary and is performed, after backwashing, with $2\,M$ hydrochloric acid as before.

---

* The resin is supplied in moist condition, and should not be allowed to dry out; particulate fracture may occur after repeated drying and re-wetting.

Weakly acidic cation exchangers (e.g. polymethylacrylic acid resins). These resins (Zerolit 226, Amberlite 50, etc.) are usually supplied in the hydrogen form. They are readily changed into the sodium form by treatment with $1M$ sodium hydroxide; an increase in volume of 80–100 per cent may be expected. The swelling is reversible and does not appear to cause any damage to the bead structure. Below a pH of about 3.5, the hydrogen form exists almost entirely in the little ionised carboxylic acid form. Exchange with metal ions will occur in solution only when these are associated in solution with anions of weak acids, i.e. pH values above about 4.

The exhausted resin is more easily regenerated than the strongly acidic exchangers; about 1.5 bed volumes of $1M$ hydrochloric acid will usually suffice.

Strongly basic anion exchangers (polystyrene quaternary ammonium resins). These resins (Duolite A113, Amberlite 400, etc.) are usually supplied in the chloride form. For conversion into the hydroxide form, treatment with $1M$ sodium hydroxide is employed, the volume used depending upon the extent of conversion desired; two bed volumes are satisfactory for most purposes. The rinsing of the resin free from alkali should be done with de-ionised water free from carbon dioxide to avoid converting the resin into the carbonate form; about 2 litres of such water will suffice for 100 g of resin. An increase in volume of about 20 per cent occurs in the conversion of the resin from the chloride to the hydroxide form.

Weakly basic anion exchangers (polystyrene tertiary amine resins). These resins (Duolite A303, Amberlite 45, etc.) are generally supplied in the free base (hydroxide) form. The salt form may be prepared by treating the resin with about four bed volumes of the appropriate acid (e.g. $1M$ hydrochloric acid) and rinsing with water to remove the excess of acid; the final effluent will not be exactly neutral, since hydrolysis occurs slowly, resulting in slightly acidic effluents. As with cation exchange, quantitative anion exchange will occur only if the anion in the resin has a lower affinity for the resin than the anion to be exchanged in the solution. When the resin is exhausted, regeneration can be accomplished by treatment with excess of $1M$ sodium hydroxide, followed by washing with de-ionised water until the effluent is neutral. If ammonia solution is used for regeneration the amount of washing required is reduced.

7.3 ION EXCHANGE CHROMATOGRAPHY

If a mixture of two or more different cations, B, C, etc., is passed through an ion exchange column, and if the quantities of these ions are small compared with the total capacity of the column for ions, then it may be possible to recover the absorbed ions separately and consecutively by using a suitable regenerating (or eluting) solution. If cation B is held more firmly by the exchange resin than cation C, all the C present will flow out of the bottom of the column before any of B is liberated, provided that the column is long enough and other experimental factors are favourable for the particular separation. This separation technique is sometimes called ion exchange chromatography. Its most spectacular success has been the separation of complex mixtures of closely related substances such as amino acids and lanthanides.

The process of removing absorbed ions is known as elution, the solution employed for elution is termed the eluant, and the solution resulting from elution is called the eluate. The liquid entering the ion exchange column may be termed the influent and the liquid leaving the column is conveniently called the effluent. If a solution of a suitable eluant is passed through a column charged with an ion B the course of the reaction may be followed by analysing continuously the effluent solution. If the concentration of B in successive portions of the eluate is plotted against the volume of the eluate, an elution curve is obtained such as is shown in Fig. 7.1. It will be seen that practically all the B is contained in a certain volume of liquid and also that the concentration of B passes through a maximum.

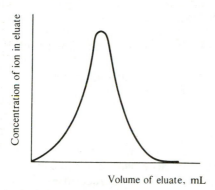

Volume of eluate, mL

Fig. 7.1

If the ion exchange column is loaded with several ions of similar charge, *B*, *C*, etc., elution curves may be obtained for each ion by the use of appropriate eluants. If the elution curves are sufficiently far apart, as in Fig. 7.2, a quantitative separation is possible; only an incomplete separation is obtained if the elution curves overlap. Ideally the curves should approach a Gaussian (normal) distribution (Section 4.9) and excessive departure from this distribution may indicate faulty technique and/or column operating conditions.

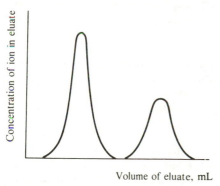

Volume of eluate, mL

Fig. 7.2

The rate at which two constituents separate in the column is determined by the ratio of the two corresponding distribution coefficients, where the distribution coefficient is given by the equation

$$K_d = \frac{\text{Amount of solute on resin}}{\text{Weight of resin, g}} \div \frac{\text{Amount of solute in solution}}{\text{Volume of solution, mL}}$$

The distribution coefficient can be determined by batch experiments in which a small known quantity of resin is shaken with a solution containing a known concentration of the solute, followed by analysis of the two phases after equilibrium has been attained. The separation factor, α, is used as a measure of the chromatographic separation possible and is given by the equation,

$$\alpha = K_{d_1}/K_{d_2}$$

where K_{d_1} and K_{d_2} are the distribution coefficients of the two constituents. The greater the deviation of α from unity the easier will be the separation. For normal laboratory practice, a useful guide is that quantitative separation should be achieved if α is above 1.2 or less than 0.8.

An important relationship exists between the weight distribution coefficient and the volume of eluant (V_{max}) required to reach the maximum concentration of an eluted ion in the effluent. This is given by the equation:

$$V_{max} = K_d V_0 + V_0$$

where V_0 is the volume of liquid in the interstices between the individual resin beads. If the latter are spheres of uniform size and close-packed in the column, V_0 is approximately 0.4 of the total bed volume, V_b. The void fraction V_0/V_b of the column may, however, be determined experimentally or calculated from density data.

The **volume distribution coefficient** is also a useful parameter for chromatographic calculations and is defined as

$$D_v = \frac{\text{Amount of ion in 1 mL of resin bed}}{\text{Amount of ion in 1 mL of interstitial volume}}$$

It is related to the weight distribution coefficient by

$$D_v = K_d \beta$$

where β is the void fraction of the settled column.

It is also related to V_{max} by the equation

$$V_{max} = V_b(D_v + \beta)$$

It should be remembered that the relationships given above are strictly applicable only when the loading of the column is less than 5 per cent of its capacity.

The application of these parameters may be illustrated by the following example.

Example. A mixture of *ca* 0.05 mmole each of chloride and bromide ions is to be separated on an anion exchange column of length 10 cm and 1 cm^2 cross-section, using 0.035M potassium nitrate as the eluant. The distribution coefficients (K_d) for the chloride and bromide ions respectively are 29 and 65.

Separation factor $\alpha = \dfrac{65}{29} = 2.24$

This value indicates that a satisfactory separation could be achieved, and this

196

is confirmed by calculation of the V_{max} values for the appearance of the chloride and bromide peaks.

From the column dimensions, the bed volume is

$$V_b = 10 \, cm \times 1.0 \, cm^2 = 10.0 \, mL$$

and the void volume (assuming $\beta = 0.4$) is

$$V_0 = 0.4 \times 10.0 \, mL = 4 \, mL$$

Hence for the chloride peak,

$$V_{max} = K_d V_0 + V_0 = (29 \times 4) + 4 = 120 \, mL$$

and for bromide,

$$V_{max} = (65 \times 4) + 4 = 264 \, mL$$

The relatively large values of V_{max} indicate, however, that the separation will be lengthy and the elution bands broad, particularly for the bromide band. The use of a more concentrated solution of eluant significantly reduces the values of V_{max} and the elution bands become much sharper. Thus the distribution coefficient for bromide using a $0.35 \, M$ potassium nitrate solution is 6.5 and using the same column, $V_{max} = (6.5 \times 4) + 4 = 30 \, mL$.

In many cases the efficient separation of a mixture by ion exchange chromatography requires that the eluant concentration be changed during the course of the elution. This may be done in a stepwise manner or by a continuous change in concentration as in gradient elution; the latter procedure can be carried out using simple laboratory equipment. A comprehensive discussion of the technique and of gradient elution devices is given in the review by L. R. Snyder.[22] A more recent development, however, is the use of microcomputer-controlled systems for gradient elution, e.g. for the separation of amino acids by ion exchange chromatography.[23]

The scope of separations by ion exchange chromatography may be extended by using for fixation or for elution a solution capable of complexing the ions exchanged. The formation of complexes may assist separations by diminishing the concentrations of free ions, and also by producing complexes of different stabilities, thus leading to significantly different behaviour with selected eluants.[24]

The results of ion exchange separations may be influenced by varying the pH, the solvent or eluant, the temperature, the nature of the ion exchange resin, the particle size, the rate of flow of eluant, and the length of the column.

7.4 ION CHROMATOGRAPHY

Ion chromatography (IC) is a relatively new technique pioneered by Small et al.[25] and which employs in a novel manner some well-established principles of ion exchange and allows electrical conductance to be used for detection and quantitative determination of ions in solution after their separation. Since electrical conductance is a property common to all ionic species in solution, a conductivity detector clearly has the potential of being a universal monitor for all ionic species.

A feature of ion exchange chromatography is, of course, that ionic solutions are used as eluants, so that the eluted species are present in an electrolyte background. The problem which now arises in the application of electrical conductance for quantifying eluted ionic species may be illustrated by the simple example of the separation and determination of sodium and potassium in a sample in which these are the only cations present. Complete separation of these two cations may be achieved using a strong acid cation exchange resin with aqueous hydrochloric acid as eluant. However, the high conductance of the hydrochloric acid in the effluent effectively 'swamps' the lower conductance due to the sodium and potassium ions, so preventing their measurement by electrical conductance.

In IC this problem of electrolyte background is overcome by means of eluant suppression. Thus in the above example of sodium and potassium analysis, if the effluent from the separating column is passed through a strong base anion exchange resin in the hydroxide form (suppressor column) the following two processes occur:

(*a*) Neutralisation of the hydrochloric acid

$$HCl + OH^- \ (resin) \rightarrow Cl^- \ (resin) + H_2O$$

(*b*) $NaCl \ (KCl) + OH^- \ (resin) \rightarrow NaOH \ (KOH) + Cl^- \ (resin)$

A consequence of these ion exchange processes is that the sample cations are presented to the conductivity detector not in a highly conducting background but in the very low conductivity of de-ionised water. It may be noted here that de-ionised water is not always the product of eluant suppression, the essential feature being that a background of low electrical conductivity is produced. Analogous schemes can be devised for anion analysis, in which case a strong acid cation exchange resin (H^+ form) is employed in the suppressor column.

Ion chromatography permits the determination of both inorganic and organic ionic species, often in concentrations of $50 \ \mu g \ L^{-1}$ (ppb) or less. Since analysis time is short (frequently less than 20 minutes) and sample volumes may be less than 1 mL, IC is a fast and economical technique. It has found increasing application in a number of different areas of chemical analysis and particularly for the quantitative determination of anions. The state-of-the-art has been reviewed.[26]

A flow scheme for the basic form of ion chromatography is shown in Fig. 7.3, which illustrates the requirements for simple anion analysis. The instrumentation used in IC does not differ significantly from that used in HPLC and the reader is referred to Chapter 8 for details of the types of pump and sample injection system employed. A brief account is given here, however, of the nature of the separator and suppressor columns and of the detectors used in ion chromatography.

Separator column. The specific capacity of the separating column is kept small by using resins of low capacity. For example, low-capacity anion exchangers have been prepared by a surface agglomeration method in which finely divided anion exchange resin is contacted with surface-sulphonated styrene–divinyl-benzene copolymer; the small particles of anion exchanger are held tenaciously on the oppositely charged surface of the sulphonated beads.[27] These resins are stable over a wide range of pH, in which respect they are superior to glass- or silica-based pellicular resins.

Suppressor column. Where electrical conductance is used for detection of sample ions in the effluent from the columns, an eluant background of low conductivity is required. The function of the suppressor column is to convert eluant ions

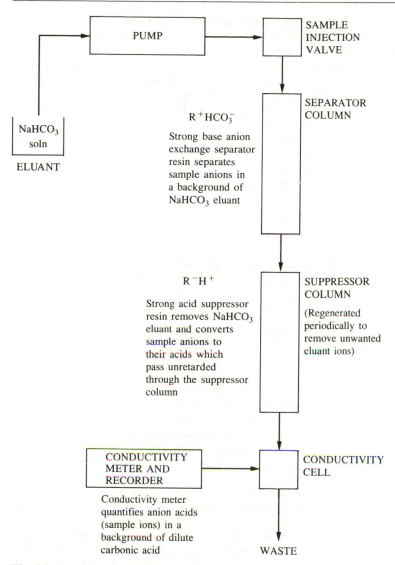

Fig. 7.3 **Ion chromatography flow scheme (anion analysis).**

into species giving low or zero conductance, e.g. where $NaHCO_3$ is used as eluant for anion analysis this is converted into a background of dilute carbonic acid by using a strong acid suppressor resin.

The low-concentration eluants used to separate the sample ions on the separator column allow a substantial number of samples (typically about 50) to be analysed before the suppressor column is completely exhausted. Clearly an important practical consideration is the need to minimise the frequency of regeneration of the suppressor column and, for this reason, the specific capacity of the column is made as large as possible by using resins of moderate to high cross-linking. Some instruments contain two suppressor columns in parallel,

199

thus allowing one column to be regenerated while the other is in use. A more convenient arrangement, however, is the use of a hollow-fibre suppressor, such as that developed by Dionex Ltd, which allows continuous operation of the ion chromatograph. In the case of anion analysis, the suppressor incorporates a tubular cation exchange membrane which is constantly regenerated by dilute sulphuric acid flowing through the outer casing.

It is appropriate to refer here to the development of non-suppressed ion chromatography. A simple chromatographic system for anions which uses a conductivity detector but requires no suppressor column has been described by Fritz and co-workers.[28] The anions are separated on a column of macroporous anion exchange resin which has a very low capacity, so that only a very dilute solution (ca $10^{-4} M$) of an aromatic organic acid salt (e.g. sodium phthalate) is required as the eluant. The low conductance of the eluant eliminates the need for a suppressor column and the separated anions can be detected by electrical conductance. In general, however, non-suppressed ion chromatography is an order of magnitude less sensitive than the suppressed mode.

Detectors.   Although electrical conductance has been widely used for detecting ions in ion chromatography, the scope of the technique has been considerably extended by the use of other types of detector. It is convenient broadly to classify detectors into two series.

1. Detectors employing electrochemical principles:
   (a) *Conductimetric detectors.* Conductance is a fundamental property of ions in solution making it an ideal technique for monitoring ion exchange separations because of its universal and linear response. It is the optimum mode of detection for strong acid anions ($pK_a < 7$), providing high sensitivity in the absence of background electrolyte.
   (b) *Amperometric detectors.* This type of detector may be used to detect ions which are electrochemically active but not readily detected by conductance measurement, e.g. weak acid anions such as CN^-, HS^- ($pK_a > 7$). The detector commonly features interchangeable silver or platinum working electrodes and may be used alone (no suppressor then being required) or simultaneously with a conductivity detector.

2. Detectors based on established optical absorption and emission techniques, are typified by:
   (a) *Spectrophotometric detectors.* The operation of spectrophotometric detectors is based on the absorbance of monochromatic light by the column effluent in accordance with the Beer–Lambert law (Section 17.2). As most organic species have significant absorption in the UV region of the spectrum, these detectors have wide application. Sensitivity clearly depends on how strongly the sample absorbs at the wavelength of maximum absorption, but detection limits in the low- (or even sub-) nanogram range may be achieved in favourable conditions.

   An alternative approach using a spectrophotometric detector has been described by Small and Miller[29] The essential feature here is that the eluting ion, which is commonly phthalate, must absorb ultraviolet light thus allowing its concentration to be monitored at a suitable wavelength as it emerges from the column. When a non-absorbing sample ion elutes from the column, the concentration of eluting ion decreases and the detector registers a negative response which is proportional to the concentration of sample ion. This procedure has contributed to the ease with which non-absorbing species can be detected, particularly sample ions with $pK_a > 7$ for which conductance detection is not appropriate.

(b) *Fluorescence detectors.* Although only a small proportion of inorganic and organic compounds are naturally fluorescent, the inherent sensitivity and selectivity of fluorescence detection offers significant advantages. The development of appropriate pre-column and post-column derivatisation procedures has furthered the application of fluorescence detection for the trace analysis of non-fluorescent or weakly fluorescing species.[30]

The reader is recommended to consult the monograph by Scott for further details of all detectors[31] used in ion chromatography.

Ion chromatography has been successfully applied to the quantitative analysis of ions in many diverse types of industrial and environmental samples. The technique has also been valuable for microelemental analysis, e.g. for the determination of sulphur, chlorine, bromine, phosphorus and iodine as heteroatoms in solid samples. Combustion in a Schöniger oxygen flask (Section 3.31)is a widely used method of degrading such samples, the products of combustion being absorbed in solution as anionic or cationic forms, and the solution then directly injected into the ion chromatograph.

A typical application of ion chromatography for the separation and determination of simple anions is illustrated by the experiment described in Section 7.15.

7.5 ION EXCHANGE IN ORGANIC AND AQUEOUS–ORGANIC SOLVENTS

Investigations in aqueous systems have established many of the fundamental principles of ion exchange as well as providing useful applications. The scope of the ion exchange process has, however, been extended by the use of both organic and mixed aqueous–organic solvent systems.[32,33]

The organic solvents generally used are oxo-compounds of the alcohol, ketone and carboxylic acid types, generally having dielectric constants below 40. Cations and anions should, therefore, pair more strongly in such solvent systems than in water and this factor may in itself be expected to alter selectivities for the resin. In addition to influencing these purely electrostatic forces, the presence of the organic solvent may enhance the tendency of a cation to complex with anionic or other ligands, thus modifying its ion exchange behaviour. In mixed aqueous–organic solvents the magnitude of such effects will clearly be dependent on the proportion of organic solvent present.

As already indicated, ion exchange resins are osmotic systems which swell owing to solvent being drawn into the resin. Where mixed solvent systems are used the possibility of preferential osmosis occurs and it has been shown that strongly acid cation and strongly basic anion resin phases tend to be predominantly aqueous with the ambient solution predominantly organic. This effect (preferential water sorption by the resin) increases as the dielectric constant of the organic solvent decreases.

An interesting consequence of selective sorption is that conditions for partition chromatography arise which may enhance the normal ion exchange separation factors. This aspect has been utilised by Korkisch[34] for separation of inorganic ions by the so-called 'combined ion exchange–solvent extraction method' (CISE).

7.6 CHELATING ION EXCHANGE RESINS

The use of complexing agents in solution in order to enhance the efficiency of separation of cation mixtures (e.g. lanthanides) using conventional cation or anion exchange resins is well established. An alternative mode of application of complex formation is, however, the use of chelating resins which are ion exchangers in which various chelating groups (e.g. dimethylglyoxime and iminodiacetic acid) have been incorporated and are attached to the resin matrix.

An important feature of chelating ion exchangers is the greater selectivity which they offer compared with the conventional type of ion exchanger. The affinity of a particular metal ion for a certain chelating resin depends mainly on the nature of the chelating group, and the selective behaviour of the resin is largely based on the different stabilities of the metal complexes formed on the resin under various pH conditions. It may be noted that the binding energy in these resins is of the order of $60-105 \, kJ \, mol^{-1}$, whereas in ordinary ion exchangers the strength of the electrostatic binding is only about $8-13 \, kJ \, mol^{-1}$.

The exchange process in a chelating resin is generally slower than in the ordinary type of exchanger, the rate apparently being controlled by a particle diffusion mechanism.

According to Gregor et al.[35] the following properties are required for a chelating agent which is to be incorporated as a functional group into an ion exchange resin:

1. the chelating agent should yield, either alone or with a cross-linking substance, a resin gel of sufficient stability to be capable of incorporation into a polymer matrix;
2. the chelating group must have sufficient chemical stability, so that during the synthesis of the resin its functional structure is not changed by polymerisation or any other reaction;
3. the steric structure of the chelating group should be compact so that the formation of the chelate rings with cations will not be hindered by the resin matrix.
4. the specific arrangements of the ligand groups should be preserved in the resin. This is particularly necessary since the complexing agents forming sufficiently stable complexes are usually at least tridentate.

These considerations indicate that many chelating agents could not be incorporated into a resin without loss of their selective complexing abilities. Ligands which do not form 1:1 complexes (e.g. 8-quinolinol) would be unsuitable, as also would molecules such as EDTA, which are insufficiently compact. In the latter case, it is improbable that the chelate configurations occurring in aqueous solution could be maintained in a cross-linked polymer. The closely related iminodiacetic acid group does, however, meet the requirements described, being compact and forming 1:1 complexes with metal cations.

The basicity of the nitrogen atom can be influenced by whether the imino group is attached directly to a benzene nucleus or whether a methylene group is interposed.

Although chelating resins containing various ligand donor atoms have been synthesised, the iminodiacetic acid resins (N and O donor atoms) undoubtedly form the largest group.[36,37] The resin based on iminodiacetic acid in a styrene divinylbenzene matrix is available commercially under the trade names of Dowex Chelating Resin A-1 and Chelex 100, and its chemical and physical properties have been fully investigated.

The starting material for the synthesis of this chelating resin is chloromethylated styrene–divinylbenzene, which undergoes an amination reaction and is then treated with monochloracetic acid:

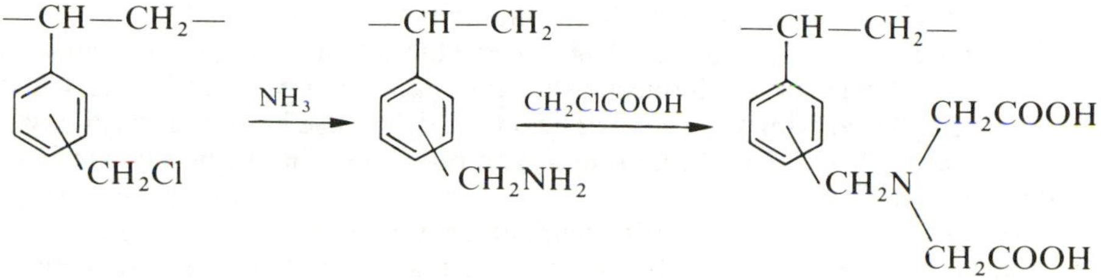

The selectivity of this type of exchange resin is illustrated by Chelex 100 which shows unusually high preference for copper, iron and other heavy metals (i.e. metals which form complexes having high stability constants with this type of ligand) over such cations as sodium, potassium and calcium; it is also much more selective for the alkaline earths than for the alkali metal cations. The resin's high affinity for these ions makes it very useful for removing, concentrating or analysing traces of them in solutions, even when large amounts of sodium and potassium are present (see the experiment described in Section 7.14).

In contrast to the above resins, the chelating resin Amberlite IRC-718 is based upon a macroreticular matrix. It is claimed to exhibit superior physical durability and adsorption kinetics when compared to chelating resins derived from gel polymers and should also be superior for use in non-aqueous solvent systems.

It is appropriate here to refer to the use of chelating resins for ligand-exchange chromatography, a useful technique for pre-concentration and separation of compounds which can form complexes or adducts with metal ions. An ion exchanger containing a complexing metal ion, e.g. Cu^{2+}, Ni^{2+}, is used as a solid sorbent; the use of chelating resins is advantageous since the successful application of ligand exchange depends on keeping the complexing metal ion in the resin. The potential ligands, e.g. amines, amino acids, polyhydric alcohols, are sorbed from solutions (or gases), separations occurring because of differences in the stabilities of the metal–ligand complexes. On this basis high capacities and selectivities can be achieved. Examples of the application of ligand-exchange chromatography are:

1. the separation and quantitative determination of amphetamine and related compounds,[38] and
2. the concentration and separation of amino acids in saline solutions.[39]

203

An extensive review of ligand-exchange chromatography has been given by Davankov.[40]

7.7 LIQUID ION EXCHANGERS

The ion exchange processes involving exchange resins occur between a solid and a liquid phase whereas in the case of liquid ion exchangers the process takes place between two immiscible solutions. Liquid ion exchangers consist of high-molecular-weight acids and bases which possess low solubility in water and high solubility in water-immiscible solvents. Thus, a solution of a base insoluble in water, in a solvent which is water-immiscible, can be used as an anion exchanger; similarly a solution of an acid insoluble in water can act as a cation exchanger for ions in aqueous solution. A comprehensive list of liquid ion exchangers has been given by Coleman et al.[41]

The liquid anion exchangers at present available are based largely on primary, secondary and tertiary aliphatic amines, e.g. the exchangers Amberlite LA.1 [N-dodecenyl(trialkylmethyl)amine] and Amberlite LA.2 [N-lauryl(trialkyl-methyl)amine], both secondary amines. These anion exchange liquids are best employed as solutions (ca 2.5 to 12.5% v/v) in an inert organic solvent such as benzene, toluene, kerosene, petroleum ether, cyclohexane, octane, etc.

The liquid exchangers Amberlite LA.1 and LA.2 may be used to remove acids from solution

$$R'R''NH + HX \rightarrow R'R''NH_2X$$

or in a salt form for various ion exchange processes

$$R'R''NH_2Cl + NaNO_3 \rightarrow R'R''NH_2NO_3 + NaCl$$

Examples of liquid cation exchangers are alkyl and dialkyl phosphoric acids, alkyl sulphonic acids and carboxylic acids, although only two appear to have been used to any extent, viz. di-(2-ethylhexyl)phosphoric(V) acid and dinonylnaphthalene sulphonic acid.

The operation of liquid ion exchangers involves the selective transfer of a solute between an aqueous phase and an immiscible organic phase containing the liquid exchangers. Thus high-molecular-weight amines in acid solution yield large cations capable of forming extractable species (e.g. ion pairs) with various anions. The technique employed for separations using liquid ion exchangers is thus identical to that used in solvent extraction separations and these exchangers offer many of the advantages of both ion exchange and solvent extraction. There are, however, certain difficulties and disadvantages associated with their use which it is important to appreciate in order to make effective use of liquid ion exchangers.

Probably the chief difficulty which arises is that due to the formation of emulsions between the organic and aqueous phases. This makes separation of the phases difficult and sometimes impossible. It is clearly important to select liquid exchangers having low surface activity and to use conditions which will minimise the formation of stable emulsions [see Section 6.7, consideration (3)].

Another disadvantage in the use of liquid ion exchangers is that it is frequently necessary to back-extract the required species from the organic phase into an aqueous phase prior to completing the determination. The organic phase may, however, sometimes be used directly for determination of the extracted species, in particular by aspirating directly into a flame and estimating extracted metal ions by flame emission or atomic absorption spectroscopy.

The extraction of metals by liquid amines has been widely investigated and depends on the formation of anionic complexes of the metals in aqueous solution. Such applications are illustrated by the use of Amberlite LA.1 for extraction of zirconium and hafnium from hydrochloric acid solutions, and the use of liquid amines for extraction of uranium from sulphuric acid solutions.[42,43]

Exhausted liquid ion exchangers may be regenerated in an analogous manner to ion exchange resins, e.g. Amberlite LA.1 saturated with nitrate ions can be converted to the chloride form by treatment with excess sodium chloride solution.

The properties and applications of liquid ion exchangers have been reviewed.[44]

APPLICATIONS IN ANALYTICAL CHEMISTRY

7.8 EXPERIMENTAL TECHNIQUES*

The simplest apparatus for ion exchange work in analysis consists of a burette provided with a glass-wool plug or sintered glass disc (porosity 0 or 1) at the lower end. Another simple column is shown in Fig. 7.4(a); the ion exchange resin is supported on a glass-wool plug or sintered-glass disc. A glass-wool pad may be placed at the top of the bed of resin and the eluting agent is added from a tap funnel supported above the column. The siphon overflow tube, attached to the column by a short length of rubber or PVC tubing, ensures that the level

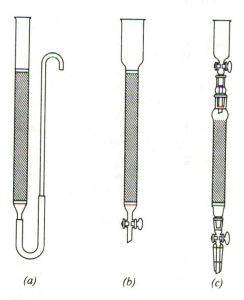

(a)   (b)   (c)

Fig. 7.4

---

* The simple techniques described in this section do not, of course, apply to ion chromatography (IC) or HPLC.

of the liquid does not fall below the top of the resin bed, so that the latter is always wholly immersed in the liquid. The ratio of the height of the column to the diameter is not very critical but is usually 10 or 20:1. Another form of column is depicted in Fig. 7.4(b) (not drawn to scale): a convenient size is 30 cm long, the lower portion of about 10 mm and the upper portion of about 25 mm internal diameter. A commercially available column, fitted with ground-glass joints is illustrated in Fig. 7.4(c).

The ion exchange resin should be of small particle size, so as to provide a large surface of contact; it should, however, not be so fine as to produce a very slow flow rate. For much laboratory work 50–100 mesh or 100–200 mesh materials are satisfactory. In all cases the diameter of the resin bead should be less than one-tenth of that of the column. Resins of medium and high cross-linking rarely show any further changes in volume, and only if subjected to large changes of ionic strength will any appreciable volume change occur. Resins of low cross-linking may change in volume appreciably even with small variations of ionic strength, and this may result in channelling and possible blocking of the column; these effects limit the use of these materials. To obtain satisfactory separations, it is essential that the solutions should pass through the column in a uniform manner. The resin particles should be packed uniformly in the column: the resin bed should be free from air bubbles so that there is no channelling.

To prepare a well-packed column, a supply of exchange resin of narrow size range is desirable. An ion exchange resin swells if the dry solid is immersed in water; no attempt should therefore be made to set up a column by pouring the dry resin into a tube and then adding water, since the expansion will probably shatter the tube. The resin should be stirred with water in an open beaker for several minutes, any fine particles removed by decantation, and the resin slurry transferred portionwise to the tube previously filled with water. The tube may be tapped gently to prevent the formation of air bubbles. To ensure the removal of entrained air bubbles, or any remaining fine particles, and also to ensure an even distribution of resin granules, it is advisable to 'backwash' the resin column before use, i.e. a stream of good-quality distilled water or of de-ionised water is run up through the bed from the bottom at a sufficient flow rate to loosen and suspend the exchanger granules. The enlarged upper portion of the exchange tube shown in Fig. 7.4(b) or (c) will hold the resin suspension during washing. If a tube of uniform bore is used the volume of resin employed must be suitably adjusted or else a tube attached by a rubber bung to the top of the column; the tube dips into an open filter flask, the side arm of which acts as the overflow and is connected by rubber tubing to waste. When the wash water is clear, the flow of water is stopped and the resin is allowed to settle in the tube. The excess of water is drained off; the water level must never fall below the surface of the resin, or else channelling will occur, with consequent incomplete contact between the resin and solutions used in subsequent operations. The apparatus with a side-arm outlet (Fig. 7.4a) has an advantage in this respect in that the resin will not run dry even if left unattended, since the outlet is above the surface of the resin.

Ion exchange resins (standard grades) as received from the manufacturers may contain unwanted ionic impurities and sometimes traces of water-soluble intermediates or incompletely polymerised material; these must be washed out before use. This is best done by passing $2M$ hydrochloric acid and $2M$ sodium hydroxide alternately through the column, with distilled-water rinsings in between, and then washing with water until the effluent is neutral and salt-free. 'Analytical grade' and/or 'chromatographic grade' ion exchange resins that have undergone this preliminary washing are available commercially.

For analytical work the exchange resin of 'analytical' grade (Amberlite) or of 'chromatographic' grade (Duolite; Amberlite, etc.) of a particle size of 100–200 mesh is preferred. However, for student work, the 'standard' grade of resin of 50–100 or 15–50 mesh, which is less expensive, is generally satisfactory. The 'standard' grade of resin must, however, be conditioned before use. Cation exchange resins must be soaked in a beaker in about twice the volume of $2M$ hydrochloric acid for 30–60 minutes with occasional stirring; the fine particles are removed by decantation or by backwashing a column with distilled or de-ionised water until the supernatant liquid is clear. Anion exchange resins may be washed with water in a beaker until the colour of the decanted wash liquid reaches a minimum intensity; they may then be transferred to a wide glass column and cycled between $1M$ hydrochloric acid and $1M$ alkali. Sodium hydroxide is used for strongly basic resins, and ammonia (preferably) or sodium carbonate for weakly basic resins. For all resins the final treatment must be with a solution leading to the resin in the desired ionic form.

A 50 mL or 100 mL burette, with Pyrex glass-wool plug or sintered-glass disc at the lower end, can generally be used for the determinations described below: alternatively, the column with side arm (Fig. 7.4a) is equally convenient in practice for student use. Reference will be made to the Duolite resins; the equivalent Amberlite or other resin (see Table 7.1 in Section 7.1) may of course be used.

7.9 DETERMINATION OF THE CAPACITY OF AN ION EXCHANGE RESIN (COLUMN METHOD)

Cation exchange resin. Dry the purified resin (e.g. Duolite C225 in the hydrogen form) by placing it in an evaporating dish, covering with a clockglass supported on two glass rods to provide protection from dust while giving access to the air, and leaving in a warm place (25–35 °C) until the resin is completely free-running (2–3 days). The capacity of the resulting resin remains constant over a long period if kept in a closed bottle. Drying at higher temperatures (say, 100 °C) is not recommended, owing to possible fracture of the resin beads.

Partly fill a small column, 15 cm × 1 cm (Fig. 7.4a) with distilled water, taking care to displace any trapped air from beneath the sintered-glass disc. Weigh out accurately about 0.5 g of the air-dried resin in a glass scoop and transfer it with the aid of a small camel-hair brush through a dry funnel into the column. Add sufficient distilled water to cover the resin. Dislodge any air bubbles that stick to the resin beads by applying intermittent pressure to the rubber tubing, thus causing the level of the liquid in the column to rise and fall slightly. Adjust the level of the outlet tube so that the liquid in the column will drain to a level about 1 cm above the resin beads.

Fill a 250 mL separatory funnel with *ca* 0.25M sodium sulphate solution. Allow this solution to drip into the column at a rate of about 2 mL per minute, and collect the effluent in a 500 mL conical flask. When all the solution has passed through the column, titrate the effluent with standard 0.1M sodium hydroxide using phenolphthalein as indicator.

The reaction may be represented as:

$$R^-H^+ + Na^+ \rightleftharpoons R^-Na^+ + H^+$$

and proceeds to completion because of the large excess and large volume of sodium sulphate solution passed through the column.

The capacity of the resin in millimoles per gram is given by av/W, where a is the molarity of the sodium hydroxide solution, v is the volume in mL, and W is the weight (g) of the resin.

Anion exchange resin. Proceed as in the previous experiment using 1.0 g, accurately weighed, of the air-dried strongly basic anion exchanger (e.g. Duolite A113, chloride form). Fill the 250 mL separatory funnel with *ca* 0.25*M* sodium nitrate solution, and allow this solution to drop into the column at the rate of about 2 mL per minute. Collect the effluent in a 500 mL conical flask, and titrate with standard 0.1 *M* silver nitrate using potassium chromate as indicator.

The reaction which occurs may be written as:

$$R^+Cl^- + NO_3^- \rightleftharpoons R^+NO_3^- + Cl^-$$

The capacity of the resin expressed as millimoles per gram is given by bv/W, where v mL of bM $AgNO_3$ are required by W g of the resin.

7.10 SEPARATION OF ZINC AND MAGNESIUM ON AN ANION EXCHANGER

Theory. Several metal ions (e.g. those of Fe, Al, Zn, Co, Mn, etc.) can be absorbed from hydrochloric acid solutions on anion exchange resins owing to the formation of negatively charged chloro-complexes. Each metal is absorbed over a well-defined range of pH, and this property can be used as the basis of a method of separation. Zinc is absorbed from 2*M* acid, while magnesium (and aluminium) are not; thus by passing a mixture of zinc and magnesium through a column of anion exchange resin a separation is effected. The zinc is subsequently eluted with dilute nitric acid.

Procedure. Prepare a column of the anion exchange resin using about 15 g of Duolite A113 in the chloride form (Section 7.9). The column should be made up in 2*M* hydrochloric acid.

Prepare separate standard zinc (about 2.5 mg Zn mL^{-1}) and standard magnesium (about 1.5 mg Mg mL^{-1}) ion solutions by dissolving accurately weighed quantities of zinc shot and magnesium (for Grignard reaction) in 2*M* hydrochloric acid and diluting each to volume in a 250 mL graduated flask. Pipette 10.0 mL of the zinc ion solution and 10.0 mL of the magnesium ion solution into a small separatory funnel supported in the top of the ion exchange column, and mix the solutions. Allow the mixed solution to flow through the column at a rate of about 5 mL per minute. Wash the funnel and column with 50 mL of 2*M* hydrochloric acid: do not permit the level of the liquid to fall below the top of the resin column. Collect all the effluent in a conical flask; this contains all the magnesium. Now change the receiver. Elute the zinc with 30 mL of water, followed by 80 mL of *ca* 0.25*M* nitric acid. Determine the magnesium and the zinc in the respective eluates by neutralisation with sodium hydroxide solution, followed by titration with standard EDTA solution using a buffer solution of pH = 10 and solochrome black indicator (Section 10.62).

The following results were obtained in a typical experiment:

Weight of zinc taken = 25.62 mg; found = 25.60 mg
Weight of magnesium taken = 14.95 mg; found = 14.89 mg

Magnesium may conveniently be determined by atomic absorption spectroscopy (Section 21.21) if a smaller amount (*ca* 4 mg) is used for the separation. Collect the magnesium effluent in a 1 L graduated flask, dilute to the mark with de-ionised water and aspirate the solution into the flame of an atomic absorption spectrometer. Calibrate the instrument using standard magnesium solutions covering the range 2 to 8 ppm.

7.11 SEPARATION OF CHLORIDE AND BROMIDE ON AN ANION EXCHANGER

Theory. The anion exchange resin, originally in the chloride form, is converted into the nitrate form by washing with sodium nitrate solution. A concentrated solution of the chloride and bromide mixture is introduced at the top of the column. The halide ions exchange rapidly with the nitrate ions in the resin, forming a band at the top of the column. Chloride ion is more rapidly eluted from this band than bromide ion by sodium nitrate solution, so that a separation is possible. The progress of elution of the halides is followed by titrating fractions of the effluents with standard silver nitrate solution.

Procedure. Prepare an anion exchange column (Section 7.8) using about 40 g of Duolite A113 (chloride form). The ion exchange tube may be 16 cm long and about 12 mm internal diameter. Wash the column with 0.6 M sodium nitrate until the effluent contains no chloride ion (silver nitrate test) and then wash with 50 mL of 0.3 M sodium nitrate.

Weigh out accurately about 0.10 g of analytical grade sodium chloride and about 0.20 g of potassium bromide, dissolve the mixture in about 2.0 mL of water and transfer quantitatively to the top of the column with the aid of 0.3 M sodium nitrate. Pass 0.3 M sodium nitrate through the column at a flow rate of about 1 mL per minute and collect the effluent in 10 mL fractions. Transfer each fraction in turn to a conical flask, dilute with an equal volume of water, add 2 drops of 0.2 M potassium chromate solution and titrate with standard 0.02 M silver nitrate.

Before commencing the elution titrate 10.0 mL of the 0.3 M sodium nitrate with the standard silver nitrate solution, and retain the product of this blank titration for comparing with the colour in the titrations of the eluates. When the titre of the eluate falls almost to zero (i.e. nearly equal to the blank titration) — *ca* 150 mL of effluent — elute the column with 0.6 M sodium nitrate. Titrate as before until no more bromide is detected (titre almost zero). A new blank titration must be made with 10.0 mL of the 0.6 M sodium nitrate.

Plot a graph of the total effluent collected against the concentration of halide in each fraction (millimoles per litre). The sum of the titres using 0.3 M sodium nitrate eluant (less blank for *each* titration) corresponds to the chloride, and the parallel figure with 0.6 M sodium nitrate corresponds to the bromide recovery.

A typical experiment gave the following results:

Weight of sodium chloride used = 0.1012 g ≡ 61.37 mg Cl⁻
Weight of potassium bromide used = 0.1934 g ≡ 129.87 mg Br⁻

Concentration of silver nitrate solution $= 0.019\,36\,M$
Cl^-: total titres (less blanks) $= 89.54\,mL \equiv 61.47\,mg$
Br^-: total titres (less blanks) $= 83.65\,mL = 129.4\,mg$

7.12 DETERMINATION OF THE TOTAL CATION CONCENTRATION IN WATER

Theory. The following procedure is a rapid one for the determination of the total cations present in water, particularly that used for industrial ion exchange plants, but may be used for all samples of water, including tap water. When water containing dissolved ionised solids is passed through a cation exchanger in the hydrogen form all cations are removed and replaced by hydrogen ions. By this means any alkalinity present in the water is destroyed, and the neutral salts present in solution are converted into the corresponding mineral acids. The effluent is titrated with $0.02\,M$ sodium hydroxide using screened methyl orange as indicator.

Procedure. Prepare a 25–30 cm column of Duolite C225 in a 14–16 mm chromatographic tube (Section 7.8). Pass 250 mL of $2\,M$ hydrochloric acid through the tube during about 30 minutes; rinse the column with distilled water until the effluent is just alkaline to screened methyl orange or until a 10 mL portion of the effluent does not require more than one drop of $0.02\,M$ sodium hydroxide to give an alkaline reaction to bromothymol blue indicator. The resin is now ready for use: the level of the water should never be permitted to drop below the upper surface of the resin in the column. Pass 50.0 mL of the sample of water under test through the column at a rate of 3–4 mL per minute, and discard the effluent. Now pass two 100.0 mL portions through the column at the same rate, collect the effluents separately, and titrate each with standard $0.02\,M$ sodium hydroxide using screened methyl orange as indicator. After the determination has been completed, pass 100–150 mL of distilled water through the column.

From the results of the titration calculate the millimoles of calcium present in the water. It may be expressed, if desired, as the equivalent mineral acidity (EMA) in terms of mg $CaCO_3$ per litre of water (i.e. parts per million of $CaCO_3$). In general, if the titre is A mL of sodium hydroxide of molarity B for an aliquot volume of V mL, the EMA is given by $(AB \times 50 \times 1000)/V$.

Commercial samples of water are frequently alkaline due to the presence of hydrogencarbonates, carbonates, or hydroxides. The alkalinity is determined by titrating a 100.0 mL sample with $0.02\,M$ hydrochloric acid using screened methyl orange as indicator (or to a pH of 3.8). To obtain the total cation content in terms of $CaCO_3$, the total methyl orange alkalinity is added to the EMA.

7.13 SEPARATION OF CADMIUM AND ZINC ON AN ANION EXCHANGER

Theory. Cadmium and zinc form negatively charged chloro-complexes which are absorbed by a strongly basic anion exchange resin, such as Duolite A113. The maximum absorption of cadmium and zinc is obtained in $0.12\,M$ hydrochloric acid containing 100 g of sodium chloride per litre. The zinc is eluted quantitatively by a $2\,M$ sodium hydroxide solution containing 20 g of sodium chloride per litre, while the cadmium is retained on the resin. Finally, the cadmium is eluted with $1 M$ nitric acid. The zinc and cadmium in their respective effluents may be determined by titration with standard EDTA.

Elements such as Fe(III), Mn, Al, Bi, Ni, Co, Cr, Cu, Ti, the alkaline-earth metals, and the lanthanides are not absorbed on the resin in the HCl–NaCl medium.

Reagents. *Anion exchange column.* Prepare an anion exchange column using 25–30 g of Duolite A113 (chloride form) following the experimental details given in Section 7.8. Allow the resin to settle in $0.5 M$ hydrochloric acid. Transfer the resin slurry to the column: after settling, the resin column should be about 20 cm in length if a 50 mL burette is used.

Reagent I. This consists of $0.12 M$ hydrochloric acid containing 100 g of analytical-grade sodium chloride per litre.

Reagent II. This consists of $2 M$ sodium hydroxide containing 20 g of sodium chloride per litre.

Zinc-ion solution. Dissolve about 7.0 g of analytical grade zinc sulphate heptahydrate in 25 mL of Reagent I.

Cadmium-ion solution. Dissolve about 6.0 g of analytical grade crystallised cadmium sulphate in 25 mL of Reagent I.

EDTA solution. 0.01 M. See Section 10.49.

Buffer solution, pH = 10. Dissolve 7.0 g of analytical grade ammonium chloride and 57 mL of concentrated ammonia solution (sp. gr. 0.88) in water and dilute to 100 mL.

Solochrome black indicator mixture. Triturate 0.20 g of the solid dyestuff with 50 g of potassium chloride.

Xylenol orange indicator. Finely grind (triturate) 0.20 g of the solid dyestuff with 50 g of potassium chloride (or nitrate). This solid mixture is used because solutions of xylenol orange are not very stable.

Nitric acid, ca $1 M$.

Procedure. Wash the anion exchange column with two 20 mL portions of Reagent I; drain the solution to about 0.5 cm above the top of the resin. Mix thoroughly equal volumes (2.00 mL each) of the zinc and cadmium ion solutions and transfer by means of a pipette 2.00 mL of the mixed solution to the top of the resin column. Allow the solution to drain to within about 0.5 cm of the top of the resin and wash down the tube above the resin with a little of Reagent I. Pass 150 mL of Reagent II through the column at a flow rate of about 4 mL min^{-1} and collect the eluate (containing the zinc) in a 250 mL graduated flask; dilute to volume with water. Wash the resin with about 50 mL of water to remove most of the sodium hydroxide solution. Now place a 250 mL graduated flask in position as receiver and pass 150 mL of $1 M$ nitric acid through the column at a rate of about 4 mL min^{-1}; the cadmium will be eluted. Dilute the effluent to 250 mL with distilled water.

The resin may be regenerated by passing Reagent I through the column, and can then be used again for analysis of another Zn–Cd sample.

Analyses. (a) *Original zinc-ion solution.* Dilute 2.00 mL (pipette) to 100 mL in a graduated flask. Pipette 10.0 mL of the diluted solution into a 250 mL conical flask, add *ca* 90 mL of water, 2 mL of the buffer solution, and sufficient of the solochrome black indicator mixture to impart a pronounced red colour to the solution. Titrate with standard $0.01 M$ EDTA to a pure blue colour (see Section 10.59).

(*b*) *Zinc ion eluate*. Pipette 50.0 mL of the solution into a 250 mL conical flask, neutralise with hydrochloric acid, and dilute to about 100 mL with water. Add 2 mL of the buffer mixture, then a little solochrome black indicator powder, and titrate with standard 0.01 M EDTA until the colour changes from red to pure blue.

(*c*) *Original cadmium ion solution*. Dilute 2.00 mL (pipette) to 100 mL in a graduated flask. Pipette 10.0 mL of the diluted solution into a 250 mL conical flask, add *ca* 40 mL of water, followed by solid hexamine and a few milligrams of xylenol orange indicator. If the pH is correct (5–6) the solution will have a pronounced red colour (see Section 10.59). Titrate with standard 0.01 M EDTA until the colour changes from red to clear orange–yellow.

(*d*) *Cadmium ion eluate*. Pipette 50.0 mL of the solution into a conical flask, and partially neutralise (to pH 3–4) with aqueous sodium hydroxide. Add solid hexamine (to give a pH of 5–6) and a little xylenol orange indicator. Titrate with standard 0.01 M EDTA to a colour change from red to clear orange–yellow.

Some typical results are given below.

0.20 mL of original Zn^{2+} solution required 17.50 mL of 0.010 38M EDTA
$\therefore$ Weight of Zn^{2+} per mL $= 17.50 \times 5 \times 0.010\,38 \times 65.38 = 59.35$ mg
50.0 mL of Zn^{2+} eluate $\equiv 17.45$ mL of 0.010 38M EDTA.
$\therefore Zn^{2+}$ recovered $= 5 \times 17.45 \times 0.010\,38 \times 65.38 = 59.21$ mg
0.200 mL of original Cd^{2+} solution required 19.27 mL of 0.010 38M EDTA.
$\therefore$ Weight of Cd^{2+} per mL $= 5 \times 19.27 \times 0.010\,38 \times 112.4 = 112.4$ mg
50.0 mL of Cd^{2+} eluate $\equiv 19.35$ mL of 0.010 38M EDTA.
$\therefore Cd^{2+}$ recovered $= 5 \times 19.35 \times 0.010\,38 \times 112.4 = 112.8$ mg.

7.14 CONCENTRATION OF COPPER(II) IONS FROM A BRINE SOLUTION USING A CHELATING ION EXCHANGE RESIN

Theory. Conventional anion and cation exchange resins appear to be of limited use for concentrating trace metals from saline solutions such as sea water. The introduction of chelating resins, particularly those based on iminodiacetic acid, makes it possible to concentrate trace metals from brine solutions and separate them from the major components of the solution. Thus the elements cadmium, copper, cobalt, nickel and zinc are selectively retained by the resin Chelex-100 and can be recovered subsequently for determination by atomic absorption spectrophotometry.[45] To enhance the sensitivity of the AAS procedure the eluate is evaporated to dryness and the residue dissolved in 90 per cent aqueous acetone.* The use of the chelating resin offers the advantage over concentration by solvent extraction that, in principle, there is no limit to the volume of sample which can be used.

Reagents. *Standard copper (II) solutions*. Dissolve 100 mg of spectroscopically pure copper metal in a slight excess of nitric acid and dilute to 1 L in a graduated flask with de-ionised water. Pipette a 10 mL aliquot into a 100 mL graduated flask and make up to the mark with acetone (analytical grade); the resultant solution contains 10 μg of copper per mL. Use this stock solution to

---

* In the illustrative experiment described here, copper(II) ions in a brine solution are concentrated from 0.1 to about 3.3 ppm prior to determination by atomic absorption spectrophotometry.

prepare a series of standard solutions containing 1.0–5.0 μg of copper per mL, each solution being 90 per cent with respect to acetone.

Sample solution. Prepare a sample solution containing 100 μg of copper(II) in 1 L of 0.5M sodium chloride solution in a graduated flask.

Ion exchange column. Prepare the Chelex-100 resin (100–500 mesh) by digesting it with excess (about 2–3 bed-volumes) of 2M nitric acid at room temperature. Repeat this process twice and then transfer sufficient resin to fill a 1.0 cm diameter column to a depth of 8 cm. Wash the resin column with several bed-volumes of de-ionised water.

Procedure. Allow the whole of the sample solution (1 L) to flow through the resin column at a rate not exceeding 5 mL min^{-1}. Wash the column with 250 mL of de-ionised water and reject the washings. Elute the copper(II) ions with 30 mL of 2M nitric acid, place the eluate in a small conical flask (100 mL, preferably silica) and evaporate carefully to dryness on a hotplate (use a low temperature setting). Dissolve the residue in 1 mL of 0.1M nitric acid introduced by pipette and then add 9 mL of acetone. Determine copper in the resulting solution using an atomic absorption spectrophotometer which has been calibrated using the standard copper(II) solutions.

Note. All glass and silica apparatus to be used should be allowed to stand overnight filled with a 1:1 mixture of concentrated nitric and sulphuric acids and then thoroughly rinsed with de-ionised water. This treatment effectively removes traces of metal ions.

7.15 DETERMINATION OF ANIONS USING ION CHROMATOGRAPHY

The experiment described illustrates the application of ion chromatography (Section 7.4) to the separation and determination of the following anions: Br^-, Cl^-, NO_3^- and NO_2^-. It may be readily extended to include other anions, such as F^-, $H_2PO_4^-$, and SO_4^{2-}. The experiment is based on the Waters ILC Series Ion/Liquid Chromatograph which does not require the use of a suppressor column.

Solutions. Weigh out accurately the following amounts of analytical-grade salts: NaCl (0.1648 g); KBr (0.1489 g); $NaNO_3$ (0.1372 g); $NaNO_2$ (0.1500 g). Dissolve each salt in 100 mL of distilled, de-ionised water in a graduated flask to give standard concentrates containing 1000 ppm of anion. Store these standard solutions in clean plastic ware (e.g. polyethylene or polypropylene); glass containers are not suitable for ion chromatography since cations tend to be leached from them.

Prepare a series of standard solutions of each anion covering the required concentration range by appropriate dilution of the standard concentrates with distilled, de-ionised water.

Borate–gluconate eluant. Prepare a buffer concentrate by dissolving the following substances in water and making up to 1 L with distilled, de-ionised water:

Sodium gluconate	16 g
Boric acid	18 g
Sodium tetraborate ($Na_2B_4O_7, 10H_2O$)	25 g
Glycerol	125 mL

The concentrate is stable for up to six months.

Prepare 1 L of the following borate/gluconate eluant* immediately prior to its use:

Buffer concentrate	20 mL
Acetonitrile	120 mL
Water	860 mL

The eluant must be filtered and de-gassed before use, e.g. using a Millipore filter system.

Column. A Waters IC-PAK A or equivalent anion exchange column, operating at ambient temperature. Equilibrate the column with the mobile phase (eluant) by allowing the latter to flow through the column for 1 hour.

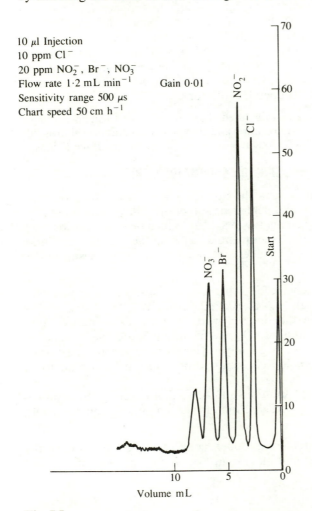

10 μl Injection
10 ppm Cl^-
20 ppm NO_2^-, Br^-, NO_3^-
Flow rate 1·2 mL min⁻¹   Gain 0·01
Sensitivity range 500 μs
Chart speed 50 cm h⁻¹

Fig. 7.5

---

* If necessary, adjust the pH to 8.4–8.5 using boric acid crystals or 0.1 *M* sodium hydroxide; the pH of the eluant is critical.

Procedure. Inject $10\,\mu L$ of standard anion solution into the column and elute with the borate/gluconate eluant at a flow rate of $1.2\,mL\,min^{-1}$. Record the chromatogram and measure the peak height (or peak area). Repeat this procedure using the range of standard anion solutions and prepare a calibration graph of peak height against anion concentration ($mg\,L^{-1}$) for each of the anions to be determined. Finally inject $10\,\mu L$ of the anion mixture (e.g. Cl^-, 10 ppm; NO_2^-, Br^-, NO_3^-, 20 ppm) and record the chromatogram; the elution sequence is Cl^-, NO_2^-, Br^- and NO_3^- as shown in the typical chromatogram (Fig. 7.5). Measure the peak height for each anion and find the anion concentration by reference to the appropriate calibration graph.

For References and Bibliography see Sections 9.9 and 9.10.

CHAPTER 8
COLUMN AND THIN–LAYER LIQUID CHROMATOGRAPHY

8.1 INTRODUCTION

The term 'liquid chromatography' covers a variety of separation techniques, such as liquid–solid, liquid–liquid, ion exchange (see Chapter 7) and size-exclusion chromatography, all of course involving a liquid mobile phase. Classical liquid column chromatography is characterised by use of relatively wide-diameter glass columns, packed with a finely divided stationary phase, with the mobile phase percolating through the column under gravity. Although many remarkable separations have been achieved, these are generally slow and examination of the recovered fractions (e.g. by chemical or spectroscopic techniques) can be tedious. Since about 1969, however, the development of modern high-performance liquid chromatography (HPLC) has enabled liquid chromatography to match the great success achieved by gas chromatography in providing the following features:

1. high resolving power;
2. speed of separation;
3. continuous monitoring of the column effluent;
4. accurate quantitative measurement;
5. repetitive and reproducible analysis using the same column; and
6. automation of the analytical procedure and data handling.

High-performance liquid chromatography is in some respects more versatile than gas chromatography since (a) it is not limited to volatile and thermally stable samples, and (b) the choice of mobile and stationary phases is wider.

The present chapter is largely concerned with HPLC, together with a summary of developments in quantitative thin-layer chromatography, but a brief account of the various types of liquid chromatography is given first together with a guide to the choice of appropriate separation mode.

8.2 TYPES OF LIQUID CHROMATOGRAPHY

There are four main types of liquid chromatography which require discussion.

1. Liquid–solid chromatography (LSC). This process, often termed adsorption chromatography, is based on interactions between the solute and fixed active sites on a finely divided solid adsorbent used as the stationary phase. The adsorbent, which may be packed in a column or spread on a plate, is generally a high surface area, active solid such as alumina, charcoal or silica gel, the last

216

of these being the most widely used. A practical consideration is that highly active adsorbents may give rise to irreversible solute adsorption; silica gel, which is slightly acidic, may strongly retain basic compounds, whilst alumina (non-acid washed) is basic and should not be used for the chromatography of base-sensitive compounds. Adsorbents of varying particle size, e.g. $20-40\ \mu m$ for TLC and down to $5\ \mu m$ for HPLC, may be purchased commercially.

The role of the solvent in LSC is clearly vital since mobile-phase (solvent) molecules compete with solute molecules for polar adsorption sites. The stronger the interaction between the mobile phase and the stationary phase, the weaker will be solute adsorption, and vice versa. The classification of solvents according to their strength of adsorption is referred to as an eluotropic series,[46] which may be used as a guide to find the optimum solvent strength for a particular separation; a trial-and-error approach may, however, be required and this is done more rapidly by TLC than by using a column technique. Solvent purity is very important in LSC since water and other polar impurities may significantly affect column performance, whilst the presence of UV-active impurities is undesirable when using UV-type detectors.

In general, the compounds best separated by LSC are those which are soluble in organic solvents and are non-ionic. Water soluble non-ionic compounds are better separated using either reverse-phase or bonded-phase chromatography.

2. Liquid–liquid (partition) chromatography (LLC). This type of chromatography is similar in principle to solvent extraction (see Chapter 6), being based upon the distribution of solute molecules between two immiscible liquid phases according to their relative solubilities. The separating medium consists of a finely divided inert support (e.g. silica gel, kieselguhr, etc.) holding a fixed (stationary) liquid phase, and separation is achieved by passing a mobile phase over the stationary phase. The latter may be in the form of a packed column, a thin layer on glass, or a paper strip.

It is convenient to divide LLC into two categories, based on the relative polarities of the stationary and mobile phases. The term 'normal LLC' is used when the stationary phase is polar and the mobile phase is **non-polar**. In this case the solute elution order is based on the principle that non-polar solutes prefer the mobile phase and elute first, while polar solutes prefer the stationary phase and elute later. In **reverse-phase chromatography** (RPC), however, the stationary phase is non-polar and the mobile phase is polar; the solute elution order is commonly the reverse of that observed in normal LLC, i.e. with polar compounds eluting first and non-polar ones later. This is a popular mode of operation due to its versatility and scope, the almost universal application of RPC arising from the fact that nearly all organic molecules have hydrophobic regions in their structure and are therefore capable of interacting with the non-polar stationary phase.* Since the mobile phase in RPC is polar, and commonly contains water, the method is particularly suited to the separation of polar substances which are either insoluble in organic solvents or bind too strongly to solid adsorbents (LSC) for successful elution. Table 8.1 shows some typical stationary and mobile phases which are used in normal and reverse phase chromatography.

---

* The reverse-phase technique is used less, however, with the advent of hydrophobic bonded phases (Section 8.2(3)).

Table 8.1   Typical stationary and mobile phases for normal and reverse phase chromatography

Stationary phases	Mobile phases
Normal	
β, β′-Oxydipropionitrile	Saturated hydrocarbons, e.g. hexane, heptane; aromatic solvents,
Carbowax (400, 600, 750, etc.)	e.g. benzene, xylene; saturated hydrocarbons mixed with up to
Glycols (ethylene, diethylene)	10 per cent dioxan, methanol, ethanol, chloroform, methylene
Cyanoethylsilicone	chloride (dichloromethane)
Reverse-phase	
Squalane	Water and alcohol–water mixtures; acetonitrile and
Zipax-HCP	acetonitrile–water mixtures
Cyanoethylsilicone	

Although the stationary and mobile phases in LLC are chosen to have as little solubility in one another as possible, even slight solubility of the stationary phase in the mobile phase may result in the slow removal of the former as the mobile phase flows over the column support. For this reason the mobile phase must be pre-saturated with stationary phase before entering the column. This is conveniently done by using a pre-column before the chromatographic column; the pre-column should contain a large-particle packing (e.g. 30–60 mesh silica gel) coated with a high percentage (30–40 per cent) of the stationary phase to be used in the chromatographic column. As the mobile phase passes through the pre-column it becomes saturated with stationary phase before entering the chromatographic column.

The support materials for the stationary phase can be relatively inactive supports, e.g. glass beads, or adsorbents similar to those used in LSC. It is important, however, that the support surface should not interact with the solute, as this can result in a mixed mechanism (partition and adsorption) rather than true partition. This complicates the chromatographic process and may give non-reproducible separations. For this reason, high loadings of liquid phase are required to cover the active sites when using high surface area porous adsorbents.

It is appropriate here to refer to **ion-pair chromatography** (IPC) which is essentially a partition-type process analogous to the ion-association systems used in solvent extraction (see Section 6.5). In this process the species of interest associates with a counter ion of opposite charge, the latter being selected to confer solubility in an organic solvent on the resulting ion pair. The technique can be used for a wide variety of ionisable compounds but particularly for those yielding large aprotic ions, e.g. quaternary ammonium compounds, and for compounds such as amino acids which are difficult to extract in the uncharged form. The stationary phase consists of an aqueous medium containing a high concentration of a counter ion and at an appropriate pH, typical support materials being cellulose, diatomaceous earth and silica gel. The mobile phase is generally an organic medium having low to moderate solvating power. The application of IPC is well illustrated by the separation of sulpha drugs on a microparticulate silica;[47] the stationary phase contains $0.1\,M$ tetrabutylammonium sulphate and is buffered at pH 9.2, and a butanol–hexane (25:75) mobile phase is used. A useful advantage of the ion-pair technique is the possibility of selecting counter ions which have a high response to specific detectors, e.g. counter ions of high molar absorptivity like bromothymol blue, or highly fluorescent anions such as anthracene sulphonate.

3. Bonded-phase chromatography (BPC). To overcome some of the problems associated with conventional LLC, such as loss of stationary phase from the support material, the stationary phase may be chemically bonded to the support material. This form of liquid chromatography, in which both monomeric and polymeric phases have been bonded to a wide range of support materials, is termed 'bonded-phase chromatography'.

Silylation reactions have been widely used to prepare bonded phases. The silanol groups ($\equiv$Si—OH) at the surface of silica gel are reacted with substituted chlorosilanes. A typical example is the reaction of silica with a dimethylchlorosilane which produces a monomeric bonded phase, since each molecule of the silylating agent can react with only one silanol group:

$$—\overset{|}{\underset{|}{Si}}—OH + Cl—\overset{CH_3}{\underset{CH_3}{\overset{|}{\underset{|}{Si}}}}—R \longrightarrow —\overset{|}{\underset{|}{Si}}—O—\overset{CH_3}{\underset{CH_3}{\overset{|}{\underset{|}{Si}}}}—R + HCl$$

The use of di- or tri-chlorosilanes in the presence of moisture can result in a polymeric layer being formed at the silica surface, i.e. a polymeric bonded phase. Monomeric bonded phases are, however, preferred since they are easier to manufacture reproducibly than the polymeric type. The nature of the main chromatographic interaction can be varied by changing the characteristics of the functional group R; in analytical HPLC the most important bonded phase is the non-polar C-18 type in which the modifying group R is an octadecyl hydrocarbon chain. Unreacted silanol groups are capable of adsorbing polar molecules and will therefore affect the chromatographic properties of the bonded phase, sometimes producing undesirable effects such as tailing in RPC. Such effects can be minimised by the process of 'end-capping' in which these silanol groups are rendered inactive by reaction with trimethylchlorosilane:

$$—\overset{|}{\underset{|}{Si}}—OH + Cl—\overset{CH_3}{\underset{CH_3}{\overset{|}{\underset{|}{Si}}}}—CH_3 \longrightarrow —\overset{|}{\underset{|}{Si}}—O—\overset{CH_3}{\underset{CH_3}{\overset{|}{\underset{|}{Si}}}}—CH_3$$

An important property of these siloxane phases is their stability under the conditions used in most chromatographic separations; the siloxane bonds are attacked only in very acidic (pH < 2) or basic (pH > 9) conditions. A large number of commercial bonded-phase packings are available in particle sizes suitable for HPLC.[48]

4. Gel permeation (exclusion) chromatography (GPC). This form of liquid chromatography permits the separation of substances largely according to their molecular size and shape. The stationary phases used in GPC are porous materials with a closely controlled pore size, the primary mechanism of retention of solute molecules being the different penetration (or permeation) by each solute molecule into the interior of the gel particles. Molecules whose size is too great will be effectively barred from certain openings into the gel network and will, therefore, pass through the column chiefly by way of the interstitial liquid volume. Smaller molecules are better able to penetrate into the interior of the gel particles, depending of course on their size and upon the distribution of pore sizes available to them, and are more strongly retained.

The materials originally used as stationary phases for GPC were the xerogels of the polyacrylamide (Bio-Gel) and cross-linked dextran (Sephadex) type. However, these semi-rigid gels are unable to withstand the high pressures used in HPLC, and modern stationary phases consist of microparticles of styrene–divinylbenzene copolymers (Ultrastyragel, manufactured by Waters Associates), silica, or porous glass.

The extensive analytical applications of GPC cover both organic and inorganic materials.[49] Although there have been many applications of GPC to simple inorganic and organic molecules, the technique has been mainly applied to studies of complex biochemical or highly polymerised molecules.

Choice of mode of separation. To select the most appropriate column type, the analyst requires some knowledge of the physical characteristics of the sample as well as the type of information required from the analysis. The diagram in Fig. 8.1 gives a general guide to the selection of a chromatographic method for separation of compounds of molecular weight < 2000; for samples of higher molecular weight (> 2000) the method of choice would be size-exclusion or gel-permeation chromatography. A prediction of the correct chromatographic system to be used for a given sample cannot be made with certainty, however, and must usually be confirmed by experiment. For a complex sample, no single method may be completely adequate for the separation and a combination of techniques may be required. Computer-aided methods for optimisation of separation conditions in HPLC have been described.[50]

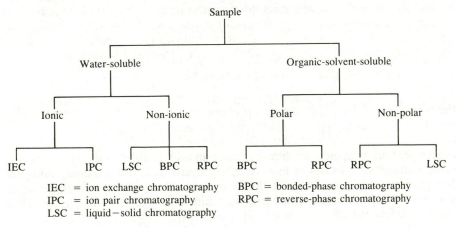

IEC = ion exchange chromatography   BPC = bonded-phase chromatography
IPC = ion pair chromatography   RPC = reverse-phase chromatography
LSC = liquid–solid chromatography

Fig. 8.1

8.3 EQUIPMENT FOR HPLC

The essential features of a modern liquid chromatograph are illustrated in the block diagram (Fig. 8.2) and comprise the following components;
1. solvent delivery system which includes a pump, associated pressure and flow controls and a filter on the inlet side;

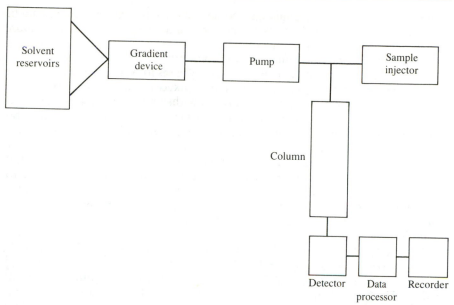

Fig. 8.2

2. sample injection system;
3. the column;
4. the detector;
5. strip chart recorder;
6. data handling device and microprocessor control.

A brief account of the individual components will be given here, but specialised texts should be consulted for a detailed description of the instrumentation available for HPLC.[51] The manufacturer's instructions must be consulted for details of the mode of operation of particular instruments.

High-pressure pumps. The pump is one of the most important components of the liquid chromatograph, since its performance directly affects retention time, reproducibility and detector sensitivity. For analytical applications in which columns 25–50 cm in length (4.0–10 mm i.d.) and packed with particles as small as 5 or 10 μm are typically used, the pump should be capable of delivering the mobile phase at flow rates of 1–5 mL min^{-1} and at pressures reaching 5000 psi. Most of the work in analytical HPLC is, however, done using pressures between about 400 and 1500 psi. Although the SI unit of pressure is the Pascal (Pa), instrument manufacturers commonly report pressures in bar or pounds per square inch (psi), the relevant conversion factors being, 1 bar = 10^5 Pa = 14.5 psi. The supply of mobile phase to the column should be constant, reproducible and pulse-free.

The types of pumps used for HPLC can be divided into two categories: constant-pressure pumps (e.g. the inexpensive gas-displacement pump) and the constant-volume type (e.g. the reciprocating and syringe pumps). The most commonly used pumps in HPLC are the single- or multi-head reciprocating type. The former delivers the flow as a series of pulses which must be damped

221

using a pulse dampener; dual- and triple-head reciprocating pumps can be operated without a pulse dampener since they minimise pulsation, but are more expensive than single-head pumps.

The choice of a suitable **mobile phase** is vital in HPLC and it is appropriate to refer to the factors influencing this choice. Thus, the eluting power of the mobile phase is determined by its overall polarity, the polarity of the stationary phase and the nature of the sample components. For 'normal-phase' separations eluting power increases with increasing polarity of the solvent, while for 'reverse-phase' separations eluting power decreases with increasing solvent polarity. Optimum separating conditions can be often achieved by using a mixture of two solvents, and gradient elution is frequently used where sample components vary widely in polarity. For gradient elution using a low-pressure mixing system, the solvents from separate reservoirs are fed to a mixing chamber and the mixed solvent is then pumped to the column; in modern instruments delivery of solvent to the pump is controlled by time-proportioning electrovalves, regulated by a microprocessor.[52]

Other properties of solvents which need to be considered are boiling point, viscosity (lower viscosity generally gives greater chromatographic efficiency), detector compatibility, flammability, and toxicity. Many of the common solvents used in HPLC are flammable and some are toxic and it is therefore advisable for HPLC instrumentation to be used in a well-ventilated laboratory, if possible under an extraction duct or hood.

Special grades of solvents are available for HPLC which have been carefully purified to remove UV-absorbing impurities and any particulate matter. If, however, other grades of solvent are used, purification may be required since impurities present would, if strongly UV-absorbing, affect the detector or, if of higher polarity than the solvent (e.g. traces of water or ethanol, commonly added as a stabiliser, in chloroform), influence the separation. It is also important to remove dissolved air or suspended air bubbles which can be a major cause of practical problems in HPLC, particularly affecting the operation of the pump and detector. These problems may be avoided by de-gassing the mobile phase before use; this can be accomplished by placing the mobile phase under vacuum, or by heating and ultrasonic stirring. Compilations of useful solvents for HPLC are available.[53]

Sample injection system. Introduction of the sample is generally achieved in one of two ways, either by using syringe injection or through a sampling valve.

Septum injectors allow sample introduction by a high-pressure syringe through a self-sealing elastomer septum. One of the problems associated with septum injectors is the leaching effect of the mobile phase in contact with the septum, which may give rise to ghost peaks. In general, syringe injection for HPLC is more troublesome than in gas chromatography.

Although the problems associated with septum injectors can be eliminated by using stop-flow septumless injection, currently the most widely used devices in commercial chromatographs are the microvolume sampling valves (Fig. 8.3) which enable samples to be introduced reproducibly into pressurised columns without significant interruption of the mobile phase flow. The sample is loaded at atmospheric pressure into an external loop in the valve and introduced into the mobile phase by an appropriate rotation of the valve. The volume of sample introduced, ranging from $2\,\mu\text{L}$ to over $100\,\mu\text{L}$, may be varied by changing

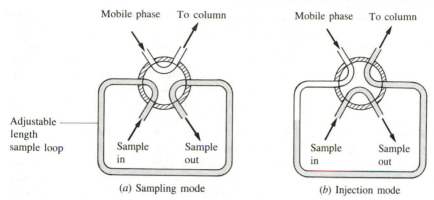

Fig. 8.3 Operation of a sample loop.

the volume of the sample loop or by using special variable-volume sample valves. Automatic sample injectors are also available which allow unattended (e.g. overnight) operation of the instrument. Valve injection is preferred for quantitative work because of its higher precision compared to syringe injection.

The column. The columns most commonly used are made from precision-bore polished stainless steel tubing, typical dimensions being 10–30 cm long and 4 or 5 mm internal diameter. The stationary phase or packing is retained at each end by thin stainless steel frits with a mesh of 2 μm or less.

The packings used in modern HPLC consist of small, rigid particles having a narrow particle-size distribution. The types of packing may conveniently be divided into the following three general categories.

(a) Porous, polymeric beads based on styrene–divinylbenzene copolymers. These are used for ion exchange (see Chapter 7) and size exclusion chromatography (Section 8.2), but have been replaced for many analytical applications by silica-based packings which are more efficient and mechanically stable.

(b) Porous-layer beads (diameter 30–55 μm) consisting of a thin shell (1–3 μm) of silica, or modified silica or other material, on an inert spherical core (e.g. glass beads). These pellicular-type packings are still used for some ion exchange applications, but their general use in HPLC has declined with the development of totally porous microparticulate packings.

(c) Totally porous silica particles (diameter < 10 μm, with narrow particle size range) are now the basis of the most commercially important column packings for analytical HPLC. Compared with the porous-layer beads, totally porous silica particles give considerable improvements in column efficiency, sample capacity, and speed of analysis.

The development of bonded phases (Section 8.2) for liquid–liquid chromatography on silica-gel columns is of major importance. For example, the widely used C-18 type permits the separation of moderately polar mixtures and is used for the analysis of pharmaceuticals, drugs and pesticides.

The procedure chosen for column packing depends chiefly on the mechanical strength of the packing and its particle size. Particles of diameter > 20 μm can usually be dry-packed, whereas for particles with diameters < 20 μm slurry

223

packing techniques are used in which the particles are suspended in a suitable solvent and the suspension (or slurry) driven into the column under pressure. The essential features for successful slurry packing of columns have been summarised.[54] Many analysts will, however, prefer to purchase the commercially available HPLC columns, for which the appropriate manufacturer's catalogues should be consulted.

Finally, the useful life of an analytical column is increased by introducing a **guard column**. This is a short column which is placed between the injector and the HPLC column to protect the latter from damage or loss of efficiency caused by particulate matter or strongly adsorbed substances in samples or solvents. It may also be used to saturate the eluting solvent with soluble stationary phase [see Section 8.2(2)]. Guard columns may be packed with microparticulate stationary phases or with porous-layer beads; the latter are cheaper and easier to pack than the microparticulates, but have lower capacities and therefore require changing more frequently.

Detectors. The function of the detector in HPLC is to monitor the mobile phase as it emerges from the column. The detection process in liquid chromatography has presented more problems than in gas chromatography; there is, for example no equivalent to the universal flame ionisation detector of gas chromatography for use in liquid chromatography. Suitable detectors can be broadly divided into the following two classes:

(a) *Bulk property detectors* which measure the difference in some physical property of the solute in the mobile phase compared to the mobile phase alone, e.g. refractive index and conductivity* detectors. They are generally universal in application but tend to have poor sensitivity and limited range. Such detectors are usually affected by even small changes in the mobile-phase composition which precludes the use of techniques such as gradient elution.

(b) *Solute property detectors*, e.g. spectrophotometric, fluorescence and electrochemical detectors. These respond to a particular physical or chemical property of the solute, being ideally independent of the mobile phase. In practice, however, complete independence of the mobile phase is rarely achieved, but the signal discrimination is usually sufficient to permit operation with solvent changes, e.g. gradient elution. They generally provide high sensitivity (about 1 in 10^9 being attainable with UV and fluorescence detectors) and a wide linear response range but, as a consequence of their more selective natures, more than one detector may be required to meet the demands of an analytical problem. Some commercially available detectors have a number of different detection modes built into a single unit, e.g. the Perkin–Elmer '3D' system which combines UV absorption, fluorescence and conductimetric detection.

Some of the important characteristics required of a detector are the following.

(a) *Sensitivity*, which is often expressed as the noise equivalent concentration, i.e. the solute concentration, C_n, which produces a signal equal to the detector noise level. The lower the value of C_n for a particular solute, the more sensitive is the detector for that solute.

---

* The conductance detector is a universal detector for ionic species and is widely used in ion chromatography (see Section 7.4).

224

(b) *A linear response.* The linear range of a detector is the concentration range over which its response is directly proportional to the concentration of solute. Quantitative analysis is more difficult outside the linear range of concentration.

(c) *Type of response*, i.e. whether the detector is universal or selective. A universal detector will sense all the constituents of the sample, whereas a selective one will only respond to certain components. Although the response of the detector will not be independent of the operating conditions, e.g. column temperature or flow rate, it is advantageous if the response does not change too much when there are small changes of these conditions.

A summary of these characteristics for different types of detectors is given in Table 8.2.

Table 8.2 Typical detector characteristics in HPLC

Type	Response	$C_n (\text{g mL}^{-1})$	Linear range*
Amperometric	Selective	10^{-10}	10^4–10^5
Conductimetric	Selective	10^{-7}	10^3–10^4
Fluorescence	Selective	10^{-12}	10^3–10^4
UV/visible absorption	Selective	10^{-8}	10^4–10^5
Refractive index	Universal	10^{-6}	10^3–10^4

* The range over which the response is essentially linear is expressed as the factor by which the lowest concentration (C_n) must be multiplied to obtain the highest concentration.

A detailed description of the various detectors available for use in HPLC is beyond the scope of the present text and the reader is recommended to consult the monograph by Scott.[55] A brief account of the principal types of detectors is given below.

Refractive index detectors. These bulk property detectors are based on the change of refractive index of the eluant from the column with respect to pure mobile phase. Although they are widely used, the refractive index detectors suffer from several disadvantages — lack of high sensitivity, lack of suitability for gradient elution, and the need for strict temperature control ($\pm 0.001\,^\circ\text{C}$) to operate at their highest sensitivity. A pulseless pump, or a reciprocating pump equipped with a pulse dampener, must also be employed. The effect of these limitations may to some extent be overcome by the use of differential systems in which the column eluant is compared with a reference flow of pure mobile phase. The two chief types of RI detector are as follows.

1. The **deflection refractometer** (Fig. 8.4), which measures the deflection of a beam of monochromatic light by a double prism in which the reference and sample cells are separated by a diagonal glass divide. When both cells contain solvent of the same composition, no deflection of the light beam occurs; if, however, the composition of the column mobile phase is changed because of the presence of a solute, then the altered refractive index causes the beam to be deflected. The magnitude of this deflection is dependent on the concentration of the solute in the mobile phase.
2. The **Fresnel refractometer** which measures the change in the fractions of reflected and transmitted light at a glass-liquid interface as the refractive

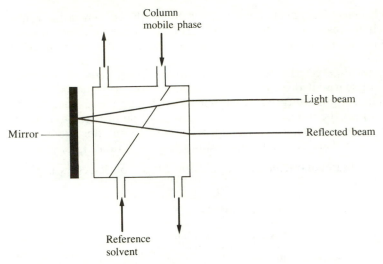

Fig. 8.4   Refractive index detector.

index of the liquid changes. In this detector both the column mobile phase and a reference flow of solvent are passed through small cells on the back surface of a prism. When the two liquids are identical there is no difference between the two beams reaching the photocell, but when the mobile phase containing solute passes through the cell there is a change in the amount of light transmitted to the photocell, and a signal is produced. The smaller cell volume (about 3 μL) in this detector makes it more suitable for high-efficiency columns but, for sensitive operation, the cell windows must be kept scrupulously clean.

Ultraviolet detectors. The UV absorption detector is the most widely used in HPLC, being based on the principle of absorption of UV visible light as the effluent from the column is passed through a small flow cell held in the radiation beam. It is characterised by high sensitivity (detection limit of about $1 \times 10^{-9}\,\mathrm{g\,mL^{-1}}$ for highly absorbing compounds) and, since it is a solute property detector, it is relatively insensitive to changes of temperature and flow rate. The detector is generally suitable for gradient elution work since many of the solvents used in HPLC do not absorb to any significant extent at the wavelengths used for monitoring the column effluent. The presence of air bubbles in the mobile phase can greatly impair the detector signal, causing spikes on the chromatogram; this effect can be minimised by degassing the mobile phase prior to use, e.g. by ultrasonic vibration. Both single and double beam (Fig. 8.5) instruments are commercially available. Although the original detectors were single- or dual-wavelength instruments (254 and/or 280 nm), some manufacturers now supply variable-wavelength detectors covering the range 210–800 nm so that more selective detection is possible.

No account of UV detectors would be complete without mention of the diode array (multichannel) detector, in which polychromatic light is passed through the flow cell. The emerging radiation is diffracted by a grating and then falls on to an array of photodiodes, each photodiode receiving a different narrow-wavelength band. A microprocessor scans the array of diodes many times a

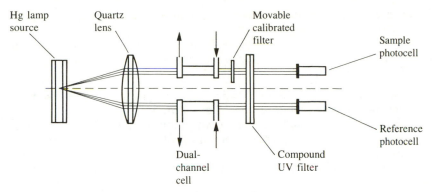

Fig. 8.5 Block diagram of a double-beam UV detector.

second and the spectrum so obtained may be displayed on the screen of a VDU or stored in the instrument for subsequent print-out. An important feature of the multichannel detector is that it can be programmed to give changes in detection wavelength at specified points in the chromatogram; this facility can be used to 'clean up' a chromatogram, e.g. by discriminating against interfering peaks due to compounds in the sample which are not of interest to the analyst.

Fluorescence detectors. These devices enable fluorescent compounds (solutes) present in the mobile phase to be detected by passing the column effluent through a cell irradiated with ultraviolet light and measuring any resultant fluorescent radiation. Although only a small proportion of inorganic and organic compounds are naturally fluorescent, many biologically active compounds (e.g. drugs) and environmental contaminants (e.g. polycyclic aromatic hydrocarbons) are fluorescent and this, together with the high sensitivity of these detectors, explains their widespread use. Because both the excitation wavelength and the detected wavelength can be varied, the detector can be made selective. The application of fluorescence detectors has been extended by means of pre- and post-column derivatisation of non-fluorescent or weakly fluorescing compounds (see Section 8.4).

Electrochemical detectors. The term 'electrochemical detector' in HPLC normally refers to amperometric or coulometric detectors, which measure the current associated with the oxidation or reduction of solutes. In practice it is difficult to use electrochemical reduction as a means of detection in HPLC because of the serious interference (large background current) caused by reduction of oxygen in the mobile phase. Complete removal of oxygen is difficult so that electrochemical detection is usually based on oxidation of the solute. Examples of compounds which can be conveniently detected in this way are phenols, aromatic amines, heterocyclic nitrogen compounds, ketones, and aldehydes. Since not all compounds undergo electrochemical oxidation, such detectors are selective and selectivity may be further increased by adjusting the potential applied to the detector to discriminate between different electroactive species. It may be noted here that an anode becomes a stronger oxidising agent as its electrode potential becomes more positive. Of course, electrochemical detection requires the use of conducting mobile phases, e.g. containing inorganic salts or mixtures of water with water-miscible organic solvents, but such conditions are often difficult to apply to techniques other than reverse phase and ion exchange chromatography.

The amperometric detector is currently the most widely used electrochemical detector, having the advantages of high sensitivity and very small internal cell volume. Three electrodes are used:

1. the working electrode, commonly made of glassy carbon, is the electrode at which the electroactive solute species is monitored;
2. the reference electrode, usually a silver–silver chloride electrode, gives a stable, reproducible voltage to which the potential of the working electrode is referred; and
3. the auxiliary electrode is the current-carrying electrode and usually made of stainless steel.

Despite their higher sensitivity and relative cheapness compared with ultraviolet detectors, amperometric detectors have a more limited range of applications, being often used for trace analyses where the ultraviolet detector does not have sufficient sensitivity.

8.4 DERIVATISATION

In liquid chromatography, in contrast to gas chromatography [see Section 9.2(2)], derivatives are almost invariably prepared to enhance the response of a particular detector to the substance of analytical interest. For example, with compounds lacking an ultraviolet chromophore in the 254 nm region but having a reactive functional group, derivatisation provides a means of introducing into the molecule a chromophore suitable for its detection. Derivative preparation can be carried out either prior to the separation (pre-column derivatisation) or afterwards (post-column derivatisation). The most commonly used techniques are pre-column off-line and post-column on-line derivatisation.

Pre-column off-line derivatisation requires no modification to the instrument and, compared with the post-column techniques, imposes fewer limitations on the reaction conditions. Disadvantages are that the presence of excess reagent and by-products may interfere with the separation, whilst the group introduced into the molecules may change the chromatographic properties of the sample.

Post-column on-line derivatisation is carried out in a special reactor situated between the column and detector. A feature of this technique is that the derivatisation reaction need not go to completion provided it can be made reproducible. The reaction, however, needs to be fairly rapid at moderate temperatures and there should be no detector response to any excess reagent present. Clearly an advantage of post-column derivatisation is that ideally the separation and detection processes can be optimised separately. A problem which may arise, however, is that the most suitable eluant for the chromatographic separation rarely provides an ideal reaction medium for derivatisation; this is particularly true for electrochemical detectors which operate correctly only within a limited range of pH, ionic strength and aqueous solvent composition.

Reagents which form a derivative that strongly absorbs UV/visible radiation are called **chromatags**; an example is the reagent ninhydrin, commonly used to obtain derivatives of amino acids which show absorption at about 570 nm. Derivatisation for fluorescence detectors is based on the reaction of non-fluorescent reagent molecules (**fluorotags**) with solutes to form fluorescent derivatives; the reagent dansyl chloride (I) is used to obtain fluorescent derivatives of proteins, amines and phenolic compounds, the excitation and emission wavelengths being 335–365 nm and 520 nm, respectively.

(I)

The reader is recommended to consult the relevant textbooks[56] (see also Bibliography, Section 9.10) for a more comprehensive account of chemical derivatisation in liquid chromatography.

8.5 QUANTITATIVE ANALYSIS

Quantitative analysis by HPLC clearly requires that a relationship is established between the magnitude of the detector signal and the concentration of a particular solute in the sample, the former being measured by either the corresponding peak area or the peak height. Peak area measurements are preferred when the column flow can be controlled precisely, since peak area is relatively independent of mobile-phase composition. Manual methods may be used for calculating peak areas (see Section 9.4) but computing integrators are preferred for data handling in chromatography and are now often a part of the instrumental package. If the peak areas are measured with an integrator, the latter prints out the retention time for each peak together with a number which is proportional to the peak area. The percentage of each compound in the mixture may then be calculated on the basis of area normalisation, i.e. by expressing each peak area as a percentage of the total area of all the peaks in the chromatogram. Since, however, the detector response is likely to differ for the various components of the mixture, it is essential to correct each peak area before using area normalisation; this is done by finding the relative response factors for the detector (see Section 8.8).

8.6 THIN–LAYER CHROMATOGRAPHY

The important difference between thin-layer chromatography (TLC) and high-performance liquid chromatography is one of practical technique rather than of the physical phenomena (adsorption, partition, etc.) on which separation is based. Thus, in TLC the stationary phase consists of a thin layer of sorbent (e.g. silica gel or cellulose powder) coated on an inert, rigid, backing material such as a glass plate or plastic foil so that the separation process occurs on a flat essentially two-dimensional surface. The analogous technique of paper chromatography has largely been superseded by TLC in analytical laboratories, especially with the advent of thin-layer plates coated with cellulose. Although TLC is widely used for qualitative analysis, it does not in general provide quantitative information of high precision and accuracy. Recent changes in the practice of TLC have, however, resulted in improved performance both in terms of separations and in quantitative measurements; these developments are referred to as high-performance thin-layer chromatography (HPTLC). A brief account of the technique of thin-layer chromatography follows together with a summary of the main features of HPTLC (Section 8.7).

Technique of thin-layer chromatography. *Preparation of the plate.* In thin-layer chromatography a variety of coating materials is available, but silica gel is most frequently used. A slurry of the adsorbent (silica gel, cellulose powder, etc.) is spread uniformly over the plate by means of one of the commercial forms of spreader, the recommended thickness of adsorbent layer being 150–250 μm. After air-drying overnight, or oven-drying at 80–90 °C for about 30 minutes, it is ready for use.

Ready to use thin-layers (i.e. pre-coated plates or plastic sheets) are commercially available; the chief advantage of plastic sheets is that they can be cut to any size or shape required, but they have the disadvantage that they bend in the chromatographic tank unless supported.

Two points of practical importance may be noted here:

1. care should be exercised in handling the plate to avoid placing fingers on the active adsorbent surface and so introducing extraneous substances;
2. pre-washing of the plate is advisable in order to remove extraneous material contained in the layer, and this may be done by running the development solvent to the top of the plate.

Sample application. The origin line, to which the sample solution is applied, is usually located 2–2.5 cm from the bottom of the plate. The accuracy and precision with which the sample 'spots' are applied is, of course, very important when quantitative analysis is required. Volumes of 1, 2 or 5 μL are applied using an appropriate measuring instrument, e.g. an Agla syringe or a Drummond micropipette (the latter is a calibrated capillary tube fitted with a small rubber teat). Care must be taken to avoid disturbing the surface of the adsorbent as this causes distorted shapes of the spots on the subsequently developed chromatogram and so hinders quantitative measurement. It may also be noted that losses may occur if the usual method of drying applied spots in a gentle current of air is used. The use of low boiling point solvents clearly aids drying and also helps to ensure compact spots ($\not> 2$–3 mm diameter).

Development of plates. The chromatogram is usually developed by the ascending technique in which the plate is immersed in the developing solvent (redistilled or chromatographic grade solvent should be used) to a depth of 0.5 cm. The tank or chamber used is preferably lined with sheets of filter paper which dip into the solvent in the base of the chamber; this ensures that the chamber is saturated with solvent vapour (Fig. 8.6). Development is allowed to proceed until the solvent front has travelled the required distance (usually 10–15 cm), the plate is then removed from the chamber and the solvent front immediately marked with a pointed object.

The plate is allowed to dry in a fume cupboard or in an oven; the drying conditions should take into account the heat- and light-sensitivity of the separated compounds.

The positions of the separated solutes can be located by various methods. Coloured substances can be seen directly when viewed against the stationary phase whilst colourless species may usually be detected by spraying the plate

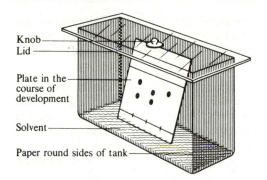

Knob
Lid
Plate in the course of development
Solvent
Paper round sides of tank

Fig. 8.6 Reproduced from D. Abbott and R. S. Andrews, *An Introduction to Chromatography*, Longman, London, 1965.

with an appropriate* reagent which produces coloured areas in the regions which they occupy. Some compounds fluoresce in ultraviolet light and may be located in this way. Alternatively if fluorescing material is incorporated in the adsorbent the solute can be observed as a dark spot on a fluorescent background when viewed under ultraviolet light. (When locating zones by this method **the eyes should be protected** by wearing special protective goggles or spectacles.) The spots located by this method can be delineated by marking with a needle.

Quantitative evaluation. Methods for the quantitative measurement of separated solutes on a thin-layer chromatogram can be divided into two categories. In the more generally used *in-situ* methods, quantitation is based on measurement of the photodensity of the spots directly on the thin-layer plate, preferably using a densitometer. The latter instrument scans the individual spots by either reflectance or absorption of a light beam, the scan usually being along the line of development of the plate. The difference in intensity of the reflected (or transmitted) light between the adsorbent and the solute spots is observed as a series of peaks plotted by a chart recorder. The areas of the peaks correspond to the quantities of the substances in the various spots. This type of procedure requires comparison with spots obtained using known amounts of standard mixtures which must be chromatographed on the same plate as the sample. The design and specifications of commercially available densitometers have been reviewed.[57]

The alternative, and cheaper, procedure is to remove the separated components by scraping off the relevant portion of the adsorbent after visualisation by a non-destructive technique. The component is conveniently extracted by placing the adsorbent in a centrifuge tube and adding a suitable solvent to dissolve the solute. When the solute has dissolved the tube is spun in a centrifuge, the supernatant liquid removed and analysed by an appropriate quantitative technique, e.g. ultraviolet, visible or fluorescence spectrometry or gas–liquid chromatography. Alternatively the solute may be extracted by transferring the adsorbent on to a short column of silica gel supported by a sinter filter and eluting with the solvent. Again the extract is analysed by a suitable quantitative technique. In each case, of course, it is necessary to obtain a calibration curve

---

* Spraying of locating reagents should always be performed in a fume cupboard.

231

for known quantities of the solute in the chosen solvent. Quantitative thin-layer elution techniques have been reviewed by Court.[58]

It should be noted that to obtain the best results in any of these quantitative TLC methods, the spots being used should have R_f values between 0.3 and 0.7; spots with R_f values <0.3 tend to be too concentrated whereas those with R_f values >0.7 are too diffuse.

8.7   HIGH PERFORMANCE THIN–LAYER CHROMATOGRAPHY (HPTLC)

Recent developments in the practice of thin-layer chromatography have resulted in a breakthrough in performance which has led to the expression 'high performance thin-layer chromatography'. These developments have not been the result of any specific advance in instrumentation (as with HPLC), but rather the culmination of improvements in the various operations involved in TLC. The three chief features of HPTLC are summarised below, but for a comprehensive account of the subject the reader is recommended to consult a more specialised text.[59]

Quality of the adsorbent layer. Layers for HPTLC are prepared using specially purified silica gel with average particle diameter of $5-15 \, \mu m$ and a narrow particle size distribution. The silica gel may be modified if necessary, e.g. chemically bonded layers are available commercially as reverse-phase plates. Layers prepared using these improved adsorbents give up to about 5000 theoretical plates and so provide a much improved performance over conventional TLC; this enables more difficult separations to be effected using HPTLC, and also enables separations to be achieved in much shorter times.

Methods of sample application. Due to the lower sample capacity of the HPTLC layer, the amount of sample applied to the layer is reduced. Typical sample volumes are $100-200 \, nL$ which give starting spots of only $1.0-1.5 \, mm$ diameter; after developing the plate for a distance of $3-6 \, cm$, compact separated spots are obtained giving detection limits about ten times better than in conventional TLC. A further advantage is that the compact starting spots allow an increase in the number of samples which may be applied to the HPTLC plate.

The introduction of the sample into the adsorbent layer is a critical process in HPTLC. For most quantitative work a platinum–iridium capillary of fixed volume (100 or 200 nL), sealed into a glass support capillary of larger bore, provides a convenient spotting device. The capillary tip is polished to provide a smooth, planar surface of small area (ca $0.05 \, mm^2$), which when used with a mechanical applicator minimises damage to the surface of the plate; spotting by manual procedures invariably damages the surface.

The availability of scanning densitometers. Commercial instruments for *in-situ* quantitative analysis based on direct photometric measurement have played an important role in modern thin-layer chromatography. Although double beam instruments are available, single beam single wavelength operation is mainly used in HPTLC since the quality and surface homogeneity of the plates are generally very good.

High performance thin-layer chromatography has found its greatest application in the areas of clinical (e.g. analysis of drugs in blood) and environmental analysis.

8.8 DETERMINATION OF ASPIRIN, PHENACETIN AND CAFFEINE IN A MIXTURE

High performance liquid chromatography is used for the separation and quantitative analysis of a wide variety of mixtures, especially those in which the components are insufficiently volatile and/or thermally stable to be separated by gas chromatography. This is illustrated by the following method which may be used for the quantitative determination of aspirin and caffeine in the common analgesic tablets, using phenacetin as internal standard; where APC tablets are available the phenacetin can also be determined by this procedure.

Sample mixture. A suitable sample mixture is obtained by weighing out accurately about 0.601 g of aspirin, 0.076 g of phenacetin and 0.092 g of caffeine. Dissolve the mixture in 10 mL absolute ethanol, add 10 mL of 0.5 M ammonium formate solution and dilute to 100 mL with de-ionised water.

Solvent (mobile phase). Ammonium formate $(0.05M)$ in 10 per cent (v/v) ethanol–water at pH 4.8. Use a flow rate of $2\,mL\,min^{-1}$ with inlet pressure of about 117 bar $(1\ bar = 10^5\ Pa)$.

Column. 15.0 cm × 4.6 mm, packed with a 5 μm silica SCX (strong cation exchanger) bonded phase.

Detector. UV absorbance at 244 nm (or 275 nm).

Procedure. Inject 1 μL of the sample solution and obtain a chromatogram. Under the above conditions the compounds are separated in about 3 minutes, the elution sequence being (1) aspirin; (2) phenacetin; (3) caffeine. Measure peak areas with an integrator and normalise the peak area for each compound (i.e. express each peak area as a percentage of the total peak area). Compare these results with the known composition of the mixture; discrepancies arise because of different detector response to the same amount of each substance.

Determine the response factors (r) for the detector relative to phenacetin $(=1)$ as internal standard by carrying out three runs, using 1 μL injection, and obtaining the average value of r.

$$\text{Relative response factor, } r = \frac{\text{Peak area of compound/Mass of compound}}{\text{Peak area of standard/Mass of standard}}$$

Correct the peak areas initially obtained by dividing by the appropriate response factor and normalise the corrected values. Compare this result with the known composition of the mixture.

8.9 THIN–LAYER CHROMATOGRAPHY — THE RECOVERY OF SEPARATED SUBSTANCES BY ELUTION TECHNIQUES

The purpose of the experiment is to illustrate the elution technique for the recovery of pure substances after their separation by thin-layer chromatography. The experiment can be readily extended to include the quantitative determination of the recovered substances.

Apparatus. Prepared silica gel plates.
Chromatographic tank (see Fig. 8.6).
Drummond (or similar) micropipette.

Chemicals. *Indicator solutions (∼0.1 per cent, aq.).* Bromophenol blue; Congo red; phenol red.
Mixture (M) of above three indicator solutions.
Developing solvent. n-Butanol–ethanol–0.2M ammonia (60:20:20 by volume). Chromatographic grade solvents should be used.

Procedure. Pour the developing solvent into the chromatographic tank to a depth of about 0.5 cm and replace the lid. Take a prepared plate and carefully 'spot' 5 μL of each indicator on the origin line (see Section 8.6, under Sample application) using a micropipette. Allow to dry, slide the plate into the tank and develop the chromatogram by the ascending solvent for about 1 h. Remove the plate, mark the solvent front and dry the plate in an oven at 60 °C for about 15 min. Evaluate the R_f value for each of the indicators using the equation

$$R_f = \frac{\text{Distance compound has moved from origin}}{\text{Distance of solvent front from origin}}$$

Take a second prepared plate and 'spot' three separate 5 μL of mixture M on to the origin line using a micropipette. Place the dry plate into the tank, replace the lid and allow the chromatogram to run for about 1 h. Remove the plate, mark the solvent front and dry the plate at 60 °C for about 15 min. Identify the separated components on the basis of their R_f values.

Carefully scrape the separated bromophenol blue 'spots' on to a sheet of clean smooth-surfaced paper using a narrow spatula (this is easier if two grooves are made down to the glass on either side of the spots). Pour the blue powder into a small centrifuge tube, add 2 mL of ethanol, 5 drops of 0.880 ammonia solution, and stir briskly until the dye is completely extracted. Centrifuge and remove the supernatant blue solution from the residual white powder. Repeat this procedure with the separated Congo red and phenol red 'spots'.

An alternative elution technique is to transfer the powder (e.g. for bromophenol blue) to a glass column fitted with a glass-wool plug or glass sinter, and elute the dye with ethanol containing a little ammonia. The eluted solution, made up to a fixed volume in a small graduated flask, may be used for colorimetric/spectrophotometric analysis of the recovered dye (see Chapter 17). A calibration curve must, of course, be constructed for each of the individual compounds.

For References and Bibliography see Sections 9.9 and 9.10.

CHAPTER 9
GAS CHROMATOGRAPHY

9.1 INTRODUCTION

Gas chromatography is a process by which a mixture is separated into its constituents by a moving gas phase passing over a stationary sorbent. The technique is thus similar to liquid–liquid chromatography except that the mobile liquid phase is replaced by a moving gas phase. Gas chromatography is divided into two major categories: **gas–liquid chromatography** (GLC), where separation occurs by partitioning a sample between a mobile gas phase and a thin layer of non-volatile liquid coated on an inert support, and **gas–solid chromatography** (GSC), which employs a solid of large surface area as the stationary phase. The present chapter deals with gas–liquid chromatography and some of its applications in the field of quantitative chemical analysis. However, before considering these applications it is appropriate to describe briefly the apparatus used in, and some of the basic principles of, gas chromatography. A comprehensive account of the various aspects of modern gas chromatography is, of course, beyond the scope of the present text and, for more detailed accounts of these topics the texts listed in the Bibliography at the end of this chapter should be consulted.

9.2 APPARATUS

A gas chromatograph (see block diagram Fig. 9.1a) consists essentially of the following parts.

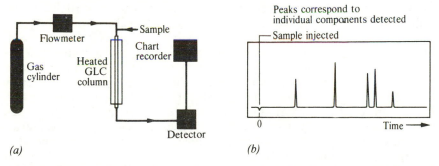

(a) *(b)*

Fig. 9.1 (*a*) Block diagram of a gas chromatograph. (*b*) Typical chart record. Reproduced by permission from R. C. Denney, *The Truth about Breath Tests*, Nelson, London, 1970.

1. A supply of carrier gas from a high-pressure cylinder. The carrier gas used is either helium, nitrogen, hydrogen or argon, the choice of gas depending on factors such as availability, purity required, consumption and the type of detector employed. Thus helium is preferred when thermal conductivity detectors are employed because of its high thermal conductivity relative to that of the vapours of most organic compounds. Associated with this high pressure supply of carrier gas are the attendant pressure regulators and flow meters to control and monitor the carrier gas flow; the operating efficiency of the apparatus is very dependent on the maintenance of a constant flow of carrier gas.

It is appropriate to emphasise here two important safety considerations:

(a) free-standing gas cylinders must always be supported by means of clamps or chains;
(b) waste gases, especially hydrogen, must be vented through an extraction hood.

2. Sample injection system and derivatisation. Numerous devices have been developed for introducing the sample, but the major applications involve liquid samples that are introduced using a microsyringe with hypodermic needle. The latter is inserted through a self-sealing silicone rubber septum and the sample injected smoothly into a heated metal block at the head of the column. Manipulation of the syringe may be regarded as an art developed with practice and the aim must be to introduce the sample in a reproducible manner. The temperature of the sample port should be such that the liquid is rapidly vaporised but without either decomposing or fractionating the sample; a useful rule of thumb is to set the sample port temperature approximately to the boiling point of the least volatile component. For greatest efficiency, the smallest possible sample size ($1-10 \mu L$) consistent with detector sensitivity should be used.

It should be noted here that the difficulty of accurately injecting small quantities of liquids imposes a significant limitation on quantitative gas chromatography. For this reason, it is essential in quantitative GLC to use a procedure, such as the use of an internal standard, which allows for any variation in size of the sample and the effectiveness with which it is applied to the column (see Sections 9.4(5) and 9.7).

Many samples are, however, unsuitable for direct injection into a gas chromatograph because, for example, of their high polarity, low volatility or thermal instability. In this respect the versatility and application of gas chromatography has been greatly extended by the formation of volatile derivatives, especially by the use of silylation reagents. The term 'silylation' is normally taken to mean the introduction of the trimethylsilyl, $-Si(CH_3)_3$, or similar group in place of active hydrogen atoms in the substance under investigation. A considerable number of such reagents is now available,[60] including some special silylating agents which give improved detector response, usually by incorporating a functional group suitable for a selective detector system. Reagents containing chlorine and bromine atoms, for example, in the silyl group are used particularly for preparing derivatives injected on to gas chromatographs fitted with electron-capture detectors. Derivatisation can also give enhanced resolution from other components in a mixture and improved peak shape for quantitative analysis.

Although inorganic compounds are generally not so volatile as are organic compounds, gas chromatography has been applied in the study of certain inorganic compounds which possess the requisite properties. If gas chromatography is to be used for metal separation and quantitative analysis, the types of compounds which can be used are limited to those that can be readily formed in virtually quantitative and easily reproducible yield. This feature, together with the requirements of sufficient volatility and thermal stability necessary for successful gas chromatography, make neutral metal chelates the most favourable compounds for use in metal analysis. β-Diketone ligands, e.g. acetylacetone and the fluorinated derivatives, trifluoroacetylacetone (TFA) and hexafluoroacetylacetone (HFA) form stable, volatile chelates with aluminium, beryllium, chromium(III) and a number of other metal ions; it is thus possible to chromatograph a wide range of metals as their β-diketone chelates.

$$\begin{array}{ccc} & O & O^- \\ & \| & | \\ CF_3-C-CH{=}C-CH_3 & & \text{TFA anion} \end{array}$$

$$\begin{array}{ccc} & O & O^- \\ & \| & | \\ CF_3-C-CH{=}C-CF_3 & & \text{HFA anion} \end{array}$$

The number of reported applications to analytical determinations at the trace level appear to be few, probably the best known being the determination of beryllium in various samples. The method generally involves the formation of the volatile beryllium trifluoroacetylacetonate chelate, its solvent extraction into benzene with subsequent separation and analysis by gas chromatography..[61]

Various types of derivatisation have now been developed for both gas and liquid chromatography. For more detailed information regarding the choice of a suitable derivative for a particular analytical problem, the appropriate works of reference should be consulted.[62,63]

For compounds of high molecular mass, however, the formation of derivatives does not help to solve the problem of involatility. This difficulty may be overcome by breaking the large molecules up into smaller and more volatile fragments which may then be analysed by gas–liquid chromatography, i.e. by using the technique known as **pyrolysis gas chromatography (PGC)**.

Pyrolysis GC is a technique in which a non-volatile sample is pyrolysed under rigidly controlled conditions, usually in the absence of oxygen, and the decomposition products separated in the gas chromatographic column. The resulting chromatogram (pyrogram) is used for both qualitative and quantitative analysis of the sample. If the latter is very complex, complete identification of the pyrolysis fragments may not be possible, but in such cases the pyrogram may be used to 'fingerprint' the sample. PGC has been applied to a wide variety of samples, but its major use has been in polymer analysis for the investigation of both synthetic and naturally occurring polymers. The various PGC systems can generally be classified into two distinct types:

(a) *Static-mode (furnace) reactors* which typically consist of a quartz reactor tube and a Pregl type of combustion furnace. Solid samples are placed in the reactor tube and the system is closed. The furnace is then placed over the combustion tube and the sample heated to the pyrolysis temperature. In this type of pyrolysis system, the time required to reach the necessary temperature is much longer (up to 30 seconds) than in dynamic pyrolysis, resulting in a greater number of secondary reactions. An important advantage of the static-mode system, however, is the usually larger sample capacity.

(b) *Dynamic (filament) reactors* in which the sample is placed on the tip of a filament or wire igniter (platinum and nichrome wires have been used) which is then sealed in a reactor chamber; in PGC the latter is typically the injection port of the gas chromatograph. As the carrier gas passes over the sample, a d.c. charge is applied, and the sample is heated rapidly to the pyrolysis temperature. As the sample decomposes, the pyrolysis products are carried away into a cooler area (reducing the possibility of secondary reactions) before entering the gas chromatographic column.

3. The column. The actual separation of sample components is effected in the column where the nature of the solid support, type and amount of liquid phase, method of packing, length and temperature are important factors in obtaining the desired resolution.

The column is enclosed in a thermostatically controlled oven so that its temperature is held constant to within $0.5\,^\circ$C, thus ensuring reproducible conditions. The operating temperature may range from ambient to over $400\,^\circ$C and for isothermal operation is kept constant during the separation process.

Although many types of column have been developed for gas chromatography, they may be divided into two major groups:

(*a*) *Packed columns*. Conventional analytical columns are usually prepared with 2–6 mm internal diameter glass tubing or 3–10 mm outer diameter metal tubing, which is normally coiled for compactness. Glass columns must be used if any of the sample components are decomposed by contact with metal.

The material chosen as the inert support should be of uniform granular size and have good handling characteristics (i.e. be strong enough not to break down in handling) and be capable of being packed into a uniform bed in a column. The surface area of the material should be large so as to promote distribution of the liquid phase as a film and ensure the rapid attainment of equilibrium between the stationary and mobile phases. The most commonly used supports (e.g. Celite) are made from diatomaceous materials which can hold liquid phases in amounts exceeding 20 per cent without becoming too sticky to flow freely and can be easily packed.

Commercial preparations of these supports are available in narrow mesh-range fractions; to obtain particles of uniform size the material should be sieved to the desired particle size range and repeatedly water floated to remove fine particles which contribute to excessive pressure drop in the final column. To a good approximation the height equivalent of a theoretical plate is proportional to the average particle diameter so that theoretically the smallest possible particles should be preferred in terms of column efficiency. Decreasing particle size will, however, rapidly increase the gas pressure necessary to achieve flow through the column and in practice the best choice is 80/100 mesh for a 3 mm i.d. column. It may be noted here that for effective packing of any column the internal diameter of the tubing should be at least eight times the diameter of the solid support particles.

Various types of **porous polymers** have also been developed as column packing material for gas chromatography, e.g. the Porapak series (Waters Associates) and the Chromosorb series (Johns Manville) which are styrene copolymers although modified with different, mainly polar, monomers. An entirely different type of porous polymer is Tenax GC which is based on 2,6-diphenylphenylene oxide; a special feature of this column packing is its high maximum operating temperature of 400 °C. Tenax GC has been used for concentrating and determining trace volatile organic constituents in gases and biological fluids.[64]

The selection of the most suitable liquid phase for a particular separation is crucial. Liquid phases can be broadly classified as follows:

1. Non-polar hydrocarbon-type liquid phases, e.g. paraffin oil (Nujol), squalane, Apiezon L grease and silicone-gum rubber; the latter is used for high-temperature work (upper limit ~400 °C).
2. Compounds of intermediate polarity which possess a polar or polarisable group attached to a large non-polar skeleton, e.g. esters of high-molecular-weight alcohols such as dinonyl phthalate.
3. Polar compounds containing a relatively large proportion of polar groups, e.g. the carbowaxes (polyglycols).
4. Hydrogen-bonding class, i.e. polar liquid phases such as glycol, glycerol, hydroxyacids, etc., which possess an appreciable number of hydrogen atoms available for hydrogen bonding.

The column packing is prepared by adding the correct amount of liquid phase dissolved in a suitable solvent (e.g. acetone or dichloromethane) to a weighed quantity of the solid support in a suitable dish. The volatile solvent is removed either by spontaneous evaporation or by careful heating, the mixture being gently agitated to ensure a uniform distribution of the liquid phase in the support. Final traces of the solvent may be removed under vacuum and the column packing re-sieved to remove any fines produced during the preparation. The relative amount of stationary liquid phase in the column packing is usually expressed on the basis of the percentage by weight of liquid phase present, e.g. 15 per cent loading indicates that 100 g column packing contains 15 g of liquid phase on 85 g of inert support. The solid should remain free flowing after being coated with the liquid phase.

Micropacked columns, sometimes referred to as packed capillary columns, have been used in gas chromatography, e.g. for the determination of trace components in complex mixtures. These columns are characterised by small internal diameters (i.d. <1.0 mm) and packing densities comparable with conventional packed columns. In general, the column packing technique requires higher pressures and constant vibration (e.g. ultrasonic) to achieve the necessary packing density. Micropacked columns give high efficiency but practical problems, especially sample injection at high back-pressures, have limited their use.

(*b*) *Open tubular columns.* These capillary columns (i.d. <1 mm) are increasingly used in GLC because of their superior resolving power for complex mixtures. This results from the high theoretical plate numbers which can be attained with long columns of this type for a relatively small pressure drop. In these capillary columns the stationary phase is coated on the inner wall of the tube, two basic types of capillary column being available:

1. wall-coated open tubular (WCOT), in which the stationary phase is coated directly on to the inner wall of the tubing;

239

2. support-coated open tubular (SCOT), which have a finely-divided layer of solid support material deposited on the inner wall on to which the stationary phase is then coated These SCOT columns are not as efficient as WCOT columns but have a higher sample capacity, which enables them to be used without a stream splitter.

Capillary columns are fabricated from thin-walled stainless steel, glass, or high-purity fused silica tubing (the last is preferred for its inertness). Typical dimensions of the columns, which are coiled, are 25–200 m long and 0.2–0.5 mm i.d.

Excellent open tubular columns may now be purchased, providing a number of stationary phases of differing polarity on WCOT and SCOT columns, and whose efficiency, greatly improved sample detectability, and thermal stability surpass those exhibited by packed columns; their chief disadvantage is that they have a lower sample capacity than packed columns.[65,66]

4. The detector. The function of the detector, which is situated at the exit of the separation column, is to sense and measure the small amounts of the separated components present in the carrier gas stream leaving the column. The output from the detector is fed to a recorder which produces a pen-trace called a **chromatogram** (Fig. 9.1*b*). The choice of detector will depend on factors such as the concentration level to be measured and the nature of the separated components. The detectors most widely used in gas chromatography are the thermal conductivity, flame-ionisation and electron-capture detectors, and a brief description of these will be given. For more detailed descriptions of these and other detectors more specialised texts should be consulted.[67–69]

Some of the important properties of a detector in gas chromatography are briefly discussed below.

(*a*) *Sensitivity.* This is usually defined as the detector response (mV) per unit concentration of analyte (mg mL^{-1}). It is closely related to the limit of detection (MDL) since high sensitivity often gives a low limit of detection. Since, however, the latter is generally defined as the amount (or concentration) of analyte which produces a signal equal to twice the baseline noise, the limit of detection will be raised if the detector produces excessive noise. The sensitivity also determines the slope of the calibration graph (slope increases with increasing sensitivity) and therefore influences the precision of the analysis [see also (*b*) below].

(*b*) *Linearity.* The linear range of a detector refers to the concentration range over which the signal is directly proportional to the amount (or concentration) of analyte. Linearity in detector response will give linearity of the calibration graph and allows the latter to be drawn with more certainty. With a convex calibration curve, the precision is reduced at the higher concentrations where the slope of the curve is much less. A large linear range is a great advantage, but detectors with a small linear range may still be used because of their other qualities, although they will need to be recalibrated over a number of different concentration ranges.

(*c*) *Stability.* An important characteristic of a detector is the extent to which the signal output remains constant with time, assuming there is a constant input. Lack of stability can be exhibited in two ways, viz. by **baseline noise** or by **drift**, both of which will limit the sensitivity of the detector. Baseline noise, caused by a rapid random variation in detector output, makes it difficult to measure small peaks against the fluctuating background. Baseline drift, a slow systematic variation in output, results in a sloping baseline which in severe cases may even go off scale during the analysis. Drift is often due to factors external to the detector, such as temperature change or column bleed, and so is controllable, whereas noise is usually due to poor contacts within the detector and imposes a more fundamental limit on its performance.

(d) *Universal or selective response*. A universal detector will respond to all the components present in a mixture. In contrast, a selective detector senses only certain components in a sample which can be advantageous if it responds only to those which are of interest, thus giving a considerably simplified chromatogram and avoiding interference.

Thermal conductivity detector. The most important of the bulk physical property detectors is the thermal conductivity detector (TCD) which is a universal, non-destructive, concentration-sensitive detector. The TCD was one of the earliest routine detectors and thermal conductivity cells or katharometers are still widely used in gas chromatography. These detectors employ a heated metal filament or a thermistor (a semiconductor of fused metal oxides) to sense changes in the thermal conductivity of the carrier gas stream. Helium and hydrogen are the best carrier gases to use in conjunction with this type of detector since their thermal conductivities are much higher than any other gases; on safety grounds helium is preferred because of its inertness.

In the detector two pairs of matched filaments are arranged in a Wheatstone bridge circuit; two filaments in opposite arms of the bridge are surrounded by the carrier gas only, while the other two filaments are surrounded by the effluent from the chromatographic column. This type of thermal conductivity cell is illustrated in Fig. 9.2(*a*) with two gas channels through the cell; a sample channel and a reference channel. When pure carrier gas passes over both the reference and sample filaments the bridge is balanced, but when a vapour emerges from the column, the rate of cooling of the sample filaments changes and the bridge becomes unbalanced. The extent of this imbalance is a measure of the concentration of vapour in the carrier gas at that instant, and the out-of-balance

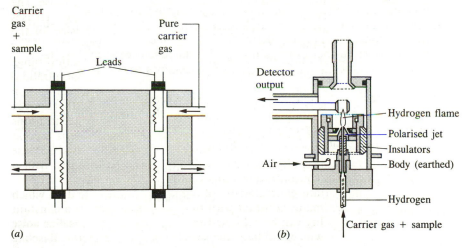

(*a*) (*b*)

Fig. 9.2 (*a*) **Thermal conductivity detector.** (*b*) **Flame ionisation detector.**

241

signal is fed to a recorder thus producing the chromatogram. The differential technique used is thus based on the measurement of the difference in thermal conductivity between the carrier gas and the carrier gas/sample mixture.

The TCD is generally used for the detection of permanent gases, light hydrocarbons and compounds which respond poorly to the flame-ionisation detector (FID). Used in conjunction with a Porapak column it is useful for the determination of water (see Section 9.5) and it has also been employed in gas chromatographic studies of metal chelates, e.g. for the quantitative determination of mixtures of beryllium, aluminium, gallium and indium trifluoroacetylacetonates (see Section 9.2(2)). For many general applications, however, it has been replaced by the FID, which is more sensitive (up to 1000-fold), has a greater linear response range, and provides a more reliable signal for quantitative analysis.

Ionisation detectors. An important characteristic of the common carrier gases is that they behave as perfect insulators at normal temperatures and pressures. The increased conductivity due to the presence of a few charged molecules in the effluent from the column thus provides the high sensitivity which is a feature of the ionisation based detectors. Ionisation detectors in current use include the flame ionisation detector (FID), thermionic ionisation detector (TID), photo-ionisation detector (PID) and electron capture detector (ECD) each, of course, employing a different method to generate an ion current. The two most widely used ionisation detectors are, however, the FID and ECD and these are described below.

Flame ionisation detector. The basis of this detector is that the effluent from the column is mixed with hydrogen and burned in air to produce a flame which has sufficient energy to ionise solute molecules having low ionisation potentials. The ions produced are collected at electrodes and the resulting ion current measured; the burner jet is the negative electrode whilst the anode is usually a wire or grid extending into the tip of the flame. This is shown diagrammatically in Fig. 9.2(*b*).

The combustion of mixtures of hydrogen and air produces very few ions so that with only the carrier gas and hydrogen burning an essentially constant signal is obtained. When, however, carbon-containing compounds are present ionisation occurs and there is a large increase in the electrical conductivity of the flame. Because the sample is destroyed in the flame a stream-splitting device is employed when further examination of the eluate is necessary; this device is inserted between the column and detector and allows the bulk of the sample to by-pass the detector.

The FID has wide applicability, being a very nearly universal detector for gas chromatography of organic compounds, and this, coupled with its high sensitivity, stability, fast response and wide linear response range ($\sim 10^7$), has made it the most popular detector in current use.[70]

Electron capture detector. Most ionisation detectors are based on measurement of the increase in current (above that due to the background ionisation of the carrier gas) which occurs when a more readily ionised molecule appears in the gas stream. The electron capture detector differs from other ionisation detectors in that it exploits the recombination phenomenon, being based on electron capture by compounds having an affinity for free electrons; the detector thus measures a decrease rather than an increase in current.

A β-ray source (commonly a foil containing ^{3}H or ^{63}Ni) is used to generate 'slow' electrons by ionisation of the carrier gas (nitrogen preferred) flowing through the detector. These slow electrons migrate to the anode under a fixed potential and give rise to a steady baseline current. When an electron-capturing gas (i.e. eluate molecules) emerges from the column and reacts with an electron, the net result is the replacement of an electron by a negative ion of much greater mass with a corresponding reduction in current flow.

The response of the detector is clearly related to the electron affinity of the eluate molecules being particularly sensitive to compounds containing halogens and sulphur, anhydrides, conjugated carbonyls, nitrites, nitrates and organo-metallic compounds. The ECD is the second most widely used ionisation detector due to its high sensitivity to a wide range of compounds. It is, for example, used in trace analysis of pesticides, herbicides, drugs, and other biologically active compounds, and is of value in detecting ultratrace amounts of metals as their chelate complexes.[71]

Compared with the flame ionisation detector, however, the ECD is more specialised and tends to be chosen for its selectivity which can simplify chromatograms. The ECD requires careful attention to obtain reliable results. Cleanliness is essential and the carrier gases must be very pure and dry. The two most likely impurities in these gases are water and oxygen which are sufficiently electronegative to produce a detector response and so give a noisy baseline.

Table 9.1 gives a summary of some important detector characteristics.

Table 9.1 Detector characteristics

Type	MDL* (g s^{-1})	Linear range	Temp. limit ($^\circ$C)	Features
TCD	10^{-6}–10^{-8}	10^4	450	Non-destructive, but temperature and flow-sensitive
FID	10^{-11}	10^7	400	Destructive, excellent stability and linearity
ECD	10^{-13}	10^2	350	Non-destructive but easily contaminated and temperature-sensitive

* The minimum detectable level is commonly given in terms of the mass flow rate in grams per second.

Element-selective detectors. Many samples, e.g. those originating from environmental studies, contain so many constituent compounds that the gas chromatogram obtained is a complex array of peaks. For the analytical chemist, who may be interested in only a few of the compounds present, the replacement of the essentially non-selective type of detector (i.e. thermal conductivity, flame ionisation, etc.) by a system which responds selectively to some property of certain of the eluted species may overcome this problem.

The most common selective detectors in use generally respond to the presence of a characteristic element or group in the eluted compound. This is well illustrated by the thermionic ionisation detector (TID) which is essentially a flame ionisation detector giving a selective response to phosphorus- and/or nitrogen-containing compounds. Typically the TID contains an electrically heated rubidium silicate bead situated a few millimetres above the detector jet tip and below the collector electrode. The temperature of the bead is maintained

243

at 600–800 °C while a plasma is sustained in the region of the bead by hydrogen and air support gases. A reaction cycle is so produced in which the rubidium is vaporised, ionised and finally recaptured by the bead. During this process an electron flow to the positive collector electrode occurs and this background current is enhanced when nitrogen or phosphorus compounds are eluted, due it is thought to the production of radicals in the flame or plasma which accelerate the rate of rubidium recycling.[72] The selectivity of this detector can facilitate otherwise difficult analyses, e.g. the determination of pesticides such as Malathion and Parathion.

Another example is the flame photometric detector (FPD) which offers simultaneous sensitivity and specificity for the determination of compounds containing sulphur and phosphorus. The operating principle of the FPD is that combustion of samples containing phosphorus or sulphur in a hydrogen-rich flame results in the formation of luminescent species that emit light characteristic of the heteroatom introduced into the flame. Selection of an appropriate filter (394 or 526 nm bandpass) allows selectivity for sulphur or phosphorus, respectively. It is advantageous to use nitrogen as the carrier gas and mix it with oxygen at the column exit; hydrogen is introduced at the burner to initiate combustion.[73]

The principles and applications of element-selective detectors have been reviewed.[74]

The element specificity of atomic absorption spectrometry has also been used in conjunction with gas chromatography to separate and determine organometallic compounds of similar chemical composition, e.g. alkyl leads in petroleum; here lead is determined by AAS for each compound as it passes from the gas chromatograph.[75]

Finally, a high degree of **specific molecular identification** can be achieved by the interfacing of the gas chromatograph with various spectroscopic instruments. Although the quantitative information obtained from a chromatogram is usually good, the certainty of identification based only on the retention parameter may be suspect. In the case of spectroscopic techniques, however, the reverse situation applies since these provide excellent qualitative information, enabling a pure substance to be identified, but less satisfactory quantitative information is often obtained from their signals. The combination of chromatographic and spectroscopic techniques thus provides more information about a sample than may be obtained from either instrument independently. The chief combined techniques are gas chromatography interfaced with mass spectrometry (GC–MS), Fourier transform infrared spectrometry (GC–FTIR), and optical emission spectroscopy (GC–OES).[76]

9.3   PROGRAMMED TEMPERATURE GAS CHROMATOGRAPHY

Gas chromatograms are usually obtained with the column kept at a constant temperature. Two important disadvantages result from this isothermal mode of operation.

1. Early peaks are sharp and closely spaced (i.e. resolution is relatively poor in this region of the chromatogram), whereas late peaks tend to be low, broad and widely spaced (i.e. resolution is excessive).
2. Compounds of high boiling point are often undetected, particularly in the

study of mixtures of unknown composition and wide boiling point range; the solubilities of the higher-boiling substances in the stationary phase are so large that they are almost completely immobilised at the inlet to the column, especially where the latter is operated at a relatively low temperature.

The above consequences of isothermal operation may be largely avoided by using the technique of programmed-temperature gas chromatography (PTGC) in which the temperature of the whole column is raised during the sample analysis.

A temperature programme consists of a series of changes in column temperatures which may be conveniently selected by a microprocessor controller. The programme commonly consists of an initial isothermal period, a linear temperature rise segment, and a final isothermal period at the temperature which has been reached, but may vary according to the separation to be effected. The rate of temperature rise, which may vary over a wide range, is a compromise between the need for a slow rate of change to obtain maximum resolution and a rapid change to minimise analysis time.

Programmed-temperature gas chromatography permits the separation of compounds of a very wide boiling range more rapidly than by isothermal operation of the column. The peaks on the chromatogram are also sharper and more uniform in shape so that, using PTGC, peak heights may be used to obtain accurate quantitative analysis.[77]

9.4 QUANTITATIVE ANALYSIS BY GLC

The quantitative determination of a component in gas chromatography using differential-type detectors of the type previously described is based upon measurement of the recorded peak area or peak height; the latter is more suitable in the case of small peaks, or peaks with narrow band width. In order that these quantities may be related to the amount of solute in the sample two conditions must prevail:

(a) the response of the detector–recorder system must be linear with respect to the concentration of the solute;

(b) factors such as the rate of carrier gas flow, column temperature, etc., must be kept constant or the effect of variation must be eliminated, e.g. by use of the internal standard method.

Peak area is commonly used as a quantitative measure of a particular component in the sample and can be measured by one of the following techniques.

1. Planimetry. The planimeter is a mechanical device which enables the peak area to be measured by tracing the perimeter of the peak. The method is slow but can give accurate results with experience in manipulation of the planimeter. Accuracy and precision, however, decrease as peak area diminishes.

2. Geometrical methods. In the so-called triangulation methods, tangents are drawn to the inflection points of the elution peak and these two lines together with the baseline form a triangle (Fig. 9.3); the area of the latter is calculated as one-half the product of the base length times the peak height, the value obtained being about 97 per cent of the actual area under the chromatographic peak when this is Gaussian in shape.

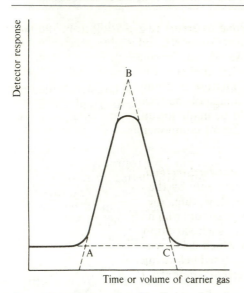

Fig. 9.3 Measurement of peak area by triangulation.

The area may also be computed as the product of the peak height times the width at half the peak height, i.e., *by the height × width at half height method.* Since the exact location of the tangents (required for the triangulation method) to the curve is not easily determined it is in general more accurate to use the method based on width at half-height.

3. Integration by weighing. The chromatographic peak is carefully cut out of the chart and the paper weighed on an analytical balance. The accuracy of the method is clearly dependent upon the constancy of the thickness and moisture content of the chart paper, and it is usually preferable (unless an automatic integrator is available) to use geometrical methods.

4. Automatic integration. The older methods for computing peak area are time-consuming and often unsatisfactory in terms of accuracy and reproducibility of results. The greater use of capillary column chromatography, with its resulting sharp, closely spaced peaks has accentuated the need for a rapid, automatic instrumental method for data processing. The instrument currently most widely used in quantitative gas chromatography is the digital integrator; real-time digital automatic integrators process the analytical signal as the analyses are being run. These systems automatically identify peaks, compute peak areas and/or peak heights, and provide the results either in printed form or in one of the various computer-compatible formats.

The measurement of individual peak areas can be difficult when the chromatogram contains overlapping peaks. However, this problem can be often overcome by the use of derivative facilities which give first- or second-derivative chromatograms (see Section 17.12, for the analogous derivative procedures used in spectroanalytical methods).

5. Data evaluation. It is, of course, necessary to correlate peak area with the amount or concentration of a particular solute in the sample. Quantitation by simple calibration with standards is prone to errors (e.g. arising from the use of a microsyringe with a conventional injection port), and is not generally used; the commonly used methods are those of area normalisation and internal standards which allow for variations in the size of sample, etc. In area normalisation, the composition of the mixture is obtained by expressing the area of each individual peak as a percentage of the total area of all the peaks in the chromatogram; correction should be made for any significant variation in sensitivity of the detector for the different components of the mixture (see Section 9.7).

Quantitative analysis using the internal standard method. The height and area of chromatographic peaks are affected not only by the amount of sample but also by fluctuations of the carrier gas flow rate, the column and detector temperatures, etc., i.e. by variations of those factors which influence the sensitivity and response of the detector. The effect of such variations can be eliminated by use of the internal standard method in which a known amount of a reference substance is added to the sample to be analysed before injection into the column. The requirements for an effective internal standard (Section 4.5) may be summarised as follows:

(*a*) it should give a completely resolved peak, but should be eluted close to the components to be measured;

(*b*) its peak height or peak area should be similar in magnitude to those of the components to be measured; and

(*c*) it should be chemically similar to but not present in the original sample.

The procedure comprises the addition of a constant amount of internal standard to a fixed volume of several synthetic mixtures which contain varying known amounts of the component to be determined. The resulting mixtures are chromatographed and a calibration curve is constructed of the percentage of component in the mixtures against the ratio of component peak area/standard peak area. The analysis of the unknown mixture is carried out by addition of the same amount of internal standard to the specified volume of the mixture; from the observed ratio of peak areas the solute concentration is read off using the calibration curve.

Provided a suitable internal standard is available, this is probably the most reliable method for quantitative GLC. For example, the concentration of ethanol in blood samples has been determined using propan-2-ol as the internal standard.

Standard addition. The sample is chromatographed before and after the addition of an accurately known amount of the pure component to be determined, and its weight in the sample is then derived from the ratio of its peak areas in the two chromatograms. Standard addition is particularly useful in the analysis of complex mixtures where it may be difficult to find an internal standard which meets the necessary requirements.

9.5 ELEMENTAL ANALYSIS USING GAS CHROMATOGRAPHY

An important application of gas chromatography is its use for determination of the elements carbon, hydrogen, nitrogen, oxygen and sulphur in organic and organometallic samples. A brief account of the procedure used is given here, based on those adopted for the Carlo Erba Elemental Analyser; further details may of course be obtained by referring to the manufacturer's manual.

Determination of C, H and N. The weighed samples (usually about 1 mg), held in a clean, dry, tin container, are dropped at pre-set intervals of time into a vertical quartz tube maintained at 1030 °C and through which flows a constant stream of helium gas. When the samples are introduced the helium stream is temporarily enriched with pure oxygen and flash combustion occurs. The mixture of gases so obtained is passed over Cr_2O_3 to obtain quantitative combustion, and then over copper at 650 °C to remove excess oxygen and reduce oxides of nitrogen to N_2. Finally the gas mixture passes through a chromatographic column (2 m long) of Porapak QS heated to approximately 100 °C. The individual components (N_2, CO_2, H_2O) are separated and eluted to a thermal conductivity detector, the detector signal being fed to a potentiometric recorder in parallel with an integrator and digital print-out. The instrument is calibrated by combustion of standard compounds, such as cyclohexanone-2,4-dinitrophenylhydrazone.

It is appropriate to mention here the determination of **Total Organic Carbon** (TOC) which is important in water analysis and water quality monitoring, giving an indication of the overall amount of organic pollutants. The water is first acidified and purged to remove carbon dioxide from any carbonate or hydrogencarbonate present. After this treatment, a small measured volume of the water is injected into a gas stream which then passes through a heated packed tube where the organic material is oxidised to carbon dioxide. The latter is determined either by infrared absorption or is converted to methane for determination by gas chromatography using a flame ionisation detector [see Section 9.2].

Determination of oxygen. The sample is weighed into a silver container which has been solvent-washed, dried at 400 °C and kept in a closed container to avoid oxidation. It is dropped into a reactor heated at 1060 °C, quantitative conversion of oxygen to carbon monoxide being achieved by a layer of nickel-coated carbon (see Note). The pyrolysis gases then flow into the chromatographic column (1 m long) of molecular sieves (5×10^{-8} cm) heated at 100 °C; the CO is separated from N_2, CH_4, and H_2, and is measured by a thermal conductivity detector.

Note. The addition of a chlorohydrocarbon vapour to the carrier gas is found to enhance the decomposition of the oxygen-containing compounds.

Determination of sulphur. The initial procedure for flash combustion of the sample is essentially that described for the determination of C, H and N. Quantitative conversion of sulphur to sulphur dioxide is then achieved by passing the combustion gases over tungsten(VI) oxide, WO_3, and excess oxygen is removed by passing the gases through a heated reduction tube containing copper. Finally the gas mixture passes through a Porapak chromatographic column heated at 80 °C in which SO_2 is separated from other combustion gases and measured by a thermal conductivity detector.

9.6 DETERMINATION OF ALUMINIUM BY GAS CHROMATOGRAPHIC ANALYSIS OF ITS TRIS(ACETYLACETONATO) COMPLEX

The purpose of this experiment is to illustrate the application of gas chromatographic analysis to the determination of trace amounts of metals as their chelate complexes. The procedure described for the determination of aluminium may be adapted for the separation and determination of aluminium and chromium(III) as their acetylacetonates.[78]

Sample. The solvent extraction of aluminium from aqueous solution using acetylacetone can provide a suitable sample solution for gas chromatographic analysis.

Take 5 mL of a solution containing about 15 mg of aluminium and adjust the pH to between 4 and 6. Equilibrate the solution for 10 minutes with two successive 5 mL portions of a solution made up of equal volumes of acetylacetone (pure, redistilled) and chloroform. Combine the organic extracts. Fluoride ion causes serious interference to the extraction and must be previously removed.

Introduce a 0.30 μL portion of the solvent extract into the gas chromatograph. It is found that solutions of concentrations greater than $0.3 M$ are unsuitable as they deposit solid and thus cause a blockage of the 1 μL microsyringe used for the injection of the sample. The syringe is flushed several times with the sample solution, filled with the sample to the required volume, excess liquid wiped from the tip of the needle and the sample injected into the chromatograph.

Apparatus. A gas chromatograph equipped with a flame-ionisation detector and data-handling system. The use of a digital integrator is particularly convenient for quantitative determinations, but other methods of measuring peak area may be used (Section 9.4).

Pure nitrogen (oxygen-free), at a flow rate of 40 mL min^{-1}, is used as carrier gas. The dimensions of the glass column are 1.6 m length and 6 mm o.d., and it is packed with 5 per cent by weight SE-30 on Chromosorb W as the stationary phase. The column is maintained at a temperature of 165 °C.

Procedure. Extract a series of aqueous aluminium solutions containing 5–25 mg aluminium in 5 mL, using the procedure described above under Sample. Calibrate the apparatus by injecting 0.30 μL of each extract into the column and recording the peak area on the chromatogram. Plot a graph of peak area against concentration.

Determine aluminium (present as its acetylacetonate) in the sample solution by injecting 0.30 μL into the column. Record the peak area obtained and read off the aluminium concentration from the calibration graph (see Note).

Note. The calibration procedure is, however, of limited accuracy and a more accurate result may be obtained using the method of standard addition (Section 9.4)

9.7 ANALYSIS OF A MIXTURE USING THE INTERNAL NORMALISATION METHOD

To obtain an accurate quantitative analysis of the composition of a mixture, a knowledge of the response of the detector to each component is required. If the detector response is not the same for each component, then the areas under the peaks clearly cannot be used as a direct measure of the proportion of the components in the mixture. The experiment described illustrates the use of an internal normalisation method for the quantitative analysis of a mixture of the following three components: ethyl acetate (ethanoate), octane, and ethyl n-propyl ketone (hexan-3-one).

Reagents and apparatus required. *Ethyl acetate (I), octane (II), ethyl* n-*propyl ketone (III) and toluene (IV), all GPR or comparable grade.*
Mixture A containing compounds (I), (II), and (III) in an unknown ratio.
Microsyringe.
Gas chromatograph. Preferably equipped with a flame ionisation detector and a digital integrator.
Column. Packed with stationary phase containing 10 per cent by weight of dinonyl phthalate.

Procedure.
1. Prepare a mixture *B* which contains equal amounts by weight of the compounds (I), (II), and (III).
2. Set the chromatograph oven to 75 °C and the carrier gas (pure nitrogen) flow rate to 40–45 mL min^{-1}.
3. When the oven temperature has stabilised, inject a 0.3 μL sample of mixture *B* and decide from the peak areas whether the detector response is the same for each component.
4. If the detector response differs, make up by weight a 1:1 mixture of each of the separate components (I, II, and III) with compound (IV). Inject a 0.1 μL sample of each mixture, measure the corresponding peak area, and hence deduce the factors which will correct the peak areas of components (I), (II), and (III) with respect to the internal standard (IV).
5. Prepare a mixture, by weight, of *A* with compound (IV). Inject a 0.3 μL sample of this mixture, measure the various peak areas and, after making appropriate corrections for differences in detector sensitivity, determine the percentage composition of A.

9.8 DETERMINATION OF SUCROSE AS ITS TRIMETHYLSILYL DERIVATIVE USING GAS–LIQUID CHROMATOGRAPHY

The purpose of the experiment is to illustrate the application of derivatisation in the analysis of sugar and related substances by gas–liquid chromatography. The silylation method described is an almost universal derivatisation procedure for carbohydrate analysis by GC.[79]

Reagents and apparatus. The reagents and solvents should be pure and dry, and should be tested in advance in the gas chromatographic system which is to be used in the experiment.
Pyridine. Purify by refluxing over potassium hydroxide, followed by distillation. Store the purified pyridine over the same reagent.
Trimethylchlorosilane (TMCS), (CH$_3$)$_3$SiCl.
Hexamethyldisilazane (HMDS), (CH$_3$)$_3$Si—NH—Si(CH$_3$)$_3$.
Reaction vessel. Use a small tube or vial fitted with a Teflon-lined screw cap.
Gas chromatograph. Operate the column isothermally at 210 °C using a flame ionisation detector.

Procedure. Treat 10 mg of sucrose with 1 mL of anhydrous pyridine, 0.2 mL of HMDS, and 0.1 mL of TMCS in the plastic-stoppered vial (or similar container). Shake the mixture vigorously for about 30 seconds and allow it to stand for 10 min at room temperature; if the carbohydrate appears to remain insoluble in the reaction mixture, warm the vial for 2–3 min at 75–85 °C. Inject 0.3 μL of the resulting mixture into the gas chromatograph.

Note. Anhydrous reaction conditions are generally essential since the silylated derivatives are sensitive to water in varying degrees.

9.9 REFERENCES FOR PART C

1. H Irving and R J P Williams, Metal complexes and partition equilibria, *J. Chem. Soc.*, 1949, 1841.
2. H Akaiwa and H Kawamoto, *Anal. Chim. Acta*, 1968, **40**, 407
3. K S Math, K S Bhatki and H Freiser, *Talanta*, 1969, **16**, 412
4. J Hála, *J. Radioanal. Chem.*, 1979, **51**, 15
5. K J Doolan and L E Smythe, *Talanta*, 1973, **20**, 241
6. D Thorburn Burns, *Anal. Proc.*, 1982, **19**, 355
7. C G Taylor and B Fryer, *Analyst*, 1969, **94**, 1106
8. R M Dagnall and T S West, *Talanta*, 1964, **11**, 1627
9. A K Babko, *Talanta*, 1968, **15**, 721
10. M Jawaid and F Ingram, *Talanta*, 1978, **25**, 91
11. K L Cheng, K Ueno and T Imamura, *Handbook of Organic Analytical Reagents*, CRC Press, Boca Raton, FL, 1982
12. T Kamada, T Shiraishi and Y Yamamoto, *Talanta*, 1978, **25**, 15
13. M S Cresser, *Solvent Extraction in Flame Spectroscopic Analysis*, Butterworth, London, 1978
14. E B Edward-Inatimi and J A W Dalziel, *Anal. Proc.*, 1980, **17**, 40
15. P D Goulden, P Brooksbank and J F Ryan, *Internat. Lab.*, Sept/Oct 1973, 31
16. J Bassett and P J Matthews, *Analyst*, 1974, **99**, 1
17. A Clearfield (Ed), *Inorganic Ion Exchange Materials*, CRC Press, Boca Raton, FL, 1982
18. R Kunin, E F Meitzner and N Bortnick, *J. Am. Chem. Soc.*, 1962, **84**, 305
19. F M Rabel, *Adv. Chromatogr.*, 1979, **17**, 53
20. R E Majors, *J. Chromatogr. Sci.*, 1977, **15**, 334
21. G J Moody and J D R Thomas, *Lab. Practice*, 1970, **19**, 487
22. L R Snyder, *Chromatogr. Rev.*, 1965, **7**, 1
23. A M C Davies, E H A Prescott and R Stansfield, *J. Chromatogr.*, 1979, **171**, 117
24. J Inczédy, *Analytical Applications of Complex Equilibria*, Ellis Horwood, Chichester, 1976
25. H Small, T S Stevens and W C Baumann, *Anal. Chem.*, 1975, **47**, 47
26. C A Pohl and E L Johnson, *J. Chromatogr. Sci.*, 1980, **18**, 442
27. T S Stevens and M A Langhorst, *Anal. Chem.*, 1982, **54**, 950
28. D T Gjerde, J S Fritz and G Schmuckler, *J. Chromatogr.*, 1979, **186**, 509
29. H Small and T E Miller, *Anal. Chem.*, 1982, **54**, 462
30. A T Rhys Williams, *Fluorescence Detection in Liquid Chromatography*, Perkin–Elmer, Beaconsfield, 1980
31. R P W Scott, *Liquid Chromatography Detectors* Elsevier, Amsterdam, 1979
32. W R Heumann, *CRC Crit. Rev. Anal. Chem.*, 1971, **2**, 425
33. G J Moody and J D R Thomas, *Analyst* 1968, **93**, 557
34. J Korkisch, *Separation Science*, 1966, **1**, 159
35. H P Gregor *et al.*, *Ind. Eng. Chem.*, 1952, **44**, 2834
36. E Blasius and B Brozio, Chelating ion-exchange resins. In *Chelates in Analytical Chemistry*, H A Flaschka and A J Barnard (Eds), Vol. 1, Marcel Dekker, New York, 1967, p 49
37. G Schmuckler, Chelating resins — Their analytical properties and applications *Talanta*, 1965, **12**, 281
38. C M de Hernandez and H F Walton, *Anal. Chem.*, 1972, **44**, 890

39. A Siegel and E T Degens, *Science*, 1966, **151**, 1098
40. V A Davankov, *Adv. Chromatogr.*, 1980, **18**, 139
41. C F Coleman *et al.*, *Talanta*, 1962, **9**, 297
42. H Green, Recent uses of liquid ion exchanges in inorganic analysis *Talanta*, 1964, **11**, 1561
43. H Green, Use of liquid ion-exchanges in inorganic analysis *Talanta*, 1973, **20**, 139
44. R Kunin and A G Winger, *Chem. Ing. Tech.*, 1962, **34**, 461
45. J P Riley and D Taylor, *Anal. Chim. Acta*, 1968, **40**, 479
46. L R Snyder, *Principles of Adsorption Chromatography*, Marcel Dekker, New York, 1968
47. B L Karger, S C Su, S Marchese and B A Persson, *J. Chromatogr. Sci.*, 1974, **12**, 678
48. R E Majors, *Am. Lab.*, 1975, **7**, 13
49. D M W Anderson, I C M Dea and A Hendrie, *Sel. Ann. Rev. Anal. Sci.*, 1971, **1**, 1
50. J C Berridge, *Anal. Proc.*, 1985, **22**, 323
51. J F K Huber (Ed), *Instrumentation for High Performance Liquid Chromatography*, Elsevier, Amsterdam, 1978
52. P A Bristow, *Anal. Chem.*, 1976, **48**, 237
53. R P W Scott, *Contemporary Liquid Chromatography*, Wiley, New York, 1976
54. J H Knox, *J. Chromatogr. Sci.*, 1977, **15**, 353
55. R P W Scott, *Liquid Chromatography Detectors*, 2nd edn, Elsevier, Amsterdam, 1986
56. J F Lawrence and R W Frei, *Chemical Derivatisation in Liquid Chromatography*, Elsevier, Amsterdam, 1976
57. J C Touchstone and J Sherma (Eds), *Densitometry in Thin-Layer Chromatography — Practice and Applications*, Wiley–Interscience, New York, 1979
58. W E Court, Quantitative thin-layer chromatography using elution techniques. In *Quantitative Paper and Thin-Layer Chromatography* E J Shellard (Ed) Academic Press, New York, 1968, p 29
59. A Zlatkis and R E Kaiser (Eds), *HPTLC: High Performance Thin-Layer Chromatography*, Elsevier, Amsterdam, 1977
60. R C Denney, Silylation reagents for chromatography *Speciality Chemicals*, 1983, **3**, 6–7, 12
61. R S Barratt, Analytical applications of gas chromatography of metal chelates *Proc. Soc. Anal. Chem.*, 1973, **10**, 167
62. K Blau and G S King (Eds), *Handbook of Derivatives for Chromatography*, London, Heyden, 1977
63. D Knapp (Ed), *Handbook of Analytical Derivatisation Reactions*, Wiley, New York, 1979
64. A Zlatkis *et al.*, *Anal. Chem.*, 1973, **45**, 763
65. W G Jennings, *Gas Chromatography with Glass Capillary Columns*, Academic Press, New York, 1980
66. M L Lee, F J Yang and K D Bartle, *Open Tubular Column Gas Chromatography, Theory and Practice*, Wiley, New York, 1984
67. D J David, *Gas Chromatographic Detectors*, Wiley, New York, 1974
68. J Sevcik, *Detectors in Gas Chromatography*, Elsevier, Amsterdam, 1976
69. E R Adlard, *CRC Crit. Rev. Anal. Chem.*, 1975, **5**, 1, 13
70. I I G McWilliam, *Chromatographia*, 1983, **17**, 241
71. A Zlatkis and C F Poole (Eds), *Electron Capture. Theory and Practice in Chromatography*, Elsevier, Amsterdam, 1981
72. R C Hall, *CRC Crit. Rev. Anal. Chem.*, 1978, **8**, 323
73. S S Brody and J E Chaney, *J. Gas Chromatogr.*, 1966, **4**, 42
74. D F S Natusch and T M Thorpe, Element selective detectors in gas chromatography, *Anal. Chem.*, 1973, **45**, 1184A

75. P R Ballinger and I M Whittemore, *Proc. Am. Chem. Soc. Div. Petroleum Chem.*, 1968, **13**, 133
76. T Hirschfield, *Anal. Chem.*, 1980, **52**, 297A
77. W E Harris and H W Habgood, *Programmed Temperature Gas Chromatography*, Wiley, New York, 1966
78. R D Hill and H Gesser, *J. Gas Chromatogr.*, 1963, **1**, 11
79. C C Sweeley *et al.*, *J. Am. Chem. Soc.*, 1963, **85**, 2497

9.10 SELECTED BIBLIOGRAPHY FOR PART C

1. A K De, S M Khopkar and R A Chalmers, *Solvent Extraction of Metals*, Van Nostrand Reinhold, London, 1970
2. J Stary, *The Solvent Extraction of Metal Chelates*, Pergamon Press, Oxford, 1964
3. Y. Marcus (Ed), *Solvent Extraction Reviews*, Vol. 1, Marcel Dekker, New York, 1971
4. H Irving and R J P Williams, Liquid–liquid extraction. In *Treatise on Analytical Chemistry*, Part 1, Vol 3, I Kolthoff and P Elving (Eds), Interscience, New York, 1961
5. R M Diamond and D G Tuck, Extraction of inorganic compounds into organic solvents. In *Progress in Inorganic Chemistry*, Vol 2, F A Cotton (Ed), Interscience, New York, 1960
6. Yu A Zolotov (trans. J Schmorak), *Extraction of Chelate Compounds*, Ann Arbor Science Publishers, Ann Arbor, MI, 1970
7. H Freiser, Some recent developments in solvent extraction *Crit. Rev. Anal. Chem.*, 1970, **1**, 47
8. H Freiser, Solvent extraction in analytical chemistry and separation science *Bunseki Kagaku*, 1981, **30**, S47–S57
9. R Paterson, *An Introduction to Ion Exchange*, Heydon–Sadtler, London, 1970
10. O Samuelson, *Ion Exchanger Separations in Analytical Chemistry*, John Wiley, New York, 1962
11. Anon., *Ion Exchange Resins*, 6th edn, British Drug Houses Ltd, Poole, Dorset, UK, 1981
12. J A Marinsky and Y Marcus (Eds), *Ion Exchange and Solvent Extraction — A series of Advances*, Marcel Dekker, New York, 1973
13. F Helfferich, *Ion Exchange*, McGraw-Hill, New York, 1962
14. M Qureshi *et al.*, Recent progress in ion-exchange studies on insoluble salts of polybasic metals *Separation Sci.* 1972, **7**, 615
15. W Riemann and H F Walton, *Ion Exchange in Analytical Chemistry*, Pergamon Press, Oxford, 1970
16. J Inczédy, Use of ion exchangers in analytical chemistry, *Rev. Anal. Chem.*, 1972, **1**, 157
17. J Inczédy, *Analytical Applications of Ion Exchangers*, Pergamon Press, Oxford, 1966
18. F C Smith, Jr, and R C Chang, *The Practice of Ion Chromatography*, John Wiley, New York, 1983
19. E Sawicki, J D Mulik and E Wittgenstein (Eds), *Ion Chromatographic Analysis of Environmental Pollutants*, Vol. 1, Ann Arbor Science, Ann Arbor, MI, 1978
20. R J Hamilton and P A Sewell, *Introduction to High Performance Liquid Chromatography*, Chapman and Hall, London, 1982
21. J H Knox (Ed), *High Performance Liquid Chromatography*, Edinburgh University Press, Edinburgh, 1982
22. J J Kirkland and L R Snyder, *Introduction to Modern Liquid Chromatography*, 2nd edn, John Wiley, New York, 1979
23. C F Simpson (Ed), *Techniques in Liquid Chromatography*, John Wiley, New York, 1982
24. E Stahl and H Jork, *Thin-Layer Chromatography: a Laboratory Handbook*, Springer-Verlag, New York, 1969

25. J C Touchstone and M F Dobbins, *Practice of Thin-Layer Chromatography*, John Wiley, New York, 1978
26. J C Touchstone (Ed), *Quantitative Thin-Layer Chromatography*, Wiley, New York, 1973
27. R C Denney, *A Dictionary of Chromatography*, 2nd edn, Macmillan, London 1982
28. J B Pattison, *A Programmed Introduction to Gas–Liquid Chromatography*, 2nd edn, Heyden, London, 1973
29. L S Ettre and A Zlatkis (Eds), *The Practice of Gas Chromatography*, Wiley, New York, 1967
30. R W Moshier and R E Sievers, *Gas Chromatography of Metal Chelates*, Pergamon Press, Oxford, 1965
31. R L Grob (Ed), *Modern Practice of Gas Chromatography*, 2nd edn, Wiley, New York, 1985
32. C F Poole and S A Schuette, *Contemporary Practice of Chromatography*, Elsevier, Amsterdam, 1984
33. J Novák, *Quantitative Analysis by Gas Chromatography*, Marcel Dekker, New York, 1975

PART D
TITRIMETRY AND GRAVIMETRY

CHAPTER 10
TITRIMETRIC ANALYSIS

THEORETICAL CONSIDERATIONS

10.1 TITRIMETRIC ANALYSIS

The term 'titrimetric analysis' refers to quantitative chemical analysis carried out by determining the volume of a solution of accurately known concentration which is required to react quantitatively with a measured volume of a solution of the substance to be determined. The solution of accurately known strength is called the **standard solution**, see Section 10.3. The weight of the substance to be determined is calculated from the volume of the standard solution used and the chemical equation and relative molecular masses of the reacting compounds.

The term 'volumetric analysis' was formerly used for this form of quantitative determination but it has now been replaced by **titrimetric analysis**. It is considered that the latter expresses the process of titration rather better, and the former is likely to be confused with measurements of volumes, such as those involving gases. In titrimetric analysis the reagent of known concentration is called the **titrant** and the substance being titrated is termed the **titrand**. The alternative name has not been extended to apparatus used in the various operations; so the terms volumetric glassware and volumetric flasks are still common, but it is better to employ the expressions graduated glassware and graduated flasks and these are used throughout this book.

The standard solution is usually added from a long graduated tube called a burette. The process of adding the standard solution until the reaction is just complete is termed a **titration**, and the substance to be determined is **titrated**. The point at which this occurs is called the **equivalence point** or the **theoretical** (or **stoichiometric**) **end point**. The completion of the titration is detected by some physical change, produced by the standard solution itself (e.g. the faint pink colour formed by potassium permanganate) or, more usually, by the addition of an auxiliary reagent, known as an indicator; alternatively some other physical measurement may be used. After the reaction between the substance and the standard solution is practically complete, the indicator should give a clear visual change (either a colour change or the formation of turbidity) in the liquid being titrated. The point at which this occurs is called the **end point of the titration**. In the ideal titration the visible end point will coincide with the stoichiometric or theoretical end point. In practice, however, a very small difference usually occurs; this represents the titration error. The indicator and the experimental conditions should be so selected that the difference between the visible end point and the equivalence point is as small as possible.

For use in titrimetric analysis a reaction must fulfil the following conditions.

1. There must be a simple reaction which can be expressed by a chemical equation; the substance to be determined should react completely with the reagent in stoichiometric or equivalent proportions.
2. The reaction should be relatively fast. (Most ionic reactions satisfy this condition.) In some cases the addition of a catalyst may be necessary to increase the speed of a reaction.
3. There must be an alteration in some physical or chemical property of the solution at the equivalence point.
4. An indicator should be available which, by a change in physical properties (colour or formation of a precipitate), should sharply define the end point of the reaction. [If no visible indicator is available, the detection of the equivalence point can often be achieved by following the course of the titration by measuring (a) the potential between an indicator electrode and a reference electrode (**potentiometric titration**, see Chapter 15); (b) the change in electrical conductivity of the solution (**conductimetric titration**, see Chapter 13); (c) the current which passes through the titration cell between an indicator electrode and a depolarised reference electrode at a suitable applied e.m.f. (**amperometric titration**, see Chapter 16); or (d) the change in absorbance of the solution (**spectrophotometric titration**, see Section 17.48).]

Titrimetric methods are normally capable of high precision (1 part in 1000) and wherever applicable possess obvious advantages over gravimetric methods. They need simpler apparatus, and are, generally, quickly performed; tedious and difficult separations can often be avoided. The following apparatus is required for titrimetric analysis: (i) calibrated measuring vessels, including burettes, pipettes, and measuring flasks (see Chapter 3); (ii) substances of known purity for the preparation of standard solutions; (iii) a visual indicator or an instrumental method for detecting the completion of the reaction.

10.2 CLASSIFICATION OF REACTIONS IN TITRIMETRIC ANALYSIS

The reactions employed in titrimetric analysis fall into four main classes. The first three of these involve no change in oxidation state as they are dependent upon the combination of ions. But the fourth class, oxidation–reduction reactions, involves a change of oxidation state or, expressed another way, a transfer of electrons.

1. Neutralisation reactions, or acidimetry and alkalimetry. These include the titration of free bases, or those formed from salts of weak acids by hydrolysis, with a standard acid (**acidimetry**), and the titration of free acids, or those formed by the hydrolysis of salts of weak bases, with a standard base (**alkalimetry**). The reactions involve the combination of hydrogen and hydroxide ions to form water.

Also under this heading must be included titrations in non-aqueous solvents, most of which involve organic compounds.

2. Complex formation reactions. These depend upon the combination of ions, other than hydrogen or hydroxide ions, to form a soluble, slightly dissociated ion or compound, as in the titration of a solution of a cyanide with silver nitrate

$(2CN^- + Ag^+ \rightleftharpoons [Ag(CN)_2]^-)$ or of chloride ion with mercury(II) nitrate solution $(2Cl^- + Hg^{2+} \rightleftharpoons HgCl_2)$.

Ethylenediaminetetra-acetic acid, largely as the disodium salt of EDTA, is a very important reagent for complex formation titrations and has become one of the most important reagents used in titrimetric analysis. Equivalence point detection by the use of metal-ion indicators has greatly enhanced its value in titrimetry.

3. Precipitation reactions. These depend upon the combination of ions to form a simple precipitate as in the titration of silver ion with a solution of a chloride (Section 10.74). No change in oxidation state occurs.

4. Oxidation–reduction reactions. Under this heading are included all reactions involving change of oxidation number or transfer of electrons among the reacting substances. The standard solutions are either oxidising or reducing agents. The principal oxidising agents are potassium permanganate, potassium dichromate, cerium(IV) sulphate, iodine, potassium iodate, and potassium bromate. Frequently used reducing agents are iron(II) and tin(II) compounds, sodium thiosulphate, arsenic(III) oxide, mercury(I) nitrate, vanadium(II) chloride or sulphate, chromium(II) chloride or sulphate, and titanium(III) chloride or sulphate.

10.3 STANDARD SOLUTIONS

The word 'concentration' is frequently used as a general term referring to a quantity of substance in a defined volume of solution. But for quantitative titrimetric analysis use is made of standard solutions in which the base unit of quantity employed is the mole. This follows the definition given by the International Union of Pure and Applied Chemistry[1] in which:

> 'The mole is the amount of substance which contains as many elementary units as there are atoms in 0.012 kilogram of carbon-12. The elementary unit must be specified and may be an atom, a molecule, an ion, a radical, an electron or other particle or a specified group of such particles.'

As a result standard solutions are now commonly expressed in terms of molar concentrations or molarity (M). Such standard solutions are specified in terms of the number of moles of solute dissolved in 1 litre of solution; for any solution,

$$\text{Molarity } (M) = \frac{\text{Moles of solute}}{\text{Volume of solution in litres}}$$

As the term 'mole' refers to an amount of substance with reference to the specified mass of carbon-12, it is possible to express the relative molecular mass (the basis for the mole) for any substance as the additive sum of the relative atomic masses (R.A.M.s) of its component elements, for example:

The relative molecular mass for sulphuric acid, H_2SO_4, is calculated from the relative atomic masses as follows:

259

Element	R.A.M.
Hydrogen	$1.0079 \times 2 = 2.0158$
Sulphur	$32.06 \quad \times 1 = 32.06$
Oxygen	$15.9994 \times 4 = 63.9986$
Relative Molecular Mass	$= 98.0744$

This approach can be used to obtain the R.A.M. of any compound, so that

1 mole of Hg_2Cl_2 has a mass of 0.472 09 kg
1 mole of $Na_2CO_3, 10H_2O$ has a mass of 0.286 141 kg
1 mole of H_2SO_4 has a mass of 0.098 074 kg

It follows from this, that a molar solution of sulphuric acid will contain 98.074 grams of sulphuric acid in 1 litre of solution, or 49.037 grams in 500 mL of solution. Similarly, a $0.1 M$ solution will contain 9.8074 grams of sulphuric acid in 1 litre of solution, and a $0.01 M$ solution will have 0.980 74 gram in the same volume. So that the concentration of any solution can be expressed in terms of the molar concentration so long as the weight of substance in any specified volume is known.

10.4 EQUIVALENTS, NORMALITIES AND OXIDATION NUMBERS

Although molar concentrations are now commonly used in determinations of reacting quantities in titrimetric analysis, it has been traditional to employ other concepts involving what are known as 'equivalent weights' and 'normalities' for this purpose. In neutralisation reactions the equivalent weight/normality concept is relatively straightforward, but for reduction–oxidation titrations it often requires an understanding of what are known as 'oxidation numbers' of the substances involved in the redox reaction. Although the modern approach is to discard this form of calculation and quantitation, the authors of this book fully appreciate that there are many scientists who do prefer to use it, and some who claim it has clear advantages over the molar concept. Because of this, a full explanation of this approach to titrimetry is retained as Appendix 17 but all other quantitative aspects in this book are in terms of moles per litre.

10.5 PREPARATION OF STANDARD SOLUTIONS

If a reagent is available in the pure state, a solution of definite molar strength is prepared simply by weighing out a mole, or a definite fraction or multiple thereof, dissolving it in an appropriate solvent, usually water, and making up the solution to a known volume. It is not essential to weigh out exactly a mole (or a multiple or sub-multiple thereof); in practice it is more convenient to prepare the solution a *little* more concentrated than is ultimately required, and then to dilute it with distilled water until the desired molar strength is obtained. If M_1 is the required molarity, V_1 the volume after dilution, M_2 the molarity originally obtained, and V_2 the original volume taken, $M_1 V_1 = M_2 V_2$, or $V_1 = M_2 V_2 / M_1$. The volume of water to be added to the volume V_2 is $(V_1 - V_2)$ mL.

The following is a list of some of the substances which can be obtained in a state of high purity and are therefore suitable for the preparation of standard solutions: sodium carbonate, potassium hydrogenphthalate, benzoic acid, sodium tetraborate, sulphamic acid, potassium hydrogeniodate, sodium oxalate, silver, silver nitrate, sodium chloride, potassium chloride, iodine, potassium bromate, potassium iodate, potassium dichromate, lead nitrate and arsenic(III) oxide.

When the reagent is not available in the pure form as in the cases of most alkali hydroxides, some inorganic acids and various deliquescent substances, solutions corresponding approximately to the molar strength required are first prepared. These are then standardised by titration against a solution of a pure substance of known concentration. It is generally best to standardise a solution by a reaction of the same type as that for which the solution is to be employed, and as nearly as possible under identical experimental conditions. The titration error and other errors are thus considerably reduced or are made to cancel out. This indirect method is employed for the preparation of, for instance, solutions of most acids (the constant boiling point mixture of definite composition of hydrochloric acid can be weighed out directly, if desired), sodium hydroxide, potassium hydroxide and barium hydroxide, potassium permanganate, ammonium and potassium thiocyanates, and sodium thiosulphate.

10.6 PRIMARY AND SECONDARY STANDARDS

In titrimetry certain chemicals are used frequently in defined concentrations as reference solutions. Such substances are referred to as **primary standards** or **secondary standards**. A primary standard is a compound of sufficient purity from which a standard solution can be prepared by direct weighing of a quantity of it, followed by dilution to give a defined volume of solution. The solution produced is then a primary standard solution. A primary standard should satisfy the following requirements.

1. It must be easy to obtain, to purify, to dry (preferably at 110–120 °C), and to preserve in a pure state. (This requirement is not usually met by hydrated substances, since it is difficult to remove surface moisture completely without effecting partial decomposition.)
2. The substance should be unaltered in air during weighing; this condition implies that it should not be hygroscopic, oxidised by air, or affected by carbon dioxide. The standard should maintain an unchanged composition during storage.
3. The substance should be capable of being tested for impurities by qualitative and other tests of known sensitivity. (The total amount of impurities should not, in general, exceed 0.01–0.02 per cent.)
4. It should have a high relative molecular mass so that the weighing errors may be negligible. (The precision in weighing is ordinarily 0.1–0.2 mg; for an accuracy of 1 part in 1000, it is necessary to employ samples weighing at least about 0.2 g.)
5. The substance should be readily soluble under the conditions in which it is employed.
6. The reaction with the standard solution should be stoichiometric and practically instantaneous. The titration error should be negligible, or easy to determine accurately by experiment.

261

In practice, an ideal primary standard is difficult to obtain, and a compromise between the above ideal requirements is usually necessary. The substances commonly employed as primary standards are indicated below:

(*a*) **Acid-base reactions** — sodium carbonate Na_2CO_3, sodium tetraborate $Na_2B_4O_7$, potassium hydrogenphthalate $KH(C_8H_4O_4)$, constant boiling point hydrochloric acid, potassium hydrogeniodate $KH(IO_3)_2$, benzoic acid (C_6H_5COOH).

(*b*) **Complex formation reactions** — silver, silver nitrate, sodium chloride, various metals (e.g. spectroscopically pure zinc, magnesium, copper, and manganese) and salts, depending upon the reaction used.

(*c*) **Precipitation reactions** — silver, silver nitrate, sodium chloride, potassium chloride, and potassium bromide (prepared from potassium bromate).

(*d*) **Oxidation–reduction reactions** — potassium dichromate $K_2Cr_2O_7$, potassium bromate $KBrO_3$, potassium iodate KIO_3, potassium hydrogen-iodate $KH(IO_3)_2$, sodium oxalate $Na_2C_2O_4$, arsenic(III) oxide As_2O_3, and pure iron.

Hydrated salts, as a rule, do not make good standards because of the difficulty of efficient drying. However, those salts which do not effloresce, such as sodium tetraborate $Na_2B_4O_7, 10H_2O$, and copper sulphate $CuSO_4, 5H_2O$, are found by experiment to be satisfactory secondary standards.[2]

A secondary standard is a substance which may be used for standardisations, and whose content of the active substance has been found by comparison against a primary standard. It follows that a secondary standard solution is a solution in which the concentration of dissolved solute has not been determined from the weight of the compound dissolved but by reaction (titration) of a volume of the solution against a measured volume of a primary standard solution.

NEUTRALISATION TITRATIONS

10.7 NEUTRALISATION INDICATORS

The object of titrating, say, an alkaline solution with a standard solution of an acid is the determination of the amount of acid which is exactly equivalent chemically to the amount of base present. The point at which this is reached is the **equivalence point, stoichiometric point**, or **theoretical end point**; the resulting aqueous solution contains the corresponding salt. If both the acid and base are strong electrolytes, the solution at the end-point will be neutral and have a pH of 7 (Section 2.17); but if either the acid or the base is a weak electrolyte, the salt will be hydrolysed to a certain degree, and the solution at the equivalence point will be either slightly alkaline or slightly acid. The exact pH of the solution at the equivalence point can readily be calculated from the ionisation constant of the weak acid or the weak base and the concentration of the solution (see Section 2.19). For any actual titration the correct end-point will be characterised by a definite value of the hydrogen-ion concentration of the solution, the value depending upon the nature of the acid and the base and the concentration of the solution.

A large number of substances, called **neutralisation** or **acid–base indicators**, change colour according to the hydrogen-ion concentration of the solution. The chief characteristic of these indicators is that the change from a predominantly 'acid' colour to a predominantly 'alkaline' colour is not sudden and abrupt, but takes place within a small interval of pH (usually about two pH units) termed the **colour-change interval** of the indicator. The position of the colour-change interval in the pH scale varies widely with different indicators. For most acid–base titrations it is possible to select an indicator which exhibits a distinct colour change at a pH close to that corresponding to the equivalence point.

The first useful theory of indicator action was suggested by W. Ostwald[3] based upon the concept that indicators in general use are very weak organic acids or bases.

The simple Ostwald theory of the colour change of indicators has been revised, and the colour changes are believed to be due to structural changes, including the production of quinonoid and resonance forms; these may be illustrated by reference to phenolphthalein, the changes of which are characteristic of all phthalein indicators: see the formulae I–IV given below. In the presence of dilute alkali the lactone ring in I opens to yield II, and the triphenylcarbinol structure (II) undergoes loss of water to produce the resonating ion III which is red. If phenolphthalein is treated with excess of concentrated alcoholic alkali the red colour first produced disappears owing to the formation of IV.

The Brønsted–Lowry concept of acids and bases[4] makes it unnecessary to distinguish between acid and base indicators: emphasis is placed upon the charge types of the acid and alkaline forms of the indicator. The equilibrium between the acidic form In_A and the basic form In_B may be expressed as:

$$In_A \rightleftharpoons H^+ + In_B \tag{1}$$

and the equilibrium constant as:

$$\frac{a_{H^+} \times a_{In_B}}{a_{In_A}} = K_{In} \tag{2}$$

The observed colour of an indicator in solution is determined by the ratio of the concentrations of the acidic and basic forms. This is given by:

$$\frac{[In_A]}{[In_B]} = \frac{a_{H^+} \times y_{In_B}}{K_{In} \times y_{In_A}} \tag{3}$$

263

where y_{In_A} and y_{In_B} are the activity coefficients of the acidic and basic forms of the indicator. Equation (3) may be written in the logarithmic form:

$$pH = -\log a_{H^+} = pK_{In} + \log\frac{[In_B]}{[In_A]} + \log\frac{y_{In_B}}{y_{In_A}} \qquad (4)$$

The pH will depend upon the ionic strength of the solution (which is, of course, related to the activity coefficient — see Section 2.5). Hence, when making a colour comparison for the determination of the pH of a solution, not only must the indicator concentration be the same in the two solutions but the ionic strength must also be equal or approximately equal. The equation incidentally provides an explanation of the so-called salt and solvent effects which are observed with indicators. The colour-change equilibrium at any particular ionic strength (constant activity-coefficient term) can be expressed by a condensed form of equation (4):

$$pH = pK'_{In} + \log\frac{[In_B]}{[In_A]} \qquad (5)$$

where pK'_{In} is termed the **apparent indicator constant**.

The value of the ratio $[In_B]/[In_A]$ (i.e. [Basic form]/[Acidic form]) can be determined by a visual colour comparison or, more accurately, by a spectrophotometric method. Both forms of the indicator are present at any hydrogen-ion concentration. It must be realised, however, that the human eye has a limited ability to detect either of two colours when one of them predominates. Experience shows that the solution will appear to have the 'acid' colour, i.e. of In_A, when the ratio of $[In_A]$ to $[In_B]$ is above approximately 10, and the 'alkaline' colour, i.e. of In_B, when the ratio of $[In_B]$ to $[In_A]$ is above approximately 10. Thus only the 'acid' colour will be visible when $[In_A]/[In_B] > 10$; the corresponding limit of pH given by equation (5) is:

$$pH = pK'_{In} - 1$$

Only the alkaline colour will be visible when $[In_B]/[In_A] > 10$, and the corresponding limit of pH is:

$$pH = pK'_{In} + 1$$

The colour-change interval is accordingly $pH = pK'_{In} \pm 1$, i.e. over approximately two pH units. Within this range the indicator will appear to change from one colour to the other. The change will be gradual, since it depends upon the ratio of the concentrations of the two coloured forms (acidic form and basic form). When the pH of the solution is equal to the apparent dissociation constant of the indicator pK'_{In}, the ratio $[In_A]$ to $[In_B]$ becomes equal to 1, and the indicator will have a colour due to an equal mixture of the 'acid' and 'alkaline' forms. This is sometimes known as the 'middle tint' of the indicator. This applies strictly only if the two colours are of equal intensity. If one form is more intensely coloured than the other or if the eye is more sensitive to one colour than the other, then the middle tint will be slightly displaced along the pH range of the indicator.

Table 10.1 contains a list of indicators suitable for titrimetric analysis and for the colorimetric determination of pH. The colour-change intervals of most of the various indicators listed in the table are represented graphically in Fig. 10.1.

264

Table 10.1 Colour changes and pH range of certain indicators

Indicator	Chemical name	pH range	Colour in acid solution	Colour in alkaline solution	pK'_{In}
Brilliant cresyl blue (acid)	Aminodiethylaminomethyldiphenazonium chloride	0.0–1.0	Red–orange	Blue	—
Cresol red (acid)	1-Cresolsulphonphthalein	0.2–1.8	Red	Yellow	—
m-Cresol purple	m-Cresolsulphonphthalein	0.5–2.5	Red	Yellow	—
Quinaldine red	1-(p-Dimethylaminophenylethylene)quinoline ethiodide	1.4–3.2	Colourless	Red	1.7
Thymol blue (acid)	Thymolsulphonphthalein	1.2–2.8	Red	Yellow	—
Tropaeolin OO	p-Anilinophenylazobenzenesulphonic acid sodium salt	1.3–2.8	Red	Yellow	—
Bromophenol blue	Tetrabromophenolsulphonphthalein	2.8–4.6	Yellow	Blue	4.1
Ethyl orange	Dimethylaminophenylazobenzenesulphonic acid sodium salt	3.0–4.5	Red	Orange	—
Methyl orange		2.9–4.6	Red	Orange	3.7
Congo red	Diphenyldiazobis-1-naphthylaminesulphonic acid disodium salt	3.0–5.0	Blue	Red	—
Bromocresol green	Tetrabromo-m-cresolsulphonphthalein	3.6–5.2	Yellow	Blue	4.7
Methyl red	1-Carboxybenzeneazodimethylaniline	4.2–6.3	Red	Yellow	5.0
Ethyl red		4.5–6.5	Red	Orange	—
Chlorophenol red	Dichlorophenolsulphonphthalein	4.6–7.0	Yellow	Red	6.1
4-Nitrophenol	4-Nitrophenol	5.0–7.0	Colourless	Yellow	7.1
Bromocresol purple	Dibromo-o-cresolsulphonphthalein	5.2–6.8	Yellow	Purple	6.1
Bromophenol red	Dibromophenolsulphonphthalein	5.2–7.0	Yellow	Red	—
Azolitmin (litmus)	—	5.0–8.0	Red	Blue	—
Bromothymol blue	Dibromothymolsulphonphthalein	6.0–7.6	Yellow	Blue	7.1
Neutral red	Aminodimethylaminotoluphenazonium chloride	6.8–8.0	Red	Orange	—
Phenol red	Phenolsulphonphthalein	6.8–8.4	Yellow	Red	7.8
Cresol red (base)	1-Cresolsulphonphthalein	7.2–8.8	Yellow	Red	8.2
1-Naphtholphthalein	1-Naphtholphthalein	7.3–8.7	Yellow	Blue	8.4
m-Cresol purple	m-Cresolsulphonphthalein	7.6–9.2	Yellow	Purple	—
Thymol blue (base)	Thymolsulphonphthalein	8.0–9.6	Yellow	Blue	8.9
o-Cresolphthalein	Di-o-cresolphthalide	8.2–9.8	Colourless	Red	—
Phenolphthalein	Phenolphthalein	8.3–10.0	Colourless	Red	9.6
Thymolphthalein	Thymolphthalein	9.3–10.5	Colourless	Blue	9.3
Alizarin yellow R	p-Nitrobenzeneazosalicylic acid	10.1–12.1	Yellow	Orange–red	—
Brilliant cresyl blue (base)	Aminodiethylaminomethyldiphenazonium chloride	10.8–12.0	Blue	Yellow	—
Tropaeolin O	p-Sulphobenzeneazoresorcinol	11.1–12.7	Yellow	Orange	—
Nitramine	2,4,6-Trinitrophenylmethylnitroamine	10.8–13.0	Colourless	Orange–brown	—

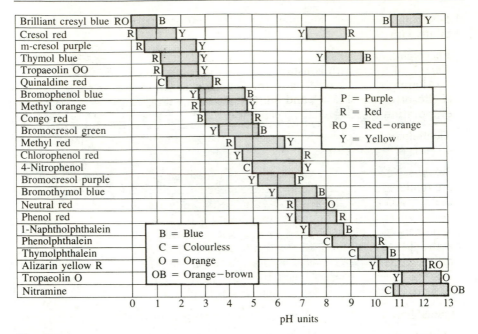

Fig. 10.1

It is necessary to draw attention to the variable pH of water which may be encountered in quantitative analysis. Water in equilibrium with the normal atmosphere which contains 0.03 per cent by volume of carbon dioxide has a pH of about 5.7; very carefully prepared conductivity water has a pH close to 7; water saturated with carbon dioxide under a pressure of one atmosphere has a pH of about 3.7 at 25 °C. The analyst may therefore be dealing, according to the conditions that prevail in the laboratory, with water having a pH between the two extremes pH 3.7 and pH 7. Hence for indicators which show their alkaline colours at pH values above 4.5, the effect of carbon dioxide introduced during a titration, either from the atmosphere or from the titrating solutions, must be seriously considered. This subject is discussed again later (Section 10.12).

10.8 PREPARATION OF INDICATOR SOLUTIONS

As a rule laboratory solutions of the indicators contain 0.5–1 g of indicator per litre of solvent. If the substance is soluble in water, e.g. a sodium salt, water is the solvent; in most other cases 70–90 per cent ethanol is employed.

Methyl orange. This indicator is available either as the free acid or as the sodium salt.

Dissolve 0.5 g of the free acid in 1 litre of water. Filter the cold solution to remove any precipitate which separates.

Dissolve 0.5 g of the sodium salt in 1 litre of water, add 15.2 mL of 0.1 M hydrochloric acid, and filter, if necessary, when cold.

Methyl red. Dissolve 1 g of the free acid in 1 litre of hot water, or dissolve in 600 mL of ethanol and dilute with 400 mL of water.

1-Naphtholphthalein. Dissolve 1 g of the indicator in 500 mL of ethanol and dilute with 500 mL of water.

Phenolphthalein. Dissolve 5 g of the reagent in 500 mL of ethanol and add 500 mL of water with constant stirring. Filter, if a precipitate forms.

Alternatively, dissolve 1 g of the dry indicator in 60 mL of 2-ethoxyethanol (Cellosolve), b.p. 135 °C, and dilute to 100 mL with distilled water: the loss by evaporation is less with this preparation.

Thymolphthalein. Dissolve 0.4 g of the reagent in 600 mL of ethanol and add 400 mL of water with stirring.

Sulphonphthaleins. These indicators are usually supplied in the acid form. They are rendered water-soluble by adding sufficient sodium hydroxide to neutralise the sulphonic acid group. One gram of the indicator is triturated in a clean glass mortar with the appropriate quantity of $0.1\,M$ sodium hydroxide solution, and then diluted with water to 1 L. The following volumes of $0.1\,M$ sodium hydroxide are required for 1 g of the indicators: bromophenol blue, 15.0 mL; bromocresol green, 14.4 mL; bromocresol purple, 18.6 mL; chlorophenol red, 23.6 mL; bromothymol blue, 16.0 mL; phenol red, 28.4 mL; thymol blue, 21.5 mL; cresol red, 26.2 mL; metacresol purple, 26.2 mL.

Quinaldine red. Dissolve 1 g in 100 mL of 80 per cent ethanol.

Methyl yellow, neutral red, and Congo red. Dissolve 1 g of the indicator in 1 L of 80 per cent ethanol. Congo red may also be dissolved in water.

4-Nitrophenol. Dissolve 2 g of the solid in 1 L of water.

Alizarin yellow R. Dissolve 0.5 g of the indicator in 1 L of 80 per cent ethanol.

Tropaeolin O and tropaeolin OO. Dissolve 1 g of the solid in 1 L of water.

Many of the indicator solutions are available from commercial suppliers already prepared for use.

10.9 MIXED INDICATORS

For some purposes it is desirable to have a sharp colour change over a narrow and selected range of pH; this is not easily seen with an ordinary acid–base indicator, since the colour change extends over two units of pH. The required result may, however, be achieved by the use of a suitable mixture of indicators; these are generally selected so that their pK'_{In} values are close together and the overlapping colours are complementary at an intermediate pH value. A few examples will be given in some detail.

(a) A mixture of equal parts of neutral red (0.1 per cent solution in ethanol) and methylene blue (0.1 per cent solution in ethanol) gives a sharp colour change from violet–blue to green in passing from acid to alkaline solution at pH 7. This indicator may be employed to titrate acetic acid (ethanoic acid) with ammonia solution or vice versa. Both acid and base are approximately of the same strength, hence the equivalence point will be at a pH ≈ 7 (Section 10.15); owing to the extensive hydrolysis and the flat nature of the titration curve, the titration cannot be performed except with an indicator of very narrow range.

267

(b) A mixture of phenolphthalein (3 parts of a 0.1 per cent solution in ethanol) and 1-naphtholphthalein (1 part of a 0.1 per cent solution in ethanol) passes from pale rose to violet at pH = 8.9. The mixed indicator is suitable for the titration of phosphoric acid to the diprotic stage ($K_2 = 6.3 \times 10^{-8}$; the equivalence point at pH ≈ 8.7).

(c) A mixture of thymol blue (3 parts of a 0.1 per cent aqueous solution of the sodium salt) and cresol red (1 part of a 0.1 per cent aqueous solution of the sodium salt) changes from yellow to violet at pH = 8.3. It has been recommended for the titration of carbonate to the hydrogencarbonate stage.

Other examples are included in Table 10.2.

Table 10.2 Some mixed indicators

Indicator mixture	pH	Colour change	Composition*
Bromocresol green; methyl orange	4.3	Orange → blue–green	1 p 0.1% (Na) in w; 1 p 0.2% in w
Bromocresol green; chlorophenol red	6.1	Pale green → blue violet	1 p 0.1% (Na) in w; 1 p 0.1% (Na) in w
Bromothymol blue; neutral red	7.2	Rose pink → green	1 p 0.1% in e; 1 p 0.1% in e
Bromothymol blue; phenol red	7.5	Yellow → violet	1 p 0.1% (Na) in w; 1 p 0.1% (Na) in w
Thymol blue; cresol red	8.3	Yellow → violet	3 p 0.1% (Na) in w; 1 p 0.1% (Na) in w
Thymol blue; phenolphthalein	9.0	Yellow → violet	1 p 0.1% in 50% e; 3 p 0.1% in 50% e
Thymolphthalein; phenolphthalein	9.9	Colourless → violet	1 p 0.1% in e; 1 p 0.1% in w

* Abbreviations: p = part, w = water, e = ethanol, Na = Na salt

The colour change of a single indicator may also be improved by the addition of a pH-sensitive dyestuff to produce the complement of one of the indicator colours. A typical example is the addition of xylene cyanol FF to methyl orange (1.0 g of methyl orange and 1.4 g of xylene cyanol FF in 500 mL of 50 per cent ethanol): here the colour change from the alkaline to the acid side is green → grey → magenta, the middle (grey) stage being at pH = 3.8. The above is an example of a **screened indicator**, and the mixed indicator solution is sometimes known as 'screened' methyl orange. Another example is the addition of methyl green (2 parts of a 0.1 per cent solution in ethanol) to phenolphthalein (1 part of a 0.1 per cent solution in ethanol); the former complements the red–violet basic colour of the latter, and at a pH of 8.4–8.8 the colour change is from grey to pale blue.

10.10 UNIVERSAL OR MULTIPLE-RANGE INDICATORS

By mixing suitable indicators together changes in colour may be obtained over a considerable portion of the pH range. Such mixtures are usually called '**universal indicators**'. They are not suitable for quantitative titrations, but may be employed for the determination of the approximate pH of a solution by the colorimetric method. One such universal indicator is prepared by dissolving 0.1 g of phenolphthalein, 0.2 g of methyl red, 0.3 g of methyl yellow, 0.4 g of bromothymol blue, and 0.5 g of thymol blue in 500 mL of absolute ethanol, and adding sodium hydroxide solution until the colour is yellow. The colour changes are: pH 2, red; pH 4, orange; pH 6, yellow; pH 8, green; pH 10, blue.

Another recipe for a universal indicator is as follows: 0.05 g of methyl orange, 0.15 g of methyl red, 0.3 g of bromothymol blue, and 0.35 g of phenolphthalein in 1 L of 66 per cent ethanol. The colour changes are: pH up to 3, red; pH 4, orange–red; pH 5, orange; pH 6, yellow; pH 7, yellowish–green; pH 8, greenish–blue; pH 9, blue; pH 10, violet; pH 11, reddish–violet. Several 'universal indicators' are available commercially as solutions and as test papers.

10.11 NEUTRALISATION CURVES

The mechanism of neutralisation processes can be understood by studying the changes in the hydrogen ion concentration during the course of the appropriate titration. The change in pH in the neighbourhood of the equivalence point is of the greatest importance, as it enables an indicator to be selected which will give the smallest titration error. The curve obtained by plotting pH as the ordinate against the percentage of acid neutralised (or the number of mL of alkali added) as abscissa is known as the neutralisation (or, more generally, the titration) curve. This may be evaluated experimentally by determination of the pH at various stages during the titration by a potentiometric method (Sections 15.15 and 15.20), or it may be calculated from theoretical principles.

10.12 NEUTRALISATION OF A STRONG ACID WITH A STRONG BASE

For this calculation it is assumed that both the acid and the base are completely dissociated and the activity coefficients of the ions are unity in order to obtain the pH values during the course of the neutralisation of the strong acid and the strong base, or vice versa, at the laboratory temperature. For simplicity of calculation consider the titration of 100 mL of $1M$ hydrochloric acid with $1M$ sodium hydroxide solution. The pH of $1M$ hydrochloric acid is 0. When 50 mL of the $1M$ base have been added, 50 mL of unneutralised $1M$ acid will be present in a total volume of 150 mL.

$[H^+]$ will therefore be $50 \times 1/150 = 3.33 \times 10^{-1}$, or pH $= 0.48$
For 75 mL of base, $[H^+] = 25 \times 1/175 = 1.43 \times 10^{-1}$, pH $= 0.84$
For 90 mL of base, $[H^+] = 10 \times 1/190 = 5.26 \times 10^{-2}$, pH $= 1.3$
For 98 mL of base, $[H^+] = 2 \times 1/198 = 1.01 \times 10^{-2}$, pH $= 2.0$
For 99 mL of base, $[H^+] = 1 \times 1/199 = 5.03 \times 10^{-3}$, pH $= 2.3$
For 99.9 mL of base, $[H^+] = 0.1 \times 1/199.9 = 5.00 \times 10^{-4}$, pH $= 3.3$

Upon the addition of 100 mL of base, the pH will change sharply to 7, i.e. the theoretical equivalence point. The resulting solution is simply one of sodium chloride. Any sodium hydroxide added beyond this will be in excess of that needed for neutralisation.

With 100.1 mL of base, $[OH^-] = 0.1/200.1 = 5.00 \times 10^{-4}$, pOH $= 3.3$ and pH $= 10.7$
With 101 mL of base, $[OH^-] = 1/201 = 5.00 \times 10^{-3}$, pOH $= 2.3$, and pH $= 11.7$

These results show that as the titration proceeds, initially the pH rises slowly, but between the addition of 99.9 and 100.1 mL of alkali, the pH of the solution

269

rises from 3.3 to 10.7, i.e. in the vicinity of the equivalence point the rate of change of pH of the solution is very rapid.

The complete results, up to the addition of 200 mL of alkali, are collected in Table 10.3; this also includes the figures for 0.1 M and 0.01 M solutions of acid and base respectively. The additions of alkali have been extended in all three cases to 200 mL; it is evident that the range from 200 to 100 mL and beyond represents the reverse titration of 100 mL of alkali with the acid in the presence of the non-hydrolysed sodium chloride solution. The data in the table are presented graphically in Fig. 10.2.

Table 10.3 pH during titration of 100 mL of HCl with NaOH of equal concentration

NaOH added (mL)	1M solution (pH)	0.1M solution (pH)	0.01M solution (pH)
0	0.0	1.0	2.0
50	0.5	1.5	2.5
75	0.8	1.8	2.8
90	1.3	2.3	3.3
98	2.0	3.0	4.0
99	2.3	3.3	4.3
99.5	2.6	3.6	4.6
99.8	3.0	4.0	5.0
99.9	3.3	4.3	5.3
100.0	7.0	7.0	7.0
100.1	10.7	9.7	8.7
100.2	11.0	10.0	9.0
100.5	11.4	10.4	9.4
101	11.7	10.7	9.7
102	12.0	11.0	10.0
110	12.7	11.7	10.7
125	13.0	12.0	11.0
150	13.3	12.3	11.3
200	13.5	12.5	11.5

In quantitative analysis it is the changes of pH near the equivalence point which are of special interest. This part of Fig. 10.2 is accordingly shown on a larger scale in Fig. 10.3, on which are also indicated the colour-change intervals of some of the common indicators.

*With 1*M *solutions,* it is evident that any indicator with an effective range between pH 3 and 10.5 may be used. The colour change will be sharp and the titration error negligible.

*With 0.1*M *solutions,* the ideal pH range for an indicator is limited to 4.5–9.5. Methyl orange will exist chiefly in the alkaline form when 99.8 mL of alkali have been added, and the titration error will be 0.2 per cent, which is negligibly small for most practical purposes; it is therefore advisable to add sodium hydroxide solution until the indicator is present completely in the alkaline form. The titration error is also negligibly small with phenolphthalein.

With 0.01M solutions, the ideal pH range is still further limited to 5.5–8.5; such indicators as methyl red, bromothymol blue, or phenol red will be suitable. The titration error for methyl orange will be 1–2 per cent.

The above considerations apply to solutions which do not contain carbon dioxide. In practice, carbon dioxide is usually present (compare Section 10.7)

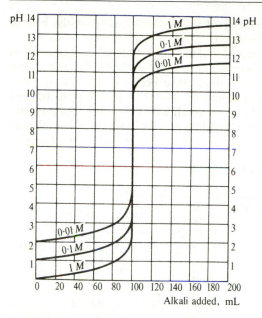

Fig. 10.2 Neutralisation curves of 100 mL of HCl with NaOH of same concentration (calculated).

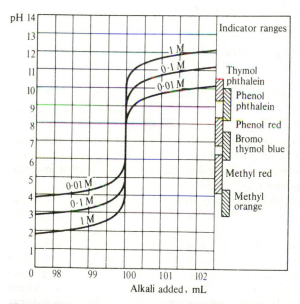

Fig. 10.3 Neutralisation curves of 100 mL of HCl with NaOH of same concentration in vicinity of equivalence point (calculated).

arising from the small quantity of carbonate in the sodium hydroxide and/or from the atmosphere. The gas is in equilibrium with carbonic acid, of which both stages of ionisation are weak. This will introduce a small error when

271

indicators of high pH range (above pH 5) are used, e.g. phenolphthalein or thymolphthalein. More acid indicators, such as methyl orange and methyl yellow, are unaffected by carbonic acid. The difference between the amounts of sodium hydroxide solution used with methyl orange and phenolphthalein is not greater than $0.15-0.2\,\text{mL}$ of $0.1\,M$ sodium hydroxide when $100\,\text{mL}$ of $0.1\,M$ hydrochloric acid are titrated. A method of eliminating this error, other than that of selecting an indicator with a pH range below pH 5, is to boil the solution while still acid to expel carbon dioxide and then to continue the titration with the cold solution. Boiling the solution is particularly efficacious when titrating dilute (e.g. $0.01\,M$) solutions.

10.13 NEUTRALISATION OF A WEAK ACID WITH A STRONG BASE

The neutralisation of $100\,\text{mL}$ of $0.1\,M$ acetic acid (ethanoic acid) with $0.1\,M$ sodium hydroxide solution will be considered here; other concentrations can be treated similarly. The pH of the solution at the equivalence point is given by (Section 2.19)

$$\text{pH} = \tfrac{1}{2}\text{p}K_w + \tfrac{1}{2}\text{p}K_a - \tfrac{1}{2}\text{p}c = 7 + 2.37 - \tfrac{1}{2}(1.3) = 8.72$$

For other concentrations, we may employ the approximate Mass Action expression:

$$[\text{H}^+] \times [\text{CH}_3\text{COO}^-]/[\text{CH}_3\text{COOH}] = K_a \tag{6}$$

or $\quad [\text{H}^+] = [\text{CH}_3\text{COOH}] \times K_a/[\text{CH}_3\text{COO}^-]$

or $\quad \text{pH} = \log[\text{Salt}]/[\text{Acid}] + \text{p}K_a \tag{7}$

The concentration of the salt (and of the acid) at any point is calculated from the volume of alkali added, due allowance being made for the total volume of the solution.

   The initial pH of $0.1M$ acetic acid is computed from equation (6); the dissociation of the acid is relatively so small that it may be neglected in expressing the concentration of acetic acid. Hence from equation (6):

$$[\text{H}^+] \times [\text{CH}_3\text{COO}^-]/[\text{CH}_3\text{COOH}] = 1.82 \times 10^{-5}$$

or $\quad [\text{H}^+]^2/0.1 = 1.82 \times 10^{-5}$

or $\quad [\text{H}^+] = \sqrt{1.82 \times 10^{-6}} = 1.35 \times 10^{-3}$

or $\quad \text{pH} = 2.87$

When $50\,\text{mL}$ of $0.1M$ alkali have been added,

$$[\text{Salt}] = 50 \times 0.1/150 = 3.33 \times 10^{-2}$$

and $\quad [\text{Acid}] = 50 \times 0.1/150 = 3.33 \times 10^{-2}$

$$\text{pH} = \log(3.33 \times 10^{-2}/3.33 \times 10^{-2}) + 4.74 = 4.74$$

The pH values at other points on the titration curve are similarly calculated. After the equivalence point has been passed, the solution contains excess of OH^- ions which will repress the hydrolysis of the salt; the pH may be assumed, with sufficient accuracy for our purpose, to be that due to the excess of base present, so that in this region the titration curve will almost coincide with that for $0.1\,M$ hydrochloric acid (Fig. 10.2 and Table 10.3). All the results are collected in Table 10.4, and are depicted graphically in Fig. 10.4. The results for the titration of 100 mL of $0.1\,M$ solution of a weaker acid ($K_a = 1 \times 10^{-7}$) with $0.1\,M$ sodium hydroxide at the laboratory temperature are also included.

Table 10.4 Neutralisation of 100 mL of 0.1M acetic acid ($K_a = 1.82 \times 10^{-5}$) and of 100 mL of 0.1M HA ($K_a = 1 \times 10^{-7}$) with 0.1M sodium hydroxide

Vol. of 0.1M NaOH used (mL)	0.1M acetic acid (pH)	0.1M HA ($K_a = 1 \times 10^{-7}$) (pH)
0	2.9	4.0
10	3.8	6.0
25	4.3	6.5
50	4.7	7.0
90	5.7	8.0
99.0	6.7	9.0
99.5	7.0	9.3
99.8	7.4	9.7
99.9	7.7	9.8
100.0	8.7	9.9
100.2	10.0	10.0
100.5	10.4	10.4
101	10.7	10.7
110	11.7	11.7
125	12.0	12.0
150	12.3	12.3
200	12.5	12.5

CH_3COOH

$C_2H_4O_2$

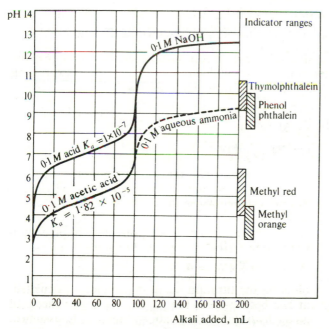

Fig. 10.4 Neutralisation curves of 100 mL of 0.1M acetic acid ($K_a = 1.82 \times 10^{-5}$) and of 0.1M acid ($K_a = 1 \times 10^{-7}$) with 0.1M sodium hydroxide (calculated).

273

For $0.1\,M$ acetic acid and $0.1\,M$ sodium hydroxide, it is evident from the titration curve that neither methyl orange nor methyl red can be used as indicators. The equivalence point is at pH 8.7, and it is necessary to use an indicator with a pH range on the slightly alkaline side, such as phenolphthalein, thymolphthalein, or thymol blue (pH range, as base, 8.0–9.6). For the acid with $K_a = 10^{-7}$, the equivalence point is at pH = 10, but here the rate of change of pH in the neighbourhood of the stoichiometric point is very much less pronounced, owing to considerable hydrolysis. Phenolphthalein will commence to change colour after about 92 mL of alkali have been added, and this change will occur to the equivalence point; thus the end point will not be sharp and the titration error will be appreciable. With thymolphthalein, however, the colour change covers the pH range 9.3–10.5; this indicator may be used, the end point will be sharper than for phenolphthalein, but nevertheless somewhat gradual, and the titration error will be about 0.2 per cent. Acids that have dissociation constants less than 10^{-7} cannot be satisfactorily titrated in $0.1\,M$ solution with a simple indicator.

In general, it may be stated that weak acids ($K_a > 5 \times 10^{-6}$) should be titrated with phenolphthalein, thymolphthalein, or thymol blue as indicators.

10.14 NEUTRALISATION OF A WEAK BASE WITH A STRONG ACID

This may be illustrated by the titration of 100 mL of $0.1M$ aqueous ammonia ($K_b = 1.85 \times 10^{-5}$) with $0.1M$ hydrochloric acid at the ordinary laboratory temperature. The pH of the solution at the equivalence point is given by the equation (Section 2.19):

$$\text{pH} = \tfrac{1}{2}\text{p}K_w - \tfrac{1}{2}\text{p}K_b + \tfrac{1}{2}\text{p}c = 7 - 2.37 + \tfrac{1}{2}(1.3) = 5.28$$

For other concentrations, the pH may be calculated with sufficient accuracy as follows (compare previous section):

$$[NH_4^+] \times [OH^-]/[NH_3] = K_b \tag{8}$$

$$\text{or}\quad [OH^-] = [NH_3] \times K_b/[NH_4^+] \tag{9}$$

$$\text{or}\quad \text{pOH} = \log[\text{Salt}]/[\text{Base}] + \text{p}K_b \tag{10}$$

$$\text{or}\quad \text{pH} = \text{p}K_w - \text{p}K_b - \log[\text{Salt}]/[\text{Base}] \tag{11}$$

After the equivalence point has been reached, the solution contains excess of H^+ ions, hydrolysis of the salt is suppressed, and the subsequent pH changes may be assumed, with sufficient accuracy, to be those due to the excess of acid present.

The results computed in the above manner are represented graphically in Fig. 10.5; the results for the titration of 100 mL of a $0.1\,M$ solution of a weaker base ($K_b = 1 \times 10^{-7}$) are also included.

It is clear that neither thymolphthalein nor phenolphthalein can be employed in the titration of $0.1\,M$ aqueous ammonia. The equivalence point is at pH 5.3, and it is necessary to use an indicator with a pH range on the slightly acid side (3–6.5), such as methyl orange, methyl red, bromophenol blue, or bromocresol green. The last-named indicators may be utilised for the titration of all weak bases ($K_b > 5 \times 10^{-6}$) with strong acids.

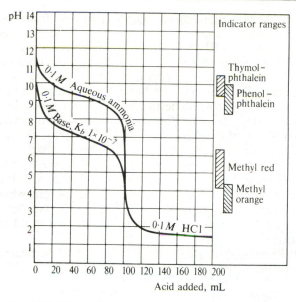

Fig. 10.5 Neutralisation curves of 100 mL 0.1*M* aqueous ammonia ($K_a = 1.8 \times 10^{-5}$) and of 0.1*M* base ($K_a = 1 \times 10^{-7}$) with 0.1*M* hydrochloric acid.

For the weak base ($K_b = 1 \times 10^{-7}$), bromophenol blue or methyl orange may be used; no sharp colour change will be obtained with bromocresol green or with methyl red, and the titration error will be considerable.

10.15 NEUTRALISATION OF A WEAK ACID WITH A WEAK BASE

This case is exemplified by the titration of 100 mL of 0.1*M* acetic acid ($K_a = 1.82 \times 10^{-5}$) with 0.1*M* aqueous ammonia ($K_b = 1.8 \times 10^{-5}$). The pH at the equivalence point is given by (Section 2.19)

$$\text{pH} = \tfrac{1}{2}\text{p}K_w + \tfrac{1}{2}\text{p}K_a - \tfrac{1}{2}\text{p}K_b = 7.0 + 2.38 - 2.37 = 7.1$$

The neutralisation curve up to the equivalence point is almost identical with that using 0.1*M* sodium hydroxide as the base; beyond this point the titration is virtually the addition of 0.1*M* aqueous ammonia solution to 0.1*M* ammonium acetate solution and equation (11) (Section 10.14) is applicable to the calculation of the pH. The titration curve for the neutralisation of 100 mL of 0.1*M* acetic acid with 0.1*M* aqueous ammonia at the laboratory temperature is shown by the broken line in Fig. 10.4. The chief feature of the curve is that the change of pH near the equivalence point and, indeed, during the whole of the neutralisation curve is very gradual. There is no sudden change in pH, and hence no sharp end point can be found with any simple indicator. A mixed indicator, which exhibits a sharp colour change over a very limited pH range, may sometimes be found which is suitable. Thus for acetic acid–ammonia solution titrations, neutral red–methylene blue mixed indicator may be used (see Section 10.9), but on the whole, it is best to avoid the use of indicators in titrations involving *both* a weak acid and a weak base.

10.16   NEUTRALISATION OF A POLYPROTIC ACID WITH A STRONG BASE

The shape of the titration curve will depend upon the relative magnitudes of the various dissociation constants. It is assumed that titrations take place at the ordinary laboratory temperature in solutions of concentration of $0.1 M$ or stronger. For a diprotic acid, if the difference between the primary and secondary dissociation constants is very large ($K_1/K_2 > 10\,000$), the solution behaves like a mixture of two acids with constants K_1 and K_2 respectively; the considerations given previously may be applied. Thus for sulphurous acid, $K_1 = 1.7 \times 10^{-2}$ and $K_2 = 1.0 \times 10^{-7}$, it is evident that there will be a sharp change of pH near the first equivalence point, but for the second stage the change will be less pronounced, yet just sufficient for the use of, say, thymolphthalein as indicator (see Fig. 10.4). For carbonic acid, however, for which $K_1 = 4.3 \times 10^{-7}$ and $K_2 = 5.6 \times 10^{-11}$, only the first stage will be just discernible in the neutralisation curve (see Fig. 10.4); the second stage is far too weak to exhibit any point of inflexion and there is no suitable indicator available for direct titration. As indicator for the primary stage, thymol blue may be used (see Section 10.13), although a mixed indicator of thymol blue (3 parts) and cresol red (1 part) (see Section 10.9) is more satisfactory; with phenolphthalein the colour change will be somewhat gradual and the titration error may be several per cent.

It can be shown that the pH at the first equivalence point for a diprotic acid is given by

$$[H^+] = \sqrt{\frac{K_1 K_2 c}{K_1 + c}}$$

Provided that the first stage of the acid is weak and that K_1 can be neglected by comparison with c, the concentration of salt present, this expression reduces to

$$[H^+] = \sqrt{K_1 K_2}, \text{ or pH} = \tfrac{1}{2}pK_1 + \tfrac{1}{2}pK_2.$$

With a knowledge of the pH at the stoichiometric point and also of the course of the neutralisation curve, it should be an easy matter to select the appropriate indicator for the titration of any diprotic acid for which K_1/K_2 is at least 10^4. For many diprotic acids, however, the two dissociation constants are too close together and it is not possible to differentiate between the two stages. If K_2 is not less than about 10^{-7}, all the replaceable hydrogen may be titrated, e.g. sulphuric acid (primary stage — a strong acid), oxalic acid, malonic, succinic, and tartaric acids.

Similar remarks apply to triprotic acids. These may be illustrated by reference to phosphoric(V) acid (orthophosphoric acid), for which $K_1 = 7.5 \times 10^{-3}$, $K_2 = 6.2 \times 10^{-8}$, and $K_3 = 5 \times 10^{-13}$. Here $K_1/K_2 = 1.2 \times 10^5$ and $K_2/K_3 = 1.2 \times 10^5$, so that the acid will behave as a mixture of three monoprotic acids with the dissociation constants given above. Neutralisation proceeds almost completely to the end of the primary stage before the secondary stage is appreciably affected, and the secondary stage proceeds almost to completion before the tertiary stage is apparent. The pH at the first equivalence point is given approximately by $(\tfrac{1}{2}pK_1 + \tfrac{1}{2}pK_2) = 4.6$, and at the second equivalence point by $(\tfrac{1}{2}pK_2 + \tfrac{1}{2}pK_3) = 9.7$; in the very weak third stage, the curve is very flat and no indicator is available for direct titration. The third equivalence point may be calculated approximately from the equation (Section 10.13):

$$pH = \tfrac{1}{2}pK_w + \tfrac{1}{2}pK_a - \tfrac{1}{2}pc = 7.0 + 6.15 - \tfrac{1}{2}(1.6) = 12.35 \text{ for } 0.1M \text{ H}_3\text{PO}_4.$$

For the primary stage (phosphoric(V) acid as a monoprotic acid), methyl orange, bromocresol green, or Congo red may be used as indicators. The secondary stage of phosphoric(V) acid is very weak (see acid $K_a = 1 \times 10^{-7}$ in Fig. 10.4) and the only suitable simple indicator is thymolphthalein (see Section 10.14); with phenolphthalein the error may be several per cent. A mixed indicator composed of phenolphthalein (3 parts) and 1-naphtholphthalein (1 part) is very satisfactory for the determination of the end point of phosphoric(V) acid as a diprotic acid (see Section 10.9). The experimental neutralisation curve of 50 mL of 0.1M phosphoric(V) acid with 0.1M potassium hydroxide, determined by potentiometric titration, is shown in Fig. 10.6.

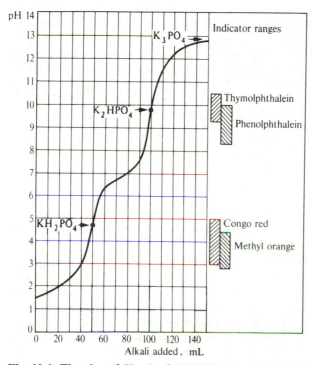

Fig. 10.6 Titration of 50 mL of 0.1M H$_3$PO$_4$ with 0.1M KOH.

There are a number of triprotic acids, e.g. citric acid with $K_1 = 9.2 \times 10^{-4}$, $K_2 = 2.7 \times 10^{-5}$, $K_3 = 1.3 \times 10^{-6}$, the three dissociation constants of which are too close together for the three stages to be differentiated easily. If $K_3 > ca\ 10^{-7}$, all the replaceable hydrogen may be titrated; the indicator will be determined by the value of K_3.

10.17 TITRATION OF ANIONS OF WEAK ACIDS WITH STRONG ACIDS: 'DISPLACEMENT TITRATIONS'

So far the titrations considered have involved a strong base, the hydroxide ion,

277

but titrations are also possible with weaker bases, such as the carbonate ion, the borate ion, the acetate ion, etc. Formerly titrations involving these ions were regarded as titrations of solutions of hydrolysed salts, and the net result was that the weak acid was displaced by the stronger acid. Thus in the titration of sodium acetate solution with hydrochloric acid the following equilibria were considered:

$$CH_3.COO^- + H_2O \rightleftharpoons CH_3.COOH + OH^- \text{ (hydrolysis)}$$

$$H^+ + OH^- = H_2O \text{ (strong acid reacts with } OH^- \text{ produced by hydrolysis)}.$$

The net result thus appeared to be:

$$H^+ + CH_3COO^- = CH_3.COOH$$

or $\quad CH_3.COONa + HCl = CH_3.COOH + NaCl$

i.e. the weak acetic acid was apparently displaced by the strong hydrochloric acid, and the process was referred to as a **displacement titration**. According to the Brønsted–Lowry theory the so-called titration of solutions of hydrolysed salts is merely the titration of a weak base with a strong (highly ionised) acid. When the anion of a weak acid is titrated with a strong acid the titration curve is identical with that observed in the reverse titration of a weak acid itself with a strong base (compare Section 10.13).

A few examples encountered in practice include the following.

Titration of borate ion with a strong acid. The titration of the tetraborate ion with hydrochloric acid is similar to that described above. The net result of the displacement titration is given by:

$$B_4O_7^{2-} + 2H^+ + 5H_2O = 4H_3BO_3$$

Boric acid behaves as a weak monoprotic acid with a dissociation constant of 6.4×10^{-10}. The pH at the equivalence point in the titration of $0.2M$ sodium tetraborate with $0.2M$ hydrochloric acid is that due to $0.1M$ boric acid, i.e. 5.6. Further addition of hydrochloric acid will cause a sharp decrease of pH and any indicator covering the pH range 3.7–5.1 (and slightly beyond this) may be used; suitable indicators are bromocresol green, methyl orange, bromophenol blue, and methyl red.

Titration of carbonate ion with a strong acid. A solution of sodium carbonate may be titrated to the hydrogencarbonate stage (i.e. with one mole of hydrogen ions), when the net reaction is:

$$CO_3^{2-} + H^+ = HCO_3^-$$

The equivalence point for the primary stage of ionisation of carbonic acid is at $pH = (\frac{1}{2}pK_1 + \frac{1}{2}pK_2) = 8.3$, and we have seen (Section 10.14) that thymol blue and, less satisfactorily, phenolphthalein, or a mixed indicator (Section 10.9) may be employed to detect the end point.

Sodium carbonate solution may also be titrated until all the carbonic acid is displaced. The net reaction is then:

$$CO_3^{2-} + 2H^+ = H_2CO_3$$

The same end point is reached by titrating sodium hydrogencarbonate solution with hydrochloric acid:

$$HCO_3^- + H^+ = H_2CO_3$$

The end point with 100 mL of 0.2M sodium hydrogencarbonate and 0.2M hydrochloric acid may be deduced as follows from the known dissociation constant and concentration of the weak acid. The end point will obviously occur when 100 mL of hydrochloric acid has been added, i.e. the solution now has a total volume of 200 mL. Consequently since the carbonic acid liberated from the sodium hydrogencarbonate (0.02 moles) is now contained in a volume of 200 mL, its concentration is 0.1M. K_1 for carbonic acid has a value of 4.3×10^{-7}, and hence we can say:

$$[H^+] \times [HCO_3^-]/[H_2CO_3] = K_1 = 4.3 \times 10^{-7} \, mol \, L^{-1}$$

and since

$$[H^+] = [HCO_3^-]$$

$$[H^+] = \sqrt{4.3 \times 10^{-7} \times 0.1} = 2.07 \times 10^{-4}$$

The pH at the equivalence point is thus approximately 3.7; the secondary ionisation and the loss of carbonic acid, due to any escape of carbon dioxide, have been neglected. Suitable indicators are therefore methyl yellow, methyl orange, Congo red, and bromophenol blue. The experimental titration curve, determined with the hydrogen electrode, for 100 mL of 0.1M sodium carbonate and 0.1M hydrochloric acid is shown in Fig. 10.7.

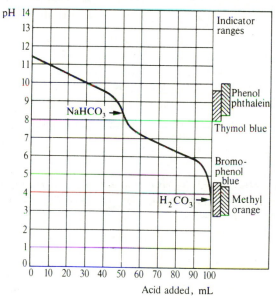

Fig. 10.7 Titration of 100 mL of 0.1M Na$_2$CO$_3$ with 0.1M HCl.

Cations of weak bases (i.e. Brønsted acids such as the phenylammonium ion $C_6H_5NH_3^+$) may be titrated with strong bases, and the treatment is similar. These were formerly regarded as salts of weak bases (e.g. aniline (phenylamine), $K_b = 4.0 \times 10^{-10}$) and strong acids: an example is aniline hydrochloride (phenylammonium chloride).

10.18 CHOICE OF INDICATORS IN NEUTRALISATION REACTIONS

As a general rule, for a titration to be feasible there should be a change of approximately two units of pH at or near the stoichiometric point produced by the addition of a small volume of the reagent. The pH at the equivalence point may be calculated by using the equations given in Section 2.19 (see also below), the pH at either side of the equivalence point (0.1–1 mL) may be calculated as described in the preceding sections, and the difference will indicate whether the change is large enough to permit a sharp end point to be observed. Alternatively, the pH change on both sides of the equivalence point may be obtained from the neutralisation curve determined by potentiometric titration (Sections 15.15 and 15.20). If the pH change is satisfactory, an indicator should be selected that changes at or near the equivalence point.

For convenience of reference, the conclusions already deduced from theoretical principles are summarised below.

Strong acid and strong base. For $0.1 M$ or more concentrated solutions, any indicator may be used which has a range between the limits pH 4.5 and pH 9.5. With $0.01 M$ solutions, the pH range is somewhat smaller (5.5–8.5). If carbon dioxide is present, either the solution should be boiled while still acid and the solution titrated when cold, or an indicator with a range below pH 5 should be employed.

Weak acid and a strong base. The pH at the equivalence point is calculated from the equation:

$$pH = \tfrac{1}{2}pK_w + \tfrac{1}{2}pK_a - \tfrac{1}{2}pc$$

The pH range for acids with $K_a > 10^{-5}$ is 7–10.5; for weaker acids ($K_a > 10^{-6}$) the range is reduced (8–10). The pH range 8–10.5 will cover most of the examples likely to be encountered; this permits the use of thymol blue, thymolphthalein, or phenolphthalein.

Weak base and strong acid. The pH at the equivalence point is computed from the equation:

$$pH = \tfrac{1}{2}pK_w - \tfrac{1}{2}pK_b + \tfrac{1}{2}pc$$

The pH range for bases with $K_b > 10^{-5}$ is 3–7, and for weaker bases ($K_b > 10^{-6}$) 3–5. Suitable indicators will be methyl red, methyl orange, methyl yellow, bromocresol green, and bromophenol blue.

Weak acid and weak base. There is no sharp rise in the neutralisation curve and, generally, no simple indicator can be used. The titration should therefore be avoided, if possible. The approximate pH at the equivalence point can be computed from the equation:

$$pH = \tfrac{1}{2}pK_w + \tfrac{1}{2}pK_a - \tfrac{1}{2}pK_b$$

It is sometimes possible to employ a mixed indicator (see Section 10.9) which exhibits a colour change over a very limited pH range, for example, neutral red–methylene blue for dilute ammonia solution and acetic (ethanoic) acid.

Polyprotic acids (or mixtures of acids, with dissociation constants K_1, K_2, and K_3) and strong bases. The first stoichiometric end point is given approximately by:

$$pH = \tfrac{1}{2}(pK_1 + pK_2)$$

The second stoichiometric end point is given approximately by:

$$pH = \tfrac{1}{2}(pK_2 + pK_3)$$

Anion of a weak acid titrated with a strong acid. The pH at the equivalence point is given by:

$$pH = \tfrac{1}{2}pK_w - \tfrac{1}{2}pK_a - \tfrac{1}{2}pc$$

Cation of a weak base titrated with a strong base. The pH at the stoichiometric end point is given by:

$$pH = \tfrac{1}{2}pK_w - \tfrac{1}{2}pK_b - \tfrac{1}{2}pc$$

As a general rule, wherever an indicator does not give a sharp end point, it is advisable to prepare an equal volume of a comparison solution containing the same quantity of indicator and of the final products and other components of the titration as in the solution under test, and to titrate to the colour shade thus obtained.

In cases where it proves impossible to find a suitable indicator (and this will occur when dealing with strongly coloured solutions) then titration may be possible by an electrometric method such as conductimetric, potentiometric or amperometric titration; see Chapters 13–16. In some instances, spectrophotometric titration (Chapter 17) may be feasible. It should also be noted that if it is possible to work in a non-aqueous solution rather than in water, then acidic and basic properties may be altered according to the solvent chosen, and titrations which are difficult in aqueous solution may then become easy to perform. This procedure is widely used for the analysis of organic materials but is of very limited application with inorganic substances and is discussed in Sections 10.19–10.21.

10.19 TITRATIONS IN NON-AQUEOUS SOLVENTS

The Brønsted–Lowry theory of acids and bases referred to in Section 10.7 can be applied equally well to reactions occurring during acid–base titrations in non-aqueous solvents. This is because their approach considers an acid as any substance which will tend to donate a proton, and a base as a substance which will accept a proton. Substances which give poor end points due to being weak acids or bases in aqueous solution will frequently give far more satisfactory end points when titrations are carried out in non-aqueous media. An additional advantage is that many substances which are insoluble in water are sufficiently soluble in organic solvents to permit their titration in these non-aqueous media.

In the Brønsted–Lowry theory, any acid (HB) is considered to dissociate in solution to give a proton (H^+) and a conjugate base (B^-); whilst any base (B) will combine with a proton to produce a conjugate acid (HB^+).

$$HB \rightleftharpoons H^+ + B^- \tag{a}$$

$$B + H^+ \rightleftharpoons HB^+ \tag{b}$$

The ability of substances to act as acids or bases will depend very much upon the nature of the solvent system which is employed. Non-aqueous solvents are classified into the four groups: aprotic solvents, protophilic solvents, protogenic solvents, and amphiprotic solvents.

Aprotic solvents include those substances which may be considered to be chemically neutral and virtually unreactive under the conditions employed. Carbon tetrachloride and benzene come in this group, they possess low dielectric constants, do not cause ionisation in solutes and do not undergo reactions with acids and bases. Aprotic solvents are frequently used to dilute reaction mixtures while taking no part in the overall process.

Protophilic solvents are substances such as liquid ammonia, amines and ketones which possess a high affinity for protons. The overall reaction taking place can be represented as:

$$HB + S \rightleftharpoons SH^+ + B^- \qquad (c)$$

The equilibrium in this reversible reaction will be greatly influenced by the nature of the acid and that of the solvent. Weak acids are normally used in the presence of strongly protophilic solvents as their acidic strengths are then enhanced and then become comparable to those of strong acids — this is referred to as the '**levelling effect**'.

Protogenic solvents are acidic in nature and readily donate protons. Anhydrous acids such as hydrogen fluoride and sulphuric acid fall in this category; because of their strength and ability to donate protons they enhance the strength of weak bases.

Amphiprotic solvents consist of liquids, such as water, alcohols and weak organic acids, which are slightly ionised and combine both protogenic and protophilic properties in being able to donate and to accept protons.

Thus, acetic (ethanoic) acid displays acidic properties in dissociating to produce protons:

$$CH_3COOH \rightleftharpoons CH_3COO^- + H^+$$

But in the presence of perchloric acid, which is a far stronger acid, acetic acid will accept a proton:

$$CH_3COOH + HClO_4 \rightleftharpoons CH_3COOH_2^+ + ClO_4^-$$

The $CH_3COOH_2^+$ ion so formed can very readily give up its proton to react with a base. A weak base will, therefore, have its basic properties enhanced, and as a consequence titrations between weak bases and perchloric acid can frequently be readily carried out using acetic acid as solvent.

In general, strongly protophilic solvents lead to the equilibrium of equation (c) being forced to the right. This effect is so powerful that in such solvents all acids act as if they were of similar strength. The converse of this occurs with strongly protogenic solvents which cause all bases to act as if they were of similar strength. Solvents which act in this way are known as 'levelling' solvents.

Determinations in non-aqueous solvents are of importance for substances which may give poor end points in normal aqueous titrations and for substances which are not soluble in water. They are also of particular value for determining the proportions of individual components in mixtures of either acids or of bases. These differential titrations are carried out in solvents which do not exert a levelling effect.

Whilst indicators may be used to establish individual end points, as in traditional acid–base titrations, potentiometric methods of end point detection are also used extensively, especially for highly coloured solutions.

Non-aqueous titrations have been used to quantify mixtures of primary, secondary and tertiary amines,[5] for studying sulphonamides, mixtures of purines and for many other organic amino compounds and salts of organic acids.

10.20 SOLVENTS FOR NON-AQUEOUS TITRATIONS

A very large number of both inorganic and organic solvents have been used for non-aqueous determinations, but a few have been used more frequently than most. Some of the most widely applied solvent systems are discussed below. In all instances pure, dry analytical reagent quality solvents should be used to assist in obtaining sharp end points.

Glacial acetic acid (ethanoic acid) is by far the most frequently employed solvent for this purpose. Before it is used it is advisable to check the water content, which may be between 0.1 and 1.0%, and to add just sufficient acetic anhydride to convert any water to the acid. The acid may be used by itself or in conjunction with other solvents such as acetic anhydride, acetonitrile and nitromethane.

Acetonitrile (methyl cyanide, cyanomethane) is frequently used with other solvents such as chloroform and phenol, and particularly with acetic acid. It enables very sharp end points to be obtained in the titration of metal acetates[6] when titrated with perchloric acid.

Alcohols: it has been found that determinations of salts of organic acids and especially of soaps are best carried out in solvent mixtures of glycols and alcohols or of glycols and hydrocarbons. The most common combinations of this type are ethylene glycol (dihydroxyethane) with propan-2-ol or butan-1-ol. The combinations provide admirable solvent power for both the polar and non-polar ends of the molecule. Another suitable solvent mixture is methanol and benzene.

Dioxan is another popular solvent which is often used in place of glacial acetic acid when mixtures of substances are to be quantified. Unlike acetic acid, dioxan is not a levelling solvent and separate end points are normally possible corresponding to the individual components in the mixtures.

Dimethylformamide (DMF) is a protophilic solvent which is frequently employed for titrations between, for instance, benzoic acid and amides, although end points may sometimes be difficult to obtain.

10.21 INDICATORS FOR NON-AQUEOUS TITRATIONS

The various relationships concerning the interconversion between un-ionised and ionised or different resonant forms of indicators referred to in Section 10.7 apply equally well to those indicators used for non-aqueous titrations. However, in this type of titration the colour change exhibited by an indicator at the end point is not always the same for different titrations as it depends upon the nature of the titrand to which it has been added. The colour corresponding to the correct end point may be established by carrying out a potentiometric titration while simultaneously observing the colour change of the indicator. The appropriate colour corresponds to the inflexion point of the titration curve (see Section 15.18).

The majority of non-aqueous titrations are carried out using a fairly limited range of indicators. Typical of those employed are the following.

(*a*) *Crystal violet* is used as a 0.5 per cent w/v solution in glacial acetic acid. Its colour change is from violet through blue, followed by green, then to greenish-yellow, in reactions in which, for instance, bases such as pyridine are titrated with perchloric acid.

(*b*) *Methyl red* is used as a 0.2 per cent w/v solution in dioxan with a yellow to red colour change.

(*c*) *1-Naphthol benzein* gives a yellow to green colour change when employed as a 0.2 per cent w/v solution in acetic acid. It gives sharp end points in nitromethane containing acetic anhydride for titrations of weak bases against perchloric acid.

(*d*) *Oracet blue B* is used as a 0.5 per cent w/v solution in acetic acid and is considered to be superior to crystal violet for titrations of bases in acetic acid with standard perchloric acid. The end point is a distinct change from blue to pink.

(*e*) *Quinaldine red* has been used as an indicator for drug determinations in dimethylformamide solution. It is used as a 0.1 per cent w/v solution in ethanol and gives a colour change from purple/red to pale green.

(*f*) *Thymol blue* is used extensively as an indicator for titrations of substances acting as acids in dimethylformamide solution. It is used as a 0.2 per cent w/v solution in methanol with a sharp colour change from yellow to blue at the end point.

10.22 PREPARATION OF A STANDARD ACID

Hydrochloric acid and sulphuric acid are widely employed in the preparation of standard solutions of acids. Both of these are commercially available as concentrated solutions; concentrated hydrochloric acid is about $10.5-12M$, and concentrated sulphuric acid is about $18M$. By suitable dilution, solutions of any desired *approximate* concentration may be readily prepared. Hydrochloric acid is generally preferred, since most chlorides are soluble in water. Sulphuric acid forms insoluble salts with calcium and barium hydroxides; for titration of hot liquids or for determinations which require boiling for some time with excess of acid, standard sulphuric acid is, however, preferable. Nitric acid is rarely employed, because it almost invariably contains a little nitrous acid, which has a destructive action upon many indicators.

For the preparation of standard solutions of hydrochloric acid, two methods are available. The first utilises the experimental fact that aqueous solutions of hydrochloric acid lose either hydrogen chloride or water upon boiling, according to whether they are more or less concentrated than the constant boiling point mixture, until they attain a practically constant composition (constant boiling point mixture), which depends upon the prevailing pressure. The composition of this constant boiling mixture, listed in Table 10.5, and its dependence upon pressure have been determined with great accuracy by Foulk and Hollingsworth.[7]

The constant boiling point acid is neither hygroscopic nor appreciably volatile, and its concentration remains unchanged if kept in a well-stoppered vessel out of direct sunlight. This acid may be employed directly in the preparation of a solution of hydrochloric acid of known concentration.

Table 10.5 Composition of constant boiling point hydrochloric acid

Pressure (Pa)	Percentage of HCl in acid (vac. wt.)	Weight of acid, weighed in air, containing 36.47 g of HCl (g)
103 991	20.173	180.621
102 658	20.197	180.407
101 325	20.221	180.193
99 992	20.245	179.979
98 659	20.269	179.766
97 325	20.293	179.555

In the second method a solution of the approximate strength required is prepared, and this is standardised against some standard alkaline substance, such as sodium tetraborate or anhydrous sodium carbonate; standard potassium iodate or pure silver may also be used (see Section 10.84). If a solution of an exact strength is required, a solution of an approximate strength somewhat greater than that desired is first prepared; this is suitably diluted with water after standardisation (for a typical calculation, see Appendix 17).

10.23 PREPARATION OF CONSTANT BOILING POINT HYDROCHLORIC ACID

Mix 400 mL of pure concentrated hydrochloric acid with 250–400 mL of distilled water so that the specific gravity of the resultant acid is 1.10 (test with a hydrometer). Insert a thermometer in the neck of a 1 L Pyrex distillation flask so that the bulb is just opposite the side tube, and attach a condenser to the side tube; use an all-glass apparatus. Place 500 mL of the diluted acid in the flask, distil the liquid at a rate of about 3–4 mL min^{-1} and collect the distillate in a small Pyrex flask. From time to time pour the distillate into a 500 mL measuring cylinder. When 375 mL has been collected in the measuring cylinder, collect a further 50 mL in the small Pyrex flask; watch the thermometer to see that the temperature remains constant. Remove the receiver and stopper it; this contains the pure constant boiling point acid. Note the barometric pressure to the nearest millimetre at intervals during the distillation and take the mean value. Interpolate the concentration of the acid from Table 10.5.

10.24 DIRECT PREPARATION OF 0.1 *M* HYDROCHLORIC ACID FROM THE CONSTANT BOILING POINT ACID

Clean and dry a small, stoppered conical flask. After weighing, do not handle the flask directly with the fingers; handle it with the aid of a tissue or a linen cloth. Add the calculated quantity of constant boiling point acid required for the preparation of 1 L of 0.1 *M* acid (see Table 10.5) with the aid of a pipette; make the final adjustment with a dropper pipette. Reweigh the flask to 0.001 g after replacing the stopper. Add an equal volume of water to prevent loss of acid, and transfer the contents to a 1 L graduated flask. Wash out the weighing flask several times with distilled water and add the washings to the original solution. Make up to the mark with distilled water. Insert the stopper and mix the solution thoroughly by shaking and inverting the flask repeatedly.

Note. Unless a solution of exact concentration is required, it is not necessary to weigh out the exact quantity of constant-boiling acid; the concentration may be calculated from the weight of acid. Thus if 18.305 g of acid, prepared at 101 325 Pa was diluted to 1 L, its concentration would be $18.305/180.193 = 0.101\,58\ \text{mol L}^{-1}$.

10.25 PREPARATION OF APPROXIMATELY 0.1 *M* HYDROCHLORIC ACID AND STANDARDISATION

Measure out 9 mL of pure concentrated hydrochloric acid by means of a burette, pour the acid into a 1 L measuring cylinder containing about 500 mL of distilled water. Make up to the litre mark with distilled water and thoroughly mix by shaking. This will give a solution that is approximately 0.1 *M*.

Note. If 1 *M* hydrochloric acid is required, use 90 mL of the concentrated acid. If 0.01 *M* acid is required, dilute two 50 mL portions of the approximately 0.1 *M* acid, removed with a 50 mL pipette, to 1 litre in a graduated flask.

Approximately 0.05 *M* sulphuric acid is similarly prepared from 3 mL of pure concentrated sulphuric acid.

Two excellent methods (utilising acid–base indicators) are available for standardisation. The first is widely employed, but the second is more convenient, less time-consuming, and equally accurate. A third, back-titration, procedure is also available.

A. Standardisation with anhydrous sodium carbonate. *Pure sodium carbonate.* Analytical reagent quality sodium carbonate of 99.9 per cent purity is obtainable commercially. This contains a little moisture and must be dehydrated by heating at 260–270 °C for half an hour and allowed to cool in a desiccator before use. Alternatively, pure sodium carbonate may be prepared by heating sodium hydrogencarbonate to 260–270 °C for 60–90 minutes; the temperature must not be allowed to exceed 270 °C, for above this temperature the sodium carbonate may lose carbon dioxide. It has been recommended[2] that the sodium hydrogencarbonate be decomposed by adding to hot (85 °C) water, and then the hydrated sodium carbonate which crystallises out is filtered off and dehydrated by heating, first at 100 °C, and finally at 260–270 °C.

In all cases the crucible is allowed to cool in a desiccator, and, before it is quite cold, the solid is transferred to a warm, dry glass-stoppered tube or bottle, out of which, when cold, it may be weighed rapidly as required. It is important to remember that anhydrous sodium carbonate is hygroscopic and has a tendency to change into the monohydrate.

Procedure. Weigh out accurately from a weighing bottle about 0.2 g of the pure sodium carbonate into a 250 mL conical flask (Note 1), dissolve it in 50–75 mL of water, and add 2 drops of methyl orange indicator (Note 2) or preferably of methyl orange–indigo carmine indicator (Section 10.9), which gives a very much more satisfactory end point (Note 3). Rinse a clean burette three times with 5 mL portions of the acid: fill the burette to a point 2–3 cm above the zero mark and open the stopcock momentarily, in order to fill the jet with liquid. Examine the jet to see that no air bubbles are enclosed. If there are, more liquid must be run out until the jet is completely filled. Re-fill, if necessary, to bring the level above the zero mark; then slowly run out the liquid until the level is between the 0.0 and 0.5 mL marks. Read the position of the meniscus to 0.01 mL (Section 3.12).

Place the conical flask containing the sodium carbonate solution upon a piece of unglazed white paper or a white tile beneath the burette, and run in the acid slowly from the burette. During the addition of the acid, the flask must be constantly rotated with one hand while the other hand controls the stopcock. Continue the addition until the methyl orange becomes a very faint yellow or the green colour commences to become paler, when the methyl orange–indigo carmine indicator is used. Wash the walls of the flask down with a little distilled water from a wash bottle, and continue the titration very carefully by adding the acid dropwise until the colour of the methyl orange becomes orange or a faint pink, or the colour of the mixed indicator is a neutral grey. This marks the end point of the titration, and the burette-reading should be taken and recorded in a note-book. The procedure is repeated with two or three other portions of sodium carbonate. The first (or preliminary) titration will indicate the location of the true end point within 0.5 mL. With experience and care, subsequent titrations can be carried out very accurately, and should yield concordant results. From the weights of sodium carbonate and the volumes of hydrochloric acid employed, the strength of the acid may be computed for each titration. The arithmetical mean is used to calculate the strength of the solution.

Notes. (1) For *elementary students*, an approximately 0.05*M* solution of sodium carbonate may be prepared by weighing out accurately about 1.3 g of pure sodium carbonate in a weighing bottle or in a small beaker, transferring it to a 250 mL graduated flask, dissolving it in water (Section 3.28), and making up to the mark. The flask is well shaken, then 25.00 mL portions are withdrawn with a pipette and titrated with the acid as described above. Individual titrations should not differ by more than 0.1 mL. Record the results as in Section 10.30.

(2) To obtain the most accurate results, a comparison solution, saturated with carbon dioxide and containing the same concentration of sodium chloride (the colour of methyl orange in a saturated aqueous solution of carbon dioxide is sensitive to the concentration of sodium chloride) and indicator as the titrated solution at the end point, should be used.

(3) This indicator is prepared by dissolving 1 g of methyl orange and 2.5 g of *purified* indigo carmine in 1 L of distilled water, and filtering the solution. The colour change on passing from alkaline to acid solution is from green to magenta with a neutral-grey colour at a pH of about 4.

The mixed indicator, bromocresol green–dimethyl yellow, may be used with advantage. The indicator consists of 4 parts of a 0.2 per cent ethanolic solution of bromocresol green and 1 part of a 0.2 per cent ethanolic solution of dimethyl yellow: about 8 drops are used for 100 mL of solution. The colour change is from blue to greenish yellow at pH = 4.0 − 4.1; the colour is yellow at pH = 3.9.

Methyl red may be used as an indicator provided the carbon dioxide in solution is expelled at the end point by boiling. This indicator gives a red colour with high concentrations of carbon dioxide such as are produced during titrations involving carbonates. Add the standard hydrochloric acid to the cold sodium carbonate solution containing 3 drops of 0.1 per cent methyl red until the indicator changes colour. Boil the solution gently (preferably with a small funnel in the mouth of the flask) for 2 minutes to expel carbon dioxide: the original colour of the indicator will return. Repeat the process until the colour no longer changes on boiling. Generally, boiling and cooling must be repeated twice. Care must be taken to avoid loss of liquid by spattering during boiling. The colour change is more easily perceived than with methyl orange.

Calculation of molarity. The molarity may be calculated by using the equation:

$$Na_2CO_3 + 2HCl = 2NaCl + CO_2 + H_2O$$

but the best method is to derive the value entirely in terms of the primary standard substance — here, sodium carbonate. The relative molecular mass of sodium carbonate is 106.00. If the weight of the sodium carbonate is divided by the number of millilitres of hydrochloric acid to which it corresponds, as found by titration, we have the weight of primary standard equivalent to 1 mL of the acid. Thus if 0.2500 g of sodium carbonate is required for the neutralisation of 45.00 mL of hydrochloric acid, 1 mL of the acid will correspond to $0.2500/45.00 = 0.005556$ g of sodium carbonate. The weight in 1 mL of $0.5M$ sodium carbonate solution is 0.05300 g. Hence the concentration of the acid is $0.005556/0.05300 = 0.1048M$.

Another method is the following: 0.2500 g of sodium carbonate required 45.00 mL of acid, hence 1 L of acid corresponds to $1000 \times 0.2500/45.00 = 5.556$ g of sodium carbonate. But 1 L of $1M$ acid corresponds to 53.00 g of sodium carbonate, hence the acid is $5.556/53.00 = 0.1048M$.

B. Standardisation against sodium tetraborate. The advantages of sodium tetraborate decahydrate (borax) are: (i) it has a large relative molecular mass, 381.44 (that of anhydrous sodium carbonate is 106.00); (ii) it is easily and economically purified by recrystallisation; (iii) heating to constant weight is not required; (iv) it is practically non-hygroscopic; and (v) a sharp end point can be obtained with methyl red at room temperatures, since this indicator is not affected by the very weak boric acid.

$$B_4O_7^{2-} + 2H^+ + 5H_2O = 4H_3BO_3$$

Pure sodium tetraborate. The salt is recrystallised from distilled water; 50 mL of water is used for every 15 g of solid. Care must be taken that the crystallisation does not take place above $55\,^\circ C$; above this temperature there is a possibility of the formation of the pentahydrate, since the transition temperature, decahydrate $\rightleftharpoons$ pentahydrate, is $61\,^\circ C$. The crystals are filtered at the pump, washed twice with water, then twice with portions of 95 per cent ethanol, followed by two portions of diethyl ether. Five-millilitre portions of water, ethanol, and ether are used for 10 g of crystals. Each washing must be followed by suction to remove the wash liquid. After the final washing, the solid is spread in a thin layer on a watch- or clock-glass and allowed to stand at room temperature for 12–18 hours. The sodium tetraborate is then dry, and may be kept in a well-stoppered tube for three to four weeks without appreciable change. An alternative method of drying is to place the recrystallised product (after having been washed twice with water) in a desiccator over a solution saturated with respect to sugar (sucrose) *and* sodium chloride. The substance is dry after about three days, and may be kept indefinitely in the desiccator without change. The latter method is more time-consuming; the product is identical with that obtained by the ethanol–ether process.

Procedure. Weigh out accurately from a weighing bottle 0.4–0.5 g of pure sodium tetraborate into a 250 mL conical flask (Note 1), dissolve it in about 50 mL of water and add a few drops of methyl red. Titrate with the hydrochloric acid contained in a burette (for details, see under A) until the colour changes to pink (Note 2). Repeat the titration with two other portions. Calculate the strength of the hydrochloric acid from the weight of sodium tetraborate and the volume of acid used. The variation of these results should not exceed $1-2$ parts per thousand. If it is greater, further titrations must be performed until the variation falls within these limits. The arithmetical mean is used to calculate the concentration of the solution.

Notes. (1) For *elementary students*, an approximately $0.05\,M$ solution of sodium tetraborate may be prepared by weighing out accurately $4.7-4.8\,g$ of analytical grade material on a watchglass or in a small beaker transferring it to a $250\,mL$ graduated flask, dissolving it in water (Section 3.28), and making up to the mark. The contents of the flask are well mixed by shaking. Portions of $25\,mL$ are withdrawn with a pipette and titrated with the acid as detailed under A. Individual titrations should not differ by more than $0.1\,mL$.

(2) For work of the highest precision a comparison solution or colour standard may be prepared for detecting the equivalence point. For $0.05\,M$ solutions, this is made by adding 5 drops of methyl red to a solution containing $1.0\,g$ of sodium chloride and $2.2\,g$ of boric acid in $500\,mL$ of water; the solution must be boiled to remove any carbon dioxide which may be present in the water. It is assumed that $20\,mL$ of wash water are used in the titration.

Calculation of the molarity. This is carried out as described in A. One mole of sodium tetraborate is $381.44\,g$.

C. Argentimetric method. This method is described in Section 10.84.

10.26 PREPARATION OF STANDARD ALKALI

Discussion. The hydroxides of sodium, potassium, and barium are generally employed for the preparation of solutions of standard alkalis; they are water-soluble strong bases. Solutions made from aqueous ammonia are undesirable, because they tend to lose ammonia, especially if the concentration exceeds $0.5\,M$; moreover, it is a weak base, and difficulties arise in titrations with weak acids (compare Section 10.15). Sodium hydroxide is most commonly used because of its cheapness. None of these solid hydroxides can be obtained pure, so that a standard solution cannot be prepared by dissolving a known weight in a definite volume of water. Both sodium hydroxide and potassium hydroxide are extremely hygroscopic; a certain amount of alkali carbonate and water are always present. Exact results cannot be obtained in the presence of carbonate with some indicators, and it is therefore necessary to discuss methods for the preparation of carbonate-free alkali solutions. For many purposes sodium hydroxide (which contains $1-2$ per cent of sodium carbonate) is sufficiently pure.

Carbonate-free sodium hydroxide solution. One of several methods of preparation may be used:

1. Rinse sodium hydroxide pellets rapidly with deionised water; this removes the carbonate from the surface. A solution prepared from the washed pellets is satisfactory for most purposes.
2. If a concentrated solution of sodium hydroxide (equal weights of pellets and water) is prepared, covered, and allowed to stand, the carbonate remains insoluble; the clear supernatant liquid may be poured or siphoned off, and suitably diluted. (Potassium carbonate is too soluble in concentrated alkali

289

for this method to be applicable for the preparation of carbonate-free potassium hydroxide solution.)

3. A method which yields a product completely free from carbonate ions consists in the electrolysis of a saturated solution of analytical grade sodium chloride with a mercury cathode and a platinum anode in the apparatus shown in Fig. 10.8. About 20–30 mL of *re-distilled* mercury are placed in a 250 mL pear-shaped Pyrex separatory funnel; 100–125 mL of an almost saturated solution of sodium chloride are then carefully introduced. Two short lengths of platinum wire are sealed into Pyrex glass tubing; one of these dips into the mercury (cathode), and the other into the salt solution (anode). A little mercury is placed in the glass tubes, and electrical contact is made by means of amalgamated copper wires dipping into the mercury in the tubes. Electrolysis is carried out using 6–8 volts and 0.5–1 amp for several hours; the funnel is shaken at intervals in order to break up the amalgam crystals that form on the surface of the mercury. The weight of the sodium dissolved in the amalgam may be roughly computed from the total current passed; the current efficiency is 75–80 per cent. When sufficient amalgam has formed, the mercury is run into a Pyrex flask containing about 100 mL of boiled-out distilled water and closed with a rubber bung carrying a soda-lime guard tube. Decomposition of the amalgam, to give the sodium hydroxide solution, is complete after several days; after 12–18 hours about 75 per cent of the amalgam is decomposed.

Fig. 10.8

4. In the anion exchange method, which is the preferred method, carbonate may be removed from either sodium hydroxide or potassium hydroxide. The solution is passed through a strong base anion exchange column (e.g. Duolite A113 or Amberlite IRA 400) in the chloride form (see Chapter 7). Initially the alkali hydroxide converts the resin into the hydroxide form; the carbonate ion has a greater affinity for the resin than has the hydroxide ion, and hence is retained on the resin: the first portions of the effluent contain chloride ion. If it is desired not to dilute the standard base appreciably and if chloride ion is objectionable, the effluent is discarded until it shows no test for chloride. Thus if a column containing one of the above resins, 35 cm long, is prepared in a 50 mL burette, about 150 mL of 4 per cent sodium hydroxide solution must be passed through the column of resin at a flow rate of 5–6 mL min^{-1} before the effluent is chloride-free; subsequently the effluent

may be collected in a 500 mL filter flask with the side arm carrying a soda-lime guard tube. When about 125 mL of liquid have been collected, 105 mL are measured out and diluted with boiled-out distilled water to 1 L. The resulting sodium hydroxide solution is carbonate-free and is about $0.1 M$. When the column becomes saturated with carbonate ion it is readily reconverted to the chloride form by passing dilute hydrochloric acid through it, followed by water to remove the excess acid.

Strong base anion exchangers in the hydroxide form may be used to prepare standard solutions of sodium or potassium hydroxide using weighed amounts of pure sodium chloride or potassium chloride. The resin, after conversion into the hydroxide form by passage of $1 M$ sodium hydroxide (prepared from $18 M$ sodium hydroxide so as to be carbonate-free) is washed with freshly boiled distilled water until the effluent contains no chloride ions and is neutral: about two litres of $1 M$ sodium hydroxide are required for 40 g of resin, and washing is with about two litres of water. About 2.92 g of sodium chloride, accurately weighed, are dissolved in 100 mL of water. The solution is passed through the column at the rate of $4 \, \text{mL min}^{-1}$; this is followed by about 300 mL of freshly boiled distilled water. The eluate is collected in a 500 mL graduated flask by means of an adapter permitting the use of a soda-lime guard tube. Towards the end the flow rate is decreased to permit careful adjustment to volume. An approximately $0.1 M$ solution of sodium hydroxide results.

A number of firms supplying laboratory chemicals offer solutions of specified concentration which can be employed for titrimetric analysis. Amongst these, some manufacturers catalogue sodium hydroxide solutions 'free from carbonate', as for example BDH Ltd 'AVS' range of solutions; if only occasional need for carbonate-free sodium hydroxide solutions arises, this is the simplest way of satisfying this need. The merits of barium hydroxide solution (Section 10.29) as a carbonate-free alkali should also be borne in mind, but this suffers from the disadvantage that the maximum concentration available is between $0.025 M$ and $0.05 M$. Whenever carbonate-free alkali is employed, it is essential that all the water used in the analyses should also be carbon-dioxide-free. With de-ionised water, there will be little cause for worry provided that the water is protected from atmospheric carbon dioxide, and with ordinary distilled water, dissolved carbon dioxide is readily removed by slowly aspirating a current of air which has been passed through a tube containing soda-asbestos or soda-lime through the water for 5–6 hours.

Attention must be directed to the fact that alkaline solutions, particularly if concentrated, attack glass. They may be preserved, if required, in polythene bottles, which are resistant to alkali. Furthermore, solutions of the strong bases absorb carbon dioxide from the air. If such solutions are exposed to the atmosphere for any appreciable time they become contaminated with carbonate. This may be prevented by the use of a storage vessel such as is shown in Fig. 3.9; the guard tube should be filled with soda-lime. A short exposure of an alkali hydroxide solution to the air will not, however, introduce any serious error. If such solutions are quickly transferred to a burette and the latter fitted with a soda-lime guard tube, the error due to contamination by carbon dioxide may be neglected.

The solution of alkali hydroxide prepared by any of the above methods must be standardised. Alkaline solutions that are subsequently to be used in the presence of carbon dioxide or with strong acids are best standardised against solutions prepared from constant boiling point hydrochloric acid or potassium hydrogeniodate or sulphamic acid, or against hydrochloric acid which has been standardised by means of sodium tetraborate or sodium carbonate. If the alkali solution is to be used in the titration of weak acids, it is best standardised against organic acids or against acid salts of organic diprotic acids, such as benzoic acid or potassium hydrogenphthalate, respectively. These substances are commercially available in a purity exceeding 99.9 per cent; potassium hydrogenphthalate is preferable, since it is more soluble in water and has a greater relative molecular mass.

Procedure A. Weigh out rapidly about 4.2 g of sodium hydroxide on a watchglass or into a small beaker, dissolve it in water, make up to 1 L with boiled-out distilled water, mix thoroughly by shaking, and pour the resultant solution into the stock bottle, which should be closed by a rubber stopper.

Procedure B (carbonate-free sodium hydroxide). Dissolve 50 g of sodium hydroxide in 50 mL of distilled water in a Pyrex flask, transfer to a 75 mL test-tube of Pyrex glass, and insert a well-fitting stopper covered with tin foil. Allow it to stand in a vertical position until the supernatant liquid is clear. For a 0.1 M sodium hydroxide solution *carefully* withdraw, using a pipette fitted with a filling device, 6.5 mL of the concentrated clear solution into a 1 L bottle or flask, and dilute quickly with 1 L of recently boiled-out water.

A clear solution can be obtained more quickly, and incidentally the transfer can be made more satisfactorily, by rapidly filtering the solution through a sintered glass funnel with exclusion of carbon dioxide with the aid of the apparatus shown in Fig. 10.9. It is advisable to calibrate the test-tube in approximately 5 mL intervals and to put the graduations on a thin slip of paper gummed to the outside of the tube.

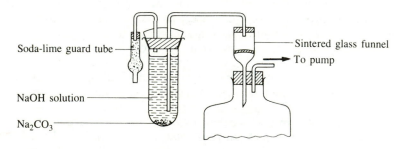

Soda-lime guard tube

Sintered glass funnel

To pump

NaOH solution

Na$_2$CO$_3$

Fig. 10.9

10.27 STANDARDISATION OF APPROXIMATELY 0.1 *M* SODIUM HYDROXIDE

If the solution contains carbonate (Procedure A), methyl orange, methyl orange–indigo carmine, or bromophenol blue must be used in standardisation against hydrochloric acid of known molar concentration. Phenolphthalein or indicators with a similar pH range, which are affected by carbon dioxide, cannot be used at the ordinary temperature (compare Section 10.7). With carbonate-free sodium hydroxide (Procedure B, above) phenolphthalein or thymol blue (Section 10.13) may be employed, and standardisation may be effected against hydrochloric acid, potassium hydrogeniodate, potassium hydrogenphthalate, benzoic acid, or other organic acids (Section 10.28).

Procedure A: with standard hydrochloric acid. Place the standardised (approx. 0.1 M) hydrochloric acid in the burette. Transfer 25 mL of the sodium hydroxide solution into a 250 mL conical flask with the aid of a pipette, dilute with a little water, add 1–2 drops of methyl orange or 3–4 drops of methyl orange–indigo carmine indicator, and titrate with the previously standardised hydrochloric acid. Repeat the titrations until duplicate determinations agree within 0.05 mL of each other.

Calculation of the molarity. In this particular case the molar concentration is readily calculated from the simple relationship:

$$V_A \times m_A = V_B \times m_B$$

where V_A and m_A refer to the volume and known molarity of the acid respectively, V_B is the volume of alkali solution required for the neutralisation, and m_B is its (unknown) molarity.

Procedure B: with potassium hydrogenphthalate. Analytical grade potassium hydrogenphthalate has a purity of at least 99.9 per cent; it is almost non-hygroscopic, but, unless a product of guaranteed purity is purchased, it is advisable to dry it at 120 °C for 2 hours, and allow it to cool in a covered vessel in a desiccator. Weigh out three 0.6–0.7-g portions of the salt into 250 mL Pyrex conical flasks (Note), add 75 mL of boiled-out water to each portion, stopper each flask and shake gently until the solid has dissolved. Titrate each solution with the sodium hydroxide solution contained in a burette, using phenolphthalein or thymol blue as indicator.

Calculation of molar concentration. This is similar to that described in Section 10.25. The R.M.M. of potassium hydrogenphthalate is 204.22. The variation in the results should not exceed 0.1–0.2 per cent.

$$HK(C_8H_4O_4) + NaOH = NaK(C_8H_4O_4) + H_2O$$

Note. For *elementary students*, an approximately 0.1 M solution is prepared by weighing out accurately about 5.1 g of the pure material, dissolving it in water, and making it up to 250 mL in a graduated flask. Volumes of 25 mL are employed in the titrations with the sodium hydroxide solution. Individual titrations should not differ by more than 0.1 mL.

10.28 OTHER STANDARD SUBSTANCES FOR ACIDIMETRY AND ALKALIMETRY

In addition to the standard substances already detailed for use in standardising acids and alkalis, numerous others have been proposed. A number of these will be briefly described.

A. Benzoic acid (C_6H_5COOH; R.M.M. = 122.12). Analytical grade material has a purity of at least 99.9 per cent. For work demanding the highest accuracy, the acid should be dried before use by careful fusion in a platinum crucible placed in an oven at about 130 °C, and then powdered in an agate mortar.

Benzoic acid is sparingly soluble in water (which is a disadvantage) and must therefore be dissolved in 95 per cent ethanol. The mode of use is similar to that already described for potassium hydrogenphthalate (Section 10.27, Procedure B). For a $0.1 M$ solution, of, say, sodium hydroxide, weigh out accurately 0.4 g portions of the acid into a 250 mL conical flask, add $10–20 \text{ mL}$ of ethanol, shake until dissolved, and then titrate the solution with the strong alkali using phenolphthalein as indicator. A blank test should be made with the same volume of ethanol and the indicator: deduct, if necessary, the volume of the alkali solution consumed in the blank test.

B. Succinic acid $\{(CH_2COOH)_2; \text{ R.M.M.} = 118.09\}$. The pure commercial product should be recrystallised from pure acetone and dried in a vacuum desiccator. The purity is checked by means of a melting-point determination $(185–185.5 \,^\circ C)$. The acid is fairly soluble in water; phenolphthalein is a suitable indicator.

C. Potassium hydrogeniodate $\{KH(IO_3)_2; \text{ R.M.M.} = 389.95\}$. Unlike the other solid standards already described, this is a strong acid and thus permits the use of any indicator having a pH range between 4.5 and 9.5 for titration with strong bases. It may be employed for the standardisation of weak bases which are subsequently to be used with strong acids; an indicator, such as methyl red, must then be used. The salt is moderately soluble in water $(1.33 \text{ g}/100 \text{ mL}$ at $15 \,^\circ C)$, it is anhydrous and non-hygroscopic, and its aqueous solution is stable for long periods; the R.M.M. is high, and a $0.01 M$ solution contains 3.8995 g per litre.

Preparation of pure potassium hydrogeniodate. Dissolve 27 g of potassium iodate in 125 mL of boiling water, and add a solution of 22 g of iodic acid in 45 mL of warm water acidified with six drops of concentrated hydrochloric acid. Potassium hydrogeniodate separates on cooling. Filter on a sintered-glass funnel, and wash with cold water. Recrystallise three times from hot water: use 3 parts of water for 1 part of the salt and stir continuously during each cooling. Dry the crystals at $100 \,^\circ C$ for several hours. The purity exceeds 99.95 per cent.

D. Sulphamic acid $(NH_2SO_2OH; \text{ R.M.M.} = 97.09)$. A product of high purity $(>99.9$ per cent$)$ is available commercially. It is a colourless, crystalline, non-hygroscopic solid, melting with decomposition at $205 \,^\circ C$. The acid is moderately soluble in water $(21.3 \text{ g}$ and 47.1 g in 100 g of water at $20 \,^\circ C$ and $80 \,^\circ C$ respectively$)$. Sulphamic acid acts as a strong acid, so that any indicator with a colour change in the pH range $4–9$ may be employed; bromothymol blue is particularly suitable for use with strong bases. It undergoes hydrolysis in aqueous solution:

$$NH_2SO_2OH + H_2O = NH_4HSO_4$$

Aqueous solutions should, preferably, not be stored; the titre does not alter on keeping if an indicator which changes in the acid range is used.

10.29 STANDARD BARIUM HYDROXIDE SOLUTION

This solution is widely employed, particularly for the titration of organic acids. Barium carbonate is insoluble, so that a clear solution is a carbonate-free strong alkali. The relative molecular mass of $Ba(OH)_2,8H_2O$ is 315.50, but a standard solution cannot be prepared by direct weighing owing to the uncertainty of the hydration and the possible presence of carbonate. To prepare an approximately $0.05M$ solution, dissolve 18 g of crystallised barium hydroxide (or of the commercial substance) in about 1 L of hot water in a large flask. Stopper the flask and allow the solution to stand for two days or until all the barium carbonate has completely settled out. Decant or siphon off the clear solution into a storage bottle of the type depicted in Fig. 3.9. A soda-lime guard tube must be provided to prevent ingress of carbon dioxide. The solution may be standardised against standard $0.1M$ hydrochloric acid, succinic acid or potassium hydrogenphthalate; phenolphthalein or thymol blue is employed as indicator.

10.30 DETERMINATION OF THE Na₂CO₃ CONTENT OF WASHING SODA

Procedure. Weigh out accurately about 3.6 g of the washing-soda crystals, dissolve in water, and make up to 250 mL in a graduated flask. Mix thoroughly. Titrate 25 mL of the solution with standard hydrochloric acid of approximately $0.1M$ concentration using methyl orange, or, better, methyl orange–indigo carmine or bromocresol green as indicator. Two consecutive titrations should agree within 0.05 mL.

Calculation. The weight of anhydrous sodium carbonate, Na_2CO_3, which has reacted with the standard hydrochloric acid can be readily calculated from the equation:

$$Na_2CO_3 + 2HCl = 2NaCl + H_2O + CO_2$$

$$106.01 \qquad 2 \times 36.46$$

The percentage of Na_2CO_3 can then be calculated from the known weight of washing soda employed.

A simpler and more general procedure is illustrated by the following example.

Weight of weighing bottle + substance = 16.7910 g

Weight of weighing bottle + residual substance = 13.0110 g

∴ Weight of sample used = 3.7800 g

This was dissolved in water and made up to 250 mL.

Titration of 25.00 mL of the carbonate solution with $0.1060M$ HCl, using methyl orange–indigo carmine as indicator.

Experiment	Reading 1 (mL)	Reading 2 (mL)	Difference = volume used (mL)
1	0.50	26.60	26.10 (preliminary)
2	0.55	26.45	25.90
3	0.50	26.45	25.95
			———
			Mean 25.93

1 mL $1M$ HCl $\equiv 0.05300$ g Na_2CO_3

$25.93 \times 0.1060 \equiv 2.749$ mL $1M$ HCl

$2.749 \times 0.05300 = 0.1457$ g Na_2CO_3 in portion titrated.

Weight of washing soda in portion titrated $= 3.7800 \times 25.0/250 = 0.3780$ g

$\therefore$ Percentage of $Na_2CO_3 = 0.1457 \times 100/0.3780 = 38.54$ per cent

Alternative method of calculation. 25.0 mL of the carbonate solution required 25.93 mL of $0.1060M$ HCl.

But 72.92 g HCl $\equiv 106.01$ g Na_2CO_3

$$25.93 \text{ mL of } 0.1060M \text{ HCl} \equiv \frac{25.93 \times 0.1060}{1000} \times 36.46 \text{ g HCl} \equiv 0.1002 \text{ g HCl}$$

$$\therefore 25 \text{ mL } Na_2CO_3 \text{ solution contain } \frac{0.1002}{72.92} \times 106.01 \text{ g } Na_2CO_3$$

$$\therefore 3.78 \text{ g washing soda contain } \frac{0.1002 \times 106.01 \times 250 \text{ g}}{72.92 \times 25} Na_2CO_3$$

$$= 1.457 \text{ g} = 38.54 \text{ per cent}$$

10.31 DETERMINATION OF THE STRENGTH OF CONCENTRATED ACIDS

Glacial acetic (ethanoic) acid. Weigh a dry, stoppered 50 mL conical flask, introduce about 2 g of glacial acetic acid and weigh again. Add about 20 mL of water and transfer the solution quantitatively to a 250 mL graduated flask. Wash the small flask several times with water and add the washings to the graduated flask. Make up to the mark with distilled, preferably boiled-out, water. Shake the flask well to ensure thorough mixing. Titrate 25 mL portions of the acid with $0.1M$ standard sodium hydroxide solution, using phenolphthalein or thymol blue as indicator.

$NaOH + CH_3COOH = CH_3COONa + H_2O$

1 mL $1M$ NaOH $\equiv 0.06005$ g CH_3COOH

Calculate the percentage of CH_3COOH in the sample of glacial acetic acid.

Note on the determination of the acetic acid content of vinegar. Vinegar usually contains 4–5 per cent acetic acid. Weigh out about 20 g vinegar as described above, and make up to 100 mL in a graduated flask. Take 25 mL with a pipette, dilute this with an equal volume of water, add a few drops of phenolphthalein, and titrate with standard $0.1M$ sodium hydroxide solution. As a result of the dilution of the vinegar, its natural colour will be so reduced that it will not interfere with the colour change of the indicator. Calculate the acetic acid content of the vinegar, and express the result in grams of acetic acid per 100 grams vinegar.

Concentrated sulphuric acid. Place about 100 mL of water in a 250 mL graduated flask, and insert a short-stemmed funnel in the neck of the flask. Charge a weight pipette with a few grams of the acid to be evaluated, and weigh. Add about 1.3–1.5 g of the acid to the flask and re-weigh the pipette. Alternatively, the acid may be weighed out in a stoppered weighing bottle, and after adding the acid to the flask, the weighing bottle is re-weighed. Rinse the funnel thoroughly, remove from the flask, and allow the flask to stand for 1–2 hours to regain the temperature of the laboratory; then the solution can be made up to the mark. Shake and mix thoroughly, and then titrate 25 mL portions with standard 0.1 M sodium hydroxide, using methyl orange or methyl orange–indigo carmine as indicator.

1 mL 1M NaOH $\equiv$ 0.4904 g H$_2$SO$_4$

Fuming sulphuric acid (oleum) should be weighed in a Lunge–Ray pipette (Fig. 3.3b).

Syrupy phosphoric acid. In this case we are dealing with a triprotic acid and theoretically three equivalence points are possible, but in practice the pH changes in the neighbourhood of the equivalence points are not very marked (see Fig. 10.6). For the first-stage neutralisation (pH 4.6) we may employ methyl orange, methyl orange–indigo carmine or bromocresol green as indicator, but it is advisable to use a comparator solution of sodium dihydrogenorthophosphate (0.03 M) containing the same amount of indicator as in the solution being titrated. For the second stage (pH 9.7), phenolphthalein is not altogether satisfactory (it changes colour on the early side), thymolphthalein is better; but the best indicator is a mixture of phenolphthalein (2 parts) with 1-naphtholphthalein (1 part) which changes from pale rose through green to violet at pH 9.6. For the third stage (pH 12.6) there is no satisfactory indicator.

Procedure. Weigh an empty stoppered weighing bottle, add about 2 g of syrupy phosphoric(V) acid and re-weigh. Transfer the acid quantitatively to a 250 mL graduated flask, and then proceed as detailed for sulphuric acid, but using the phenolphthalein–1-naphtholphthalein mixed indicator.

H$_3$PO$_4$ + 2OH$^-$ = HPO$_4^{2-}$ + 2H$_2$O

1 mL 1M NaOH $\equiv$ 0.04902 g H$_3$PO$_4$

10.32 DETERMINATION OF A MIXTURE OF CARBONATE AND HYDROXIDE (ANALYSIS OF COMMERCIAL CAUSTIC SODA)

Discussion. Two methods may be used for this analysis. In the first method the total alkali (carbonate + hydroxide) is determined by titration with standard acid, using methyl orange, methyl orange–indigo carmine, or bromophenol blue as indicator. In a second portion of solution the carbonate is precipitated with a slight excess of barium chloride solution, and, without filtering, the solution is titrated with standard acid using thymol blue or phenolphthalein as indicator. The latter titration gives the hydroxide content, and by subtracting this from the first titration, the volume of acid required for the carbonate is obtained.

Na$_2$CO$_3$ + BaCl$_2$ = BaCO$_3$ (insoluble) + 2NaCl

The second method utilises two indicators. It has been stated in Section 10.17 that the pH of half-neutralised sodium carbonate, i.e. at the sodium hydrogencarbonate stage, is about 8.3, but the pH changes comparatively slowly in the neighbourhood of the equivalence point; consequently the indicator colour-change with phenolphthalein (pH range 8.3–10.0) or thymol blue [pH range (base) 8.0–9.6] is not too sharp. This difficulty may be overcome by using a comparison solution containing sodium hydrogencarbonate of approximately the same concentration as the unknown and the same volume of indicator. A simpler method is to employ a mixed indicator (Section 10.9) composed of

6 parts of thymol blue and 1 part of cresol red; this mixture is violet at pH 8.4, blue at pH 8.3 and rose at pH 8.2. With this mixed indicator the mixture has a violet colour in alkaline solution and changes to blue in the vicinity of the equivalence point; in making the titration the acid is added slowly until the solution assumes a rose colour. At this stage all the hydroxide has been neutralised and the carbonate converted into hydrogencarbonate. Let the volume of standard acid consumed be v mL.

$$OH^- + H^+ = H_2O$$

$$CO_3^{2-} + H^+ = HCO_3^-$$

Another titration is performed with methyl orange, methyl orange–indigo carmine or bromophenol blue as indicator. Let the volume of acid be V mL.

$$OH^- + H^+ = H_2O$$

$$CO_3^{2-} + 2H^+ = H_2CO_3$$

$$H_2CO_3 \rightleftharpoons H_2O + CO_2$$

Then $V - 2(V - v)$ corresponds to the hydroxide, $2(V - v)$ to the carbonate, and V to the total alkali. To obtain satisfactory results by this method the solution titrated must be cold (as near 0 °C as is practicable), and loss of carbon dioxide must be prevented as far as possible by keeping the tip of the burette immersed in the liquid.

Procedure A. Weigh out accurately in a glass-stoppered weighing bottle about 2.5 g of commercial sodium hydroxide (e.g. in flake form). Transfer quantitatively to a 500 mL graduated flask and make up to the mark. Shake the flask well. Titrate 25 or 50 mL of this solution with standard 0.1 M hydrochloric acid, using methyl orange or methyl orange–indigo carmine as indicator. Carry out two or three titrations: these should not differ by more than 0.1 mL. This gives the total alkalinity (hydroxide + carbonate). Warm another 25 or 50 mL portion of the solution to 70 °C and add 1 per cent barium chloride solution slowly from a burette or pipette in *slight* excess, i.e. until no further precipitate is produced. Cool to room temperature, add a few drops of phenolphthalein to the solution, and titrate very slowly and with constant stirring with standard 0.1 M hydrochloric acid; the end point is reached when the colour just changes from pink to colourless. If thymol blue is used as indicator, the colour change is from blue to yellow. The amount of acid used corresponds to the hydroxide present.

This method yields only approximate results because of the precipitation of basic barium carbonate in the presence of hydroxide. More accurate results are obtained by considering the above titration as a preliminary one in order to ascertain the approximate hydroxide content, and then carrying out another titration as follows. Treat 25–50 mL of the solution with sufficient standard hydrochloric acid to neutralise most of the hydroxide, then heat and precipitate as before. Under these conditions, practically pure barium carbonate is precipitated.

1 mL 1 M HCl ≡ 0.0401 g NaOH

1 mL 1 M HCl ≡ 0.053 00 g Na$_2$CO$_3$

Procedure B. The experimental details for the preparation of the initial solution are similar to those given under Procedure A. Titrate 25 or 50 mL of the cold solution with standard $0.1 M$ hydrochloric acid and methyl orange, methyl orange–indigo carmine, or bromophenol blue as indicator. Titrate another 25 or 50 mL of the cold solution, diluted with an equal volume of water, slowly with the standard acid using phenolphthalein or, better, the thymol–blue cresol red mixed indicator; in the latter case, the colour at the end point is rose.

Calculate the result as described in the Discussion above.

10.33 DETERMINATION OF A MIXTURE OF CARBONATE AND HYDROGENCARBONATE

The two methods available for this determination are modifications of those described in Section 10.32 for hydroxide/carbonate mixtures. In the first procedure, which is particularly valuable when the sample contains relatively large amounts of carbonate and small amounts of hydrogencarbonate, the total alkali is first determined in one portion of the solution by titration with standard $0.1 M$ hydrochloric acid using methyl orange, methyl orange–indigo carmine, or bromophenol blue as indicator:

$$CO_3^{2-} + 2H^+ = H_2CO_3$$

$$HCO_3^- + H^+ = H_2CO_3$$

$$H_2CO_3 \rightleftharpoons H_2O + CO_2$$

Let this volume correspond to V mL $1 M$ HCl. To another sample, a measured excess of standard $0.1 M$ sodium hydroxide (free from carbonate) over that required to transform the hydrogencarbonate to carbonate is added:

$$HCO_3^- + OH^- = CO_3^{2-} + H_2O$$

A slight excess of 10 per cent barium chloride solution is added to the hot solution to precipitate the carbonate as barium carbonate, and the excess of sodium hydroxide solution immediately determined, without filtering off the precipitate, by titration with the same standard acid; phenolphthalein or thymol blue is used as indicator. If the volume of excess of sodium hydroxide solution added corresponds to v mL of $1 M$ sodium hydroxide and v' mL $1 M$ acid corresponds to the excess of the latter, then $v - v' =$ hydrogencarbonate, and $V - (v - v') =$ carbonate.

In the second procedure a portion of the cold solution is slowly titrated with standard $0.1 M$ hydrochloric acid, using phenolphthalein, or better, the thymol blue–cresol red mixed indicator. This (say, Y mL) corresponds to half the carbonate (compare Section 10.32):

$$CO_3^{2-} + H^+ \rightleftharpoons HCO_3^-$$

Another sample of equal volume is then titrated with the same standard acid using methyl orange, methyl orange–indigo carmine or bromophenol blue as indicator. The volume of acid used (say, y mL) corresponds to carbonate + hydrogencarbonate. Hence $2Y =$ carbonate, and $y - 2Y =$ hydrogencarbonate.

10.34 DETERMINATION OF BORIC ACID

Discussion. Boric acid acts as a weak monoprotic acid ($K_a = 6.4 \times 10^{-10}$); it

299

cannot therefore be titrated accurately with $0.1 M$ standard alkali (compare Section 10.13). However, by the addition of certain organic polyhydroxy compounds, such as mannitol, glucose, sorbitol, or glycerol, it acts as a much stronger acid (for mannitol $K_a \simeq 1.5 \times 10^{-4}$) and can be titrated to a phenolphthalein end point.

The effect of polyhydroxy compounds has been explained on the basis of the formation of 1:1 and 1:2-mole ratio complexes between the hydrated borate ion and 1,2- or 1,3-diols:

$$
2 \begin{array}{c} >C(OH) \\ | \\ >C(OH) \end{array} + H_3BO_3 = \left[\begin{array}{c} >C-O \quad O-C< \\ | \quad B \quad | \\ >C-O \quad O-C< \end{array} \right]^- H^+ + 3H_2O
$$

Glycerol has been widely employed for this purpose but mannitol and sorbitol are more effective, and have the advantage that being solids they do not materially increase the volume of the solution being titrated: 0.5–0.7 g of mannitol or sorbitol in 10 mL of solution is a convenient quantity.

The method may be applied to commercial boric acid, but as this material may contain ammonium salts it is necessary to add a slight excess of sodium carbonate solution and then to boil down to half-bulk to expel ammonia. Any precipitate which separates is filtered off and washed thoroughly, then the filtrate is neutralised to methyl red, and after boiling, mannitol is added, and the solution titrated with standard $0.1 M$ sodium hydroxide solution:

H[boric acid complex] + NaOH = Na[boric acid complex] + H_2O

$1 \text{ mL } 1 M \text{ NaOH} \equiv 0.06184 \text{ g } H_3BO_3$

A mixture of boric acid and a strong acid can be analysed by first titrating the strong acid using methyl red indicator, and then after adding mannitol or sorbitol, the titration is continued using phenolphthalein as indicator. Mixtures of sodium tetraborate and boric acid can be similarly analysed by titrating the salt with standard hydrochloric acid (Section 10.25, B), and then adding mannitol and continuing the titration with standard sodium hydroxide solution: it must of course be borne in mind that in this second titration the boric acid liberated in the first titration will also react.

Procedure.   To determine the purity of a sample of boric acid, weigh accurately about 0.8 g of the acid, transfer quantitatively to a 250 mL graduated flask and make up to the mark. Pipette 25 mL of the solution into a 250 mL conical flask, add an equal volume of distilled water, 2.5–3 g of mannitol or sorbitol, and titrate with standard $0.1 M$ sodium hydroxide solution using phenolphthalein as indicator. It is advisable to check whether any blank correction must be made: dissolve a similar weight of mannitol (sorbitol) in 50 mL of distilled water, add phenolphthalein, and ascertain how much sodium hydroxide solution must be added to produce the characteristic end point colour.

10.35 DETERMINATION OF AMMONIA IN AN AMMONIUM SALT

Discussion.   Two methods, the direct and indirect, may be used for this determination.

In the **direct method**, a solution of the ammonium salt is treated with a solution of a strong base (e.g. sodium hydroxide) and the mixture distilled. Ammonia is quantitatively expelled, and is absorbed in an excess of standard acid. The excess of acid is back-titrated in the presence of methyl red (or methyl orange, methyl orange–indigo carmine, bromophenol blue, or bromocresol green). Each millilitre of $1M$ monoprotic acid consumed in the reaction is equivalent to $0.017\,032\,g$ NH_3:

$$NH_4^+ + OH^- \rightarrow NH_3\uparrow + H_2O$$

In the **indirect method**, the ammonium salt (other than the carbonate or bicarbonate) is boiled with a known excess of standard sodium hydroxide solution. The boiling is continued until no more ammonia escapes with the steam. The excess of sodium hydroxide is titrated with standard acid, using methyl red (or methyl orange–indigo carmine) as indicator.

Procedure (direct method). Fit up the apparatus shown in Fig. 10.10; note that in order to provide some flexibility, the spray trap is joined to the condenser by a hemispherical ground joint and this makes it easier to clamp both the flask and the condenser without introducing any strain into the assembly. The flask may be of round-bottom form (capacity 500–1000 mL), or (as shown in the diagram), a Kjeldahl flask. The latter is particularly suitable when nitrogen in organic compounds is determined by the Kjeldahl method: upon completion of the digestion with concentrated sulphuric acid, cooling, and dilution of the contents, the digestion (Kjeldahl) flask is attached to the apparatus as shown in Fig. 10.10. The purpose of the spray trap is to prevent droplets of sodium

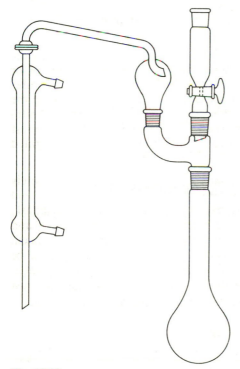

Fig. 10.10

hydroxide solution being driven over during the distillation process. The lower end of the condenser is allowed to dip into a known volume of standard acid contained in a suitable receiver, e.g. a conical flask. A commercial distillation assembly is available in which the tap funnel shown is replaced by a special liquid addition unit: this is similar in form to the tap funnel, but the tap and barrel are replaced by a small vertical ground-glass joint which can be closed with a tapered glass rod. This modification is especially useful when numerous determinations have to be made as it obviates the tendency of glass taps to 'stick' after prolonged contact with concentrated solutions of sodium hydroxide.

For practice, weigh out accurately about 1.5 g of ammonium chloride, dissolve it in water, and make up to 250 mL in a graduated flask. Shake thoroughly. Transfer 50.0 mL of the solution into the distillation flask and dilute with 200 mL of water: add a few anti-bumping granules (fused alumina) to promote regular ebullition in the subsequent distillation. Place 100.0 mL of standard 0.1 M hydrochloric acid in the receiver and adjust the flask so that the end of the condenser just dips into the acid. Make sure that all the joints are fitting tightly. Place 100 mL of 10 per cent sodium hydroxide solution in the funnel. Run the sodium hydroxide solution into the flask by opening the tap, close the tap as soon as the alkali has entered. Heat the flask so that the contents boil gently. Continue the distillation for 30–40 minutes, by which time all the ammonia should have passed over into the receiver; open the tap before removing the flame. Disconnect the trap from the top of the condenser. Lower the receiver and rinse the condenser with a little water. Add a few drops of methyl red* and titrate the excess of acid in the solution with standard 0.1 M sodium hydroxide. Repeat the determination.

Calculate the percentage of NH_3 in the solid ammonium salt employed.

1 mL 0.1 M HCl ≡ 1.703 mg NH_3

Procedure (indirect method). Weigh out accurately 0.1–0.2 g of the ammonium salt into a 500 mL Pyrex conical flask, and add 100 mL of standard 0.1 M sodium hydroxide. Place a small funnel in the neck of the flask in order to prevent mechanical loss, and boil the mixture until a piece of filter paper moistened with mercury(I) nitrate solution and held in the escaping steam is no longer turned black. Cool the solution, add a few drops of methyl red, and titrate with standard 0.1 M hydrochloric acid. Repeat the determination.

10.36 DETERMINATION OF ORGANIC NITROGEN (THE KJELDAHL PROCEDURE)

Discussion. Although other chemical and physical methods now exist for the determination of organic nitrogen, the Kjeldahl procedure is still used very extensively as it remains a highly reliable technique with well-established routines. The basic concept of the method is the digestion of organic material, e.g. proteins, using sulphuric acid and a catalyst to convert any organic nitrogen to ammonium sulphate in solution. By making the mixture alkaline any ammonia can be steam-distilled off and the resulting alkaline distillate titrated with standard acid (see also Section 10.35).

---

*A sharper colour change is obtained with the mixed indicator methyl red–bromocresol green (prepared from 1 part of 0.2 per cent methyl red in ethanol and 3 parts of 0.1 per cent bromocresol green in ethanol).

Procedure. Weigh out accurately part of the organic sample, sufficient to contain about 0.04 g of nitrogen, and place it in the long-necked Kjeldahl digestion flask. Add 0.7 g of mercury(II) oxide, 15 g of potassium sulphate and 40 mL of concentrated sulphuric acid. Heat the flask gently in a slightly inclined position. Some frothing is likely to occur and may be controlled by the use of an anti-foaming agent. When foaming ceases boil the reactants for 2 h. After cooling, add 200 mL of water and 25 mL of $0.5 M$ sodium thiosulphate solution and mix well. To the mixture add a few anti-bumping granules, then carefully pour sufficient $11 M$ sodium hydroxide solution down the inside of the flask to make the mixture strongly alkaline (approximately 115 mL). Before mixing the reagents, connect the flask to a distillation apparatus (Fig. 10.10) in which the tip of the delivery tube is submerged just below the surface of a measured volume of $0.1 M$ hydrochloric acid. Ensure that the contents of the distillation flask are well mixed, then boil until at least 150 mL of liquid have been distilled into the receiver. Add methyl red indicator to the hydrochloric acid solution and titrate with $0.1 M$ sodium hydroxide (titration a mL). Carry out a blank titration on an equal measured volume of the $0.1 M$ hydrochloric acid (titration b mL).

Using the quantities and concentrations given above, the percentage of nitrogen in the sample is given by:

$$N = \frac{(b - a) \times 0.1 \times 14 \times 100}{\text{Weight of sample (g)}} \text{ per cent}$$

10.37 DETERMINATION OF NITRATES

Discussion. Nitrates are reduced to ammonia by means of aluminium, zinc or, most conveniently, by Devarda's alloy (50 per cent Cu, 45 per cent Al, 5 per cent Zn) in strongly alkaline solution:

$$3NO_3^- + 8Al + 5OH^- + 2H_2O = 8AlO_2^- + 3NH_3$$

The ammonia is distilled into excess of standard acid as in Section 10.35.

Nitrites are similarly reduced, and must be allowed for if nitrate alone is to be determined.

Procedure. Weigh out accurately about 1.0 g of the nitrate. Dissolve it in water and transfer the solution quantitatively to the distillation flask of Fig. 10.10. Dilute to about 240 mL. Add 3 g of pure, finely divided Devarda's alloy (it should all pass a 20 mesh sieve). Fit up the apparatus completely and place $75-100$ mL standard $0.2 M$ hydrochloric acid in the receiver (500 mL Pyrex conical flask). Introduce 10 mL of 20 per cent $(0.5 M)$ sodium hydroxide solution through the funnel, and immediately close the trap. Warm *gently* to start the reaction, and allow the apparatus to stand for an hour, by which time the evolution of hydrogen should have practically ceased and the reduction of nitrate to ammonia be complete. Then boil the liquid gently and continue the distillation until $40-50$ mL of liquid remain in the distillation flask. Open the tap before removing the flame. Wash the condenser with a little distilled water, and titrate the contents of the receiver plus the washings with standard $0.2 M$ sodium hydroxide, using methyl red as indicator. Repeat the determination. For very accurate work, it is recommended that a blank test be carried out with distilled water.

1 mL $1 M$ HCl $\equiv 0.062\,01$ g NO_3^-

10.38 DETERMINATION OF PHOSPHATE (PRECIPITATION AS QUINOLINE MOLYBDOPHOSPHATE)

Discussion. When a solution of an orthophosphate is treated with a large excess of ammonium molybdate solution in the presence of nitric acid at a temperature of 20–45 °C, a precipitate is obtained, which after washing is converted into ammonium molybdophosphate with the composition $(NH_4)_3[PO_4,12MoO_3]$. This may be titrated with standard sodium hydroxide solution using phenolphthalein as indicator, but the end point is rather poor due to the liberation of ammonia. If, however, the ammonium molybdate is replaced by a reagent containing sodium molybdate and quinoline, then quinoline molybdophosphate is precipitated which can be isolated and titrated with standard sodium hydroxide:

$$(C_9H_7N)_3[PO_4,12MoO_3] + 26NaOH$$
$$= Na_2HPO_4 + 12Na_2MoO_4 + 3C_9H_7N + 14H_2O$$

The main advantages over the ammonium molybdophosphate method are: (1) quinoline molybdophosphate is less soluble and has a constant composition, and (2) quinoline is a sufficiently weak base not to interfere in the titration.

Calcium, iron, magnesium, alkali metals, and citrates do not affect the analysis. Ammonium salts interfere and must be eliminated by means of sodium nitrite or sodium hypobromite. The hydrochloric acid normally used in the analysis may be replaced by an equivalent amount of nitric acid without any influence on the course of the reaction. Sulphuric acid leads to high and erratic results and its use should be avoided.

The method may be standardised, if desired, with pure potassium dihydrogenorthophosphate (see below): sufficient 1:1 hydrochloric acid must be present to prevent precipitation of quinoline molybdate; the molybdophosphate complex is readily formed at a concentration of 20 mL of concentrated hydrochloric acid per 100 mL of solution especially when warm, and precipitation of the quinoline salt should take place *slowly* from boiling solution. A 'blank' determination should always be made; it is mostly due to silica.

Solutions required. *Sodium molybdate solution.* Prepare a 15 per cent solution of sodium molybdate, $Na_2MoO_4,2H_2O$. Store in a polythene bottle.

Quinoline hydrochloride solution. Add 20 mL of redistilled quinoline to 800 mL of hot water containing 25 mL of pure concentrated hydrochloric acid, and stir well. Cool to room temperature, add a little filter paper pulp ('accelerator'), and again stir well. Filter with suction through a paper-pulp pad, but do not wash. Dilute to 1 L with water.

Mixed indicator solution. Mix two volumes of 0.1 per cent phenolphthalein solution and three volumes of 0.1 per cent thymol blue solution (both in ethanol).

To standardise the procedure, potassium dihydrogenphosphate(V) which has been dried at 105 °C is usually suitable; if necessary it may be further purified by dissolving 100 g in 200 mL of boiling distilled water, keeping on a boiling water bath for several hours, filtering through a fluted filter paper from any turbidity which may appear, and cooling rapidly with constant stirring. The crystals are filtered with suction on hardened filter paper, washed twice with ice-cold water and once with 50 per cent ethanol, and dried at 105 °C.

Procedure. This will be described by reference to the standardisation with potassium dihydrogenphosphate(V). Weigh accurately 0.08–0.09 g of the pure salt into a 250 mL conical flask and dissolve in about 50 mL of distilled water. Add 20 mL of concentrated hydrochloric acid, then 30 mL of the sodium molybdate solution. Heat to boiling, and add a few drops of the quinoline reagent from a burette while swirling the solution in the flask. Again heat to boiling and add the quinoline reagent drop by drop with constant swirling until 1 or 2 mL have been added. Boil again, and to the gently boiling solution add the reagent a few millilitres at a time, with swirling, until 60 mL in all have been introduced. A coarsely crystalline precipitate is thus produced. Allow the suspension to stand in a boiling-water bath for 15 minutes, and then cool to room temperature. Prepare a paper-pulp filter in a funnel fitted with a porcelain cone, and tamp well down. Decant the clear solution through the filter and wash the precipitate twice by decantation with about 20 mL of hydrochloric acid (1:9); this removes most of the excess of quinoline and of molybdate. Transfer the precipitate to the pad with cold water, washing the flask well; wash the filter and precipitate with 30 mL portions of water, letting each washing run through before applying the next, until the washings are acid-free (test for acidity with pH test-paper; about six washings are usually required). Transfer the filter pad and precipitate back to the original flask: insert the funnel into the flask and wash with about 50 mL of water to ensure the transfer of all traces of precipitate. Shake the flask well so that filter paper and precipitate are completely broken up. Run in 50.0 mL of standard (carbonate-free) $0.5M$ sodium hydroxide, swirling during the addition. Shake until the precipitate is *completely* dissolved. Add a few drops of the mixed indicator solution and titrate with standard $0.5M$ hydrochloric acid to an end point which changes sharply from pale green to pale yellow.

Run a blank on the reagents, but use $0.1M$ acid and alkali solutions for the titrations; calculate the blank to $0.5M$ sodium hydroxide. Subtract the blank (which should not exceed 0.5 mL) from the volume neutralised by the original precipitate.

$$1 \text{ mL } 0.5M \text{ NaOH} \equiv 1.830 \text{ mg } PO_4^{3-}$$

Wilson[8] has recommended that the hydrochloric acid added before precipitation be replaced by citric acid, and the subsequent washing of the precipitate is then carried out solely with distilled water.

The method can be applied to the determination of phosphorus in a wide variety of materials, e.g. phosphate rock, phosphatic fertilisers and metals, and is suitable for use in conjunction with the oxygen-flask procedure (Section 3.31). In all cases it is essential to ensure that the material is so treated that the phosphorus is converted to orthophosphate; this may usually be done by dissolution in an oxidising medium such as concentrated nitric acid or in 60 per cent perchloric acid.

10.39 DETERMINATION OF THE RELATIVE MOLECULAR MASS OF AN ORGANIC ACID

Discussion. Many of the common carboxylic acids are readily soluble in water and can be titrated with sodium hydroxide or potassium hydroxide solutions. For sparingly soluble organic acids the necessary solution can be achieved by using a mixture of ethanol and water as solvent.

The theory of titrations between weak acids and strong bases is dealt with in Section 10.13, and is usually applicable to both monoprotic and polyprotic acids (Section 10.16). But for determinations carried out in aqueous solutions it is not normally possible to differentiate easily between the end points for the individual carboxylic acid groups in diprotic acids, such as succinic acid, as the dissociation constants are too close together. In these cases the end points for titrations with sodium hydroxide correspond to neutralisation of all the acidic groups. As some organic acids can be obtained in very high states of purity, sufficiently sharp end points can be obtained to justify their use as standards, e.g. benzoic acid and succinic acid (Section 10.28). The titration procedure described in this section can be used to determine the relative molecular mass (R.M.M.) of a pure carboxylic acid (if the number of acidic groups is known) or the purity of an acid of known R.M.M.

Procedure. Weigh out accurately about 4 g of the pure organic acid, dissolve it in the minimum volume of water (Note 1), or 1:1 (v/v) ethanol/water mixture, and transfer the solution to a 250 mL graduated flask. Ensure the solution is homogeneous and make up to the required volume. Use a pipette to measure out accurately a 25 mL aliquot and transfer to a 250 mL conical flask. Using two drops of phenolphthalein solution as indicator, titrate with standard $0.2M$ (approx.) sodium hydroxide solution (Note 2) until the colourless solution becomes faintly pink. Repeat with further 25 mL volumes of the acid solution until two results in agreement are obtained.

The relative molecular mass is given by:

$$\text{R.M.M.} = \frac{1000\ W \times P}{10\ V \times M}$$

where

W is the weight of the acid taken,
P is the number of carboxylic acid groups,
V is the volume of sodium hydroxide used, and
M is the molarity of the sodium hydroxide.

Notes. (1) In order to obtain sharp end points all de-ionised water used should be carbon-dioxide-free, as far as is possible.

(2) Volumes of $0.2M$ sodium hydroxide required will normally be in the range from about 15 mL to 30 mL depending upon the nature of the organic acid being determined.

10.40 DETERMINATION OF HYDROXYL GROUPS IN CARBOHYDRATES

Discussion. Hydroxyl groups present in carbohydrates can be readily acetylated by acetic (ethanoic) anhydride in ethyl acetate containing some perchloric acid. This reaction can be used as a basis for determining the number of hydroxyl groups in the carbohydrate molecule by carrying out the reaction with excess acetic anhydride followed by titration of the excess using sodium hydroxide in methyl cellosolve.

Solutions required. *Acetic anhydride.* Prepare 250 mL of a $2.0M$ solution in ethyl acetate containing 4.0 g of 72 per cent perchloric acid. The solution is made by adding 4.0 g (2.35 mL) of 72 per cent perchloric acid to 150 mL of ethyl acetate in a 250 mL graduated flask. Pipette 8.0 mL of acetic anhydride into the flask, allow to stand for half an hour. Cool the flask to 5 °C, add 42 mL of cold acetic anhydride. Keep the mixture at 5 °C for 1 h and then allow it to attain room temperature (Note 1).

Sodium hydroxide. Prepare a solution of approximately 0.5 M sodium hydroxide in methylcellosolve. This should be standardised by titration with potassium hydrogenphthalate using the mixed indicator given below.

Pyridine/water. Make up 100 mL of a mixture formed from pyridine and water in the ratio of 3 parts to 1 part by volume.

Mixed indicator. This should be prepared from 1 part of 0.1 per cent neutralised aqueous cresol red and 3 parts of 0.1 per cent neutralised thymol blue.

Procedure. Weigh out accurately between 0.15 g and 0.20 g of the carbohydrate into a 100 mL stoppered conical flask. Pipette into the flask exactly 5.0 mL of the acetic anhydride/ethyl acetate solution. Carefully swirl the contents until the solid has fully dissolved, or mix using a magnetic stirrer. **Do not heat the solution**. Add 1.5 mL of water and again swirl to mix the contents, then add 10 mL of the 3:1 pyridine/water solution, mix by swirling and allow the mixture to stand for 5 min. Titrate the excess acetic anhydride with the standardised 0.5 M sodium hydroxide using the mixed indicator to give a colour change from yellow to violet at the end point.

Carry out a blank determination on the acetic anhydride/ethyl acetate solution following the above procedure without adding the carbohydrate. Use the difference between the blank titration, V_b, and the sample titration, V_s, to calculate the number of hydroxyl groups in the sugar (Note 2).

Calculation. The volume of 0.05 M NaOH used is given by $V_b - V_s$, so the number of moles of acetic anhydride used in reacting with hydroxyl group is:

$$\frac{(V_b - V_s) \times 0.5}{2 \times 1000}$$

But each acetic anhydride molecule reacts with two hydroxyl groups, so the number of moles of hydroxyl groups is:

$$N = \frac{2(V_b - V_s) \times 0.5}{2 \times 1000} = \frac{(V_b - V_s)}{2000}$$

If the relative molecular mass (R.M.M.) of the carbohydrate is known, then the number of hydroxyl groups per molecule is given by:

$$\frac{N \times \text{R.M.M.}}{G}$$

where G is the mass of carbohydrate taken.

Notes. (1) All solutions should be freshly prepared before use. Perchloric acid solutions must not be exposed to sunlight or elevated temperatures as they can be **EXPLOSIVE**.

(2) The solutions from the titrations should be disposed of promptly after the determination has been carried out.

10.41 DETERMINATION OF A MIXTURE OF ANILINE AND ETHANOLAMINE USING A NON–AQUEOUS TITRATION

Discussion. Conventional acid–base titrations are not usually suitable for establishing the individual proportions of amines of different basicity when they are mixed together. But various non-aqueous procedures have been shown to be admirable for this purpose. In one well-established method[9] the mixture of amines in acetonitrile is titrated with perchloric acid. The two individual end-points are best obtained using a potentiometric procedure employing a glass electrode and a saturated calomel electrode (Sections 15.3 and 15.6).

Solutions required. *Perchloric acid.* Prepare an approximately $0.1\,M$ solution by adding 2.13 mL of 72 per cent perchloric acid to dioxane and making up to 250 mL in a graduated flask. This solution should be standardised by titrating against 25 mL aliquots of a standard $0.1\,M$ solution of pure potassium hydrogenphthalate in glacial acetic acid (made by dissolving 2 g of potassium hydrogenphthalate in glacial acetic acid and making up to 100 mL), using crystal violet indicator (Section 10.20).

Mixture of amines. A suitable mixture for analysis can be prepared by accurately weighing roughly equal amounts of aniline and ethanolamine. The determination is best carried out using a solution made from about 4 g of each amine diluted to 100 mL with acetonitrile in a graduated flask.

Procedure. Use 5 mL aliquots of the amine mixture diluted with an additional 20 mL of acetonitrile contained in a 100 mL beaker. Carry out the titration using the $0.1\,M$ perchloric acid, following its progress by means of meter readings (in millivolts) obtained from the two electrodes dipping into the amine solution. Constant stirring is required throughout. The ethanolamine gives rise to the first end-point, and the second end point corresponds to the aniline.

As the acetonitrile may contain basic impurities which also react with the perchloric acid, it is desirable to carry out a blank determination on this solvent. Subtract any value for this blank from the titration values of the amines before calculating the percentages of the two amines in the mixture.

10.42 DETERMINATION OF THE SAPONIFICATION VALUE OF OILS AND FATS

Discussion. For oils and fats, which are esters of long-chain fatty acids, the saponification value (or number) is defined as the number of milligrams of potassium hydroxide which will neutralise the free fatty acids obtained from the hydrolysis of 1 g of the oil or fat. This means that the saponification number is inversely proportional to the relative molecular masses of the fatty acids obtained from the esters. A typical reaction from the hydrolysis of a glyceride is:

$$
\begin{array}{lll}
CH_2-O-CO-C_{17}H_{35} & CH_2OH & \\
| & | & \\
CH-O-CO-C_{17}H_{35} + 3KOH \longrightarrow & CHOH + & 3C_{17}H_{35}COOK \\
| & | & \\
CH_2-O-CO-C_{17}H_{35} & CH_2OH & \\
\end{array}
$$

Stearin   Glycerol   Potassium stearate

Procedure. Prepare an approximately $0.5\,M$ solution of potassium hydroxide by dissolving 30 g potassium hydroxide in 20 mL of water and make the final volume to 1 L using 95 per cent ethanol. Leave the solution to stand for 24 h before decanting and filtering the solution. Using 25 mL aliquots, titrate the potassium hydroxide solution with $0.5M$ hydrochloric acid using phenolphthalein indicator (record as titration a mL).

For the hydrolysis, accurately weigh approximately 2 g of the fat or oil into a 250 mL conical flask with a ground-glass joint and add 25 mL of the potassium hydroxide solution. Attach a reflux condenser and heat the flask contents on a steam bath for 1 h with occasional shaking. While the solution is still hot add phenolphthalein indicator and titrate the excess potassium hydroxide with the $0.5M$ hydrochloric acid (record as titration b mL).

$$\text{The saponification value} = \frac{(a-b) \times 0.5 \times 56.1}{\text{Weight of sample (mg)}}$$

COMPLEXATION TITRATIONS

10.43 INTRODUCTION

The nature of complexes, their stabilities and the chemical characteristics of complexones have been dealt with in some detail in Sections 2.21 to 2.27. This particular section is concerned with the way in which complexation reactions can be employed in titrimetry, especially for determining the proportions of individual cations in mixtures.

The vast majority of complexation titrations are carried out using multidentate ligands such as EDTA or similar substances as the complexone. However, there are other more simple processes which also involve complexation using monodentate or bidentate ligands and which also serve to exemplify the nature of this type of titration. This is demonstrated in the determination outlined in Section 10.44.

10.44 A SIMPLE COMPLEXATION TITRATION

A simple example of the application of a complexation reaction to a titration procedure is the titration of cyanides with silver nitrate solution. When a solution of silver nitrate is added to a solution containing cyanide ions (e.g. an alkali cyanide) a white precipitate is formed when the two liquids first come into contact with one another, but on stirring it re-dissolves owing to the formation of a stable complex cyanide, the alkali salt of which is soluble:

$$Ag^+ + 2CN^- \rightleftharpoons [Ag(CN)_2]^-$$

When the above reaction is complete, further addition of silver nitrate solution yields the insoluble silver cyanoargentate (sometimes termed insoluble silver cyanide); the end point of the reaction is therefore indicated by the formation of a permanent precipitate or turbidity.

The only difficulty in obtaining a sharp end point lies in the fact that silver cyanide, precipitated by local excess concentration of silver ion somewhat prior to the equivalence point, is very slow to re-dissolve and the titration is time-consuming. In the Déniges modification, iodide ion (usually as KI, ca $0.01M$) is used as the indicator and aqueous ammonia (ca $0.2M$) is introduced to dissolve the silver cyanide.

The iodide ion and ammonia solution are added before the titration is commenced; the formation of silver iodide (as a turbidity) will indicate the

309

end point:

$$[Ag(NH_3)_2]^+ + I^- \rightleftharpoons AgI + 2NH_3$$

During the titration any silver iodide which would tend to form will be kept in solution by the excess of cyanide ion always present until the equivalence point is reached:

$$AgI + 2CN^- \rightleftharpoons [Ag(CN)_2]^- + I^-$$

The method may also be applied to the analysis of silver halides by dissolution in excess of cyanide solution and back-titration with standard silver nitrate. It can also be utilised indirectly for the determination of several metals, notably nickel, cobalt, and zinc, which form stable stoichiometric complexes with cyanide ion. Thus if a Ni(II) salt in ammoniacal solution is heated with excess of cyanide ion, the $[Ni(CN)_4]^{2-}$ ion is formed quantitatively; since it is more stable than the $[Ag(CN)_2]^-$ ion, the excess of cyanide may be determined by the Liebig–Denigès method. The metal ion determinations are, however, more conveniently made by titration with EDTA: see the following sections.

10.45 TITRATION CURVES

If, in the titration of a strong acid, pH is plotted against the volume of the solution of the strong base added, a point of inflexion occurs at the equivalence point (compare Section 10.12). Similarly, in the EDTA titration, if pM (negative logarithm of the 'free' metal ion concentration: $pM = -\log[M^{n+}]$) is plotted against the volume of EDTA solution added, a point of inflexion occurs at the equivalence point; in some instances this sudden increase may exceed 10 pM units. The general shape of titration curves obtained by titrating 10.0 mL of a $0.01M$ solution of a metal ion M with a $0.01M$ EDTA solution is shown in Fig. 10.11. The apparent stability constants (see Sections 2.21, 2.23 and 2.27) of various metal–EDTA complexes are indicated at the extreme right of the

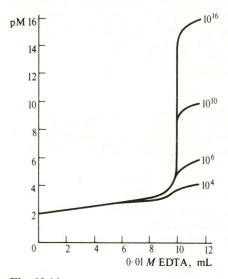

Fig. 10.11

curves. It is evident that the greater the stability constant, the sharper is the end point provided the pH is maintained constant.

In acid–base titrations the end point is generally detected by a pH-sensitive indicator. In the EDTA titration a metal ion-sensitive indicator (abbreviated, to **metal indicator** or **metal-ion indicator**) is often employed to detect changes of pM. Such indicators (which contain types of chelate groupings and generally possess resonance systems typical of dyestuffs) form complexes with specific metal ions, which differ in colour from the free indicator and produce a sudden colour change at the equivalence point. The end point of the titration can also be evaluated by other methods including potentiometric, amperometric, and spectrophotometric techniques.

10.46 TYPES OF EDTA TITRATIONS

The most important procedures for the titration of metal ions with EDTA are the following.

A. Direct titration. The solution containing the metal ion to be determined is buffered to the desired pH (e.g. to pH $= 10$ with NH_4^+–aq. NH_3) and titrated directly with the standard EDTA solution. It may be necessary to prevent precipitation of the hydroxide of the metal (or a basic salt) by the addition of some auxiliary complexing agent, such as tartrate or citrate or triethanolamine. At the equivalence point the magnitude of the concentration of the metal ion being determined decreases abruptly. This is generally determined by the change in colour of a metal indicator or by amperometric, spectrophotometric, or potentiometric methods.

B. Back-titration. Many metals cannot, for various reasons, be titrated directly; thus they may precipitate from the solution in the pH range necessary for the titration, or they may form inert complexes, or a suitable metal indicator is not available. In such cases an excess of standard EDTA solution is added, the resulting solution is buffered to the desired pH, and the excess of the EDTA is back-titrated with a standard metal ion solution; a solution of zinc chloride or sulphate or of magnesium chloride or sulphate is often used for this purpose. The end point is detected with the aid of the metal indicator which responds to the zinc or magnesium ions introduced in the back-titration.

C. Replacement or substitution titration. Substitution titrations may be used for metal ions that do not react (or react unsatisfactorily) with a metal indicator, or for metal ions which form EDTA complexes that are more stable than those of other metals such as magnesium and calcium. The metal cation M^{n+} to be determined may be treated with the magnesium complex of EDTA, when the following reaction occurs:

$$M^{n+} + MgY^{2-} \rightleftharpoons (MY)^{(n-4)+} + Mg^{2+}$$

The amount of magnesium ion set free is equivalent to the cation present and can be titrated with a standard solution of EDTA and a suitable metal indicator.

An interesting application is the titration of calcium. In the direct titration of calcium ions, solochrome black gives a poor end point; if magnesium is present, it is displaced from its EDTA complex by calcium and an improved end point results (compare Section 10.51).

D. Alkalimetric titration. When a solution of disodium ethylenediaminetetra-acetate, Na_2H_2Y, is added to a solution containing metallic ions, complexes are formed with the liberation of two equivalents of hydrogen ion:

$$M^{n+} + H_2Y^{2-} \rightleftharpoons (MY)^{(n-4)+} + 2H^+$$

The hydrogen ions thus set free can be titrated with a standard solution of sodium hydroxide using an acid–base indicator or a potentiometric end point; alternatively, an iodate–iodide mixture is added as well as the EDTA solution and the liberated iodine is titrated with a standard thiosulphate solution.

The solution of the metal to be determined must be accurately neutralised before titration; this is often a difficult matter on account of the hydrolysis of many salts, and constitutes a weak feature of alkalimetric titration.

E. Miscellaneous methods. Exchange reactions between the tetracyano-nickelate(II) ion $[Ni(CN)_4]^{2-}$ (the potassium salt is readily prepared) and the element to be determined, whereby nickel ions are set free, have a limited application. Thus silver and gold, which themselves cannot be titrated complexometrically, can be determined in this way.

$$[Ni(CN)_4]^{2-} + 2Ag^+ \rightleftharpoons 2[Ag(CN)_2]^- + Ni^{2+}$$

These reactions take place with sparingly soluble silver salts, and hence provide a method for the determination of the halide ions Cl^-, Br^-, I^-, and the thiocyanate ion SCN^-. The anion is first precipitated as the silver salt, the latter dissolved in a solution of $[Ni(CN)_4]^{2-}$, and the equivalent amount of nickel thereby set free is determined by rapid titration with EDTA using an appropriate indicator (murexide, bromopyrogallol red).

Fluoride may be determined by precipitation as lead chlorofluoride, the precipitate being dissolved in dilute nitric acid and, after adjusting the pH to 5–6, the lead is titrated with EDTA using xylenol orange indicator.[10]

Sulphate may be determined by precipitation as barium sulphate or as lead sulphate. The precipitate is dissolved in an excess of standard EDTA solution, and the excess of EDTA is back-titrated with a standard magnesium or zinc solution using solochrome black as indicator.

Phosphate may be determined by precipitating as $Mg(NH_4)PO_4,6H_2O$, dissolving the precipitate in dilute hydrochloric acid, adding an excess of standard EDTA solution, buffering at pH = 10, and back-titrating with standard magnesium ion solution in the presence of solochrome black.

10.47 TITRATION OF MIXTURES, SELECTIVITY, MASKING AND DEMASKING AGENTS

EDTA is a very unselective reagent because it complexes with numerous doubly, triply and quadruply charged cations. When a solution containing two cations which complex with EDTA is titrated without the addition of a complex-forming indicator, and if a titration error of 0.1 per cent is permissible, then the ratio of the stability constants of the EDTA complexes of the two metals M and N must be such that $K_M/K_N \geqq 10^6$ if N is not to interfere with the titration of M. Strictly, of course, the constants K_M and K_N considered in the above expression should be the apparent stability constants of the complexes. If complex-forming indicators are used, then for a similar titration error $K_M/K_N \geqq 10^8$.

The following procedures will help to increase the selectivity:

(*a*) **Suitable control of the pH of the solution.** This, of course, makes use of the different stabilities of metal–EDTA complexes. Thus bismuth and thorium can be titrated in an acidic solution (pH = 2) with xylenol orange or methylthymol blue as indicator and most divalent cations do not interfere. A mixture of bismuth and lead ions can be successfully titrated by first titrating the bismuth at pH 2 with xylenol orange as indicator, and then adding hexamine to raise the pH to about 5, and titrating the lead (see Section 10.70).

(*b*) **Use of masking agents.** **Masking** may be defined as the process in which a substance, without physical separation of it or its reaction products, is so transformed that it does not enter into a particular reaction. **Demasking** is the process in which the masked substance regains its ability to enter into a particular reaction.

By the use of masking agents, some of the cations in a mixture can often be 'masked' so that they can no longer react with EDTA or with the indicator. An effective masking agent is the cyanide ion; this forms stable cyanide complexes with the cations of Cd, Zn, Hg(II), Cu, Co, Ni, Ag, and the platinum metals, but not with the alkaline earths, manganese, and lead:

$$M^{2+} + 4CN^- \rightarrow [M(CN)_4]^{2-}$$

It is therefore possible to determine cations such as Ca^{2+}, Mg^{2+}, Pb^{2+}, and Mn^{2+} in the presence of the above-mentioned metals by masking with an excess of potassium or sodium cyanide. A small amount of iron may be masked by cyanide if it is first reduced to the iron(II) state by the addition of ascorbic acid. Titanium(IV), iron(III), and aluminium can be masked with triethanolamine; mercury with iodide ions; and aluminium, iron(III), titanium(IV), and tin(II) with ammonium fluoride (the cations of the alkaline-earth metals yield slightly soluble fluorides).

Sometimes the metal may be transformed into a different oxidation state: thus copper(II) may be reduced in acid solution by hydroxylamine or ascorbic acid. After rendering ammoniacal, nickel or cobalt can be titrated using, for example, murexide as indicator without interference from the copper, which is now present as Cu(I). Iron(III) can often be similarly masked by reduction with ascorbic acid.

(*c*) **Selective demasking.** The cyanide complexes of zinc and cadmium may be demasked with formaldehyde–acetic acid solution or, better, with chloral hydrate:

$$[Zn(CN)_4]^{2-} + 4H^+ + 4HCHO \rightarrow Zn^{2+} + 4HO \cdot CH_2 \cdot CN$$

The use of masking and selective demasking agents permits the successive titration of many metals. Thus a solution containing Mg, Zn, and Cu can be titrated as follows:

1. Add excess of standard EDTA and back-titrate with standard Mg solution using solochrome black as indicator. This gives the sum of all the metals present.
2. Treat an aliquot portion with excess of KCN (**CARE!**)* and titrate as before. This gives Mg only.

---

*Considerable care must be taken at **all** times when using potassium cyanide to avoid any form of physical contact and chemical antidotes must be kept permanently and easily available.

3. Add excess of chloral hydrate (or of formaldehyde–acetic acid solution, 3:1) to the titrated solution in order to liberate the Zn from the cyanide complex, and titrate until the indicator turns blue. This gives the Zn only. The Cu content may then be found by difference.

(*d*) **Classical separation.** These may be applied if they are not tedious; thus the following precipitates may be used for separations in which, after being re-dissolved, the cations can be determined complexometrically: CaC_2O_4, nickel dimethylglyoximate, $Mg(NH_4)PO_4,6H_2O$, and CuSCN.

(*e*) **Solvent extraction.** This is occasionally of value. Thus zinc can be separated from copper and lead by adding excess of ammonium thiocyanate solution and extracting the resulting zinc thiocyanate with 4-methylpentan-2-one (isobutyl methyl ketone); the extract is diluted with water and the zinc content determined with EDTA solution.

(*f*) **Choice of indicators.** The indicator chosen should be one for which the formation of the metal–indicator complex is sufficiently rapid to permit establishment of the end point without undue waiting, and should preferably be reversible.

(*g*) **Removal of anions.** Anions, such as orthophosphate, which can interfere in complexometric titrations may be removed using ion exchange resins. For the use of ion exchange resins in the separation of cations and their subsequent EDTA titration, see Chapter 7.

(*h*) **'Kinetic masking'.** This is a special case in which a metal ion does not effectively enter into the complexation reaction because of its kinetic inertness (see Section 2.25). Thus the slow reaction of chromium(III) with EDTA makes it possible to titrate other metal ions which react rapidly, without interference from Cr(III); this is illustrated by the determination of iron(III) and chromium(III) in a mixture (Section 10.66).

10.48 METAL ION INDICATORS

General properties. The success of an EDTA titration depends upon the precise determination of the end point. The most common procedure utilises metal ion indicators. The requisites of a metal ion indicator for use in the visual detection of end points include:

(*a*) The colour reaction must be such that before the end point, when nearly all the metal ion is complexed with EDTA, the solution is strongly coloured.
(*b*) The colour reaction should be specific or at least selective.
(*c*) The metal–indicator complex must possess sufficient stability, otherwise, because of dissociation, a sharp colour change is not obtained. The metal–indicator complex must, however, be less stable than the metal–EDTA complex to ensure that, at the end point, EDTA removes metal ions from the metal indicator–complex. The change in equilibrium from the metal–indicator complex to the metal–EDTA complex should be sharp and rapid.
(*d*) The colour contrast between the free indicator and the metal–indicator complex should be such as to be readily observed.
(*e*) The indicator must be very sensitive to metal ions (i.e. to pM) so that the colour change occurs as near to the equivalence point as possible.

(f) The above requirements must be fulfilled within the pH range at which the titration is performed.

Dyestuffs which form complexes with specific metal cations can serve as indicators of pM values; 1:1-complexes (metal: dyestuff = 1:1) are common, but 1:2-complexes and 2:1-complexes also occur. The metal ion indicators, like EDTA itself, are chelating agents; this implies that the dyestuff molecule possesses several ligand atoms suitably disposed for coordination with a metal atom. They can, of course, equally take up protons, which also produces a colour change; metal ion indicators are therefore not only pM but also pH indicators.

Theory of the visual use of metal ion indicators. Discussion will be confined to the more common 1:1-complexes. The use of a metal ion indicator in an EDTA titration may be written as:

$$M-In + EDTA \rightarrow M-EDTA + In$$

This reaction will proceed if the metal–indicator complex M–In is less stable than the metal–EDTA complex M–EDTA. The former dissociates to a limited extent, and during the titration the free metal ions are progressively complexed by the EDTA until ultimately the metal is displaced from the complex M–In to leave the free indicator (In). The stability of the metal–indicator complex may be expressed in terms of the formation constant (or indicator constant) K_{In}:

$$K_{In} = [M-In]/[M][In]$$

The indicator colour change is affected by the hydrogen ion concentration of the solution, and no account of this has been taken in the above expression for the formation constant. Thus solochrome black, which may be written as H_2In^-, exhibits the following acid–base behaviour:

$$H_2In^- \xrightarrow[\text{5.3–7.3}]{\text{pH}} HIn^{2-} \xrightarrow[\text{10.5–12.5}]{\text{pH}} In^{3-}$$

Red   Blue   Yellow–orange

In the pH range 7–11, in which the dye itself exhibits a blue colour, many metal ions form red complexes; these colours are extremely sensitive, as is shown, for example, by the fact that $10^{-6} - 10^{-7}$ molar solutions of magnesium ion give a distinct red colour with the indicator. From the practical viewpoint, it is more convenient to define the apparent indicator constant K'_{In}, which varies with pH, as:

$$K'_{In} = [MIn^-]/[M^{n+}][In]$$

where

$[MIn^-]$ = concentration of metal–indicator complex,
$[M^{n+}]$ = concentration of metallic ion, and
$[In]$ = concentration of indicator not complexed with metallic ion.

(This, for the above indicator, is equal to $[H_2In^-] + [HIn^{2-}] + [In^{3-}]$.)
The equation may be expressed as:

$$\log K'_{In} = pM + \log[MIn^-]/[In];$$

315

$\log K'_{In}$ gives the value of pM when half the total indicator is present as the metal ion complex. Some values for $\log K'_{In}$ for $CaIn^-$ and $MgIn^-$ respectively (where H_2In^- is the anion of solochrome black) are: 0.8 and 2.4 at pH = 7; 1.9 and 3.4 at pH = 8; 2.8 and 4.4 at pH = 9; 3.8 and 5.4 at pH = 10; 4.7 and 6.3 at pH = 11; 5.3 and 6.8 at pH = 12. For a small titration error K'_{In} should be large ($> 10^4$), the ratio of the apparent stability constant of the metal–EDTA complex K'_{MY} to that of the metal–indicator complex K'_{In} should be large ($> 10^4$), and the ratio of the indicator concentration to the metal ion concentration should be small ($< 10^{-2}$).

The visual metallochromic indicators discussed above form by far the most important group of indicators for EDTA titrations and the operations subsequently described will be confined to the use of indicators of this type; nevertheless there are certain other substances which can be used as indicators.[11]

Some examples of metal ion indicators. Numerous compounds have been proposed for use as pM indicators; a selected few of these will be described. Where applicable, *Colour Index* (C.I.) references are given.[12] It has been pointed out by West,[11] that apart from a few miscellaneous compounds, the important visual metallochromic indicators fall into three main groups: (*a*) hydroxyazo compounds; (*b*) phenolic compounds and hydroxy-substituted triphenylmethane compounds; (*c*) compounds containing an aminomethyldicarboxymethyl group: many of these are also triphenylmethane compounds.

Note. In view of the varying stability of solutions of these indicators, and the possible variation in sharpness of the end point with the age of the solution, it is generally advisable (if the stability of the indicator solution is suspect), to dilute the solid indicator with 100–200 parts of potassium (or sodium) chloride, nitrate or sulphate (potassium nitrate is usually preferred) and grind the mixture well in a glass mortar. The resultant mixture is usually stable indefinitely if kept dry and in a tightly stoppered bottle.

Murexide (C.I. 56085). This is the ammonium salt of purpuric acid, and is of interest because it was probably the first metal-ion indicator to be employed in the EDTA titration. Murexide solutions are reddish violet up to pH = $9(H_4D^-)$, violet from pH 9 to pH 11 (H_3D^{2-}), and blue–violet (or blue) above pH 11 (H_2D^{3-}). These colour changes are due to the progressive displacement of protons from imido groups; since there are four such groups, murexide may be represented as H_4D^-. Only two of these four acidic hydrogens can be removed by adding an alkali hydroxide, so that only two pK values need be considered; these are $pK_4 = 9.2$ $(H_2D^- \rightarrow H_3D^{2-})$ and $pK_3 = 10.5$ $(H_3D^{2-} \rightarrow H_2D^{3-})$. The anion H_4D^- can also take up a proton to yield the yellow and unstable purpuric acid, but this requires a pH of about 0.

Murexide forms complexes with many metal ions: only those with Cu, Ni, Co, Ca and the lanthanides are sufficiently stable to find application in analysis. Their colours in alkaline solution are orange (copper), yellow (nickel and cobalt), and red (calcium); the colours vary somewhat with the pH of the solution.

Murexide may be employed for the direct EDTA titration of calcium at pH = 11; the colour change at the end-point is from red to blue–violet, but is far from ideal. The colour change in the direct titration of nickel at pH 10–11 is from yellow to blue–violet.

Aqueous solutions of murexide are unstable and must be prepared each day. The indicator solution may be prepared by suspending 0.5 g of the powdered dyestuff in water, shaking thoroughly, and allowing the undissolved portion to settle. The saturated supernatant liquid is used for titrations. Every day the old supernatant liquid is decanted and the residue treated with water as before to provide a fresh solution of the indicator. Normally it is better to prepare a mixture of the indicator with pure sodium chloride in the ratio 1:500, and employ 0.2–0.4 g in each titration.

Solochrome black (eriochrome black T). This substance is sodium 1-(1-hydroxy-2-naphthylazo)-6-nitro-2-naphthol-4-sulphonate, and has the *Colour Index* reference C.I. 14645. In strongly acidic solutions the dye tends to polymerise to a red–brown product, and consequently the indicator is rarely applied in titrations of solutions more acidic than pH = 6.5.

The sulphonic acid group gives up its proton long before the pH range of 7–12, which is of immediate interest for metal-ion indicator use. Only the dissociation of the two hydrogen atoms of the phenolic groups need therefore be considered, and so the dyestuff may be represented by the formula H_2D^-. The two pK values for these hydrogen atoms are 6.3 and 11.5 respectively. Below pH = 5.5, the solution of solochrome black is red (due to H_2D^-), between pH 7 and 11 it is blue (due to HD^{2-}), and above pH = 11.5 it is yellowish–orange (due to D^{3-}). In the pH range 7–11 the addition of metallic salts produces a brilliant change in colour from blue to red:

$$M^{2+} + HD^{2-} \text{ (blue)} \rightarrow MD^- \text{ (red)} + H^+$$

This colour change can be observed with the ions of Mg, Mn, Zn, Cd, Hg, Pb, Cu, Al, Fe, Ti, Co, Ni, and the Pt metals. To maintain the pH constant (*ca* 10) a buffer mixture is added, and most of the above metals must be kept in solution with the aid of a weak complexing reagent such as ammonia or tartrate. The cations of Cu, Co, Ni, Al, Fe(III), Ti(IV), and certain of the Pt metals form such stable indicator complexes that the dyestuff can no longer be liberated by adding EDTA: direct titration of these ions using solochrome black as indicator is therefore impracticable, and the metallic ions are said to 'block' the indicator. However, with Cu, Co, Ni, and Al a back-titration can be carried out, for the rate of reaction of their EDTA complexes with the indicator is extremely slow and it is possible to titrate the excess of EDTA with standard zinc or magnesium ion solution.

Cu, Ni, Co, Cr, Fe, or Al, even in traces, must be absent when conducting a direct titration of the other metals listed above; if the metal ion to be titrated does not react with the cyanide ion or with triethanolamine, these substances can be used as masking reagents. It has been stated that the addition of 0.5–1 mL of 0.001 *M o*-phenanthroline prior to the EDTA titration eliminates the 'blocking effect' of these metals with solochrome black and also with xylenol orange (see below).

The indicator solution is prepared by dissolving 0.2 g of the dyestuff in 15 mL of triethanolamine with the addition of 5 mL of absolute ethanol to reduce the viscosity; the reagent is stable for several months. A 0.4 per cent solution of the pure dyestuff in methanol remains serviceable for at least a month.

Patton and Reeder's indicator. The indicator is 2-hydroxy-1-(2-hydroxy-4-sulpho-1-naphthylazo)-3-naphthoic acid; the name may be abbreviated to HHSNNA. Its main use is in the direct titration of calcium, particularly in the presence of magnesium. A sharp colour change from wine red to pure blue is obtained when calcium ions are titrated with EDTA at pH values between 12

317

and 14. Interferences are similar to those observed with solochrome black, and can be obviated similarly. This indicator may be used as an alternative to murexide for the determination of calcium.

The dyestuff is thoroughly mixed with 100 times its weight of sodium sulphate, and 1 g of the mixture is used in each titration. The indicator is not very stable in alkaline solution.

Solochrome dark blue or calcon (C.I. 15705). This is sometimes referred to as eriochrome blue black RC; it is in fact sodium 1-(2-hydroxy-1-naphthylazo)-2-naphthol-4-sulphonate. The dyestuff has two ionisable phenolic hydrogen atoms; the protons ionise stepwise with pK values of 7.4 and 13.5 respectively. An important application of the indicator is in the complexometric titration of calcium in the presence of magnesium; this must be carried out at a pH of about 12.3 (obtained, for example, with a diethylamine buffer: 5 mL for every 100 mL of solution) in order to avoid the interference of magnesium. Under these conditions magnesium is precipitated quantitatively as the hydroxide. The colour change is from pink to pure blue.

The indicator solution is prepared by dissolving 0.2 g of the dyestuff in 50 mL of methanol.

Calmagite. This indicator, 1-(1-hydroxyl-4-methyl-2-phenylazo)-2-naphthol-4-sulphonic acid, has the same colour change as solochrome black, but the colour change is somewhat clearer and sharper. An important advantage is that aqueous solutions of the indicator are stable almost indefinitely. It may be substituted for solochrome black without change in the experimental procedures for the titration of calcium plus magnesium (see Sections 10.54 and 10.62).

Calmagite functions as an acid–base indicator:

$$H_3D \xrightleftharpoons[\text{low pH}]{} H_2D^- \xrightleftharpoons[\text{pH 7.1–9.1}]{} HD^{2-} \xrightleftharpoons[\text{pH 11.4–13.3}]{} D^{3-}$$

Bright red	Bright red	Clear blue	Reddish-orange

The hydrogen of the sulphonic acid group plays no part in the functioning of the dye as a metal ion indicator. The acid properties of the hydroxyl groups are expressed by p$K_1 = 8.14$ and p$K_2 = 12.35$. The blue colour of calmagite at pH $= 10$ is changed to red by the addition of magnesium ions, the change being reversible:

$$HD^{2-} \xrightleftharpoons[\text{}]{Mg^{2+}} MgD^{2-}$$

Clear blue   Red

This is the basis of the indicator action in the EDTA titration. The pH of 10 is attained by the use of an aqueous ammonia–ammonium chloride buffer mixture.

The combining ratio between calcium or magnesium and the indicator is 1:1; the magnesium compound is the more stable. Calmagite is similar to solochrome black in that small amounts of copper, iron, and aluminium interfere seriously in the titration of calcium and magnesium, and similar masking agents may be used. Potassium hydroxide should be employed for the neutralisation of large amounts of acid since sodium ions in high concentration cause difficulty.

The indicator solution is prepared by dissolving 0.05 g of calmagite in 100 mL of water. It is stable for at least 12 months when stored in a polythene bottle out of sunlight.

Calcichrome. This indicator, cyclotris-7-(1-azo-8-hydroxynaphthalene-3,6-disulphonic acid), is very selective for calcium. It is in fact not very suitable as an indicator for EDTA titrations because the colour change is not particularly sharp, but if EDTA is replaced by CDTA (see Section 2.26), then the indicator gives good results for calcium in the presence of large amounts of barium and small amounts of strontium.[13]

Fast sulphon black F (C.I. 26990). This dyestuff is the sodium salt of 1-hydroxy-8-(2-hydroxynaphthylazo)-2-(sulphonaphthylazo)-3,6-disulphonic acid. The colour reaction seems virtually specific for copper ions. In ammoniacal solution it forms complexes with only copper and nickel; the presence of ammonia or pyridine is required for colour formation. In the direct titration of copper in ammoniacal solution the colour change at the end point is from magenta or [depending upon the concentration of copper(II) ions] pale blue to bright green. The indicator action with nickel is poor. Metal ions, such as those of Cd, Pb, Ni, Zn, Ca, and Ba, may be titrated using this indicator by the prior addition of a reasonable excess of standard copper(II) solution.

The indicator solution consists of a 0.5 per cent aqueous solution.

Bromopyrogallol red. This metal ion indicator is dibromopyrogallol sulphonphthalein and is resistant to oxidation; it also possesses acid–base indicator properties. The indicator is coloured orange–yellow in strongly acidic solution, claret red in nearly neutral solution, and violet to blue in basic solution. The dyestuff forms coloured complexes with many cations. It is valuable for the determination, for example, of bismuth (pH = 2–3. nitric acid solution; end-point blue to claret red).

The indicator solution is prepared by dissolving 0.05 g of the solid reagent in 100 mL of 50 per cent ethanol.

Xylenol orange. This indicator is 3,3'-bis[N,N-di(carboxymethyl)aminomethyl]-o-cresolsulphonphthalein; it retains the acid–base properties of cresol red and displays metal indicator properties even in acid solution (pH = 3–5). Acidic solutions of the indicator are coloured lemon-yellow and those of the metal complexes intensely red.

Direct EDTA titrations of Bi, Th, Zn, Cd, Pb, Co, etc., are readily carried out and the colour change is sharp. Iron(III) and, to a lesser extent, aluminium interfere. By appropriate pH adjustment certain pairs of metals may be titrated successfully in a single sample solution. Thus bismuth may be titrated at pH = 1–2, and zinc or lead after adjustment to pH = 5 by addition of hexamine.

The indicator solution is prepared by dissolving 0.5 g of xylenol orange in 100 mL of water. For storage it is best kept as a solid mixture with potassium nitrate (page 316).

Thymolphthalein complexone (thymolphthalexone). This is thymolphthalein di(methyliminediacetic acid); it contains a stable lactone ring and reacts only in an alkaline medium. The indicator may be used for the titration of calcium; the colour change is from blue to colourless (or a slight pink). Manganese and also nickel may be determined by adding an excess of standard EDTA solution,

319

and titrating the excess with standard calcium chloride solution; the colour change is from very pale blue to deep blue.

The indicator solution consists of a 0.5 per cent solution in ethanol. Alternatively, a finely ground mixture (1:100) with potassium nitrate may be used.

Methylthymol blue (methylthymol blue complexone). This compound is very similar in structure to the preceding one from which it is derived by replacement of the lactone grouping by a sulphonic acid group. By contrast, however, it will function in both acidic and alkaline media, ranging from pH = 0, under which condition bismuth may be titrated with a colour change from blue to yellow, to pH = 12; where the alkaline earths may be titrated with a colour change from blue to colourless. At intermediate pH values a wide variety of doubly charged metal ions may be titrated; of particular interest is its use as an indicator for the titration of Hg(II), an ion for which very few indicators are available. It is also suitable for determining calcium in the presence of magnesium provided that the proportion of the latter is not too high, and is therefore of value in determining the hardness of water. The indicator does not keep well in solution and is used as a solid mixture: 1 part to 100 of potassium nitrate.

Zincon. This is 1-(2-hydroxy-5-sulphophenyl)-3-phenyl-5-(2-carboxyphenyl) formazan, which is a specific indicator for zinc at pH 9–10. Its most important use, however, is as indicator for titration of calcium in the presence of magnesium, using the complexone EGTA (Section 2.26); the magnesium–EGTA complex is relatively weak and does not interfere with the calcium titration. Calcium and magnesium do not give coloured complexes with the indicator, and the procedure is to add a little of the zinc complex of EGTA. The titration is carried out in a buffer at pH 10, and under these conditions calcium ions decompose the Zn–EGTA complex, liberating zinc ions which give a blue colour with the indicator. As soon as all the calcium has been titrated, excess EGTA reconverts the zinc ions to the EGTA complex, and the solution acquires the orange colour of the metal-free indicator.

Variamine blue (C.I. 37255). The end point in an EDTA titration may sometimes be detected by changes in redox potential, and hence by the use of appropriate redox indicators. An excellent example is variamine blue (4-methoxy-4'-aminodiphenylamine), which may be employed in the complexometric titration of iron(III). When a mixture of iron(II) and (III) is titrated with EDTA the latter disappears first. As soon as an amount of the complexing agent equivalent to the concentration of iron(III) has been added, pFe(III) increases abruptly and consequently there is a sudden decrease in the redox potential (compare Section 2.33); the end point can therefore be detected either potentiometrically or with a redox indicator (10.91). The stability constant of the iron(III) complex FeY^- (EDTA = Na_2H_2Y) is about 10^{25} and that of the iron(II) complex FeY^{2-} is 10^{14}; approximate calculations show that the change of redox potential is about 600 millivolts at pH = 2 and that this will be almost independent of the concentration of iron(II) present. The jump in redox potential will also be obtained if no iron(II) salt is actually added, since the extremely minute amount of iron(II) necessary is always present in any 'pure' iron(III) salt.

The visual detection of the sharp change in redox potential in the titration of an iron(III) salt with EDTA is readily made with variamine blue as indicator.

The almost colourless leuco form of the base passes upon oxidation into the strongly coloured indamine. When titrating iron(III) at a pH of about 3 and the colourless hydrochloride of the leuco base is added, oxidation to the violet–blue indamine occurs with the formation of an equivalent amount of iron(II). At the end point of the EDTA titration, the small amount of iron(II) formed when the indicator was introduced is also transformed into the Fe(III)–EDTA complex FeY^-, whereupon the blue indamine is reduced back to the leuco base.

The indicator solution is a 1 per cent solution of the base in water.

10.49 STANDARD EDTA SOLUTIONS

Disodium dihydrogenethylenediaminetetra-acetate of analytical reagent quality is available commercially but this may contain a trace of moisture. After drying the reagent at $80\,^{\circ}C$ its composition agrees with the formula $Na_2H_2C_{10}H_{12}O_8N_2, 2H_2O$ (relative molar mass 372.24), but it should not be used as a primary standard. If necessary, the commercial material may be purified by preparing a saturated solution at room temperature: this requires about 20 g of the salt per 200 mL of water. Add ethanol slowly until a permanent precipitate appears; filter. Dilute the filtrate with an equal volume of ethanol, filter the resulting precipitate through a sintered glass funnel, wash with acetone and then with diethyl ether. Air-dry at room temperature overnight and then dry in an oven at $80\,^{\circ}C$ for at least 24 hours.

Solutions of EDTA of the following concentrations are suitable for most experimental work: $0.1\,M$, $0.05\,M$, and $0.01\,M$. These contain respectively 37.224 g, 18.612 g, and 3.7224 g of the dihydrate per litre of solution. As already indicated, the dry analytical grade salt cannot be regarded as a primary standard and the solution must be standardised; this can be done by titration of nearly neutralised zinc chloride or zinc sulphate solution prepared from a known weight of zinc pellets, or by titration with a solution made from specially dried lead nitrate.

The water employed in making up solutions, particularly dilute solutions, of EDTA should contain no traces of multicharged ions. The distilled water normally used in the laboratory may require distillation in an all-Pyrex glass apparatus or, better, passage through a column of cation exchange resin in the sodium form — the latter procedure will remove all traces of heavy metals. De-ionised water is also satisfactory; it should be prepared from distilled water since tap water sometimes contains non-ionic impurities not removed by an ion exchange column. The solution may be kept in Pyrex (or similar borosilicate glass) vessels, which have been thoroughly steamed out before use. For prolonged storage in borosilicate vessels, the latter should be boiled with a strongly alkaline, 2 per cent EDTA solution for several hours and then repeatedly rinsed with de-ionised water. Polythene bottles are the most satisfactory, and should always be employed for the storage of very dilute (e.g. $0.001\,M$) solutions of EDTA. Vessels of ordinary (soda) glass should not be used; in the course of time such soft glass containers will yield appreciable amounts of cations (including calcium and magnesium) and anions to solutions of EDTA.

Water purified or prepared as described above should be used for the preparation of *all* solutions required for EDTA or similar titrations.

10.50 SOME PRACTICAL CONSIDERATIONS

The following points should be borne in mind when carrying out complexometric titrations.

A. Adjustment of pH. **For many EDTA titrations the pH of the solution is extremely critical**; often limits of ± 1 unit of pH, and frequently limits of ± 0.5 unit of pH must be achieved for a successful titration to be carried out. To achieve such narrow limits of control it is necessary to make use of a pH meter while adjusting the pH value of the solution, and even for those cases where the latitude is such that a pH test-paper can be used to control the adjustment of pH, only a paper of the narrow range variety should be used.

In some of the details which follow, reference is made to the addition of a buffer solution, and in all such cases, to ensure that the requisite buffering action is in fact achieved, it is necessary to make certain that the original solution has first been made almost neutral by the cautious addition of sodium hydroxide or ammonium hydroxide, or of dilute acid, before adding the buffer solution. When an acid solution containing a metallic ion is neutralised by the addition of alkali care must be taken to ensure that the metal hydroxide is not precipitated.

B. Concentration of the metal ion to be titrated. Most titrations are successful with 0.25 millimole of the metal ion concerned in a volume of 50–150 mL of solution. If the metal ion concentration is too high, then the end point may be very difficult to discern, and if difficulty is experienced with an end point then it is advisable to start with a smaller portion of the test solution, and to dilute this to 100–150 mL before adding the buffering medium and the indicator, and then repeating the titration.

C. Amount of indicator. The addition of too much indicator is a fault which must be guarded against: in many cases the colour due to the indicator intensifies considerably during the course of the titration, and further, many indicators exhibit dichroism, i.e. there is an intermediate colour change one to two drops before the real end-point. Thus, for example, in the titration of lead using xylenol orange as indicator at $pH = 6$, the initial reddish–purple colour becomes orange–red, and then with the addition of one or two further drops of reagent, the solution acquires the final lemon yellow colour. This **end point anticipation**, which is of great practical value, may be virtually lost if too much of the indicator is added so that the colour is too intense. In general, a satisfactory colour is achieved by the use of 30–50 mg of a 1 per cent solid mixture of the indicator in potassium nitrate.

D. Attainment of the end point. In many EDTA titrations the colour change in the neighbourhood of the end point may be slow. In such cases, cautious addition of the titrant coupled with continuous stirring of the solution is advisable; the use of a magnetic stirrer is recommended. Frequently, a sharper end point may be achieved if the solution is warmed to about $40\,^{\circ}\mathrm{C}$. Titrations with CDTA (see Section 2.26) are always slower in the region of the end point than the corresponding EDTA titrations.

E. Detection of the colour change. With all of the metal ion indicators used in complexometric titrations, detection of the end point of the titration is dependent upon the recognition of a specified change in colour; for many observers this can be a difficult task, and for those affected by colour blindness it may be virtually impossible. These difficulties may be overcome by replacing the eye by a photocell which is much more sensitive, and eliminates the human element. To carry out the requisite operations it is necessary to have available a colorimeter or a spectrometer in which the cell compartment is large enough to accommodate the titration vessel (a conical flask or a tall form beaker). A simple apparatus may be readily constructed in which light passing through the solution is first allowed to strike a suitable filter and then a photocell; the current generated in the latter is measured with a galvanometer. Whatever form of instrument is used, the wavelength of the incident light is selected (either by an optical filter or by the controls on the instrument) so that the titration solution (including the indicator) shows a maximum transmittance. The titration is then carried out stepwise, taking readings of the transmittance after each addition of EDTA; these readings are then plotted against volume of EDTA solution added, and at the end point (where the indicator changes colour), there will be an abrupt alteration in the transmittance, i.e. a break in the curve, from which the end point may be assessed accurately.

F. Alternative methods of detecting the end point. In addition to the visual and spectrophotometric detection of end points in EDTA titrations with the aid of metal ion indicators, the following methods are also available for end point detection.

1. Potentiometric titration using a mercury electrode (see Section 15.24).
2. Potentiometric titration using a selective ion electrode (Sections 15.7, 15.8) responsive to the ion being titrated.
3. Potentiometric titration using a bright platinum-saturated calomel electrode system; this can be used when the reaction involves two different oxidation states of a given metal.
4. By amperometric titration (Chapter 16).
5. By coulometric analysis (Chapter 14).
6. By conductimetric titration (Chapter 13).

The procedure for the simple complexation titration described in Section 10.44, i.e. the determination of cyanide by titration with standard silver nitrate solution which involves formation of the complex cyanoargentate ion $[Ag(CN)_2]^-$, is most conveniently included with the use of standard silver nitrate solutions under 'Precipitation titrations' (Section 10.87). The following sections are devoted to applications of ethylenediaminetetra-acetic (EDTA) and its congeners. These reagents possess great versatility arising from their inherent potency as complexing agents and from the availability of numerous metal ion indicators (Section 10.48), each effective over a limited range of pH, but together covering a wide range of pH values: to these factors must be added the additional refinements offered by 'masking' and 'demasking' techniques (Section 10.47).

It is clearly impossible, within the scope of the present volume, to give details for all the cations (and anions) which can be determined by EDTA or similar types of titration. Accordingly, details of a few typical determinations are given which serve to illustrate the general procedures to be followed and the use of various buffering agents and of some different indicators. A conspectus of some selected procedures for the commoner cations is then given, followed by some examples of the uses of EDTA for the determination of the components of mixtures; finally, some examples of the determination of anions are given. The relevant theoretical sections (2.21–2.27) should be consulted before commencing the determinations.

DETERMINATION OF INDIVIDUAL CATIONS

10.51 DETERMINATION OF ALUMINIUM: BACK TITRATION

Pipette 25 mL of an aluminium ion solution (approximately 0.01 M) into a conical flask and from a burette add a slight excess of 0.01 M EDTA solution; adjust the pH to between 7 and 8 by the addition of ammonia solution (test drops on phenol red paper or use a pH meter). Boil the solution for a few minutes to ensure complete complexation of the aluminium; cool to room temperature and adjust the pH to 7–8. Add 50 mg of solochrome black/potassium nitrate mixture [see Section 10.50(C)] and titrate rapidly with standard 0.01 M zinc sulphate solution until the colour changes from blue to wine red.

After standing for a few minutes the fully titrated solution acquires a reddish-violet colour due to the transformation of the zinc dye complex into the aluminium–solochrome black complex; this change is irreversible, so that over-titrated solutions are lost.

Every millilitre difference between the volume of 0.01 M EDTA added and the 0.01 M zinc sulphate solution used in the back-titration corresponds to 0.2698 mg of aluminium.

The standard zinc sulphate solution required is best prepared by dissolving about 1.63 g (accurately weighed) of granulated zinc in dilute sulphuric acid, nearly neutralising with sodium hydroxide solution, and then making up to 250 mL in a graduated flask; alternatively, the requisite quantity of zinc sulphate may be used. In either case, de-ionised water must be used.

10.52 DETERMINATION OF BARIUM: DIRECT TITRATION

Pipette 25 mL barium ion solution (ca 0.01 M) into a 250 mL conical flask and dilute to about 100 mL with de-ionised water. Adjust the pH of the solution to 12 by the addition of 3–6 mL of 1 M sodium hydroxide solution; the pH *must* be checked with a pH meter as it must lie between 11.5 and 12.7. Add 50 mg of methyl thymol blue/potassium nitrate mixture [see Section 10.50(C)] and titrate with standard (0.01 M) EDTA solution until the colour changes from blue to grey.

1 mole EDTA ≡ 1 mole Ba^{2+}

10.53 DETERMINATION OF BISMUTH: DIRECT TITRATION

Pipette 25 mL of the bismuth solution (approx. 0.01 M) into a 500 mL conical flask and dilute with de-ionised water to about 150 mL. If necessary, adjust the pH to about 1 by the cautious addition of dilute aqueous ammonia or of dilute nitric acid; use a pH meter. Add 30 mg of the xylenol orange/potassium nitrate mixture (see Section 10.50) and then titrate with standard 0.01 M EDTA solution until the red colour starts to fade. From this point add the titrant slowly until the end point is reached and the indicator changes to yellow.

1 mole EDTA ≡ 1 mole Bi^{3+}

10.54 DETERMINATION OF CALCIUM: SUBSTITUTION TITRATION

Discussion. When calcium ions are titrated with EDTA a relatively stable calcium complex is formed:

$$Ca^{2+} + H_2Y^{2-} \rightleftharpoons CaY^{2-} + 2H^+$$

With calcium ions alone, no sharp end point can be obtained with solochrome black indicator and the transition from red to pure blue is not observed. With magnesium ions, a somewhat less stable complex is formed:

$$Mg^{2+} + H_2Y^{2-} \rightleftharpoons MgY^{2-} + 2H^+$$

and the magnesium indicator complex is more stable than the calcium–indicator complex but less stable than the magnesium–EDTA complex. Consequently, during the titration of a solution containing magnesium and calcium ions with EDTA in the presence of solochrome black the EDTA reacts first with the free calcium ions, then with the free magnesium ions, and finally with the magnesium–indicator complex. Since the magnesium–indicator complex is wine red in colour and the free indicator is blue between pH 7 and 11, the colour of the solution changes from wine red to blue at the end point:

$$MgD^- \text{ (red)} + H_2Y^{2-} = MgY^{2-} + HD^{2-} \text{ (blue)} + H^+$$

If magnesium ions are not present in the solution containing calcium ions they must be added, since they are required for the colour change of the indicator. A common procedure is to add a small amount of magnesium chloride to the EDTA solution before it is standardised. Another procedure, which permits the EDTA solution to be used for other titrations, is to incorporate a little magnesium–EDTA (MgY^{2-}) $(1–10$ per cent) in the buffer solution or to add a little $0.1\,M$ magnesium–EDTA (Na_2MgY) to the calcium-ion solution:

$$MgY^{2-} + Ca^{2+} = CaY^{2-} + Mg^{2+}$$

Traces of many metals interfere in the determination of calcium and magnesium using solochrome black indicator, e.g. Co, Ni, Cu, Zn, Hg, and Mn. Their interference can be overcome by the addition of a little hydroxylammonium chloride (which reduces some of the metals to their lower oxidation states), or also of sodium cyanide or potassium cyanide which form very stable cyanide complexes ('masking'). Iron may be rendered harmless by the addition of a little sodium sulphide.

The titration with EDTA, using solochrome black as indicator, will yield the calcium content of the sample (if no magnesium is present) or the total calcium and magnesium content if both metals are present. To determine the individual elements, calcium may be evaluated by titration using a suitable indicator, e.g., Patton and Reeder's indicator or calcon — see Sections 10.48 and 10.60, or by titration with EGTA using zincon as indicator — see Section 10.61. The difference between the two titrations is a measure of the magnesium content.

Procedure. Prepare an ammonia–ammonium chloride buffer solution (pH 10), by adding 142 mL concentrated ammonia solution (sp. gr. 0.88–0.90) to 17.5 g ammonium chloride and diluting to 250 mL with de-ionised water.

Prepare the magnesium complex of EDTA, Na_2MgY, by mixing equal volumes of $0.2\,M$ solutions of EDTA and of magnesium sulphate. Neutralise with sodium hydroxide solution to a pH between 8 and 9 (phenolphthalein just reddened). Take a portion of the solution, add a few drops of the buffer solution (pH 10), and a few milligrams of the solochrome black–potassium nitrate indicator mixture [see Section 10.50(C)]. A violet colour should be produced which turns blue on the addition of a drop of 0.01 M EDTA solution and red on the addition of a single drop of 0.01 M magnesium sulphate solution; this confirms the equimolarity of magnesium and EDTA. If the solution does not pass this test, it may be treated with more EDTA or with more magnesium sulphate solution until the required condition of equimolarity is attained; this gives an approximately 0.1 M solution.

Pipette 25.0 mL of the 0.01 M calcium ion solution into a 250 mL conical flask, dilute it with about 25 mL of distilled water, add 2 mL buffer solution, 1 mL 0.1 M Mg–EDTA, and 30–40 mg solochrome black/potassium nitrate mixture. Titrate with the EDTA solution until the colour changes from wine red to clear blue. No tinge of reddish hue should remain at the equivalence point. Titrate slowly near the end point.

1 mole EDTA $\equiv$ 1 mole Ca^{2+}

10.55 DETERMINATION OF COPPER: DIRECT TITRATION

The indicator used is fast sulphon black F which is virtually specific in its colour reaction with copper in ammoniacal solution; it forms coloured (red) complexes with only copper and nickel, but the indicator action with nickel is poor.

Procedure. Prepare an indicator solution by dissolving 0.5 g of fast sulphon black F in 100 mL of de-ionised water.

Pipette 25 mL of the copper solution (0.01 M) into a conical flask, add 100 mL de-ionised water, 5 mL concentrated ammonia solution and 5 drops of the indicator solution. Titrate with standard EDTA solution (0.01 M) until the colour changes from purple to dark green.

1 mole EDTA $\equiv$ 1 mole Cu^{2+}

It should be noted that this method is only applicable to solutions containing up to 25 mg copper ions in 100 mL of water; if the concentration of Cu^{2+} ions is too high, the intense blue colour of the copper(II) ammine complex masks the colour change at the end point. The indicator solution must be freshly prepared.

10.56 DETERMINATION OF IRON(III): DIRECT TITRATION

Procedure. Prepare the indicator solution by dissolving 1 g variamine blue in 100 mL de-ionised water: as already pointed out (Section 10.48), variamine blue acts as a redox indicator.

Pipette 25 mL iron(III) solution (0.05 M) into a conical flask and dilute to 100 mL with de-ionised water. Adjust the pH to 2–3; Congo red paper may be used — to the first perceptible colour change. Add 5 drops of the indicator solution, warm the contents of the flask to 40 °C, and titrate with standard (0.05 M) EDTA solution until the initial blue colour of the solution turns grey just before the end point, and with the final drop of reagent changes to yellow.

This particular titration is well adapted to be carried out potentiometrically (Section 15.24).

1 mole EDTA $\equiv$ 1 mole Fe^{3+}

10.57 DETERMINATION OF NICKEL: DIRECT TITRATIONS

Procedure (a). Prepare the indicator by grinding 0.1 g murexide with 10 g of potassium nitrate; use about 50 mg of the mixture for each titration.

Also prepare a $1M$ solution of ammonium chloride by dissolving 26.75 g of the analytical grade solid in de-ionised water and making up to 500 mL in a graduated flask.

Pipette 25 mL nickel solution ($0.01 M$) into a conical flask and dilute to 100 mL with de-ionised water. Add the solid indicator mixture (50 mg) and 10 mL of the $1M$ ammonium chloride solution, and then add concentrated ammonia solution dropwise until the pH is about 7 as shown by the yellow colour of the solution. Titrate with standard ($0.01 M$) EDTA solution until the end point is approached, then render the solution strongly alkaline by the addition of 10 mL of concentrated ammonia solution, and continue the titration until the colour changes from yellow to violet. The pH of the final solution must be 10; at lower pH values an orange–yellow colour develops and more ammonia solution must be added until the colour is clear yellow. Nickel complexes rather slowly with EDTA, and consequently the EDTA solution must be added dropwise near the end point.

Procedure (b). Prepare the indicator by dissolving 0.05 g bromopyrogallol red in 100 mL of 50 per cent ethanol, and a buffer solution by mixing 100 mL of $1M$ ammonium chloride solution with 100 mL of $1M$ aqueous ammonia solution.

Pipette 25 mL nickel solution ($0.01 M$) into a conical flask and dilute to 150 mL with de-ionised water. Add about 15 drops of the indicator solution, 10 mL of the buffer solution and titrate with standard EDTA solution ($0.01 M$) until the colour changes from blue to claret red.

1 mole EDTA $\equiv$ 1 mole Ni^{2+}

10.58 DETERMINATION OF SILVER: INDIRECT METHOD

Silver halides can be dissolved in a solution of potassium tetracyanonickelate(II) in the presence of an ammonia–ammonium chloride buffer, and the nickel ion set free may be titrated with standard EDTA using murexide as indicator.

$$2Ag^+ + [Ni(CN)_4]^{2-} \rightleftharpoons 2[Ag(CN)_2]^- + Ni^{2+}$$

It can be shown from a consideration of the overall stability constants of the ions $[Ni(CN)_4]^{2-}$ (10^{27}) and $[Ag(CN)_2]^-$ (10^{21}) that the equilibrium constant for the above ionic reaction is 10^{15}, i.e. the reaction proceeds practically completely to the right. An interesting exercise is the analysis of a solid silver halide, e.g. silver chloride.

Procedure. Prepare the murexide indicator as described in Section 10.57(a), and an ammonium chloride solution ($1M$) by dissolving 26.75 g ammonium chloride in de-ionised water in a 500 mL graduated flask.

327

The **potassium tetracyanonickelate(II)** which is required is prepared as follows. Dissolve 25 g of analytical grade $NiSO_4, 7H_2O$ in 50 mL distilled water and add portionwise, with agitation, 25 g potassium cyanide. (**Caution:** use a fume cupboard.) A yellow solution forms and a white precipitate of potassium sulphate separates. Gradually add, with stirring, 100 mL of 95 per cent ethanol, filter off the precipitated potassium sulphate with suction, and wash twice with 2 mL ethanol. Concentrate the filtrate at about 70 °C — an infrared heater is convenient for this purpose. When crystals commence to separate, stir frequently. When the crystalline mass becomes thick (without evaporating completely to dryness), allow to cool and mix the crystals with 50 mL ethanol. Separate the crystals by suction filtration and wash twice with 5 mL portions ethanol. Spread the fine yellow crystals in thin layers upon absorbent paper, and allow to stand for 2–3 days in the air, adequately protected from dust. During this period the excess of potassium cyanide is converted into potassium carbonate. The preparation is then ready for use; it should be kept in a stoppered bottle.

Treat an aqueous suspension of about 0.072 g (accurately weighed) silver chloride with a mixture of 10 mL of concentrated ammonia solution and 10 mL of $1M$ ammonium chloride solution, then add about 0.2 g of potassium cyanonickelate and warm gently. Dilute to 100 mL with de-ionised water, add 50 mg of the indicator mixture and titrate with standard (0.01 M) EDTA solution, adding the reagent dropwise in the neighbourhood of the end point, until the colour changes from yellow to violet.

1 mole EDTA $\equiv$ 2 moles Ag^+

Palladium(II) compounds can be determined by a similar procedure, but in this case, after addition of the cyanonickelate, excess of standard (0.01 M) EDTA solution is added, and the excess is back-titrated with standard (0.01 M) manganese(II) sulphate solution using solochrome black indicator.

Gold may be titrated similarly.

10.59 DETAILS FOR THE DETERMINATION OF A SELECTION OF METAL IONS BY EDTA TITRATION

With the detailed instructions given in Sections 10.51–10.58 it should be possible to carry out any of the following determinations in Table 10.6 without serious problems arising. In all cases it is recommended that the requisite pH value for the titration should be established by use of a pH meter, but in the light of experience the colour of the indicator at the required pH may, in some cases, be a satisfactory guide. Where no actual buffering agent is specified, the solution should be brought to the required pH value by the cautious addition of dilute acid or of dilute sodium hydroxide solution or aqueous ammonia solution as required.

ANALYSIS OF MIXTURES OF CATIONS

10.60 DETERMINATION OF CALCIUM AND MAGNESIUM

Discussion. Patton and Reeder's indicator (HHSNNA), see Section 10.48, makes it possible to determine calcium in the presence of magnesium, and is

Table 10.6 Summarised procedures for EDTA titrations of some selected cations

Metal	Titration type	pH	Buffer	Indicator (Note 1)	Colour change (Note 2)		Notes
Aluminium*	Back	7–8	Aq. NH_3	SB	B	R	
Barium*	Direct	12		MTB	B	Gr	(3)
Bismuth	Direct*	1		XO	R	Y	
	Direct	0–1		MTB	B	Y	
Cadmium	Direct	5	Hexamine	XO	R	Y	
Calcium	Direct	12		MTB	B	Gr	
	Substn.*	7–11	Aq. NH_3/NH_4Cl	SB	R	B	
Cobalt	Direct	6	Hexamine	XO	R	Y	(4)
Copper*	Direct			FSB	P	G	(6)
Gold	See silver						
Iron(III)*	Direct	2–3		VB	B	Y	(4)
Lead	Direct	6	Hexamine	XO	R	Y	
Magnesium	Direct	10	Aq. NH_3/NH_4Cl	SB	R	B	(4a)
Manganese	Direct	10	Aq. NH_3/NH_4Cl	SB	R	B	(5)
	Direct	10	Aq. NH_3	TPX	B	PP	(5)
Mercury	Direct	6	Hexamine	XO	R	Y	
	Direct	6	Hexamine	MTB	B	Y	
Nickel	Direct*	7–10	Aq. NH_3/NH_4Cl	M	Y	V	
	Direct*	7–10	Aq. NH_3/NH_4Cl	BPR	B	R	
	Back	10	Aq. NH_3/NH_4Cl	SB	B	R	
Palladium	See silver						
Silver*	Indirect			M	Y	V	
Strontium	Direct	12		MTB	B	Gr	
	Direct	10–11		TPX	B	PP	
Thorium	Direct	2–3		XO	R	Y	
	Direct	2–3		MTB	B	Y	
Tin(II)	Direct	6	Hexamine	XO	R	Y	
Zinc	Direct	10	Aq. NH_3/NH_4Cl	SB	R	B	
	Direct	6	Hexamine	XO	R	Y	
	Direct	6	Hexamine	MTB	B	Y	

* Details in Sections 10.51–10.58.

Notes to Table 10.6

(1) BPR = bromopyrogallol red; FSB = fast sulphone black F; M = murexide; MTB = methylthymol blue; SB = solochrome black; TPX = thymolphthalexone; VB = variamine blue; XO = xylenol orange.

(2) B = Blue; G = Green; Gr = Grey; P = Purple; PP = Pale pink; R = Red; V = Violet; Y = Yellow.

(3) Can also be determined by precipitation as $BaSO_4$ and dissolution in excess EDTA (Section 10.73).

(4) Temperature 40 °C. (4a) Warming optional.

(5) Add 0.5 g hydroxylammonium chloride (to prevent oxidation), and 3 mL triethanolamine (to prevent precipitation in alkaline solution); use boiled-out (air-free) water.

(6) In presence of concentrated aqueous ammonia.

used in the determination of the hardness of water and in the analysis of limestone and dolomite. Titration using solochrome black gives calcium and magnesium together, and the difference between the two titrations gives the magnesium content of the mixture (Note 1).

Calcon may also be used for the titration of calcium in the presence of magnesium (compare Section 10.48). The neutral solution (say, 50 mL) is treated with 5 mL of diethylamine (giving a pH of about 12.5, which is sufficiently high to precipitate the magnesium quantitatively as the hydroxide) and four drops of calcon indicator are added. The solution is stirred magnetically and titrated with standard EDTA solution until the colour changes from pink to a pure blue.

A sharper end point may be obtained by adding 2–3 drops of 1 per cent aqueous poly(vinyl alcohol) to the sample solution, then adjusting the pH to 12.5 with sodium hydroxide, adding 2–3 drops of 10 per cent aqueous potassium cyanide solution, warming to 60 °C (**Caution:** use a fume cupboard), and treating the warm solution with 3–4 drops of calcon indicator. The solution is titrated with 0.01 M EDTA to a red–blue end-point. The poly(vinyl alcohol) reduces the adsorption of the dye on the surface of the precipitate. The solution is prepared by mixing 1.0 g of medium-viscosity poly(vinyl alcohol) with 100 mL of boiling water in a mechanical homogenizer.

Procedure. Prepare the indicators by grinding (*a*) 0.5 g HHSNNA with 50 g potassium chloride, and (*b*) 0.2 g solochrome black with 50 g potassium chloride. The following solutions will also be required:

*Magnesium chloride solution (0.01*M*).* Dissolve 0.608 g pure magnesium turnings in dilute hydrochloric acid, nearly neutralise with sodium hydroxide solution (1 M) and make up to 250 mL in a graduated flask with de-ionised water. Pipette 25 mL of the resulting 0.1 M solution into a 250 mL graduated flask and make up to the mark with de-ionised water.

*Potassium hydroxide solution (ca 8*M*).* Dissolve 112 g potassium hydroxide pellets in 250 mL of de-ionised water.

Buffer solution. Add 55 mL of concentrated hydrochloric acid to 400 mL de-ionised water and mix thoroughly. Slowly pour 310 mL of redistilled monoethanolamine with stirring into the mixture and cool to room temperature (Note 2). Titrate 50.0 mL of the standard magnesium chloride solution with standard (0.01 M) EDTA solution using 1 mL of the monoethanolamine–hydrochloric acid solution as the buffer and solochrome black as the indicator. Add 50.0 mL of the magnesium chloride solution to the volume of EDTA solution required to complex the magnesium exactly (as determined in the last titration), pour the mixture into the monoethanolamine–hydrochloric acid solution, and mix well. Dilute to 1 litre (Note 3).

Determination of calcium. Pipette two 25.0 mL portions of the mixed calcium and magnesium ion solution (not more than 0.01M with respect to either ion) into two separate 250 mL conical flasks and dilute each with about 25 mL of de-ionised water. To the first flask add 4 mL 8M potassium hydroxide solution (a precipitate of magnesium hydroxide may be noted here), and allow to stand for 3–5 minutes with occasional swirling. Add about 30 mg each of potassium cyanide (**Caution:** poison) and hydroxylammonium chloride and swirl the contents of the flask until the solids dissolve. Add about 50 mg of the HHSNNA indicator mixture and titrate with 0.01 M EDTA until the colour changes from red to blue. Run into the second flask from a burette a volume of EDTA solution equal to that required to reach the end point less 1 mL. Now add 4 mL of the potassium hydroxide solution, mix well and complete the titration as with the first sample; record the exact volume of EDTA solution used. Perform a blank titration, replacing the sample with de-ionised water.

Determination of total calcium and magnesium. Pipette 25.0 mL of the mixed calcium and magnesium ion solution into a 250 mL conical flask, dilute to about 50 mL with distilled water, add 5 mL of the buffer solution, and mix by swirling. Add about 30 mg of potassium cyanide (**Caution!**) and 30 mg of hydroxylammonium chloride, shake gently until the solids dissolve, and then add about 50 mg of the solochrome black indicator mixture. Titrate with the EDTA solution to a pure blue end point. Perform a blank titration, replacing the 25 mL sample solution with de-ionised water.

Calculate the volume of standard EDTA solution equivalent to the magnesium by subtracting the total volume required for the calcium from the volume required for the total calcium and magnesium for equal amounts of the test sample.

Notes. (1) The usefulness of the HHSNNA indicator for the titration of calcium depends upon the fact that the pH of the solution is sufficiently high to ensure the quantitative precipitation of the magnesium as hydroxide and that calcium forms a more stable complex with EDTA than does magnesium. The EDTA does not react with magnesium [present as $Mg(OH)_2$] until all the free calcium and the calcium–indicator complex have been complexed by the EDTA. If the indicator is added before the potassium hydroxide, a satisfactory end-point is not obtained because magnesium salts form a lake with the indicator as the pH increases and the magnesium indicator-lake is co-precipitated with the magnesium hydroxide.

(2) The monoethanolamine–hydrochloric acid buffer has a buffering capacity equal to the ammonia–ammonium chloride buffer commonly employed for the titration of calcium and magnesium with EDTA and solochrome black (compare Section 10.54). The buffer has excellent keeping qualities, sharp end points are obtainable, and the strong ammonia solution is completely eliminated.

(3) When relatively pure samples of calcium are titrated using solochrome black as indicator, magnesium must be added to obtain a sharp end point, hence magnesium is usually added to the buffer solution (compare Section 10.54). The addition of magnesium to the EDTA solution prevents a sharp end point when calcium is titrated using HHSNNA as indicator. The introduction of complexed magnesium into the buffer eliminates the need for two EDTA solutions and ensures an adequate amount of magnesium, even when small amounts of this element are titrated.

10.61 DETERMINATION OF CALCIUM IN THE PRESENCE OF MAGNESIUM USING EGTA AS TITRANT

Discussion. Calcium may be determined in the presence of magnesium by using EGTA as titrant, because whereas the stability constant for the calcium–EGTA complex is about 1×10^{11}, that of the magnesium–EGTA complex is only about 1×10^5, and thus magnesium does not interfere with the reagent. The method described in the preceding section, which involves precipitation of magnesium hydroxide, is not satisfactory if the magnesium content of the mixture is much greater than about 10 per cent of the calcium content, since co-precipitation of calcium hydroxide may occur. Titration with EGTA is therefore to be recommended for the determination of small amounts of calcium in the presence of larger amounts of magnesium.

The indicator used in the titration is zincon (Section 10.48) which gives rise to an indirect end point with calcium. Detection of the end point is dependent upon the reaction

$$ZnEGTA^{2-} + Ca^{2+} = Zn^{2+} + CaEGTA^{2-}$$

and the zinc ions liberated form a blue complex with the indicator. At the end point, the zinc–indicator complex is decomposed:

$$ZnIn^- + H_2EGTA^{2-} \rightleftharpoons ZnEGTA^{2-} + HIn^-$$

and the solution acquires the orange–red colour of the indicator.

Procedure. Prepare an EGTA solution $(0.05\,M)$ by dissolving 19.01 g in 100 mL sodium hydroxide solution $(1\,M)$ and diluting to 1 L in a graduated flask with de-ionised water. Prepare the indicator by dissolving 0.065 g zincon in 2 mL sodium hydroxide solution $(0.1\,M)$ and diluting to 100 mL with de-ionised water, and a buffer solution (pH 10) by dissolving 25 g sodium tetraborate, 3.5 g ammonium chloride, and 5.7 g sodium hydroxide in 1 L of de-ionised water.

Prepare 100 mL of **Zn–EGTA complex solution** by taking 50 mL of $0.05\,M$ zinc sulphate solution and adding an equivalent volume of $0.05\,M$ EGTA solution; exact equality of zinc and EGTA is best achieved by titrating a 10 mL portion of the zinc sulphate solution with the EGTA solution using zincon indicator, and from this result the exact volume of EGTA solution required for the 50 mL portion of zinc sulphate solution may be calculated.

The EGTA solution may be standardised by titration of a standard $(0.05\,M)$ calcium solution, prepared by dissolving 5.00 g calcium carbonate in dilute hydrochloric acid contained in a 1 L graduated flask, and then after neutralising with sodium hydroxide solution diluting to the mark with de-ionised water: use zincon indicator in the presence of Zn–EGTA solution (see below).

To determine the calcium in the calcium–magnesium mixture, pipette 25 mL of the solution into a 250 mL conical flask, add 25 mL of the buffer solution and check that the resulting solution has a pH of 9.5–10.0. Add 2 mL of the Zn–EGTA solution and 2–3 drops of the indicator solution. Titrate slowly with the standard EGTA solution until the blue colour changes to orange–red.

10.62 DETERMINATION OF THE TOTAL HARDNESS (PERMANENT AND TEMPORARY) OF WATER

The hardness of water is generally due to dissolved calcium and magnesium salts and may be determined by complexometric titration.

Procedure. To a 50 mL sample of the water to be tested add 1 mL buffer solution (ammonium hydroxide/ammonium chloride, pH 10, Section 10.54) and 30–40 mg solochrome black indicator mixture. Titrate with standard EDTA solution $(0.01\,M)$ until the colour changes from red to pure blue. Should there be no magnesium present in the sample of water it is necessary to add 0.1 mL magnesium–EDTA solution $(0.1\,M)$ before adding the indicator (see Section 10.54). The total hardness is expressed in parts of $CaCO_3$ per million of water.

If the water contains traces of interfering ions, then 4 mL of buffer solution should be added, followed by 30 mg of hydroxylammonium chloride and then 50 mg analytical-grade potassium cyanide (**Caution**) before adding the indicator.

Notes. (1) Somewhat sharper end points may be obtained if the sample of water is first acidified with dilute hydrochloric acid, boiled for about a minute to drive off carbon dioxide, cooled, neutralised with sodium hydroxide solution, buffer and indicator solution added, and then titrated with EDTA as above.

(2) **The permanent hardness** of a sample of water may be determined as follows. Place 250 mL of the sample of water in a 600 mL beaker and boil gently for 20–30 minutes. Cool and filter it directly into a 250 mL graduated flask: do not wash the filter paper,

but dilute the filtrate to volume with de-ionised water and mix well. Titrate 50.0 mL of the filtrate by the same procedure as was used for the total hardness. This titration measures the permanent hardness of the water. Calculate this hardness as parts per million of $CaCO_3$.

Calculate the **temporary hardness** of the water by subtracting the permanent hardness from the total hardness.

(3) If it is desired to determine both the calcium and the magnesium in a sample of water, determine first the total calcium and magnesium content as above, and calculate the result as parts per million of $CaCO_3$.

The calcium content may then be determined by titration wth EDTA using either Patton and Reeder's indicator or calcon (Section 10.60), or alternatively by titration with EGTA (see Section 10.61).

10.63 DETERMINATION OF CALCIUM IN THE PRESENCE OF BARIUM USING CDTA AS TITRANT

There is an appreciable difference between the stability constants of the CDTA complexes of barium ($\log K = 7.99$) and calcium ($\log K = 12.50$), with the result that calcium may be titrated with CDTA in the presence of barium; the stability constants of the EDTA complexes of these two metals are too close together to permit independent titration of calcium in the present of barium.

The indicator calcichrome (see Section 10.48) is specific for calcium at pH 11–12 in the presence of barium.

Procedure. Prepare the CDTA solution ($0.02 M$) by dissolving 6.880 g of the solid reagent in 50 mL of sodium hydroxide solution ($1 M$) and making up to 1 L with de-ionised water; the solution may be standardised against a standard calcium solution prepared from 2.00 g of calcium carbonate (see Section 10.61). The indicator is prepared by dissolving 0.5 g of the solid in 100 mL of water.

Pipette 25 mL of the solution to be analysed into a 250 mL conical flask and dilute to 100 mL with de-ionised water: the original solution should be about $0.02 M$ with respect to calcium and may contain barium to a concentration of up to $0.2 M$. Add 10 mL sodium hydroxide solution ($1 M$) and check that the pH of the solution lies between 11 and 12; then add three drops of the indicator solution. Titrate with the standard CDTA solution until the pink colour changes to blue.

10.64 DETERMINATION OF CALCIUM AND LEAD IN A MIXTURE

With methylthymol blue, lead may be titrated at a pH of 6 without interference by calcium; the calcium is subsequently titrated at pH 12.

Procedure. Pipette 25 mL of the test solution (which may contain both calcium and lead at concentrations of up to $0.01 M$) into a 250 mL conical flask and dilute to 100 mL with de-ionised water. Add about 50 mg of methylthymol blue/potassium nitrate mixture followed by dilute nitric acid until the solution is yellow, and then add powdered hexamine until the solution has an intense blue colour (pH *ca* 6). Titrate with standard ($0.01 M$) EDTA solution until the colour turns to yellow; this gives the titration value for lead.

Now carefully add sodium hydroxide solution ($1 M$) until the pH of the solution has risen to 12 (pH meter); 3–6 mL of the sodium hydroxide solution will be required. Continue the titration of the bright blue solution with the

EDTA solution until the colour changes to grey; this gives the titration value for calcium.

10.65 DETERMINATION OF MAGNESIUM, MANGANESE AND ZINC IN A MIXTURE: USE OF FLUORIDE ION AS A DEMASKING AGENT

Discussion. In mixtures of magnesium and manganese the sum of both ion concentrations may be determined by direct EDTA titration. Fluoride ion will demask magnesium selectively from its EDTA complex, and if excess of a standard solution of manganese ion is also added, the following reaction occurs at room temperature:

$$MgY^{2-} + 2F^- + Mn^{2+} = MgF_2 + MnY^{2-}$$

The excess of manganese ion is evaluated by back-titration with EDTA. The amount of standard manganese ion solution consumed is equivalent to the EDTA 'liberated' by the fluoride ion, which is in turn equivalent to the magnesium in the sample.

Mixtures of manganese, magnesium, and zinc can be similarly analysed. The first EDTA end point gives the sum of the three ions. Fluoride ion is added and the EDTA liberated from the magnesium–EDTA complex is titrated with manganese ion as detailed above. Following the second end point cyanide ion is added to displace zinc from its EDTA chelate and to form the stable cyanozincate complex $[Zn(CN)_4]^{2-}$; the liberated EDTA (equivalent to the zinc) is titrated with standard manganese-ion solution.

Details for analysis of Mn–Mg–Zn mixtures are given below.

Procedure. Prepare a manganese(II) sulphate solution (approx. 0.05 M) by dissolving 11.15 g of the analytical-grade solid in 1 L of de-ionised water; standardise the solution by titration with 0.05 M EDTA solution using solochrome black indicator after the addition of 0.25 g of hydroxylammonium chloride — see below.

Prepare a buffer solution (pH 10) by dissolving 8.0 g ammonium nitrate in 65 mL of de-ionised water and adding 35 mL of concentrated ammonia solution (sp. gr. 0.88).

Pipette 25 mL of the solution containing magnesium, manganese and zinc ions (each approx. 0.02 M), into a 250 mL conical flask and dilute to 100 mL with de-ionised water. Add 0.25 g hydroxylammonium chloride [this is to prevent oxidation of Mn(II) ions], followed by 10 mL of the buffer solution and 30–40 mg of the indicator/potassium nitrate mixture. Warm to 40 °C and titrate (preferably stirring magnetically) with the standard EDTA solution to a pure blue colour.

After the end point, add 2.5 g of sodium fluoride, stir (or agitate) for 1 minute. Now introduce the standard manganese(II) sulphate solution from a burette in 1 mL portions until a *permanent* red colour is obtained; note the exact volume added. Stir for 1 minute. Titrate the excess of manganese ion with EDTA until the colour changes to pure blue.

After the second end point, add 4–5 mL of 15 per cent aqueous potassium cyanide solution (**CARE**), and run in the standard manganese ion solution from a burette until the colour changes sharply from blue to red. Record the exact volume of manganese(II) sulphate solution added.

Calculate the weights of magnesium, zinc, and manganese in the sample solution.

Example of calculation. In the standardisation of the Mn(II) solution, 25.0 mL of the solution required 30.30 mL of 0.0459 M EDTA solution.

$\therefore$ Molarity of Mn(II) solution $= (30.30 \times 0.0459)/25.0 = 0.0556\,M$

First titration of mixture with EDTA (Mg + Mn + Zn) = 33.05 mL

Second titration (after adding NaF; gives Mg) = 9.85 mL of Mn(II) solution and the excess of Mn(II) required 1.26 mL of standard EDTA solution.

$\therefore$ EDTA liberated by NaF $= (9.85 \times 0.0556) - (1.26 \times 0.0459)$ mmol
$$= 0.4899 \text{ mmol}$$

and weight of magnesium per mL $= (0.4899 \times 24.31)/1000$ g
$$= 11.91 \text{ mg}$$

Third titration (after adding KCN; gives Zn) = 8.46 mL of Mn(II) solution.

$\therefore$ EDTA liberated by KCN $= (8.46 \times 0.0556)$ mmol $= 0.4703$ mmol

and weight of zinc per mL $= (0.4703 \times 65.38)/1000$ g $= 30.75$ mg mL^{-1}

In the first titration, (33.05×0.0459) millimoles of EDTA were used $= 1.5170$ mmol,

which represents the total amount of metal ion titrated (Mn + Mg + Zn). Hence amount of Mn $= 1.5170 - (0.4899 + 0.4703) = 0.5568$ mmol

and weight of manganese $= (0.5568 \times 54.94)/1000$ g mL^{-1} = 30.60 mg mL^{-1}

10.66 DETERMINATION OF CHROMIUM(III) AND IRON(III) IN A MIXTURE: AN EXAMPLE OF KINETIC MASKING

Iron (and nickel, if present) can be determined by adding an excess of standard EDTA to the cold solution, and then back-titrating the solution with lead nitrate solution using xylenol orange as indicator; provided the solution is kept cold, chromium does not react. The solution from the back-titration is then acidified, excess of standard EDTA solution added and the solution boiled for 15 minutes when the red–violet Cr(III)–EDTA complex is produced. After cooling and buffering to pH 6, the excess EDTA is then titrated with the lead nitrate solution.

Procedure. Place 10 mL of the solution containing the two metals (the concentration of neither of which should exceed 0.01 M) in a 600 mL beaker fitted with a magnetic stirrer, and dilute to 100 mL with de-ionised water. Add 20 mL of standard (approx. 0.01 M) EDTA solution and add hexamine to adjust the pH to 5–6. Then add a few drops of the indicator solution (0.5 g xylenol orange dissolved in 100 mL of water) and titrate the excess EDTA with a standard lead nitrate solution (0.01 M), i.e. to the formation of a red–violet colour.

To the resulting solution now add a further 20 mL portion of the standard EDTA solution, add nitric acid (1 M) to adjust the pH to 1–2, and then boil the solution for 15 minutes. Cool, dilute to 400 mL by the addition of de-ionised water, add hexamine to bring the pH to 5–6, add more of the indicator solution, and titrate the excess EDTA with the standard lead nitrate solution.

The first titration determines the amount of EDTA used by the iron, and the second, the amount of EDTA used by the chromium.

10.67 DETERMINATION OF MANGANESE IN PRESENCE OF IRON: ANALYSIS OF FERRO-MANGANESE

After dissolution of the alloy in a mixture of concentrated nitric and hydrochloric acids the iron is masked with triethanolamine in an alkaline medium, and the manganese titrated with standard EDTA solution using thymolphthalexone as indicator. The amount of iron(III) present must not exceed 25 mg per 100 mL of solution, otherwise the colour of the iron(III)–triethanolamine complex is so intense that the colour change of the indicator is obscured. Consequently, the procedure can only be used for samples of ferro-manganese containing more than about 40 per cent manganese.

Procedure. Dissolve a weighed amount of ferro-manganese (about 0.40 g) in concentrated nitric acid and then add concentrated hydrochloric acid (or use a mixture of the two concentrated acids); prolonged boiling may be necessary. Evaporate to a small volume on a water bath. Dilute with water and filter directly into a 100 mL graduated flask, wash with distilled water and finally dilute to the mark. Pipette 25.0 mL of the solution into a 500 mL conical flask, add 5 mL of 10 per cent aqueous hydroxylammonium chloride solution, 10 mL of 20 per cent aqueous triethanolamine solution, 10–35 mL of concentrated ammonia solution, about 100 mL of water, and 6 drops of thymolphthalexone indicator solution. Titrate with standard 0.05 M EDTA until the colour changes from blue to colourless (or a very pale pink).

10.68 DETERMINATION OF NICKEL IN PRESENCE OF IRON: ANALYSIS OF NICKEL STEEL

Nickel may be determined in the presence of a large excess of iron(III) in weakly acidic solution by adding EDTA and triethanolamine; the intense brown precipitate dissolves upon the addition of aqueous sodium hydroxide to yield a colourless solution. The iron(III) is present as the triethanolamine complex and only the nickel is complexed by the EDTA. The excess of EDTA is back-titrated with standard calcium chloride solution in the presence of thymolphthalexone indicator. The colour change is from colourless or very pale blue to an intense blue. The nickel–EDTA complex has a faint blue colour; the solution should contain less than 35 mg of nickel per 100 mL.

In the back-titration small amounts of copper and zinc and trace amounts of manganese are quantitatively displaced from the EDTA and are complexed by the triethanolamine: small quantities of cobalt are converted into a triethanolamine complex during the titration. Relatively high concentrations of copper can be masked in the alkaline medium by the addition of thioglycollic acid until colourless. Manganese, if present in quantities of more than 1 mg, may be oxidised by air and forms a manganese(III)–triethanolamine complex, which is intensely green in colour; this does not occur if a little hydroxylammonium chloride solution is added.

Procedure. Prepare a standard calcium chloride solution (0.01 M) by dissolving 1.000 g of calcium carbonate in the minimum volume of dilute hydrochloric acid and diluting to 1 L with de-ionised water in a graduated flask. Also prepare a 20 per cent aqueous solution of triethanolamine.

Weigh out accurately a 1.0 g sample of the nickel steel and dissolve it in the minimum volume of concentrated hydrochloric acid (about 15 mL) to which a little concentrated nitric acid (*ca* 1 mL) has been added. Dilute to 250 mL in a graduated flask. Pipette 25.0 mL of this solution into a conical flask, add 25.0 mL of 0.01 *M* EDTA and 10 mL of triethanolamine solution. Introduce 1 *M* sodium hydroxide solution, with stirring, until the pH of the solution is 11.6 (use a pH meter). Dilute to about 250 mL. Add about 0.05 g of the thymolphthalexone/potassium nitrate mixture; the solution acquires a very pale blue colour. Titrate with 0.01 *M* calcium chloride solution until the colour changes to an intense blue. If it is felt that the end point colour change is not sufficiently distinct, add a further small amount of the indicator, a known volume of 0.01 *M* EDTA and titrate again with 0.01 *M* calcium chloride.

10.69 DETERMINATION OF LEAD AND TIN IN A MIXTURE: ANALYSIS OF SOLDER

A mixture of tin(IV) and lead(II) ions may be complexed by adding an excess of standard EDTA solution, the excess EDTA being determined by titration with a standard solution of lead nitrate; the total lead-plus-tin content of the solution is thus determined. Sodium fluoride is then added and this displaces the EDTA from the tin(IV)–EDTA complex; the liberated EDTA is determined by titration with a standard lead solution.

Procedure. Prepare a standard EDTA solution (0.2 *M*), a standard lead solution (0.01 *M*), a 30 per cent aqueous solution of hexamine, and a 0.2 per cent aqueous solution of xylenol orange.

Dissolve a weighed amount (about 0.4 g) of solder in 10 mL of concentrated hydrochloric acid and 2 mL of concentrated nitric acid; gentle warming is necessary. Boil the solution gently for about 5 minutes to expel nitrous fumes and chlorine, and allow to cool slightly, whereupon some lead chloride may separate. Add 25.0 mL of standard 0.2 *M* EDTA and boil for 1 minute; the lead chloride dissolves and a clear solution is obtained. Dilute with 100 mL of de-ionised water, cool and dilute to 250 mL in a graduated flask. Without delay, pipette two or three 25.0 mL portions into separate conical flasks. To each flask add 15 mL hexamine solution, 110 mL de-ionised water, and a few drops of xylenol orange indicator. Titrate with the standard lead nitrate solution until the colour changes from yellow to red. Now add 2.0 g sodium fluoride; the solution acquires a yellow colour owing to the liberation of EDTA from its tin complex. Titrate again with the standard lead nitrate solution until a permanent (i.e. stable for 1 minute) red colour is obtained. Add the titrant dropwise near the end point; a temporary pink or red colour gradually reverting to yellow signals the approach of the end point.

10.70 DETERMINATION OF BISMUTH, CADMIUM AND LEAD IN A MIXTURE: ANALYSIS OF A LOW–MELTING ALLOY

The analysis of low-melting alloys such as Wood's metal is greatly simplified by complexometric titration, and tedious gravimetric separations are avoided. The alloy is treated with concentrated nitric acid, evaporated to a small volume, and after dilution the precipitated tin(IV) oxide is filtered off; heavy metals adsorbed by the precipitate are removed by washing with a known volume of standard EDTA solution previously made slightly alkaline with aqueous ammonia. The hydrated tin(IV) oxide is ignited and weighed. The Bi, Pb, and Cd are determined in the combined filtrate and washings from the tin separation: these are diluted to a known volume and aliquots used in the subsequent titrations. The bismuth content is determined by titration with standard EDTA at pH 1–2 using xylenol orange as indicator; then, after adjustment of the pH to 5–6 with hexamine, the combined Pb plus Cd can be titrated with EDTA. 1,10-Phenanthroline is then added to mask the cadmium, and the liberated EDTA is titrated with standard lead nitrate solution; this gives the cadmium content and thence the lead content is obtained by difference.

Procedure. Prepare a standard solution of lead nitrate ($0.05 M$), a 0.05 per cent aqueous solution of xylenol orange indicator, and a 1,10-phenanthroline solution ($0.05 M$) by dissolving 0.90 g of pure 1,10-phenanthroline in 1.5 mL of concentrated nitric acid and 100 mL of water.

Weigh out accurately 2.0–2.5 g of Wood's metal and dissolve it in hot concentrated nitric acid (*ca* 50 mL). Evaporate the resulting solution to a small volume, dilute to about 150 mL with water and boil for 1–2 minutes. Filter off the precipitate of hydrated tin(IV) oxide through a quantitative filter paper (Whatman No. 542) and keep the filtrate. Make a known volume (say 50.0 mL) of a $0.05 M$ EDTA solution slightly basic with aqueous ammonia. Wash the precipitate on the filter with this solution and then with 50 mL of water. The final wash liquid should give no precipitate with 5 per cent sodium sulphide solution. Transfer the filtrate and washings (containing the other metals as well as the excess of EDTA) quantitatively to a 500 mL graduated flask, and dilute to the mark with de-ionised water. Char the filter paper in the usual way and, after ignition, weigh the tin(IV) oxide (see Section 11.9).

Into a conical flask, pipette a 50.0 or 100.0 mL aliquot of the solution and adjust the pH to 1–2 with aqueous ammonia solution (use pH test-paper). Add five drops of xylenol orange indicator and titrate with additional $0.05 M$ EDTA until the colour changes sharply from red to yellow. This gives the bismuth content. Record the total (combined) volume of EDTA solution used. Now add small amounts of hexamine (*ca* 5 g) until an intense red–violet coloration persists, and titrate with the standard EDTA to a yellow end point; the further consumption of EDTA corresponds to the lead-plus-cadmium content.

To determine the cadmium content, add 20–25 mL of the 1,10-phenanthroline solution and titrate the liberated EDTA with the $0.05 M$ lead nitrate solution until the colour change from yellow to red–violet occurs — a little practice is required to discern the end point precisely. Introduce further 2–5 mL portions of the 1,10-phenanthroline solution and note whether the indicator colour changes: if so, continue the titration with the lead nitrate solution. The consumption of lead nitrate solution corresponds to the cadmium content.

DETERMINATION OF ANIONS

Anions do not complex directly with EDTA, but methods can be devised for the determination of appropriate anions which involve either (i) adding an excess of a solution containing a cation which reacts with the anion to be determined, and then using EDTA to measure the excess of cation added; or (ii) the anion is precipitated with a suitable cation, the precipitate is collected, dissolved in excess EDTA solution and then the excess EDTA is titrated with a standard solution of an appropriate cation. The procedure involved in the first method will be self-evident but some details are given for determinations carried out by the second method.

10.71 DETERMINATION OF HALIDES (EXCLUDING FLUORIDE) AND THIOCYANATES

The procedure involved in the determination of these anions is virtually that discussed in Section 10.58 for the indirect determination of silver. The anion to be determined is precipitated as the silver salt; the precipitate is collected and dissolved in a solution of potassium tetracyanonickelate(II) in the presence of an ammonia/ammonium chloride buffer. Nickel ions are liberated and titrated with standard EDTA solution using murexide as indicator:

$$2Ag^+ + [Ni(CN)_4]^{2-} = Ni^{2+} + 2[Ag(CN)_2]^-$$

The method may be illustrated by the determination of bromide; details for the preparation of the potassium tetracyanonickelate are given in Section 10.58.

Pipette 25.0 mL of the bromide ion solution (0.01–0.02M) into a 400 mL beaker, add excess of dilute silver nitrate solution, filter off the precipitated silver bromide on a sintered glass filtering crucible, and wash it with cold water. Dissolve the precipitate in a warm solution prepared from 15 mL of concentrated ammonia solution, 15 mL of 1M ammonium chloride, and 0.3 g of potassium tetracyanonickelate. Dilute to 100–200 mL, add three drops of murexide indicator, and titrate with standard EDTA (0.01M) (slowly near the end point) until the colour changes from yellow to violet.

1 mole EDTA $\equiv$ 2 moles Br$^-$

10.72 DETERMINATION OF PHOSPHATES

The phosphate is precipitated as $Mg(NH_4)PO_4,6H_2O$, the precipitate is filtered off, washed, dissolved in dilute hydrochloric acid, an excess of standard EDTA solution added, the pH adjusted to 10, and the excess of EDTA titrated with standard magnesium chloride or magnesium sulphate solution using solochrome black as indicator. The initial precipitation may be carried out in the presence of a variety of metals by first adding sufficient EDTA solution (1M) to form complexes with all the multicharged metal cations, then adding excess of magnesium sulphate solution, followed by ammonia solution: alternatively, the cations may be removed by passing the solution through a cation exchange resin in the hydrogen form.

Procedure. Prepare a standard (0.05M) solution of magnesium sulphate or chloride from pure magnesium (Section 10.60), an ammonia–ammonium chloride buffer solution (pH 10) (Section 10.54), and a standard (0.05M) solution of EDTA.

Pipette 25.0 mL of the phosphate solution (approx. 0.05M) into a 250 mL beaker and dilute to 50 mL with de-ionised water; add 1 mL of concentrated hydrochloric and a few drops of methyl red indicator. Treat with an excess of 1M magnesium sulphate solution (*ca* 2 mL), heat the solution to boiling, and add concentrated ammonia solution dropwise and with vigorous stirring until the indicator turns yellow, followed by a further 2 mL. Allow to stand for several hours or overnight. Filter the precipitate through a sintered-glass crucible (porosity G4) and wash thoroughly with 1 M ammonia solution (about 100 mL). Rinse the beaker (in which the precipitation was made) with 25 mL of hot 1 M hydrochloric acid and allow the liquid to percolate through the filter crucible, thus dissolving the precipitate. Wash the beaker and crucible with a further 10 mL of 1 M hydrochloric acid and then with about 75 mL of water. To the filtrate and washings in the filter flask add 35.0 mL of 0.05 M EDTA, neutralise the solution with 1 M sodium hydroxide, add 4 mL of buffer solution and a few drops of solochrome black indicator. Back-titrate with standard 0.05 M magnesium chloride until the colour changes from blue to wine red.

10.73 DETERMINATION OF SULPHATES

The sulphate is precipitated as barium sulphate from acid solution, the precipitate is filtered off and dissolved in a measured excess of standard EDTA solution in the presence of aqueous ammonia. The excess of EDTA is then titrated with standard magnesium chloride solution using solochrome black as indicator.

Procedure. Prepare a standard magnesium chloride solution (0.05 M) and a buffer solution (pH 10); see Section 10.72. Standard EDTA (0.05 M) will also be required.

Pipette 25.0 mL of the sulphate solution (0.02–0.03 M) into a 250 mL beaker, dilute to 50 mL, and adjust the pH to 1 with 2 M hydrochloric acid; heat nearly to boiling. Add 15 mL of a nearly boiling barium chloride solution (ca 0.05 M) fairly rapidly and with vigorous stirring: heat on a steam bath for 1 hour. Filter with suction through a filter-paper disc (Whatman filter paper No. 42) supported upon a porcelain filter disc or a Gooch crucible, wash the precipitate thoroughly with cold water, and drain. Transfer the filter-paper disc and precipitate quantitatively to the original beaker, add 35.0 mL standard 0.05 M EDTA solution and 5 mL concentrated ammonia solution and boil gently for 15–20 minutes; add a further 2 mL concentrated ammonia solution after 10–15 minutes to facilitate the dissolution of the precipitate. Cool the resulting clear solution, add 10 mL of the buffer solution (pH = 10), a few drops of solochrome black indicator, and titrate the excess of EDTA with the standard magnesium chloride solution to a clear red colour.

Sulphate can also be determined by an exactly similar procedure by precipitation as lead sulphate from a solution containing 50 per cent (by volume) of propan-2-ol (to reduce the solubility of the lead sulphate), separation of the precipitate, dissolution in excess of standard EDTA solution, and back-titration of the excess EDTA with a standard zinc solution using solochrome black as indicator.

PRECIPITATION TITRATIONS

10.74 PRECIPITATION REACTIONS

The most important precipitation processes in titrimetric analysis utilise silver nitrate as the reagent (argentimetric processes). Discussion of the theory will, therefore, be confined to argentimetric processes; the same principles can, of course, be applied to other precipitation reactions. Consider the changes in ionic concentration which occur during the titration of 100 mL of $0.1 M$ sodium chloride with $0.1 M$ silver nitrate. The solubility product of silver chloride at the laboratory temperature is 1.2×10^{-10}. The initial concentration of chloride ions, $[Cl^-]$, is 0.1 mol L^{-1}, or $pCl^- = 1$ (see Section 2.17). When 50 mL of $0.1 M$ silver nitrate have been added, 50 mL of $0.1 M$ sodium chloride remain in a total volume of 150 mL: thus $[Cl^-] = 50 \times 0.1/150 = 3.33 \times 10^{-2}$, or $pCl^- = 1.48$. With 90 mL of silver nitrate solution $[Cl^-] = 10 \times 0.1/190 = 5.3 \times 10^{-3}$, or $pCl^- = 2.28$.

Now

$$a_{Ag^+} \times a_{Cl^-} \approx [Ag^+] \times [Cl^-] = 1.2 \times 10^{-10} = K_{sol. \ AgCl}$$

or

$$pAg^+ + pCl^- = 9.92 = pAgCl$$

In the last calculation, $pCl^- = 1.48$, hence $pAg^+ = 9.92 - 1.48 = 8.44$. In this manner, the various concentrations of chloride and silver ions can be computed up to the equivalence point. At the equivalence point:

$$Ag^+ = Cl^- = \sqrt{K_{sol. \ AgCl}}$$

$$pAg^+ = pCl^- = \tfrac{1}{2}pAgCl = 9.92/2 = 4.96$$

and a saturated solution of silver chloride with no excess of silver or chloride ions is present.

With 100.1 mL of silver nitrate solution, $[Ag^+] = 0.1 \times 0.1/200.1 = 5 \times 10^{-5}$, or $pAg^+ = 4.30$; $pCl^- = pAgCl - pAg^+ = 9.92 - 4.30 = 5.62$.*

The values calculated in this way up to the addition of 110 mL of $0.1 M$ silver nitrate are collected in Table 10.7. Similar values for the titration of 100 mL of $0.1 M$ potassium iodide with $0.1 M$ silver nitrate are included in the same table ($K_{sol. \ AgI} = 1.7 \times 10^{-16}$).

It will be seen by inspecting the silver-ion exponents in the neighbourhood of the equivalence point (say, between 99.8 and 100.2 mL) that there is a marked change in the silver-ion concentration, and the change is more pronounced for silver iodide than for silver chloride, since the solubility product of the latter is about 10^6 larger than for the former. This is shown more clearly in the titration curve in Fig. 10.12, which represents the change of pAg^+ in the range between 10 per cent before and 10 per cent after the stoichiometric point in the titration of $0.1 M$ chloride and $0.1 M$ iodide respectively with $0.1 M$ silver nitrate. An almost identical curve is obtained by potentiometric titration using a silver electrode (see Section 15.20); the pAg^+ values may be calculated from the e.m.f. figures as in the calculation of pH.

---

* This is not strictly true, since the dissolved silver chloride will contribute silver and chloride ions to the solution; the actual concentration is ca 1×10^{-5} g ions L^{-1}. If the excess of silver ions added is greater than 10 times this value, i.e. $> 10\sqrt{K_{sol. \ AgCl}}$, the error introduced by neglecting the ionic concentration produced by the dissolved salt may be taken as negligible for the purpose of the ensuing discussion.

Table 10.7 Titration of 100 mL of 0.1M NaCl and 100 mL of 0.1M KI respectively with 0.1M AgNO$_3$ ($K_{sol. AgCl} = 1.2 \times 10^{-10}$; $K_{sol. AgI} = 1.7 \times 10^{-16}$)

Vol. of 0.1M AgNO$_3$ (mL)	Titration of chloride		Titration of iodide	
	pCl$^-$	pAg$^+$	pI$^-$	pAg$^+$
0	1.0	—	1.0	—
50	1.5	8.4	1.5	14.3
90	2.3	7.6	2.3	13.5
95	2.6	7.3	2.6	13.2
98	3.0	6.9	3.0	12.8
99	3.3	6.6	3.3	12.5
99.5	3.7	6.2	3.7	12.1
99.8	4.0	5.9	4.0	11.8
99.9	4.3	5.6	4.3	11.5
100.0	5.0	5.0	7.9	7.9
100.1	5.6	4.3	11.5	4.3
100.2	5.9	4.0	11.8	4.0
100.5	6.3	3.6	12.2	3.6
101	6.6	3.3	12.5	3.3
102	6.9	3.0	12.8	3.0
105	7.3	2.6	13.2	2.6
110	7.6	2.3	13.5	2.4

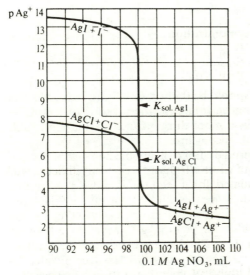

Fig. 10.12 Titration curves of 100 mL of 0.1M NaCl and of 100 mL of 0.1M KI respectively with 0.1M AgNO$_3$ (calculated).

10.75 DETERMINATION OF END POINTS IN PRECIPITATION REACTIONS

Many methods are utilised in determining end points in these reactions, but only the most important will be mentioned here.

(*a*) **Formation of a coloured precipitate.** This may be illustrated by the Mohr procedure for the determination of chloride and bromide. In the titration of a neutral solution of, say, chloride ions with silver nitrate solution, a small quantity of potassium chromate solution is added to serve as indicator. At the end point the chromate ions combine with silver ions to form the sparingly soluble, red, silver chromate.

The theory of the process is as follows. This is a case of fractional precipitation (Section 2.8), the two sparingly soluble salts being silver chloride (K_{sol} 1.2×10^{-10}) and silver chromate (K_{sol} 1.7×10^{-12}). It is best studied by considering an actual example encountered in practice, viz. the titration of, say, $0.1\,M$ sodium chloride with $0.1\,M$ silver nitrate in the presence of a few millilitres of dilute potassium chromate solution. Silver chloride is the less soluble salt and the initial chloride concentration is high; hence silver chloride will be precipitated. At the first point where red silver chromate is just precipitated both salts will be in equilibrium with the solution. Hence:

$$[Ag^+] \times [Cl^-] = K_{sol.\ AgCl} = 1.2 \times 10^{-10}$$

$$[Ag^+]^2 \times [CrO_4^{2-}] = K_{sol.\ Ag_2CrO_4} = 1.7 \times 10^{-12}$$

$$[Ag^+] = \frac{K_{sol.\ AgCl}}{[Cl^-]} = \sqrt{\frac{K_{sol.\ Ag_2CrO_4}}{[CrO_4^{2-}]}}$$

$$\frac{[Cl^-]}{\sqrt{[CrO_4^{2-}]}} = \frac{K_{sol.\ AgCl}}{\sqrt{K_{sol.\ Ag_2CrO_4}}} = \frac{1.2 \times 10^{-10}}{\sqrt{1.7 \times 10^{-12}}} = 9.2 \times 10^{-5}$$

At the equivalence point $[Cl^-] = \sqrt{K_{sol.\ AgCl}} = 1.1 \times 10^{-5}$. If silver chromate is to precipitate at this chloride-ion concentration:

$$[CrO_4^{2-}] = \left(\frac{[Cl^-]}{9.2 \times 10^{-5}}\right)^2 = \left(\frac{1.1 \times 10^{-5}}{9.2 \times 10^{-5}}\right)^2 = 1.4 \times 10^{-2}$$

or the potassium chromate solution should be $0.014\,M$. It should be noted that a slight excess of silver nitrate solution must be added before the red colour of silver chromate is visible. In practice, a more dilute solution ($0.003–0.005\,M$) of potassium chromate is generally used, since a chromate solution of concentration $0.01–0.02\,M$ imparts a distinct deep orange colour to the solution, which renders the detection of the first appearance of silver chromate somewhat difficult. The error introduced can be readily calculated, for if $[CrO_4^{2-}] = $ (say) 0.003, silver chromate will be precipitated when:

$$[Ag^+] = \sqrt{\frac{K_{sol.\ Ag_2CrO_4}}{CrO_4^{2-}}} = \sqrt{\frac{1.7 \times 10^{-12}}{3 \times 10^{-3}}} = 2.4 \times 10^{-5}$$

If the theoretical concentration of indicator is used:

$$[Ag^+] = \sqrt{\frac{1.7 \times 10^{-12}}{1.4 \times 10^{-2}}} = 1.1 \times 10^{-5}$$

The difference is $1.3 \times 10^{-5}\,mol\,L^{-1}$. If the volume of the solution at the equivalence point is $150\,mL$, then this corresponds to $1.3 \times 10^{-5} \times 150 \times 10^4/1000 = 0.02\,mL$ of $0.1\,M$ silver nitrate. This is the theoretical titration error, and is therefore negligible. In actual practice another factor must be considered, viz. the small excess of silver nitrate solution which must be added before the eye

343

can detect the colour change in the solution; this is of the order of one drop or *ca* 0.05 mL of 0.1 M silver nitrate.

The titration error will increase with increasing dilution of the solution being titrated and is quite appreciable (*ca* 0.4* per cent) in dilute, say 0.01 M, solutions when the chromate concentration is of the order 0.003–0.005 M. This is most simply allowed for by means of an indicator blank determination, e.g. by measuring the volume of standard silver nitrate solution required to give a perceptible coloration when added to distilled water containing the same quantity of indicator as is employed in the titration. This volume is subtracted from the volume of standard solution used.

It must be mentioned that the titration should be carried out in neutral solution or in very faintly alkaline solution, i.e. within the pH range 6.5–9. In acid solution, the following reaction occurs:

$$2CrO_4^{2-} + 2H^+ \rightleftharpoons 2HCrO_4^- \rightleftharpoons Cr_2O_7^{2-} + H_2O$$

$HCrO_4^-$ is a weak acid; consequently the chromate-ion concentration is reduced and the solubility product of silver chromate may not be exceeded. In markedly alkaline solutions, silver hydroxide ($K_{sol.}$ 2.3×10^{-8}) might be precipitated. A simple method of making an acid solution neutral is to add an excess of pure calcium carbonate or sodium hydrogencarbonate. An alkaline solution may be acidified with acetic acid and then a slight excess of calcium carbonate is added. The solubility product of silver chromate increases with rising temperature; the titration should therefore be performed at room temperature. By using a mixture of potassium chromate and potassium dichromate in proportions such as to give a neutral solution, the danger of accidentally raising the pH of an unbuffered solution beyond the acceptable limits is minimised; the mixed indicator has a buffering effect and adjusts the pH of the solution to 7.0 ± 0.1. In the presence of ammonium salts, the pH must not exceed 7.2 because of the effect of appreciable concentrations of ammonia upon the solubility of silver salts. Titration of iodide and of thiocyanate is not successful because silver iodide and silver thiocyanate adsorb chromate ions so strongly that a false and somewhat indistinct end point is obtained.

(b) Formation of a soluble coloured compound. This procedure is exemplified by Volhard's method for the titration of silver in the presence of free nitric acid with standard potassium thiocyanate or ammonium thiocyanate solution. The indicator is a solution of iron(III) nitrate or of iron(III) ammonium sulphate. The addition of thiocyanate solution produces first a precipitate of silver thiocyanate ($K_{sol.}$ 7.1×10^{-13}):

$$Ag^+ + SCN^- \rightleftharpoons AgSCN$$

When this reaction is complete, the slightest excess of thiocyanate produces a reddish-brown coloration, due to the formation of a complex ion:†

$$Fe^{3+} + SCN^- \rightleftharpoons [FeSCN]^{2+}$$

This method may be applied to the determination of chlorides, bromides,

---

* The errors for 0.1 M and 0.01 M bromide may be calculated to be 0.04 and 0.4 per cent respectively.
† This is the complex formed when the ratio of thiocyanate ion to iron(III) ion is low; higher complexes, $[Fe(SCN)_2]^+$, etc., are important only at higher concentrations of thiocyanate ion.

and iodides in acid solution. Excess of standard silver nitrate solution is added, and the excess is back-titrated with standard thiocyanate solution. For the chloride estimation, we have the following two equilibria during the titration of excess of silver ions:

$$Ag^+ + Cl^- \rightleftharpoons AgCl$$

$$Ag^+ + SCN^- \rightleftharpoons AgSCN$$

The two sparingly soluble salts will be in equilibrium with the solution, hence:

$$\frac{[Cl^-]}{[SCN^-]} = \frac{K_{sol.\ AgCl}}{K_{sol.\ AgSCN}} = \frac{1.2 \times 10^{-10}}{7.1 \times 10^{-13}} = 169$$

When the excess of silver has reacted, the thiocyanate may react with the silver chloride, since silver thiocyanate is the less soluble salt, until the ratio $[Cl^-]/[SCN^-]$ in the solution is 169:

$$AgCl + SCN^- \rightleftharpoons AgSCN + Cl^-$$

This will take place before reaction occurs with the iron(III) ions in the solution, and there will consequently be a considerable titration error. It is therefore absolutely necessary to prevent the reaction between the thiocyanate and the silver chloride. This may be effected in several ways, of which the first is probably the most reliable:

1. The silver chloride is filtered off before back-titrating. Since at this stage the precipitate will be contaminated with adsorbed silver ions, the suspension should be boiled for a few minutes to coagulate the silver chloride and thus remove most of the adsorbed silver ions from its surface before filtration. The cold filtrate is titrated.
2. After the addition of silver nitrate, potassium nitrate is added as coagulant, the suspension is boiled for about 3 minutes, cooled and then titrated immediately. Desorption of silver ions occurs and, on cooling, re-adsorption is largely prevented by the presence of potassium nitrate.
3. An immiscible liquid is added to 'coat' the silver chloride particles and thereby protect them from interaction with the thiocyanate. The most successful liquid is nitrobenzene (ca 1.0 mL for each 50 mg of chloride): the suspension is well shaken to coagulate the precipitate before back-titration.

With bromides, we have the equilibrium:

$$\frac{[Br^-]}{[SCN^-]} = \frac{K_{sol.\ AgBr}}{K_{sol.\ AgSCN}} = \frac{3.5 \times 10^{-13}}{7.1 \times 10^{-13}} = 0.5$$

The titration error is small, and no difficulties arise in the determination of the end point. Silver iodide ($K_{sol.}$ 1.7×10^{-16}) is less soluble than the bromide; the titration error is negligible, but the iron(III) indicator should not be added until excess of silver is present, since the dissolved iodide reacts with Fe^{3+} ions:

$$2Fe^{3+} + 2I^- \rightleftharpoons 2Fe^{2+} + I_2$$

(*c*) **Use of adsorption indicators.** K. Fajans introduced a useful type of indicator for precipitation reactions as a result of his studies on the nature of adsorption. The action of these indicators is due to the fact that at the equivalence point the indicator is adsorbed by the precipitate, and during the process of adsorption a change occurs in the indicator which leads to a substance of different colour; they have therefore been termed **adsorption indicators**. The substances employed are either acid dyes, such as those of the fluorescein series, e.g. fluorescein and eosin which are utilised as the sodium salts, or basic dyes, such as those of the rhodamine series (e.g. rhodamine 6G), which are used as the halogen salts.

The theory of the action of these indicators is based upon the properties of colloids, Section 11.3. When a chloride solution is titrated with a solution of silver nitrate, the precipitated silver chloride adsorbs chloride ions (a precipitate has a tendency to adsorb its own ions); this may be termed the primary adsorbed layer. By a process known as secondary adsorption, oppositely charged ions present in solution are held around it (shown diagrammically in Fig. 10.13a). As soon as the stoichiometric point is reached, silver ions are present in excess and these now become primarily adsorbed; nitrate ions will be held by secondary adsorption (Fig. 10.13b). If fluorescein is also present in the solution, the negative fluorescein ion, which is much more strongly adsorbed than is the nitrate ion, is immediately adsorbed, and will reveal its presence on the precipitate, not by its own colour, which is that of the solution, but by the formation of a pink complex of silver and a modified fluorescein ion on the surface with the first trace of excess of silver ions. An alternative view is that during the adsorption of the fluorescein ion a rearrangement of the structure of the ion occurs with the formation of a coloured substance. It is important to notice that the colour change takes place at the *surface* of the precipitate. If chloride is now added, the suspension remains pink until chloride ions are present in excess, the adsorbed silver ions are converted into silver chloride, which primarily adsorbs chloride ions. The secondary adsorbed fluorescein ions pass back into solution, to which they impart a greenish-yellow colour.

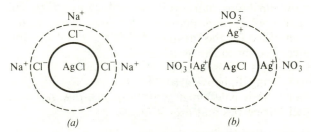

(a)   *(b)*

Fig. 10.13 (*a*) AgCl precipitated in the presence of excess of Cl$^-$; (*b*) AgCl precipitated in the presence of excess of Ag$^+$.

The following conditions will govern the choice of a suitable adsorption indicator:

(*a*) The precipitate should separate as far as possible in the colloidal condition. Large quantities of neutral salts, particularly of multicharged ions, should be avoided owing to their coagulating effect. The solution should not be too dilute, as the amount of precipitate formed will be small and the colour change far from sharp with certain indicators.

(*b*) The indicator ion must be of opposite charge to the ion of the precipitating agent.

(*c*) The indicator ion should not be adsorbed before the particular compound has been completely precipitated, but it should be strongly adsorbed immediately after the equivalence point. The indicator ion should not be too strongly adsorbed by the precipitate; if this occurs, e.g. eosin (tetrabromofluorescein) in the chloride–silver titration, the adsorption of the indicator ion may be a primary process and will take place before the equivalence point.

A disadvantage of adsorption indicators is that silver halides are sensitised to the action of light by a layer of adsorbed dyestuff. For this reason, titrations should be carried out with a minimum exposure to sunlight. When using adsorption indicators, only 2×10^{-4} to 3×10^{-3} mol of dye per mol of silver halide is added; this small concentration is used so that an appreciable fraction of the added indicator is actually adsorbed on the precipitate.

For the titration of chlorides, fluorescein may be used. This indicator is a very weak acid ($K_a = ca\ 1 \times 10^{-8}$); hence even a small amount of other acids reduces the already minute ionisation, thus rendering the detection of the end point (which depends essentially upon the adsorption of the free anion) either impossible or difficult to observe. The optimum pH range is between 7 and 10. Dichlorofluorescein is a stronger acid and may be utilised in slightly acid solutions of pH greater than 4.4; this indicator has the further advantage that it is applicable in more dilute solutions.

Eosin (tetrabromofluorescein) is a stronger acid than dichlorofluorescein and can be used down to a pH of 1–2; the colour change is sharpest in an acetic acid solution (pH < 3). Eosin is so strongly adsorbed on silver halides that it cannot be used for chloride titrations; this is because the eosin ion can compete with chloride ion before the equivalence point and thereby gives a premature indication of the end point. With the more strongly adsorbing ions, Br^-, I^- and SCN^-, the competition is not serious and a very sharp end point is obtained in the titration of these ions, even in dilute solutions. The colour of the precipitate is magenta. Rose Bengal (dichlorotetraiodofluorescein) and dimethyldiiodofluorescein have been recommended for the titration of iodides.

Other dyestuffs have been recommended as adsorption indicators for the titration of halides and other ions. Thus cyanide ion may be titrated with standard silver nitrate solution using diphenylcarbazide as adsorption indicator (see Section 10.44): the precipitate is pale violet at the end point. A selection of adsorption indicators, their properties and uses, is given in Table 10.8.

(*d*) Turbidity method. The appearance of a turbidity is sometimes utilised to mark the end-point of a reaction, as in Liebig's method for cyanides (see Section 10.44). A method which should be included here is the **turbidity procedure for the determination of silver with chloride**; first introduced by Gay Lussac. A standard solution of sodium chloride is titrated with a solution of silver nitrate or vice versa. Under certain conditions the addition of an indicator is unnecessary, because the presence of a turbidity caused by the addition of a few drops of one of the solutions to the other will show that the end point has not been reached. The titration is continued until the addition of the appropriate solution produces no turbidity. Accurate results are obtained.

The procedure may be illustrated by the following simple experiment, which is a modification of the Gay Lussac–Stas method. The sodium chloride solution is added to the silver solution in the presence of free nitric acid and a small quantity of pure barium nitrate (the latter to assist coagulation of the precipitate).

Table 10.8 Selected adsorption indicators: properties and uses

Indicator	Use
Fluorescein	Cl^-, Br^-, I^-, with Ag^+
Dichlorofluorescein	Cl^-, Br^-, BO_3^{3-}, with Ag^+
Tetrabromofluorescein (eosin)	Br^-, I^-, SCN^-, with Ag^+
Dichloro-tetraiodofluorescein (Rose Bengal)	I^- in presence of Cl^-, with Ag^+
Di-iodo-dimethylfluorescein	I^-, with Ag^+
Tartrazine	Ag^+, with I^- or SCN^-; $I^- + Cl^-$, with excess Ag^+, back-titration with I^-
Sodium alizarin sulphonate (alizarin red S)	$[Fe(CN)_6]^{4-}$, $[MoO_4]^{2-}$ with Pb^{2+}
Rhodamine 6G	Ag^+ with Br^-
Phenosafranine	Cl^-, Br^-, with Ag^+
	Ag^+, with Br^-

* The colour change is as indicator passes from solution to precipitate, unless otherwise stated.

Weigh out accurately about 0.40 g of silver nitrate into a well-stoppered 200 mL bottle. Add about 100 mL of water, a few drops of concentrated nitric acid, and a small crystal of barium nitrate. Titrate with standard 0.1 M sodium chloride by adding 20 mL at once, stoppering the bottle, and shaking it vigorously until the precipitate of silver chloride has coagulated and settled, leaving a clear solution. The volume of sodium chloride solution taken should leave the silver still in excess. Continue to add the chloride solution, 1 mL at a time, stoppering and shaking after each addition, until no turbidity is produced: note the total volume of sodium chloride solution. Repeat the determination, using a fresh sample of silver nitrate of about the same weight, and run in initially that volume of the 0.1 M sodium chloride, less 1 mL, which the first titration has indicated will be required, and thereafter add the chloride solution dropwise (i.e. in about 0.05 mL portions). It will be found that the end point can be determined within one drop.

The following sections are concerned with the use of standard solutions of reagents such as silver nitrate, sodium chloride, potassium (or ammonium) thiocyanate, and potassium cyanide. Some of the determinations which will be considered strictly involve complex formation rather than precipitation reactions, but it is convenient to group them here as reactions involving the use of standard silver nitrate solutions. Before commencing the experimental work, the theoretical Sections 10.74 and 10.75 should be studied.

10.76 PREPARATION OF 0.1 M SILVER NITRATE

Discussion. Very pure silver can be obtained commercially, and a standard solution can be prepared by dissolving a known weight (say, 10.787 g) in nitric acid in a conical flask having a funnel in the neck to prevent mechanical loss, and making up to a known volume (say, 1 L for a 0.1 M solution). The presence of acid must, however, be avoided in determinations with potassium chromate as indicator or in determinations employing adsorption indicators. It is therefore preferable to employ a neutral solution prepared by dissolving silver nitrate (relative molecular mass, 169.87) in water.

Analytical grade silver nitrate has a purity of at least 99.9 per cent, so that a standard solution can be prepared by direct weighing. If, however, commercial

Colour change at end-point*	Further data of interest
Yellowish-green → pink	Solution must be neutral or weakly basic
Yellowish-green → red	Useful pH range 4.4–7
Pink → reddish-violet	Best in acetic (ethanoic) acid solution; useful down to pH 1–2
Red → purple	Accurate if $(NH_4)_2CO_3$ added
Orange–red → blue–red	Useful pH range 4–7
Colourless *solution* → green *solution*	Sharp colour change in $I^- + Cl^-$ back-titration
Yellow → pink	Neutral solution
Orange–pink → reddish-violet	Best in dilute (up to 0.3M) HNO_3
Red ppt. → blue ppt.	Sharp, reversible colour change on ppt., but only if NO_3^- is present.
Blue ppt. → red ppt.	Tolerance up to 0.2M HNO_3

recrystallised silver nitrate be employed, or if an additional check of the molarity of the silver nitrate solution is required, standardisation may be effected with pure sodium chloride. Sodium chloride has a purity of 99.9–100.0 per cent and may be used as a primary standard. Sodium chloride is very slightly hygroscopic, and for accurate work it is best to dry the finely powdered solid in an electric oven at 250–350 °C for 1–2 hours, and allow it to cool in a desiccator.

Procedure. From silver nitrate. Dry some finely powdered analytical grade silver nitrate at 120 °C for 2 hours and allow it to cool in a covered vessel in a desiccator. Weigh out accurately 8.494 g, dissolve it in water and make up to 500 mL in a graduated flask. This gives a 0.1000M solution. Alternatively, about 8.5 g of pure, dry silver nitrate may be weighed out accurately, dissolved in 500 mL of water in a graduated flask, and the molar concentration calculated from the weight of silver nitrate employed.

In many cases the analytical grade material may be replaced by 'pure recrystallised' silver nitrate, but in that case it is advisable to standardise the solution against sodium chloride. Solutions of silver nitrate should be protected from light and are best stored in amber-coloured glass bottles.

10.77 STANDARDISATION OF SILVER NITRATE SOLUTION

Sodium chloride has a relative molecular mass of 58.44. A 0.1000M solution is prepared by weighing out 2.922 g of the pure dry salt (see Section 10.74) and dissolving it in 500 mL of water in a graduated flask. Alternatively about 2.9 g of the pure salt is accurately weighed out, dissolved in 500 mL of water in a graduated flask and the molar concentration calculated from the weight of sodium chloride employed.

(a) With potassium chromate as indicator: The Mohr titration. The reader is referred to Section 10.75 for the detailed theory of the titration. Prepare the indicator solution by dissolving 5 g potassium chromate in 100 mL of water. The final volume of the solution in the titration is 50–100 mL, and 1 mL of the indicator solution is used, so that the indicator concentration in the actual titration is 0.005–0.0025M.

Alternatively, and preferably, dissolve 4.2 g potassium chromate and 0.7 g

349

potassium dichromate in 100 mL of water; use 1 mL of indicator solution for each 50 mL of the final volume of the test solution.

Pipette 25 mL of the standard 0.1 M sodium chloride into a 250 mL conical flask resting upon a white tile, and add 1 mL of the indicator solution (preferably with a 1 mL pipette). Add the silver nitrate solution slowly from a burette, swirling the liquid constantly, until the red colour formed by the addition of each drop begins to disappear more slowly: this is an indication that most of the chloride has been precipitated. Continue the addition dropwise until a faint but distinct change in colour occurs. This *faint* reddish-brown colour should persist after brisk shaking. If the end point is overstepped (production of a deep reddish-brown colour), add more of the chloride solution and titrate again. Determine the indicator blank correction by adding 1 mL of the indicator to a volume of water equal to the final volume in the titration (Note), and then 0.01 M silver nitrate solution until the colour of the blank matches that of the solution titrated. The indicator blank correction, which should not amount to more than 0.03–0.10 mL of silver nitrate, is deducted from the volume of silver nitrate used in the titration. Repeat the titration with two further 25 mL portions of the sodium chloride solution. The various titrations should agree within 0.1 mL.

Note. A better blank is obtained by adding about 0.5 g of calcium carbonate before determining the correction. This gives an inert white precipitate similar to that obtained in the titration of chlorides and materially assists in matching the colour tints.

(*b*) **With an adsorption indicator: Discussion.** The detailed theory of the process is given in Section 10.75. Both fluorescein and dichlorofluorescein are suitable for the titration of chlorides. In both cases the end point is reached when the white precipitate in the greenish-yellow solution suddenly assumes a pronounced reddish tint. The change is reversible upon the addition of chloride. With fluorescein the solution must be neutral or only faintly acidic with acetic acid; acid solutions should be treated with a slight excess of sodium acetate. The chloride solution should be diluted to about 0.01–0.05 M, for if it is more concentrated the precipitate coagulates too soon and interferes. Fluorescein cannot be used in solutions more dilute than 0.005 M. With more dilute solutions resort must be made to dichlorofluorescein, which possesses other advantages over fluorescein. Dichlorofluorescein gives good results in very dilute solutions (e.g. for drinking water) and is applicable in the presence of acetic (ethanoic) acid and in weakly acid solutions. For this reason the chlorides of copper, nickel, manganese, zinc, aluminium, and magnesium, which cannot be titrated according to Mohr's method, can be determined by a direct titration when dichlorofluorescein is used as indicator.

For the reverse titration (chloride into silver nitrate), tartrazine (four drops of a 0.2 per cent solution per 100 mL) is a good indicator. At the end point, the almost colourless liquid assumes a blue colour.

The indicator solutions are prepared as follows:

Fluorescein. Dissolve 0.2 g fluorescein in 100 mL of 70 per cent ethanol, or dissolve 0.2 g sodium fluoresceinate in 100 mL of water.

Dichlorofluorescein. Dissolve 0.1 g dichlorofluorescein in 100 mL of 60–70 per cent ethanol, or dissolve 0.1 g sodium dichlorofluoresceinate in 100 mL of water.

Procedure. Pipette 25 mL of the standard 0.1 M sodium chloride into a 250 mL conical flask. Add 10 drops of either fluorescein or dichlorofluorescein indicator, and titrate with the silver nitrate solution in a diffuse light, while rotating the flask constantly. As the end point is approached, the silver chloride coagulates appreciably, and the local development of a pink colour upon the addition of a drop of the silver nitrate solution becomes more and more pronounced. Continue the addition of the silver nitrate solution until the precipitate suddenly assumes a pronounced pink or red colour. Repeat the titration with two other 25 mL portions of the chloride solution. Individual titrations should agree within 0.1 mL.

Calculate the molar concentration of the silver nitrate solution.

10.78 DETERMINATION OF CHLORIDES AND BROMIDES

Either the Mohr titration or the adsorption indicator method may be used for the determination of chlorides in neutral solution by titration with standard 0.1 M silver nitrate. If the solution is acid, neutralisation may be effected with chloride-free calcium carbonate, sodium tetraborate, or sodium hydrogencarbonate. Mineral acid may also be removed by neutralising most of the acid with ammonia solution and then adding an excess of ammonium acetate. Titration of the neutral solution, prepared with calcium carbonate, by the adsorption indicator method is rendered easier by the addition of 5 mL of 2 per cent dextrin solution; this offsets the coagulating effect of the calcium ion. If the solution is basic, it may be neutralised with chloride-free nitric acid, using phenolphthalein as indicator.

Similar remarks apply to the **determination of bromides**; the Mohr titration can be used, and the most suitable adsorption indicator is **eosin** which can be used in dilute solutions and even in the presence of 0.1 M nitric acid, but in general, acetic (ethanoic) acid solutions are preferred. Fluorescein may be used but is subject to the same limitations as experienced with chlorides [Section 10.77(**b**)]. With eosin indicator, the silver bromide flocculates approximately 1 per cent before the equivalence point and the local development of a red colour becomes more and more pronounced with the addition of silver nitrate solution: at the end point the precipitate assumes a magenta colour.

The indicator is prepared by dissolving 0.1 g eosin in 100 mL of 70 per cent ethanol, or by dissolving 0.1 g of the sodium salt in 100 mL of water.

For the **reverse titration** (bromide into silver nitrate), rhodamine 6G (10 drops of a 0.05 per cent aqueous solution) is an excellent indicator. The solution is best adjusted to 0.05 M with respect to silver ion. The precipitate acquires a violet colour at the end point.

Thiocyanates may also be determined using adsorption indicators in exactly similar manner to chlorides and bromides, but an iron(III) salt indicator is usually preferred (Section 10.82).

10.79 DETERMINATION OF IODIDES

Discussion. The Mohr method cannot be applied to the titration of iodides (or of thiocyanates), because of adsorption phenomena and the difficulty of distinguishing the colour change of the potassium chromate. Eosin is a suitable

351

adsorption indicator, but di-iododimethylfluorescein is better. Eosin is employed as described under bromides (Section 10.78).

The di-iododimethylfluorescein indicator is prepared by dissolving 1.0 g in 100 mL of 70 per cent ethanol. The colour change is from an orange–red to a blue–red on the precipitate.

10.80 DETERMINATION OF MIXTURES OF HALIDES WITH ADSORPTION INDICATORS

(*a*) **Chloride and iodide in a mixture.** These two ions differ considerably in the ease with which they are adsorbed on the corresponding silver halide. This makes it possible to select adsorption indicators which will permit the determination of chloride and iodide in the presence of one another. Thus the iodide may be determined by titration with standard $0.1 M$ silver nitrate using di-iododimethylfluorescein and the iodide + chloride by a similar titration using fluorescein. Chloride is obtained by difference. If a large excess of chloride is present, the result for iodide may be as much as 1 per cent high. If, however, rose Bengal (dichlorotetraiodofluorescein) is used as indicator (colour change, carmine red to blue–red) in the presence of ammonium carbonate, the iodide titration is exact.

(*b*) **Bromide and iodide in a mixture.** The total halide (bromide + iodide) is determined by titration with standard $0.1 M$ silver nitrate using eosin or fluorescein as indicator. The iodide is determined by titration with $0.01–0.2 M$ silver nitrate, using di-iododimethylfluorescein as indicator. Bromide is obtained by difference.

Numerous adsorption indicators have been suggested for various purposes, but a full treatment is outside the scope of this work.

10.81 DETERMINATION OF MIXTURES OF HALIDES BY AN INDIRECT METHOD

Discussion. The method is applicable to the determination of a mixture of two salts having the same anion (e.g. sodium chloride and potassium chloride) or the same cation (e.g. potassium chloride and potassium bromide). For example, to determine the amount of sodium and potassium chlorides in a mixture of the two salts, a known weight (w_1 g) of the solid mixture is taken, and the total chloride is determined with standard $0.1 M$ silver nitrate, using Mohr's method or an adsorption indicator. Let w_2 g of silver nitrate be required for the complete precipitation of w_1 g of the mixture, which contains x g of NaCl and y g of KCl. Then:

$$x + y = w_1$$

$$\frac{169.87 x}{58.44} + \frac{169.87 y}{74.55} = w_2$$

Upon solving these two simultaneous equations, the values for x and y are deduced.

Now suppose that the determination of potassium chloride and potassium bromide in a mixture is desired. The total halide is determined by Mohr's method or with an adsorption indicator. Let the weight of the mixture be w_3 g and w_4 g, be the weight of silver nitrate required for complete precipitation,

p g be the weight of the potassium chloride, and q g be the weight of the potassium bromide. Then:

$$p + q = w_3$$

$$\frac{169.87p}{74.55} + \frac{169.87q}{119.00} = w_4$$

The values of p and q can be obtained by solving the simultaneous equations.

It can be shown that the method depends upon the difference between the relative molecular masses of the two components of the mixture and that it is most satisfactory when the two constituents are present in approximately equal proportions.

10.82 PREPARATION AND USE OF 0.1*M* AMMONIUM OR POTASSIUM THIOCYANATE: TITRATIONS ACCORDING TO VOLHARD'S METHOD

Discussion. Volhard's original method for the determination of silver in dilute nitric acid solution by titration with standard thiocyanate solution in the presence of an iron(III) salt as indicator has proved of great value not only for silver determinations, but also in numerous indirect analyses. The theory of the Volhard process has been given in Section 10.75. In this connection it must be pointed out that the concentration of the nitric acid should be from 0.5 to 1.5 M (strong nitric acid retards the formation of the thiocyanato–iron(III) complex ($[FeSCN]^{2+}$) and at a temperature not exceeding 25 °C (higher temperatures tend to bleach the colour of the indicator). The solutions must be free from nitrous acid, which gives a red colour with thiocyanic acid, and may be mistaken for 'iron(III) thiocyanate'. Pure nitric acid is prepared by diluting the usual pure acid with about one-quarter of its volume of water and boiling until perfectly colourless; this eliminates any lower oxides of nitrogen which may be present.

The method may be applied to those anions (e.g. chloride, bromide, and iodide) which are completely precipitated by silver and are sparingly soluble in dilute nitric acid. Excess of standard silver nitrate solution is added to the solution containing free nitric acid, and the residual silver nitrate solution is titrated with standard thiocyanate solution. This is sometimes termed **the residual process.** Anions whose silver salts are slightly soluble in water, but which are soluble in nitric acid, such as phosphate, arsenate, chromate, sulphide, and oxalate, may be precipitated in neutral solution with an excess of standard silver nitrate solution. The precipitate is filtered off, thoroughly washed, dissolved in dilute nitric acid, and the silver titrated with thiocyanate solution. Alternatively, the residual silver nitrate in the filtrate from the precipitation may be determined with thiocyanate solution after acidification with dilute nitric acid.

Both ammonium and potassium thiocyanates are usually available as deliquescent solids; the analytical-grade products are, however, free from chlorides and other interfering substances. An approximately 0.1 M solution is, therefore, first prepared, and this is standardised by titration against standard 0.1 M silver nitrate.

Procedure. Weigh out about 8.5 g ammonium thiocyanate, or 10.5 g potassium thiocyanate, and dissolve it in 1 L of water in a graduated flask. Shake well.

Standardisation. Use 0.1 M silver nitrate, which has been prepared and standardised as described in Section 10.77.

The iron(III) indicator solution consists of a cold, saturated solution of ammonium iron(III) sulphate in water (about 40 per cent) to which a few drops of 6 M nitric acid have been added. One millilitre of this solution is employed for each titration.

Pipette 25 mL of the standard 0.1 M silver nitrate into a 250 mL conical flask, add 5 mL of 6 M nitric acid and 1 mL of the iron(III) indicator solution. Run in the potassium or ammonium thiocyanate solution from a burette. At first a white precipitate is produced, rendering the liquid of a milky appearance, and as each drop of thiocyanate falls in, it produces a reddish-brown cloud, which quickly disappears on shaking. As the end point approaches, the precipitate becomes flocculent and settles easily; finally one drop of the thiocyanate solution produces a faint brown colour, which no longer disappears upon shaking. This is the end point. The indicator blank amounts to 0.01 mL of 0.1 M silver nitrate. It is essential to shake vigorously during the titration in order to obtain correct results.*

The standard solution thus prepared is stable for a very long period if evaporation is prevented.

Use of tartrazine as indicator. Satisfactory results may be obtained by the use of tartrazine as indicator. Proceed as above, but add 4 drops of tartrazine (0.5 per cent aqueous solution) in lieu of the iron(III) indicator. The precipitate will appear pale yellow during the titration, but the supernatant liquid (best viewed by placing the eye at the level of the liquid and looking through it) is colourless. At the end point, the supernatant liquid assumes a bright lemon-yellow colour. The titration is sharp to one drop of 0.1 M thiocyanate solution.

10.83 DETERMINATION OF SILVER IN A SILVER ALLOY

A commercial silver alloy in the form of wire or foil is suitable for this determination. Clean the alloy with emery cloth and weigh it accurately. Place it in a 250 mL conical flask, add 5 mL water and 10 mL concentrated nitric acid; place a funnel in the mouth of the flask to avoid mechanical loss. Warm the flask gently until the alloy has dissolved. Add a little water and boil for 5 minutes in order to expel oxides of nitrogen. Transfer the cold solution quantitatively to a 100 mL graduated flask and make up to the mark with distilled water. Titrate 25 mL portions of the solution with standard 0.1 M thiocyanate.

1 mole KSCN $\equiv$ 1 mole Ag$^+$

Note. The presence of metals whose salts are colourless does not influence the accuracy of the determination, except that mercury and palladium must be absent since their thiocyanates are insoluble. Salts of metals (e.g. nickel and cobalt) which are coloured must not be present to any considerable extent. Copper does not interfere, provided it does not form more than about 40 per cent of the alloy.

---

* The freshly precipitated silver thiocyanate adsorbs silver ions, thereby causing a false end point which, however, disappears with vigorous shaking.

10.84 DETERMINATION OF CHLORIDES (VOLHARD'S METHOD)

Discussion. The chloride solution is treated with excess of standard silver nitrate solution, and the residual silver nitrate determined by titration with standard thiocyanate solution. Now silver chloride is more soluble than silver thiocyanate, and would react with the thiocyanate thus:

$$AgCl \text{ (solid)} + SCN^- \rightleftharpoons AgSCN \text{ (solid)} + Cl^-$$

It is therefore necessary to remove the silver chloride by filtration. The filtration may be avoided by the addition of a little nitrobenzene (about 1 mL for each 0.05 g of chloride); the silver chloride particles are probably surrounded by a film of nitrobenzene. Another method, applicable to chlorides, in which filtration of the silver chloride is unnecessary, is to employ tartrazine as indicator (Section 10.82).

Procedure A (HCl content of concentrated hydrochloric acid). Ordinary concentrated hydrochloric acid is usually 10–11 M, and must be diluted first. Measure out accurately 10 mL of the concentrated acid from a burette into a 1 L graduated flask and make up to the mark with distilled water. Shake well. Pipette 25 mL into a 250 mL conical flask, add 5 mL 6 M nitric acid and then add 30 mL standard 0.1 M silver nitrate (or sufficient to give 2–5 mL excess). Shake to coagulate the precipitate,* filter through a quantitative filter paper (or through a porous porcelain or sintered-glass crucible), and wash thoroughly with very dilute nitric acid (1:100). Add 1 mL of the iron(III) indicator solution to the combined filtrate and washings, and titrate the residual silver nitrate with standard 0.1 M thiocyanate.

Calculate the volume of standard 0.1 M silver nitrate that has reacted with the hydrochloric acid, and therefrom the percentage of HCl in the sample employed.

Procedure B. Pipette 25 mL of the diluted solution into a 250 mL conical flask containing 5 mL 6 M nitric acid. Add a slight excess of standard 0.1 M silver nitrate (about 30 mL in all) from a burette. Then add 2–3 mL pure nitrobenzene and 1 mL of the iron(III) indicator, and shake vigorously to coagulate the precipitate. Titrate the residual silver nitrate with standard 0.1 M thiocyanate until a permanent faint reddish-brown coloration appears.

From the volume of silver nitrate solution added, subtract the volume of silver nitrate solution that is equivalent to the volume of standard thiocyanate required. Then calculate the percentage of HCl in the sample.

Procedure C. Pipette 25 mL of the diluted solution into a 250 mL conical flask containing 5 mL of 6 M nitric acid, add a slight excess of 0.1 M silver nitrate (30–35 mL) from a burette, and four drops of tartrazine indicator (0.5 per cent aqueous solution). Shake the suspension for about a minute in order to ensure that the indicator is adsorbed on the precipitate as far as possible. Titrate the residual silver nitrate with standard 0.1 M ammonium or potassium thiocyanate with swirling of the suspension until the very pale yellow supernatant liquid (viewed with the eye at the level of the liquid) assumes a rich lemon-yellow colour.

---

*It is better to boil the suspension for a few minutes to coagulate the silver chloride and thus remove most of the adsorbed silver ions from its surface before filtration.

Bromides can also be determined by the Volhard method, but as silver bromide is less soluble than silver thiocyanate it is not necessary to filter off the silver bromide (compare chloride). The bromide solution is acidified with dilute nitric acid, an excess of standard 0.1 M silver nitrate added, the mixture thoroughly shaken, and the residual silver nitrate determined with standard 0.1 M ammonium or potassium thiocyanate, using ammonium iron(III) sulphate as indicator.

Iodides can also be determined by this method, and in this case too there is no need to filter off the silver halide, since silver iodide is very much less soluble than silver thiocyanate. In this determination the iodide solution must be very dilute in order to reduce adsorption effects. The dilute iodide solution (*ca* 300 mL), acidified with dilute nitric acid, is treated very slowly and with vigorous stirring or shaking with standard 0.1 M silver nitrate until the yellow precipitate coagulates and the supernatant liquid appears colourless. Silver nitrate is then present in excess. One millilitre of iron(III) indicator solution is added, and the residual silver nitrate is titrated with standard 0.1 M ammonium or potassium thiocyanate.

10.85 DETERMINATION OF FLUORIDE: PRECIPITATION AS LEAD CHLOROFLUORIDE COUPLED WITH VOLHARD TITRATION

Discussion. This method is based upon the precipitation of lead chlorofluoride, in which the chlorine is determined by Volhard's method, and from this result the fluorine content can be calculated. The advantages of the method are, the precipitate is granular, settles readily, and is easily filtered; the factor for conversion to fluorine is low; the procedure is carried out at pH 3.6–5.6, so that substances which might be co-precipitated, such as phosphates, sulphates, chromates, and carbonates, do not interfere. Aluminium must be entirely absent, since even very small quantities cause low results; a similar effect is produced by boron (>0.05 g), ammonium (>0.5 g), and sodium or potassium (>10 g) in the presence of about 0.1 g of fluoride. Iron must be removed, but zinc is without effect. Silica does not vitiate the method, but causes difficulties in filtration.

Procedure. Pipette 25.0 mL of the solution containing between 0.01 and 0.1 g fluoride into a 400 mL beaker, add two drops of bromophenol blue indicator, 3 mL of 10 per cent sodium chloride, and dilute the mixture to 250 mL. Add dilute nitric acid until the colour just changes to yellow, and then add dilute sodium hydroxide solution until the colour just changes to blue. Treat with 1 mL of concentrated hydrochloric acid, then with 5.0 g of lead nitrate, and heat on a water bath. Stir gently until the lead nitrate has dissolved, and then immediately add 5.0 g of crystallised sodium acetate and stir vigorously. Digest on the water bath for 30 minutes, with occasional stirring, and allow to stand overnight.

Meanwhile, a washing solution of lead chlorofluoride is prepared as follows. Add a solution of 10 g of lead nitrate in 200 mL of water to 100 mL of a solution containing 1.0 g of sodium fluoride and 2 mL of concentrated hydrochloric acid, mix it thoroughly, and allow the precipitate to settle. Decant the supernatant liquid, wash the precipitate by decantation with five portions of water, each of about 200 mL. Finally add 1 L of water to the precipitate, shake the mixture at intervals during an hour, allow the precipitate to settle, and filter the liquid. Further quantities of wash liquid may be prepared as needed by treating the precipitate with fresh portions of water. The solubility of lead chlorofluoride in water is $0.325\,g\,L^{-1}$ at $25\,°C$.

Separate the original precipitate by decantation through a Whatman No. 542 or No. 42 paper. Transfer the precipitate to the filter, wash once with cold water, four or five times with the saturated solution of lead chlorofluoride, and finally once more with cold water. Transfer the precipitate and paper to the beaker in which precipitation was made, stir the paper to a pulp in 100 mL of 5 per cent nitric acid, and heat on the water bath until the precipitate has dissolved (5 minutes). Add a slight excess of standard $0.1\,M$ silver nitrate, digest on the bath for a further 30 minutes, and allow to cool to room temperature whilst protected from the light. Filter the precipitate of silver chloride through a sintered-glass crucible, wash with a little cold water, and titrate the residual silver nitrate in the filtrate and washings with standard $0.1\,M$ thiocyanate. Subtract the amount of silver found in the filtrate from that originally added. The difference represents the amount of silver that was required to combine with the chlorine in the lead chlorofluoride precipitate.

1 mole $AgNO_3 \equiv$ 1 mole F^-

10.86 DETERMINATION OF ARSENATES

Discussion. Arsenates in solution are precipitated as silver arsenate, Ag_3AsO_4, by the addition of neutral silver nitrate solution: the solution must be neutral, or if slightly acid, an excess of sodium acetate must be present to reduce the acidity; if strongly acid, most of the acid should be neutralised by aqueous sodium hydroxide. The silver arsenate is dissolved in dilute nitric acid, and the silver titrated with standard thiocyanate solution. The silver arsenate has nearly six times the weight of the arsenic, hence quite small amounts of arsenic may be determined by this procedure.

Arsenites may also be determined by this procedure but must first be oxidised by treatment with nitric acid. Small amounts of antimony and tin do not interfere, but chromates, phosphates, molybdates, tungstates, and vanadates, which precipitate as the silver salts, should be absent. An excessive amount of ammonium salts has a solvent action on the silver arsenate.

Procedure. Place 25 mL of the arsenate solution in a 250 mL beaker, add an equal volume of distilled water and a few drops of phenolphthalein solution. Add sufficient sodium hydroxide solution to give an alkaline reaction, and then discharge the red colour from the solution by just acidifying with acetic (ethanoic) acid. Add a slight excess of silver nitrate solution with vigorous stirring, and allow the precipitate to settle in the dark. Pour off the supernatant liquid through a sintered-glass crucible, wash the precipitate by decantation with cold distilled water, transfer the precipitate to the crucible, and wash it free from silver nitrate solution. Wash out the receiver thoroughly. Dissolve the silver arsenate in dilute nitric acid ($ca\ 1\,M$) (which leaves any silver chloride undissolved), wash with very dilute nitric acid, and make up the filtrate and washings to 250 mL in a graduated flask. Titrate a convenient aliquot portion

357

with standard ammonium (or potassium) thiocyanate solution in the presence of ammonium iron(III) sulphate as indicator.

3 moles KSCN $\equiv$ 1 mole AsO_4^{3-}

10.87 DETERMINATION OF CYANIDES

Discussion. The theory of the titration of cyanides with silver nitrate solution has been given in Section 10.44. All silver salts except the sulphide are readily soluble in excess of a solution of an alkali cyanide, hence chloride, bromide, and iodide do not interfere. The only difficulty in obtaining a sharp end point lies in the fact that silver cyanide is often precipitated in a curdy form which does not readily re-dissolve, and, moreover, the end point is not easy to detect with accuracy.

There are two methods for overcoming these disadvantages. In the first the precipitation of silver cyanoargentate at the end point can be avoided by the addition of ammonia solution, in which it is readily soluble, and if a little potassium iodide solution is added before the titration is commenced, sparingly soluble silver iodide, which is insoluble in ammonia solution, will be precipitated at the end point. The precipitation is best seen by viewing against a black background.

In the second method diphenylcarbazide is employed as an adsorption indicator. The end-point is marked by the pink colour becoming pale violet (almost colourless) on the colloidal precipitate in dilute solution (*ca* 0.01 *M*) before the opalescence is visible. In 0.1 *M* solutions, the colour change is observed on the precipitated particles of silver cyanoargentate.

Procedure. NOTE: Potassium cyanide and all other cyanides are deadly poisons, and extreme care must be taken in their use. Details for the disposal of cyanides and other dangerous and toxic chemicals may be found in Refs 14 and 15.

For practice in the method, the cyanide content of potassium cyanide (laboratory reagent grade) may be determined.

Method A. Weigh out accurately about 3.5 g of potassium cyanide from a glass-stoppered weighing bottle, dissolve it in water and make up to 250 mL in a graduated flask. Shake well. Transfer 25.0 mL of this solution by **means of a burette and NOT a pipette** to a 250 mL conical flask, add 75 mL water, 5–6 mL 6*M* ammonia solution, and 2 mL 10 per cent potassium iodide solution. Place the flask on a sheet of black paper, and titrate with standard 0.1 *M* silver nitrate. Add the silver nitrate solution dropwise as soon as the yellow colour of silver iodide shows any signs of persisting. When one drop produces a permanent turbidity, the end-point has been reached.

Method B. Prepare the solution and transfer 25 mL of it to a 250 mL conical flask as detailed under Method A. Add two to three drops of diphenylcarbazide indicator and titrate with standard 0.1 *M* silver nitrate solution until a permanent violet colour is just produced.

The diphenylcarbazide indicator is prepared by dissolving 0.1 g of the solid in 100 mL of ethanol.

1 mole $AgNO_3$ $\equiv$ 2 moles CN^-

10.88 DETERMINATION OF POTASSIUM

Discussion. Potassium may be precipitated with excess of sodium tetraphenylborate solution as potassium tetraphenylborate. The excess of reagent is determined by titration with mercury(II) nitrate solution. The indicator consists of a mixture of iron(III) nitrate and dilute sodium thiocyanate solution. The end-point is revealed by the decolorisation of the iron(III)–thiocyanate complex due to the formation of the colourless mercury(II) thiocyanate. The reaction between mercury(II) nitrate and sodium tetraphenylborate under the experimental conditions used is not quite stoichiometric; hence it is necessary to determine the volume in mL of $Hg(NO_3)_2$ solution equivalent to 1 mL of a $NaB(C_6H_5)_4$ solution. Halides must be absent.

Procedure. Prepare the sodium tetraphenylborate solution by dissolving 6.0 g of the solid in about 200 mL of distilled water in a glass-stoppered bottle. Add about 1 g of moist aluminium hydroxide gel, and shake well at five-minute intervals for about 20 minutes. Filter through a Whatman No. 40 filter paper, pouring the first runnings back through the filter if necessary, to ensure a clear filtrate. Add 15 mL of $0.1 M$ sodium hydroxide to the solution to give a pH of about 9, then make up to 1 L and store the solution in a polythene bottle.

Prepare a mercury(II) nitrate solution ($0.03 M$) by dissolving 10.3 g recrystallised mercury(II) nitrate, $Hg(NO_3)_2, H_2O$, in 800 mL distilled water containing 20 mL $2 M$ nitric acid. Dilute to 1 L in a graduated flask and then standardise by titrating with a standard thiocyanate solution using iron(III) indicator solution. Prepare the indicator solutions for the main titration by dissolving separately 5 g hydrated iron(III) nitrate in 100 mL of distilled water and filtering, and 0.08 g sodium thiocyanate in 100 mL of distilled water.

Standardisation. Pipette 10.0 mL of the sodium tetraphenylborate solution into a 250 mL beaker and add 90 mL water, 2.5 mL $0.1 M$ nitric acid, 1.0 mL iron(III) nitrate solution, and 10.0 mL sodium thiocyanate solution. Without delay stir the solution mechanically, then slowly add from a burette 10 drops of mercury(II) nitrate solution. Continue the titration by adding the mercury(II) nitrate solution at a rate of 1–2 drops per second until the colour of the indicator is temporarily discharged. Continue the titration more slowly, but maintain the rapid state of stirring. The end point is arbitrarily defined as the point when the indicator colour is discharged and fails to reappear for 1 minute. Perform at least three titrations, and calculate the mean volume of mercury(II) nitrate solution equivalent to 10.0 mL of the sodium tetraphenylborate solution.

Pipette 25.0 mL of the potassium ion solution (about 10 mg K^+) into a 50 mL graduated flask, add 0.5 mL $1 M$ nitric acid and mix. Introduce 20.0 mL of the sodium tetraphenylborate solution, dilute to the mark, mix, then pour the mixture into a 150 mL flask provided with a ground stopper. Shake the stoppered flask for 5 minutes on a mechanical shaker to coagulate the precipitate, then filter most of the solution through a dry Whatman No. 40 filter paper into a dry beaker. Transfer 25.0 mL of the filtrate into a 250 mL conical flask and add 75 mL of water, 1.0 mL of iron(III) nitrate solution, and 1.0 mL of sodium thiocyanate solution. Titrate with the mercury(II) nitrate solution as described above.

Note. This determination is only suitable for students with analytical experience and should not be attempted by beginners.

OXIDATION–REDUCTION TITRATIONS

10.89 CHANGE OF THE ELECTRODE POTENTIAL DURING THE TITRATION OF A REDUCTANT WITH AN OXIDANT

In Sections 10.11–10.16 it is shown how the change in pH during acid–base titrations may be calculated, and how the titration curves thus obtained can be used (a) to ascertain the most suitable indicator to be used in a given titration, and (b) to determine the titration error. Similar procedures may be carried out for oxidation–reduction titrations. Consider first a simple case which involves only change in ionic charge, and is theoretically independent of the hydrogen-ion concentration. A suitable example, for purposes of illustration, is the titration of 100 mL of $0.1 M$ iron(II) with $0.1 M$ cerium(IV) in the presence of dilute sulphuric acid:

$$Ce^{4+} + Fe^{2+} \rightleftharpoons Ce^{3+} + Fe^{3+}$$

The quantity corresponding to $[H^+]$ in acid–base titrations is the ratio $[Ox]/[Red]$. We are concerned here with two systems, the Fe^{3+}/Fe^{2+} ion electrode (1), and the Ce^{4+}/Ce^{3+} ion electrode (2).

For (1) at 25 °C:

$$E_1 = E_1^{\ominus} + \frac{0.0591}{1} \log\frac{[Fe^{3+}]}{[Fe^{2+}]} = +0.75 + 0.0591 \log\frac{[Fe^{3+}]}{[Fe^{2+}]}$$

For (2), at 25 °C:

$$E_2 = E_2^{\ominus} + \frac{0.0591}{1} \log\frac{[Ce^{4+}]}{[Ce^{3+}]} = +1.45 + 0.0591 \log\frac{[Ce^{4+}]}{[Ce^{3+}]}$$

The equilibrium constant of the reaction is given by (Section 2.33):

$$\log K = \log\frac{[Ce^{3+}] \times [Fe^{3+}]}{[Ce^{4+}] \times [Fe^{2+}]} = \frac{1}{0.0591}(1.45 - 0.75) = 11.84$$

or

$$K = 7 \times 10^{11}$$

The reaction is therefore virtually complete.

During the addition of the cerium(IV) solution up to the equivalence point, its only effect will be to oxidise the iron(II) (since K is large) and consequently change the ratio $[Fe^{3+}]/[Fe^{2+}]$. When 10 mL of the oxidising agent have been added, $[Fe^{3+}]/[Fe^{2+}] = 10/90$ (approx.) and

$$E_1 = 0.75 + 0.0591 \log 10/90 = 0.75 - 0.056 = 0.69 \text{ volt}$$

With 50 mL of the oxidising agent, $E_1 = E_1^{\ominus} = 0.75$ volt
With 90 mL, $E_1 = 0.75 + 0.0591 \log 90/10 = 0.81$ volt
With 99 mL, $E_1 = 0.75 = 0.0591 \log 99/1 = 0.87$ volt
With 99.9 mL, $E_1 = 0.75 + 0.0591 \log 99.9/0.1 = 0.93$ volt

At the equivalence point (100.0 mL) $[Fe^{3+}] = [Ce^{3+}]$ and $[Ce^{4+}] = [Fe^{2+}]$, and the electrode potential is given by:*

$$\frac{E_1^{\ominus} + E_2^{\ominus}}{2} = \frac{0.75 + 1.45}{2} = 1.10 \text{ volts}$$

The subsequent addition of cerium(IV) solution will merely increase the ratio $[Ce^{4+}]/[Ce^{3+}]$. Thus:

With 100.1 mL, $E_2 = 1.45 + 0.059\ 1 \log 0.1/100 = 1.27$ volts
With 101 mL, $E_2 = 1.45 + 0.059\ 1 \log 1/100 = 1.33$ volts
With 110 mL, $E_2 = 1.45 + 0.059\ 1 \log 10/100 = 1.39$ volts
With 190 mL, $E_2 = 1.45 + 0.059\ 1 \log 90/100 = 1.45$ volts

These results are shown in Fig. 10.14.

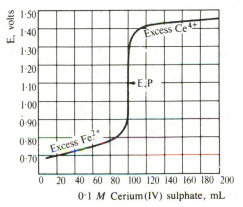

Fig. 10.14 Titration of 100 mL of 0.1M iron(II) with 0.1M cerium(IV) sulphate (calculated).

It is of interest to calculate the iron(II) concentration in the neighbourhood of the equivalence point. When 99.9 mL of the cerium(IV) solution have been added, $[Fe^{2+}] = 0.1 \times 0.1/199.9 = 5 \times 10^{-5}$, or $pFe^{2+} = 4.3$. The concentration at the equivalence point is given by (Section 2.33):

$$[Fe^{3+}]/[Fe^{2+}] = \sqrt{K} = \sqrt{7 \times 10^{11}} = 8.4 \times 10^5$$

Now $[Fe^{3+}] = 0.05M$, hence $[Fe^{2+}] = 5 \times 10^{-2}/8.4 \times 10^5 = 6 \times 10^{-8}M$, or $pFe^{2+} = 7.2$. Upon the addition of 100.1 mL of cerium(IV) solution, the reduction potential (see above) is 1.27 volts. The $[Fe^{3+}]$ is practically unchanged at $5 \times 10^{-2}M$, and we may calculate $[Fe^{2+}]$ with sufficient accuracy for our

---

* For a deduction of this expression and a discussion of the approximations involved, see a textbook of electrochemistry. It can similarly be shown that for the reaction:

$$a\,Ox_I + b\,Red_{II} \rightleftharpoons b\,Ox_{II} + a\,Red_I$$

the potential at the equivalence point is given by:

$$E_0 = \frac{b\,E_1^{\ominus} + a\,E_2^{\ominus}}{a+b}$$

where $E_1^{\ominus}$ refers to Ox_I, Red_I, and $E_2^{\ominus}$ to Ox_{II}, Red_{II}.

purpose from the equations:

$$E = E_1^{\ominus} + 0.0591 \log \frac{[Fe^{3+}]}{[Fe^{2+}]}$$

$$1.27 = 0.75 + 0.0591 \log \frac{5 \times 10^{-2}}{[Fe^{2+}]}$$

$$[Fe^{2+}] = 1 \times 10^{-10}$$

or

$$pFe^{2+} = 10$$

Thus pFe^{2+} changes from 4.3 to 10 between 0.1 per cent before and 0.1 per cent after the stoichiometric end point. These quantities are of importance in connection with the use of indicators for the detection of the equivalence point.

It is evident that the abrupt change of the potential in the neighbourhood of the equivalence point is dependent upon the standard potentials of the two oxidation–reduction systems that are involved, and therefore upon the equilibrium constant of the reaction; it is independent of the concentrations unless these are extremely small. The change in redox potential for a number of typical oxidation–reduction systems is exhibited graphically in Fig. 10.15. For the MnO_4^-, Mn^{2+} system and others which are dependent upon the pH of the

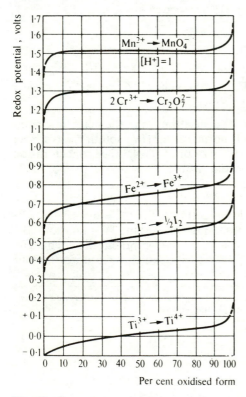

Fig. 10.15 Variation of redox potentials with oxidant/reductant ratio.

solution, the hydrogen-ion concentration is assumed to be molar: lower acidities give lower potentials. The value at 50 per cent oxidised form will, of course, correspond to the standard redox potential. As an indication of the application of the curves, consider the titration of iron(II) with potassium dichromate. The titration curve would follow that of the Fe(II)/Fe(III) system until the end-point was reached, then it would rise steeply and continue along the curve for the $Cr_2O_7^{2-}/Cr^{3+}$ system: the potential at the equivalence point can be determined as already described.

It is possible to titrate two substances by the same titrant provided that the standard potentials of the substances being titrated, and their oxidation or reduction products, differ by about 0.2 V. Stepwise titration curves are obtained in the titration of mixtures or of substances having several oxidation states. Thus the titration of a solution containing Cr(VI), Fe(III) and V(V) by an acid titanium(III) chloride solution is an example of such a mixture: in the first step Cr(VI) is reduced to Cr(III) and V(V) to V(IV); in the second step Fe(III) is reduced to Fe(II); in the third step V(IV) is reduced to V(III); chromium is evaluated by difference of the volumes of titrant used in the first and third steps. Another example is the titration of a mixture of Fe(II) and V(IV) sulphates with Ce(IV) sulphate in dilute sulphuric acid: in the first step Fe(II) is oxidised to Fe(III) and in the second 'jump' V(IV) is oxidised to V(V) the latter change is accelerated by heating the solution after oxidation of the Fe(II) ion is complete. The titration of a substance having several oxidation states is exemplified by the stepwise reduction by acid chromium(II) chloride of Cu(II) ion to the Cu(I) state and then to the metal.

10.90 FORMAL POTENTIALS

Standard potentials $E^{\ominus}$ are evaluated with full regard to activity effects and with all ions present in simple form: they are really limiting or ideal values and are rarely observed in a potentiometric measurement. In practice, the solutions may be quite concentrated and frequently contain other electrolytes; under these conditions the activities of the pertinent species are much smaller than the concentrations, and consequently the use of the latter may lead to unreliable conclusions. Also, the actual active species present (see example below) may differ from those to which the ideal standard potentials apply. For these reasons 'formal potentials' have been proposed to supplement standard potentials. The formal potential is the potential observed experimentally in a solution containing one mole each of the oxidised and reduced substances together with other specified substances at specified concentrations. It is found that formal potentials vary appreciably, for example, with the nature and concentration of the acid that is present. The formal potential incorporates in one value the effects resulting from variation of activity coefficients with ionic strength, acid–base dissociation, complexation, liquid-junction potentials, etc., and thus has a real practical value. Formal potentials do not have the theoretical significance of standard potentials, but they are observed values in actual potentiometric measurements. In dilute solutions they usually obey the Nernst equation fairly closely in the form:

$$E = E^{\ominus\prime} + \frac{0.0591}{n} \log \frac{[Ox]}{[Red]} \text{ at } 25\,^{\circ}C$$

where $E^{\ominus\prime}$ is the formal potential and corresponds to the value of E at *unit* concentrations of oxidant and reductant, and the quantities in square brackets refer to molar concentrations. It is useful to determine and to tabulate $E^{\ominus\prime}$ with equivalent amounts of various oxidants and their conjugate reductants at various concentrations of different acids. If one is dealing with solutions whose composition is identical with or similar to that to which the formal potential pertains, more trustworthy conclusions can be derived from formal potentials than from standard potentials.

To illustrate how the use of standard potentials may occasionally lead to erroneous conclusions, consider the hexacyanoferrate(II)–hexacyano-ferrate(III) and the iodide–iodine systems. The standard potentials are:

$$[Fe(CN)_6]^{3-} + e \rightleftharpoons [Fe(CN)_6]^{4-}; \ E^{\ominus} = +0.36 \ \text{volt}$$

$$I_2 + 2e \rightleftharpoons 2I^-; \ E^{\ominus} = +0.54 \ \text{volt}$$

It would be expected that iodine would quantitatively oxidise hexacyanoferrate(II) ions:

$$2[Fe(CN)_6]^{4-} + I_2 = 2[Fe(CN)_6]^{3-} + 2I^-$$

In fact $[Fe(CN)_6]^{4-}$ ion oxidises iodide ion quantitatively in media containing $1M$ hydrochloric, sulphuric, or perchloric acid. This is because in solutions of low pH, protonation occurs and the species derived from $H_4Fe(CN)_6$ are weaker than those derived from $H_3Fe(CN)_6$; the activity of the $[Fe(CN)_6]^{4-}$ ion is decreased to a greater extent than that of the $[Fe(CN)_6]^{3-}$ ion, and therefore the reduction potential is increased. The actual redox potential of a solution containing equal concentrations of both cyanoferrates in $1M$ HCl, H_2SO_4 or $HClO_4$ is $+0.71$ volt, a value that is greater than the potential of the iodine–iodide couple.

Some results of formal potential measurements may now be mentioned. If there is no great difference in complexation of either the oxidant or its conjugate reductant in various acids, the formal potentials lie close together in these acids. Thus for the Fe(II)–Fe(III) system $E^{\ominus} = +0.77$ volt, $E^{\ominus\prime} = +0.73$ volt in $1M$ $HClO_4$, $+0.70$ volt in $1M$ HCl, $+0.68$ volt in $1M$ H_2SO_4, and $+0.61$ volt in $0.5M$ $H_3PO_4 + 1M$ H_2SO_4. It would seem that complexation is least in perchloric acid and greatest in phosphoric(V) acid.

For the Ce(III)–Ce(IV) system $E^{\ominus\prime} = +1.44$ volts in $1M$ H_2SO_4, $+1.61$ volts in $1M$ HNO_3, and $+1.70$ volts in $1M$ $HClO_4$. Perchloric acid solutions of cerium(IV) perchlorate, although unstable on standing, react rapidly and quantitatively with many inorganic compounds and have greater oxidising power than cerium(IV) sulphate–sulphuric acid or cerium(IV) nitrate–nitric acid solutions.

10.91 DETECTION OF THE END POINT IN OXIDATION–REDUCTION TITRATIONS

A. Internal oxidation–reduction indicators. As discussed in Sections 10.10–10.16, acid–base indicators are employed to mark the sudden change in pH during acid–base titrations. Similarly an oxidation–reduction indicator should mark the sudden change in the oxidation potential in the neighbourhood of the equivalence point in an oxidation–reduction titration. The ideal oxidation–reduction indicator will be one with an oxidation potential intermediate between that of the solution titrated and that of the titrant, and which exhibits a sharp, readily detectable colour change.

An oxidation–reduction indicator (redox indicator) is a compound which exhibits different colours in the oxidised and reduced forms:

$$In_{Ox} + ne \rightleftharpoons In_{Red}$$

The oxidation and reduction should be reversible. At a potential E the ratio of the concentrations of the two forms is given by the Nernst equation:

$$E = E_{In}^{\ominus} + \frac{RT}{nF} \ln a_{In.ox}/a_{In.Red}$$

$$E \approx E_{In}^{\ominus} + \frac{RT}{nF} \ln \frac{[In_{Ox}]}{[In_{Red}]}$$

where $E_{In}^{\ominus}$ is the standard (strictly the formal) potential of the indicator. If the colour intensities of the two forms are comparable a practical estimate of the colour-change interval corresponds to the change in the ratio $[In_{Ox}]/[In_{Red}]$ from 10 to $\frac{1}{10}$, this leads to an interval of potential of:

$$E = E_{In}^{\ominus} \pm \frac{0.0591}{1} \text{ volts at } 25\,°C$$

If the colour intensities of the two forms differ considerably the intermediate colour is attained at potential somewhat removed from $E_{In}^{\ominus}$, but the error is unlikely to exceed 0.06 volt. For a sharp colour change at the end point, $E_{In}^{\ominus}$ should differ by about at least 0.15 volt from the standard (formal) potentials of the other systems involved in the reaction.

One of the best oxidation–reduction indicators is the 1,10-phenanthroline–iron(II) complex. The base 1,10-phenanthroline combines readily in solution with iron(II) salts in the molecular ratio 3 base:1 iron(II) ion forming the intensely red 1,10-phenanthroline–iron(II) complex ion; with strong oxidising agents the iron(III) complex ion is formed, which has a pale blue colour. The colour change is a very striking one:

$$[Fe(C_{12}H_8N_2)_3]^{3+} + e \rightleftharpoons [Fe(C_{12}H_8N_2)_3]^{2+}$$
$$\text{Pale blue} \qquad\qquad\qquad\quad \text{Deep red}$$

The standard redox potential is 1.14 volts; the formal potential is 1.06 volts in $1\,M$ hydrochloric acid solution. The colour change, however, occurs at about 1.12 volts, because the colour of the reduced form (deep red) is so much more intense than that of the oxidised form (pale blue). The indicator is of great value in the titration of iron(II) salts and other substances with cerium(IV) sulphate solutions. It is prepared by dissolving 1,10-phenanthroline hydrate (relative molecular mass = 198.1) in the calculated quantity of $0.02\,M$ acid-free iron(II) sulphate, and is therefore 1,10-phenanthroline–iron(II) complex sulphate (known as **ferroin**). One drop is usually sufficient in a titration: this is equivalent to less than 0.01 mL of $0.05\,M$ oxidising agent, and hence the indicator blank is negligible at this or higher concentrations.

It has been shown (Section 10.89) that the potential at the equivalence point is the mean of the two standard redox potentials. In Fig. 10.14, the curve shows the variation of the potential during the titration of $0.1\,M$ iron(II) ion with

0.1 M cerium(IV) solution, and the equivalence point is at 1.10 volts. Ferroin changes from deep red to pale blue at a redox potential of 1.12 volts: the indicator will therefore be present in the red form. After the addition of, say, a 0.1 per cent excess of cerium(IV) sulphate solution the potential rises to 1.27 volts, and the indicator is oxidised to the pale blue form. It is evident that the titration error is negligibly small.

The standard or formal potential of ferroin can be modified considerably by the introduction of various substituents in the 1,10-phenanthroline nucleus. The most important substituted ferroin is 5-nitro-1,10-phenanthroline iron(II) sulphate (nitroferroin) and 4,7-dimethyl-1,10-phenanthroline iron(II) sulphate (dimethylferroin). The former ($E^{\ominus} = 1.25$ volts) is especially suitable for titrations using Ce(IV) in nitric or perchloric acid solution where the formal potential of the oxidant is high. The 4,7-dimethylferroin has a sufficiently low formal potential ($E^{\ominus} = 0.88$ volt) to render it useful for the titration of Fe(II) with dichromate in 0.5 M sulphuric acid.

Mention should be made of one of the earliest internal indicators. This is a 1 per cent solution of diphenylamine in concentrated sulphuric acid, and was introduced for the titration of iron(II) with potassium dichromate solution. An intense blue–violet coloration is produced at the end point. The addition of phosphoric(V) acid is desirable, for it lowers the formal potential of the Fe(III)–Fe(II) system so that the equivalence point potential coincides more nearly with that of the indicator. The action of diphenylamine (I) as an indicator depends upon its oxidation first into colourless diphenylbenzidine (II), which is the real indicator and is reversibly further oxidised to diphenylbenzidine violet (III). Diphenylbenzidine violet undergoes further oxidation if it is allowed to stand with excess of dichromate solution; this further oxidation is irreversible, and red or yellow products of unknown composition are produced.

A solution of diphenylbenzidine in concentrated sulphuric acid acts similarly to diphenylamine. The reduction potential of the system II, III is 0.76 volt in 0.5–1 M sulphuric acid. It is therefore evident that a lowering of the potential of the Fe(III)–Fe(II) system is desirable, as already mentioned, in order to obtain a sharp colour change. The disadvantage of diphenylamine and of diphenylbenzidine is their slight solubility in water. This has been overcome by the use of the soluble barium or sodium diphenylaminesulphonate, which is employed in 0.2 per cent aqueous solution. The redox potential ($E^{\ominus}_{\text{In}}$) is slightly higher (0.85 volt in 0.5 M sulphuric acid), and the oxidised form has a reddish-violet colour resembling that of potassium permanganate, but the colour slowly disappears on standing; the presence of phosphoric(V) acid is desirable in order to lower the redox potential of the system.

A list of selected redox indicators, together with their colour changes and reduction potentials in an acidic medium, is given in Table 10.9.

Table 10.9 Some oxidation–reduction indicators

Indicator	Colour change		Formal potential at pH = 0 (volts)
	Oxidised form	Reduced form	
5-Nitro-1,10-phenanthroline iron(II) sulphate (nitroferroin)	Pale blue	Red	1.25
1,10-Phenanthroline iron(II) sulphate (ferroin)	Pale blue	Red	1.06
2,2'-Bipyridyl iron(II) sulphate	Faint blue	Red	1.02
5,6-Dimethylferroin	Pale blue	Red	0.97
N-Phenylanthranilic acid,	Purple red	Colourless	0.89
4,7-Dimethyl-1,10-phenanthroline iron(II) sulphate (4,7-dimethylferroin)	Pale blue	Red	0.88
Diphenylaminesulphonic acid	Red–violet	Colourless	0.85
Diphenylbenzidine	Violet	Colourless	0.76
Diphenylamine	Violet	Colourless	0.76
3,3'-Dimethylnaphthidine	Purplish-red	Colourless	0.71
Starch–I_3^-,KI	Blue	Colourless	0.53
Methylene blue	Blue	Colourless	0.52

At this stage reference may be made to **potential mediators**, i.e. substances which undergo reversible oxidation–reduction and reach equilibrium *rapidly*. If we have a mixture of two ions, say M^{2+} and M^+, which reaches equilibrium slowly with an inert electrode, and a very small quantity of cerium(IV) salt is added, then the reaction:

$$M^+ + Ce^{4+} \rightarrow M^{2+} + Ce^{3+}$$

takes place until the tendency of M^+ to be oxidised to M^{2+} is exactly balanced by the tendency of Ce^{3+} to be oxidised to Ce^{4+}, that is, until the M^{2+}, M^+ and Ce^{4+}, Ce^{3+} potentials are equal. A platinum or other inert electrode rapidly attains equilibrium with the Ce(III) and Ce(IV) ions, and will soon register a stable potential which is also that due to the $M^{2+} + e \rightleftharpoons M^+$ system. If the potential mediator is employed in small amount, then a negligible quantity of M^+ is converted into M^{2+} when equilibrium is reached, and the measured potential may be regarded as that of the original system. Potential mediators are, of course, useful in the measurement of the oxidation–reduction potentials of redox systems; in this connection mention may be made of the use of potassium iodide ($\equiv$ iodide–iodine system) in the arsenate–arsenite system in acid solution. It is evident that redox indicators (e.g. 1,10-phenanthroline–iron(II) ion) may act as potential mediators.

B. Self-indicating reagents. This is well illustrated by potassium permanganate, one drop of which will impart a visible pink coloration to several hundred millilitres of solution, even in the presence of slightly coloured ions, such as iron(III). The colours of cerium(IV) sulphate and of iodine solutions have also been employed in the detection of end points, but the colour change is not so marked as for potassium permanganate; here, however, sensitive internal

367

indicators (1,10-phenanthroline–iron(II) ion or N-phenylanthranilic acid and starch respectively) are available.

This method has the drawback that an excess of oxidising agent is always present at the end point. For work of the highest accuracy, the indicator blank may be determined and allowed for, or the error may be considerably reduced by performing the standardisation and determination under similar experimental conditions.

C. Potentiometric methods. This is a procedure which depends upon measurement of the e.m.f. between a reference electrode and an indicator (redox) electrode at suitable intervals during the titration, i.e. a potentiometric titration is carried out. The procedure is discussed fully in Chapter 15; let it suffice at this stage to point out that the procedure is applicable not only to those cases where suitable indicators are available, but also to those cases, e.g. coloured or very dilute solutions, where the indicator method is inapplicable, or of limited accuracy.

OXIDATIONS WITH POTASSIUM PERMANGANATE

10.92 DISCUSSION

This valuable and powerful oxidising agent was first introduced into titrimetric analysis by F. Margueritte for the titration of iron(II). In acid solutions, the reduction can be represented by the following equation:

$$MnO_4^- + 8H^+ + 5e \rightleftharpoons Mn^{2+} + 4H_2O$$

The standard potential in acid solution, $E^{\ominus}$, has been calculated to be 1.51 volts; hence the permanganate ion in acid solution is a strong oxidising agent.

Sulphuric acid is the most suitable acid, as it has no action upon permanganate in dilute solution. With hydrochloric acid, there is a likelihood of the reaction:

$$2MnO_4^- + 10Cl^- + 16H^+ = 2Mn^{2+} + 5Cl_2 + 8H_2O$$

taking place, and some permanganate may be consumed in the formation of chlorine. This reaction is particularly liable to occur with iron salts unless special precautions are adopted (see below). With a small excess of free acid, a very dilute solution, low temperature and slow titration with constant shaking, the danger from this cause is minimised. There are, however, some titrations, such as those with arsenic(III) oxide, antimony(III), and hydrogen peroxide, which can be carried out in the presence of hydrochloric acid.

In the analysis of iron ores, solution is frequently effected in concentrated hydrochloric acid; the iron(III) is reduced and the iron(II) is then determined in the resultant solution. To do this, it is best to add about 25 mL of **Zimmermann and Reinhardt's solution** (this is sometimes termed **preventive solution**), which is prepared by dissolving 50 g of crystallised manganese(II) sulphate ($MnSO_4,4H_2O$) in 250 mL water, adding a cooled mixture of 100 mL concentrated sulphuric acid and 300 mL water, followed by 100 mL syrupy orthophosphoric acid. The manganese(II) sulphate lowers the reduction potential of the $MnO_4^- - Mn(II)$ couple (compare Section 2.31) and thereby makes it a weaker oxidising agent; the tendency of the permanganate ion to oxidise chloride ion is thus reduced. It has been stated that a further function of the manganese(II) sulphate is to supply an adequate concentration of Mn^{2+} ions to react with any local excess of permanganate ion. Mn(III) is probably formed in the reduction of permanganate ion to manganese(II); the Mn(II), and also the orthophosphoric acid, exert a depressant effect upon the potential of the Mn(III)–Mn(II) couple, so that Mn(III) is reduced by Fe^{2+} ion rather than by chloride ion. The phosphoric(V) acid combines with the yellow Fe^{3+} ion to form the complex ion $[Fe(HPO_4)]^+$, thus rendering the end point more clearly visible. The phosphoric(V) acid lowers the reduction potential of the Fe(III)–Fe(II) system by complexation, and thus tends to increase the reducing power of the Fe^{2+} ion. Under these conditions permanganate ion oxidises iron(II) rapidly and reacts only slowly with chloride ion.

For the titration of colourless or slightly coloured solutions, the use of an indicator is unnecessary, since as little as 0.01 mL of 0.02 M potassium permanganate imparts a pale-pink colour to 100 mL of water. The intensity of the colour in dilute solutions may be enhanced, if desired, by the addition of a redox indicator (such as sodium diphenylamine sulphonate, N-phenylanthranilic acid, or ferroin) just before the end point of the reaction; this is usually not required, but is advantageous if more dilute solutions of permanganate are used.

Potassium permanganate may also be used in *strongly* alkaline solutions. Here two consecutive partial reactions take place:
(1) the relatively rapid reaction:

$$MnO_4^- + e \rightleftharpoons MnO_4^{2-}$$

and (2) the relatively slow reaction:

$$MnO_4^{2-} + 2H_2O + 2e \rightleftharpoons MnO_2 + 4OH^-$$

The standard potential $E^\ominus$ of reaction (1) is 0.56 volt and of reaction (2) 0.60 volt. By suitably controlling the experimental conditions (e.g. by the addition of barium ions, which form the sparingly soluble barium manganate as a fine, granular precipitate), reaction (1) occurs almost exclusively. In *moderately alkaline* solutions permanganate is reduced quantitatively to manganese dioxide. The half-cell reaction is:

$$MnO_4^- + 2H_2O + 3e \rightleftharpoons MnO_2 + 4OH^-$$

and the standard potential $E^\ominus$ is 0.59 volt.

Potassium permanganate is not a primary standard. It is difficult to obtain the substance perfectly pure and completely free from manganese dioxide. Moreover, ordinary distilled water is likely to contain reducing substances (traces of organic matter, etc.) which will react with the potassium permanganate to form manganese dioxide. The presence of the latter is very objectionable because it catalyses the auto-decomposition of the permanganate solution on standing:

$$4MnO_4^- + 2H_2O = 4MnO_2 + 3O_2 + 4OH^-$$

Permanganate is inherently unstable in the presence of manganese(II) ions:

$$2MnO_4^- + 3Mn^{2+} + 2H_2O = 5MnO_2 + 4H^+$$

This reaction is slow in acid solution, but it is very rapid in neutral solution. For these reasons, potassium permanganate solution is rarely made up by dissolving weighed amounts of the purified solid in water; it is more usual to heat a freshly prepared solution to boiling and keep it on the steam bath for an hour or so, and then filter the solution through a non-reducing filtering medium, such as purified glass wool or a sintered-glass filtering crucible (porosity No. 4). Alternatively, the solution may be allowed to stand for 2–3 days at room temperature before filtration. The glass-stoppered bottle or flask should be carefully freed from grease and prior deposits of manganese dioxide: this may be done by rinsing with dichromate–sulphuric acid cleaning mixture* and then thoroughly with distilled water. Acidic and alkaline solutions are less stable than neutral ones. Solutions of permanganate should be protected from unnecessary exposure to light: a dark-coloured bottle is recommended. Diffuse daylight causes no appreciable decomposition, but bright sunlight slowly decomposes even pure solutions.

Potassium permanganate solutions may be standardised using arsenic(III) oxide or sodium oxalate as primary standards: secondary standards include metallic iron, and iron(II) ethylenediammonium sulphate (or ethylenediamine iron(II) sulphate), $FeSO_4, C_2H_4(NH_3)_2SO_4, 4H_2O$. Full details of the first two methods are given in Section 10.93 below. Standardisation using metallic iron is similar to that for potassium dichromate given in Section 10.100.

10.93 PREPARATION OF 0.02 *M* POTASSIUM PERMANGANATE

Weigh out about 3.2–3.25 g potassium permanganate on a watchglass, transfer it to a 1500 mL beaker, add 1 L water, cover the beaker with a clockglass, heat the solution to boiling, boil gently for 15–30 minutes and allow the solution to cool to the laboratory temperature. Filter the solution through a funnel containing a plug of purified glass wool, or, more simply, through a sintered-glass or porcelain filtering crucible or funnel. Collect the filtrate in a vessel which has been cleaned with chromic acid mixture* and then thoroughly washed with distilled water. The filtered solution should be stored in a clean, glass-stoppered bottle, and kept in the dark or in diffuse light except when in use: alternatively, it may be kept in a dark brown glass bottle.

10.94 STANDARDISATION OF PERMANGANATE SOLUTIONS

Method A: With arsenic(III) oxide. This procedure, which utilises arsenic(III) oxide as a primary standard and potassium iodide or potassium iodate as a catalyst for the reaction, is convenient in practice and is a trustworthy method for the standardisation of permanganate solutions. Analytical grade arsenic(III) oxide has a purity of at least 99.8 per cent, and the results by this method agree to within 1 part in 3000 with the sodium oxalate procedure (Method B, below).

$$As_2O_3 + 4OH^- = 2HAsO_3^{2-} + H_2O$$

$$5H_3AsO_3 + 2MnO_4^- + 6H^+ = 5H_3AsO_4 + 2Mn^{2+} + 3H_2O$$

Procedure. Dry some arsenic(III) oxide at 105–110 °C for 1–2 hours, cover the container, and allow to cool in a desiccator. Accurately weigh approximately

---

* **Caution:** This is a very powerful reagent and should only be used by experienced chemists.

0.25 g of the dry oxide, and transfer it to a 400 mL beaker. Add 10 mL of a cool solution of sodium hydroxide, prepared from 20 g sodium hydroxide and 100 mL water (Note 1). Allow to stand for 8–10 minutes, stirring occasionally. When solution is complete, add 100 mL water, 10 mL pure concentrated hydrochloric acid, and 1 drop $0.0025M$ potassium iodide or potassium iodate (Note 2). Add the permanganate solution from a burette until a faint pink colour persists for 30 seconds. Add the last 1–1.5 mL dropwise, allowing each drop to become decolorised before the next drop is introduced. For the most accurate work it is necessary to determine the volume of permanganate solution required to duplicate the pink colour at the end point. This is done by adding permanganate solution to a solution containing the same amounts of alkali, acid, and catalyst as were used in the test. The correction should not be more than 0.03 mL. Repeat the determination with two other similar quantities of oxide. Calculate the concentration of the potassium permanganate solution. Duplicate determinations should agree within 0.1 per cent.

Notes. (1) For *elementary students*, it is sufficient to weigh out accurately about 1.25 g of arsenic(III) oxide, dissolve this in 50 mL of a cool 20 per cent solution of sodium hydroxide, and make up to 250 mL in a graduated flask. Shake well. Measure 25.0 mL of this solution by means of a burette and **not** with a pipette (**caution — the solution is highly poisonous**) into a 500 mL conical flask, add 100 mL water, 10 mL pure concentrated hydrochloric acid, one drop potassium iodide solution, and titrate with the permanganate solution to the first permanent pink colour as detailed above. Repeat with two other 25 mL portions of the solution. Successive titrations should agree within 0.1 mL.

(2) $0.0025M$ Potassium iodide = 0.41 g KI L^{-1}. $0.0025M$ Potassium iodate = 0.54 g $KIO_3 L^{-1}$.

Calculation. It is evident from the equation given above that if the weight of arsenic(III) oxide is divided by the number of millilitres of potassium permanganate solution to which it is equivalent, as found by titration, we have the weight of primary standard equivalent to 1 mL of the permanganate solution.

Method B: With sodium oxalate. This reagent is readily obtained pure and anhydrous, and the ordinary material has a purity of at least 99.9 per cent. In the experimental procedure originally employed a solution of the oxalate, acidified with dilute sulphuric acid and warmed to 80–90 °C, was titrated with the permanganate solution slowly ($10–15$ mL min^{-1}) and with constant stirring until the first permanent faint pink colour was obtained; the temperature near the end-point was not allowed to fall below 60 °C. However with this procedure the results may be 0.1–0.45 per cent high; the titre depends upon the acidity, the temperature, the rate of addition of the permanganate solution, and the speed of stirring. Because of this it is best to make *a more rapid* addition of 90–95 per cent of the permanganate solution (about $25–35$ mL min^{-1}) to a solution of sodium oxalate in $1M$ sulphuric acid at 25–30 °C, the solution is then warmed to 55–60 °C and the titration completed, the last 0.5–1 mL portion being added dropwise. The method is accurate to 0.06 per cent. Full experimental details are given below.

$$2Na^+ + C_2O_4^{2-} + 2H^+ \rightleftharpoons H_2C_2O_4 + 2Na^+$$

$$2MnO_4^- + 5H_2C_2O_4 + 6H^+ = 2Mn^{2+} + 10CO_2 + 8H_2O$$

It should be mentioned that if oxalate is to be determined it is often not convenient to use the room temperature technique for unknown amounts of oxalate. The permanganate solution may then be standardised against sodium oxalate at about 80 °C using the same procedure in the standardisation as in the analysis.

Procedure. Dry some analytical grade sodium oxalate at 105–110 °C for 2 hours, and allow it to cool in a covered vessel in a desiccator. Weigh out accurately from a weighing bottle about 0.3 g of the dry sodium oxalate into a 600 mL beaker, add 240 mL of recently prepared distilled water, and 12.5 mL of concentrated sulphuric acid (**caution**) or 250 mL of 1 M sulphuric acid. Cool to 25–30 °C and stir until the oxalate has dissolved (Note 1). Add 90–95 per cent of the required quantity of permanganate solution from a burette at a rate of 25–35 mL min^{-1} while stirring slowly (Note 2). Heat to 55–60 °C (use a thermometer as stirring rod), and complete the titration by adding permanganate solution until a faint pink colour persists for 30 seconds. Add the last 0.5–1 mL dropwise, with particular care to allow each drop to become decolorised before the next is introduced. For the most exact work, it is necessary to determine the excess of permanganate solution required to impart a pink colour to the solution. This is done by matching the colour produced by adding permanganate solution to the same volume of boiled and cooled dilute sulphuric acid at 55–60 °C. This correction usually amounts to 0.03–0.05 mL. Repeat the determination with two other similar quantities of sodium oxalate.

Notes. (1) For *elementary students*, it is sufficient to weigh out accurately about 1.7 g of sodium oxalate, transfer it to a 250 mL graduated flask, and make up to the mark. Shake well, Use 25 mL of this solution per titration and add 150 mL of *ca* 1 M sulphuric acid. Carry out the titration rapidly at the ordinary temperature until the *first* pink colour appears throughout the solution, and allow to stand until the solution is colourless. Warm the solution to 50–60 °C and continue the titration to a permanent faint pink colour. It must be remembered that oxalate solutions attack glass, so that the solution should not be stored more than a few days.

(2) An approximate value of the volume of permanganate solution required can be computed from the weight of sodium oxalate employed. In the first titration about 75 per cent of this volume is added, and the determination is completed at 55–60 °C. Thereafter, about 90–95 per cent of the volume of permanganate solution is added at the laboratory temperature.

Provided that it is stored with due regard to the precautions referred to in Section 10.92 the standardised permanganate solution will keep for a long time, but it is advisable to re-standardise the solution frequently to confirm that no decomposition has set in.

10.95 ANALYSIS OF HYDROGEN PEROXIDE

Hydrogen peroxide is usually encountered in the form of an aqueous solution containing about 6 per cent, 12 per cent or 30 per cent hydrogen peroxide, and frequently referred to as '20-volume', '40-volume', and '100-volume' hydrogen peroxide respectively; this terminology is based upon the volume of oxygen liberated when the solution is decomposed by boiling. Thus 1 mL of '100-volume' hydrogen peroxide will yield 100 mL of oxygen measured at standard temperature and pressure.

The following reaction occurs when potassium permanganate solution is added to hydrogen peroxide solution acidified with dilute sulphuric acid:

$$2MnO_4^- + 5H_2O_2 + 6H^+ = 2Mn^{2+} + 5O_2 + 8H_2O$$

This forms the basis of the method of analysis given below.

It is good practice to use a fairly high concentration of acid and a reasonably low rate of addition in order to reduce the danger of forming manganese dioxide, which is an active catalyst for the decomposition of hydrogen peroxide. For slightly coloured solutions or for titrations with dilute permanganate, the use of ferroin as indicator is recommended. Organic substances may interfere. A fading end point indicates the presence of organic matter or other reducing agents, in which case the iodometric method is better (Section 10.118).

Procedure. Transfer 25.0 mL of the '20-volume' solution by means of a burette to a 500 mL graduated flask, and dilute with water to the mark. Shake thoroughly. Pipette 25.0 mL of this solution to a conical flask, dilute with 200 mL water, add 20 mL dilute sulphuric acid (1:5), and titrate with standard 0.02 M potassium permanganate to the first permanent, faint pink, colour. Repeat the titration; two consecutive determinations should agree within 0.1 mL.

Calculate: (i) the weight of hydrogen peroxide per L of the original solution and (ii) the 'volume strength', i.e. the number of millilitres of oxygen at s.t.p. that can be obtained from 1 mL of the original solution.

Analysis of metallic peroxides. A metallic peroxide, such as **sodium peroxide**, can be analysed in similar manner, provided that care is taken to avoid loss of oxygen during the dissolution of the peroxide. This may be done by working in a medium containing boric acid which is converted to the relatively stable 'perboric acid' upon the addition of the peroxide.

Procedure. To 100 mL of distilled water, add 5 mL of concentrated sulphuric acid, cool and then add 5 g of pure boric acid; when this has dissolved cool the mixture in ice. Transfer gradually from a weighing bottle about 0.5 g (accurately weighed) of the sodium peroxide sample (**handle with care**) to the well-stirred, ice-cold solution. When the addition is complete, transfer the solution to a 250 mL graduated flask, make up to the mark, and then titrate 50 mL portions of the solution with standard 0.02 M permanganate solution.

10.96 DETERMINATION OF NITRITES

Discussion. Nitrites react in warm acid solution (ca 40 °C) with permanganate solution in accordance with the equation:

$$2MnO_4^- + 5NO_2^- + 6H^+ = 2Mn^{2+} + 5NO_3^- + 3H_2O$$

If a solution of a nitrite is titrated in the ordinary way with potassium permanganate, poor results are obtained, because the nitrite solution has first to be acidified with dilute sulphuric acid. Nitrous acid is liberated, which being volatile and unstable, is partially lost. If, however, a measured volume of standard potassium permanganate solution, acidified with dilute sulphuric acid, is treated with the nitrite solution, added from a burette, until the permanganate is just decolorised, results accurate to 0.5–1 per cent may be obtained. This is due to the fact that nitrous acid does not react instantaneously with the permanganate. This method may be used to determine the purity of commercial potassium nitrite.

373

Procedure. Weigh out accurately about 1.1 g of commercial potassium nitrite, dissolve it in cold water, and dilute to 250 mL in a graduated flask. Shake well. Measure out 25.0 mL of standard $0.02 M$ potassium permanganate into a 500 mL flask, add 225 mL of $0.5 M$ sulphuric acid, and heat to 40 °C. Place the nitrite solution in the burette, and add it slowly and with constant stirring until the permanganate solution is just decolorised. Better results are obtained by allowing the tip of the burette to dip under the surface of the diluted permanganate solution. Towards the end the reaction is sluggish, so that the nitrite solution must be added very slowly.

More accurate results may be secured by adding the nitrite to an acidified solution in which permanganate is present in excess (the tip of the pipette containing the nitrite solution should be below the surface of the liquid during the addition), and back-titrating the excess potassium permanganate with a solution of ammonium iron(II) sulphate which has recently been compared with the permanganate solution.

10.97 DETERMINATION OF PERSULPHATES

Discussion. Alkali persulphates (peroxydisulphates) can readily be evaluated by adding to their solutions a known excess of an acidified iron(II) salt solution, and determining the excess of iron(II) by titration with standard potassium permanganate solution.

$$S_2O_8^{2-} + 2Fe^{2+} + 2H^+ = 2Fe^{3+} + 2HSO_4^-$$

By adding phosphoric acid or hydrofluoric acid, the reduction is complete in a few minutes at room temperature. Many organic compounds interfere.

Another procedure utilises standard oxalic acid solution. When a sulphuric acid solution of a persulphate is treated with excess of standard oxalic acid solution in the presence of a little silver sulphate as catalyst, the following reaction occurs:

$$H_2S_2O_8 + H_2C_2O_4 = 2H_2SO_4 + 2CO_2$$

The excess of oxalic acid is titrated with standard potassium permanganate solution.

Procedure A. Prepare an approximately $0.1 M$ solution of ammonium iron(II) sulphate by dissolving about 9.8 g of the solid in 200 mL of sulphuric acid ($0.5 M$) in a 250 mL graduated flask, and then making up to the mark with freshly boiled and cooled distilled water. Standardise the solution by titrating 25 mL portions with standard potassium permanganate solution ($0.02 M$) after the addition of 25 mL sulphuric acid ($0.5 M$).

Weigh out accurately about 0.3 g potassium persulphate into a conical flask and dissolve it in 50 mL of water. Add 5 mL syrupy phosphoric(V) acid or 2.5 mL 35–40 per cent hydrofluoric acid (**CARE!**), 10 mL $2.5 M$ sulphuric acid, and 50.0 mL of the *ca* $0.1 M$ iron(II) solution. After 5 minutes, titrate the excess of Fe^{2+} ion with standard $0.02 M$ potassium permanganate.

From the difference between the volume of $0.02 M$ permanganate required to oxidise 50 mL of the iron(II) solution and that required to oxidise the iron(II) salt remaining after the addition of the persulphate, calculate the percentage purity of the sample.

Procedure B. Prepare an approximately $0.05 M$ solution of oxalic acid by dissolving about 1.6 g of the compound and making up to 250 mL in a graduated flask. Standardise the solution with standard $(0.02 M)$ potassium permanganate solution using the procedure described in Section 10.94 (Method B).

Weigh out accurately 0.3–0.4 g potassium persulphate into a 500 mL conical flask, add 50 mL of $0.05 M$-oxalic acid, followed by 0.2 g of silver sulphate dissolved in 20 mL of 10 per cent sulphuric acid. Heat the mixture in a water bath until no more carbon dioxide is evolved (15–20 minutes), dilute the solution to about 100 mL with water at about 40 °C, and titrate the excess of oxalic acid with standard $0.02 M$ potassium permanganate.

OXIDATIONS WITH POTASSIUM DICHROMATE

10.98 DISCUSSION

Potassium dichromate is not such a powerful oxidising agent as potassium permanganate (compare reduction potentials in Table 2.6 in Section 2.31), but it has several advantages over the latter substance. It can be obtained pure, it is stable up to its fusion point, and it is therefore an excellent primary standard. Standard solutions of exactly known concentration can be prepared by weighing out the pure dry salt and dissolving it in the proper volume of water. Furthermore, the aqueous solutions are stable indefinitely if adequately protected from evaporation. Potassium dichromate is used only in acid solution, and is reduced rapidly at the ordinary temperature to a green chromium(III) salt. It is not reduced by cold hydrochloric acid, provided the acid concentration does not exceed 1 or $2 M$. Dichromate solutions are less easily reduced by organic matter than are those of permanganate and are also stable towards light. Potassium dichromate is therefore of particular value in the determination of iron in iron ores: the ore is usually dissolved in hydrochloric acid, the iron(III) reduced to iron(II), and the solution then titrated with standard dichromate solution:

$$Cr_2O_7^{2-} + 6Fe^{2+} + 14H^+ = 2Cr^{3+} + 6Fe^{3+} + 7H_2O$$

In acid solution, the reduction of potassium dichromate may be represented as:

$$Cr_2O_7^{2-} + 14H^+ + 6e \rightleftharpoons 2Cr^{3+} + 7H_2O$$

The green colour due to the Cr^{3+} ions formed by the reduction of potassium dichromate makes it impossible to ascertain the end-point of a dichromate titration by simple visual inspection of the solution and so a redox indicator must be employed which gives a strong and unmistakable colour change; this procedure has rendered obsolete the external indicator method which was formerly widely used. Suitable indicators for use with dichromate titrations include N-phenylanthranilic acid (0.1 per cent solution in $0.005 M$ NaOH) and sodium diphenylamine sulphonate (0.2 per cent aqueous solution); the latter must be used in presence of phosphoric(V) acid.

10.99 PREPARATION OF 0.02 *M* POTASSIUM DICHROMATE

Analytical grade potassium dichromate has a purity of not less than 99.9 per cent and is satisfactory for most purposes.* Powder finely about 6 g of the analytical grade material in a glass or agate mortar, and heat for 30–60 minutes in an air oven at 140–150 °C. Allow to cool in a closed vessel in a desiccator. Weigh out accurately about 5.88 g of the dry potassium dichromate into a weighing bottle and transfer the salt quantitatively to a 1 L graduated flask, using a small funnel to avoid loss. Dissolve the salt in the flask in water and make up to the mark; shake well. Alternatively, place a little over 5.88 g of potassium dichromate in a weighing bottle, and weigh accurately. Empty the salt into a 1 L graduated flask, and weigh the bottle again. Dissolve the salt in water, and make up to the mark.

The molarity of the solution can be calculated directly from the weight of salt taken, but if the salt has only been weighed out approximately, then the solution must be standardised as described in the following section.

10.100 STANDARDISATION OF POTASSIUM DICHROMATE SOLUTION AGAINST IRON

With metallic iron. Use iron wire of 99.9 per cent assay value (Note 1). Insert a well-fitting rubber stopper provided with a bent delivery tube into a 500 mL conical flask and clamp the flask in a retort stand in an inclined position, the tube being so bent as to dip into a small beaker containing saturated sodium hydrogencarbonate solution or 20 per cent potassium hydrogencarbonate solution (prepared from the solids) (Fig. 10.16). Place 100 mL 1.5 M sulphuric acid (from 92 mL water and 8 mL concentrated sulphuric acid) in the flask, and add 0.5–1 g sodium hydrogencarbonate in two portions; the carbon dioxide produced will drive out the air. Meanwhile, weigh out accurately about 0.2 g of iron wire, place it quickly into the flask, replace the stopper and bent tube, and warm gently until the iron has dissolved completely. Cool the flask rapidly

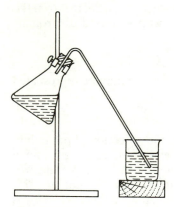

Fig. 10.16

---

* If only a 'pure' grade (as distinct from analytical grade) of commercial salt is available, or if there is some doubt as to the purity of the salt, the following method of purification should be used. A concentrated solution of the salt in hot water is prepared and filtered. The crystals which separate on cooling are filtered on a sintered-glass filter funnel and sucked dry. The resultant crystals are recrystallised again. The purified crystals are then dried at 180–200 °C, ground to a fine powder in a glass or agate mortar, and again dried at 140–150 °C to constant weight.

under a stream of cold water, with the delivery tube still dipping into the solution in the beaker (Note 2).

Titrate the cooled solution immediately with the dichromate solution, using either sodium diphenylamine sulphonate or N-phenylanthranilic acid as indicator. If the former is selected, add 6–8 drops of the indicator, followed by 5 mL of syrupy phosphoric(V) acid: titrate slowly with the dichromate solution, stirring well, until the pure green colour changes to a grey–green. Then add the dichromate solution dropwise until the first tinge of blue–violet, which remains permanent on shaking, appears. If the latter indicator is selected, add 200 mL of 1M sulphuric acid, then 0.5 mL of the indicator; add the dichromate solution, with shaking until the colour changes from green to violet–red (Note 3).

1 mole $K_2Cr_2O_7 \equiv 6$ moles Fe

Notes. (1) Iron wire of 99.9 per cent purity is available commercially and is a suitable analytical standard. If the wire exhibits any sign of rust, it should be drawn between two pieces of fine emery cloth, and then wiped with a clean, dry cloth before use. The general reaction which occurs has been given in Section 10.92.

(2) As the flask cools, the hydrogencarbonate solution is automatically drawn in until the pressure of the carbon dioxide inside the flask is equal to the atmospheric pressure.

(3) The standardisation may also be effected with ethylenediammonium iron(II) sulphate.

10.101 DETERMINATION OF CHROMIUM IN A CHROMIUM(III) SALT

Discussion. Chromium(III) salts are oxidised to dichromate by boiling with excess of a persulphate solution in the presence of a little silver nitrate (catalyst). The excess of persulphate remaining after the oxidation is complete is destroyed by boiling the solution for a short time. The dichromate content of the resultant solution is determined by the addition of excess of a standard iron(II) solution and titration of the excess of the latter with standard 0.02M potassium dichromate.

$$2Cr^{3+} + 3S_2O_8^{2-} + 7H_2O \xrightarrow{(AgNO_3)} Cr_2O_7^{2-} + 6HSO_4^- + 8H^+$$

$$2S_2O_8^{2-} + 2H_2O = O_2\uparrow + 4HSO_4^-$$

Procedure. Weigh out accurately an amount of the salt which will contain about 0.25 g of chromium, and dissolve it in 50 mL distilled water. Add 20 mL of ca 0.1 M silver nitrate solution, followed by 50 mL of a 10 per cent solution of ammonium or potassium persulphate. Boil the liquid gently for 20 minutes. Cool, and dilute to 250 mL in a graduated flask. Remove 50 mL of the solution with a pipette, add 50 mL of a 0.1 M ammonium iron(II) sulphate solution (Section 10.97, Procedure A), 200 mL of 1M sulphuric acid, and 0.5 mL of N-phenylanthranilic acid indicator. Titrate the excess of the iron(II) salt with standard 0.02 M potassium dichromate until the colour changes from green to violet–red.

Standardise the ammonium iron(II) sulphate solution against the 0.02M potassium dichromate, using N-phenylanthranilic acid as indicator. Calculate the volume of the iron(II) solution which was oxidised by the dichromate originating from the chromium salt, and from this the percentage of chromium in the sample.

Note. Lead or barium can be determined by precipitating the sparingly soluble chromate, dissolving the washed precipitate in dilute sulphuric acid, adding a known excess of ammonium iron(II) sulphate solution, and titrating the excess of Fe^{2+} ion with $0.02M$ potassium dichromate in the usual way.

$$2PbCrO_4 + 2H^+ = 2Pb^{2+} + Cr_2O_7^{2-} + H_2O$$

10.102 DETERMINATION OF CHLORATE

Discussion. Chlorate ion is reduced by warming with excess of iron(II) in the presence of a relatively high concentration of sulphuric acid:

$$ClO_3^- + 6Fe^{2+} + 6H^+ = Cl^- + 6Fe^{3+} + 3H_2O$$

The excess Fe^{2+} ion is determined by titration with standard dichromate solution in the usual way.

Procedure. To obtain experience in the method, the purity of analytical-grade potassium chlorate may be determined. Prepare a $0.02M$ potassium chlorate solution. Into a 250 mL conical flask, place 25.0 mL of the potassium chlorate solution, 25.0 mL of $0.2M$ ammonium iron(II) sulphate solution in $2M$ sulphuric acid and add **cautiously** 12 mL concentrated sulphuric acid. Heat the mixture to boiling (in order to ensure completion of the reduction), and cool to room temperature by placing the flask in running tap water. Add 20 mL 1:1 water/phosphoric(V) acid, followed by 0.5 mL sodium diphenyl-amine-sulphonate indicator. Titrate the excess Fe^{2+} ion with standard $0.02M$ potassium dichromate to a first tinge of purple coloration which remains on stirring.

Standardise the ammonium iron(II) sulphate solution by repeating the procedure but using 25 mL distilled water in place of the chlorate solution. The difference in titres is equivalent to the amount of potassium chlorate added.

10.103 DETERMINATION OF CHEMICAL OXYGEN DEMAND

Discussion. One very important application of potassium dichromate is in a back-titration for the environmental determination[16] of the amount of oxygen required to oxidise all the organic material in a sample of impure water, such as sewage effluent. This is known as the chemical oxygen demand (C.O.D.) and is expressed in terms of milligrams of oxygen required per litre of water, $mg\,L^{-1}$. The analysis of the impure water sample is carried out in parallel with a blank determination on pure, double-distilled water.

Procedure. Place a 50 mL volume of the water sample in a 250 mL conical flask with a ground-glass neck which can be fitted with a water condenser for refluxing. Add 1 g of mercury(II) sulphate, followed by 80 mL of a silver sulphate/sulphuric acid solution (Note 1). Then add 10 mL of approximately $0.00833M$ standard potassium dichromate solution (Note 2), fit the flask with the reflux condenser and boil the mixture for 15 minutes. On cooling rinse the inside of the condenser with 50 mL of water into the flask contents. Add either diphenylamine indicator (1 mL) or ferroin indicator and titrate with $0.025M$ ammonium iron(II) sulphate solution (Note 3). Diphenylamine gives a colour change from blue to green at the end-point, whilst that for ferroin is blue–green to red–brown. Call this titration A mL. Repeat the back-titration for the blank (titration B mL). The difference between the two values is the amount of potassium dichromate used up in the oxidation. The C.O.D. is calculated from the relationship:

$$C.O.D. = (A - B) \times 0.2 \times 20 \text{ mg L}^{-1}$$

as a 1 mL difference between the titrations corresponds to 0.2 mg of oxygen required by the 50 mL sample (a correction must, of course, be made if solutions of slightly different molarities are employed); see Note 4.

Notes. (1) This solution is prepared by dissolving 5 g of silver sulphate in 500 mL of concentrated sulphuric acid.

(2) The required concentration is obtained by weighing out 1.225 g of potassium dichromate and diluting to 500 mL with de-ionised water in a graduated flask.

(3) Dissolve 4.9 g of ammonium iron(II) sulphate heptahydrate in 150 mL of water and add 2.5 mL of concentrated sulphuric acid. Dilute the solution to 500 mL in a graduated flask.

(4) This method gives high results with samples possessing a high chloride content due to reaction between the mercury(II) sulphate and the chloride ions. In these cases the problem can be overcome by following a procedure using chromium(III) potassium sulphate, $Cr(III)K(SO_4)_2,12H_2O$.[17]

OXIDATIONS WITH CERIUM(IV) SULPHATE SOLUTION

10.104 GENERAL DISCUSSION

Cerium(IV) sulphate is a powerful oxidising agent; its reduction potential in $0.5-4.0M$ sulphuric acid at $25\,^{\circ}C$ is 1.43 ± 0.05 volts. It can be used only in acid solution, best in $0.5M$ or higher concentrations: as the solution is neutralised, cerium(IV) hydroxide [hydrated cerium(IV) oxide] or basic salts precipitate. The solution has an intense yellow colour, and in hot solutions which are not too dilute the end point may be detected without an indicator; this procedure, however, necessitates the application of a blank correction, and it is therefore preferable to add a suitable indicator.

The advantages of cerium(IV) sulphate as a standard oxidising agent are:

1. Cerium(IV) sulphate solutions are remarkably stable over prolonged periods. They need not be protected from light, and may even be boiled for a short time without appreciable change in concentration. The stability of sulphuric acid solutions covers the wide range of 10–40 mL of concentrated sulphuric acid per litre. It is evident, therefore, that an acid solution of cerium(IV) sulphate surpasses a permanganate solution in stability.
2. Cerium(IV) sulphate may be employed in the determination of reducing agents in the presence of a high concentration of hydrochloric acid (contrast potassium permanganate, Section 10.92).
3. Cerium(IV) solutions in $0.1\,M$ solution are not too highly coloured to obstruct vision when reading the meniscus in burettes and other titrimetric apparatus.
4. In the reaction of cerium(IV) salts in acid solution with reducing agents, the simple change

$$Ce^{4+} + e \rightleftharpoons Ce^{3+}$$

is assumed to take place. With permanganate, of course, a number of reduction products are produced according to the experimental conditions.

Solutions of cerium(IV) sulphate in dilute sulphuric acid are stable even at boiling temperatures. Hydrochloric acid solutions of the salt are unstable because of reduction to cerium(III) by the acid with the simultaneous liberation of chlorine:

$$2Ce^{4+} + 2Cl^- = 2Ce^{3+} + Cl_2$$

This reaction takes place quite rapidly on boiling, and hence hydrochloric acid cannot be used in oxidations which necessitate boiling with excess of cerium(IV) sulphate in acid solution: sulphuric acid must be used in such oxidations. However, direct titration with cerium(IV) sulphate in a dilute hydrochloric acid medium, e.g. for iron(II) may be accurately performed at room temperature, and in this respect cerium(IV) sulphate is superior to potassium permanganate [cf. (2) above]. The presence of hydrofluoric acid is harmful, since fluoride ion forms a stable complex with Ce(IV) and decolorises the yellow solution.

Formal potential measurements show that the redox potential of the Ce(IV)–Ce(III) system is greatly dependent upon the nature and the concentration of the acid present; thus the following values are recorded for the acids named in molar solution: H_2SO_4 1.44 V, HNO_3 1.61 V, $HClO_4$ 1.70 V, and in 8M perchloric acid solution the value is 1.87 V.

It has been postulated on the basis of the formal potential measurements that Ce(IV) exists as anionic complexes $[Ce(SO_4)_4]^{4-}$ or $[Ce(SO_4)_3]^{2-}$, $[Ce(NO_3)_6]^{2-}$, and $[Ce(ClO_4)_6]^{2-}$; in consequence, solid salts such as ammonium cerium(IV) sulphate, $2(NH_4)_2SO_4.Ce(SO_4)_2,2H_2O$, and ammonium cerium(IV) nitrate, $2NH_4NO_3.Ce(NO_3)_4,4H_2O$, have been formulated as ammonium tetrasulphatocerate(IV), $(NH_4)_4[Ce(SO_4)_4],2H_2O$, and ammonium hexanitratocerate(IV), $(NH_4)_2[Ce(NO_3)_6],4H_2O$, respectively. For convenience, the term cerium(IV) sulphate will be retained.

Solutions of cerium(IV) sulphate may be prepared by dissolving cerium(IV) sulphate or the more soluble ammonium cerium(IV) sulphate in dilute (0.5–1.0M) sulphuric acid. Ammonium cerium(IV) nitrate may be purchased of analytical grade, and a solution of this in 1M sulphuric acid may be used for many of the purposes for which cerium(IV) solutions are employed, but in some cases the presence of nitrate ion is undesirable. The nitrate ion may be removed by evaporating the solid reagent which concentrated sulphuric acid, or alternatively a solution of the nitrate may be precipitated with aqueous ammonia and the resulting cerium(IV) hydroxide filtered off and dissolved in sulphuric acid.

Internal indicators suitable for use with cerium(IV) sulphate solutions include N-phenylanthranilic acid, ferroin [1,10-phenanthroline iron(II)], and 5,6-dimethylferroin.

10.105   PREPARATION OF 0.1 M CERIUM(IV) SULPHATE

Method A. Evaporate 55.0 g of ammonium cerium(IV) nitrate almost to dryness with excess (48 mL) of concentrated sulphuric acid in a Pyrex evaporating dish (**FUME CUPBOARD**). Dissolve the resulting cerium (IV) sulphate in 1M sulphuric acid (28 mL concentrated sulphuric acid to 500 mL water), transfer to a 1 L graduated flask, add 1M sulphuric acid until near the graduation mark, and make up to the mark with distilled water. Shake well.

Method B. Weigh out 35–36 g of pure cerium(IV) sulphate into a 500 mL beaker, add 56 mL of 1:1-sulphuric acid/water and stir, with frequent additions of water and gentle warming, until the salt is dissolved. Transfer to a 1 L glass-stoppered graduated flask, and when cold, dilute to the mark with distilled water. Shake well.

Alternatively, weigh out 64–66 g of ammonium cerium(IV) sulphate into a solution prepared by adding 28 mL of concentrated sulphuric acid to 500 mL of water: stir the mixture until the solid has dissolved. Transfer to a 1 L graduated flask, and make up to the mark with distilled water.

The relative molecular masses of cerium(IV) sulphate $Ce(SO_4)_2$ and ammonium cerium(IV) sulphate $(NH_4)_4[Ce(SO_4)_4],2H_2O$ are 333.25 and 632.56 respectively.

Method C. Place about 21 g of cerium(IV) hydroxide in a 1500 mL beaker, and add, with stirring, 100 mL of concentrated sulphuric acid. Continue the stirring and introduce 300 mL of distilled water slowly and cautiously. Allow to stand overnight, and if any residue remains, filter the solution into a 1 L graduated flask and dilute to the mark.

10.106 STANDARDISATION OF CERIUM(IV) SULPHATE SOLUTIONS

Method A: Standardisation with arsenic(III) oxide. *Discussion.* The most trustworthy method for standardising cerium(IV) sulphate solutions is with pure arsenic(III) oxide. The reaction between cerium(IV) sulphate solution and arsenic(III) oxide is very slow at the ambient temperature; it is necessary to add a trace of osmium tetroxide as catalyst. The arsenic(III) oxide is dissolved in sodium hydroxide solution, the solution acidified with dilute sulphuric acid, and after adding 2 drops of an 'osmic acid' solution prepared by dissolving 0.1 g osmium tetroxide in 40 mL of 0.05M sulphuric acid, and the indicator (1–2 drops ferroin or 0.5 mL N-phenylanthranilic acid), it is titrated with the cerium(IV) sulphate solution to the first sharp colour change: orange–red to very pale blue or yellowish-green to purple respectively.

$$2Ce^{4+} + H_3AsO_3 + H_2O = 2Ce^{3+} + H_3AsO_4 + 2H^+$$

Procedure. Weigh out accurately about 0.2 g of arsenic(III) oxide, previously dried at 105–110 °C for 1–2 hours, and transfer to a 500 mL beaker or to a 500 mL conical flask. Add 20 mL of approx. 2M sodium hydroxide solution, and warm the mixture gently until the arsenic(III) oxide has *completely* dissolved. Cool to room temperature, and add 100 mL water, followed by 25 mL 2.5M sulphuric acid. Then add 3 drops 0.01 M osmium tetroxide solution (0.25 g osmium tetroxide (**CARE! FUME CUPBOARD**) dissolved in 100 mL 0.05M sulphuric acid) and 0.5 mL N-phenylanthranilic acid indicator (or 1–2 drops of ferroin). Titrate with the 0.1 M cerium(IV) sulphate solution until the first sharp colour change occurs (see Discussion above). Repeat with two other samples of approximately equal weight of arsenic(III) oxide.

Method B: Standardisation with sodium oxalate. Standardisation may also be carried out with sodium oxalate; in this case, an indirect procedure must be used as the redox indicators are themselves oxidised at the elevated temperatures which are necessary. The procedure, therefore, is to add an excess of the cerium(IV) solution, and then, after cooling, the excess is determined by back-titration with an iron(II) solution. It is possible to carry out a direct titration of the sodium oxalate if a potentiometric procedure is used (Chapter 15).

Procedure. Prepare an approximately $0.1 M$ solution of ammonium iron(II) sulphate in dilute sulphuric acid and titrate with the cerium(IV) sulphate solution using ferroin indicator.

Weigh out accurately about 0.2 g sodium oxalate into a 250 mL conical flask and add 25–30 mL $1 M$ sulphuric acid. Heat the solution to about 60 °C and then add about 30 mL of the cerium(IV) solution to be standardised dropwise, adding the solution as rapidly as possible consistent with drop formation. Re-heat the solution to 60 °C, and then add a further 10 mL of the cerium(IV) solution. Allow to stand for three minutes, then cool and back-titrate the excess cerium(IV) with the iron(II) solution using ferroin as indicator.

Practically all the determinations described under potassium permanganate and potassium dichromate may be carried out with cerium(IV) sulphate. Use is made of the various indicators already detailed and also, in some cases where great accuracy is not required, of the pale yellow colour produced by the cerium(IV) sulphate itself. Only a few determinations will, therefore, be considered in some detail.

10.107 DETERMINATION OF COPPER

Discussion. Copper(II) ions are quantitatively reduced in $2 M$ hydrochloric acid solution by means of the silver reductor (Section 10.140) to the copper(I) state. The solution, after reduction, is collected in a solution of ammonium iron(III) sulphate, and the Fe^{2+} ion formed is titrated with standard cerium(IV) sulphate solution using ferroin or N-phenylanthranilic acid as indicator.

Comparatively large amounts of nitric acid, and also zinc, cadmium, bismuth, tin, and arsenate have no effect upon the determination; the method may therefore be applied to determine copper in brass.

Procedure (copper in crystallised copper sulphate). Weigh out accurately about 3.1 g of copper sulphate crystals, dissolve in water, and make up to 250 mL in a graduated flask. Shake well. Pipette 50 mL of this solution into a small beaker, add an equal volume of *ca* $4 M$ hydrochloric acid. Pass this solution through a silver reductor at the rate of 25 mL min^{-1}, and collect the filtrate in a 500 mL conical flask charged with 20 mL $0.5 M$ iron(III) ammonium sulphate solution (prepared by dissolving the appropriate quantity of the analytical grade iron(III) salt in $0.5 M$ sulphuric acid). Wash the reductor column with six 25 mL portions of $2 M$ hydrochloric acid. Add 1 drop of ferroin indicator or 0.5 mL N-phenylanthranilic acid, and titrate with $0.1 M$ cerium(IV) sulphate solution. The end point is sharp, and the colour imparted by the Cu^{2+} ions does not interfere with the detection of the equivalence point.

Procedure (copper in copper(I) chloride). Prepare an ammonium iron(III) sulphate solution by dissolving 10.0 g of the salt in about 80 mL of $3 M$ sulphuric acid and dilute to 100 mL with acid of the same strength. Weigh out accurately about 0.3 g of the sample of copper(I) chloride into a dry 250 mL conical flask and add 25.0 mL of the iron(III) solution. Swirl the contents of the flask until the copper(I) chloride dissolves, add a drop or two of ferroin indicator, and titrate with standard $0.1 M$ cerium(IV) sulphate.

Repeat the titration with 25.0 mL of the iron solution, omitting the addition of the copper(I) chloride. The difference in the two titrations gives the volume of $0.1 M$ cerium(IV) sulphate which has reacted with the known weight of copper(I) chloride.

10.108 DETERMINATION OF MOLYBDATE

Discussion. Molybdates [Mo(VI)] are quantitatively reduced in $2M$ hydrochloric acid solution at 60–80 °C by the silver reductor to Mo(V). The reduced molybdenum solution is sufficiently stable over short periods of time in air to be titrated with standard cerium(IV) sulphate solution using ferroin or N-phenylanthranilic acid as indicator. Nitric acid must be completely absent; the presence of a little phosphoric(V) acid during the reduction of the molybdenum(VI) is not harmful and, indeed, appears to increase the rapidity of the subsequent oxidation with cerium(IV) sulphate. Elements such as iron, copper, and vanadium interfere; nitrate interferes, since its reduction is catalysed by the presence of molybdates.

Procedure.   Weigh out accurately about 2.5 g ammonium molybdate $(NH_4)_6Mo_7O_{24},4H_2O$, dissolve in water and make up to 250 mL in a graduated flask. Pipette 50 mL of this solution into a small beaker, add an equal volume of $4M$ hydrochloric acid, then 3 mL of 85 per cent phosphoric(V) acid, and heat the solution to 60–80 °C. Pour hot $2M$ hydrochloric acid through a silver reductor, and then pass the molybdate solution through the hot reductor at the rate of about $10\,\text{mL min}^{-1}$. Collect the reduced solution in a 500 mL beaker or 500 mL conical flask, and wash the reductor with six 25 mL portions of $2M$ hydrochloric acid; the first two washings should be made with the hot acid (rate: $10\,\text{mL min}^{-1}$) and the last four washings with the cold acid (rate: 20–$25\,\text{mL min}^{-1}$). Cool the solution, add one drop of ferroin or 0.5 mL N-phenylanthranilic acid, and titrate with standard $0.1 M$ cerium(IV) sulphate. The precipitate of cerium(IV) phosphate, which is initially formed, dissolves on shaking. Add the last 0.5 mL of the reagent dropwise and with vigorous stirring or shaking.

10.109 DETERMINATION OF NITRITES

Discussion. Satisfactory results are obtained by adding the nitrite solution to excess of standard $0.1 M$ cerium(IV) sulphate, and determining the excess of cerium(IV) sulphate with a standard iron(II) solution (compare Section 10.96).

$$2Ce^{4+} + NO_2^- + H_2O = 2Ce^{3+} + NO_3^- + 2H^+$$

For practice, determine the percentage of NO_2^- in potassium nitrite, or the purity of sodium nitrite, preferably of analytical-grade quality.

Procedure. Weigh out accurately about 1.5 g of sodium nitrite and dissolve it in 500 mL of boiled-out water in a graduated flask. Shake thoroughly. Place 50 mL of standard $0.1 M$ cerium(IV) sulphate in a conical flask, and add 10 mL of $2M$ sulphuric acid. Transfer 25 mL of the nitrite solution to this flask by means of a pipette, and keep the tip of the pipette below the surface of the liquid during the addition. Allow to stand for 5 minutes, and titrate the excess of cerium(IV) sulphate with standard $0.1 M$ ammonium iron(II) sulphate, using ferroin or N-phenylanthranilic acid as indicator. Repeat the titration with two further portions of the nitrite solution. Standardise the iron solution by titrating 25 mL of it with the cerium(IV) solution in the presence of dilute sulphuric acid.

Determine the volume of the standard cerium(IV) sulphate solution which has reacted with the nitrite solution, and therefrom calculate the purity of the sodium nitrite employed.

Note. Cerium(IV) sulphate may also be used for the following analyses.

Hydrogen peroxide. The diluted solution, which may contain nitric or hydrochloric acid in any concentration between 0.5 and $3M$ or sulphuric acid in the concentration range 0.25 to $1.5M$, is titrated directly with standard cerium(IV) sulphate solution, using ferroin or N-phenylanthranilic acid as indicator. The reaction is:

$$2Ce^{4+} + H_2O_2 = 2Ce^{3+} + O_2 + 2H^+$$

Persulphate (peroxydisulphate). Persulphate cannot be determined directly by reduction with iron(II) because the reaction is too slow:

$$S_2O_8^{2-} + 2Fe^{2+} = 2SO_4^{2-} + 2Fe^{3+}$$

An excess of a standard solution of iron(II) must therefore be added and the excess back-titrated with standard cerium(IV) sulphate solution. Erratic results are obtained, depending upon the exact experimental conditions, because of induced reactions leading to oxidation by air of iron(II) ion or to decomposition of the persulphate; these induced reactions are inhibited by bromide ion in concentrations not exceeding $1M$ and, under these conditions, the determination may be carried out in the presence of organic matter.

To 25.0 mL of $0.01–0.015M$ persulphate solution in a 150 mL conical flask, add 7 mL of $5M$ sodium bromide solution and 2 mL of $3M$ sulphuric acid. Stopper the flask. Swirl the contents, then add excess of $0.05M$ ammonium iron(II) sulphate (15.0 mL), and allow to stand for 20 minutes. Add 1 mL of $0.001M$ ferroin indicator, and titrate the excess of Fe^{2+} ion with $0.02M$ cerium(IV) sulphate in $0.5M$ sulphuric acid to the *first* colour change from orange to yellow.

Oxalates. Oxalates can be determined by means of the indirect method described in Section 10.106.

Hexacyanoferrate(II). This can be determined by titration in $1M$ H_2SO_4 using N-phenylanthranilic acid.

OXIDATION AND REDUCTION PROCESSES INVOLVING IODINE: IODOMETRIC TITRATIONS

10.110 GENERAL DISCUSSION

The direct iodometric titration method (sometimes termed **iodimetry**) refers to titrations with a standard solution of iodine. The indirect iodometric titration method (sometimes termed **iodometry**) deals with the titration of iodine liberated in chemical reactions. The normal reduction potential of the reversible system:

$$I_2 \text{ (solid)} + 2e \rightleftharpoons 2I^-$$

is 0.5345 volt. The above equation refers to a saturated aqueous solution in the presence of solid iodine; this half-cell reaction will occur, for example, towards the end of a titration of iodide with an oxidising agent such as potassium permanganate, when the iodide ion concentration becomes relatively low. Near the beginning, or in most iodometric titrations, when an excess of iodide ion is present, the tri-iodide ion is formed

$$I_2 (aq.) + I^- \rightleftharpoons I_3^-$$

since iodine is readily soluble in a solution of iodide. The half-cell reaction is better written:

$$I_3^- + 2e \rightleftharpoons 3I^-$$

and the standard reduction potential is 0.5355 volt. Iodine or the tri-iodide ion is therefore a much weaker oxidising agent than potassium permanganate, potassium dichromate, and cerium(IV) sulphate.

In most direct titrations with iodine (iodimetry) a solution of iodine in potassium iodide is employed, and the reactive species is therefore the tri-iodide ion I_3^-. Strictly speaking, all equations involving reactions of iodine should be written with I_3^- rather than with I_2, e.g.

$$I_3^- + 2S_2O_3^{2-} = 3I^- + S_4O_6^{2-}$$

is more accurate than

$$I_2 + 2S_2O_3^{2-} = 2I^- + S_4O_6^{2-}$$

For the sake of simplicity, however, the equations in this book will usually be written in terms of molecular iodine rather than the tri-iodide ion.

Strong reducing agents (substances with a much lower reduction potential), such as tin(II) chloride, sulphurous acid, hydrogen sulphide, and sodium thiosulphate, react completely and rapidly with iodine even in acid solution. With somewhat weaker reducing agents, e.g. arsenic(III), or antimony(III), complete reaction occurs only when the solution is kept neutral or very faintly acid; under these conditions the reduction potential of the reducing agent is a minimum, or its reducing power is a maximum.

If a strong oxidising agent is treated in neutral or (more usually) acid solution with a large excess of iodide ion, the latter reacts as a reducing agent and the oxidant will be quantitatively reduced. In such cases, an equivalent amount of iodine is liberated, and is then titrated with a standard solution of a reducing agent, which is usually sodium thiosulphate.

The normal reduction potential of the iodine–iodide system is independent of the pH of the solution so long as the latter is less than about 8; at higher values iodine reacts with hydroxide ions to form iodide and the extremely unstable hypoiodite, the latter being transformed rapidly into iodate and iodide by self-oxidation and reduction:

$$I_2 + 2OH^- = I^- + IO^- + H_2O$$

$$3IO^- = 2I^- + IO_3^-$$

The reduction potentials of certain substances increase considerably with increasing hydrogen ion concentration of the solution. This is the case with systems containing permanganate, dichromate, arsenate, antimonate, bromate, etc., i.e. with anions which contain oxygen and therefore require hydrogen for complete reduction. Many weak oxidising anions are completely reduced by iodide ions if their reduction potentials are raised considerably by the presence in solution of a large amount of acid.

By suitable control of the pH of the solution, it is sometimes possible to titrate the reduced form of a substance with iodine, and the oxidised form, after the addition of iodide, with sodium thiosulphate. Thus with the arsenite–arsenate system:

$$H_3AsO_3 + I_2 + H_2O \rightleftharpoons H_3AsO_4 + 2H^+ + 2I^-$$

the reaction is completely reversible. At pH values between 4 and 9, arsenite can be titrated with iodine solution. In strongly acid solutions, however, arsenate is reduced to arsenite and iodine is liberated. Upon titration with sodium thiosulphate solution, the iodine is removed and the reaction proceeds from right to left.

Two important **sources of error in titrations involving iodine** are: (*a*) loss of iodine owing to its appreciable volatility; and (*b*) acid solutions of iodide are oxidised by oxygen from the air:

$$4I^- + O_2 + 4H^+ = 2I_2 + 2H_2O$$

In the presence of excess of iodide, the volatility is decreased markedly through the formation of the tri-iodide ion; at room temperature the loss of iodine by volatilisation from a solution containing at least 4 per cent of potassium iodide is negligible provided the titration is not prolonged unduly. Titrations should be performed in cold solutions in conical flasks and not in open beakers. If a solution is to stand it should be kept in a glass-stoppered vessel. The atmospheric oxidation of iodide is negligible in neutral solution in the absence of catalysts, but the rate of oxidation increases rapidly with decreasing pH. The reaction is catalysed by certain metal ions of variable charge value (particularly copper), by nitrite ion, and also by strong light. For this reason titrations should not be performed in direct sunlight, and solutions containing iodide should be stored in amber glass bottles. Furthermore, the air oxidation of iodide ion may be induced by the reaction between iodide and the oxidising agent, especially when the main reaction is slow. Solutions containing an excess of iodide and acid must therefore not be allowed to stand longer than necessary before titration of the iodine. If prolonged standing is necessary (as in the titration of vanadate or Fe^{3+} ions) the solution should be free from air before the addition of iodide and the air displaced from the titration vessel by carbon dioxide [e.g. by adding small portions (0.2–0.5 g) of pure sodium hydrogencarbonate to the acid solution, or a little solid carbon dioxide, dry ice]; potassium iodide is then introduced and the glass stopper replaced immediately.

It seems appropriate to refer at this point to the uses of a standard solution containing **potassium iodide and potassium iodate**. This solution is quite stable and yields iodine when treated with acid:

$$IO_3^- + 5I^- + 6H^+ = 3I_2 + 3H_2O$$

The standard solution is prepared by dissolving a weighed amount of pure potassium iodate in a solution containing a slight excess of pure potassium iodide, and diluting to a definite volume. This solution has two important uses. The first is as a source of a known quantity of iodine in titrations [compare Section 10.115(A)]; it must be added to a solution containing strong acid; it cannot be employed in a medium which is neutral or possesses a low acidity.

The second use is in the determination of the acid content of solutions iodometrically or in the **standardisation of solutions of strong acids**. It is evident from the above equation that the amount of iodine liberated is equivalent to the acid content of the solution. Thus if, say, 25 mL of an approximately $0.1\,M$ solution of a strong acid is treated with a slight excess of potassium iodate (say, 25 mL of $0.02\,M$ potassium iodate solution, Section 10.126) and a slight excess of potassium iodide solution (say, 10 mL of a 10 per cent solution), and the liberated iodine titrated with standard $0.1\,M$ sodium thiosulphate with the aid of starch as an indicator, the concentration of the acid may be readily evaluated.

10.111 DETECTION OF THE END POINT

A solution of iodine in aqueous iodide has an intense yellow to brown colour. One drop of $0.05\,M$ iodine solution imparts a perceptible pale yellow colour to 100 mL of water, so that in otherwise colourless solutions iodine can serve as its own indicator. The test is made much more sensitive by the use of a solution of starch as indicator. Starch reacts with iodine in the presence of iodide to form an intensely blue-coloured complex, which is visible at very low concentrations of iodine. The sensitivity of the colour reaction is such that a blue colour is visible when the iodine concentration is $2 \times 10^{-5}\,M$ and the iodide concentration is greater than $4 \times 10^{-4}\,M$ at $20\,°C$. The colour sensitivity decreases with increasing temperature of the solution; thus at $50\,°C$ it is about ten times less sensitive than at $25\,°C$. The sensitivity decreases upon the addition of solvents, such as ethanol: no colour is obtained in solutions containing 50 per cent ethanol or more. It cannot be used in a strongly acid medium because hydrolysis of the starch occurs.

Starches can be separated into two major components, amylose and amylopectin, which exist in different proportions in various plants. Amylose, which is a straight-chain compound and is abundant in potato starch, gives a blue colour with iodine and the chain assumes a spiral form. Amylopectin, which has a branched-chain structure, forms a red–purple product, probably by adsorption.

The great merit of starch is that it is inexpensive. It possesses the following disadvantages: (1) insolubility in cold water; (2) instability of suspensions in water; (3) it gives a water-insoluble complex with iodine, the formation of which precludes the addition of the indicator early in the titration (for this reason, in titrations of iodine, the starch solution should not be added until just prior to the end point when the colour begins to fade); and (4) there is sometimes a 'drift' end point, which is marked when the solutions are dilute.

Most of the shortcomings of starch as an indicator are absent in **sodium starch glycollate**. This is a white, non-hygroscopic powder, readily soluble in hot water to give a faintly opalescent solution, which is stable for many months; it does not form a water-insoluble complex with iodine, and hence the indicator may be added at any stage of the reaction. With excess of iodine (e.g. at the beginning of a titration with sodium thiosulphate) the colour of the solution containing 1 mL of the indicator (0.1 per cent aqueous solution) is green; as the iodine concentration diminishes the colour changes to blue, which becomes intense just before the end point is reached. The end point is very sharp and reproducible and there is no 'drift' in dilute solution.

Carbon tetrachloride has been used in certain reactions instead of starch solution. One litre of water at $25\,°C$ will dissolve 0.335 g of iodine, but the same volume of carbon tetrachloride will dissolve about 28.5 g. Iodine is therefore about 85 times as soluble in carbon tetrachloride as it is in water, and the carbon tetrachloride solution is highly coloured. When a little carbon tetrachloride is added to an aqueous solution containing iodine and the solution well shaken, the great part of the iodine will dissolve in the carbon tetrachloride; the latter will fall to the bottom since it is immiscible with water, and the colour of the organic layer will be much deeper than that of the original aqueous solution. The reddish-violet colour of iodine in carbon tetrachloride is visible in very low concentrations of iodine; thus on shaking 10 mL of carbon tetrachloride with 50 mL of $10^{-5}M$ iodine, a distinct violet coloration is produced in the organic layer. This enables many iodometric determinations to be carried out with comparative ease. The titrations are performed in 250 mL glass-stoppered bottles or flasks with accurately ground stoppers. After adding the excess of potassium iodide solution and 5–10 mL of carbon tetrachloride to the reaction mixture, the titration with sodium thiosulphate is commenced. At first the presence of iodine in the aqueous solution will be apparent, and gentle rotation of the liquid causes sufficient mixing. Towards the end of the titration the bottle or flask is stoppered and shaken after each addition of sodium thiosulphate solution; the end point is reached when the carbon tetrachloride just becomes colourless. Equally satisfactory results can be obtained with chloroform.

Preparation and use of starch solution. Make a paste of 0.1 g of soluble starch with a little water, and pour the paste, with constant stirring, into 100 mL of boiling water, and boil for 1 minute. Allow the solution to cool and add 2–3 g of potassium iodide. Keep the solution in a stoppered bottle.

Only freshly prepared starch solution should be used. Two millilitres of a 1 per cent solution per 100 mL of the solution to be titrated is a satisfactory amount; the same volume of starch solution should always be added in a titration. In the titration of iodine, starch must not be added until just before the end point is reached. Apart from the fact that the fading of the iodine colour is a good indication of the approach at the end point, if the starch solution is added when the iodine concentration is high, some iodine may remain adsorbed even at the end point. The indicator blank is negligibly small in iodimetric and iodometric titrations of $0.05M$ solutions; with more dilute solutions, it must be determined in a liquid having the same composition as the solution titrated has at the end point.

A solid solution of starch in urea may also be employed. Reflux 1 g of soluble starch and 19 g of urea with xylene. At the boiling point of the organic solvent the urea melts with little decomposition, and the starch dissolves in the molten urea. Allow to cool, then remove the solid mass and powder it; store the product in a stoppered bottle. A few milligrams of this solid added to an aqueous solution containing iodine then behaves like the usual starch indicator.

Preparation and use of sodium starch glycollate indicator. Sodium starch glycollate, prepared as described below, dissolves slowly in cold but rapidly in hot water. It is best dissolved by mixing, say, 5.0 g of the finely powdered solid with 1–2 mL ethanol, adding 100 mL cold water, and boiling for a few minutes with vigorous stirring: a faintly opalescent solution results. This 5 per cent stock solution is diluted to 1 per cent concentration as required. The most convenient concentration for use as an indicator is 0.1 mg mL^{-1}, i.e. 1 mL of the 1 per cent aqueous solution is added to 100 mL of the solution being titrated.

10.112 PREPARATION OF 0.05*M* IODINE SOLUTION

Discussion. In addition to a small solubility (0.335 g of iodine dissolves in 1 L of water at 25 °C), aqueous solutions of iodine have an appreciable vapour pressure of iodine, and therefore decrease slightly in concentration on account of volatilisation when handled. Both difficulties are overcome by dissolving the iodine in an aqueous solution of potassium iodide. Iodine dissolves readily in aqueous potassium iodide: the more concentrated the solution, the greater is the solubility of the iodine. The increased solubility is due to the formation of a tri-iodide ion:

$$I_2 + I^- \rightleftharpoons I_3^-$$

The resulting solution has a much lower vapour pressure than a solution of iodine in pure water, and consequently the loss by volatilisation is considerably diminished. Nevertheless, the vapour pressure is still appreciable so that **precautions should always be taken to keep vessels containing iodine closed except during the actual titrations**. When an iodide solution of iodine is titrated with a reductant, the free iodine reacts with the reducing agent, this displaces the equilibrium to the left, and eventually all the tri-iodide is decomposed; the solution therefore behaves as though it were a solution of free iodine.

For the preparation of standard iodine solutions, resublimed iodine and iodate-free potassium iodide should be employed. The solution may be standardised against pure arsenic(III) oxide or with a sodium thiosulphate solution which has been recently standardised against potassium iodate.

The equation for the ionic reaction is:

$$I_2 + 2e \rightleftharpoons 2I^-$$

Procedure: Preparation of 0.05*M* iodine. Dissolve 20 g of iodate-free potassium iodide in 30–40 mL of water in a glass-stoppered 1 L graduated flask. Weigh out about 12.7 g of resublimed iodine on a watchglass on a rough balance (never on an analytical balance on account of the iodine vapour), and transfer it by means of a small dry funnel into the concentrated potassium iodide solution. Insert the glass stopper into the flask, and shake in the cold until all the iodine has dissolved. Allow the solution to acquire room temperature, and make up to the mark with distilled water.

The iodine solution is best preserved in small glass-stoppered bottles. These should be filled completely and kept in a cool, dark place.

10.113 STANDARDISATION OF IODINE SOLUTIONS

(A) With arsenic(III) oxide: Discussion. As already indicated (Section 10.94), arsenic(III) oxide which has been dried at 105–110 °C for two hours is an excellent primary standard. The reaction between this substance and iodine is a reversible one:

$$H_3AsO_3 + I_2 + H_2O \rightleftharpoons H_3AsO_4 + 2H^+ + 2I^-$$

and only proceeds quantitatively from left to right if the hydrogen iodide is removed from the solution as fast as it is formed. This may be done by the addition of sodium hydrogencarbonate: sodium carbonate and sodium hydroxide cannot be used, since they react with the iodine, forming iodide,

389

hypoiodite, and iodate. Actually it has been shown that complete oxidation of the arsenite occurs when the pH of the solution lies between 4 and 9, the best value being 6.5, which is very close to the neutral point. Buffer solutions are employed to maintain the correct pH. A 0.12 M solution of sodium hydrogencarbonate saturated with carbon dioxide has a pH of 7; a solution saturated with both sodium tetraborate and boric acid has a pH of about 6.2, whilst a $Na_2HPO_4-NaH_2PO_4$ solution is almost neutral. Any of these three buffer solutions is suitable, but as already stated the first-named is generally employed.

Procedure. Weigh out accurately about 2.5 g of finely powdered arsenic(III) oxide, transfer to a 500 mL beaker, and dissolve it in a concentrated solution of sodium hydroxide, prepared from 2 g of iron-free sodium hydroxide and 20 mL of water. Dilute to about 200 mL, and neutralise the solution with 1M hydrochloric acid, using a pH meter. When the solution is faintly acid transfer the contents of the beaker quantitatively to a 500 mL graduated flask, add 2 g of pure sodium hydrogencarbonate, and, when all the salt has dissolved, dilute to the mark and shake well.

Using a burette or a pipette with a safety pump (this is necessary owing to the poisonous properties of the solution) measure out 25.0 mL of the arsenite solution into a 250 mL conical flask, add 25–50 mL of water, 5 g of sodium hydrogencarbonate, and 2 mL of starch solution. Swirl the solution carefully until the hydrogencarbonate has dissolved. Then titrate slowly with the iodine solution, contained in a burette, to the first blue colour.

Alternatively, the arsenite solution may be placed in the burette, and titrated against 25.0 mL of the iodine solution contained in a conical flask. When the solution has a pale yellow colour, add 2 mL of starch solution, and continue the titration slowly until the blue colour is just destroyed.

If it is desired to base the standardisation directly upon arsenic(III) oxide, proceed as follows. Weigh out accurately about 0.20 g of pure arsenic(III) oxide into a conical flask, dissolve it in 10 mL of 1M sodium hydroxide, and add a small excess of dilute sulphuric acid (say, 12–15 mL of 0.5M acid). Mix thoroughly and cautiously. Then add carefully a solution of 2 g of sodium hydrogencarbonate in 50 mL of water, followed by 2 mL of starch solution. Titrate slowly with the iodine solution to the first blue colour. Repeat with two other similar quantities of the oxide.

(B) With standard sodium thiosulphate solution. Sodium thiosulphate solution, which has been recently standardised, preferably against pure potassium iodate, is employed. Transfer 25 mL of the iodine solution to a 250 mL conical flask, dilute to 100 mL and add the standard thiosulphate solution from a burette until the solution has a pale yellow colour. Add 2 mL of starch solution, and continue the addition of the thiosulphate solution slowly until the solution is just colourless.

10.114 PREPARATION OF 0.1M SODIUM THIOSULPHATE

Discussion. Sodium thiosulphate ($Na_2S_2O_3,5H_2O$) is readily obtainable in a state of high purity, but there is always some uncertainty as to the exact water content because of the efflorescent nature of the salt and for other reasons. The substance is therefore unsuitable as a primary standard. It is a reducing agent by virtue of the half-cell reaction:

$$2S_2O_3^{2-} \rightleftharpoons S_4O_6^{2-} + 2e$$

An approximately $0.1M$ solution is prepared by dissolving about 25 g crystallised sodium thiosulphate in 1 L of water in a graduated flask. The solution is standardised by any of the methods described below.

Before dealing with these, it is necessary to refer briefly to the stability of thiosulphate solutions. Solutions prepared with conductivity (equilibrium) water are perfectly stable. However, ordinary distilled water usually contains an excess of carbon dioxide; this may cause a slow decomposition to take place with the formation of sulphur:

$$S_2O_3^{2-} + H^+ = HSO_3^- + S$$

Moreover, decomposition may also be caused by bacterial action (e.g. *Thiobacillus thioparus*), particularly if the solution has been standing for some time. For these reasons, the following recommendations are made:

1. Prepare the solution with recently boiled distilled water.
2. Add 3 drops of chloroform or $10\,mg\,L^{-1}$ of mercury(II) iodide; these compounds improve the keeping qualities of the solution.

    Bacterial activity is least when the pH lies between 9 and 10. The addition of a *small* amount, $0.1\,g\,L^{-1}$, of sodium carbonate is advantageous to ensure the correct pH. In general, alkali hydroxides, sodium carbonate ($> 0.1\,g\,L^{-1}$), and sodium tetraborate should not be added, since they tend to accelerate the decomposition:

$$S_2O_3^{2-} + 2O_2 + H_2O \rightleftharpoons 2SO_4^{2-} + 2H^+$$

3. Avoid exposure to light, as this tends to hasten the decomposition.

The standardisation of thiosulphate solutions may be effected with potassium iodate, potassium dichromate, copper and iodine as primary standards, or with potassium permanganate or cerium(IV) sulphate as secondary standards. Owing to the volatility of iodine and the difficulty of preparation of perfectly pure iodine, this method is not a suitable one for beginners. If, however, a standard solution of iodine (see Sections 10.112 and 10.113) is available, this may be used for the standardisation of thiosulphate solutions.

Procedure. Weigh out 25 g of sodium thiosulphate crystals ($Na_2S_2O_3,5H_2O$), dissolve in boiled-out distilled water, and make up to 1 L in a graduated flask with boiled-out water. If the solution is to be kept for more than a few days, add 0.1 g sodium carbonate or three drops of chloroform.

10.115 STANDARDISATION OF SODIUM THIOSULPHATE SOLUTIONS

(A) With potassium iodate. Potassium iodate has a purity of at least 99.9 per cent: it can be dried at 120 °C. This reacts with potassium iodide in acid solution to liberate iodine:

$$IO_3^- + 5I^- + 6H^+ = 3I_2 + 3H_2O$$

Its relative molecular mass is 214.00; a $0.02M$ solution therefore contains 4.28 g of potassium iodate per litre.

Weigh out accurately 0.14–0.15 g of pure dry potassium iodate, dissolve it in 25 mL of cold, boiled-out distilled water, add 2 g of iodate-free potassium iodide (Note 1) and 5 mL of 1M sulphuric acid (Note 2). Titrate the liberated iodine with the thiosulphate solution with constant shaking. When the colour of the liquid has become a pale yellow, dilute to ca 200 mL with distilled water, add 2 mL of starch solution, and continue the titration until the colour changes from blue to colourless. Repeat with two other similar portions of potassium iodate.

Notes. (1) The absence of iodate is indicated by adding dilute sulphuric acid when no immediate yellow coloration should be obtained. If starch is added, no immediate blue coloration should be produced.

(2) Only a small amount of potassium iodate is needed so that the error in weighing 0.14–0.15 g may be appreciable. In this case it is better to weigh out accurately 4.28 g of the salt (if a slightly different weight is used, the exact molarity is calculated), dissolve it in water, and make up to 1 L in a graduated flask. Twenty-five millilitres of this solution are treated with excess of pure potassium iodide (1 g of the solid or 10 mL of 10 per cent solution), followed by 3 mL of 1M sulphuric acid, and the liberated iodine is titrated as detailed above.

(B) With potassium dichromate. Potassium dichromate is reduced by an acid solution of potassium iodide, and iodine is set free:

$$Cr_2O_7^{2-} + 6I^- + 14H^+ = 2Cr^{3+} + 3I_2 + 7H_2O$$

This reaction is subject to a number of errors: (1) the hydriodic acid (from excess of iodide and acid) is readily oxidised by air, especially in the presence of chromium(III) salts, and (2) it is not instantaneous. It is accordingly best to pass a current of carbon dioxide through the reaction flask before and during the titration (a more convenient but less efficient method is to add some solid sodium hydrogencarbonate to the acid solution, and to keep the flask covered as much as possible), and to allow 5 minutes for its completion.

Place 100 mL of cold, recently boiled distilled water in a 500 mL conical, preferably glass-stoppered, flask, add 3 g of iodate-free potassium iodide and 2 g of pure sodium hydrogencarbonate, and shake until the salts dissolve. Add 6 mL of concentrated hydrochloric acid slowly while gently rotating the flask in order to mix the liquids; run in 25.0 mL of standard 0.017M potassium dichromate (see Note), mix the solutions well, and wash the sides of the flask with a little boiled-out water from the wash bottle. Stopper the flask (or cover it with a small watchglass), and allow to stand in the dark for 5 minutes in order to complete the reaction. Rinse the stopper or watchglass, and dilute the solution with 300 mL of cold, boiled-out water. Titrate the liberated iodine with the sodium thiosulphate solution contained in a burette, while constantly rotating the liquid so as to mix the solutions thoroughly. When most of the iodine has reacted as indicated by the solution acquiring a yellowish-green colour, add 2 mL of starch solution and rinse down the sides of the flask; the colour should change to blue. Continue the addition of the thiosulphate solution dropwise, and swirling the liquid constantly, until one drop changes the colour from greenish-blue to light green. The end-point is sharp, and is readily observed in a good light against a white background. Carry out a blank determination, substituting distilled water for the potassium dichromate solution; if the potassium iodide is iodate-free, this should be negligible.

Note. If preferred, about 0.20 g of potassium dichromate may be accurately weighed out, dissolved in 50 mL of cold, boiled-out water, and the titration carried out as detailed above.

Alternative procedure. The following method utilises a trace of copper sulphate as a catalyst to increase the speed of the reaction; in consequence, a weaker acid (acetic acid) may be employed and the extent of atmospheric oxidation of hydriodic acid reduced. Place 25.0 mL of 0.017M potassium dichromate in a 250 mL conical flask, add 5.0 mL of glacial acetic acid, 5 mL of 0.001M copper sulphate, and wash the sides of the flask with distilled water. Add 30 mL of 10 per cent potassium iodide solution, and titrate the iodine as liberated with the approximately 0.1M thiosulphate solution, introducing a little starch indicator towards the end. The titration may be completed in 3–4 minutes after the addition of the potassium iodide solution. Subtract 0.05 mL to allow for the iodine liberated by the copper sulphate catalyst.

(C) With a standard solution of iodine. If a standard solution of iodine is available (see Section 10.112), this may be used to standardise the thiosulphate solution. Measure a 25.0 mL portion of the standard iodine solution into a 250 mL conical flask, add about 150 mL distilled water and titrate with the thiosulphate solution, adding 2 mL of starch solution when the liquid is pale yellow in colour.

When thiosulphate solution is added to a solution containing iodine the overall reaction, which occurs rapidly and stoichiometrically under the usual experimental conditions (pH < 5), is:

$$2S_2O_3^{2-} + I_2 = S_4O_6^{2-} + 2I^-$$

or

$$2S_2O_3^{2-} + I_3^- = S_4O_6^{2-} + 3I^-$$

It has been shown that the colourless intermediate $S_2O_3I^-$ is formed by a rapid reversible reaction:

$$S_2O_3^{2-} + I_2 \rightleftharpoons S_2O_3I^- + I^-$$

The intermediate reacts with thiosulphate ion to provide the main course of the overall reaction:

$$S_2O_3I^- + S_2O_3^{2-} = S_4O_6^{2-} + I^-$$

The intermediate also reacts with iodide ion:

$$2S_2O_3I^- + I^- = S_4O_6^{2-} + I_3^-;$$

This explains the reappearance of iodine after the end point in the titration of very dilute iodine solutions by thiosulphate.

10.116 DETERMINATION OF COPPER IN CRYSTALLISED COPPER SULPHATE

Procedure. Weigh out accurately about 3.0 g of the salt, dissolve it in water, and make up to 250 mL in a graduated flask. Shake well. Pipette 50.0 mL of this solution into a 250 mL conical flask, add 1 g potassium iodide (or 10 mL of a 10 per cent solution) (Note 1), and titrate the liberated iodine with standard

0.1M sodium thiosulphate (Note 2). Repeat the titration with two other 50 mL portions of the copper sulphate solution.

The reaction, written in molecular form, is:

$$2CuSO_4 + 4KI = 2CuI + I_2 + 2K_2SO_4$$

from which it follows that:

$$2CuSO_4 \equiv I_2 \equiv 2Na_2S_2O_3$$

Notes. (1) If in a similar determination, free mineral acid is present, a few drops of dilute sodium carbonate solution must be added until a *faint* permanent precipitate remains, and this is removed by means of a drop or two of acetic acid. The potassium iodide is then added and the titration continued. For accurate results, the solution should have a pH of 4–5.5.

(2) After the addition of the potassium iodide solution, run in standard 0.1M sodium thiosulphate until the brown colour of the iodine fades, then add 2 mL of starch solution, and continue the addition of the thiosulphate solution until the blue colour commences to fade. Then add about 1 g of potassium thiocyanate or ammonium thiocyanate, preferably as a 10 per cent aqueous solution: the blue colour will instantly become more intense. Complete the titration as quickly as possible. The precipitate possesses a pale pink colour, and a distinct permanent end point is readily obtained.

10.117 DETERMINATION OF CHLORATES

Discussion. One procedure is based upon the reaction between chlorate and iodide in the presence of concentrated hydrochloric acid:

$$ClO_3^- + 6I^- + 6H^+ = Cl^- + 3I_2 + 3H_2O$$

The liberated iodine is titrated with standard sodium thiosulphate solution.

In another method the chlorate is reduced with bromide in the presence of *ca* 8M hydrochloric acid, and and the bromine liberated is determined iodimetrically:

$$ClO_3^- + 6Br^- + 6H^+ = Cl^- + 3Br_2 + 3H_2O$$

Procedure. (*a*) Place 25 mL of the chlorate solution (approx. 0.02M) in a glass-stoppered conical flask and add 3 mL of concentrated hydrochloric acid followed by two portions of about 0.3 g each of pure sodium hydrogencarbonate to remove air. Add immediately about 1.0 g of iodate-free potassium iodide and 22 mL of concentrated hydrochloric acid. Stopper the flask, shake the contents, and allow it to stand for 5–10 minutes. Titrate the solution with standard 0.1M sodium thiosulphate in the usual manner.

(*b*) Place 10.0 mL of the chlorate solution in a glass-stoppered flask, add *ca* 1.0 g potassium bromide and 20 mL concentrated hydrochloric acid (the final concentration of acid should be about 8M). Stopper the flask, shake well, and allow to stand for 5–10 minutes. Add 100 mL of 1 per cent potassium iodide solution, and titrate the liberated iodine with standard 0.1M sodium thiosulphate.

10.118 ANALYSIS OF HYDROGEN PEROXIDE

Discussion. Hydrogen peroxide reacts with iodide in acid solution in accordance with the equation:

394

$$H_2O_2 + 2H^+ + 2I^- = I_2 + 2H_2O$$

The reaction velocity is comparatively slow, but increases with increasing concentration of acid. The addition of three drops of a neutral 20 per cent ammonium molybdate solution renders the reaction almost instantaneous, but as it also accelerates the atmospheric oxidation of the hydriodic acid, the titration is best conducted in an inert atmosphere (nitrogen or carbon dioxide).

The iodometric method has the advantage over the permanganate method (Section 10.95) that it is less affected by stabilisers which are sometimes added to commercial hydrogen peroxide solutions. These preservatives are often boric acid, salicylic acid, and glycerol, and render the results obtained by the permanganate procedure less accurate.

Procedure. Dilute the hydrogen peroxide solution to *ca* 0.3 per cent H_2O_2. Thus, if a '20-volume' hydrogen peroxide is used, transfer 10.0 mL by means of a burette or pipette to a 250 mL graduated flask, and make up to the mark. Shake well. Remove 25.0 mL of this diluted solution, and add it gradually and with constant stirring to a solution of 1 g of pure potassium iodide in 100 mL of 1M sulphuric acid (1:20) contained in a stoppered bottle. Allow the mixture to stand for 15 minutes, and titrate the liberated iodine with standard 0.1M sodium thiosulphate, adding 2 mL starch solution when the colour of the iodine has been nearly discharged. Run a blank determination at the same time.

Better results are obtained by transferring 25.0 mL of the diluted hydrogen peroxide solution to a conical flask, and adding 100 mL 1M(1:20) sulphuric acid. Pass a slow stream of carbon dioxide or nitrogen through the flask, add 10 mL of 10 per cent potassium iodide solution, followed by three drops of 3 per cent ammonium molybdate solution. Titrate the liberated iodine immediately with standard 0.1M sodium thiosulphate in the usual way.

Note. The above method may also be used for all per-salts.

10.119 DETERMINATION OF DISSOLVED OXYGEN

Discussion. One of the most useful titrations involving iodine is that originally developed by Winkler[18] to determine the amount of oxygen in samples of water. The dissolved oxygen content is not only important with respect to the species of aquatic life which can survive in the water, but is also a measure of its ability to oxidise organic impurities in the water (see also Section 10.103). Despite the advent of the oxygen-selective electrode (Section 16.36) direct titrations on water samples are still used extensively.[19]

In order to avoid loss of oxygen from the water sample it is 'fixed' by its reaction with manganese(II) hydroxide which is converted rapidly and quantitatively to manganese(III) hydroxide:

$$4Mn(OH)_2 + O_2 + 2H_2O \rightarrow 4Mn(OH)_3$$

The brown precipitate obtained dissolves on acidification and oxidises iodide ions to iodine:

$$Mn(OH)_3 + I^- + 3H^+ \rightarrow Mn^{2+} + \tfrac{1}{2}I_2 + 3H_2O$$

The free iodine may then be determined by titration with sodium thiosulphate (Section 10.113).

$$2S_2O_3^{2-} + I_2 \rightarrow S_4O_6^{2-} + 2I^-$$

This means that 4 moles of thiosulphate correspond to 1 mole of dissolved oxygen.

The main interference in this process is due to the presence of nitrites (especially in waters from sewage treatment). This is overcome by treating the original water sample with sodium azide, which destroys any nitrite when the sample is acidified:

$$HNO_2 + HN_3 \rightarrow N_2 + N_2O + H_2O$$

Procedure. The water sample should be collected by carefully filling a 200–250 mL bottle to the very top and stoppering it while it is below the water surface. This should eliminate any further dissolution of atmospheric oxygen. By using a dropping pipette placed below the surface of the water sample, add 1 mL of a 50 per cent manganese(II) solution (Note 1) and in a similar way add 1 mL of alkaline iodide–azide solution (Note 2). Re-stopper the water sample and shake the mixture well. The manganese(III) hydroxide forms as a brown precipitate. Allow the precipitate to settle completely for 15 minutes and add 2 mL of concentrated phosphoric(V) acid (85 per cent). Replace the stopper and turn the bottle upside-down two or three times in order to mix the contents. The brown precipitate will dissolve and release iodine in the solution (Note 3).

Measure out a 100 mL portion of the solution with a pipette and titrate the iodine with approximately $M/80$ standard sodium thiosulphate solution adding 2 mL of starch solution as indicator as the titration proceeds and *after* the titration liquid has become pale yellow in colour.

Calculate the dissolved oxygen content and express it as mg L^{-1}; 1 mL of $M/80$ thiosulphate $\equiv$ 1 mg dissolved oxygen.

Notes. (1) Prepared by dissolving 50 g of manganese(II) sulphate pentahydrate in water and making up to 100 mL.

(2) Prepared from 40 g of sodium hydroxide, 20 g of potassium iodide and 0.5 g of sodium azide made up to 100 mL with water.

(3) If the brown precipitate has not completely dissolved then add a little more (a few drops) phosphoric(V) acid.

10.120 DETERMINATION OF THE AVAILABLE CHLORINE IN HYPOCHLORITES

Discussion. Most hypochlorites are normally obtained only in solution, but calcium hypochlorite exists in the solid form in commercial bleaching powder which consists essentially of a mixture of calcium hypochlorite $Ca(OCl)_2$ and the basic chloride $CaCl_2,Ca(OH)_2,H_2O$; some free slaked lime is usually present. The active constituent is the hypochlorite, which is responsible for the bleaching action. Upon treating bleaching powder with hydrochloric acid, chlorine is liberated:

$$OCl^- + Cl^- + 2H^+ = Cl_2 + H_2O$$

The **available chlorine** refers to the chlorine liberated by the action of dilute acids on the hypochlorite, and is expressed as the percentage by weight in the case of bleaching powder. Commercial bleaching powder contains 36–38 per cent of available chlorine.

Two methods are in common use for the determination of the available chlorine. In the first, the hypochlorite solution or suspension is treated with an excess of a solution of potassium iodide, and strongly acidified with acetic acid:

$$OCl^- + 2I^- + 2H^+ \rightleftharpoons Cl^- + I_2 + H_2O$$

The liberated iodine is titrated with standard sodium thiosulphate solution. The solution should not be strongly acidified with hydrochloric acid, for the little calcium chlorate which is usually present, by virtue of the decomposition of the hypochlorite, will react slowly with the potassium iodide and liberate iodine:

$$ClO_3^- + 6I^- + 6H^+ = Cl^- + 3I_2 + 3H_2O$$

In the second method, the hypochlorite solution or suspension is titrated against standard sodium arsenite solution; this is best done by adding an excess of the arsenite solution and then back-titrating with standard iodine solution.

Procedure (iodometric method). Weigh out accurately about 5.0 g of the bleaching powder into a clean glass mortar. Add a little water, and rub the mixture to a smooth paste. Add a little more water, triturate with the pestle, allow the mixture to settle, and pour off the milky liquid into a 500 mL graduated flask. Grind the residue with a little more water, and repeat the operation until the whole of the sample has been transferred to the flask either in solution or in a state of very fine suspension, and the mortar washed quite clean. The flask is then filled to the mark with distilled water, well shaken, and 50.0 mL of the turbid liquid immediately withdrawn with a pipette. This is transferred to a 250 mL conical flask, 25 mL of water added, followed by 2 g of iodate-free potassium iodide (or 20 mL of a 10 per cent solution) and 10 mL of glacial acetic acid. Titrate the liberated iodine with standard $0.1M$ sodium thiosulphate.

10.121 DETERMINATION OF ARSENIC(V)

The reaction is the reverse of that employed in the standardisation of iodine with sodium arsenite solution (Section 10.113):

$$As_2O_5 + 4H^+ + 4I^- \rightleftharpoons As_2O_3 + 2I_2 + 2H_2O$$

or

$$H_3AsO_4 + 2H^+ + 2I^- \rightleftharpoons H_3AsO_3 + I_2 + H_2O$$

For good results, the following experimental conditions must be observed: (1) the hydrochloric acid concentration in the final solution should be at least $4M$; (2) air should be displaced from the titration mixture by adding a little solid sodium hydrogencarbonate; (3) the solution must be allowed to stand for at least 5 minutes before the liberated iodine is titrated; and (4) constant stirring is essential during the titration to prevent decomposition of the thiosulphate in the strongly acid solution.

Treat the arsenate solution (say, 20.0 mL of *ca* 0.025*M*) in a glass-stoppered conical flask with concentrated hydrochloric acid to give a solution in $4M$ hydrochloric acid. Displace the air by introducing two 0.4 g portions of pure sodium hydrogencarbonate into the flask. Add 1.0 g of pure potassium iodide, replace the stopper, mix the solution, and allow to stand for at least 5 minutes. Titrate the solution, whilst stirring vigorously, with standard 0.1*M* sodium thiosulphate.

A similar procedure may also be used for the determination of **antimony(V)**, whilst **antimony(III)** may be determined like arsenic(III) by direct titration with standard iodine solution (Section 10.113), but in the antimony titration it is necessary to include some tartaric acid in the solution; this acts as complexing agent and prevents precipitation of antimony as hydroxide or as basic salt in alkaline solution. On the whole, however, the most satisfactory method for determining antimony is by titration with potassium bromate (Section 10.133).

10.122 DETERMINATION OF SULPHUROUS ACID AND OF SULPHITES

Discussion. The iodimetric determination is based upon the equations:

$$SO_3^{2-} + I_2 + H_2O = SO_4^{2-} + 2H^+ + 2I^-$$

$$HSO_3^- + I_2 + H_2O = SO_4^{2-} + 3H^+ + 2I^-$$

For accurate results, the following experimental conditions must be observed:

(a) the solutions should be very dilute;
(b) the sulphite must be added slowly and with constant stirring to the iodine solution, and not conversely; and
(c) exposure of the sulphite to the air should be minimised.

In determinations of sulphurous acid and sulphites, excess of standard $0.05M$ iodine is diluted with several volumes of water, acidified with hydrochloric or sulphuric acid, and a known volume of the sulphite or sulphurous acid solution is added slowly and with constant stirring from a burette, with the jet close to the surface of the liquid. The excess of iodine is then titrated with standard $0.1M$ sodium thiosulphate. Solid soluble sulphites are finely powdered and added directly to the iodine solution. Insoluble sulphites (e.g. calcium sulphite) react very slowly, and must be in a very fine state of division.

Procedure. Pipette 25.0 mL standard $(0.05M)$ iodine solution into a 500 mL conical flask and add 5 mL $2M$ hydrochloric acid and 150 mL distilled water. Weigh accurately sufficient solid sulphite to react with about 20 mL $0.05M$ iodine solution and add this to the contents of the flask; swirl the liquid until all the solid has dissolved and then titrate the excess iodine with standard $(0.1M)$ sodium thiosulphate using starch indicator. If the sulphite is in solution, then a volume of this equivalent to about 20 mL of $0.05M$ iodine should be pipetted into the contents of the flask in place of the weighed amount of solid.

10.123 DETERMINATION OF HYDROGEN SULPHIDE AND OF SULPHIDES

Discussion. The iodimetric method utilises the reversible reaction

$$H_2S + I_2 \rightleftharpoons 2H^+ + 2I^- + S$$

For reasonably satisfactory results, the sulphide solution must be dilute (concentration not greater than 0.04 per cent or $0.01M$), and the sulphide solution added to excess of acidified $0.005M$ or $0.05M$ iodine and not conversely. Loss of hydrogen sulphide is thus avoided, and side reactions are almost entirely eliminated. (With solutions more concentrated than about $0.01M$, the precipitated sulphur encloses a portion of the iodine, and this escapes the subsequent titration with the standard sodium thiosulphate solution.) The excess of iodine is then titrated with standard thiosulphate solution, using starch as indicator.

Excellent results are obtained by the following method, which is of wider applicability. When excess of standard sodium arsenite solution is treated with hydrogen sulphide solution and then acidified with hydrochloric acid, arsenic(III) sulphide is precipitated:

$$As_2O_3 + 3H_2S = As_2S_3 + 3H_2O$$

The excess of arsenic(III) oxide is determined with $0.05M$ iodine and starch.

The procedure is illustrated by determination of the concentration of hydrogen sulphide water.

Procedure. Prepare a saturated solution of hydrogen sulphide by bubbling the gas through distilled water (**CARE! FUME CUPBOARD**). Place 50.0 mL of standard $0.05M$ sodium arsenite in a 250 mL graduated flask, add 20 mL of the hydrogen sulphide water, mix well, and add sufficient hydrochloric acid to render the solution distinctly acid. A yellow precipitate of arsenic(III) sulphide is formed, but the liquid itself is colourless. Make up to the mark with distilled water, and shake thoroughly. Filter the mixture through a dry filter paper into a dry vessel. Remove 100 mL of the filtrate, neutralise it with sodium hydrogencarbonate, and titrate with standard $0.05M$ iodine to the first blue colour with starch. The quantity of residual arsenic(III) oxide is thus determined, and is deducted from the original 50 mL employed.

Note. If certain sulphides are treated with hydrochloric acid, hydrogen sulphide is evolved and can be absorbed in an ammoniacal cadmium chloride solution: upon acidification hydrogen sulphide is released.

10.124 DETERMINATION OF HEXACYANOFERRATES(III)

Discussion. The reaction between hexacyanoferrates(III) (ferricyanides) and soluble iodides is a reversible one:

$$2[Fe(CN)_6]^{3-} + 2I^- \rightleftharpoons 2[Fe(CN)_6]^{4-} + I_2$$

In strongly acid solution the reaction proceeds from left to right, but is reversed in almost neutral solution. Oxidation also proceeds quantitatively in a slightly acid medium in the presence of a zinc salt. The very sparingly soluble potassium zinc hexacyanoferrate(II) is formed, and the hexacyanoferrate(II) ions are removed from the sphere of action:

$$2[Fe(CN)_6]^{4-} + 2K^+ + 3Zn^{2+} = K_2Zn_3[Fe(CN)_6]_2$$

The procedure may be used to determine the purity of potassium hexacyanoferrate(III).

Procedure. Weigh out accurately about 10 g of the salt and dissolve it in 250 mL of water in a graduated flask. Pipette 25 mL of this solution into a 250 mL conical flask, add about 20 mL of 10 per cent potassium iodide solution, 2 mL of $1M$ sulphuric acid, and 15 mL of a solution containing 2.0 g crystallised zinc sulphate. Titrate the liberated iodine immediately with standard $0.1M$ sodium thiosulphate and starch; add the starch solution (2 mL) after the colour has faded to a pale yellow. The titration is complete when the blue colour has just disappeared. When great accuracy is required, the process should be conducted in an atmosphere of carbon dioxide.

OXIDATIONS WITH POTASSIUM IODATE

10.125 GENERAL DISCUSSION

Potassium iodate is a powerful oxidising agent, but the course of the reaction is governed by the conditions under which it is employed. The reaction between potassium iodate and reducing agents such as iodide ion or arsenic(III) oxide in solutions of moderate acidity ($0.1-2.0M$ hydrochloric acid) stops at the stage when the iodate is reduced to iodine:

$$IO_3^- + 5I^- + 6H^+ = 3I_2 + 3H_2O$$

$$2IO_3^- + 5H_3AsO_3 + 2H^+ = I_2 + 5H_3AsO_4 + H_2O$$

As already indicated, the first of these reactions is very useful for the generation of known amounts of iodine, and it also serves as the basis of a method for standardising solutions of acids (Section 10.110).

With a more powerful reductant, e.g. titanium(III) chloride, the iodate is reduced to iodide:

$$IO_3^- + 6Ti^{3+} + 6H^+ = I^- + 6Ti^{4+} + 3H_2O$$

In more strongly acid solutions ($3-6M$ hydrochloric acid) reduction occurs to iodine monochloride, and it is under these conditions that it is most widely used.[20,21]

$$IO_3^- + 6H^+ + Cl^- + 4e \rightleftharpoons ICl + 3H_2O$$

In hydrochloric acid solution, iodine monochloride forms a stable complex ion with chloride ion:

$$ICl + Cl^- \rightleftharpoons ICl_2^-$$

The overall half-cell reaction may therefore be written as:

$$IO_3^- + 6H^+ + 2Cl^- + 4e \rightleftharpoons ICl_2^- + 3H_2O$$

The reduction potential is 1.23 volts; hence under these conditions potassium iodate acts as a very powerful oxidising agent.

Oxidation by iodate ion in a strong hydrochloric acid medium proceeds through several stages:

$$IO_3^- + 6H^+ + 6e \rightleftharpoons I^- + 3H_2O$$

$$IO_3^- + 5I^- + 6H^+ = 3I_2 + 3H_2O$$

$$IO_3^- + 2I_2 + 6H^+ = 5I^+ + 3H_2O$$

In the initial stages of the reaction free iodine is liberated[22]: as more titrant is added, oxidation proceeds to iodine monochloride, and the dark colour of the solution gradually disappears. The overall reaction may be written as:

$$IO_3^- + 6H^+ + 4e \rightleftharpoons I^+ + 3H_2O$$

The reaction has been used for the determination of many reducing agents: the optimum acidity for reasonably rapid reaction varies from one reductant to another within the range 2.5–9M hydrochloric acid; in many cases the concentration of acid is not critical, but for Sb(III) it is 2.5–3.5M.

Under these conditions starch cannot be used as indicator because the characteristic blue colour of the starch–iodine complex is not formed at high concentrations of acid. In the original procedure, a few millilitres of an immiscible solvent (carbon tetrachloride or chloroform) were added to the solution being titrated contained in a glass-stoppered bottle or conical flask. The end point is marked by the disappearance of the last trace of violet colour, due to iodine, from the solvent: iodine monochloride is not extracted and imparts a pale-yellowish colour to the aqueous phase. The extraction end point is very sharp. The main disadvantage is the inconvenience of vigorous shaking with the extraction solvent in a stoppered vessel after each addition of the reagent near the end point.

The immiscible solvent may be replaced by certain dyes, e.g. amaranth (C.I. 16185), colour change red to colourless; xylidine ponceau (C.I. 16255), colour change orange to colourless; naphthalene black 12B (C.I. 20470), colour change green to faint pink; the first two of these are generally preferred. The indicators are used as 0.2 per cent aqueous solutions and about 0.5 mL per titration is added near the end point. The dyes are destroyed by the first excess of iodate, and hence the indicator action is irreversible. The indicator blank is equivalent to 0.05 mL of 0.025M potassium iodate per 1.0 mL of indicator solution, and is therefore virtually negligible.

p-Ethoxychrysoidine is a moderately satisfactory reversible indicator. It is used as a 0.1 per cent solution in ethanol (about 12 drops per titration), and the colour change is from red to orange; the colour is red–purple just before the end point. The indicator is added after the colour of the iodine commences to fade. A blank determination should be made for each new batch of indicator.

10.126 PREPARATION OF 0.025M POTASSIUM IODATE

Dry some potassium iodate at 120 °C for 1 hour and allow it to cool in a covered vessel in a desiccator. Weigh out exactly 5.350 g of the finely powdered potassium iodate on a watchglass, and transfer it by means of a clean camel-hair brush directly into a dry 1 L graduated flask. Add about 400–500 mL of water, and gently rotate the flask until the salt is completely dissolved. Make up to the mark with distilled water. Shake well. The solution will keep indefinitely.

It must be emphasised that this 0.025M solution is intended for the reaction:

$$IO_3^- + 6H^+ + Cl^- + 4e \rightleftharpoons ICl + 3H_2O$$

but when used in solutions of moderate acidity leading to the liberation of free iodine (Section 10.115), ideally the solution then requires 4.28 g L^{-1} potassium iodate; the method of preparation will be as described above with suitable adjustment of the weight of salt taken.

10.127 DETERMINATION OF ARSENIC OR OF ANTIMONY

Discussion. The determination of arsenic in arsenic(III) compounds is based upon the following reaction:

$$IO_3^- + 2H_3AsO_3 + 2H^+ + Cl^- = ICl + 2H_3AsO_4 + H_2O$$

A similar reaction occurs with antimony(III) compounds. The determination of antimony(III) in the presence of tartrate is not very satisfactory with an immiscible solvent to assist in indicating the end point; amaranth, however, gives excellent results.

$$IO_3^- + 2[SbCl_4]^- + 6H^+ + 5Cl^- = ICl + 2[SbCl_6]^- + 3H_2O$$

To assay a sample of arsenic(III) oxide the following procedure may be used.

Procedure. Weigh out accurately about 1.1 g of the oxide sample, dissolve in a small quantity of warm 10 per cent sodium hydroxide solution, and make up to 250 mL in a graduated flask. Use a burette to measure 25.0 mL of this solution into a stoppered reagent bottle of about 250 mL capacity, add 25 mL water, 60 mL concentrated hydrochloric acid and about 5 mL carbon tetrachloride or chloroform. Cool to room temperature. Run in the standard 0.025M potassium iodate from a burette until the solution, which at first is strongly coloured with iodine, becomes pale brown. The bottle is then stoppered and vigorously shaken, and the organic solvent layer acquires the purple colour due to iodine. Continue to add small volumes of the iodate solution, shaking vigorously after each addition, until the organic layer is only very faintly violet. Continue the addition dropwise, with shaking after each drop, until the solvent loses the last trace of violet and has only a very pale yellow colour (due to iodine chloride). The end point is very sharp and, after a little experience, is rarely overshot. If this should occur, a small volume of the oxide solution is added from a graduated pipette, and the end point re-determined. Allow to stand for ten minutes and observe whether the organic layer shows any purple colour; the absence of colour confirms that the titration is complete.

The acidity of the mixture at the end of the titration should be not less than 3M and not more than 5M; if the acidity is too high the reaction takes place slowly.

Note. A 250 mL graduated flask with a short neck and a well-fitting ground-glass stopper may also be used. The colour of the organic layer is readily seen by inverting the flask so that the layer of solvent indicator collects in the neck.

Alternatively, in this and all subsequent titrations with 0.025M potassium iodate, a 250 mL conical flask may be used and the carbon tetrachloride or chloroform indicator replaced by 0.5 mL amaranth or xylidine ponceau indicator, which is added after most of the iodine colour has disappeared from the reaction mixture (see Section 10.125).

10.128 DETERMINATION OF HYDRAZINE

Discussion. Hydrazine reacts with potassium iodate under the usual Andrews[20] conditions, thus:

$$IO_3^- + N_2H_4 + 2H^+ + Cl^- = ICl + N_2 + 3H_2O$$

Thus

$$KIO_3 \equiv N_2H_4$$

To determine the N_2H_4, H_2SO_4 content of hydrazinium sulphate, use the following method.

Procedure. Weigh out accurately 0.08–0.1 g of hydrazinium sulphate into a 250 mL reagent bottle, add a mixture of 30 mL of concentrated hydrochloric acid, 20 mL of water, and 5 mL of chloroform or carbon tetrachloride. Run in the standard 0.025M potassium iodate slowly from a burette, with shaking the stoppered bottle between the additions, until the organic layer is just decolorised.

10.129 DETERMINATION OF MERCURY

Discussion. The mercury is precipitated as mercury(I) chloride and the latter is reacted with standard potassium iodate solution:

$$IO_3^- + 2Hg_2Cl_2 + 6H^+ + 13Cl^- = ICl + 4[HgCl_4]^{2-} + 3H_2O$$

Thus

$$KIO_3 \equiv 4Hg \equiv 2Hg_2Cl_2$$

To determine the purity of a sample of a mercury(II) salt, the following procedure in which the compound is reduced with phosphorous (phosphonic) acid may be used; to assay a sample of a mercury(I) salt, the reduction with phosphorous acid is omitted.

Procedure. Weigh out accurately about 2.5 g of finely powdered mercury(II) chloride, and dissolve it in 100 mL of water in a graduated flask. Shake well. Transfer 25.0 mL of the solution to a conical flask, add 25 mL water, 2 mL 1M hydrochloric acid, and excess of 50 per cent phosphorous(III) acid solution. Stir thoroughly and allow to stand for 12 hours or more. Filter the precipitated mercury(I) chloride through a quantitative filter paper and wash the precipitate moderately with cold water. Transfer the precipitate with the filter paper quantitatively to a 250 mL reagent bottle, add 30 mL concentrated hydrochloric acid, 20 mL water, and 5 mL carbon tetrachloride or chloroform. Titrate the mixture with standard 0.025M potassium iodate in the usual manner (Section 11.127).

$$2HgCl_2 + H_3PO_3 + H_2O = Hg_2Cl_2 + 2HCl + H_3PO_4$$

10.130 DETERMINATIONS OF OTHER IONS

Copper(II) compounds. Many other metallic ions which are capable of undergoing oxidation by potassium iodate can also be determined. Thus, for example, copper(II) compounds can be analysed by precipitation of copper(I) thiocyanate which is titrated with potassium iodate:

$$7IO_3^- + 4CuSCN + 18H^+ + 7Cl^- = 7ICl + 4Cu^{2+} + 4HSO_4^- + 4HCN + 5H_2O$$

As a typical example, 0.8 g of copper(II) sulphate ($CuSO_4,5H_2O$) is dissolved in water, 5 mL of 0.5M sulphuric acid added, and the solution made up to 250 mL in a graduated flask; 25.0 mL of the resulting solution are pipetted into a 250 mL conical flask, 10–15 mL of freshly prepared sulphurous acid solution added, and then after heating to boiling, 10 per cent ammonium thiocyanate solution is added slowly from a burette with constant stirring until there is no further change in colour, and then 4 mL of reagent is added in excess. After allowing the precipitate to settle for 10–15 minutes, it is filtered through a fine filter paper and then washed with cold 1 per cent ammonium sulphate solution until free from thiocyanate. It is then transferred quantitatively into the vessel in which the titration is to be performed, and after adding 30 mL of concentrated hydrochloric acid, followed by 20 mL of water, the titration is carried out in the usual manner either with an organic solvent present, or with an internal indicator being added as the end point is approached.

Thallium(I) salts. These are oxidised in accordance with the equation:

$$IO_3^- + 2Tl^+ + 6H^+ + Cl^- = ICl + 2Tl^{3+} + 3H_2O$$

so that

$$KIO_3 \equiv 2Tl$$

The solution should contain 0.25–0.30 g Tl^+ in 20 mL plus 60 mL of concentrated hydrochloric acid and is titrated as usual with $0.025M$ KIO_3 solution.

Tin(II) salts. Tin(II) salts are likewise oxidised in accordance with the equation

$$IO_3^- + 2Sn^{2+} + 6H^+ + Cl^- = ICl + 2Sn^{4+} + 3H_2O$$

so that

$$KIO_3 \equiv 2Sn$$

If the bulk of the iodate solution is added rapidly, atmospheric oxidation does not present a serious problem, but the method cannot be used in the presence of salts of antimony(III), copper(I), or iron(II). The solution, which should contain for example 0.15 g $SnCl_2,2H_2O$ in 25 mL, is treated with 30 mL of concentrated hydrochloric acid and 20 mL of water and is then titrated in the usual manner with standard potassium iodate solution.

Vanadates. Vanadates are reduced by iodides in strongly acid (hydrochloric) solution in an atmosphere of carbon dioxide to the vanadium(IV) condition:

$$2VO_4^{3-} + 2I^- + 12H^+ = 2VO^{2+} + I_2 + 6H_2O$$

The liberated iodine and the excess of iodide is determined by titration with standard potassium iodate solution; the hydrochloric acid concentration must not be allowed to fall below $7M$ in order to prevent re-oxidation of the vanadium compound by iodine chloride.

Place 25.0 mL of the solution containing 0.05–0.10 g of vanadium (as vanadate) in a 250 mL glass-stoppered reagent bottle, and pass a rapid current of carbon dioxide for 2–3 minutes into the bottle, but not through the solution. Then add sufficient concentrated hydrochloric acid through a funnel to make the solution 6–$8M$ during the titration. Introduce a known volume (excess) of approximately $0.05M$ potassium iodide, which has been titrated against the standard iodate solution. Mix the contents of the bottle, allow to stand for 1–2 minutes, add 5 mL of carbon tetrachloride, and then titrate as rapidly as possible with standard $0.025M$ potassium iodate until no more iodine colour can be detected in the organic layer. Add concentrated hydrochloric acid as needed during the titration so that the concentration does not fall below $7M$.

Soluble sulphides. Hydrogen sulphide and soluble sulphides can also be determined by oxidation with potassium iodate in an alkaline medium. Mix 10.0 mL of the sulphide solution containing about 2.5 mg sulphide with 15.0 mL $0.025M$ potassium iodate (Section 10.126) and 10 mL of $10M$ sodium hydroxide. Boil gently for 10 minutes, cool, add 5 mL of 5 per cent potassium iodide solution and 20 mL of $4M$ sulphuric acid. Titrate the liberated iodine, which is equivalent to the unused iodate, with standard $0.1M$ sodium thiosulphate in the usual manner.

OXIDATIONS WITH POTASSIUM BROMATE

10.131 GENERAL DISCUSSION

Potassium bromate is a powerful oxidising agent which is reduced smoothly to bromide:

$$BrO_3^- + 6H^+ + 6e \rightleftharpoons Br^- + 3H_2O$$

The relative molecular mass is 167.00, and a $0.02M$ solution contains 3.34 g L^{-1} potassium bromate. At the end of the titration free bromine appears:

$$BrO_3^- + 5Br^- + 6H^+ = 3Br_2 + 3H_2O$$

The presence of free bromine, and consequently the end-point, can be detected by its yellow colour, but it is better to use indicators such as methyl orange, methyl red, naphthalene black 12B, xylidine ponceau, and fuchsine. These indicators have their usual colour in acid solution, but are destroyed by the first excess of bromine. With all irreversible oxidation indicators the destruction of the indicator is often premature to a slight extent: a little additional indicator is usually required near the end point. The quantity of bromate solution consumed by the indicator is exceedingly small, and the 'blank' can be neglected for $0.02M$ solutions. Direct titrations with bromate solution in the presence of irreversible dyestuff indicators are usually made in hydrochloric acid solution, the concentration of which should be at least $1.5-2M$. At the end of the titration some chlorine may appear by virtue of the reaction:

$$10Cl^- + 2BrO_3^- + 12H^+ = 5Cl_2 + Br_2 + 6H_2O$$

This immediately bleaches the indicator.

The titrations should be carried out slowly so that the indicator change, which is a time reaction, may be readily detected. If the determinations are to be executed rapidly, the volume of the bromate solution to be used must be known approximately, since ordinarily with irreversible dyestuff indicators there is no simple way of ascertaining when the end point is close at hand. With the highly coloured indicators (xylidine ponceau, fuchsine, or naphthalene black 12B), the colour fades as the end point is approached (owing to local excess of bromate) and another drop of indicator can be added. At the end point the indicator is irreversibly destroyed and the solution becomes colourless or almost so. If the fading of the indicator is confused with the equivalence point, another drop of the indicator may be added. If the indicator has faded, the additional drop will colour the solution; if the end point has been reached, the additional drop of indicator will be destroyed by the slight excess of bromate present in the solution.

The introduction of reversible redox indicators for the determination of arsenic(III) and antimony(III) has considerably simplified the procedure; those at present available include 1-naphthoflavone, and p-ethoxychrysoidine. The addition of a little tartaric acid or potassium sodium tartrate is recommended when antimony(III) is titrated with bromate in the presence of the reversible

405

indicators; this will prevent hydrolysis at the lower acid concentrations. The end point may be determined with high precision by potentiometric titration (see Chapter 15).

Examples of determinations utilising direct titration with bromate solutions are expressed in the following equations:

$$BrO_3^- + 3H_3AsO_3 \xrightarrow{(HCl)} Br^- + 3H_3AsO_4$$

$$2BrO_3^- + 3N_2H_4 \xrightarrow{(HCl)} 2Br^- + 3N_2 + 6H_2O$$

$$BrO_3^- + NH_2OH \xrightarrow{(HCl)} Br^- + NO_3^- + H^+ + H_2O$$

$$BrO_3^- + 6[Fe(CN)_6]^{4-} + 6H^+ \longrightarrow Br^- + 6[Fe(CN)_6]^{3-} + 3H_2O$$

Various substances cannot be oxidised directly with potassium bromate, but react quantitatively with an excess of bromine. Acid solutions of bromine of exactly known concentration are readily obtainable from a standard potassium bromate solution by adding acid and an excess of bromide:

$$BrO_3^- + 5Br^- + 6H^+ = 3Br_2 + 3H_2O$$

In this reaction 1 mole of bromate yields six atoms of bromine. Bromine is very volatile, and hence such operations should be conducted at as low a temperature as possible and in conical flasks fitted with ground-glass stoppers. The excess of bromine may be determined iodometrically by the addition of excess of potassium iodide and titration of the liberated iodine with standard thiosulphate solution:

$$2I^- + Br_2 = I_2 + 2Br^-$$

Potassium bromate is readily available in a high state of purity; the product has an assay value of at least 99.9 per cent. The substance can be dried at $120-150\,°C$, is anhydrous, and the aqueous solution keeps indefinitely. It can therefore be employed as a primary standard. Its only disadvantage is that one-sixth of the relative molecular mass is a comparatively small quantity.

10.132 PREPARATION OF 0.02M POTASSIUM BROMATE

Dry some finely powdered potassium bromate for $1-2$ hours at $120\,°C$, and allow to cool in a closed vessel in a desiccator. Weigh out accurately 3.34 g of the pure potassium bromate, and dissolve in 1 L of water in a graduated flask.

10.133 DETERMINATION OF ANTIMONY OR OF ARSENIC

Discussion. The antimony or the arsenic must be present as antimony(III) or arsenic(III). The reaction of arsenic(III) or antimony(III) with potassium bromate may be written:

$$2KBrO_3 + 3M_2O_3 + 2HCl = 2KCl + 3M_2O_5 + 2HBr \ (M = As \ or \ Sb)$$

The presence of tin and of considerable quantities of iron and copper interfere with the determinations.

To determine the purity of a sample of arsenic(III) oxide follow the general procedure outlined in Section 10.127 but when the 25 mL sample of solution is being prepared for titration, add 25 mL water, 15 mL of concentrated hydrochloric acid and then two drops of indicator solution (xylidine ponceau or naphthalene black 12B; see Section 10.125). Titrate slowly with the standard 0.02M potassium bromate with constant swirling of the solution. As the end point approaches, add the bromate solution dropwise with intervals of 2–3 seconds between the drops until the solution is colourless or very pale yellow. If the colour of the indicator fades, add another drop of indicator solution. (The immediate discharge of the colour indicates that the equivalence point has been passed and the titration is of little value.)

As an alternative, a reversible indicator may be employed, either (a) 1-naphthoflavone (0.5% solution in ethanol, which gives an orange-coloured solution at the end-point), or (b) p-ethoxychrysoidine (0.1% aqueous solution, colour change pink to pale yellow). Under these conditions, the measured 25 mL portion of the arsenic solution is treated with 10 mL of 10 per cent potassium bromide solution, 6 mL of concentrated hydrochloric acid, 10 mL of water and either 0.5 mL of indicator (a) or two drops of indicator (b).

10.134 DETERMINATION OF METALS BY MEANS OF 8-HYDROXYQUINOLINE ('OXINE')

Discussion. Various metals (e.g. aluminium, iron, copper, zinc, cadmium, nickel, cobalt, manganese, and magnesium) under specified conditions of pH yield well-defined crystalline precipitates with 8-hydroxyquinoline. These precipitates have the general formula $M(C_9H_6ON)_n$, where n is the charge on the metal M ion [see, however, Section 11.11(c)]. Upon treatment of the oxinates with dilute hydrochloric acid, the oxine is liberated. One molecule of oxine reacts with two molecules of bromine to give 5,7-dibromo-8-hydroxyquinoline:

$$C_9H_7ON + 2Br_2 = C_9H_5ONBr_2 + 2H^+ + 2Br^-$$

Hence 1 mole of the oxinate of a double-charged metal corresponds to 4 moles of bromine, whilst that of a triple-charged metal corresponds to 6 moles. The bromine is derived by the addition of standard 0.02M potassium bromate and excess of potassium bromide to the acid solution.

$$BrO_3^- + 5Br^- + 6H^+ = 3Br_2 + 3H_2O$$

Full details are given for the determination of aluminium by this method. Many other metals may be determined by this same procedure, but in many cases complexometric titration offers a simpler method of determination. In cases where the oxine method offers advantages, the experimental procedure may be readily adapted from the details given for aluminium.

Determination of aluminium. Prepare a 2 per cent solution of 8-hydroxyquinoline [see Section 11.11(c)] in 2M acetic acid; add ammonia solution until a *slight* precipitate persists, then re-dissolve it by warming the solution.

Transfer 25 mL of the solution to be analysed, containing about 0.02 g of aluminium, to a conical flask, add 125 mL of water and warm to 50–60 °C. Then add a 20 per cent excess of the oxine solution (1 mL will precipitate 0.001 g of Al), when the complex $Al(C_9H_6ON)_3$ will be formed. Complete the precipitation by the addition of a solution of 4.0 g of ammonium acetate in the minimum quantity of water, stir the mixture, and allow to cool. Filter the granular precipitate through a sintered-glass crucible of porosity No. 4 and wash with warm water (Note 1). Dissolve the complex in warm concentrated hydrochloric acid, collect the solution in a 250 mL reagent bottle, add a few drops of indicator (0.1 per cent solution of the sodium salt of methyl red or 0.1 per cent methyl orange solution), and 0.5–1 g of pure potassium bromide. Titrate slowly with standard 0.02M potassium bromate until the colour becomes pure yellow (with either indicator). The exact end point is not easy to detect, and the best procedure is to add an excess of potassium bromate solution, i.e., a further 2 mL beyond the estimated end point, so that the solution now contains free bromine. Dilute the solution considerably with 2M hydrochloric acid (to prevent the precipitation of 5,7-dibromo-8-hydroxyquinoline during the titration), then add (after 5 minutes) 10 mL of 10 per cent potassium iodide solution, and titrate the liberated iodine with standard 0.1M sodium thiosulphate, using starch as indicator (Note 2) to determine the excess bromate (see Section 10.131).

From the above discussion, it is evident that $Al \equiv 12Br$.

Notes. (1) This will remove the excess of oxine. Complications due to adsorption of iodine will thus be avoided.

(2) A brown additive compound of iodine with the dibromo compound may separate during the titration; this compound usually dissolves during the subsequent titration with thiosulphate, yielding a yellow solution so that the end point with starch may be found in the usual manner. Occasionally, the dark-coloured compound, which contains adsorbed iodine, may not dissolve readily and thus introduces an uncertainty in the end-point: this difficulty may be avoided by adding 10 mL of carbon disulphide before introducing the potassium iodide solution.

10.135 DETERMINATION OF HYDROXYLAMINE

The method based upon the reduction of iron(III) solutions in the presence of sulphuric acid, boiling, and subsequent titration in the cold with standard 0.02M potassium permanganate frequently yields high results unless the experimental conditions are closely controlled:

$$2NH_2OH + 4Fe^{3+} = N_2O + 4Fe^{2+} + 4H^+ + H_2O$$

Better results are obtained by oxidation with potassium bromate in the presence of hydrochloric acid:

$$NH_2OH + BrO_3^- = NO_3^- + Br^- + H^+ + H_2O$$

The hydroxylamine solution is treated with a measured volume of 0.02M potassium bromate so as to give 10–30 mL excess, followed by 40 mL of 5M hydrochloric acid. After 15 minutes the excess of bromate is determined by the addition of potassium iodide solution and titration with standard 0.1M sodium thiosulphate (compare Section 10.134).

10.136 DETERMINATION OF PHENOL

Discussion. A number of phenols can be substituted rapidly and quantitatively with bromine produced from bromate and bromide[23] in acid solution (Section 10.131). The determination involves treating phenol (Note 1) with an excess of potassium bromate and potassium bromide; when bromination of the phenol is complete the unreacted bromine is then determined by adding excess potassium iodide and back-titrating the liberated iodine with standard sodium thiosulphate.

$$+ 3Br_2 \longrightarrow + 3HBr$$

Procedure. Prepare an approx. 0.02M standard solution of potassium bromate by weighing accurately about 1.65 g of the analytical grade reagent, dissolving it in water and making it up to 500 mL in a graduated flask (Note 2).

To determine the purity of a sample of phenol, weigh out accurately approximately 0.3 g of phenol, dissolve it in water and make the volume to 250 mL in a graduated flask. Pipette 25 mL volumes of this solution into 250 mL stoppered (ground-glass) conical flasks. To each flask pipette 25 mL of the standard potassium bromate solution and add 0.5 g of potassium bromide and 5 mL of 3M sulphuric acid (Note 3). Mix the reagents and let them stand for 15 minutes, then rapidly add about 2.5 g of potassium iodide to each flask, immediately re-stoppering and swirling the contents to dissolve the solid. Titrate the liberated iodine with the standard 0.1M sodium thiosulphate until the solution is only slightly yellow, then add 5 mL of starch indicator solution and continue the titration until the blue colour disappears.

Calculate the amount of excess bromate from the amount of thiosulphate needed for the back-titration of the free iodine, and hence the quantity of bromate which reacted with the phenol.

Notes. (1) Other phenols which undergo this type of reaction include 4-chlorophenol, m-cresol (3-methylphenol) and 2-naphthol.

(2) The concentration of the potassium bromate can be checked by the following method: pipette 25 mL of the solution into a 250 mL conical flask, add 2.5 g of potassium iodide and 5 mL of 3M sulphuric acid. Titrate the liberated iodine with standard 0.1M sodium thiosulphate (Section 10.114) until the solution is faintly yellow. Add 5 mL of starch indicator solution and continue the titration until the blue colour disappears.

(3) The flasks must be stoppered at all times after the addition of reagents to prevent the loss of bromine.

THE REDUCTION OF HIGHER OXIDATION STATES

10.137 GENERAL DISCUSSION

It has already been indicated that before titration with an oxidising agent can be carried out, it may in some cases be necessary to reduce the compound supplied to a lower state of oxidation. Such a situation is frequently encountered with the determination of iron; iron(III) compounds must be reduced to iron(II) before titration with potassium permanganate or potassium dichromate can be performed. It is possible to carry out such determinations directly as a **reductimetric titration** by the use of solutions of powerful reducing agents such as chromium(II) chloride, titanium(III) chloride or vanadium(II) sulphate, but the problems associated with the preparation, storage and handling of these reagents have militated against their widespread use. Titanium(III) sulphate has found application in the analysis of certain types of organic compounds,[24] but is of limited application in the inorganic field. An apparatus suitable for the preparation, storage and manipulation of chromium(II) and vanadium(II) solutions is described in Ref. 25; with both these reagents it is necessary [and it is also advisable with Ti(III) solutions], to carry out titrations in an atmosphere of hydrogen, nitrogen or carbon dioxide, and in view of the instability of most indicators in the presence of these powerful reducing agents, it is frequently necessary to determine the end-point potentiometrically.

The most important method for reduction of compounds to an oxidation state suitable for titration with one of the common oxidising titrants is based upon the use of metal amalgams, but there are various other methods which can be used, and these will be discussed in the following sections.

10.138   REDUCTION WITH AMALGAMATED ZINC: THE JONES REDUCTOR

Amalgamated zinc is an excellent reducing agent for many metallic ions. Zinc reacts rather slowly with acids, but upon treatment with a dilute solution of a mercury(II) salt, the metal is covered with a thin layer of mercury; the amalgamated metal reacts quite readily. Reduction with amalgamated zinc is usually carried out in the 'reductor', due to C. Jones. This consists of a column of amalgamated zinc contained in a long glass tube provided with a stopcock, through which the solution to be reduced may be drawn. A large surface is exposed, and consequently such a zinc column is much more efficient than pieces of zinc placed in the solution.

A suitable form of the Jones reductor, with approximate dimensions, is shown in Fig. 10.17. A perforated porcelain plate, covered with glass wool, supports

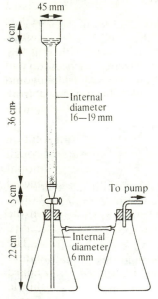

Fig. 10.17

410

the zinc column. The tube below the tap passes through a tightly fitting one-holed rubber stopper into a 750 mL filter flask. It is advisable to connect another filter flask in series with the water-pump, so that if any water 'sucks back' it will not spoil the determination. The amalgamated zinc is prepared as follows. About 300 g of granulated zinc (or zinc shavings, or pure 20–30-mesh zinc) are covered with 2 per cent mercury(II) chloride solution in a beaker. The mixture is stirred for 5–10 minutes, then the solution is decanted from the zinc, which is washed three times with water by decantation. The resulting amalgamated zinc should have a bright silvery lustre. The porcelain plate is placed in position, covered with a layer of glass wool and then the amalgamated zinc added: the latter should reach to the shoulder of the tube. The zinc is washed with distilled water (500 mL), using gentle suction. If the reductor is not to be used immediately, it must be left full of water in order to prevent the formation of basic salts by atmospheric oxidation, which impair the reducing surface. If the moist amalgam is exposed to the moisture of the atmosphere, hydrogen peroxide may be generated:

$$Zn + O_2 + 2H_2O = Zn(OH)_2 + H_2O_2$$

but no hydrogen peroxide is formed if acid is present.

To use the reductor for the reduction of iron(III). Proceed as follows. The zinc is activated by filling the cup (which holds about 50 mL) with $1M$ (*ca* 5 per cent) sulphuric acid, the tap being closed. The flask is connected to a filter pump, the tap opened, and the acid *slowly* drawn through the column until it has fallen to *just above* the level of the zinc; the tap is then closed and the process repeated twice. The tap is shut, and the flask detached, cleaned, and replaced. The reductor is now ready for use. It is important to note that during use the level of the liquid should always be just above the top of the zinc column. The solution to be reduced should have a volume of 100–150 mL, contain not more than 0.25 g of iron, and be about $1M$ in sulphuric acid. The cold iron solution is passed through the reductor, using gentle suction, at a flow rate not exceeding 75–100 mL min^{-1}. As soon as the reservoir is nearly emptied of the solution, 100 mL of 2.5 per cent (or *ca* $0.5M$) sulphuric acid is passed through in two portions, followed by 100–150 mL of water. The last washing is necessary in order to wash out all of the reduced compound and also the acid, which would otherwise cause unnecessary consumption of the zinc. Disconnect the flask from the reductor, wash the end of the delivery tube, and titrate immediately with standard $0.02M$ potassium permanganate.

Carry out a blank determination, preferably before passing the iron solution through the reductor, by running the same volumes of acid and water through the apparatus as are used in the actual determination. This should not amount to more than about 0.1 mL of $0.02M$ permanganate, and should be deducted from the volume of permanganate solution used in the subsequent titration.

It must be emphasised that if hydrochloric acid has been employed in the original solution of the iron-bearing material, the volume should be reduced to *ca* 25 mL and then diluted to *ca* 150 mL with 5 per cent sulphuric acid. The determination is carried out as detailed above, but 25 mL of Zimmermann–Reinhardt or 'preventive solution' must be added before titration with standard potassium permanganate solution. For the determination of iron in hydrochloric acid solution, it is more convenient to reduce the solution in a silver reductor

(Section 10.140) and to titrate the reduced solution with either standard potassium dichromate or standard cerium(IV) sulphate solution.

Applications and limitations of the Jones reductor. Solutions containing 1–10 per cent by volume of sulphuric acid or 3–15 per cent by volume of concentrated hydrochloric acid can be used in the reductor. Sulphuric acid is, however, generally used, as hydrochloric acid may interfere in the subsequent titration, e.g. with potassium permanganate.

Nitric acid must be absent, for this is reduced to hydroxylamine and other compounds which react with permanganate. If nitric acid is present, evaporate the solution just to dryness, wash the sides of the vessel with about 3 mL of water, carefully add 3–4 mL of concentrated sulphuric acid, and evaporate until fumes of the latter are evolved. Repeat this operation twice to ensure complete removal of the nitric acid, dilute to 100 mL with water, add 5 mL of concentrated sulphuric acid, and proceed with the reduction.

Organic matter (acetates, etc.) must be absent. It is removed by heating to fumes of sulphuric acid in a covered beaker, then carefully adding drops of a saturated solution of potassium permanganate until a permanent colour is obtained, and finally continuing the fuming for a few minutes.

Solutions containing compounds of copper, tin, arsenic, antimony, and other reducible metals must never be used. These must be removed before the reduction by treatment with hydrogen sulphide.

Other ions which are reduced in the reductor to a definite lower oxidation state are those of titanium to Ti^{3+}, chromium to Cr^{2+}, molybdenum to Mo^{3+}, niobium to Nb^{3+}, and vanadium to V^{2+}. Uranium is reduced to a mixture of U^{3+} and U^{4+}, but by bubbling a stream of air through the solution in the filter flask for a few minutes, the dirty dark-green colour changes to the bright apple-green colour characteristic of pure uranium(IV) salts. Tungsten is reduced, but not to any definite lower oxidation state.

With the exception of iron(II) and uranium(IV), the reduced solutions are extremely unstable and readily re-oxidise upon exposure to air. They are best stabilised in a five-fold excess of a solution of 150 g of ammonium iron(III) sulphate and 150 mL of concentrated sulphuric acid per litre [approximately 0.3M with respect to iron] contained in the filter flask. The iron(II) formed is then titrated with a standard solution of a suitable oxidising agent. Titanium and chromium are completely oxidised and produce an equivalent amount of iron(II) sulphate; molybdenum is re-oxidised to the Mo(V) (red) stage, which is fairly stable in air, and complete oxidation is effected by the permanganate, but the net result is the same, viz. Mo(III) → Mo(VI); vanadium is re-oxidised to the V(IV), condition, which is stable in air, and the final oxidation is completed by slow titration with potassium permanganate solution or with cerium(IV) sulphate solution.

10.139   REDUCTION WITH LIQUID AMALGAMS

Makazono introduced liquid zinc amalgam as a reducing agent and subsequent Japanese workers have used liquid amalgams of cadmium, bismuth, and lead. The advantages claimed for liquid amalgam reductions are: (a) complete reduction is achieved in a few minutes; (b) the amalgam can be used repeatedly;

and (c) no blank correction is required as in the Jones reductor. The reduction potentials of the saturated metal amalgams are as follows:

$$Zn^{2+} + 2e \rightleftharpoons Zn; \qquad\qquad -0.76 \text{ volt}$$

$$Cd^{2+} + 2e \rightleftharpoons Cd; \qquad\qquad -0.40 \text{ volt}$$

$$Pb^{2+} + 2e \rightleftharpoons Pb; \qquad\qquad -0.13 \text{ volt}$$

$$BiO^+ + 2H^+ + 3e \rightleftharpoons Bi + H_2O; \quad +0.32 \text{ volt}$$

The most powerful reductant is therefore zinc amalgam, whilst bismuth amalgam is the least reducing. The final reduction products obtained with these amalgams for a few elements are collected in Table 10.10.

Table 10.10

Liquid amalgam	Iron	Titanium	Molybdenum	Vanadium	Uranium	Tungsten
Zinc	Fe^{2+}	Ti^{3+}	Mo^{3+}	V^{2+}	U^{4+}†	W^{3+}
Cadmium	Fe^{2+}	Ti^{3+}	Mo^{3+}	V^{2+}	U^{4+}	—
Lead	Fe^{2+}	Ti^{3+}	Mo^{3+}	V^{2+}	U^{4+}	W^{3+}
Bismuth	Fe^{2+}	Ti^{3+}	Mo^{3+} or Mo^{5+}*	VO^{2+}	U^{4+}	W^{5+}

* The exact product depends upon the pH of the solution.
† Some U^{3+} is also formed.

The zinc amalgam is prepared by washing 15 g of pure, fine-mesh zinc shot with dilute sulphuric acid, and then heating for 1 hour on the water bath with 300 g of mercury plus 5 mL of 1:4 sulphuric acid. (**CAUTION: Mercury vapour is highly poisonous**; the operation must therefore be performed in a fume cupboard with a good draught.) The whole is allowed to cool, the amalgam washed several times with dilute sulphuric acid, and the liquid portion separated from the solid by means of a separatory funnel. The solid is reserved for another preparation of the amalgam. The liquid amalgam is preserved under dilute sulphuric acid; reaction with the latter is very slow, and the same sample of amalgam may be employed for several reductions. The other amalgams are prepared in similar manner, except that for bismuth and lead, hydrochloric acid is used in place of sulphuric acid. All the amalgams can also be prepared electrolytically using a mercury cathode.

Reductions with liquid amalgams are usually carried out in a separatory funnel in an atmosphere of carbon dioxide, so that when the reduction is complete the remaining amalgam can be easily removed before the titration of the reduced solution is attempted. In a simplified procedure, 30–40 mL of carbon tetrachloride are added to the flask in which reduction has been carried out, at the stage when the amalgam is normally separated: this produces three layers in the flask, the amalgam, surmounted by the carbon tetrachloride, which thus separates the aqueous solution from the amalgam. A mechanically operated stirrer is then inserted so that it will agitate the aqueous layer without disturbing the layers below, and the aqueous solution can then be titrated with the appropriate reagent.

10.140 THE SILVER REDUCTOR

The silver reductor has a relatively low reduction potential (the Ag/AgCl electrode potential in $1M$ hydrochloric acid is 0.2245 volt), and consequently it is not able to effect many of the reductions which can be made with amalgamated zinc. The silver reductor is preferably used with hydrochloric acid solutions, and this is frequently an advantage. The various reductions which can be effected with the silver and the amalgamated zinc reductors are summarised in Table 10.11.*

Table 10.11

Silver reductor: hydrochloric acid solution	Amalgamated zinc (Jones) reductor: sulphuric acid solution
$Fe^{3+} \rightarrow Fe^{2+}$	$Fe^{3+} \rightarrow Fe^{2+}$
Ti^{4+} not reduced	$Ti^{4+} \rightarrow Ti^{3+}$
$Mo^{6+} \rightarrow Mo^{5+}$ (2M HCl; 60–80 °C)	$Mo^{6+} \rightarrow Mo^{3+}$
Cr^{3+} not reduced	$Cr^{3+} \rightarrow Cr^{2+}$
$UO_2^{2+} \rightarrow U^{4+}$ (4M HCl; 60–90 °C)	$UO_2^{2+} \rightarrow U^{3+} + U^{4+}$
$V^{5+} \rightarrow V^{4+}$	$V^{5+} \rightarrow V^{2+}$
$Cu^{2+} \rightarrow Cu^+$ (2M HCl)	$Cu^{2+} \rightarrow Cu^0$

The silver reductor (shaped like a short, squat Jones reductor tube) may be constructed from a tube 12 cm long and 2 cm internal diameter fused to a reservoir bulb of 50–75 mL capacity. It is not always necessary to use suction. The silver is conveniently prepared as follows on a large scale; for preparations on a smaller scale the procedure must be appropriately adapted. A solution of 500 g of silver nitrate in 2500 mL of water, slightly acidified with dilute nitric acid, is placed in a 4 L beaker. Cathodes consisting of two heavy-gauge platinum plates, each 10 cm square, are suspended in the electrolyte by the use of a heavy copper bus-bar connection to a source of current. The anode consists of either a silver rod 200 mm long and 10–25 mm in diameter or a similar weight of silver as a heavy-gauge rectangular sheet; it is suspended in the centre of the electrolyte with the platinum cathodes placed at the outer edges of the deposition cell. Silver is deposited as granular crystals with high ratio of surface to mass by a current of 60–70 A at 5–6 volts. These crystals, obtained in excellent yield, are deposited on the four outside edges of the cathodes; they should be dislodged by gentle tapping, and washed by decantation with dilute sulphuric acid. About 30 g of silver in this form occupy a volume of 40–50 mL — sufficient to fill one reductor tube.

The necessary quantity of silver is introduced into the reductor above a small plug of glass wool: by means of a glass rod flattened at one end, it is compressed to as great an extent as necessary without restricting the free flow of solution through the column. The reductor is rinsed with 100 mL of $1M$ hydrochloric acid, added in five equal portions, each consecutive portion being allowed to pass through the reductor to just above the level of the silver.

The dark silver chloride coating which covers the silver of the upper part of the reductor when hydrochloric acid solutions are employed moves further down

---

* Strictly speaking the higher charge states 6+ and 5+ should be presented as oxidation states (VI) and (V).

the column in use, and when it extends to about three-quarters of the length of the column, the reductor must be regenerated by the following method. The reductor is rinsed with water and filled completely with 1:3-ammonia solution. The silver chloride dissolves; after 10 minutes, the solution is rinsed out of the reductor tube with water, followed by $1M$ hydrochloric acid and is then ready for re-use. As a precautionary measure, the ammoniacal solution of silver chloride should be immediately acidified. The wastage of silver associated with this method of regeneration may be avoided by filling the tube with sulphuric acid ($0.1M$) and then inserting a rod of zinc with its lower end well buried in the silver; when the reduction is complete (as evidenced by loss of the dark colour), the column is well washed with water and is then ready for use.

Examples of the use of the silver reductor are given in Sections 10.107 and 10.108.

10.141 OTHER METHODS OF REDUCTION

Although as already stated the use of metal amalgams, and in particular use of the Jones reductor or of the related silver reductor, is the best method of reducing solutions in preparation for titration with an oxidant, it may happen that for occasional use there is no Jones reductor available, and a simpler procedure will commend itself. In practical terms, the need is most likely to arise in connection with the determination of iron, for which the reduction of iron(III) to iron(II) may be necessary.

Tin(II) chloride solution. Many iron ores are brought into solution with concentrated hydrochloric acid and the resulting solution may be readily reduced with tin(II) chloride:

$$2Fe^{3+} + Sn^{2+} = 2Fe^{2+} + Sn^{4+}$$

The hot solution (70–90 °C) from about 0.3 g of iron ore, which should occupy a volume of 25–30 mL and be 5–6M with respect to hydrochloric acid, is reduced by adding concentrated tin(II) chloride solution dropwise from a separatory funnel or a burette, with stirring, until the yellow colour of the solution has *nearly* disappeared. The reduction is then completed by diluting the concentrated solution of tin(II) chloride with 2 volumes of dilute hydrochloric acid, and adding the dilute solution dropwise, with agitation after each addition, until the liquid has a faint green colour, quite free from any tinge of yellow. The solution is then rapidly cooled under the tap to about 20 °C, with protection from the air, and the slight excess of tin(II) chloride present removed by adding 10 mL of a saturated solution (*ca* 5 per cent) of mercury(II) chloride rapidly in one portion and with thorough mixing; a *slight* silky white precipitate of mercury(I) chloride should be obtained.

The small amount of mercury(I) chloride in suspension has no appreciable effect upon the oxidising agent used in the subsequent titration, but if a heavy precipitate forms, or a grey or black precipitate is obtained, too much tin(II) solution has been used; the results are inaccurate and the reduction must be repeated. Finely divided mercury reduces permanganate or dichromate ions and also slowly reduces Fe^{3+} ions in the presence of chloride ion.

After the addition of the mercury(II) chloride solution, the whole is allowed to stand for five minutes, then diluted to about 400 mL and titrated with standard potassium dichromate solution (Section 10.99), or with standard permanganate solution in the presence of 'preventive solution' (Section 10.92).

Blank runs on the reagents should be carried through all the operations, and corrections made, if necessary.

The concentrated solution of tin(II) chloride is prepared by dissolving 12 g of pure tin or 30 g of crystallised tin(II) chloride ($SnCl_2,2H_2O$) in 100 mL of concentrated hydrochloric acid and diluting to 200 mL with water.

Other reducing agents which have been used include hydrogen sulphide and sulphurous acid. The former reagent is passed through an acidic solution of the ion to be reduced. The latter is employed either by passing sulphur dioxide from a siphon of the liquid gas or by adding a freshly prepared sulphurous acid solution or ammonium hydrogensulphite solution to an acidic solution of the ion.

The reagents listed above can be applied to the reduction of many other ions in addition to Fe^{3+}, and there are also a number of other substances which can be employed as reducing agents; thus for example hydroxylammonium salts are frequently added to solutions to ensure that reagents do not undergo atmospheric oxidation, and as an example of an unusual reducing agent, phosphorous(III) acid may be used to reduce mercury(II) to mercury(I); see Section 10.129.

For References and Bibliography see Sections 11.79 and 11.80.

CHAPTER 11
GRAVIMETRY

11.1 INTRODUCTION TO GRAVIMETRIC ANALYSIS

Gravimetric analysis or quantitative analysis by weight is the process of isolating and weighing an element or a definite compound of the element in as pure a form as possible. The element or compound is separated from a weighed portion of the substance being examined. A large proportion of the determinations in gravimetric analysis is concerned with the transformation of the element or radical to be determined into a pure stable compound which can be readily converted into a form suitable for weighing. The weight of the element or radical may then be readily calculated from a knowledge of the formula of the compound and the relative atomic masses of the constituent elements.

The separation of the element or of the compound containing it may be effected in a number of ways, the most important of which are: (*a*) precipitation methods; (*b*) volatilisation or evolution methods; (*c*) electroanalytical methods; and (*d*) extraction and chromatographic methods. Only (*a*) and (*b*) will be discussed in this chapter: (*c*) is considered in Part E, and (*d*) in Part C.

It is appropriate to mention at this stage the reasons for the continuing use of gravimetric analysis despite the disadvantage that it is generally somewhat time-consuming. The advantages offered by gravimetric analysis are:

(*a*) it is accurate and precise when using modern analytical balances;
(*b*) possible sources of error are readily checked, since filtrates can be tested for completeness of precipitation and precipitates may be examined for the presence of impurities;
(*c*) it has the important advantage of being an absolute method, i.e. one involving direct measurement without any form of calibration being required;
(*d*) determinations can be carried out with relatively inexpensive apparatus, the most expensive requirements being a muffle furnace and, in some cases, platinum crucibles.

Two general applications of gravimetric analysis are:

(*a*) the analysis of standards which are to be used for the testing and/or calibration of instrumental techniques;
(*b*) analyses requiring high accuracy, although the time-consuming nature of gravimetry limits this application to small numbers of determinations.

11.2 PRECIPITATION METHODS

These are perhaps the most important with which we are concerned in gravimetric analysis. The constituent being determined is precipitated from solution in a form which is so slightly soluble that no appreciable loss occurs when the precipitate is separated by filtration and weighed. Thus, in the determination of silver, a solution of the substance is treated with an excess of sodium chloride or potassium chloride solution, the precipitate is filtered off, well washed to remove soluble salts, dried at 130–150 °C, and weighed as silver chloride. Frequently the constituent being determined is weighed in a form other than that in which it was precipitated. Thus magnesium is precipitated, as ammonium magnesium phosphate $Mg(NH_4)PO_4,6H_2O$, but is weighed, after ignition, as the pyrophosphate $Mg_2P_2O_7$. The following factors determine a successful analysis by precipitation.

1. The precipitate must be so insoluble that no appreciable loss occurs when it is collected by filtration. In practice this usually means that the quantity remaining in solution does not exceed the minimum detectable by the ordinary analytical balance, viz. 0.1 mg.
2. The physical nature of the precipitate must be such that it can be readily separated from the solution by filtration, and can be washed free of soluble impurities. These conditions require that the particles are of such size that they do not pass through the filtering medium, and that the particle size is unaffected (or, at least, not diminished) by the washing process.
3. The precipitate must be convertible into a pure substance of definite chemical composition; this may be effected either by ignition or by a simple chemical operation, such as evaporation, with a suitable liquid.

Factor 1, which is concerned with the completeness of precipitation, has already been dealt with in connection with the solubility-product principle, and the influence upon the solubility of the precipitate of (i) a salt with a common ion, (ii) salts with no common ion, (iii) acids and bases, and (iv) temperature (Sections 2.6–2.11).

It was assumed throughout that the compound which separated out from the solution was chemically pure, but this is not always the case. The purity of the precipitate depends *inter alia* upon the substances present in solution both before and after the addition of the reagent, and also upon the exact experimental conditions of precipitation. In order to understand the influence of these and other factors, it will be necessary to give a short account of the properties of colloids.

Problems which arise with certain precipitates include the coagulation or flocculation of a colloidal dispersion of a finely divided solid to permit its filtration and to prevent its re-peptisation upon washing the precipitate. It is therefore desirable to understand the basic principles of the colloid chemistry of precipitates, for which an appropriate textbook should be consulted (see the Bibliography, Section 11.80). However, some aspects of the colloidal state relevant to quantitative analysis are indicated below.

11.3 THE COLLOIDAL STATE

The colloidal state of matter is distinguished by a certain range of particle size, as a consequence of which certain characteristic properties become apparent.

Before discussing these, mention must be made of the various units which are employed in expressing small dimensions. The most important of these are:

$1 \mu m = 10^{-3}$ mm; $\qquad$ 1 nm $= 10^{-6}$ mm;

1 Ångström unit $= Å = 10^{-10}$ metre $= 10^{-7}$ mm $= 0.1$ nm.

Colloidal properties are, in general, exhibited by substances of particle size ranging between 0.1 μm and 1 nm. Ordinary quantitative filter paper will retain particles down to a diameter of about 10^{-2} mm or 10 μm, so that colloidal solutions in this respect behave like true solutions and are not filterable (size of molecules is of the order of 0.1 nm or 10^{-8} cm). The limit of vision under the microscope is about 0.2 μm. If a powerful beam of light is passed through a colloidal solution and the solution viewed at right angles to the incident light, a scattering of light is observed. This is the so-called **Tyndall effect**, which is not exhibited by true solutions.

An important consequence of the smallness of the size of colloidal particles is that the ratio of surface area to weight is extremely large. Phenomena, such as adsorption, which depend upon the size of the surface will therefore play an important part with substances in the colloidal state.

For convenience, colloids are divided into two main groups, designated as **lyophobic** and **lyophilic colloids**. The chief properties of each class are summarised in Table 11.1, although it must be emphasised that the distinction is not an absolute one, since some gelatinous precipitates (e.g. aluminium hydroxide and other metallic hydroxides) have properties intermediate between those of lyophobic and lyophilic colloids.

Table 11.1

Lyophobic colloids	Lyophilic colloids
(1) The dispersion (or sols) are only slightly viscous. Examples: sols of metals, silver halides, metallic sulphides, etc.	(1) The dispersions are very viscous; they set to jelly-like masses known as gels. Examples: sols of silicic acid, tin(IV) oxide, gelatin.
(2) A comparatively minute concentration of an electrolyte results in flocculation. The change is, in general, irreversible; water has no effect upon the flocculated solid.	(2) Comparatively large concentrations of electrolytes are required to cause precipitation ('salting out'). The change is, in general, reversible, and reversal is effected by the addition of a solvent (water).
(3) Lyophobic colloids, ordinarily, have an electric charge of definite sign, which can be changed only by special methods.	(3) Most lyophilic colloids change their charge readily, e.g. they are positively charged in an acid medium and negatively charged in an alkaline medium.
(4) The ultra-microscope reveals bright particles in vigorous motion (Brownian movement).	(4) Only a diffuse light cone is exhibited under the ultra-microscope.

The process of dispersing a gel or a flocculated solid to form a sol is called **peptisation**, and is briefly dealth with on page 421 and in Section 11.8.

The stability of lyophobic colloids is intimately associated with the electrical charge on the particles.* Thus in the formation of an arsenic(III) sulphide sol

---

* Lyophilic colloids are mainly stabilised by solvation.

by precipitation with hydrogen sulphide in acid solution, sulphide ions are primarily adsorbed (since every precipitate has a tendency to adsorb its own ions), and some hydrogen ions are secondarily adsorbed. The hydrogen ions or other ions which are secondarily adsorbed have been termed 'counter-ions'. Thus the so-called electrical double layer is set up between the particles and the solution. An arsenic(III) sulphide particle is represented diagrammatically in Fig. 11.1. The colloidal particle of arsenic(III) sulphide has a negatively charged surface, with positively charged counter-ions which impart a positive charge to the liquid immediately surrounding it. If an electric current is passed through the solution, the negative particles will move towards the anode; the speed is comparable with that of electrolytic ions. The electrical conductivity of a sol is, however, quite low because the number of current-carrying particles is small compared with that in a solution of an electrolyte at an appreciable concentration; the large charge carried by the colloidal particles is not sufficient to compensate for their smaller number.

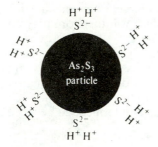

Fig. 11.1

If the electrical double layer is destroyed, the sol is no longer stable, and the particles will flocculate, thereby reducing the large surface area. Thus if barium chloride solution is added, barium ions are preferentially adsorbed by the particles; the charge distribution on the surface is disturbed and the particles flocculate. After flocculation, it is found that the dispersion medium is acid owing to the liberation of the hydrogen counter-ions. It appears that ions of opposite charge to those primarily adsorbed on the surface are necessary for coagulation. The minimum amount of electrolyte necessary to cause flocculation of the colloid is called the **flocculation** or **coagulation value**. It has been found that the latter depends primarily upon the charge numbers of the ions of the opposite charge to that on the colloidal particles: the nature of the ions has some influence also.

If two sols of opposite sign are mixed, mutual coagulation usually occurs owing to the neutralisation of charges. The above remarks apply largely to lyophobic colloids. Lyophilic colloids are generally much more difficult to coagulate than lyophobic colloids. If a lyophilic colloid, e.g. of gelatin, is added to a lyophobic colloid, e.g. of gold, then the lyophobic colloid appears to be strongly protected against the flocculating action of electrolytes. It is probable that the particles of the lyophilic colloid are adsorbed by the lyophobic colloid and impart their own properties to the latter. The lyophilic colloid is known as a **protective colloid**. This explains the relative stability produced by the addition of a little gelatin to the otherwise unstable gold sols. For this reason also, organic matter, which might form a protective colloid, is generally destroyed before proceeding with an inorganic analysis.

During the flocculation of a colloid by an electrolyte, the ions of opposite sign to that of the colloid are adsorbed to a varying degree on the surface; the higher the charge of the ion, the more strongly is it adsorbed. In all cases, the precipitate will be contaminated by surface adsorption. Upon washing the precipitate with water, part of the adsorbed electrolyte is removed, and a new difficulty may arise. The electrolyte concentration in the supernatant liquid may fall below the coagulation value, and the precipitate may pass into colloidal solution again. This phenomenon, which is known as **peptisation**, is of great importance in quantitative analysis. By way of illustration, consider the precipitation of silver by excess of chloride ions in acid solution and the subsequent washing of the coagulated silver chloride with water; the adsorbed hydrogen ions will be removed by the washing process and a portion of the precipitate may pass through the filter. If, however, washing is carried out with dilute nitric acid, no peptisation occurs. For this reason, precipitates are always washed with a suitable solution of an electrolyte which does not interfere with the subsequent steps in the determination.

11.4 SUPERSATURATION AND PRECIPITATE FORMATION

The solubility of a substance at any given temperature in a given solvent is the amount of the substance dissolved by a known weight of that solvent when the substance is in equilibrium with the solvent. The solubility depends upon the particle size, when these are smaller than about 0.01 mm in diameter; the solubility increases greatly the smaller the particles, owing to the increasing role played by surface effects. (The definition of solubility given above refers to particles larger than 0.01 mm.) A supersaturated solution is one that contains a greater concentration of solute than corresponds to the equilibrium solubility at the temperature under consideration. Supersaturation is therefore an unstable state which may be brought to a state of stable equilibrium by the addition of a crystal of the solute ('seeding' the solution) or of some other substance, or by mechanical means such as shaking or stirring. The difficulty of precipitation of ammonium magnesium phosphate will at once come to mind as an example of supersaturation.

According to von Weimarn, supersaturation plays an important part in determining the particle size of a precipitate. He deduced that the initial velocity of precipitation is proportional to $(Q - S)/S$, where Q is the total concentration of the substance that is to precipitate, and S is the equilibrium solubility; $(Q - S)$ will denote the supersaturation at the moment precipitation commences. The expression applies approximately only when Q is large compared with S. The influence of the degree of supersaturation is well illustrated by von Weimarn's results for the precipitation of barium sulphate from solutions of barium thiocyanate and manganese sulphate respectively. When the concentrations of these solutions were greater than molar, a gelatinous precipitate was instantly obtained, whereas with very dilute solutions ($< 10^{-3}M$) the precipitate appeared after about one month and contained relatively large crystal particles (e.g. 0.03 mm × 0.015 mm).

These results indicated that the particle size of a precipitate decreases with increasing concentration of the reactants. For the production of a crystalline precipitate, for which the adsorption errors will be least and filtration will be easiest, $(Q - S)/S$ should be as small as possible. There is obviously a practical limit to reducing $(Q - S)/S$ by making Q very small, since for a precipitation to be of value in analysis, it must be complete in a comparatively short time and the volumes of solutions involved must not be too large. There is, however, another method which may be used, viz. that of increasing S. For example, barium sulphate is about 50 times more soluble in $2M$ hydrochloric acid than in water: if $0.05M$ solutions of barium chloride and sulphuric acid are prepared in boiling $2M$ hydrochloric acid and the solutions mixed, a typical crystalline precipitate of barium sulphate is slowly formed.[26,27]

Applications of the above conceptions are to be found in the following recognised procedures in gravimetric analysis.

1. Precipitation is usually carried out in hot solutions, since the solubility generally increases with rise in temperature.
2. Precipitation is effected in dilute solution and the reagent is added slowly and with thorough stirring. The slow addition results in the first particles precipitated acting as nuclei which grow as further material precipitates.
3. A suitable reagent is often added to increase the solubility of the precipitate and thus lead to larger primary particles.
4. A procedure which is commonly employed to prevent supersaturation from occurring is that of precipitation from homogeneous solution. This is achieved by generating the precipitating agent within the solution by means of a homogeneous reaction at a similar rate to that required for precipitation of the species.

11.5 THE PURITY OF THE PRECIPITATE: CO–PRECIPITATION

When a precipitate separates from a solution, it is not always perfectly pure: it may contain varying amounts of impurities dependent upon the nature of the precipitate and the conditions of precipitation. The contamination of the precipitate by substances which are normally soluble in the mother liquor is termed **co-precipitation**. We must distinguish between two important types of co-precipitation. The first is concerned with adsorption at the *surface* of the particles exposed to the solution, and the second relates to the occlusion of foreign substances during the process of crystal growth from the primary particles.

With regard to surface adsorption, this will, in general, be greatest for gelatinous precipitates and least for those of pronounced macrocrystalline character. Precipitates with ionic lattices appear to conform to the Paneth–Fajans–Hahn adsorption rule, which states that the ion that is most strongly adsorbed by an ionic substance (crystal lattice) is that ion which forms the least soluble salt. Thus on sparingly soluble sulphates, it is found that calcium ions are adsorbed preferentially over magnesium ions because calcium sulphate is less soluble than magnesium sulphate. Also silver iodide adsorbs silver acetate much more strongly than it does silver nitrate under comparable conditions, since the former is the less soluble. The deformability of the adsorbed ions and the electrolytic dissociation of the adsorbed compound also have a considerable influence; the smaller the dissociation of the compound, the greater is the adsorption. Thus hydrogen sulphide, a weak electrolyte, is strongly adsorbed by metallic sulphides.

The second type of co-precipitation may be visualised as occurring during the building up of the precipitate from the primary particles. The latter will be subject to a certain amount of surface adsorption, and during their coalescence the impurities will either be partially eliminated if large single crystals are formed and the process takes place slowly, or, if coalescence is rapid, large crystals composed of loosely bound small crystals may be produced and some of the impurities may be entrained within the walls of the large crystals. If the impurity is isomorphous or forms a solid solution with the precipitate, the amount of co-precipitation may be very large, since there will be no tendency for elimination during the 'ageing' process. The latter actually occurs during the precipitation of barium sulphate in the presence of alkali nitrates; in this particular case X-ray studies have shown that the abnormally large co-precipitation (which may be as high as 3.5 per cent if precipitation occurs in the presence of high concentrations of nitrate) is due to the formation of solid solutions. Fortunately, however, such cases are comparatively rare in analysis.

Appreciable errors may also be introduced by **post-precipitation**. This is the precipitation which occurs on the surface of the first precipitate *after* its formation. It occurs with sparingly soluble substances which form supersaturated solutions; they usually have an ion in common with the primary precipitate. Thus in the precipitation of calcium as oxalate in the presence of magnesium, magnesium oxalate separates out gradually upon the calcium oxalate; the longer the precipitate is allowed to stand in contact with the solution, the greater is the error due to this cause. A similar effect is observed in the precipitation of copper or mercury(II) sulphide in $0.3M$ hydrochloric acid in the presence of zinc ions; zinc sulphide is slowly post-precipitated.

Post-precipitation differs from co-precipitation in several respects:

(*a*) The contamination increases with the time that the precipitate is left in contact with the mother liquor in post-precipitation, but usually decreases in co-precipitation.

(*b*) With post-precipitation, contamination increases the faster the solution is agitated by either mechanical or thermal means. The reverse is usually true with co-precipitation.

(*c*) The magnitude of contamination by post-precipitation may be much greater than in co-precipitation.

It is convenient to consider now the influence of **digestion**. This is usually carried out by allowing the precipitate to stand for 12–24 hours at room temperature, or sometimes by warming the precipitate for some time in contact with the liquid from which it was formed: the object is, of course, to obtain complete precipitation in a form which can be readily filtered. During the process of digestion or of the ageing of precipitates, at least two changes occur. The very small particles, which have a greater solubility than the larger ones, will, after precipitation has occurred, tend to pass into solution, and will ultimately re-deposit upon the larger particles; co-precipitation on the minute particles is thus eliminated and the total co-precipitation on the ultimate precipitate reduced. The rapidly formed crystals are probably of irregular shape and possess a comparatively large surface; upon digestion these tend to become more regular in character and also more dense, thus resulting in a decrease in the area of the

423

surface and a consequent reduction of adsorption. The net result of digestion is usually to reduce the extent of co-precipitation and to increase the size of the particles, rendering filtration easier.

11.6 CONDITIONS OF PRECIPITATION

No universal rules can be given which are applicable to all cases of precipitation, but, by the intelligent application of the principles enumerated in the foregoing paragraphs, a number of fairly general rules may be stated:

1. Precipitation should be carried out in dilute solution, due regard being paid to the solubility of the precipitate, the time required for filtration, and the subsequent operations to be carried out with the filtrate. This will minimise the errors due to co-precipitation.
2. The reagents should be mixed slowly and with constant stirring. This will keep the degree of supersaturation small and will assist the growth of large crystals. A slight excess of the reagent is all that is generally required; in exceptional cases a large excess may be necessary. In some instances the order of mixing the reagents may be important. Precipitation may be effected under conditions which increase the solubility of the precipitate, thus further reducing the degree of supersaturation (compare Section 11.5).
3. Precipitation is effected in hot solutions, provided the solubility and the stability of the precipitate permit. Either one or both of the solutions should be heated to just below the boiling point or other more favourable temperature. At the higher temperature: (*a*) the solubility is increased with a consequent reduction in the degree of supersaturation, (*b*) coagulation is assisted and sol formation decreased, and (*c*) the velocity of crystallisation is increased, thus leading to better-formed crystals.
4. Crystalline precipitates should be digested for as long as practical, preferably overnight, except in those cases where post-precipitation may occur. As a rule, digestion on the steam bath is desirable. This process decreases the effect of co-precipitation and gives more readily filterable precipitates. Digestion has little effect upon amorphous or gelatinous precipitates.
5. The precipitate should be washed with the appropriate dilute solution of an electrolyte. Pure water may tend to cause peptisation. (For theory of washing, see Section 11.8 below.)
6. If the precipitate is still appreciably contaminated as a result of co-precipitation or other causes, the error may often be reduced by dissolving it in a suitable solvent and then re-precipitating it. The amount of foreign substance present in the second precipitation will be small, and consequently the amount of the entrainment by the precipitate will also be small.

11.7 PRECIPITATION FROM HOMOGENEOUS SOLUTION

The major objective of a precipitation reaction is the separation of a pure solid phase in a compact and dense form which can be filtered easily. The importance of a small degree of supersaturation has long been appreciated, and it is for this reason that a dilute solution of a precipitating agent is added slowly and with stirring. In the technique known as precipitation from homogeneous solution, the precipitant is not added as such but is slowly generated by a homogeneous chemical reaction within the solution. The precipitate is thus formed under conditions which eliminate the undesirable concentration effects which are inevitably associated with the conventional precipitation process. The precipitate is dense and readily filterable; co-precipitation is reduced to a minimum. Moreover, by varying the rate of the chemical reaction producing the precipitant in homogeneous solution, it is possible to alter further the physical appearance of the precipitate — the slower the reaction, the larger (in general) are the crystals formed.

Many different anions can be generated at a slow rate; the nature of the anion is important in the formation of compact precipitates. It is convenient to deal with the subject under separate headings.

Hydroxides and basic salts. The necessity for careful control of the pH has long been recognised. This is accomplished by making use of the hydrolysis of urea, which decomposes into ammonia and carbon dioxide as follows:

$$CO(NH_2)_2 + H_2O = 2NH_3 + CO_2$$

Urea possesses negligible basic properties ($K_b = 1.5 \times 10^{-14}$), is soluble in water and its hydrolysis rate can be easily controlled. It hydrolyses rapidly at 90–100 °C, and hydrolysis can be quickly terminated at a desired pH by cooling the reaction mixture to room temperature. The use of a hydrolytic reagent *alone* does not result in the formation of a compact precipitate; the physical character of the precipitate will be very much affected by the presence of certain anions. Thus in the precipitation of aluminium by the urea process, a dense precipitate is obtained in the presence of succinate, sulphate, formate, oxalate, and benzoate ions, but not in the presence of chloride, chlorate, perchlorate, nitrate, sulphate, chromate, and acetate ions. The preferred anion for the precipitation of aluminium is succinate. It would appear that the main function of the 'suitable anion' is the formation of a basic salt which seems responsible for the production of a compact precipitate. The pH of the initial solution must be appropriately adjusted.

The following are suitable anions for urea precipitations of some metals: sulphate for gallium, tin, and titanium; formate for iron, thorium, and bismuth; succinate for aluminium and zirconium.

The urea method generally results in the deposition on the surface of the beaker of a thin, tenacious, and somewhat transparent film of the basic salt. This film cannot be removed by scraping with a 'policeman'. It is dissolved by adding a few millilitres of hydrochloric acid, covering the beaker with a clockglass, and refluxing for 5–10 minutes; the small amount of metallic ion is precipitated by ammonia solution and filters readily through the same filter containing the previously precipitated basic salt.

The urea hydrolysis method may be applied also to:

1. the precipitation of barium as barium chromate in the presence of ammonium acetate;
2. the precipitation of large amounts of nickel as the dimethylglyoximate; and
3. the precipitation of aluminium as the oxinate.

Phosphates. Insoluble orthophosphates may be precipitated with phosphate ion derived from trimethyl or triethyl phosphate by stepwise hydrolysis. Thus 1.8M sulphuric acid containing zirconyl ions and trimethyl phosphate on

425

heating gives a dense precipitate of variable composition, which is ignited to and weighed as the dipolyphosphate (pyrophosphate) ZrP_2O_7.

Metaphosphoric acid may also be used; it hydrolyses in warm acid solution forming phosphoric(V) acid. Thus bismuth may be precipitated as bismuth phosphate in a dense, crystalline form.

Oxalates. Urea may be employed to raise the pH of an acid solution containing hydrogenoxalate ion $HC_2O_4^-$, thus affording a method for the slow generation of oxalate ion. Calcium oxalate may thus be precipitated in a dense form:

$$CO(NH_2)_2 + 2HC_2O_4^- + H_2O = 2NH_4^+ + CO_2 + 2C_2O_4^{2-}$$

Dimethyl oxalate and diethyl oxalate can also be hydrolysed to yield oxalate ion:

$$(C_2H_5)_2C_2O_4 + 2H_2O = 2C_2H_5OH + 2H^+ + C_2O_4^{2-}$$

Diethyl oxalate is usually preferred because of its slower rate of hydrolysis. Satisfactory results are obtained in the precipitation of calcium, magnesium, and zinc: thorium is precipitated using dimethyl oxalate.

Calcium can be determined as the oxalate by precipitation from homogeneous solution by cation release from the EDTA complex in the presence of oxalate ion.[28]

Sulphates. Sulphate ion may be generated by the hydrolysis of aminosulphonic (sulphamic) acid:

$$NH_2SO_3H + H_2O = NH_4^+ + H^+ + SO_4^{2-}$$

The reaction has been used to produce barium sulphate in a coarsely crystalline form.

The hydrolysis of dimethyl sulphate also provides a source of sulphate ion, and the reaction has been used for the precipitation of barium, strontium, and calcium as well as lead:

$$(CH_3)_2SO_4 + 2H_2O = 2CH_3OH + 2H^+ + SO_4^{2-}$$

11.8 WASHING THE PRECIPITATE

The experimental aspect of this important subject is dealt with in Section 3.36. Only some general theoretical considerations will be given here. Most precipitates are produced in the presence of one or more soluble compounds, and it is the object of the washing process to remove these as completely as possible. It is evident that only surface impurities will be removed in this way. The composition of the wash solution will depend upon the solubility and chemical properties of the precipitate and upon its tendency to undergo peptisation, the impurities to be removed, and the influence of traces of the wash liquid upon the subsequent treatment of the precipitate before weighing. Pure water cannot, in general, be employed owing to the possibility of producing partial peptisation of the precipitate and, in many cases, the occurrence of small losses as a consequence of the slight solubility of the precipitate: a solution of some electrolyte is employed. This should possess a common ion with the precipitate in order to reduce solubility errors, and should easily be volatilised in the preparation of the precipitate for weighing. For these reasons, ammonium salts, ammonia solution, and dilute acids are commonly employed. If the filtrate is required in

426

a subsequent determination, the selection is limited to substances which will not interfere in the sequel. Also, hydrolysable substances will necessitate the use of solutions containing an electrolyte which will depress the hydrolysis (compare Section 2.19). Whether the wash liquid is employed hot or at some other temperature will depend primarily upon the solubility of the precipitate; if permissible, hot solutions are to be preferred because of the greater solubility of the foreign substances and the increased speed of filtration.

It is convenient to divide wash solutions into three classes:

1. *Solutions which prevent the precipitate from becoming colloidal and passing through the filter.* This tendency is frequently observed with gelatinous or flocculated precipitates but rarely with well-defined crystalline precipitates. The wash solution should contain an electrolyte. The nature of the electrolyte is immaterial, provided it is without action upon the precipitate either during washing or ignition. Ammonium salts are therefore widely used. Thus dilute ammonium nitrate solution is employed for washing iron(III) hydroxide [hydrated iron(III) oxide], and 1 per cent nitric acid for washing silver chloride.

2. *Solutions which reduce the solubility of the precipitate.* The wash solution may contain a moderate concentration of a compound with one ion in common with the precipitate, use being made of the fact that substances tend to be less soluble in the presence of a slight excess of a common ion. Most salts are insoluble in ethanol and similar solvents, so that organic solvents can sometimes be used for washing precipitates. Sometimes a mixture of an organic solvent (e.g. ethanol) and water or a dilute electrolyte is effective in reducing the solubility to negligible proportions. Thus 100 mL of water at 25 °C will dissolve 0.7 mg of calcium oxalate, but the same volume of dilute ammonium oxalate solution dissolves only a negligible weight of the salt. Also 100 mL of water at room temperature will dissolve 4.2 mg of lead sulphate, but dilute sulphuric acid or 50 per cent aqueous ethanol has practically no solvent action on the compound.

3. *Solutions which prevent the hydrolysis of salts of weak acids and bases.* If the precipitate is a salt of weak acid and is slightly soluble it may exhibit a tendency to hydrolyse, and the soluble product of hydrolysis will be a base; the wash liquid must therefore be basic. Thus $Mg(NH_4)PO_4$ may hydrolyse appreciably to give the hydrogenphosphate ion HPO_4^{2-} and hydroxide ion, and should accordingly be washed with dilute aqueous ammonia. If salts of weak bases, such as hydrated iron(III), chromium(III), or aluminium ion, are to be separated from a precipitate, e.g. silica, by washing with water, the salts may be hydrolysed and their insoluble basic salts or hydroxides may be produced together with an acid:

$$[Fe(H_2O_6)]^{3+} \rightleftharpoons [Fe(OH)(H_2O)_5]^{2+} + H^+$$

The addition of an acid to the wash solution will prevent the hydrolysis of iron(III) or similar salts: thus dilute hydrochloric acid will serve to remove iron(III) and aluminium salts from precipitates that are insoluble in this acid.

Solubility losses are reduced by employing the minimum quantity of wash solution consistent with the removal of impurities. It can be readily shown that washing is more efficiently carried out by the use of many small portions of liquid than with a few large portions, the total volume being the same in both instances. Under ideal conditions, where the foreign substance is simply

427

mechanically associated with the particles of the precipitate, the following expression may be shown to hold:

$$x_n = x_0 \left(\frac{u}{u+v} \right)^n$$

where x_0 is the concentration of impurity before washing, x_n is the concentration of impurity after n washings, u is the volume in millilitres of the liquid remaining with the precipitate after draining, and v is the volume in millilitres of the solution used in each washing. It follows from this expression that it is best: (a) to allow the liquid to drain as far as possible in order to maintain u at a minimum; and (b) to use a relatively small volume of liquid and to increase the number of washings. Thus if $u = 1$ mL and $v = 9$ mL, five washings would reduce the surface impurity to 10^{-6} of its original value; one washing with the same volume of liquid, viz. 45 mL, would only reduce the concentration to $1/46$ or 2.2×10^{-2} of its initial concentration.

In practice, the washing process is not quite so efficient as the above simple theory would indicate, since the impurities are not merely mechanically associated with the surface. Furthermore, solubility losses are not so great as one would expect from the solubility data because the wash solution passing through the filter is not saturated with respect to the precipitate. Frequent qualitative tests must be made upon portions of the filtrate for some foreign ion which is known to be present in the original solution; as soon as these tests are negative, the washing is discontinued.

11.9 IGNITION OF THE PRECIPITATE: THERMOGRAVIMETRIC METHOD OF ANALYSIS

In addition to superficially adherent water, precipitates may contain:

(a) adsorbed water, present on all solid surfaces in amount dependent on the humidity of the atmosphere;
(b) occluded water, present in solid solution or in cavities within crystals;
(c) sorbed water, associated with substances having a large internal surface development, e.g. hydrous oxides; and
(d) essential water, present as water of hydration or crystallisation [e.g. CaC_2O_4,H_2O or $Mg(NH_4)PO_4,6H_2O$] or as water of constitution [the water is not present as such but is formed on heating, e.g. $Ca(OH)_2 \rightarrow CaO + H_2O$].

In addition to the evolution of water, the ignition of precipitates often results in thermal decomposition reactions involving the dissociation of salts into acidic and basic components, e.g. the decomposition of carbonates and sulphates; the decomposition temperatures will obviously be related to the thermal stabilities.

The temperatures at which precipitates may be dried, or ignited to the required chemical form, can be determined from a study of the **thermogravimetric curves** for the individual substances. It should be emphasised, however, that thermogravimetric curves must be interpreted with care, paying due regard to the different experimental conditions which apply in thermogravimetry (temperature is usually changing at a regular rate) and in routine gravimetric analysis (the precipitate is brought to a specified temperature and maintained at that temperature for a definite time). A concise account of the principles and

428

applications of thermogravimetry follows and a small number of illustrative experiments are described in Section 11.78.

Thermogravimetry is a technique in which a change in the weight of a substance is recorded as a function of temperature or time. The basic instrumental requirement for thermogravimetry is a precision balance with a furnace programmed for a linear rise of temperature with time. The results may be presented (1) as a thermogravimetric (TG) curve, in which the weight change is recorded as a function of temperature or time, or (2) as a derivative thermogravimetric (DTG) curve where the first derivative of the TG curve is plotted with respect to either temperature or time.

A typical thermogravimetric curve, for copper sulphate pentahydrate $CuSO_4,5H_2O$, is given in Fig. 11.2.

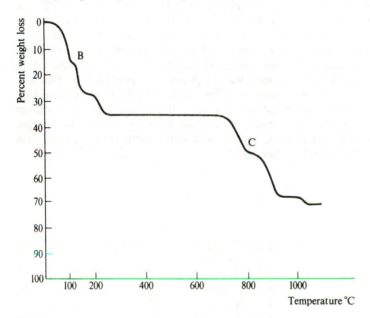

Fig. 11.2

The following features of the TG curve should be noted:

(a) the horizontal portions (plateaus) indicate the regions where there is no weight change;
(b) the curved portions are indicative of weight losses;
(c) since the TG curve is quantitative, calculations on compound stoichiometry can be made at any given temperature.

As Fig. 11.2 shows, copper sulphate pentahydrate has four distinct regions of decomposition:

	Approximate temperature region
$CuSO_4,5H_2O \rightarrow CuSO_4,H_2O$	90–150 °C
$CuSO_4,H_2O \rightarrow CuSO_4$	200–275 °C
$CuSO_4 \rightarrow CuO + SO_2 + \frac{1}{2}O_2$	700–900 °C
$2CuO \rightarrow Cu_2O + \frac{1}{2}O_2$	1000–1100 °C

429

The precise temperature regions for each of the reactions are dependent upon the experimental conditions (see Section 11.78). Although in Fig. 11.2 the ordinate is shown as the percentage weight loss, the scale on this axis may take other forms:

1. as a true weight scale;
2. as a percentage of the total weight;
3. in terms of relative molecular mass units.

An additional feature of the TG curve (Fig. 11.2) should now be examined, namely the two regions B and C where there are changes in the slope of the weight loss curve. If the rate of change of weight with time dW/dt is plotted against temperature, a derivative thermogravimetric (DTG) curve is obtained (Fig. 11.3). In the DTG curve when there is no weight loss then $dW/dt = 0$. The peak on the derivative curve corresponds to a maximum slope on the TG curve. When dW/dt is a minimum but not zero there is an inflexion, i.e. a change of slope on the TG curve. Inflexions B and C on Fig. 11.2 may imply the formation of intermediate compounds. In fact the inflexion at B arises from the formation of the trihydrate $CuSO_4,3H_2O$, and that at point C is reported by Duval[29] to be due to formation of a golden-yellow basic sulphate of composition $2CuO \cdot SO_3$. Derivative thermogravimetry is useful for many complicated determinations and any change in the rate of weight loss may be readily identified as a trough indicating consecutive reactions; hence weight changes occurring at close temperatures may be ascertained.

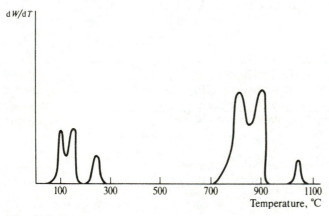

Fig. 11.3

Experimental factors. In the previous section it was stated that the precise temperature regions for each reaction of the thermal decomposition of copper sulphate pentahydrate is dependent upon experimental conditions. When a variety of commercial thermobalances became available in the early 1960s it was soon realised that a wide range of factors could influence the results obtained. Reviews of these factors have been made by Simons and Newkirk[30] and by Coats and Redfern[31] as a basis for establishing criteria necessary to obtain meaningful and reproducible results.

The factors which may affect the results can be classified into the two main groups of instrumental effects and the characteristics of the sample.

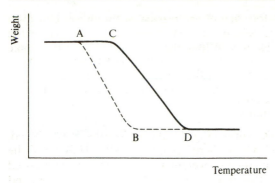

Fig. 11.4

Instrumental factors. *Heating rate.* When a substance is heated at a fast rate, the temperature of decomposition will be higher than that obtained at a slower rate of heating. The effect is shown for a single-step reaction in Fig. 11.4. The curve AB represents the decomposition curve at a slow heating rate, whereas the curve CD is that due to the faster heating rate. If T_A and T_C are the decomposition temperatures at the start of the reaction and the final temperatures on completion of the decomposition are T_B and T_D, the following features can be noted:

$$T_A < T_C$$

$$T_B < T_D$$

$$T_B - T_A < T_D - T_C$$

The heating rate has only a small effect when a fast reversible reaction is considered. The points of inflexion B and C obtained on the thermogravimetric curve for copper sulphate pentahydrate (Fig. 11.2) may be resolved into a plateau if a slower heating rate is used. Hence the detection of intermediate compounds by thermogravimetry is very dependent upon the heating rate employed.

Furnace atmosphere. The nature of the surrounding atmosphere can have a profound effect upon the temperature of a decomposition stage. For example, the decomposition of calcium carbonate occurs at a much higher temperature if carbon dioxide rather than nitrogen is employed as the surrounding atmosphere. Normally the function of the atmosphere is to remove the gaseous products evolved during thermogravimetry, in order to ensure that the nature of the surrounding gas remains as constant as possible throughout the experiment. This condition is achieved in many modern thermobalances by heating the test sample *in vacuo*.

The most common atmospheres employed in thermogravimetry are:

1. 'static air' (air from the surroundings flows through the furnace);
2. 'dynamic air', where compressed air from a cylinder is passed through the furnace at a measured flow rate;
3. nitrogen gas (oxygen-free) which provides an inert environment.

Atmospheres that take part in the reaction — for example, humidified air —

have been used in the study of the decomposition of such compounds as hydrated metal salts.

Since thermogravimetry is a dynamic technique, convection currents arising in a furnace will cause a continuous change in the gas atmosphere. The exact nature of this change further depends upon the furnace characteristics so that widely differing thermogravimetric data may be obtained from different designs of thermobalance.

Crucible geometry. The geometry of the crucible can alter the slope of the thermogravimetric curve. Generally, a flat, plate-shaped crucible is preferred to a 'high-form' cone shape because the diffusion of any evolved gases is easier with the former type.

Sample characteristics. The weight, particle size, and the mode of preparation (the pre-history) of a sample all govern the thermogravimetric results. A large sample can often create a deviation from linearity in the temperature rise. This is particularly true when a fast exothermic reaction is studied, for example the evolution of carbon monoxide during the decomposition of calcium oxalate to calcium carbonate. A large volume of sample in a crucible can impede the diffusion of evolved gases through the bulk of the solid large crystals, especially those of certain metallic nitrates which may undergo decrepitation ('spitting' or 'spattering') when heated. Other samples may swell, or foam and even bubble. In practice a small sample weight with as small a particle size as practicable is desirable for thermogravimetry.

Diverse thermogravimetric results can be obtained from samples with different pre-histories; for example, TG and DTG curves showed that magnesium hydroxide prepared by precipitation methods has a different temperature of decomposition from that for the naturally occurring material.[32] It follows that the source and/or the method of formation of the sample should be ascertained.

Applications. Some of the applications of thermogravimetry are of particular importance to the analyst. These are:

1. the determination of the purity and thermal stability of both primary and secondary standards;
2. the investigation of correct drying temperatures and the suitability of various weighing forms for gravimetric analysis;
3. direct application to analytical problems (automatic thermogravimetric analysis);
4. the determination of the composition of complex mixtures.

Thermogravimetry is a valuable technique for the assessment of the purity of materials. Analytical reagents, especially those used in titrimetric analysis as primary standards, e.g. sodium carbonate, sodium tetraborate, and potassium hydrogenphthalate, have been examined. Many primary standards absorb appreciable amounts of water when exposed to moist atmospheres. TG data can show the extent of this absorption and hence the most suitable drying temperature for a given reagent may be determined.

The thermal stability of EDTA as the free acid and also as the more widely used disodium salt, $Na_2EDTA,2H_2O$, has been reported by Wendlandt.[33] He showed that the dehydration of the disodium salt commences at between 110 and 125 °C, which served to confirm the view of Blaedel and Knight[34] that $Na_2EDTA,2H_2O$ could be safely heated to constant weight at 80 °C.

Undoubtedly the most widespread application of thermogravimetry in analytical chemistry has been in the study of the recommended drying temperatures of gravimetric precipitates. Duval studied over a thousand gravimetric precipitates by this method and gave the recommended drying temperatures. He further concluded that only a fraction of these precipitates are suitable weighing forms for the elements. The results recorded by Duval were obtained with materials prepared under specified conditions of precipitation and this must be borne in mind when assessing the value of a given precipitate as a weighing form, since conditions of precipitation can have a profound effect on the pyrolysis curve. It must be stressed that the rejection of a precipitate because it does not give a stable plateau on the pyrolysis curve at one given rate is unjustified. Further, the limits of the plateau should not be taken as indicative of thermal stability within the complete temperature range. The weighing form is not necessarily isothermally stable at all temperatures that lie on the horizontal portion of a thermogravimetric curve. A slow rate of heating is to be preferred, especially with a large sample weight, over the temperature ranges in which chemical changes take place. Thermogravimetric curves must be interpreted with due regard to the fact that while they are being obtained the temperature is changing at a uniform rate, whereas in routine gravimetric analysis the precipitate is often brought rapidly to a specified temperature and maintained at that temperature for a definite time.

Thermogravimetry may be used to determine the composition of binary mixtures. If each component possesses a characteristic unique pyrolysis curve, then a resultant curve for the mixture will afford a basis for the determination of its composition. In such an automatic gravimetric determination the initial weight of the sample need not be known. A simple example is given by the automatic determination of a mixture of calcium and strontium as their carbonates.

Both carbonates decompose to their oxides with the evolution of carbon dioxide. The decomposition temperature for calcium carbonate is in the temperature range 650–850 °C, whilst strontium carbonate decomposes between 950 and 1150 °C. Hence the amount of calcium and strontium present in a mixture may be calculated from the weight losses due to the evolution of carbon dioxide at the lower and higher temperature ranges respectively. This method could be extended to the analysis of a three-component mixture, as barium carbonate is reported to decompose at an even higher temperature (~ 1300 °C) than strontium carbonate.

Thermogravimetric analysis has also been used in conjunction with other techniques, such as differential thermal analysis (DTA), gas chromatography, and mass spectrometry, for the study and characterisation of complex materials such as clays, soils and polymers.[35]

QUANTITATIVE SEPARATIONS BASED UPON PRECIPITATION METHODS

11.10 FRACTIONAL PRECIPITATION

The simple theory of fractional precipitation has been given in Section 2.8. It was shown that when the solubility products of two sparingly soluble salts having an ion in common differ sufficiently, then one salt will precipitate almost completely before the other commences to separate. This separation is actually possible for a mixture of chloride and iodide, but in other cases the theoretical predictions must be verified experimentally because of the possibility of co-precipitation (Section 11.5) affecting the results. Some separations based upon fractional precipitation, which are of practical importance, will now be considered.

A. Precipitation of sulphides. In order to understand fully the separations dependent upon the sulphide ion, we shall consider first the quantitative relationships involved in a saturated solution of hydrogen sulphide. The following equilibria are present:

$$H_2S \rightleftharpoons H^+ + HS^-$$

$$HS^- \rightleftharpoons H^+ + S^{2-}$$

$$[H^+] \times [HS^-]/[H_2S] = K_1 = 1.0 \times 10^{-7} \tag{1}$$

$$[H^+] \times [S^{2-}]/[HS^-] = K_2 = 1.0 \times 10^{-14} \tag{2}$$

The very small value of K_2 indicates that the secondary dissociation and consequently $[S^{2-}]$ is exceedingly small. It follows therefore that only the primary ionisation is of importance, and $[H^+]$ and $[HS^-]$ are practically equal in value. A saturated aqueous solution of hydrogen sulphide at 25 °C, at atmospheric pressure, is approximately $0.1M$, and calculation shows (see Section 2.14) that in this solution

$$[H^+] = [HS^-] = 1 \times 10^{-4} \, mol \, L^{-1}$$

$$[S^{2-}] = 1 \times 10^{-14} \, mol \, L^{-1}$$

and $[S^{2-}]$ is inversely proportional to the square of the hydrogen ion concentration. Clearly, by varying the pH of the solution the sulphide ion concentration may be controlled, and in this way, separations of metallic sulphides may be effected.

As shown in Section 2.15, in a solution of $0.25M$ hydrochloric acid saturated with hydrogen sulphide (this is the solution employed for the precipitation of the sulphides of the Group II metals in qualitative analysis),

$$[HS^-] = 4 \times 10^{-8} \, mol \, L^{-1}$$

and

$$[S^{2-}] = 1.6 \times 10^{-21} \, mol \, L^{-1}$$

Thus by changing the acidity from $9.5 \times 10^{-5}M$ (that present in saturated hydrogen sulphide water) to $0.25M$, the sulphide ion concentration is reduced from 1×10^{-14} to 1.6×10^{-21}.

With the aid of a table of solubility products of metallic sulphides, we can calculate whether certain sulphides will precipitate under any given conditions of acidity and also the concentration of the metallic ions remaining in solution. Precipitation of a metallic sulphide MS will occur when $[M^{2+}] \times [S^{2-}]$ exceeds the solubility product, and the concentration of metallic ions remaining in the solution may be calculated from the equation:

$$[M^{2+}] = \frac{S_{MS}}{[S^{2-}]} = \frac{S_{MS} \times [H^+]^2}{1.0 \times 10^{-21} \times [H_2S]} \qquad (3)$$

As an example, consider the precipitation of copper(II) sulphide ($K_{S,CuS} = 8.5 \times 10^{-45}$) and iron(II) sulphide ($K_{S,FeS} = 1.5 \times 10^{-19}$) from $0.01M$ solutions of the metallic ions in the presence of $0.25M$ hydrochloric acid. For copper(II) sulphide, the solubility product is readily exceeded;

$$[S^{2-}] = 1.6 \times 10^{-21}, [Cu^{2+}] = 0.01$$

and precipitation will occur until

$$[Cu^{2+}] = 8.5 \times 10^{-45} \times [H^+]^2 / 1.0 \times 10^{-21} \times [H_2S]$$
$$= 8.5 \times 10^{-45} \times (0.25)^2 / 1.0 \times 1^{-21} \times 0.1$$
$$= 5 \times 10^{-24}$$

i.e. precipitation is virtually complete. With iron(II) sulphide, the solubility product cannot be exceeded and precipitation will not occur under these conditions. If, however, the acidity is sufficiently decreased, and consequently $[S^{2-}]$ increased, iron(II) sulphide will be precipitated.

It must be pointed out that the above calculation is approximate only, and may be regarded merely as an illustration of the principles involved in considering the precipitation of sulphides under various experimental conditions; the solubility products of most metallic sulphides are not known with any great accuracy. It is by no means certain that the sulphide ion S^{2-} is the most important reactant in acidified solutions; it may well be that in many cases the active precipitant is the hydrogensulphide ion HS^-, the concentration of which is considerable, and that intermediate products are formed. Also much co-precipitation and post-precipitation occur in sulphide precipitations unless the experimental conditions are rigorously controlled.

B. Precipitation and separation of hydroxides at controlled hydrogen ion concentration or pH. The underlying theory is very similar to that just given for sulphides. Precipitation will depend largely upon the solubility product of the metallic hydroxide and the hydroxide ion concentration, or since $pH + pOH = pK_w$ (Section 2.16), upon the hydrogen ion concentration of the solution.

As shown above the sulphide ion concentration of a saturated aqueous solution of hydrogen sulphide may be controlled within wide limits by suitably changing the concentration of hydrogen ions — a common ion — of the solution. In a like manner the hydroxide ion concentration of a solution of a weak base, such as aqueous ammonia ($K_b = 1.8 \times 10^{-5}$), may be regulated by the addition of a common ion, e.g. ammonium ions in the form of the completely dissociated ammonium chloride. The magnitude of the effect is best illustrated by means of an example. In a $0.1M$ ammonia solution, the degree of dissociation is given (Section 2.13) approximately by:

$$\alpha = \sqrt{KV} = \sqrt{1.8 \times 10^{-5} \times 10} = 0.0013$$

Hence $[OH^-] = 0.0013$, $[NH_4^+] = 0.0013$, and $[NH_3] = 0.0987$. As shown in Section 2.15, by the addition of 0.5 mole of ammonium chloride to 1 L of this solution, $[OH^-]$ is reduced to 3.6×10^{-6} mol L^{-1}.

Thus the addition of half a mole of ammonium chloride to a $0.1M$ solution of aqueous ammonia has decreased the hydroxide ion concentration from 1.3×10^{-3} to $3.6 \times 10^{-6}\,\text{mol L}^{-1}$, or has changed pOH from 2.9 to 5.4, i.e. the pH has changed from 11.1 to 8.6.

An immediate application of the use of the aqueous ammonia–ammonium chloride mixture may be made to the familiar example of the prevention of precipitation of magnesium hydroxide (S.P. 1.5×10^{-11}). We can first calculate the minimum hydroxide ion concentration necessary to prevent precipitation in, say, $0.1M$ magnesium solution.

$$[\text{OH}^-] = \sqrt{\frac{K_{s,\text{Mg(OH)}_2}}{[\text{Mg}^{2+}]}} = \sqrt{\frac{1.5 \times 10^{-11}}{0.1}} = 1.22 \times 10^{-5}M$$

or

$$\text{pOH} = 4.9 \quad \text{and} \quad \text{pH} = 14.0 - 4.9 = 9.1$$

If an aqueous ammonia solution which is $0.1M$ is employed, the concentration of NH_4^+ ion as ammonium chloride or other ammonium salt necessary to prevent the precipitation of magnesium hydroxide can be readily calculated as follows. Substituting in the mass action equation:

$$\frac{[\text{NH}_4^+] \times [\text{OH}^-]}{[\text{NH}_3]} = 1.8 \times 10^{-5}$$

$$\frac{[\text{NH}_4^+) \times 1.22 \times 10^{-5}}{0.1} = 1.8 \times 10^{-5}$$

or

$$[\text{NH}_4^+] = 1.48 \times 10^{-1}M$$

This corresponds to an ammonium chloride concentration of $1.48 \times 10^{-1} \times 53.5 = 7.9\,\text{g L}^{-1}$.

Now consider the conditions necessary for the practically complete precipitation of magnesium hydroxide from a $0.1M$ solution of, say, magnesium chloride. A pOH slightly in excess of 4.9 (i.e. pH = 9.1) might fail to precipitate the hydroxide owing to supersaturation. Let us suppose the hydroxide ion concentration is increased ten-fold, i.e. to pOH 3.9 or pH 10.1, then, provided no supersaturation is present:

$$[\text{Mg}^{2+}] = \frac{K_{s,\text{Mg(OH)}_2}}{[\text{OH}^-]^2} = \frac{1.5 \times 10^{-11}}{(1.22 \times 10^{-4})^2} = 0.001M$$

i.e. the concentration of the magnesium ions remaining in solution is $0.001M$, or 1 per cent of the magnesium ions would remain unprecipitated. If pOH is changed to 2.9 or pH to 11.1, it can be shown in a similar way that the concentration of the magnesium ions left in solution is $ca\ 1 \times 10^{-5}M$, so that the precipitation error is 0.1 per cent, a negligible quantity; in other words, magnesium is precipitated quantitatively at a pH of 11.1.

Values for the solubility products of metallic hydroxides are, however, not very precise, so that it is not always possible to make exact theoretical calculations. The approximate pH values at which various hydroxides begin to precipitate from dilute solution are collected in Table 11.2.

Table 11.2   pH values at which various hydroxides are precipitated

pH	Metal ion	pH	Metal ion
3	Sn^{2+}, Fe^{3+}, Zr^{4+}	7	Fe^{2+}
4	Th^{4+}	8	Co^{2+}, Ni^{2+}, Cd^{2+}
5	Al^{3+}	9	Ag^+, Mn^{2+}, Hg^{2+}
6	Zn^{2+}, Cu^{2+}, Cr^{3+}	11	Mg^{2+}

The precipitated metallic hydroxides or hydrated oxides are gelatinous in character, and they tend to be contaminated with anions by adsorption and occlusion, and sometimes with basic salts. The values presented in Table 11.2 suggest that many separations should be possible by fractional precipitation of the hydroxides, but such separations are not always practical owing to high local concentrations of base when the solution is treated with alkali. Such unequal concentrations of base result in regions of high local pH and lead to the precipitation of more soluble hydroxides, which may be occluded in the desired precipitate. Slow, or preferably homogeneous, precipitation overcomes this difficulty, and much sharper separations may be achieved.

The common tripositive cations may be separated from many dipositive cations by the 'basic acetate' or 'basic benzoate' method. These separations are based upon the fact that the equilibria for the first dissociation of the typical ions are:

$$[M(H_2O)_x]^{3+} + H_2O \rightleftharpoons [M(H_2O)_{x-1}(OH)]^{2+} + H_3O^+$$
$$(K = 5 \times 10^{-3} - 1 \times 10^{-5})$$

$$[M(H_2O)_y]^{2+} + H_2O \rightleftharpoons [M(H_2O)_{y-1}(OH)]^+ + H_3O^+$$
$$(K = 10^{-7} - 10^{-12})$$

Any strong acid that may be present is first neutralised. Then, by selecting an appropriate base, whose conjugate acid has a K_a of about 10^{-5}, the equilibrium for the tripositive cations will be forced to the right; the base is too weak, however, to remove the hydroxonium ions from the equilibrium of the dipositive cations. Since a large excess of the basic ion is added, a basic salt of the tripositive metal usually precipitates instead of the normal hydroxide. Acetate or benzoate ions (in the form of the sodium salts) are the most common bases that are employed for this procedure. The precipitation of basic salts may be combined with precipitation from homogeneous solution, and thus very satisfactory separations may be obtained.

11.11 ORGANIC PRECIPITANTS

Separation of one or more metal ions from mixtures may be made with the aid of organic reagents, with which they yield sparingly soluble and often coloured compounds. These compounds usually have high relative molecular masses, so that a small amount of the ions will yield a relatively large amount of the precipitate. The ideal organic precipitant should be *specific* in character, i.e. it should give a precipitate with only one particular ion. In few cases, however, has this ideal been attained; it is more usual to find that the organic reagent will react with a group of ions, but frequently by a rigorous control of the experimental conditions it is possible to precipitate only one of the ions of the group. Sometimes the precipitated compound may be weighed after drying at a suitable temperature; in other cases the composition is not quite definite and the substance is converted by ignition to the oxide of the metal; in a few instances, a titrimetric method is employed which utilises the quantitatively precipitated organic complex.

The original work in this field was largely empirical in character and was directed towards a search for specific, or at least highly selective, reagents for particular metal ions. A more fundamental approach is to consider the factors which lead to selectivity and to give quantitative consideration to the equilibria involved. Frequently sufficient selectivity can be achieved for a particular purpose by controlling such variables as the concentration of the reagent and the pH, and also by taking advantage of secondary complexing agents (masking agents — see Section 10.47).

It is difficult to give a rigid classification of the numerous organic reagents. The most important, however, are those which form chelate complexes, which involve the formation of one or more (usually five- or six-membered) rings incorporating the metal ion; ring formation leads to a relatively great stability. One classification of such reagents is concerned with the number of hydrogen ions displaced from a neutral molecule in forming one chelate ring. A guide to the applicability of organic reagents for analytical purposes may be obtained from a study of the formation constant of the coordination compound (which is a measure of its stability), the effect of the nature of the metallic ion and of the ligand on the stability of complexes, and of the precipitation equilibria involved, particularly in the production of uncharged chelates. For further details, the reader is referred to Sections 2.22–2.25 and to the books on chelate compounds listed in the Bibliography at the end of Part D. Selected examples of precipitation reagents follow, those of a similar chemical type being grouped together as far as possible (see Ref. 11, Part C, for a more comprehensive account of organic analytical reagents).

A. Dimethylglyoxime. This reagent (I) was discovered by L. Tschugaeff and was applied by O. Brunck for the determination of nickel in steel. It gives a bright red precipitate (II) of $Ni(C_4H_7O_2N_2)_2$ with nickel salt solutions; precipitation is usually carried out in ammoniacal solution or in a buffer solution containing ammonium acetate and acetic acid. The complex is weighed after drying at 110–120 °C. A slight excess of the reagent exerts no action on the precipitate, but a large excess should be avoided because of: (a) the possible precipitation of the dimethylglyoxime itself due to its low solubility in water (it is used in ethanolic solution),* and (b) the increased solubility of the precipitate in water–ethanol mixtures.* The interference of iron(III), aluminium, or bismuth is prevented by the addition of a soluble tartrate or citrate; when much cobalt, zinc, or manganese is present, precipitation should take place in a sodium acetate, rather than an ammonium acetate, buffer.

Solutions of palladium(II) salts give a characteristic yellow precipitate in dilute hydrochloric or sulphuric acid solution; the composition is similar to that of nickel, viz. $Pd(C_4H_7O_2N_2)_2$, and the precipitate can be dried at

---

* These possible errors may be avoided by employing disodium dimethylglyoxime, which is soluble in water — see below.

110–120 °C and weighed. The precipitate is almost insoluble in hot water, but dissolves readily in ammonia and cyanide solutions. Gold is reduced to the metal by the reagent, and platinum (if present in appreciable quantity) is partially precipitated either as a greenish complex compound or as the metal, upon boiling the solution. The precipitation of palladium is not complete in the presence of nitrates.

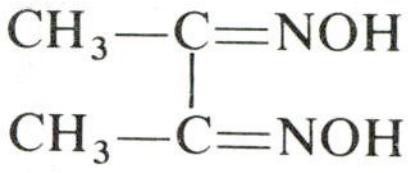

$$CH_3-\underset{|}{C}=NOH$$
$$CH_3-C=NOH$$
$$(I)$$

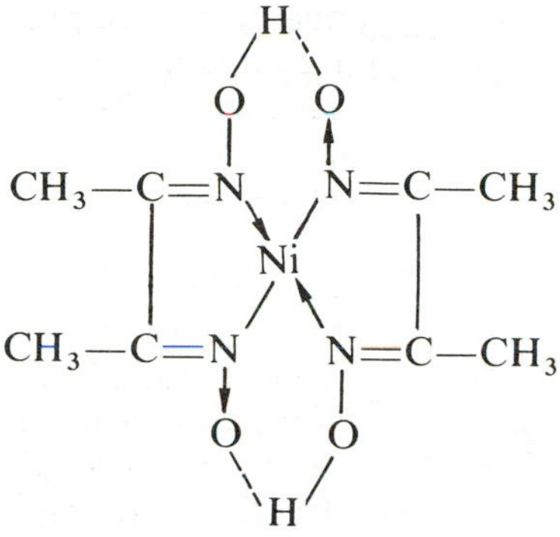

(II)

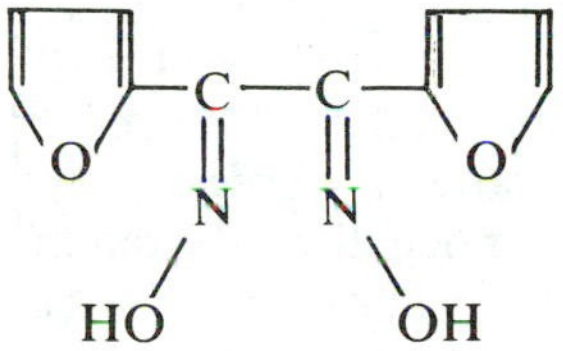

(III)

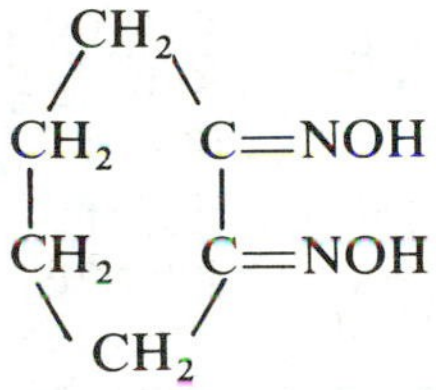

(IV)

Dimethylglyoxime is only slightly soluble in water ($0.40\ \text{g L}^{-1}$); consequently it is employed as a 1 per cent solution in ethanol. The sodium salt of dimethylglyoxime $Na_2C_4H_6O_2N_2,8H_2O$ is available commercially: this is soluble in water and may be employed as 2–3 per cent aqueous solution.

Furil-α-dioxime (III) has also been proposed for the determination of nickel. It gives a red precipitate with nickel salts in ammoniacal solution. The complex is less soluble than nickel dimethylglyoxime, and has a smaller nickel content, thus giving a larger weight of precipitate for a given weight of nickel. The great advantage of furil-α-dioxime is its solubility in water, which precludes the possibility of contaminating the precipitate of the nickel derivative with the free reagent. A 2 per cent aqueous solution is normally used. The reagent is, however, expensive.

Cyclohexane-1,2-dione dioxime (nioxime) (IV) is more soluble in water ($8.2\ \text{g L}^{-1}$ at 21 °C) than is dimethylglyoxime: it is an excellent reagent for the gravimetric determination of palladium.

B. Cupferron (ammonium salt of *N*-nitroso-*N*-phenylhydroxylamine) (V).* This reagent, the ammonium salt of nitrosophenylhydroxylamine, forms insoluble compounds with a number of metals in both weakly acid and strongly acid solutions. It is most useful when employed in strongly acid solutions (5–10 per cent by volume of hydrochloric or sulphuric acid) and then precipitates iron(III), vanadium(V), titanium(IV), zirconium(IV), cerium(IV), niobium(V), tantalum(V), tungsten(VI), gallium(III) and tin(IV), separating these elements from aluminium, beryllium, chromium, manganese, nickel, cobalt, zinc, uranium(VI), calcium, strontium and barium. The presence of tartrate and oxalate has no effect upon the precipitation of metals by cupferron.

(V)

The cupferron method is very satisfactory for the separation of iron, titanium, zirconium, vanadium and, in special cases, tin, tantalum, uranium, and gallium.

The reagent is usually employed as a 6 per cent aqueous solution; this should be freshly prepared, since it does not keep satisfactorily for more than a few days. The solid reagent should be stored in amber bottles containing a few lumps of ammonium carbonate. Precipitation is always carried out in the cold, since cupferron is decomposed into nitrosobenzene on heating. Sufficient reagent is added to form the curdy precipitate of the metallic derivative of cupferron and to give a white flocculent precipitate of free nitrosophenylhydroxylamine (needles). Precipitates should be filtered as soon after their formation as possible, since excess of cupferron is not very stable in acid solution. Nitric acid solutions cannot be used for the precipitation, since oxidising agents destroy the reagent. The addition of macerated filter paper assists the filtration of the precipitate and also the subsequent gradual ignition. A disadvantage of the reagent is that the precipitates cannot be weighed after drying, but must be ignited to the corresponding oxide and weighed in this form. The ignition must be done cautiously in a *large* crucible with a gradual increase in temperature to avoid mechanical loss.

N-Benzoyl-*N*-phenylhydroxylamine, $C_6H_5CO(C_6H_5)NOH$, has been employed as a reagent similar to cupferron in its reactions, but it is more stable. The reagent is moderately soluble in hot water but easily soluble in ethanol and other organic solvents. The Cu(II), Fe(III) and Al complexes can be weighed as such [e.g. as $Cu(C_{13}H_{10}O_2N)_2$] but the Ti compound must be ignited to the oxide.

---

* The name cupferron was assigned to the compound by O. Baudisch, and is derived from the fact that the reagent precipitates both copper and iron. Cupferron precipitates iron completely in strong mineral-acid solution, and copper is only quantitatively precipitated in faintly acid solution. The selectivity of the reagent is greatest in strongly acid solution.

C. 8-Hydroxyquinoline (oxine) (VI). Oxine (C_9H_7ON) forms sparingly soluble derivatives with metallic ions, which have the composition $M(C_9H_6ON)_2$ if the coordination number of the metal is four (e.g. magnesium, zinc, copper, cadmium, lead, and indium), $M(C_9H_6ON)_3$ if the coordination number is six (e.g. aluminium, iron, bismuth, and gallium), and $M(C_9H_6ON)_4$ if the coordination number is eight (e.g. thorium and zirconium). There are, however, some exceptions, for example $TiO(C_9H_6ON)_2$, $MnO_2(C_9H_6ON)_2$, $WO_2(C_9H_6ON)_2$, and $UO_2(C_9H_6ON)_2$. By proper control of the pH of the solution, by the use of complex-forming reagents and by other methods, numerous separations may be carried out: thus aluminium may be separated from beryllium in an ammonium acetate–acetic acid buffer, and magnesium from the alkaline-earth metals in ammoniacal buffer. The pH values, extracted from the literature, for the quantitative precipitation of metal oxinates are collected in Table 11.3.

(VI)

8-Hydroxyquinoline is an almost colourless, crystalline solid, m.p. 75–76 °C; it is almost insoluble in water. The reagent is prepared for use in either of the following ways.

(a) Two grams of oxine are dissolved in 100 mL of 2M acetic acid, and ammonia solution is added dropwise until a turbidity begins to form; the solution is clarified by the addition of a little acetic (ethanoic) acid. This solution is stable for long periods, particularly if it is kept in an amber bottle.

(b) Two grams of oxine are dissolved in 100 mL of methanol or ethanol (this reagent cannot be used for the determination of aluminium) or in acetone. The solution is stable for about ten days if protected from light. It is stated that the alcoholic solution may be employed in cases where precipitation occurs at a high pH, and the acetic acid solution for precipitations at low pH.

Table 11.3 pH range for precipitation of metal oxinates

Metal	pH		Metal	pH	
	Initial precipitation	Complete precipitation		Initial precipitation	Complete precipitation
Aluminium	2.9	4.7– 9.8	Manganese	4.3	5.9– 9.5
Bismuth	3.7	5.2– 9.4	Molybdenum	2.0	3.6– 7.3
Cadmium	4.5	5.5–13.2	Nickel	3.5	4.6–10.0
Calcium	6.8	9.2–12.7	Thorium	3.9	4.4– 8.8
Cobalt	3.6	4.9–11.6	Titanium	3.6	4.8– 8.6
Copper	3.0	> 3.3	Tungsten	3.5	5.0– 5.7
Iron(III)	2.5	4.1–11.2	Uranium	3.7	4.9– 9.3
Lead	4.8	8.4–12.3	Vanadium	1.4	2.7– 6.1
Magnesium	7.0	> 8.7	Zinc	3.3	> 4.4

The following general conditions for conducting precipitations with 8-hydroxyquinoline may be given.

1. The reagent is added to the cold solution (or frequently at 50–60 °C) until the yellow or orange–yellow colour of the supernatant liquid indicates that a small but definite excess is present.
2. The precipitate is coagulated by a short period of heating at a temperature not exceeding 70 °C.
3. The precipitate may be filtered through paper or any variety of filtering crucible.
4. The filtrate should possess a yellow or orange colour, indicating the presence of excess of precipitant. If a turbidity appears, a portion should be heated; if the turbidity disappears, it may be assumed to be due to excess of reagent crystallising out, and is harmless. Otherwise, more reagent should be added, and the solution filtered again.
5. Washing of the precipitate may often be effected with hot or cold water (according to the solubility of the metal 'oxinate') and is continued until the filtrates become colourless. The use of ethanol is permissible if it is known to have no effect upon the precipitate.
6. The washed precipitate may be dried at 105–110 °C (usually hydrated 'oxinate') or at 130–140 °C (anhydrous 'oxinate'). In cases where prolonged heating at 130–140 °C is required, slight decomposition may occur. Frequently, ignition to the oxide yields a more suitable form for weighing, but care must be exercised to prevent loss, since many 'oxinates' are appreciably volatile; it is usually best to cover the complex with oxalic acid (1–3 g) and heat gradually. The determination may also be completed titrimetrically by dissolving the precipitate in dilute hydrochloric acid and titrating with a standard solution of potassium bromate as detailed in Section 10.134.

D. Benzoin-α-oxime (cupron) (VII). This compound yields a green precipitate, $CuC_{14}H_{11}O_2N$, with copper in dilute ammoniacal solution, which may be dried to constant weight at 100 °C. Ions which are precipitated by aqueous ammonia are kept in solution by the addition of tartrate; the reagent is then specific for copper. Copper may thus be separated from cadmium, lead, nickel, cobalt, zinc, aluminium, and small amounts of iron.

From strongly acidic solutions benzoin-α-oxime precipitates molybdate and tungstate ions quantitatively; chromate, vanadate, niobate, tantalate, and palladium(II) are partially precipitated. The molybdate complex is best ignited at 500–525 °C to MoO_3 before weighing; alternatively, the precipitate may be dissolved in ammonia solution and the molybdenum precipitated as lead molybdate, in which form it is conveniently weighed.

Benzoin-α-oxime is a white, crystalline solid, m.p. 152 °C, which is sparingly soluble in water but fairly soluble in ethanol. The reagent is employed as a 2 per cent solution in ethanol.

C_6H_5—CH—OH

$\quad\quad\quad$ |

C_6H_5—C=NOH

(VII)

E. Nitron (VIII). The strong organic base 1,4-diphenyl-3-phenylamino-1*H*-

1,2,4-triazolium hydroxide, which is named nitron, yields a sparingly soluble crystalline nitrate $C_{20}H_{16}N_4,HNO_3$ in solutions acidified with acetic (ethanoic) acid or sulphuric acid. Perchlorate, perrhenate, tetrafluoroborate, and tungstate also form insoluble salts, and can be determined in a similar manner. Numerous other anions, including bromide, iodide, chlorate, thiocyanate, nitrite, and chromate, interfere, but may easily be removed by preliminary treatment. The results in the presence of chloride are generally high, possibly because of co-precipitation.

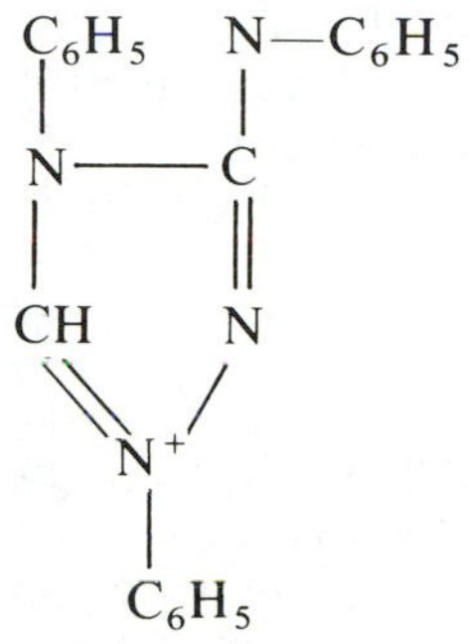

(VIII)

Nitron is a yellow, crystalline solid, m.p. 189 °C, which is insoluble in water. The reagent consists of a 10 per cent solution in 5 per cent acetic acid; it should be filtered, if necessary, and the clear solution protected from light.

F. Anthranilic acid (IX). The sodium salt of anthranilic acid precipitates, in neutral or weakly acid solution, zinc, cadmium, cobalt, nickel, copper, lead, silver, and mercury. Several of these salts, including the anthranilates of cadmium, zinc, nickel, cobalt, and copper, are suitable for the quantitative precipitation and gravimetric determination of these elements; the salts have the general formula $M(C_7H_6O_2N)_2$, and may be dried at 105–110 °C. The precipitations must be carried out at a controlled pH range: in too strongly acidic solutions the precipitates will not form, whilst in too strongly basic solutions the organometallic complexes undergo decomposition. In general, sodium anthranilate is limited in use to the precipitation of a single listed cation from a relatively pure solution in which small amounts of ammonium, alkaline earths, and alkali-metal salts may be present.

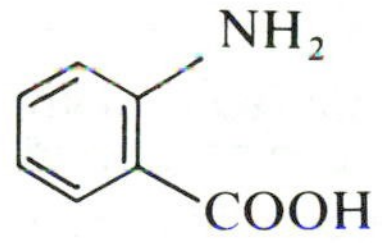

(IX)

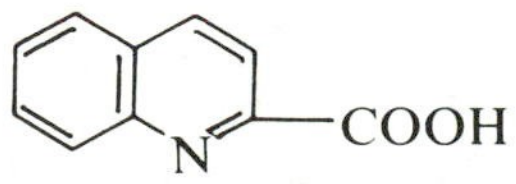

(X)

443

The reagent consists of a 3 per cent aqueous solution of pure sodium anthranilate.

G. Quinaldic acid (X). This organic reagent gives insoluble complexes with copper, cadmium, zinc, manganese, silver, cobalt, nickel, lead, mercury, iron(II), palladium(II), and platinum(II), and insoluble basic salts with iron(III), aluminium, chromium, beryllium, and titanium. The formation of insoluble quinaldates is influenced by the pH of the solution. Thus copper quinaldate $Cu(C_{10}H_6NO_2)_2, H_2O$ (after drying at 110–115 °C) may be precipitated from relatively acidic solutions, whilst under the same conditions the more soluble cadmium and zinc quinaldates remain in solution. Complexing reagents may also assist in rendering the reagent more selective. The quinaldates of copper, cadmium, and zinc are well-defined crystalline salts, which are readily filtered, washed, and dried.

The reagent consists of a 2 per cent aqueous solution of the acid or its sodium salt.

H. 8-Hydroxyquinaldine (XI). The reactions of 8-hydroxyquinaldine are, in general, similar to 8-hydroxyquinoline described under (C) above, but unlike the latter it does not produce an insoluble complex with aluminium. In acetic acid–acetate solution precipitates are formed with bismuth, cadmium, copper, iron(II) and iron(III), chromium, manganese, nickel, silver, zinc, titanium (TiO^{2+}), molybdate, tungstate, and vanadate. The same ions are precipitated in ammoniacal solution with the exception of molybdate, tungstate, and vanadate, but with the addition of lead, calcium, strontium, and magnesium; aluminium is not precipitated, but tartrate must be added to prevent the separation of aluminium hydroxide.

(XI)

8-Hydroxyquinaldine (2-methyloxine) is a pale yellow, crystalline solid, m.p. 72 °C; it is insoluble in water, but readily soluble in hot ethanol, benzene, and diethyl ether. The reagent is prepared by dissolving 5 g of 8-hydroxyquinaldine in 12 g of glacial acetic acid and diluting to 100 mL with water: the solution is stable for about a week.

I. Sodium tetraphenylborate $Na^+[B(C_6H_5)_4]^-$. This is a useful reagent for potassium: the solubility product of the potassium salt is 2.25×10^{-8}. Precipitation is usually effected at pH 2 or at pH 6.5 in the presence of EDTA. Rubidium and caesium interfere; ammonium ion forms a slightly soluble salt and can be removed by ignition; mercury(II) interferes in acid solution but does not do so at pH 6.5 in the presence of EDTA.

11.12 VOLATILISATION OR EVOLUTION METHODS

Evolution or volatilisation methods depend essentially upon the removal of volatile constituents. This may be effected in several ways: (1) by simple ignition in air or in a current of an indifferent gas; (2) by treatment with some chemical reagent whereby the desired constituent is rendered volatile; and (3) by treatment with a chemical reagent whereby the desired constituent is rendered non-volatile. The volatilised substance may be absorbed in a weighed quantity of a suitable medium when the estimation is a *direct* one, or the weight of the residue remaining after the volatilisation of a component is determined, and the proportion of the constituent calculated from the loss in weight; the latter is the *indirect* method. Examples of each of these procedures are given in the following paragraphs; full experimental details will be found later in this chapter.

The determination of superficially bound moisture or of water of crystallisation in hydrated compounds may be carried out simply by heating the substance to a suitable temperature and weighing the residue (see Section 16.35, for a method involving the use of the Karl Fischer reagent). Substances that decompose upon heating can be studied more fully by thermal analysis (Section 11.9). The water may also be absorbed in a weighed quantity of an appropriate drying agent, such as anhydrous calcium chloride or magnesium perchlorate.

The determination of carbon dioxide in carbonate-containing materials may be effected by treating the sample with excess of acid and absorbing the carbon dioxide in an alkaline absorbent, such as soda-lime or 'Carbosorb'.* The gas is completely expelled by heating the solution and by passing a current of purified air through the apparatus; it is, of course, led through a drying agent to remove water vapour before passing to the carbon dioxide absorption apparatus. The gain in weight of the latter is due to carbon dioxide.

In the determination of carbon in steels and alloys, the substance is burnt in pure oxygen in the presence of catalysts and the carbon dioxide absorbed as in the previous example. Precautions are taken to remove other volatile constituents such as sulphur dioxide. This method may be employed in the determination of carbon and hydrogen in organic compounds; the sample is burnt in a controlled stream of oxygen, and the water and carbon dioxide are absorbed separately in an appropriate absorbent, e.g. in calcium chloride saturated with carbon dioxide and in soda-lime. Elemental analysis of organic compounds is, however, now generally carried out using gas chromatography (see Section 9.5).

Another example is the determination of pure silica in an impure ignited silica residue. The latter is treated in a platinum crucible with a mixture of sulphuric and hydrofluoric acids; the silica is converted into the volatile silicon tetrafluoride:

$$SiO_2 + 4HF \rightleftharpoons SiF_4\uparrow + 2H_2O$$

The residue consists of the impurities, and the loss in weight of the crucible gives the amount of pure silica present, provided that the contaminants are in the same form before and after the hydrofluoric acid treatment and are not volatilised in the operation. Although silicon is not the only element that forms a volatile fluoride, it is by far the most abundant and most often encountered element; consequently the volatilisation method of separation is generally satisfactory.

---

* The self-indicating granules of 'Carbosorb' show when the absorbent is exhausted.

PRACTICAL GRAVIMETRIC ANALYSIS

11.13   INTRODUCTION

The student should, of course, become familiar with the general technique of gravimetric analysis before commencing experimental work. Sections 3.32–3.37 in particular should be carefully studied, along with Sections 3.1 for an introduction to laboratory working and 3.5 for care and use of balances. It is, however, essential to emphasise here the importance of careful working at all stages in a gravimetric determination and of performing each determination in duplicate.

For all gravimetric determinations described in this chapter, the phrase 'Allow to cool in a desiccator' should be interpreted as cooling the crucible, etc., **provided with a well-fitting cover** in a desiccator. The crucible, etc., should be weighed as soon as it has acquired the laboratory temperature (for a detailed discussion, see Section 3.22).

In the succeeding sections an account of the gravimetric determination of cations, arranged in alphabetical order, is followed by a similar arrangement for anions. One selected procedure is given in detail for each cation and anion, the essential features of other procedures being summarised in Sections 11.52 and 11.77 respectively. The determinations listed below are particularly suitable for students seeking to gain experience in the technique of gravimetric analysis.

1. Determination of aluminium as the 8-hydroxyquinolate (Section 11.14).
2. Determination of bismuth as oxyiodide (Section 11.20).
3. Determination of copper as copper(I) thiocyanate (Section 11.26).
4. Determination of lead as chromate (Section 11.29).
5. Determination of nickel as the dimethylglyoximate (Section 11.35).
6. Determination of strontium as strontium hydrogenphosphate (Section 11.42).
7. Determination of zinc as 8-hydroxyquinaldinate (Section 11.50).
8. Determination of chloride as silver chloride (Section 11.57).
9. Determination of fluoride as lead chlorofluoride (Section 11.59).
10. Determination of sulphate as barium sulphate (Section 11.72).

CATIONS

11.14   DETERMINATION OF ALUMINIUM AS THE 8-HYDROXYQUINOLATE, $Al(C_9H_6ON)_3$, WITH PRECIPITATION FROM HOMOGENEOUS SOLUTION

Discussion. Some of the details of this method have already been given in Section 11.11(C). This procedure separates aluminium from beryllium, the alkaline earths, magnesium, and phosphate. For the gravimetric determination a 2 per cent or 5 per cent solution of oxine in $2M$ acetic acid may be used: 1 mL of the latter solution is sufficient to precipitate 3 mg of aluminium. For practice in this determination, use about 0.40 g, accurately weighed, of aluminium ammonium sulphate. Dissolve it in 100 mL of water, heat to 70–80 °C, add the appropriate volume of the oxine reagent, and (if a precipitate has not already formed) slowly introduce $2M$ ammonium acetate solution until a precipitate just appears, heat to boiling, and then add 25 mL of $2M$ ammonium acetate solution dropwise and with constant stirring (to ensure complete precipitation).

If the supernatant liquid is yellow, enough oxine reagent has been added. Allow to cool, and collect the precipitated aluminium 'oxinate' on a weighed sintered glass (porosity No. 4) or porous porcelain filtering crucible, and wash well with cold water. Dry to constant weight at 130–140 °C. Weigh as $Al(C_9H_6ON)_3$.

Precipitation may also be effected from homogeneous solution. The experimental conditions must be carefully controlled. The solution containing 25–50 mg of aluminium should also contain 1.25–2.0 mL of concentrated hydrochloric acid in a total volume of 150–200 mL. After addition of excess of the oxine reagent, 5 g of urea is added for each 25 mg of aluminium present, and the solution is heated to boiling. The beaker is covered with a clockglass and heated for 2–3 hours at 95 °C. Precipitation is complete when the supernatant liquid, originally greenish-yellow, acquires an orange–yellow colour. The cold solution is filtered through a sintered-glass filtering crucible (porosity No. 3 or 4), the precipitate washed well with cold water, and dried to constant weight at 130 °C.

Procedure. The solution should contain 25–50 mg of aluminium and 1.0–2.0 mL of concentrated hydrochloric acid in a volume of 150–200 mL. For practice in this determination, weigh out accurately about 0.45 g of aluminium ammonium sulphate, dissolve it in water containing about 1.0 mL of concentrated hydrochloric acid, and dilute to about 200 mL. Add 5–6 mL of oxine reagent (a 10 per cent solution in 20 per cent acetic acid) and 5 g of urea. Cover the beaker with a clockglass and heat on an electric hotplate at 95 °C for 2.5 hours. Precipitation is complete when the supernatant liquid, originally greenish-yellow, acquires a pale orange–yellow colour. The precipitate is compact and filters easily. Allow to cool and collect the precipitate in a sintered-glass filtering crucible (porosity No. 3 or 4), wash with a little hot water and finally with cold water. Dry at 130 °C. Weigh as $Al(C_9H_6ON)_3$.

11.15 AMMONIUM

Ammonium may be determined by precipitation with sodium tetraphenylborate as the sparingly soluble ammonium tetraphenylborate $NH_4[B(C_6H_5)_4]$, using a similar procedure to that described for potassium; it is dried at 100 °C. For further details of the reagent, including interferences, notably potassium, rubidium, and caesium, see Section 11.38.

If the ammonium salt is present with other cations and anions, a titrimetric procedure (see Chapter 10) is usually employed.

11.16 ANTIMONY

Antimony pyrogallate, $Sb(C_6H_5O_3)$. Antimony(III) salts in the presence of tartrate ions may be quantitatively precipitated with a large excess of aqueous pyrogallol as the dense antimony pyrogallate. The method facilitates a simple separation from arsenic; the latter element may be determined in the filtrate from the precipitation of antimony by direct treatment with hydrogen sulphide.

Procedure. The solution should contain antimony(III) (0.1–0.2 g). Add a slight excess over the calculated quantity of potassium sodium tartrate to avoid the formation of basic salts upon dilution. Dissolve approximately five times the theoretical quantity of pure pyrogallol in 100 mL of air-free water, add this all at once to the antimony solution, and dilute to 250 mL. After 30–60 seconds the clear mixture becomes turbid, and then a dense, cloudy precipitate forms which separates out rapidly. Allow to stand for 2 hours, filter through a weighed sintered glass or porcelain filtering crucible, wash several times with cold water to remove the excess of pyrogallol (50 mL is usually sufficient), dry at 100–105 °C to steady weight. To confirm that no pyrogallol remains wash again with cold water, dry at 100–105 °C, and weigh; repeat the operation until the weight is constant. Weigh as $Sb(C_6H_5O_3)$.

It should be pointed out that the titrimetric methods described for the determination of antimony (Chapter 10) are to be preferred to the gravimetric methods as they are simpler, more rapid, and quite as accurate.

11.17 ARSENIC

Arsenic(III) sulphide, As_2S_3: Discussion. The arsenic must be present as arsenic(III). In this condition [ensured by the addition of, for example, iron(II) sulphate, copper(I) chloride, pyrogallol, or phosphorous(III) acid] arsenic may be separated from other elements by distillation from a hydrochloric acid solution, the temperature of the vapour being held below 108 °C; arsenic trichloride (also germanium chloride, if present) volatilises and is collected in water or in hydrochloric acid.

Procedure. Pass a rapid stream of washed hydrogen sulphide through a solution of the arsenic(III) (Note 1) in $9M$ hydrochloric acid at 15–20 °C. Allow to stand for an hour or two, and filter through a weighed filtering crucible (sintered glass or porcelain) (Note 2). Wash the precipitate with $8M$ hydrochloric acid saturated with hydrogen sulphide, then successively with ethanol, carbon disulphide (to remove any free sulphur which may be present), and ethanol. Dry at 105 °C to constant weight, and weigh as As_2S_3.

Notes. (1) A suitable solution for practice in this determination is prepared by dissolving about 0.3 g of arsenic(III) oxide, accurately weighed, in $9M$ hydrochloric acid.

(2) Sometimes a film of the sulphide adheres to the glass vessel in which precipitation was carried out; this can be dissolved in a little ammonia solution and the sulphide re-precipitated with the acid washing liquor.

11.18 BARIUM

Determination of barium as sulphate: Discussion. This method is most widely employed. The effect of various interfering ions (e.g. calcium, strontium, lead, nitrate, etc., which contaminate the precipitate) is dealt with in Section 11.72. The solubility of barium sulphate in cold water is about $2.5\,mg\,L^{-1}$; it is, however, greater in hot water or in dilute hydrochloric or nitric acid, and less in solutions containing a common ion.

The barium sulphate may be precipitated either by the use of sulphuric acid, or from homogeneous solution by the use of sulphamic acid solution which produces sulphate ions on boiling:

$$NH_2SO_3H + H_2O = NH_4^+ + SO_4^{2-} + H^+$$

Procedure: Precipitation with sulphuric acid. The solution (100 mL) should contain not more than 0.15 g of barium (see Note) and not more than 1 per cent by volume of concentrated hydrochloric acid. Heat to boiling, add a slight excess of hot 0.5M sulphuric acid slowly and with constant stirring. Digest on the steam bath until the precipitate has settled, filter, wash with hot water containing two drops of sulphuric acid per litre, and then with a little water until the acid is removed. Full experimental details of the filtration, washing, and ignition processes (900–1000 °C) are given in Section 11.72. Weigh as $BaSO_4$.

Note. A suitable solution for practice may be prepared by dissolving about 0.3 g of accurately weighed barium chloride in 100 mL water and adding 1 mL of concentrated hydrochloric acid.

Procedure: Precipitation from homogeneous solution (sulphamic acid method). The sample solution may contain up to 100 mg of barium, preferably present as the chloride. A solution prepared from about 0.18 g of accurately weighed barium chloride may be used to obtain experience in the determination. Dilute the solution to about 100 mL; add 1.0 g sulphamic acid. Heat the covered beaker on an electric hotplate at 97–98 °C; continue the heating for 30 minutes after the first turbidity appears. Filter through a weighed porcelain filtering crucible and wash with warm distilled water. Ignite to constant weight at 900 °C (preferably in an electric muffle furnace). Weigh as $BaSO_4$.

11.19 BERYLLIUM

Determination of beryllium by precipitation with ammonia solution and subsequent ignition to beryllium oxide:* Discussion. Beryllium may be determined by precipitation with aqueous ammonia solution in the presence of ammonium chloride or nitrate, and subsequently igniting and weighing as the oxide BeO. The method is not entirely satisfactory owing to the gelatinous nature of the precipitate, its tendency to adhere to the sides of the vessel, and the possibility of adsorption effects.

Beryllium is sometimes precipitated together with aluminium hydroxide, which it resembles in many respects. Separation from aluminium (and also from iron) may be effected by means of oxine. An acetic (ethanoic) acid solution containing ammonium acetate is used; the aluminium and iron are precipitated as oxinates, and the beryllium in the filtrate is then precipitated with ammonia solution. Phosphate must be absent in the initial precipitation of beryllium and aluminium hydroxides.

The precipitation by ammonia solution of such elements as Al, Bi, Cd, Cr, Ca, Cu, Fe, Pb, Mn, Ni, and Zn may be prevented by complexation with EDTA: upon boiling the ammoniacal solution, beryllium hydroxide is precipitated quantitatively.

In all the above methods the element is weighed as the oxide, BeO, which is somewhat hygroscopic [compare aluminium(III) oxide]. The ignited residue, contained in a covered crucible, must be cooled in a desiccator containing concentrated sulphuric acid or phosphorus(V) pentoxide, and weighed immediately it has acquired the laboratory temperature.

Procedure. The beryllium solution (200 mL), prepared with nitric acid or hydrochloric acid and containing about 0.1 g of Be, must be almost neutral and contain no other substance precipitable by ammonia solution. Heat to boiling,

---

* Beryllium and its compounds are toxic and care should be taken to avoid inhalation of dusts or contact with eyes and skin.

449

and add dilute ammonia solution slowly and with constant stirring until present in *very slight* excess. Add a Whatman accelerator or one-half of a Whatman ashless tablet, boil for 1 or 2 minutes, and filter on a Whatman No. 41 or 541 filter paper. Transfer as much of the precipitate as possible by rinsing with hot 2 per cent ammonium nitrate solution. Remove any precipitate adhering to the walls of the beaker by dissolving in the minimum volume of hot, very dilute, nitric acid, heating to boiling, and precipitating as before. Filter through the same paper, and wash thoroughly with the ammonium nitrate solution. Place the paper and precipitate in a weighed silica or platinum crucible, dry, and then slowly decompose the hydroxide by raising the temperature gradually to 700 °C, and finally ignite at about 1000 °C for at least 1 hour. Cool in a covered crucible in a desiccator charged with concentrated sulphuric acid, phosphorus(V) pentoxide, or anhydrous magnesium perchlorate, and weigh immediately when cold as BeO.

In the presence of interfering elements, proceed as follows. Neutralise 80–120 mL of the solution containing 15–25 mg of beryllium with ammonia solution until the hydroxides commence to precipitate. Re-dissolve the precipitate by the addition of a few drops of dilute hydrochloric acid. Add 0.5 g of ammonium chloride and sufficient 0.5M EDTA solution to complex all the heavy elements present. Add a slight excess of dilute ammonia solution, with stirring, boil for 2–3 minutes, add a little ashless filter pulp, filter, and complete the determination as above.

11.20 BISMUTH

Determination of bismuth as oxyiodide: Discussion. The cold bismuth solution, weakly acid with nitric acid, is treated with an excess of potassium iodide when BiI_3 and some $K[BiI_4]$ are formed:

$$Bi(NO_3)_3 + 3KI = BiI_3 + 3KNO_3$$

$$BiI_3 + KI = K[BiI_4]$$

Upon dilution and boiling, bismuth oxyiodide is formed, and is weighed as such after suitable drying.

$$BiI_3 + H_2O = BiOI + 2HI$$
(black)

$$K[BiI_4] + H_2O = BiOI + KI + 2HI$$
(yellow)

A large excess of potassium iodide should be avoided, since the complex salt is not so readily hydrolysed as the tri-iodide. This is an excellent method, because the oxyiodide is precipitated in a form which is very convenient for filtration and weighing.

Procedure. The cold bismuth nitrate solution, containing 0.1–0.15 g of Bi (Note 1), must be slightly acid with nitric acid (Note 2), and occupy a volume of about 20 mL. Add finely powdered solid potassium iodide, slowly and with stirring, until the supernatant liquid above the black precipitate of bismuth tri-iodide is just coloured yellow (due to $K[BiI_4]$). Dilute to 200 mL with boiling water, and boil for a few minutes. The black tri-iodide is converted into the copper-coloured precipitate of the oxyiodide. The supernatant liquid should be colourless; if this is yellow, a further 100 mL of water should be added, and the boiling continued until colourless. Add a few drops of methyl orange indicator, and then sodium acetate solution (25 g L^{-1}) from a burette until the solution is neutral. Filter off the precipitate through a weighed sintered-glass or porcelain filtering crucible, wash with hot water, and dry at 105–110 °C to constant weight. Weigh as BiOI.

Notes. (1) A suitable solution for practice can be obtained by dissolving about 0.15 g of pure bismuth, accurately weighed, in the minimum volume of 1:4 nitric acid. Alternatively, a corresponding amount of bismuth nitrate may be used.

(2) Chloride and bromide should be absent. If the solution is strongly acid with nitric acid, it should be evaporated to dryness on the water bath, and the residue dissolved in a little dilute nitric acid.

11.21 CADMIUM

Determination of cadmium as quinaldate: Discussion. Quinaldic acid or its sodium salt precipitates cadmium quantitatively from acetic (ethanoic) acid or neutral solutions. The precipitate is collected on a sintered-glass crucible, and dried at 125 °C. A determination may be completed in about 90 minutes. For the limitations of the method, see Section 11.11(G).

Procedure. The solution (150 mL) should be neutral or weakly acid with acetic acid, and should contain 0.1–0.15 g of Cd. Heat the solution to boiling, and remove the source of heat. Add the reagent (a 3.3 per cent solution of quinaldic acid or of the sodium salt in water) dropwise with vigorous stirring until present in slight excess. Then neutralise carefully with dilute ammonia solution, and allow the white curdy precipitate to settle. When cold, wash with cold water by decantation, filter through a sintered-glass or porcelain filtering crucible, wash thoroughly with cold water, and dry at 125 °C to constant weight. Weigh as $Cd(C_{10}H_6O_2N)_2$.

11.22 CALCIUM

Determination of calcium as oxalate. Discussion. The calcium is precipitated as calcium oxalate CaC_2O_4,H_2O by treating a hot hydrochloric acid solution with ammonium oxalate, and slowly neutralising with aqueous ammonia solution:

$$Ca^{2+} + C_2O_4^{2-} + H_2O = CaC_2O_4,H_2O$$

The precipitate is washed with dilute ammonium oxalate solution and then weighed in one of the following forms:

1. As CaC_2O_4,H_2O by drying at 100–105 °C for 1–2 hours. This method is not recommended for accurate work, because of the hygroscopic nature of the oxalate and the difficulty of removing the co-precipitated ammonium oxalate at this low temperature. The results are usually 0.5–1 per cent high.
2. As $CaCO_3$ by heating at 475–525 °C in an electric muffle furnace. This is the most satisfactory method, since calcium carbonate is non-hygroscopic.

$$CaC_2O_4 = CaCO_3 + CO$$

3. As CaO by igniting at 1200 °C. This method is widely used, but the resulting calcium oxide has a comparatively small relative molecular mass and is hygroscopic; precautions must therefore be taken to prevent absorption of moisture (and of carbon dioxide).

$$CaCO_3 = CaO + CO_2$$

Calcium oxalate monohydrate has a solubility of 0.0067 g and 0.0140 g L^{-1} at 25° and 95 °C respectively. The solubility is less in neutral solutions containing moderate concentrations of ammonium oxalate owing to the common-ion effect (Section 2.7); hence a dilute solution of ammonium oxalate is employed as the wash liquid in the gravimetric determination.

Procedure. Weigh out accurately sufficient of the sample to contain 0.2 g of calcium* into a 500 mL beaker covered with a clockglass and provided with a stirring rod. Add 10 mL of water, followed by about 15 mL of dilute hydrochloric acid (1:1). Heat the mixture until the solid has dissolved, and boil gently for several minutes in order to expel carbon dioxide. Rinse down the sides of the beaker and the clockglass, and dilute to 200 mL: add 2 drops of methyl red indicator. Heat the solution to boiling, and add very slowly a warm solution of 2.0 g of ammonium oxalate in 50 mL of water. Add to the resultant hot solution (about 80 °C) filtered dilute ammonia solution (1:1) dropwise and with stirring until the mixture is neutral or faintly alkaline (colour change from red to yellow). Allow the solution to stand without further heating for at least an hour. After the precipitate has settled, test the solution for complete precipitation with a few drops of ammonium oxalate solution. The subsequent procedure will depend on whether the calcium oxalate is to be weighed as the carbonate or as the oxide.

Weighing as calcium carbonate. Decant the clear supernatant liquid through a weighed silica or porcelain filtering crucible. Transfer the precipitate to the crucible with a jet of water from the wash bottle; any precipitate adhering to the beaker or to the stirring rod is transferred with the aid of a rubber-tipped rod ('policeman'). Wash the precipitate with a cold, very dilute, ammonium oxalate solution (0.1–0.2 per cent) at least five times, or until the washings give no test for chloride ion (add dilute nitric acid and a few drops of silver nitrate solution to 5 mL of the washings). Dry the precipitate in the steam oven or at 100–120 °C for 1 hour, and then transfer to an electrically heated muffle furnace, maintained at 500 ± 25 °C for 2 hours. Cool the crucible and contents in a desiccator, and weigh. Further heating at 500 °C should not affect the weight. As a final precaution, moisten the precipitate with a few drops of saturated ammonium carbonate solution, dry at 110 °C, and weigh again. A gain in weight indicates that some oxide was present; this should not occur.

$$CaO + (NH_4)_2CO_3 = CaCO_3 + 2NH_3 + H_2O$$

Weighing as calcium oxide. Decant the clear supernatant liquid through a Whatman No. 40 or 540 filter paper, transfer the precipitate to the filter, and wash with a cold 0.1–0.2 per cent ammonium oxalate solution until free from chloride. Transfer the moist precipitate to a previously ignited and weighed

---

* 0.5 gram of calcium carbonate, or of calcite, which has been finely powdered in an agate mortar and dried at 110–130 °C for 1 hour, is suitable.

platinum crucible, and ignite gently at first over a Bunsen flame and finally for 10–15 minutes with a Meker or Fisher high-temperature burner until two successive weighings do not differ by more than 0.0003 g. The covered crucible and contents are placed in a desiccator containing pure concentrated sulphuric acid or phosphorus(V) pentoxide (but not calcium chloride), and weighed as soon as cold.

Precipitation of calcium oxalate may also be made from *homogeneous solution* either by use of urea as reagent or by the use of dimethyl oxalate. Both procedures lead to satisfactory separations from magnesium.

Procedure. To obtain experience in this method, weigh out accurately about 0.25 g calcium carbonate, dissolve it in 5 mL dilute hydrochloric acid (1:5), and dilute to 150 mL. Adjust the pH to 4.7 with dilute ammonia solution (use a pH meter). Add 100 mL ammonium acetate–acetic (ethanoic) acid buffer (2.5M with respect to each) and 5.0 g pure dimethyl oxalate (see Note). Cover the beaker and heat on a temperature-controlled hotplate at 90 °C for 2.5 hours: stir occasionally. Precipitation usually commences after 10 minutes. As a precautionary measure add, 10 minutes before filtration, 5 mL of a solution containing 0.25 g ammonium oxalate. Cool the solution rapidly to room temperature, filter through a weighed porcelain filtering crucible of medium porosity, and wash with 1 per cent ammonium oxalate solution. Dry the precipitate for 1 hour at 120 °C and then ignite in an electric muffle furnace at 500 °C for 2 hours. Weigh as $CaCO_3$.

Note. Unless pure dimethyl oxalate is used, immediate precipitation of some fine calcium oxalate will occur. Impure dimethyl oxalate should be recrystallised from ethanol and stored in a desiccator.

11.23 CERIUM

Determination of cerium as cerium(IV) iodate and subsequent ignition to cerium(IV) oxide: Discussion. Cerium may be determined as cerium(IV) iodate, $Ce(IO_3)_4$, which is ignited to and weighed as the oxide, CeO_2. Thorium (also titanium and zirconium) must, however, be first removed (see Section 11.44); the method is then applicable in the presence of relatively large quantities of lanthanides. Titrimetric methods (see Section 10.104 to Section 10.109) are generally preferred.

Procedure. The solution should not exceed 50 mL in volume, all metallic elements should be present as nitrates, and the cerium content should not exceed 0.10 g. Treat the solution with half its volume of concentrated nitric acid, and add 0.5 g potassium bromate (to oxidise the cerium). When the latter has dissolved, add ten to fifteen times the theoretical quantity of potassium iodate in nitric acid solution (see Note) slowly and with constant stirring, and allow the precipitated cerium(IV) iodate to settle. When cold, filter the precipitate through a fine filter paper (e.g. Whatman No. 42 or 542), allow to drain, rinse once, and then wash back into the beaker in which precipitation took place by means of a solution containing 0.8 g potassium iodate and 5 mL concentrated nitric acid in 100 mL. Mix thoroughly, collect the precipitate on the same paper, drain, wash back into the beaker with hot water, boil, and treat at once with concentrated nitric acid dropwise until the precipitate just dissolves (20–25 mL of acid are required per 0.1 g of cerium). Add 0.25 g potassium bromate and as much potassium iodate–nitric acid solution as before. When cold, collect the cerium(IV) iodate upon the same filter paper, wash once with the washing solution, return to the beaker, stir with the washing solution, filter again, and wash three times with the same solution. Place the filter paper and precipitate in the same beaker, add 5–8 g oxalic acid and 50 mL water, and heat to boiling. After all the iodine has been expelled, set aside for several hours, filter, wash with cold water, dry, and ignite (at 500–600 °C) to constant weight in a platinum crucible. Weight as CeO_2.

Note. This is prepared by dissolving 5 g of potassium iodate in 167 mL of concentrated nitric acid, and diluting to 500 mL.

11.24 CHROMIUM

Determination of chromium as lead chromate (precipitation from homogeneous solution): Discussion. Use is made of the homogeneous generation of chromate ion produced by the slow oxidation of chromium(III) by bromate at 90–95 °C in the presence of excess of lead nitrate solution and an acetate buffer. The crystals of lead chromate produced are relatively large and easily filtered; the volume of the precipitate is about half that produced by the standard method of precipitation.

$$2Cr^{3+} + BrO_3^- + 5H_2O = 2CrO_4^{2-} + Br^- + 10H^+$$

$$BrO_3^- + 5Br^- + 6H^+ = 3Br_2 + 3H_2O$$

$$Pb^{2+} + CrO_4^{2-} = PbCrO_4$$

Cations forming insoluble chromates, such as those of silver, barium, mercury(I), mercury(II), and bismuth, do not interfere because the acidity is sufficiently high to prevent their precipitation. Bromide ion from the generation may be expected to form insoluble silver bromide, and so it is preferable to separate silver prior to the precipitation. Ammonium salts interfere, owing to competitive oxidation by bromate, and should be removed by treatment with sodium hydroxide.

$$BrO_3^- + 2NH_4^+ = Br^- + N_2 + 2H^+ + 3H_2O$$

Procedure. Use a sample solution containing about 50 mg of chromium(III). Neutralise the solution by the addition of sodium hydroxide solution until a precipitate just begins to form. Add 10 mL acetate buffer solution (6M in acetic acid and 0.6M in sodium acetate), 10 mL lead nitrate solution (3.5 g per 100 mL), and 10 mL potassium bromate solution (2.0 g per 100 mL). Heat to 90–95 °C: after generation (of chromate) and precipitation are complete (about 45 minutes), as shown by the clear supernatant liquid, cool, filter on a weighed sintered-glass or porcelain filtering crucible, wash with a little 0.1 per cent nitric acid, and dry to constant weight at 120 °C. Weigh as $PbCrO_4$.

11.25 COBALT

Determination of cobalt as cobalt tetrathiocyanatomercurate(II) (mercurithiocyanate): Discussion. This method is based upon the fact that cobalt(II) in almost neutral solution forms a blue complex salt $Co[Hg(SCN)_4]$ with a reagent prepared by dissolving 1 mole of mercury(II) chloride and 5 moles of ammonium thiocyanate in water. The precipitate is sparingly soluble in water, soluble in acids and in a large excess of the reagent, soluble in diethyl ether, chloroform, and carbon tetrachloride, and sparingly soluble in absolute ethanol. It may be dried at 100–110 °C. The following elements interfere: copper, cadmium, zinc, iron(II), iron(III), nickel, manganese(II), bismuth, silver and mercury(II); iron(III) may be rendered innocuous by the addition of phosphate.

Procedure The almost neutral sample solution may conveniently contain 35–40 mg of cobalt in 25 mL (Note 1) and be free from the interfering elements mentioned above. Add, with constant stirring, 4.8 mL of the mercury(II) chloride solution (Note 2) followed by 5.2 mL of the ammonium thiocyanate reagent (Note 2). Do not scratch the sides of the beaker with the stirring rod. A dark blue precipitate forms after stirring for 1–3 minutes; continue the stirring for a further 2–3 minutes and allow to stand for 2 hours at room temperature. Collect the precipitate in a weighed sintered-glass (porosity No. 4) or porcelain filtering crucible; use the filtrate to assist the transfer of any residual precipitate in the beaker. Wash the precipitate with 2–3 mL of a dilute solution of the precipitating reagent (Note 3) and finally with 5 mL ice-cold water. Dry at 100 °C. Weigh as $Co[Hg(SCN)_4]$.

Notes. (1) A suitable solution for practice analysis may be prepared by dissolving about 5.0 g of accurately weighed, pure, ammonium cobalt sulphate in water and diluting to 500 mL in a graduated flask. Use 25 mL for each determination.

The cobalt content may be rapidly checked by titration with standard EDTA solution in the presence of xylenol orange as indicator (see Section 10.59).

(2) Solution (i): dissolve 5.4 g finely powdered mercury(II) chloride in 100 mL of distilled water; slight warming may be necessary.

Solution (ii): dissolve 6.0 g ammonium thiocyanate in 100 mL of distilled water.

It is preferable to add the solutions separately to the cobalt solution: a slight excess (up to 10 per cent) of solution (ii) is not harmful. About 1.0 mL of each solution is required for the precipitation of 10 mg of cobalt. The excess of the reagent should not be more than 10–15 per cent owing to the solubility of the precipitate in the ammonium mercurithiocyanate solution.

(3) The washing solution is prepared by adding 1.0 mL each of solutions (i) and (ii) to 100 mL of water.

11.26 COPPER

Determination of copper as copper(I) thiocyanate: Discussion. This is an excellent method, since most thiocyanates of other metals are soluble. Separation may thus be effected from bismuth, cadmium, arsenic, antimony, tin, iron, nickel, cobalt, manganese, and zinc. The addition of 2–3 g of tartaric acid is desirable for the prevention of hydrolysis when bismuth, antimony, or tin is present. Excessive amounts of ammonium salts or of the thiocyanate precipitant should be absent, as should also oxidising agents; the solution should only be slightly acidic, since the solubility of the precipitate increases with decreasing pH. Lead, mercury, the precious metals, selenium, and tellurium interfere and contaminate the precipitate.

The essential experimental conditions are:

1. *slight acidity* of the solution with respect to hydrochloric acid or sulphuric

acid, since the solubility of the precipitate increases appreciably with decreasing pH;

2. the presence of a *reducing agent*, such as sulphurous acid or ammonium hydrogensulphite, to reduce copper(II) to copper(I);

3. a *slight excess of ammonium thiocyanate*, since a large excess increases the solubility of the copper(I) thiocyanate due to the formation of a complex thiocyanate ion;

4. the *absence of oxidising agents*.

The reaction may be represented as:

$$2Cu^{2+} + HSO_3^- + H_2O = 2Cu^+ + HSO_4^- + 2H^+$$

$$Cu^+ + SCN^- = CuSCN$$

The precipitate is curdy (compare silver chloride) and is readily coagulated by boiling. It is washed with dilute ammonium thiocyanate solution: a little sulphurous acid or ammonium hydrogensulphite is added to the wash solution to prevent any oxidation of the copper(I) salt.

Procedure. Weigh out accurately about 0.4 g of the copper salt (Note 1) into a 250 mL beaker, and dissolve it in 50 mL of water. Add a few drops of dilute hydrochloric acid, and then a slight excess (about 20–30 mL are required) of freshly prepared saturated sulphurous acid solution. Alternatively, add 25 mL ammonium hydrogensulphite solution: the latter is prepared by diluting to ten times its volume the commercial concentrated solution, which has a specific gravity of 1.33 and contains about 54 per cent sulphur dioxide. Dilute the cold liquid to 150–200 mL, heat nearly to boiling, and add freshly prepared 10 per cent ammonium thiocyanate solution, slowly and with constant stirring, from a burette until present in *slight* excess. The precipitate of copper(I) thiocyanate should be white; the mother liquor should be colourless and smell of sulphur dioxide. Allow to stand for two hours, but preferably overnight. Filter through a weighed filtering crucible (sintered glass, or porcelain), and wash the precipitate 10 to 15 times with a cold solution prepared by adding to every 100 mL of water 1 mL of a 10 per cent solution of ammonium thiocyanate and 5–6 drops of saturated sulphurous acid solution, and finally several times with 20 per cent ethanol to remove ammonium thiocyanate (Note 2). Dry the precipitate to constant weight at 110–120 °C (Note 3). Weigh as CuSCN.

Notes. (1) Copper sulphate pentahydrate is suitable for practice in this determination: 0.4 g of this contains about 0.1 g of Cu.

(2) Alternatively, but less desirably, the precipitate may be washed with cold water until the filtrate gives only a slight reddish coloration with iron(III) chloride, and finally with 20 per cent ethanol.

(3) The precipitate, collected in a sintered-glass (porosity No. 4) or porcelain filtering crucible, may be weighed more rapidly as follows. Wash the copper(I) thiocyanate five or six times with ethanol, followed by a similar treatment with small volumes of anhydrous diethyl ether, then suck the precipitate dry at the pump for 10 minutes, wipe the outside of the crucible with a clean linen cloth and leave it in a vacuum desiccator for 10 minutes. Weigh as CuSCN.

11.27 GOLD

Determination of gold as the metal: Discussion. Gold is nearly always determined as the metal. The reducing agents generally employed are sulphur dioxide, oxalic acid, and iron(II) sulphate. If nitric acid is present it must be removed by repeated evaporation with concentrated hydrochloric acid, and the solution diluted with water. With sulphurous acid, small amounts of the platinum metals (primarily platinum) may be carried down with the precipitate. It is therefore usually necessary to re-dissolve the solid in dilute *aqua regia* and to re-precipitate the gold; oxalic acid gives a better separation from the platinum metals in the second precipitation, although the precipitate is somewhat finely divided. Iron(II) sulphate gives satisfactory results for gold alone, but difficulties are introduced if the platinum metals are subsequently to be determined. Oxalic acid is slow in its action, and yields a precipitate which is difficult to filter.

The best results are obtained with quinol (1,4-dihydroxybenzene) as the reducing agent. Precipitation in hot 1.2M hydrochloric acid solution is rapid, the gold is readily filtered, and occlusion of the platinum metals is negligible. Precipitation in the cold is complete in 2 hours. Palladium in the filtrate can be precipitated directly with dimethylglyoxime, whilst platinum in the filtrate may be determined either by evaporating to dryness in order to destroy organic matter and then digesting with a little *aqua regia* or by reduction with sodium formate and formic acid.

Gold may also be separated from hydrochloric acid solutions of the platinum metals by extraction with diethyl ether or with ethyl acetate (compare Chapter 6); except in special cases these methods do not offer any special advantages over the reduction to the metal.

Procedure. The solution must be free from nitric acid, be about 1.2M with respect to hydrochloric acid (*ca* 5 mL of concentrated hydrochloric acid in 50 mL of water), and contain up to 0.2 g of Au in 50 mL. Heat the solution to boiling, add excess of 5 per cent aqueous quinol solution (3 mL for every 25 mg of Au), and boil for 20 minutes. Allow to cool, and filter either through a weighed porcelain filtering crucible or through a Whatman No. 42 or 542 filter paper; wash thoroughly with hot water. The small particles of gold remaining in the bottom of the beaker (easily seen with a small flash lamp) are best removed with pieces of ashless filter paper. Ignite the porcelain filtering crucible to constant weight. If filter paper is used, transfer to a weighed porcelain or silica crucible, burn off the paper carefully, and ignite to constant weight. Weigh as Au.

11.28 IRON

Determination of iron as iron(III) oxide by initial formation of basic iron(III) formate: Discussion. The precipitation of iron as iron(III) hydroxide by ammonia solution yields a gelatinous precipitate which is rather difficult to wash and to filter. Iron(III) can, however, be precipitated from homogeneous solution as a dense basic formate by the urea hydrolysis method. The precipitate obtained is more readily filtered and washed and adsorbs fewer impurities than that formed by other hydrolytic procedures. Ignition yields iron(III) oxide.

The pH at which basic iron(III) formate begins to precipitate depends upon several factors, which include the initial iron and chloride concentration: a high concentration of ammonium chloride is essential to prevent colloid formation. It is important to use an optimum initial pH to avoid a large excess of free acid, which would have to be neutralised by urea hydrolysis, and yet there must be present sufficient acid to prevent the formation of a gelatinous precipitate prior to boiling the solution: ideally, a turbidity should appear about 5–10 minutes after the solution has begun to boil. For iron contents of 5 mg to 55 mg per 100 mL, the optimum initial pH is between 2.00 and 1.70. Some reduction occurs during the precipitation (due to the presence of both formate and chloride): re-oxidation of iron(II) to iron(III) is easily effected by the addition of hydrogen peroxide towards the end of the procedure. Precipitation as basic formate enables iron to be separated from manganese(II), cobalt, nickel, copper, zinc, cadmium, magnesium, calcium, and barium. When copper is present the solution must be cooled before hydrogen peroxide is added, otherwise the vigorous decomposition of the hydrogen peroxide may result in loss of some of the solution.

Attention is directed to the fact that if ignition is carried out in a platinum crucible at a temperature above 1100 °C some reduction to the oxide Fe_3O_4 may occur, and at temperatures above 1200 °C some of the oxide may be reduced to the metal and alloy with the platinum. This accounts in part for the contamination of the platinum crucible by iron which sometimes occurs in analytical work. This oxide is not produced if silica crucibles are employed for the ignitions.

Procedure. For practice in this determination ammonium iron(III) sulphate (iron alum) may be used. Weigh out accurately about 1.0 g iron alum, dissolve it in dilute hydrochloric acid, add 2.0 mL formic acid (sp. gr. 1.20; *ca* 90 per cent), 10 g ammonium chloride, and 4.5 g urea (as a 10 per cent aqueous solution); mix well. Dilute to about 350 mL. Adjust the pH to 1.80 (use a pH meter) by the addition of hydrochloric acid.* Dilute to 400 mL, insert a boiling rod, and boil gently for about 90 minutes or until a pH of about 4.0 is reached. Add 5 mL of 3 per cent hydrogen peroxide solution and boil for a further 5 minutes. Then add 10 mL of 0.02 per cent gelatin solution: the latter will improve the filtering and washing properties of the precipitate. Filter on a Whatman No. 40 or 540 filter paper and wash the precipitate 15 times with hot 1 per cent ammonium nitrate solution adjusted to pH 4. Remove as much as possible of the precipitate adhering to the walls of the beaker with the aid of a rubber-tipped stirring rod.

Dissolve any film adhering tenaciously to the walls of the beaker by adding 4–5 mL of concentrated hydrochloric acid, cover with a clockglass, and reflux gently for a few minutes. Then wash the beaker, clockglass, and stirring rod with 25 mL distilled water, add a few drops of methyl red indicator and then dilute ammonia solution dropwise until the colour of the solution is a distinct yellow. Boil for 2–3 minutes to coagulate the precipitate, filter, and wash on a separate filter paper. Place both filter papers in a weighed porcelain or silica crucible, char the filter papers over a small flame, and then ignite at a red heat or in an electric muffle furnace at 850 °C. Heating for 1 hour is usually sufficient. Weigh as Fe_2O_3.

11.29 LEAD

Determination of lead as chromate: Discussion. Although this method is limited in its applicability because of the general insolubility of chromates, it is a useful

---

* Alternatively, add pure aqueous ammonia solution (1:1) by means of a dropper pipette until a definite precipitate of iron(III) hydroxide just begins to form. Now add concentrated hydrochloric acid dropwise, stirring and allowing to stand for a minute or two after each 2–3 drops, until a clear solution is obtained: then add 1.5 mL of concentrated hydrochloric acid and mix well.

procedure for gaining experience in gravimetric analysis. The best results are obtained by precipitating from homogeneous solution utilising the homogeneous generation of chromate ion produced by slow oxidation of chromium(III) by bromate at 90–95 °C in the presence of an acetate buffer. For further details see Section 11.24.

Procedure. Use a sample solution containing 0.1–0.2 g lead. Neutralise the solution by adding sodium hydroxide until a precipitate just begins to form. Add 10 mL acetate buffer solution [6M in acetic (ethanoic) acid and 0.6M in sodium acetate], 10 mL chromium nitrate solution (2.4 g per 100 mL), and 10 mL potassium bromate solution (2.0 g per 100 mL). Heat to 90–95 °C. After generation (of chromate) and precipitation are complete (about 45 minutes) as shown by a clear supernatant liquid, cool, filter through a weighed sintered-glass or porcelain filtering crucible, wash with a little 1 per cent nitric acid, and dry at 120 °C. Weigh as $PbCrO_4$.

11.30 LITHIUM

Determination of lithium as lithium aluminate: Discussion. Lithium may be determined as lithium aluminate by precipitation with excess of sodium aluminate solution in the cold, the final pH of the solution being adjusted to 12.6–13.0. The precipitate is washed with water until free from alkali and weighed as $2Li_2O \cdot 5Al_2O_3$ after heating at 500–550 °C. The solubility in water is 0.008 g L^{-1} at room temperature; it is 0.09 g L^{-1} at pH 12.6.

Procedure. The sample solution (20 mL) may contain up to 10 mg of lithium, and the pH should be about 3.0. Add 40 mL of the cold reagent (see Note) for each 10 mg of lithium. Adjust the pH to 12.6 by the addition of 1M sodium hydroxide solution: use a pH meter. Allow to stand for 30 minutes, and collect the voluminous precipitate in a porcelain filtering crucible. Wash with small volumes of ice-cold water until the washings are no longer alkaline to phenolphthalein. Ignite at 500–550 °C in an electric muffle furnace. Weigh as $2Li_2O \cdot 5Al_2O_3$.

Note. Prepare the precipitating reagent by dissolving 5.0 g aluminium potassium sulphate (potash alum) in 90 mL warm water. Cool and add dropwise with stirring, while cooling in ice, a solution of 2.0 g sodium hydroxide in 5.0 mL water until the initially formed precipitate re-dissolves. After standing for 12 hours, filter, adjust the pH to 12.6, and dilute to 100 mL with water.

11.31 MAGNESIUM

Determination of magnesium as the 8-hydroxyquinolate: Procedure. (Note 1) To a solution containing about 20 mg of magnesium ion in 100 mL of water, add 2 g of ammonium chloride and 0.5 mL of *o*-cresolphthalein indicator (0.2 per cent solution in ethanol). Neutralise with 6M aqueous ammonia solution until a violet colour is obtained (pH about 9.5) and then add 2 mL of the ammonia solution in excess. Heat to 70–80 °C, add very slowly and with constant stirring a 1 per cent solution of 8-hydroxyquinoline (oxine) in 2M acetic acid until a small excess is present as shown by the yellow colour of the supernatant liquid (Note 2). Digest the precipitate on the steam bath for 10 minutes with frequent stirring, and collect the precipitate on a weighed sintered-glass filtering

459

crucible. Wash the precipitate with dilute ammonia solution (1:40) until the washings are colourless, dry to constant weight at 100–110 °C and weigh as Mg $(C_9H_6ON)_2 \cdot 2H_2O$. Alternatively the precipitate may be dried at 155–160 °C and weighed as the anhydrous compound $Mg(C_9H_6ON)_2$.

Notes. (1) The procedure is applicable if only alkali metals are present with the magnesium. Numerous ions interfere [Section 11.11(C)].

(2) A large excess of the reagent should be avoided as there is a danger that the 'oxine' itself may precipitate.

11.32 MANGANESE

Determination of manganese as the ammonium phosphate or as the pyrophosphate: Discussion. The only method which is at all widely used for the gravimetric determination of manganese is the precipitation as ammonium manganese phosphate, $MnNH_4PO_4,H_2O$, in slightly ammoniacal solution containing excess of ammonium salts. The precipitate may be weighed in this form after drying at 100–105 °C, or it may be ignited and subsequently weighed as manganese pyrophosphate, $Mn_2P_2O_7$. The latter procedure is by far the better one. The method is, however, of limited application because of the interfering influence of numerous other elements. Titrimetric methods are generally preferred (see Chapter 10); the potentiometric determination of manganese (see Section 15.23) may also be recommended.

Procedure. The solution (200 mL) should be slightly acid, contain not more than 0.2 g of Mn(II) in 200 mL, and no other cations except those of the alkali metals (see Note). Almost neutralise the solution with dilute ammonia solution, add 20 g of ammonium chloride and a considerable excess of diammonium hydrogenphosphate $(NH_4)_2HPO_4$ (say, 2 g of the solid). If a precipitate forms at this point, dissolve it by the addition of a few drops of 1:3 hydrochloric acid. Heat the solution almost to boiling (90–95 °C), and add dilute ammonia solution (1:3) dropwise and with constant stirring until a precipitate $\{Mn_3(PO_4)_2\}$ begins to form; immediately suspend the addition of the alkali. Continue the heating and stirring until the precipitate becomes crystalline $(MnNH_4PO_4,H_2O)$. Then add another drop or two of ammonia solution, stir as before, etc., and so continue until no more precipitate is produced and its silky appearance remains unchanged. The precipitate must be maintained at 90–95 °C throughout; a large excess of ammonia solution must be avoided. Allow the solution to stand at room temperature (or, better, at 0 °C) for 2 hours. Filter through a weighed porcelain filtering crucible, and wash the precipitate with cold, 1 per cent ammonium nitrate solution until free from chloride. Dry at a gentle heat, and then heat at 700–800 °C (in an electric crucible furnace or within a larger nickel crucible) to constant weight. Weigh as $Mn_2P_2O_7$. Alternatively, but less desirably, the precipitate in the porcelain filtering crucible may be dried at 100–105 °C to constant weight and weighed as $MnNH_4PO_4,H_2O$; in this case, a sintered-glass filtering crucible may also be used.

Note. A suitable solution for practice may be prepared by one of the following methods:

(a) Dissolve 0.7 g of accurately weighed manganese(II) sulphate $MnSO_4,4H_2O$ in 200 mL of water.

460

(*b*) Dissolve 0.5 g of accurately weighed potassium permanganate in very dilute sulphuric acid, and reduce the solution with sulphur dioxide or with ethanol. Remove the excess of sulphur dioxide or of acetaldehyde (and ethanol) by boiling. Dilute to 200 mL.

11.33 MERCURY

Determination of mercury as mercury(II) thionalide: Discussion. Thionalide $C_{10}H_7 \cdot NH \cdot CO \cdot CH_2 \cdot SH$ may be used for the quantitative precipitation of mercury(II) as $Hg(C_{12}H_{10}ONS)_2$. Sulphate does not interfere. Attention is drawn to the following experimental points.

1. The chloride ion concentration of the solution should not exceed $0.1M$; the results are high if the chloride ion concentration is excessive.
2. If nitric acid solutions of mercury(II) nitrate are used, the latter must be converted into mercury(II) chloride by the addition of at least an equivalent amount of chloride ion.
3. A three-fold excess of reagent should be employed.

Procedure. The sample solution may contain 5 to 75 mg of mercury(II). A solution, prepared from mercury(II) chloride and containing, say, 20 mg mercury in 150 mL of water, may be used for practice in this determination. Heat the solution to 80–85 °C and add, with constant stirring, a three-fold excess of a 1 per cent solution of thionalide in acetic acid. The precipitate coagulates upon stirring. Filter the hot solution through a sintered-glass filtering crucible (porosity No. 3) which has been preheated by pouring hot water through it. (The use of a warm filtering crucible is essential; the separation of thionalide in the pores of the sintered plate of the crucible, which would render filtration difficult, is thus avoided.) Wash with hot water until free from acid, and dry to constant weight at 105 °C. Weigh as $Hg(C_{12}H_{10}ONS)_2$.

11.34 MOLYBDENUM

Determination of molybdenum with 8-hydroxyquinoline (oxine): Discussion. Molybdates yield sparingly soluble orange–yellow molybdyl 'oxinate' with oxine solution; the pH of the solution should be between the limits 3.3–7.6. The complex differs from other 'oxinates' in being insoluble in organic solvents and in many concentrated inorganic acids. The freshly precipitated compound dissolves only in concentrated sulphuric acid and in hot solutions of caustic alkalis. This determination is of particular interest, as it allows a complete separation of molybdenum from rhenium.

Procedure. Neutralise the solution of alkali molybdate, containing up to 0.1 g of Mo, to methyl red, and then acidify with a few drops of $1M$ sulphuric acid. Add 5 mL $2M$ ammonium acetate, dilute to 50–100 mL, and heat to boiling. Precipitate the molybdenum by the addition of a 3 per cent solution of oxine in dilute acetic acid (see Note), until the supernatant liquid becomes perceptibly yellow. Boil gently and stir for 3 minutes, filter through a filtering crucible (sintered-glass or porcelain), wash with hot water until free from the reagent, and dry to constant weight at 130–140 °C. Weigh as $MoO_2(C_9H_6ON)_2$.

Note. The oxine reagent may be prepared by dissolving 3 g oxine in 8.5 mL of warm glacial acetic acid, pouring into 80 mL water, and diluting to 100 mL.

461

11.35 NICKEL

Determination of nickel as the dimethylglyoximate: Discussion. Nickel is precipitated by the addition of an ethanolic solution of dimethylglyoxime $\{CH_3 \cdot C(:NOH) \cdot C(:NOH) \cdot CH_3$, referred to in what follows as $H_2DMG\}$ to a hot, faintly acid solution of the nickel salt, and then adding a slight excess of aqueous ammonia solution (free from carbonate). The precipitate is washed with cold water and then weighed as nickel dimethylglyoximate after drying at $110-120\,^\circ C$. With large precipitates, or in work of high accuracy, a temperature of $150\,^\circ C$ should be used: any reagent that may have been carried down by the precipitate is volatilised.

$$Ni^{2+} + 2H_2DMG = Ni(HDMG)_2 + 2H^+$$

For the structure of the complex and further details about the reagent, see Section 11.11(A).

The precipitate is soluble in free mineral acids (even as little as is liberated by reaction in neutral solution), in solutions containing more than 50 per cent of ethanol by volume, in hot water (0.6 mg per 100 mL), and in concentrated ammoniacal solutions of cobalt salts, but is insoluble in dilute ammonia solution, in solutions of ammonium salts, and in dilute acetic (ethanoic) acid–sodium acetate solutions. Large amounts of aqueous ammonia and of cobalt, zinc, or copper retard the precipitation; extra reagent must be added, for these elements consume dimethylglyoxime to form various soluble compounds. Better results are obtained in the presence of cobalt, manganese, or zinc by adding sodium or ammonium acetate to precipitate the complex; iron(III), aluminium, and chromium(III) must, however, be absent.

Dimethylglyoxime forms sparingly soluble compounds with palladium, platinum, and bismuth. Palladium and gold are partially precipitated in weakly ammoniacal solution; in weakly acid solution palladium is quantitatively precipitated and gold partially. Bismuth is precipitated in strongly basic solution. These elements, and indeed all the elements of the hydrogen sulphide group, should be absent. Iron(II) yields a red-coloured soluble complex in ammoniacal solution and leads to high results if much of it is present. Silicon and tungsten interfere only when present in amounts of more than a few milligrams. Iron(III), aluminium, and chromium(III) are rendered inactive by the addition of a soluble tartrate or citrate, with which these elements form complex ions.

Dimethylglyoxime is almost insoluble in water, and is added in the form of a 1 per cent solution in 90% ethanol (rectified spirit) or absolute ethanol; 1 mL of this solution is sufficient for the precipitation of 0.0025 g of nickel. As already pointed out, the reagent is added to a hot feebly acid solution of a nickel salt, and the solution is then rendered faintly ammoniacal. This procedure gives a more easily filterable precipitate than does direct precipitation from cold or from ammoniacal solutions. Only a slight excess of the reagent should be used, since dimethylglyoxime is not very soluble in water or in very dilute ethanol and may precipitate; if a very large excess is added (such that the alcohol content of the solution exceeds 50 per cent), some of the precipitate may dissolve.

Procedure. *A. Nickel in a nickel salt.* Weigh out accurately 0.3–0.4 g of pure ammonium nickel sulphate $(NH_4)_2SO_4 \cdot NiSO_4, 6H_2O$ into a 500 mL beaker provided with a clockglass cover and stirring rod. Dissolve it in water, add 5 mL of dilute hydrochloric acid (1:1) and dilute to 200 mL. Heat to $70-80\,^\circ C$, add a slight excess of the dimethylglyoxime reagent (at least 5 mL for every 10 mg of Ni present), and immediately add dilute ammonia solution dropwise, directly to the solution and not down the beaker wall, and with constant stirring until precipitation takes place, and then in slight excess. Allow to stand on the steam bath for 20–30 minutes, and test the solution for complete precipitation when the red precipitate has settled out. Allow the precipitate to stand for 1 hour, cooling at the same time. Filter the cold solution through a sintered-glass or porcelain filtering crucible, previously heated to 110–120 °C and weighed after cooling in a desiccator. Wash the precipitate with cold water until free from chloride, and dry it at 110–120 °C for 45–50 minutes. Allow to cool in a desiccator and weigh. Repeat the drying until constant weight is attained. Weigh as $Ni(C_4H_7O_2N_2)_2$, which contains 20.32 per cent of Ni.

B. Nickel in nickel steel. Weigh out accurately about 1 g of the drillings or borings of the nickel steel* (or sufficient of the sample to contain 0.03–0.04 g of nickel) into a 100–150 mL beaker or porcelain basin, dissolve it in the minimum volume of concentrated hydrochloric acid (about 20 mL should suffice), and boil with successive additions of concentrated nitric acid (*ca* 5 mL) to ensure complete oxidation of the iron to the iron(III) state. Dilute somewhat, filter, if necessary, from any solid material, and wash the paper with hot water; dilute the filtrate (or solution) to 250 mL in a 400 mL beaker. Add 5 g of citric or tartaric acid, neutralise the solution with dilute aqueous ammonia solution,† and then barely acidify (litmus) with dilute hydrochloric acid. Warm the solution to 60–80 °C, add a slight excess of a 1 per cent ethanolic solution of dimethylglyoxime (20–25 mL), immediately followed by dilute ammonia solution dropwise until the liquid is slightly ammoniacal, stir well, and allow to stand on the steam bath for 20–30 minutes. Allow the solution to stand at least 1 hour and cool to room temperature during this time. Filter off the precipitate through a weighed filtering crucible; test the filtrate for complete precipitation with a little dimethylglyoxime solution, and wash the precipitate with cold water until free from chloride. Dry the precipitate at 100–120 °C for 45–60 minutes, and weigh as $Ni(C_4H_7O_2N_2)_2$.

Calculate the percentage of nickel in the steel.

11.36 PALLADIUM

Determination of palladium with dimethylglyoxime: Discussion. This is one of the best methods for the determination of the element. Gold must be absent, for it precipitates as the metal even from cold solutions. The platinum metals do not, in general, interfere but moderate amounts of platinum may cause a little contamination of the precipitate, and with large amounts a second precipitation is desirable. The precipitate is decomposed by digestion on the water bath with a little *aqua regia*, and diluted with an equal volume of

---

* Bureau of Analysed Samples 'Nickel Steel, No. 222' is suitable: this steel contains about 3.5 per cent of nickel (BCS-CRM 222/1, ECRM 154/1).

† If a precipitate appears or if the solution is not clear when it is rendered ammoniacal, more tartaric or citric acid must be added until a perfectly clear solution is obtained upon adding dilute ammonia solution. Any insoluble matter should be filtered off and washed with hot water containing a little ammonia solution.

water; the resulting solution is largely diluted with water, and the palladium re-precipitated with dimethylglyoxime.

An objection to the precipitation of palladium with dimethylglyoxime is the voluminous character of the precipitate. Hence if much palladium is present, an aliquot part of the solution should be used.

Procedure. The solution should contain not more than 0.1 g Pd in 250 mL, be 0.25M with respect to hydrochloric or nitric acid, and be free from nickel and gold. Add, at room temperature, a 1 per cent solution of dimethylglyoxime in 95 per cent ethanol. Use 2–5 mL of the reagent for every 10 mg of palladium. Allow the solution to stand for 1 hour, and then filter through a weighed filtering crucible (sintered-glass, or porcelain). Test the filtrate with a little of the reagent to make sure that precipitation is complete. Wash the orange–yellow precipitate of palladium dimethylglyoximate thoroughly, first with cold water and then with hot water. Dry at 110 °C to constant weight. Weigh as $Pd(C_4H_7O_2N_2)_2$.

11.37 PLATINUM

Determination as metallic platinum: Discussion. The platinum solution is treated with formic acid, best at pH 6, and the precipitated platinum weighed. A 2 per cent solution of hypophosphorous (phosphinic) acid may also be used as the reducing agent.

Procedure. In this determination any excess of nitric and/or hydrochloric acid present must be removed. Evaporate the solution of platinum, containing no other platinum metals (ruthenium, rhodium, palladium, osmium, and iridium) or gold, to a syrup on the steam bath so as to remove as much hydrochloric acid as possible. If nitric acid was present, dissolve the residue in 5 mL of water, heat on the water bath for a few minutes, add 5 mL of concentrated hydrochloric acid, and again evaporate to a syrupy consistency. Dissolve the residue in water, and dilute so that the solution does not contain more than 0.5 g of Pt in 100 mL. For each 100 mL of solution, add 3 g of anhydrous sodium acetate and 1 mL of formic acid. Heat on the boiling water bath for several hours. Filter through a quantitative filter paper. Add a little more sodium acetate and formic acid to the filtrate and digest in order to ensure complete precipitation. Wash the precipitate with water until free from chloride, dry and ignite the filter paper in contact with the precipitate to constant weight. Weigh as metallic Pt.

11.38 POTASSIUM

Determination of potassium as potassium tetraphenylborate:* Discussion. A solution of sodium tetraphenylborate $Na[B(C_6H_5)_4]$ is probably the best precipitant for potassium, but it is expensive. Precipitation may be effected at a temperature below 20 °C in dilute mineral acid solution (pH 2), in which interference from most foreign ions is negligible. The precipitate is granular and settles readily; it is washed with a saturated aqueous solution of the precipitate (prepared independently), and the potassium tetraphenylborate (TPB) is dried at 120 °C and weighed. The compound decomposes at temperatures above 265 °C. The precipitate is of constant composition $K[B(C_6H_5)_4]$, and is

---

* This method is particularly useful where a high degree of accuracy is not required.

sparingly soluble in water ($\equiv 5.1$ mg L^{-1} of potassium at 20 °C). Very few elements interfere with the determination: these include the ions of silver, mercury(II), thallium(I), rubidium, and caesium; ammonium ion, which forms a slightly soluble salt, can be removed by ignition prior to the addition of the reagent.

Procedure. For practice in this determination, weigh out accurately about 0.10 g potassium chloride and dissolve it in 50 mL distilled water. Add 10 mL of $0.1M$ hydrochloric acid. Then introduce from a burette 40 mL of the sodium TPB reagent (Note 1) slowly (5–10 minutes) and stir continuously. The temperature throughout must be below 20 °C. Allow the precipitate to settle during 1 hour. Collect the precipitate on a sintered-glass filtering crucible (porosity No. 4), wash the precipitate with a small volume (5–10 mL in small portions) of saturated potassium TPB solution (Note 2), and finally with 1–2 mL ice-cold distilled water (Note 3). Dry at 120 °C and cool the covered crucible in a desiccator. Weigh as $K[B(C_6H_5)_4]$.

Notes. (1) Prepare the sodium TPB reagent by dissolving 3.0 g of the solid reagent in 500 mL distilled water in a glass-stoppered bottle. Add about 1 g moist aluminium hydroxide gel, break up the gel if necessary, and shake the suspension for 15 minutes. Filter through a Whatman No. 40 filter paper. Re-filter the first part of the filtrate, if necessary, to ensure a clear filtrate.

(2) Precipitate about 0.1 g potassium (present as potassium chloride in 50 mL water) with 40 mL of the sodium TPB solution added slowly and with constant stirring. Allow to stand for 30 minutes, filter through a sintered-glass filtering crucible, wash with distilled water, and dry for 1 hour at 120 °C. Shake 20–25 mg of the dry precipitate with 200 mL distilled water in a stoppered bottle at 5-minute intervals during 1 hour. Filter through a Whatman No. 40 filter paper, and use the filtrate as the wash liquid.

(3) The reagent is expensive; it is therefore desirable to recover it from potassium TPB precipitates remaining from gravimetric determinations or obtained by adding potassium chloride to filtrates, wash liquids, etc. The potassium TPB is dissolved in acetone and the acetone solution is passed through a strongly acidic cation exchange resin (sodium form); the effluent contains sodium TPB, and is evaporated to dryness on a water bath. The resulting sodium TPB is recrystallised from acetone.

11.39 SELENIUM AND TELLURIUM

Discussion. This gravimetric determination depends upon the separation and weighing as elementary selenium or tellurium (or as tellurium dioxide). Alkali selenites and selenious acid are reduced in hydrochloric acid solution with sulphur dioxide, hydroxylammonium chloride, hydrazinium sulphate or hydrazine hydrate. Alkali selenates and selenic acid are not reduced by sulphur dioxide alone, but are readily reduced by a saturated solution of sulphur dioxide in concentrated hydrochloric acid. In working with selenium it must be remembered that appreciable amounts of the element may be lost on warming strong hydrochloric acid solutions of its compounds: if dilute acid solutions (concentration $<6M$) are heated at temperatures below 100 °C the loss is negligible.

With tellurium, precipitation of the element with sulphur dioxide is slow in dilute hydrochloric acid solution and does not take place at all in the presence of excess of acid; moreover, the precipitated element is so finely divided that it oxidises readily in the subsequent washing process. Satisfactory results are obtained by the use of a mixture of sulphur dioxide and hydrazinium chloride as the reducing agent, and the method is applicable to both tellurites and tellurates. Another method utilises excess of sodium hypophosphite (phosphinate) in the presence of dilute sulphuric acid as the reducing agent.

A process for the gravimetric determination of mixtures of selenium and tellurium is also described. Selenium and tellurium* occur in practice either as the impure elements or as selenides or tellurides. They may be brought into solution by mixing intimately with 2 parts of sodium carbonate and 1 part of potassium nitrate in a nickel crucible, covering with a layer of the mixture, and then heating gradually to fusion. The cold melt is extracted with water, and filtered. The elements are then determined in the filtrate.

A. Determination of selenium: Procedure. The selenium must be present as selenium(IV), and the selenium content of the solution must not exceed 0.25 g per 150 mL. Take an amount of the oxide, selenite, etc., that will contain not more than 0.25 g selenium, and dissolve it in 100 mL concentrated hydrochloric acid. Add, with constant stirring and at not over 25 °C, 50 mL cold concentrated hydrochloric acid that has been saturated with sulphur dioxide at room temperature, allow the solution to stand until the red selenium settles out, filter through a weighed filtering crucible (sintered-glass, or porcelain), wash well successively with cold concentrated hydrochloric acid, cold water until free from chloride, then with ethanol, and finally with diethyl ether. Dry the precipitate for 3–4 hours at 30–40 °C to remove ether, and then to constant weight at 100–110 °C. Weigh as Se.

B. Determination of tellurium: Procedure. The solution should contain not more than 0.2 g tellurium in 50 mL of 3M hydrochloric acid (*ca* 25 per cent by volume of hydrochloric acid). Heat to boiling, add 15 mL of a freshly prepared, saturated solution of sulphur dioxide, then 10 mL of a 15 per cent aqueous solution of hydrazinium chloride, and finally 25 mL more of the saturated solution of sulphur dioxide. Boil until the precipitate settles in an easily filterable form; this should require not more than 5 minutes. Allow to settle, filter through a weighed filtering crucible (sintered-glass, or porcelain), and immediately wash with hot water until free from chloride. Finally wash with ethanol (to remove all water and prevent oxidation), and dry to constant weight at 105 °C. Weigh as Te.

In the alternative method of reduction, which is particularly valuable for the determination of small amounts of tellurium, the procedure is as follows. Treat the solution containing, say, up to about 0.01 g Te in 90 mL with 10 mL of 1:3-sulphuric acid, then add 10 g sodium hypophosphite (phosphinate), and heat on a steam bath for 3 hours. Collect and weigh the precipitated tellurium as above.

C. Determination of mixtures of selenium and tellurium: Procedure. Dissolve the mixed oxides (not exceeding 0.25 g of each) in 100 mL of concentrated hydrochloric acid, and add with constant stirring 50 mL cool concentrated hydrochloric acid which has been saturated with sulphur dioxide at the ordinary temperature. Allow the solution to stand until the *red* selenium has settled, filter through a weighed filtering crucible (sintered-glass or porcelain) and complete

---

* Tellurium and its compounds are toxic and cause irritation to eyes and skin: contact and inhalation should be avoided.

the determination as described in (A). Preserve the filtrate and washings and concentrate to 50 mL on a water bath below 100 °C (above 100 °C tellurium is lost as chloride). Determine tellurium as described under (B).

11.40 SILVER

Determination of silver as chloride: Discussion. The theory of the process is given under Chloride (Section 11.57). Lead, copper(I), palladium(II), mercury(I), and thallium(I) ions interfere, as do cyanides and thiosulphates. If a mercury(I) [or copper(I) or thallium(I)] salt is present, it must be oxidised with concentrated nitric acid before the precipitation of silver; this process also destroys cyanides and thiosulphates. If lead is present, the solution must be diluted so that it contains not more than 0.25 g of the substance in 200 mL, and the hydrochloric acid must be added very slowly. Compounds of bismuth and antimony that hydrolyse in the dilute acid medium used for the complete precipitation of silver must be absent. For possible errors in the weight of silver chloride due to the action of light, see Section 11.57.

Procedure. The solution (200 mL) should contain about 0.1 g of silver and about 1 per cent by volume of nitric acid. Heat to about 70 °C, and add approximately 0.2M pure hydrochloric acid slowly and with constant stirring until no further precipitation occurs; avoid a large excess of the acid. Do not expose the precipitate to too much bright light. Warm until the precipitate settles, allow to cool to about 25 °C, and test the supernatant liquid with a few drops of the acid to be sure that precipitation is complete. Allow the precipitate to settle in a dark place for several hours or, preferably, overnight. Pour the supernatant liquid through a weighed sintered-glass or porcelain filtering crucible, wash the precipitate by decantation with 0.1M nitric acid, transfer the precipitate to the crucible, and wash again with 0.01M nitric acid until free from chloride. Dry the precipitate first at 100 °C and then at 130–150 °C, allow to cool in a desiccator and weigh. Repeat the heating, etc., until constant weight is obtained. Weigh as AgCl.

11.41 SODIUM

Determination of sodium as sodium zinc uranyl acetate:* Discussion. Treatment of a concentrated solution of a sodium salt with a large excess of zinc uranyl acetate reagent results in the precipitation of sodium zinc uranyl acetate. This substance is moderately soluble in water (5.88 g in 100 mL of water at 20 °C) so that a special washing technique must be used. The solubility in a solution containing excess of the reagent is less. About 10 volumes of the reagent are added for each volume of the sample solution, which should not contain more than 8 mg of sodium per mL; precipitation of the triple acetate is usually complete in 1 hour. One milligram of sodium yields 66.88 mg of the triple salt; the latter is relatively bulky, so that the amount of sodium that can be handled in a single determination is limited.

Lithium interferes, since it forms a sparingly soluble triple acetate. Potassium has no effect provided not more than 50 mg mL^{-1} are present. Sulphate must

---

* This method is only of use where a high degree of accuracy is not required.

be absent when potassium is present, for potassium sulphate is sparingly soluble in the reagent. Moderate amounts of ammonium salts, calcium, barium, and magnesium may be tolerated; for larger amounts, a double precipitation is necessary. Phosphates, arsenates, molybdates, oxalates, tartrates, sulphates (in the presence of potassium), and strontium interfere.

Procedure. The neutral or feebly acid sample solution, free from the interfering substances mentioned above, should contain not more than 8 mg of sodium per mL, preferably as chloride. Treat the sample solution (say, 1.5 mL) (Note 1) with 15 mL of zinc uranyl acetate reagent (Note 2), and stir vigorously, preferably mechanically for at least 30 minutes. Allow to stand for 1 hour, and filter through a weighed sintered-glass or porcelain filtering crucible (porosity No. 4). Wash the precipitate four times with 2 mL portions of the precipitating reagent (allow the wash liquid to drain completely before adding the next portion), then ten times with 95 per cent ethanol saturated with sodium zinc uranyl acetate at room temperature (2 mL portions), and finally with a little dry diethyl ether or acetone. Dry for 30 minutes only at $55-60\,°C$. Weigh as $NaZn(UO_2)_3(C_2H_3O_2)_9,6H_2O$.

Notes. (1) A suitable solution for practice may be prepared by evaporating 20.0 mL of $0.02M$ sodium chloride, prepared from the salt, to about 1.5 mL on a water bath.

(2) The reagent is prepared by mixing equal volumes of Solutions A and B and filtering after standing overnight.

Solution A: dissolve 20 g crystallised uranyl acetate $UO_2(C_2H_3O_2)_2,2H_2O$ in 4 mL glacial acetic acid and 100 mL water (warming may be necessary).

Solution B: dissolve 60 g crystallised zinc acetate $Zn(C_2H_3O_2)_2,3H_2O$ in 3 mL glacial acetic acid and 100 mL of water.

11.42 STRONTIUM

Determination of strontium as strontium hydrogenphosphate, $SrHPO_4$: Discussion. Strontium (30–200 mg) may be precipitated as $SrHPO_4$ using potassium dihydrogenphosphate; precipitation commences at pH 4 and is quantitative at pH 5.7–6. The flocculent precipitate soon becomes crystalline. It may be weighed as $SrHPO_4$ after drying at $120\,°C$; alternatively, it may be ignited and weighed as $Sr_2P_2O_7$. Ions which yield insoluble phosphates should be absent; the sodium, potassium, or ammonium ion concentration should not exceed $0.2M$.

Procedure. Treat the sample solution (60 mL, say, prepared by weighing out accurately about 0.15 g pure $SrCl_2,6H_2O$ and dissolving in water) with 10–20 mL $0.5M$ potassium dihydrogenphosphate and heat to the boiling point. Add $1M$ potassium hydroxide from a dropper pipette until an appreciable precipitate is formed. The final pH should be about 6; this can be detected by adding the base until bromocresol purple indicator in the solution just turns purple, or a pH meter may be used. Boil until the initial flocculent precipitate becomes crystalline (30–60 minutes). Allow to stand for 1 hour. Collect the precipitate in a sintered-glass (porosity No. 4) or porcelain filtering crucible; remove any precipitate adhering to the walls of the beaker with a rubber-tipped stirring rod, and wash the precipitate with a little cold water. Dry at $120\,°C$. Weigh as $SrHPO_4$.

11.43 THALLIUM

Determination of thallium as chromate: Discussion. The thallium* must be present in the thallium(I) state. If present as a thallium(III) salt, reduction must be effected (before precipitation) with sulphur dioxide; the excess of sulphur dioxide is boiled off.

Procedure. The solution (100 mL) should contain about 0.1 g of Tl, no excessive amounts of ammonium salts, and no substances that form precipitates with ammonia solution, or reduce potassium chromate, or react with potassium or thallium(I) chromate in ammoniacal solution. Neutralise the thallium solution with dilute ammonia solution (2:1), and add 3 mL in excess. Heat to about 80 °C, and add 2 g of potassium chromate in the form of a 10 per cent solution slowly and with constant stirring. Allow to stand at the laboratory temperature for at least 12 hours. Filter through a weighed filtering crucible (sintered-glass, or porcelain), wash with 1 per cent potassium chromate solution, then sparingly with 50 per cent ethanol, and dry at 120 °C to constant weight. Weigh as Tl_2CrO_4.

11.44 THORIUM

Determination of thorium as sebacate and subsequent ignition to the oxide, ThO_2: Discussion. This procedure permits of the separation by a single precipitation of thorium from relatively large amounts of the lanthanides (Ce, La, Pr, Nd, Sm, Gd) and also from cerium(IV).

Procedure. The solution (100 mL) should be neutral or faintly acid, and contain not more than 0.1 g Th. Heat the solution to boiling and add slowly and with constant stirring a hot, almost saturated, solution of pure sebacic acid in slight excess. The precipitate is voluminous, but granular, and therefore easily manipulated. Filter off immediately, wash thoroughly with hot water, dry, and ignite (use either a Meker or Fisher burner or an electric muffle furnace at 700–800 °C) in a weighed platinum, porcelain, or silica crucible to constant weight. Weigh as ThO_2.

11.45 TIN

Determination of tin with *N*-benzoyl-*N*-phenylhydroxylamine: Discussion. *N*-Benzoyl-*N*-phenylhydroxylamine $C_6H_5CON(OH)C_6H_5$, as a 1 per cent solution in ethanol, precipitates a complex $(C_{13}H_{11}O_2N)_2SnCl_2$, m.p. 171 °C, from tin(IV) solutions containing 1–8 per cent concentrated hydrochloric acid: the complex can be dried at 110 °C. Apparently the reagent reduces tin(IV) to tin(II) and then forms the addition compound. Copper can be quantitatively precipitated by the reagent at pH 3.6–6.0; no interference is encountered from copper, lead, or zinc in precipitating tin from, for example, brass solutions containing 7 per cent by volume of concentrated hydrochloric acid.

N-Benzoyl-*N*-phenylhydroxylamine is a white crystalline solid, m.p. 121 °C:

---

* **Thallium and its compounds are toxic** and cause irritation to eyes and the skin: contact and inhalation should be avoided.

its solubility in water is 0.04 g per 100 mL at 25 °C and 0.5 g per 100 mL at about 80 °C.

The reagent has been used for the determination of copper, iron, and aluminium. The pH ranges for quantitative precipitation are: Cu, 3.6–6.0; Fe, 3.0–5.5; and Al, 3.6–6.4. Incomplete precipitation occurs at a lower pH, and high results are obtained at higher pH values. Titanium must be precipitated below 25 °C and ignited to, and weighed as, the dioxide. Zirconium is also precipitated. Iron and aluminium cannot be precipitated in the presence of phosphate; chromium(III) interferes with the precipitation of iron(III). The following elements do not give precipitates with the reagent at pH 4: bismuth, cadmium, cobalt, manganese, nickel, uranium(IV), and zinc.

Procedure. To the sample solution of tin(IV) chloride (containing 5–20 mg of tin), add 10 mL concentrated hydrochloric acid and dilute to *ca* 600 mL with distilled water. Add from a separatory funnel, dropwise and with constant stirring, 5 mL of a 1 per cent solution of the reagent in ethanol for each 10 mg of tin present plus 8 mL in excess. Cool in an ice bath for 4 hours, filter on a filtering crucible (sintered-glass or porcelain), wash with a few millilitres of ice water, and dry at 110 °C. Weigh as $(C_{13}H_{11}O_2N)_2SnCl_2$.

11.46 TITANIUM

Determination of titanium with tannic acid and phenazone: Discussion. This method affords a separation from iron, aluminium, chromium, manganese, nickel, cobalt, and zinc, and is applicable in the presence of phosphates and silicates. Small quantities of titanium (2–50 mg) may be readily determined.

Procedure. The titanium content of the solution should not exceed 0.1 g of TiO_2, and the titanium should be present as the sulphate or chloride. Add dilute ammonia to the solution until the odour persists, then (cautiously) 10 mL of concentrated sulphuric acid and 40 mL of 10 per cent tannic acid solution. Dilute to 400 mL, stir thoroughly, and cool. Introduce a 20 per cent aqueous solution of 'phenazone' (antipyrine; 2,3-dimethyl-1-phenyl-5-pyrazolone) with constant stirring until an orange–red flocculent precipitate is obtained. Stop the stirring, and continue the addition of the phenazone solution until a white, cheese-like precipitate (produced by the interaction of tannic acid and phenazone) is formed in addition to the red precipitate. Boil the mixture, remove the flame, add 40 g ammonium sulphate, and allow to cool with occasional stirring. Filter the bulky precipitate through a Whatman No. 41 or 541 filter paper, supported on a Whatman filter cone (hardened, No. 51), with slight suction, and wash with a solution of 100 mL water, 3 mL concentrated sulphuric acid, 10 g ammonium sulphate, and 1 g phenazone. Dry the precipitate at 100 °C, transfer to a weighed crucible, heat gently at first, and then ignite at 700–800 °C to constant weight. Weigh as TiO_2.

Note. If the wet precipitate is heated directly, caking occurs which renders the complete oxidation of the carbonaceous matter very slow. If alkali metals were originally present, the ignited oxide must be washed with hot water, filtered, and re-ignited to constant weight.

470

11.47 TUNGSTEN

Determination of tungsten as the trioxide (tannic acid–phenazone method): Discussion. Tungstic acid is incompletely precipitated from solutions of tungstates by tannic acid. If, however, phenazone (2,3-dimethyl-1-phenyl-5-pyrazolone) is added to the cold solution after treatment with excess of tannic acid, precipitation is quantitative. This process effects a separation from aluminium, and also from iron, chromium, manganese, zinc, cobalt, and nickel if a double precipitation is used.

Procedure. The solution of tungstate (200–250 mL) should contain not more than 0.15 g of WO_3, and be faintly ammoniacal. Add 6–7 mL of concentrated sulphuric acid and 7–8 g of ammonium sulphate, and heat to boiling. Treat with 6 mL of 10 per cent aqueous tannic acid solution, keep the mixture on the water bath for a few minutes, and allow to cool to room temperature. A flocculent dark-brown precipitate separates. When cold, stir in 10 mL of a 10 per cent aqueous solution of phenazone. Filter the precipitate through a weighed silica, or porcelain, filtering crucible (Note 1), wash with the special wash liquid (Note 2), and ignite to constant weight at 800–900 °C. Weigh as WO_3.

Notes. (1) The filtrate must be colourless. If it is yellow, insufficient phenazone has been added.

(2) The special wash liquid contains 1 mL concentrated sulphuric acid, 10 g ammonium sulphate, and 0.4 g phenazone in 200 mL of water.

11.48 URANIUM

Determination of uranium with cupferron: Discussion. Cupferron does not react with uranium(VI), but uranium(IV) is quantitatively precipitated. These facts are utilised in the separation of iron, vanadium, titanium, and zirconium from uranium(VI). After precipitation of these elements in acid solution with cupferron, the uranium in the filtrate is reduced to uranium(IV) by means of a Jones reductor and then precipitated with cupferron (thus separating it from aluminium, chromium, manganese, zinc, and phosphate). Ignition of the uranium(IV) cupferron complex affords U_3O_8.

Procedure. If uranium is to be determined in the filtrate from the precipitation of the iron group by cupferron, concentrate the solution to 50 mL, add 20 mL of concentrated nitric acid and 10 mL of concentrated sulphuric acid (if not already present) and evaporate until fumes of sulphur trioxide appear. If organic matter still remains (as shown by the appearance of a dark colour upon evaporation), repeat the treatment with nitric acid. Finally, expel the nitric acid by evaporating to strong fuming, after the addition of a little water. Dilute the solution so that it contains about 6 mL of concentrated sulphuric acid per 100 mL. Cool to room temperature and pass the solution through a Jones reductor (Section 10.138); wash the reductor with 5 per cent sulphuric acid, cool the combined reduced solution and washings to 5–10 °C, and add excess of a freshly prepared 6 per cent solution of cupferron. The precipitate does not usually form until about 5 mL cupferron solution has been added. Introduce a Whatman 'accelerator' or one-quarter of an 'ashless tablet', allow to settle for a few minutes, and filter through a quantitative filter paper. Wash with cold 4 per cent sulphuric acid containing 1.5 g L^{-1} of cupferron. Dry the precipitate

471

at 100 °C, ignite cautiously in a platinum crucible, first at a low temperature and then at 1000 °C, to constant weight. Weigh as U_3O_8.

11.49 VANADIUM

Determination of vanadium as silver vanadate: Discussion. Vanadates are precipitated by excess of silver nitrate solution in the presence of sodium acetate; after boiling, the precipitate consists of silver orthovanadate. The following reactions occur with a solution of a metavanadate:

$$2NaVO_3 + 2CH_3COONa + H_2O \rightleftharpoons Na_4V_2O_7 + 2CH_3COOH$$

$$Na_4V_2O_7 + 4AgNO_3 \rightleftharpoons Ag_4V_2O_7 + 4NaNO_3$$

$$Ag_4V_2O_7 + 2AgNO_3 + 2CH_3COONa + H_2O \rightleftharpoons$$
$$2Ag_3VO_4 + 2CH_3COOH + 2NaNO_3$$

Titrimetric methods (see Chapter 10) are, however, more convenient and less influenced by interfering elements, and are generally preferred.

Procedure. Neutralise the solution (200 mL), containing not more than 0.2 g of alkali vanadate, if acid by aqueous sodium hydroxide, or if alkaline by the addition of nitric acid to the boiling solution until it becomes yellow, followed by decolorisation with dilute ammonia solution. Add 3 g of ammonium acetate, 0.5 mL of concentrated ammonia solution, and then excess of silver nitrate solution, heat to boiling and then keep on a steam bath for 30 minutes. Test for complete precipitation with more silver nitrate solution; if a turbidity is produced, boil the liquid until it becomes clear. Allow the dense brown precipitate of silver vanadate to settle, and collect it on a weighed filtering crucible (sintered-glass, or porcelain), wash with hot water, and dry at 110 °C. Weigh as Ag_3VO_4.

It has been stated that the results obtained by precipitation of vanadate as silver orthovanadate Ag_3VO_4 are not altogether satisfactory. Better results are obtained by precipitation at pH 4.5 as silver metavanadate $AgVO_3$; the precipitate is weighed after drying at 100–105 °C.

11.50 ZINC

Determination of zinc as 8-hydroxyquinaldinate: Discussion. Zinc may be precipitated by 8-hydroxyquinaldine (2-methyloxine) in acetic acid–acetate solution: it can thus be separated from aluminium and magnesium [see Section 11.11(G)]. It can be weighed as $Zn(C_{10}H_8ON)_2$ after drying at 130–140 °C. The co-precipitated reagent is volatile at 130 °C.

Procedure. The solution may contain up to 0.05 g Zn in 200 mL. Add dilute aqueous ammonia solution until a white precipitate of zinc hydroxide just appears. Re-dissolve the zinc hydroxide with a drop of acetic acid. Add a slight excess of the reagent (see Note) (2 mL for each 10 mg of Zn present) and then 2–3 drops of concentrated ammonia solution; the pH should be at least 5.5. Digest the precipitate at 60–80 °C for 15 minutes, allow to stand for 10–20 minutes, and filter through a sintered-glass or porcelain filtering crucible. Dry to constant weight at 130–140 °C. Weigh as $Zn(C_{10}H_8ON)_2$.

If aluminium is present, add 1 g of ammonium tartrate to the clear, slightly acid solution. Introduce the reagent (2 mL for each 10 mg of Zn present), dilute the solution to 200 mL, and heat to 60–80 °C. Neutralise the excess of acid by adding dilute ammonia solution (1:5) dropwise until the complex salt which forms on the addition of each drop just re-dissolves on stirring. Add, with stirring, 45 mL of 2M ammonium acetate solution. The pH should be at least 5.5. Allow the solution to stand for 10–20 minutes, and complete the determination as above.

Note. The reagent is prepared by dissolving 5 g 8-hydroxyquinaldine in 12 g glacial acetic (ethanoic) acid and diluting to 100 mL with water.

11.51 ZIRCONIUM

Determination of zirconium with mandelic acid and subsequent ignition to the dioxide, ZrO_2: Discussion. Zirconium may be precipitated from a hydrochloric acid solution with mandelic acid $(C_6H_5 \cdot CH(OH) \cdot COOH)$ as zirconium mandelate, $Zr(C_8H_7O_3)_4$, which is ignited to and weighed as the dioxide. Quantitative separation is thus made from titanium, iron, vanadium, aluminium, chromium, thorium, cerium, tin, barium, calcium, copper, bismuth, antimony, and cadmium. If sulphuric acid is employed, the concentration should not exceed 5 per cent: higher concentrations give low results.

Procedure. The solution (20–30 mL) may contain 0.05–0.2 g Zr, and should possess a hydrochloric acid content of about 20 per cent by volume. Add 50 mL of 16 per cent aqueous mandelic acid solution and dilute to 100 mL. Raise the temperature slowly to 85 °C and maintain this temperature for 20 minutes. Filter off the resulting precipitate through a quantitative filter paper, wash it with a hot solution containing 2 per cent hydrochloric acid and 5 per cent mandelic acid. Ignite the filter and precipitate to the oxide in the usual manner: a temperature of 900–1000 °C is satisfactory. Weigh as ZrO_2.

Note. Bromomandelic acid is a superior reagent for this determination, but is more expensive. A similar procedure to that above is employed.

11.52 ALTERNATIVE PROCEDURES FOR CATIONS

A summary of some alternative procedures for cations is given here under the headings of the various precipitating reagents used. Section references, given in parentheses, relate to the other procedure already given for that cation.

1. Anthranilic acid. *Cadmium (≯ 100 mg in 150 mL).* Add 3 per cent sodium anthranilate solution to the boiling solution (neutral or weakly acid). Filter and wash the precipitate with 50 mL of diluted reagent (1:20), then with ethanol. Dry at 105–110 °C and weigh as $Cd(C_7H_6O_2N)_2$ (Section 11.21).

Cobalt (II) and zinc. These may be determined in a similar manner (see also Sections 11.25 and 11.50 respectively).

2. Benzoin-α-oxime (cupron). *Copper (II) (≯ 50 mg in ammoniacal solution).* Add 2 per cent ethanolic solution of the reagent to the boiling solution, filter and wash the precipitate with hot dilute ammonia solution (1:100), then with hot water and finally with warm ethanol. Dry at 105–115 °C and weigh as $Cu(C_{14}H_{11}O_2N)$ (Section 11.26).

473

3. Chromate. *Barium (≯ 400 mg in 200 mL of neutral solution).* Add 1 mL of $6M$ acetic (ethanoic) acid and 10 mL of $3M$ ammonium acetate. To the boiling solution add a slight excess of hot 10 per cent ammonium chromate solution. Cool the solution, filter and wash the precipitate with hot water until washings give a negligible reaction with silver nitrate. Dry at 120 °C and weigh as $BaCrO_4$ (Section 11.18).

4. Cupferron. *Tin (IV) (ca 150 mg in 300 mL).* Add 3 g of boric acid, 2.5 mL of concentrated sulphuric acid (**CARE**) and finally an excess of 10 per cent aqueous solution of cupferron. Stir vigorously, filter (Whatman No. 41 or 541 filter paper) and wash the precipitate with cold water. Dry the precipitate and filter paper in a weighed crucible, ignite carefully and weigh as SnO_2 (Section 11.45).

5. Cyclohexane-1,2-dionedioxime (nioxime). *Palladium (II) (5–30 mg in about 200 mL at pH 1–5).* To the hot solution (60 °C) add slowly, with stirring, 0.50 mL of a 0.8 per cent aqueous solution of nioxime for each milligram of palladium. Keep at 60 °C for 30 minutes, filter and wash the precipitate with hot water. Dry at 110 °C and weigh as $Pd(C_6H_9O_2N_2)_2$ (Section 11.36).

6. Hydroxide. *Aluminium (ca 100 mg in 200 mL).* Add 5 g ammonium chloride and a few drops of methyl red indicator, and heat just to boiling. Add pure dilute ammonia solution (1:1) slowly until the solution has a distinct yellow colour. Boil the solution for 1–2 minutes and filter at once through a quantitative filter paper. Wash the precipitate with hot 2 per cent ammonium chloride solution (neutral to methyl red). Dry and char the filter paper in a silica crucible (previously ignited, cooled and weighed), ignite for about 15 minutes, cool and weigh as Al_2O_3 (Section 11.14).

7. 8-Hydroxyquinaldine (2-methyloxine). *Magnesium (≯ 50 mg in 150–200 mL).* Add 3 mL of reagent solution (dissolve 5 g reagent in 12 g glacial acetic acid and dilute to 100 mL with water) for every 10 mg of magnesium present and then add concentrated ammonia solution until the pH is at least 9.3 (or no further precipitate forms). Heat at 60–80 °C for 20 minutes, filter and wash the precipitate with hot water. Dry at 130–140 °C and weigh as $Mg(C_{10}H_8ON)_2$ (Section 11.31).

8. 8-Hydroxyquinoline (oxine). *Iron (III) {ca 30 mg in 25 mL of weakly acidic (HCl) solution}.* Add a solution of 3 g ammonium acetate in 125 mL water, followed by 12–15 mL of reagent solution {2 per cent in $2M$ acetic (ethanoic) acid}. Heat the solution at 80–90 °C for 30 minutes, filter and wash the precipitate with 1 per cent acetic acid and with water. Dry at 130–140 °C and weigh as $Fe(C_9H_6ON)_3$ (Section 11.28).

Uranium (VI) (≯ 300 mg in about 200 mL of 1–2 per cent acetic acid). Add 5 g analytical grade ammonium acetate, heat to boiling and slowly add 4 per cent reagent solution. Heat on a water bath for 5–10 minutes, cool and filter. Wash the precipitate with hot water and then with cold water. Dry at 105–110 °C and weigh as $UO_2(C_9H_6ON)_2 \cdot C_9H_7ON$ (or ignite and weigh as U_3O_8) (Section 11.48).

9. Phosphate. *Bismuth (≯ 100 mg in 100 mL, free from Cl^- and SO_4^{2-}).* Add dilute ammonia solution until a slight permanent precipitate is formed and dissolve this in 2 mL of nitric acid (1:1). To the boiling solution add 30 mL of 10 per cent diammonium hydrogenphosphate slowly and with stirring. Dilute to 400 mL with boiling water and allow to stand for 10–15 minutes on a hotplate. Filter through a weighed porcelain filter crucible and wash the precipitate with 3 per cent ammonium nitrate solution (acidified with a few drops of nitric acid per litre). Dry the precipitate, then heat at 800 °C and weigh as $BiPO_4$ (Section 11.20).

Magnesium (≯ 100 mg in neutral or slightly acid solution). Add 5 mL of concentrated hydrochloric acid, dilute to 150 mL and then add 10 mL of freshly prepared diammonium hydrogenphosphate solution (25 per cent, w/v). Add concentrated ammonia solution slowly with stirring until methyl red indicator turns yellow and then add 5 mL in excess. Allow the solution to stand for at least 4 hours, filter through a porcelain filter crucible, and wash with cold 0.8M aqueous ammonia (in small portions). Dry the crucible and precipitate at about 120 °C, then heat it at 1000–1100 °C and weigh as $Mg_2P_2O_7$ (Section 11.31).

Zirconium (≯ 100 mg in ca 1.8M sulphuric acid solution). Add freshly prepared 10 per cent aqueous diammonium hydrogenphosphate solution in 50–100-fold excess. Dilute to 300 mL, boil for a few minutes, allow to digest on a water bath for 15–30 minutes and cool to about 60 °C. Filter through a quantitative filter paper, wash first with 150 mL of 1M sulphuric acid containing 2.5 g diammonium hydrogenphosphate and then with cold 5 per cent ammonium nitrate solution until the filtrate is sulphate-free. Dry the filter paper and precipitate at 110 °C, place in a platinum crucible and carefully burn off the filter paper. Finally heat at 1000 °C for 1–3 hours and weigh as ZrP_2O_7 (Section 11.51).

10. Pyrogallol. *Bismuth (100–200 mg in 150 mL of weakly nitric acid solution).* Add dilute ammonia solution until a faint permanent turbidity is obtained and then clear the solution by careful addition of a little dilute nitric acid. To the boiling solution add a slight excess of a solution of pure pyrogallol in air-free water. Continue boiling for a short time, test for completeness of precipitation and filter. Wash the precipitate with 0.05M nitric acid and then with water. Dry at 105 °C and weigh as $Bi(C_6H_3O_3)$ (Section 11.20).

11. Quinaldinic acid. *Zinc { ≯ 100 mg in a solution acidified with acetic (ethanoic) acid to pH 3–4}.* Heat to boiling and add 3 per cent sodium quinaldate solution with stirring until precipitation is complete (25 per cent excess should be used). Cool to room temperature, filter and wash the precipitate with cold water and then a little ethanol. Dry at 105–110 °C and weigh as $Zn(C_{10}H_6O_2N)_2,H_2O$ (Section 11.50).

12. Salicylaldoxime. *Copper (50–100 mg in 100 mL solution).* Add 2M sodium hydroxide solution until a slight permanent precipitate is formed and dissolve this by addition of a little dilute acetic acid. Add freshly prepared reagent solution* in slight excess at room temperature, filter and wash the precipitate with water until the washings give no colour with iron(III) chloride. Dry at 100–105 °C and weigh as $Cu(C_7H_6O_2N)_2$ (Section 11.26).

Lead (ca 100 mg in 25 mL solution). Add 10 mL of freshly prepared reagent solution,* dilute to 50 mL and add 12.5 mL of concentrated ammonia solution.

---

* See footnote on page 476.

Stir for 1 hour and then allow the precipitate to settle. Filter by decantation, wash the precipitate until free from excess reagent, dry at 105 °C and weigh as $Pb(C_7H_5O_2N)$ (Section 11.29).

Nickel ($\not> 30$ mg in 100 mL of neutral or very faintly acid solution). Add a slight excess of reagent,* stir thoroughly and filter. Wash the precipitate with cold water until free from excess reagent, dry at 100 °C and weigh as $Ni(C_7H_6O_2N)_2$ (Section 11.35).

13. Sulphide. *Antimony(III) in 100 mL of hydrochloric acid solution (1:4).* Heat to boiling and immediately pass in a rapid stream of washed hydrogen sulphide. Maintain the solution at 90–100 °C and continue passage of the gas until the precipitate is crystalline and black in colour (30–35 minutes). Dilute with an equal volume of water, mix and heat again while the gas is slowly passed into the suspension for some minutes. When the solution is clear, cool and filter (sintered-glass, or porcelain, filter crucible). Wash the precipitate a few times with water, then with ethanol, and draw air through the crucible to dry the precipitate as far as possible. Heat for 2 hours at 100–130 °C in a current of carbon dioxide (crucible and contents placed in a wide glass tube passing through an electric furnace) and a further 2 hours at 280–300 °C. Cool in a slow stream of carbon dioxide, place in a desiccator for 20–30 minutes and weigh as Sb_2S_3 (Section 11.16).

Mercury(II) ($\not> 100$ mg in 100 mL solution). Add a few millilitres of dilute hydrochloric acid and saturate the cold solution with washed hydrogen sulphide. Allow the black precipitate to settle, filter and wash with cold water (if the presence of sulphur is suspected, wash the precipitate with hot water, ethanol, or carbon disulphide). Dry at 105–110 °C and weigh as HgS (Section 11.33).

ANIONS

11.53 BORATE

Determination of borate as nitron tetrafluoroborate: Discussion. Boric acid (100–250 mg) in aqueous solution may be determined by conversion into tetrafluoroboric acid and precipitation of the latter with a large excess of nitron [see Section 11.11(E)] as nitron tetrafluoroborate, which is weighed after drying at 110 °C. The accuracy is about 1 per cent.

$$H_3BO_3 + 4HF \rightleftharpoons HBF_4 + 3H_2O$$

$$HBF_4 + C_{20}H_{16}N_4 \rightleftharpoons C_{20}H_{16}N_4 \cdot HBF_4$$

Fluoride ion, and weak acids and bases do not interfere, but nitrate, nitrite, perchlorate, thiocyanate, chromate, chlorate, iodide, and bromide do. Since analysis of almost all boron-containing compounds requires a preliminary treatment which ultimately results in an aqueous boric acid sample, this procedure may be regarded as a gravimetric determination of boron.

Procedure. Place the aqueous sample solution of boric acid (containing 100–250 mg of H_3BO_3) in a 250 mL polythene beaker and dilute to about

---

* Dissolve 1.0 g of salicylaldehyde oxime in 5 mL of 95 per cent ethanol and pour the solution slowly into 95 mL of water (temperature $\not> 80$ °C). Shake the mixture until clear and filter if necessary.

60 mL with distilled water. Add 15.0 mL of the nitron solution (Note 1) and 1.0–1.3 g of 48 per cent hydrofluoric acid (**CARE!**). Allow the solution to stand for 10–20 hours, and cool in an ice bath for 2 hours. Collect the precipitate in a porcelain filtering crucible, and wash it with five 10 mL portions of saturated nitron tetrafluoroborate solution (Note 2); drain the precipitate after each washing. Dry at 105–110 °C for 2 hours, and weigh as $C_{20}H_{16}N_4 \cdot HBF_4$.

Notes. (1) Prepare the nitron reagent by dissolving 3.75 g nitron in 25 mL 5 per cent acetic (ethanoic) acid (by volume). Store in a dark bottle.

(2) Prepare the wash solution by adding an excess of solid nitron tetrafluoroborate to 100 mL of water and shaking mechanically for 2 hours.

11.54 BROMATE AND BROMIDE

Discussion. These anions are both determined as silver bromide, AgBr, by precipitation with silver nitrate solution in the presence of dilute nitric acid. With the bromate, initial reduction to the bromide is achieved by the procedures described for the chlorate (Section 11.56) and the iodate (Section 11.63). Silver bromide is less soluble in water than is the chloride. The solubility of the former is 0.11 mg L^{-1} at 21 °C as compared with 1.54 mg L^{-1} for the latter; hence the procedure for the determination of bromide is practically the same as that for chloride. Protection from light is even more essential with the bromide than with the chloride because of its greater sensitivity (see Section 11.57).

11.55 CARBONATE

Determination of carbonate by the evolution of carbon dioxide: Discussion. The carbonate is decomposed by dilute acid, and either the loss in weight due to the escape of carbon dioxide determined (indirect method) or the carbon dioxide evolved is absorbed in a suitable medium and the increase in weight of the absorbent determined (**direct method**). The direct method gives more satisfactory results, and will therefore be described. The indirect method is often employed, however, for samples containing relatively large amounts of carbonate.

The decomposition of the carbonate may be effected with dilute hydrochloric acid, dilute perchloric acid, or phosphoric(V) acid. The last-named acid is perhaps the most convenient because of its comparative non-volatility and the fact that the reaction can be more easily controlled than with the other acids. If dilute hydrochloric acid is employed, a short, water-cooled condenser should be inserted between the decomposition flask and the absorption train (see below).

Two absorbents are required, one for water vapour, the other for carbon dioxide. The absorbents for water vapour which are generally employed are: (*a*) anhydrous calcium chloride (14–20 mesh), (*b*) anhydrous calcium sulphate ('Drierite' or 'Anhydrocel'), and (*c*) anhydrous magnesium perchlorate ('Anhydrone'). Both (*b*) and (*c*) are preferable to (*a*); (*c*) absorbs about 50 per cent of its weight of water, but is expensive. Anhydrous calcium chloride usually contains a little free lime, which will absorb carbon dioxide also; it is essential to saturate the **U**-tube containing calcium chloride with dry carbon dioxide for several hours and then to displace the carbon dioxide by a current of pure dry air before use.

The absorbent for carbon dioxide in general use is: (*d*) soda-lime (this is

477

available also in the form of self-indicating granules, 'Carbosorb', which indicate when the absorbent is exhausted). In all cases the carbon dioxide is absorbed in accordance with the following equation:

$$2NaOH + CO_2 = Na_2CO_3 + H_2O$$

Water is formed in the reaction; hence it is essential to fill one-quarter or one-third of the tube with any of the desiccants referred to above (Fig. 11.5).

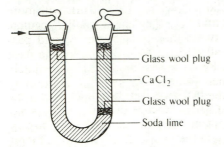

Glass wool plug

$CaCl_2$

Glass wool plug

Soda lime

Fig. 11.5

Procedure. Fit up the apparatus shown in Fig. 11.6. A is a flask of about 100 mL capacity, B is a dropping funnel containing 20–25 mL of phosphoric(V) acid, C is soda-lime guard tube, D is a bubbler containing phosphoric(V) acid, E is a **U**-tube containing calcium chloride which has been saturated with carbon dioxide and the residual carbon dioxide displaced by air (this may be replaced by anhydrous calcium sulphate, or by anhydrous magnesium perchlorate, if available).* F and G are **U**-tubes containing soda-lime, and H is a guard **U**-tube containing the same desiccant as in E. The **U**-tubes may be suspended by silver wires attached to hooks on the glass or metal rod I, or by some other means. All joints are made with short lengths of stout-walled rubber tubing, and the two ends of the glass tubing should be in contact. Rubber bungs are employed in A, B, and C. Before proceeding with the actual determination, make sure that the apparatus is gas-tight.

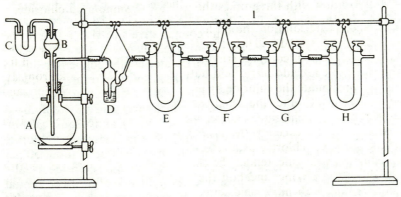

Fig. 11.6

---

* The first one-third of this tube may be filled with anhydrous copper sulphate to remove any hydrogen sulphide or hydrogen chloride present from sulphides or chlorides in the limestone.

Weigh out accurately 0.5–0.6 g of the carbonate (Note 1) into the flask A, which should be clean and dry. Remove the two soda-lime tubes F and G, wipe them with a clean linen handkerchief or cloth, and leave them in the balance case for 45 minutes. Open the taps of the **U**-tubes momentarily to the air in the balance case, and weigh them separately. Replace them on the drying train; place 25 mL of phosphoric(V) acid in B, and see that the apparatus is connected up as in Fig. 11.6. Open the taps of the **U**-tubes. Run in sufficient phosphoric(V) acid from the tap funnel to cover the solid in the flask (the 25 mL will more than suffice). Close the tap of the funnel and heat the flask carefully; regulate the temperature so that not more than two bubbles of gas per second pass through the bubbler D. After about 30–40 minutes, the contents of the flask should be boiling; boil for 2–3 minutes. Remove the flame, and immediately attach a filter pump and large bubbler [similar to D, and containing phosphoric(V) acid] to the end of the tube H. Open the tap funnel, and draw air through the apparatus at the rate of about two bubbles per second for 20 minutes. Remove the tubes F and G, close the taps, treat them as before, and weigh them. From the increase in weight, calculate the percentage of CO_2 in the sample (Note 2).

Notes. (1) For practice in this determination, the student may employ calcium carbonate or 'Limestone, 15e' (Analysed Samples for Students) from the Bureau of Analysed Samples.

(2) For the most accurate work, and particularly when the amount of carbon dioxide is small, a 'blank' experiment must be run with the reagents alone before the determination proper is carried out.

11.56 CHLORATE

Determination of chlorate as silver chloride: Discussion. The chlorate is reduced to chloride, and the latter is determined as silver chloride, AgCl. The reduction may be performed with iron(II) sulphate solution, sulphur dioxide, or by zinc powder and acetic (ethanoic) acid. Alkali chlorates may be quantitatively converted into chlorides by three evaporations with concentrated hydrochloric acid, or by evaporation with three times the weight of ammonium chloride.

Procedure. The chlorate solution should have a volume of about 100 mL, and contain about 0.2 g ClO_3. Add 50 mL of a 10 per cent solution of crystallised iron(II) sulphate, heat with constant stirring to the boiling point, and boil for 15 minutes. Allow to cool, add nitric acid until the precipitated basic iron(III) salt is dissolved, precipitate the chloride by means of silver nitrate solution, and collect and weigh as AgCl after the usual treatment (Section 11.57).

Alternatively, treat the chlorate solution with excess of sulphur dioxide, boil the solution to remove the excess of the gas, render slightly acid with nitric acid, and precipitate the silver chloride as above.

For the reduction with zinc, render the chlorate solution strongly acid with acetic acid, add excess of zinc, and boil the mixture for 1 hour. Dissolve the excess of unused zinc with nitric acid, filter, and treat the filtrate with silver nitrate in the usual manner.

Note. Hypochlorites and chlorites may be reduced to chlorides with sulphur dioxide, and determined in the same way.

479

11.57 CHLORIDE

Determination of chloride as silver chloride: Discussion. The aqueous solution of the chloride is acidified with dilute nitric acid in order to prevent the precipitation of other silver salts, such as the phosphate and carbonate, which might form in neutral solution, and also to produce a more readily filterable precipitate. A slight excess of silver nitrate solution is added, whereupon silver chloride is precipitated:

$$Cl^- + Ag^+ = AgCl$$

The precipitate, which is initially colloidal, is coagulated into curds by heating the solution and stirring the suspension vigorously; the supernatant liquid becomes almost clear. The precipitate is collected in a filtering crucible, washed with very dilute nitric acid, in order to prevent it from becoming colloidal (Section 11.8), dried at $130-150\,°C$, and finally weighed as AgCl.

Silver chloride has a solubility in water of $1.4\,mg\,L^{-1}$ at $20\,°C$, and $21.7\,mg\,L^{-1}$ at $100\,°C$. The solubility is less in the presence of very dilute nitric acid (up to 1 per cent), and is very much less in the presence of moderate concentrations of silver nitrate (see Section 2.7; the optimum concentration of silver nitrate is $0.05\,g\,L^{-1}$, but the solubility is negligibly small up to about $1.7\,g\,L^{-1}$). The solubility is increased by the presence of ammonium and of alkali-metal salts, and by large concentrations of acids. Under the conditions of the precipitation, very little occlusion occurs. If silver chloride is washed with pure water, it may become colloidal and run through the filter. For this reason the wash solution should contain an electrolyte (compare Sections 11.3 and 11.8). Nitric acid is generally employed because it is without action on the precipitate and is readily volatile; its concentration need not be greater than $0.01M$. Completeness of washing of the precipitate is tested for by determining whether the excess of the precipitating agent, silver nitrate, has been removed. This may be done by adding one or two drops of $0.1M$ hydrochloric acid to $3-5\,mL$ of the washings collected after the washing process has been continued for some time; if the solution remains clear or exhibits only a very slight opalescence, all the silver nitrate has been removed.

Silver chloride is light-sensitive; decomposition occurs into silver and chlorine, and the silver remains colloidally dispersed in the silver chloride and thereby imparts a purple colour to it. The decomposition by light is only superficial, and is negligible unless the precipitate is exposed to direct sunlight and is stirred frequently. Hence the determination must be carried out in as subdued a light as possible, and when the solution containing the precipitate is set aside, it should be placed in the dark (e.g. in a locker), or the vessel containing it should be covered with thick brown paper.

It has been found that in a solution containing silver chloride and $1-2$ per cent excess of $0.2M$ silver nitrate, exposure to direct sunlight for 5 hours with occasional stirring leads to a positive error of about 2.1 per cent whilst exposure in a bright laboratory, with no direct or reflected sunlight and occasional stirring, gives a positive error of about 0.2 per cent. This positive error is due to the liberation of chlorine during exposure to light: the chlorine is largely changed back to chloride ions, which cause further precipitation of silver chloride. A possible reaction is:

$$3Cl_2 + 5Ag^+ + 3H_2O = 5AgCl + ClO_3^- + 6H^+$$

On the other hand, in the determination of silver by precipitation with a slight excess of 0.2M hydrochloric acid (Section 11.40), the error is negative, e.g. 0.4 per cent after 2 hours' exposure in direct sunlight with no stirring, and 0.1 per cent after 2 hours' exposure in a bright laboratory, with no direct or reflected sunlight and occasional stirring. This arises from the loss of chlorine which escapes from the precipitate. The weight of the precipitate may be brought to the correct value by treatment with nitric acid, followed by hydrochloric acid.

Procedure. Weigh out accurately about 0.2 g of the solid chloride (or an amount containing approximately 0.1 g of chlorine)* into a 250 mL beaker provided with a stirring rod and covered with a clockglass. Add about 150 mL of water, stir until the solid has dissolved, and add 0.5 mL of concentrated nitric acid. To the cold solution add 0.1M silver nitrate slowly and with constant stirring. Only a slight excess should be added; this is readily detected by allowing the precipitate to settle and adding a few drops of silver nitrate solution, when no further precipitate should be obtained. **Carry out the determination in subdued light**. Heat the suspension nearly to boiling, while stirring constantly, and maintain it at this temperature until the precipitate coagulates and the supernatant liquid is clear (2–3 minutes). Make certain that precipitation is complete by adding a few drops of silver nitrate solution to the supernatant liquid. If no further precipitate apppears, set the beaker aside in the dark, and allow the solution to stand for about 1 hour before filtration. In the meantime prepare a sintered-glass filtering crucible; the crucible must be dried at the same temperature as is employed in heating the precipitate (130–150 °C) and allowed to cool in a desiccator. Collect the precipitate in the weighed filtering crucible. Wash the precipitate two or three times by decantation with about 10 mL of cold very dilute nitric acid (say, 0.5 mL of the concentrated acid added to 200 mL of water) before transferring the precipitate to the crucible. Remove the last small particles of silver chloride adhering to the beaker with a 'policeman' (Section 3.23). Wash the precipitate in the crucible with very dilute nitric acid added in small portions until 3–5 mL of the washings, collected in a test-tube, give no turbidity with one or two drops of 0.1M hydrochloric acid. Place the crucible and contents in an oven at 130–150 °C for 1 hour, allow to cool in a desiccator, and weigh. Repeat the heating and cooling until constant weight is attained.

Calculate the percentage of chlorine in the sample.

Note on the gravimetric standardisation of hydrochloric acid. The gravimetric standard-isation of hydrochloric acid by precipitation as silver chloride is a convenient and accurate method, which has the additional advantage of being independent of the purity of any primary standard (compare Section 10.38). Measure out from a burette 30–40 mL of the, say, 0.1M hydrochloric acid which is to be standardised. Dilute to 150 mL, precipitate (but omit the addition of nitric acid), and weigh the silver chloride. From the weight of the precipitate, calculate the chloride concentration of the solution, and thence the concentration of the hydrochloric acid.

11.58 CYANIDE

Note. Great care must be taken in the use and determination of cyanides owing to their highly poisonous nature.

---

*Analytical grade potassium chloride or sodium chloride, dried at 110–120 °C, is suitable.

Discussion. This anion may be determined as silver cyanide, AgCN, the experimental details are similar to those given for chloride, except that, owing to the volatility of hydrocyanic acid, **the solution must not be heated**. The **cold** solution of alkali cyanide is treated with a slight excess of silver nitrate solution, **faintly** acidified with nitric acid, the precipitate allowed to settle, collected on a weighed filtering crucible, and weighed as AgCN after drying at 100 °C.

11.59 FLUORIDE

Determination as lead chlorofluoride: Discussion. This method, which is based upon the precipitation of lead chlorofluoride, has been used extensively for the determination of macro and semimicro amounts of fluoride. The procedure for the precipitation is described in Section 10.85 but for a gravimetric finish the precipitate is filtered through a weighed sintered-glass filtering crucible, washed several times with a saturated solution of lead chlorofluoride and finally once with cold water. It is weighed as PbClF after drying to constant weight at 140–150 °C.

11.60 FLUOROSILICATE

Discussion. The determination of this anion is of little practical importance. The methods available for its determination will, however, be outlined. Alkali fluorosilicates are decomposed by heating with sodium carbonate solution into a fluoride and silicic acid:

$$Na_2[SiF_6] + 2Na_2CO_3 + H_2O = 6NaF + H_2SiO_3 + 2CO_2$$

Insoluble fluorosilicates are brought into solution by fusion with four times the bulk of fusion mixture, and extracting the melt with water. In either case, the solution is treated with a considerable excess of ammonium carbonate, warmed to 40 °C, and, after standing for 12 hours, the precipitated silicic acid is filtered off, and washed with 2 per cent ammonium carbonate solution. The filtrate contains a little silicic acid, which may be removed by shaking with a little freshly precipitated cadmium oxide. The fluoride in the filtrate is determined as described in Section 11.59.

If an acid solution of a fluorosilicate is rendered faintly alkaline with aqueous sodium hydroxide and then shaken with freshly precipitated cadmium oxide, all the silicic acid is adsorbed by the suspension. The alkali fluoride is then determined in the filtrate.

11.61 HEXAFLUOROPHOSPHATE

Determination as the tetraphenylarsonium compound: Discussion. Any ions forming a precipitate with tetraphenylarsonium chloride (e.g. MnO_4^-, ClO_4^-, Br^-, I^-, IO_3^- and SCN^-) will interfere. Difluorophosphates give slight interference which can be overcome by boiling the solution for a few minutes to hydrolyse difluorophosphate; the hexafluorophosphate ion is not affected by this treatment.

Procedure. To an aliquot containing 36–55 mg of potassium hexafluorophosphate add aqueous ammonia until the solution is basic. A final concentration of ammonia in the range $5-10M$ is satisfactory and the total volume of solution should be about 50 mL. Warm to about 50 °C and add about twice the equivalent amount of $0.015M$ tetraphenylarsonium chloride slowly and with stirring. Allow to stand for 30 minutes, filter through a medium-porosity sintered-glass filter and wash with five 10 mL portions of dilute aqueous ammonia. Dry the precipitate to constant weight at 105–115 °C and weigh as $(C_6H_5)_4AsPF_6$.

11.62 HYPOPHOSPHITE (PHOSPHINATE)

This anion is determined similarly to phosphite (Section 11.70), i.e. indirectly as mercury(I) chloride, Hg_2Cl_2. In this case the reaction with mercury(II) chloride solution is:

$$4HgCl_2 + H_3PO_2 + 2H_2O = 2Hg_2Cl_2 + H_3PO_4 + 4HCl$$

so that

$$2Hg_2Cl_2 \equiv H_3PO_2$$

11.63 IODATE

Determination of iodate as silver iodide: Discussion. Iodates are readily reduced by sulphurous acid to iodides; the latter are determined by precipitation with silver nitrate solution as silver iodide, AgI. Iodates cannot be converted quantitatively into iodides by ignition, for the decomposition takes place at a temperature at which the iodide is appreciably volatile.

Periodates are also reduced by sulphurous acid, and may therefore be similarly determined. Similar remarks apply to bromates; these are ultimately weighed as silver bromide, AgBr.

Procedure. Acidify the iodate solution (100 mL containing *ca* 0.3 g of IO_3) (see Note) with sulphuric acid, and pass in sulphur dioxide (or add a freshly prepared saturated solution of sulphurous acid) until the solution, which at first becomes yellow, on account of the separation of iodine, is again colourless. Boil off the excess of sulphur dioxide, and precipitate the iodide with dilute silver nitrate solution as described in Section 11.64. Weigh as AgI.

Note. For practice in this determination, potassium iodate may be employed.

11.64 IODIDE

Determination of iodide as silver iodide: Discussion. This anion is usually determined by precipitation as silver iodide, AgI. Silver iodide is the least soluble of the silver halides; 1 litre of water dissolves 0.0035 mg at 21 °C. Co-precipitation and similar errors are more likely to occur with iodide than with the other halides.

Procedure. Precipitation is therefore made by adding a very dilute solution, say $0.05M$, of silver nitrate slowly and with constant stirring to a dilute ammoniacal solution of the iodide until precipitation is complete, and then adding excess of nitric acid (1 per cent by volume). The precipitate is collected in the usual manner, washed with 1 per cent nitric acid, and finally with a *little* water to remove nitric acid. Peptisation tends to occur with excess of water. Other details of the determination will be found in Section 11.57.

11.65 NITRATE

Determination of nitrate as nitron nitrate: Discussion. The mono-acid base nitron, $C_{20}H_{16}N_4$, forms a fairly insoluble crystalline nitrate, $C_{20}H_{16}N_4,HNO_3$ (solubility is 0.099 g L^{-1} at about 20 °C), which can be used for the quantitative determination of nitrates [see Section 11.11(E)]. The sulphate and acetate are soluble so that precipitation may be made in sulphuric or acetic (ethanoic) acid solution. Perchlorates (0.08 g), iodides (0.17 g), thiocyanates (0.4 g), chromates (0.6 g), chlorates (1.2 g), nitrites (1.9 g), bromides (6.1 g), hexacyanoferrate(II), hexacyanoferrate(III), oxalates, and considerable quantities of chlorides interfere, and should be absent. The figures in parentheses are the approximate solubilities of the nitron salts in g L^{-1} at about 20 °C.

Procedure. The solution (75–100 mL) should be neutral and contain about 0.1 g NO_3. Add 1 mL glacial acetic acid or 0.5 mL $1M$ sulphuric acid and heat the solution nearly to the boiling point. Then introduce in one portion 10–12 mL of the nitron reagent (see Note), stir, and cool in ice–water for 2 hours. Filter through a weighed filtering crucible (sintered-glass, or porcelain). Wash with 10–15 mL of a cold saturated solution of nitron nitrate, added in several portions, and drain the precipitate well after each washing. Finally, wash twice with 3 mL portions of ice-cold water. Dry at 105 °C (1 hour is usually required), and weigh as $C_{20}H_{16}N_4,HNO_3$.

Note. Prepare the reagent by dissolving 5 g of nitron in 50 mL of 5 per cent acetic acid. Store in an amber bottle.

11.66 NITRITE

No satisfactory direct gravimetric procedure is available but nitrite can be oxidised to nitrate by permanganate or cerium(IV) and then determined in that form. The determination of total nitrate + nitrite is an important analysis, e.g. for soil samples. Nitrite may be destroyed using urea, sulphamic acid or hydrazine sulphate; the reaction with the former is:

$$CO(NH_2)_2 + 2HNO_2 \rightarrow 2N_2 + CO_2 + 3H_2O$$

11.67 OXALATE

Determination of oxalate as calcium oxalate and as calcium carbonate or calcium oxide: Discussion. The neutral solution of alkali oxalate is acidified with acetic (ethanoic) acid, heated to boiling, and precipitated with boiling calcium chloride solution. After standing for 12 hours, the precipitate is filtered off, washed with hot water, and weighed either as calcium oxalate, or after heating, as calcium carbonate, $CaCO_3$, or as calcium oxide, CaO. Further details are given in Section 11.22.

11.68 PERCHLORATE

Determination of perchlorate as silver chloride: Discussion. Perchlorates are not reduced by iron(II) sulphate solution, sulphurous acid, or by repeated evaporation with concentrated hydrochloric acid; reduction occurs, however, with titanium(III) sulphate solution. Ignition of perchlorates with ammonium chloride in a platinum crucible or in a porcelain crucible in the presence of a little platinum powder results in reduction to the chlorides (the platinum acts as a catalyst), which may be determined in the usual manner. Losses occur when perchlorates are ignited alone.

Procedure. The perchlorate, if supplied as a solution, is evaporated to dryness on the water bath (see Note); otherwise the solid perchlorate is used directly. Intimately mix about 0.4 g of the perchlorate with 1.5 g of ammonium chloride in a *platinum* crucible covered with a watchglass or lid, ignite gently until fuming ceases and continue the heating for 1 hour. Do not fuse the resulting chloride, as the crucible may be attacked. Repeat the ignition with another 1.5 g of ammonium chloride. Dissolve the residue in a little water, filter through a small quantitative filter paper to remove any platinum powder which may be present, and determine the chloride in the filtrate as silver chloride (Section 10.57).

Note. **Evaporation with perchloric acid or perchlorates should be carried out in a fume cupboard which is kept clean and free from combustible materials.** In the presence of carbon and easily oxidisable organic compounds a violent explosion may occur on heating. The determination is not suitable for the beginner and should only be carried out by an experienced analyst.

11.69 PHOSPHATE

Determination of phosphate as ammonium molybdophosphate. This may be readily effected by precipitation with excess of ammonium molybdate in warm nitric acid solution; arsenic, vanadium, titanium, zirconium, silica and excessive amounts of ammonium salts interfere. The yellow precipitate obtained may be weighed as either ammonium molybdophosphate, $(NH_4)_3[PMo_{12}O_{40}]$, after drying at 200–400 °C, or as $P_2O_5,24MoO_3$, after heating at 800–825 °C for about 30 minutes.

Procedure. Prepare a solution of anhydrous disodium hydrogenphosphate or of analytical-grade potassium dihydrogenphosphate containing about 125 mg of P_2O_5 in 150 mL. Warm to 60 °C, and run in 100 mL of ammonium molybdate reagent (see Note) also warmed to 60 °C: use a fast-flowing pipette for the addition and stir well. Heat to 60 °C for about 1 hour with frequent stirring. Collect the precipitate in a weighed porcelain filtering crucible using two 20 mL portions of 2 per cent ammonium nitrate solution to transfer it from the beaker (remove any precipitate adhering to the walls of the beaker with a rubber-tipped glass rod); wash the precipitate in the crucible with five 10 mL portions of 2 per cent ammonium nitrate solution. Dry the precipitate at 280 °C, and weigh as $(NH_4)_3[PMo_{12}O_{40}]$. As an additional check, ignite the precipitate at 800–825 °C in an electric muffle furnace; weigh as $P_2O_5,24MoO_3$. Both solids are appreciably hygroscopic; the covered crucible, after cooling in a desiccator, should be weighed as soon as it has acquired the laboratory temperature.

Note. Prepare the ammonium molybdate reagent as follows. Dissolve 125 g ammonium nitrate in 125 mL water in a flask and add 175 mL nitric acid, sp. gr. 1.42. Dissolve 12.5 g ammonium molybdate* in 75 mL of water and add this slowly and with constant shaking to the nitrate solution. Dilute to 500 mL with water, heat the flask in a water bath at

---

* This is the heptamolybdate $(NH_4)_6Mo_7O_{24},4H_2O$.

60 °C for 6 hours, and allow the solution to stand for 24 hours. If a precipitate forms, filter through a Whatman No. 42 filter paper. This reagent has good keeping qualities and no precipitation should occur for at least 3 months.

11.70 PHOSPHITE (PHOSPHONATE)

Determination of phosphite as mercury(I) chloride: Discussion. The acid solution of phosphite reduces mercury(II) chloride solution to mercury(I) chloride which is weighed. The reaction is:

$$2HgCl_2 + H_3PO_3 + H_2O = Hg_2Cl_2 + H_3PO_4 + 2HCl$$

whence

$$Hg_2Cl_2 \equiv H_3PO_3$$

Procedure. The phosphite solution (30 mL) should contain about 0.1 g HPO_3^{2-}. Place 50 mL of 3 per cent mercury(II) chloride solution, 20 mL of 10 per cent sodium acetate, and 5 mL of glacial acetic (ethanoic) acid in a 250 mL beaker, and add the phosphite solution dropwise, and with stirring, in the cold. Allow to stand on a water bath at 40–50 °C for 2 hours. When cold, filter, through a weighed filtering crucible (sintered-glass, or porcelain), wash two or three times with 1 per cent hydrochloric acid, and then four times with warm water. Dry at 105–110 °C, and weigh as Hg_2Cl_2.

11.71 SILICATE

For analytical purposes silicates may be conveniently divided into the following two classes: (*a*) those ('soluble' silicates) which are decomposed by acids, such as hydrochloric acid, to form silicic acid and the salts (e.g. chlorides) of the metals present; and (*b*) those ('insoluble' silicates) which are not decomposed by any acid, except hydrofluoric acid. There are also many silicates which are partially decomposed by acids; for analytical purposes these will be included in class (*b*).

A. Determination of silica in a 'soluble' silicate: Discussion. Most of the silicates which come within the classification of 'soluble' silicates are the orthosilicates formed from SiO_4^{4-} units in combination with just one or two cations. More highly condensed silicate structures give rise to the 'insoluble' silicates.

Procedure. Weigh out accurately about 0.4 g of the finely powdered silicate (Note 1) into a platinum or porcelain dish, add 10–15 mL water, and stir until the silicate is thoroughly wet. Place the dish, covered with a clockglass, on the water bath, and add gradually 25 mL 1:1 hydrochloric acid. The contents of the dish must be continuously stirred with a glass rod; when no gritty particles remain, the powder will have been completely decomposed. Evaporate the liquid to dryness: stir the residue continuously and break up any lumps with the glass rod. When the powder appears to be dry, place the basin in an air oven at 100–110 °C for 1 hour in order to dehydrate the silica. Moisten the residue with 5 mL of concentrated hydrochloric acid, and bring the acid into contact with the solid with the aid of a stirring rod. Add 75 mL of water, rinse down the sides of the dish, and heat on a steam bath for 10–20 minutes to assist in the solution of the soluble salts. Filter off the separated silica on a Whatman

No. 41 or 541 filter paper. Wash the precipitate first with warm, dilute hydrochloric acid (approx. $0.5M$), and then with hot water until free from chlorides. Pour the filtrate and washings into the original dish, evaporate to dryness on the steam bath, and heat in an air oven at 100–110 °C for 1 hour. Moisten the residue with 5 mL concentrated hydrochloric acid, add 75 mL water, warm to extract soluble salts, and filter through a fresh, but smaller, filter paper. Wash with warm dilute hydrochloric acid (approx. $0.1M$), and finally with a little hot water. Fold up the moist filters, and place them in a weighed platinum crucible. Dry the paper with a small flame, char the paper, and burn off the carbon over a low flame; take care that none of the fine powder is blown away. When all the carbon has been oxidised, cover the crucible, and heat for an hour at the full temperature of a Meker-type burner in order to complete the dehydration. Allow to cool in a desiccator, and weigh. Repeat the ignition, etc., until the weight is constant.

To determine the exact SiO_2 content of the residue, moisten it with 1 mL water, add two or three drops of concentrated sulphuric acid and about 5 mL of the purest available hydrofluoric acid. (**CARE!**) Place the crucible in an air bath (Section 3.21) and evaporate the hydrofluoric acid in a fume cupboard (hood) with a small flame until the acid is completely expelled; the liquid should not be boiled. (The crucible may also be directly heated with a small non-luminous flame.) Then increase the heat to volatilise the sulphuric acid, and finally heat with a Meker-type burner for 15 minutes. Allow to cool in a desiccator and weigh. Re-heat to constant weight. The loss in weight represents the weight of the silica (Note 2).

Notes. (1) For practice in this determination, powdered, fused sodium silicate may be used.

(2) It is advisable to carry out a blank determination with the hydrofluoric acid, and to allow for any non-volatile substances, if necessary.

B. Determination of silica in an 'insoluble' silicate, and ultimate weighing as silica, SiO_2: Discussion.

Insoluble silicates are generally fused with sodium carbonate, and the melt, which contains the silicate in acid-decomposable form, is then treated with hydrochloric acid. The acid solution of the decomposed silicate is evaporated to dryness on the water bath to separate the gelatinous silicic acid SiO_2,xH_2O as insoluble silica SiO_2,yH_2O; the residue is heated at 110–120 °C to dehydrate the silica partially and render it as insoluble as possible. The residue is extracted with hot dilute hydrochloric acid to remove salts of iron, aluminium, and other metals which may be present. The greater portion of the silica remains undissolved, and is filtered off. The filtrate is evaporated to dryness, and the residue heated at 110–120 °C as before in order to render insoluble the small amount of silicic acid that has escaped dehydration. The residue is again treated with dilute hydrochloric acid, and the second portion of silica is filtered off on a fresh filter. The two washed precipitates are combined, and ignited in a platinum crucible at about 1050 °C to silicon dioxide, SiO_2, and the latter is weighed. The ignited residue is not usually pure silicon dioxide; it will generally contain small amounts of the oxides of iron, aluminium, titanium, etc. The amount of impurity may be determined, if desired, by treating the weighed residue in a platinum crucible with an excess of hydrofluoric acid and a little concentrated sulphuric acid. The silica is expelled as the volatile silicon tetrafluoride; the impurities (e.g. Al_2O_3 and Fe_2O_3) are first converted into the fluorides, which pass into the sulphates in contact with the less-volatile sulphuric acid, whilst the subsequent brief ignition (at 1050–1100 °C for a few minutes) converts the sulphates back into oxides. Thus, for example:

$$SiO_2 + 6HF = H_2[SiF_6] + 2H_2O$$

$$H_2[SiF_6] = SiF_4 + 2HF$$

$$Al_2O_3 + 6HF = 2AlF_3 + 3H_2O$$

$$2AlF_3 + 3H_2SO_4 = Al_2(SO_4)_3 + 6HF$$

$$Al_2(SO_4)_3 = Al_2O_3 + 3SO_3$$

The loss in weight, therefore, represents the amount of pure silicon dioxide present.

Procedure. Weigh out accurately into a platinum crucible about 1.0 g of the finely powdered dry silicate (see Note), add six times the weight of anhydrous sodium carbonate (or, better, of fusion mixture), and mix the solids thoroughly by stirring with a thin, rounded glass rod. Cover the mixture with a little more of the carbonate, and then cover the crucible. Heat the mixture gradually until after about 20 minutes a tranquil melt is obtained; the cover of the crucible is lifted occasionally to examine the contents. Maintain the temperature of a quiet liquid fusion for about 30 minutes. Allow to cool. Place the crucible and lid in a covered deep porcelain or platinum basin (or in a large casserole), cover it with water, and leave overnight, or warm on the water bath until the contents are well disintegrated. Introduce very slowly by means of a pipette or a bent funnel about 25 mL of concentrated hydrochloric acid into the covered vessel. Warm on the steam bath until the evolution of carbon dioxide has ceased. Remove and rinse the coverglass, crucible, and lid, and evaporate the contents of the dish to *complete* dryness on the steam bath, crushing all lumps with a glass rod. Heat the residue for an hour at 100–110 °C to dehydrate the silica. Complete the determination as described in (A).

Note. 'Feldspar (Potash), No. 29dG' available from the Bureau of Analysed Samples (BCS-CRM 376).

(C) Determination of silica in an 'insoluble' silicate as quinoline molybdosilicate: Discussion. Silica may also be determined gravimetrically as quinoline molybdosilicate. The solution of silicic acid is treated with ammonium molybdate to form molybdosilicic acid, $H_4[SiO_4,12MoO_3]$, which is then precipitated as quinoline molybdosilicate, $(C_9H_7)_4H_4[SiO_4,12MoO_3]$. The latter is weighed after drying at 150 °C. The experimental conditions lead to quinoline molybdosilicate in a pure form suitable for weighing.

Phosphate, arsenate, and vanadate interfere. Borate, fluoride, and large amounts of aluminium, calcium, magnesium, and the alkali metals have no effect in the determination, but large amounts of iron (> 5 per cent) appear to produce slightly low results.

Procedure. The method to be described is especially suitable for ceramic materials such as fireclay, firebrick, or silica brick. The finely ground sample should be dried at 110 °C. The weight of sample to be employed depends largely upon the silica content of the material, since not more than 35–40 mg of silica should be present in the aliquot employed for the determination. For samples containing more than 65 per cent SiO_2 (Note 1), use 0.25 g; for samples containing less than 65 per cent SiO_2, use 0.50 g (Note 2).

Place 7 g of sodium hydroxide pellets in a nickel crucible (4.5 cm × 4.5 cm) and fuse gently until the water is expelled and a clear melt results. Allow to cool, introduce the weighed sample evenly on to the solidified melt, moisten with a little ethanol and gently evaporate the ethanol on a hot plate; this reduces the tendency to spurting in the subsequent fusion. Heat gently over a Bunsen burner, with occasional rotation of the crucible, until the sodium hydroxide is just molten, after which raise the temperature to a dull red heat for 2–5 minutes; the sample should then have dissolved completely. Carefully cool the crucible by partial immersion in cold water; when the melt has just solidified transfer the hot crucible to a 400 mL nickel beaker and cover with a clock glass. Raise the clock glass slightly, fill the crucible with boiling water and replace the cover. This should suffice to dissolve the fused mass; otherwise add a little more boiling water. When the vigorous reaction has subsided, wash the clock glass and sides of the beaker with hot water; remove the crucible with clean tongs, carefully rinsing it inside and out with hot water. Dilute the suspension to 175 mL; do not exceed this volume. Place 20 mL of concentrated hydrochloric acid in a 500 mL conical flask; pour the fusion extract, with swirling, into the acid, rinse the beaker with a little hot water, and add the rinsings to the flask. Cool rapidly to room temperature and dilute to 250 mL in a graduated flask.

Withdraw an aliquot part containing about 35 mg of silica, dilute it to about 250 mL in an 800 mL beaker, add 3 g of sodium hydroxide pellets and swirl until dissolved. Add 10 drops of thymol blue indicator (0.04 per cent solution in dilute ethanol, 1:4) followed by concentrated hydrochloric acid dropwise, swirling constantly, until the colour of the indicator changes from blue, through yellow, just to red; do not allow the solution to become too hot (Note 3). Now add 10 mL of dilute hydrochloric acid (1:9) and dilute to 400 mL. Add 50 mL of 10 per cent ammonium molybdate solution (Note 4) from a burette: stir vigorously during the addition and for 1 minute afterwards. Allow to stand for 10 minutes, add 50 mL of concentrated hydrochloric acid, and immediately precipitate the yellow molybdosilicate by introducing 50 mL of the quinoline reagent (Note 5) from a burette, stirring constantly. A cream-coloured, finely divided precipitate of quinoline molybdosilicate forms. Warm the suspension to about 80 °C during about 10 minutes and maintain this temperature for 5 minutes in order to coagulate the precipitate. Cool in running water to below 20 °C and collect the precipitate in a sintered glass filtering crucible (porosity No. 4); wash the precipitate six times with the special wash solution (Note 6), taking care not to allow the precipitate to run dry during the filtration and washing. Dry at 150 °C for 2 hours and cool the covered crucible in a desiccator. Weigh as $(C_9H_7)_4H_4[SiO_4,12MoO_3]$.

Notes. (1) 'Silica brick, No. 267' (a British Chemical Standard) may be used.

(2) 'Firebrick ECRM No. 269' may be used.

(3) This process is to ensure that the silica is in the correct form for reaction with ammonium molybdate. If the solution is too hot, the red colour may not develop.

(4) Prepare the 10 per cent ammonium molybdate solution by dissolving 25 g ammonium molybdate in water and diluting to 250 mL in a polythene bottle. It keeps for about 4 weeks.

(5) Prepare the 2 per cent quinoline hydrochloride solution by adding 10 mL pure quinoline to about 400 mL hot water containing 12.5 mL of concentrated hydrochloric

acid, and stirring constantly. Cool the solution, add a little ashless filter pulp, and leave to settle. Filter the solution through a paper pulp pad, but do not wash. Dilute the filtrate with water to 500 mL.

(6) Prepare the wash solution by diluting 5 mL of the 2 per cent quinoline hydrochloride solution with water to 200 mL.

11.72 SULPHATE

Determination of sulphate as barium sulphate: Discussion. The method consists in slowly adding a dilute solution of barium chloride to a hot solution of the sulphate slightly acidified with hydrochloric acid:

$$Ba^{2+} + SO_4^{2-} \rightarrow BaSO_4$$

The precipitate is filtered off, washed with water, carefully ignited at a red heat, and weighed as barium sulphate.

The reaction upon which the determination depends appears to be a simple one, but is in reality subject to numerous possible errors; satisfactory results can be obtained only if the experimental conditions are carefully controlled. Before some of these are discussed, the student is recommended to read Sections 11.3–11.6.

Barium sulphate has a solubility in water of about $3 \, mg \, L^{-1}$ at the ordinary temperature. The solubility is increased in the presence of mineral acids, because of the formation of the hydrogensulphate ion ($SO_4^{2-} + H^+ \rightleftharpoons HSO_4^-$); thus the solubilities at room temperature in the presence of 0.1, 0.5, 1.0, and $2.0M$ hydrochloric acid are 10, 47, 87, and $101 \, mg \, L^{-1}$ respectively, but the solubility is less in the presence of a moderate excess of barium ions. Nevertheless, it is customary to carry out the precipitation in weakly acid solution in order to prevent the possible formation of the barium salts of such anions as chromate, carbonate, and phosphate, which are insoluble in neutral solutions; moreover, the precipitate so obtained consists of large crystals, and is therefore more easily filtered (compare Section 11.4). It is also of great importance to carry out the precipitation at boiling temperature, for the relative supersaturation is less at higher temperatures (compare Section 11.4). The concentration of hydrochloric acid is, of course, limited by the solubility of the barium sulphate, but it has been found that a concentration of $0.05M$ is suitable; the solubility of the precipitate in the presence of barium chloride at this acidity is negligible. The precipitate may be washed with cold water, and losses, owing to solubility influences, may be neglected except for the most accurate work.

Barium sulphate exhibits a marked tendency to carry down other salts (see co-precipitation, Section 11.5). Whether the results will be low or high will depend upon the nature of the co-precipitated salt. Thus barium chloride and barium nitrate are readily co-precipitated. These salts will be an addition to the true weight of the barium sulphate, hence the results will be high, since the chloride is unchanged upon ignition and the nitrate will yield barium oxide. The error due to the chloride will be considerably reduced by the very slow addition of hot dilute barium chloride solution to the hot sulphate solution, which is constantly stirred; that due to the nitrate cannot be avoided, and hence nitrate ion must always be removed by evaporation with a large excess of hydrochloric acid before precipitation. Chlorate has a similar effect to nitrate, and is similarly removed.

In the presence of certain cations [sodium, potassium, lithium, calcium, aluminium, chromium, and iron(III)], co-precipitation of the sulphates of these metals occurs, and the results will accordingly be low. This error cannot be entirely avoided except by the removal of the interfering ions. Aluminium, chromium, and iron may be removed by precipitation, and the influence of the other ions, if present, is reduced by considerably diluting the solution and by digesting the precipitate (Section 11.5). It must be pointed out that the general method of re-precipitation, in order to obtain a purer precipitate, cannot be employed, because no simple solvent (other than concentrated sulphuric acid) is available in which the precipitate may be easily dissolved.

Positively charged barium sulphate, which is obtained when sulphate is precipitated by excess of barium ions, can be coagulated by the addition of a trace of agar-agar. About 1 mg of agar-agar as a 1 per cent aqueous solution will cause the flocculation of about 0.1 g of barium sulphate, but in practice somewhat larger quantities are generally used. The resulting precipitate does not creep up the sides of the vessel.

Negatively charged barium sulphate, obtained in the determination of barium, is not appreciably improved by agar-agar; this precipitate as a rule presents little difficulty in filtration.

Pure barium sulphate is not decomposed when heated in dry air until a temperature of about 1400 °C is reached:

$$BaSO_4 = BaO + SO_3$$

The precipitate is, however, easily reduced to sulphide at temperatures above 600 °C by the carbon of the filter paper:

$$BaSO_4 + 4C = BaS + 4CO$$

The reduction is avoided by first charring the paper without inflaming, and then burning off the carbon slowly at a low temperature with free access of air. If a reduced precipitate is obtained, it may be re-oxidised by treatment with sulphuric acid, followed by volatilisation of the acid and re-heating. The final ignition of the barium sulphate need not be made at a higher temperature than 600–800 °C (dull red heat). A Vitreosil or porcelain filtering crucible may be used, and the difficulty of reduction by carbon is entirely avoided.

Procedure. Weigh out accurately about 0.3 g of the solid*(or a sufficient amount to contain 0.05–0.06 g of sulphur) into a 400 mL beaker, provided with a stirring rod and clockglass cover. Dissolve the solid† in about 25 mL of water, add 0.3–0.6 mL of concentrated hydrochloric acid, and dilute to 200–255 mL. Heat the solution to boiling, add dropwise from a burette or pipette 10–12 mL of warm 5 per cent barium chloride solution (5 g $BaCl_2,2H_2O$ in 100 mL of water — ca 0.2M). Stir the solution constantly during the addition. Allow the precipitate to settle for a minute or two. Then test the supernatant liquid for

---

* Potassium sulphate may be employed.

† For sulphates which are insoluble in water and acids, it is best to mix the finely powdered solid with six to twelve times its bulk of anhydrous sodium carbonate in a platinum crucible (Sections 3.20 and 3.31), heat the covered crucible slowly to fusion, and maintain in the fused state for 15 minutes. The melt is extracted with water, the solution filtered, the residue washed with hot 1 per cent sodium carbonate solution, and the cold filtrate carefully acidified with hydrochloric acid (to methyl orange). The sulphate is determined as above.

complete precipitation by adding a few drops of barium chloride solution. If a precipitate is formed, add slowly a further 3 mL of the reagent, allow the precipitate to settle as before, and test again; repeat this operation until an excess of barium chloride is present. When an excess of the precipitating agent has been added, keep the covered solution hot, but not boiling, for an hour (steam bath, low-temperature hotplate, or small flame) in order to allow time for complete precipitation.* The volume of the solution should not be allowed to fall below 150 mL; if the clockglass covering the beaker is removed, the underside must be rinsed off into the beaker by means of a stream of water from a wash bottle. The precipitate should settle readily, and a clear supernatant liquid should be obtained. Test the latter with a few drops of barium chloride solution for complete precipitation. If no precipitate is obtained, the barium sulphate is ready for filtration. The determination may be completed by either of the following procedures.

(1) Filter paper method. Decant the clear solution through an ashless filter paper (Whatman No. 40 or 540), and collect the filtrate in a clean beaker. Test the filtrate with a few drops of barium chloride: if a precipitate forms, the entire sample must be discarded and a new determination commenced. If no precipitate forms discard the liquid, rinse out the beaker, and place it under the funnel; this is in order to avoid the necessity of re-filtering the whole solution if any precipitate should pass through the filter. Transfer the precipitate to the filter with the aid of a jet of hot water from the wash bottle. Use a rubber-tipped rod ('policeman') to remove any precipitate adhering to the walls of the beaker or to the stirring rod, and transfer the precipitate to the filter paper. Wash the precipitate with small portions of hot water. Direct the jet as near the top of the filter paper as possible, and let each portion of the wash solution run through before adding the next. Continue the washing until about 5 mL of the wash solution gives no opalescence with a drop or two of silver nitrate solution. Eight or ten washings are usually necessary.

Fold the moist paper around the precipitate and place it in a porcelain or silica (Vitreosil) crucible, previously ignited to redness, cooled in a desiccator and weighed. Dry the paper by placing the loosely covered crucible upon a triangle several centimetres above a small flame. Then gradually increase the heat until the paper chars and volatile matter is expelled. Do not allow the paper to burst into flame, as mechanical loss may thus ensue. When the charring is complete, raise the temperature of the crucible to dull redness and burn off the carbon with free acess of air† (crucible slightly inclined with cover displaced, Section 3.37). When the precipitate is white,‡ ignite the crucible at a red heat for 10–15 minutes. Then allow the crucible to cool somewhat in the air, transfer it to a desiccator, and, when cold, weigh the crucible and contents. Repeat the ignition with 10-minute periods of heating, subsequent cooling in a desiccator, etc., until constant weight (± 0.0002 g) is attained.

Calculate the percentage of SO_4 in the sample.

---

* An equivalent result is obtained by allowing the solution to stand at the laboratory temperature for about 18 hours.
† Any dark matter on the crucible cover may be removed by placing it, clean side down, on a triangle, and heating it for some time.
‡ If the precipitate is slightly discoloured, add a drop or two of dilute sulphuric acid, evaporate gently, etc.

(2) Filtering crucible method. Clean, ignite, and weigh either a porcelain filtering crucible or a Vitreosil filtering crucible (porosity, No. 4). Carry out the ignition either upon a crucible ignition-dish or by placing the crucible inside a nickel crucible at a red heat (or, if available, in an electric muffle furnace at 600–800 °C), allow to cool in a desiccator and weigh. Filter the supernatant liquid, after digestion of the precipitate, through the weighed crucible, using gentle suction. Reject the filtrate, after testing for complete precipitation with a little barium chloride solution. Transfer the precipitate to the crucible and wash with warm water until 3–5 mL of the filtrate give no precipitate with a few drops of silver nitrate solution. Dry the crucible and precipitate in an oven at 100–110 °C, and then ignite in a manner similar to that used for the empty crucible for periods of 15 minutes until constant weight is attained (see Note).

Note. A rapid method for weighing the precipitate is as follows. (This procedure should *not* be employed by elementary students or beginners in the study of quantitative analysis.) Filter off the precipitated barium sulphate through a weighed filtering crucible (sintered-glass, or porcelain) and wash it with hot water until the chloride reaction of the washings is negative. Then wash five or six times with small volumes of ethanol, followed by five or six small volumes of anhydrous diethyl ether. Suck the precipitate dry on the pump for 10 minutes, wipe the outside of the crucible dry with a clean linen cloth, leave in a vacuum desiccator for 10 minutes (or until constant in weight), and weigh as $BaSO_4$. The result is of a moderate order of accuracy.

11.73 SULPHIDE

Determination of sulphur in mineral sulphides: Introduction. The methods to be described apply to most insoluble sulphides. In these the sulphur is oxidised to sulphuric acid, and determined as barium sulphate. Two procedures are available for effecting the oxidation.

A. Dry Process: Discussion. The oxidation is carried out by fusion with sodium peroxide, or, less efficiently, with sodium carbonate and potassium nitrate:

$$2FeS_2 + 15Na_2O_2 = Fe_2O_3 + 4Na_2SO_4 + 11Na_2O$$

The sulphide is fused with the sodium peroxide in an iron or nickel crucible (platinum is strongly attacked — Sections 3.20 and 3.31), the fused mass treated with water, filtered, and acidified. The excess peroxide is removed by boiling, and the sulphate ion precipitated with barium chloride. The decomposition of the sulphide is rapid, but the method has several disadvantages. Amongst these may be mentioned: the slight attack on the metal crucible, thus preventing the subsequent determination of the metal content of the sample; the introduction of appreciable quantities of sodium salts, thus increasing the error due to co-precipitation (Sections 11.5 and 11.72); and the possible contamination by sulphur from the flame gases, since sulphur dioxide is rapidly absorbed by the alkaline melt. The last error may be minimised by keeping the crucible covered during the ignition.

Procedure. Dry some finely powdered pyrites* at 100 °C for 1 hour. Place about 1 g of anhydrous sodium carbonate into the crucible, and weigh accurately into it 0.4–0.5 g of the pyrites. Add 5–6 g of sodium peroxide, and mix well with a

---

* Burnt Pyrites, No. 45AG; available from the Bureau of Analysed Samples, is suitable.

stout copper or nickel wire or with a thin glass rod. Wipe the wire or rod, if necessary, with a small piece of quantitative filter paper, and add the latter to the crucible; cover the mixture with a thin layer of peroxide. Place the covered crucible on a triangle, and heat it with a very small flame. Increase the temperature gradually until after 10–15 minutes the crucible is at a *dull* red heat (the lower the temperature, the less is the crucible attacked) and just sufficiently hot to keep the mass completely fused. Remove the cover occasionally and examine the contents; be sure that the whole mass is fluid. Maintain the mass fluid for 15 minutes to complete the oxidation. Allow to cool, extract the crucible with water in a covered 600 mL beaker, rinse off the crucible-cover into the beaker, remove the crucible with a glass rod and wash it well; dilute to 300 mL. Boil the solution for 15 minutes in order to destroy the excess of peroxide ($Na_2O_2 + 2H_2O = 2NaOH + H_2O_2$), neutralise part of the alkali by adding 5–6 mL of concentrated hydrochloric acid with stirring, add a Whatman 'accelerator' or a quarter of an 'ashless tablet', and filter through a Whatman No. 541 filter paper. Wash the residue at least ten times with hot 1 per cent sodium carbonate solution (10–20 mL portions). Acidify the combined filtrate and washings contained in an 800–1000 mL beaker with concentrated hydrochloric acid, using methyl red or methyl orange as indicator, and add 2 mL of acid in excess. Dilute, if necessary, to 600 mL, and heat to boiling. Precipitate the sulphate by the slow addition with stirring of a boiling 5 per cent solution of barium chloride; the latter is added in slight excess of the calculated amount required, assuming the pyrites to be pure FeS_2. Complete the determination as in Section 11.72.

Calculate the percentage of sulphur in the sample.

B. Wet process: Discussion. The sulphide is oxidised (1) by bromine in carbon tetrachloride solution, followed by nitric acid, (2) by sodium chlorate and hydrochloric acid, or (3) by a mixture of nitric and hydrochloric acids and a little bromine. The use of the first-named oxidising agent will be described; the reaction may be represented by:

$$2FeS_2 + 6HNO_3 + 15Br_2 + 16H_2O = 2Fe(NO_3)_3 + 4H_2SO_4 + 30HBr$$

The method has the advantage of not introducing any metallic ions, but it is essential to remove the excess of nitric acid (see Discussion in Section 11.72). The action is slower than by the fusion method.

Procedure. Dry some finely powdered iron pyrites (Note 1) at 100 °C for 1 hour. Weigh out accurately 0.4–0.5 g of the pyrites into a dry 400 mL beaker, add 6 mL of a mixture of 2 volumes of pure liquid bromine and 3 volumes of pure carbon tetrachloride (**fume cupboard!**), and cover with a clockglass. Allow to stand in the fume cupboard for 15–20 minutes and swirl the contents of the beaker occasionally during this period. Then add 10 mL of concentrated nitric acid down the side of the beaker, and allow to stand for another 15–20 minutes, swirling occasionally as before. Heat the covered beaker below 100 °C by placing it on a hotplate or thermostatically controlled water bath until all action has ceased and most of the bromine has been expelled (about 1 hour). Raise the clockglass cover slightly, or displace it to one side, and evaporate the liquid to dryness on the steam bath. Add 10 mL concentrated hydrochloric acid, mix well, and again evaporate to dryness to eliminate most of the nitric acid. Place the beaker in an oven or air bath at 95–100 °C for 30–60 minutes in order to dehydrate any silica which may be present (Note 2). If the dry residue is heated at a temperature above 100 °C, loss of sulphuric acid may occur and the determination will be rendered useless. Moisten the cold, dry residue with 1–2 mL of concentrated hydrochloric acid and, after an interval of 3–5 minutes, dilute with 50 mL of hot water, and rinse the sides of the beaker and the coverglass with water. Digest the contents of the beaker at 100 °C for 10 minutes in order to dissolve all soluble salts. Allow the solution to cool for 5 minutes, and add 0.2–0.3 g of aluminium powder to reduce the iron(III). Gently swirl (or stir) until the solution becomes colourless. Allow to cool, add a Whatman 'accelerator', stir, and rinse down the coverglass and the sides of the beaker. Filter through a Whatman No. 540 paper, and collect the filtrate in an 800 mL beaker; wash the filter thoroughly with hot water. Dilute the combined filtrate and washings to 600 mL and add 2 mL of concentrated hydrochloric acid. Precipitate the sulphate in the cold (Note 3) by running in from a burette, *without stirring*, a 5 per cent solution of barium chloride at a rate not exceeding $5 \, mL \, min^{-1}$ until an excess of 5–10 mL is present (Note 4). When all the precipitant has been added, stir gently and allow the precipitate to settle for 2 hours, but preferably overnight. Filter through a No. 540 filter paper or, preferably, through a porcelain filtering crucible, wash with warm water until free from chloride, and ignite to constant weight.

Calculate the percentage of sulphur in the sample.

Notes. (1) The procedure is applicable to most **mineral sulphides**; many of these contain silica, and provision is made for the removal of this impurity in the experimental details.

(2) If the iron pyrites or the sample of sulphide contains no appreciable proportion of silica, the heating at 95–100 °C may be omitted.

(3) If a drop or two of tin(II) chloride solution is added to prevent re-oxidation of the Fe(II) salt by air, precipitation of the barium sulphate may be made in boiling solution according to the usual procedure (Section 11.72).

(4) Calculate the volume of 5 per cent barium chloride solution which must be added from the approximate sulphur content of the iron pyrites FeS_2 or of the mineral sulphide.

11.74 SULPHITE

Determination of sulphite by oxidation to sulphate and precipitation as barium sulphate: Discussion. Sulphites may be readily converted into sulphates by boiling with excess of bromine water, sodium hypochlorite, sodium hypobromite, or ammoniacal hydrogen peroxide (equal volumes of 20-volume hydrogen peroxide and 1:1 ammonia solution). The excess of the reagent is decomposed by boiling, the solution acidified with hydrochloric acid, precipitated with barium chloride solution, and the barium sulphate collected and weighed in the usual manner (Section 11.72).

11.75 THIOCYANATE

Determination as copper(I) thiocyanate, CuSCN. The solution (100 mL) should be neutral or slightly acid (hydrochloric or sulphuric acid), and contain not more than 0.1 g SCN^-. It is saturated with sulphur dioxide in the cold (or 50 mL of freshly prepared saturated sulphurous acid solution added), and then treated dropwise and with constant stirring with about 60 mL of 0.1M copper sulphate solution. The mixture is again saturated with sulphur dioxide (or 10 mL of saturated sulphurous acid solution added), allowed to stand for a few hours, the precipitate collected on a weighed filtering crucible (sintered-glass, or porcelain), washed several times with cold water containing sulphurous acid until the copper is removed [potassium hexacyanoferrate(II) test], and finally once with ethanol. The precipitate is dried at 110–120 °C to constant weight, and weighed as CuSCN.

11.76 THIOSULPHATE

Conversion of thiosulphate to sulphate and determination as barium sulphate: Discussion. Thiosulphates are oxidised to sulphates by methods similar to those described for sulphites (Section 11.74), e.g. by heating on a water bath with an ammoniacal solution of hydrogen peroxide, followed by boiling to expel the excess of the reagent. The sulphate is then determined as barium sulphate, $BaSO_4$.

$$S_2O_3^{2-} \equiv 2SO_4^{2-}$$

11.77 ALTERNATIVE PROCEDURES FOR ANIONS

Alternative procedures for a few anions are summarised here. Section references are to the gravimetric procedure already given for the particular anion.

Fluoride as triphenyltin fluoride ($\not> 40$ mg in about 25 mL almost neutral solution). Add 95 per cent ethanol to the aqueous solution so that it comprises 60–70 per cent of the final volume. Heat to boiling and slowly add twice the calculated quantity of the hot reagent (see Note) with vigorous stirring. Again heat to boiling, then allow to cool while continuing to stir. Allow to stand overnight and then cool for 1 hour in ice. Filter and wash the precipitate with about 50 mL of 95 per cent ethanol saturated with triphenyltin fluoride. Dry at 110 °C and weigh as $(C_6H_5)_3SnF$ (Section 11.59).

Note. Prepare the reagent by shaking vigorously 4.0 g of triphenyltin chloride with 200 mL of 95 per cent ethanol. Filter any undissolved residue and dilute with an equal volume of 95 per cent ethanol.

Perchlorate as potassium perchlorate (*ca* 400 mg as the sodium salt in 25 mL solution). Warm the solution to 80–90 °C and treat with a slight excess of a cold saturated solution of potassium acetate. Allow to cool and after 1 hour collect the precipitate on a weighed porcelain filtering crucible. Wash twice with $0.05M$ potassium chloride and then with four 5 mL portions of equal volumes of anhydrous ethyl acetate and anhydrous butan-1-ol. Dry the precipitate at 110 °C for 30–60 minutes and then heat at 350 °C in an electric furnace for 15 minutes. Weigh as $KClO_4$ (Section 11.68).

Phosphate as $Mg_2P_2O_7$ ($\not> 100$ mg of P_2O_5 in 50–100 mL of neutral or weakly acid solution). Add 3 mL of concentrated hydrochloric acid and a few drops of methyl red indicator. Introduce 25 mL of magnesia mixture (see Note), then slowly add pure concentrated ammonia solution, while stirring vigorously, until the indicator turns yellow. The procedure from this stage is the same as that described for the determination of magnesium as $Mg_2P_2O_7$ (Section 11.52); see also Section 11.69.

Note. Prepare by dissolving 25 g magnesium chloride ($MgCl_2,6H_2O$) and 50 g ammonium chloride in 250 mL of water. Add a slight excess of ammonia solution, allow to stand overnight and filter if necessary. Acidify with dilute hydrochloric acid, add 1 mL concentrated hydrochloric acid and dilute to 500 mL.

Phosphite as $Mg_2P_2O_7$. Treat the aqueous phosphite (100 mL) with 5 mL concentrated nitric acid, evaporate to a small volume on a water bath, add 1 mL fuming nitric acid and heat again. Dilute the solution and carry out the determination as just described for phosphate, weighing as $Mg_2P_2O_7$ (Section 11.70).

Thiocyanate as $BaSO_4$. Treat the alkali thiocyanate solution with excess bromine water and heat for 1 hour on a water bath. Acidify the solution with hydrochloric acid and determine the sulphate formed by precipitating and weighing as $BaSO_4$ (Section 11.72); see also Section 11.75.

$$SCN^- + 4Br_2 + 4H_2O = SO_4^{2-} + 7Br^- + 8H^+ + BrCN$$

Thiosulphate as Ag_2S. Add a slight excess of $0.1M$ silver nitrate solution to the cold, almost neutral, thiosulphate solution. Heat at 60 °C in a covered vessel and, after cooling, filter and wash the silver sulphide precipitate with ammonium nitrate solution, water and finally with ethanol. Dry at 110 °C and weigh as Ag_2S (Section 11.76).

11.78 THERMOGRAVIMETRIC EXPERIMENTS

A limited number of thermogravimetric experiments are outlined below. For more detailed information on these and other studies the reader is referred to the publications listed in the Bibliography at the end of Part D, page 499.

When using a modern thermobalance which incorporates an electronic microbalance requiring small sample weights, the following operating precautions should be noted.

(a) The weight of sample selected is dependent upon the actual weight loss anticipated.

(b) The crucible should not be handled, because of the danger of transferring grease or moisture to the crucible. A platinum crucible may be cleaned by placing it in dilute nitric acid. If the platinum crucible is heavily contaminated it may be cleaned by heating with a sodium carbonate/sodium nitrate fusion mixture.

(c) A representative sample should be taken from the original batch. If the material is thought to be inhomogeneous, several samples should be run and different results will confirm the inhomogeneity. The sample particle size should be smaller than 100 mesh ($< 150 \mu$m) to ensure that an even layer is distributed in the crucible.

(d) The method of obtaining a sample depends upon the nature of the material; thus a circular disc may be cut from a film of material by the use of an appropriate cork borer or leather punch. Fibrous material, which does not pack easily, may be squeezed between metal foil before being transferred to the crucible. Liquid samples may be transferred to the crucible by means of a hypodermic syringe. Air-sensitive samples should be loaded into the crucible in a glove box and transferred rapidly to the thermobalance which should be set up all ready for a dry inert gas flow.

497

(*e*) **Materials which creep or froth should not be used in a thermobalance.** It is always sound practice to heat the test material in a small crucible in an oven or muffle furnace to ascertain whether or not there is any creeping or frothing *before* using the sample for thermogravimetry. Considerable damage can occur to a thermobalance if samples are not monitored in this way prior to analysis in the apparatus.

The following experiments are designed to make the operator familiar with the use of the thermobalance.

A. The thermal decomposition of calcium oxalate monohydrate. This determination may be carried out on any standard thermobalance. In all cases the manufacturer's handbook should be consulted for full detailed instructions for operating the instrument.

Initially, zero the balance on the 10 mg range with an empty crucible in position and use an air flow of 10 mL min^{-1}. Weigh accurately about 2 mg of the calcium oxalate monohydrate directly into the crucible and record the weight on the chart. The recorder variable range may now be used to expand the sample weight to 100 per cent of full scale. Select a suitable heating rate (30 °C min^{-1}) and record the pyrolysis curve of calcium oxalate monohydrate from ambient temperature to 1000 °C in terms of percentage sample weight loss. From the TG curve estimate the purity of the calcium oxalate (see Section 11.9).

As additional experiments, investigate the decomposition of calcium oxalate in a static air atmosphere and in a nitrogen atmosphere at a flow rate of 10 mL min^{-1}. Compare the final stage of the decomposition, i.e. the conversion of calcium carbonate to calcium oxide, using different furnace atmospheres.

B. The thermal decomposition of copper sulphate pentahydrate. Follow the procedure outlined in (A) above, but in this case weigh out accurately about 6 mg of the copper sulphate. Record the thermal decomposition of copper sulphate from ambient temperature to 1000 °C using a heating rate of 10 °C min^{-1} and an air atmosphere with a flow rate of 10 mL min^{-1}. Examine the effect of varying the heating rate on the dehydration reactions by selecting rates of 2, 20 and 100 °C min^{-1} in addition to 10 °C min^{-1} as used previously. Further experiments may be designed to study the effect of differing particle size on the dehydration reactions of copper sulphate pentahydrate.

C. Other useful substances to study. The following substances show interesting pyrolysis curves and an assessment of the purity of these materials may be investigated:

cadmium sulphate, $3CdSO_4,8H_2O$;

ammonium magnesium phosphate, $MgNH_4PO_4,6H_2O$; and

the disodium salt of ethylenediaminetetra-acetic acid, $Na_2EDTA,2H_2O$.

11.79   REFERENCES FOR PART D

1. Anon, *Information Bulletin* No. 36, International Union of Pure and Applied Chemistry, Aug. 1974
2. C Woodward and H N Redman *High-Precision Titrimetry*, Society for Analytical Chemistry, London, 1973

3. W Ostwald, *Scientific Foundations of Analytical Chemistry*, 1895, p 118; see also A R Hantzsch, *Ber. Dtsch. Chem. Gesell.*, 1908, **61**, 1171, 1187
4. J N Brønsted, *Rec. Trav. Chim.*, 1923, **42**, 718; T M Lowry, *Trans. Farad. Soc.*, 1924, **20**, 13
5. S Siggia, J G Hanna and I R Kervenski, *Anal. Chem.*, 1950, **22**, 1295
6. J S Fritz, *Anal. Chem.*, 1954, **26**, 1701
7. C W Foulk and M Hollingsworth, *J Am. Chem. Soc.*, 1923, **45**, 1223
8. H N Wilson, *Analyst*, 1951, **76**, 65
9. J S Fritz, *Anal. Chem.* 1953, **25**, 407
10. W J Williams, *Handbook of Anion Determinations*, Butterworths, London, 1979, p 350
11. T S West, *Complexometry*, 3rd edn, BDH Chemicals Ltd, Poole, 1969
12. Anon, *Colour Index*, 2nd edn, Society of Dyers and Colourists, Bradford, 1956
13. R A Close and T S West, *Talanta*, 1960, **5**, 221
14. Anon, *Laboratory Waste Disposal Manual*, 2nd edn, Manufacturing Chemists Association, Washington, DC, 1969 (revised edition 1974)
15. P J Gaston, *The Care, Handling and Disposal of Dangerous Chemicals*, Northern Publishers Ltd, Aberdeen, 1964
16. I L Marr and M S Cresser, *Environmental Chemical Analysis*, International Textbook Company, Glasgow, 1983, p 121
17. K C Thompson, D Mendham, D Best and K E de Casseres, *Analyst*, 1986, **111**, 483
18. L W Winkler, *Ber. Dtsch. Chem. Gesell.*, 1888, **21**, 2843
19. Ref. 16, pp 116–117
20. L W Andrews, *J. Am. Chem. Soc.*, 1903, **25**, 76
21. G S Jamieson, *Volumetric Iodate Methods*, Reinhold, New York, 1926
22. G J Moody and J D R Thomas, *J. Chem. Educ.*, 1963, **40**, 151, and *Education in Chemistry*, 1964, **1**, 214
23. D A Skoog, D M West and F J Holler, *Fundamentals of Analytical Chemistry*, 5th edn, Holt, Rinehart and Winston, New York, 1987
24. A I Vogel, *Elementary Practical Organic Chemistry*, Part III, Quantitative Organic Analysis, Longmans Green & Co., London, 1958
25. C M Ellis and A I Vogel, *Analyst*, 1956, **81**, 693
26. T B Smith, *Analytical Processes*, 2nd edn, Arnold, London, 1940
27. H A Laitinen and W E Harris, *Chemical Analysis*, 2nd edn, McGraw-Hill, New York, 1975
28. R Grzeskowiak and T A Turner, *Talanta*, 1973, **20**, 351
29. C Duval and M de Clercq, *Anal. Chim. Acta.*, 1951, **5**, 282
30. E K Simons and A E Newkirk, *Talanta*, 1964, **11**, 549
31. A W Coats and J P Redfern, *Analyst*, 1963, **88**, 906
32. R C Turner, I Hoffman and D Chen, *Can. J. Chem.*, 1963, **41**, 243
33. W W Wendlandt, *Anal. Chem.*, 1960, **32**, 848
34. W J Blaedel and H T Knight, *Anal. Chem.*, 1954, **26**, 741
35. C J Keattch and D Dollimore, *An Introduction to Thermogravimetry*, 2nd edn, London, Heyden, 1975

11.80 SELECTED BIBLIOGRAPHY FOR PART D

1. D Cooper and C Doran, *Classical Methods of Chemical Analysis*, Vol. I, ACOL–Wiley, Chichester, 1987
2. J Mendham, D Dodd and D Cooper, *Classical Methods of Chemical Analysis*, Vol II, ACOL–Wiley, Chichester, 1987
3. D A Skoog and D M West, *Analytical Chemistry — An Introduction*, 4th edn, Holt, Rinehart & Winston, New York, 1986
4. E Harns and B Kratchvil, *An Introduction to Chemical Analysis*, Holt, Rinehart & Winston, New York, 1982
5. O Budevsky, *Foundations of Chemical Analysis*, Wiley, New York, 1979

6. A Vincent, *Oxidation and Reduction In Inorganic and Analytical Chemistry*, Wiley, Chichester, 1985
7. L Meites, *Handbook of Analytical Chemistry*, McGraw-Hill, New York, 1963
8. R Pribil, *Analytical Applications of EDTA and Related Compounds*, Pergamon Press, Oxford, 1972
9. G Schwarzenbach and H Flaschka, *Complexometric Titrations*, 2nd edn, Methuen and Co., London, 1969
10. W Wagner and C J Hull, *Inorganic Titrimetric Analysis*, Marcel Dekker Inc., New York, 1971
11. C L Wilson and D W Wilson, *Comprehensive Analytical Chemistry*, Elsevier, Amsterdam, 1962
12. L F Hamilton, S G Simpson and D W Ellis, *Calculations of Analytical Chemistry*, 7th edn, McGraw-Hill, New York, 1969
13. L Gordon, M L Salutsky and H H Willard, *Precipitation from Homogeneous Solution*, John Wiley, New York, 1959
14. H Flaschka and A J Barnard, Jr, Tetraphenylboron (TPB) as an analytical reagent. In: *Advances in Analytical Chemistry and Instrumentation*, Vol I, C N Reilley (Ed), Interscience Publishers, New York, 1960
15. F E Beamish and J A Page, Inorganic gravimetric and volumetric analysis, *Anal. Chem.*, 1956, **28**, 694; 1958, **30**, 805
16. F E Beamish and A D Westland, Volumetric and gravimetric analytical methods for inorganic compounds, *Anal. Chem.*, 1960, **32**, 249R
17. C L Wilson and D W Wilson (Eds), *Comprehensive Analytical Chemistry*, Vol 1A, Classical Analysis, Elsevier, Amsterdam and London, 1960
18. L Erdey, *Gravimetric Analysis*, Part 1 (1963) and Parts 2 and 3 (1965), Pergamon Press, Oxford
19. N H Furman (Ed), *Standard Methods of Chemical Analysis*, 6th edn, Vol. 1, The Elements, Van Nostrand, Princeton, NJ, 1962
20. Anon, *Organic Reagents for Metals and for Certain Radicals*, Vol. II, Hopkin and Williams, London, 1964
21. C Duval, *Inorganic Thermogravimetric Analysis*, 2nd edn, Elsevier, Amsterdam, 1963
22. D J Shaw, *Introduction to Colloid and Surface Chemistry*, 3rd edn, Butterworths, London, 1980
23. I M Kolthoff and P J Elving (Eds), *Treatise on Analytical Chemistry*, Part II (17 volumes), dealing with Analytical methods for the elements, Wiley, New York, 1959–1981; some volumes in 2nd edn from 1978
24. C J Keatch, *An Introduction to Thermogravimetry*, 2nd edn, Heyden, London, 1975

PART E
ELECTROANALYTICAL METHODS

CHAPTER 12
ELECTRO–GRAVIMETRY

12.1 THEORY OF ELECTRO–GRAVIMETRIC ANALYSIS

In electro-gravimetric analysis the element to be determined is deposited electrolytically upon a suitable electrode. Filtration is not required, and provided the experimental conditions are carefully controlled, the co-deposition of two metals can often be avoided. Although this procedure has to a large extent been superseded by potentiometric methods based upon the use of ion-selective electrodes (see Chapter 15), the method, when applicable has many advantages. The theory of the process is briefly discussed below in order to understand how and when it may be applied: for a more detailed treatment see Refs 1–9.

Electro-deposition is governed by Ohm's Law and by Faraday's two Laws of Electrolysis (1833–1834). The latter state:

1. The amounts of substances liberated (or dissolved) at the electrodes of a cell are directly proportional to the quantity of electricity which passes through the solution.
2. The amounts of different substances liberated or dissolved by the same quantity of electricity are proportional to their relative atomic (or molar) masses divided by the number of electrons involved in the respective electrode processes.

It follows from the Second Law that when a given current is passed in series through solutions containing copper(II) sulphate and silver nitrate respectively, then the weights of copper and silver deposited in a given time will be in the ratio of 63.55/2 to 107.87.

Ohm's Law expresses the relation between the three fundamental quantities, current, electromotive force, and resistance:

The current I is directly proportional to the electromotive force E and indirectly proportional to the resistance R, i.e.

$$I = E/R$$

Electrical units. The fundamental SI unit is the unit of current which is called the **ampere** (A), and which is defined as the constant current which, if maintained in two parallel rectilinear conductors of negligible cross-section and of infinite length and placed one metre apart in a vacuum, would produce between these conductors a force equal to 2×10^{-7} newton per metre length.

The unit of electrical potential is the **volt** (V) which is the difference of potential between two points of a conducting wire which carries a constant current of one ampere, when the power dissipated between these two points is one joule per second.

The unit of electrical resistance is the **ohm** (Ω) which is the resistance between two points of a conductor when a constant difference of potential of one volt applied between these two points produces a current of one ampere.

The unit quantity of electricity is the **coulomb** (C), and is defined as the quantity of electricity passing when a current of one ampere flows for one second.

To liberate one mole of electrons, or of a singly charged ion, will require $L \times e$ coulombs, where L is the Avogadro constant $(6.022 \times 10^{23}\,\mathrm{mol}^{-1})$ and e is the elementary charge $(1.602 \times 10^{-10}\,\mathrm{C})$; the resultant quantity $(9.647 \times 10^{4}\,\mathrm{C\,mol}^{-1})$ is termed the Faraday constant (F).

12.2 SOME TERMS USED IN ELECTRO-GRAVIMETRIC ANALYSIS

Voltaic (galvanic) and electrolytic cells. A cell consists of two electrodes and one or more solutions in an appropriate container. If the cell can furnish electrical energy to an external system it is called a voltaic (or galvanic) cell. The chemical energy is converted more or less completely into electrical energy, but some of the energy may be dissipated as heat. If the electrical energy is supplied from an external source the cell through which it flows is termed an electrolytic cell and Faraday's Laws account for the material changes at the electrodes. A given cell may function at one time as a galvanic cell and at another as an electrolytic cell: a typical example is the lead accumulator or storage cell.

Cathode. The cathode is the electrode at which reduction occurs. In an electrolytic cell it is the electrode attached to the negative terminal of the source, since electrons leave the source and enter the electrolysis cell at that terminal. The cathode is the positive terminal of a galvanic cell, because such a cell accepts electrons at this terminal.

Anode. The anode is the electrode at which oxidation occurs. It is the positive terminal of an electrolysis cell or the negative terminal of a voltaic cell.

Polarised electrode. An electrode is polarised if its potential deviates from the reversible or equilibrium value. An electrode is said to be 'depolarised' by a substance if that substance lowers the amount of polarisation.

Current density. The current density is defined as the current per unit area of electrode surface. It is generally expressed in amperes per square centimetre (or per square decimetre) of the electrode surface.

Current efficiency. By measuring the amount of a particular substance that is deposited and comparing this with the theoretical quantity (calculated by Faraday's Laws), the actual current efficiency may be obtained.

Decomposition potential. If a small potential of, say, 0.5 volt is applied to two smooth platinum electrodes immersed in a solution of $1\,M$ sulphuric acid, then an ammeter placed in the circuit will at first show that an appreciable current is flowing, but its strength decreases rapidly, and after a short time it becomes virtually equal to zero. If the applied potential is gradually increased, there is a slight increase in the current until, when the applied potential reaches a certain value, the current suddenly increases rapidly with increase in the e.m.f. It will be observed, in general, that at the point at which there is a sudden increase in

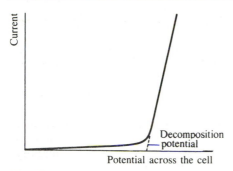

Fig. 12.1

current, bubbles of gas commence to be freely evolved at the electrodes. Upon plotting the current against the applied potential, a curve similar to that shown in Fig. 12.1 is obtained; the point at which the current suddenly increases is evident, and in the instance under consideration is at about 1.7 volts. The potential at this point is termed the 'decomposition potential' and it is at this point that the evolution of both hydrogen and oxygen in the form of bubbles is first observed. We may define the **decomposition potential** of an electrolyte as the minimum external potential that must be applied in order to bring about continuous electrolysis.

If the circuit is broken after the e.m.f. has been applied, it will be observed that the reading on the voltmeter is at first fairly steady, and then decreases, more or less rapidly, to zero. The cell is now clearly behaving as a source of current, and is said to exert a '**back**' or '**counter**' or '**polarisation**' e.m.f., since the latter acts in a direction opposite to that of the applied e.m.f. This back e.m.f. arises from the accumulation of oxygen and hydrogen at the anode and cathode respectively; two gas electrodes are consequently formed, and the potential difference between them opposes the applied e.m.f. When the primary current from the battery is shut off, the cell produces a moderately steady current until the gases at the electrodes are either used up or have diffused away; the voltage then falls to zero. This back e.m.f. is present even when the current from the battery passes through the cell and accounts for the shape of the curve in Fig. 12.1.

The back e.m.f. is usually regarded as being made up of three components: (*a*) the reversible back e.m.f.; (*b*) a concentration polarisation e.m.f.; and (*c*) an activation overpotential.

(a) Reversible back e.m.f. This is the e.m.f. of the voltaic cell set up by the passage of the electrolysis current. Consider the electrolysis of a molar solution* of zinc bromide between smooth platinum electrodes. The application of a potential will result in the deposition of zinc on the cathode (thus producing a zinc electrode) and liberation of bromine at the anode (thus producing a bromine electrode). The reaction at the cathode is:

$$Zn^{2+} + 2e \rightleftharpoons Zn; \text{ i.e. a reduction}$$

---

* i.e. Molar with respect to Zn^{2+}. Ideally, concentration should be replaced by activity; the former is, however, sufficiently accurate for our purpose.

505

and that at the anode is:

$2Br^- \rightleftharpoons Br_2 + 2e$; i.e. an oxidation

The potential of the cathode at 25 °C can be calculated from the expression (Section 2.28):

$$E_{cathode} = E_{Zn}^{\ominus} + \frac{0.0591}{2} \log [Zn^{2+}] = E_{Zn}^{\ominus}$$

since $[Zn^{2+}] = 1 \text{ mol L}^{-1}$.

At the anode:

$$E_{anode} = E_{Br_2}^{\ominus} - \frac{0.0591}{1} \log [Br^-] = E_{Br_2}^{\ominus} - \frac{0.0591}{1} \log 2$$

since $[Br^-] = 2 \text{ mol L}^{-1}$.

The e.m.f. of the resulting cell will therefore be:

$$E_{Zn}^{\ominus} - \left\{ E_{Br_2}^{\ominus} - \frac{0.0591}{1} \log 2 \right\} = 0.76 - (-1.07 - 0.02) = 1.85 \text{ volts}$$

In general, it may be stated that the theoretical back or polarisation e.m.f. E_{back} is given by:

$$E_{back} = E_{cathode} - E_{anode}$$

where $E_{cathode}$ and E_{anode} are the values established by the presence of the electrolysis products furnished by the solution which is electrolysed.

(b) Concentration polarisation e.m.f. (concentration overpotential). In the electrolysis of an acidic solution of copper(II) sulphate between platinum electrodes, concentration changes take place in the solution in the neighbourhood of the electrodes. At the cathode depletion of copper ions occurs near the surface; the reversible potential of the copper electrode therefore shifts in the negative direction. At the anode accumulation of hydrogen ions ($2H_2O \rightarrow O_2 + 4H^+ + 4e$) and perhaps of oxygen (if the solution is not already saturated with it) causes the reversible potential of the oxygen electrode to shift in the positive direction. Both effects tend to increase the back e.m.f. The concentration overpotential is increased by increased current density and decreased by stirring.

(c) Activation overpotential. This is due to the effect of the potential applied to the electrode on the activation energy of the electrode reaction; for details a textbook of physical chemistry should be consulted (see Section 3.39). The effect is most marked when gases are liberated at the electrode, and varies according to the nature of the electrode material.

Overpotential. It has been found by experiment that the decomposition voltage of an electrolyte varies with the nature of the electrodes employed for the electrolysis and is, in many instances, higher than that calculated from the difference of the *reversible* electrode potentials. The excess voltage over the calculated back e.m.f. is termed the overpotential. Overpotential may occur at the anode as well as at the cathode. The decomposition voltage E_D is therefore:

$$E_D = E_{cathode} + E_{o.c.} - (E_{anode} + E_{o.a.})$$

where $E_{o.c.}$ and $E_{o.a.}$ are the overpotentials at the cathode and anode respectively.

The overpotential at the anode or cathode is a function of the following variables.

1. The nature and the physical state of the metal employed for the electrodes. The fact that reactions involving gas evolution usually require less overpotential at platinised than at polished platinum electrodes is due to the much larger effective area of the platinised electrode and thus the smaller current density at a given electrolysis current.
2. The physical state of the substance deposited. If it is a metal, the overpotential is usually small; if it is a gas, such as oxygen or hydrogen, the overpotential is relatively great.
3. The current density employed. For current densities up to $0.01 \, A \, cm^{-2}$, the increase in overpotential is very rapid; above this figure the increase in overpotential continues, but less rapidly.
4. The change in concentration, or the concentration gradient, existing in the immediate vicinity of the electrodes; as this increases, the overpotential rises.

The overpotential of hydrogen is of great importance in electrolytic determinations and separations. It is greatest with the relatively soft metals, such as bismuth (0.4 V), lead (0.4 V), tin (0.5 V), zinc (0.7 V), cadmium (1.1 V) and mercury (1.2 V): the overvoltage values given refer to the electrolysis of $0.05 \, M$ sulphuric acid with a current density of $0.01 \, A \, cm^{-2}$, and can be compared with the value (0.09 V) for a bright platinum electrode under similar conditions. The existence of hydrogen overpotential renders possible the electro-gravimetric determination of metals, such as cadmium and zinc, which otherwise would not be deposited before the reduction of hydrogen ion. In alkaline solution, the hydrogen overpotential is slightly higher (0.05–0.03 volt) than in acid solution.

Oxygen overpotential is about 0.4–0.5 volt at a polished platinum anode in acid solution, and is of the order of 1 volt in alkaline solution with current densities of $0.02–0.03 \, A \, cm^{-2}$. As a rule the overpotential associated with the deposition of metals on the cathode is quite small (about 0.1–0.3 volt) because the depositions proceed nearly reversibly.

12.3 COMPLETENESS OF DEPOSITION

For the electrolysis of a solution to be maintained, the potential applied to the electrodes of the cell ($E_{app.}$) must overcome the decomposition potential of the electrolyte (E_D) (which as shown above includes the back e.m.f. and also any overpotential effects), as well as the electrical resistance of the solution. Thus, $E_{app.}$ must be equal to or greater than ($E_D + IR$), where I is the electrolysis current, and R the cell resistance. As electrolysis proceeds, the concentration of the cation which is being deposited decreases, and consequently the cathode potential changes.

If the relevant ionic concentration in the solution is c_i, and the ion concerned has a charge number of 2, then at a temperature of 25 °C, the cathode potential

507

will have a value given by:

$$E_1 = E_M^\ominus + \frac{0.0591}{2} \log c_i = E_M + 0.0296 \log c_i \text{ volts}$$

If the ionic concentration is reduced by deposition to one ten-thousandth of its original value, thus giving an accuracy of 0.01 per cent in the determination, the new cathode potential will be:

$$E_2 = E_M + 0.926 \log(c_i \times 10^{-4}) = E_M + 0.0296 \log c_i + 0.0296 \log 10^{-4}$$

$$= (E_M + 0.0296 \log c_i) - 4 \times 0.0296 = E_l - 0.118 \text{ volts}$$

It follows that if the original solution contains two cations whose deposition potentials differ by about 0.25 V, then the cation of higher deposition potential should be deposited without any contamination by the ion of lower deposition potential. In practice, in may be necessary to take steps to ensure that the cathode potential is unable to fall to a level where deposition of the second ion may occur (see Section 12.6).

12.4 ELECTROLYTIC SEPARATION OF METALS

If a current is passed through a solution containing copper(II), hydrogen and cadmium(II) ions, copper will be deposited first at the cathode. As the copper deposits, the electrode potential decreases, and when the potential becomes equal to that given by the hydrogen ions, hydrogen gas will form at the cathode. The potential at the cathode will remain virtually constant as long as hydrogen is evolved, which would mean as long as any water remains, and it is therefore unable to become sufficiently negative to permit the deposition of cadmium. Thus, metal ions with positive reduction potentials may be separated, without external control of the cathode potential, from metal ions having negative reduction potentials.

Silver can be readily separated from copper, even though they both have positive reduction potentials, because the difference between the two values is large (silver, $+0.779$ V; copper, $+0.337$ V), but as indicated above, when the standard potentials of the two metals differ only slightly, the electro-separation is more difficult. An obvious solution to this problem is to decrease the concentration of one of the ions being discharged by incorporating it in a complex ion of large stability constant (Section 2.23). As an illustration of the kind of result achieved, the deposition potentials for $0.1 M$ solutions of ions M^{2+} of the following metals have the values indicated: zinc, $+0.79$ V; cadmium, $+0.44$ V; copper, $+0.34$ V. When 0.1 mole of the corresponding cyanides are dissolved in potassium cyanide to give an excess concentration of potassium cyanide of $0.4 M$, the deposition potentials become: zinc, $+1.18$ V; cadmium, $+0.87$ V; copper, $+0.96$ V.

An interesting application of these results is to the direct quantitative separation of copper and cadmium. The copper is first deposited in acid solution; the solution is then made slightly alkaline with pure aqueous sodium hydroxide, potassium cyanide is added until the initial precipitate just re-dissolves, and the cadmium is deposited electrolytically.

508

12.5 CHARACTER OF THE DEPOSIT

The ideal deposit for analytical purposes is adherent, dense, and smooth; in this form it is readily washed without loss. Flaky, spongy, powdery, or granular deposits adhere only loosely to the electrode, and for this and other reasons should be avoided.

As a rule, more satisfactory deposits are obtained when the metal is deposited from a solution in which it is present as complex ions rather than as simple ions. Thus silver is obtained in a more adherent form from a solution containing the $[Ag(CN_2)]^-$ ion than from silver nitrate solution. Nickel when deposited from solutions containing the complex ion $[Ni(NH_3)_6]^{2+}$ is in a very satisfactory state for drying and weighing. Mechanical stirring often improves the character of the deposit, since large changes of concentration at the electrode are reduced, i.e. concentration polarisation is brought to a minimum.

Increased current density up to a certain critical value leads to a diminution of grain size of the deposit. Beyond this value, which depends upon the nature of the electrolyte, the rate of stirring, and the temperature, the deposits tend to become unsatisfactory. At sufficiently high values of the current density, evolution of hydrogen may occur owing to the depletion of metal ions near the cathode. If appreciable evolution of hydrogen occurs, the deposit will usually become broken up and irregular; spongy and poorly adherent deposits are generally obtained under such conditions. For this reason the addition of nitric acid or ammonium nitrate is often recommended in the determination of certain metals, such as copper; bubble formation is thus considerably reduced. The action of the nitrate ion at the copper cathode can be represented by:

$$NO_3^- + 10H^+ + 8e \rightleftharpoons NH_4^+ + 3H_2O$$

The nitrate ion is reduced to ammonium ion at a lower (i.e. less negative) cathode potential than that at which hydrogen ion is discharged, and, therefore, acts to decrease hydrogen evolution. The nitrate ion acts as a **cathodic depolariser**.

Raising the temperature, say to between 70 and 80 °C, often improves the physical properties of the deposit. This is due to several factors, which include the decrease in resistance of the solution, increased rate of stirring and of diffusion.

12.6 ELECTROLYTIC SEPARATION OF METALS WITH CONTROLLED CATHODE POTENTIAL

In the common method of electro-gravimetric analysis, a potential slightly in excess of the decomposition potential of the electrolyte under investigation is applied, and the electrolysis allowed to proceed without further attention, except perhaps occasionally to increase the applied potential to keep the current at approximately the same value. This procedure, termed constant-current electrolysis, is (as explained in Section 12.4) of limited value for the separation of mixtures of metallic ions. The separation of the components of a mixture where the decomposition potentials are not widely separated may be effected by the application of **controlled cathode potential electrolysis**. An auxiliary standard electrode (which may be a saturated calomel electrode with the tip of the salt bridge very close to the cathode or working electrode) is inserted in the solution, and thus the voltage between the cathode and the reference half-cell may be measured. As shown in Section 12.3, every ten-fold decrease in metal ion concentration makes the cathode potential $0.0591/n$ volt more negative at 25 °C (n is the charge number of the ion). For an accuracy of 0.1 per cent in the determination of the metal M, the concentration of the ions is reduced to 10^{-3} of the original value, and consequently the cathode potential (initially E) will decrease by $3 \times 0.0591/2$ volt $= 0.088$ volt. However, provided the cathode potential does not fall below E_D volts, where E_D is the deposition potential of the ion M^{2+}, the metal M will be deposited free from contamination by any other metal which may be present. Manual control of the cathode potential is possible, but is a tedious operation; however the procedure can be automated by including a potentiostat in the circuit. This is an electronic device which controls the potential difference between the cathode and the reference electrode. A diagram of a suitable circuit for controlled potential electro-deposition is shown in Fig. 12.2(a); Fig. 12.2(b) shows how the reference (saturated calomel) electrode is located in close proximity to the electrode whose potential is being controlled. In this diagram, deposition is occurring at the cathode, which is therefore referred to as the working electrode; the anode may be termed the auxiliary electrode, or the secondary electrode.

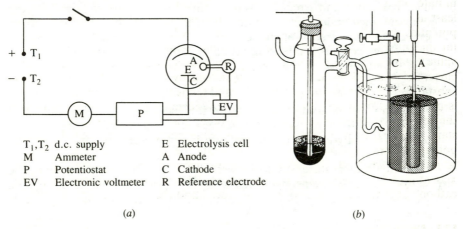

T_1, T_2	d.c. supply	E	Electrolysis cell
M	Ammeter	A	Anode
P	Potentiostat	C	Cathode
EV	Electronic voltmeter	R	Reference electrode

(a)   (b)

Fig. 12.2

At all stages of the electrolysis the electronic voltmeter reads the potential difference between the cathode and the reference electrode, and the required limiting value of this potential difference is entered into the potentiostat. If the measured potential at any instant differs from the pre-set value, the potentiostat will adjust the current flowing to restore the required potential difference.

It must be emphasised that in evaluating the limiting cathode potential to be applied in the separation of two given metals, simple calculation of the equilbrium potentials from the Nernst Equation is insufficient: due account must be taken of any overpotential effects. If we carry out, for each metal, the procedure described in Section 12.2 for determination of decomposition potentials, but include a reference electrode (calomel electrode) in the circuit, then we can ascertain the value of the cathode potential for each current setting and plot the current–potential curves. Schematic current–cathode potential

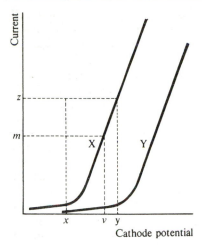

Fig. 12.3

curves for the reduction of two substances X and Y are shown in Fig. 12.3. To initiate the deposition of the substance X, the cathode potential E_c must be at least as large as the value of the deposition potential indicated by x, but the potential should not exceed y, for if it does Y will deposit also. Consequently, for the complete deposition of X, the cathode potential should be limited to a value (v) slightly less than that which corresponds to the potential y. The initial current should not exceed the value indicated by m. As the deposition of the substance X proceeds, the cathode potential tends to become more negative but is prevented from exceeding the value y by decreasing the voltage applied to the cell. To deal with this situation the potentiostat must be programmed to start with a pre-set potential difference value corresponding to the cathode potential v, and then at a pre-determined rate the current is gradually reduced so that ultimately the potential difference corresponding to cathode potential x is attained: at all stages of the operation the potential difference between cathode and reference electrode is monitored and the requisite control exerted.

12.7 APPARATUS

The essential requirements for a constant-current electrolytic determination — a source of direct current (which may be a mains-operated unit producing a rectified smoothed output of 3–15 volts), a variable resistance, an ammeter (reading up to 10 amperes), a voltmeter (10–15 volts), and a pair of platinum electrodes — can be readily assembled in most laboratories, but if a number of determinations are to be performed a commercial electrolysis unit will doubtless be preferred. This will be equipped with rectifier, a motor drive for a paddle-type stirrer or with a magnetic stirrer, and a hotplate.

The electrodes are made of platinum gauze as the open construction assists the circulation of the solution. It is possible to use one of the electrodes as stirrer for the solution, but special arrangements must then be made for connection of the electrolysis current to this electrode, and an independent glass-paddle stirrer or a magnetic stirrer offer a simple alternative.Typical electrodes are the Fischer type depicted in Fig. 12.4; a glass tube is slid into

Fig. 12.4

the loops on the wire of the outer electrode, and the wire to the inner electrode passes through this tube. When it is necessary to know the current density for a particular determination, the area of a gauze electrode may be regarded as approximately equal to twice the area of a foil electrode of the same dimensions.

Use and care of electrodes. Electrodes must be free from grease, otherwise an adherent deposit may not be obtained. For this reason an electrode should never be touched on the deposition surface with the fingers; it should always be handled by the platinum connecting wire attached to the electrode. Platinum electrodes are easily rendered grease-free by heating them to redness in a flame.

Before use, electrodes must be carefully cleaned to remove any previous deposits. Deposits of copper, silver, cadmium, mercury, and many other metals can be removed by immersion in dilute nitric acid (1:1), rinsing with water, then boiling with fresh 1:1 nitric acid for 5–10 minutes, followed by a final washing with water. Deposits of lead dioxide are best removed by means of 1:1 nitric acid containing a little hydrogen peroxide to reduce the lead to the Pb(II) condition; ethanol or oxalic acid may replace the hydrogen peroxide.

After thorough rinsing with distilled water, the electrode is rinsed with 15–20 mL of pure acetone delivered from a small all-glass wash bottle, and is then placed on a clock glass and dried for 3–4 minutes in an electric oven at 110 °C. After cooling for about 5 minutes at the laboratory temperature, it is weighed and is then ready for the determination.

Electrolysis of chloride solutions may be carried out provided that a sufficient amount (1–5 g) of either hydrazinium chloride or of hydroxylammonium chloride is added as an anodic depolariser:

$$N_2H_5^+ \rightleftharpoons N_2 + 5H^+ + 4e; \qquad 2NH_2OH \rightleftharpoons N_2O + 4H^+ + H_2O + 4e$$

If no depolariser is added to an acidic chloride solution, corrosion of the anode occurs and the dissolved platinum is deposited on the cathode, leading to erroneous results and to destruction of the anode. A number of metals (for example, zinc and bismuth) should not be deposited on a platinum surface.

These metals, particularly zinc, appear to react with the platinum in some way, for when they are dissolved off with nitric acid the platinum surface is dulled or blackened. Injury to the platinum can be prevented in these cases by first plating it with copper, and then depositing the metal on this.

A **mercury cathode** finds widespread application for separations by constant current electrolysis. The most important use is the separation of the alkali and alkaline-earth metals, Al, Be, Mg, Ta, V, Zr, W, U, and the lanthanides from such elements as Fe, Cr, Ni, Co, Zn, Mo, Cd, Cu, Sn, Bi, Ag, Ge, Pd, Pt, Au, Rh, Ir, and Tl, which can, under suitable conditions, be deposited on a mercury cathode. The method is therefore of particular value for the determination of Al, etc., in steels and alloys: it is also applied in the separation of iron from such elements as titanium, vanadium, and uranium. In an uncontrolled constant-current electrolysis in an acid medium the cathode potential is limited by the potential at which hydrogen ion is reduced: the overpotential of hydrogen on mercury is high (about 0.8 volt), and consequently more metals are deposited from an acid solution at a mercury cathode than with a platinum cathode.[10]

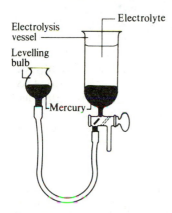

Electrolysis vessel

Electrolyte

Levelling bulb

Mercury

Fig. 12.5

A convenient type of mercury electrode vessel is shown in Fig. 12.5, the electrolysis vessel contains the platinum anode which is inserted into the electrolyte. Electrical contact to the mercury cathode is made by inserting into the mercury pool a short piece of platinum wire sealed into the end of a glass tube; the circuit is completed by dipping a copper connecting wire into mercury contained in the tube. A stirrer must also be included and this must be arranged so that it agitates the surface of the mercury as well as the aqueous solution. When electrolysis is complete, the levelling-bulb is lowered until the mercury reaches the upper end of the stopcock bore, keeping the circuit closed at all times; the stopcock is then turned through 180° and the electrolyte is collected in a suitable vessel.

Controlled-potential separation of many metals can be effected with the aid of the mercury cathode. This is because the optimum control potential and the most favourable solution conditions for a given separation can be deduced from polarograms recorded with the dropping mercury electrode: see Chapter 16.

An important use of a mercury cathode is in the purification of electrolyte solutions, for example the removal of traces of heavy metals from potassium chloride solutions. All such impurities have much more positive deposition

513

potentials than the potassium ion, and hence by carrying out electrolysis of the solution with the cathode potential controlled to a few millivolts more positive than the value for the deposition potential of potassium ions, the heavy metals will be transferred to the mercury. This procedure is of particular significance when specially pure reagents are required for trace analyses.

The electrolysis vessel usually consists of a tall form beaker without a lip: the outer electrode should fit easily into the beaker, but the volume of liquid between electrode and beaker should be kept as small as possible. To prevent loss by spray, the beaker should be covered by a watchglass: this should be drilled to accommodate the electrode connections, and then split into two.

When electrolysis is complete, stirring is stopped, and the electrolysis vessel carefully lowered away (or the electrodes may be carefully raised), **without breaking the electrolysis circuit**; this latter procedure is necessary, since otherwise the electrolyte in contact with the electrode may dissolve some of the deposit. The electrodes are washed **as they emerge** from the solution with a fine stream of distilled water directed uniformly around the upper rims of both electrodes from a wash bottle: the first 10–15 mL will then contain virtually all the electrolyte adhering to the electrodes. It is unnecessary to save subsequent washings, thus avoiding excessive dilution of the residual solution. The electrode is then disconnected and rinsed with pure acetone, and then dried at 100–110 °C for 3–4 minutes. The electrode with its deposit is weighed after cooling for about 5 minutes at the laboratory temperature.

If a **controlled-potential determination** is to be carried out, additional equipment will be required, namely an electronic voltmeter, a potentiostat and a reference electrode. The latter is most commonly a saturated calomel electrode, the construction of which is described in Chapter 14.

12.8 DETERMINATION OF COPPER (CONSTANT CURRENT PROCEDURE)

Discussion. Copper may be deposited from either sulphuric or nitric acid solution, but, usually, a mixture of the two acids is employed. If such a solution is electrolysed with an e.m.f. of 2–3 volts the following reactions occur:

Cathode: $Cu^{2+} + 2e \rightleftharpoons Cu$

$2H^+ + 2e \rightleftharpoons H_2$

Anode: $4OH^- \rightleftharpoons O_2 + 2H_2O + 4e$

The acid concentration of the solution must not be too great, otherwise the deposition of the copper may be incomplete or the deposit will not adhere satisfactorily to the cathode. The beneficial effect of nitrate ion is due to its depolarising action at the cathode:

$NO_3^- + 10H^+ + 8e \rightleftharpoons NH_4^+ + 3H_2O$

The reduction potential of the nitrate ion is lower than the discharge potential of hydrogen, and therefore hydrogen is not liberated. The nitric acid must be free from nitrous acid, as the nitrite ion hinders complete deposition and introduces other complications. The nitrous acid may be removed (*a*) by boiling the nitric acid before adding it, (*b*) by the addition of urea to the solution:

$2H^+ + 2NO_2^- + CO(NH_2)_2 = 2N_2 + CO_2 + 3H_2O$

or most efficiently (c) by the addition of a little sulphamic acid:

$$H^+ + NO_2^- + {}^-O \cdot SO_2 \cdot NH_2 = N_2 + HSO_4^- + H_2O$$

The action is rapid, and the acidity of the electrolyte is unaffected. The error due to nitrous acid is increased by the presence of a large amount of iron; iron is reduced by the current to the iron(II) state, whereupon the nitric acid is reduced. This error may be minimised by the proper regulation of the pH and by the addition of ammonium nitrate instead of nitric acid, or, best, by the removal of the iron prior to the electrolysis, or by complexation with phosphate or fluoride.

The solution should be free from the following, which either interfere or lead to an unsatisfactory deposit: silver, mercury, bismuth, selenium, tellurium, arsenic, antimony, tin, molybdenum, gold and the platinum metals, thiocyanate, chloride, oxidising agents such as oxides of nitrogen, or excessive amounts of iron(III), nitrate or nitric acid. Chloride ion is avoided because Cu(I) is stabilised as a chloro-complex and remains in solution to be re-oxidised at the anode unless hydrazinium chloride is added as depolariser.

The electrolytic deposit should be salmon-pink in colour, silky in texture, and adherent. If it is dark, the presence of foreign elements and/or oxidation is indicated. Spongy or coarsely crystalline deposits are likely to yield high results; they arise from the use of too high current densities or improper acidity and absence of nitrate ion.

Procedure. The solution (100 mL) may contain 0.2–0.3 g of Cu (see Note). Add cautiously 2 mL of concentrated sulphuric acid, 1 mL of concentrated nitric acid (free from nitrous acid by boiling or by the addition of a little urea, or, better, 0.5 g of sulphamic acid), and transfer to, unless already present in, the electrolysis vessel. Clean, dry and weigh the platinum gauze cathode as described in Section 12.7, then assemble the apparatus (complete with a magnetic stirrer) on the electro-deposition apparatus; the electrodes should be clamped so that they are about 80–90 per cent covered by the solution. Cover the beaker with the split clock glass, and following the instructions pertaining to the electrolysis unit, switch on the stirrer, and then (at a minimum setting) the electrolysis current; finally adjust the latter so that a potential difference of 3–4 volts is applied to the cell and the current is 2–4 A. Continue the electrolysis until the blue colour of the solution has entirely disappeared (usually somewhat less than 1 hour), reduce the current to 0.5–1 A, and test for completeness of deposition by rinsing the split clock glass, raising the level of the liquid by about 0.5 cm by the addition of distilled water, and continuing the electrolysis for 15–20 minutes. If no copper plates out on the fresh surface of the cathode, electrolysis may be regarded as complete.

Lower the beaker very slowly, or raise the electrodes, and at the same time direct a continuous stream of distilled water from a wash bottle against the upper edge of the cathode. This washing must be done immediately the cathode is removed out of the solution, and the circuit must not be broken during the process. When the cathode has been thoroughly washed, break the circuit, dip the cathode into a beaker of distilled water, and then rinse it with analytical grade acetone. Dry at 100–110 °C for 3–4 minutes, and weigh after cooling in air for about 5 minutes.

From the increase in weight of the cathode, calculate the copper content of the solution. After the cathode has been weighed, it should be cleaned with nitric acid as described in Section 12.7, and re-weighed; the loss in weight will serve as a check.

Note. Larger quantities of copper may be present, particularly if rapid electrolysis is employed; the quantity given is, however, convenient for instructional purposes. For practice in the determination, prepare the solution *either* by weighing out accurately about 1.0 g of copper sulphate pentahydrate *or* by dissolving about 0.25 g, accurately weighed, of copper in 1:1 nitric acid, boiling to remove nitrous fumes, just neutralising with ammonia solution, and then just acidifying with dilute sulphuric acid and diluting to 100 mL.

12.9 SOME EXAMPLES OF METALS WHICH CAN BE DETERMINED BY ELECTRO–GRAVIMETRY

Constant current procedure. With the exception of lead, which from nitric acid solutions is deposited on the *anode* as PbO_2, the ions listed in Table 12.1 are deposited as metal on the cathode. With the ions indicated by an asterisk in Table 12.1, it is advisable to use a platinum cathode which has been plated with copper before the initial weighing: this is because, in these cases, the deposited metals cannot be readily distinguished on a platinum surface and it is difficult to be certain when deposition is complete.

Table 12.1 Conditions for the determination of metals by electrogravimetry

Ion	Electrolyte	Electrical details
Cd^{2+} *	Potassium cyanide forming $K_2[Cd(CN)_4]$	1.5–2 A; 2.5–3 V
Co^{2+} *	Ammoniacal sulphate	4 A; 3–4 V
Cu^{2+}	Sulphuric acid/nitric acid	2–4 A; 3–4 V
Pb^{2+} *	Tartrate buffer or chloride solution (solubility limits the amount of lead to less than 50 mg per 100 mL)	2 A; 2–3 V
Pb^{2+}	Nitric acid. PbO_2 deposited on anode; use empirical conversion factor of 0.864	5 A; 2–3 V
Ni^{2+} *	Ammoniacal sulphate	4 A; 3–4 V
Ag^+	Potassium cyanide forming $K[Ag(CN)_2]$	0.5–1.0 A; 2.5–3 V
Zn^{2+} *	Potassium hydroxide solution	4 A; 3.5–4.5 V

* Advisable to use a platinum cathode plated with copper before initial weighing: see text.

The separations summarised in Table 12.2 can be readily accomplished electrolytically; the first is dependent upon a large difference in deposition potentials (Section 12.4), and the second upon the fact that Pb^{2+} can be anodically deposited as PbO_2.

Table 12.2 Simple electrolytic separations

Ions	Electrolyte	Electrical details
Cu/Ni	Deposit Cu from H_2SO_4 solution;	2–4 A; 3–4 V
	neutralise with ammonia, add 15 mL concentrated aqueous ammonia and deposit Ni	4 A; 3–4 V
Cu/Pb	Nitric acid solution; deposit Cu on cathode and PbO_2 on anode	1.5–2 A; 2 V

12.10 DETERMINATION OF ANTIMONY, COPPER, LEAD AND TIN IN BEARING METAL (CONTROLLED–POTENTIAL PROCEDURE)

The principles of electrolysis with controlled cathode potential have been discussed in Section 12.6, and the details given below serve to illustrate the procedure. In this case the amounts of copper and antimony (which are deposited simultaneously) are small, and so the cathode potential can be set immediately to the limiting value, but with the higher proportion of tin it can be set initially to a value which is more positive than the limiting value so as to speed up the deposition process.

Procedure. Weigh accurately 0.2–0.4 g of the alloy (as drillings or fine filings) into a small beaker. Dissolve the alloy by warming with a mixture of 10 mL concentrated hydrochloric acid, 10 mL water, and 1 g ammonium chloride (the last-named to minimise the loss of tin as tetrachloride). Solution may be hastened by the addition, drop by drop, of concentrated nitric acid. When all the alloy has dissolved, boil off the excess of chlorine and nitrous fumes, add 5 mL concentrated hydrochloric acid, dilute to 150 mL, and then add 1 g of hydrazinium chloride. Stir the solution efficiently and electrolyse, limiting the cathode potential to −0.36 volts *vs* S.C.E. (saturated calomel electrode); copper and antimony are deposited together. After 30–45 minutes the current becomes constant (usually at about 20 mA): remove the saturated calomel electrode, stop the stirrer, withdraw the electrodes from the solution while washing them with distilled water, and then break the electrolysis circuit. After the normal procedure, weigh the dried cathode.

Separate the copper and antimony by dissolving the deposit in a mixture of 5 mL concentrated nitric acid, 5 mL 40 per cent hydrofluoric acid (**CARE**), and 10 mL water: boil off the nitrogen oxides, dilute to 150 mL, and add dropwise a solution of potassium dichromate until the liquid is distinctly yellow. Deposit the copper by electrolysing the solution at room temperature and limiting the cathode *vs* S.C.E. potential to −0.36 volt. Evaluate the weight of antimony by difference.

To the solution from which the copper and antimony have been separated as above, add 5 mL concentrated hydrochloric acid and 1 g hydrazinium chloride. Electrolyse using a weighed copper-coated cathode after adding sufficient distilled water to cover the electrode. Set the potentiostat to give a cathode potential of −0.6 volts *vs* the S.C.E. changing to −0.7 volt *vs* the S.C.E. over a period of 20 minutes; continue the electrolysis for a further 25 minutes to complete the deposition of lead and tin. Neutralise the electrolyte by adding dilute ammonia solution (1:1), otherwise some tin may re-dissolve during the washing of the electrodes, then remove the cathode, wash, dry and weigh to determine the weight of tin and lead.

Dissolve the deposit from the cathode in 15 mL nitric acid, sp. gr. 1.20, in a 400 mL beaker, and finally wash the cathode with water. Evaporate the resulting solution almost to dryness, cool, and add a further 15 mL nitric acid, sp. gr. 1.2. Digest hot for a time and then filter the hydrated tin(IV) oxide on a paper-pulp pack, and wash it four times with hot water. Dilute the resulting filtrate and washings to 100 mL, and heat to boiling. Electrolyse the hot solution with a small platinum gauze anode at 4–5 A until the deposition of PbO_2 is complete (about 5 minutes). Remove the anode, dry, and weigh as before. Calculate the percentage of lead from the weight of PbO_2 using the empirical factor of 0.864. Evaluate the tin content by subtraction from the combined weight of tin and lead.

Calculate the percentages of antimony, copper, lead, and tin in the alloy.

In a similar determination described by Lingane and Jones,[11] an alloy containing copper, bismuth, lead, and tin is dissolved in hydrochloric acid as described above, and then 100 mL of sodium tartrate solution ($0.1 M$) is added, followed by sufficient sodium hydroxide solution ($5 M$) to adjust the pH to 5.0. After the addition of hydrazinium chloride (4 g), the solution is warmed to 70 °C and then electrolysed. Copper is deposited at -0.3 volt, and then sequentially, bismuth at -0.4 volt, and lead at -0.6 volt; all cathode potentials quoted are *vs* the S.C.E. After deposition of the lead, the solution is acidified with hydrochloric acid and the tin then deposited at a cathode potential of -0.65 volt *vs* the S.C.E.

Some examples of controlled cathode potential determinations are shown in Table 12.3.

Table 12.3 Examples of controlled cathode potential determinations

Metal	Electrolyte	$E_{cathode}$ *vs* S.C.E. (V)	Separated from
Antimony	Hydrazine/HCl	-0.3	Pb, Sn
Cadmium	Acetate buffer	-0.8	Zn
Copper	Tartrate/hydrazine/Cl$^-$	-0.3	Bi, Cd, Pb, Ni, Sn, Zn
Lead	Tartrate/hydrazine/Cl$^-$	-0.6	Cd, Fe, Mn, Ni, Sn, Zn
Nickel	Tartrate/NH$_4$OH	-1.1	Al, Fe, Zn
Silver	Acetate buffer	$+0.1$	Cu, heavy metals

For References and Bibliography see Sections 16.38 and 16.39.

CHAPTER 13
CONDUCTIMETRY

13.1 GENERAL CONSIDERATIONS

Ohm's Law states that the current I (amperes) flowing in a conductor is directly proportional to the applied electromotive force E (volts) and inversely proportional to the resistance R (ohms) of the conductor:

$$I = E/R$$

The reciprocal of the resistance is termed the **conductance** (G): this is measured in reciprocal ohms (or Ω^{-1}), for which the name siemens (S) is used. The resistance of a sample of homogeneous material, length l, and cross-section area a, is given by:

$$R = \rho.l/a$$

where ρ is a characteristic property of the material termed the **resistivity** (formerly called specific resistance). In SI units, l and a will be measured respectively in metres and square metres, so that ρ refers to a metre cube of the material, and

$$\rho = R.a/l$$

is measured in Ω metres. Formerly, resistivity measurements were made in terms of a centimetre cube of substance, giving ρ the units Ω cm. The reciprocal of resistivity is the **conductivity**, κ (formerly specific conductance), which in SI units is the conductance of a one metre cube of substance and has the units $\Omega^{-1} m^{-1}$, but if ρ is measured in Ω cm, then κ will be measured in $\Omega^{-1} cm^{-1}$. Virtually all the data at present recorded in the literature are expressed in terms of κ measured in $\Omega^{-1} cm^{-1}$ units, and these values will therefore be adopted in this book. Furthermore, most of the existing data are expressed in terms of the 'international ohm' and not the SI 'absolute' unit introduced in 1948; the relationship between these quantities is given by

1 mean international ohm $= 1.000\,49\,\Omega$

or for instruments calibrated in the U.S.A.,

1 US international ohm $= 1.000\,495\,\Omega$

(see Section 12.1).

The conductance of an electrolytic solution at any temperature depends only on the ions present, and their concentration. When a solution of an electrolyte is diluted, the conductance will decrease, since fewer ions are present per millilitre

519

of solution to carry the current. If all the solution be placed between two electrodes 1 cm apart and large enough to contain the whole of the solution, the conductance will increase as the solution is diluted. This is due largely to a decrease in inter-ionic effects for strong electrolytes and to an increase in the degree of dissociation for weak electrolytes.

The **molar conductivity** (Λ) of an electrolyte is defined as the conductivity due to one mole and is given by:

$$\Lambda = 1000 \, \kappa/C = \kappa.1000 \, V$$

where C is the concentration of the solution in mol L^{-1}, and V is the dilution in L (i.e. the number of litres containing one mole). Clearly, since κ has the dimensions $\Omega^{-1} cm^{-1}$, the units of Λ are $\Omega^{-1} cm^2 mol^{-1}$, or in SI units, $\Omega^{-1} m^2 mol^{-1}$.

For strong electrolytes the molar conductivity increases as the dilution is increased, but it appears to approach a limiting value known as the **molar conductivity at infinite dilution**. The quantity Λ^{∞} can be determined by graphical extrapolation for dilute solutions of strong electrolytes. For weak electrolytes the extrapolation method cannot be used for the determination of Λ^{∞} but it may be calculated from the molar conductivities at infinite dilution of the respective ions, use being made of the 'Law of Independent Migration of Ions'. At infinite dilution the ions are independent of each other, and each contributes its part of the total conductivity, thus:

$$\Lambda^{\infty} = \Lambda^{\infty} (\text{cat}) + \Lambda^{\infty} (\text{an})$$

where $\Lambda^{\infty} (\text{cat})$ and $\Lambda^{\infty} (\text{an})$ are the ionic molar conductivities at infinite dilution of the cation and anion respectively. The values for the limiting ionic molar conductivities for some ions in water at 25 °C are collected in Table 13.1.

Table 13.1 Limiting ionic molar conductivities at 25 °C ($\Omega^{-1} cm^2 mol^{-1}$)

Cations				Anions			
M^+		M^{2+}		X^-		X^{2-}	X^{3-}
H^+	349.8	Ca^{2+}	119.0	OH^-	198.3	CO_3^{2-} 138.6	PO_4^{3-} 240
Na^+	50.1	Mg^{2+}	106.2	F^-	55.4	SO_4^{2-} 160.0	
K^+	73.5	Cu^{2+}	107.2	Cl^-	76.3		
Li^+	38.7	Zn^{2+}	105.6	Br^-	78.1		
NH_4^+	73.5			NO_3^-	71.5		
Ag^+	61.9			HCO_3^-	44.5		
$N(CH_3)_4^+$	44.9			CH_3COO^-	40.9		

13.2 THE MEASUREMENT OF CONDUCTIVITY

To measure the conductivity of a solution it is placed in a cell carrying a pair of platinum electrodes which are firmly fixed in position. It is usually very difficult to measure precisely the area of the electrodes and their distance apart, and so if accurate conductivity values are to be determined, the **cell constant** must be evaluated by calibration with a solution of accurately known conductivity, e.g. a standard potassium chloride solution.* It is now a common practice for the cell constant to be determined by the manufacturer, and to allow for measurements with solutions of widely differing conductivities, different cells can be obtained offering a range of cell constants.

The measurements are made by connecting the cell to a conductivity meter which supplies alternating current at a frequency of about 1000 Hz to the cell. The use of alternating current reduces the possibility of electrolysis occurring and causing polarisation at the electrodes, but it introduces the complication that the cell has a capacitance in addition to its resistance. The modern conductivity meter has a sophisticated electronic circuit which eliminates capacitance effects, can measure a wide range of conductivities (e.g. $0.001\ \mu\Omega^{-1}$ cm^{-1}–$1300\ m\Omega^{-1}\ cm^{-1}$), and with provision for automatic range switching. By operation of a balancing control, the instrument is adjusted so that the constant of the cell in use is displayed on the digital meter which records the conductivity values. A temperature sensor (which may be a platinum resistance thermometer which is sometimes incorporated in the conductivity cell) is also connected to the meter: this will automatically correct conductivities measured over a range of temperatures to value at $25\,°C$, the temperature for which the apparatus is calibrated. The clean conductivity cell is rinsed with, and then charged with, the solution whose conductivity is to be measured, and the required result is immediately displayed on the meter. Details of a circuit for a 'conductance bridge' are given in Ref. 13.

13.3 CONDUCTIMETRY AS AN ANALYTICAL TOOL

Direct measurement of conductivity is potentially a very sensitive procedure for measuring ionic concentrations, but it must be used with caution since any charged species present in a solution will contribute to the total conductance.

Conductimetric measurements can also be used to ascertain the end-point in many titrations, but such use is limited to comparatively simple systems in which there are no excessive amounts of reagents present. Thus, many oxidation titrations which require the presence of relatively large amounts of acid are not suited to conductimetric titration. Conductimetric titrations have been largely superseded by potentiometric procedures (see Chapter 15), but there are occasions when the conductimetric method can be advantageous.[14]

13.4 APPLICATIONS OF DIRECT CONDUCTIMETRIC MEASUREMENTS

Purity of water. The purity of distilled or de-ionised water is commonly checked by conductimetric measurements. The conductivity of pure water is about $5 \times 10^{-8}\ \Omega^{-1}\ cm^{-1}$, and the smallest trace of ionic impurity leads to a large increase in conductivity. Conductimetric monitoring is employed in laboratories to check the operation of ion exchange units producing de-ionised water, and finds similar industrial application where processes requiring the use of very pure water (e.g. manufacture of semiconductors) are carried on.

There are important industrial applications, such as the control of boiler feed water and of boiler blow-down in large steam-generating plants; to check the

---

*Solutions containing (1) 7.419 138 g, (2) 0.745 263 g of potassium chloride in 1000 g of solution have conductivities at $25\,°C$ of (1) $0.012\ 8560\ \Omega^{-1}\ cm^{-1}$, (2) $0.001\ 4088\ \Omega^{-1}\ cm^{-1}$ (Ref. 12).

concentrations of acid pickling baths, of alkaline degreasing baths, and for the completion of rinsing and washing operations. Conductimetric monitoring of rivers and lakes is used to control pollution, and in oceanography, conductimetry measurements are performed to determine salinity values. In such measurements it is frequently possible to adjust the meter to correct for background currents so that the conductivity meter readings immediately indicate the level of pollution (or of salinity). The meter is calibrated against solutions of known concentration of suitable electrolytes, e.g. sodium chloride for salinity readings.

Ion chromatography (see Section 7.4). Conductivity cells can be coupled to ion chromatographic systems to provide a sensitive method for measuring ionic concentrations in the eluate. To achieve this end, special micro-conductivity cells have been developed of a flow-through pattern and placed in a thermostatted enclosure: a typical cell may contain a volume of about $1.5\,\mu L$ and have a cell constant of approximately $15\,cm^{-1}$. It is claimed[15] that sensitivity is improved by use of a bipolar square-wave pulsed current which reduces polarisation and capacitance effects, and the changes in conductivity caused by the heating effect of the current (see Refs 16, 17).

13.5   THE BASIS OF CONDUCTIMETRIC TITRATIONS

The addition of an electrolyte to a solution of another electrolyte under conditions producing no appreciable change in volume will affect the conductance of the solution according to whether or not ionic reactions occur. If no ionic reaction takes place, such as in the addition of one simple salt to another (e.g. potassium chloride to sodium nitrate), the conductance will simply rise. If ionic reaction occurs, the conductance may either increase or decrease; thus in the addition of a base to a strong acid, the conductance decreases owing to the replacement of the hydrogen ion of high conductivity by another cation of lower conductivity. This is the principle underlying conductimetric titrations, i.e. the substitution of ions of one conductivity by ions of another conductivity.

Consider how the conductance of a solution of a strong electrolyte A^+B^- will change upon the addition of a reagent C^+D^-, assuming that the cation A^+ (which is the ion to be determined) reacts with the ion D^- of the reagent. If the product of the reaction AD is relatively insoluble or only slightly ionised, the reaction may be written:

$$A^+B^- + C^+D^- = AD + C^+B^-$$

Thus in the reaction between A^+ ions and D^- ions, the A^+ ions are replaced by C^+ ions during the titration. As the titration proceeds the conductance increases or decreases, depending upon whether the conductivity of the C^+ ions is greater or less than that of the A^+ ions.

During the progress of neutralisation, precipitation, etc., changes in conductance may, in general, be expected, and these may therefore be employed in determining the end points as well as the progress of the reactions. The conductance is measured after each addition of a small volume of the reagent, and the points thus obtained are plotted to give a graph which ideally consists of two straight lines intersecting at the equivalence point. The accuracy of the method is greater the more acute the angle of intersection and the more nearly the points of the graph lie on a straight line. The volume of the solution should not change appreciably; this may be achieved by employing a titrating reagent which is 20 to 100 times more concentrated than the solution being titrated, and the latter should be as dilute as practicable. Thus if the conductivity cell contains about 100 mL of the solution at the beginning of the titration, the reagent (say, of 50 times the concentration of the solution being analysed) may be placed in a 5 mL microburette, graduated in 0.01 or 0.02 mL. A correction for the dilution effect may, however, be made by multiplying the values of the conductance by the factor $(V + v)/V$, in which V is the original volume of the solution and v is the volume of reagent added.

In contrast to potentiometric titration methods (see Chapter 15), but similar to amperometric titration methods (see Chapter 16), measurements near the equivalence point have no special significance. Indeed, owing to hydrolysis, dissociation, or solubility of the reaction product, the values of the conductance measured in the vicinity of the equivalence point are usually worthless in the construction of the graph, since one or both curves will give a rounded portion at this point. Even if the conductance of the reaction product at the equivalence point is appreciable, the reaction may frequently be employed for conductimetric titration if the conductivity of the reaction product AD is practically completely suppressed by a reasonable excess of A^+ and D^-. Thus conductimetric methods may be applied where visual or potentiometric methods fail to give results owing to considerable solubility or hydrolysis at the equivalence point, for example, in many precipitation reactions producing moderately soluble substances, in the direct titration of weak acids by weak bases, and in the displacement titration of salts of moderately weak acids or bases by strong acids or bases. A further important advantage is that the method is as accurate in dilute as in more concentrated solutions; it can also be employed with coloured solutions.

It may be noted that very weak acids, such as boric acid and phenol, which cannot be titrated potentiometrically in aqueous solution, can be titrated conductimetrically with relative ease. Mixtures of certain acids can be titrated more accurately by conductimetric than by potentiometric (pH) methods. Thus mixtures of hydrochloric acid (or any other strong acid) and acetic (ethanoic) acid (or any other weak acid of comparable strength) can be titrated with a weak base (e.g. aqueous ammonia) or with a strong base (e.g. sodium hydroxide): reasonably satisfactory end points are obtained.

Attention is directed to the importance of temperature control in conductance measurements. While the use of a thermostat is not essential in conductimetric titrations, constancy of temperature is required but it is usually only necessary to place the conductivity cell in a large vessel of water at the laboratory temperature.

The relative change of conductance of the solution during the reaction and upon the addition of an excess of reagent largely determines the accuracy of the titration; under optimum conditions this is about 0.5 per cent. Large amounts of foreign electrolytes, which do not take part in the reaction, must be absent, since these have a considerable effect upon the accuracy. In consequence, the conductimetric method has much more limited application than visual, potentiometric, or amperometric procedures.

13.6 APPARATUS AND MEASUREMENTS

A conductivity cell for conductimetric titrations may be of any kind that lends itself to thorough stirring of the contents (preferably by mechanical means), and permits the periodical addition of reagents. As explained above, it may be necessary to place the cell in a large vessel of water in order to maintain constancy of temperature, but in most circumstances the cell may be used at the ambient temperature of the laboratory. The cell should be constructed of Pyrex or other resistance glass and fitted with platinised platinum electrodes; the platinising helps to minimise polarisation effects. The size and separation of the electrodes will be governed by the change of conductance during the titration: for low-conductance solutions (e.g. when extremely dilute), the electrodes should be large and close together; for precipitation reactions the electrodes must be vertical. The cell constant need not be known.

The following procedure may be used for platinising the electrodes. The conductivity vessel and electrodes are thoroughly cleaned by immersion in a warm solution of potassium dichromate in concentrated sulphuric acid. After washing with distilled water until free from acid, the electrodes are plated from a solution containing 3 g chloroplatinic acid and 0.025 g lead acetate per 100 mL. The current may be obtained from two accumulators (4 volts), the poles of which are connected to the ends of a suitable sliding resistance. The current is adjusted so as to produce a moderate evolution of hydrogen. Each electrode should be used alternately as anode and cathode (i.e. the current should be reversed every half-minute) and electrolysis should be continued until both electrodes are covered with a jet-black deposit. The time may vary from about two to about five minutes. After platinising, the electrodes must be freed from traces of chlorine; dilute sulphuric acid is electrolysed during 15 minutes using the two platinised electrodes (connected together) as cathode and another platinum electrode as anode. The electrodes are then washed with distilled water and afterwards kept immersed in distilled water until required for use.

For most purposes a special cell is not required and good results are obtained by clamping a commercially available dip cell (shown diagrammatically in Fig. 13.1) inside a beaker which is placed on a magnetic stirrer. With this arrangement, the dipping cell should be lifted clear of the solution after each addition from the burette to ensure that the liquid between the electrodes becomes thoroughly mixed. The cell is connected to the conductivity meter, and in many cases a pen recorder can also be attached to give a plot of the change in conductance during the titration: when the recorder is in use, the automatic range change of the conductivity meter can be overridden.

Fig. 13.1

After measuring the solution to be titrated (up to 25 mL) into the beaker, it is diluted with distilled water to at least 100 mL, and the stirrer set in motion. The titrating agent (concentration at least 10 times that of the solution being titrated) should be placed in a 5 or 10 mL microburette; the reagent is added in small portions, and the solution is stirred or shaken after each addition. The conductance is measured after the well-mixed solution has been allowed to stand for a minute or two. The addition of the titrating reagent is continued until at least five readings beyond the equivalence point have been made. It is often advisable to carry out a preliminary titration; this will provide information as to the increments of the reagent best suited for the particular titration, e.g. in increments of 0.5 mL, etc. The conductance is plotted as ordinate against the volume of the titrating reagent as abscissa; the two straight portions of the curve are extrapolated until they intersect and the point of intersection is taken as the equivalence point of the reaction. If a large volume of titrant has been added, the volume correction factor (see Section 13.5) must be applied.

13.7 APPLICATIONS OF CONDUCTIMETRIC TITRATIONS

Some typical conductimetric titration curves are collected in Fig. 13.2 (a)–(d).

Strong acid with a strong base. The conductance first falls, due to the replacement of the hydrogen ion (Λ^∞ 350, Table 13.1) by the added cation (Λ^∞ 40–80) and then, after the equivalence point has been reached, rapidly rises with further additions of strong alkali due to the large Λ^∞ value of the hydroxyl ion (198). The two branches of the curve are straight lines provided the volume of the reagent added is negligible, and their intersection gives the end point; curves 1 and 2, Fig. 13.2(a).

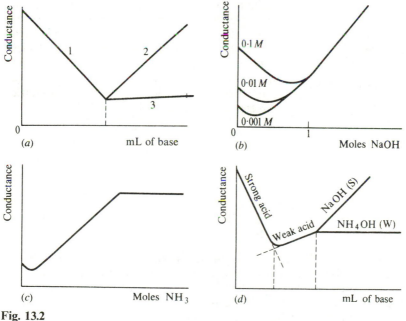

Fig. 13.2

Strong acid with a weak base. The titration of a strong acid with a moderately weak base ($K_b \approx 10^{-5}$) may be illustrated by the neutralisation of dilute sulphuric acid by dilute ammonia solution [curves 1 and 3, Fig. 13.2(a)]. The first branch of the graph reflects the disappearance of the hydrogen ions during the neutralisation, but after the end point has been reached the graph becomes almost horizontal, since the excess aqueous ammonia is not appreciably ionised in the presence of ammonium sulphate.

Weak acid with a strong base. In the titration of a weak acid with a strong base, the shape of the curve will depend upon the concentration and the dissociation constant K_a of the acid. Thus in the neutralisation of acetic acid ($K_a \simeq 1.8 \times 10^{-5}$) with sodium hydroxide solution, the salt (sodium acetate) which is formed during the first part of the titration tends to repress the ionisation of the acetic acid still present so that its conductance decreases. The rising salt concentration will, however, tend to produce an increase in conductance. In consequence of these opposing influences the titration curves may have minima, the position of which will depend upon the concentration and upon the strength of the weak acid. As the titration proceeds, a somewhat indefinite break will occur at the end point, and the graph will become linear after all the acid has been neutralised. Some curves for acetic acid–sodium hydroxide titrations are shown in Fig. 13.2(b): clearly it is not possible to fix an accurate end point.

For moderately strong acids (K_a ca 10^{-3}) the influence of the rising salt concentration is less pronounced, but, nevertheless, difficulty is also experienced in locating the end point accurately and generally titrations of weak and moderately strong acids with a strong base are not suitable for conductimetric techniques.

For very weak acids however, e.g. boric acid [trioxoboric(III) acid], the initial conductance is very small but increases as the neutralisation proceeds owing to the salt formed. The conductance values near the equivalence point are high because of hydrolysis; beyond the equivalence point the hydrolysis is considerably reduced by the excess alkali. To determine the end point, values of the conductance considerably removed from the equivalence point must therefore be used for extrapolation.

Weak acids with weak bases. The titration of a weak acid and a weak base can be readily carried out, and frequently it is preferable to employ this procedure rather than use a strong base. Curve (c) in Fig. 13.2 is the titration curve of 0.003 M acetic acid with 0.0973 M aqueous ammonia solution. The neutralisation curve up to the equivalence point is similar to that obtained with sodium hydroxide solution, since both sodium and ammonium acetates are strong electrolytes; after the equivalence point an excess of aqueous ammonia solution has little effect upon the conductance, as its dissociation is depressed by the ammonium salt present in the solution. The advantages over the use of strong alkali are that the end point is easier to detect, and in dilute solution the influence of carbon dioxide may be neglected.

Mixture of a strong acid and a weak acid with a strong base. Upon adding a strong base to a mixture of a strong acid and a weak acid (e.g. hydrochloric and acetic acids), the conductance falls until the strong acid is neutralised, then rises as the weak acid is converted into its salt, and finally rises more steeply as excess alkali is introduced. Such a titration curve is shown as S in Fig. 13.2(d).

The three branches of the curve will be straight lines except in so far as: (a) increasing dissociation of the weak acid results in a rounding-off at the first end-point, and (b) hydrolysis of the salt of the weak acid causes a rounding-off at the second end point. Usually, extrapolation of the straight portions of the three branches leads to definite location of the end points. Here also titration with a weak base, such as aqueous ammonia solution, is frequently preferable to strong alkali for reasons already mentioned in discussing weak acids: curve W in Fig. 13.2(d) is obtained by substituting aqueous ammonia solution for the strong alkali. The procedure may be applied to the determination of mineral acid in vinegar or other weak organic acids ($K \not< 10^{-5}$) and can be used to analyse 'aspirin' tablets.

Other types of titrations which can be performed conductimetrically include (1) displacement titrations, in which, for example, a salt of a weak acid (sodium acetate) is treated with a strong acid (hydrochloric acid), or a salt of a weak base (ammonium chloride) is treated with a strong base (sodium hydroxide); (2) precipitation titrations (silver nitrate–sodium chloride); (3) complexation titrations (EDTA with many metallic ions). In all these cases satisfactory results are dependent upon a large difference in ion conductance values between the replaced and the replacing ions, and they are generally less satisfactory than acid–base titrations. For precipitation titrations there are also restrictions imposed by the solubility and the rate of precipitation of the relevant substance.

13.8 HIGH-FREQUENCY TITRATIONS (OSCILLOMETRY)

As already indicated conductimetric measurements are normally made with alternating current of frequency 10^3 Hz, and this leads to the existence of capacitance as well as resistance in the conductivity cell. If the frequency of the current is increased further to $10^6 - 10^7$ Hz, the capacitance effect becomes even more marked, and the normal conductivity meter is no longer suitable for measuring the conductance.

The cell employed must be re-designed, and in a typical form the metal electrodes (which need not be platinum) encircle the outside of the glass container and are situated about 2.5 cm apart. Thus they are not in contact with the liquid which can be advantageous for dealing with corrosive materials. Various forms of apparatus suitable for use with such cells have been devised.[18,19]

At these high frequencies, the retarding effect of the ion-atmosphere on the movement of a central ion is greatly decreased and conductance tends to be increased. The capacitance effect is related to the absorption of energy due to induced polarisation and the continuous re-alignment of electrically unsymmetrical molecules in the oscillating field. With electrolyte solutions of low dielectric constant, it is the conductance which is mainly affected, whilst in solutions of low conductance and high dielectric constant, the effect is mostly in relation to capacitance.

Clearly for titration purposes, it is low-dielectric constant conducting solutions which will be important, and addition of a suitable reagent to such a solution permits the plotting of a titration curve from which the end point can be deduced as described in Section 13.7. It should be noted that in view of the enhanced conductance in the high-frequency field, the maximum concentration of reagents is much smaller than with normal conductimetric titrations, and the maximum concentration will depend on the frequency chosen. It is found that many titrations which are unsatisfactory as normal conductimetric titrations can be performed satisfactorily under high-frequency conditions. This applies in particular to many EDTA titrations, the determination of thorium (as Th^{4+}) with sodium carbonate, and of beryllium (as Be^{2+}) with sodium hydroxide.

Since the capacitance response of the measuring instrument is mainly affected by solutions of low conductance and high dielectric constant, it follows that changes in dielectric constant will be detected by the apparatus, and this gives rise to a potential analytical procedure. Most hydrocarbons have a low dielectric constant (e.g. benzene 2.27), whilst water has a high value (78.5). Hence, traces of water (or of other highly polar material) in benzene can be determined by noting the capacitance reading with the contaminated material, and comparing the result with a standard curve which has been obtained by taking readings with a number of benzene–water mixtures of known composition. A similar procedure can be used to analyse mixtures of acetone and benzene, and of various alcohols and water.

For References and Bibliography see Sections 16.38 and 16.39.

CHAPTER 14
COULOMETRY

14.1 GENERAL DISCUSSION

Coulometric analysis is an application of Faraday's First Law of Electrolysis which may be expressed in the form that the extent of chemical reaction at an electrode is directly proportional to the quantity of electricity passing through the electrode. For each mole of chemical change at an electrode ($96\,487 \times n$) coulombs are required; i.e. the Faraday constant multiplied by the number of electrons involved in the electrode reaction. The weight of substance produced or consumed in an electrolysis involving Q coulombs is therefore given by the expression

$$W = \frac{M_r \times Q}{96\,487\,n}$$

where M_r is the relative atomic (or molecular) mass of the substance liberated or consumed. Analytical methods based upon the measurement of a quantity of electricity and the application of the above equation are termed **coulometric methods** — a term derived from 'coulomb'.

The fundamental requirement of a coulometric analysis is that the electrode reaction used for the determination proceeds with 100 per cent efficiency so that the quantity of substance reacted can be expressed by means of Faraday's Law from the measured quantity of electricity (coulombs) passed. The substance being determined may directly undergo reaction at one of the electrodes (**primary coulometric analysis**), or it may react in solution with another substance generated by an electrode reaction (**secondary coulometric analysis**).

Two distinctly different coulometric techniques are available: (1) coulometric analysis with controlled potential of the working electrode, and (2) coulometric analysis with constant current. In the former method the substance being determined reacts with 100 per cent current efficiency at a working electrode, the potential of which is controlled. The completion of the reaction is indicated by the current decreasing to practically zero, and the quantity of the substance reacted is obtained from the reading of a coulometer in series with the cell or by means of a current–time integrating device. In method (2) a solution of the substance to be determined is electrolysed with constant current until the reaction is completed (as detected by a visual indicator in the solution or by amperometric, potentiometric, or spectrophotometric methods) and the circuit is then opened. The total quantity of electricity passed is derived from the product current (amperes) × time (seconds): the present practice is to include an electronic integrator in the circuit.

14.2 COULOMETRY AT CONTROLLED POTENTIAL

In a controlled-potential coulometric analysis, the current generally decreases exponentially with time according to the equation

$$I_t = I_0 e^{-k't} \quad \text{or} \quad I_t = I_0 10^{-kt}$$

where I_0 is the initial current, I_t the current at time t, and k (k') are constants. A typical time vs current curve is shown in Fig. 14.1; the current decreases more or less exponentially to almost zero. In many cases an appreciable 'background current' is observed with the supporting electrolyte alone, and in such instances the current finally decays to the background current rather than to zero; a correction can be applied by assuming that the background current is constant during the electrolysis.

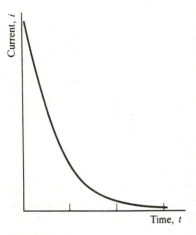

Fig. 14.1

In electrolysis at controlled potential, the quantity of electricity Q (coulombs) passed from the beginning of the determination to time t is given by

$$Q = \int_0^t I_t \, dt$$

where I_t is the current at time t. The above equations relating the variation of current with time, can also be expressed in terms of the concentration of electrolyte at time t (C_t) and the initial concentration (C_0), i.e.

$$C_t = C_0 e^{-k't}$$

This equation is that of a first-order reaction process, and thus the fraction of material electrolysed at any instant is independent of the initial concentration. It follows that if the limit of accuracy of the determination is set at $C_t = 0.001 C_0$, the time t required to achieve this result will be independent of the initial concentration. The constant k' in the above equation can be shown to be equal to Am/V, where A is the area of the pertinent electrode, V the volume of the solution and m the mass transfer coefficient of the electrolyte.[20] It follows that to make t small A and m must be large, and V small, and this leads to the subdivision of controlled potential coulometry into (1) stirred solutions, (2) flowing streams, and (3) thin-layer cavity cells.

14.3 APPARATUS AND GENERAL TECHNIQUE

Originally, the number of coulombs passed was determined by including a coulometer in the circuit, e.g. a silver, an iodine or a hydrogen–oxygen coulometer. The amount of chemical change taking place in the coulometer can be ascertained, and from this result the number of coulombs passed can be calculated, but with modern equipment an electronic integrator is used to measure the quantity of electricity passed.

Apparatus. The source of current is a potentiostat which is used in conjunction with a reference electrode (commonly a saturated calomel electrode) to control the potential of the working electrode. The circuit will be essentially that shown in Fig. 12.2(a) but with the addition of the integrator or of a coulometer.

For stirred-solution coulometry a mercury cathode is commonly used (see Fig. 12.5), suitably equipped for coulometric determinations. The cell has a capacity of about 100 mL and is closed with a Bakelite cover (fitted on to the top of the glass cell, which is ground flat) and a gas delivery tube is provided for removing dissolved air from the solution with nitrogen or other inert gas; the excess nitrogen escapes through the loosely fitting glass sleeve through which the shaft of a glass stirrer passes. Removal of the air is necessary, because oxygen is reduced at the mercury cathode at about -0.05 volt vs S.C.E., and this would interfere with the determination of most substances. The area of the mercury cathode is about 20 cm^2. Two kinds of anode, immersed directly in the test solution, may be used, viz. a large helical silver wire (ca 2.6 mm diameter, helix 5 cm long, and 3 cm diameter; area about 100 cm^2) or a platinum gauze cylinder (area 75 cm^2) mounted vertically and co-axially with the stirrer shaft. The silver anode is employed when the solution contains metals, such as bismuth, which tend to be oxidised to insoluble higher oxides at a platinum anode: chloride ion, at least equivalent to the quantity of the cathode reaction (but preferably in 50–100 per cent excess) is added. The reaction at the silver anode is

$$Ag + Cl^- \rightleftharpoons AgCl + e$$

Hydrazine is used as depolariser at the platinum anode for metals which are not reduced by this compound:

$$N_2H_5^+ \rightleftharpoons N_2 + 5H^+ + 4e$$

The evolution of nitrogen aids in removing dissolved air. A salt bridge (4 mm tube) attached to the saturated calomel electrode is filled with 3 per cent agar gel saturated with potassium chloride and its tip is placed within 1 mm of the mercury cathode when the mercury is not being stirred; this ensures that the tip trails in the mercury surface when the latter is stirred. It is essential that the mercury–solution interface (not merely the solution) be vigorously stirred, and for this purpose the propeller blades of the glass stirrer are partially immersed in the mercury.

One of the outstanding advantages of the mercury cathode is that the optimum control potential for a given separation is easily determinable from polarograms recorded with the dropping mercury electrode. This potential corresponds to the beginning of the polarographic diffusion current plateau (Chapter 16); there is usually no advantage in employing a control potential more than about 0.15 volt greater than the half-wave potential. Some values for the half-wave potential ($E_{1/2}$) and suitable values for the cathode potential are collected in Table 14.1.

Table 14.1 Deposition of metals at controlled potential of the mercury cathode

Element	Supporting electrolyte	Volts vs S.C.E.	
		$E_{1/2}$	$E_{cathode}$
Cu	0.5M acid sodium tartrate, pH 4.5	-0.09	-0.16
Bi	0.5M acid sodium tartrate, pH 4.5	-0.23	-0.40
Pb	0.5M acid sodium tartrate, pH 4.5	-0.48	-0.56
Cd	1M NH_4Cl + 1M aq. NH_3	-0.81	-0.85
Zn	1M NH_4Cl + 1M aq. NH_3	-1.33	-1.45
Ni	1M pyridine + HCl, pH 7.0	-0.78	-0.95
Co	1M pyridine + HCl, pH 7.0	-1.06	-1.20

Details have been collected for the determination of some 50 elements by this technique[21,22] and it is possible to effect many difficult separations, such as Cu and Bi, Cd and Zn, Ni and Co; it has been widely used in the nuclear energy industry. A number of organic compounds can also be determined by this procedure, e.g. trichloroacetic acid and 2,4,6-trinitrophenol are reduced at a mercury cathode in accordance with the equations

$$Cl_3CCOO^- + H^+ + 2e \rightleftharpoons CHCl_2COO^- + Cl^-$$

$$C_6H_2(NO_2)_3(OH) + 18H^+ + 18e \rightleftharpoons C_6H_2(NH_2)_3(OH) + 6H_2O$$

and so can be determined coulometrically[23].

General technique. A coulometric determination at controlled potential of the mercury cathode is performed by the following general method. The supporting electrolyte (50–60 mL) is first placed in the cell and the air is removed by passing a rapid stream of nitrogen through the solution for about 5 minutes. The cathode mercury is then introduced through the stopcock at the bottom of the cell by raising the mercury reservoir. The stirrer is started and the tip of the bridge from the reference electrode is adjusted so that it just touches, or trails slightly in, the stirred mercury cathode. The potentiostat is adjusted to maintain the desired control potential and the solution is electrolysed, with nitrogen passing continuously, until the current decreases to a very small constant value (the 'background current'). This preliminary electrolysis removes traces of reducible impurities; the current usually decreases to 1 mA or less after about 10 minutes. A known volume (say, 10–40 mL) of the sample solution is then pipetted into the cell, and the electrolysis is allowed to proceed until the current decreases to the same small value observed with the supporting electrolyte alone. Electrolysis is usually complete within an hour. The electronic integrator is read and the weight of metal deposited is calculated as explained in Section 14.1.

14.4 SEPARATION OF NICKEL AND COBALT BY COULOMETRIC ANALYSIS AT CONTROLLED POTENTIAL

Reagents. *Standard nickel and cobalt ion solutions.* Prepare solutions of nickel and cobalt ion (*ca* 10 mg mL^{-1}) from pure ammonium nickel sulphate and pure ammonium cobalt sulphate respectively.

Pyridine. Re-distil pyridine and collect the middle fraction boiling within a 2°C range, viz. 113–115°C.

Supporting electrolyte. Prepare a supporting electrolyte composed of 1.00 M pyridine and 0.50 M chloride ion, adjusted to a pH of 7.0 ± 0.2 for use with a silver anode, or 1.00 M pyridine, 0.30 M chloride ion and 0.20 M hydrazinium sulphate, adjusted to a pH of 7.0 ± 0.2, for use with a platinum cathode. A small background current is obtained with the latter.

Procedure. Place 90 mL of the supporting electrolyte in the cell, remove dissolved air with pure nitrogen, and subject the solution to a preliminary electrolysis with the potential of the mercury cathode −1.20 volt *vs* S.C.E. to remove traces of reducible impurities; stop the electrolysis when the background current (*ca* 2 mA) has decreased to a constant value (30–60 minutes). Prepare the coulometer, adjust the potentiostat to maintain the potential of the cathode at the value to be used in the determination (−0.95 volt *vs* S.C.E. for nickel) and add 10 mL of each of the prepared solutions to the cell. Electrolyse until the current has decreased to the value of the background current. Note the number of coulombs passed and calculate the weight of nickel deposited. Now adjust the potential to −1.20 volt and continue the electrolysis until the current again falls to the background current value, and from the number of coulombs passed in this second electrolysis, calculate the weight of cobalt present. If necessary, for each determination, correction for the background current can be carried out by subtracting the quantity $I_b t$ from Q (the number of coulombs recorded), where I_b is the base current, and t the duration of the electrolysis in seconds.

14.5 FLOWING STREAM COULOMETRY

As indicated in Section 14.2, the time required for a coulometric determination can be reduced by using a flowing stream technique. In this technique, the electrode is a cylinder of **reticular vitreous carbon**[24] which is a material of porous structure having a porosity factor of 0.95 which indicates a large open internal volume. This means that if a solution is caused to flow through the electrode, the effective cross-sectional area of the stream of liquid is comparatively large. This electrode (which is the working electrode) is surrounded by a metal cylinder (usually of stainless steel) which acts as the auxiliary electrode. For a reasonable flow rate (about 1 mL s^{-1}), the solution must be pumped through the electrode and the assembly is mounted inside a small chamber in which connection to the reference electrode is made. With a suitable flow rate and a small concentration of the electrolyte under determination entering the electrode, the effluent will be virtually free of this material, which means that the electrolysis is completed inside the electrode.[25] The usual procedure is to take a small known volume (say 20 μL) of the solution to be analysed and to inject this into a stream of supporting electrolyte flowing in a tube of about 0.5 mm diameter. Mixing takes place, and at the entry to the working electrode the diluted test solution may occupy a volume of about 200 μL. At the specified flow rate, all the material to be determined will pass through the electrode in about 12–15 seconds, and thus a rapid rate of determination has been achieved.

The electrode will of course be incorporated in a circuit similar to that previously described in which a potentiostat controls the potential of the working electrode and may also provide a counter-current facility to nullify the background current: an electronic integrator will also be included.

In the thin-layer cavity cell technique, a cell is constructed to give a thin cavity on one wall of which the metal-plate working electrode is mounted. This wall is separated by a Teflon sheet in which a central aperture has been cut out, from the opposite wall of the cavity: this wall contains entry and exit tubes for the test solution which is caused to flow past the working electrode; provision is made for connections to the other electrodes. If the Teflon sheet is thin enough (about 0.05 mm), the distance between the two walls of the cavity is less than the normal thickness of the diffusion layer of the electrolyte when undergoing electrolysis, and so electrolysis within the cavity is rapid.[26]

It is also possible to reduce the time required for conventional controlled-potential coulometry by adopting the procedure of **predictive coulometry**.[27] A given determination will need a certain number of coulombs (Q_∞) for completion, and if at time t, Q_t coulombs have been passed, then Q_R further coulombs will be required to complete the determination, and $Q_R = Q_\infty - Q_t$. By choosing a number of times t_1, t_2, t_3 separated by a common interval (say 10 seconds) and measuring the corresponding numbers of coulombs passed Q_1, Q_2, Q_3, it can be shown that

$$Q_\infty = Q_3 + \frac{(Q_2 - Q_3)^2}{2Q_2 - (Q_1 + Q_3)}$$

A computer is programmed to calculate Q_∞ from the value of Q_t at successive intervals of 10 seconds until a constant value is obtained for Q_∞, thus completing the determination.

COULOMETRY AT CONSTANT CURRENT: COULOMETRIC TITRATIONS

14.6 GENERAL DISCUSSION

Coulometry at controlled potential is applicable only to the limited number of substances which undergo quantitative reaction at an electrode during electrolysis. By using coulometry at controlled or constant current, the range of substances that can be determined may be extended considerably, and includes many which do not react quantitatively at an electrode. Constant-current electrolysis is employed to generate a reagent which reacts stoichiometrically with the substance to be determined. The quantity of substance reacted is calculated with the aid of Faraday's Law, and the quantity of electricity passed can be evaluated simply by timing the electrolysis at constant current. Since the current can be varied from, say, 0.1 to 100 mA, amounts of material corresponding to 1×10^{-9} to 1×10^{-6} mole per second of electrolysis time can be determined. In titrimetric analysis the reagent is added from a burette; in **coulometric**

titrations the reagent is generated electrically and its amount is evaluated from a knowledge of the current and the generating time. The electron becomes the standard reagent. In many respects, e.g. detection of end points, the procedure differs only slightly from ordinary titrations.

The fundamental requirements of a coulometric titration are: (1) that the reagent-generating electrode reaction proceeds with 100 per cent efficiency, and (2) that the generated reagent reacts stoichiometrically and, preferably, rapidly with the substance being determined. The reagent may be generated directly within the test solution or, less frequently, it may be generated in an external solution which is allowed to run continuously into the test solution.

Since a small quantity of electricity can be readily measured with a high degree of accuracy, the method has high sensitivity. Coulometric titrimetry has several important advantages.

1. Standard solutions are not required and in their place the coulomb becomes the primary working standard.
2. Unstable reagents, such as bromine, chlorine, silver(II) ion (Ag^{2+}), and titanium(III) ion, can be utilised, since they are generated and consumed immediately; there is no loss on storage or change in titre.
3. When necessary very small amounts of titrants may be generated: this dispenses with the difficulties involved in the standardisation and storage of dilute solutions and the procedure is ideally adapted for use on a micro or semimicro scale.
4. The sample solution is not diluted in the internal generation procedure.
5. By pre-titration of the generating solution before the addition of the sample, more accurate results can be obtained. Since the effect of impurities in the generating solution is minimised.
6. The method (which is largely electrical in nature) is readily adapted to remote control: this is significant in the titration of radioactive or dangerous materials. It may also be adapted to automatic control because of the relative ease of the automatic control of current.

Several methods are available for the detection of end points in coulometric titrations. These are the following.

(*a*) Use of chemical indicators: these must not be electro-active. Examples include methyl orange for bromine, starch for iodine, dichlorofluorescein for chloride, and eosin for bromide and iodide.
(*b*) Potentiometric observations. Electrolytic generation is continued until the e.m.f. of a reference electrode–indicating electrode assembly placed in the test solution attains a pre-determined value corresponding to the equivalence point.
(*c*) Amperometric procedures. These are based upon the establishment of conditions such that either the substance being determined or, more usually, the titrant undergoes reaction at an indicator electrode to produce a current which is proportional to the concentration of the electro-active substance. With the potential of the indicator electrode maintained constant, or nearly so, the end point can be established from the course of the current change during the titration. The voltage impressed upon the indicator electrode is well below the 'decomposition voltage' of the pure supporting electrolyte but close to or above the 'decomposition voltage' of the supporting

535

electrolyte plus free titrant; consequently, as long as any of the substance being determined remains to react with the titrant, the indicator current remains very small but increases as soon as the end point is passed and free titrant is present. There is a relatively inexhaustible supply of titrant ion (e.g. bromide ion in coulometric titrations with bromine), and the indicator current beyond the equivalence point is therefore governed largely by the rate of diffusion of the free titrant (e.g. bromine) to the surface of the indicator electrode. The indicator current is consequently proportional to the concentration of the free titrant (bromine) in the bulk of the solution and to the area of the indicator electrode (cathode for bromine). The indicator current will increase with increasing rate of stirring, since this decreases the thickness of the diffusion layer at the electrode; it is also somewhat temperature-dependent. The generation time at which the equivalence point is reached may be determined by calibrating the indicator electrode system with the supporting electrolyte alone by generating the titrant (e.g. bromine) for various times (say, 10–50 seconds) to evaluate the constant in the relation $I_i = Kt$, where I_i is the indicator current and t is the time. The generating time to the equivalence point may then be obtained from the observed final value of the indicator current in the actual titration, calculating the excess generating time and subtracting this from the total generating time in the titration. Alternatively, and more simply, the equivalence point time may be located by measuring three values of the indicator current at three measured times beyond the equivalence point and extrapolating to zero current.

(*d*) Application of the biamperometric (dead-stop) method (compare Section 16.32).

(*e*) Spectrophotometric observations (compare Sections 17.53 to 17.57). The titration cell consists of a spectrophotometer cuvette (2 cm light path). The motor-driven glass propeller stirrer and the working platinum electrode are placed in the cell in such a way as to be out of the light path: a platinum electrode in dilute sulphuric acid in an adjacent cuvette also placed in the cell-holder serves as an auxiliary electrode and is connected with the titration cell by an inverted **U**-tube salt bridge. The appropriate wavelength is set on the instrument. Before the end point the absorbance changes only very slowly, but a rapid and linear response occurs beyond the equivalence point. Examples are: the titration of Fe(II) in dilute sulphuric acid with electro-generated Ce(IV) at 400 nm, and the titration of arsenic(III) with electro-generated iodine at 342 nm.

The principle of coulometric titration. This involves the generation of a titrant by electrolysis and may be illustrated by reference to the titration of iron(II) with electro-generated cerium(IV). A large excess of Ce(III) is added to the solution containing the Fe(II) ion in the presence of, say $1M$ sulphuric acid. Consider what happens at a platinum anode when a solution containing Fe(II) ions alone is electrolysed at constant current. Initially the reaction

$$Fe^{2+} \rightleftharpoons Fe^{3+} + e$$

will proceed with 100 per cent current efficiency. At the anode surface the concentration of Fe(III) ions formed is relatively large, while that of the Fe(II) ions, which is governed by the rate of transfer from the bulk of the solution, is very small: the potential of the anode gradually acquires a value which is much more positive (more oxidising) than the standard potential of the Fe(III)/Fe(II) couple (0.77 volt). As electrolysis proceeds, the anode potential becomes more and more positive (oxidising) at a rate that depends on the current density, and ultimately it becomes so positive (ca 1.23 volt) that oxygen evolution from the oxidation of water begins ($2H_2O \rightleftharpoons O_2 + 4H^+ + 4e$), and this occurs before all the Fe(II) ions in the bulk of the solution are oxidised. As soon as oxygen evolution commences, the current efficiency for the oxidation of Fe(II) falls below 100 per cent and the quantity of Fe(II) initially present cannot be computed from Faraday's Law. If the electrolysis is conducted in the presence of a relatively large concentration of Ce(III) ions the following reactions will take place at the anode. At a certain potential of the anode, which is considerably less than that required for oxygen evolution, oxidation of Ce(III) to Ce(IV) sets in, and the Ce(IV) thus produced is transferred to the bulk of the solution, where it oxidises Fe(II):

$$Ce^{4+} + Fe^{2+} = Ce^{3+} + Fe^{3+}$$

The potential of the working electrode is thus stabilised by the reagent-generating reaction, and hence is prevented from drifting to a value such that an interfering reaction may result.

Stoichiometrically, the total quantity of electricity passed is exactly the same as it would have been if the Fe(II) ions had been directly oxidised at the anode and the oxidation of Fe(II) proceeds with 100 per cent efficiency. The equivalence point is marked by the first persistence of excess Ce(IV) in the solution, and may be detected by any of the methods described above. The Ce^{3+} ions added to the Fe(II) solution undergo no net change and are said to act as a **mediator**.

Side reactions are avoided at the generating electrode provided there is not complete depletion (at the electrode surface) of the substance involved in the generation of the titrant. The concentration of the titrant depends upon the current through the cell, the area of the generating electrode, and the rate of stirring; the concentration of the generating substance is usually between 0.01 M and 0.1 M.

14.7 INSTRUMENTATION

A number of coulometric titrators are available commercially and are simple to operate. Suitable apparatus can also be assembled from readily available equipment. The essential requirements are:

(a) *A constant current source.* This can be a large-capacity storage battery, but is preferably an amperostat which is an electronically controlled instrument providing a constant current; such an instrument can often provide a constant voltage (act as a potentiostat) in addition to its constant-current function.

(b) An *integrator.* This is an electronic device for measuring the product (current × time), i.e. the number of coulombs. If necessary this may be replaced by an accurately calibrated milliammeter which is coupled with a quartz crystal clock to record the duration of the electrolysis.

Figure 14.2(a) is a schematic diagram of a suitable circuit for coulometric titration with internal generation of titrant and using the dead-stop or

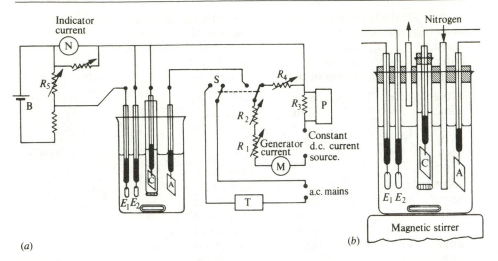

Fig. 14.2

amperometric end point technique. The high-wattage resistances R_1 (large) and R_2 (small) permit the current to be varied. The calibrated milliammeter M records the generating current; a more accurate value of the current is obtained by measuring the voltage drop across a standard resistance R_3 (say, 100 ohms) with a potentiometer P. The variable resistance R_4 (high-wattage type) is so connected that the electrolysis current flows through it whenever the electrolysis cell is disconnected from the circuit; its value (*ca* 20 ohms) should not differ greatly from that of the cell, and it is adjusted so that the current will have nearly the same magnitude as in the titration. This arrangement ensures that the resistances R_1 and R_2 are at constant temperature, and so minimises the variations in their resistance which would occur if the current through them were interrupted periodically. S is a double-pole toggle switch which permits simultaneous operation of the cell and timer T and also allows the current to pass through R_4 when the cell is disconnected.

The titration cell is shown in detail in Fig. 14.2(*b*). It consists of a tall-form beaker (without lip) of about 150 or 200 mL capacity. Provision is made for magnetic stirring and for passing a stream of inert gas (e.g. nitrogen) through the solution. The main generator electrode A may consist of platinum foil (1 cm × 1 cm or 4 cm × 2.5 cm) and the auxiliary electrode C may consist of platinum foil (1 cm × 1 cm or 4 cm × 2.5 cm) bent into a half-cylinder so as to fit into a wide glass tube (*ca* 1 cm diameter). The isolation of the auxiliary generator electrode C within the glass cylinder (closed by a sintered-glass disc) from the bulk of the solution avoids any effects arising from undesirable reactions at this electrode.

E_1 and E_2 are the indicator electrodes. These may consist of a tungsten pair for a biamperometric end point; for an amperometric end point they may both be of platinum foil or one can be platinum and the other a saturated calomel reference electrode. The voltage impressed upon the indicator electrodes is supplied by battery B (*ca* 1.5 volts) via a variable resistance R_5; N records the indicator current. For a potentiometric end point E_1 and E_2 may consist of either platinum–tungsten bimetallic electrodes. or E_1 may be an S.C.E. and E_2

a glass electrode. These are connected directly to a pH meter with a subsidiary scale calibrated in millivolts. The indicator electrodes should be positioned outside the electric field (current path) between the generator electrodes; otherwise, spurious indicator currents may be produced, particularly in the amperometric detection of the equivalence point.

General procedure. The electrolysis cell is set up with both generator and indicator electrodes in position and provision is made, if necessary, for passing an inert gas (e.g., nitrogen) through the solution. The titration cell is charged with the solution from which the titrant will be generated electrolytically, together with the solution to be titrated. The auxiliary electrode compartment is filled with a solution of the appropriate electrolyte at a higher level than the solution in the titration cell. The indicator electrodes are connected to a suitable apparatus for the detection of the end point, e.g. a pH meter with additional millivolt scale, or a galvanometer. Stirring is effected with a magnetic stirrer. The reading of the digital indicator instrument is taken. The current, previously adjusted to a suitable value, is then switched on and reaction between the internally generated titrant and the test solution allowed to proceed. Readings are taken periodically (more frequently as the end point is approached) of the integrating counter and of the indicating instrument (e.g. pH meter); it is usually necessary to switch off the electrolysis current while the readings of the indicating instrument are recorded. The end point of the titration is readily evaluated from the plot of the reading of the indicating instrument (e.g. millivolts) against the counter reading; the first- or second-derivative curve is drawn to locate the equivalence point accurately. It is possible to repeat the titration with a fresh volume of the test solution, if the end point is determined potentiometrically, subsequent determinations may be stopped at the potential found for the equivalence point in the initial titration.

14.8 EXTERNAL GENERATION OF TITRANT

The limitations of coulometric titration with internal generation of the titrant include the following.

1. No substance may be present which undergoes reaction at the generator electrodes; for example, in acidimetric titrations the test solutions must not contain substances which are reduced at the generator cathode.
2. When applied on a macro scale — samples of 1–5 millimoles — generation rates of 100–500 milliamps are required: parasitic currents may be induced in the indicator electrodes at currents in excess of about 10–20 mA; consequently precise location of the equivalence point by amperometric methods is not trustworthy.

To overcome these limitations, the reagent can be generated at constant current with 100 per cent efficiency in an external generator cell and subsequently delivered to the titration cell. This technique is identical with an ordinary titration except that the reagent is generated electrolytically. A double-arm electrolytic cell for external generation of the titrant is shown in Fig. 14.3. The generator electrodes consist of two platinum spirals near the centre of the inverted **U**-tube. The space between the electrodes is packed with glass wool to prevent turbulent mixing; the downward legs of the generator tube are

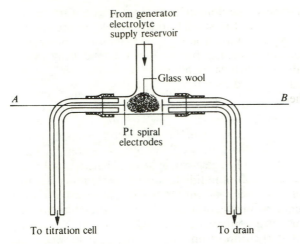

From generator
electrolyte
supply reservoir

Glass wool

A   *B*

Pt spiral
electrodes

To titration cell   To drain

Fig. 14.3

constructed of 1 mm capillary tubing to reduce the inconvenience due to hold-up.
The solution of the electrolyte, which upon electrolysis will yield the desired
titrant, is fed continuously into the top of the generator cell. The solution is
divided at the **T**-joint so that about equal quantities flow through each of the
arms of the cell. As these portions of the solution flow past the electrodes,
electrolysis occurs: the products of electrolysis are swept along by the flow of
the solution through the arms and emerge from the delivery tips. A beaker
containing the substance to be titrated is placed beneath the appropriate delivery
tip, and the solution from the other tip is run to waste. Thus for the titration
of acids, electrode *A* functions as a cathode in sodium sulphate generator
electrolyte and the hydroxide ion generated by the reaction

$$2H_2O + 2e \rightleftharpoons 2OH^- + H_2$$

flows into the test solution. The hydrogen ion and oxygen generated at the other
electrode by the reaction

$$2H_2O \rightleftharpoons 4H^+ + O_2 + 4e$$

are swept out of the other arm into the drain. For the titration of bases, the
generator electrode which delivers to the titration cell is employed as the anode.
For titrations with electrically generated iodine, the generator electrolyte consists
of potassium iodide solution, and the iodine solution formed at the anode flows
into the titration vessel.

   A minor disadvantage of external generation of titrant is the dilution of the
contents of the titration cell; care is therefore necessary in suitably adjusting
the rate of flow and the concentration of the generator solution. The procedure
is, however, admirably suited for automatic control.

14.9 EXPERIMENTAL DETAILS FOR TYPICAL COULOMETRIC TITRATIONS AT CONSTANT CURRENT

In the following pages experimental details are given for some typical coulometric
titrations at constant current.

By virtue of its inherent accuracy, coulometric titration is very suitable for the determination of substances present in small amount, and quantities of the order of 10^{-7}–10^{-5} mole are typical. Larger amounts of material require very long electrolysis times unless an amperostat capable of delivering relatively large currents (up to 2 A) is available. In such cases, a common procedure is to start the electrolysis with a large current, and then to switch to a much lower output as the end point is approached.

14.10 ANTIMONY(III)

Discussion. Iodine (or tri-iodide ion $I_3^- \rightleftharpoons I_2 + I^-$) is readily generated with 100 per cent efficiency by the oxidation of iodide ion at a platinum anode, and can be used for the coulometric titration of antimony(III). The optimum pH is between 7.5 and 8.5, and a complexing agent (e.g. tartrate ion) must be present to prevent hydrolysis and precipitation of the antimony. In solutions more alkaline than pH of about 8.5, disproportionation of iodine to iodide and iodate(I) (hypoiodite) occurs. The reversible character of the iodine–iodide complex renders equivalence point detection easy by both potentiometric and amperometric techniques; for macro titrations, the usual visual detection of the end point with starch is possible.

Apparatus. Use the apparatus shown in Figs. 14.2(*a*) and (*b*). The generator cathode (isolated auxiliary electrode) consists of platinum foil (4 cm × 2.5 cm, bent into a half cylinder) and the generator anode (working electrode) is a rectangular platinum foil (4 cm × 2.5 cm). For potentiometric end point detection, use a platinum-foil electrode 1.25 cm × 1.25 cm (or a silver-rod electrode) in combination with an S.C.E. connected to the cell by a potassium chloride– or potassium nitrate–agar bridge.

Reagents. *Supporting electrolyte.* Prepare a 0.1 *M* phosphate buffer of pH = 8 containing 0.1 *M* potassium iodide and 0.025 *M* potassium tartrate (0.17 g $Na_2HPO_4, 12H_2O$, 3.38 g $NaH_2PO_4, 2H_2O$, 4.15 g KI and 1.5 g potassium tartrate, in 250 mL of water).

Procedure. Place 45 mL of the supporting electrolyte in the cell and fill the isolated cathode compartment with the same solution to a level well above that in the cell. Pipette 5.00, 10.00, or 15.00 mL of the 0.01 *M* antimony solution into the cell and titrate coulometrically with a current of 40 milliamps. Stir the solution continuously by means of the magnetic stirrer and take e.m.f. readings of the Pt–S.C.E. electrode combination at suitable time intervals: the readings may be somewhat erratic initially, but become steady and reproducible after about 3 minutes. Evaluate the end point of the titration from the graph of e.m.f. *vs* counter reading: this shows a marked change of e.m.f. at the end point. If it proves difficult to locate the end point precisely, recourse may be made to the first- and second-differential plots.

If it is desired to use the biamperometric method for detecting the end point, then the calomel electrode and also the silver rod (if used) must be removed and replaced by two platinum plates 1.25 cm × 1.25 cm. The potentiometer (or pH meter) used to measure the e.m.f. must also be removed, and one of the indicator electrodes is then joined to a sensitive galvanometer fitted with a variable shunt. The indicator circuit is completed through a potential divider placed across a 1.5 V dry battery (see Fig. 16.17, Section 16.32). Charge the electrolysis cell as described above, adjust the potential across the indicator electrodes to about 150 mV, and set the galvanometer shunt to give maximum deflection on the galvanometer. Switch on the electrolysis current and read the indicator current from time to time. Plot indicator current against counter reading and extrapolate to zero current to locate the end point.

For determination of the end point by a visual method, add 1–2 mL of 1 per cent starch solution, and stop the titration immediately the solution has acquired a uniform blue colour.

14.11 8-HYDROXYQUINOLINE (OXINE)

Discussion. Bromine may be electro-generated with 100 per cent current efficiency by the oxidation of bromide ion at a platinum anode. Bromination of oxine proceeds according to the equation:

$$C_9H_7ON + 2Br_2 = C_9H_5ONBr_2 + 2H^+ + 2Br^-$$

and thus four Faraday constants are required per mole of oxine. The end point is detected amperometrically.

Apparatus. Set up the apparatus as in Section 14.10 with two small platinum plates connected to apparatus for the amperometric detection of the end point.

Reagents. *Supporting electrolyte.* Prepare $0.2 M$ potassium bromide from the analytical grade salt.

Oxine solution. $0.003 M$ oxine (use the analytical-grade material) in $0.0025 M$ hydrochloric acid.

Procedure. Place 40 mL of the supporting electrolyte in the coulometric cell and pipette 10.00 mL of the oxine solution into it. Charge the cathode compartment with the $0.2 M$ potassium bromide. Pass a current of 30 mA while stirring the solution magnetically. Adjust the sensitivity of the indicating apparatus to a suitable value. Near the end point transient deflections occur and serve to give warning of its approach. The end point is at the first permanent deflection and the reading of the counter is taken.

14.12 CHLORIDE, BROMIDE AND IODIDE

(*a*) **With Hg(I) ions: Discussion.** Mercury(I) ions can be generated at 100 per cent efficiency from mercury-coated gold or from mercury pool anodes, and employed for the coulometric-titration of halides. The end point is conveniently determined potentiometrically. In titrations of chloride ion, the addition of methanol (up to 70–80 per cent) is desirable in order to reduce the solubility of the mercury(I) chloride.

The standard potentials [*vs* N.H.E. (see Section 2.28)] of the fundamental couples involving uncomplexed mercury(I) and mercury(II) ions are:

$$Hg_2^{2+} + 2e = 2Hg; \qquad E^\ominus = +0.80 \text{ volt}$$

$$Hg^{2+} + 2e = Hg; \qquad E^\ominus = +0.88 \text{ volt}$$

$$2Hg^{2+} + 2e = Hg_2^{2+}; \qquad E^\ominus = +0.91 \text{ volt}$$

The oxidation of Hg to Hg_2^{2+} requires a smaller (less oxidising) potential than to Hg^{2+}; mercury(I) ions are the main product when a mercury electrode is subjected to anodic polarisation in a non-complexing medium. From a stoichiometric standpoint it does not matter whether oxidation of a mercury anode produces the mercury(I) or mercury(II) salt of a given anion, because the same quantity of electricity per mole of the anion is involved in either case: thus the same number of coulombs per mole of the anion are required to form either Hg_2Cl_2 or $HgCl_2$.

Apparatus. The apparatus is similar to that described in Section 14.7. The generator anode A now consists of a mercury pool, 0.5–1 cm deep, at the bottom of the cell; electrical connection is made by means of a platinum wire sealed through glass tubing and dipping into the mercury. For titrations of chloride and bromide, the mercury-pool generator anode serves also as the indicator electrode and is used in conjunction with a saturated calomel reference electrode; the latter is connected to the cell through a saturated potassium nitrate salt bridge. For titrations of iodide, the indicator electrode consists of a silver rod fitted through glass tubing and held by the cover of the cell. During the titration the contents of the electrolysis cell are stirred vigorously with a magnetic stirrer; the stirrer bar floats on the surface of the mercury pool anode.

Reagents. *Supporting electrolyte.* For chloride and bromide, use $0.5M$ perchloric acid. For iodide, use $0.1M$ perchloric acid plus $0.4M$ potassium nitrate. It is recommended that a stock solution of about five times the above concentrations be prepared ($2.5M$ perchloric acid for chloride and bromide; $0.5M$ perchloric acid $+ 2.0M$ potassium nitrate for iodide), and dilution to be effected in the cell according to the volume of test solution used. The reagents must be chloride-free.

Catholyte. The electrolyte in the isolated cathode compartment may be either the same supporting electrolyte as in the cell or $0.1M$ sulphuric acid: the formation of mercury(I) sulphate causes no difficulty.

Chloride. Experience in this determination may be obtained by the titration of, say, carefully standardised *ca* $0.005M$ hydrochloric acid.

Pipette 5.00 or 10.00 mL of the hydrochloric acid into the cell, add 35–40 mL of methanol and 10 mL of the stock solution of the supporting electrolyte. Fill the isolated cathode compartment with supporting electrolyte of the same concentration as that in the main body of the solution or with $0.1M$ sulphuric acid; the level of the liquid must be kept above that in the titration cell. Note the counter reading, stir magnetically, and commence the electrolysis at about 50 mA. Stop the generating current periodically, record the counter reading, and observe the potential between the mercury pool and the S.C.E. Plot a potential–counter reading curve and evaluate the equivalence point from the first or second differential graph. The approach of the equivalence point is readily detected in practice: successive small increments of 0.05 or 0.1 counter unit result in a relatively large change of potential (*ca* 30 millivolts per 0.1 counter unit).

Bromide. A $0.01M$ solution of potassium bromide, prepared from the pure salt previously dried at $110\,°C$, is suitable for practice in this determination. The experimental details are similar to those given above for chloride except that no methanol need be added. The titration cell may contain 10.00 mL of the bromide solution, 30 mL of water, and 10 mL of the stock solution of supporting electrolyte.

Iodide. A 0.01 M solution of potassium iodide, prepared from the dry salt with boiled-out water, is suitable for practice in this determination. The experimental details are similar to those given for bromide, except that the indicator electrode consists of a silver rod immersed in the solution. The titration cell may be charged with 10.00 mL of the iodide solution, 30 mL of water, and 10 mL of the stock solution of perchloric acid + potassium nitrate. In the neighbourhood of the equivalence point it is necessary to allow at least 30–60 seconds to elapse before steady potentials are established.

(b) With Ag(I) ions: Discussion. Silver ions can be electro-generated with 100 per cent efficiency at a silver anode and can be applied to precipitation titrations. The end points can be determined potentiometrically.

The supporting electrolyte may be 0.5 M potassium nitrate for bromide and iodide; for chloride, 0.5 M potassium nitrate in 25–50 per cent ethanol must be used because of the appreciable solubility of silver chloride in water.

Apparatus. Use the apparatus of Section 14.7. The generator anode is of *pure* silver foil (3 cm × 3 cm); the cathode in the isolated compartment is a platinum foil (3 cm × 3 cm) bent into a half-cylinder. For the potentiometric end point detection, use a short length of silver wire as the indicator electrode; the electrical connection to the saturated calomel reference electrode is made by means of an agar–potassium nitrate bridge.

Reagent. *Supporting electrolyte.* Prepare a 0.5 M potassium nitrate from the pure salt; for chloride determinations the solution must be prepared with a mixture of equal volumes of distilled water and ethanol.

Procedure. The determinations are carried out as described above.

14.13 TITRATION OF ACIDS

General discussion. The limiting reactions in aqueous solution at platinum electrodes are:

$$2H_2O \rightleftharpoons O_2 + 4H^+ + 4e \text{ (anode)}$$

$$2H_2O + 2e \rightleftharpoons H_2 + 2OH^- \text{ (cathode)}$$

Consequently anodic electro-generation of hydrogen ion for the titration of bases and cathodic electro-generation of hydroxide ion for the titration of acids is readily accomplished. One of the many advantages of coulometric titration of acids is that difficulties associated with the presence of carbon dioxide in the test solution or of carbonate in the standard titrant base are easily avoided: carbon dioxide can be removed completely by passing nitrogen through the original acid solution before the titration is commenced. The presence of any substance that is reduced more easily than hydrogen ion or water at a platinum cathode, or which is oxidised more easily than water at a platinum anode, will, of course, interfere.

When internal generation is used in association with a platinum auxiliary electrode the latter must be placed in a separate compartment; contact between the auxiliary electrode compartment and the sample solution is made through some sort of a diaphragm, e.g. a tube with a sintered-glass disc or an agar–salt bridge. For the titration of acids a silver anode may be used in combination with a platinum cathode in presence of bromide ions; the silver electrode is placed inside a straight tube closed by a sintered disc at its lower end and this can be inserted directly into the test solution. A bromide ion concentration of about $0.05 M$ is satisfactory.

(a) **With isolated platinum auxiliary generating electrode.** *Apparatus.* Use the cell (*ca* 150 mL capacity) shown in Fig. 14.2(*b*) but modified by placing the auxiliary electrode in a small beaker which is connected to the cell by means of an inverted **U**-tube salt bridge containing a gel consisting of saturated potassium chloride solution with 3 per cent agar. The end point detection system consists of a pH meter in conjunction with the glass electrode–saturated calomel electrode assembly supplied with the instrument.

Reagents. *Supporting electrolyte.* $0.1 M$ sodium chloride solution.
Catholyte. This consists of $0.1 M$ sodium chloride solution to which a little dilute sodium hydroxide solution is added.
Hydrochloric acid, $0.01 M$ and $0.001 M$. Prepare with boiled-out water and concentrated hydrochloric acid, and standardise.

Procedure. Place 50 mL of the supporting electrolyte in the coulometric cell, and pass nitrogen through the solution until a pH of 7.0 is attained: thenceforth pass nitrogen over the surface of the solution. Pipette 10.00 mL of the acid into the cell. Adjust the current to a suitable value (40 or 20 mA), then start the electrolysis: stop the titration when the equivalence point pH (7.00) is reached.

(b) **With silver auxiliary electrode.** *Apparatus.* The titration cell is shown diagrammatically in Fig. 14.2, but note that the silver anode is placed inside the glass tube with a sintered disc at the lower end. The platinum cathode and the silver anode consist of stout wires coiled into helices. The silver anode may be used repeatedly before the silver bromide coating becomes so thick that it must be removed — about 30 successive titrations of 0.1 mmole samples at 20 mA. When finally necessary, the silver bromide coating may be removed by dissolution in potassium cyanide solution (**CARE**).

Reagent. *Supporting electrolyte.* Prepare a $0.05 M$ sodium bromide solution.

Procedure. Place 50 mL of the supporting electrolyte in the beaker and add some of the same solution to the tube carrying the silver electrode so that the liquid level in this tube is just above the beaker. Pass nitrogen into the solution until the pH is 7.0. Pipette 10.00 mL of either $0.01 M$ or $0.001 M$ hydrochloric acid into the cell. Continue the passage of nitrogen. Proceed with the titration as described under (*a*) above.

Several successive samples may be titrated without renewing the supporting electrolyte.

Note. The above techniques are generally applicable to many other acids, both strong and weak. The only limitation is that the anion must not be reducible at the platinum cathode and must not react in any way with the silver anode or with silver bromide (e.g. by complexation).

Titration of bases. The titration of a base with electro-generated hydrogen ions

Table 14.2 Some typical coulometric titrations

Reagent generated	Electrolyte composition	Notes*	Substances titrated	End-point detection†
Neutralisation reactions				
H^+	Sodium sulphate (0.2M)		OH^-; organic bases	P
OH^-	Sodium sulphate (0.2M)		H^+; organic acids	P
Redox reactions				
Cl_2	HCl (2.0M)		As(III); I^-	A
Br_2	KBr (0.2M)		Sb(III); Tl(I); U(IV); I^-; SCN^-; NH_3; N_2H_4; NH_2OH	A
I_2	KI (0.1M), phosphate buffer, pH 8		As(III); Sb(III); $S_2O_3^{2-}$; ascorbic acid	A, P, I
Ce(IV)	$Ce_2(SO_4)_3$ (0.1M)		Fe(II); Ti(III); U(IV); As(III); $Fe(CN)_6^{4-}$	P
Mn(III)	$MnSO_4$ (0.5M), H_2SO_4 (1.8M)		Fe(II); As(III); oxalic acid	P
Ag(II)	$AgNO_3$ (0.1M), HNO_3 (5M)	1	As(III); Ce(III); V(IV)	P
Cu(I)	$CuSO_4$ (0.1M)		V(V); Cr(VI); IO_3^-; Br_2	P
Fe(II)	$Fe_2(SO_4)_3(NH_4)_2SO_4$ (0.3M), H_2SO_4 (2M)		V(V); Cr(VI); MnO_4^-	P
Ti(III)	Ti(IV) sulphate (0.6M), H_2SO_4 (6M)		Fe(III); Ce(IV); V(V); U(VI)	P
Precipitation reactions				
Ag(I)	KNO_3 (0.5M)	2	Cl^-; Br^-; I^-; mercaptans	P
Hg(I)	$HClO_4$ (0.5M)	3	Cl^-; Br^-	P
	$HClO_4$ (0.1M), KNO_3 (0.4M)		I^-	P
$Fe(CN)_6^{4-}$	$K_3Fe(CN)_6$ (0.2M), H_2SO_4 (0.1M)		Zn	P
Complexation reactions				
EDTA	$Hg(NH_3)Y^{2-}$ (0.1M), NH_4NO_3 (0.1M), pH 8.3	4‡	Ca(II); Cu(II); Zn(II); Pb(II)	P
Miscellaneous reactions				
Substitution				
Br_2	KBr (0.2M)		Aromatic amines; phenols; 'oxine'	A
Addition				
Br_2	KBr (0.2M)	5	Unsaturated hydrocarbons, e.g. alkenes; cyclohexene	A

* Notes: 1, Gold anode; 2, silver anode; 3, mercury anode; 4, mercury cathode: for reagent see footnote ‡; 5, trace of mercury(II) acetate dissolved in acetic acid/methanol mixture added as catalyst.

† A, amperometric; I, indicator; P, potentiometric.

‡ Mercury–EDTA reagent: Prepare a stock solution containing 8.4 g mercury(II) nitrate and 9.3 g disodium-EDTA in 250 mL (each reagent 0.1M). Mix 25 mL stock solution with 75 mL ammonium nitrate solution (0.1M) and adjust to pH 8.3 with concentrated ammonia solution.

at a platinum cathode $(2H_2I \rightleftharpoons O_2 + 4H^+ + 4e)$ may be carried out as described above for the titration of acids using an isolated auxiliary electrode: the connections to the electrodes must be reversed because it is now required to generate hydrogen ions in the coulometric cell.

Reagent. *Supporting electrolyte.* Prepare a $0.2 M$ sodium sulphate solution.

Procedure. Experience in this titration may be acquired by titration of, say, 5.00 mL of accurately standardised $0.01 M$ sodium hydroxide solution. Use 50 mL of supporting electrolyte and a current of 30 mA.

14.14 SOME FURTHER EXAMPLES OF COULOMETRIC TITRATIONS

A selection of coulometric titrations of different types is collected in Table 14.2. It may be noted that the Karl Fischer method for determining water was first developed as an amperometric titration procedure (Section 16.35), but modern instrumentation treats it as a coulometric procedure with electrolytic generation of I_2. The reagents referred to in the table are generated at a platinum cathode unless otherwise indicated in the Notes.

Extensive summaries of the applications of coulometric analysis will be found in Refs 21 and 22.

For References and Bibliography see Sections 16.38 and 16.39.

CHAPTER 15
POTENTIOMETRY

15.1 INTRODUCTION

As shown in Section 2.28, when a metal M is immersed in a solution containing its own ions M^{n+}, then an electrode potential is established, the value of which is given by the **Nernst equation**:

$$E = E^{\ominus} + (RT/nF)\ln a_{M^{n+}}$$

where $E^{\ominus}$ is a constant, the standard electrode potential of the metal. E can be measured by combining the electrode with a reference electrode (commonly a saturated calomel electrode: see Section 15.3), and measuring the e.m.f. of the resultant cell. It follows that knowing the potential E_r of the reference electrode, we can deduce the value of the electrode potential E, and provided the standard electrode potential $E^{\ominus}$ of the given metal is known, we can then proceed to calculate the metal ion activity $a_{M^{n+}}$ in the solution. For a dilute solution the measured ionic activity will be virtually the same as the ionic concentration, and for stronger solutions, given the value of the activity coefficient, we can convert the measured ionic activity into the corresponding concentration.

This procedure of using a single measurement of electrode potential to determine the concentration of an ionic species in solution is referred to as **direct potentiometry**. The electrode whose potential is dependent upon the concentration of the ion to be determined is termed the **indicator electrode**, and when, as in the case above, the ion to be determined is directly involved in the electrode reaction, we are said to be dealing with an '**electrode of the first kind**'.

It is also possible in appropriate cases to measure by direct potentiometry the concentration of an ion which is not directly concerned in the electrode reaction. This involves the use of an '**electrode of the second kind**', an example of which is the silver–silver chloride electrode which is formed by coating a silver wire with silver chloride; this electrode can be used to measure the concentration of chloride ions in solution.

The silver wire can be regarded as a silver electrode with a potential given by the Nernst equation as

$$E = E^{\ominus}_{Ag} + (RT/nF)\ln a_{Ag^+}$$

The silver ions involved are derived from the silver chloride, and by the solubility product principle (Section 2.6), the activity of these ions will be governed by the chloride-ion activity

$$a_{Ag^+} = K_{s(AgCl)}/a_{Cl^-}$$

548

Hence the electrode potential can be expressed as

$$E = E_{Ag}^{\ominus} + (RT/nF)\ln K_s - (RT/nF)\ln a_{Cl^-}$$

and is clearly governed by the activity of the chloride ions, so that the value of the latter can be deduced from the measured electrode potential.

In a similar manner, in a solution containing the species Hg^{2+}, HgY^{2-}, $MY^{(n-4)+}$ and M^{n+}, where Y is the complexing agent EDTA and M^{n+} is a metallic ion which forms complexes with it, the concentration of the mercury ion is controlled by the stability constants of the complex ions $MY^{(n-4)+}$ and HgY^{2-} (the latter has a very high stability constant), and the concentration of the metal ions M^{n+}. Hence, a mercury electrode placed in this solution will acquire a potential which is determined by the concentration of the ion M^{n+}.

In the Nernst equation the term RT/nF involves known constants, and introducing the factor for converting natural logarithms to logarithms to base 10, the term has a value at a temperature of $25\,°C$ of 0.0591 V when n is equal to 1. Hence, for an ion M^+, a ten-fold change in ionic activity will alter the electrode potential by about 60 millivolts, whilst for an ion M^{2+}, a similar change in activity will alter the electrode potential by approximately 30 millivolts, and it follows that to achieve an accuracy of 1 per cent in the value determined for the ionic concentration by direct potentiometry, the electrode potential must be capable of measurement to within 0.26 mV for the ion M^+, and to within 0.13 mV for the ion M^{2+}.

An element of uncertainty is introduced into the e.m.f. measurement by the **liquid junction potential** which is established at the interface between the two solutions, one pertaining to the reference electrode and the other to the indicator electrode. This liquid junction potential can be largely eliminated, however, if one solution contains a high concentration of potassium chloride or of ammonium nitrate, electrolytes in which the ionic conductivities of the cation and the anion have very similar values.

One way of overcoming the liquid junction potential problem is to replace the reference electrode by an electrode composed of a solution containing the same cation as in the solution under test, but at a known concentration, together with a rod of the same metal as that used in the indicator electrode: in other words we set up a concentration cell (Section 2.29). The activity of the metal ion in the solution under test is given by

$$E_{cell} = (RT/nF)\ln\frac{(activity)_{known}}{(activity)_{unknown}}$$

In view of the problems referred to above in connection with direct potentiometry, much attention has been directed to the procedure of **potentiometric titration** as an analytical method. As the name implies, it is a titrimetric procedure in which potentiometric measurements are carried out in order to fix the end point. In this procedure we are concerned with *changes* in electrode potential rather than in an accurate value for the electrode potential with a given solution, and under these circumstances the effect of the liquid junction potential may be ignored. In such a titration, the change in cell e.m.f. occurs most rapidly in the neighbourhood of the end point, and as will be explained later (Section 15.18), various methods can be used to ascertain the point at which the rate of potential change is at a maximum: this is at the end point of the titration.

549

In the present chapter consideration is given to various types of indicator and reference electrodes, to the procedures and instrumentation for measuring cell e.m.f., to some selected examples of determinations carried out by direct potentiometry, and to some typical examples of potentiometric titrations.

REFERENCE ELECTRODES

15.2 THE HYDROGEN ELECTRODE

All electrode potentials are quoted with reference to the standard hydrogen electrode (Section 2.28), and hence this must be regarded as the primary reference electrode. A typical hydrogen electrode has already been described (Section 2.28), and the electrode shown in Fig. 2.2 is the Hildebrand bell-type electrode. The platinum electrode is surrounded by an outer tube into which hydrogen enters through a side inlet, escaping at the bottom through the test solution. There are several small holes near the bottom of the bell; when the speed of the gas is suitably adjusted, the hydrogen escapes through the small openings only. Because of the periodic formation of bubbles, the level of the liquid inside the tube fluctuates, and a part of the foil is alternately exposed to the solution and to hydrogen. The lower end of the foil is continuously immersed in the solution to avoid interruption of the electric current. It should be noted that although in Fig. 2.2 an open vessel is shown, in practice the electrode will be used in a stoppered flask with a suitable exit for the hydrogen, so that an oxygen-free atmosphere can be maintained in the flask.

The preparation and the use of the hydrogen electrode. The hydrogen ions of the solution are brought into equilibrium with the gaseous hydrogen by means of platinum black; the latter adsorbs hydrogen and acts catalytically. The platinum black may be supported on platinum foil of about 1 cm^2 total area but a platinum wire, 1 cm long and 0.3 mm in diameter, is often satisfactory. The platinum electrode is first carefully cleaned with hot chromic acid mixture (Section 3.8; **CAUTION!**) and thoroughly washed with distilled water. Then it is plated from a solution containing 3.0 g of chloroplatinic acid and 25 mg of lead acetate per 100 mL distilled water with platinum foil as an anode. The current may be obtained from a 4 V battery connected to a suitable sliding resistance; the current is adjusted to produce a moderate evolution of hydrogen, and the process is complete in about 2 minutes. It is important that only a *thin*, jet-black deposit be made; thick deposits lead to unsatisfactory hydrogen electrodes. After platinising, the electrode must be freed from traces of chlorine: it is washed thoroughly with water, electrolysed in *ca* 0.25 M sulphuric acid as cathode for about 30 minutes, and again well washed with water. Hydrogen electrodes should be stored in distilled water; they should never be touched with the fingers. It is advisable to have two hydrogen electrodes so that the readings obtained with one can be periodically checked against the other. In operation, hydrogen is supplied to the electrode from a cylinder of the compressed gas.

When used as a standard electrode, the hydrogen electrode operates in a solution containing hydrogen ions at constant (unit) activity based usually on hydrochloric acid, and the hydrogen gas must be at 1 atmosphere pressure: the effect of change in gas pressure is discussed in Ref. 28.

Although the hydrogen electrode is the primary reference electrode, it suffers from various disadvantages including some difficulty in preparing and operating satisfactorily. The platinum black coating of the electrode, in common with many catalysts, is susceptible to 'poisoning' by a variety of substances including arsenic, mercury, hydrogen sulphide, cyanide ions, and surface-active substances such as proteins. Furthermore, the electrode cannot be used in the presence of oxidising or reducing agents. In practice, therefore, subsidiary standard electrodes which can be kept permanently set up and which are available for immediate use are preferred for most purposes. The most common of these subsidiary standard electrodes are the calomel and the silver–silver chloride electrode.

15.3 THE CALOMEL ELECTRODE

The most widely used reference electrode, due to its ease of preparation and constancy of potential, is the **calomel electrode**. A calomel half-cell is one in which mercury and calomel [mercury(I) chloride] are covered with potassium chloride solution of definite concentration; this may be $0.1 M$, $1 M$, or saturated. These electrodes are referred to as the decimolar, the molar and the saturated calomel electrode (S.C.E.) and have the potentials, relative to the standard hydrogen electrode at 25 °C, of 0.3358, 0.2824 and 0.2444 volt.* Of these electrodes the S.C.E. is most commonly used, largely because of the suppressive effect of saturated potassium chloride solution on liquid junction potentials. However, this electrode suffers from the drawback that its potential varies rapidly with alteration in temperature owing to changes in the solubility of potassium chloride, and restoration of a stable potential may be slow owing to the disturbance of the calomel–potassium chloride equilibrium. The potentials of the decimolar and molar electrodes are less affected by change in temperature and are to be preferred in cases where accurate values of electrode potentials are required. The electrode reaction is

$$Hg_2Cl_2(s) + 2e \rightleftharpoons 2Hg \text{ (liq)} + 2Cl^-$$

and the electrode potential is governed by the chloride-ion concentration of the solution.

One form of calomel electrode is shown in Fig. 15.1(a). It consists of a stoppered glass vessel provided with a bent side tube fitted with a three-way tap which carries a short upper and a long lower tube; the latter is drawn out to a constriction at the bottom end. A short platinum wire is fused into the bottom of the vessel so that it protrudes into the interior, and a narrow glass tube sealed to the bottom of the vessel is bent round parallel to the vessel. A little mercury placed in the bottom of this tube provides electrical connection with the interior of the vessel through the sealed-in platinum wire. **Mercury and mercury compounds must be handled with care** (see Section 16.8).

To set up a saturated calomel electrode, a saturated solution of potassium chloride is first prepared from pure potassium chloride and de-ionised water, and this is then shaken for some hours with analytical grade mercury(I) chloride so that the solution is also saturated with this substance. Pure mercury is placed in the electrode vessel to a depth of about 1 cm: the platinum contact must be

---

* These figures include the liquid junction potential.[29]

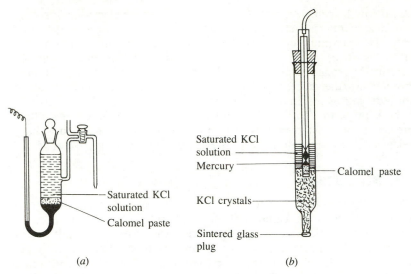

Saturated KCl
solution

Mercury

Calomel paste

KCl crystals

Sintered glass
plug

Saturated KCl
solution

Calomel paste

(a)   (b)

Fig. 15.1

totally covered with mercury. A calomel paste is prepared by grinding a few grams of mercury(I) chloride with pure mercury and a few millilitres of the saturated potassium chloride in a glass mortar; the supernatant liquid is poured off and the grinding repeated twice more with fresh quantities of potassium chloride solution. Some of this paste is transferred to the electrode vessel to form a 1 cm layer on top of the mercury, and the vessel is then filled with the saturated potassium chloride solution. By judiciously tilting the vessel, the side arm up to and including the bore of the three-way tap is filled with potassium chloride solution and the vessel is then stoppered. To put the electrode into service, the three-way tap is turned so that contact is made between the upper and lower external tubes, and the lower tube is placed in pure saturated potassium chloride solution (no calomel present); by applying suction to the upper tube, liquid is drawn up until the bore of the tap is filled. The tap is then turned so that contact is *almost* made between the side tube and the contents of the electrode vessel: in this position there will be ample conductivity round the barrel of the tap for the electrode to function satisfactorily, and there is no danger that the contents of the electrode vessel may be transferred to the test solution. In use, the constricted lower end of the side tube is dipped into the test solution, and after use the solution in the side tube is drained away and the tube washed out with de-ionised water. If the test solution is affected by chloride ions, the side arm may be filled with $3M$ potassium nitrate solution.

Compact, ready-prepared calomel electrodes are available commercially and find wide application especially in conjunction with pH meters and ion-selective meters. A typical electrode is shown in Fig. 15.1(b). With time, the porous contact disc at the base of the electrode may become clogged, thus giving rise to a very high resistance. In some forms of the electrode the sintered disc may be removed and a new porous plate inserted, and in some modern electrodes an ion exchange membrane is incorporated in the lower part of the electrode which prevents any migration of mercury(I) ions to the sintered disc and thus to the test solution. These commercially available electrodes are normally supplied with saturated potassium chloride solution.

Some commercial electrodes are supplied with a double junction. In such arrangements, the electrode depicted in Fig. 15.1(b) is mounted in a wider vessel of similar shape which also carries a porous disc at the lower end. This outer vessel may be filled with the same solution (e.g. saturated potassium chloride solution) as is contained in the electrode vessel: in this case the main function of the double junction is to prevent the ingress of ions from the test solution which may interfere with the electrode. Alternatively, the outer vessel may contain a different solution from that involved in the electrode (e.g. 3 M potassium nitrate or 3 M ammonium nitrate solution), thus preventing chloride ions from the electrode entering the test solution. This last arrangement has the disadvantage that a second liquid junction potential is introduced into the system, and on the whole it is preferable wherever possible to choose a reference electrode which will not introduce interferences.

For some purposes, modifications of the calomel electrode may be preferred. Thus, if it is necessary to avoid the presence of potassium ions, the electrode may be prepared with sodium chloride solution replacing the potassium chloride. In some cases the presence of chloride ions may be inimical and a mercury(I) sulphate electrode may then be used: this is prepared in similar manner to a calomel electrode using mercury(I) sulphate and potassium sulphate or sodium sulphate solution.

15.4 THE SILVER–SILVER CHLORIDE ELECTRODE

This electrode is perhaps next in importance to the calomel electrode as a reference electrode. It consists of a silver wire or a silver-plated platinum wire, coated electrolytically with a thin layer of silver chloride, dipping into a potassium chloride solution of known concentration which is saturated with silver chloride: this is achieved by the addition of two or three drops of 0.1 M silver nitrate solution. Saturated potassium chloride solution is most commonly employed in the electrode, but 1 M or 0.1 M solutions can equally well be used: as explained in Section 15.1, the potential of the electrode is governed by the activity of the chloride ions in the potassium chloride solution.

Commercial forms of the electrode are available and in general are similar to the calomel electrode depicted in Fig. 15.1(b) with the replacement of the mercury by a silver electrode, and calomel by silver chloride. The remarks concerning clogging of the sintered disc, and the use of ion exchange membranes and double junctions to reduce this are equally applicable to the silver–silver chloride electrode.

Values of the electrode potentials for the more common reference electrodes are collected in Table 15.1 together with an indication of the effect of temperature for the most important electrodes.

INDICATOR ELECTRODES

15.5 GENERAL DISCUSSION

As already stated, the indicator electrode of a cell is one whose potential is dependent upon the activity (and therefore the concentration) of a particular

Table 15.1 Potentials of common reference electrodes

Electrode	Potential *vs* standard hydrogen electrode (volts)			
	15 °C	20 °C	25 °C	30 °C
Calomel				
KCl(sat) (S.C.E.)	0.2512	0.2477	0.2444	0.2409
1.0*M* KCl	0.2852	0.2838	0.2824	0.2810
0.1*M* KCl	0.3365	0.3360	0.3358	0.3356
Mercury(I) sulphate				
K_2SO_4(sat)	—	—	0.656	—
0.05*M* H_2SO_4	—	—	0.680	—
Silver–silver chloride				
KCl(sat)	0.2091	0.2040	0.1989	0.1939
1.0*M* KCl	—	—	0.2272	—
0.1*M* KCl	—	—	0.2901	—

ionic species whose concentration is to be determined. In direct potentiometry or the potentiometric titration of a metal ion, a simple indicator electrode will usually consist of a carefully cleaned rod or wire of the appropriate metal: it is most important that the surface of the metal to be dipped into the solution is free from oxide films or any corrosion products. In some cases a more satisfactory electrode can be prepared by using a platinum wire which has been coated with a thin film of the appropriate metal by electro-deposition.

When hydrogen ions are involved, a hydrogen electrode can obviously be used as indicator electrode, but its function can also be performed by other electrodes, foremost among which is the glass electrode. This is an example of a membrane electrode in which the potential developed between the surface of a glass membrane and a solution is a linear function of the pH of the solution, and so can be used to measure the hydrogen ion concentration of the solution. Since the glass membrane contains alkali-metal ions, it is also possible to develop glass electrodes which can be used to determine the concentration of these ions in solution, and from this development (which is based upon an ion exchange mechanism), a whole range of membrane electrodes have evolved based upon both solid-state and liquid membrane ion exchange materials; these electrodes constitute the important series of **ion-selective electrodes**[30] (sometimes called ion-sensitive electrodes) which are now available for many different ions (Sections 15.7–15.13).

Indicator electrodes for anions may take the form of a gas electrode (e.g. oxygen electrode for OH^-; chlorine electrode for Cl^-), but in many instances consist of an appropriate electrode of the second kind: thus as shown in Section 15.1, the potential of a silver–silver chloride electrode is governed by the chloride-ion activity of the solution. Selective-ion electrodes are also available for many anions.

The indicator electrode employed in a potentiometric titration will, of course, be dependent upon the type of reaction which is under investigation. Thus, for an acid–base titration, the indicator electrode is usually a glass electrode (Section 15.6); for a precipitation titration (halide with silver nitrate, or silver with chloride) a silver electrode will be used, and for a redox titration [e.g. iron(II) with dichromate] a plain platinum wire is used as the redox electrode.

For some industrial operations an antimony electrode is used to measure hydrogen-ion concentrations. The electrode consists of a rod of antimony which invariably has a coating of oxide, and placed in an aqueous solution the equilibrium

$$Sb_2O_3(s) + 6H^+ + 6e \rightleftharpoons 2Sb(s) + 3H_2O$$

is established. This gives rise to an electrode potential

$$E = E^{\ominus}_{Sb_2O_3,Sb} + \frac{RT}{6F} \ln[H^+]^6 = E^{\ominus}_{Sb_2O_3,Sb} - 0.0591 \text{ pH}$$

since the activities of the solids and of liquid water are constant. This electrode can be used in the pH range 3–8, but each electrode must be calibrated in a series of solutions of known pH. It fails in the presence of (a) strong oxidising agents, (b) complexing agents such as tartrates, and (c) solutions of pH lower than 3 or greater than 8 because the oxide dissolves in acidic and in alkaline solutions. Its great attraction is that it is cheap, simple to use, and is rugged: it is particularly useful for making measurements on slurries and gels.

15.6 THE GLASS ELECTRODE

The glass electrode is the most widely used hydrogen ion responsive electrode, and its use is dependent upon the fact that when a glass membrane is immersed in a solution, a potential is developed which is a linear function of the hydrogen ion concentration of the solution. The basic arrangement of a glass electrode is shown in Fig. 15.2(a); the bulb B is immersed in the solution of which it is required to measure the hydrogen ion concentration, and the electrical circuit is completed by filling the bulb with a solution of hydrochloric acid (usually

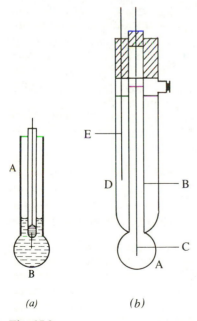

(a)   (b)

Fig. 15.2

0.1 M), and inserting a silver–silver chloride electrode. Provided that the internal hydrochloric acid solution is maintained at constant concentration, the potential of the silver–silver chloride electrode inserted into it will be constant, and so too will the potential between the hydrochloric acid solution and the inner surface of the glass bulb. Hence the only potential which can vary is that existing between the outer surface of the glass bulb and the test solution in which it is immersed, and so the overall potential of the electrode is governed by the hydrogen ion concentration of the test solution.

Glass electrodes are now available as **combination electrodes** which contain the indicator electrode (a thin glass bulb) and a reference electrode (silver–silver chloride) combined in a single unit as depicted in Fig. 15.2(b). The thin glass bulb A and the narrow tube B to which it is attached are filled with hydrochloric acid and carry a silver–silver chloride electrode C. The wide tube D is fused to the lower end of tube B and contains saturated potassium chloride solution which is also saturated with silver chloride; it carries a silver–silver chloride electrode E. The assembly is sealed with an insulating cap.

The nature of the glass used for construction of the glass electrode is very important. Hard glasses of the Pyrex type are not suitable, and for many years a lime–soda glass (Corning 015) of the approximate composition SiO_2 72 per cent, Na_2O 22 per cent, CaO 6 per cent was universally used for the manufacture of glass electrodes. Such electrodes were extremely satisfactory over the pH range 1–9, but in solutions of higher alkalinity the electrode was subject to an 'alkaline error' and tended to give low values for the pH. The error increased with the concentration of alkali-metal ions in solution, and for example at pH 12 in the presence of sodium ions, the error varied from -1.0 pH ($[Na^+] = 1M$) to -0.4 pH ($[Na^+] = 0.1M$); the errors were smaller in solutions containing lithium, potassium, barium, or calcium ions. Attempts were therefore made to discover glasses which would give electrodes free from this alkaline error, and it was found that the required result could be achieved by replacing most or all of the sodium content of the glass by lithium. An electrode constructed of a glass having a composition SiO_2 63 per cent, Li_2O 28 per cent, Cs_2O 2 per cent, BaO 4 per cent, La_2O_3 3 per cent has an error of only -0.12 pH at pH 12.8 in the presence of sodium ions at a concentration of $2M$. Lithium-based glasses are now exclusively used for hydrogen ion responsive glass electrodes required for use at high pH values.

To measure the hydrogen ion concentration of a solution the glass electrode must be combined with a reference electrode, for which purpose the saturated calomel electrode is most commonly used, thus giving the cell:

Ag,AgCl(s)|HCl($0.1M$)|Glass|Test solution $\vdots$ KCl(sat'd),Hg_2Cl_2(s)|Hg

Owing to the high resistance of the glass membrane, a simple potentiometer cannot be employed for measuring the cell e.m.f. and specialised instrumentation (Section 15.14) must be used. The e.m.f. of the cell may be expressed by the equation:

$$E = K + (RT/F) \ln a_{H^+}$$

or at a temperature of 25 °C by the expression:

$$E = K + 0.0591 \text{ pH}$$

In these equations K is a constant partly dependent upon the nature of the glass used in the construction of the membrane, and partly upon the individual character of each electrode; its value may vary slightly with time. This variation of K with time is related to the existence of an **asymmetry potential** in a glass electrode which is determined by the differing responses of the inner and outer surfaces of the glass bulb to changes in hydrogen ion activity; this may originate as a result of differing conditions of strain in the two glass surfaces. Owing to the asymmetry potential, if a glass electrode is inserted into a test solution which is in fact identical with the internal hydrochloric acid solution, then the electrode has a small potential which is found to vary with time. On account of the existence of this asymmetry potential of time-dependent magnitude, a constant value cannot be assigned to K, and every glass electrode must be standardised frequently by placing in a solution of known hydrogen ion activity (a buffer solution).

So-called combination electrodes may be purchased in which the glass electrode and the saturated calomel reference electrode are combined into a single unit, thus giving a more robust piece of equipment, and the convenience of having to insert and support a single probe in the test solution instead of the two separate components.

As will be apparent from the above discussion, the operation of a glass electrode is related to the situations existing at the inner and outer surfaces of the glass membrane. Glass electrodes require soaking in water for some hours before use and it is concluded that a hydrated layer is formed on the glass surface, where an ion exchange process can take place. If the glass contains sodium, the exchange process can be represented by the equilibrium

$$H_{soln}^+ + Na_{glass}^+ \rightleftharpoons H_{glass}^+ + Na_{soln}^+$$

The concentration of the solution within the glass bulb is fixed, and hence on the inner side of the bulb an equilibrium condition leading to a constant potential is established. On the outside of the bulb, the potential developed will be dependent upon the hydrogen ion concentration of the solution in which the bulb is immersed. Within the layer of 'dry' glass which exists between the inner and outer hydrated layers, the conductivity is due to the interstitial migration of sodium ions within the silicate lattice. For a detailed account of the theory of the glass electrode a textbook of electrochemistry should be consulted.

In view of the equilibrium shown in the equation above it is not surprising that if the solution to be measured contains a high concentration of sodium ions, say a sodium hydroxide solution, the pH determined is too low. Under these conditions sodium ions from a solution pass into the hydrated layer in preference to hydrogen ions, and consequently the measured e.m.f. (and hence the pH) are too low. This is the reason for the 'alkaline error' encountered with the glass electrode constructed from a lime–soda glass. Likewise in strongly acid solutions (hydrogen ion concentration in excess of $1 M$), errors also arise but to a much smaller degree; this effect is related to the fact that in the relatively concentrated solutions involved, the activity of the water in the solution is reduced and this can affect the hydrated layer of the electrode which is involved in the ion exchange reaction.

The glass electrode can be used in the presence of strong oxidants and reductants, in viscous media, and in the presence of proteins and similar substances which seriously interfere with other electrodes. It can also be adapted for measurements with small volumes of solutions. It may give erroneous results when used with very poorly buffered solutions which are nearly neutral.

The glass electrode should be thoroughly washed with distilled water after each measurement and then rinsed with several portions of the next test solution before making the following measurement. The glass electrode should not be allowed to become dry, except during long periods of storage: it will return to its responsive condition when immersed in distilled water for at least 12 hours prior to use (see Ref. 31).

ION–SELECTIVE ELECTRODES

15.7 ALKALI METAL ION–SELECTIVE GLASS ELECTRODES

If the preference for hydrogen ion exchange shown by lime–soda glasses can be reduced, then other cations will become involved in the ion exchange process and the possibility of an electrode responsive to metallic ions such as sodium and potassium exists. The required effect can be achieved by the introduction of aluminium oxide into the glass, and as shown in Table 15.2, this approach has led to new glass electrodes of great importance to the analyst.

Table 15.2 Composition of glasses for cation-sensitive glass electrodes

Composition	For determination of
Na_2O 22%, CaO 6%, SiO_2 72%	H^+ (subject to alkaline error)
Li_2O 28%, Cs_2O 2%, BaO 4%, La_2O_3 3%, SiO_2 63%	H^+ (alkaline error reduced)
Li_2O 15%, Al_2O_3 25%, SiO_2 60%	Li^+
Na_2O 11%, Al_2O_3 18%, SiO_2 71%	Na^+, Ag^+
Na_2O 27%, Al_2O_3 5%, SiO_2 68%	K^+, NH_4^+

In all cases some sensitivity to hydrogen ions remains; in any potentiometric determination with these modified glass electrodes the hydrogen ion concentration of the solution must be reduced so as to be not more than 1 per cent of the concentration of the ion being determined, and in a solution containing more than one kind of alkali metal cation, some interference will be encountered.

The construction of these electrodes is exactly similar to that already described for the pH responsive glass electrode. They must of course be used in conjunction with a reference electrode and for this purpose a silver–silver chloride electrode is usually preferred. A 'double junction' reference electrode is often used. The electrode response to the activity of the appropriate cation is given by the usual Nernst equation:

$$E = k + (RT/nF) \log a_{M^{n+}}$$

and for a singly charged cation, since $-\log a_{M^+} = pM$ (cf. pH)

$$E = k - 0.0591 \, pM \text{ (at } 25\,°C)$$

Such an electrode may, however, also show a response to certain other singly charged cations, and when an interfering cation C^+ is present in the test solution, an equilibrium is established between ions M^+ in the glass surface in contact with the solution, and the ions C^+ in the solution:

$$M_{gl}^+ + C_{soln}^+ \rightleftharpoons C_{gl}^+ + M_{soln}^+$$

The equilibrium constant (exchange constant) for this equilibrium is given by

$$K_{ex} = \frac{a_{M^+} \times a'_{C^+}}{a'_{M^+} \times a_{C^+}}$$

where a_{M^+} and a_{C^+} are the activities of the ions in the test solution and the corresponding a' values are the activities of those ions in the surface layer of glass.

The electrode potential under these conditions is given by

$$E = K_M + \frac{2.303\,RT}{nF}\ln(a_M + k^{pot}_{M,C}a_C)$$

where k_M is the asymmetry potential of the electrode in presence of the ion M^+, and $k^{pot}_{M,C}$ is termed the '**selectivity coefficient**' of the electrode for M over C. Both k_M and $k^{pot}_{M,C}$ can be evaluated by making e.m.f. measurements with two solutions containing different known amounts of the two ions. The selectivity coefficient is a measure of the interference of the ion C^+ in the determination of the ion M^+, but the value is dependent upon several variables, such as the total ion concentration of the solution and the ratio of the activity of the ion being determined to that of the interfering ion. A small selectivity coefficient indicates an electrode which is not greatly susceptible to interference by the specified ion:

$$k_{M,C} = \frac{\text{Response to } C^+}{\text{Response to } M^+}$$

15.8 OTHER SOLID MEMBRANE ELECTRODES

The glass membrane of the electrodes discussed above may be replaced by other materials such as a single crystal or a disc pressed from finely divided crystalline material; it may be advantageous to incorporate the crystalline material into an inert carrier such as a suitable polymer thus producing a heterogeneous-membrane electrode.

Pungor and co-workers[32] developed an iodide ion selective electrode by incorporating finely dispersed silver iodide into a silicone rubber monomer and then carrying out polymerisation. A circular portion of the resultant silver-iodide-impregnated polymer was used to seal the lower end of a glass tube which was then partly filled with potassium iodide solution ($0.1\,M$), and a silver wire was inserted to dip into the potassium iodide solution. When the membrane end of the assembly is inserted into a solution containing iodide ions, we have a situation exactly similar to that encountered with glass membrane electrodes. The silver iodide particles in the membrane set up an exchange equilibrium with the solutions on either side of the membrane. Inside the electrode, the iodide ion concentration is fixed and a stable situation results. Outside the electrode, the position of equilibrium will be governed by the iodide ion concentration of the external solution, and a potential will, therefore, be established across the membrane and this potential will vary according to the iodide ion concentration of the test solution.

A single crystal electrode is exemplified by the lanthanum fluoride electrode in which a crystal of lanthanum fluoride is sealed into the bottom of a plastic container to produce a fluoride ion electrode. The container is charged with a solution containing potassium chloride and potassium fluoride and carries a silver wire which is coated with silver chloride at its lower end: it thus constitutes a silver–silver chloride reference electrode.

The lanthanum fluoride crystal is a conductor for fluoride ions which being small can move through the crystal from one lattice defect to another, and equilibrium is established between the crystal face inside the electrode and the internal solution. Likewise, when the electrode is placed in a solution containing fluoride ions, equilibrium is established at the external surface of the crystal. In general, the fluoride ion activities at the two faces of the crystal are different and so a potential is established, and since the conditions at the internal face are constant, the resultant potential is proportional to the fluoride ion activity of the test solution.

The pressed disc (or pellet) type of crystalline membrane electrode is illustrated by silver sulphide, in which substance silver ions can migrate. The pellet is sealed into the base of a plastic container as in the case of the lanthanum fluoride electrode, and contact is made by means of a silver wire with its lower end embedded in the pellet: this wire establishes equilibrium with silver ions in the pellet and thus functions as an internal reference electrode. Placed in a solution containing silver ions the electrode acquires a potential which is dictated by the activity of the silver ions in the test solution. Placed in a solution containing sulphide ions, the electrode acquires a potential which is governed by the silver ion activity in the solution, and this is itself dictated by the activity of the sulphide ions in the test solution and the solubility product of silver sulphide — i.e. it is an electrode of the second kind (Section 15.1).

If the pellet contains a mixture of silver sulphide and silver chloride (or bromide or iodide), the electrode acquires a potential which is determined by the activity of the appropriate halide ion in the test solution. Likewise, if the pellet contains silver sulphide together with the insoluble sulphide of copper(II), cadmium(II), or lead(II), we produce electrodes which respond to the activity of the appropriate metal ion in a test solution.

15.9   ION EXCHANGE ELECTRODES

Ion exchange electrodes can be prepared using an organic liquid ion exchanger which is immiscible with water, or an ion-sensing material is dissolved in an organic solvent which is immiscible with water, and placed in a tube sealed at the lower end by a thin hydrophobic membrane such as 'millipore' cellulose acetate filter: aqueous solutions will not penetrate this film. The basis of the construction of such an electrode is indicated in Fig. 15.3. The membrane (A) seals the bottom of the electrode vessel, which is divided by the central tube into an inner (B) and outer compartment (D). Compartment B contains an aqueous solution of known concentration of the chloride of the metal ion to be determined: this solution is also saturated with silver chloride and carries a silver electrode (C), which thus forms a reference electrode. The liquid ion exchange material is placed in reservoir D and the pores of the membrane become impregnated with the organic liquid which thus makes contact with the aqueous test solution in which the electrode is placed; this solution also carries a suitable reference electrode, e.g. a calomel electrode. This kind of electrode is referred to as a liquid-membrane electrode.

Following a design by Thomas and co-workers[33a,b] it is now usual to prepare

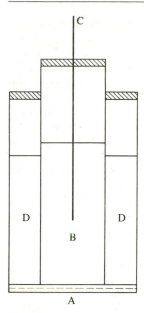

Fig. 15.3

solid ion exchange membranes by dissolving the liquid ion exchange material together with poly(vinyl chloride) (PVC) in a suitable organic solvent such as tetrahydrofuran and then allowing the solvent to evaporate. A disc is cut from the flexible residue and cemented to a PVC tube to produce an electrode vessel, in which the PVC membrane replaces the cellulose acetate and reservoir material previously used, so that only a single compartment is needed. Clearly it is no longer possible to refer to a liquid membrane electrode; most ion exchange electrodes are now of this type.

A typical example of a PVC matrix membrane electrode is the calcium ion electrode in which the cation exchange material is based upon a dialkyl phosphate such as didecyl hydrogenphosphate or better still dioctylphenyl hydrogenphosphate dissolved in dioctyl phenyl phosphonate. In contact with an aqueous solution containing calcium ions, reaction involving the loss of a proton from each of two molecules of ester occurs to form a calcium (dialkyl or dialkylphenyl) phosphate at the surface of the membrane, which thus acquires calcium ions which can equilibrate with any other solution containing calcium ions in which it may be placed. At the internal face of the membrane, a solution of some specified calcium ion concentration is present and so a definite potential is established. On the outer side of the membrane, the potential established will be determined by the calcium ion activity of the test solution, and thus the overall electrode potential can be related to the activity of calcium ions in the test solution. The electrode fails in acid solution because the reaction producing the calcium (dialkyl) phosphate is reversed. If the solvent used is changed to decanol, the electrode becomes responsive to other ions similar to calcium, including magnesium, and can be used as a 'water hardness' electrode.

An electrode sensitive to nitrate ions can be prepared by using the salt hexadecyl-(tridodecyl)-ammonium nitrate in the membrane, and a perchlorate

561

[chlorate(VII) ion] electrode can be produced based upon a membrane containing tris(o-phenanthroline) iron(II) perchlorate.

The ionic organic ion exchangers used in the electrodes described above can in some cases be replaced by neutral organic ligands. A typical example is the potassium ion-selective electrode based upon the antibiotic valinomycin: this substance contains a number of oxygen atoms that can form a ring and coordinate with a potassium ion by displacing its hydration shell. The electrode is set up with an internal potassium chloride solution of definite concentration from which some potassium ions are extracted by the valinomycin into the inner surface of the membrane. When the electrode is placed in an aqueous solution containing potassium ions, some of these are extracted into the external surface of the membrane, and the resultant electrode potential will be dependent upon the potassium ion activity of the test solution. Many synthetic neutral organic ligands are now available which can be used as sensors in ion-selective electrodes for a large range of cations.

15.10 ENZYME–BASED ELECTRODES

Such electrodes make use of an enzyme to convert the substance to be determined into an ionic product which can itself be detected by a known ion-selective electrode. A typical example is the **urea electrode**, in which the enzyme urease is employed to hydrolyse urea:

$$CO(NH_2)_2 + H_2O + 2H^+ \xrightarrow{\text{urease}} 2NH_4^+ + CO_2$$

and the progress of the reaction can be followed by means of a glass electrode which is sensitive to ammonium ions. The final concentration of ammonium ions determined can be related to the urea present.

The urease is incorporated into a polyacrylamide gel which is allowed to set on the bulb of the glass electrode and may be held in position by nylon gauze. Preferably, the urease can be chemically immobilised on to bovine serum albumin or even on to nylon. When the electrode is inserted into a solution containing urea, ammonium ions are produced, diffuse through the gel and cause a response by the ammonium ion probe:

$$E_{cell} = k + 0.0295 \log a_{urea}$$

Penicillin can likewise be determined by using the enzyme penicillinase to destroy the penicillin with production of hydrogen ions which can be determined using a normal glass pH electrode. Many other organic materials can be determined by similar procedures.[34,35]

15.11 GAS SENSING ELECTRODES

Such electrodes can be used to analyse solutions of gases such as ammonia, carbon dioxide, nitrogen dioxide, sulphur dioxide and hydrogen sulphide. For the last named, a sulphide ion responsive electrode is used, and for nitrogen dioxide a nitrate ion responsive electrode is employed, whilst for the other gases named a glass pH electrode is used. To determine the proportion of any of these gases in a stream of gas, the gaseous mixture is passed through a scrubber where the gas is dissolved in water and the resultant liquid is then examined with the appropriate gas sensing electrode.

The essential features of a gas sensing electrode can be appreciated by reference to Fig. 15.3; only the central portion of this diagram, namely vessel B and its attachments, are now relevant. The membrane A is permeable to the dissolved gas in the test solution and may be a microporous membrane manufactured from either polytetrafluoroethylene or from polypropylene, both of which materials are water repellent and are not penetrated by aqueous solutions, but they allow gas molecules to pass through: this kind of membrane is used with ammonia, carbon dioxide, and nitrogen dioxide. Alternatively, the membrane is a very thin homogeneous film, commonly of silicone rubber, through which the gas diffuses (sulphur dioxide, hydrogen sulphide). Electrode C is now a glass pH electrode or other suitable ion-selective electrode and a silver–silver chloride reference electrode is also incorporated in B. The internal solution in B contains sodium chloride and an electrolyte appropriate to the gas which is being determined: NH_3, NH_4Cl; CO_2, $NaHCO_3$; NO_2, $NaNO_2$; SO_2, $K_2S_2O_5$; H_2S, a citrate buffer. Membrane A is small in area, and the volume of liquid in B is also small so that it rapidly equilibrates with the test solution.

15.12 ION–SELECTIVE FIELD EFFECT TRANSISTORS (ISFET)

A novel development of the use of ion-selective electrodes is the incorporation of a very thin ion-selective membrane (C) into a modified metal oxide semiconductor field effect transistor (A) which is encased in a non-conducting shield (B) (Fig. 15.4). When the membrane is placed in contact with a test solution containing an appropriate ion, a potential is developed, and this potential affects the current flowing through the transistor between terminals T_1 and T_2.

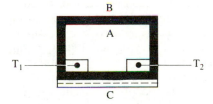

Fig. 15.4

By calibration against solutions containing known activities of the ion being determined, measurement of the current can be used to ascertain the activity of the ion in the test solution. Such measurements can be carried out with very small volumes of liquid, and find application in biochemical analyses.[36,37] However, the simpler ion-selective electrodes discussed above can be readily adapted for dealing with small volumes, and even for intracellular measurements.

In an alternative procedure designed to deal with minute volumes of liquid, Walter[38] set up a 'layer cell' based upon the technique employed in 'instant colour' photographic films, Such a cell designed to determine potassium ions made use of two layer assemblies terminating in valinomycin electrodes, so that with a standard potassium chloride solution added to one assembly, and the solution under test to the other, with the two assemblies joined by a salt bridge, a concentration cell was set up, measurement of the e.m.f. of which made possible the calculation of the potassium ion concentration in the test solution.

15.13 COMMERCIALLY AVAILABLE ION–SELECTIVE ELECTRODES

A number of ion-selective electrodes are available from laboratory supply houses: whilst not intended to be an exhaustive list, Table 15.3 serves to indicate the variety of determinations for which electrodes are available. An indication is also given of the lower limit of detection of the electrodes; this figure may vary somewhat according to the source of the electrode but full details are furnished by the manufacturer of the effective range of use of each electrode and of likely interferences.

Table 15.3 A selection of commercially available ion-selective electrodes

Ion	Lower limit of detection, pX*	Ion	Lower limit of detection, pX*
Glass membrane		*Crystalline solid-state membrane*	
H^+	13	Ag^+	8
Na^+	6	Cd^{2+}	7
K^+	4	Cu^{2+}	8
Ag^+	5	Pb^{2+}	6
NH_4^+	5	F^-	6
Ion-exchange membrane		Cl^-	4.3
Ca^{2+}	6.3	Br^-	5.3
K^+	6	I^-	7.3
$Ca^{2+} + Mg^{2+}$	5.3	CN^-	6
NO_3^-	5.3	SCN^-	5.3
ClO_3^-	5.3	S^{2-}	8

* $pX = \log_{10} 1/[X]$; (X expressed in mol L^{-1}).

Considerations of importance relating to the use of ion-selective electrodes are (1) the concentration range over which the electrode may be used, and (2) the response time. If the e.m.f. of a given ion-selective electrode is measured in a series of solutions containing the relevant ion at varying activities, then on plotting the e.m.f. against the logarithm of the ionic activity, a graph is obtained as in Fig. 15.5. The curve falls into three distinct parts: (*a*) a straight line portion AB; (*b*) a curved portion BC; (*c*) a nearly horizontal portion CD. The straight line AB has a slope equal to $2.303 RT/nF = 59.1$ millivolt at $25 \,°C$, when $n = 1$, and in this region the electrode is said to show a Nernstian response; point B

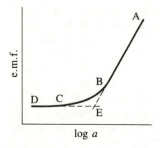

Fig. 15.5

may be regarded as the lower limit of measurement for practical purposes. Nevertheless, measurements can be made over the curved portion of the curve (BC) by making a series of readings with solutions of known activities falling within the required range and then plotting a calibration curve. IUPAC defines the lower detection limit as 'The concentration of the ion at which the extrapolated linear portion of the graph intersects the extrapolated Nernstian portion of the graph', i.e. point E in Fig. 15.5. The detection limit will, of course, be influenced by the presence of interfering ions.

The response time of an electrode is defined as the time taken for the cell e.m.f. to a reach a value which is 1 millivolt from the final equilibrium value. The response time is obviously affected by the type of electrode, particularly with regard to the nature of the membrane, and is also affected by the presence of interfering ions, and by change in temperature.

A further matter of importance is the influence of ionic strength on activity values in view of which it is important that the test and the standard solutions should be of comparable ionic strength. When dealing with a test solution containing a single electrolyte little difficulty will be encountered in arranging for the test and standard solutions to be of similar ionic strength, but this may not be the case when the test solution has arisen from an involved analytical procedure. In such a case, a 'total ionic strength adjuster buffer' (TISAB) is added to both test and standard solutions so as to achieve comparable values for the ionic strength in all solutions, the value being governed by the ionic strength adjuster buffer rather than by the sample itself. A number of electrolytes may be used in this fashion, due attention being paid to ensure that errors do not arise owing to complexation, or to poor selectivity of the electrode in use with respect to the added ion in relation to the ion whose concentration is to be determined.

General details for the care and maintenance of ion-selective electrodes are given in Ref. 31.

INSTRUMENTATION AND MEASUREMENT OF CELL e.m.f.

15.14 pH METERS AND SELECTIVE ION METERS

The most satisfactory method for measurement of the e.m.f. of a cell was, for many years, that known as the Poggendorff's compensation method. The principle of the method is to balance the unknown e.m.f. against a known e.m.f. which can be easily varied. When these two e.m.f.s are exactly equal, no current flows through a galvanometer in the circuit, and *no current is taken from the cell under test*. Such a simple procedure is, however, not applicable to circuits in which a high resistance exists, as, for example, with a glass electrode. With the introduction of solid-state circuitry which has simplified the measurement of small d.c. potentials in circuits of high impedance, the production of the direct-reading type of pH meter became possible. The modern pH meter is an electronic digital voltmeter, scaled to read pH directly, and may range from a comparatively simple hand-held instrument, suitable for use in the field, to more elaborate bench models, often provided with a scale-expansion facility, with a resolution of 0.001 pH unit and an accuracy of ± 0.001 unit.

As already explained, a glass electrode has an 'asymmetry potential' which makes it impossible to relate a measured electrode potential directly to the pH of the solution, and makes it necessary to calibrate the electrode. A pH meter therefore always includes a control ('Set buffer', 'Standardise' or 'Calibrate') so that with the electrode assembly (glass plus reference electrode or a combination electrode) placed in a buffer solution of known pH, the scale reading of the instrument can be adjusted to the correct value.

The Nernst equation shows that the glass electrode potential for a given pH value will be dependent upon the temperature of the solution. A pH meter, therefore, includes a biasing control so that the scale of the meter can be adjusted to correspond to the temperature of the solution under test. This may take the form of a manual control, calibrated in °C, and which is set to the temperature of the solution as determined with an ordinary mercury thermometer. In some instruments, arrangements are made for automatic temperature compensation by inserting a temperature probe (a resistance thermometer) into the solution, and the output from this is fed into the pH meter circuit.

Some instruments also include what is known as a 'Slope control'. This is to allow for the fact that in some cases, if a meter is calibrated at a certain pH (say pH 4.00), then when the electrode assembly is placed in a new buffer solution of different pH (say 9.20), the meter reading may not agree exactly with the known pH of the solution. In this event, the slope control is adjusted so that the meter reading in the second solution agrees with the known pH value. The meter is again checked in the first buffer solution, and provided the scale reading is correct (4.00), it is assumed that the meter will give accurate readings for all pH values falling within the limits of the two buffer solutions.

Using a given glass electrode–reference electrode assembly, if we measure the cell e.m.f. over a range of pH, all measurements being at the same temperature, and if the readings are then repeated for a series of different temperatures, then on plotting the results as a series of isothermal curves, we find that at some pH value (pH_i), the cell e.m.f. is independent of temperature; pH_i is referred to as the 'isopotential pH'. If the composition of the solution surrounding the inner silver–silver chloride electrode is altered, or if an entirely different external reference electrode is used, then the value of pH_i changes, and some pH meters include an 'Isopotential' control which can be used to take account of such changes in the electrode system.

Mode of operation. Before use, it is obviously necessary to become familiar with the instruction manual issued with the pH meter it is proposed to employ, but the general procedure for making a pH measurement is similar for all instruments, and will follow a pattern such as that detailed below.

1. Switch on and allow the instrument to warm up; the time for this will be quite short if the circuit is of the solid-state type. While this is taking place, make certain that the requisite buffer solutions for calibration of the meter are available, and if necessary prepare any required solutions: this is most conveniently done by dissolving an appropriate 'buffer tablet' (these are obtainable from many suppliers of pH meters and from laboratory supply houses) in the specified volume of distilled water.

2. If the instrument is equipped with a manual temperature control, take the temperature of the solutions and set the control to this value; if automatic control is available, then place the temperature probe into some of the first

standard buffer solution contained in a small beaker which has been previously rinsed with a little of the solution.

3. Insert the electrode assembly into the same beaker, and if available, set the selector switch of the instrument to read pH.

4. Adjust the 'Set buffer' control until the meter reading agrees with the known pH of the buffer solution.

5. Remove the electrode assembly (and the thermometer probe if used), rinse in distilled water, and place into a small beaker containing a little of the second buffer solution. If the meter reading does not agree exactly with the known pH, adjust the 'Slope' control until the required reading is obtained.

6. Remove the electrode assembly, rinse in distilled water, place in the first buffer solution and confirm that the correct pH reading is shown on the meter: if not, repeat the calibration procedure.

7. If the calibration is satisfactory, rinse the electrodes, etc., with distilled water, and introduce into the test solution contained in a smaller beaker. Read off the pH of the solution.

8. Remove the electrodes, etc., rinse in distilled water, and leave standing in distilled water.

Direct-reading meters suitable for use with ion-selective electrodes are available from a number of manufacturers; they are sometimes referred to as ion activity meters. They are very similar in construction to pH meters, and most can in fact be used as a pH meter, but by virtue of the extended range of measurements for which they must be used (anions as well as cations, and doubly charged as well as singly charged ions), the circuitry is necessarily more complex and scale expansion facilities are included. They are commonly used in the millivolt mode.

As with a pH meter, the electrode appropriate to the measurement to be undertaken must be calibrated in solutions of known concentration of the chosen ion; at least two reference solutions should be used, differing in concentration by 2–5 units of pM according to the particular determination to be made. The general procedure for carrying out a determination with one of these instruments is outlined in Section 15.16.

DIRECT POTENTIOMETRY

The use of a pH meter or an ion activity meter to measure the concentration of hydrogen ions or of some other ionic species in a solution is clearly an example of direct potentiometry. In view of the discussion in the preceding sections the procedure involved will be evident, and two examples will suffice to illustrate the experimental method.

15.15 DETERMINATION OF pH

At this stage it should be pointed out that the original definition of $pH = -\log c_H$ (due to Sørensen, 1909; and which may be written as pcH) is not exact, and cannot be determined exactly by electrometric methods. It is realised that the activity rather than the concentration of an ion determines the e.m.f. of a galvanic cell of the type commonly used to measure pH, and hence pH may be defined as

$$pH = -\log a_H.$$

where a_{H^+} is the activity of the hydrogen ion. This quantity as defined is also not capable of precise measurement, since any cell of the type

$H_2, Pt|H^+$ (unknown)$\|$Salt bridge$\|$Reference electrode

used for the measurement inevitably involves a liquid junction potential of more or less uncertain magnitude. Nevertheless, the measurement of pH by the e.m.f. method gives values corresponding more closely to the activity than the concentration of hydrogen ion. It can be shown that the pcH value is nearly equal to $-\log 1.1\, a_{H^+}$, hence:

$$pH = pcH + 0.04$$

This equation is a useful practical formula for converting tables of pH based on the Sørensen scale to an approximate activity basis, in line with the practical definition of pH given below.

The modern definition of pH is an operational one and is based on the work of standardisation and the recommendations of the US National Bureau of Standards (NBS). In the 1987 IUPAC definition[39] the *difference* in pH between two solutions S (a standard) and X (an unknown) at the same temperature with the same reference electrode and with hydrogen electrodes at the same hydrogen pressure is given by:

$$pH\,(X) - pH\,(S) = \frac{E_X - E_S}{2.3026\, RT/F}$$

where E_X is the e.m.f. of the cell

$H_2, Pt|$Solution $X \| 3.5M$ KCl$|$Reference electrode

and E_S is the e.m.f. of the cell

$H_2, Pt|$Solution $S \| 3.5M$ KCl$|$Reference electrode

The two hydrogen electrodes may be replaced by a *single* glass electrode which is transferred from one cell to the other. The pH difference thus determined is a pure number. The pH scale is defined by specifying the nature of the standard solution and assigning a pH value to it.

The **IUPAC definition of pH**[39] is based upon a $0.05M$ solution of potassium hydrogenphthalate as the 'reference value pH standard' (RVS). In addition, six further 'primary standard solutions' are also defined which between them cover a range of pH values lying between 3.5 and 10.3 at room temperature, and these are further supplemented by a number of 'operational standard solutions' which extend the pH range covered to 1.5–12.6 at room temperature. The composition of the RVS solution, of three of the primary standard solutions and of two of the operational standard solutions is detailed below, and their pH values at various temperatures are given in Table 15.4. It should be noted that the concentrations are expressed on a *molal* basis, i.e. moles of solute per kilogram of solution.

The British standard (BS 1647:1984, Parts 1 and 2) is also based upon potassium hydrogenphthalate and a number of reference solutions of a range of substances, and leads to results which are very similar to the figures given in Table 15.4. When applied to dilute solutions ($<0.1M$) at pH between 2 and

Table 15.4 pH of IUPAC standards from 0 to 90 °C

| Temperature (°C) | RVS | Primary | | | Operational | |
		P1	P2	P3	O1	O2
0	4.000	—	3.863	6.984	—	13.360
5	3.998	—	3.840	6.951	—	13.159
10	3.997	—	3.820	6.923	1.638	12.965
15	3.998	—	3.802	6.900	1.642	12.780
20	4.001	—	3.788	6.881	1.644	12.602
25	4.005	3.557	3.776	6.865	1.646	12.431
30	4.011	3.552	3.766	6.853	1.648	12.267
35	4.018	3.549	3.759	6.844	1.649	12.049
40	4.027	3.547	3.754	6.838	1.650	11.959
50	4.050	3.549	3.749	6.833	1.653	11.678
60	4.060	3.560	—	6.836	1.660	11.423
70	4.116	3.580	—	6.845	1.671	11.192
80	4.159	3.610	—	6.859	1.689	10.984
90	4.21	3.650	—	6.876	1.720	10.800

12 it conforms approximately to the equation

$$pH = -\log\{c_{H^+} y_{1:1}\} \pm 0.02$$

where $y_{1:1}$ is the mean activity coefficient which a typical 1:1 electrolyte would have in that solution.

Details for the preparation of the solutions referred to in the table are as follows (*note that concentrations are expressed in molalities*): all reagents must be of the highest purity. Freshly distilled water protected from carbon dioxide during cooling, having a pH of 6.7–7.3, should be used, and is essential for basic standards. De-ionised water is also suitable. Standard buffer solutions may be stored in well-closed Pyrex or polythene bottles. If the formation of mould or sediment is visible the solution must be discarded.

RVS. 0.05m Potassium hydrogenphthalate. Dissolve 10.21 g of the solid (dried below 130 °C) in water and dilute to 1 kg. The pH is not affected by atmospheric carbon dioxide: the buffer capacity is rather low. The solution should be replaced after 5–6 weeks, or earlier if mould-growth is apparent.

P1. Saturated potassium hydrogentartrate solution. The pH is insensitive to changes of concentration and the temperature of saturation may vary from 22 to 28 °C: the excess of solid must be removed. The solution does not keep for more than a few days unless a preservative (crystal of thymol) is added.

P2. 0.025m Phosphate buffer. Dissolve 3.40 g of KH_2PO_4 and 3.55 g of Na_2HPO_4 (dried for 2 hours at 110–113 °C) in carbon-dioxide-free water and dilute to 1 kg. The solution is stable when protected from undue exposure to the atmosphere.

P3. 0.01m Borax. Dissolve 3.81 g of sodium tetraborate $Na_2B_4O_7,10H_2O$ in carbon dioxide-free water and dilute to 1 kg. The solution should be protected from exposure to atmospheric carbon dioxide, and replaced about a month after preparation.

O1. 0.05m Potassium tetroxalate. Dissolve 12.70 g of the dihydrate in water and dilute to 1 kg. The salt $KHC_2O_4,H_2C_2O_4,2H_2O$ must not be dried above 50 °C. The solution is stable and the buffer capacity is relatively high.

O2. Saturated calcium hydroxide solution. Shake a large excess of finely divided calcium hydroxide vigorously with water at 25 °C, filter through a sintered glass filter (porosity 3) and store in a polythene bottle. Entrance of carbon dioxide into the solution should be avoided. The solution should be replaced if a turbidity develops. The solution is $0.0203\,M$ at 25 °C, $0.0211\,M$ at 20 °C, and $0.0195\,M$ at 30 °C.

For most purposes it is not necessary to follow the procedures given above for the preparation of standard buffer solutions: the buffer tablets which are available from laboratory suppliers, when dissolved in the specified volume of distilled (de-ionised) water, produce buffer solutions suitable for the calibration of pH meters.

Measurement of the pH of a given solution. The normal procedure is to use a glass electrode together with a saturated calomel reference electrode and to measure the e.m.f. of the cell with a pH meter. The procedure for use of a pH meter has already been described in Section 15.14; the instruction manual of the instrument available for use should however be consulted for details of minor variations in the controls supplied. The glass electrode supplied with the instrument should be standing in distilled water: if for any reason it is necessary to make use of a new electrode, then this must be left soaking in distilled water for at least 12 hours before measurements are attempted. Never handle the bulb of the electrode: the glass is approximately 0.1 mm thick. Remember that the assembly is necessarily somewhat fragile and treat it with great care: in particular the electrode must always be supported within the measuring vessel (special electrode stands are usually supplied with pH meters) and not allowed to stand on the base of the vessel.

Prepare the buffer solutions for calibration of the pH meter if these are not already available; the potassium hydrogenphthalate buffer (pH 4), and the sodium tetraborate buffer (pH 9.2) are the most commonly used for calibration purposes.

Check whether the instrument supplied is equipped for automatic temperature compensation, and, if so, that the temperature probe (resistance thermometer) is available. If it is not so equipped, then the temperature of the solutions to be used must be measured, and the appropriate setting made on the manual temperature control of the instrument.

Proceed to measure the pH of the given solution, following the steps outlined in Section 15.14. On completion of the determination, remember to wash down the electrodes with distilled water, and to leave them standing in distilled water.

15.16 DETERMINATION OF FLUORIDE

This determination involves the use of an ion-selective electrode in conjunction with an ion activity meter. The electrode must (as with the glass electrode used for pH measurements) be calibrated, using solutions containing the appropriate ion at known concentrations. Whereas for pH measurements it suffices to calibrate the glass electrode at two pH values, for ion-selective electrodes it is advisable to plot a calibration graph by making measurements with five to six standard solutions of varying concentration. This calibration graph can be used to ascertain the fluoride ion concentration in a test solution by measuring the e.m.f. of the calibrated electrode system when placed in the test solution.

As an alternative to plotting a calibration curve, the method of standard addition may be used. The appropriate ion-selective electrode is first set up, together with a suitable reference electrode in a known volume (V_t) of the test solution, and then the resultant e.m.f. (E_t) is measured. Applying the usual Nernst equation, we can say

$$E_t = k_e + k \log y_t C_t$$

where k_e is the electrode constant, k is theoretically $2.303\,RT/nF$ but in practice is the experimentally determined slope of the E *vs* $\log C$ plot for the given electrode, y_t and C_t are the activity coefficient and the concentration respectively of the ion to be determined in the test solution. A known volume V_2 of a standard solution (concentration C_s) of the ion to be determined is added to the test solution, and the new e.m.f. E_2 is measured; C_s should be 50–100 times greater than the value of C_t. For the new e.m.f. E_2 we can write:

$$E_2 = k_e + k \log y_t (V_t C_t + V_2 C_s)/(V_t + V_2)$$

where V_t is the original volume of the test solution.

Provided that the first and second solutions are of similar ionic strength, the activity coefficients will be the same in each solution, and the difference between the two e.m.f. values can be expressed as

$$\Delta E = (E_2 - E_t) = k \log(V_t C_t + V_2 C_s)/C_t (V_t + V_2)$$

whence

$$C_t = \frac{C_s}{10^{\Delta E/k}(1 + V_t/V_2) - V_t/V_2}$$

Hence provided the value of the slope constant k is known, the unknown concentration C_t can be calculated.

Procedure. Set up the ion activity meter in accordance with the manual supplied with the instrument.

The electrodes required are a fluoride ion selective electrode and a calomel reference electrode of the type supplied for use with pH meters.

Prepare the following solutions.

Sodium fluoride standards. Using analytical grade sodium fluoride and de-ionised water, prepare a standard solution which is approximately 0.05 M ($2.1\,\text{g L}^{-1}$), and of accurately known concentration (solution A). Take 10 mL of solution A and dilute to 1 L in a graduated flask to obtain solution B which contains approximately $10\,\text{mg L}^{-1}$ fluoride ion. A 20 mL volume of solution B further diluted (graduated flask) to 100 mL gives a standard (solution C), containing approximately $2\,\text{mg L}^{-1}$ fluoride ion, and by diluting 10 mL and 5 mL portions of solution B to 100 mL, we obtain standards D and E, containing respectively 1 and $0.5\,\text{mg L}^{-1}$.

Total Ionic Strength Adjustment Buffer (TISAB). Dissolve 57 mL acetic acid, 58 g sodium chloride and 4 g cyclohexane diaminotetra-acetic acid (CDTA) in 500 mL of de-ionised water contained in a large beaker. Stand the beaker inside a water bath fitted with a constant-level device, and place a rubber tube connected to the cold water tap *inside* the bath. Allow water to flow slowly into the bath and discharge through the constant level: this will ensure that in the subsequent treatment the solution in the beaker will remain at constant temperature.

Insert into the beaker a calibrated glass electrode–calomel electrode assembly which is joined to a pH meter, then with constant stirring and continuous monitoring of the pH, add slowly sodium hydroxide solution (5 M), until the solution acquires a pH of 5.0–5.5. Pour into a 1 L graduated flask and make up to mark with de-ionised water. The buffering procedure is necessary because OH$^-$ ions having the same charge and similar size to the F$^-$ ion act as an interference with the LaF$_3$ electrode.

The resulting solution will exert a buffering action in the region pH 5–6, the CDTA will complex any polyvalent ions which may interact with fluoride, and by virtue of its relatively high concentration the solution will furnish a medium of high total ionic strength, thus obviating the possibility of variation of e.m.f. owing to varying ionic strength of the test solutions.

Pipette 25 mL of solution B into a 100 mL beaker mounted on a magnetic stirrer and add an equal volume of TISAB from a pipette. Stir the solution to ensure thorough mixing, stop the stirrer, insert the fluoride ion–calomel electrode system and measure the e.m.f. The electrode rapidly comes to equilibrium, and a stable e.m.f. reading is obtained immediately. Wash down the electrodes and then insert into a second beaker containing a solution prepared from 25 mL each of standard solution C and TISAB; read the e.m.f. Carry out further determinations using the standards D and E.

Plot the observed e.m.f. values against the concentrations of the standard solutions, using a semi-log graph paper which covers four cycles (i.e. spans four decades on the log scale): use the log axis for the concentrations, which should be in terms of fluoride ion concentration. A straight line plot (calibration curve) will be obtained. With increasing dilution of the solutions there tends to be a departure from the straight line: with the electrode combination and measuring system referred to above, this becomes apparent when the fluoride ion concentration is reduced to ca 0.2 mg L^{-1}.

Now take 25 mL of the test solution, add 25 mL TISAB and proceed to measure the e.m.f. as above. Using the calibration curve, the fluoride ion concentration of the test solution may be deduced. The procedure described is suitable for measuring the fluoride ion concentration of tap water in areas where fluoridation of the supply is undertaken.

The result may be checked by adding four successive portions (2 mL) of standard solution C to the test solution of which the e.m.f. has already been determined, and measuring the e.m.f. after each addition; the calculation for this standard addition procedure is as described above.

As an alternative to the calculations described, recourse may be made to the **Gran's Plot** procedure to evaluate the initial concentration of the test solution. It was shown by Gran[40] that if antilog ($E_{cell} = nF/2.303RT$) is plotted against volume of reagent added, a straight line is obtained which when extrapolated cuts the abscissa at a point corresponding to the concentration of the test solution. Special graph paper is available (Gran's plot paper) which is a semi-antilog paper, in which the vertical axis is scaled to antilogarithm values, and the horizontal axis is a normal linear scale; using this paper the observed cell e.m.f. is plotted against the volume of reagent added. The Gran plot is particularly useful for determining end-points in potentiometric titrations.

POTENTIOMETRIC TITRATIONS

15.17 PRINCIPLES OF POTENTIOMETRIC TITRATIONS

In the previous sections dealing with direct potentiometry the procedure involved measurement of the e.m.f. between two electrodes; an indicator electrode, the potential of which is a function of the concentration of the ion to be determined, and a reference electrode of constant potential: accurate determination of the e.m.f. is crucial. In potentiometric titrations absolute potentials or potentials with respect to a standard half-cell are not usually required, and measurements are made while the titration is in progress. The equivalence point of the reaction will be revealed by a sudden change in potential in the plot of e.m.f. readings against the volume of the titrating solution; any method which will detect this abrupt change of potential may be used. One electrode must maintain a constant, but not necessarily known, potential; the other electrode must serve as an indicator of the changes in ion concentration, and must respond rapidly. The solution must, of course, be stirred during the titration. A simple arrangement for potentiometric titration is given in Fig. 15.6. A is a reference electrode (e.g. a saturated calomel half-cell), B is the indicator electrode, and C is a mechanical stirrer (it may be replaced, with advantage, by a magnetic stirrer); the solution to be titrated is contained in the beaker. When basic or other solutions requiring the exclusion of atmospheric carbon dioxide or of air are titrated it is advisable to use either a three- or four-necked flask or a tall lipless beaker equipped so that nitrogen may be bubbled through the solution before and, if necessary, during the titration.

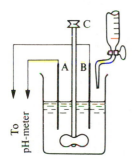

Fig. 15.6

The e.m.f. of the cell containing the initial solution is determined, and relatively large increments (1–5 mL) of the titrant solution are added until the equivalence point* is approached; the e.m.f. is determined after each addition. The approach of the e.p. is indicated by a somewhat more rapid change of the e.m.f. In the vicinity of the equivalence point, equal increments (e.g. 0.1 or 0.05 mL) should be added; the equal additions in the region of the e.p. are particularly important when the equivalence point is to be determined by the analytical method described below. Sufficient time should be allowed after each addition for the

---

* The abbreviation e.p. will be used for equivalence point.

indicator electrode to reach a reasonably constant potential (to *ca* $+1-2$ millivolts) before the next increment is introduced. Several points should be obtained well beyond the e.p.

To measure the e.m.f. the electrode system must be connected to a potentiometer or to an electronic voltmeter; if the indicator electrode is a membrane electrode (e.g. a glass electrode), then a simple potentiometer is unsuitable and either a pH meter or a selective-ion meter must be employed: the meter readings may give directly the varying pH (or pM) values as titration proceeds, or the meter may be used in the millivoltmeter mode, so that e.m.f. values are recorded. Used as a millivoltmeter, such meters can be used with almost any electrode assembly to record the results of many different types of potentiometric titrations, and in many cases the instruments have provision for connection to a recorder so that a continuous record of the titration results can be obtained, i.e. a titration curve is produced.

A number of commercial titrators are available in which the electrical measuring unit is coupled to a chart recorder to produce directly a titration curve, and by linking the delivery of titrant from the burette to the movement of the recorder chart, an auto-titrator is produced. It is possible to stop the delivery of the titrant when the indicator electrode attains the potential corresponding to the equivalence point of the particular titration: this is a feature of some importance when a number of repetitive titrations have to be performed. Many such instruments are controlled by a microprocessor so that the whole titration procedure is, to a large extent, automated. In addition to the normal titration curve, such instruments will also plot the first-derivative curve ($\Delta E / \Delta V$), the second-derivative curve ($\Delta^2 E / \Delta V^2$), and will provide a Gran's plot (Section 15.18).

15.18 LOCATION OF END POINTS

Generally speaking, the end point of a titration can be most easily fixed by examination of the titration curve including the derivative curves to which this gives rise, or by examining a Gran's Plot. There are, however, occasions when a simple and independent electrical method of detection may be useful, especially when dealing with routine determinations. In these procedures, the normal reference electrode–indicator electrode pair are no longer used, and there is no continuous recording of e.m.f. values during the titration.

One such procedure involves the use of a simple bimetallic electrode system. A tungsten electrode does not respond readily to changes in potential in certain oxidation–reduction systems (e.g. $Cr_2O_7^{2-}$, Fe^{2+}) whereas a platinum electrode does. Hence a platinum–tungsten couple can be used instead of the usual combination of a platinum electrode and reference (e.g. calomel) electrode to indicate the end-point. With the Pt–W couple the tungsten appears to undergo anodic oxidation and acts as a kind of 'attackable reference electrode'. The use of such bimetallic systems is empirical, and the optimum conditions must be established by trial. With the Pt–W pair the potential is small at first, remains at this value until very near the equivalence point, when it usually increases slightly, and then there is an abrupt change at the equivalence point; a sensitive millivoltmeter is used to monitor the change in e.m.f. The method is unsuitable for the titration of very dilute solutions ($<0.001M$).

Another procedure involves the use of **polarised indicator electrodes**. Some redox couples (e.g. $Cr_2O_7^{2-}$, Cr^{3+}; MnO_4^-, Mn^{2+}; and $S_2O_3^{2-}$, $S_4O_6^{2-}$) encountered in titrimetric analysis are somewhat slow in establishing steady potentials at a platinum electrode when the measurement is made in the ordinary way (i.e. with zero current). To eliminate long waiting periods for the attainment of steady potentials in cases of this kind, polarised indicator electrodes, at which electrolysis is forced to occur at a low rate, may be used; polarised monometallic and polarised bimetallic systems have been employed. The former consists of a polarised metallic electrode and an unpolarised electrode (e.g. a calomel electrode). A bimetallic system consists of two identical pure platinum wires, one polarised anodically and the other cathodically with a polarising current of the order of a few microamperes; these appear to behave as two dissimilar metals, and their single electrode potentials respond in a different manner. At all events a distinct change in behaviour is apparent at the equivalence point, and if the potential difference between the electrodes is plotted against the volume of reagent added the usual derivative type of curve is obtained. The potential difference developed at the end point may be of the order of 100–200 millivolts.

In potentiometric titrations with polarised electrodes the measured quantity is the change in e.m.f. at constant current (compare Amperometry, Sections 16.23–16.27, in which the change in current at constant applied e.m.f. is measured). The use of polarised electrodes can entail an error corresponding to the amount of electrolysis that occurs at the electrode; this is always present with a single polarised electrode, and also occurs with two identical polarised electrodes whenever the titration couple behaves irreversibly. The error can be reduced to negligible proportions by using small electrodes and a small electrolysis current. The main value of polarised electrodes is for titrations involving irreversible couples where an unpolarised electrode is often very slow in acquiring a constant potential whereas a polarised indicator electrode reaches a steady potential quickly at constant current, and large variations of potential are observed at the end point. The polarising current is supplied from a dry cell in series with a suitable variable resistance (see Fig. 16.1), and the potential change at the end point is recorded by a sensitive millivoltmeter.

When a titration curve has been obtained, i.e. a plot of e.m.f. readings obtained with the normal reference electrode–indicator electrode pair against volume of titrant added, either by manual plotting of the experimental readings, or with suitable equipment, plotted automatically during the course of the titration; it will in general be of the same form as the neutralisation curve for an acid, i.e. an S-shaped curve as shown in Fig. 10.2 (Section 10.12). The central portion of such a curve is shown in Fig. 15.7(a), and clearly the end point will be located on the steeply rising portion of the curve; it will in fact occur at the point of inflexion. Although when the curve shows a very clearly marked steep portion one can give an approximate value of the end point as being mid-way along the steep part of the curve, it is generally necessary to carry out some geometrical construction in order to fix the end point exactly. Three procedures may be adopted for this purpose: (a) the method of bisection; (b) the method of parallel tangents; (c) the method of circle fitting.

Unless the curve has been plotted automatically, the accuracy of the results obtained by any of the above procedures will be dependent upon the skill with which the titration curve has been drawn through the points plotted on the

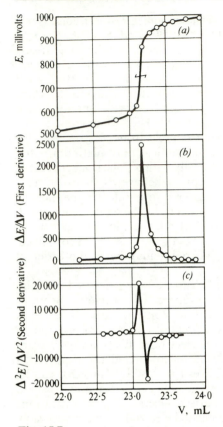

Fig. 15.7

graph from the experimental observations. It is therefore usually considered preferable to employ **analytical (or derivative) methods** of locating the end point, these consist in plotting the first derivative curve ($\Delta E/\Delta V$ against V), or the second derivative curve ($\Delta^2 E/\Delta V^2$ against V). The first derivative curve gives a maximum at the point of inflexion of the titration curve, i.e. at the end point, whilst the second derivative curve ($\Delta^2 E/\Delta V^2$) is zero at the point where the slope of the $\Delta E/\Delta V$ curve is a maximum.

The procedure may be illustrated by the actual results obtained for the potentiometric titration of 25.0 mL of *ca* 0.1 *M* ammonium iron(II) sulphate with standard (0.1095 *M*) cerium(IV) sulphate solution using platinum and saturated calomel electrodes:

$$Fe^{2+} + Ce^{4+} = Fe^{3+} + Ce^{3+}$$

The relevant results are collected in Table 15.5, as are also the calculated values for the first derivative $\Delta E/\Delta V$ (millivolt mL^{-1}) and the second derivative $\Delta^2 E/\Delta V^2$; it is clear that for locating the end point, only the experimental figures in the vicinity of the equivalence point are required. It is convenient, and simplifies the calculations, if small equal volumes of titrant are added in the neighbourhood of the end point, but this is not essential.

Table 15.5 Potentiometric titration of Fe^{2+} solution with 0.1095M Ce^{4+} solution, using platinum and calomel electrodes

Ce^{4+} solution added, V (mL)	E (mV)	$\Delta E/\Delta V$ (mV mL^{-1})	$\Delta^2 E/\Delta V^2$ (mV mL^{-2})
22.00	523		
		40	
22.50	543		
		70	
22.60	550		0
		70	
22.70	557		100
		80	
22.80	565		200
		100	
22.90	575		500
		150	
23.00	590		1 500
		300	
23.10	620		21 000
		2400	
23.20	860		$-18\,500$
		550	
23.30	915		$-2\,600$
		290	
23.40	944		$-1\,500$
		140	
23.50	958		-840
		56	
24.00	986		-300

In Fig. 15.7 are presented: (a) the part of the experimental titration curve in the vicinity of the equivalence point; (b) the first derivative curve, i.e. the slope of the titration curve as a function of V (the equivalence point is indicated by the maximum, which corresponds to the inflexion in the titration curve); and (c) the second derivative curve, i.e. the slope of curve (b) as a function of V (the second derivative becomes zero at the inflexion point and provides a more exact measurement of the equivalence point).

The optimum volume increment ΔV depends upon the magnitude of the slope of the titration curve at the equivalence point and this can easily be estimated from a preliminary titration. In general, the greater the slope at the e.p., the smaller should ΔV be, but it should also be large enough so that the successive values of ΔE exhibit a significant difference.

When the titration curve is symmetrical about the equivalence point the end point, defined by the maximum value of $\Delta E/\Delta V$, is identical with the true stoichiometrical equivalence point. A symmetrical titration curve is obtained when the indicator electrode is reversible and when in the titration reaction one mole or ion of the titrant reagent reacts with one mole or ion of the substance titrated. Asymmetrical titration curves result when the number of molecules or ions of the reagent and the substance titrated are unequal in the titration reaction, e.g. in the reaction

$$5Fe^{2+} + MnO_4^- + 8H^+ = 5Fe^{3+} + Mn^{2+} + 4H_2O$$

In such reactions, even though the indicator electrode functions reversibly, the maximum value of $\Delta E/\Delta V$ will not occur exactly at the stoichiometric equivalence point. The resulting **titration error** (difference between end point and equivalence point) can be calculated or can be determined by experiment and a correction applied. The titration error is small when the potential change at the equivalence point is large. With most of the reactions used in potentiometric analysis, the titration error is usually small enough to be neglected. It is assumed that sufficient time is allowed for the electrodes to reach equilibrium before a reading is recorded.

As has been indicated, if suitable automatic titrators are used, then the derivative curve may be plotted directly and there is no need to undertake the calculations described above.

When the potential of the indicator electrode at the equivalence point is known, either from a previous experiment or from calculations, the end point can be determined simply by adding the titrant solution until this equivalence-point potential is reached. This technique is analogous to ordinary titrations with indicators and is very convenient and rapid; this procedure can be very readily followed when an auto-titrator is employed.

The Gran's Plot procedure is a relatively simple method for fixing an end point. As explained in Section 15.16, if a series of additions of reagent are made in a potentiometric titration, and the cell e.m.f. (E) is read after each addition, then if antilog$(E \times nF/2.303RT)$ is plotted against the volume of reagent added, a straight line is obtained which when extrapolated cuts the volume axis at a point corresponding to the equivalence point volume of the reagent; plotting is simplified if the special semi-antilog Gran's Plot paper is used. The particular advantage of this method is that the titration need not be pursued to the actual end point; it is only necessary to have the requisite number of observations before the end point to permit a straight line to be drawn, and the greatest accuracy is achieved by using results over the last 20 per cent of the equivalence point volume.

15.19 SOME GENERAL CONSIDERATIONS

In this and succeeding sections experimental details are given for some typical titrations carried out in aqueous solutions; with this information it should be possible to deduce the procedure to be followed in other cases. The procedure when non-aqueous solutions are employed is considered in Section 15.25.

The majority of potentiometric titrations involve chemical reactions which can be classified as (*a*) neutralisation reactions, (*b*) oxidation–reduction reactions, (*c*) precipitation reactions or (*d*) complexation reactions, and for each of these different types of reaction, certain general principles can be enunciated.

(*a*) **Neutralisation reactions.** The indicator electrode may be a hydrogen, glass, or antimony electrode; a calomel electrode is generally employed as the reference electrode.

The accuracy with which the end point can be found potentiometrically depends upon the magnitude of the change in e.m.f. in the neighbourhood of the equivalence point, and this depends upon the concentration and the strength of the acid and alkali (compare Sections 10.13–10.16). Satisfactory results are obtained in all cases except: (*a*) those in which either the acid or the base is very weak ($K < 10^{-8}$) and the solutions are dilute, and (*b*) those in which both the acid and the base are weak. In the latter case an accuracy of about 1 per cent may be obtained in $0.1 M$ solution.

The method may be used to titrate a mixture of acids which differ greatly in their strengths, e.g. acetic (ethanoic) and hydrochloric acids; the first break in the titration curve occurs when the stronger of the two acids is neutralised, and the second when neutralisation is complete. For this method to be successful, the two acids or bases should differ in strength by at least 10^5 to 1.

(*b*) **Oxidation–reduction reactions.** The theory of oxidation–reduction reactions is given in Section 2.31. The determining factor is the ratio of the concentrations of the oxidised and reduced forms of certain ion species. For the reaction:

Oxidised form + *n* electrons $\rightleftharpoons$ Reduced form

the potential *E* acquired by the indicator electrode at 25 °C is given by:

$$E = E^{\ominus} + \frac{0.0591}{n} \log\frac{[Ox]}{[Red]}$$

where $E^{\ominus}$ is the standard potential of the system. The potential of the immersed electrode is thus controlled by the *ratio* of these concentrations. During the oxidation of a reducing agent or the reduction of an oxidising agent the ratio, and therefore the potential, changes more rapidly in the vicinity of the end point of the reaction. Thus titrations involving such reactions [e.g. iron(II) with potassium permanganate or potassium dichromate or cerium(IV) sulphate] may be followed potentiometrically and afford titration curves characterised by a sudden change of potential at the equivalence point. The indicator electrode is usually a bright platinum wire or foil, and the oxidising agent is generally placed in the burette.

(*c*) **Precipitation reactions.** The theory of precipitation reactions is given in Sections 10.74–10.75. The ion concentration at the equivalence point is determined by the solubility product of the sparingly soluble material formed during the titration. In the precipitation of an ion *I* from solution by the addition of a suitable reagent, the concentration of *I* in the solution will clearly change most rapidly in the region of the end point. The potential of an indicator electrode responsive to the concentration of *I* will undergo a like change, and hence the change can be followed potentiometrically. Here one electrode may be a saturated calomel or silver–silver chloride electrode, and the other must be an electrode which will readily come into equilibrium with one of the ions of the precipitate. For example, in the titration of silver ions with a halide (chloride, bromide, or iodide) this must be a silver electrode. It may consist of a silver wire, or of a platinum wire or gauze plated with silver and sealed into a glass tube. Since a halide is to be determined, the salt bridge must be a saturated solution of potassium nitrate. Excellent results are obtained by titrating, for example, silver nitrate solutions with thiocyanate ions. In many cases, the use of appropriate ion-selective electrodes will be possible.

(*d*) **Complexation reactions.** In many cases of this type of titration, complex formation results from the interaction of a sparingly soluble precipitate with an excess of reagent; this occurs for example when a solution of potassium cyanide is titrated with silver nitrate, where silver cyanide initially produced dissolves in excess potassium cyanide to give the complex ion $[Ag(CN)_2]^-$ and consequently only a very small concentration of silver ions. This situation continues up to the point where all the cyanide ion has been converted to the complex ion, the increasing concentration of which also means a gradually increasing concentration of free silver ions and consequently a gradual rise in the potential of a silver electrode in the solution. At the end point, there is a marked rise in potential which enables the end point to be determined, but if the addition of silver nitrate is continued past this point, the e.m.f. changes only very gradually and silver cyanide is precipitated. Finally, a second rapid change in potential is observed at the point where all the cyanide ion has been precipitated as silver cyanide. For this particular titration a silver electrode is the obvious indicator electrode, and as reference electrode either a mercury–mercury(I) sulphate electrode, or a calomel electrode which is isolated from the solution to be titrated by means of a potassium nitrate or potassium sulphate salt bridge.

For complexation titrations involving the use of EDTA, an indicator electrode can be set up by using a mercury electrode in the presence of mercury(II)–EDTA complex (see Section 15.24).

15.20 SOME EXPERIMENTAL DETAILS FOR POTENTIOMETRIC TITRATIONS

A few simple experiments will be briefly described, the performance of which will provide experience of the technique. Experiment 1 will require the use of an electronic millivoltmeter or of a pH meter which should be used in the millivolt mode; this same apparatus can be used for the other experiments. If available, a commercial titrator (or auto-titrator) may be used after experience has been gained with the simple equipment, and apart from Experiment 1, many of the determinations can be performed using a potentiometer.

Experiment 1 NEUTRALISATION REACTIONS
Prepare solutions of acetic (ethanoic) acid and of sodium hydroxide, each approximately $0.1\,M$, and set up a pH meter as described in Section 15.14.

The following general instructions are applicable to most potentiometric titrations and are given in detail here to avoid subsequent repetition.

(a) Fit up the apparatus shown in Fig. 15.6 with the electrode assembly (or combination electrode) supplied with the pH meter supported inside the beaker. The beaker has a capacity of about 400 mL and contains 50 mL of the solution to be titrated (the acetic acid).

(b) Select a burette, and by means of a piece of polythene tubing attach to the jet a piece of glass capillary tubing about 8–10 cm in length. Charge the burette with the sodium hydroxide solution taking care to remove all air bubbles from the capillary extension, and then clamp the burette so that the end of the capillary is immersed in the solutions to be titrated. This procedure ensures that all additions recorded on the burette have in fact been added to the solution, and no drops have been left adhering to the tip of the burette, a factor which can be of some significance for e.m.f. readings made near the end point of the titration.

(c) Stir the solution in the beaker gently. Read the potential difference between the electrodes with the aid of the meter. Record the reading and also the volume of alkali in the burette.

(d) Add 2–3 mL of solution from the burette, stir for about 30 seconds, and, after waiting for a further half-minute, measure the e.m.f. of the cell.

(e) Repeat the addition of 1 mL portions of the base, stirring and measuring the e.m.f. after each addition until a point is reached within about 1 mL of the expected end point. Henceforth, add the solution in portions of 0.1 mL or less, and record the potentiometer readings after each addition. Continue the additions until the equivalence point has been passed by 0.5–1.0 mL.

(f) Plot potentials as ordinates and volumes of reagent added as abscissae; draw a smooth curve through the points. The equivalence point is the volume corresponding to the steepest portion of the curve. In some cases the curve is practically vertical, one drop of solution causing a change of 100–200 millivolts in the e.m.f. of the cell; in other cases the slope is more gradual.

(g) Locate the end point of the titration by plotting $\Delta E/\Delta V$ for small increments of the titrant in the vicinity of the equivalence point ($\Delta V = 0.1$ mL or 0.05 mL) against V. There is a maximum in the plot at the end point [compare Fig. 15.7(b)].

(h) Plot the second derivative curve, $\Delta^2 E/\Delta V^2$, against V: the second derivative becomes zero at the end point [compare Fig. 15.7(c)]. This method, although laborious, gives the most exact evaluation of the end point; a Gran's Plot may also be made.

Other suggested experiments include titration of $0.05 M$ Na_2CO_3 with $0.1 M$ HCl, and of $0.1 M$ boric acid in the presence of 4 g of mannitol with $0.1 M$ NaOH.

Experiment 2 OXIDATION–REDUCTION REACTION

Experience in this kind of titration may be obtained by determining the iron(II) content of a solution by titration with a standard potassium dichromate solution.

Prepare 250 mL of $0.02 M$ potassium dichromate solution and an equal volume of *ca* $0.1 M$ ammonium iron(II) sulphate solution; the latter must contain sufficient dilute sulphuric acid to produce a clear solution, and the exact weight of ammonium iron(II) sulphate employed should be noted. Place 25 mL of the ammonium iron(II) sulphate solution in the beaker, add 25 mL of *ca* $2.5 M$ sulphuric acid and 50 mL of water. Charge the burette with the $0.02 M$ potassium dichromate solution, and add a capillary extension tube. Use a bright platinum electrode as indicator electrode and an S.C.E. reference electrode. Set the stirrer in motion. Proceed with the titration as directed in Experiment 1. After each addition of the dichromate solution measure the e.m.f. of the cell. Determine the end point: (1) from the potential–volume curve and (2) by the derivative method. Calculate the molarity of the ammonium iron(II) sulphate solution, and compare this with the value calculated from the actual weight of solid employed in preparing the solution.

Repeat the experiment using another 25 mL of the ammonium iron(II) sulphate solution but with a pair of polarised platinum electrodes. Set up two small platinum plate electrodes (0.5 cm square) in the titration beaker and remove the two electrodes previously in use. Connect the platinum plates to a polarising circuit consisting of a 3.0 volt dry battery joined to a 21 megohm resistor so that a minute current will flow between the electrodes when they are placed in solution. Also join the electrodes to the circuit used for measuring the cell e.m.f.; a simple potentiometer which is excellent for the first part of the experiment cannot be used with the polarised electrodes, and the most satisfactory procedure is to use the millivolt scale of a pH meter. Some commercial potentiometric titration units make provision for titration with polarised electrodes. The end point of the titration is indicated by the large jump in e.m.f.

The experiment may also be repeated using a platinum (indicator) electrode and a tungsten wire reference electrode. If the tungsten electrode has been left idle for more than a few days, the surface must be cleaned by dipping into just molten sodium nitrate (**CARE!**). The salt should be only just at the melting point or the tungsten will be rapidly attacked; it should remain in the melt for a few seconds only and is then thoroughly washed with distilled water.

Experiment 3 PRECIPITATION REACTIONS

The indicator electrode must be reversible to one or the other of the ions which is being precipitated. Thus in the titration of a potassium iodide solution with standard silver nitrate solution, the electrode must be either a silver electrode or a platinum electrode in the presence of a little iodine (best introduced by adding a little of a freshly prepared alcoholic solution of iodine), i.e. an iodine electrode (reversible to I^-). The exercise recommended is the standardisation of silver nitrate solution with pure sodium chloride.

Prepare an approximately $0.1 M$ silver nitrate solution. Place 0.1169 g of dry sodium chloride in the beaker, add 100 mL of water, and stir until dissolved. Use a silver wire electrode (or a silver-plated platinum wire), and a silver–silver chloride or a saturated calomel reference electrode separated from the solution by a potassium nitrate–agar bridge (see below). Titrate the sodium chloride solution with the silver nitrate solution following the general procedure described in Experiment 1; it is important to have efficient stirring and to wait long enough after each addition of titrant for the e.m.f. to become steady. Continue the titration 5 mL beyond the end point. Determine the end point and thence the molarity of the silver nitrate solution.

The **salt bridge** which is required in this experiment is prepared from a piece of narrow glass tubing which is first bent at right angles giving a limb long enough to reach to near the bottom of the titration vessel. The tube is then given a second right angle bend in such a position that the horizontal limb will extend from the titration vessel to a suitable position in which a small beaker can be supported; the two vertical limbs of the bridge should be of equal length. Clean the tube thoroughly and then clamp with the two vertical limbs extending upwards. Dissolve 3 g of agar in 100 mL of hot (almost boiling) distilled water, and then add 40 g of potassium nitrate. As soon as the salt has dissolved, allow to cool for a few minutes, and then carefully pour the hot liquid into the inverted bridge tube so that it is filled *completely* and with no air bubbles entrained in the liquid; a drawn-out thistle funnel will be found useful for this operation. Allow the tube to cool completely still in the inverted position, and when cold it may be found that at the ends of the tube the gel has contracted somewhat, so that when the tube is placed in a liquid, an air bubble is trapped at the bottom of the tube; if this has happened, the extreme ends of the tube should be carefully cut off.

For this particular titration, a three- or four-necked flat bottom flask is conveniently used as titration vessel; the salt bridge can then be inserted into one of the necks of the flask and held in position by means of a bung. The free end of the bridge is allowed to dip into a small beaker containing potassium nitrate solution ($3M$), and the side arm of the reference electrode is then inserted into the beaker. When not in use, the salt bridge should be stored with the two ends immersed in potassium nitrate solution contained in two test-tubes. A potassium chloride–agar bridge is obtained by replacing the potassium nitrate by 40 g of potassium chloride.

An interesting extension of the above experiment is the titration of a mixture of halides (chloride/iodide) with silver nitrate solution. Prepare a solution (100 mL) containing both potassium chloride and potassium iodide; weigh each substance accurately and arrange for the solution to be about $0.025M$ with respect to each salt. A silver nitrate solution of known concentration (about $0.05M$) will also be required.

Pipette 10 mL of the halide solution into the titration vessel and dilute to about 100 mL with distilled water. Insert a silver electrode, an agar–potassium nitrate salt bridge and complete the cell with a saturated calomel electrode dipping into the small beaker containing potassium nitrate solution and the second arm of the bridge. Fit a 10 mL microburette with a capillary tube extension, and fill with the silver nitrate solution. Add 1 mL of the silver nitrate solution to the contents of the titration vessel and read the cell e.m.f. after allowing adequate time for the value to become stable; complete the titration in accordance with the details previously given, but remember that there will be two end points, one in the neighbourhood of 5 mL of silver nitrate (I^-), and the other in the neighbourhood of 10 mL (Cl^-).

A comment on the polarity of the electrodes of the silver–calomel electrode cell may be helpful at this point. With the respective values of electrode potentials (calomel 0.245 V, $E^{\ominus}_{Ag}$ 0.799 V) one would normally expect the silver electrode to be the positive electrode of the cell, but at the start of the above titration, the concentration of silver ions in solution is so minute that the log term in the Nernst equation in fact has a large negative value, and the potential of the silver electrode actually becomes smaller than that of the calomel electrode. With continued addition of silver nitrate, the concentration of silver ions in solution gradually rises and the potential of the silver electrode increases, and at a point which occurs near the first end point of the titration it becomes equal to, and subsequently greater than, the potential of the calomel electrode. When this point is reached, it is necessary to reverse the connections from the cell to the potentiometer, and in order to plot a satisfactory titration curve the subsequent readings must be regarded as negative: conversely, of course, the initial readings may be regarded as negative, and those after the change over point as positive.

15.21 DETERMINATION OF COPPER

Prepare a solution of the sample, containing about 0.1 g copper and no interfering elements, by any of the usual methods; any large excess of nitric acid and all traces of nitrous acid must be removed. Boil the solution to expel most of the acid, add about 0.5 g urea (to destroy the nitrous acid) and boil again. Treat the cooled solution with concentrated ammonia solution dropwise until the deep-blue cuprammonium compound is formed, and then add a further two drops. Decompose the cuprammonium complex with glacial acetic acid and add 0.2 mL in excess. Too great a dilution of the final solution should be avoided, otherwise the reaction between the copper(II) acetate and the potassium iodide may not be complete.

Place the prepared copper acetate solution in the beaker and add 10 mL of 20 per cent potassium iodide solution. Set the stirrer in motion and add distilled water, if necessary, until the platinum plate electrode is fully immersed. Use a saturated calomel reference electrode, and carry out the normal potentiometric titration procedure using a standard sodium thiosulphate solution as titrant.

15.22 DETERMINATION OF CHROMIUM

The chromium in the substance is converted into chromate or dichromate by any of the usual methods. A platinum indicator electrode and a saturated calomel electrode are used. Place a known volume of the dichromate solution in the titration beaker, add 10 mL of 10 per cent sulphuric acid or hydrochloric acid per 100 mL of the final volume of the solution and also 2.5 mL of 10 per cent phosphorus(V) acid. Insert the electrodes, stir, and after adding 1 mL of a standard ammonium iron(II) sulphate solution, the e.m.f. is measured. Continue to add the iron solution, reading the e.m.f. after each addition, then plot the titration curve and determine the end point.

15.23 DETERMINATION OF MANGANESE

The method is based upon the titration of manganese(II) ions with permanganate in neutral pyrophosphate solution:

$$4Mn^{2+} + MnO_4^- + 8H^+ + 15H_2P_2O_7^{2-} = 5Mn(H_2P_2O_7)_3^{3-} + 4H_2O$$

The manganese(III) pyrophosphate complex has an intense reddish-violet colour; consequently the titration must be performed potentiometrically. A bright platinum indicator electrode and a saturated calomel reference electrode may be used. The change in potential at the equivalence point at a pH between 6 and 7 is large (about 300 millivolts); the potential of the platinum electrode becomes constant rapidly after each addition of the potassium permanganate solution, thus permitting direct tiration to almost the equivalence point and reducing the time required for a determination to less than 10 minutes. With relatively pure manganese solutions, a sodium pyrophosphate concentration of $0.2-0.3M$, a pH between 6 and 7, the equivalence-point potential is $+0.47 \pm 0.02$ volt vs the saturated calomel electrode. At a pH above 8 the pyrophosphate complex is unstable and the method cannot be used.

Large amounts of chloride, cobalt(II), and chromium(III) do not interfere; iron(III), nickel, molybdenum(VI), tungsten(VI), and uranium(VI) are innocuous; nitrate, sulphate, and perchlorate ions are harmless. Large quantities of magnesium, cadmium, and aluminium yield precipitates which may co-precipitate manganese and should therefore be absent. Vanadium causes difficulties only when the amount is equal to or larger than the amount of manganese; when it is present originally in the +4 state, it is oxidised slowly in the titration to the +5 state along with the manganese. Small amounts of vanadium (up to about one-fifth of the amount of manganese) cause little error. The interference of large amounts of vanadium(V) can be circumvented by performing the titration at a pH of 3–3.5. Oxides of nitrogen interfere because of their reaction with potassium permanganate: hence when nitric acid is used to dissolve the sample, the resulting solution must be boiled thoroughly and a small amount of urea or sulphamic acid must be added to the acid solution to remove the last traces of oxides of nitrogen before introducing the sodium pyrophosphate solution.

For initial practice in the method determine the manganese content of anhydrous manganese(II) sulphate. Heat manganese(II) sulphate crystals to 280 °C, allow to cool, grind to a fine powder, re-heat at 280 °C for 30 minutes, and allow to cool in a desiccator. Weigh accurately about 2.2 g of the anhydrous manganese(II) sulphate, dissolve it in water and make up to 250 mL in a graduated flask.

Prepare a 0.02 M solution of potassium permanganate and standardise it against arsenic(III) oxide.

Prepare 5 M sodium hydroxide solution using analytical grade material: test a 10 mL sample for reducing agents by adding a drop of the permanganate solution; no green coloration should develop.

Prepare also a saturated solution of the purest available sodium pyrophosphate (do not heat above 25 °C, otherwise appreciable hydrolysis may occur); 12 g of the hydrated solid $Na_4P_2O_7,10H_2O$ will dissolve in 100–150 mL of water according to the purity of the compound. It is essential to employ freshly made sodium pyrophosphate solution in the determination.

Place 150 mL of the sodium pyrophosphate solution in a 250–400 mL beaker, adjust the pH to 6–7 by the addition of concentrated sulphuric acid from a 1 mL graduated pipette (use a pH meter). Add 25 mL of the manganese(II) sulphate solution and adjust the pH again to 6–7 by the addition of 5 M sodium hydroxide solution. Introduce a bright platinum electrode into the solution, and connect the latter through a saturated potassium chloride bridge to a standard calomel electrode; complete the assembly for potentiometric titrations. Stir the mixture, add the potassium permanganate solution in 2 mL portions at first, reduce this to 0.1 mL portions in the vicinity of the end point: determine the potential after each addition. Plot the e.m.f. values (ordinates) against the volume of potassium permanganate solution added (abscissae), and determine the equivalence point. From the curve read off the potential at the equivalence point; this should be +0.47 volt. Calculate the percentage of Mn in the sample.

The method can be readily adapted to the determination of manganese in steel and in manganese ores.

Pyrolusite. Dissolve 1.5–2 g, accurately weighed, pyrolusite in a mixture of 25 mL of 1:1 hydrochloric acid and 6 mL concentrated sulphuric acid, and dilute to 250 mL. Filtration is unnecessary. Titrate an aliquot part containing 80–100 mg manganese: add 200 mL freshly prepared, saturated sodium pyrophosphate solution, adjust the pH to a value between 6 and 7, and perform the potentiometric titration as described above.

Steel. Dissolve 5 g, accurately weighed, of a steel in 1:1 nitric acid with the aid of the minimum volume of hydrochloric acid in a Kjeldahl flask. Boil the solution down to a small volume with excess of concentrated nitric acid to re-oxidise any vanadium present reduced by the hydrochloric acid: this step is unnecessary if vanadium is known to be absent. Dilute, boil to remove gaseous oxidation products, allow to cool, add 1 g of urea or sulphamic acid, and dilute to 250 mL. Titrate 50 mL portions as above.

15.24 POTENTIOMETRIC EDTA TITRATIONS WITH THE MERCURY ELECTRODE

Discussion. The indicator electrode employed is a mercury–mercury(II)–EDTA complex electrode. A mercury electrode in contact with a solution containing metal ions M^{n+} (to be titrated) and a small added quantity of a mercury(II)–EDTA complex HgY^{2-} (EDTA $= Na_2H_2Y$) exhibits a potential corresponding to the half-cell:

$$Hg \mid Hg^{2+}, HgY^{2-}, MY^{(n-4)+}, M^{n+}$$

It can be shown that the potential at equilibrium is given by:

$$E = E^{\ominus}_{Hg^{2+},Hg} + \frac{RT}{2F} \ln \frac{[HgY^{2-}]}{[MY^{(n-4)+}]} \cdot \frac{K_{MY}}{K_{HgY}} + \frac{RT}{2F} \ln [M^{n+}]$$

where K_{MY} and K_{HgY} are the stability (or formation) constants of the metal–EDTA and mercury–EDTA complexes respectively. The first two terms on the right-hand side of this equation are essentially constant during a potentiometric titration, especially in the region of the end point, hence the measured potential of the electrode becomes a linear function of pM. The mercury–mercury(II)–EDTA complex electrode will, for convenience, be subsequently described as a mercury indicator electrode; it is clearly a pM indicator electrode.

The potential of the mercury indicator electrode depends upon the total mercury concentration in the solution. In practice it is found that the addition of 1 drop of a $0.001-0.01 M$ solution of the mercury–EDTA complex HgY^{2-} is sufficient to establish a reasonably constant value for the mercury content so that trace additions of this metal do not seriously alter the shape of the titration curve. In complexometric titrations of metal ions with the mercury electrode, the experimental conditions, such as pH, the kind and nature of the buffer solution, must be carefully controlled. The buffer should be present in an amount sufficient to prevent pH changes during the titration: a large excess of buffer should be avoided, as this may decrease the extent of the potential break at the end point. Halide ions must not be present in appreciable concentrations because they may interfere with the electrode reaction, especially for titrations performed under acid conditions (for example chloride interferes at a pH less than about 6.5). At a pH lower than 2, the mercury(II)–EDTA complex dissociates to such an extent that a poorly defined titration curve results. At a pH above about 11, oxygen reacts with mercury, leading to a distorted titration curve; this may be often avoided by bubbling nitrogen through the solution before and during the titration. Direct and also back-titration procedures have been used for the determination of numerous metal ions and a selection of these is given below.

Apparatus. *Mercury electrode.* The electrode, together with the essential dimensions, is shown in Fig. 15.8; it is easily constructed from Pyrex tubing.

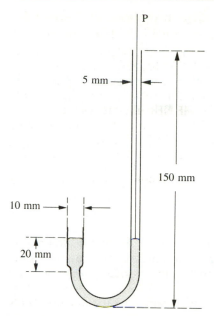

Fig. 15.8

The platinum wire P dipping into the mercury may be welded to a copper wire, but it is preferable to use a platinum wire sufficiently long to protrude at the top of the electrode tube. The mercury must be pure and clean; in case of doubt, the mercury should be washed with dilute nitric acid and then thoroughly rinsed with distilled water. The electrode is filled with mercury so that the wide portion is half-full: it is most important that no mercury is spilled into the titration vessel during the titration. After each titration the electrode is repeatedly washed with distilled water.

Titration assembly. The electrode system consists of a mercury electrode and a saturated calomel [or, in some cases, a mercury–mercury(I) sulphate] reference electrode, both supported in a 250 mL Pyrex beaker. Provision is made for magnetic stirring and the potential is followed by means of an electronic millivoltmeter or an auto-titrator.

Reagents required. *Standard EDTA solution, 0.05 M.* See Section 10.49.

Mercury–EDTA solution. Mix small equal volumes of $0.05 M$ mercury(II) nitrate and $0.05 M$ EDTA; neutralise the liberated acid by the addition of a few drops of $3 M$ ammonia solution. (In acid solution an insoluble precipitate, probably HgH_2Y, forms after a few days.) Dilute 10.0 mL of this solution to 100 mL with distilled water. The resulting *ca* $0.0025 M$ mercury–EDTA solution is used for most titrations.

Ammonia buffer solution. Mix 20 g ammonium nitrate and 35 mL concentrated ammonia solution, and make up to 100 mL with distilled water. Dilute 80 mL to 1 L with distilled water. The pH is about 10.1.

Acetate buffer solution. Mix equal volumes of $0.5 M$ sodium acetate solution and $0.5 M$ acetic (ethanoic) acid solution. The resulting solution has a pH of about 4.7.

Triethanolamine buffer solution. Prepare a *ca* 0.5 *M* aqueous solution of triethanolamine and add 2.5 *M* nitric acid until the pH is 8.5 (use a pH meter).

Procedure. The general procedure is as follows. Place 25.0 mL of the metal ion solution (approximately 0.05 *M*) in a 250 mL Pyrex beaker, add 25 mL of the appropriate buffer solution and one drop of 0.0025 *M* mercury–EDTA solution (one drop of 0.025 *M* mercury–EDTA solution for calcium and magnesium). Use the titration assembly described above. Stir magnetically. Titrate potentiometrically with standard 0.05 *M* EDTA solution added from a burette supported over the beaker. Reduce the volume of EDTA solution added to 0.1 mL or less as soon as the potential begins to rise; wait for a steady potential to be established after each addition. Soon after the end point the change of potential with each addition of EDTA becomes smaller and only a few large additions need be made. Care must be taken that mercury is not spilled into the solution either during the insertion of the mercury electrode or during the titration.

Plot the titration curve (potential in millivolts *vs* S.C.E. against volume of standard EDTA solution) and evaluate the end point. In general, results accurate to better than 0.1 per cent are obtained. Brief notes on determinations with various metal ion solutions follow.

Calcium. 25.0 mL calcium ion solution + 25 mL ammonia or triethanolamine buffer.

Magnesium. 25.0 mL magnesium ion solution + 25 mL ammonia buffer.

Nickel. 25.0 mL nickel ion solution + 25 mL ammonia buffer.

Cobalt. 20.0 mL cobalt ion solution + 25 mL ammonia buffer.*

Copper. 25.0 mL of copper(II) ion solution + 25 mL of acetate buffer.

Mercury. 25.0 mL mercury(II) ion solution + 25 mL acetate buffer.

Zinc. 25.0 mL zinc ion solution + 25 mL acetate buffer.

Bismuth. 25.0 mL bismuth ion solution + solid hexamine to pH about 4.6; the precipitate of basic bismuth salt dissolves as the EDTA solution is added but the titration is slow.

Lead. 25.0 mL lead ion solution + solid hexamine to pH about 4.6.

Thorium.

(1) 25.0 mL thorium ion solution + 90 mL 0.001 *M* nitric acid, 3 *M* ammonia solution added until pH about 3.2; mercury–mercury(I) sulphate reference electrode.

(2) 25.0 mL thorium ion solution + solid hexamine to pH about 4.6; 40.0 mL 0.05 *M* EDTA added, and excess back-titrated with standard lead nitrate solution.

Chromium. 25.0 mL chromium(III) ion solution (0.02 *M*, prepared by dilution of stock solution) + 50.0 mL 0.02 *M* EDTA + 50 mL acetate buffer, boiled for 10 minutes, solution cooled, pH adjusted to 4.6 with hexamine, 1 drop of mercury–EDTA solution added, and then back-titrated with standard zinc ion solution.

Aluminium. 25.0 mL aluminium ion solution, acidified with a few drops of 2.5 *M* nitric acid (to pH 1–2), boiled for 1 minute, 50.0 mL 0.05 *M* EDTA added to hot solution, solution cooled, 50 mL acetate buffer and 1 drop of 0.0025 *M* mercury–EDTA added, excess of EDTA back-titrated with standard zinc ion solution.

---

* A larger excess of ammonia buffer is required to ensure the formation of the cobalt-ammine.

15.25 POTENTIOMETRIC TITRATIONS IN NON-AQUEOUS SOLVENTS

As indicated in Section 2.4 the strength of an acid (and of a base) is dependent upon the solvent in which it has been dissolved, and in Sections 10.19–10.21 it has been shown how this modification of strength can be used to carry out titrations in non-aqueous solvents which are impossible to perform in aqueous solution. Potentiometric methods can be used to determine the end point of such non-aqueous titrations, which are mainly of the acid–base type and offer very valuable methods for the determination of many organic compounds.

The general procedure for performing a potentiometric titration with non-aqueous solutions is in essence the same as that employed for aqueous solutions, but there are some important points of difference.

1. An electronic millivoltmeter or a pH meter can be used to measure the e.m.f. of the titration cell, but in the latter case, the instrument must be used in the millivolt mode, since 'pH' has no significance in non-aqueous solutions.
2. Many of the non-aqueous solvents used must be protected from exposure to the air, and titrations with such materials must be conducted in a closed vessel such as a three- or four-necked flask. It must also be noted that organic solvents have much greater coefficients of thermal expansion than has water, and every effort must therefore be made to ensure that all solutions are kept as nearly as possible at constant temperature.
3. The reagents commonly employed are:
    (a) Potassium methoxide dissolved in a mixture of benzene and methanol; in some cases potassium may be replaced by lithium or by sodium, and the methanol by ethanol or propan-1-ol. Solvents for the titrand are commonly benzene–methanol, dimethylformamide, ethane-1,2-diamine or 1-aminobutane.
    (b) Quaternary ammonium hydroxides dissolved in methanol–benzene or in propan-2-ol, and with acetonitrile or pyridine as solvent for the titrand; the reagents in both (a) and (b) are used for the titration of acids.
    (c) Perchloric acid dissolved in glacial acetic acid, which is used for the titration of basic substances which are themselves dissolved in glacial acetic acid.
4. As indicator electrodes glass and antimony electrodes are commonly used, but it must be noted that in benzene–methanol solutions, a glass–antimony electrode pair may be used in which the glass electrode functions as reference electrode. Glass electrodes should not be maintained in non-aqueous solvents for long periods, as the hydration layer of the glass bulb may be impaired and the electrode will then cease to function satisfactorily.
5. Reference electrodes are usually a calomel or a silver–silver chloride electrode. It is advisable that these be of the double-junction pattern so that potassium chloride solution from the electrode does not contaminate the test solution. Thus, for example, in titrations involving glacial acetic acid as solvent, the outer vessel of the double junction calomel electrode may be filled with glacial acetic acid containing a little lithium perchlorate to improve the conductance.

Table 15.6 indicates common reagents and solvents and the appropriate electrode combination for a variety of acid–base titrations.

Table 15.6 Some common potentiometric non-aqueous titrations

Substances determined: acids, enols, imides, phenols, sulphonamides

Reagent	Solvent	Electrodes*	
		Ref.	Ind.
CH_3OK/benzene–methanol	Benzene–methanol	Gl.	Sb
		Cal.	Sb
	Dimethylformamide	Gl.	Sb
	Ethane-1,2-diamine	Gl.	Sb
	1-Aminobutane	Gl.	Sb
R_4NOH/benzene–methanol	Acetonitrile	Cal.	Gl.
	Pyridine	Cal.	Sb

Substances determined: amines, amine salts, amino acids, salts of acids

Reagent	Solvent	Electrodes*	
		Ref.	Ind.
$HClO_4$/CH_3COOH	Glacial acetic acid	Cal.	Gl.
		Ag, AgCl	Gl.

* Abbreviations; Gl., glass; Cal., calomel.

For References and Bibliography see Sections 16.37 and 16.38.

CHAPTER 16
VOLTAMMETRY

16.1 INTRODUCTION

Voltammetry is concerned with the study of voltage–current–time relationships during electrolysis carried out in a cell. The technique commonly involves studying the influence of changes in applied voltage on the current flowing in the cell, but in some circumstances, the variation of current with time may be investigated. The procedure normally involves the use of a cell with an assembly of three electrodes: (1) a working electrode at which the electrolysis under investigation takes place; (2) a reference electrode which is used to measure the potential of the working electrode; and (3) an auxiliary electrode which, together with the working electrode, carries the electrolysis current. In some circumstances the working electrode may be a dropping mercury electrode (D.M.E.), and the auxiliary electrode is a pool of mercury at the base of the cell: for this special case the technique is referred to as **polarography**.

Techniques which come under the general heading of voltammetry are:

(*a*) polarography (d.c. and a.c.);
(*b*) anodic stripping voltammetry;
(*c*) chronopotentiometry.

Amperometry refers to measurement of current under a constant applied voltage and under these conditions it is the concentration of the analyte which determines the magnitude of the current. Such measurements may be used to follow the change in concentration of a given ion during a titration, and thus to fix the end point: this procedure is referred to as **amperometric titration**.

In a polarographic cell, in view of the relative surface areas of the two electrodes, it follows that at the large auxiliary or counter-electrode the current density will be very small, whilst at the working electrode it may be high. In consequence, the counter-electrode is not readily polarised, and when small currents flow through the cell, the concentration of the ions in the electrode layer (i.e. the layer of solution immediately adjacent to the electrode) remains virtually equal to the concentration in the bulk solution, and the potential of the electrode is maintained at a constant value. By contrast, at the dropping electrode, the electrode layer tends to become depleted of the ions being discharged at the electrode, and if the solution is not stirred, then the diffusion of ions across the resultant concentration gradient becomes an important factor in determining the magnitude of the current flowing.

The total current flowing will in fact be equal to the current carried by the ions undergoing normal electrolytic migration, plus the current due to the diffusion of ions:

$$I = I_d + I_m$$

where I is the total current, I_d the diffusion current, and I_m the migration current. There is, however, a complicating factor in that, in dilute solution, the depletion of the electrode layer leads to an increase in the resistance of the solution, and thus to a change in the Ohm's Law potential drop $(I \times R)$ in the cell; consequently the exact potential operative at the electrode is open to doubt. To overcome this, it is usual to add an excess of an indifferent electrolyte to the system (e.g. $0.1 M$ KCl), and under these conditions the solution is maintained at a low and constant resistance, whilst the migration current of the species under investigation virtually disappears, i.e. $I = I_d$.

The rate of diffusion of the ion to the electrode surface is given by Fick's Second Law as

$$\frac{\partial c}{\partial t} = \frac{D \partial^2 c}{\partial x^2}$$

where D is the diffusion coefficient, c = concentration, t = time, and x = distance from the electrode surface, and the potential of the electrode is controlled by the Nernst equation:

$$E = E^\ominus + \frac{RT}{nF} \ln \frac{a_{Ox}}{a_{Red}}$$

POLAROGRAPHY

16.2 BASIC PRINCIPLES

If a steadily increasing voltage is applied to a cell incorporating a relatively large quiescent mercury anode and a minute mercury cathode (composed of a succession of small mercury drops falling slowly from a fine capillary tube), it is frequently possible to construct a reproducible current voltage curve. The electrolyte is a dilute solution of the material under examination (which must be electro-active) in a suitable medium containing an excess of an indifferent electrolyte (**base or ground solution**, or **supporting electrolyte**) to carry the bulk of the current and raise the conductivity of the solution, thus ensuring that the material to be determined, if charged, does not migrate to the dropping mercury cathode. From an examination of the current–voltage curve, information as to the nature and concentration of the material may be obtained (Heyrovsky).[41] Heyrovsky and Shikata[42] developed an apparatus which increased the applied voltage at a steady rate and simultaneously recorded photographically the current–voltage curve. Since the curves obtained with this instrument are a graphical representation of the polarisation of the dropping electrode, the apparatus was called a **polarograph**, and the records obtained with it, **polarograms**; the photographic recorder is now replaced by a pen recorder, and in some circumstances, by an oscilloscope.

The basic apparatus for polarographic analysis is depicted in Fig. 16.1. The dropping mercury electrode is here shown as the cathode (its most common function); it is sometimes referred to as the working or micro-electrode. The anode is a pool of mercury, and its area is correspondingly large, so that it may be regarded as incapable of becoming polarised, i.e. its potential remains almost constant in a medium containing anions capable of forming insoluble salts with mercury (Cl^-, SO_4^{2-}, etc.); it acts as a convenient non-standardised reference electrode, the exact potential of which will depend upon the nature and the concentration of the supporting electrolyte. The polarisation of the cell is therefore governed by the reactions occurring at the dropping mercury cathode. Inlet and outlet tubes are provided to the cell for expelling dissolved oxygen from the solution by the passage of an inert gas (hydrogen or nitrogen) before, but not during, an actual measurement — otherwise the polarogram of the dissolved oxygen will appear in the current–voltage curve. P is a potentiometer by which any e.m.f. up to 3 volts may be gradually applied to the cell, and R is a pen recorder. It may be mentioned that under these conditions the current–voltage curve is really a current–cathode potential curve, but displaced by a constant voltage corresponding to the potential of the anode. For some purposes it is advisable to employ an external anode of known potential (e.g. a saturated calomel electrode): an internal electrode is more convenient for most analytical work, since absolute values of the cathode potential are not usually required.

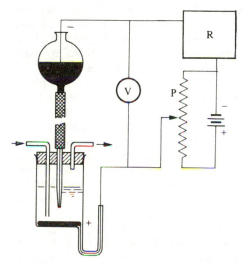

Fig. 16.1

Consider what will occur if an external e.m.f. is applied to the cell shown in Fig. 16.1 charged with, say, a dilute, oxygen-free solution of cadmium chloride. All the positively charged ions present in the solution will be attracted to the negative working electrode by (a) an electrical force, due to the attraction of oppositely charged bodies to each other, and by (b) a diffusive force, arising from the concentration gradient produced at the electrode surface. The total current passing through the cell can be regarded as the sum of these two factors. A typical simple current–voltage curve is shown in Fig. 16.2. The working electrode, being perfectly polarisable, assumes the correspondingly increasing negative potential applied to it; from A to B practically no current will pass through the cell. At B, where the potential of the micro-electrode is equal to the deposition potential of the cadmium ions with respect to a metallic cadmium

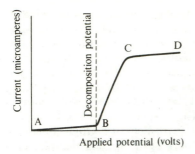

Fig. 16.2

electrode, the current suddenly commences to increase and the working electrode becomes depolarised by the cadmium ions, which are then discharged upon the electrode surface to form metallic cadmium; consequently a rapid increase in the current flowing through the cell will be observed. At the point C the current no longer increases linearly with applied potential but approaches a steady limiting value at the point D: no increase in current is observed at higher cathode potentials unless a second compound able to depolarise the working electrode is present in the solution. At any point on the curve between B and C (usually spoken of as the **polarographic wave**) the number of cadmium ions reaching the micro-electrode surface as a result of migration and diffusion from the main bulk of the solution always exceeds the number of cadmium ions which react at and are deposited upon the electrode. At the point C the rate of supply of the cadmium ions from the main bulk of the solution to the working electrode surface has become equal to the rate of their deposition. Hence, at potentials more negative than point D, the concentration of undischarged cadmium ions at the micro-electrode surface is negligibly small relative to the cadmium ion concentration in the bulk of the solution; no further increase in current passing through the electrolytic cell can be expected, since the limiting current is now fixed by the rate at which cadmium ions can reach the electrode surface.

A number of **polarisable micro-electrodes** (e.g. a rotating platinum wire, *ca* 3 mm long and 0.5 mm diameter, suitably mounted, or stationary noble metal electrodes) have been used in determining current–voltage curves, but the most satisfactory is a slowly growing drop of mercury issuing, under a head of 40–60 cm of mercury, from a resistance-glass capillary (0.05–0.08 mm in diameter and 5–9 cm long) in small, uniform drops. The dropping mercury electrode has the following advantages.

(*a*) Its surface is reproducible, smooth, and continuously renewed; this is conducive to good reproducibility of the current potential curve and eliminates passivity or poisoning effects.

(*b*) Mercury forms amalgams (solid solutions) with many metals.

(*c*) The diffusion current assumes a steady value immediately after each change of applied potential, and is reproducible.

(*d*) The large hydrogen overpotential on mercury renders possible the deposition of substances difficult to reduce, e.g. the alkali metal ions, aluminium ion and manganese(II) ion. (The current–potential curves of these ions are inaccessible with a platinum micro-electrode.)

(*e*) The surface area can be calculated from the weight of the drops.

The dropping mercury electrode may be applied over the range $+0.4$ to about -2.0 volts with reference to the S.C.E. Above $+0.4$ volt mercury dissolves and gives an anodic wave; it begins to oxidise to mercury(I) ion. At potentials more negative than about -1.8 volts *vs* S.C.E., visible hydrogen evolution occurs in acid solutions and the usual supporting electrolytes commence to discharge. The range may be extended to about -2.6 volts *vs* S.C.E. by using supporting electrolytes having higher reduction potentials than the alkali metals; tetra-alkyl ammonium hydroxides or their salts are satisfactory for this purpose.

Reference has already been made to the convenience for routine analytical work of using a mercury pool covering the bottom of the electrolysis cell as the **non-polarisable reference electrode**; the mercury pool is connected to an external circuit via a platinum wire sealed through the wall of the cell. If the solution covering the mercury pool contains chloride ion, the mercury pool acts as a calomel electrode of the particular chloride ion concentration. Whilst convenient for routine polarographic determinations, the mercury pool never possesses a definite, known potential and the potential does not attain a constant value in the absence of chloride ions or other depolarising ions: further the internal reference electrode cannot be used in the presence of powerful oxidising or reducing agents. With the mercury pool anode, the *apparent* half-wave potential of a given substance will be in terms of the total e.m.f. applied to the cell and varies with the potential of the mercury anode. If absolute values are required it is best to use a cell which incorporates a reference electrode (usually a saturated calomel electrode); see Section 16.8. External reference electrodes have potentials which are accurately known and may be used with solutions containing strong oxidising or reducing agents, and the sample solution need not contain a depolarising anion.

Clearly, with the apparatus described above we are dealing with direct current, and the technique as thus carried out is termed conventional d.c. polarography to distinguish it from various modifications (see later) and from the use of alternating current (a.c. polarography).

DIRECT CURRENT POLAROGRAPHY

16.3 THEORETICAL PRINCIPLES

It is necessary to consider the factors which affect the limiting current with a dropping mercury cathode.

Residual (or condenser) current. Mercury is unique in remaining electrically uncharged when it is dropping freely into a solution containing an indifferent electrolyte, such as potassium chloride or potassium nitrate. If a current–voltage curve is determined for a solution containing ions with a strongly negative reduction potential (e.g. potassium ions), a small current will flow before the decomposition of the solution begins. This current increases almost linearly with the applied voltage, and it is observed even when the purest, air-free solutions are used, so that it cannot be due to the reduction of impurities. It must therefore be considered a non-faradaic or condenser current, made appreciable by the continual charging of new mercury drops to the applied potential. It is known that metals, when submerged in an electrolyte, are covered with an electrical double layer of positively and negatively charged ions. The capacity of the double layer and hence the charging current vary, depending upon the potential which is imposed upon the metal.

In practice, one often finds that the indifferent electrolyte contains traces of impurities so that small, almost imperceptible currents are superimposed upon the condenser current. It is customary to include all these in the residual current, and this must be subtracted from the total observed current.

Migration current. Electro-active material reaches the surface of the electrode largely by two processes. One is the migration of charged particles in the electric field caused by the potential difference existing between the electrode surface and the solution; the other is concerned with the diffusion of particles, and will be discussed in a succeeding paragraph. Heyrovsky showed that the migration current can be practically eliminated if an indifferent electrolyte is added to the solution in a concentration so large that its ions carry essentially all the current. (An indifferent electrolyte is one which conducts the current but does not react with the material under investigation, nor at the electrodes within the potential range studied.) In practice, this means that the concentration of the added electrolyte ('supporting electrolyte') must be at least 100-fold that of the electro-active material.

An example will make this conception of supporting electrolyte clear. Imagine that an electrolytic solution is composed of potassium ions $(0.10 M)$ and copper(II) ions $(0.005 M)$. If we assume that the molar conductivities of K^+ and $\frac{1}{2}Cu^{2+}$ are approximately equal, then it follows that *ca* 90 per cent of the current will be transported to the cathode by the potassium ions and only 10 per cent by the copper ions. Both ions will tend to diffuse towards any portion of the solution where a concentration gradient exists, but the rate of diffusion will be slow. If the concentration of the potassium ions be increased until it represents 99 per cent of the total cations present, practically all the current passing through the cell will be transported by the potassium ions. Under such conditions the electro-active material can reach the electrode surface only by diffusion. It must be emphasised that the supporting electrolyte must be composed of ions which are discharged at higher potentials than, and which will not interfere or react chemically with, the ions under investigation.

Diffusion current. When an excess of supporting electrolyte is present in the solution the electrical force on the reducible ions is nullified; this is because the ions of the added salt carry practically all the current and the potential gradient is compressed or shortened to a region so very close to the electrode surface that it is no longer operative to attract electro-reducible ions. Under these conditions the limiting current is almost solely a diffusion current. Ilkovič[43] examined the various factors which govern the diffusion current and deduced the following equation:

$$I_d = 607\, n\, D^{1/2}\, C\, m^{2/3}\, t^{1/6}$$

where

I_d = the average diffusion current in microamperes during the life of the drop;
n = the number of faradays of electricity required per mole of the electrode reaction (or the number of electrons consumed in the reduction of one molecule of the electro-active species);

D = the diffusion coefficient of the reducible or oxidisible substance expressed as $cm^2 s^{-1}$;

C = its concentration in $mmol L^{-1}$;

m = the rate of flow of mercury from the dropping electrode expressed in $mg s^{-1}$; and

t = drop time in s.

The constant 607 is a combination of natural constants, including the Faraday constant; it is slightly temperature-dependent and the value 607 is for 25 °C. The Ilkovič equation is important because it accounts quantitatively for the many factors which influence the diffusion current: in particular, the linear dependence of the diffusion current upon n and C. Thus, with all the other factors remaining constant, the diffusion current is directly proportional to the concentration of the electro-active material — this is of great importance in quantitative polarographic analysis.

The original Ilkovič equation neglects the effect on the diffusion current of the curvature of the mercury surface. This may be allowed for by multiplying the right-hand side of the equation by $(1 + AD^{1/2} t^{1/6} m^{-1/3})$, where A is a constant and has a value of 39. The correction is not large (the expression in parentheses usually has a value between 1.05 and 1.15) and account need only be taken of it in very accurate work.

The diffusion current I_d depends upon several factors, such as temperature, the viscosity of the medium, the composition of the base electrolyte, the molecular or ionic state of the electro-active species, the dimensions of the capillary, and the pressure on the dropping mercury. The temperature coefficient is about 1.5–2 per cent °C^{-1}; precise measurements of the diffusion current require temperature control to about 0.2 °C, which is generally achieved by immersing the cell in a water thermostat (preferably at 25 °C). A metal ion complex usually yields a different diffusion current from the simple (hydrated) metal ion. The drop time t depends largely upon the pressure on the dropping mercury and to a smaller extent upon the interfacial tension at the mercury–solution interface; the latter is dependent upon the potential of the electrode. Fortunately t appears only as the sixth root in the Ilkovič equation, so that variation in this quantity will have a relatively small effect upon the diffusion current. The product $m^{2/3} t^{1/6}$ is important because it permits results with different capillaries under otherwise identical conditions to be compared: the ratio of the diffusion currents is simply the ratio of the $m^{2/3} t^{1/6}$ values.

Unless the individual drops fall under their own weight when they are completely formed, the diffusion currents are not reproducible: stirring of the solution under investigation is therefore not permissible.

Polarographic maxima. Current–voltage curves obtained with the dropping mercury cathode frequently exhibit pronounced maxima, which are reproducible and which can be usually eliminated by the addition of certain appropriate 'maximum suppressors'. These maxima vary in shape from sharp peaks to rounded humps, which gradually decrease to the normal diffusion-current curve as the applied voltage is increased. A typical example is shown in Fig. 16.3. Curve A is that for copper ions in 0.1M potassium hydrogencitrate solution, and curve B is the same polarogram in the presence of 0.005 per cent acid fuchsine solution.

To measure the true diffusion current, the maxima must be eliminated or

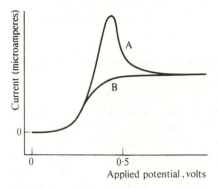

Fig. 16.3

suppressed. Fortunately this can be done easily by the addition of a very small quantity of a surface-active substance, such as a dyestuff (as in the example above), gelatin, or other colloids. The function of any such maximum suppressor is probably to form an adsorbed layer on the aqueous side of the mercury–solution interface which resists compression; this prevents the streaming movement of the diffusion layer (which is believed to be responsible for the current maximum) at the interface. Gelatin is widely used: the amount present in the solution should lie between 0.002 and 0.01 per cent; higher concentrations will usually suppress the diffusion current. Other maximum suppressors which are very effective are Triton X-100 (a non-ionic detergent), which is used at a concentration of 0.002–0.004 per cent, added in the form of a 0.1 per cent stock solution in water, and methyl cellulose (0.005 %).

Half-wave potentials. The salient features of a typical current–applied voltage curve (polarogram) are shown in Fig. 16.4.

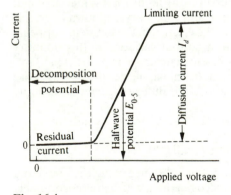

Fig. 16.4

The conventional method of drawing the current–voltage curves is to plot the applied e.m.f. as abscissae reading in increasing negative values on the right: current is plotted as ordinates, cathodic currents (resulting from reduction) being regarded as positive and anodic currents negative. The height of the curve (**wave height**) is the diffusion current, and is a function of the concentration of the reacting material; the potential corresponding to the point of inflexion of the curve (**half-wave potential**) is characteristic of the nature of the reacting material. This is the essential basis of quantitative and qualitative polarographic analysis. It should be noted that since the current flowing increases as the mercury drop becomes larger and then abruptly decreases when the drop falls, the recorder trace in fact has a saw-tooth appearance, especially at the upper end. The curve shown in Fig. 16.4 represents the mean value of the oscillations in current.

The underlying theory may be simplified as follows. Polarography is concerned with electrode reactions at the indicator or micro-electrode, i.e. with reactions involving a transfer of electrons between the electrode and the components of the solution. These components are called oxidants when they can accept electrons, and reductants when they can lose electrons. The electrode is a cathode when a reduction can take place at its surface, and an anode when oxidation occurs at its surface. During the reduction of an oxidant at the cathode, electrons leave the electrode with the formation of an equivalent amount of the reductant in solution:

Oxidant $+ n$ Electrons $\rightleftharpoons$ Reductant

or

$$Ox + ne \rightleftharpoons Red \tag{1}$$

The reductant differs from the oxidant merely by n electrons, and together they form an oxidation–reduction system. Consider the reversible reduction of an oxidant to a reductant at a dropping mercury cathode. The electrode potential is given by:

$$E = E^\ominus + \frac{RT}{nF} \ln \frac{a_{ox.}}{a_{red.}} \tag{2}$$

where $a_{ox.}$ and $a_{red.}$ are the activities of the oxidant and reductant respectively *as they exist at the electrode surface* (henceforth called the electrode–solution interface and denoted by the subscript 's'). Substitution of the concentrations for activities will not introduce any appreciable error and equation (2) may therefore be written as:

$$E = E^\ominus + \frac{RT}{nF} \ln \frac{[Ox]_s}{[Red]_s} \tag{3}$$

Here $E^\ominus$ is the standard potential of the reaction against the reference electrode used to measure the potential of the dropping electrode, and the potential E refers to the average value during the life of a mercury drop. Before the commencement of the polarographic wave only a small residual current flows, and the concentration of any electro-active substance must be the same at the electrode interface as in the bulk of the solution. As soon as the decomposition potential is exceeded, some of the reducible substance (oxidant) at the interface is reduced, and must be replenished from the body of the solution by means of diffusion. The reduction product (reductant) does not accumulate at the interface, but diffuses away from it into the solution or into the electrode material. If the applied potential is increased to a value at which all the oxidant reaching the interface is reduced, only the newly formed reductant will be present; the current then flowing will be the diffusion current. The current I at any point on the wave is determined by the rate of diffusion of the oxidant from the bulk of the solution to the electrode surface under a concentration gradient $[Ox]$ to $[Ox]_s$.

$$I = K([Ox] - [Ox]_s)^* \qquad (4)$$

When $[Ox]_s$ is reduced to almost zero, equation (4) may be written:

$$I = K[Ox] = I_d \qquad (5)$$

where I_d is the diffusion current. From equations (4) and (5), it follows that:

$$[Ox]_s = (I_d - I)/K \qquad (6)$$

If the reductant (Red) is soluble in water and none was originally present with the oxidant, it will diffuse from the surface of the electrode to the bulk of the solution. The concentration of $[Red]_s$ at the surface at any value of I will be proportional to the rate of diffusion of the reductant from the surface of the electrode to the solution under a concentration gradient $[Red]_s$ and hence also the current:

$$I = k[Red]_s^* \qquad (7)$$

If the reductant is insoluble in water but soluble in the mercury phase (amalgam formation), equation (7) still holds. Substituting in equation (3), we have:

$$E = E^\ominus - \frac{RT}{nF} \ln \frac{K}{k} + \frac{RT}{nF} \ln \frac{I_d - I}{I} = E^{\ominus\prime} + \frac{RT}{nF} \ln \frac{I_d - I}{I} \qquad (8)$$

where $E^{\ominus\prime} = (E^\ominus - K')$ and $K' = \frac{RT}{nF} \ln \frac{K}{k}$.

When I is equal to $I_d/2$, equation (8) reduces to:

$$E = E_{1/2} = E^{\ominus\prime} + \frac{RT}{nF} \ln \frac{I_d/2}{I_d/2} = E'_0 \qquad (9)$$

The potential at the point on the polarographic wave where the current is equal to one-half the diffusion current is termed the **half-wave potential** and is designated by $E_{1/2}$. It is quite clear from equation (9) that $E_{1/2}$ is a characteristic constant for a reversible oxidation–reduction system and that its value is independent of the concentration of the oxidant $[Ox]$ in the bulk of the solution. It follows from equations (8) and (9) that at $25\,^\circ$C:

$$E = E_{1/2} + \frac{0.0591}{n} \log \frac{I_d - I}{I} \qquad (10)$$

This equation represents the potential as a function of the current at any point on the polarographic wave, it is sometimes termed the equation of the polarographic wave.

The half-wave potential is also independent of the electrode characteristics, and can, therefore, serve for the qualitative identification of an unknown substance. Owing to the proximity of many different half-wave potentials, its use for qualitative analysis is of limited application unless the number of

---

*The constants K and k may be evaluated from the Ilkovič equation.

possibilities is strictly limited by the nature of the unknown: polarographic detectors find application in liquid chromatography. The theoretical treatment for anodic waves is similar to the above.

It follows from equation (10) that when $\log(I_d - I)/I$ (where I is the current at any point on the polarographic wave minus the residual current) is plotted against the corresponding potential of the micro-electrode (ordinates), a straight line should be obtained with a slope of $0.0591/n$ for a reversible reaction; the intercept of the graph upon the vertical axis gives the half-wave potential of the system. Hence, n, the number of electrons taking part in the reversible reaction, may be determined. In applying equation (10), it is necessary to correct both I and I_d for the residual currents at the corresponding values of the applied potential and to correct the applied potential itself for any IR drop in the cell circuit. After these corrections have been made, both $E_{1/2}$ and the slope of the log plot are found to be independent of the concentration of the electro-active ion. Because it is concentration-independent, the half-wave potential is generally preferred to the somewhat vague 'decomposition potential'. It also follows from equation (10) that the range of potentials over which the polarographic wave extends decreases with increasing values of n; thus, the wave is steeper in the reduction of aluminium or lanthanum ions than that for the lead or cadmium ions, which in turn is steeper than that of an alkali metal ion or thallium(I) ion.

If the reaction at the indicator electrode involves complex ions, satisfactory polarograms can be obtained only if the dissociation of the complex ion is very rapid as compared with the diffusion rate, so that the concentration of the simple ion is maintained constant at the electrode interface. Consider the general case of the dissociation of a complex ion:

$$MX_p^{(n-pb)+} \rightleftharpoons M^{n+} + pX^{b-} \tag{11}$$

The instability constant may be written:

$$K_{\text{instab.}} = \frac{[M^{n+}][X^{b-}]^p}{[MX_p^{(n-pb)+}]} \tag{12}$$

(strictly, activities should replace concentrations).

The electrode reaction, assuming amalgam formation, can be represented:

$$M^{n+} + ne + \text{Hg} \rightleftharpoons M(\text{Hg}) \tag{13}$$

Combining equations (11) and (13), we have:

$$MX_p^{(n-pb)+} + ne + \text{Hg} \rightleftharpoons M(\text{Hg}) + pX^{b-} \tag{14}$$

It can be shown[44] that the expression for the electrode potential can be written:

$$E_{1/2} = E^{\ominus} + \frac{0.0591}{n} \log K_{\text{instab.}} - \frac{0.0591}{n} \log[X^{b-}]^p \tag{15}$$

Here p is the coordination number of the complex ion formed, X^{b-} is the ligand and n is the number of electrons involved in the electrode reaction. The concentration of the complex ion does not enter into equation (15), so that the observed half-wave potential will be constant and independent of the concentration of the complex metal ion. Furthermore, the half-wave potential is more negative the smaller value of $K_{\text{instab.}}$, i.e. the more stable the complex ion. The half-wave potential will also shift with a change in the concentration of the ligand, and if the former is determined at two different concentrations of the complex-forming agent, we have:

$$\Delta E_{1/2} = -p \cdot \frac{0.0591}{n} \times \Delta \log[X^{b-}] \tag{16}$$

This relationship enables one to determine the coordination number p of the complex ion and thus its formula.

It can also be shown that:

$$(E_{1/2})_c - (E_{1/2})_s = \frac{0.0591}{n} \log K_{i_n \text{stab.}} - p \cdot \frac{0.0591}{n} \log[X^{b-}] \tag{17}$$

where $(E_{1/2})_c$ and $(E_{1/2})_b$ are the half-wave potentials of the complex and simple ions respectively at 25 °C, $K_{\text{instab.}}$ is the instability (or disssociation) constant and $[X^{b-}]$ is the concentration of the complexing agent X^{b-} in the body of the solution. It is assumed that the ligand is present in sufficiently large amount so that its concentration is practically the same at the surface of the dropping electrode as in the bulk of the solution. This equation may be employed to evaluate the instability constant of the complex ion: it involves merely the comparison of the half-wave potential at a given concentration of the complexing agent with that of the simple metal ion.

The shift of the half-wave potentials of metal ions by complexation is of value in polarographic analysis to eliminate the interfering effect of one metal upon another, and to promote sufficient separation of the waves of metals in mixtures to make possible their simultaneous determination. Thus, in the analysis of copper-base alloys for nickel, lead, etc., the reduction wave of copper(II) ions in most supporting electrolytes precedes that of the other metals and swamps those of the other metals present; by using a cyanide supporting electrolyte, the copper is converted into the difficultly reducible cyanocuprate(I) ion and, in such a medium, nickel, lead, etc., can be determined.

16.4 QUANTITATIVE TECHNIQUE

General considerations. Conventional d.c. polarographic analysis is most conveniently carried out if the concentration of the electro-active substance is $10^{-4} - 10^{-3}$ molar and the volume of the solution is between 2 and 25 mL. It is, however, possible to deal with concentrations as high as 10^{-2} molar or as low as 10^{-5} molar and to employ volumes appreciably less than 1 mL.

Oxygen dissolved in electrolytic solutions is easily reduced at the dropping mercury electrode, and produces a polarogram consisting of two waves of approximately equal height and extending over a considerable voltage range; their position depends upon the pH of the solution, being displaced to higher voltages by alkali. The concentration of oxygen in aqueous solutions that are saturated with air at room temperature is about $2.5 \times 10^{-4} M$, consequently its polarographic behaviour is of considerable practical importance. A typical polarogram for air-saturated $1 M$ potassium chloride solution (in the presence of 0.01 per cent methyl red) is given in Fig. 16.5 (curve A). It has been stated that the first wave (starting at about -0.1 volt relative to S.C.E.) is due to the reduction of oxygen to hydrogen peroxide:

$$O_2 + 2H_2O + 2e = H_2O_2 + 2OH^- \text{ (neutral or alkaline solution)}$$

$O_2 + 2H^+ + 2e = H_2O_2$ (acid solution)

The second wave is ascribed to the reduction of the hydrogen peroxide either to hydroxyl ions or to water:

$H_2O_2 + 2e = 2OH^-$ (alkaline solution)

$H_2O_2 + 2H^+ + 2e = 2H_2O$ (acid solution)

It is, therefore, necessary to remove any dissolved oxygen from the electrolytic solution whenever cathodic regions are being investigated in which oxygen interferes. This is easily accomplished by bubbling an inert gas (nitrogen or hydrogen) through the solution for about 10–15 minutes before determining the current–voltage curve. Curve B in Fig. 16.5 was obtained after the removal of the oxygen by oxygen-free nitrogen from a cylinder of the compressed gas. The gas stream must be discontinued during the actual measurements to prevent its stirring effect interfering with the normal formation of drops of mercury or with the diffusion process near the micro-electrode.

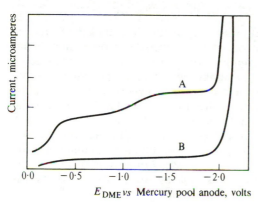

E_{DME} *vs* Mercury pool anode, volts

Fig. 16.5

Measurement of the oxygen polarographic wave has been used as a method for determining dissolved oxygen: the given sample is first examined after adding some potassium chloride solution as supporting electrolyte, and the procedure is then repeated after subjecting the sample to de-oxygenation by passing oxygen-free nitrogen through it for five minutes.

The effect of temperature has already been referred to; the electrolysis cell should be immersed in a thermostat bath maintained within $\pm 0.2\,°C$; for many purposes a temperature variation of $\pm 0.5\,°C$ is permissible. A temperature of $25\,°C$ is usually employed.

As a precautionary measure to prevent the appearance of maxima, sufficient gelatin to give a final concentration of 0.005 per cent should be added. The gelatin should preferably be prepared fresh each day; bacterial action usually appears after a few days. Other maximum suppressors (e.g. Triton X-100 and methyl cellulose) are sometimes used.

Two or more electro-active ions may be determined successively if their half-wave potentials differ by at least 0.4 volt for singly charged ions and 0.2 volt for doubly charged ions provided that the ions are present in approximately equal concentrations. If the concentrations differ considerably, the difference between the half-wave potentials must be correspondingly larger. If the waves of two ions overlap or interfere, various experimental devices may be employed. The half-wave potential of one of the ions may be displaced to more negative potentials by the use of suitable complexing agents which are incorporated in the supporting electrolyte; for example, Cu^{2+} ions may be complexed by the addition of potassium cyanide. Sometimes one ion may be removed by precipitation (e.g. with a mixture of lead and zinc, the lead will not interfere if it is precipitated as sulphate; the lead sulphate formed need not be removed by filtration); the possibilities of adsorption or co-precipitation of part of the other ions must, however, be borne in mind; electrolytic separations are also very useful. The need for using complexation or separation procedures can often be avoided if one of the alternative techniques (Section 16.9) is used instead of conventional d.c. polarography.

16.5 EVALUATION OF QUANTITATIVE RESULTS

Three methods which have been widely used in practice are described.

Wave height–concentration plots. Solutions of several different concentrations of the ion under investigation are prepared, the composition of the supporting electrolyte and the amount of maximum suppressor added being the same for the comparison standards and for the unknown. The heights of the waves obtained are measured in any convenient manner and plotted as a function of the concentration. The polarogram of the unknown is produced exactly as the standards, and the concentration is read from the graph. The method is strictly empirical, and no assumptions, except correspondence with the conditions of the calibration, are made. The wave height need not be a linear function of the concentration, although this is frequently the case. For results of the highest precision the unknown should be bracketed by standard solutions run consecutively.

Internal standard (pilot ion) method. The relative diffusion currents of ions in the same supporting electrolyte are independent of the characteristics of the capillary electrode and, to a close approximation, of the temperature. Hence upon determining the relative wave heights with the unknown ion and with some standard or 'pilot' ion added to the solution in known amount, and comparing these with the ratio for known amounts of the same two ions, previously determined, the concentration of the unknown ion may be established. This procedure has limited application, primarily because only a small number of ions are available to act as pilot or reference ions. The main requirement for such an ion (if singly charged) is that its half-wave potential should differ by at least 0.4 volt from the unknown or any other ion in the solution with which it might interfere. When a single unknown is present, this condition can usually be satisfied, but in complex mixtures there is seldom sufficient difference between the half-wave potentials to introduce additional waves.

Method of standard addition. The polarogram of the unknown solution is first recorded, after which a known volume of a standard solution of the same ion is added to the cell and a second polarogram is taken. From the magnitude of the heights of the two waves, the known concentration of ion added, and the volume of the solution after the addition, the concentration of the unknown may be readily calculated as follows. If I_1 is the observed diffusion current ($\equiv$ wave height) of the unknown solution of volume V mL and of concentration C_u, and I_2 is the observed diffusion current after v mL of a standard solution of concentration C_s have been added, then according to the Ilkovič equation we have:

$$I_1 = kC_u$$

and

$$I_2 = k(VC_u + vC_s)(V + v)$$

Thus

$$k = I_2(V + v)/(VC_u + vC_s)$$

whence

$$C_u = \frac{I_1 vC_s}{(I_2 - I_1)(V + v) + I_1 v}$$

The accuracy of the method depends upon the precision with which the two volumes of solution and the corresponding diffusion currents are measured. The material added should be contained in a medium of the same composition as the supporting electrolyte, so that the latter is not altered by the addition. The assumption is made that the wave height is a linear function of the concentration in the range of concentration employed. The best results would appear to be obtained when the wave height is about doubled by the addition of the known amount of standard solution. This procedure is sometimes referred to as **spiking**.

16.6 MEASUREMENT OF WAVE HEIGHTS

With a well-defined polarographic wave where the limiting current plateau is parallel to the residual current curve, the measurement of the diffusion current is relatively simple. In the exact procedure, illustrated in Fig. 16.6(a), the actual

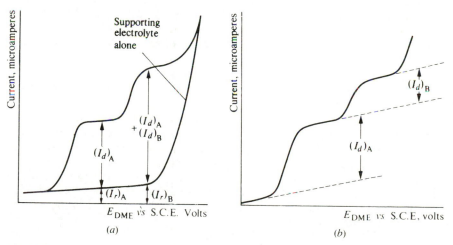

Fig. 16.6

residual current curve is determined separately with the supporting electrolyte alone; by subtracting the residual current from the value of the current at the diffusion-current plateau (both measured at the same applied voltage), the diffusion current is obtained. It may be noted that when employing polarograms produced with a pen recorder, a line is drawn through the midpoints of the recorder oscillations. For subsequent electro-active substances, the diffusion current would be evaluated by subtracting both the residual current and all preceding diffusion currents.

It is simpler, though less exact, to apply the extrapolation method. The part of the residual current curve preceding the initial rise of the wave is extrapolated: a line parallel to it is drawn through the diffusion current plateau as shown in Fig. 16.6(b). For succeeding waves, the diffusion current plateau of the preceding wave is used as a pseudo-residual current curve.

16.7   POLAROGRAPHS FOR CLASSICAL d.c. POLAROGRAPHY

The essential requirements for producing polarographic current–voltage curves are:

1. a means of applying a variable known d.c. voltage ranging from 0 to 2–3 volts to the electrolysis cell, and
2. a method for recording the resultant current.

The apparatus depicted in Fig. 16.1 represents a manual polarograph which can be usefully employed to study the basic techniques of polarography.

Commercial polarographs are also available in which the voltage scan is carried out automatically while a chart recorder plots the current–voltage curve. A counter-current control is incorporated which applies a small opposing current to the cell which can be adjusted to compensate for the residual current: this leads to polarograms which are better defined. Most of these instruments also incorporate circuits which permit the performance of alternative, more sensitive types of polarography as discussed in Section 16.9

A useful feature of many polarographs is the facility of plotting **derivative polarograms**, i.e. curves obtained by plotting dI/dE against E. Such curves show a peak at the half-wave potential, and by measuring the height of the peak it is possible to obtain quantitative data on the reducible substance; the height of the peak is proportional to the concentration of the ion being discharged.

A typical conventional polarogram for $0.003 M$-cadmium sulphate in $1 M$ potassium chloride in the presence of 0.001 per cent gelatin, and the corresponding derivative curve, are shown in Fig. 16.7 (I_{max} is the maximum current recorded in the derivative mode).

A derivative plot can be used to measure half-wave potentials which are closer than 0.15 V; this is not possible with a normal polarogram owing to the disturbing effect of the diffusion current of the first ion to be discharged on the second step in the polarogram. Also in cases where the element with the lower half-wave potential is present in much higher concentration, e.g. in the analysis of copper for cadmium content, the analysis is almost impossible without previous chemical separation, but if a derivative polarogram is prepared, the series of peaks obtained in positions approximating to the half-wave potentials enable one to identify and quantify the individual elements. Figure 16.8 illustrates the polarogram (a) obtained with copper and cadmium ions in the ratio of 40

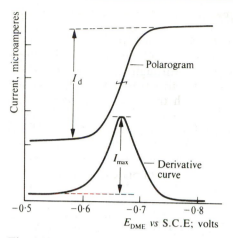

Fig. 16.7

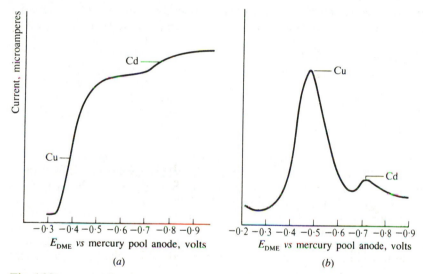

Fig. 16.8

to 1, and the corresponding derivative polarogram (*b*): the two peaks are clearly visible.

Many modern polarographs provide for **potentiostatic control** of the dropping electrode potential: this is particularly valuable when solutions of high resistance are involved, as for example when non-aqueous solvents (or mixtures of water and organic solvents) are used. A high resistance in the polarographic cell leads to a large Ohm's Law voltage drop ($I \times R$) across the cell, which not only influences the measured electrode potential but may also distort the polarogram: in extreme cases with some non-aqueous solvents, a straightforward *I vs E* plot may appear to be virtually a straight line, and only after correction of the ohmic voltage drop is a normal polarogram obtained. Potentiostatic control requires the introduction of a third electrode (a counter-electrode) into the polarographic cell.

607

16.8 ANCILLARY EQUIPMENT FOR POLAROGRAPHY

Mercury. Double distilled mercury is usually recommended for polarographic work. The re-distilled mercury of commerce is generally satisfactory for most determinations; it should be filtered through a filter-paper cone with a small pin-hole in the tip (or through a sintered-glass funnel) before use in order to remove any surface oxides or dust.

Used mercury should be washed with water, thoroughly agitated for about 12 hours in contact with 10 per cent nitric acid (a filter flask, arranged to admit air through the bottom of the mercury and connected to a water pump, is satisfactory), then thoroughly washed with distilled water, dried with filter paper, and re-distilled under reduced pressure.

CAUTION. **Mercury vapour is a cumulative poison.** All vessels containing mercury should be stoppered. Any spilled mercury should be immediately collected and placed in a flask containing water, and the bench (floor) dusted with powdered sulphur. Employ a tray under all vessels containing mercury and for all operations involving the transfer of mercury.

Dropping mercury electrode assembly. The assembly consists of a mercury reservoir (e.g. a 100 mL levelling bulb), a connecting tube between the reservoir and the capillary tube, and a small glass electrolysis cell in which the unknown solution is placed (Fig. 16.1). The heavy-walled rubber tubing, 80–100 cm long, should be sulphur-free; Neoprene tubing is frequently employed. The inside surface should be steamed out for 30 minutes before use, followed by drying with air filtered through a cotton filter-plug. Electrical connection to the mercury in the reservoir is effected by a platinum wire sealed into the end of a soft-glass tube, which is partly filled with mercury and held in place by the stopper of the reservoir.

The capillary tube has a length of 5–10 cm and a bore diameter of about 0.05 mm (range 0.04 – 0.07 mm); the outside diameter is usually about 6–7 mm: the delivery tip is cut accurately horizontal. At a given pressure the drop time (which is the time that elapses between the fall of two successive drops) is directly proportional to the length of the capillary, but inversely to the third power of its internal radius; it is also inversely proportional to the pressure on the drop. A capillary suitable for use in polarography should have a length and bore such that the application of a pressure of about 50 cm of mercury will cause a drop weighing 6–10 mg to fall every 3–6 seconds when the tip of the capillary is immersed in distilled water. The dropping mercury electrode must be mounted so that it is within $\pm 5°$ of the vertical; deviation from this angle produces erratic dropping.

With careful treatment, a capillary should remain serviceable for many months. It is absolutely essential that no solid matter of any kind should be allowed to reach the inside of the capillary. The electrode must never be allowed to stand in a solution when the mercury is not flowing.

The following procedure is recommended. The sample solution is de-aerated, then, with the tip of the capillary in the air, the mercury pressure is raised at least 10 cm above the previously found equilibrium height, the capillary is inserted into the cell, and the mercury level is finally adjusted to the desired value. After completion of the measurements the capillary is withdrawn from the cell and washed thoroughly with a stream of water from a wash bottle while the mercury is still issuing from the tip and is being collected in a microbeaker.

The mercury reservoir is then lowered until the mercury flow *just ceases* (not further) and the electrode is allowed to stand in the air. It is good technique, at the beginning of each period of use, to immerse the capillary for *ca* 1 min. in 1:1-nitric acid while mercury is flowing, then wash it well with distilled water: a further precaution is to allow the mercury drops to form in distilled water for about 15 minutes.

If the capillary becomes partly or completely blocked it is sometimes possible to clear it by carefully drawing strong nitric acid through it until the foreign matter has been completely dissolved, followed by distilled water to remove all traces of acid: the capillary is finally dried by drawing a stream of warm air (filtered through a cotton-wool plug) through it.

For reproducible results with the same dropping mercury electrode, it is important that the height of the mercury in the reservoir above the capillary tip should be constant, i.e. the same pressure on the dropping mercury tip be maintained; the small quantity of mercury passing through the capillary does not appreciably affect the volume in the reservoir. The stand supporting the electrode should permit the rapid immersion and removal of the capillary from the solution in the polarographic cell, particularly when the latter is in position in a thermostat. In many cases the dropping mercury electrode can, with advantage, be replaced by a Hanging (or Static) Mercury Drop Electrode (H.M.D.E. or S.M.D.E.) in which drops of mercury are produced at the tip of a rather wider capillary than that normally used in the D.M.E., by operation of a solenoid-controlled plunger. The solenoid is adjusted so that development of the mercury drop ceases before the point at which the drop would fall under its own weight. In this way an electrode surface of fixed area is established, so that once the surface has been charged by the condenser current, it is only the diffusion current which continues to flow, and this can be measured with greater accuracy than is possible with the normal dropping electrode. At pre-determined intervals the drop is dislodged from the capillary by operation of a 'drop knocker', and a new mercury drop is then dispensed so that a further reading can be made.

Polarographic cells. Numerous types of polarographic cells have been described and various forms are available commercially; the choice may well be dictated by the electrode stand in use.

The **H**-type cell devised by Lingane and Laitinen and shown in Fig. 16.9 will be found satisfactory for many purposes; a particular feature is the built-in reference electrode. Usually a saturated calomel electrode is employed, but if the presence of chloride ion is harmful a mercury(I) sulphate electrode (Hg/Hg_2SO_4 in potassium sulphate solution: potential *ca* $+0.40$ volts *vs* S.C.E.) may be used. It is usually designed to contain $10-50$ mL of the sample solution in the left-hand compartment, but it can be constructed to accommodate a smaller volume down to $1-2$ mL. To avoid polarisation of the reference electrode the latter should be made of tubing at least 20 mm in diameter, but the dimensions of the solution compartment can be varied over wide limits. The compartments are separated by a cross-member filled with a 4 per cent agar-saturated potassium chloride gel, which is held in position by a medium-porosity sintered Pyrex glass disc (diameter at least 10 mm) placed as near the solution compartment as possible in order to facilitate de-aeration of the test solution. By clamping the cell so that the cross-member is vertical, the molten

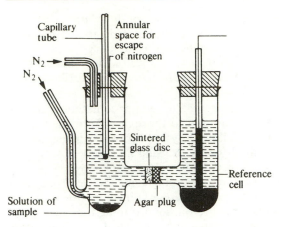

Capillary tube

Annular space for escape of nitrogen

N_2 →

N_2

Sintered glass disc

Reference cell

Solution of sample

Agar plug

Fig. 16.9

agar gel is pipetted into the cross-member and the cell is allowed to stand undisturbed until the gel has solidified.

In use, the solution compartment (either dried by aspiration of air through it or rinsed with several portions of the test solution) is charged with at least enough test solution to cover the entire sintered-glass disc. Dissolved air is removed by bubbling pure nitrogen through the solution via the side arm: by means of a two-way tap in the gas stream, the gas is then diverted over the surface of the solution. Measurements should not be attempted while gas is bubbling through the solution, for the stirring causes high and erratic currents. Finally, the dropping electrode is inserted through another hole in the stopper (which should be large enough for ease of insertion and removal of the capillary) and the measurements are made. When the **H**-cell is not in use the left-hand compartment should be kept filled either with water or with saturated potassium chloride solution (or other electrolyte appropriate to the reference electrode being used) to prevent the agar plug from drying out.

When potentiometric control is exercised, then, as already explained, an additional (counter-) electrode must be used, and under these conditions a suitably sized four-necked flask makes a convenient electrolysis vessel; if the flask is pear-shaped the mercury is readily collected at the bottom of the flask. One neck of the flask is fitted with a nitrogen delivery tube provided with a two-way tap so that the gas can be bubbled through the solution, or alternatively allowed to flow over the surface of the liquid [see Fig. 16.15(a)]. The dropping electrode is inserted through a loosely bored stopper (so that nitrogen can escape) into a second neck, the side arm of a calomel electrode (or other suitable reference electrode) is introduced through the stopper of a third neck, whilst a tube carrying a platinum plate is inserted through the stopper of the fourth neck to act as auxiliary electrode.

The reference electrode must be sited as close as possible to the D.M.E. so that the resistance of the solution between the two electrodes is reduced to a minimum, and the potentiostat then maintains the e.m.f. of the D.M.E.–reference electrode combination at the correct value. This arrangement has the further advantage that no sensible current is passed through the reference electrode

610

and hence there is no possibility of polarisation of the latter with consequent variation in potential.

Maximum suppressors. Gelatin is widely used as a maximum suppressor in spite of the fact that its aqueous solution deteriorates fairly rapidly, and must therefore be prepared afresh every few days as needed. Usually a 0.2 per cent stock solution is prepared as follows. Allow 0.2 g of pure powdered gelatin (the grade sold for bacteriological work is very satisfactory) to stand in 100 mL of boiled-out distilled water for about 30 minutes with occasional swirling: warm the flask containing the mixture to about 70 °C on a water bath for about 15 minutes or until all the solid has dissolved. The solution must not be boiled or heated with a free flame. Stopper the flask firmly. This solution does not usually keep for more than about 48 hours. Its stability may be increased to a few days by adding a few drops of sulphur-free toluene or a small crystal of thymol, but the addition is rarely worth while and is not recommended.

A gelatin concentration of 0.005 per cent, which corresponds to 0.25 mL of the stock 0.2 per cent solution in each 10 mL of the solution being analysed, is usually sufficient to eliminate maxima. Higher concentrations (certainly not above 0.01 per cent) should not be used, since these will distort the wave form and decrease the diffusion current markedly.

Triton X-100, like gelatin, suppresses both positive and negative maxima, but, unlike gelatin, its aqueous solution is stable. A stock 0.2 per cent solution is prepared by shaking 0.20 g of Triton X-100 thoroughly with 100 mL of water. About 0.1 mL of this solution should be added to each 10 mL of the sample solution to give a Triton X-100 concentration of 0.002 per cent.

16.9 MODIFIED VOLTAMMETRIC PROCEDURES

As already indicated, quantitative conventional d.c. polarography is limited at best to solutions with electrolytes at concentrations greater than $10^{-5}M$, and two different ions can only be investigated when their half-wave potentials differ by at least 0.2 V. These limitations are largely due to the condenser current associated with the charging of each mercury drop as it forms, and various procedures have been devised to overcome this problem. These include:

(a) pulse polarography (Section 16.10);
(b) rapid scan polarography (Section 16.11);
(c) sinusoidal a.c. polarography (Section 16.12);
(d) stripping analysis (Section 16.20).

These modified procedures involve the use of specially constructed polarographs.

16.10 PULSE POLAROGRAPHY

Barker and Jenkins[45] attempted to solve the problem by application of the polarising current in a series of pulses; one pulse of approximately 0.05 second duration being applied during the growth of a mercury drop, and at a fixed point near the end of the life of the drop. Two different procedures may, however, be employed: (a) pulses of increasing amplitude may be superimposed upon a constant d.c. potential, or (b) pulses of constant amplitude may be applied to a steadily increasing d.c. potential.

In method (a) the onset of the pulse is marked by a sudden rise in the total current passing: this is largely due to the condenser (charging) current which, however, soon decays to zero. The faradaic current also decays, but only to the level of the diffusion current, and if the current measurement is made in the last stages of the pulse (in the final 17 ms of its duration), it gives the faradaic current alone. The resulting polarogram is similar to a conventional d.c. polarogram except that the characteristic saw-tooth pattern of the latter is replaced by a stepped curve.

In method (b) the applied potential varies with time as shown in Fig. 16.10. The sloping line which is periodically interrupted by the pulses of pulse height B (which may be from 5 to 100 millivolts in magnitude) represents the normal steadily increasing d.c. voltage. The time interval A between the termination of two succeeding mercury drops represents the drop time. The current is measured *twice* during the lifetime of each mercury drop: once just before the application of the pulse (points corresponding to C′ in Fig. 16.10) as well as at the usual point C near the end of the pulse. The current at C′ is that which would be observed in normal d.c. polarography; its value is stored in the polarograph. The onset of the pulse is then marked by a sudden rise in current, which as in method (a) soon settles down as the condenser current decays, and near the end of the pulse (point C), the current is again read. This value is then compared with that stored in the instrument (the value for point C′), and the difference between them is amplified and recorded. Clearly, if the measurements of current are made on the residual current curve, or on a plateau of the d.c. polarogram, the difference in current between points C and C′ will be small, but if the measurements are made on a polarographic wave, an appreciable difference in current will be recorded, and it will of course reach a maximum value when the applied d.c. potential is equal to the half-wave potential. The difference in current plotted against the applied d.c. potential will therefore be a peaked curve with the height of the peak proportional to the concentration of the reducible substance in the solution, just as with a derivative polarogram. Differential pulse polarography is a very satisfactory method for the determination of many substances with a detection limit of about $10^{-8} M$, and a resolution of about 0.05 V.

An important feature of pulse polarography is the sampling of the current at definite points in the lifetime of the mercury drop, and it is essential to

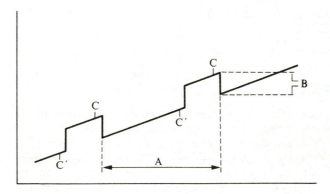

Fig. 16.10

establish an exact timing procedure. Various methods have been adopted to achieve the desired result: these include mechanical tapping of the capillary to dislodge the mercury drop at precise time intervals, or alternatively the natural drop time of the capillary is utilised, with the fall of the mercury drop serving to actuate the timing circuits which control the measurement of the current.

16.11 RAPID SCAN POLAROGRAPHY

In this technique the applied potential is swept rapidly through a range of up to 2 V during part of the lifetime of a single mercury drop: typically the voltage sweep occurs during the last 2 seconds of the lifetime of a mercury drop which has a drop time of about 6 seconds. The resultant current–voltage curve has a peak in it reminiscent of a derivative polarogram. It can be shown that the peak potential is related to the half-wave potential of the ion being discharged by the expression

$$E_s = E_{1/2} = 1.1\frac{RT}{nF}$$

The resultant peak current is greater than the diffusion current recorded with a conventional d.c. polarograph by a factor of ten or even more. The method thus shows enhanced sensitivity and it can be used to make measurements with solutions having concentrations as low as $10^{-6} - 10^{-7} M$ and with a resolution of the order of 0.05 V.

The D.M.E. can with advantage be replaced by an S.M.D.E. (Section 16.8), and it is possible to use platinum, graphite, or glassy carbon electrodes, in which case the procedure should be termed voltammetry rather than polarography.

16.12 SINUSOIDAL a.c. POLAROGRAPHY

In this procedure, a constant sine wave a.c. potential of a few millivolts is superimposed upon the usual d.c. potential sweep. The applied d.c. potential is measured in the usual way and these results are coupled with measurements of the alternating current.

If the values of the a.c. current are plotted against the potential applied by the potentiometer, a series of peaks are obtained as illustrated in Fig. 16.11(a): the normal d.c. polarogram of the same solution is also shown (b).

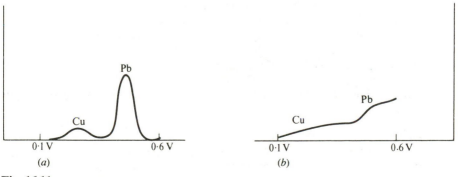

Fig. 16.11

The a.c. curve is seen to be similar in character to a derivative polarogram (Figs 16.7; 16.8b; Section 16.7) but must not be confused with this type of curve. Each peak in the a.c. curve corresponds to a step in the normal polarographic record. The voltage of the peak is the same as that of the midpoint of the step, and the height of a peak above the baseline is proportional to the concentration of the depolariser, and thus corresponds to the step height: it will be apparent that with closely separated waves, measurements are much more readily made from the a.c. polarogram than from the d.c. polarogram, and it is considered that peaks separated by 40 mV can be resolved as compared with the separation of 200 mV required in d.c. polarography: the limit of sensitivity $(10^{-5}M)$ is however not greatly different from that achieved in d.c. polarography, and this is related to the fact that the residual current is rather large. This arises because the condenser current is relatively large as compared with the diffusion or faradaic current: the curve is thus similar to that produced by Rapid Scan Polarography (see Section 16.11).

16.13 QUANTITATIVE APPLICATIONS OF POLAROGRAPHY

(*a*) **Inorganic.** Many inorganic anions, cations, and some molecules can be determined polarographically. Cathodic reduction waves are particularly valuable for the determination of cations of the transition metals. The ease of determination is frequently dependent upon selection of the appropriate supporting electrolyte: for example, with potassium chloride as supporting electrolyte the waves for copper(II) and iron(III) interfere with each other, but with potassium fluoride as supporting electrolyte, the iron(III) forms a complex with fluoride ions and the iron(III) wave becomes 0.5 V more negative whilst the copper wave is practically unaffected; thus interference is no longer experienced.

The polarographic determination of metal ions such as Al^{3+} which are readily hydrolysed can present problems in aqueous solution, but these can often be overcome by the use of non-aqueous solvents. Typical non-aqueous solvents, with appropriate supporting electrolytes shown in parentheses, include acetic acid (CH_3CO_2Na), acetonitrile $(LiClO_4)$, dimethylformamide (tetrabutyl-ammonium perchlorate), methanol (KCN or KOH), and pyridine (tetraethyl-ammonium perchlorate). In these media a platinum micro-electrode is employed in place of the dropping mercury electrode.

The polarographic method is applicable to the determination of inorganic anions such as bromate, iodate, dichromate, vanadate, etc. Hydrogen ions are involved in many of these reduction processes, and the supporting electrolyte must therefore be adequately buffered.

Typical applications in the inorganic field are the analysis of minerals, metals (including alloys), fertilisers, natural waters, industrial effluents and polluted atmospheres. The technique can also be used to establish the formulae of various complexes.

(*b*) **Organic.** Many organic functional groups will undergo reduction or oxidation at a dropping electrode and thus lend themselves to polarographic determination. In general, the reactions of organic compounds at the dropping electrode are slower and are often more complex than those of inorganic ions;

nevertheless polarographic investigations can be useful for structure determination and for qualitative and quantitative analysis.

Reactions of organic substances at the dropping electrode usually involve hydrogen ions; a typical reaction can be represented by the equation

$$R + nH^+ + ne \rightleftharpoons RH_n$$

where RH_n is the reduced form of the reducible organic compound R. As hydrogen ions (supplied from the solution) are involved in the reaction, the supporting electrolyte must be well buffered. Change in the pH of the supporting electrolyte may even lead to the formation of different reaction products. Thus, in slightly alkaline solution, benzaldehyde is reduced at -1.4 V with formation of benzyl alcohol, but in acid solution (pH < 2), reduction takes place at -1.0 V with formation of hydrobenzoin:

$$2C_6H_5CHO + 2H^+ + 2e \rightleftharpoons C_6H_5CH(OH)CH(OH)C_6H_5$$

Whilst some organic compounds can be investigated in aqueous solution, it is frequently necessary to add an organic solvent to improve the solubility; suitable water-miscible solvents include ethanol, methanol, ethane-1,2-diol, dioxan, acetonitrile and acetic (ethanoic) acid. In some cases a purely organic solvent must be used and anhydrous materials such as acetic acid, formamide and diethylamine have been employed; suitable supporting electrolytes in these solvents include lithium perchlorate and tetra-alkylammonium salts R_4NX (R = ethyl or butyl; X = iodide or perchlorate).

The following functional groups can be expected to react at the dropping electrode.

C=C when conjugated with another double bond or an aromatic ring;
C≡C when conjugated with an aromatic ring;
C—X (X = halogen), but with some exceptions;
C=O (aldehydes, ketones, quinones);
COOH simple monocarboxylic acids do not react, but dicarboxylic acids in which the carboxyl groups are conjugated with each other do;
Peroxides and epoxides;
C=N;
Nitro, nitroso, and azo groups;
Heterocycles with two or more nitrogen atoms in the ring;
C=S;
S—S, S—H (mercaptans give an *anodic* wave).

Polarographic methods can be used to examine food and food products; biological materials; herbicides, insecticides and pesticides; petroleum and petroleum products; pharmaceuticals. The examination of blood and urine samples is frequently carried out to establish the presence of drugs and to obtain quantitative results.

Polarographic detectors find application in monitoring the effluent from chromatography columns, including those used in HPLC. Applications of polarography are discussed in Refs 46–48 (Section 16.38).

When information regarding the polarographic behaviour of a substance is not available, cognisance must be taken of the fact that in addition to the normal polarographic wave associated with the reduction and the diffusion of the species under investigation, there may be other factors which influence the current observed. These include the following.

(*a*) *Kinetic currents*, where the rate of a chemical reaction exerts a controlling influence. Thus, a species *S* may not itself be electro-active, but may be capable of conversion to a substance *O* which undergoes reduction at the dropping electrode:

$$S \rightleftharpoons O + ne \rightleftharpoons R$$

If the rate of formation of *O* is slower than the rate of diffusion, the diffusion current will be controlled by the rate at which *O* is formed.

(*b*) *Catalytic currents*. These are observed when the product of reduction at the electrode is reconverted to the original species by interaction with another substance in solution which acts as an oxidising agent:

$$S + ne \rightleftharpoons R; R + X \rightleftharpoons S$$

This is observed for example with a solution containing Fe(III) ions and hydrogen peroxide: Fe(II) ions formed at the electrode are converted back to Fe(III) ions by the hydrogen peroxide: such behaviour results in an enhanced diffusion wave.

(*c*) *Adsorption currents*. If either the oxidised or the reduced form of an electro-active substance is adsorbed on the surface of the electrode, the behaviour of the system is altered. If the oxidised form is adsorbed, reduction takes place at a more negative potential; if the reduced form is adsorbed, then the reduction is favoured and a 'pre-wave' is observed in the polarogram. When the electrode surface is completely covered by reductant, the current is controlled by a normal diffusion process and a normal polarographic wave is obtained.

Various tests can be applied, including the influence of the height of the mercury column of the dropping electrode on the diffusion current to decide if any of the above non-diffusion processes are operative.[49]

A development of the study of adsorption currents is referred to as Tensammetry, and a useful review[50] refers to the application of this technique to the examination of natural waters as an aid to 'speciation', the procedure for deciding how a metal ion is distributed between the various species in which the metal may be present.

16.14 DETERMINATION OF THE HALF-WAVE POTENTIAL OF THE CADMIUM ION IN 1 *M* POTASSIUM CHLORIDE SOLUTION

The following experiments (Section 16.14–16.16), which can well be performed with a manual polarograph, serve to illustrate the general procedure to be followed in d.c. polarography: it is advantageous to use a chart recorder to produce the polarogram.

If a commercial polarograph which includes a potentiostat is employed, then the three-electrode procedure (Sections 16.7 and 16.8) is conveniently used with the controlled potential supplied by the potentiostat applied between the dropping electrode and the calomel reference electrode, while the electrolysis current flows between the working (mercury) electrode and the auxiliary platinum electrode. Follow the operating instructions for the particular apparatus in use. Such an instrument will normally make it possible to perform the pulsed techniques referred to in Sections 16.10–16.12, and it will be instructive to repeat the experiment using these alternative techniques and to compare the results obtained.

Make sure that the reservoir of the dropping electrode contains an adequate supply of mercury, and that mercury drops freely from the capillary when the tip is immersed in distilled water while the reservoir is raised to near the maximum height of the stand: allow the mercury to drop for 5–10 minutes. Replace the beaker of water by one containing 1 *M* potassium chloride solution and adjust the rate of dropping by varying the height of the mercury reservoir until the dropping rate is 20–24 per minute: then note the height of the mercury column.

When the adjustment has been completed, rinse the capillary well with a stream of distilled water from a wash bottle and then dry by blotting with filter paper. Insert the capillary through an inverted cone of quantitative filter paper and clamp vertically over a small beaker. Lower the levelling bulb until the mercury drops just cease to flow.

Pipette 10 mL of a cadmium sulphate solution (1.0 g Cd^{2+} L^{-1}) into a 100 mL graduated flask, add 2.5 mL of 0.2 per cent gelatin solution, 50 mL of 2 *M* potassium chloride solution and dilute to the mark. The resulting solution (*A*) will contain 0.100 g Cd^{2+} L^{-1} in a base solution (supporting electrolyte) of 1 *M* potassium chloride with 0.005 per cent gelatin solution as suppressor.

Measurements. Place 5.0 mL of the solution *A* in a polarographic cell equipped with an external reference electrode (saturated calomel electrode). Pass pure nitrogen through the solution at a rate of about two bubbles per second for 10–15 minutes in order to remove dissolved oxygen. Raise the mercury reservoir to the previously determined height and insert the capillary into the cell so that the capillary tip is immersed in the solution. Connect the S.C.E. to the positive terminal and the mercury in the reservoir to the negative terminal of the polarograph. After about 15 minutes stop the passage of inert gas through the solution; the electrical measurements may now be commenced.

Carry out a preliminary test, adjusting the sensitivity of the instrument so that the trace on the recorder uses as much of the width of the chart paper as possible at the maximum applied potential; this of course must not exceed the decomposition potential of the supporting electrolyte.

Now commence the voltage sweep using a scan rate of 5 mV per second, or with a manual polarograph, increase the voltage in steps of 0.05 V. The recorder plot will take the form shown in Fig. 16.4: if a manual instrument is used, then since the current oscillates as mercury drops grow and then fall away, the plot will have a saw-tooth appearance, and for measurement purposes a smooth curve must be drawn through the midpoint of the peaks of the plot.

When the experiment has been completed, clean the capillary as described above and then store it by inserting through a bored cork (or silicone rubber bung — normal rubber bungs which contain sulphur must be avoided) which is then placed in a test-tube containing a little pure mercury. Lower the mercury reservoir until drops no longer issue from the capillary, then push the end of the capillary into the mercury pool.

The mercury at the bottom of the electrolysis vessel must be carefully recovered, washed by shaking with distilled water, and then stored under water in a 'Recovered Mercury' bottle.

Determine the half-wave potential from the current–voltage curve as described in Section 16.6; the value in $1\,M$ potassium chloride should be about -0.60 *vs* S.C.E. Measure the maximum height of the diffusion wave after correction has been made for the residual current; this is the diffusion current I_d, and is proportional to the total concentration of cadmium ions in the solution.

Measure the height of the diffusion wave I, after correcting for the residual current at each increment of the applied voltage. Plot the values of $\log(I_d - I)/I$ as abscissae against applied voltage as ordinates (Fig. 16.12; strictly speaking, the values of the applied voltage are negative). Determine the slope of the graph, which should be equal to about 0.030, and read off the intercept on the voltage axis. The latter is the desired half-wave potential of the cadmium ion *vs* S.C.E.

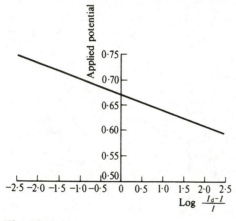

Fig. 16.12

As an additional exercise, the current–voltage curve of the supporting electrolyte ($1\,M$ potassium chloride) may be evaluated; this gives the residual current directly and no extrapolation is required for the determination of I and I_d.

16.15 INVESTIGATION OF THE INFLUENCE OF DISSOLVED OXYGEN

The solubility of oxygen in water at the ordinary laboratory temperature is about 8 mg (or 2.5×10^{-4} mol) per litre. Oxygen gives two polarographic waves ($O_2 \to H_2O_2 \to H_2O$) which occupy a considerable voltage range, and their positions depend upon the pH of the solution. Unless the test solution contains a substance which yields a large wave or waves compared with which those due to oxygen are negligible, dissolved oxygen will interfere. In general, particularly in dilute solution, dissolved oxygen must be removed by passing pure nitrogen or hydrogen through the solution.

Place some $1\,M$ potassium chloride solution containing 0.005 per cent gelatin in a polarographic cell immersed in a thermostat. Make the usual preliminary adjustments with regard to sensitivity of the recorder, then record the current–applied voltage curve. Now pass oxygen-free nitrogen through the solution for 10–15 minutes. Record the polarogram using the same sensitivity. It will be observed that the two oxygen waves are absent in the new polarogram (compare Fig. 16.5).

16.16 DETERMINATION OF CADMIUM IN SOLUTION

Two procedures may be employed: (1) that dependent upon wave height–concentration plots, and (2) the method of standard additions. The theory has been given in Section 16.5.

(1) **Wave height–concentration plots**. Prepare from the stock solution containing $1.000\,g\,Cd^{2+}\,L^{-1}$ solutions containing respectively 0.1, 0.05, 0.025, and $0.01\,g\,Cd^{2+}\,L^{-1}$ by transferring to $100\,mL$ graduated flasks 10, 5.0, 2.5, and $1.0\,mL$ of the stock solution, adding $50\,mL$ of $2M$ potassium chloride solution and $2.5\,mL$ of 0.2 per cent gelatin solution to each flask and then diluting to the mark with distilled water. Mix $10\,mL$ of the unknown solution (which may contain, say, about $0.04\,g$ of cadmium L^{-1}) in a $100\,mL$ graduated flask with $50\,mL$ of $2M$ potassium chloride solution and $2.5\,mL$ of 0.2 per cent gelatin solution and dilute to the mark. Record the polarograms of the four standard solutions and of the unknown solution following the procedure described in Section 16.14, and determine the wave heights from each polarogram. Draw a calibration curve (wave heights as ordinates, concentrations as abscissae) for the four standard solutions: read off from the curve the concentration corresponding to the wave height of the unknown solution.

(2) **Method of standard addition**. The polarogram of the unknown solution will have been determined under (1). A new polarogram must now be recorded after the addition of a known volume of a standard solution containing the same ion, care being taken that in the resulting solution the concentrations of the supporting electrolyte and the suppressor are maintained constant.

Place $10\,mL$ of the unknown solution, $5\,mL$ of the stock solution $(1.000\,g\,Cd^{2+}\,L^{-1})$, $50\,mL$ of $2M$ potassium chloride solution, and $2.5\,mL$ of 0.2 per cent gelatin solution in a $100\,mL$ graduated flask, and dilute to the mark with distilled water. Transfer a suitable volume to the polarographic cell in a thermostat, remove the dissolved air with nitrogen, and record the polarogram in the usual way: it is important that the sensitivity be kept at the previous value. Determine the new wave height. Calculate the concentration of the unknown solution as described in Section 16.5. Compare this value of the concentration with that found by method (1).

16.17 DETERMINATION OF LEAD AND COPPER IN STEEL

In the application of the polarographic method of analysis to steel a serious difficulty arises owing to the reduction of iron(III) ions at or near zero potential in many base electrolytes. One method of surmounting the difficulty is to reduce iron(III) to iron(II) with hydrazinium chloride in a hydrochloric acid medium. The current near zero potential is eliminated, but that due to the reduction of iron(II) ions at about -1.4 volts *vs* S.C.E. still occurs. Other metals (including copper and lead) which are reduced at potentials less negative than this can then be determined without interference from the iron. Alternatively, the Fe^{3+} to Fe^{2+} reduction step may be shifted to more negative potentials by complex ion formation.

The following procedure may be used for the simultaneous determination of copper and lead in plain carbon steels. Dissolve 5.0 g of the steel, accurately weighed, in a mixture of 25 mL of water and 25 mL of concentrated hydrochloric acid: heat gently to minimise the loss of acid. Add a few drops of saturated potassium chlorate solution to dissolve carbides, etc., and boil the mixture until the solution is clear. Cool and dilute to 50 mL with water in a graduated flask. Pipette 2.00 mL of this solution into a polarographic cell and add: 1.0 mL of 20 per cent hydrazinium chloride solution to reduce any iron(III) to the iron(II) state, 1.0 mL of 0.2 per cent methyl cellulose to act as a maximum suppressor, and 5.5 mL of 2.0 M sodium formate solution to adjust the pH of the solution to that at which reduction of Fe(III) and Cu(II) ions takes place. Place the cell in a nearly boiling water bath for 10 minutes in order to complete the reduction. Cool. Analyse the solution polarographically: use a saturated calomel reference electrode. The first step in the polarogram is due to the reduction of copper(I) ions to the metal and has a half-wave potential of -0.25 volt vs S.C.E. The second step, which is due to lead, has a half-wave potential of -0.45 vs S.C.E. Carry out a calibration by adding known amounts of copper and lead to a solution of steel of low copper and lead content, and determine the increase in wave heights due to the additions.

Calculate the percentage of copper and of lead in the sample of steel.

16.18 DETERMINATION OF NITROBENZENE IN ANILINE

Aniline usually contains traces of nitrobenzene which can be determined polarographically.

Procedure. Prepare a buffer solution (pH 7) by dissolving 3.7 g pure citric acid and 23.6 g of potassium dihydrogenphosphate in 1 L of distilled water.

Prepare a stock solution of nitrobenzene by dissolving about 0.25 g (accurately weighed in a weighing bottle) in 50 mL of 95 per cent ethanol, and adding this to 450 mL of the buffer solution contained in a 500 mL graduated flask. Rinse the weighing bottle with two successive 5 mL portions of 95 per cent ethanol and add each rinsing to the 500 mL flask; finally make the flask up to the mark with buffer solution.

Weigh 1–2 g of the aniline sample into a 50 mL graduated flask and dissolve the aniline by adding 5 mL concentrated hydrochloric acid. Allow the fumes to subside, then make the flask up to the mark with buffer solution, adding 1.5 mL 0.2 per cent gelatin solution just before the final adjustment. Prepare a series of solutions for polarography by placing (1) 30 mL, (2) 40 mL, (3) 50 mL, (4) 60 mL of the stock nitrobenzene solution in a series of 100 mL graduated flasks. Add to each flask 2.5 mL of 0.2 per cent gelatin solution and make up to the mark with buffer solution. Examine each solution in turn in the polarograph and then plot wave heights against nitrobenzene concentration to give a calibration curve.

Finally, determine the polarogram of the aniline solution, and from the wave height deduce the nitrobenzene content of the sample; $E_{1/2}$ ca -0.4 V.

16.19 DETERMINATION OF ASCORBIC ACID IN FRUIT JUICE

The juice of many fruits and particularly those of the citrus family contain appreciable quantities of ascorbic acid (vitamin C). It is not possible to examine the undiluted juice polarographically as it contains many diverse compounds, some of which are surface-active and, therefore, interfere with the development of a polarographic wave. If, however, the juice is well diluted, the interference due to the surface-active substance is virtually eliminated and polarography becomes possible. Ascorbic acid is prone to oxidation by dissolved oxygen, and solutions containing it should be prepared with water which has been boiled and then allowed to cool in an atmosphere of nitrogen. Ascorbic acid gives an anodic wave, and the dropping electrode must be made the anode of the cell; start with a potential of $+0.2$ V (just below the oxidation potential of mercury) and sweep through zero voltage to an increasing negative potential.

Procedure. Prepare a standard solution of ascorbic acid (ca $10^{-3}M$) by dissolving about 0.18 g (accurately weighed) in a 1 L graduated flask using oxygen-free water.

Prepare a buffer solution (pH 4.5) by dissolving 6 g acetic (ethanoic) acid and 13.6 g sodium acetate in 1 L of distilled water. Pipette 10 mL of a commercial sample of citrus fruit juice into a 1 L graduated flask and make up to the mark with oxygen-free water.

The polarographic investigation is best carried out using the differential pulse technique. Place 10 mL of the buffer solution in the polarographic cell and de-oxygenate by bubbling oxygen-free nitrogen through it for 10 minutes. Use a micropipette to add a 10 μL sample of the diluted fruit juice to the cell and record the polarogram. Add two successive 5 μL portions of the standard ascorbic acid solution to the cell and record the polarogram after each addition, thus using the standard addition method (Section 16.5) to determine the unknown concentration; $E_{1/2}$ ca -0.05 V.

STRIPPING VOLTAMMETRY

16.20 BASIC PRINCIPLES

Consider a conventional d.c. polarographic system set up with an oxygen-free solution of supporting electrolyte containing one or more ions reducible at a mercury cathode, but fitted with a Hanging (or Static) Mercury Drop Electrode (Section 16.8). If the potentiostat of the polarograph is then set to a *fixed value* which is chosen to be 0.2–0.4 V more negative than the highest reduction potential encountered among the reducible ions, then electrolysis will occur, deposition of metals will take place on the H.M.D.E. cathode and, usually, amalgam formation will take place. The rate of amalgam formation will be governed by the magnitude of the current flowing, by the concentrations of the reducible ions, and by the rate at which the ions travel to the electrode; the latter can be controlled by stirring the solution. Given sufficient time, virtually the whole of the reducible ion content of the solution may be transferred to the mercury cathode, but complete exhaustion of the solution is not really necessary for the present procedure, and in practice, electrolysis is carried out for a carefully controlled time interval so that a fraction (say 10 per cent) of the reducible ions are discharged. This operation is often referred to as a **concentration step**, the metals become concentrated into the relatively small volume of the mercury drop.

The electrolysis current is not stopped, the stirrer switched off, and the cell allowed to stand for about 30 seconds to allow the solution to become quiescent.

The potentiostat is then caused to make a voltage sweep in reverse, starting from the potential used in the electrolysis. This means that a gradually increasing positive potential is applied to the H.M.D.E., which is now the anode of the cell. If the current is measured and plotted against the voltage (a recorder is used), then initially a gradually increasing current corresponding to the residual current of conventional polarography, and due mainly to the ground solution, is observed. As the potential approaches the oxidation potential of one of the metals dissolved in the mercury, then ions of that metal pass into solution from the amalgam and the current increases rapidly and attains a maximum value when the potential has a value approximating to the appropriate oxidation potential. The metal is said to be 'stripped' from the amalgam, and if the potential were held at the value corresponding to the maximum current, all of the metal would eventually be returned to the solution. In actual fact, however, the potential is not held stationary, and as the potential sweep continues, the current declines from its maximum value and settles down to a new approximately steady value. In other words, the curve shows a peak. With continuing rise in the anodic potential, fresh peaks will be produced in the curve as the oxidation potentials of the different metals contained in the amalgam are reached: by analogy with polarogram, the resulting curve is termed a **voltammogram** (or **stripping voltammogram**). A typical result with a single metal concentrated in the mercury is shown in Fig. 16.13.

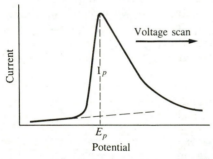

Fig. 16.13

The peaks are characterised by the **peak potential** E_p, by the **peak current** I_p (i.e. the height of the peak) and by the **breadth** b (i.e. the voltage span of the peak at the point where the current is $0.5I_p$): these parameters are, however, dependent upon characteristics of the electrode and upon the rate of the voltage sweep during the stripping process. The magnitude of the peak current is proportional to the concentration in the amalgam of the metal being stripped, and is, therefore, proportional to its concentration in the original solution.

From the nature of the process described above it has been referred to as 'stripping polarography', but the term **'anodic stripping voltammetry'** is preferred. It is also possible to reverse the polarity of the two electrodes of the cell, thus leading to the technique of cathodic stripping voltammetry.

In just the same way as differential pulse polarography represents a vast improvement over conventional polarography (see Section 16.10), the application of a pulsed procedure leads to the greatly improved technique of differential pulsed anodic (cathodic) stripping volammetry. A particular feature of this technique is that owing to the much better resolution and greater sensitivity which is achieved, the concentration of metals in the H.M.D.E. can be reduced and consequently, the time needed for the concentration step can be cut considerably.

The technique can be used to measure concentrations in the range $10^{-6}-10^{-9}M$ and as such is eminently suitable for the determination of trace metal impurities; of recent years it has found application in the analysis of semiconductor materials, in the investigation of pollution problems, and in speciation studies.

16.21 SOME FUNDAMENTAL FEATURES

In view of the limitations referred to above, and particularly the influence of electrode characteristics upon the peaks in the voltammogram, some care must be exercised in setting up an apparatus for stripping voltammetry. The optimum conditions require:

(a) in the concentration step a small mercury volume by comparison with the volume of solution to be electrolysed; efficient stirring of the solution during the electrolysis, otherwise the deposition procedure may be unduly prolonged;

(b) in the stripping operation, as fast a voltage scan as possible consistent with the avoidance of peak tailing.

Electrodes. The Hanging Mercury Drop Electrode is traditionally associated with the technique of stripping voltammetry and its capabilities were investigated by Kemula and Kublik.[51] In view of the importance of drop size it is essential to be able to set up *exactly* reproducible drops, and this can be done as explained in Section 16.8 for the S.M.D.E.

In an alternative technique use is made of a platinum wire sealed into a glass tube. The wire is first thoroughly cleaned and is then used as anode for the electrolysis of pure perchloric acid: this treatment exerts a polishing effect. The current is then reversed, and the electrode used as cathode to ensure that no oxide or adsorbed oxygen films remain on the surface of the electrode. Still using it as cathode, the electrode is now used for the electrolysis of mercury(II) nitrate solution and it thus becomes plated with mercury to give a Mercury Film Electrode (M.F.E.). This is claimed to possess advantages related to its rigidity and also to the greater surface area/volume ratio as compared with a mercury drop. With the electrode based on platinum, however, should there be any bare platinum areas which have escaped plating in contact with the solution, then complications may arise owing to the smaller hydrogen overpotential on platinum compared with that on mercury, and metals having high positive electrode potentials may fail to deposit when the electrode is used for the concentration step.

Of recent years the use of mercury film electrodes based on substrates other than platinum has been explored, and increased sensitivity is claimed for electrodes based on wax-impregnated graphite, on carbon paste and on vitreous carbon: a technique of simultaneous deposition of mercury and of the metals to be determined has also been developed.

Cells. The cell employed can be a suitable polarographic cell, or can be specially constructed to fulfil the following requirements. Efficient *reproducible* stirring of the solution is essential, and for this purpose a magnetic stirrer is usually suitable. Exclusion of oxygen is important, and so the cell must be provided with a cover, and provision made for passing pure (oxygen-free) nitrogen through the solution before commencing the experiment, and over the surface of the liquid during the determination. The cover of the cell must provide a firm seating for the H.M.D.E. (or other type of electrode used), and must also have openings for the reference electrode (usually an S.C.E.) and for a platinum counter-electrode if it is required to operate under conditions of controlled potential. If the solution under investigation is to be analysed for mercury, then the reference electrode should be isolated from the solution by use of a salt bridge.

Reagents. In view of the sensitivity of the method, the reagents employed for preparing the ground solutions must be very pure, and the water used should be re-distilled in an all-glass, or better, an all-silica apparatus: the traces of organic material sometimes encountered in demineralised water (Section 3.17) make such water unsuitable for this technique unless it is distilled. The common supporting electrolytes include potassium chloride, sodium acetate–acetic acid buffer solutions, ammonia–ammonium chloride buffer solutions, hydrochloric acid and potassium nitrate.

The normal analytical grade chemicals often contain trace impurities which are quite unimportant for most analytical purposes, but in terms of stripping voltammetry may represent serious contamination: this is especially true if heavy metals are involved. It is therefore necessary to employ reagents of very high purity (e.g. the B.D.H. 'Aristar' reagents or similar grade), or alternatively to subject the purest material available to an electrolytic purification process. The stirred solution is electrolysed with a small current (10 mA) for 24 hours, using a pool of mercury at the bottom of a beaker as cathode and a platinum anode; pure nitrogen is passed through the solution before commencing the electrolysis so as to remove dissolved oxygen, and during the purification process, a current of pure nitrogen is maintained over the surface of the solution. It will usually be necessary to use some form of potentiostatic control during the electrolysis process.

In view of the foregoing remarks, it is clear that all glassware used in the preliminary treatment of samples to be subjected to stripping voltammetry, as well as the apparatus to be used in the actual determination, must be scrupulously cleaned. It is usually recommended that glassware be soaked for some hours in *pure* nitric acid ($6M$), or in a 10 per cent solution of *pure* 70 per cent perchloric acid, followed by washing with de-ionised water.

Concentration process. In the concentration process a fraction of the amount of a metal ion in solution is deposited at the mercury electrode, and the concentration of the metal in the mercury may become from 10 to 1000 times greater than the original ionic concentration in the solution; it is to this pre-concentration process that the great sensitivity of the stripping method is due. Since the deposition is not exhaustive it is important that, in a given investigation, the same fraction of metal is incorporated in the mercury for each voltammogram recorded, and to achieve this, the electrode surface area, the deposition time and the rate of stirring must be carefully duplicated. Electrolysis times vary from 30 seconds to 30 minutes depending on the concentration of the analyte in the solution under investigation, the electrode in use and the stripping procedure to be employed. Clearly, the weaker (in analyte) the solution being analysed, the longer must be the electrolysis time. Electrolysis at a mercury film electrode is quicker than at a H.M.D.E., and the pre-concentration step may be shortened if differential pulse stripping is employed rather than direct current stripping.

Certain materials may, during the electrolysis process, give rise to species which form an insoluble salt with Hg(I) ions, thus giving rise to an insoluble film on the mercury surface. Substances in this category include halides (other than fluorides), sulphides, thiocyanates, mercaptans and many organic thio compounds. In the subsequent stripping process, the voltage scan must be in the *negative* direction so that the procedure is then referred to as Cathodic Stripping Voltammetry.

Many 'polarographic analysers' are available commercially which can be used for the pre-concentration electrolysis step, and will then apply the requisite stripping procedure and produce a recorder chart of the result. In many modern instruments, microprocessor control is incorporated so that the requisite series of operations take place automatically.

16.22 DETERMINATION OF LEAD IN TAP WATER

In the following determination use of the S.M.D.E. and of differential pulse stripping is described. All the glass apparatus used must be rigorously cleaned; vessels should be filled with pure $6M$ nitric and left standing overnight and then thoroughly cleaned with re-distilled water.

Procedure. Prepare (1) a standard $(0.01 M)$ lead solution by dissolving 1.65 g (accurately weighed) analytical grade lead nitrate in re-distilled water and adjusting to the mark in a 500 mL graduated flask; (2) a supporting electrolyte solution $(0.02 M$ potassium nitrate) by dissolving 1.0 g of 'Aristar' (or similar highly purified material) in 500 mL of redistilled water. Confirm that this solution does not contain significant amounts of impurities (especially lead) by carrying out the pre-concentration and stripping procedure described in detail below using 10 mL of the solution and 10 mL of re-distilled water. If a significant lead peak is obtained, the solution must be discarded.

Place 10 mL of the tap water and 10 mL of the supporting electrolyte in the electrolysis vessel, add a magnetic stirring bar and mount the vessel on a magnetic stirrer. Insert an S.C.E., a platinum plate auxiliary electrode, a nitrogen circulating tube and the capillary of the mercury electrode: operate the dispensing mechanism to form a mercury drop at the end of the capillary. Pass oxygen-free nitrogen through the solution for 5 minutes, and then adjust the nitrogen to pass over the surface of the liquid.

Make the connections to the polarographic analyser and adjust the applied voltage to -0.8 V, i.e. a value well in excess of the deposition potential of lead ions. Set the stirrer in motion noting the setting of the speed controller, and after 15–20 seconds, switch on the electrolysis current and at the same time start a stopclock; allow electrolysis to proceed for 5 minutes. On completion of the electrolysis time, turn off the stirrer, but leave the electrolysis potential applied to the cell. After 30 seconds to allow the liquid to become quiescent, replace the electrolysis current by the pulsed stripping potential and set the chart recorder in motion. When the lead peak at *ca* 0.5 V has been passed, turn off the stripping current and the recorder. A suitable scan rate for stripping is $5 \, \text{mV s}^{-1}$.

Clean out the electrolysis cell and charge it with 10 mL of supporting electrolyte, 10 mL of re-distilled water and 1 µL of the standard lead solution. Set up a new mercury drop, pass nitrogen for 5 minutes and then record a new stripping voltammogram by repeating the above procedure with timings and stirring rate repeated exactly. Repeat with three further solutions containing 2, 3 and 4 µL of the standard lead solution. Measure the peak heights of the five voltammograms and thus deduce the lead content of the tap water.

AMPEROMETRY

16.23 AMPEROMETRIC TITRATIONS

It has been shown (Section 16.3; Fig. 16.4) that the limiting current is independent of the applied voltage impressed upon a dropping mercury electrode (or other indicator micro-electrode). The only factor affecting the limiting current, if the migration current is almost eliminated by the addition of sufficient supporting electrolyte, is the rate of diffusion of electro-active material from the bulk of the solution to the electrode surface. Hence, the diffusion current (= limiting current − residual current) is proportional to the concentration of the electro-active material in the solution. If some of the electro-active material is removed by interaction with reagent, the diffusion current will decrease. This is the fundamental principle of **amperometric titrations**. The observed diffusion current at a suitable applied voltage is measured as a function of the volume of the titrating solution: the end point is the point of intersection of two lines giving the change of current before and after the equivalence point.

Some advantages of amperometric titrations may be mentioned:

1. The titration can usually be carried out rapidly, since the end point is found graphically; a few current measurements at constant applied voltage before and after the end point suffice.
2. Titrations can be carried out in cases in which the solubility relations are such that potentiometric or visual indicator methods are unsatisfactory; for example, when the reaction product is markedly soluble (precipitation titration) or appreciably hydrolysed (acid–base titration). This is because the readings near the equivalence point have no special significance in amperometric titrations. Readings are recorded in regions where there is excess of titrant, or of reagent, at which points the solubility or hydrolysis is suppressed by the Mass Action effect; the point of intersection of these lines gives the equivalence point.
3. A number of amperometric titrations can be carried out at dilutions ($ca \ 10^{-4}M$) at which many visual or potentiometric titrations no longer yield accurate results.
4. 'Foreign' salts may frequently be present without interference and are, indeed, usually added as the supporting electrolyte in order to eliminate the migration current.

If the current–voltage curves of the reagent and of the substance being titrated are not known, the polarograms must first be determined in the supporting

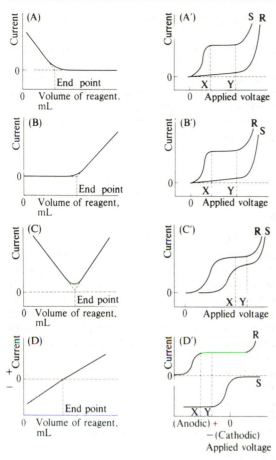

Fig. 16.14

electrolyte in which the titration is to be carried out. The voltage applied at the beginning of the titration must be such that the total diffusion current of the substance to be titrated, or of the reagent, or of both, is obtained. In Fig. 16.14 are collected the most common types of curves encountered in amperometric titrations together with the corresponding hypothetical polarograms of each individual substance: S refers to the solute to be titrated and R to the titrating reagent. The slight 'rounding off' in the vicinity of the equivalence point is due to solubility of precipitate, salt hydrolysis, or dissociation of a complex; this curvature does not usually interfere, since the end-point is located by extending the linear branches to the point of intersection. For each amperometric titration the applied voltage is adjusted to a value between X and Y shown in Fig. 16.14, (A')–(D'). In (A) only the material being titrated gives a diffusion current [see (A')], i.e. the electro-active material is removed from the solution by precipitation with an inactive substance (for example, lead ions titrated with oxalate or sulphate ions). In (B) the solute gives no diffusion current but the reagent does [see (B')], i.e. an electro-active precipitating reagent is added to an inactive substance (for example, sulphate ions titrated with barium or lead ions). In (C)

627

both the solute and the titrating reagent give diffusion currents [see (C′)] and a sharp **V**-shaped curve is obtained (for example, lead ion titrated with dichromate ion, nickel ion with dimethylglyoxime, and copper ion with benzoin α-oxime). Finally, in (D) the solute gives an anodic diffusion current (i.e. is, oxidised at the dropping mercury cathode) at the same potential as the titrating reagent gives a cathodic diffusion current [see (D′)]; here the current changes from anodic to cathodic or vice versa and the end point of the titration is indicated by a zero current. Examples of (D) include the titration of iodide ion with mercury(II)(as nitrate), of chloride ion with silver ion, and of titanium(III) in an acidified tartrate medium with iron(III). Because the diffusion coefficient of the reagent is usually slightly different from the substance being titrated, the slope of the line before the end point differs slightly from that after the end point [compare (D)]; in practice, it is easy to add the reagent until the current acquires a zero value or, more accurately, the value of the residual current for the supporting electrolyte.

To take into account the change in volume of the solution during the titration, the observed currents should be multiplied by the factor $(V + v)/V$, where V is the initial volume of the solution and v is the volume of the titrating reagent added. Alternatively, this correction may be avoided (or considerably reduced) by adding the reagent from a semimicro burette in a concentration 10 to 20 times that of the solute. The use of concentrated reagents has the additional advantage that comparatively little dissolved oxygen is introduced into the system, thus rendering unnecessary prolonged bubbling with inert gas after each addition of the reagent. The migration current is eliminated by adding sufficient supporting electrolyte; if necessary, a suitable maximum suppressor is also introduced.

16.24 TECHNIQUE OF AMPEROMETRIC TITRATIONS WITH THE DROPPING MERCURY ELECTRODE

An excellent and inexpensive titration cell consists of a commercial resistance glass (e.g. Pyrex), 100 mL, three-necked, flat or round-bottomed flask to which a fourth neck is sealed. The complete assembly is depicted schematically in Fig. 16.15(a). The burette (preferably of the semimicro type and graduated in 0.01 mL), dropping electrode, a two-way gas-inlet tube (thus permitting nitrogen to be passed either through the solution or over its surface), and an agar–potassium salt bridge (*not* shown in the figure) are fitted into the four necks by means of rubber stoppers. The agar–salt bridge is connected through an intermediate vessel (a weighing bottle may be used) containing saturated potassium chloride solution to a large saturated calomel electrode. The agar–salt bridge is made from a gel which is 3 per cent in agar and contains sufficient potassium chloride to saturate the solution at the room temperature, when chloride ions interfere with the titrations, the connection is made with an agar–potassium nitrate bridge.

The simple electrical circuit shown in Fig. 16.15(b) is suitable for this procedure. The voltage applied to the titration cell is supplied by two 1.5 V dry cells and is controlled by the potential divider R (a 50–100 ohm variable resistance); it can be measured on the digital voltmeter V. The current flowing is read on the micro-ammeter M.

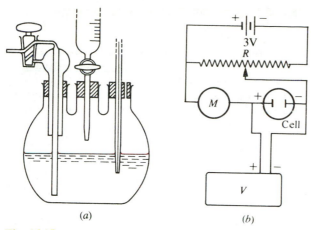

(a)   (b)

Fig. 16.15

Thermostatic control is not essential provided the cell is maintained at a fairly constant temperature during the titration. It is advantageous to store the reagent beneath an atmosphere of inert gas: this precaution is not absolutely necessary if the reagent solution has 10 to 20 times the concentration of the solution being titrated and is added from a semimicro burette. If the solute is electro-reducible, sufficient electrolyte should be added to eliminate the migration current; if the reagent is electro-reducible and the solute is not, the addition of a supporting electrolyte is usually not required, since sufficient electrolyte is formed during the titration to eliminate the migration current beyond the end point. It may be necessary to add a suitable maximum suppressor, such as gelatin. If the polarographic characteristics of the solute and the reagent are not known, the current–voltage curve of each must be determined in the medium in which the titration is being carried out. The applied voltage is then adjusted at the beginning of the titration to such a value that the diffusion current of the unknown solute or of the reagent, or of both, is obtained. Frequently the voltage range is comparatively large and, in consequence, great accuracy is not required in adjusting the applied voltage.

The general procedure is as follows. A known volume of the solution under test is placed in the titration cell, which is then assembled as in Fig. 16.15(a): the electrical connections are completed (dropping mercury electrode as cathode; saturated calomel half-cell as anode), and dissolved oxygen is removed by passing a slow stream of pure nitrogen for about 15 minutes. The applied voltage is then adjusted to the desired value, and the initial diffusion current is noted. A known volume of the reagent is run in from a semimicro burette, nitrogen is bubbled through the solution for about 2 minutes to eliminate traces of oxygen from the added liquid and to ensure complete mixing. The flow of gas through the solution is then stopped, but is allowed to pass over the surface of the solution (thus maintaining an inert, oxygen-free atmosphere). The current and burette readings are both noted. This procedure is repeated until sufficient readings have been obtained to permit the end point to be determined as the intersection of the two linear parts of the graph.

16.25 DETERMINATION OF LEAD WITH STANDARD POTASSIUM DICHROMATE SOLUTION

Both lead ion and dichromate ion yield a diffusion current at an applied potential to a dropping mercury electrode of -1.0 volt against the saturated calomel electrode (S.C.E.). Amperometric titration gives a **V**-shaped curve [Fig. 16.14 (C)]. The exercise described refers to the determination of lead in lead nitrate; the application to the determination of lead in dilute aqueous solutions ($10^{-3} - 10^{-4} M$) is self-evident.

Reagents. *Lead nitrate solution.* Dissolve an accurately weighed amount of lead nitrate in 250 mL water in a graduated flask to give an approximately 0.01 M solution. For the titration, dilute 10 mL of this solution (use a pipette) to 100 mL in a graduated flask, thus yielding a *ca* 0.001 M solution of known strength.

Potassium dichromate, ca 0.05 M solution. Use the appropriate quantity, accurately weighed, of the dry solid.

Potassium nitrate, ca 0.01 M solution. For use as the supporting electrolyte.

Procedure. Use the electrical equipment of Fig. 16.15(*b*). Set up the dropping mercury electrode assembly and allow the mercury to drop into distilled water for at least 5 minutes. Meanwhile, place 25.0 mL of the *ca* 0.001 M lead nitrate solution in the titration cell, add 25 mL 0.01 M potassium nitrate solution, complete the cell assembly, and bubble nitrogen slowly through the solution for 15 minutes. Make the necessary electrical connections. Apply a potential of -1.0 volt *vs* S.C.E.: at this potential both the lead and the dichromate ions yield diffusion currents. Turn the three-way tap so that the nitrogen now passes over the surface of the solution. Adjust the microammeter range so that the reading is at the 'high' end of the scale. Do not alter the applied voltage during the determination. Add the *ca* 0.05 M dichromate solution in 0.05 mL portions until within 1 mL of the end point, and henceforth in 0.01 mL portions until about 1 mL beyond the end point, and continue with additions of 0.05 mL. After each addition pass nitrogen through the solution for 1 minute to ensure thorough mixing and also de-oxygenation, turn the tap so that the nitrogen passes over the surface of the solution, and observe the current. It will be seen that a large initial current will decrease as the titration proceeds to a small value at the equivalence point, and then increase again beyond the equivalence point. Plot the values of the current as ordinates against the volume of reagent added as abscissae: draw two straight lines through the branches of the 'curve'. The point of intersection is the equivalence point. Calculate the percentage of lead in the sample of lead nitrate.

16.26 DETERMINATION OF SULPHATE WITH STANDARD LEAD NITRATE SOLUTION

Solutions as dilute as 0.001 M with respect to sulphate may be titrated with 0.01 M lead nitrate solution in a medium containing 30 per cent ethanol with reasonable accuracy. For solutions 0.01 M or higher in sulphate the best results are obtained in a medium containing about 20 per cent ethanol. The object of the alcohol is to reduce the solubility of the lead sulphate and thus minimise the magnitude of the rounded portion of the titration curve in the vicinity of the equivalence point. The titration is performed in the absence of oxygen at a potential of -1.2 volts (*vs* S.C.E.), at which potential lead ions yield a diffusion current. A 'reversed **L**' graph [compare Fig. 16.14 (B)] is obtained: the intersection of the two branches gives the end-point. A supporting electrolyte need not be added, since the current does not increase appreciably until an excess of lead is present in the solution, and the amount of salt formed during the titration suffices to suppress completely the migration current of lead ions.

Reagents. *Potassium sulphate.* Prepare an approximately $0.01\,M$ solution in a 100 mL graduated flask using an accurately weighed quantity of the solid.

*Lead nitrate solution, approx. 0.1*M. Prepare similarly in a 100 mL graduated flask from a known weight of the dry solid.

Procedure. Use the apparatus and technique described in Section 16.25. Introduce 25.0 mL of the potassium sulphate solution into the cell, add two to three drops of thymol blue followed by a few drops of concentrated nitric acid until the colour is just red (pH 1.2); finally, add 25 mL of 95 per cent ethanol. Connect the saturated calomel electrode through an agar–potassium nitrate bridge to the cell. Fill the semimicro burette with the standard lead nitrate solution. Pass nitrogen through the solution in the cell for 15 minutes and then over the surface of the solution. Meanwhile adjust the applied voltage to -1.2 V. Add the lead nitrate solution from the burette using the same sequence of additions as in the previous Section 16.25. After each addition of lead nitrate pass nitrogen through the solution for about 1 minute to allow time for the lead sulphate to precipitate: for more dilute solutions the nitrogen should be passed for 3 minutes before reading the current. Plot the titration curve, deduce the end point and calculate the percentage of SO_4 in the sample of potassium sulphate.

16.27 TITRATION OF AN IODIDE SOLUTION WITH MERCURY(II) NITRATE SOLUTION

This experiment illustrates the titration of a substance yielding an anodic step (iodide ion) with a solution of an oxidant [mercury(II) nitrate] giving a cathodic diffusion current at the same applied voltage. The magnitude of the anodic diffusion current decreases up to the end point: upon adding an excess of titrant, the diffusion current increases, but in the opposite direction. The type of graph obtained is similar to that in Fig. 16.14 (D). The end point of the titration is given by the intersection of the two linear portions of the graph with the volume (of titrant) axis: the diffusion current is then approximately zero. The two linear parts do not usually have the same slope, because the titrant and the substance being titrated have different diffusion currents for equivalent concentrations.

Reagents. *Potassium iodide solution,* ca *0.004*M. Dissolve 0.68 g potassium iodide, accurately weighed, in 1 L water.

Mercury(II) nitrate solution. Dissolve 1.713 g pure mercury(II) nitrate monohydrate in 500 mL of $0.05\,M$ nitric acid.

*Nitric acid, 0.1*M.

Procedure. Equip the titration vessel with a dropping mercury electrode, an agar–KNO_3 bridge connected to an S.C.E. through saturated potassium chloride solution contained in a 10 mL beaker, a nitrogen gas inlet, and a magnetic stirrer. Charge the flask with 25.0 mL of the iodide solution, add 25 mL $0.1\,M$ nitric acid, and 2.5 mL warm 1 per cent gelatin solution. Connect the dropping mercury electrode to the negative terminal of the polarising unit and the positive terminal to the S.C.E. Set the applied potential at zero and adjust the micro-ammeter reading at the centre of the scale. Pass nitrogen through the solution for at least 5 minutes while stirring magnetically. Run in the mercury(II) nitrate solution from a semimicro burette and take readings of the current at 0.10 mL intervals. The end point corresponds to zero current, but continue the titration beyond this point to obtain the cathodic current due to excess of mercury(II) nitrate. Plot current against volume of mercury(II) nitrate solution, and evaluate the exact end point from the graph.

Calculate the concentration of the mercury(II) nitrate solution from the known concentration of the potassium iodide solution.

It may be noted that it is not always necessary to use a D.M.E. when performing an amperometric titration; in some cases a graphite rod may be used. Use of a platinum electrode is considered in Section 16.28.

TITRATIONS WITH THE ROTATING PLATINUM MICRO-ELECTRODE

16.28 DISCUSSION AND APPARATUS

The dropping mercury electrode cannot be used at markedly positive potentials (say, above about 0.4 volt *vs* S.C.E.) because of the oxidation of the mercury. By replacing the dropping mercury electrode by an inert platinum electrode, it was hoped to extend the range of polarographic work in the positive direction to the voltage approaching that at which oxygen is evolved, namely, 1.1 volts. The attainment of a steady diffusion current is slow with a stationary platinum electrode, but the difficulty may be overcome by rotating the platinum electrode at constant speed: the diffusion layer thickness is considerably reduced, thus increasing the sensitivity and the rate of attainment of equilibrium. Difficulties, however, arise in obtaining reproducible values for the diffusion currents from day to day, but nevertheless, it is suitable as an indicator electrode in amperometric titrations. The larger currents (about 20 times those at the dropping mercury electrode) attained with the rotating platinum electrode allow correspondingly smaller currents to be measured without loss of accuracy and thus very dilute solutions (up to $10^{-4}M$) may be titrated. In order to obtain a linear relation between current and amount of reagent added, the speed of stirring must be kept constant during the titration: a speed of about 600 revolutions per minute is generally suitable.

The construction of a simple rotating platinum micro-electrode will be evident from Fig. 16.16. The electrode is constructed from a standard 'mercury seal'. About 5 mm of platinum wire (0.5 mm diameter) protrudes from the wall of a length of 6 mm glass tubing; the latter is bent at an angle approaching a right angle a short distance from the lower end. Electrical connection is made to the electrode by a stout amalgamated copper wire passing through the tubing to the mercury covering the sealed-in platinum wire; the upper end of the copper wire passes through a small hole blown in the stem of the stirrer and dips into mercury contained in the 'mercury seal'. A wire from the latter is connected to the source of applied voltage. The tubing forms the stem of the electrode, which is rotated at a *constant* speed of 600 r.p.m.

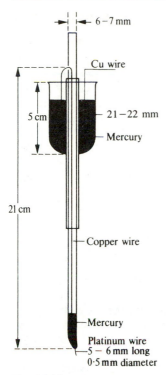

6 − 7 mm

Cu wire

5 cm

21 − 22 mm

Mercury

21 cm

Copper wire

Mercury

Platinum wire
5 − 6 mm long
0·5 mm diameter

Fig. 16.16

16.29 DETERMINATION OF THIOSULPHATE WITH IODINE

Dilute solutions of sodium thiosulphate (e.g. $0.001 M$) may be titrated with dilute iodine solutions (e.g. $0.005 M$) at zero applied voltage. For satisfactory results, the thiosulphate solution should be present in a supporting electrolyte which is $0.1 M$ in potassium chloride and $0.004 M$ in potassium iodide. Under these conditions no diffusion current is detected until after the equivalence point when excess of iodine is reduced at the electrode; a reversed **L**-type of titration graph is obtained.

Dilute solutions of iodine, e.g. $0.0001 M$, may be titrated similarly with standard thiosulphate. The supporting electrolyte consists of $1.0 M$ hydrochloric acid and $0.004 M$ potassium iodide. No external e.m.f. is required when an S.C.E. is employed as reference electrode.

Reagents. ca *0.001* M *Sodium thiosulphate solution,* $0.1 M$ with respect to potassium chloride and $0.004 M$ with respect to potassium iodide.
Standard 0.005 M *iodine solution in 0.004* M *potassium iodide.*

Procedure. Place 25.0 mL of the thiosulphate solution in the titration cell. Set the applied voltage to zero with respect to the S.C.E. after connecting the rotating platinum micro-electrode to the polarising unit. Adjust the range of the micro-ammeter. Titrate with the standard $0.005 M$ iodine solution in the usual manner.

Plot the titration graph, evaluate the end point, and calculate the exact concentration of the thiosulphate solution. As a check, repeat the titration using freshly-prepared starch indicator solution.

16.30 DETERMINATION OF ANTIMONY WITH STANDARD POTASSIUM BROMATE SOLUTION

Dilute solutions of antimony(III) and arsenic(III) (*ca* 0.0005 *M*) may be titrated with standard 0.002 *M* potassium bromate in a supporting electrolyte of 1 *M* hydrochloric acid containing 0.05 *M* potassium bromide. The two electrodes are a rotating platinum micro-electrode and an S.C.E.: the former is polarised to +0.2 volt. A reversed **L**-type of titration graph is obtained.

Reagents. *0.005* M *Potassium antimonyl tartrate solution.* Dissolve 1.625 g of the solid in 1 L of distilled water. Dilute 25.0 mL of this solution to 250 mL with 1 *M* hydrochloric acid which is 0.05 *M* in potassium bromide.
   Standard 0.002 M *potassium bromate solution.* From the pure solid.

Procedure. Pipette 25.0 mL of the antimony solution into the titration cell. Set the applied voltage at 0.2 volt *vs* S.C.E., and adjust the range of the micro-ammeter. Titrate in the usual manner, and calculate the concentration of the antimony solution.

16.31 EXAMPLES OF AMPEROMETRIC TITRATIONS USING A SINGLE POLARISED ELECTRODE

In addition to the titrations described in detail in Sections 16.25–16.27, 16.29 and 16.30, many other titrations may be performed amperometrically. An indication of the scope of this method is given by the procedures listed in Table 16.1.

Table 16.1 Examples of amperometric titrations

Titrant	Electrode	Species determined
Complexation reactions		
EDTA	DME	Many metallic ions; cf. Chapter 10
Precipitation reactions		
Dimethylglyoxime	DME	Ni^{2+}
Lead nitrate	DME	SO_4^{2-}, MoO_4^{2-}, F^-
Mercury(II) nitrate	DME	I^-
Silver nitrate	Rotating Pt	Cl^-, Br^-, I^-, CN^-, thiols
Sodium tetraphenylborate	Graphite	K^+
Thorium(IV) nitrate	DME	F^-
Potassium dichromate	DME	Pb^{2+}, Ba^{2+}
Oxidation reactions		
Iodine	Rotating Pt	As(III), $Na_2S_2O_3$
Potassium bromate/KBr	Rotating Pt	As(III), Sb(III), N_2H_4
Additions	Rotating Pt	Alkenes
Substitutions	Rotating Pt	Some phenols, aromatic amines

BIAMPEROMETRIC TITRATIONS

16.32 GENERAL DISCUSSION

The titrations so far discussed in this chapter have been concerned with the use of a reference electrode (usually S.C.E.), in conjunction with a polarised electrode (dropping mercury electrode or rotating platinum micro-electrode). Titrations may also be performed in a uniformly stirred solution by using two small but similar platinum electrodes to which a small e.m.f. (1–100 millivolts) is applied: the end point is usually shown by either the disappearance or the appearance of a current flowing between the two electrodes. For the method to be applicable the only requirement is that a reversible oxidation–reduction system be present either before or after the end point.

A simple apparatus suitable for this procedure is shown in Fig. 16.17. B is a 3 volt torch battery or 2 volt accumulator, M is a micro-ammeter, R is a 500 ohm, 0.5 watt potentiometer, and E,E are platinum electrodes. The potentiometer is set so that there is a potential drop of about 80–100 millivolts across the electrodes.

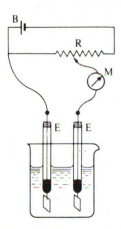

Fig. 16.17

In a titration with two indicator electrodes and when the reactant involves a reversible system (e.g. $I_2 + 2e \rightleftharpoons 2I^-$), an appreciable current flows through the cell. The amount of oxidised form reduced at the cathode is equal to that formed by oxidation of the reduced form at the anode. Both electrodes are depolarised until the oxidised component or the reduced component of the system has been consumed by a titrant. After the end point, only one electrode remains depolarised if the titrant (e.g. thiosulphate ion, $2S_2O_3^{2-} \rightarrow S_4O_6^{2-} + 2e$) does not involve a reversible system. Current thus flows until the end point: at or after the end point the current is zero or virtually zero. In the determination of iodine by titration with thiosulphate a rapid decrease in current is observed in the neighbourhood of the end point and this has led to the name '**dead-stop end point**'. The complementary type of end point, which resembles a reversed **L**-type amperometric graph, is probably more desirable in practice, and is obtained in the titration of an irreversible couple (say, thiosulphate) by a

635

reversible couple (say, iodine): the current is very low before the end point, and a very rapid increase in current signals the end point. When both systems are reversible [e.g. iron(II) ions with cerium(IV) or permanganate ions; applied potential 100 millivolts], the current is zero or close to zero at the equivalence point and a **V**-shaped titration graph results.

16.33 TITRATION OF THIOSULPHATE WITH IODINE ('DEAD-STOP END POINT')

Reagents required. Prepare a *ca* 0.001 *M* sodium thiosulphate solution and also a standard 0.005 *M* iodine solution.

Procedure. Pipette 25.0 mL of the thiosulphate solution into the titration cell e.g. a 150 mL Pyrex beaker. Insert two similar platinum wire or foil electrodes* into the cell and connect to the apparatus of Fig. 16.17. Apply 0.10 volt across the electrodes. Adjust the range of the micro-ammeter to obtain full-scale deflection for a current of 10–25 milliamperes. Stir the solution with a magnetic stirrer. Add the iodine solution from a 5 mL semimicro burette slowly in the usual manner and read the current (galvanometer deflection) after each addition of the titrant. When the current begins to increase, stop the addition; then add the titrant by small increments of 0.05 or 0.10 mL. Plot the titration graph, evaluate the end point, and calculate the concentration of the thiosulphate solution. It will be found that the current is fairly constant until the end point is approached and increases rapidly beyond.

16.34 DETERMINATION OF NITRATE

Discussion. 'Dead-stop' end point titrimetry may be applied to the determination of nitrate ion by titration with ammonium iron(II) sulphate solution in a strong sulphuric acid medium:

$$4FeSO_4 + 2HNO_3 + 2H_2SO_4 = 2Fe_2(SO_4)_3 + N_2O_3 + 3H_2O$$

Two platinum electrodes are immersed in sulphuric acid of suitable concentration containing the nitrate ion to be determined and a potential of about 100 millivolts is applied. Upon titration with 0.4 *M* ammonium iron(II) sulphate solution there is an initial rise in current followed by a gradual fall, with a marked increase at the end point: the latter is easily determined from a plot of current against volume of iron solution added. The concentration of water should not be allowed to rise above 25 per cent (w/w). The temperature of the solution should not exceed 40 °C.

Reagents. *Sulphuric acid, ca 25 per cent v/v (Solution A)*. Add cautiously 250 mL of concentrated sulphuric acid to 750 mL of water, cool, and dilute to 1 L. **(Take care with this addition.)**

  Ammonium iron(II) sulphate solution, ca 0.4M. Dissolve about 15.6 g, accurately weighed, of ammonium iron(II) sulphate in 100 mL of Solution A.

  Potassium nitrate solution, ca 0.3M. Dissolve about 3.0 g, accurately weighed,

---

* A length of 6–7 mm of platinum wire of 0.5 mm diameter sealed into a glass tube is satisfactory; electrical connection is made by means of a copper wire dipping into a little mercury in contact with the platinum wire.

of potassium nitrate in a small volume of Solution A and dilute to 100 mL in a graduated flask with concentrated sulphuric acid.

Determine the end point from the plot of current against volume of iron reagent. Calculate the weight of nitrate ion equivalent to 1.0 mL of the $0.4M$ iron solution.

When dealing with small amounts of nitrate ion it is advisable to pass a current of pure nitrogen through the solution before commencing the titration, and to maintain an atmosphere of nitrogen in the flask throughout the titration.

If chloride is present, saturated aqueous silver acetate solution should be added in amounts slightly more than the calculated quantity prior to the addition of concentrated sulphuric acid. The procedure may be applied to the routine analysis of mixtures of nitric and sulphuric acids, and to the determination of nitrogen in esters such as nitroglycerine and nitrocellulose; the latter are easily hydrolysed by strong sulphuric acid after dispersal in glacial acetic acid.

16.35 DETERMINATION OF WATER WITH THE KARL FISCHER REAGENT

For the determination of small amounts of water, Karl Fischer (1935) proposed a reagent prepared by the action of sulphur dioxide upon a solution of iodine in a mixture of anhydrous pyridine and anhydrous methanol. Water reacts with this reagent in a two-stage process in which one molecule of iodine disappears for each molecule of water present:

$$3C_5H_5N + I_2 + SO_2 + H_2O = 2C_5H_5NH^+I^- + C_5H_5\overset{+}{N}\diagup\underset{^-O}{\overset{SO_2}{|}} \qquad (a)$$

$$C_5H_5\overset{+}{N}\diagup\underset{^-O}{\overset{SO_2}{|}} + CH_3OH = C_5H_5N\diagdown\diagup\underset{H}{\overset{OSO_2OCH_3}{}} \qquad (b)$$

The end point of the reaction is conveniently determined electrometrically using the dead-stop end point procedure. If a small e.m.f. is applied across two platinum electrodes immersed in the reaction mixture a current will flow as long as free iodine is present, to remove hydrogen and depolarise the cathode. When the last trace of iodine has reacted the current will decrease to zero or very close to zero. Conversely, the technique may be combined with a direct titration of the sample with the Karl Fischer reagent: here the current in the electrode circuit suddenly increases at the first appearance of unused iodine in the solution.

The original Karl Fischer reagent prepared with an excess of methanol was somewhat unstable and required frequent standardisation. It was found that the stability was improved by replacing the methanol by 2-methoxyethanol.

The method is clearly confined to those cases where the test substance does not react with either of the components of the reagent, nor with the hydrogen iodide which is formed during the reaction with water: the following compounds interfere in the Karl Fischer titration.

637

(1) Oxidising agents, such as chromates, dichromates, copper(II) and iron(III) salts, higher oxides, and peroxides

$$MnO_2 + 4C_5H_5NH^+ + 2I^- = Mn^{2+} + 4C_5H_5N + I_2 + 2H_2O$$

(2) Reducing agents, such as thiosulphates, tin(II) salts and sulphides.
(3) Compounds which can be regarded as forming water with the components of the Karl Fischer reagent, for example:
   (a) basic oxides —

$$ZnO + 2C_5H_5NH^+ = Zn^{2+} + 2C_5H_5N + H_2O;$$

   (b) salts of weak oxy-acids —

$$NaHCO_3 + C_5H_5NH^+ = Na^+ + H_2O + CO_2 + C_5H_5N$$

The Karl Fischer procedure was applied to the determination of water present in hydrated salts or adsorbed on the surface of solids. The procedure, where applicable, was more rapid and direct than the commonly used drying process. A sample of the finely powdered solid, containing 5–10 millimoles (90–180 mg) of water, was dissolved or suspended in 25 mL of dry methanol in a 250-mL glass-stoppered graduated flask. The mixture was titrated with standard Karl Fischer reagent to the usual electrometric end point. A blank titration was also carried out on a 25 mL sample of the methanol used to determine what correction (if any) needed to be applied to the titre obtained with the salt.

The Karl Fischer procedure has now been simplified and the accuracy improved by modification to a coulometric method (Chapter 14). In this procedure the sample under test is added to a pyridine–methanol solution containing sulphur dioxide and a soluble iodide. Upon electrolysis, iodine is liberated at the anode and reactions (a) and (b) then follow: the end point is detected by a pair of electrodes which function as a biamperometric detection system and indicate the presence of free iodine. Since one mole of iodine reacts with one mole of water it follows that 1 mg of water is equivalent to 10.71 coulombs.

With this simplified procedure it is much easier to protect the system from atmospheric moisture, which must obviously be excluded. Modern K-F Titrators are equipped with special titration vessels which are designed to prevent the ingress of atmospheric moisture. Many also have microprocessors attached which will carry out the requisite operations automatically and will often provide a print-out of the results including the percentage moisture content.

The determination of adsorbed water on solids, and of the water of crystallisation of many salt hydrates, can be effected as in the original Karl Fischer method by adding a known weight of the solid to the reagent, and then carrying out the coulometric titration. To deal with materials which may contain substances which interfere with the Karl Fischer reaction, many modern instruments are provided with ancillary water vaporisation units in which the material can be heated in an atmosphere of carefully dried nitrogen, thus vaporising the water. The gas stream leaving the vaporiser is passed through the titration cell containing K-F reagent where the water is absorbed and then determined by coulometric titration.

16.36   THE DETERMINATION OF GLUCOSE USING AN ENZYME ELECTRODE

In presence of the enzyme glucose oxidase, an aqueous solution of glucose undergoes oxidation to gluconic acid with formation of hydrogen peroxide which can be determined by anodic oxidation at a fixed potential.

$$C_6H_{12}O_6 + O_2 \xrightarrow[\text{Phosphate buffer}]{\text{Enzyme}} \text{Gluconic acid} + H_2O_2$$

The enzyme is used in an enzyme electrode in which a tube is sealed at its lower end with a cellulose acetate membrane. An outer membrane of collagen is also attached to the end of the electrode tube and glucose oxidase enzyme is contained in the space between the two diaphragms.

When the electrode is placed in an aqueous solution of glucose which has been suitably diluted with a phosphate buffer solution (pH 7.3), solution passes through the outer membrane into the enzyme where hydroxen peroxide is produced. Hydrogen peroxide can diffuse through the inner membrane which, however, is impermeable to other components of the solution. The electrode vessel contains phosphate buffer, a platinum wire and a silver wire which act as electrodes. A potential of 0.7 volts is applied to the electrodes (the apparatus shown in Fig. 16.17 is suitable) with the platinum wire as anode. At this electrode the reaction $H_2O_2 \rightarrow O_2 + 2H^+ + 2e$ takes place, and the oxygen produced is reduced at the silver cathode:

$$\tfrac{1}{2}O_2 + 2H^+ + 2e \rightarrow H_2O$$

After a short time to allow for equilibration, the current settles down to a steady value, the magnitude of which is governed by the concentration of hydrogen peroxide in the electrode, and this is proportional to the glucose concentration of the test solution. By making readings with a series of standard glucose solutions (prepared in the same phosphate buffer solution), and then plotting the observed steady currents against concentration of glucose, the unknown concentration can be deduced (see Ref. 52, and Section 15.10).

16.37   THE CLARK CELL FOR DETERMINATION OF OXYGEN

A gold disc to which a gold wire is attached is placed inside a glass tube about 1.5 cm in diameter. A thin plastic film (Teflon is suitable) is stretched tightly over the end of the tube and fixed in position by an O-ring. A conducting solution is placed in the tube (0.1 M potassium chloride), and a silver–silver chloride electrode inserted, the latter consisting of a silver wire coated with silver chloride by electrolysis of a chloride solution. The lower end of the silver wire is coiled into a helix which is placed around (but does not touch) the gold disc and lower end of the gold wire: both wires are inserted through a plastic closure which serves to seal the top end of the tube.

When the tube is placed in a solution which contains oxygen, oxygen will pass through the membrane to the internal solution, and upon application of a voltage (0.6–0.8 V) to the two electrodes, the oxygen undergoes reduction at the gold cathode:

$$\tfrac{1}{2}O_2 + 2H^+ + 2e \rightarrow H_2O$$

The amperometric current thus produced is read on a micro-ammeter: when a

639

steady current is obtained it is controlled by the rate of diffusion of oxygen to the cathode and this is determined by the concentration of dissolved oxygen inside the cell, which in turn is related to the oxygen content of the test solution. Calibration is carried out by making readings with solutions which have been saturated with oxygen at varying partial pressures.

The apparatus is sometimes referred to as an 'oxygen electrode', but it is actually a cell. Although the Teflon membrane is impermeable to water and, therefore, to most substances dissolved in water, dissolved gases can pass through, and gases, such as chlorine, sulphur dioxide and hydrogen sulphide, can affect the electrode. The apparatus can be made readily portable and it is, therefore, of value for use in the field and can be used to monitor the oxygen content of rivers and lakes (see Ref. 53).

16.38   REFERENCES FOR PART E

1. D A Skoog, *Principles of Instrumental Analysis*, 3rd edn, Saunders College Publishing, Philadelphia, 1985
2. J J Lingane, *Electroanalytical Chemistry*, 2nd edn, Interscience, New York, 1958
3. I M Kolthoff and P J Elving, *Treatise on Analytical Chemistry*, Part I, Vol 4, Wiley, New York, 1959
4. *Wilson and Wilson's Comprehensive Analytical Chemistry*, Vol IIA, Elsevier, Amsterdam, 1964
5. H H Willard, L L Merrit, Jr, J A Dean and F A Settle, Jr, *Instrumental Methods of Analysis*, 6th edn, Van Nostrand, Princeton, 1981
6. A J Bard and L R Faulkner *Electrochemical Methods: Fundamentals and Applications*, J. Wiley and Sons, New York, 1980
7. G D Christian and J E O'Reilly, *Instrumental Analysis*, 2nd edn, Allyn and Bacon, Boston, 1986
8. R Greef, R Peat, L M Peter, D Pletcher and J Robinson, *Instrumental Methods in Electrochemistry*, Ellis Horwood, Chichester, 1985
9. B H Vassos and G W Ewing *Electroanalytical Chemistry*, *John Wiley and Sons*, New York, 1983
10. J A Maxwell and R P Graham, The mercury cathode, *Chem. Rev.* 1950, **46**, 471
11. J J Lingane and S Jones, *Anal. Chem.*, 1951, **23**, 1804
12. G Jones and B C Bradshaw, *J. Am. Chem. Soc.*, 1933, **55**, 1780
13. J D Mann and P D Jamieson, *Lab. Pract.*, 1981, **30**, 1229
14. *Wilson and Wilson's Comprehensive Analytical Chemistry*, Vol. XXII, Elsevier Science, Amsterdam, 1983
15. D E Johnson and C G Enke, *Anal. Chem.*, 1970, **42**, 329
16. J S Fritz, D T Gjerde and C Phlandt, *Ion Chromatography*, Huthig, Heidelberg, 1982
17. R J Anderson and R C Hall, *Am. Lab.*, 1980, **12**, 108
18. W J Blaedel and D L Petitjean. In *Physical Methods of Chemical Analysis*, Vol 3, W C Berl (Ed), Academic Press, New York, 1956
19. E Pungor, *Oscillometry and Conductometry*, Pergamon Press, Oxford, 1965
20. R M Fuoss and F Accascina, *Electrolytic Conductance*, Van Nostrand, New York, 1959
21. J E Harrar. In *Electroanalytical Chemistry*, Vol 8, A J Bard (Ed), Marcel Dekker, New York, 1975
22. *Wilson and Wilson's Comprehensive Analytical Chemistry*, Vol 11D, Elsevier, Amsterdam, 1975
23. T Meites and L Meites, *Anal. Chem.*, 1955, **27**, 1531; 1956, **28**, 103
24. A N Strohl and D J Curran, *Anal. Chem.*, 1979, **51**, 353, 1050
25. J Ruzicka and E H Hansen, *Flow Injection Analysis*, John Wiley, New York, 1981

26. C N Reilley, *Rev. Pure Appl. Chem.*, 1968, **18**, 137
27. F B Stephens, F Jakob, R Rigdon and J E Harrar, *Anal. Chem.*, 1970, **42**, 764
28. G Mattock, *pH Measurement and Titration*, Heywood, London, 1961
29. I M Kolthoff and P J Elving, *Treatise on Analytical Chemistry*, Pt I, Vol 1, Wiley, New York, 1978
30. *Wilson and Wilson's Comprehensive Analytical Chemistry*, Vol. XXII, Elsevier Science, Amsterdam, 1986
31. J Marlow, (*a*) Care and Maintenance of pH and redox electrodes; (*b*) Care and Maintenance of Ion-Selective Electrodes. In *International Laboratory*, Vol XII, Part 2, 1987
32. E Pungor, J Havas and K Tóth, *Zeit. Chem.*, 1965, **5**, 9
33. (*a*) G J Moody, R B Oke and J D R Thomas, *Analyst*, 1970, **75**, 910; (*b*) A Craggs, G J Moody and J D R Thomas, *J. Chem. Educ.*, 1974, **51**, 541
34. J Koryta (Ed), *Use of Enzyme Electrodes in Biomedical Investigations*, Wiley, Chichester, 1980
35. H Freiser (Ed), *Ion-Selective Electrodes in Analytical Chemistry*, Vol 2, Plenum, New York, 1980
36. J Janata and R J Huber, *Ion Selective Electrode Rev.*, 1979, **1**, 31
37. A Haemmerli, J Janata and A M Brown, *Anal. Chem.*, 1980, **52**, 1179
38. B Walter, *Anal. Chem.*, 1983, **55**, 508A
39. IUPAC *Manual of Symbols and Terminology for Physicochemical Quantities and Units*, Butterworth, London, 1969
40. G Gran, *Analyst*, 1952, **77**, 661
41. J Heyrovsky, *Chemicke Listy*, 1922, **16**, 256
42. J Heyrovsky and M Shikata, *Rec. Trav. Chim.*, 1925, **44**, 496
43. D Ilkovič, *Coll. Czech. Chem. Comm.*, 1934, **6**, 498
44. D R Crow, *Polarography of Metal Complexes*, Academic Press, London, 1969
45. G C Barker and I L Jenkins, *Analyst*, 1952, **77**, 685
46. A M Bond, *Modern Polarographic Methods in Analytical Chemistry*, Marcel Dekker, New York, 1980
47. P Zuman, *Topics in Organic Polarography*, Plenum, New York, 1970
48. W F Smyth, *Polarography of Molecules of Biological Significance*, Academic Press, New York, 1979
49. Ref. 7, p 60
50. P M Bersier and J Bersier, *Analyst*, 1988, **113**, 3
51. W Kemula and Z Kublik, *Anal. Chim. Acta*, 1958, **18**, 104
52. G Sittapalam and G S Wilson, *J. Chem. Ed.*, 1982, **59**, 70
53. I Fatt, *Polarographic Oxygen Sensors*, CRC Press, Cleveland, OH, 1976

16.39 SELECTED BIBLIOGRAPHY FOR PART E

1. W J Albery, *Electrode Processes*, Clarendon, Oxford, 1975
2. P L Bailey, *Analysis with Ion-Selective Electrodes*, 2nd edn, Heyden, London, 1980
3. A J Bard (Ed), *Electroanalytical Chemistry: A Series of Advances*, Marcel Dekker, New York, 1966–
4. A J Bard and L R Faulkner, *Electrochemical Methods*, Wiley–Interscience, New York, 1980
5. A M Bond, *Modern Polarographic Methods in Analytical Chemistry*, Marcel Dekker, New York, 1980
6. H Freiser (Ed), *Ion-Selective Electrodes in Analytical Chemistry*, Vols 1, 2, Plenum, New York, 1980

7. R Kalvoda, *Electroanalytical Methods in Chemical and Environmental Analysis*, Plenum, New York, 1987
8. P T Kissinger and W R Heineman, *Laboratory Techniques in Electroanalytical Chemistry*, Marcel Dekker, New York, 1984
9. J A Plambeck, *Electroanalytical Chemistry*, Wiley, New York, 1982
10. J Robbins, *Ions in Solution*, Clarendon, Oxford, 1972
11. D T Sawyer and J L Roberts, *Experimental Electrochemistry for Chemists*, Wiley–Interscience, New York, 1984
12. Southampton Electrochemistry Group, *Instrumental Methods in Electrochemistry*, Ellis Horwood, Chichester, 1985
13. B H Vassos and G W Ewing, *Electroanalytical Chemistry*, Wiley, New York, 1983
14. J Wang, *Stripping Analysis: Principles, Instrumentation and Applications*, Verlag Chemie, Weinham, 1985
15. H H Willard, L I Merritt Jr, J A Dean and F A Settle, Jr *Instrumental Methods of Analysis*, 6th edn, Van Nostrand, Princeton, 1981

The following ACOL texts (John Wiley & Sons, Chichester and New York, 1987) are useful for students.

16. T Riley and C Tomlinson, *Principles of Electroanalytical Methods*
17. A Evans, *Potentiometry and Ion-Selective Electrodes*
18. T Riley and A Watson, *Polarography and Other Voltammetric Methods*

PART F
SPECTROANALYTICAL METHODS

CHAPTER 17
COLORIMETRY AND SPECTROPHOTOMETRY

17.1 GENERAL DISCUSSION

The variation of the colour of a system with change in concentration of some component forms the basis of what the chemist commonly terms **colorimetric analysis**. The colour is usually due to the formation of a coloured compound by the addition of an appropriate reagent, or it may be inherent in the desired constituent itself. The intensity of the colour may then be compared with that obtained by treating a known amount of the substance in the same manner.

Colorimetry is concerned with the determination of the concentration of a substance by measurement of the relative absorption of light with respect to a known concentration of the substance. In **visual colorimetry**, natural or artificial white light is generally used as a light source, and determinations are usually made with a simple instrument termed a **colorimeter**, or colour comparator. When the eye is replaced by a photoelectric cell (thus largely eliminating the errors due to the personal characteristics of each observer) the instrument is termed a **photoelectric colorimeter**. The latter is usually employed with light contained within a comparatively narrow range of wavelengths furnished by passing white light through filters, i.e. materials in the form of plates of coloured glass, gelatin, etc., transmitting only a limited spectral region: the name **'filter photometer'** is sometimes applied to such an instrument.

In **spectrophotometric analysis** a source of radiation is used that extends into the ultraviolet region of the spectrum. From this, definite wavelengths of radiation are chosen possessing a bandwidth of less than 1 nm. This process necessitates the use of a more complicated and consequently more expensive instrument. The instrument employed for this purpose is a **spectrophotometer**.

An optical spectrometer is an instrument possessing an optical system which can produce dispersion of incident electromagnetic radiation, and with which measurements can be made of the quantity of transmitted radiation at selected wavelengths of the spectral range. A photometer is a device for measuring the intensity of transmitted radiation or a function of this quantity. When combined in the spectrophotometer the spectrometer and photometer are employed conjointly to produce a signal corresponding to the difference between the transmitted radiation of a reference material and that of a sample at selected wavelengths.

The chief advantage of colorimetric and spectrophotometric methods is that they provide a simple means for determining minute quantities of substances. The upper limit of colorimetric methods is, in general, the determination of constituents which are present in quantities of less than 1 or 2 per cent.

The sensitivity can, however, be improved if the technique of **derivative spectrophotometry** (Section 17.12) is employed. The development of inexpensive photoelectric colorimeters has placed this branch of instrumental chemical analysis within the means of even the smallest teaching institution.

In this chapter we are concerned with analytical methods that are based upon the absorption of electromagnetic radiation. Light consists of radiation to which the human eye is sensitive, waves of different wavelengths giving rise to light of different colours, while a mixture of light of these wavelengths constitutes white light. White light covers the entire visible spectrum 400–760 nm. The approximate wavelength ranges of colours are given in Table 17.1. The visual perception of colour arises from the selective absorption of certain wavelengths of incident light by the coloured object. The other wavelengths are either reflected or transmitted, according to the nature of the object, and are perceived by the eye as the colour of the object. If a solid opaque object appears white, all wavelengths are reflected equally; if the object appears black, very little light of any wavelength is reflected; if it appears blue, the wavelengths that give the blue stimulus are reflected, etc.

Table 17.1   Approximate wavelengths of colours

Ultraviolet	<400 nm	Yellow	570–590 nm
Violet	400–450 nm	Orange	590–620 nm
Blue	450–500 nm	Red	620–760 nm
Green	500–570 nm	Infrared	>760 nm

It must be emphasised that the range of electromagnetic radiation extends considerably beyond the visible region. The approximate limits of wavelength and frequency for the various types of radiation, including the frequency range of sound waves, are shown in Fig. 17.1 (not drawn to scale); this may be regarded as an electromagnetic spectrum. It will be seen that γ-rays and X-rays have very short wavelengths, while ultraviolet, visible, infrared and radio waves have progressively longer wavelengths. For colorimetry and spectrophotometry, the visible region and the adjacent ultraviolet region are of major importance.

Electromagnetic waves are usually described in terms of (a) wavelength λ (distance between the peaks of waves in cm, unless otherwise specified), (b) wavenumber $\tilde{v}$ (number of waves per cm), and (c) the frequency v (number of waves per second). The three quantities are related as follows:

$$\frac{1}{\text{Wavelength}} = \text{Wavenumber} = \frac{\text{Frequency}}{\text{Velocity of light}}$$

$$\frac{1}{\lambda} = \tilde{v} = \frac{v}{c}$$

The units in common use are:

1 Ångstrom unit $= 1 \text{ Å} = 10^{-10} \text{ metre} = 10^{-8} \text{ cm}$

1 nanometre $= 1 \text{ nm} = 10 \text{ Å} = 10^{-7} \text{ cm}$

1 micrometre $= 1 \text{ } \mu\text{m} = 10^4 \text{ Å} = 10^{-4} \text{ cm}$

Velocity of light $= c = 2.99793 \times 10^8 \text{ m s}^{-1}$

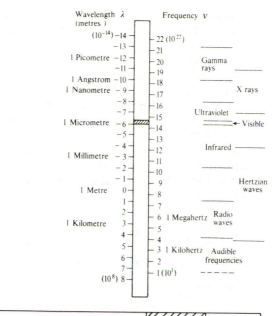

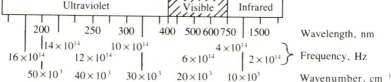

Fig. 17.1

Wavenumber $\tilde{v} = 1/\lambda$ waves per cm

Frequency $v = c/\lambda \approx 3 \times 10^{10}/\lambda$ waves per second

To comply completely with SI units these functions should be calculated using the metre as the basic unit. It is, however, still common practice to use centimetres for this purpose.

17.2 THEORY OF SPECTROPHOTOMETRY* AND COLORIMETRY

When light (monochromatic or heterogeneous) falls upon a homogeneous medium, a portion of the incident light is reflected, a portion is absorbed within the medium, and the remainder is transmitted. If the intensity of the incident light is expressed by I_0, that of the absorbed light by I_a, that of the transmitted light by I_t, and that of the reflected light by I_r, then:

$$I_0 = I_a + I_t + I_r$$

---

* Spectrophotometry proper is mainly concerned with the following regions of the spectrum: ultraviolet, 185–400 nm; visible 400–760 nm; and infrared, 0.76–15 μm. Colorimetry is concerned with the visible region of the spectrum. In this chapter attention will be confined largely to the visible and near ultraviolet region of the spectrum.

For air–glass interfaces arising from the use of glass cells, it may be stated that about 4 per cent of the incident light is reflected. I_r is usually eliminated by the use of a control, such as a comparison cell, hence:

$$I_0 = I_a + I_t \qquad (1)$$

Credit for investigating the change of absorption of light with the thickness of the medium is frequently given to Lambert,[1] although he really extended concepts originally developed by Bouguer.[2] Beer[3] later applied similar experiments to solutions of different concentrations and published his results just prior to those of Bernard.[4] This very confusing story has been explained by Malinin and Yoe.[5] The two separate laws governing absorption are usually known as Lambert's Law and Beer's Law. In the combined form[6] they are referred to as the Beer–Lambert Law.

Lambert's Law. This law states that when monochromatic light passes through a transparent medium, the rate of decrease in intensity with the thickness of the medium is proportional to the intensity of the light. This is equivalent to stating that the intensity of the emitted light decreases exponentially as the thickness of the absorbing medium increases arithmetically, or that any layer of given thickness of the medium absorbs the same fraction of the light incident upon it. We may express the law by the differential equation:

$$-\frac{dI}{dl} = kI \qquad (2)$$

where I is the intensity of the incident light of wavelength λ, l is the thickness of the medium, and k is a proportionality factor. Integrating equation (2) and putting $I = I_0$ when $l = 0$, we obtain:

$$\ln \frac{I_0}{I_t} = kl$$

or, stated in other terms,

$$I_t = I_0 \cdot e^{-kl} \qquad (3)$$

where I_0 is the intensity of the incident light falling upon an absorbing medium of thickness l, I_t is the intensity of the transmitted light, and k is a constant for the wavelength and the absorbing medium used. By changing from natural to common logarithms we obtain:

$$I_t = I_0 \cdot 10^{-0.4343kl} = I_0 \cdot 10^{-Kl} \qquad (4)$$

where $K = k/2.3026$ and is usually termed the **absorption coefficient**. The absorption coefficient is generally defined as the reciprocal of the thickness (l cm) required to reduce the light to $\frac{1}{10}$ of its intensity. This follows from equation (4), since:

$$I_t/I_0 = 0.1 = 10^{-Kl} \quad \text{or} \quad Kl = 1 \quad \text{and} \quad K = 1/l$$

The ratio I_t/I_0 is the fraction of the incident light transmitted by a thickness l of the medium and is termed the **transmittance** T. Its reciprocal I_0/I_t is the **opacity**, and the **absorbance** A of the medium (formerly called the **optical density**

D or **extinction** E) is given by:*

$$A = \log I_0/I_t \tag{5}$$

Thus a medium with absorbance 1 for a given wavelength transmits 10 per cent of the incident light at the wavelength in question.

Beer's Law. We have so far considered the light absorption and the light transmission for monochromatic light as a function of the thickness of the absorbing layer only. In quantitative analysis, however, we are mainly concerned with solutions. Beer studied the effect of concentration of the coloured constituent in solution upon the light transmission or absorption. He found the same relation between transmission and concentration as Lambert had discovered between transmission and thickness of the layer [equation (3)], i.e. the intensity of a beam of monochromatic light decreases exponentially as the concentration of the absorbing substance increases arithmetically. This may be written in the form:

$$I_t = I_0 \cdot e^{-k'c}$$
$$= I_0 \cdot 10^{-0.4343k'c} = I_0 \cdot 10^{-K'c} \tag{6}$$

where c is the concentration, and k' and K' are constants. Combining equations (4) and (5), we have:[6]

$$I_t = I_0 \cdot 10^{-acl} \tag{7}$$

or

$$\log I_0/I_t = acl \tag{8}$$

This is the fundamental equation of colorimetry and spectrophotometry, and is often spoken of as the **Beer–Lambert Law**. The value of a will clearly depend upon the method of expression of the concentration. If c is expressed in mole L^{-1} and l in centimetres then a is given the symbol ε and is called the **molar absorption coefficient** or molar absorptivity (formerly the molar extinction coefficient).

The specific absorption (or extinction) coefficient E_s (sometimes termed absorbancy index) may be defined as the absorption per unit thickness (path length) and unit concentration.

Where the molecular weight of a substance is not definitely known, it is obviously not possible to write down the molecular absorption coefficient, and in such cases it is usual to write the unit of concentration as a superscript, and the unit of length as a subscript. Thus

$$E_{1\,cm}^{1\%}\ 325\,nm = 30$$

means that for the substance in question, at a wavelength of 325 nm, a solution of length 1 cm, and concentration 1 per cent (1 per cent by weight of solute *or* 1 g of solid per 100 mL of solution) $\log I_0/I_t$ has a value of 30.

It will be apparent that there is a relationship between the absorbance A, the transmittance T, and the molar absorption coefficient, since:

$$A = \varepsilon cl = \log \frac{I_0}{I_t} = \log \frac{1}{T} = -\log T \tag{9}$$

---

* The term 'internal transmission density' (D_i) has been proposed.

The scales of spectrophotometers are often calibrated to read directly in absorbances, and frequently also in percentage transmittance. It may be mentioned that for colorimetric measurements I_0 is usually understood as the intensity of the light transmitted by the pure solvent, or the intensity of the light entering the solution; I_t is the intensity of the light emerging from the solution, or transmitted by the solution. It will be noted that:

(a) The **absorption coefficient** (or extinction coefficient) is the absorbance for unit path length:

$$K = A/t \text{ or } I_t = I_0 \cdot 10^{-Kt}$$

(b) The **specific absorption coefficient** (or absorbancy index) is the absorbance per unit path length and unit concentration:

$$E_s = A/cl \text{ or } I_t = I_0 \cdot 10^{-E_s cl}$$

(c) The **molar absorption coefficient** is the specific absorption coefficient for a concentration of 1 mol L^{-1} and a path length of 1 cm.

$$\varepsilon = A/cl$$

Application of Beer's Law. Consider the case of two solutions of a coloured substance with concentrations c_1 and c_2. These are placed in an instrument in which the thickness of the layers can be altered and measured easily, and which also allows a comparison of the transmitted light (e.g. a Duboscq colorimeter, Section 17.5). When the two layers have the same colour intensity:

$$I_{t_1} = I_0 \cdot 10^{-\varepsilon l_1 c_1} = I_{t_2} = I_0 \cdot 10^{-\varepsilon l_2 c_2} \tag{10}$$

Here l_1 and l_2 are the lengths of the columns of solutions with concentrations c_1 and c_2 respectively when the system is optically balanced. Hence, under these conditions, and when Beer's law holds:

$$l_1 c_1 = l_2 c_2 \tag{11}$$

A colorimeter can, therefore, be employed in a dual capacity: (a) to investigate the validity of Beer's Law by varying c_1 and c_2 and noting whether equation (11) applies, and (b) for the determination of an unknown concentration c_2 of a coloured solution by comparison with a solution of known concentration c_1. It must be emphasised that equation (11) is valid only if Beer's Law is obeyed over the concentration range employed and the instrument has no optical defects.

When a spectrophotometer is used it is unnecessary to make comparison with solutions of known concentration. With such an instrument the intensity of the transmitted light or, better, the ratio I_t/I_0 (the transmittance) is found directly at a known thickness l. By varying l and c the validity of the Beer–Lambert Law, equation (9), can be tested and the value of ε may be evaluated. When the latter is known, the concentration c_x of an unknown solution can be calculated from the formula:

$$c_x = \frac{\log I_0/I_t}{\varepsilon l} \tag{12}$$

Attention is directed to the fact that the molar absorption coefficient ε depends upon the wavelength of the incident light, the temperature, and the solvent employed. In general, it is best to work with light of wavelength approximating to that for which the solution exhibits a maximum selective absorption (or minimum selective transmittance): the maximum sensitivity is thus attained.

For matched cells (i.e. l constant) the Beer–Lambert Law may be written:

$$c \propto \log \frac{I_0}{I_t}$$

$$c \propto \log \frac{1}{T}$$

or

$$c \propto A \tag{13}$$

Hence, by plotting A [or $\log(1/T)$] as ordinate, against concentration as abscissa, a straight line will be obtained and this will pass through the point $c = 0$, $A = 0$ ($T = 100$ per cent). This calibration line may then be used to determine unknown concentrations of solutions of the same material after measurement of absorbances.

Deviation from Beer's Law. Beer's Law will generally hold over a wide range of concentration if the structure of the coloured ion or of the coloured non-electrolyte in the dissolved state does not change with concentration. Small amounts of electrolytes, which do not react chemically with the coloured components, do not usually affect the light absorption; large amounts of electrolytes may result in a shift of the maximum absorption, and may also change the value of the extinction coefficient. Discrepancies are usually found when the coloured solute ionises, dissociates, or associates in solution, since the nature of the species in solution will vary with the concentration. The law does not hold when the coloured solute forms complexes, the composition of which depends upon the concentration. Also, discrepancies may occur when monochromatic light is not used. The behaviour of a substance can always be tested by plotting $\log I_0/I_t$, or $\log 1/T$ against the concentration: a straight line passing through the origin indicates conformity to the law.

For solutions which do not follow Beer's Law, it is best to prepare a calibration curve using a series of standards of known concentration. Instrumental readings are plotted as ordinates against concentrations in, say, mg per 100 mL or 1000 mL as abscissae. For the most precise work each calibration curve should cover the dilution range likely to be met with in the actual comparison.

17.3 CLASSIFICATION OF METHODS OF 'COLOUR' MEASUREMENT OR COMPARISON

The basic principle of most colorimetric measurements consists in comparing under well-defined conditions the colour produced by the substance in unknown amount with the same colour produced by a known amount of the material being determined. The quantitative comparison of these two solutions may, in general, be carried out by one or more of six methods. It is not essential to prepare a series of standards with the spectrophotometer; the molar absorption coefficient can be calculated from one measurement of the absorbance or transmittance of a standard solution, and the unknown concentration can then be calculated with the aid of the molar absorption coefficient and the observed value of the absorbance or transmittance [cf. Section 17.2, equations (12) and (13)].

A. Standard series method (Section 17.4). The test solution contained in a Nessler tube is diluted to a definite volume, thoroughly mixed, and its colour compared with a series of standards similarly prepared. The concentration of the unknown is then, of course, equal to that of the known solution whose colour it matches exactly. The accuracy of the method will depend upon the concentrations of the standard series; the probable error is of the order of ± 3 per cent, but may be as high as ± 8 per cent.

For convenience, artificial standards, e.g. Lovibond glasses or salt solutions, such as iron(III) chloride in aqueous hydrochloric acid (yellow), aqueous cobalt chloride (pink), aqueous copper sulphate (blue), and aqueous potassium dichromate (orange) are sometimes used. It is essential to standardise the artificial standards against known amounts of the substance being determined, the latter always being treated under exactly similar conditions. The disadvantage of this method is that the spectral absorption curves of the test solutions and of the sub-standard glasses or solutions may be far from identical; the error due to this cause is greatly magnified in the case of observers suffering from partial colour blindness.

B. Duplication method. This is usually applied as the so-called **colorimetric titration** in which a known volume (x mL) of the test solution is treated in a Nessler cylinder with a measured volume (y mL) of appropriate reagent so that a colour is developed. Distilled water (x mL) is placed in a second Nessler cylinder together with y mL of reagent. A standard solution of the substance under test is now added to the second cylinder from a microburette until the colour developed matches that in the first tube; the concentration of the test solution can then be calculated. The standard solution should be of such concentration that it amounts to no more than 2 per cent of the final solution. This method is only approximate but has the merit that only the simplest apparatus is required: it will not be discussed further.

C. Dilution method. The sample and standard solution are contained in glass tubes of the same diameter, and are observed *horizontally* through the tubes. The more concentrated solution is diluted until the colours are identical in intensity when observed horizontally through the same thickness of solution. The relative concentrations of the original solutions are then proportional to the heights of the matched solutions in the tubes. This is the least accurate method of all, and will not be discussed further.

D. Balancing method (Section 17.5). This method forms the basis of all colorimeters of the plunger type, e.g. in the Duboscq colorimeter. The comparison is made in two tubes, and the height of the liquid in one tube is adjusted so that when both tubes are observed vertically the colour intensities in the tubes are equal. The concentration in one of the tubes being known, that in the other may be calculated from the respective lengths of the two columns of liquid and the relation [equation (11)]:

$$c_1 l_1 = c_2 l_2$$

It must be emphasised again that this simple proportionality holds only if Beer's law is applicable, and that the relation holds with greater exactness if a beam of monochromatic light (obtained with the aid of a suitable colour filter) rather than white light is employed. As a general rule, it is preferable that the solutions under comparison should not differ greatly in concentration, and for the most accurate work an empirically constructed calibration curve should be used. As usually employed with white light, the accuracy obtainable with a Duboscq colorimeter is of the order of ± 7 per cent; the accuracy is increased appreciably if monochromatic light is employed.

E. Photoelectric photometer method (Section 17.6). In this method the human eye is replaced by a suitable photoelectric cell; the latter is employed to afford a direct measure of the light intensity, and hence of the absorption. Instruments incorporating photoelectric cells measure the light absorption and not the colour of the substance: for this reason the term 'photoelectric colorimeters' is a misnomer; better names are photoelectric comparators, photometers, or, best, **absorptiometers**.

Essentially most such instruments consist of a light source, a suitable light filter to secure an approximation to monochromatic light (hence the name photoelectric filter photometer), a glass cell for the solution, a photoelectric cell to receive the radiation transmitted by the solution, and a measuring device to determine the response of the photoelectric cell. The comparator is first calibrated in terms of a series of solutions of known concentration, and the results plotted in the form of a curve connecting concentrations and readings of the measuring device employed. The concentration of the unknown solution is then determined by noting the response of the cell and referring to the calibration curve.

These instruments are available in a number of different forms incorporating one or two photocells. With the one-cell type, the absorption of light by the solution is usually measured directly by determining the current output of the photoelectric cell in relation to the value obtained with the pure solvent. It is of the utmost importance to use a light source of constant intensity, and if the photocells exhibit a 'fatigue effect' it is necessary to allow them to attain their equilibrium current after each change of light intensity. The two-cell type of filter photometer is usually regarded as the more trustworthy (provided the electrical circuit is appropriately designed) in that any fluctuation of the intensity of the light source will affect both cells alike if they are matched for their spectral response. Here the two photocells, illuminated by the same source of light, are balanced against each other through a galvanometer: the test solution is placed before one cell and the pure solvent before the other, and the current output difference is measured.

F. Spectrophotometric method (Section 17.7). This is undoubtedly the most accurate method for determining *inter alia* the concentration of substances in solution, but the instruments are, of necessity, more expensive. A spectrophotometer may be regarded as a refined filter photoelectric photometer which permits the use of continuously variable and more nearly monochromatic bands of light. The essential parts of a spectrophotometer are: (1) a source of radiant energy, (2) a monochromator, i.e. a device for isolating monochromatic light or, more accurately expressed, narrow bands of radiant energy from the light source, (3) glass or silica cells for the solvent and for the solution under test, and (4) a device to receive or measure the beam or beams of radiant energy passing through the solvent or solution.

In the following sections it is proposed to discuss the more important of the above methods in somewhat greater detail. For a more complete treatment the reader is referred to the special treatises on the subject (see the selected Bibliography in Section 21.27).

17.4 STANDARD SERIES METHOD

In this method colourless glass tubes of uniform cross-section and with flat bottoms are usually employed. These are termed **Nessler tubes**. The best variety have polished, flat bottoms. They are made in either the 'low' form with a height of 175–200 mm and a diameter of 25–32 mm or as a 'high' form with a height of 300–375 mm and a diameter of 21–24 mm. The solution of the substance being determined is made up to a definite volume, and the colour is compared with that of a series of standards prepared in the same way from known amounts of the component being determined. Volumes of 50 or 100 mL of the unknown and standard solutions are placed in Nessler tubes, and the solutions are viewed *vertically* through the length of the columns of the liquid. The concentration of the unknown is equal to that of the standard having the same colour.* As a general rule, it will be found that the colour intensity of the unknown lies between two successive standards. Another series of standards may then be prepared covering this range over smaller concentration intervals.

For the comparison of colours in Nessler tubes, the simplest apparatus consists of a modified wooden test-tube rack finished dull black, and provided with an inclined opal glass reflector or mirror, arranged to reflect light up through the tubes. The Nessler tubes rest on a narrow ledge, and do not come into contact with the reflector. The unknown and standards are compared by placing them adjacent to each other and looking vertically down through them.

This procedure serves as the basis for the colorimetric determination of pH by employing a series of buffer solutions and suitable indicators. A series of appropriate buffer solutions is selected, differing successively in pH by about 0.2, covering the pH range of the solutions under investigation; the range of the buffer solutions required will be indicated by a preliminary pH determination. Equal volumes, say 10 mL, of the buffer solutions differing successively in pH by about 0.2 are placed in test-tubes of colourless glass and having approximately the same dimensions, and a small equal quantity of a suitable indicator for the particular pH range is added to each tube. A series of different colours corresponding to the different pH values is thus obtained. An equal volume (say 10 mL) of the test solution is treated with an equal volume of indicator to that used for the buffer solutions, and the resulting colour is compared with that of the coloured standard buffer solutions. When a complete match is found, the test solution and the corresponding buffer solution have the same pH. This

---

* It is advisable, wherever possible, to make a preliminary determination of the concentration of the unknown solution by adding from a burette a solution of the component in known concentration to a Nessler tube containing the reagents diluted with a suitable amount of water until the depth of colour obtained is practically the same as that of an equal volume of the unknown solution also contained in a Nessler cylinder and standing at its side. A series of standards on either side of this concentration is then prepared.

procedure for determining the pH of solutions is now rarely used, having been supplanted by the potentiometric method using a pH meter (Chapter 16). Small compact units, 'stick meters' are available which make it possible to measure the pH of a solution very easily and with a minimum of preparation.

For turbid or slightly coloured solutions, the direct-comparison method given above can no longer be applied. The interference due to the coloured substance can be eliminated in a simple way by the following device, suggested by Walpole. In Fig. 17.2, A, B, C, and D are glass cylinders with plane bases standing in a box which is painted dull black on the inside. A contains the coloured solution to be tested (here the test solution + indicator), B contains an equal volume of water, C contains a solution of known strength for comparison (here the standard buffer solution + indicator), while D contains the same volume of the solution to be tested as was originally added to A. The colour of the unknown solution is thus compensated for.

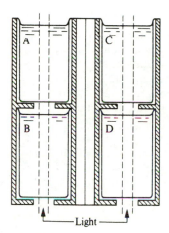

Fig. 17.2

Standard series using glass comparators. A number of devices are manufactured which employ permanent tinted glass standards mounted in special viewers. An example of such an instrument is the Lovibond '2000' Comparator* (Fig. 17.3). The discs containing the nine glass colour standards fit into the 'comparator', which is furnished with four compartments to receive small test-tubes or rectangular cells, and is also provided with an opal glass screen. The disc can revolve in the comparator, and each colour standard passes in turn in front of an aperture through which the solution in the cell (or cells) can be observed. As the disc revolves, the value of the colour standard visible in the aperture appears in a special recess.

The instrument is equipped with a prism so that the colour of the liquid under examination and that of the standard which has been selected on the disc are brought into juxtaposition, thus making it easier to compare the two. The 'Nesslerimeter' is a similar type of instrument, made in a tall form to accommodate Nessler tubes.

---

* Manufactured by Tintometer Ltd, Salisbury, UK.

 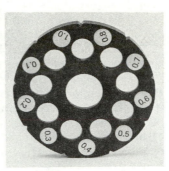

Fig. 17.3

Standards are available which cater for the performance of over 400 chemical analyses: these include the determination of several metals and of a number of organic substances, including many of biochemical significance.

The same firm also manufacture 'Tintometers' which enable the colour of a material to be expressed in terms of the proportion of each of the three primary colours red, yellow and blue which combine to give a matching hue. The instrument has a range of filters giving 700 gradations in yellow, a similar number in red, and 500 gradations in blue, so that a very precise assessment can be made of the synthesis of a particular shade. Colour is an important factor in the presentation of many products, both natural and manufactured. Uniformity of colour is regarded as an indication of quality in products as diverse as beverages, cosmetics, foods and paints.

17.5 BALANCING METHOD

Plunger-type colorimeters. The plunger-type of colorimeter with two halves of the field of view illuminated by the light passing through the unknown and standard solutions respectively was invented by J. Duboscq of Paris in 1854. Various improved modifications of the instrument were subsequently developed by manufacturers of optical apparatus.

The essential principles of a **Duboscq colorimeter** are illustrated in Fig. 17.4. Light from an even source of illumination concealed in the base of the instrument passes through the windows (matt white screens) in the top of the base through the solutions to be tested and through the plungers. Some of the light is absorbed in passing through the liquids, the amount of absorption being dependent upon the concentration and the depth of the solution. The two beams of light from the plungers are then brought to a common axis by a prism system. On looking through the eyepiece, a wide, circular field is visible, light from one cup illuminating one half, and light from the second cup illuminating the other half of the field. The depths of the columns of liquids are adjusted by rotating the milled heads on either side of the instrument, which raises and lowers the cups, until the two halves of the field are identical in intensity, i.e. until the dividing line practically disappears. When this condition holds and Beer's law is applicable, the concentrations of the two solutions are inversely proportional

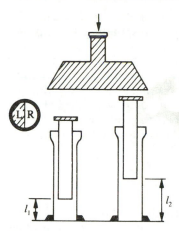

Fig. 17.4

to their depths, which are normally read on the scales attached to the cup carriers. The two scales are 60 cm long and are engraved on metal: they are divided into millimetres, and a vernier scale enables readings to be taken to within 0.1 mm.

Use of Duboscq-type colorimeter. The colorimeter must be kept scrupulously clean. The cups and plungers are rinsed with distilled water and either dried with soft lens-polishing material or rinsed with the solution to be measured.

Make sure that the readings are zero when the plungers just touch the bottoms of the cups. Place the standard solution in one cup, and an equal volume of the unknown solution in the other; do not fill the cups above the shoulder. Set the unknown solution at a scale reading of 10.0 mm and adjust the standard until the fields are matched. Carry out at least six adjustments with the cup containing the standard solution, and calculate the mean value.

The plungers should always remain below the surface of the liquid. Since the eye may become fatigued and unable to detect small differences, it is recommended after making an adjustment to close the eyes for a moment or to look at something else, and then see if the adjustment still appears satisfactory. It is advisable to approach the match point both from above and below.

If l_1 and l_2 are the average readings for the cups containing the solutions of known and unknown concentration respectively, and c_1 and c_2 are the corresponding concentrations, then if Beer's Law holds:

$$c_1 l_1 = c_2 l_2 \quad \text{or} \quad c_2 = c_1 \frac{l_1}{l_2}$$

It will be noted that if $l_2 = 10.0$, the standard scale when multiplied by 10 will give the percentage concentration of the sample in terms of the standard.

Owing to optical and mechanical imperfections of some makes of colorimeters, it is sometimes found that the same reading cannot be obtained in the adjustment for illumination when the cups are filled with the same solution and balanced. In such a case one of the cups (say, the left one) is filled with a reference solution (which may be a solution containing the component to be determined) of the same colour and approximately the same intensity as the unknown and the

657

plunger set at some convenient point (about the middle) of the scale. Fill the other cup with a solution having a colour corresponding to a known concentration of the component to be determined, and adjust this cup to colour balance. Take the reading and repeat the adjustment, say ten times, in such a way that the balancing point is approached five times from the lower and five times from the higher side. Calculate the average reading (l_1). Remove the cup, rinse it thoroughly, and fill it with the unknown solution. Repeat the balancing exactly as for the standard solution, and find the average of, say, ten readings (l_2). If c_1 is the concentration in the standard solution, then the concentration of the unknown solution is given by:

$$c_2 = c_1 \frac{l_1}{l_2}$$

(This method is comparable in many respects with the method of weighing by substitution.) If Beer's Law is not valid for the solution, it is best to arrange matters so that the colour intensity of the standard lies close to that of the unknown.

Immediately the determination has been completed, empty the cups and rinse both the cups and plungers with distilled water. Leave the colorimeter in a scrupulously clean condition.

17.6 PHOTOELECTRIC PHOTOMETER METHOD

Photoelectric colorimeters (absorptiometers). One of the greatest advances in the design of colorimeters has been the use of photoelectric cells to measure the intensity of the light, thus eliminating the errors due to the personal characteristics of each observer.

The photovoltaic or barrier-layer cell, in which light striking the surface of a semiconductor such as selenium mounted upon a base plate (usually iron) leads to the generation of an electric current, the magnitude of which is governed by the intensity of light beam, has been widely used in many absorptiometers. It, however, suffers from the defects that (1) amplification of the current produced by the cell is difficult to achieve which means that the cell does not have a very high sensitivity for low levels of light and (2) the cell tends to become fatigued. For these reasons it has now been largely superseded by the **photomultiplier** and by **silicon diode detectors**.

Photo-emissive cells. In the simplest form of photo-emissive cell (also called phototube) a glass bulb is coated internally with a thin, sensitive layer, such as caesium or potassium oxide and silver oxide (i.e. one which emits electrons when illuminated), a free space being left to permit the entry of the light. This layer is the cathode. A metal ring inserted near the centre of the bulb forms the anode, and is maintained at a high voltage by means of a battery. The interior of the bulb may be either evacuated or, less desirably, filled with an inert gas at low pressure (e.g. argon at about 0.2 mm). When light, penetrating the bulb, falls on the sensitive layer, electrons are emitted, thereby causing a current to flow through an outside circuit; this current may be amplified by electronic means, and is taken as a measure of the amount of light striking the photosensitive surface. Otherwise expressed, the emission of electrons leads to a potential drop across a high resistance (R) in series with the cell and the battery;

the fall in potential may be measured by a suitable potentiometer (M), and is related to the amount of light falling on the cathode. The action of the photo-emissive cell is shown diagrammatically in Fig. 17.5.

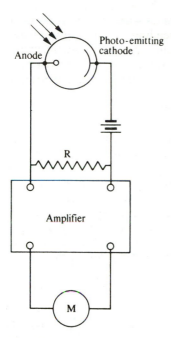

Fig. 17.5

The sensitivity of a photo-emissive cell (phototube) may be considerably increased by means of the so-called **photomultiplier tube**. The latter consists of an electrode covered with a photo-emissive material and a series of positively charged plates, each charged at a successively higher potential. The plates are covered with a material which emits several (2–5) electrons for each electron collected on its surface. When the electrons hit the first plate, secondary electrons are emitted in greater number than initially struck the plate, with the net result of a large amplification (up to 10^6) in the current output of the cell. The output of a photomultiplier tube is limited to several milliamperes, and for this reason only low incident radiant energy intensities can be employed. It can measure intensities about 200 times weaker than those measurable with an ordinary photoelectric cell and amplifier.

Phototube detectors are normally sensitive either to radiation of wavelength 200 nm to 600–650 nm, or of wavelength 600–1000 nm. To scan a complete spectral range an instrument must therefore contain two photocells; a 'red' sensitive cell (600–800 nm) and a 'blue' cell (200–600 nm).

The **silicon diode** (photodiode) detector consists of a strip of p-type silicon on the surface of a silicon chip (n-type silicon). By application of a biasing potential with the silicon chip connected to the positive pole of the biasing source, electrons and holes are caused to move away from the $p–n$ junction. This creates a depletion region in the neighbourhood of the junction which in effect becomes a capacitor. When light strikes the surface of the chip, free

659

electrons and holes are created which migrate to discharge the capacitor: the magnitude of the resultant current is a measure of the intensity of the light. Such a detector has a greater sensitivity than a single phototube, but less than that shown by a photomultiplier.

With modern technology it is possible to form a large number of such photodiodes on the surface of a single silicon chip. This chip also contains an integrating circuit which can scan each photodiode in turn to give a signal which is transmitted to a microprocessor. Each photodiode can be programmed to respond to a certain small band of wavelengths so that the complete spectrum can be scanned virtually instantaneously.[7]

When using a spectrophotometer which is equipped with a diode array, an absorption spectrum is obtained by electronic scanning rather than the mechanical scanning which occurs in a conventional spectrophotometer. This results in what is virtually instantaneous recording of the absorption curve; it may be accomplished in 1–5 seconds. In consequence, samples are exposed to radiation for such short periods of time that there is little possibility of photochemical reactions taking place, and the effects of fluorescence in the samples are minimised. The speed of operation makes such instruments useful for the investigation of fast chemical reactions, and for monitoring the eluate from liquid chromatographs. Diode detectors do however suffer from somewhat limited resolution; about 1 nm in the ultraviolet, and about 2 nm in the visible region.

17.7 WAVELENGTH SELECTION

Bearing in mind that the colour of a substance is related to its ability to absorb selectively in the visible region of the electromagnetic spectrum, it follows that having achieved the ability to measure the intensity of light with a high degree of accuracy, if we wish to analyse a solution by measuring the extent to which some coloured component absorbs light, the accuracy will be improved if measurements are made at the wavelength which is being absorbed. In this connection, it must be borne in mind that the observed colour is due to the radiation which is *not* absorbed, or in other words by the radiation which is transmitted by the coloured solution: the colour corresponding to this radiation is said to be **complementary** to the colour corresponding to that of the radiation which is being absorbed. Complementary colours are listed in Table 17.2. Procedures which can be used to select specified regions of the visible part of

Table 17.2 Complementary colours

Wavelength (nm)	Hue (transmitted)	Complementary hue
400–435	Violet	Yellowish-green
435–480	Blue	Yellow
480–490	Greenish-blue	Orange
490–500	Bluish-green	Red
500–560	Green	Purple
560–580	Yellowish-green	Violet
580–595	Yellow	Blue
595–610	Orange	Greenish-blue
610–750	Red	Bluish-green

the electromagnetic spectrum include the use of (1) filters, (2) prisms and (3) diffraction gratings.

Light filters. Optical filters are used in colorimeters (absorptiometers) for isolating any desired spectral region. They consist of either thin films of gelatin containing different dyes or of coloured glass.

Interference filters (transmission type). These have somewhat narrower transmitted bands than coloured filters and are essentially composed of two highly reflecting but partially transmitting films of metal (usually silver separated by a spacer film of transparent material). The amount of separation of the metal films governs the wavelength position of the pass band, and hence the colour of the light that the filter will transmit. This is the result of an optical interference effect which produces a high transmission of light when the optical separation of the metal films is effectively a half wavelength or a multiple of a half-wavelength. Light which is not transmitted is for the most part reflected. The wavelength region covered is either 253–390 nm or 380–1100 nm, peak transmission is between 25 and 50 per cent and the bandwidth is less than 18 nm for the narrow-band filters suitable for colorimetry.

Absorptiometers equipped with filters are now rarely used but they are inexpensive, and for certain specified measurements can be very satisfactory.

Prisms. To obtain improved resolution of spectra in both the visible and ultraviolet regions of the spectrum it is necessary to employ a better optical system than that possible with filters. In many instruments, both manual and automatic, this is achieved by using prisms to disperse the radiation obtained from incandescent tungsten or deuterium sources. The dispersion is dependent upon the fact that the refractive index, n, of the prism material varies with wavelength, λ, the dispersive power being given by $dn/d\lambda$. The separation achieved between different wavelengths is dependent upon both the dispersive power and the apical angle of the prism.

In instruments in which the radiation is only passed through the prism in a single direction it is common to use a 60° prism. In some cases double dispersion is achieved by reflecting the radiation back through the prism by placing a mirrored surface behind the prism, as in the Littrow mounting, Fig. 17.6. Monochromatic radiation of different wavelengths is brought to focus on the instrument slit by rotation of the prism.

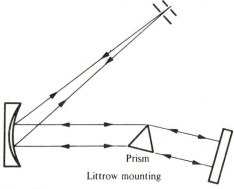

Prism

Littrow mounting

Fig. 17.6

Unfortunately no single material is entirely suitable for use over the full range of 200–1000 nm, although fused silica is the favourite compromise material. Glass prisms can be employed between 400 and 1000 nm for the visible region, but are not transparent to ultraviolet radiation. For the region below 400 nm quartz or fused silica prisms are required. If quartz is employed for a 60° single pass prism it is necessary to make the prism in two halves, one half from right-handed quartz and the other from left-handed quartz in order that polarisation effects introduced by one will be reversed by the other.

Prisms have the advantage that, unlike the diffraction gratings described below, they only produce a single-order spectrum.

Diffraction gratings. This alternative method of dispersion uses the principle of diffraction of radiation from a series of closely spaced lines marked on a surface. Early diffraction gratings were made of glass through which the radiation passed and became diffracted; these are known as transmission gratings. To achieve the diffraction of ultraviolet radiation, however, modern grating spectrophotometers employ metal reflection gratings with which the radiation is reflected from the surfaces of a series of parallel grooves. These are often known as echelette gratings.

The principle of diffraction is dependent upon the differences in path length experienced by a wavefront incident at an angle to the individual surfaces of the grooves of the grating. If i is the angle of incidence and r the angle of reflection the path difference between rays from adjacent grooves is given by

$$d \sin i - d \sin r$$

where d is the distance between the grooves, Fig. 17.7. Because of the path difference that is created, the new wavefronts interfere with each other except

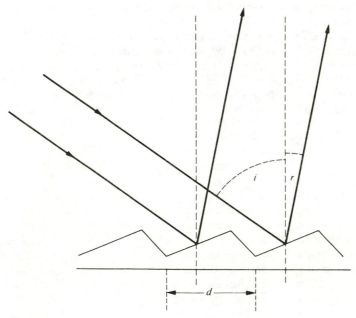

Fig. 17.7

when the path difference is an integral number of wavelengths, i.e. when

$$n\lambda = d(\sin i \pm \sin r) \tag{14}$$

When polychromatic radiation is incident upon the diffraction grating this equation can usually only be satisfied for a single wavelength at a time. Rotation of the grating to change the angle of incidence i will bring each wavelength in turn to a position to satisfy the equation, thus serving as a method of monochromation.

Diffraction gratings suffer from the disadvantage that they produce second-order and higher-order spectra which can overlap the desired first-order spectrum. This overlap is most commonly seen between the long-wavelength region of the first-order spectrum and the shorter-wavelength region of the second-order spectrum. The difficulty is overcome by using carefully positioned filters in the instrument to block the undesired wavelengths.

For ultraviolet/visible spectrophotometers the gratings employed have between 10 000 and 30 000 lines cm^{-1}. This very fine ruling means that the value of d in equation (14) is small and produces high dispersion between wavelengths in the first-order spectrum. Only a single grating is required to cover the region between 200 and 900 nm.

A recent development is the **holographic grating**. These can be produced by allowing an interference pattern produced by two monochromatic laser beams to impinge on a layer of photoresist material: subsequent photographic development produces a series of parallel grooves which constitute the grating. A reflective coating is applied, and if the grating is mounted on a flexible base, it can be produced in curved forms which serve to collimate light beams, thus making it possible to dispense with some lenses in the spectrophotometer. This method of production gives smoother lines than those made by a mechanical ruling engine and so less light is scattered and the resultant beam of light is of higher intensity.

Monochromation and bandwidth. From the foregoing discussion it is clear that for simple absorptiometers (colorimeters) where only the visual spectrum is involved, filters or prisms will suffice for selecting the appropriate spectral region for a given determination. On the other hand, with diffraction gratings, which are employed in spectrophotometers, a much wider range of wavelengths extending into the ultraviolet may be examined, and the instrument is of much greater sensitivity. The portion of the instrument which enables selection of an appropriate spectral region is referred to as the **monochromator**: it contains, in addition to the prism or diffraction grating, an entrance slit which reduces the incident beam of radiation to a suitable area, and also an exit slit which selects the wavelength of the radiation which is to be presented to the sample under investigation. An important feature is the range of wavelengths present in the beam to which the sample is subjected; this is measured by what is termed the **spectral half-bandwidth**, which is the wavelength range embraced by the beam at the point where the intensity of the beam is one-half of its maximum value (Fig. 17.8). The width of the slits in the monochromator can be adjusted, and the narrower the slit the better the resolution of an absorption band, but decrease in slit width necessarily reduces the intensity of the beam reaching the detector. In practice, a compromise may need to be made between resolution and adequate intensity of radiation to permit accurate absorption readings.

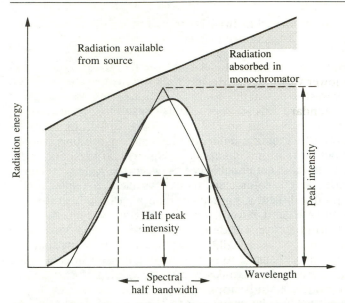

Fig. 17.8 Reproduced with permission from J. E. Steward (Ed.), *Introduction to Ultraviolet and Visible Spectrophotometry*, **2nd edn, Philips/Pye Unicam, Cambridge, 1985.**

The focusing of radiation within the instrument was formerly done by means of lenses, but these suffer from chromatic aberration and particularly in respect of the relationship between the visible and ultraviolet parts of the spectrum. Focusing is now usually carried out by means of suitably curved mirrors having a reflecting surface coated with aluminium which is protected by a silica film.

17.8 RADIATION SOURCES

For simple absorptiometers a **tungsten lamp** is the usual source of illumination. With spectrophotometers, however, to cover the full wavelength range of the instrument, two lamps are provided. The first is usually a tungsten–halogen (or quartz–iodine) lamp which covers wavelengths ranging from the red end of the visible spectrum (750–800 nm) to the near-ultraviolet (300–320 nm): the lamp is provided with a quartz outer sheath to permit use of the ultraviolet part of the emission. For measurements in the far-ultraviolet (down to 200 nm), a **hydrogen or a deuterium lamp** is used: the latter is preferred on account of the greater intensity of the radiation. This must also be provided with a quartz envelope. Xenon arc lamps with a spectral range of 250–600 nm may also be used.

17.9 CELLS

To investigate the absorption of radiation by a given solution, the solution must be placed in a suitable container called a cell (or cuvette) which can be accurately located in the beam of radiation. The instrument is provided with a cell-carrier which serves to site the cells correctly. Standard cells are of rectangular form with a 1 cm light path, but larger cells are available when solutions of low absorbance are to be examined, and likewise for solutions of high absorbance, cells of short path length can be obtained (semimicro or micro cells).

For aqueous solutions it is possible to obtain comparatively cheap cells made of polystyrene. Standard cells are made of glass to cover the wavelength range 340–1000 nm, but for lower wavelengths (down to 220 nm) they must be made of silica, and for the lowest wavelength (down to 185 nm) a special grade of silica must be used. All standard cells are supplied with a lid to prevent spillage, but if volatile solvents are to be used, special cells with a well-fitting stopper should be employed. In addition to the usual rectangular-pattern cell, continuous flow cells (which are useful as chromatographic detectors) and sampling cells (which are fitted with tubes making it possible to empty and fill the cells without removing them from the spectrophotometer) are also available.

Standard cells are produced in three grades: Grade A have a path length tolerance of 0.1 per cent; Grade B have a tolerance of 0.5 per cent and are regarded as suitable for routine use; Grade C can have a tolerance of up to 3 per cent. Even the highest quality cells differ slightly from each other and in work of the highest accuracy it is usual to select a matching pair of cells, one to hold the test solution and the other, a blank or reference solution. If unmatched cells are used, then a correction must be applied for the differing transmissions, but with many modern spectrophotometers which embody microprocessor control, the necessary correction can be done automatically.

17.10 DATA PRESENTATION

The most common method of recording the absorbance of a solution was to use a micro-ammeter to measure the output of the photoelectric cell, and this method is still applied to the simplest absorptiometers. The meter is usually provided with a dual scale calibrated to read (1) percentage absorbance and (2) transmission. For quantitative measurements it is more convenient to work in terms of absorbance rather than transmittance; this is emphasised by reference to the two graphs shown in Fig. 17.9.

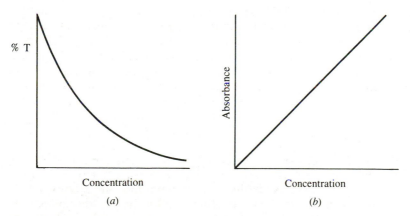

Fig. 17.9 Reproduced with permission from J. E. Steward (Ed.), *Introduction to Ultraviolet and Visible Spectrophotometry*, 2nd edn, Philips/Pye Unicam, Cambridge, 1985.

In more sophisticated instruments, the modern tendency is to replace the micro-ammeter by a digital read-out, and there is an increasing trend to use visual display units to show the results. Such instruments are controlled by microprocessors which may either show sequentially the successive operations which must be performed to measure the absorbance of a solution at a fixed wavelength or to observe the absorption spectrum of a sample; alternatively the whole procedure may be automated. Such instruments will display the absorption spectrum on the VDU screen, and by linking to a printer, a permanent record is produced.

17.11   LAYOUT OF INSTRUMENTS

The essential features of a simple filter absorptiometer — light source, filter to select the appropriate wavelength of light, a container for the solution, a photocell to receive the transmitted light, and a meter for measuring the response of the photocell — are assembled as shown in Fig. 17.10.

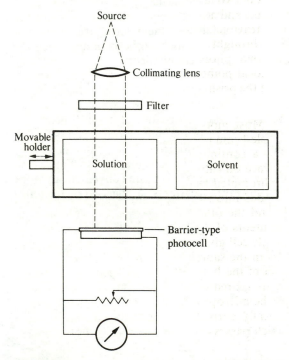

Fig. 17.10

The make-up of a **single-beam spectrophotometer** is shown in Fig. 17.11. An image of the light source A is focused by the condensing mirror B and the diagonal mirror C on the entrance slit at D: the entrance slit is the lower of two slits placed vertically, one above the other. Light falling on the collimating mirror E is rendered parallel and reflected to the quartz prism F. The back surface of the prism is aluminised, so that light refracted at the first surface is reflected back through the prism, undergoing further refraction as it emerges

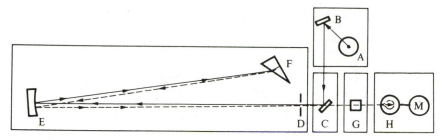

Fig. 17.11

from the prism. The collimating mirror focuses the spectrum in the plane of the slits D, and light of the wavelength for which the prism is set passes out of the monochromator through the exit (upper) slit, through the absorption cell G to the photocell H. The photocell response is amplified and is registered on the meter M.

The light source A in fact consists of two lamps as explained in Section 17.8; a tungsten–halogen lamp for the visible and near ultraviolet part of the spectrum and a deuterium lamp for the far ultraviolet. They are mounted on a movable arm which allows each lamp to be brought to the correct operating position as required. Likewise there are two phototubes (Section 17.6) and the appropriate one is brought to the focal point as the lamps are changed over. In modern versions of the instrument the prism F will be replaced by a diffraction grating.

Double-beam spectrophotometers. Most modern general-purpose ultraviolet/ visible spectrophotometers are double-beam instruments which cover the range between about 200 and 800 nm by a continuous automatic scanning process producing the spectrum as a pen trace on calibrated chart paper.

In these instruments the monochromated beam of radiation, from tungsten and deuterium lamp sources, is divided into two identical beams, one of which passes through the reference cell and the other through the sample cell. The signal for the absorption of the contents of the reference cell is automatically subtracted from that from the sample cell giving a net signal corresponding to the absorption for the components in the sample solution.

The splitting and recombination of the beam is accomplished by means of two rotating sector mirrors which are geared to the same electric motor so that they work in unison (Fig. 17.12). The microprocessor which is used to operate such an instrument will automatically correct for the 'dark current' of the photocell, i.e. the small current which passes even when the cell is not exposed to radiation.

A further factor which must be taken into consideration is that of stray light within the instrument. As indicated in Section 17.7, diffraction gratings can give rise to spectra of different orders, some of which may overlap the main or first-order beam, and the effect of these can be overcome by suitable filters correctly sited within the instrument. Although the interior of the mono-chromator is blackened, some stray light arising from reflections within the monochromator may pass through the exit slit, and when the photodetector reading is small, the stray light, which can contain wavelengths which are not absorbed by the solution under test, may well account for a significant proportion

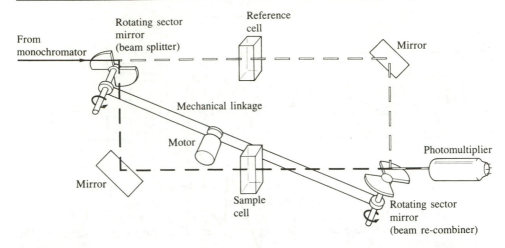

Fig. 17.12 Reproduced with permission from J. E. Steward (Ed.), *Introduction to Ultraviolet and Visible Spectrophotometry*, 2nd edn, Philips/Pye Unicam, Cambridge, 1985.

of the reading. When this happens, a calibration plot (Section 17.2) departs from the expected straight line, and with increasing absorbance of the solution, becomes curved towards the concentration axis. This complicates quantitative determinations and every effort is therefore made to reduce the amount of stray light. One method of achieving this is by the inclusion of an additional monochromator through which the radiation passes before entering the main monochromator.

17.12 DERIVATIVE SPECTROPHOTOMETRY

As shown in Section 15.17, the location of the end point of a potentiometric titration can often be accomplished more exactly from the first or second derivative of the titration curve. than from the titration curve itself. Similarly, absorption observations will often yield more information from derivative plots than from the original absorption curve. This technique was used as long ago as 1955,[8] but with the development of microcomputers which permit rapid generation of derivative curves, the method has acquired great impetus.[9,10]

If we consider an absorption band showing a normal (Gaussian) distribution [Fig. 17.13(*a*)], we find [Figs. (*b*) and (*d*)] that the first- and third-derivative plots are disperse functions that are unlike the original curve, but they can be used to fix accurately the wavelength of maximum absorption, λ_{max} (point M in the diagram).

The second- and fourth-order derivatives [Figs. 17.13(*c*) and (*e*)] have a central peak which is sharper than the original band but of the same height; its sign alternates with increasing order. It is clear that resolution is improved in the even-order spectra, and this offers the possibility of separating two absorption bands which may in fact merge in the zero-order spectrum. Thus, a mixture of two substances C and D gave a zero-order spectrum as indicated in Fig. 17.14(*a*) showing no well-defined absorption bands, but the second-order

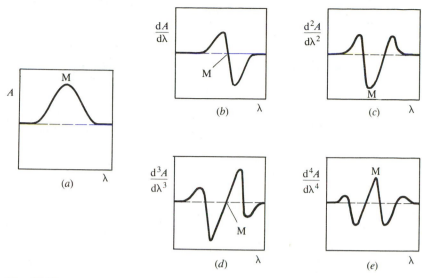

Fig. 17.13

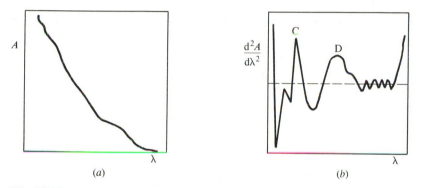

Fig. 17.14

spectrum deduced from this curve showed peaks at 280 nm and 330 nm [Fig. 17.14(b)].

The influence of an impurity (Y) on the absorption spectrum of a substance (X) can often be eliminated by considering derivative curves as shown in Fig. 17.15: the second-order plot of the mixture is identical with that of pure X. When the interference spectrum can be described by an nth-order polynomial, the interference is eliminated in the $(n + 1)$ derivative.

For quantitative measurements peak heights (expressed in mm) are usually measured of the long-wave peak satellite of either the second- or fourth-order derivative curves, or for the short-wave peak satellite of the same curves. This is illustrated in Fig. 17.16(a) for a second-order derivative: D_L is the long-wave peak height and D_S the short-wave peak height. Some workers[11] have preferred to use the peak tangent baseline (D_B) or the derivative peak zero (D_Z) measurements [Fig. 17.16(b)].

Derivative spectra can be recorded by means of a wavelength modulation

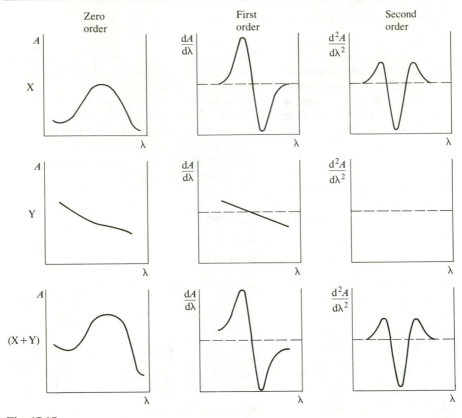

Fig. 17.15

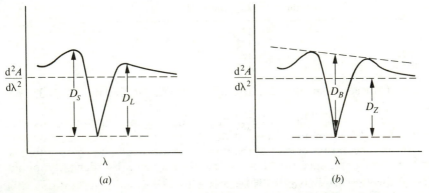

Fig. 17.16

device in which beams of radiation differing in wavelength by a small amount (1–2 nm) fall alternately on the sample cell and the difference between the two readings is recorded. In an alternative procedure, a derivative unit involving resistance/capacitance circuits, filters and operational amplifier are attached to the spectrophotometer, but as already indicated, derivative curves are most readily obtained by computer-based calculations.

670

17.13 THE ORIGINS OF ABSORPTION SPECTRA

The absorption of radiation is due to the fact that molecules contain electrons which can be raised to higher energy levels by the absorption of energy. The requisite energy can in some cases be supplied by radiation of visible wavelengths, thus producing an absorption spectrum in the visible region, but in other cases the higher energy associated with ultraviolet radiation is required. In addition to the change in electronic energy which follows the absorption of radiation, there are also changes associated with variation in the vibrational energy of the atoms in the molecule and changes in rotational energy. This means that many different amounts of energy will be absorbed depending upon the varying vibrational levels to which the electrons may be raised and the result is that we do not observe a sharp absorption line, but instead a comparatively broad absorption band.

Electrons in a molecule can be classified in three different types.

1. Electrons in a single covalent bond (σ-bond): these are tightly bound and radiation of high energy (short wavelength) is required to excite them.
2. Electrons attached to atoms such as chlorine, oxygen or nitrogen as 'lone pairs': these non-bonding electrons can be excited at a lower energy (longer wavelength) than tightly bound bonding electrons.
3. Electrons in double or triple bonds (π-orbitals) which can be excited relatively easily. In molecules containing a series of alternating double bonds (conjugated systems), the π-electrons are delocalised and require less energy for excitation so that the absorption rises to higher wavelengths. These statements are illustrated by the data in Table 17.3, which also includes the molar absorption coefficients for the λ_{max} absorption.

Table 17.3 Some electronic transitions in selected organic molecules

Compound	Transition	λ_{max} (nm)	ε ($m^2 mol^{-1}$)
CH_4	$\sigma \to \sigma^*$	122	—
CH_3Cl	$n \to \sigma^*$	173	200
$CH_2{=}CH_2$	$\pi \to \pi^*$	162	1500
$(CH_3)_2C{=}O$	$\pi \to \pi^*$	185	95
	$n \to \pi^*$	277	2
$H_2C{=}CH{-}CH{=}CH_2$	$\pi \to \pi^*$	180	2100
	$\pi \to \pi^*$	200	800
	$\pi \to \pi^*$	255	22

* Indicates excited orbitals.

The absorption of a given substance is greatly affected if it contains what is referred to as a **chromophore**. This is a functional group which has a characteristic absorption spectrum in the visible or ultraviolet region. Such groups invariably contain double or triple bonds and include the C=C linkage (and, therefore, the benzene ring), the C≡C bond, the nitro and nitroso groups, the azo group, the carbonyl and thiocarbonyl groups. If the chromophore is conjugated with another of the same or a different kind, then the absorption is enhanced and a new absorption band appears at a higher wavelength. Many of these features can be identified in Table 17.3.

The absorption of a given molecule may also be enhanced by the presence of groups called **auxochromes**. These do not absorb significantly in the ultraviolet region but may have a profound effect on the absorption of the molecule to which they are attached. Important auxochromes include OH, NH$_2$, CH$_3$ and NO$_2$ groups, and their effect, which is to displace the absorption maximum to a longer wavelength, is referred to as a **bathochromic shift**: it is related to the electron-donating properties of the auxochromes.

Many complexes of metals with organic ligands absorb in the visible part of the spectrum and are important in quantitative analysis. The colours arise from (i) $d \rightarrow d$ transitions within the metal ion (these usually produce absorptions of low intensity) and (ii) $n \rightarrow \pi^*$ and $\pi \rightarrow \pi^*$ transitions within the ligand. Another type of transition referred to as 'charge-transfer' may also be operative in which an electron is transferred between an orbital in the ligand and an unfilled orbital of the metal or vice versa. These give rise to more intense absorption bands which are of analytical importance.

For detailed consideration of the relationships between chemical constitution and the absorption of visible/ultraviolet radiation, textbooks of physical chemistry or of spectroscopy should be consulted.[12-17] A table of λ_{max} and ε_{max} values is given in Appendix 10.

EXPERIMENTAL COLORIMETRIC DETERMINATIONS

17.14 SOME GENERAL REMARKS UPON COLORIMETRIC DETERMINATIONS

Visual methods have been virtually displaced for most determinations by methods depending upon the use of photoelectric cells (filter photometers or absorptiometers, and spectrophotometers), thus leading to reduction of the experimental errors of colorimetric determinations. The so-called photoelectric colorimeter is a comparatively inexpensive instrument, and should be available in every laboratory. The use of spectrophotometers has enabled determinations to be extended into the ultraviolet region of the spectrum, whilst the use of chart recorders means that the analyst is not limited to working at a single fixed wavelength.

The choice of a colorimetric procedure for the determination of a substance will depend upon such considerations as the following.

(a) A colorimetric method will often give more accurate results at low concentrations than the corresponding titrimetric or gravimetric procedure. It may also be simpler to carry out.

(b) A colorimetric method may frequently be applied under conditions where no satisfactory gravimetric or titrimetric procedure exists, e.g. for certain biological substances.

(c) Colorimetric procedures possess advantages for the routine determination of some of the components of a number of similar samples by virtue of the rapidity with which they may be made: there is often no serious sacrifice of accuracy over the corresponding gravimetric or titrimetric procedures provided the experimental conditions are rigidly controlled.

The criteria for a satisfactory colorimetric analysis are:

1. *Specificity of the colour reaction.* Very few reactions are specific for a

particular substance, but many give colours for a small group of related substances only, i.e. they are selective. By utilising such devices as the introduction of other complex-forming compounds, by altering the oxidation states, and control of pH, close approximation to specificity may often be obtained. This subject is discussed in detail below.

2. *Proportionality between colour and concentration.* For visual colorimeters it is important that the colour intensity should increase linearly with the concentration of the substance to be determined. This is not essential for photoelectric instruments, since a calibration curve may be constructed relating the instrumental reading of the colour with the concentration of the solution. Otherwise expressed, it is desirable that the system follows Beer's law even when photoelectric colorimeters are used.

3. *Stability of the colour.* The colour produced should be sufficiently stable to permit an accurate reading to be taken. This applies also to those reactions in which colours tend to reach a maximum after a time: the period of maximum colour must be long enough for precise measurements to be made. In this connection the influence of other substances and of experimental conditions (temperature, pH, stability in air, etc.) must be known.

4. *Reproducibility.* The colorimetric procedure must give reproducible results under specific experimental conditions. The reaction need not necessarily represent a stoichiometrically quantitative chemical change.

5. *Clarity of the solution.* The solution must be free from precipitate if comparison is to be made with a clear standard. Turbidity scatters as well as absorbing the light.

6. *High sensitivity.* It is desirable, particularly when minute amounts of substances are to be determined, that the colour reaction be highly sensitive. It is also desirable that the reaction product absorbs strongly in the visible rather than in the ultraviolet; the interfering effect of other substances in the ultraviolet is usually more pronounced.

In view of the selective character of many colorimetric reactions, it is important to control the operational procedure so that the colour is specific for the component being determined. This may be achieved by isolating the substance by the ordinary methods of inorganic analysis; double precipitation is frequently necessary to avoid errors due to occlusion and co-precipitation. Such methods of chemical separation may be tedious and lengthy and if minute quantities are under consideration, appreciable loss may occur owing to solubility, supersaturation, and peptisation effects. Use may be made of any of the following processes in order to render colour reactions specific and/or to separate the individual substances.

(*a*) Suppression of the action of interfering substances by the formation of complex ions or of non-reactive complexes.

(*b*) Adjustment of the pH; many reactions take place within well-defined limits of pH.

(*c*) Removal of the interfering substance by extraction with an organic solvent, sometimes after suitable chemical treatment.

(*d*) Isolation of the substance to be determined by the formation of an organic complex, which is then removed by extraction with an organic solvent. This method may be combined with (*a*) in which an interfering ion is prevented from forming a soluble organic complex by converting it into a complex ion which remains in the aqueous layer (see Chapter 6).

(e) Separation by volatilisation. This method is of limited application, but gives good results, e.g. distillation of arsenic as the trichloride in the presence of hydrochloric acid.

(f) Electrolysis with a mercury cathode or with controlled cathode potential.

(g) Application of physical methods utilising selective absorption, chromatographic separations, and ion exchange separations.

Some remarks concerning **standard curves** seem appropriate at this point. The usual method of use of a filter photometer or a spectrophotometer requires the construction of a standard curve (also termed the reference or calibration curve) for the constituent being determined. Suitable quantities of the constituent are taken and treated in the same way as the sample solution for the development of colour and the measurement of the transmission (or absorbance) at the optimum wavelength. The absorbance ($\log I_0/I_t$) is plotted against the concentration: a straight-line plot is obtained if Beer's Law is obeyed. The curve may then be used for future determinations of the constituent under the same experimental conditions. When the absorbance is directly proportional to the concentration, only a few points are required to establish the line: when the relation is not linear, a greater number of points will generally be necessary. The standard curve should be checked at intervals. When a filter photometer is used, the characteristics of the filter and the light source may change with time.

When plotting the standard curve it is customary to assign a transmission of 100 per cent to the blank solution (reagent solution plus water); this represents zero concentration of the constituent. It may be mentioned that some coloured solutions have an appreciable temperature coefficient of transmission, and the temperature of the determination should not differ appreciably from that at which the calibration curve was prepared.

17.15 CHOICE OF SOLVENT

Clearly the requirements which a solvent to be used in colorimetric or spectrophotometric determinations must meet are that it (1) must be a good solvent for the substance under determination, (2) does not interact with the solute, and (3) must not show significant absorption at the wavelength to be employed in the determination. For inorganic compounds water usually meets these requirements, but for the majority of organic compounds it is necessary to use an organic solvent. An immediate complication which arises is that polar solvents such as alcohols, esters and ketones (water also falls into this category) tend to obliterate the fine structure of absorption spectra which are related to vibrational effects. To preserve these details, the absorption measurements must be carried out in a hydrocarbon (non-polar) solvent. Thus, for example, a solution of phenol in cyclohexane gives an absorption curve showing three sharply defined peaks in the ultraviolet, but an aqueous solution of the same concentration gives a single broad absorption band covering the same wavelength range as that observed in the hydrocarbon solvent.

There is the further complication as far as spectrophotometry is concerned, i.e. that all solvents show absorption at some point in the ultraviolet, and care must be taken to choose a solvent for a particular determination which is transparent in the requisite wavelength region. Organic solvents are listed in order of 'cut-off wavelengths', which is the wavelength at which the transmittance is reduced to 25 per cent when measured in a 10 cm cell against water as reference material. Values for some typical solvents are collected in Table 17.4.

Table 17.4 Cut-off wavelengths for some common solvents

Solvent	(nm)	Solvent	(nm)
Water	190	Dichloromethane	233
Hexane	199	Trichloromethane (chloroform)	247
Heptane	200	Tetrachloromethane (carbon tetrachloride)	257
Diethyl ether	205	Benzene	280
Ethanol	207	Pyridine	306
Methanol	210	Propanone (acetone)	331
Cyclohexane	212		

Any impurities present in the solvents may affect the 'cut-off' value, and it is therefore essential to employ materials of the highest purity. Most major suppliers of laboratory chemicals offer products which have been specially purified and carefully tested to ensure that they are of the requisite standard for use in spectrophotometric determinations. Such chemicals are usually identified by a special name as for example the 'Spectrosol', materials supplied by BDH Ltd. In many cases, however, it suffices to subject the purest material available to spectrophotometric examination, and if there is no appreciable absorption over the spectral range required for the proposed determination, the solvent may be used: otherwise careful purification will be needed.[18]

17.16 GENERAL PROCEDURE FOR COLORIMETRIC DETERMINATIONS

In any colorimetric determination the exact procedure will be governed partly by the particular instrument to be employed and partly by the nature of the substance which is being determined. There are nevertheless certain general principles which are universally applicable and these are also relevant to spectrophotometric determinations in general.

It is obviously important to ensure that the colorimeter (or spectrophotometer) is functioning correctly and that no adjustments or replacements are necessary. The exact procedures to be followed in such an instrument check will be detailed in the operating manual supplied by the manufacturer, and in many modern microprocessor-controlled instruments the instrument can be programmed for example to check, and if necessary to correct, the wavelength setting. In the absence of such automated facilities, matters which should be checked include the following four items. For instruments in regular use only items (1) and (4) are likely to require attention, but items (2) and (3) should be checked periodically and after any apparent misuse of the instrument, or if it has been left unused for a long period.

1. *Illumination.* The emission of any lamp tends to decrease with age and must be replaced when the time of maximum usage is reached: a log should be kept of the burning time of each lamp.
2. *Wavelength scale.* It is most important that this is correctly adjusted. Many instruments are supplied with test filters and by measuring the λ_{max} values

675

of the absorption peaks of the filters, the scale can be checked. It can be done more precisely for an instrument containing a hydrogen or deuterium lamp by checking the red (656.1 nm) and blue–green (486.0 nm) lines in the hydrogen spectrum. Alternatively, a neon or a mercury lamp can be substituted for the normal illumination of the instrument and the resultant spectra contain a number of lines of accurately known wavelength with which to test the scale.

The colour filters used with absorptiometers should be examined from time to time by measuring the absorbance in a spectrophotometer; if the results depart markedly from those expected for the filter, it should be replaced. Normally a filter as supplied with the instrument will be used, but if necessary, filters from the 'Wratten' range supplied by Kodak Ltd or from the 'Spectrum Filter' or 'Bright Spectrum Filter' series supplied by Ilford Ltd may be used.

3. *Absorbance scale.* This can be checked by using one or more standard solutions which have been carefully prepared; examples include potassium dichromate in either acid or alkaline solution, and potassium nitrate solution. Full details of recommended standard solutions and of their standard absorption values are given in Ref. 19.

4. *Cells.* Unless matched cells (Section 17.9) are to be employed, cells to be used must be tested to ensure that under identical conditions, two chosen cells show the same transmission within very narrow limits. This is done by setting up the instrument at a selected wavelength, preferably one to be used in the determination on hand, and with no cells in position, adjusting the controls so that the scale reads 0 per cent absorbance (or 100 per cent transmittance). The chosen cells, which must be clean and dry, are filled with the solvent to be used and the outsides dried with paper tissue; care must be taken to handle them by the ground surfaces of the cells only and to avoid touching the polished faces. Check the transmission of one cell using the second cell as a blank and, after noting the reading, empty the cell, refill with solvent and repeat the reading, which should be consistent with the first. Now replace the solvent in the second cell and again repeat the reading. If the difference in absorbance between the two cells is less than 0.02, the cells can be accepted for use; many modern instruments are capable of correcting automatically for a small difference between the cells. If a double-beam instrument is in use, then, of course, the difference in the absorbance of the two cells is obtained immediately, and the checking procedure is more straightforward.

After use the cells must be carefully washed with distilled water, or if an organic solvent which is not miscible with water was employed, they must be rinsed with a solvent which is miscible with both the solvent used and with water, and then well washed with distilled water. Finally, they are rinsed with ethanol followed by drying, which may be conveniently done in a vacuum desiccator. Cells which have become contaminated can usually be cleaned by soaking in a solution of a detergent, such as Teepol. For obstinate contamination recourse may be made to the use of sulphuric acid–sodium dichromate cleaning mixture (**CAUTION!**) (Section 3.8). After soaking overnight the cells are well washed with water and finally dried.

The determination itself will normally require the preparation of solutions of

(i) a weighed quantity of the material under investigation in an appropriate solvent: (ii) a standard solution of the compound being determined in the same solvent; (iii) the requisite reagent; and (iv) any ancillary reagents such as buffers, acids or alkalis necessary to establish the correct conditions for formation of the required coloured product. If ultraviolet measurements are to be carried out, it is often possible to make the determination on the solution of the substance without the need for adding other reagents. In view of the sensitivity of colorimetric and spectrophotometric methods, the solutions of which the absorbance is measured are usually very dilute. In order to take sufficient material to permit an accurate weight to be achieved when preparing the original solution of analyte and of the corresponding standard solution, it is commonly necessary to prepare solutions which are too concentrated for the absorbance measurements, and these must then be diluted accurately to the appropriate strength.

In connection with colour-producing reagents, it must be recognised that solutions of such reagents are frequently unstable and normally should not be stored for more than a day or so. Even in the solid state, many of these materials tend to deteriorate slowly and it is advisable that, in general, only small quantities should be stored so that fresh supplies are obtained at frequent intervals. A little-used reagent which may have been in stock for some months should be subjected to a trial run with the appropriate substance before commencing the actual determination.

Measurement of the absorbance of the test and standard solutions will be carried out in the manner described above for comparing cells, but, of course, choosing the wavelength appropriate to the substance being determined. The blank solution will be of similar composition to the test solution, but without any of the determinand. The prepared solution of the test material must be diluted if necessary so that the absorbance lies in the 0.2–1.5 region.

If the determination is a routine procedure, then a calibration plot will be available and it will be a simple matter to ascertain the concentration of the test solution; in this case it is unnecessary to prepare a standard solution of the compound being determined. If a calibration plot is not available, then a series of solutions containing say 5.0, 7.5, 10.0 and 15.0 mL of the standard solution are diluted to, say, 100 mL in graduated flasks. The absorbance of each of these solutions is measured and the results plotted against concentration.

For measurements in the ultraviolet, concentrations are frequently calculated by using the relationship (Section 17.2)

Molar absorption coefficient $(\varepsilon) = A/cl$

where A is the measured absorbance of a solution of concentration c, in a cell of path-length l. It follows that if ε for the compound being determined is known, then the concentration of the solution can be calculated. In some cases (especially with natural products) it is not possible to give ε values because relative molecular masses are not known; recourse may then be made to $E_{1\%}^{1\,\text{cm}}$ values which is the absorptivity representing the absorbance of a 1 per cent solution in a cell of 1 cm path length (see Section 17.9): such values are recorded for many materials, for example, numerous pharmaceutical preparations.

The following procedures are arranged in alphabetical order with cations

677

first (Sections 17.17–17.35), followed by anions (Sections 17.36–17.41)* and a few typical examples relating to organic compounds (Section 17.41–17.44). Although the methods can be mostly carried out using absorptiometers, spectrophotometers can be used if desired. With this in mind, the appropriate wavelength for each determination is given and, where applicable, reference is also made to a suitable colour filter.

CATIONS

17.17 ALUMINIUM

Discussion. Among the reagents that have been used for the colorimetric determination of aluminium are ammonium aurintricarboxylate (aluminon) and eriochrome cyanine R. The latter appears to be somewhat superior, and its use will therefore be described. At a pH of 5.9–6.1, zinc, nickel, manganese, and cadmium interfere negligibly, but iron and copper must be absent. One procedure for removing interfering elements, e.g. in the analysis of steels, is to pass the solution through a cellulose column; iron and other elements are separated by elution with a mixture of concentrated hydrochloric acid and freshly distilled butanone (ethyl methyl ketone) (8:192 v/v). The aluminium, and any nickel present, are recovered by passing dilute hydrochloric acid (1:5 v/v) through the column.

Reagents. *Eriochrome cyanine R solution.* Dissolve 0.1 g of the solid reagent in water, dilute to 100 mL, and filter through a Whatman No. 541 filter paper if necessary. This solution should be prepared daily.

Standard aluminium solution. Dissolve 1.319 g aluminium potassium sulphate in water and dilute to 1 L in a graduated flask; 1 mL $\equiv$ 75 μg Al.

Buffer solution, concentrated. Dissolve 27.5 g ammonium acetate and 11.0 g hydrated sodium acetate in 100 mL water: add 1.0 mL glacial acetic (ethanoic) acid and mix well.

Buffer solution, dilute. To one volume of concentrated buffer solution, add five volumes water and adjust the pH to 6.1 by adding acetic acid or sodium hydroxide solution.

Procedure. Transfer an aliquot of the solution (say, 20.0 mL), containing 2–70 μg Al and free from interfering elements, to a 250 mL beaker, add 5 mL of five volume hydrogen peroxide and mix well. Adjust the pH of the solution to 6.0 (using either 0.2M sodium hydroxide or 0.2M hydrochloric acid), add 5.0 mL of eriochrome cyanine R solution, and mix. Introduce 50 mL of the dilute buffer solution, and dilute without delay to 100 mL in a graduated flask. Measure the absorbance after 30 minutes with a spectrophotometer at 535 nm against a reagent blank in a 5 mm cell. For an absorptiometer, use a yellow–green filter and 1 cm cells.

Construct the calibration curve using 0, 1, 2, 3, 4, and 5 mL of the standard aluminium solution.

---

* A large number of reagents for metals are available and in some instances a reagent will produce a reaction suitable for colorimetric determination with several metals. The table reproduced in Appendix 2 gives a clearer idea of the wide range of reagents available.

17.18 AMMONIA

Discussion. J. Nessler in 1856 first proposed an alkaline solution of mercury(II) iodide in potassium iodide as a reagent for the colorimetric determination of ammonia. Various modifications of the reagent have since been made. When Nessler's reagent is added to a dilute ammonium salt solution, the liberated ammonia reacts with the reagent fairly rapidly but not instantaneously to form an orange–brown product, which remains in colloidal solution, but flocculates on long standing. The colorimetric comparison must be made before flocculation occurs.

The reaction with Nessler's reagent [an alkaline solution of potassium tetraiodomercurate(II)] may be represented as:

$$2K_2[HgI_4] + 2NH_3 = NH_2Hg_2I_3 + 4KI + NH_4I$$

The reagent is employed for the determination of ammonia in very dilute ammonia solutions and in water. In the presence of interfering substances, it is best to separate the ammonia first by distillation under suitable conditions. The method is also applicable to the determination of nitrates and nitrites: these are reduced in alkaline solution by Devarda's alloy to ammonia, which is removed by distillation. The procedure is applicable to concentrations of ammonia as low as $0.1 \, \text{mg L}^{-1}$.

Reagents. Nessler's reagent is prepared as follows. Dissolve 35 g potassium iodide in 100 mL water, and add 4 per cent mercury(II) chloride solution, with stirring or shaking, until a slight red precipitate remains (about 325 mL are required). Then introduce with stirring, a solution of 120 g sodium hydroxide in 250 mL water and make up to 1 L with distilled water. Add a little more mercury(II) chloride solution until there is a permanent turbidity. Allow the mixture to stand for one day and decant from the sediment. Keep the solution stoppered in a dark-coloured bottle.

The following is an alternative method of preparation. Dissolve 100 g mercury(II) iodide and 70 g potassium iodide in 100 mL ammonia-free water. Add slowly, and with stirring, to a cooled solution of 160 g sodium hydroxide pellets (or 224 g potassium hydroxide) in 700 mL ammonia-free water, and dilute to 1 L with ammonia-free distilled water. Allow the precipitate to settle, preferably for a few days, before using the pale yellow supernatant liquid.

Ammonia-free water may be prepared in a conductivity-water still, or by means of a column charged with a mixed cation and anion exchange resin (e.g. Permutit Bio-Deminrolit or Amberlite MB-1), or as follows. Redistil 500 mL of distilled water in a Pyrex apparatus from a solution containing 1 g potassium permanganate and 1 g anhydrous sodium carbonate; reject the first 100 mL portion of the distillate and then collect about 300 mL.

Procedure. For practice in this determination, employ either a very dilute ammonium chloride solution or ordinary distilled water which usually contains sufficient ammonia for the exercise.

Prepare a standard ammonium chloride solution as follows. Dissolve 3.141 g ammonium chloride, dried at 100 °C, in ammonia-free water and dilute to 1 L with the same water. This stock solution is too concentrated for most purposes. A standard solution is made by diluting 10 mL of this solution to 1 L with ammonia-free water: 1 mL contains 0.01 mg of NH_3.

679

If necessary, dilute the sample to give an ammonia concentration of 1 mg L^{-1} and fill a 50 mL Nessler tube to the mark. Prepare a series of Nessler tubes containing the following volumes of standard ammonium chloride solution diluted to 50 mL: 1.0, 2.0, 3.0, 4.0, 5.0, and 6.0 mL. The standards contain 0.01 mg NH$_3$ for each mL of the standard solution. Add 1 mL of Nessler's reagent to each tube, allow to stand for 10 minutes, and compare the unknown with the standards in a Nessler stand (Section 17.4) or in a BDH Nesslerimeter. This will give an approximate figure which will enable another series of standards to be prepared and more accurate results to be obtained.

A photoelectric colorimeter or a spectrophotometer may, of course, be used. When 1 mL of the Nessler reagent is added to 50 mL of the sample, a blue colour filter in the wavelength region 400–425 nm allows measurements with a 1 cm path in the nitrogen range 20–250 μg. Nitrogen concentrations approaching up to 1 mg can be determined with a green colour filter or in the wavelength range near 525 nm. The calibration curve should be prepared under exactly the same conditions of temperature and reaction time adopted for the sample.

17.19 ANTIMONY

Discussion. The procedure is based on the formation of yellow tetraiodo-antimonate(III) acid (HSbI$_4$) when antimony(III) in sulphuric acid solution is treated with excess of potassium iodide solution. Spectrophotometric measurements may be made at 425 nm in the visible region or, more precisely, at 330 nm in the ultraviolet region. Appreciable amounts of bismuth, copper, lead, nickel, tin, tungstate, and molybdate interfere.

Reagents. *Potassium iodide solution.* Dissolve 14.0 g potassium iodide and 1.0 g crystallised ascorbic acid in redistilled water and dilute to 100 mL.

Standard antimony solution. Dissolve 0.2668 g antimonyl potassium tartrate in redistilled water, add 160 mL concentrated sulphuric acid, and dilute to 1 L with water in a graduated flask.

Procedure. Use a solution containing 0.15–1.8 mg antimony per 100 mL; it should be slightly acidic with sulphuric acid (1.2–1.5M). Transfer a 10 mL aliquot to a 50 mL graduated flask, add 25 mL of the potassium iodide–ascorbic acid reagent, and dilute to the mark with 25 per cent v/v sulphuric acid. Mix thoroughly and measure the absorbance at 425 nm or at 330 nm using a reagent blank as reference solution.

Construct a calibration curve using appropriate volumes of the standard antimony solution treated in the same way as for the sample solution.

17.20 ARSENIC

Discussion. Of the numerous procedures available for the determination of minute amounts of arsenic,* only one will be described, viz. the molybdenum blue method. It possesses great sensitivity and precision, and is readily applied colorimetrically or spectrophotometrically.

Molybdenum blue method.   When arsenic, as arsenate, is treated with ammonium molybdate solution and the resulting heteropolymolybdoarsenate (arseno-molybdate) is reduced with hydrazinium sulphate or with tin(II) chloride, a blue soluble complex 'molybdenum blue' is formed. The constitution is uncertain, but it is evident that the molybdenum is present in a lower oxidation state. The stable blue colour has a maximum absorption at about 840 nm and shows no appreciable change in 24 hours. Various techniques for carrying out the determination are available, but only one can be given here. Phosphate reacts in the same manner as arsenate (and with about the same sensitivity) and must be absent.

Both macro and micro quantities of arsenic may be isolated by distillation of arsenic(III) chloride from hydrochloric acid solution in an all-glass apparatus in a stream of carbon dioxide or nitrogen: a reducing agent, such as hydrazinium sulphate, is used to reduce arsenic(V) to arsenic(III). The distillate may be collected in cold water. Germanium accompanies arsenic in the distillation; if phosphate is present in large amounts the distillate should be re-distilled under the same conditions. Another method of isolation involves volatilisation of arsenic as arsine by the action of zinc in hydrochloric or sulphuric acid solution. Appreciable amounts of certain reducible heavy metals, such as copper, nickel, and cobalt, slow down the evolution of arsine, as do also large amounts of metals that are precipitated by zinc. Copper in more than small quantities prevents complete evolution of arsine; the error amounts to 20 per cent (for 5–10 μg As) with 50 mg of copper. The arsine which is evolved may be absorbed in a sodium hydrogencarbonate solution of iodine. The absorption apparatus should be so designed that the arsine is completely absorbed.

Reagents.†   *Potassium iodide solution.*   Dissolve 15 g of solid in 100 mL water.
Tin(II) chloride solution.   Dissolve 40 g hydrated tin(II) chloride in 100 mL concentrated hydrochloric acid.
Zinc.   Use 20–30 mesh or granulated; arsenic-free.
Iodine–potassium iodide solution.   Dissolve 0.25 g iodine in a small volume of water containing 0.4 g potassium iodide, and dilute to 100 mL.
Sodium disulphite solution (sodium metabisulphite).   Dissolve 0.5 g of the solid reagent ($Na_2S_2O_5$) in 10 mL water. Prepare fresh daily.

---

* Very small amounts of arsenic (0.001–0.1 mg) may be determined by volatilising the element as arsine AsH_3 and comparing the coloration formed upon discs of dry paper impregnated with mercury(II) chloride with that obtained by the use of known amounts of arsenic (**Gutzeit test**). Although the method is still used in practice and suitable apparatus is available from most laboratory supply houses, it is doubtful whether the accuracy exceeds 10 per cent of the true value. Too much dependence is placed upon the rate of evolution of arsine, which is not necessarily the same as the rate of evolution of hydrogen in the reduction apparatus. On the whole, the spectrophotometric procedure, based upon molybdenum blue or the silver diethyldithiocarbamate complex, is far superior. In all evolution methods arsenic must be in the arsenic(III) state.
† Special pure, arsenic-free reagents are available from chemical supply houses (e.g. BDH Ltd) and are symbolised by 'AST' after the name of the compound, these should be used as far as possible in the determination and for the preparation of the above reagents.

Sodium hydrogencarbonate solution. Dissolve 4.2 g of the solid in 100 mL water.

Ammonium molybdate–hydrazinium sulphate reagent. Solution (*a*): dissolve 1.0 g ammonium molybdate in 10 mL water and add 90 mL of $3M$ sulphuric acid. Solution (*b*): dissolve 0.15 g pure hydrazinium sulphate in 100 mL water. Mix 10.0 mL each of solutions (*a*) and (*b*) just before use.

Hydrochloric acid. This must be arsenic-free.

Standard arsenic solution. Dissolve 1.320 g arsenic(III) oxide in the minimum volume of $1M$ sodium hydroxide solution, acidify with dilute hydrochloric acid, and make up to 1 L in a graduated flask: 1 mL contains 1 mg of As. A solution containing 0.001 mg As per mL is prepared by dilution.

Procedure. The arsenic must be in the arsenic(III) state; this may be secured by first distilling in an all-glass apparatus with concentrated hydrochloric acid and hydrazinium sulphate, preferably in a stream of carbon dioxide or nitrogen. Another method consists in reducing the arsenate (obtained by the wet oxidation of a sample) with potassium iodide and tin(II) chloride: the acid concentration of the solution after dilution to 100 mL must not exceed 0.2–$0.5M$; 1 mL of 50 per cent potassium iodide solution and 1 mL of a 40 per cent solution of tin(II) chloride in concentrated hydrochloric acid are added, and the mixture heated to boiling.

Transfer an aliquot portion of the arsenate solution, having a volume of 25 mL and containing not more than 20 µg of arsenic, to the 50 mL Pyrex evolution vessel A shown in Fig. 17.17, and add sufficient concentrated hydrochloric acid to make the total volume present in the solution 5–6 mL, followed by 2 mL of the potassium iodide solution and 0.5 mL of the tin(II) chloride solution. Allow to stand at room temperature for 20–30 minutes to permit the complete reduction of the arsenate.

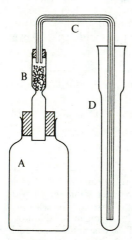

Fig. 17.17

The tube B is loosely packed with purified glass wool soaked in lead acetate solution (to remove hydrogen sulphide and trap acid spray), and C is a capillary tube (4 mm external and 0.5 mm internal diameter). Place 1.0 mL iodine–potassium iodide solution and 0.2 mL of the sodium hydrogencarbonate solution in the narrow absorption tube D. Mix with the end of the delivery tube.

Rapidly add 2.0 g of zinc to the vessel A, immediately insert the stopper, and allow the gases to bubble through the solution for 30 minutes. At the end of this time the solution in D should still contain some iodine. Disconnect the delivery tube C and leave it in the absorption tube. Add 5.0 mL of the ammonium molybdate–hydrazine reagent and a drop or two of sodium disulphite solution. Heat the resulting colourless solution in a water bath at 95–100 °C, cool, transfer to a 10 mL graduated flask, and make up to volume with water.

Measure the transmittance of the solution at 840 nm or with a red filter with maximum transmission above 700 nm. Charge the reference cell with a solution obtained by taking the iodine–iodide–hydrogencarbonate mixture and treating it with molybdate–hydrazinium sulphate–disulphite as in the actual procedure.

Construct the calibration curve by taking, say, 0, 2.5, 5.0, 7.5, and 10.0 μg As (for a final volume of 10 mL), mixing with iodine–iodide–hydrogencarbonate solution, adding molybdate–hydrazinium sulphate–disulphite, and heating to 95–100 °C.

The following procedure has been recommended by the Analytical Methods Committee of the Society for Analytical Chemistry for the determination of small amounts of arsenic in organic matter.[20] Organic matter is destroyed by wet oxidation, and the arsenic, after extraction with diethylammonium diethyldithiocarbamate in chloroform, is converted into the arsenomolybdate complex: the latter is reduced by means of hydrazinium sulphate to a molybdenum blue complex and determined spectrophotometrically at 840 nm and referred to a calibration graph in the usual manner.

17.21 BERYLLIUM

Discussion. Minute amounts of beryllium may be readily determined spectrophotometrically by reaction under alkaline conditions with 4-nitrobenzeneazoorcinol. The reagent is yellow in a basic medium; in the presence of beryllium the colour changes to reddish-brown. The zone of optimum alkalinity is rather critical and narrow; buffering with boric acid increases the reproducibility. Aluminium, up to about 240 mg per 25 mL, has little influence provided an excess of 1 mole of sodium hydroxide is added for each mole of aluminium present. Other elements which might interfere are removed by preliminary treatment with sodium hydroxide solution, but the possible co-precipitation of beryllium must be considered. Zinc interferes very slightly but can be removed by precipitation as sulphide. Copper interferes seriously, even in such small amounts as are soluble in sodium hydroxide solution. The interference of small amounts of copper, nickel, iron and calcium can be prevented by complexing with EDTA and triethanolamine.

Procedure. Transfer the almost neutral sample solution of beryllium (containing 5 to 80 μg of the element in a volume of about 10 mL) to a 25 mL graduated flask, add 2.8 mL of 2.0M sodium hydroxide (or more if much aluminium is present), 5.0 mL of 0.64M boric acid solution, and 6.0 mL of the dye solution (see Note), dilute to the mark with distilled water, and mix well. Measure the transmittance at 520 nm, or with a green filter preferably using a 2 cm cell.

Construct a calibration curve (for details, see Sections 17.14 and 17.16) using beryllium sulphate and the experimental conditions given above: cover the range 5–80 μg of beryllium. Evaluate the concentration of the sample solution of beryllium with the aid of the calibration curve.

Note. Prepare the dye solution by stirring 0.025 g of 4-nitrobenzeneazo-orcinol mechanically for several hours with 0.1M sodium hydroxide; filter before use.

17.22 BISMUTH

Discussion. When potassium iodide solution is added to a dilute sulphuric acid solution containing a small amount of bismuth a yellow to orange coloration, due to the formation of an iodobismuthate(III) ion, is produced. The colour intensity increases with iodide concentration up to about 1 per cent potassium iodide and then remains practically constant.

The reaction is a sensitive one, but is subject to a number of interferences. The solution must be free from large amounts of lead, thallium(I), copper, tin, arsenic, antimony, gold, silver, platinum, and palladium, and from elements in sufficient quantity to colour the solution, e.g. nickel. Metals giving insoluble iodides must be absent, or present in amounts not yielding a precipitate. Substances which liberate iodine from potassium iodide interfere, for example iron(III); the latter should be reduced with sulphurous acid and the excess of gas boiled off, or by a 30 per cent solution of hypophosphorous acid. Chloride ion reduces the intensity of the bismuth colour. Separation of bismuth from copper can be effected by extraction of the bismuth as dithizonate by treatment in ammoniacal potassium cyanide solution with a 0.1 per cent solution of dithizone in chloroform; if lead is present, shaking of the chloroform solution of lead and bismuth dithizonates with a buffer solution of pH 3.4 results in the lead alone passing into the aqueous phase. The bismuth complex is soluble in a pentan-1-ol–ethyl acetate mixture, and this fact can be utilised for the determination in the presence of coloured ions, such as nickel, cobalt, chromium, and uranium.

Procedure. Prepare a standard solution of bismuth by dissolving 0.100 g pure bismuth in 20 mL concentrated sulphuric acid, and diluting to 1 L with water: 1 mL contains 0.1 mg Bi. Other standard solutions may be obtained by dilution.

Treat the colourless solution (*ca* 15 mL), free from interfering substances and about 1M in sulphuric acid, with 1 mL of 30 per cent hypophosphorous acid solution and 1 mL of 10 per cent aqueous potassium iodide solution. Dilute to 25 mL and match the yellow colour produced against standards containing the same concentrations of sulphuric acid and hypophosphorous acid. Alternatively, measure the absorbance at or near 460 nm or with a blue filter.

In the extraction procedure the yellow solution is allowed to stand for 10 minutes, and then extracted with 3 mL portions of a 3:1 mixture by volume of pentan-1-ol and ethyl acetate until the last extract is colourless. Make up the combined extracts to a definite volume (10 mL or 25 mL) with the organic solvent, and determine the transmittance (460 nm) at once. Construct the calibration curve by extracting known amounts of bismuth under the same conditions as the sample.

Bismuth in lead: Discussion. This method is based upon the extraction of bismuth as cupferrate by chloroform from $0.1M$ acid solution: as little as 1 μg of bismuth can be separated from 10 g of lead.

Procedure. Dissolve a suitable weight of the sample of lead in $6M$ nitric acid: add a little 50 per cent aqueous tartaric acid to clear the solution if antimony or tin is present. Cool, transfer to a separatory funnel, and dilute to about 25 mL. Add concentrated ammonia solution to the point where the slight precipitate will no longer dissolve on shaking, then adjust the pH to 1, using nitric acid or ammonia solution. Add 1 mL freshly prepared 1 per cent cupferron solution, mix, and extract with 5 mL chloroform. Separate the chloroform layer, and repeat the extraction twice with 1 mL portions of cupferron solution + 5 mL of chloroform. Wash the combined chloroform extracts with 5 mL of water. Extract the bismuth from the chloroform by shaking with two 10 mL portions of $1M$ sulphuric acid. Run the sulphuric acid solution into a 25 mL graduated flask. Add 3 drops saturated sulphur dioxide solution and 4 mL of 20 per cent aqueous potassium iodide. Dilute to volume and measure the transmission at 460 nm.

17.23 BORON

Discussion. Minute amounts of boron are usually separated by distillation from an acid solution as methyl borate. Borosilicate glass should be avoided, even for the storage of chemicals. The apparatus should be constructed of fused silica;* a platinum dish receiver may also be used. Distillation may be made from a strong acid solution [sulphuric or phosphoric(V) acid]. In the simplest apparatus methanol vapour is passed through a flask containing the solution of the sample and is condensed and collected in an excess of either calcium hydroxide or sodium hydroxide solution in a silica or platinum dish. In a more efficient apparatus the methanol is made to cycle between the sample dissolved in the acid medium and a flask containing calcium or sodium hydroxide solution: distillation can thus be continued for several hours with only a small amount of methanol. At the end of the distillation the contents of the receiver in which the methyl borate was collected (which must be strongly alkaline — a minimum of four times the theoretical amount of base) are evaporated to dryness. The residue is used for the colorimetric determination. Most of the reagents, e.g. quinalizarin (1,2,5,8-tetrahydroxyanthraquinone) or 1,1'-dianthrimide (1,1'-iminodianthraquinone) react only in concentrated sulphuric acid solution. With the former the absorption maxima for the reagent and its boron complex lie close together, while with the latter the maximum absorption for the reagent is below 400 nm and for the boron complex is at 620 nm. The use of dianthrimide will accordingly be described. The colour change of 1,1'-dianthrimide from greenish-yellow to blue in the presence of borates in concentrated sulphuric acid is the basis of a trustworthy method for the determination of micro amounts of boron; the effective range of the reagent is 0.5–6 μg and the colour is stable for several hours.

Interferences in the distillation method are fluoride and large amounts of gelatinous silica. Fluoride interference may be overcome by the addition of calcium chloride. Strong oxidising agents, such as chromate and nitrate, interfere,

---

* Corning Vycor glass, containing 96 per cent of silica, is usually suitable.

since they destroy the reagent. Boron in natural waters can be determined without separation; the residue obtained after evaporation to dryness with a little calcium hydroxide solution may be used directly in the colour formation. In the analysis of steel by dissolution in sulphuric acid no oxidising compounds are formed which can interfere with the reaction.

Reagents. *Dianthrimide reagent solution.* Dissolve 150 mg of 1,1'-dianthrimide in 1 L concentrated sulphuric acid (*ca* 96 per cent w/w). Keep in the dark and protected from moisture.

Standard boron solution. Dissolve 0.7621 g boric acid in water and dilute to 1 L. Take 50 mL of this solution and dilute to 1 L: the resulting solution contains 6.667 μg B per mL.

Dilute sulphuric acid. Prepare a 1:3 v/v solution.

Procedure (boron in steel). Dissolve about 3 g of the steel (B content $\not> 0.02$ per cent), accurately weighed, in 40 mL dilute sulphuric acid in a 150 mL Vycor or silica flask fitted with a reflux condenser. Heat until dissolved. Filter through a quantitative filter paper into a 100 mL graduated flask. Wash with hot water, cool to room temperature, and dilute to the mark with water. This flask (A) contains the acid-soluble boron.

Ignite the filter in a platinum crucible, fuse with 2.0 g of anhydrous sodium carbonate, dissolve the melt in 40 mL of dilute sulphuric acid, and add 1 mL of sulphurous acid solution (about 6 per cent) to reduce any iron(III) salt, etc., formed in the fusion, and filter if necessary. Transfer the solution to a 100 mL graduated flask, dilute to the mark, and mix. This flask (B) contains the acid-insoluble boron.

Transfer 3.0 mL of solutions A and B to two dry, glass-stoppered conical flasks (Vycor or silica). Add 25 mL of dianthrimide reagent solution to each with shaking, and insert the glass stoppers loosely. For the blank use 3.0 mL of solutions A and B in two similar 50 mL conical flasks and add 25 mL concentrated sulphuric acid (98 per cent w/w). Heat all four flasks in a boiling water bath for 60 minutes. Cool to room temperature and measure the absorbance of each of the solutions at 620 nm (orange filter) against pure concentrated sulphuric acid in 1 cm or 2 cm cells. Correct for the blanks.

To construct the calibration curve, run 5–50 mL of the standard boron solution by means of a burette into 100 mL graduated flasks, add 30 mL of dilute sulphuric acid, and make up to volume. These solutions contain 1–10 μg of B per 3 mL. Use 3 mL of each solution and of a boron-free comparison solution and proceed as above. Plot a calibration curve relating absorbance and boron content.

Calculate the total boron content of the steel (i.e. acid-soluble plus acid-insoluble boron).

An alternative method for the determination of boron is given under Section 6.9.

17.24 CHROMIUM

Discussion. Small amounts of chromium (up to 0.5 per cent) may be determined colorimetrically in alkaline solution as chromate; uranium and cerium interfere, but vanadium has little influence. The transmittance of the solution is measured at 365–370 nm or with the aid of a filter having maximum transmission in the violet portion of the spectrum. The standard solution used for the preparation of the reference curve should have the same alkalinity as the sample solution, and should preferably have the same concentration of foreign salts. Standards may be prepared from analytical grade potassium chromate.

A more sensitive method is to employ 1,5-diphenylcarbazide $CO(NH \cdot NHC_6H_5)_2$; in acid solution (ca $0.2M$), chromates give a soluble violet compound with this reagent.

Molybdenum(VI), vanadium(V), mercury, and iron interfere; permanganates, if present, may be removed by boiling with a little ethanol. If the ratio of vanadium to chromium does not exceed 10:1, nearly correct results may be obtained by allowing the solution to stand for 10–15 minutes after the addition of the reagent, since the vanadium–diphenylcarbazide colour fades fairly rapidly. Vanadate can be separated from chromate by adding oxine to the solution and extracting at a pH of about 4 with chloroform; chromate remains in the aqueous solution. Vanadium as well as iron can be precipitated in acid solution with cupferron and thus separated from chromium(III).

Procedure. Prepare a 0.25 per cent solution of diphenylcarbazide in 50 per cent acetone as required. The test solution may contain from 0.2 to 0.5 part per million of chromate. To about 15 mL of this solution add sufficient $3M$ sulphuric acid to make the concentration about $0.1M$ when subsequently diluted to 25 mL, add 1 mL of the diphenylcarbazide reagent and make up to 25 mL with water. Match the colour produced against standards prepared from $0.0002M$ potassium dichromate solution. A green filter having the transmission maximum at about 540 nm may be used.

Chromium in steel: Discussion. The chromium in the steel is oxidised by perchloric acid to the dichromate ion, the colour of which is intensified by iron(III) perchlorate which is itself colourless. The coloured solution is compared with a blank in which the dichromate is reduced with ammonium iron(II) sulphate. The method is not subject to interference by iron or by moderate amounts of alloying elements usually present in steel.

Procedure. Place a 1.000 g sample of the steel (Cr content < 0.1 per cent) in a 100 mL beaker and dissolve it in 10 mL of dilute nitric acid (1:1) and 20 mL of perchloric acid (sp. gr. 1.70; 70–72 per cent). [If the Cr content is 0.1–1 per cent, dissolve a 0.5000 g sample in 10 mL of dilute nitric acid (1:1) and 15 mL of perchloric acid (sp. gr. 1.70).] Evaporate to dense fumes of perchloric acid and boil gently for 5 minutes to oxidise the chromium. Cool the beaker and contents rapidly, dissolve soluble salts by adding 20 mL of water, transfer the solution quantitatively to a glass-stoppered 50 mL graduated flask, and dilute to the mark. Remove an aliquot portion to an absorption cell, reduce it with a little (ca 20 mg) ammonium iron(II) sulphate, and adjust the colorimeter or spectrophotometer so that the reading with this solution is zero; a violet filter having a maximum transmission between 410 and 480 nm may be used in the colorimeter. Discard the solution in the absorption cell, and re-fill it with an equal volume of the oxidised solution: the reading is a measure of the colour due to the dichromate.

Standardisation may be carried out by the use of solutions prepared from a chromium-free standard steel and standard potassium dichromate solution. After dissolution of the standard steel, the solution is boiled with perchloric acid, potassium dichromate added and the resulting solution is diluted to volume, and measurements are carried out as above. The chromium content of any unknown steel may then be deduced from the colorimeter reading.

17.25 COBALT

Discussion. An excellent method for the colorimetric determination of minute amounts of cobalt is based upon the soluble red complex salt formed when cobalt ions react with an aqueous solution of nitroso-R-salt (sodium 1-nitroso-2-hydroxynaphthalene-3,6-disulphonate). Three moles of the reagent combine with 1 mole of cobalt.

The cobalt complex is usually formed in a hot acetate–acetic acid medium. After the formation of the cobalt colour, hydrochloric acid or nitric acid is added to decompose the complexes of most of the other heavy metals present. Iron, copper, cerium(IV), chromium(III and VI), nickel, vanadyl vanadium, and copper interfere when present in appreciable quantities. Excess of the reagent minimises the interference of iron(II); iron(III) can be removed by diethyl ether extraction from a hydrochloric acid solution. Most of the interferences can be eliminated by treatment with potassium bromate, followed by the addition of an alkali fluoride. Cobalt may also be isolated by dithizone extraction from a basic medium after copper has been removed (if necessary) from acidic solution. An alumina column may also be used to adsorb the cobalt nitroso-R-chelate anion in the presence of perchloric acid, the other elements are eluted with warm $1M$ nitric acid, and finally the cobalt complex with $1M$ sulphuric acid, and the absorbance measured at 500 nm.

Procedure. The test solution should contain between 0.001 and 0.02 mg of cobalt. Evaporate almost to dryness, add 1 mL of concentrated nitric acid, and continue the evaporation just to dryness to oxidise any iron(II) which may be present. Dissolve the residue in 10 mL of water containing 0.5 mL each of 1:1 hydrochloric acid and 1:10 nitric acid. Boil for a few minutes to dissolve any solid material. Add 2.0 mL of a 0.2 per cent aqueous solution of nitroso-R-salt and also 2.0 g of hydrated sodium acetate. The pH of the solution should be close to 5.5; check with bromocresol green indicator or with a pH meter. Boil for 1 minute, add 1.0 mL of concentrated hydrochloric acid, and boil again for 1 minute. Cool to room temperature, dilute to 25 mL in a graduated flask, and compare the colour with standards or use a spectrophotometer. Determine the absorbance at 425 nm (green filter) against a reagent blank. In the presence of 2 mg or more of iron it is best to make the measurement at 500 nm (violet filter) to reduce the error resulting from the absorption of light by the yellow solution.

Standard solutions may be conveniently prepared with spectroscopically pure cobalt.

Cobalt in steel: Discussion. An alternative, but less sensitive, method utilises 2-nitroso-1-naphthol, and this can be used for the determination of cobalt in steel. The pink cobalt(III) complex is formed in a citrate medium at pH 2.5–5. Citrate serves as a buffer, prevents the precipitation of metallic hydroxides, and complexes iron(III) so that it does not form an extractable nitrosonaphtholate complex. The cobalt complex forms slowly (*ca* 30 minutes) and is extracted with chloroform.

Procedure. Prepare the 2-nitroso-1-naphthol reagent by dissolving 1.0 g of the solid in 100 mL of glacial acetic acid. Add 1 g of activated carbon: shake the solution before use and filter off the required volume.

Dissolve a known weight (*ca* 0.5 g) of the steel by any suitable procedure. Treat the acidic sample solution (< 200 μg Co), containing iron in the iron(II) state, with 10–15 mL of 40 per cent (w/v) sodium citrate solution, dilute to 50–75 mL and adjust the pH to 3–4 (indicator paper) with 2*M* hydrochloric acid or sodium hydroxide. Cool to room temperature, add 10 mL of 3 per cent (10-volume) hydrogen peroxide and, after 3 minutes, 2 mL of the reagent solution. Allow to stand for at least 30 minutes at room temperature. Extract the solution in a separatory funnel by shaking vigorously for 1 minute with 25 mL of chloroform: repeat the extraction twice with 10 mL portions of chloroform. Dilute the combined extracts to 50 mL with chloroform and transfer to a clean separatory funnel. Add 20 mL of 2*M* hydrochloric acid, shake for 1 minute, run the chloroform layer into another separatory funnel, and shake for 1 minute with 20 mL of 2*M* sodium hydroxide. Determine the absorbance of the clear chloroform phase in a 1 cm cell at 530 nm.

For the preparation of standard cobalt solutions, use analytical grade cobalt(II) chloride or spectroscopically pure cobalt dissolved in hydrochloric acid; subject solutions containing 0, 5, 10, 25, 50, 100, 150, and 200 μg of Co to the whole procedure.

17.26 COPPER

Discussion. Small quantities of copper may be determined by the diethyldithio-carbamate method (Section 6.10) or by the 'neo-cuproin' method (Section 6.11), an extraction being necessary in both cases. In another somewhat simpler procedure, the copper is complexed with biscyclohexanone oxalyldihydrazone and the resulting blue colour is measured by a suitable spectrophotometer within the range 570–600 nm (orange filter). The solution measured should contain not more than 100 μg of copper.

Reagents. *Bicyclohexanone oxalyldihydrazone solution* (copper reagent). Dissolve 0.1 g of the solid reagent in 10 mL ethanol (or industrial methylated spirit) and 10 mL hot water, and dilute to 200 mL. Filter, if necessary.

Synthetic standard solution (for analysis of steel). Dissolve an appropriate weight of pure iron (Johnson Matthey) in a mixture of equal volumes of concentrated hydrochloric acid and concentrated nitric acid; with this solution as base, add a suitable amount of copper nitrate solution containing 0.01 g copper per L.

Procedure (copper in steel). Weigh out accurately a 0.1 g sample of the steel* into a 150 mL conical beaker, add 5 mL concentrated hydrochloric acid and 5 mL concentrated nitric acid, and warm gently. In the presence of interfering amounts of chromium, add 5 mL perchloric acid, sp. gr. 1.70, and evaporate until strong fuming occurs. When the sample has dissolved or after the fuming with perchloric acid, cool, add 50 mL cold distilled water, followed by 10 mL acid solution (1:1 HCl/HNO_3). Carefully add 10 mL concentrated ammonia

---

* The following British Chemical Standard or Euronorm Certified Reference Materials may be used for practice in this determination: BCS No. 402; ECRM 084-1.

solution, sp. gr. 0.88, cool to room temperature, and dilute to 100 mL in a graduated flask. Return the solution to the original beaker and transfer a 10 mL aliquot to a 100 mL graduated flask. Add 20 mL of the copper reagent, dilute to 100 mL with distilled water, and transfer to a 100 mL dry beaker. Allow to stand for 10–15 minutes, and then measure the absorbance with a spectrophotometer.

Construct a calibration curve using the synthetic standard solution: add the standard copper solution immediately before the reagent.

17.27 IRON

Two procedures will be described — the thiocyanate and the 1,10-phenanthroline methods.

A. Thiocyanate method: Discussion. Iron(III) reacts with thiocyanate to give a series of intensely red-coloured compounds, which remain in true solution: iron(II) does not react. Depending upon the thiocyanate concentration, a series of complexes can be obtained; these complexes are red and can be formulated as $[Fe(SCN)_n]^{3-n}$, where $n = 1, \ldots, 6$. At low thiocyanate concentration the predominant coloured species is $[Fe(SCN)]^{2+}$ $\{Fe^{3+} + SCN^- \rightarrow [Fe(SCN)]^{2+}\}$, at $0.1M$ thiocyanate concentration it is largely $[Fe(SCN)_2]^+$, and at very high thiocyanate concentration it is $[Fe(SCN)_6]^{3-}$. In the colorimetric determination a large excess of thiocyanate should be used, since this increases the intensity and also the stability of the colour. Strong acids (hydrochloric or nitric acid — concentration 0.05–$0.5M$) should be present to suppress the hydrolysis:

$$Fe^{3+} + 3H_2O \rightleftharpoons Fe(OH)_3 + 3H^+$$

Sulphuric acid is not recommended, because sulphate ions have a certain tendency to form complexes with iron(III) ions. Silver, copper, nickel, cobalt, titanium, uranium, molybdenum, mercury ($> 1\,g\,L^{-1}$), zinc, cadmium, and bismuth interfere. Mercury(I) and tin(II) salts, if present, should be converted into the mercury(II) and tin(IV) salts, otherwise the colour is destroyed. Phosphates, arsenates, fluorides, oxalates, and tartrates interfere, since they form fairly stable complexes with iron(III) ions; the influence of phosphates and arsenates is reduced by the presence of a comparatively high concentration of acid.

When large quantities of interfering substances are present, it is usually best to proceed in either of the following ways: (1) remove the iron by precipitation with a slight excess of ammonia solution, and dissolve the precipitate in dilute hydrochloric acid; (2) extract the 'iron(III) thiocyanate' three times either with pure diethyl ether or, better, with a mixture of pentan-1-ol and pure diethyl ether (5:2) and employ the organic layer for the colour comparison.

Reagents. Prepare the following solutions.

Standard solution of iron(III). Use method (*a*), (*b*) or (*c*). (*a*) Dissolve 0.7022 g ammonium iron(II) sulphate in 100 mL water, add 5 mL of 1:5 sulphuric acid, and run in cautiously a dilute solution of potassium permanganate ($2\,g\,L^{-1}$) until a slight pink coloration remains after stirring well. Dilute to 1 L and mix thoroughly. 1 mL $\equiv$ 0.1 mg of Fe. (*b*) Dissolve 0.864 g ammonium iron(III) sulphate in water, add 10 mL concentrated hydrochloric acid and dilute to 1 L. 1 mL $\equiv$ 0.1 mg of Fe. (*c*) Dissolve 0.1000 g of electrolytic iron or pure iron wire in 50 mL of 1:3 nitric acid, boil to expel oxides of nitrogen, and dilute to 1 L with de-ionised water.

Potassium thiocyanate solution. Dissolve 20 g potassium thiocyanate in 100 mL water: the solution is *ca 2M*.

Procedure. Dissolve a weighed portion of the substance in which the amount of iron is to be determined in a suitable acid, and evaporate nearly to dryness to expel excess of acid. Dilute slightly with water, oxidise the iron to the iron(III) state with dilute potassium permanganate solution or with a little bromine water, and make up the liquid to 500 mL or other suitable volume. Take 40 mL of this solution and place in a 50 mL graduated flask, add 5 mL of the thiocyanate solution and 3 mL of 4*M* nitric acid. Add de-ionised water to dilute to the mark. Prepare a blank using the same quantities of reagents. Measure the absorbance of the sample solution in a spectrophotometer at 480 nm (blue–green filter). Determine the concentration of this solution by comparison with values on a reference curve obtained in the same way from different concentrations of the standard iron solution.

B. 1,10-Phenanthroline method: Discussion. Iron(II) reacts with 1,10-phenanthroline to form an orange–red complex $[(C_{12}H_8N_2)_3Fe]^{2+}$. The colour intensity is independent of the acidity in the pH range 2–9, and is stable for long periods. Iron(III) may be reduced with hydroxylammonium chloride or with benzene-1,4-diol (quinol). Silver, bismuth, copper, nickel, and cobalt interfere seriously, as do perchlorate, cyanide, molybdate, and tungstate. The iron–phenanthroline complex (as the perchlorate) may be extracted with nitrobenzene and measured at 515 nm (blue–green filter) against a reagent blank.

Both iron(II) and iron(III) can be determined spectrophotometrically: the reddish-orange iron(II) complex absorbs at 515 nm, and both the iron(II) and the yellow iron(III) complex have identical absorption at 396 nm, the amount being additive. The solution, slightly acid with sulphuric acid, is treated with 1,10-phenanthroline, and buffered with potassium hydrogenphthalate at a pH of 3.9: the reading at 396 nm gives the total iron and that at 515 nm the iron(II).

Reagents. Prepare the following solutions.
1,10-Phenanthroline. 0.25 per cent solution of the monohydrate in water.
Sodium acetate. 0.2*M* and 2*M*.
Hydroxylammonium chloride. 10 per cent aqueous solution, or benzene-1,4-diol (quinol), 1 per cent solution in an acetic acid buffer of pH *ca* 4.5 (mix 65 mL of 0.1*M* acetic acid and 35 mL of 0.1*M* sodium acetate solution). Prepare when required.

Procedure. Take an aliquot portion of the unknown slightly acid solution containing 0.1–0.5 mg iron and transfer it to a 50 mL graduated flask. Determine, by the use of a similar aliquot portion containing a few drops of bromophenol blue, the volume of sodium acetate solution required to bring the pH to 3.5 ± 1.0. Add the same volume of acetate solution to the original aliquot part and then 4 mL each of the quinol and 1,10-phenanthroline solutions. Make up to the mark with distilled water, mix well, and allow to stand for 1 hour to complete the reduction of the iron. Compare the intensity of the colour produced with standards, similarly prepared, in any convenient way. If a colorimeter is employed, use a filter showing maximum transmission at 480–520 nm; for a spectrophotometer, use a wavelength of 515 nm.

The iron may also be reduced with hydroxylammonium chloride. Add 5 mL of the 10 per cent hydroxylammonium solution, adjust the pH of the slightly acid solution to 3–6 with sodium acetate, then add 4 mL of the 1,10-phenanthroline solution, dilute to 50 mL, mix, and measure the absorbance after 5–10 minutes.

17.28 LEAD

Discussion. For the determination of small amounts of lead (0.005–0.25 mg) advantage is taken of the fact that when a sulphide is added to a solution containing lead *ions* a brown colour, due to the formation of colloidal lead sulphide, is produced. However, for general use the dithizone method (see Section 6.13) is to be preferred and this will be described.

Reagents. *Standard lead solution.* Dissolve 0.160 g of analytical grade lead nitrate in 1 L of distilled water; 10.0 mL of this solution, diluted to 250 mL gives a working solution containing 4 μg of lead mL^{-1}.

Dithizone reagent. Dissolve 5 mg of the solid in 100 mL of chloroform (Section 6.6).

Ammonia–cyanide–sulphite solution. Prepare by diluting 35 mL of concentrated ammonia solution and 3.0 mL of 10 per cent potassium cyanide solution (**CAUTION**) to 100 mL and adding 0.15 g of sodium sulphite.

Procedure. Place 10.0 mL of the working lead solution in a 250 mL separatory funnel, add 75 mL of the ammonia–cyanide–sulphite solution and then by the cautious addition of dilute hydrochloric acid adjust the pH of the solution to 9.5 (pH-meter). This operation must be carried out slowly: if the pH of the solution falls even temporarily below 9.5, HCN may be liberated and so use of a fume cupboard is necessary. Now add 7.5 mL of the dithizone reagent to the separatory funnel, followed by a further 17.5 mL of chloroform. Shake for 1 minute, allow the layers to separate, then remove the chloroform layer. Measure the absorbance of this against a blank solution, using a 1 cm cell and a wavelength of 510 nm (green filter).

Repeat the procedure with 5.0 mL, 7.5 mL and 15.0 mL portions of the working lead solution and then with 10 mL of the test solution.

17.29 MAGNESIUM

Two methods are commonly used for the determination of magnesium. Titan yellow may be used to obtain a coloured colloidal suspension, or solochrome black to give a red soluble complex. In most cases the second of these is to be preferred.

Solochrome black method: Discussion. The difficulties inherent in the colloidal systems involved in 'lake' methods may be avoided by the use of organic reagents which form soluble coloured complexes with magnesium in basic solution. One such reagent is solochrome black, which forms red soluble complexes with magnesium. The colour is not stable: calcium, copper, manganese, iron, aluminium, cobalt, nickel, etc., interfere. By buffering at pH 10.1, a single complex is formed by one magnesium ion and two molecules of the dye. Calcium may be separated from magnesium by precipitation as sulphate in the presence of a large excess of methanol.

Reagents. *Buffer solution (pH = 10.1).* This consists of a 0.75 per cent w/v solution of ammonium chloride in dilute ammonia solution, prepared by mixing 5 volumes of concentrated ammonia solution (sp. gr. 0.88) and 95 volumes of water.

Solochrome black solution. Prepare a 0.1 per cent solution in methanol; warm to speed solution and filter.

Procedure. Transfer the neutral sample solution ($<100\,\mu$g Mg), free from calcium and other metals, to a 100 mL graduated flask with calibrated neck. Add 25 mL of the buffer solution, dilute to just below the 90 mL graduation mark, and shake. Add 10.0 mL of the solochrome black solution carefully. Shake to mix and dilute to the 100 mL mark with water. Measure the absorbance immediately at 520 nm (green filter) against that of a blank solution, similarly prepared but containing no magnesium.

17.30 MOLYBDENUM

Discussion. Molybdenum may be determined colorimetrically by the thiocyanate–tin(II) chloride method (for details, see Section 6.14) or by the dithiol method described here.

Toluene-3,4-dithiol, usually called 'dithiol', yields a slightly soluble, dark-green complex, $(CH_3.C_6H_3.S_2)_3Mo(VI)$, with molybdenum(VI) in a mineral acid medium, which can be extracted by organic solvents. The resulting green solution is used for the colorimetric determination of molybdenum.

Dithiol is a less selective reagent than thiocyanate for molybdenum. Tungsten interferes most seriously but does not do so in the presence of tartaric acid or citric acid (see Section 17.34). Tin does not interfere if the absorbance is read at 680 nm. Strong oxidants oxidise the reagent; iron(III) salts should be reduced with potassium iodide solution and the liberated iodine removed with thiosulphate.

Procedure. Prepare the dithiol reagent by adding 0.1–0.2 g of dithiol to 100 mL of 0.25 M sodium hydroxide solution, followed by 0.5 mL of thioglycollic acid (to inhibit oxidation of the reagent); keep at 5 °C and prepare fresh daily.

Add to the sample solution (containing 1–25 μg of Mo) 4 mL of 1:3 sulphuric acid, 3 drops of 85 per cent phosphoric(V) acid, and 0.5 g of citric acid. Dilute with water to 20 mL and add 2 mL of dithiol solution. Allow to stand at room temperature for 2 hours. Extract the molybdenum complex with 13 mL and 10 mL portions respectively of re-distilled butyl acetate, and make up to 25.0 mL with this solvent in a graduated flask; filter through glass wool if not entirely clear. Determine the absorbance of the solution at 670 nm. Prepare a calibration curve as detailed in Section 6.14.

17.31 NICKEL

Discussion. When dimethylglyoxime is added to an alkaline solution of a nickel salt which has been treated with an oxidising agent (such as bromine), a red coloration is obtained. The red soluble complex contains nickel in a higher oxidation state.* The nickel complex formed absorbs at about 445 nm provided absorbance readings are made within 10 minutes of mixing. The dimethylglyoxime–oxidising agent method must be distinguished from the divalent nickel–dimethylglyoxime procedure which yields a nickel(II) dimethylglyoximate soluble in chloroform: full details of this solvent extraction method are given in Section 6.15.

Cobalt(II), gold(III), and dichromate ions interfere under the conditions of the test. Metals which precipitate in ammoniacal solution can be removed by double precipitation, or by taking advantage of the solubility of nickel(II) dimethylglyoxime in chloroform [the nickel(III) complex is insoluble, as is also the brown cobalt dimethylglyoxime]. Copper may accompany the nickel in the extraction; most of the copper is removed from the chloroform extract when it is shaken with dilute ammonia solution, whereas the nickel remains in the organic solvent. The nickel(II) dimethylglyoxime in the chloroform layer may be decomposed by shaking with dilute hydrochloric acid; most of the dimethylglyoxime remains in the chloroform; the nickel is transferred to the aqueous phase and may be determined colorimetrically. Citrate or tartrate may be added to prevent the precipitation of iron, aluminium, etc. Much manganese may interfere, but this is prevented by adding hydroxylammonium chloride to maintain it as Mn(II).

Procedure (nickel in steel). Dissolve 0.50 g, accurately weighed,† of the steel in 10 mL of warm 1:1 nitric acid, boil to expel oxides of nitrogen, cool, and make up to 250 mL with water in a graduated flask. Mix well, and transfer 5 mL of the solution to a 50 mL graduated flask. To 5 mL of this solution add 5 mL of 10 per cent citric acid solution, neutralise with concentrated ammonia solution and add a few drops in excess (pH > 7.5). Add 2 mL of a 1 per cent dimethylglyoxime solution in ethanol (or more if copper or cobalt is present). Extract with three 3 mL portions of chloroform, shaking for 30 seconds each time. Shake the combined chloroform extracts with 6 mL of 0.5 M ammonia solution (1:30); shake the ammonia washing with 2 mL of chloroform and add the latter to the main chloroform extract. Return the nickel to the ionic state by shaking the chloroform extract vigorously for 1 minute with two 5 mL portions of 0.5 M hydrochloric acid. Transfer the hydrochloric acid solution to a 25 mL graduated flask, dilute to about 20 mL, add 1 mL of saturated bromine water, followed by 2 mL of concentrated ammonia solution. Cool below 30 °C if necessary, add 1 mL of 1 per cent dimethylglyoxime solution, and dilute to volume. Measure the absorbance at 445 nm (blue filter) after 5 minutes. The standard solutions for the construction of the calibration curve should contain approximately the same concentration of iron (nickel-free) as the sample solution.

Prepare the standard nickel solution by dissolving 0.673 g pure ammonium nickel sulphate in water and diluting to 1 L: 1 mL contains 0.1 mg of Ni. The solution may be further diluted to a basis of 0.01 mg of Ni per mL, if necessary. Pure nickel may also be employed for the preparation of the standard solution.

---

* This has been regarded as nickel(III) and also as nickel(IV) dimethylglyoxime.

† The weight of steel to be taken will naturally depend upon the nickel content. The final nickel concentration should not exceed 0.6 mg per 100 mL because a precipitate may form above this concentration.

17.32 TIN

Discussion. In acid solution, toluene-3,4-dithiol (dithiol) forms a red compound when warmed with tin(II) salts (compare molybdenum, Section 17.30). Tin(IV) also reacts, but more slowly than tin(II); thioglycollic acid may be employed to reduce tin(IV) to tin(II). The reagent is not stable, being easily reduced, and hence should be prepared as required. A dispersant is generally added to the solution under test.

Many heavy metals react with dithiol to give coloured precipitates, e.g. bismuth, iron(III), copper, nickel, cobalt, silver, mercury, lead, cadmium, arsenic, etc.; molybdate and tungstate also react. Of the various interfering elements, only arsenic distils over with the tin when a mixture is distilled from a medium of concentrated sulphuric acid and concentrated hydrobromic acid in a current of carbon dioxide. If arsenic is present in quantities larger than that of the tin it should be removed.

Reagents. *Dithiol reagent.* Dissolve 0.1 g dithiol in 2.5 mL 5M sodium hydroxide solution. Add 0.5 mL thioglycollic acid, and dilute to 50 mL with water. Prepare fresh daily.

Dispersant solution. Prepare a 1 per cent aqueous solution of sodium lauryl sulphate.

Standard tin solution. Dissolve 1.000 g tin in 100 mL of 1:1 hydrochloric acid and dilute with the same concentration of acid to 1 L:1 mL contains 1 mg Sn. Prepare more dilute solutions as required (e.g. 0.01 mg Sn per mL) by dilution with 1:1 hydrochloric acid.

Procedure. Transfer a 10 mL aliquot of the sample solution, which should be 0.5M in hydrochloric acid and contain not more than 0.25 mg of tin, to a 25 mL graduated flask, and add in the order given 1 drop thioglycollic acid, 2.0 mL concentrated hydrochloric acid, 0.5 mL of the dispersant solution, and 1.0 mL of the dithiol reagent with thorough mixing after each addition. Place the flask in a water bath at 60 °C for 10 minutes, cool, and dilute the contents to the mark. Measure the absorbance at 545 nm (yellow–green filter) against a reagent blank.

Construct a concentration–absorbance curve with the aid of the standard tin solution.

Procedure (tin in canned foods). The procedure provides for the removal of interfering copper by the addition of diethylammonium diethyldithiocarbamate in chloroform reagent.*

Weigh 5 or 10 g of the sample, depending on the expected tin content, into a small porcelain crucible. Dry and char the sample on a hotplate; heat to ash in a muffle furnace at 600 °C. Add 1 g of fusion mixture (3 parts Na_2CO_3 + 1 part KCN by weight) and fuse this with the ash by holding the crucible with nickel tongs over a Bunsen or Meker burner. Cool the crucible, place it in a small beaker, and cover the latter with a watchglass. Add 10 mL water, and run 10 mL dilute hydrochloric acid (1:1) cautiously into the crucible (**FUME CUPBOARD!**). Boil the contents of the beaker gently for 30 minutes. Cool and filter: wash the beaker, crucible, and filter with water.

---

* This reagent is prepared from 3.0 mL of diethylamine in chloroform and 1 mL of carbon disulphide in 9 mL of chloroform. Mix carefully and store in a dark bottle in a refrigerator.

If copper is known to be absent or present only in negligible proportions, dilute the solution with water to 50 mL in a graduated flask, and continue as detailed below. Otherwise, transfer the solution to a small separatory funnel and add 5 mL of the diethylammonium diethyldithiocarbamate in chloroform reagent (diluted 1:20 with chloroform when required). Shake and run off the chloroform layer, extract the aqueous layer with successive 1 mL portions of the reagent until the chloroform layer is colourless; finally, wash the aqueous layer with a few mL of chloroform. Dilute the aqueous solution with water to 50 mL in a graduated flask.

To 10.0 mL of the solution thus prepared add 0.5 mL of dilute hydrochloric acid (1:1) and proceed as above. Measure the absorbance at 545 nm, or use a green filter with an absorptiometer.

17.33 TITANIUM

Discussion. With an acidic titanium(IV) solution hydrogen peroxide produces a yellow colour:* with small amounts of titanium (up to 0.5 mg of TiO_2 per mL), the intensity of the colour is proportional to the amount of the element present. Comparison is usually made with standard titanium(IV) sulphate solutions; a method for their preparation from potassium titanyl oxalate is described below. The hydrogen peroxide solution should be about 3 per cent strength (ten volume) and the final solution should contain sulphuric acid having a concentration from about 0.75 to 1.75 M in order to prevent hydrolysis to a basic sulphate and to prevent condensation to metatitanic acid. The colour intensity increases slightly with rise of temperature; hence the solutions to be compared should have the same temperature, preferably 20–25 °C.

Elements which interfere are: (*a*) iron, nickel, chromium, etc., because of the colour of their solutions; (*b*) vanadium, molybdenum, and, under some conditions, chromium, because they form coloured compounds with hydrogen peroxide; (*c*) fluorine (even in minute amount) and large quantities of phosphates, sulphates, and alkali salts (the influence of the last two is progressively reduced the greater the concentration of sulphuric acid present — up to 10 per cent). The influence of elements of class (*a*) is overcome, if present in small amount, by matching the colour by the addition of like quantities of the coloured elements to the standard before hydrogen peroxide is added. When large amounts of iron are present, as in the analysis of cast irons and steels, two methods may be adopted: (1) phosphoric(V) acid can be added in like amount to both unknown and standard, after the addition of hydrogen peroxide; (2) the iron content of the unknown solution is determined, and a quantity of standard ammonium iron(III) sulphate solution, containing the same amount of iron, is added to the standard solution. Large quantities of nickel, chromium, etc., must be removed. Elements of class (*b*) must also be removed; vanadium and molybdenum are most easily separated by precipitation of the titanium with sodium hydroxide solution in the presence of a little iron. Fluoride has the most powerful effect in bleaching the colour; it must be removed by repeated evaporation with concentrated sulphuric acid. The bleaching effect of phosphoric acid is overcome

---

* The coloured species formed has been stated to be $[TiO(SO_4)_2]^{2-}$ or a similar ion; and it has also been formulated as $[Ti(H_2O_2)]^{4+}$ or an analogous complex.

by adding a like amount to the standard, or by adding 1 mL of 0.1 per cent uranyl acetate solution for each 0.1 mg of Ti present.

Preparation of standard titanium solution. Weigh out 3.68 g potassium titanyl oxalate $K_2TiO(C_2O_4)_2,2H_2O$ into a Kjeldahl flask; add 8 g ammonium sulphate and 100 mL concentrated sulphuric acid. Gradually heat the mixture to boiling and boil for 10 minutes. Cool, pour the solution into 750 mL of water, and dilute to 1 L in a graduated flask; 1 mL $\equiv$ 0.50 mg of Ti.

If there is any doubt concerning the purity of the potassium titanyl oxalate, standardise the solution by precipitating the titanium with ammonia solution or with cupferron solution, and igniting the precipitate to TiO_2.

Procedure. The sample solution should preferably contain titanium as sulphate in sulphuric acid solution, and be free from the interfering constituents mentioned above. The final acidity may vary from 0.75 to 1.75 M. If iron is present in appreciable amounts, add dilute phosphoric(V) acid from a burette until the yellow colour of the iron(III) is eliminated: the same amount of phosphoric(V) acid must be added to the standards. If alkali sulphates are present in the test solution in appreciable quantity, add a like amount to the standards. Add 10 mL of 3 per cent hydrogen peroxide solution and dilute the solution to 100 mL in a graduated flask; the final concentration of Ti may conveniently be 2–25 parts per million. Compare the colour produced by the unknown solution with that of standards of similar composition by any of the usual methods.

For a filter colorimeter use a blue filter (maximum transmission 400–420 nm); a wavelength of 410 nm is employed for a spectrophotometer. In the latter case, the effect of iron, nickel, chromium(III), and other coloured ions not reacting with hydrogen peroxide may be compensated by using a solution of the sample, not treated with hydrogen peroxide, in the reference cell.

17.34 TUNGSTEN

Discussion. Toluene-3,4-dithiol (dithiol) may be used for the colorimetric determination of tungsten; it forms a slightly soluble coloured complex with tungsten(VI) which can be extracted with butyl or pentyl acetate and other organic solvents. Molybdenum reacts similarly (see Section 17.30) and must be removed before tungsten can be determined. The molybdenum complex can be preferentially developed in cold weak acid solution and selectively extracted with pentyl acetate before developing the tungsten colour in a hot solution of increased acidity. The procedure will be illustrated by describing the determination of tungsten in steel.

Reagents. *Dithiol reagent solution.* Dissolve 1 g toluene-3,4-dithiol in 100 mL pentyl acetate. This should be prepared immediately before use.

Standard tungsten solution. Dissolve 0.1794 g sodium tungstate $Na_2WO_4,2H_2O$ in water and dilute to 1 L:1 mL $\equiv$ 0.1 g W. For use, dilute 100 mL of this solution to 1 L:1 mL $\equiv$ 0.01 mg W.

'Mixed acid'. Mix 15.0 mL concentrated sulphuric acid and 15.0 mL ortho-phosphoric acid (sp. gr. 1.75), and dilute to 100 mL with distilled water.

Procedure (tungsten in steel). Dissolve 0.5 g of the steel, accurately weighed, in 30 mL of the 'mixed acid' by heating, oxidise with concentrated nitric acid, and evaporate to fuming. Extract with 100 mL water, boil, transfer to a 500 mL graduated flask, cool, dilute to the mark with water, and mix. Pipette a 15 mL aliquot into a 50 mL flask, evaporate to fuming, cool, add 5 mL dilute hydrochloric acid (sp. gr. 1.06), warm until the salts dissolve, and cool to room temperature. Add 5 drops of 10 per cent aqueous hydroxylammonium sulphate solution, 10 mL of the dithiol reagent, and allow to stand in a bath at 20–25 °C for 15 minutes with periodic shaking. Transfer the contents quantitatively to a 25 mL separatory funnel, using 3–4 mL portions of pentyl acetate for washing. Shake and allow the layers to separate. Run off the lower acid layer containing the tungsten and reserve it in the original 50 mL flask. Wash the pentyl acetate layer twice consecutively with 5 mL portions of hydrochloric acid (sp. gr. 1.06) and combine the acid washings with the original acid layer. Discard the molybdenum-containing pentyl acetate layer. Evaporate the acid tungsten solution carefully to fuming (to expel dissolved pentyl acetate), then add a few drops of concentrated nitric acid during fuming to clear up any charred organic matter. Add 5 mL of 10 per cent tin(II) chloride solution (in concentrated hydrochloric acid) and heat to 100 °C for 4 minutes: add 10 mL of the dithiol reagent and heat at 100 °C for 10 minutes longer with periodic shaking. Transfer to a 25 mL stoppered separatory funnel, and rinse three times with 2 mL portions of pentyl acetate. Shake, separate, and draw off the lower acid layer and discard it. Add 5 mL concentrated hydrochloric acid to the organic layer, repeat the extraction and again discard the lower layer. Draw off the pentyl acetate layer containing the tungsten complex into a 50 mL graduated flask and dilute to volume with pentyl acetate. Measure the absorbance with a spectrophotometer at 630 nm in 4 cm cells, or use an absorptiometer and a red filter.

Refer the readings to a calibration curve prepared from a solution containing spectrographically pure iron to which suitable amounts of standard sodium tungstate solution have been added.

17.35 VANADIUM

Of the two methods commonly used for the determination of vanadium the second, in which phosphotungstovanadic acid is formed, is employed most frequently.

A. Vanadyl sulphate method: Discussion. When hydrogen peroxide is added to a solution containing small quantities of vanadium(V) (up to 0.1 mg of V per mL) in sulphuric acid solution, a reddish-brown coloration is produced: this is thought to be due to the formation of a compound of the type $(VO)_2(SO_4)_3$. A large excess of hydrogen peroxide tends to reduce the colour intensity and to change the colour from red–brown to yellow. With a hydrogen peroxide concentration of 0.03 per cent, the sulphuric acid concentration can vary between 0.3 and $3M$ without any appreciable effect on the colour: with higher concentrations of hydrogen peroxide, the acidity must be increased to permit development of the maximum colour intensity.

The colour is unaffected by the presence of phosphate or fluoride. Titanium and molybdenum(VI) (which give colours with hydrogen peroxide) and tungsten interfere. Titanium may be removed by adding fluoride or hydrofluoric acid, which simultaneously remove the yellow colour due to iron(III). If titanium is absent, phosphate may be used to decolorise any iron(III) salt present. Oxalic acid eliminates the interference due to tungsten. In the presence of elements which yield coloured solutions, such as chromium or nickel, it is best to add equal amounts of these elements to the standard solution. If steel is being analysed, the most convenient procedure is to use a like steel as standard.

B. Phosphotungstovanadic acid method: Discussion. Vanadium may also be determined by making use of the yellow, soluble phosphotungstovanadic acid formed upon adding phosphoric(V) acid and sodium tungstate to an acid vanadate solution. The most intense colour is obtained when the molecular ratio of phosphoric(V) acid to sodium tungstate is in the range 3:1 to 20:1, and the tungstate concentration in the test solution is 0.01 to 0.1 M; the preferred concentrations are 0.5 M in phosphoric(V) acid and 0.025 M in sodium tungstate.

The following interfere: (a) coloured ions, such as chromate, copper, and cobalt; (b) titanium, zirconium, bismuth, antimony, and tin yield slightly soluble phosphates or basic salts except in very low concentrations; (c) potassium and ammonium ions give sparingly soluble phosphotungstates; (d) molybdenum(VI) in relatively high concentration (>0.5 mg mL^{-1}); (e) iodide, thiocyanate, etc., reduce phosphotungstic acid; and (f) iron in concentrations greater than 1 mg mL^{-1} (slight interference even in the presence of phosphoric acid).

Procedure. Render the solution *ca* 0.5 M in mineral acid, and add 1.0 mL of 1:2 phosphoric(V) acid and 0.5 mL of 0.5 M sodium tungstate solution (prepared by dissolving 16.5 g sodium tungstate $Na_2WO_4,2H_2O$ in 100 mL water) for each 10 mL of test solution. Heat to boiling, cool, dilute to volume, and determine the absorbance of the resulting solution* at 400 nm (violet filter). If small amounts of coloured ions (nickel, cobalt, dichromate, etc.) are present, these should be incorporated in the comparison solution, preferably by employing an aliquot portion of the original sample solution.

Vanadium in steel. Dissolve 1.0 g, accurately weighed, of the steel in 50 mL of 1:4 sulphuric acid. When solution is complete, introduce 10 mL of concentrated nitric acid, and boil until nitrous fumes are no longer evolved. Dilute the solution to 100 mL with hot water, heat to boiling, and add saturated potassium permanganate solution until a pink colour persists or a precipitate is formed. Boil for 5 minutes. Filter off any tungsten(VI) oxide or manganese oxide which may be precipitated. Add a slight excess of freshly prepared sulphurous acid, and boil off the excess. Cool, add 5 mL phosphoric(V) acid and 5 mL of ten volume hydrogen peroxide.

Simultaneously with the main determination prepare in an analogous manner a comparison solution from a standard steel which contains no vanadium but is otherwise similar; add a standard solution of vanadium to the control, followed by hydrogen peroxide, etc., and compare this colorimetrically or spectrophotometrically with the solution obtained from the unknown steel.

ANIONS

17.36 CHLORIDE

Two procedures are commonly employed for the colorimetric determination of chloride.

---

* The yellow colour may also be extracted with 2-methylpropanol, and read at 400 nm against a reagent blank.

A. Mercury(II) chloranilate method: Discussion. The mercury(II) salt of chloranilic acid (2,5-dichloro-3,6-dihydroxy-p-benzoquinone) may be used for the determination of small amounts of chloride ion. The reaction is:

$$HgC_6Cl_2O_4 + 2Cl^- + H^+ = HgCl_2 + HC_6Cl_2O_4^-$$

The amount of reddish-purple acid-chloranilate ion liberated is proportional to the chloride ion concentration. Methyl cellosolve (2-methoxyethanol) is added to lower the solubility of mercury(II) chloranilate and to suppress the dissociation of the mercury(II) chloride; nitric acid is added (concentration $0.05 M$) to give the maximum absorption. Measurements are made at 530 nm in the visible or 305 nm in the ultraviolet region. Bromide, iodide, iodate, thiocyanate, fluoride, and phosphate interfere, but sulphate, acetate, oxalate, and citrate have little effect at the 25 mg L^{-1} level. The limit of detection is 0.2 mg L^{-1} of chloride ion; the upper limit is about 120 mg L^{-1}. Most cations, but not ammonium ion, interfere and must be removed.

Silver chloranilate cannot be used in the determination because it produces colloidal silver chloride.

Procedure. Remove interfering cations by passing the aqueous solution containing the chloride ion through a strongly acidic ion exchange resin in the hydrogen form (e.g. Zerolit 225 or Amberlite 120) contained in a tube 15 cm long and 1.5 cm in diameter. Adjust the pH of the effluent to 7 with dilute nitric acid or aqueous ammonia and pH paper. To an aliquot containing not more than 1 mg of chloride ion in less than 45 mL of water in a 100 mL graduated flask, add 5 mL 1 M nitric acid and 50 mL methylcellosolve. Dilute the mixture to volume with distilled water, add 0.2 g mercury(II) chloranilate, and shake the flask intermittently for 15 minutes. Separate the excess of mercury(II) chloranilate by filtration through a fine ashless filter paper or by centrifugation. Measure the absorbance of the clear solution with a spectrophotometer at 530 nm (yellow–green filter) against a blank prepared in the same manner.

Construct a calibration curve using standard ammonium chloride solution (1–100 mg L^{-1} Cl$^-$) and deduce the chloride ion concentration of the test solution with its aid.

Mercury(II) chloranilate may be prepared by adding dropwise a 5 per cent solution of mercury(II) nitrate in 2 per cent nitric acid to a stirred solution of chloranilic acid at 50 °C until no further precipitate forms. Decant the supernatant liquid, wash the precipitate three times by decantation with ethanol and once with diethyl ether, and dry in a vacuum oven at 60 °C. The compound is available commercially.

Mercury(II) thiocyanate method: Discussion. This second procedure for the determination of trace amounts of chloride ion depends upon the displacement of thiocyanate ion from mercury(II) thiocyanate by chloride ion; in the presence of iron(III) ion a highly coloured iron(III) thiocyanate complex is formed, and the intensity of its colour is proportional to the original chloride ion concentration:

$$2Cl^- + Hg(SCN)_2 + 2Fe^{3+} = HgCl_2 + 2[Fe(SCN)]^{2+}$$

The method is applicable to the range 0.5–100 μg of chloride ion.

Procedure. Place a 20 mL aliquot of the chloride solution in a 25 mL graduated flask, add 2.0 mL of 0.25 M ammonium iron(III) sulphate [$Fe(NH_4)(SO_4)_2$, $12H_2O$] in 9 M nitric acid, followed by 2.0 mL of a saturated solution of mercury(II) thiocyanate in ethanol. After 10 minutes measure the absorbance of the sample solution and also of the blank in 5 cm cells in a spectrophotometer at 460 nm (blue filter) against water in the reference cell. The amount of chloride ion in the sample corresponds to the difference between the two absorbances and is obtained from a calibration curve.

Construct a calibration curve using a standard sodium chloride solution containing 10 μg Cl^- mL^{-1}: cover the range 0–50 μg as above. Plot absorbance against micrograms of chloride ion.

17.37 FLUORIDE

Fluoride, in the absence of interfering anions (including phosphate, molybdate, citrate, and tartrate) and interfering cations (including cadmium, tin, strontium, iron, and particularly zirconium, cobalt, lead, nickel, zinc, copper, and aluminium), may be determined with thorium chloranilate in aqueous 2-methoxyethanol at pH 4.5; the absorbance is measured at 540 nm or, for small concentrations 0–2.0 mg L^{-1} at 330 nm.

In water as solvent, the reaction is:

$$Th(C_6Cl_2O_4)_2 + 6F^- + 2H^+ \rightleftharpoons ThF_6^{2-} + 2HC_6Cl_2O_4^-$$

In aqueous 2-methoxyethanol, the main reaction is stated to be:

$$Th(C_6Cl_2O_4)_2 + 2F^- + H^+ \rightleftharpoons ThF_2C_6O_4 + HC_6Cl_2O_4^-$$

Interfering cations, except aluminium and zirconium, can be removed by passage through an ion exchange column. In the presence of interfering anions and also aluminium and zirconium, fluoride may be separated as hydrofluorosilicic acid by distilling with dilute perchloric acid at 135 °C (temperature maintained by the addition of water) in the presence of a few glass beads.

A calibration curve for the range 0.2–10 mg fluoride ion per 100 mL is constructed as follows. Add the appropriate amount of standard sodium fluoride solution, 25 mL of 2-methoxyethanol, and 10 mg of a buffer [0.1 M in both sodium acetate and acetic (ethanoic) acid] to a 100 mL graduated flask. Dilute to volume with distilled water and add about 0.05 g of thorium chloranilate. Shake the flask intermittently for 30 minutes (the reaction in the presence of 2-methoxyethanol is about 90 per cent complete after 30 minutes and almost complete after 1 hour) and filter about 10 mL of the solution through a dry Whatman No. 42 filter paper. Measure the absorbance of the filtrate in a 1 cm cell at 540 nm (yellow–green filter) against a blank, prepared in the same manner, using a suitable spectrophotometer. Prepare a calibration curve for the concentration range 0.0–0.2 mg fluoride ion per 100 mL in the same way, but add only 10.0 mL of 2-methoxyethanol; measure the absorbance of the filtrate in a 1 cm silica cell at 330 nm.

Treat the fluoride sample solution in the same manner as described for the calibration curve after removing interfering ions and adjusting the pH to about 5 with dilute nitric acid or sodium hydroxide solution. Read off the fluoride concentration from the calibration curve and the observed value of the absorbance.

17.38 NITRITE

Discussion. General procedures for the determination of nitrites are usually based upon some form of diazotisation reaction, often involving carcinogenic materials such as the naphthylamines. In the following method these compounds are avoided.

In this case the nitrite ion, under acidic conditions, causes diazotisation of sulphanilamide (4-aminobenzenesulphonamide) to occur, and the product is coupled with *N*-(1-naphthyl)ethylenediamine dihydrochloride.

Reagents. *Sulphanilamide solution (A).* Dissolve 0.5 g sulphanilamide in 100 mL of 20 per cent v/v hydrochloric acid.

N- (1-naphthyl)-ethylenediamine dihydrochloride solution (B). Dissolve 0.3 g of the solid reagent in 100 mL of 1 per cent v/v hydrochloric acid.

Procedure. To 100 mL of the neutral sample solution (containing not more than 0.4 mg nitrite) add 2.0 mL of solution A and, after 5 minutes, 2.0 mL of solution B. The pH at this point should be about 1.5. Measure the absorbance after 10 minutes in the wavelength region of 550 nm (yellow–green filter), in a spectrophotometer against a blank solution prepared in the same manner. Calculate the concentration of the nitrite from a calibration plot prepared from a series of standard nitrite solutions.

17.39 PHOSPHATE

Two methods are commonly used for the determination of phosphate.

A. Molybdenum blue method: Discussion. Orthophosphate and molybdate ions condense in acidic solution to give molybdophosphoric acid (phosphomolybdic acid), which upon selective reduction (say, with hydrazinium sulphate) produces a blue colour, due to molybdenum blue of uncertain composition. The intensity of the blue colour is proportional to the amount of phosphate initially incorporated in the heteropoly acid. If the acidity at the time of reduction is $0.5\,M$ in sulphuric acid and hydrazinium sulphate is the reductant, the resulting blue complex exhibits maximum absorption at 820–830 nm.

B. Phosphovanadomolybdate method: Discussion. This second method is considered to be slightly less sensitive than the previous molybdenum blue method, but it has been particularly useful for phosphorus determinations carried out by means of the Schöniger oxygen flask method (Section 3.31). The phosphovanadomolybdate complex formed between the phosphate, ammonium vanadate, and ammonium molybdate is bright yellow in colour and its absorbance can be measured between 460 and 480 nm.

Reagents. *Ammonium vanadate solution.* Dissolve 2.5 g ammonium vanadate (NH_4VO_3) in 500 mL hot water, add 20 mL concentrated nitric acid and dilute with water to 1 mL in a graduated flask.

Ammonium molybdate solution. Dissolve 50 g ammonium molybdate, $(NH_4)_6$ $Mo_7O_{24},4H_2O$, in warm water and dilute to 1 L in a graduated flask. Filter the solution before use.

Procedure. Dissolve 0.4 g of the phosphate sample in $2.5\,M$ nitric acid to give 1 L in a graduated flask. Place a 10 mL aliquot of this solution in a 100 mL graduated flask, add 50 mL water, 10 mL of the ammonium vanadate solution, 10 mL of the ammonium molybdate solution and dilute to the mark. Determine the absorbance of this solution at 465 nm against a blank prepared in the same manner, using 1 cm cells.

Prepare a series of standards from potassium dihydrogenphosphate covering the range 0–2 mg phosphorus per 100 mL and containing the same concentration of acid, ammonium vanadate, and ammonium molybdate as the previous solution. Construct a calibration curve and use it to calculate the concentration of phosphorus in the sample.

17.40 SILICATE

Discussion. Small quantities of dissolved silicic acid react with a solution of a molybdate in an acid medium to give an intense yellow coloration, due probably to the complex molybdosilicic acid $H_4[SiMo_{12}O_{40}]$. The latter may be employed as a basis for the colorimetric determination of silicate (absorbance measurements at 400 nm). It is usually better to reduce the complex acid to molybdenum blue (the composition is uncertain); a solution of a mixture of 1-amino-2-naphthol-4-sulphonic acid and sodium hydrogensulphite solution is a satisfactory reducing agent.

Phosphates, arsenates, and germanates give similar colorations and either they must be removed or their interferences must be eliminated by the addition of suitable reagents: arsenic and germanium can be removed by evaporation with hydrochloric acid, and phosphate by precipitation as ammonium magnesium phosphate in acetic (ethanoic) acid solution, or it may be rendered innocuous by the addition of ammonium citrate. Elements such as barium, bismuth, lead, and antimony give precipitates or turbidities, and must be absent. Water used for dilution should be freshly distilled in an all-Pyrex apparatus or passed through a mixed bed ion exchange column, and stored in polythene containers. Water tends to dissolve significant traces of silica on standing in glass, particularly soda-glass, vessels.

Reagents. *Ammonium molybdate solution.* Dissolve 8.0 g ammonium molybdate crystals in water, add 9 mL concentrated sulphuric acid, and dilute to 100 mL.

Reducing agent. Solution A: dissolve 10 g sodium hydrogensulphite in 70 mL water. Solution B: dissolve 0.8 g anhydrous sodium hydrogensulphite in 20 mL water, and add 0.16 g 1-amino-2-naphthol-4-sulphonic acid. Mix solution A with solution B, and dilute to 100 mL.

Tartaric acid solution. Prepare a 10 per cent aqueous solution.

Standard solution of silica. Fuse 0.107 g of pure, dry precipitated silica with 1.0 g of anhydrous sodium carbonate in a platinum crucible. Cool the melt, dissolve it in de-ionised water, dilute to 500 mL, and store in a polythene bottle. 1 mL $\equiv$ 0.1 mg Si. Dilute as appropriate, say, to 1 mL $\equiv$ 0.01 mg Si.

Procedure. The sample solution, free from interfering elements and radicals, may conveniently occupy a volume of about 50 mL and contain between 0.01 and 0.1 mg of silica; the pH should be 4.5–5.0 Add 1 mL of the ammonium molybdate solution and, after 5 minutes, add 5 mL of the tartaric acid solution and mix. Introduce 1.0 mL of the reducing agent and dilute to 100 mL in a graduated flask. Measure the absorbance at *ca* 815 nm after 20 minutes against de-ionised water.

Construct a calibration curve using 0, 1.0, 2.5, 5.0, 7.5, and 10.0 mL of the standard silica solution (1 mL $\equiv$ 0.01 mg Si) which have been treated similarly.

17.41 SULPHATE

Discussion. The barium salt of chloranilic acid (2,5-dichloro-3,6-dihydroxy-*p*-benzoquinone) illustrates the principle of a method which may find wide application in the colorimetric determination of various anions. In the reaction

$$Y^- + MA \text{ (solid)} = A^- + MY \text{ (solid)}$$

where Y^- is the anion to be determined and A^- is the coloured anion of an organic acid, MY must be so much less soluble than MA that the reaction is quantitative. MA must be only sparingly soluble so that the blanks will not be too high. Sulphate ion in the range 2–400 mg L^{-1} may be readily determined by utilising the reaction between barium chloranilate with sulphate ion in acid solution to give barium sulphate and the acid-chloranilate ion:

$$SO_4^{2-} + BaC_6Cl_2O_4 + H^+ = BaSO_4 + HC_6Cl_2O_4^-$$

The amount of acid chloranilate ion liberated is proportional to the sulphate-ion concentration. The reaction is carried out in 50 per cent aqueous ethanol buffered at an apparent pH of 4. Most cations must be removed because they form insoluble chloranilates: this is simply effected by passage of the solution through a strongly acidic ion exchange resin in the hydrogen form (see Section 7.2). Chloride, nitrate, hydrogencarbonate, phosphate, and oxalate do not interfere at the 100 mg L^{-1} level. The pH of the solution governs the absorbance of chloranilic acid solutions at a particular wavelength; chloranilic acid is yellow, acid-chloranilate ion is dark purple, and chloranilate ion is light purple. At pH 4 the acid-chloranilate ion gives a broad peak at 530 nm, and this wavelength is employed for measurements in the visible region. A much more intense absorption occurs in the ultraviolet: a sharp band at 332 nm enables the limit of detection of sulphate ion to be extended to 0.06 mg L^{-1}.

Procedure. Pass the aqueous solution containing sulphate ion (2–400 mg L^{-1}) through a column 1.5 cm in diameter and 15 cm long of Zerolit 225 or equivalent cation exchange resin in the hydrogen form. Adjust the effluent to pH 4 with dilute hydrochloric acid or ammonia solution. Make up to volume in a graduated flask. To an aliquot containing up to 40 mg of sulphate ion in less than 40 mL in a 100 mL graduated flask, add 10 mL of a buffer (pH = 4; a 0.05 M solution of potassium hydrogenphthalate) and 50 mL of 95 per cent ethanol. Dilute to the mark with distilled water, add 0.3 g of barium chloranilate and shake the flask for 10 minutes. Remove the precipitated barium sulphate and the excess of barium chloranilate by filtering or centrifuging. Measure the absorbance of the filtrate with a filter colorimeter or a spectrophotometer at 530 nm (green filter) against a blank prepared in the same manner. Construct a calibration curve using standard potassium sulphate solutions prepared from the analytical grade salt.

SOME TYPICAL ORGANIC COMPOUNDS

17.42 PRIMARY AMINES

The determination of primary amines on the macro scale is most conveniently carried out by titration in non-aqueous solution (Section 10.41), but for small quantities of amines spectroscopic methods of determination are very valuable. In some cases the procedure is applicable to aromatic amines only, and the diazotisation method described for determination of nitrite (Section 17.38) can be adapted as a method for the determination of aromatic primary amines. On the other hand, the naphthaquinone method can be applied to both aliphatic and aromatic primary amines.

A. Diazotisation method: Discussion. In this procedure the amine is diazotised and then coupled with *N*-(1-naphthyl)ethylenediamine. This leads to formation of a coloured product whose concentration can be determined with an absorptiometer or a spectrophotometer.

Reagents. N-*(1-naphthyl) ethylenediamine dihydrochloride*. Dissolve 0.3 g of the solid in 100 mL of 1 per cent v/v hydrochloric acid (solution A).
  Sodium nitrite. Dissolve 0.7 g sodium nitrite in 100 mL distilled water (solution B).
  1M Hydrochloric acid.
  90 per cent Ethanol (rectified spirit).

Procedure. Weigh out 10–15 mg of the amine sample and dissolve in 1*M* hydrochloric acid in a 50 mL graduated flask. Place 2.0 mL of this solution in a small conical flask clamped in a 400 mL beaker containing tap water, and then add 1 mL of solution B; allow to stand for 5 minutes. Now add 5 mL of 90 per cent ethanol and after waiting a further 3 minutes, add 2 mL of solution A. A red coloration develops rapidly, the absorbance of which can be measured against a blank solution containing all the reagents except the amine. The measurement should be made at a wavelength of about 550 nm (yellow–green filter); the exact value for λ_{max} varies slightly with the nature of the amine. A calibration curve can be prepared using a series of solutions of the pure amine of appropriate concentrations, which are treated in the manner described above.

B. Naphthaquinone method: Discussion. Many primary amines develop a blue colour when treated with ortho-quinones; the preferred reagent is the sodium salt of 1,2-naphthaquinone-4-sulphonic acid.

Reagents. *1,2-Naphthaquinone-4-sulphonic acid sodium salt.* Dissolve 0.4 g of the solid in 100 mL distilled water (solution C).
  Buffer solution. Dissolve 4.5 g of disodium hydrogenphosphate in 1 L of distilled water and carefully add 0.1*M* sodium hydroxide solution to give a pH of 10.2–10.4 (pH meter).

Procedure. Place 25 mL of an aqueous solution of the amine, 10 mL of solution C and 1 mL of buffer solution in a 100 mL stoppered conical flask: the amount of amine should not exceed 10 µg. After 1 minute add 10 mL chloroform and stir on a magnetic stirrer for 15–20 minutes. Transfer to a

705

separatory funnel, and after the phases have separated, run off the chloroform layer. Measure the absorbance at a wavelength of 450 nm (violet filter) against a chloroform blank.

17.43 CARBONYL GROUP

When present in macro quantities, aldehydes and ketones can be determined by conversion to the 2,4-dinitrophenylhydrazone which can be collected and weighed. When present in smaller quantities ($10^{-3}M$ or less), although hydrazone formation takes place, it does not separate from methanol solution, but if alkali is added an intense red coloration develops: the reagent itself only produces a slight yellow colour. Measurement of the absorbance of the red solution thus provides a method for quantitative determination.

Reagents. *Methanol.* This must be free from aldehydes and ketones: if necessary reflux 1 L of the purest material available for 2 hours with 5 g of 2,4-dinitrophenylhydrazine and five drops of concentrated hydrochloric acid. Then distil the methanol through a fractionating column and collect the fraction boiling at 64.5–65.5 °C.

2,4-Dinitrophenylhydrazine. Prepare a saturated solution in 20 mL of the purified methanol. This solution should be discarded after one week (solution A).

Potassium hydroxide. Dissolve 10 g of solid in 20 mL distilled water and then make up to 100 mL with purified methanol.

Procedure. Dissolve 0.1 g of the sample in 10 mL purified methanol and transfer 1.0 mL of this solution to a stoppered test-tube. Add 1.0 mL of solution A and one drop of concentrated hydrochloric acid, then place the stoppered tube in a beaker of boiling water for 5 minutes. Cool, and then add 5.0 mL of the potassium hydroxide solution. Measure the absorbance of the solution at 480 nm (blue–green filter) against a blank obtained by subjecting 1.0 mL of purified methanol to the above procedure.

17.44 ANIONIC DETERGENTS

An early method developed for the assay of detergents based upon the sodium salts of the higher homologues of the alkanesulphonic acids,[21] involved treatment of an aqueous solution of the detergent with methylene blue in the presence of chloroform. Reaction takes place between the ionic dyestuff (which is a chloride) and the detergent:

$$(MB^+)Cl^- + RSO_3Na \rightarrow (MB^+)(RSO_3^-) + NaCl$$

where MB^+ indicates the cation of methylene blue.

The reaction product can be extracted by chloroform, whilst the original dyestuff is insoluble in this medium, and the intensity of the colour in the chloroform layer is proportional to the concentration of the detergent. The method is specially valuable for the determination of small concentrations of detergents and is therefore useful in pollution studies.

Reagents. *Methylene blue solution.* Dissolve 0.1 g of the solid (use redox indicator quality) in 100 mL distilled water.

Hydrochloric acid (6M).

Chloroform.

Procedure. Weigh sufficient solid material to contain 0.001–0.004 mmol detergent and dissolve in 100 mL distilled water. Place 20 mL of this solution in a 150 mL separatory funnel and neutralise by adding 6 M hydrochloric acid dropwise; when neutrality is achieved (use a test-paper), add 3–4 further drops of acid. Add 20 mL chloroform (trichloromethane) and 1 mL of the methylene blue solution. Shake for 1 minute, allow to stand for 5 minutes, then run the chloroform layer into a 100 mL separatory funnel; retain the aqueous layer (solution A). Add 20 mL distilled water to the chloroform solution, shake for one minute, allow to stand for 5 minutes, then transfer the chloroform layer through a small filter funnel with a cotton-wool plug at the apex, into a 100 mL graduated flask. Repeat the extraction of solution A three more times following the above procedure exactly. After the final extraction rinse the filter funnel and plug with a little chloroform, and finally make the solution in the graduated flask up to the mark with chloroform.

Measure the absorbance of this solution at 650 nm (red filter) in a 1 cm cell against a blank of distilled water, and also carry out the above procedure on the reagents, using distilled water in place of the sample solution: use the resultant absorbance reading to correct that given by the sample if necessary. If not already available, a calibration curve must be constructed for the detergent being measured, using four appropriate dilutions of a standard solution and carrying out the procedure detailed above on each.

An obvious modification of the above procedure will permit the determination of long-chain amines or quaternary ammonium salts (cationic surfactants):

$$R_3NH^+X^- + R'SO_3^- \rightarrow (R_3NH^+)(R'SO_3^-) + X^-$$

In this case the sulphonic acid group is present in a sulphon-phthalein dye: namely the indicator bromophenol blue. As in the previous example, the species $(R_3NH^+)(R'SO_3^-)$ can be extracted into chloroform whilst the indicator itself is not extracted, and the colour of the extract is proportional to the quantity of surfactant in the material under test.

17.45 PHENOLS

Discussion. The formation of coloured compounds by coupling phenols with diazotised primary aromatic amines has long been recognised as a method of determining phenols, and procedures have been evolved whereby the phenol solution is titrated with a diazonium solution which has been calibrated against known concentrations of the phenol. The resultant reaction products are coloured, but many are only sparingly soluble in water and organic solvents and do not therefore lend themselves to colorimetric determination.

However, 4-aminophenazone when treated with a mild oxidising agent such as hexacyanoferrate(III) ions in alkaline solution and in the presence of a phenol, reacts to form a dye which can be extracted with chloroform. The absorbance of this solution can be used to ascertain the amount of phenol present and this provides a good method for the determination of traces of phenols.

The method can be applied to most phenols substituted in the *ortho* or *meta* position and to phenols substituted in the *para* position by OH or OCH$_3$ groups; most other substituents in the *para* position inhibit the reaction. Aromatic primary amines unsubstituted in the *para* position interfere and must be removed if possible before commencing the determination by extraction with acid.

Reagents. *4-aminophenazone.* Dissolve 3 g of the compound in 100 mL of distilled water. (Solution A).

Potassium hexacyanoferrate(III) (potassium ferricyanide). Dissolve 2 g of the solid in 100 mL distilled water. (Solution B).

Aqueous ammonia (4M).

Procedure. Dissolve the sample in distilled water and take an aliquot which should contain not more than 50 μg of phenolic compound. Use the aqueous ammonia to adjust the pH of the solution to 9.7–10.3 (pH meter), and then dilute to 500 mL with distilled water. Transfer the solution to a large separatory funnel, add 1.0 mL of solution A followed by 10 mL of solution B. Shake well to ensure thorough mixing, and then carry out three extractions with successive portions of 15 mL, 10 mL and 5 mL of chloroform (trichloromethane). Combine the chloroform extracts and make up the volume to 30 mL. Measure the absorbance of the extract against a blank of chloroform at a wavelength of 460 nm (blue filter), using 1 cm cells. The colour may tend to fade after 10 minutes and so speed is essential.

EXPERIMENTAL DETERMINATIONS WITH ULTRAVIOLET/VISIBLE SPECTROPHOTOMETERS

17.46 DETERMINATION OF THE ABSORPTION CURVE AND CONCENTRATION OF A SUBSTANCE (POTASSIUM NITRATE)

Discussion. Potassium nitrate is an example of an inorganic compound which absorbs mainly in the ultraviolet, and can be employed to obtain experience in the use of a manually operated ultraviolet/visible spectrophotometer. Some of the exercise can also be carried out employing an automatic recording spectrophotometer (see Section 17.16).

The absorbance and the percentage transmission of an approximately 0.1 M potassium nitrate solution is measured over the wavelength range 240–360 nm at 5 nm intervals and at smaller intervals in the vicinity of the maxima or minima. Manual spectrophotometers are calibrated to read both absorbance and percentage transmission on the dial settings, whilst the automatic recording double beam spectrophotometers usually use chart paper printed with both scales. The linear conversion chart, Fig. 17.18, is useful for visualising the relationship between these two quantities.

The three normal means of presenting the spectrophotometric data are described below: by far the most common procedure is to plot absorbance against wavelength (measured in nanometres). The wavelength corresponding to the absorbance maximum (or minimum transmission) is read from the plot and is used for the preparation of the calibration curve. This point is chosen

Fig. 17.18

for two reasons: (1) it is the region in which the greatest difference in absorbance between any two different concentrations will be obtained, thus giving the maximum sensitivity for concentration studies, and (2) as it is a turning point on the curve it gives the least alteration in absorbance value for any slight variation in wavelength.

No general rule can be given concerning the strength of the solution to be prepared, as this will depend upon the spectrophotometer used for the study. Usually a $0.01-0.001\,M$ solution is sufficiently concentrated for the highest absorbances, and other concentrations are prepared by dilution. The concentrations should be selected such that the absorbance lies between about 0.3 and 1.5.

For the determination of the concentration of a substance, select the wavelength of maximum absorption for the compound (e.g. 302.5–305 nm for potassium nitrate) and construct a calibration curve by measuring the absorbances of four or five concentrations of the substance (e.g. 2, 4, 6, 8 and $10\,\mathrm{g}\,KNO_3\,L^{-1}$) at the selected wavelength. Plot absorbance (ordinates) against concentration (abiscissae). If the compound obeys Beer's Law a linear calibration curve, passing through the origin, will be obtained. If the absorbance of the unknown solution is measured the concentration can be obtained from the calibration curve.

If it is known that the compound obeys Beer's Law the molar absorption coefficient ε can be determined from one measurement of the absorbance of a

709

standard solution. The unknown concentration is then calculated using the value of the constant ε and the measured value of the absorbance under the same conditions.

Procedure. Dry some pure potassium nitrate at 110 °C for 2–3 hours and cool in a desiccator. Prepare an aqueous solution containing $10.000\,\mathrm{g\,L^{-1}}$. With the aid of a precision spectrophotometer* and matched 1 cm rectangular cells, measure the absorbance and the percentage transmission over a series of wavelengths covering the range 240–350 nm. Plot the data in three different ways: (1) absorbance against wavelength; (2) percentage transmission against wavelength; and (3) $\log \varepsilon$ (molecular decadic absorption coefficient) against wavelength. The curves obtained for potassium nitrate are shown in Figs. 17.19(a)–(c). From the curves, evaluate the wavelength of maximum absorption (or minimum transmission). Use this value of the wavelength to determine the absorbance of solutions of potassium nitrate containing 2.000, 4.000, 6.000, and 8.000 g of potassium nitrate $\mathrm{L^{-1}}$. Run a blank on the two cells, filling both the blank cell and the sample cell with distilled water; if the cells are correctly matched no difference in absorbance should be discernible. Plot the absorbances (ordinates) against concentration.

Determine the absorbance of an unknown solution of potassium nitrate and read the concentration from the calibration curve.

17.47 THE EFFECT OF SUBSTITUENTS ON THE ABSORPTION SPECTRUM OF BENZOIC ACID

In Section 17.13 reference has been made to the influence of various substituents in the benzene ring on the absorption of ultraviolet radiation, and the purpose of this exercise is to examine the effect in the case of benzoic acid by comparing the absorption spectrum of benzoic acid with those given by 4-hydroxybenzoic acid and 4-aminobenzoic acid.

Reagents. *Hydrochloric acid (0.1 M).*
Sodium hydroxide solution (0.1 M).
Benzoic acid (1).
4-Hydroxybenzoic acid (2).
4-Aminobenzoic acid (3).

Procedure. Prepare a solution of benzoic acid in distilled water by dissolving 0.100 g in a 100 mL graduated flask and making up to the mark (solution A_1). Prepare similar solutions of the other two acids giving solutions A_2 and A_3. Now take 10.0 mL of solution A_1 and dilute to the mark in a 100 mL graduated flask with distilled water giving a solution B_1 containing 0.01 mg of benzoic acid $\mathrm{mL^{-1}}$. Prepare similar solutions B_2 and B_3 from solutions A_2 and A_3.

Use a recording spectrophotometer to plot the absorption curves of the three separate solutions, in each case using distilled water as the blank. Use silica cells and record the spectra over the range 210–310 nm.

Now prepare a new solution of 4-hydroxybenzoic acid (solution C_2) by

---

*When reporting spectrophotometric measurements, details should be given of the concentration used, the solvent employed, the make and model of the instrument, as well as the slit widths employed, together with any other pertinent information.

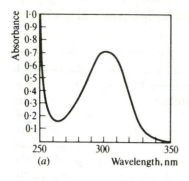

(a)   Wavelength, nm

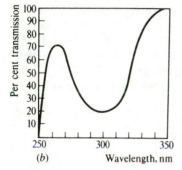

(b)   Wavelength, nm

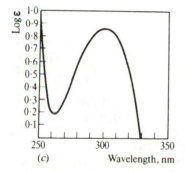

(c)   Wavelength, nm

Fig. 17.19

placing 10.0 mL of solution A_2 in a 100 mL graduated flask and making up to the mark with 0.1 M hydrochloric acid solution. Prepare a similar solution C_3 for the 4-aminobenzoic acid and plot the absorption curves of these two solutions.

Prepare another 4-hydroxybenzoic acid solution (D_2) by placing 10.0 mL of solution A_2 in a 100 mL graduated flask and making up to the mark with 0.1 M sodium hydroxide solution. Prepare a similar solution D_3 from solution A_3 and record the absorption spectra of these two new solutions.

Examine the seven absorption spectra, record the λ_{max} values of absorption peaks; comment upon the effect of —OH and —NH$_2$ groups upon the absorption spectrum of benzoic acid, and of hydrochloric acid and sodium hydroxide upon the spectra of the two substituted benzoic acids.

17.48   SIMULTANEOUS SPECTROPHOTOMETRIC DETERMINATION (CHROMIUM AND MANGANESE)

Discussion.   This section is concerned with the simultaneous spectrophotometric determination of two solutes in a solution. The absorbances are additive, provided there is no reaction between the two solutes. We may write:

$$A_{\lambda_1} = {}_{\lambda_1}A_1 + {}_{\lambda_1}A_2 \tag{14}$$

$$A_{\lambda_2} = {}_{\lambda_2}A_1 + {}_{\lambda_2}A_2 \tag{15}$$

where A_{λ_1} and A_{λ_2} are the *measured* absorbances at the two wavelengths λ_1 and λ_2; the subscripts 1 and 2 refer to the two different substances, and the subscripts λ_1 and λ_2 refer to the different wavelengths. The wavelengths are selected to coincide with the absorption maxima of the two solutes: the absorption spectra of the two solutes should not overlap appreciably (compare Fig. 17.20), so that substance 1 absorbs strongly at wavelength λ_1 and weakly at wavelength λ_2, and substance 2 absorbs strongly at λ_2 and weakly at λ_1. Now $A = \varepsilon cl$, where ε is the molar absorption coefficient at any particular wavelength, c is the concentration expressed in $mol\,L^{-1}$, and l is the thickness (length) of the absorbing solution expressed in cm. If l is 1 cm:

$$A_{\lambda_1} = {}_{\lambda_1}\varepsilon_1 . c_1 + {}_{\lambda_1}\varepsilon_2 . c_2 \tag{16}$$

$$A_{\lambda_2} = {}_{\lambda_2}\varepsilon_1 . c_1 + {}_{\lambda_2}\varepsilon_2 . c_2 \tag{17}$$

Solution of these simultaneous equations gives:

$$c_1 = \frac{{}_{\lambda_2}\varepsilon_2 . A_{\lambda_1} - {}_{\lambda_1}\varepsilon_2 . A_{\lambda_2}}{{}_{\lambda_1}\varepsilon_1 . {}_{\lambda_2}\varepsilon_2 - {}_{\lambda_1}\varepsilon_2 . {}_{\lambda_2}\varepsilon_1} \tag{18}$$

$$c_2 = \frac{{}_{\lambda_1}\varepsilon_1 . A_{\lambda_2} - {}_{\lambda_2}\varepsilon_1 . A_{\lambda_1}}{{}_{\lambda_1}\varepsilon_1 . {}_{\lambda_2}\varepsilon_2 - {}_{\lambda_1}\varepsilon_2 . {}_{\lambda_2}\varepsilon_1} \tag{19}$$

The values of the molar absorption coefficients ε_1 and ε_2 can be deduced from measurements of the absorbances of pure solutions of substances 1 and 2. By measuring the absorbance of the mixture at wavelengths λ_1 and λ_2, the concentrations of the two components can be calculated.

The above considerations will be illustrated by the simultaneous determination of manganese and chromium in steel and other ferro-alloys. The absorption spectra of 0.001 M permanganate and dichromate ions in 1 M sulphuric acid, determined with a spectrophotometer and against 1 M sulphuric acid in the reference cell, are shown in Fig. 17.20. For permanganate, the absorption maximum is at 545 nm, and a small correction must be applied for dichromate absorption. Similarly the peak dichromate absorption is at 440 nm, at which permanganate only absorbs weakly. Absorbances for these two ions, individually and in mixtures, obey Beer's Law provided the concentration of sulphuric acid is at least 0.5 M. Iron(III), nickel, cobalt, and vanadium absorb at 425 nm and 545 nm, and should be absent or corrections must be made.

Reagents.   *Potassium dichromate.* 0.002 M, 0.001 M, and 0.0005 M in 1 M sulphuric acid and 0.7 M phosphoric(V) acid, prepared from the analytical grade reagents.

Potassium permanganate. 0.002 M, 0.001 M, and 0.0005 M in 1 M sulphuric

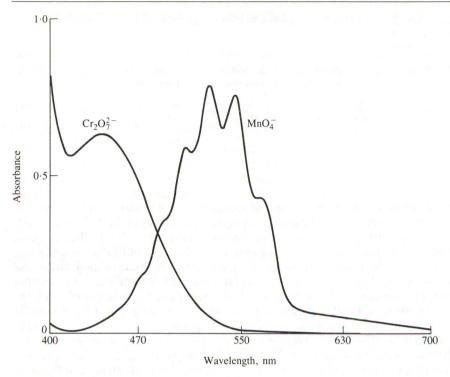

Fig. 17.20

acid and $0.7\,M$ phosphoric(V) acid, prepared from the analytical grade reagents. All flasks must be scrupulously clean.

Procedure. (*a*) *Determination of molar absorption coefficients and verification of additivity of absorbances.* The molar absorption coefficients must be determined for the particular set of cells and the spectrophotometer employed. For the present purpose we may write:

$$A = \varepsilon\, cl$$

where ε is the molar absorption coefficient, c is the concentration (mol L^{-1}), and l is the cell thickness or length (cm).

Measure the absorbance A of the above three solutions of potassium dichromate and of potassium permanganate, each solution separately, at both 440 nm and 545 nm in 1 cm cells. Calculate ε in each case and record the mean values for $Cr_2O_7^{2-}$ and MnO_4^- at the two wavelengths.

Mix $0.001\,M$ potassium dichromate and $0.0005\,M$ potassium permanganate in the following amounts (plus 1.0 mL of concentrated sulphuric acid), and prepare a set of results similar to those in Table 17.5, which is a set of typical results included for guidance only. Measure the absorbance of each of the mixtures at 440 nm. Calculate the absorbance of the mixtures from:

$$A_{440} = {}_{440}\varepsilon_{Cr} \cdot c_{Cr} + {}_{440}\varepsilon_{Mn} \cdot c_{Mn}$$

713

Table 17.5   Test of additivity principle with $Cr_2O_7^{2-}$ and MnO_4^- mixtures at 440 nm

$K_2Cr_2O_7$ solution (mL)	$KMnO_4$ solution (mL)	A observed	A calculated
50	0	0.371	—
45	5	0.338	0.340
40	10	0.307	0.308
35	15	0.277	0.277
25	25	0.211	0.214
15	35	0.147	0.151
5	45	0.086	0.088
0	50	0.057	—

(b) *Determination of chromium and manganese in an alloy steel.** Weigh out accurately about 1.0 g of the alloy steel into a 300 mL Kjeldahl flask, add 30 mL of water and 10 mL of concentrated sulphuric acid [also 10 mL of 85 per cent phosphoric(V) acid if tungsten is present]. Boil gently until decomposition is complete or the reaction subsides. Then add 5 mL of concentrated nitric acid in several small portions. If much carbonaceous residue persists, add a further 5 mL of concentrated nitric acid, and boil down to copious fumes of sulphuric acid. Dilute to about 100 mL and boil until all salts have dissolved. Cool, transfer to a 250 mL graduated flask, and dilute to the mark.

Pipette a 25 mL or 50 mL aliquot of the clear sample solution into a 250 mL conical flask, add 5 mL concentrated sulphuric acid, 5 mL 85 per cent phosphoric(V) acid, and 1–2 mL of 0.1 M silver nitrate solution, and dilute to about 80 mL. Add 5 g potassium persulphate, swirl the contents of the flask until most of the salt has dissolved, and heat to boiling. Keep at the boiling point for 5–7 minutes. Cool slightly, and add 0.5 g pure potassium periodate. Again heat to boiling and maintain at the boiling point for about 5 minutes. Cool, transfer to a 100 mL graduated flask, and measure the absorbances at 440 nm and 545 nm in 1 cm cells.

Calculate the percentage of chromium and manganese in the sample. Use equations (18) and (19) and values of the molar absorption coefficients ε determined above: these will give concentrations expressed in mol L^{-1}, from which values the percentages can readily be calculated. Each value will require correction for the amounts of vanadium, cobalt, nickel, and iron which may be present, using Table 17.6. The values listed are the equivalent percentages of the respective constituent to be subtracted from the apparent Cr and Mn percentages for each 1 per cent of the element in question. It can be shown that utilising the known (or determined) molar absorption coefficients ($_{545}\varepsilon_{Mn}$ 2.35; $_{545}\varepsilon_{Cr}$ 0.011; $_{440}\varepsilon_{Cr}$ 0.369; $_{440}\varepsilon_{Mn}$ 0.095):

$$\text{Mn, per cent} = \frac{0.005\,49\,V}{W}(0.426\,A_{545} - 0.013\,A_{440})$$

$$\text{Cr, per cent} = \frac{0.010\,40\,V}{W}(2.71\,A_{440} - 0.110\,A_{545})$$

for a sample of W grams in a volume of V mL.

---

* British Chemical Standard BCS–CRM No. 225/2 Ni–Cr–Mo steel is suitable for practice in this determination.

Table 17.6 Corrections for interfering substances

Substance	Cr correction at 440 nm (per cent)	Mn correction at 545 nm (per cent)
$Cr_2O_7^{2-}$	—	0.0025
MnO_4^-	0.490	—
VO_2^+	0.0266	—
Co^{2+}	0.0072	0.0011
Ni^{2+}	0.0039	0.0001
Fe^{3+}	0.0005	—

17.49 DETERMINATION OF THE ABSORPTION CURVES OF AROMATIC HYDROCARBONS AND THE ANALYSIS OF BINARY MIXTURES

This exercise provides the opportunity of examining the absorption spectra of typical aromatic hydrocarbons and of investigating the possibility of analysing mixtures of hydrocarbons by ultraviolet spectrophotometry.

Reagents. *Methanol, benzene.* Both of 'Spectrosol' or equivalent purity.
  Toluene (methylbenzene). Analytical grade.

Procedure. Using a 0.1 mL capillary micropipette with 0.005 mL graduations, place 0.05 mL of the benzene in a 25 mL graduated flask and prepare a stock solution by diluting to the mark with methanol. Prepare a series of five dilutions of the stock solution by using a 2 mL graduated pipette to transfer respectively 0.25, 0.50, 0.75, 1.00, and 1.50 mL of the solution into a series of 10 mL graduated flasks, and making up to the mark with methanol.

Using stoppered quartz cells, use solution 5 (i.e. the most concentrated of the test solutions) to plot an absorption curve using pure methanol as the blank. Take absorbance readings over the wavelength range 200–300 nm, but preferably a spectrophotometer equipped with a chart recorder should be employed. Make a note of the λ_{max} values for the peaks observed in the curve. There is a well-developed peak at approximately 250 nm, and using each of the test solutions in turn, measure the absorption of each of them at the observed peak wavelength and test the validity of Beer's Law.

Now, starting with 0.05 mL toluene, repeat the procedure to obtain five working solutions 1′–5′ and use solution 5′ to plot the absorption curve of toluene: again record the λ_{max} values for the peaks of the curve. There is a well-developed peak at approximately 270 nm, and using the five test solutions, measure the absorbance of each at the observed peak wavelength and test the application of Beer's Law. Measure solution 5′ also at the wavelength used for benzene, and solution 5 at the wavelength used for toluene.

Prepare a benzene–toluene mixture by placing 0.05 mL of each liquid in a 25 mL graduated flask and making up to the mark with methanol. Take 1.5 mL of this solution, place in a 10 mL graduated flask and dilute to the mark with methanol; this solution contains benzene at the same concentration as solution 5, and toluene at the same concentration as solution 5′. Measure the absorbances of this solution at the two wavelengths selected for the Beer's Law plots of both benzene and toluene. Then use the procedure detailed in Section 17.48 to evaluate the composition of the solution and compare the result with that calculated from the amounts of benzene and toluene taken.

A similar procedure can be applied to mixtures of 1,2-dimethylbenzene (*o*-xylene) and 1,4-dimethylbenzene (*p*-xylene).

17.50 DETERMINATION OF PHENOLS IN WATER

Phenols show an ultraviolet absorption spectrum with a band between 270 and 280 nm, the intensity of which is greatly increased by working in alkaline solution so that the phenol is predominantly in the form of the phenoxide ion; at the same time there is a shift in the absorption band and many phenols under these conditions show a well-developed peak at a wavelength of 287–296 nm. Using an average value of λ_{max} of 293 nm, the molar absorption coefficients of a number of common phenols in alkaline solution have been determined at this wavelength,[22] and these values can be used for quantitative measurements. Alternatively, calibration curves may be prepared using a pure sample of the phenol to be determined. The ultraviolet spectrophotometric method is of wider application than the colorimetric method for determining phenols described in Section 17.45.

Water samples showing contamination by phenols are best examined by extracting the phenol into an organic solvent; tri-*n*-butyl phosphate is very suitable for this purpose. Photometric measurements can be carried out on the extract, and the requisite alkaline conditions are achieved by the addition of tetra-*n*-butylammonium hydroxide.

Reagents. *Stock phenol solution.* Weigh out 0.5 g phenol, dissolve in distilled water and make up to the mark in a 500 mL graduated flask: it is recommended that freshly boiled and cooled distilled water be used.

Standard phenol solution (0.025 mg L^{-1}). Dilute 25.0 mL of the stock solution to 1.0 L using freshly boiled and cooled distilled water. This solution must be freshly prepared.

*Tetra-*n-*butylammonium hydroxide (0.1 M solution in methanol).* Prepare an anion exchange column using an anion exchange resin such as Duolite A113 or Amberlite IRA-400, convert to the hydroxide form and after washing with water, pass 300–400 mL of methanol through the column to remove water (see Section 7.2).

Dissolve 20 g of tetra-*n*-butylammonium iodide in 100 mL of dry methanol and pass this solution through the column at a rate of about 5 mL min^{-1}: the effluent must be collected in a vessel fitted with a 'Carbosorb' guard tube to protect it from atmospheric carbon dioxide. Then pass 200 mL of dry methanol through the column. Standardise the methanolic solution by carrying out a potentiometric titration of an accurately weighed portion (about 0.3 g) of benzoic acid. Calculate the molarity of the solution and add sufficient dry methanol to make it approximately 0.1 *M*.

Hydrochloric acid (5 M).

*Tri-*n-*butyl phosphate.*

Procedure. Prepare four test solutions of phenol by placing 200 mL of boiled and cooled distilled water in each of four stoppered, 500 mL bottles, and adding to each 5 g of sodium chloride; this assists the extraction procedure by 'salting out' the phenol. Add respectively 5.0, 10.0, 15.0 and 20.0 mL of the standard phenol solution to the four bottles, then adjust the pH of each solution to about 5 by the careful addition of 5 *M* hydrochloric acid (use a test-paper). Add distilled water to each bottle to make a total volume of 250 mL and then add 20.0 mL tri-*n*-butyl phosphate. Stopper each bottle securely (the stoppers should be wired on), and then shake in a mechanical shaker for 30 minutes. Transfer to separatory funnels, and when the phases have settled, run off and discard the aqueous layers.

Prepare an alkaline solution of the phenol concentrate by placing 4.0 mL of a tri-*n*-butyl phosphate layer in a 5 mL graduated flask and then adding 1.0 mL of the tetra-*n*-butylammonium hydroxide: do this for each of the four solutions. The reference solution consists of 4 mL of the organic layer (in which the phenol is undissociated) plus 1 mL of methanol. Measure the absorbance of each of the extracts from the four test solutions and plot a calibration curve.

The unknown solution (which should contain between 0.5 and 2.0 mg phenol L^{-1}) is treated in the manner described above, and by reference to the calibration curve the absorbance reading will determine the phenol content of the unknown sample.

If the sample supplied may contain organic substances which can be extracted into tri-*n*-butyl phosphate, a preliminary extraction with carbon tetrachloride must be carried out. Solid potassium hydroxide is added to a 600–700 mL portion of the sample to raise the pH to about 12; use a test-paper. A 20 mL portion of carbon tetrachloride is added and shaking for 30 minutes is then carried out using a mechanical shaker. Using a separatory funnel, the organic layer is discarded and a 200 mL portion of the aqueous layer is transferred to a 500 mL bottle with a well-fitting stopper. Hydrochloric acid (5*M*) is added to bring the pH of the solution to about 5, and then the above procedure is followed.

17.51 DETERMINATION OF THE ACTIVE CONSTITUENTS IN A MEDICINAL PREPARATION BY DERIVATIVE SPECTROSCOPY

For this determination a spectrophotometer which is equipped to produce derivative curves is essential.

'Actifed' is a medicinal preparation in which the effective components are the two drugs pseudoephedrine hydrochloride and triprolidine hydrochloride. The absorption spectrum of 'Actifed' tablets dissolved in 0.1*M* hydrochloric acid is similar to that shown in Fig. 17.14(*a*) which is clearly of no value for quantitative determinations. A second derivative spectrum however is similar in character to that shown in Fig. 17.14(*b*) in which peak C corresponds to the pseudoephedrine hydrochloride and D to the triprolidine hydrochloride and from which it is possible to make quantitative measurements. Experience showed that it is advisable to use different response times for the two peaks: with the instrument used a response setting of 3 was found to give the best results for pseudoephedrine hydrochloride, whilst a setting of 4 was best for the triprolidine hydrochloride.

Reagents. *Pseudoephedrine hydrochloride.*
   Triprolidine hydrochloride.
   *Hydrochloric acid (0.1*M*).*

Procedure. Prepare standard solutions of pseudoephedrine hydrochloride by weighing out accurately about 60 mg into a 500 mL graduated flask. Add about 50 mL 0.1*M* hydrochloric acid to dissolve the solid and then make up to the mark with 0.1*M* hydrochloric acid. In a series of 50 mL graduated flasks place respectively 25, 30, and 40 mL of this solution and dilute to the mark with the hydrochloric acid; with the original (undiluted) solution, this gives four standard solutions.

Prepare standard solutions of triprolidine hydrochloride by weighing accurately about 0.1 g of the solid into a 100 mL graduated flask, adding about 50 mL of hydrochloric acid (0.1 M) and swirling the flask until the solid has dissolved, and then making to the mark with the hydrochloric acid. Dilute 10 mL of this solution in a 100 mL graduated flask using the 0.1 M hydrochloric acid to make up the volume. In a series of 50 mL graduated flasks pipette respectively 25, 30, and 40 mL of this diluted solution, and make up to the mark with the hydrochloric acid, thus giving four standard solutions.

Weigh eight to ten 'Actifed' tablets, grind to a fine powder using a pestle and mortar, and then weigh accurately an amount of the powder equivalent to the weight of one tablet into a 500 mL graduated flask. Add about 200 mL of the 0.1 M hydrochloric acid, stopper the flask, and shake for 5 minutes to dissolve the tablets: dilute to the mark with the hydrochloric acid. Filter the turbid liquid through a dry filter paper and reject the first 20 mL of filtrate. Collect the succeeding filtrate in a dry vessel as the test solution.

Set up the spectrophotometer to record in the second-derivative mode and record the spectra of the four standard triprolidine hydrochloride solutions. Use quartz cells, with 0.1 M hydrochloric acid in the reference cell and scan between 210 and 350 nm. For each spectrum measure the long-wave peak heights D_L [see Fig. 17.16(a)] covering the wavelength range 290–310 nm; plot the results against the concentrations of the solutions and confirm that a straight line results.

Likewise, record the second derivative spectra of the four standard pseudo-ephedrine hydrochloride solutions and record the peak heights D_L at 258–259 nm: plot the results against concentration and confirm that a straight line is obtained.

Now record the second derivative spectrum of the 'Actifed' solution, determine the long-wave peak heights for both components, and by comparison with the calibration plots of the individual components, deduce their proportions in the tablets.

17.52 SPECTROPHOTOMETRIC DETERMINATION OF THE pK VALUE OF AN INDICATOR (THE ACID DISSOCIATION CONSTANT OF METHYL RED)

Discussion. The dissociation of an acid–base indicator is well suited to spectrophotometric study; the procedure involved will be illustrated by the determination of the acid dissociation constant of methyl red (MR). The acidic (HMR) and basic (MR⁻) forms of methyl red are shown below.

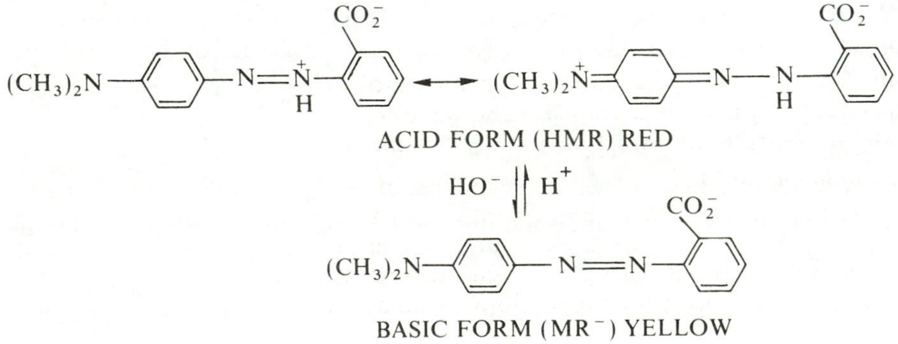

ACID FORM (HMR) RED

BASIC FORM (MR⁻) YELLOW

The acid dissociation constant K is given by the equation:

$$K = \frac{[H^+][MR^-]}{[HMR]} \tag{20}$$

$$pK = pH - \log\frac{[MR^-]}{[HMR]} \tag{21}$$

Both HMR and MR$^-$ have strong absorption peaks in the visible portion of the spectrum; the colour change interval from pH 4 to pH 6 can be conveniently obtained with a sodium acetate–acetic acid buffer system.

The determination of pK involves three steps:

(a) Evaluation of the wavelengths at which HMR (λ_A) and MR$^-$ (λ_B) exhibit maximum absorption.

(b) Verification of Beer's Law for both HMR and MR$^-$ at wavelengths λ_A and λ_B.

(c) Determination of the relative amounts of HMR and MR$^-$ present in solution as a function of pH.

By using the same concentration of indicator in each of the measurements at different values of pH and measuring the absorbance for each solution at λ_A and at λ_B, the relative amounts of HMR and MR$^-$ in solution can be calculated from the two equations:

$$A_A = d_{A.HMR}[HMR] + d_{A.MR^-}[MR^-] \tag{22}$$

$$A_B = d_{B.HMR}[HMR] + d_{B.MR^-}[MR^-] \tag{23}$$

where $d_{A.HMR}$, $d_{A.MR^-}$, $d_{B.HMR}$ and $d_{B.MR^-}$ are derived from the graphs plotted in (b). By solving these two simultaneous equations, the ratio [MR$^-$]/[HMR] can be obtained and thence pK with the aid of equation (21). Equations (22) and (23) imply that the observed absorbances (A) at λ_A and λ_B are the simple additive sums of the absorbances (d) due to HMR and MR$^-$.

Reagents. *Methyl red solution.* Dissolve 0.10 g pure crystalline methyl red in 30 mL 95 per cent ethanol and dilute to 50 mL with water. The solution required in the experiment (standard solution) is prepared by transferring 5.0 mL of the above stock solution to 50 mL of 95 per cent ethanol contained in a 100 mL graduated flask and diluting to 100 mL with water.

Sodium acetate, 0.04 M and 0.01 M.

Acetic acid, 0.02 M.

Hydrochloric acid, 0.1 M and 0.01 M. The exact concentrations of these two solutions are not critical.

Procedure. The study can be carried out using either a manually operated single-beam spectrophotometer, or an automatic recording double-beam spectrophotometer. In both cases the wavelengths at which HMR and MR$^-$ exhibit absorption maxima are readily obtained from the spectra.

(a) Prepare solution A by diluting a mixture of 10.0 mL of the standard solution of the indicator (MR) and 1.0 mL of 0.1 M hydrochloric acid to 100 mL; the pH of this solution is about 2, so that the indicator MR is present entirely as HMR. Using 1 cm cells, determine the absorption spectrum of this solution over the range 350–600 nm against a blank of distilled water.

For manual plotting cover the range in increments of 25 nm except for the portion between 500 and 550 nm which should be covered in 10 nm increments. From the spectrum of absorbance against wavelength determine the wavelength λ_A at which the maximum absorbance occurs: this is about 520 nm.

Prepare solution B by diluting a mixture of 10.0 mL of the standard solution of the indicator and 25.0 mL of $0.04\,M$ sodium acetate to 100 mL. The pH of this solution is about 8, so that the indicator MR is present entirely as MR$^-$. Measure the absorbance of solution B over the range 350–600 nm as detailed for solution A: with a manual spectrophotometer use 25 nm steps except for 400–450 nm, where 10 nm steps are recommended. Determine the wavelength λ_B of maximum absorbance as above: this is about 430 nm. The type of plots obtained for solutions A and B is shown in Fig. 17.21. The absorption peaks are not completely separated, but cross at a wavelength of about 460 nm. This point is known as the 'isobestic point'. If the absorbance of a solution containing both HMR and MR$^-$ is measured at this wavelength, the observed absorbance is independent of the relative amounts of HMR and MR$^-$ present and depends only on the total amount of the indicator MR in the solution.

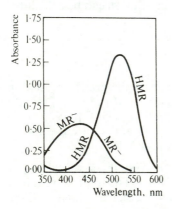

Fig. 17.21

(*b*) Using solution A, measure out 40.0 mL, 25.0 mL, and 10.0 mL into separate 50 mL graduated flasks, and dilute in each case to the mark with $0.01\,M$ hydrochloric acid. The resulting solutions will contain 0.8, 0.5, and 0.2 times (respectively) the initial concentration of HMR. Similarly, using solution B and diluting with $0.01\,M$ sodium acetate, prepare three solutions containing respectively 0.8, 0.5, and 0.2 times the initial concentration of MR$^-$. Measure the absorbance of each of the six solutions versus water at wavelengths of λ_A and λ_B. It is important in obtaining the experimental absorbance to be sure that all the measurements are made at constant temperature, say at the temperature of the room housing the spectrophotometer. Plot absorbance against relative concentration of the indicator MR: in each case straight-line plots should be obtained, as in Fig.17.22.

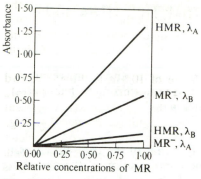

Fig. 17.22

(c) Prepare solutions in four 100 mL graduated flasks as listed in Table 17.7. Determine the pH values of each of the solutions (typical values are incorporated in the table) and measure the absorbance of each solution at wavelengths λ_A and λ_B. All these solutions contain the same concentration of indicator as solutions A and B used in (a). For each prepared solution, obtain the values of the absorbances $d_{A.HMR}$, $d_{A.MR^-}$, $d_{B.HMR}$, and $d_{B.MR^-}$ from the plots in Fig. 17.22, at relative concentrations of 1.0, and solve the simultaneous equations (22) and (23) in order to evaluate the relative amounts of HMR and MR^- in solution. From the relative amounts of HMR and MR^- present as a function of pH, calculate the value of pK for methyl red using equation (21). Some typical results are collected in Table 17.8.

Table 17.7

	Flask number			
	1	2	3	4
Standard indicator solution MR (mL)	10.0	10.0	10.0	10.0
0.04M Sodium acetate (mL)	25.0	25.0	25.0	25.0
0.02M Acetic card (mL)	50.0	25.0	10.0	5.0
Water (to mark)	15.0	40.0	55.0	60.0
pH	4.84	5.15	5.53	5.81

Table 17.8

Solution number	Observed pH	Absorbance at λ_A	Absorbance at λ_B	$\dfrac{[MR^-]}{[HMR]}$	pK
1	4.84	0.605	0.204	0.679	5.01
2	5.15	0.442	0.263	1.403	5.00
3	5.53	0.254	0.317	3.436	4.99
4	5.81	0.168	0.348	6.740	4.98
				Mean	5.00

DETERMINATIONS BY SPECTROPHOTOMETRIC TITRATIONS

17.53 SPECTROPHOTOMETRIC TITRATIONS

General discussion. In a spectrophotometric titration the end point is evaluated from data on the absorbance of the solution. For monochromatic light passing through a solution, Beer's Law may be written as:

$$\text{Absorbance} = \log I_0 / I_t = \varepsilon c l$$

where I_0 is the intensity of the incident light, I_t that of the transmitted light, ε is the molar absorption coefficient, c is the concentration of the absorbing species, and l is the thickness or length of the light path through the absorbing medium. Since spectrophotometric titrations are carried out in a vessel for which the light path is constant, the absorbance is proportional to the concentration. Thus in a titration in which the titrant, the reactant, or a reactive product absorbs radiation, the plot of absorbance versus volume of titrant added will consist, if the reaction is complete and the volume change is small, of two straight lines intersecting at the end-point.

The shape of a photometric titration curve will be dependent upon the optical properties of the reactant, titrant, and products of the reaction at the wavelength used. Some typical titration plots are given in Fig. 17.23.

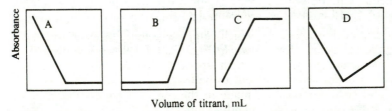

Volume of titrant, mL

Fig. 17.23

Diagram A is characteristic of systems where the substance titrated is converted into a non-absorbing product.

Diagram B is typical of the titration where the titrant alone absorbs.

Diagram C corresponds to systems where the substance titrated and the titrant are colourless and the product alone absorbs.

Diagram D is obtained when a coloured reactant is converted into a colourless product by a coloured titrant.

Owing to the linear response of absorbance to concentration, an appreciable break will often be obtained in a photometric titration, even though the changes in concentration are insufficient to give a clearly defined inflexion point in a potentiometric titration. Photometric titrations have several advantages over direct colorimetric determinations. The presence of other substances absorbing at the same wavelength does not necessarily cause interference, since only the change in absorbance is significant. The precision of locating the titration line (required for the evaluation of the equivalence point) by pooling the information derived from several points is greater than the precision of any single point; furthermore, the procedure may be useful for reactions which tend to be appreciably incomplete near the equivalence point. An accuracy and precision of a few tenths per cent are attainable with comparative ease by spectro-photometric titration. The optimum concentration of the solution to be analysed depends upon the molar absorption coefficient of the absorbing species involved, and is usually of the order of $10^{-4} - 10^{-5} M$. The effect of dilution can be made negligible by the use of a sufficiently concentrated titrant. If relatively large volumes of titrant are added the effect of dilution may be corrected by multiplying the observed absorbances by the factor $(V+v)/V$, where V is the initial volume and v is the volume added; if the dilution is of the order of only a few per cent the lines in the titration plots appear straight. The operating wavelength is selected on the basis of two considerations: avoidance of interference by other absorbing substances and need for an absorption coefficient which will cause the change in absorbance to fall within a convenient range. The latter is particularly important, because serious photometric error is possible in high-absorbance regions. Light leakage must, of course, be avoided.

The experimental technique is simple. The cell containing the solution to be titrated is placed in the light path of a spectrophotometer, a wavelength appropriate to the particular titration is selected, and the absorption is adjusted to some convenient value by means of the sensitivity and slit-width controls. A measured volume of the titrant is added to the stirred solution, and the absorbance is read again. This is repeated at several points before the end point and several more points after the end point. The latter is found graphically.

17.54 APPARATUS FOR SPECTROPHOTOMETRIC TITRATIONS

A special titration cell is necessary which completely fills the cell compartment of the spectrophotometer. One shown in Fig. 17.24 can be made from 5 mm Perspex sheet, cemented together with special Perspex cement, and with dimensions suitable for the instrument to be used. Since Perspex is opaque to ultraviolet light, two openings are made in the cell to accommodate circular quartz windows* 23 mm in diameter and 1.5 mm thick: the windows are inserted in such a way that the beam of monochromatic light passes through their centres

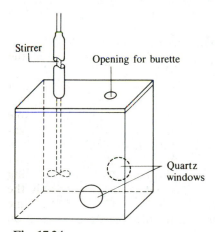

Fig. 17.24

* These were actually fused silica polarimeter end plates.

to the photoelectric cell. The Perspex cover of the cell has two small openings for the tip of a 5 mL microburette and for a microstirrer, respectively, held by means of rubber bungs: the stirrer is 'sleeved'. The whole of the cell, with the exception of the quartz windows, is covered with black paper and, as a further precaution, the top of the cell is covered with a black cloth: it is most important to exclude all extraneous light. In some circumstances, a probe-type photometer based upon the use of fibre optics may be employed.

17.55 SIMULTANEOUS DETERMINATION OF ARSENIC(III) AND ANTIMONY(III) IN A MIXTURE

Discussion. In acid solution arsenic(III) can be oxidised to arsenic(V) and antimony(III) to antimony(V) by the well-established titration with a solution of potassium bromate and potassium bromide (Section 10.133). The end point for such determinations is usually observed indirectly, and very good results have been obtained by the spectrophotometric method of Sweetser and Bricker.[23] No change in absorbance at 326 nm is obtained until all the arsenic(III) has been oxidised, the absorbance then decreases to a minimum at the antimony(III) end point at which it rises again as excess titrant is added.

Reagents. *Bromate/bromide solution.* Prepare a standard bromate/bromide solution by dissolving 2.78 g potassium bromate and 9.9 g potassium bromide in water and diluting to 1 L in a graduated flask. This solution is $0.017 M$ potassium bromate with a slight excess of the theoretical amount of potassium bromide. Analytical grade reagents should be employed.

Arsenic/antimony solution. Prepare a mixed solution containing approximately 115 mg arsenic and 160 mg antimony in 1 L by dissolving about 150 mg arsenic(III) oxide and 300 mg antimony(III) chloride in $6 M$ hydrochloric acid.

Procedure. Place 80 mL of the arsenic/antimony solution in the titration cell of the spectrophotometer. Titrate with standard bromate/bromide solution at 326 nm taking an absorbance reading at least every 0.2 mL. From the curve obtained calculate the concentration of arsenic and antimony in the solution.

17.56 DETERMINATION OF COPPER(II) WITH EDTA

Discussion. The titration of a copper ion solution with EDTA may be carried out photometrically at a wavelength of 745 nm. At this wavelength the copper–EDTA complex has a considerably greater molar absorption coefficient than the copper solution alone. The pH of the solution should be about 2.4.

The effect of different ions upon the titration is similar to that given under iron(III) (Section 17.57). Iron(III) interferes (small amounts may be precipitated with sodium fluoride solution): tin(IV) should be masked with 20 per cent aqueous tartaric acid solution. The procedure may be employed for the determination of copper in brass, bronze, and bell metal without any previous separations except the removal of insoluble lead sulphate when present.

Reagents. *Copper ion solution, 0.04 M.* Wash analytical grade copper with petroleum spirit (b.p. 40–60 °C) to remove any surface grease and dry at 100 °C. Weigh accurately about 1.25 g of the pure copper, dissolve it in 5 mL of concentrated nitric acid, and dilute to 1 L in a graduated flask. Titrate this standard copper solution with the EDTA solution using fast sulphon black as indicator (Section 10.55), and thus obtain a further check on the molarity of the EDTA.

EDTA solution, 0.10 M, and buffer solution. pH 2.2. See Section 17.57.

Procedure. Charge the titration cell (Fig. 17.24) with 10.00 mL of the copper ion solution, 20 mL of the acetate buffer (pH = 2.2), and about 120 mL of water. Position the cell in the spectrophotometer and set the wavelength scale at 745 nm. Adjust the slit width so that the reading on the absorbance scale is zero. Stir the solution and titrate with the standard EDTA: record the absorbance every 0.50 mL until the value is about 0.20 and subsequently every 0.20 mL. Continue the titration until about 1.0 mL after the end point; the latter occurs when the absorbance readings become fairly constant. Plot absorbance against mL of titrant added; the intersection of the two straight lines (see Fig. 17.23 C) is the end point.

Calculate the concentration of copper ion (mg mL^{-1}) in the solution and compare this with the true value.

17.57 DETERMINATION OF IRON(III) WITH EDTA

Discussion. Salicylic acid and iron(III) ions form a deep-coloured complex with a maximum absorption at about 525 nm: this complex is used as the basis for the photometric titration of iron(III) ion with standard EDTA solution. At a pH of *ca* 2.4 the EDTA–iron complex is much more stable (higher stability constant) than the iron–salicylic acid complex. In the titration of an iron–salicylic acid solution with EDTA the iron–salicylic acid colour will therefore gradually disappear as the end point is approached. The spectrophotometric end point at 525 nm is very sharp.

Considerable amounts of zinc, cadmium, tin(IV), manganese(II), chromium(III), and smaller amounts of aluminium cause little or no interference at pH 2.4: the main interferences are lead(II), bismuth, cobalt(II), nickel, and copper(II).

Reagents. *EDTA solution, 0.10 M.* See Section 10.49. Standardise accurately (Section 10.49).

Iron (III) solution, 0.05 M. Dissolve about 12.0 g, accurately weighed, of ammonium iron(III) sulphate in water to which a little dilute sulphuric acid is added, and dilute the resulting solution to 500 mL in a graduated flask. Standardise the solution with standard EDTA using variamine blue B as indicator (Section 10.56).

Sodium acetate–acetic acid buffer. Prepare a solution which is 0.2 M in sodium acetate and 0.8 M in acetic acid. The pH is 4.0.

Sodium acetate–hydrochloric acid buffer. Add 1 M hydrochloric acid to 350 mL of 1 M sodium acetate until the pH of the mixture is 2.2 (pH meter).

Salicylic acid solution. Prepare a 6 per cent solution of salicylic acid in acetone.

Procedure. Transfer 10.00 mL of the iron(III) solution to the titration cell (Fig. 17.24), add about 10 mL of the buffer solution of pH = 4.0 and about 120 mL of water: the pH of the resulting solution should be 1.7–2.3. Insert the titration cell into the spectrophotometer; immerse the stirrer and the tip of the 5 mL microburette (graduated in 0.02 mL) in the solution. Switch on the tungsten lamp and allow the spectrophotometer to 'warm up' for about 20 minutes. Stir the solution. Add about 4.0 mL of the standard EDTA (note the volume accurately). Set the wavelength at 525 nm, and adjust the slit width of the instrument so that the reading on the absorbance scale is 0.2–0.3. Now add 1.0 mL of the salicylic acid solution; the absorbance immediately increases to a very large value (> 2). Continue the stirring. Add the EDTA solution slowly from the microburette until the absorbance approaches 1.8; record the volume of titrant. Introduce the EDTA solution in 0.05 mL aliquots and record the absorbance after each addition. Continue the titration until at least four readings are taken beyond the end point (fairly constant absorbance). Plot absorbance against mL of titrant added: the intersection of the two straight lines (see Fig. 17.23 A) gives the true end point.

Calculate the concentration of iron(III) (mg mL^{-1}) in the solution and compare this with the true value.

Determination of iron(III) in the presence of aluminium. Iron(III) (concentration *ca* 50 mg per 100 mL) can be determined in the presence of up to twice the amount of aluminium by photometric titration with EDTA in the presence of 5-sulphosalicylic acid (2 per cent aqueous solution) as indicator at pH 1.0 at a wavelength of 510 nm. The pH of a strongly acidic solution may be adjusted to the desired value with a concentrated solution of sodium acetate: about 8–10 drops of the indicator solution are required. The spectrophotometric titration curve is of the form shown in Fig. 17.23.

Determination of organic compounds. The application of photometric titrimetry to organic compounds may be exemplified by the **titration of phenols**. This can be carried out by working at the λ_{max} value (in the ultraviolet) for the phenol being determined (see Section 17.50). It has been shown that by titrating with tetra-*n*-butylammonium hydroxide and using propan-2-ol as solvent, it is possible to differentiate between substituted phenols.[24]

Further details of photometric titrations will be found in Refs 25 and 26.

SOME NEPHELOMETRIC DETERMINATIONS

17.58 GENERAL DISCUSSION

Small amounts of some insoluble compounds may be prepared in a state of aggregation such that moderately stable suspensions are obtained. The optical properties of each suspension will vary with the concentration of the dispersed phase. When light is passed through the suspension, part of the incident radiant energy is dissipated by absorption, reflection, and refraction, while the remainder is transmitted. Measurement of the intensity of the transmitted light as a function of the concentration of the dispersed phase is the basis of **turbidimetric analysis**. When the suspension is viewed at right angles to the direction of the incident light the system appears opalescent due to the reflection of light from the particles of the suspension (Tyndall effect). The light is reflected irregularly and diffusely, and consequently the term 'scattered light' is used to account for this opalescence or cloudiness. The measurement of the intensity of the scattered light (at right angles to the direction of the incident light) as a function of the concentration of the dispersed phase is the basis of **nephelometric analysis** (Gr. *nephele* = a cloud). Nephelometric analysis is most sensitive for very dilute suspensions

($\not> 100$ mg L^{-1}). Techniques for turbidimetric analysis and nephelometric analysis resemble those of filter photometry and fluorimetry respectively.

The construction of calibration curves is recommended in nephelometric and turbidimetric determinations, since the relationship between the optical properties of the suspension and the concentration of the disperse phase is, at best, semi-empirical. If the cloudiness or turbidity is to be reproducible, the utmost care must be taken in its preparation. The precipitate must be very fine, so as not to settle rapidly. The intensity of the scattered light depends upon the number and the size of the particles in suspension, and provided that the average size of particles is fairly reproducible, analytical applications are possible.

The following conditions should be carefully controlled in order to produce suspensions of reasonably uniform character:

1. the concentrations of the two ions which combine to produce the precipitate as well as the ratio of the concentrations in the solutions which are mixed;
2. the manner, the order, and the rate of mixing;
3. the amounts of other salts and substances present, especially protective colloids (gelatin, gum arabic, dextrin, etc.);
4. the temperature.

17.59 INSTRUMENTS FOR NEPHELOMETRY AND TURBIDIMETRY

Visual and photoelectric colorimeters may be used as turbidimeters: a blue filter usually results in greater sensitivity. A calibration curve must be constructed using several standard solutions, since the light transmitted by a turbid solution does not generally obey the Beer–Lambert Law precisely.

'Visual' nephelometers (comparator type) have been superseded by the photoelectric type. It is, however, possible to adapt a good Duboscq colorimeter (Section 17.5) for nephelometric work. Since the instrument is to measure scattered light, the light path must be so arranged that the light enters the side of the cups at right angles to the plungers instead of through the bottoms. The usual cups are therefore replaced by clear glass tubes with opaque bottoms; the glass plungers are accurately fitted with opaque sleeves. The light, which enters at right angles to the cups, must be regulated so that equal illumination is obtained on both sides. A standard suspension is placed in one cup, and the unknown solution is treated in an identical manner and placed in the other cup. The dividing line between the two fields in the eyepiece must be thin and sharp, and seem to disappear when the fields are matched.

Most fluorimeters (see Chapter 18) may be adapted for use in nephelometry and instruments (nephelometers) for such measurements are also available. The essential feature of such an instrument is the reflector designed to collect the light which has been scattered by the particles in a cloudy or turbid solution. In a typical arrangement, the solution is placed in a test-tube which is supported above a light source as shown in Fig. 17.25, and the scattered light is directed by the reflector on to an annular photocell. The current thus generated is fed to a sensitive micro-ammeter. A series of colour filters may be included so that coloured solutions can be dealt with; the filter selected should be similar in colour to that of the solution. A metal test-tube cover excludes extraneous light, and sensitivity and zero-setting controls are provided. A number of test-tubes will be needed, and these should be selected so that as far as possible they match each other.

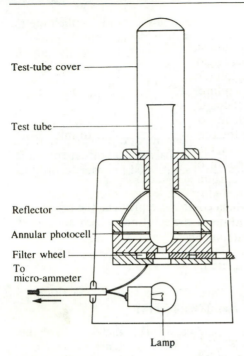

Test-tube cover

Test tube

Reflector

Annular photocell

Filter wheel

To
micro-ammeter

Lamp

Fig. 17.25

The general procedure for operating such an instrument is as follows.

1. Adjust the micro-ammeter to the zero of the scale.
2. Remove the cap cover, place the standard in position, and replace the cap. The standard is generally the matched test-tube containing the most concentrated suspension of the substance being determined: the concentration must, of course, be known.
3. Select the filter required. This should be similar in colour to that of the solution.
4. Adjust the sensitivity control of the micro-ammeter to obtain a reading of 100 divisions on the scale.
5. Remove the standard.
6. Fill a test-tube with distilled water or with a 'blank' solution to a depth of not less than 3 cm, and set to zero by means of the zero control.
7. Check the reading of the standard (100 divisions).
8. Repeat procedures (6) and (7) until full-scale deflection and zero settings are obtained.
9. Replace the standard suspension with more dilute suspensions, and note the various scale readings. Draw a calibration curve relating meter readings and the concentrations of the substance being determined.
10. Fill a test-tube with the sample to be determined to a depth similar to that used for the standards, and insert into the instrument. Note the galvanometer deflection: evaluate the concentration from the calibration curve.

17.60 DETERMINATION OF SULPHATE

Discussion. The turbidity of a dilute barium sulphate suspension is difficult to reproduce; it is therefore essential to adhere rigidly to the experimental procedure detailed below. The velocity of the precipitation, as well as the concentration of the reactants, must be controlled by adding (after all the other components are present) pure solid barium chloride of definite grain size. The rate of solution of the barium chloride controls the velocity of the reaction. Sodium chloride and hydrochloric acid are added before the precipitation in order to inhibit the growth of microcrystals of barium sulphate: the optimum pH is maintained and minimises the effect of variable amounts of other electrolytes present in the sample upon the size of the suspended barium sulphate particles. A glycerol–ethanol solution helps to stabilise the turbidity. The reaction vessel is shaken gently in order to obtain a uniform particle size: each vessel should be shaken at the same rate and the same number of times. The unknown must be treated exactly like the standard solution. The interval between the time of precipitation and measurement must be kept constant.

Reagents. *Standard sulphate solution.* Dissolve 1.814 g dry potassium sulphate in distilled water and dilute to 1 L in a graduated flask. This solution contains 1.000 mg of sulphate ion per mL.

Sodium chloride–hydrochloric acid reagent. Dissolve 60 g sodium chloride in 200 mL distilled water, add 5 mL pure concentrated hydrochloric acid, and dilute to 250 mL.

Barium chloride. Use crystals of barium chloride that pass through a 20 mesh sieve and are retained by a 39 mesh sieve.

Glycerol–ethanol solution. Dissolve 1 volume of pure glycerol in 2 volumes of absolute ethanol.

Procedure. Run 0.5, 1.0, 1.5, 2.0, 2.5, and 3.0 mL of the standard potassium sulphate solution from a calibrated burette into separate 100 mL graduated flasks. To each flask add 10 mL of the sodium chloride–hydrochloric acid reagent and 20 mL of the glycerol–ethanol solution, and dilute to 100 mL with distilled water. Add 0.3 g of the sieved barium chloride to each flask, stopper each flask, and shake for 1 minute by inverting each flask once per second: all the barium chloride should dissolve. Allow each flask to stand for 2–3 minutes and measure the turbidity in the nephelometer: take care to avoid small air bubbles adhering to the walls of the matched test-tubes. Use the most concentrated solution as standard and, by means of the sensitivity control, adjust the micro-ammeter reading to 100 divisions. Prepare a 'blank' solution, repeat the above sequence of operations, but do not add any sulphate solution. Place the 'blank' solution in the nephelometer and adjust to zero reading of the scale by means of the zero control. Check the reading of the most turbid solution, and adjust any deviation from 100 by means of the sensitivity control. Repeat the measurements with the five other standard sulphate solutions. Plot the reading against the sulphate-ion content per mL.

Determine the sulphate-ion content of an unknown solution, say *ca* 0.3 mg mL^{-1}, using the calibration curve.

17.61 DETERMINATION OF PHOSPHATE

Discussion. Phosphate ion is determined nephelometrically following the formation of strychnine molybdophosphate. This turbidity is white in colour and consists of extremely fine particles (compare ammonium molybdophosphate, which is yellow and is composed of rather large grains). The precipitate must not be agitated, as it tends to agglomerate easily; it is somewhat sensitive to temperature changes.

Reagents. *Standard phosphate solution.* Dissolve 1.721 g potassium dihydrogen-phosphate (dried at 110 °C) in 1 L of water in a graduated flask. Pipette 10.0 mL of this solution into a 1 L graduated flask and dilute to the mark. The resulting dilute solution contains 0.01 mg phosphorus pentoxide per mL.

Molybdate–strychnine reagent.[*] This reagent is prepared in two parts; these are mixed just before use, since the addition of the acid molybdate solution to the strychnine sulphate solution produces a precipitate after 24 hours. Solution A (acid molybdate solution): place 30 g molybdenum trioxide in a 500 mL conical flask, add 10 g sodium carbonate and 200 mL water. Boil the mixture until a clear solution is obtained. Filter the hot solution, if necessary. Add 200 mL 5 M sulphuric acid, allow to cool, and dilute to 500 mL.

Solution B (strychnine sulphate solution): dissolve 1.6 g strychnine sulphate in 100 mL warm distilled water, cool and dilute to 500 mL.

Prepare the reagent by adding solution B rapidly to an equal volume of solution A, and shake the resulting mixture thoroughly; filter off the bluish-white precipitate through a Whatman No. 42 filter paper. The resulting clear solution will keep for about 20 hours. Solutions A and B may be kept indefinitely.

Saturated sodium sulphate solution. Prepare a saturated aqueous solution at 50 °C and cool to room temperature. Filter before use.

Sulphuric acid, 1 M. Dilute 27 mL of concentrated sulphuric acid to 500 mL in a graduated flask.

Procedure. Run in 1.0, 2.0, 4.0, 6.0, 8.0, and 10.0 mL of the standard phosphate solution from a calibrated burette into separate 100 mL graduated flasks. To each flask add 18 mL 1 M sulphuric acid and 16 mL saturated sodium sulphate solution, and dilute to approximately 95 mL with distilled water. Now add 2.0 mL of the molybdate–strychnine reagent and dilute to 100 mL; mix the contents of the flask by gently inverting several times, but do not shake. Allow the flasks to stand for 20 minutes to permit the turbidities to develop before making the measurements. Prepare a 'blank' solution by repeating the above sequence of operations, but omit the addition of the phosphate solution. Use the most concentrated solution as the initial standard and adjust the micro-ammeter reading to 100 divisions. Introduce the 'blank' solution into the matched test-tube of the nephelometer and adjust the reading to zero. Check the standard solution for a galvanometer reading of 100. Repeat the above with the five other phosphate solutions. Plot galvanometer reading against mg P_2O_5 per mL.

Determine the phosphate content of an unknown solution, containing say *ca* 0.005 mg P_2O_5 per mL: use the calibration graph.

---

[*] Strychnine is a toxic alkaloid. **It should only be handled with gloves** and under no circumstances should it be ingested.
For References and Bibliography see Sections 21.26 and 21.27.

CHAPTER 18
SPECTROFLUORIMETRY

18.1 GENERAL DISCUSSION

Fluorescence is caused by the absorption of radiant energy and the re-emission of some of this energy in the form of light. The light emitted is almost always of higher wavelength than that absorbed (Stokes' Law). In true fluorescence the absorption and emission take place in a short but measurable time — of the order of $10^{-12} - 10^{-9}$ second. If the light is emitted with a time delay ($> 10^{-8}$ second) the phenomenon is known as **phosphorescence**; this time delay may range from a fraction of a second to several weeks, so that the difference between the two phenomena may be regarded as one of degree only. Both fluorescence and phosphorescence are designated by the term **photoluminescence**; the latter is therefore the general term applied to the process of absorption and re-emission of light energy.

At present the most widely used type of photoluminescence in analytical chemistry is fluorescence, which is distinguished from other forms of photo-luminescence by the fact that the excited molecule returns to the ground state immediately after excitation.

When a molecule absorbs a photon of ultraviolet radiation it undergoes a transition to an excited electronic state and one of its electrons is promoted to an orbital of higher energy. There are two important types of transition for organic molecules:

(a) $n \rightarrow \pi^*$, in which an electron in a non-bonding orbital is promoted to a π-antibonding orbital;
(b) $\pi \rightarrow \pi^*$, in which an electron in a π-bonding orbital is raised to a π-antibonding orbital.

It is the $\pi \rightarrow \pi^*$ type of excitation which leads to significant fluorescence, the $n \rightarrow \pi^*$ type producing only a weak fluorescence. The electronic transitions corresponding to charge-transfer bands also lead to strong fluorescence.

The electronic energy is not, however, the only type of energy affected when a molecule absorbs a photon of UV radiation. Organic molecules have a large number of vibrations and each of these contributes a series of nearly equally spaced vibrational levels to each of the electronic states. The various energy states available to a molecule may be represented by means of an energy-level diagram. For details of such diagrams and the allowed energy transitions an appropriate textbook should be consulted (see Bibliography, Section 21.27; see also Section 17.13 for the origins of absorption spectra).

It is clear that before a molecule can emit radiation by the mechanism of fluorescence it must first be able to absorb radiation. Not all molecules which absorb UV or visible radiation are, however, fluorescent and it is useful to quantify the extent to which a particular molecule fluoresces. This is done by means of the **quantum yield,** ϕ_f, which is defined as the fraction of the incident radiation which is re-emitted as fluorescence.

$$\phi_f(\leqslant 1) = \frac{\text{No. of photons emitted}}{\text{No. of photons absorbed}} = \frac{\text{Quantity of light emitted}}{\text{Quantity of light absorbed}}$$

A proportion of the excited molecules may lose their excess energy by undergoing bond dissociation leading to a photochemical reaction or may return to the ground state by other mechanisms. Clearly the quantum efficiency will then be less than unity and may be extremely small. The value of ϕ_f is an inherent property of a molecule and is determined to a large extent by its structure. In general a high value of ϕ_f is associated with molecules possessing an extensive system of conjugated double bonds with a relatively rigid structure due to ring formation. This is illustrated by the intense fluorescence exhibited by organic molecules such as anthracene, fluorescein and other condensed-ring aromatic structures. The number of simple inorganic species which are fluorescent is much more limited, the chief examples being lanthanide and actinide compounds. In the case of metal ions this limitation may be overcome, however, by the formation of a complex with an appropriate organic ligand; for example, many of the metal-ion complexes formed with the complexing agent 8-hydroxyquinoline are fluorescent.

Quantitative aspects. The total fluorescence intensity, F, is given by the equation $F = I_a\phi_f$ where I_a is the intensity of light absorption and ϕ_f the quantum efficiency of fluorescence. Since $I_0 = I_a + I_t$ where $I_0 =$ intensity of incident light and $I_t =$ intensity of transmitted light, then

$$F = (I_0 - I_t)\phi_f$$

and since $I_t = I_0 e^{-\varepsilon cl}$ (Beer–Lambert Law)

$$F = I_0(1 - e^{-\varepsilon cl})\phi_f \tag{1}$$

For weakly absorbing solutions, when εcl is small, the equation becomes

$$F = I_0 . 2.3\, \varepsilon cl\, \phi_f \tag{2}$$

so that for very dilute solutions (a few ppm, or less) the total fluorescence intensity F is proportional to both the concentration of the sample and the intensity of the excitation energy.

It is instructive to compare the sensitivity which may be achieved by absorption and fluorescence methods. The overall precision with which absorbance can be measured is certainly not better than 0.001 units using a 1 cm cell. Since for most molecules the value of $\varepsilon_{\max}$ is rarely greater than 10^5, then on the basis of the Beer–Lambert Law the minimum detectable concentration is given by $c_{\min} > 10^{-3}/10^5 = 10^{-8} M$.

With fluorescence, however, the sensitivity is limited in principle only by the maximum intensity of the exciting light source so that under ideal conditions,

$$c_{\min} = 10^{-12} M$$

In general the limit of detection of the fluorescence technique is of the order of 10^3 times lower than that for UV absorption spectrometry.

Selectivity may also be superior using fluorescence methods since (a) not all absorbing species fluoresce, and (b) the analyst can select two wavelengths (excitation and emission) as compared with one for absorption methods. This inherent selectivity may be inadequate, however, and must often be enhanced by chemical separation, e.g. solvent extraction (see Chapter 6). Improved selectivity can also be achieved using derivative techniques, i.e. by basing the measurement of a sample component on the derivative spectrum rather than the original fluorescence emission spectrum; thus weak shoulders on the latter convert into easily quantified peaks in the derivative spectrum. It is important to distinguish between the **emission** and **excitation fluorescence spectra**. In the former, excitation is carried out at a fixed wavelength and the emission intensity recorded as a function of wavelength. Excitation spectra, on the other hand, are obtained by measuring the fluorescence intensity at a fixed wavelength while the wavelength of the exciting radiation is varied.

Factors such as dissociation, association, or solvation, which result in deviation from the Beer–Lambert law, can be expected to have a similar effect in fluorescence. Any material that causes the intensity of fluorescence to be less than the expected value given by equation (2) is known as a **quencher**, and the effect is termed quenching; it is normally caused by the presence of foreign ions or molecules. Fluorescence is affected by the pH of the solution, by the nature of the solvent, the concentration of the reagent which is added in the determination of inorganic ions, and, in some cases, by temperature. The time taken to reach the maximum intensity of fluorescence varies considerably with the reaction.

An important aspect of quenching in analysis is that the fluorescence exhibited by the analyte may be quenched by the molecules of some compound present in the sample, i.e. this is an example of a matrix effect. If the concentration of the quenching species is constant this may be allowed for by using suitable standards (i.e. containing the same concentration of quenching species), but difficulties occur when there is an unpredictable variation in the concentration of quenching species.

18.2 INSTRUMENTS FOR FLUORIMETRIC ANALYSIS

Instruments for the measurement of fluorescence are known as fluorimeters or spectrofluorimeters. The essential parts of a simple fluorimeter are shown in Fig. 18.1. The light from a mercury-vapour lamp (or other source of ultraviolet light)* is passed through a condensing lens, a primary filter (to permit the light band required for excitation to pass), a sample container,† a secondary filter (selected to absorb the primary radiant energy but transmit the fluorescent

---

* Since the intense UV radiation emitted by these lamps is damaging to the eyes, it is essential never to look at the unshielded lamp when it is on. Care must also be exercised in the handling and use of the commonly employed high-pressure xenon lamps, which may shatter and explode if dropped.

† Fluorescence cells are usually made of glass or silica with all four faces polished. For precise quantitative work it is important that these cells are always inserted into the cell-holder the same way round. Cells must always be washed out and carefully stored after use.

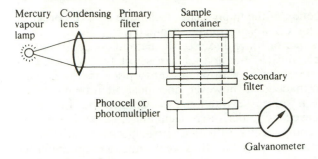

Mercury vapour lamp — Condensing lens — Primary filter — Sample container — Secondary filter — Photocell or photomultiplier — Galvanometer

Fig. 18.1

radiation), a receiving photocell placed in a position at right angles to the incident beam (in order that it may not be affected by the primary radiation), and a sensitive galvanometer or other device for measuring the output of the photocell. Since fluorescence intensity is proportional to the intensity of irradiation, the light source must be very stable if fluctuations in its intensity are not compensated for. It is usual, therefore, to employ a two-cell instrument; the galvanometer is used as a null instrument, and readings are taken on a potentiometer used in balancing the photocells against each other. Since the two photocells are selected so as to be similar in spectral response, it is assumed that fluctuations in the intensity of the light source are minimised.

The simpler fluorimeters are manual instruments operating only at a single selected wavelength at any one time. Despite this they are perfectly suitable for quantitative measurements, as these are almost always carried out at a fixed wavelength. The experiments listed at the end of this chapter have all been carried out at single fixed wavelengths.

The more advanced spectrofluorimeters are capable of automatically scanning fluorescent spectra between about 200 and 900 nm and produce a chart record of the spectrum obtained. These can also operate at a fixed wavelength and are equally suitable for carrying out quantitative work; their main application tends to be for the detection and determination of small concentrations of organic substances.

Some commercial spectrophotometers have fluorescence attachments which allow the sample to be irradiated from an ancillary source and the resulting fluoresence to pass through the monochromator for spectral analysis.

18.3 SOME APPLICATIONS OF FLUORIMETRY

Fluorimetry is generally used if there is no colorimetric method sufficiently sensitive or selective for the substance to be determined. In inorganic analysis the most frequent applications are for the determination of metal ions as fluorescent organic complexes. Many of the complexes of oxine fluoresce strongly: aluminium, zinc, magnesium, and gallium are sometimes determined at low concentrations by this method. Aluminium forms fluorescent complexes with the dyestuff eriochrome blue black RC (pontachrome blue black R), whilst beryllium forms a fluorescent complex with quinizarin.

The analysis of non-metallic elements and anionic species may present a problem since many do not readily form suitable derivatives for fluorimetric analysis. The best-known fluorimetric methods for non-metals are those for boron and selenium which both involve derivatisation reactions leading to ring closure, e.g. the derivative (I) formed by the condensation reaction between boric acid and benzoin.

Important organic applications are to the determination of quinine and the vitamins riboflavin (vitamin B_2) and thiamine (vitamin B_1). Riboflavin fluoresces in aqueous solution; thiamine must first be oxidised with alkaline hexacyanoferrate(III) solution to thiochrome, which gives a blue fluorescence in butanol solution. Under standard conditions, the net fluorescence of the thiochrome produced by oxidation of the vitamin B_1 is directly proportional to its concentration over a given range. The fluorescence can be measured either by reference to a standard quinine solution in a null-point instrument or directly in a spectrofluorimeter.[27]

The phenomenon of quenching has already been mentioned (Section 18.1) and the application of quenching methods will be briefly indicated here. The principle of such methods is that the emission of a fluorescent species is quenched by the analyte so that the intensity of fluorescence decreases as the analyte concentration increases. A major limitation of such procedures, however, is that since quenching is completely non-specific, applications are restricted to analyses in which only the analyte is able to quench the fluorescence. Probably the most important application of quenching is in the determination of molecular oxygen, a paramagnetic species which is a particularly effective quenching agent for molecules with relatively long fluorescent lifetimes; with eosin ($\tau \approx 10^{-3}$ s), 50 per cent quenching is produced by about $10 \, \text{mg L}^{-1}$ oxygen. Fluorescence quenching thus provides a useful method for monitoring low levels of oxygen, e.g. in a supply of 'oxygen-free' nitrogen.

The intensity and colour of the fluorescence of many substances depend upon the pH of the solution; indeed, some substances are so sensitive to pH that they can be used as pH indicators. These are termed '**fluorescent**' or '**luminescent**' indicators. Those substances which fluoresce in ultraviolet light and change in colour or have their fluorescence quenched with change in pH can be used as fluorescent indicators in acid–base titrations. The merit of such indicators is that they can be employed in the titration of coloured (and sometimes of intensely coloured) solutions in which the colour changes of the usual indicators would

735

Table 18.1 Some fluorescent indicators

Name of indicator	Approx. pH range	Colour change
Acridine	5.2– 6.6	Green to violet–blue
Chromotropic acid	3.0– 4.5	Colourless to blue
2-Hydroxycinnamic acid	7.2– 9.0	Colourless to green
3,6-Dihydroxyphthalimide	0.0– 2.5	Colourless to yellowish-green
	6.0– 8.0	Yellowish-green to green
Eosin	3.0– 4.0	Colourless to green
Erythrosin-B	2.5– 4.0	Colourless to green
Fluorescein	4.0– 6.0	Colourless to green
4-Methylaesculetin	4.0– 6.2	Colourless to blue
	9.0–10.0	Blue to light green
2-Naphthoquinoline	4.4– 6.3	Blue to colourless
Quinine sulphate	3.0– 5.0	Blue to violet
	9.5–10.0	Violet to colourless
Quininic acid	4.0– 5.0	Yellow to blue
Umbelliferone	6.5– 8.0	Faint blue to bright blue

be masked. Titrations are best performed in a silica flask. Examples of fluorescent indicators are given in Table 18.1.

EXPERIMENTAL

18.4 QUININE

Discussion. This determination is an ideal experiment with which to gain experience in quantitative fluorimetry. It can be employed particularly for the determination of the amount of quinine in samples of tonic water.

Reagents. *Dilute sulphuric acid*, ca *0.05*M. Add 3.0 mL concentrated sulphuric acid to 100 mL water, and dilute to 1 L with distilled water.

Standard solution of quinine. Weigh out accurately 0.100 g quinine and dissolve it in 1 L 0.05 M sulphuric acid in a graduated flask. Dilute 10.0 mL of the above solution to 1 L with 0.05M sulphuric acid. The resulting solution contains 0.001 00 mg quinine per mL.

With the aid of a calibrated burette, run 10.0, 17.0, 24.0, 31.0, 38.0, 45.0, 52.0, and 62.0 mL of the above dilute standard solution into separate 100 mL graduated flasks and dilute each to the mark with 0.05M sulphuric acid.

Procedure. Measure the fluorescence of each of the above solutions at 445 nm, using that containing 62.0 mL of the dilute quinine solution as standard for the fluorimeter. Use LF2 or an equivalent primary filter ($\lambda_{ex} = 350$ nm) and gelatin as the secondary filter if using a simple fluorimeter.

Now prepare test solutions containing, say, 0.000 25 and 0.000 45 mg quinine per mL. Determine their concentrations by measuring the fluorescence on the instrument and using the calibration curve (see Note).

To determine the quinine content of tonic water it is first necessary to de-gas the sample either by leaving the bottle open to the atmosphere for a prolonged period or by stirring it vigorously in a beaker for several minutes. Take 12.5 mL of the de-gassed tonic water and make up to 25 mL in a graduated flask with 0.1 M sulphuric acid. From this solution prepare other dilutions with 0.05 M sulphuric acid until a fluorimeter reading is obtained that falls on the calibration line previously prepared. From the value obtained calculate the concentration of quinine in the original tonic water.

Note. It is good practice to make the fluorescence measurements for samples and standards as close together as possible to minimise any 'drift' in instrument response.

18.5 ALUMINIUM

The procedure utilises eriochrome blue black RC (also called pontachrome blue black R; Colour Index No. 15705) at a pH of 4.8 in a buffer solution. Beryllium gives no fluorescence and does not interfere; iron, chromium, copper, nickel, and cobalt mask the fluorescence; fluoride must be removed if present. The method may be adapted for the determination of aluminium in steel.

Reagents. *Standard solution of aluminium.* Dissolve 1.760 g aluminium potassium sulphate crystals in distilled water, add 3 mL concentrated sulphuric acid, and dilute to 1 L in a graduated flask. Pipette 10.0 mL of this solution into a little water, add 2.0 mL concentrated sulphuric acid, and dilute to 1 L with distilled water. This solution contains 0.00100 mg aluminium per L.
   Ammonium acetate solution, 10 per cent. Dissolve 25 g of the pure salt in water and dilute to 250 mL.
   Dilute sulphuric acid. Add 25 mL concentrated sulphuric acid to 200 mL water, cool, and dilute to 500 mL in a graduated flask.
   Eriochrome blue black RC, 0.1 per cent. Prepare a 0.1 per cent solution in 90 per cent ethanol.

Procedure. Into 100 mL graduated flasks, each containing 10 mL of the ammonium acetate solution, 1 mL of the dilute sulphuric acid, and 3 mL of the eriochrome blue black RC solution, run in from a burette 15.0, 20.0, 25.0, 30.0, 35.0, 40.0, 45.0, and 50.0 mL of the standard aluminium solution. Dilute each of the above solutions with distilled water, and adjust to a pH of 4.6 ± 0.2 if necessary before making up to the 100 mL mark. Allow the solutions to stand for at least 1 hour.
   Measure the fluorescence of each of the above solutions at 590 nm, using that containing $0.005 \, mg \, mL^{-1}$ Al as standard. The use of a primary filter (Corning 5543 or 5874 has been recommended) will depend upon the quality of the eriochrome blue black RC: it can often be dispensed with. The secondary filter may be a Chance OR2 or Corning 2408, or LF7 for a Locart instrument. Draw a calibration curve, plotting instrument readings against concentration of aluminium. Determine the number of mg of Al per L in an unknown solution (say *ca* $0.25 \, mg \, L^{-1}$), utilising the calibration curve.

18.6 CADMIUM

Discussion. Cadmium may be precipitated quantitatively in alkaline solution in the presence of tartrate by 2-(2-hydroxyphenyl)benzoxazole. The complex dissolves readily in glacial acetic acid, giving a solution with an orange tint and a bright blue fluorescence in ultraviolet light. The acetic (ethanoic) acid solution is used as a basis for the fluorimetric determination of cadmium.[28]

Reagents. *2-(2-Hydroxyphenyl) benzoxazole solution.* Dissolve 1.0 g of the solid reagent in 1 L of 95 per cent ethanol.

Standard cadmium ion solution. Prepare a standard cadmium ion solution containing *ca* 0.04 mg mL^{-1} Cd using hydrated cadmium sulphate.

Solutions for calibration curve with fluorimeter. Prepare the cadmium complex of the reagent by precipitating it from a solution of a pure cadmium salt as follows. Introduce a large excess of sodium tartrate, warm to 60 °C, adjust the pH to 9–10 by the addition of 0.5 M sodium hydroxide, add a slight excess of the reagent, and digest at 60 °C for 15 minutes. Filter on a sintered-glass crucible (medium porosity), wash with 50 per cent ethanol (rendered faintly ammoniacal) to remove excess of the reagent, and dry at 130–140 °C for 1–2 hours. Weigh out 0.2371 g of the complex ($\equiv$ 0.0500 g Cd) and dissolve it in 1 L of glacial acetic acid. Remove volumes of the acetic acid solution equivalent to 2.5, 2.0, 1.0, 0.5, and 0.10 mg Cd, and dilute each to exactly 50 mL with glacial acetic acid. Measure the fluorescence of each of the above solutions, using the appropriate filters (e.g. a yellow filter such as Corning 3-74 between the sample and the photocell). Plot fluorimeter readings against concentration of Cd per 50 mL.

Procedure. Use an aqueous solution of the sample (25–30 mL) containing from 0.1–2.0 mg of Cd and about 0.1 g of ammonium tartrate. Add an equal volume of 95 per cent ethanol, warm to 60 °C, treat with a slight excess of the reagent solution (4 mL $\equiv$ *ca* 1 mg Cd), adjust the pH to 9–11, digest at 60 °C for 15 minutes, filter on a medium-porosity glass crucible, wash with 20–25 mL of 95 per cent ethanol containing a trace of ammonia, and dry the precipitate at 130 °C for 30–45 minutes. Dissolve the precipitate in 50.0 mL of glacial acetic acid, and measure the fluorescence of the solution as in the calibration procedure. Evaluate the cadmium content from the calibration curve.

18.7 CALCIUM

Discussion. This method is based upon the formation of a fluorescent chelate between calcium ions and calcein [fluorescein bis(methyliminodiacetic acid)] in alkaline solution.[29] The procedure described below[30] has been employed for the determination of calcium in biological materials.[31] *

Reagents. *Standard calcium solution.* Prepare a standard solution containing 40.0 mg L^{-1} calcium by dissolving the calculated quantity of calcium carbonate in the minimum amount of hydrochloric acid and diluting to 1 L in a graduated flask.

Calcein solution. Dissolve sufficient calcein, or its disodium salt, in the minimum amount of 0.40 M potassium hydroxide solution and dilute with water to give a concentration of 60 mg L^{-1} in a graduated flask. A small amount of EDTA solution (about 1.0 mL of 0.03 M for every 100 mL calcein solution) may be needed in the calcein solution to achieve balancing of the blank on the fluorimeter. This is only necessary in those cases in which the potassium hydroxide used is found to contain a small amount of calcium impurity.

---

* Calcium in the 10–500 ng range can be determined by using the selective spectrofluorimetric reagent 1,5-bis(dicarboxymethyl-aminoethyl)2,6-dihydroxynaphthalene at pH 11.7 (Ref. 32).

Aqueous solutions of calcein are not stable for longer than 24 hours and should be kept in the dark as much as possible.

Potassium hydroxide solution.   Prepare a 0.4M potassium hydroxide solution by dissolving solid potassium hydroxide (preferably calcium-free) in de-ionised water and make up to 1 L in a graduated flask.

Procedure.   Prepare a series of calcium ion solutions covering the concentration range 0–4 μg per 25 mL by adding sufficient of the 40 mg L^{-1} calcium standard to 25 mL graduated flasks each containing 5.0 mL of 0.4 M potassium hydroxide solution and 1 mL of calcein solution. Dilute each to 25 mL using de-ionised water. Determine the fluorescence for each solution at 540 nm with excitation at either 330 nm or 480 nm, and plot a calibration curve.

Prepare the sample solution in a similar manner to give a fluorescence value falling within the range of the calibration curve, and hence obtain the original calcium concentration in the sample.

18.8   ZINC

The zinc complex of oxine fluoresces in ultraviolet light, and this forms the basis of the following method.

Reagents.   *Standard zinc solution.* Dissolve about 4.0 g, accurately weighed zinc shot in 35 mL concentrated hydrochloric acid, and dilute with distilled water to 1 L in a graduated flask. Pipette 10.0 mL of this solution into a 1litre graduated flask and dilute to the mark with distilled water.

8-Hydroxyquinoline (oxine) solution, 5 per cent. Dissolve 5.0 g oxine in 12 g glacial acetic acid and dilute to 100 mL with distilled water.

Standard dichlorofluorescein solution. Add a 0.1 per cent ethanolic solution of dichlorofluorescein dropwise to 1 L of distilled water until the resulting solution has a fluorescence slightly greater than that produced by the most concentrated zinc solution to be investigated (see below). About 0.8–1.0 mL of dichlorofluorescein solution is required.

Gum arabic solution, 2 per cent. Grind finely 2.0 g gum arabic in a glass mortar, dissolve it in water, and dilute to 100 mL; filter, if necessary.

Ammonium acetate solution, ca 2M. Dissolve 15.5 g crystallised ammonium acetate in water and dilute to 100 mL.

Procedure.   By means of a calibrated burette, run 5.0, 10.0, 15.0, 20.0, and 25.0 mL of the standard zinc solution into separate 100 mL graduated flasks. To each flask add 10 mL of the ammonium acetate solution, 4 mL of the gum arabic solution, dilute to about 45 mL with distilled water, and mix by swirling. Now add exactly 0.40 mL of the oxine solution (use, for example, a micrometer syringe or a micropipette), dilute to the mark with distilled water, shake gently, and transfer immediately to the cell of a fluorimeter for measurement. Employ the dichlorofluorescein solution as standard. Use a Chance OB2 as primary filter and OY2 as the secondary filter. Commence measurements with the most concentrated zinc solution. It is important that the fluorescence of the zinc–oxine mixtures be determined immediately after they are prepared, since the fine suspension of zinc oxinate slowly settles to the bottom of the cell, Plot instrument readings against zinc content (mg mL^{-1}). Use the calibration curve for determining the zinc content of test solutions containing, say, 4.5 and 6.5 mg of zinc L^{-1}.

18.9 DETERMINATION OF CODEINE AND MORPHINE IN A MIXTURE

This experiment[33] illustrates how adjustment of pH may be used to control fluorescence and so make the determination more specific. The alkaloids codeine and morphine can be determined independently because whilst both fluoresce strongly at the same wavelength in dilute sulphuric acid solution, morphine gives a generally negligible fluorescence in dilute sodium hydroxide. The fluorescence intensities of the two compounds are assumed to be additive.

Solutions. Prepare the following series of standard solutions of codeine and morphine, each of which should cover the range $5-20\,\text{mg L}^{-1}$:

(a) codeine in H_2SO_4 $(0.05M)$;
(b) codeine in NaOH $(0.1M)$;
(c) morphine in H_2SO_4 $(0.05M)$;
(d) morphine in NaOH $(0.1M)$.

Prepare solutions of the accurately weighed sample (codeine–morphine mixture) in H_2SO_4 $(0.05M)$ and in NaOH $(0.1M)$.

Procedure. Measure the fluorescence intensities of each of the series of standard solutions at 345 nm, with excitation at 285 nm. Construct a calibration graph for each of the four series (a), (b), (c), and (d) above.

Measure the fluorescence intensity of the sample in NaOH solution using the above emission and excitation wavelengths. Read off the codeine concentration from the appropriate calibration graph (b). Calculate the fluorescence intensity which corresponds to this concentration of codeine in H_2SO_4 using calibration graph (a). Now measure the fluorescence intensity of the sample in H_2SO_4 solution and subtract the fluorescence intensity due to codeine. The value obtained gives the fluorescence intensity due to morphine in H_2SO_4 and its concentration can be deduced from calibration graph (c).

Calibration graph (d) may be used to correct for the small fluorescence intensity due to morphine in NaOH; this is not negligible when the morphine concentration is high and the codeine concentration is low.

For References and Bibliography see Sections 21.26 and 21.27.

CHAPTER 19
INFRARED SPECTROPHOTOMETRY

19.1 INTRODUCTION

The infrared region of the electromagnetic spectrum may be divided into three main sections:[34]

1. near-infrared (overtone region) $0.8-2.5\,\mu m$ ($12\,500-4000\,cm^{-1}$);
2. middle infrared (vibration–rotation region) $2.5-50\,\mu m$ ($4000-200\,cm^{-1}$);
3. far-infrared (rotation region) $50-1\,000\,\mu m$ ($200-10\,cm^{-1}$).

The main region of interest for analytical purposes is from 2.5 to $25\,\mu m$ (micrometres), i.e. 4000 to 400 wavenumbers (waves per centimetre, cm^{-1}). Normal optical materials such as glass or quartz absorb strongly in the infrared, so instruments for carrying our measurements in this region differ from those used for the electronic (visible/ultraviolet) region.

Infrared spectra originate from the different modes of vibration and rotation of a molecule. At wavelengths below $25\,\mu m$ the radiation has sufficient energy to cause changes in the vibrational energy levels of the molecule, and these are accompanied by changes in the rotational energy levels. The pure rotational spectra of molecules occur in the far-infrared region and are used for determining molecular dimensions.

In the case of simple diatomic molecules it is possible to calculate the vibrational frequencies by treating the molecule as a harmonic oscillator. The frequency of vibration is given by:

$$v = \frac{1}{2\pi}\sqrt{\frac{f}{\mu}}\,s^{-1}$$

where $v =$ the frequency (vibrations per second), $f =$ force constant (the stretching or restoring force between two atoms) in newtons per metre, and μ is the reduced mass per molecule (in kilograms) defined by the relationship:

$$\mu = \frac{m_1 \times m_2}{m_1 + m_2} = \frac{A_{r_1} \times A_{r_2}}{[A_{r_1} + A_{r_2}]L \times 1000}\,kg$$

where m_1 and m_2 are masses of the individual atoms, and A_{r_1} and A_{r_2} are the relative atomic masses, L is Avogadro's constant.

It is, however, customary to quote absorption bands in units of wavenumber ($\tilde{v}$) which are expressed in reciprocal centimetres, cm^{-1}. In some instances wavelengths (λ) measured in micrometres (μm) are also used. The relationship

741

between these units is given by:

$$\tilde{v} = \frac{1}{\lambda} = \frac{v}{c}$$

so

$$\tilde{v} = \frac{1}{2\pi c}\left(\frac{f}{\mu}\right)^{1/2} \text{cm}^{-1}$$

There is usually good agreement between calculated and experimental values for wavenumbers. As an example, we may take the C—O bond in methanol (CH_3OH). For this, $f = 5 \times 10^2 \, \text{N m}^{-1}$, $\mu = 6.85 \times m_u$ kg (m_u is the unified atomic mass constant $= 1.660 \times 10^{-27}$ kg), and the velocity of light $c = 2.998 \times 10^{10} \text{ cm s}^{-1}$.

$$\tilde{v} = \frac{1}{2 \times \pi \times 2.998 \times 10^{10}}\left(\frac{5 \times 10^2}{6.85 \times 1.66 \times 10^{-27}}\right)^{1/2}$$

$$= \frac{20.97 \times 10^{13}}{18.84 \times 10^{10}} = 1113 \, \text{cm}^{-1}$$

The observed C—O band for methanol is at $1034 \, \text{cm}^{-1}$.

This simple calculation has not taken into consideration any possible effects arising from other atoms in the molecule. More sophisticated methods of calculation which take account of these interactions have been developed but are outside the scope of this book and students should consult other appropriate texts[35] if they wish to study the theory of the subject further.

For a vibrational mode* to appear in the infrared spectrum, and therefore absorb energy from the incident radiation, it is essential that a change in dipole moment occurs during the vibration. Vibration of two similar atoms against each other, for example oxygen or nitrogen molecules, will not result in a change of electrical symmetry or dipole moment of the molecule, and such molecules will not absorb in the infrared region.

In many of the normal modes of vibration of a molecule the main participants in the vibration will be two atoms held together by a chemical bond. These vibrations have frequencies which depend primarily on the masses of the two vibrating atoms and on the force constant of the bond between them. The frequencies are also slightly affected by other atoms attached to the two atoms concerned. These vibrational modes are characteristic of the groups in the molecule and are useful in the identification of a compound, particularly in establishing the structure of an unknown substance.

Some of these group frequencies are listed in Table 19.1, and a more complete correlation table is provided in Appendix 11.

The above, of course, is a very simplified picture, as many bands of much weaker intensities occur at shorter wavelengths (these are known as overtone bands and combination bands), but these are unlikely to be confused with the

---

* Bond vibration modes are divisible into two distinct types, stretching and bending (deformation) vibrations. The former constitute the periodic stretchings of the bond along the bond axis. The latter are displacements occurring at right angles to the bond axis. For further information, a textbook on infrared spectroscopy should be consulted.

Table 19.1 Approximate positions of some infrared absorption bands

Group	Wavenumber (cm^{-1})	Wavelength (μm)
C—H (aliphatic)	2700–3000	3.33– 3.70
C—H (aromatic)	3000–3100	3.23– 3.33
O—H (phenolic)	3700	2.70
O—H (phenolic, hydrogen bonding)	3300–3700	2.70– 3.03
S—H	2570–2600	3.85– 3.89
N—H	3300–3370	2.97– 3.03
C—O	1000–1050	9.52–10.00
C=O (aldehyde)	1720–1740	5.75– 5.8
C=O (ketone)	1705–1725	5.80– 5.86
C=O (acid)	1650	6.06
C=O (ester)	1700–1750	5.71– 5.88
C=N	1590–1660	6.02– 6.23
C—C	750–1100	9.09–13.33
C=C	1620–1670	5.99– 6.17
C≡C	2100–2250	4.44– 4.76
C≡N	2100–2250	4.44– 4.76
CH$_3$—, —CH$_2$—	1350–1480	6.76– 7.41
C—F	1000–1400	7.14–10.00
C—Cl	600– 800	12.50–16.67
C—Br	500– 600	16.67–20.00
C—I	500	20.00

much more intense fundamental bands originating from normal modes of vibration.

Infrared absorption spectra can be employed for the identification of pure compounds or for the detection and identification of impurities. Most of the applications are concerned with organic compounds, primarily because water, the chief solvent for inorganic compounds, absorbs strongly beyond 1.5 μm. Moreover, inorganic compounds often have broad absorption bands, whereas organic substances may give rise to numerous narrower bands.

The infrared absorption spectrum of a compound may be regarded as a sort of 'finger-print' of that compound; see Fig. 19.1. Thus for the identification of

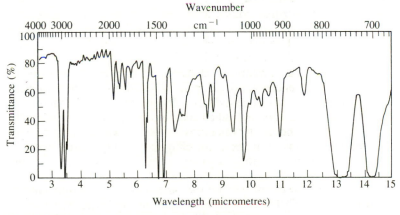

Fig. 19.1 The infrared spectrum of polystyrene.

a pure compound the spectrum of the unknown substance is compared with the spectra of a limited number of possible substances suggested by other properties. When a match between spectra is obtained, identification is complete. This procedure is especially valuable for distinguishing between structural isomers[36] (but not optical isomers).

The spectrum of a mixture of compounds is essentially that of the sum of the spectra of the individual components, provided association, dissociation, polymerisation, or compound formation does not take place. In order to detect an impurity in a substance, comparison can be made of the spectrum of the substance with that of the pure compound: impurities will cause extra absorption bands to appear in the spectrum. The most favourable case will occur when the impurities present possess characteristic groupings not present in the main constituent.

19.2 APPARATUS AND INSTRUMENTS

For measurements in the middle-infrared region, 2.5–50 μm, there are several differences between the instruments used for UV/visible spectrophotometry and those designed for infrared determinations. These changes are mainly dictated by the fact that glass and quartz absorb strongly in the infrared region and photomultipliers are insensitive to the radiation. Front-surfaced mirrors are largely employed to avoid the necessity of radiation passing through glass or quartz layers as reflection from metallic surfaces is generally very efficient in the infrared region. But absorption cells and windows must be fabricated from infrared transparent materials. The substances most commonly used with infrared radiation and their useful transmission ranges are given in Table 19.2.

Table 19.2 Transmission ranges of materials for cells and windows

Material	Transmission range	
	(μm)	(cm^{-1})
Lithium fluoride	2.5– 5.9	4000–1695
Calcium fluoride	2.4– 7.7	4167–1299
Sodium chloride	2.0–15.4	5000– 649
Potassium bromide	9.0–26.0	1111– 385
Caesium bromide	9.0–26.0	1111– 385
KR-5 (TlBr + TlI)	25.0–40.0	400– 250

The main sources of infrared radiation used in spectrophotometers are (1) a nichrome wire wound on a ceramic support, (2) the Nernst glower, which is a filament containing zirconium, thorium and cerium oxides held together by a binder, (3) the Globar, a bonded silicon carbide rod. These are heated electrically to temperatures within the range 1200–2000 °C when they will glow and produce the infrared radiation approximating to that of a black body.

Traditional infrared spectrophotometers were constructed with mono-chromation being carried out using sodium chloride or potassium bromide prisms, but these had the disadvantage that the prisms are hygroscopic and the middle-infrared region normally necessitated the use of two different prisms in order to obtain adequate dispersion over the whole range.

For these reasons diffraction gratings have displaced prisms as the main means of monochromation in the infrared region. Gratings provide higher resolving powers than do prisms and can be designed to operate effectively over a wider spectral range. Even so, most grating instruments operate with two gratings with an automatic change of grating occurring around 2000 cm^{-1}. The layout of a typical grating infrared spectrophotometer is shown in Fig. 19.2.

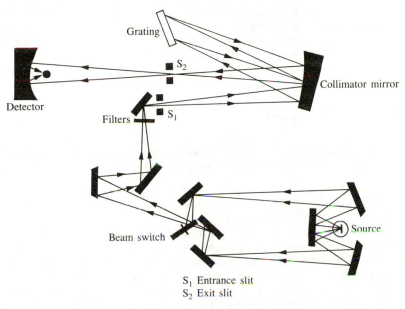

Fig. 19.2 **Layout of an infrared spectrophotometer employing a diffraction grating for monochromation. Reproduced by permission from R. C. J. Osland,** *Principles and Practices of Infrared Spectroscopy,* **2nd edn, Philips Ltd, 1985.**

More advanced infrared spectrophotometers produce the infrared spectra by a procedure based upon interferometry. This is known as FT-IR, from the name 'Fourier transform infrared spectroscopy'.[37] The instruments are normally based upon the Michelson interferometer in which the radiation from an infrared source is split into two beams by a half-silvered 45° mirror so that the resulting beams are at right angles to each other. If an absorbing material is placed in one of the beams, then the resulting interferogram will carry the spectral characteristics of the sample in the beam. The actual conversion of the information from the interferogram into an infrared spectrum is very complex and has only been possible by the development of computers. There are, however, great advantages in the use of FT-IR. All the frequencies are recorded simultaneously, there is an improvement in signal-to-noise ratios, and it is easier to study small samples or materials with weak absorptions. In addition to this the time taken for a full spectral scan is less than 1 second and this makes it possible to obtain improved spectra by carrying out repetitive scans and averaging the collected signals. This is because the signal-to-noise ratio is directly related to $\sqrt{n}$, where n is the number of scans. Thus 16 repeat scans give a four-fold enhancement of the S/N ratio. Quantitative infrared analysis has

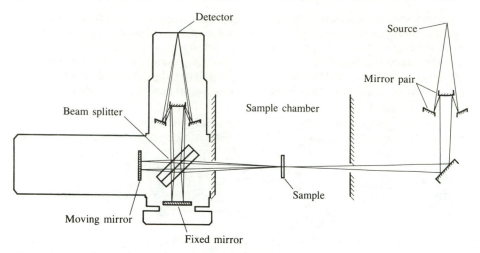

Detector

Source

Mirror pair

Sample chamber

Beam splitter

Sample

Moving mirror

Fixed mirror

Fig. 19.3   Layout of a Fourier-transform infrared spectrometer. Reproduced by permission of Lloyd Instruments PLC, Warsash, Southampton SO3 6HP.

benefited greatly from the development of FT-IR. The layout of a typical FT-IR spectrophotometer is shown in Fig. 19.3.

Detection of the infrared signal is, of course, of prime importance. A range of detectors is available for this purpose, the type used in any particular instrument depending upon the type and quality of the spectrophotometer.

The **thermocouple** is made by welding together two wires of metals 1 and 2 in such a manner that a segment of metal 1 is connected to two terminal wires of metal 2. One junction between metals 1 and 2 is heated by the infrared beam, and the other junction is kept at constant temperature: small changes in ambient temperature are thus minimised. To avoid losses of energy by convection, the couples are enclosed in an evacuated vessel with a window transparent to infrared radiation. The metallic junctions are also covered with a black deposit to decrease reflection of the incident beam

A **bolometer** is essentially a thin blackened platinum strip in an evacuated glass vessel with a window transparent to the infrared rays: it is connected as one arm of a Wheatstone bridge, and any radiation absorbed raises the temperature of the strip and changes its resistance. Two identical elements are usually placed in the opposite arms of a bridge; one of the elements is in the path of the infrared beam and the other compensates for variations in ambient temperature. Both the above receptors give a very small direct current, which may be amplified by special methods to drive a recorder.

The **Golay pneumatic detector**, which is sometimes used, consists of a gas-filled chamber which undergoes a pressure rise when heated by radiant energy. Small pressure changes cause deflections of one wall of the chamber; this movable wall also functions as a mirror and reflects a light beam directed upon it to a photocell, the amount of light reflected bearing a direct relation to the gas-chamber expansion, and hence to the radiant energy of the light from the monochromator. This detector responds to the total light energy received as distinct from energy received per unit area (thermocouples and bolometers).

The **pyroelectric detectors** fitted in many modern instruments use ferroelectric materials operating below their Curie-point temperatures. When infrared radiation is incident on the detector there is a change in polarisation which can be employed to produce an electrical signal. The detector will only produce a signal when the intensity of the incident radiation changes. These detectors are of especial value in FT-IR where rapid response times are needed and for this purpose they use deuterium triglycine sulphate as the detecting medium in an evacuated chamber.

All infrared spectrophotometers are provided with chart recorders which will present the complete infrared spectrum on a single continuous sheet, usually with wavelength and wavenumber scales shown for the abscissa and with absorbance and percentage transmittance as the ordinates. More advanced instruments also possess visual display units on which the spectra can be displayed as they are recorded and on which they can be compared with earlier spectra previously obtained or with spectra drawn from an extensive library held in a computer memory. These modern developments have all led to quantitative infrared spectrophotometry being a much more viable and useful analytical procedure than it was just a few years ago.

19.3 DEDICATED PROCESS ANALYSERS

One very important group of infrared instruments consists of spectrometers used for quantitative measurements either as part of a continuous industrial monitoring process or for environmental studies. These instruments are normally purpose-made, dedicated machines designed to run virtually automatically, and are normally intended only to measure a single compound or family of compounds.

This type of instrument is typified by the non-dispersive continuous stream analyser used for the detection of carbon monoxide. In this device identical infrared beams are passed through the reference cell and the sample cell with the two signals being balanced against each other after detection on a diaphragm detector. This detector consists of two small compartments of equal volume filled with the pure gas which has to be determined. When an increase in the carbon monoxide level in the sample flow chamber occurs, the infrared radiation at 4.2 μm is absorbed and the strength of the infrared beam on the detector is diminished. As a result the diaphragm becomes distended due to the unbalanced heating effect on the cell and a signal corresponding in magnitude to the quantity of carbon monoxide in the sample is recorded, usually on a continuous sheet or tape. This type of process analyser is illustrated in Fig. 19.4.

Another dedicated application of quantitative infrared spectrometry which has aroused much interest in recent years is its application to the measurement of ethanol in the breath of motorists who are suspected of having had alcoholic drinks prior to driving motor vehicles. It is not the purpose of this book to discuss the justification or arguments surrounding this application and this has been well documented elsewhere.[38] However, these analysers are used in many countries throughout the world and are greatly relied upon in the battle against the combination of drink–driving. Typical of the infrared analysers used for this purpose is the Lion Intoximeter 3000, the operating system of which is shown in Fig. 19.5.

Unlike the carbon monoxide measuring instrument discussed above, the Lion Intoximeter 3000 uses an interference filter to produce monochromatic radiation

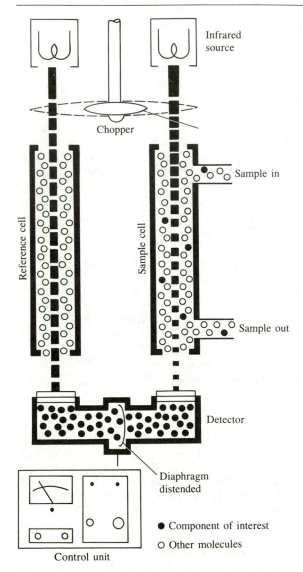

Fig. 19.4 Non-dispersive continuous stream infrared analyser. Reproduced by permission of Beckman Instrument Co.

of 3.39 μm which corresponds to the C–H stretching frequency for ethanol. The infrared radiation source is a nichrome filament helix wound around a ceramic rod and heated to 800 °C. The beam from the nichrome source is divided into two before passing through the fixed path length, double chambered gas sample cell. Under normal circumstances the composition of the atmosphere in the two cells will be identical; the resulting signals will be balanced, and the ratios of the two energy levels can be used to set and establish the baseline conditions. When ethanol is passed into the machine, either from a simulator or from someone with ethanol in their breath, the infrared beam in the sample cell is

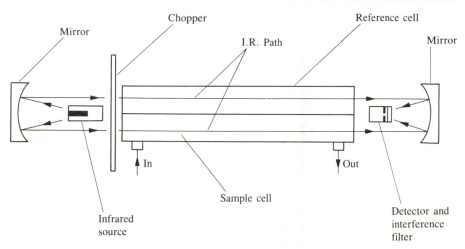

Fig. 19.5 Diagram of the infrared analyser unit of the Lion Intoximeter 3000 breath alcohol analyser. Reproduced by permission of Lion Laboratories Ltd, Ty Verlon Industrial Estate, Barry, Wales CF6 3BE, UK.

partially absorbed by the alcohol. The amount of radiation reaching the detector from the sample cell will depend upon the concentration of ethanol in the cell. The infrared detector again measures the ratio of the reference and sample chamber beams and the resulting signal is converted into a corresponding breath alcohol value. In this instrument the monochromation takes place after the beams have passed through the cells and before they are measured on the solid-state photoconductive detector.[39] The Intoximeter 3000 is designed to take a 70 mL sample of deep lung air only after at least an initial 1.5 L volume has been expelled through the sample tube by the subject. Condensation of alcohol and water from the breath is prevented by maintaining the sample flow unit at 45 °C.

So long as a compound has a fairly intense absorption which is unlikely to overlap with those of other substances with which it is likely to be mixed, then it is possible to monitor that substance on a continuous basis with a dedicated infrared detector. Gases such as carbon dioxide, nitrogen oxides, ethylene oxide and ammonia can now be measured and regulated using these devices.

19.4 INFRARED CELLS FOR LIQUID SAMPLES

As many solutions for infrared spectrophotometry involve organic solvents, it is necessary to use sample cells which can not only be stoppered to prevent solvent evaporation, but also be readily dismantled for cleaning and polishing. Cells used for accurate quantitative work must have smooth, polished, parallel window surfaces and be of fixed path length. Cells are available commercially with fixed path lengths from 0.025 mm to 1.0 mm, and with variable path lengths to 6.0 mm. For this purpose cells are made with transparent potassium bromide (or less commonly sodium chloride) plate windows, cut from large crystals, held in a stainless-steel former with a lead or polytetrafluoroethylene spacer providing the fixed separation between the plates. An accurate measurement of the path length can be made by the procedure given in Section 19.5.

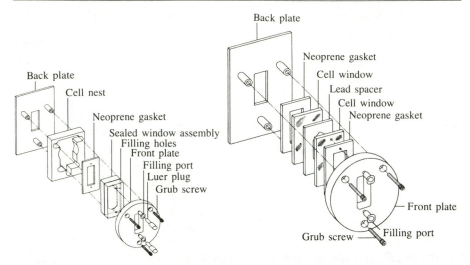

Fig. 19.6 Exploded views of fixed-path-length cells. Reproduced by permission of Specac Ltd, Lagoon Road, Orpington, Kent BR5 3QX.

The construction of two typical infrared cells is shown in Fig. 19.6. Such cells have to be carefully filled by using a syringe or Pasteur pipette to ensure that no air is trapped inside. To prevent evaporation the ports should be plugged with small plastic stoppers once the cell has been filled with the solution.

19.5 MEASUREMENT OF CELL PATH LENGTH

When a beam of monochromatic radiation is passed through the windows of an infrared cell some reflection occurs on the window surfaces and interference takes place between radiation passing from the internal surface of the first window and that reflected back from the internal surface of the second window. This interference is at a maximum when $2d = (n + 1/2)\lambda$, where d is the distance in μm between the inner surfaces of the two cell windows, λ is the wavelength in μm, and n is any integral number. If the wavelength λ of the monochromatic radiation is varied continuously an interference pattern consisting of a series of waves (Fig. 19.7) is obtained.

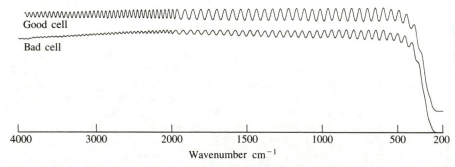

Fig. 19.7 Interference patterns from empty fixed-path-length cell. Reproduced by permission from R. C. J. Osland, *Principles and Practices of Infrared Spectroscopy*, 2nd edn, Philips Ltd, 1985.

A value for the cell path-length, d, can be calculated from the formula:

$$d = \frac{\Delta n}{2(\tilde{v}_1 - \tilde{v}_2)} \, \text{cm}$$

where n is the number of complete interference fringes between wavenumbers $\tilde{v}_1$ and $\tilde{v}_2$.

19.6 MEASUREMENT OF INFRARED ABSORPTION BANDS

As with electronic spectra, the use of infrared spectra for quantitative determinations depends upon the measurement of the intensity of either the transmission or absorption of the infrared radiation at a specific wavelength, usually the maximum of a strong, sharp, narrow, well-resolved absorption band. Most organic compounds will possess several peaks in their spectra which satisfy these criteria and which can be used so long as there is no substantial overlap with the absorption peaks from other substances in the sample matrix.

The background to any spectrum does not normally correspond to a 100 per cent transmittance at all wavelengths and measurements are best made by what is known as the baseline method.[40] This involves selecting an absorption peak to which a tangential line can be drawn, as shown in Fig. 19.8. This is then used to establish a value for I_0 by measuring vertically from the tangent through the peak to the wavenumber scale. Similarly, a value for I is obtained by measuring the corresponding distance from the absorption peak maximum. So that for any peak the absorbance will not be the value corresponding to the height of the absorption from the chart paper abscissa, but the value of A_{calc} obtained from the equation:

$$A_{\text{calc}} = \log \frac{1}{T} = \log \frac{I_0}{I}$$

where I_0 and I are values measured using the tangential baseline.

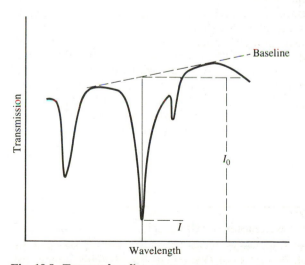

Fig. 19.8 Tangent baseline measurement.

This procedure has the great advantage that some potential sources of error are eliminated. The measurements do not depend upon accurate wavelength positions as they are made with respect to the spectrum itself and any cell errors are avoided by using the same fixed-path-length cell. Also, by employing this measured ratio any variations in the source intensity, the instrument optics or sensitivity are eliminated.

19.7 THE BEER–LAMBERT RELATIONSHIP IN QUANTITATIVE INFRARED SPECTROPHOTOMETRY

Infrared spectra are recorded using either or both absorbance and percentage transmission just as they are in visible/ultraviolet electronic spectra, and the Beer–Lambert relationship:

$$A = \varepsilon c l = \log \frac{I_0}{I} = \log \frac{1}{T}$$

as given in Section 17.2 applies as well to infrared spectra as it does to electronic spectra. Similarly, for a mixture of compounds the observed absorbance at a particular wavelength (or frequency) will be the sum of the absorbances for the individual constituents of the mixture at the wavelength:

$$A_{\text{observed}} = A_1 + A_2 + A_3 = \varepsilon_1 c_1 l + \varepsilon_2 c_2 l + \varepsilon_3 c_3 l$$

as the path length l is constant for the mixture.

It has taken a long time for quantitative infrared spectrophotometry to become a commonly used procedure for several reasons.

(a) The molar absorption coefficients are usually smaller by a factor of 10 than are those in the electronic region. So the infrared procedure is usually less sensitive.

(b) The most accurate range for quantitative measurements is between 55 per cent and 20 per cent T (0.26 A to 0.70 A), with the accuracy diminishing rapidly outside these values.

(c) Older null-balance spectrophotometers possessed an instrumental error of ± 1 per cent T.

Whilst nothing can improve upon the disadvantage of low molar absorption coefficients, instrumental designs and improvements with ratio recording and FT-IR instruments have virtually overcome the accuracy and instrumental limitations referred to in (b) and (c) above. As a result, quantitative infrared procedures are now much more widely used and are frequently applied in quality control and materials investigations. Applications fall into several distinct groups:

(a) direct application of the Beer–Lambert relationship;
(b) use of a calibration graph;
(c) standard addition methods.

(a) **Measurements using the Beer–Lambert relationship.** Where a compound gives a strong, narrow, well-defined absorption band, this can be used as the basis for quantitative measurements simply by comparing the magnitude of the absorbance (A_u) of the unknown concentration (c_u) with the corresponding absorbance (A_s) of a standard solution of known concentration (c_s) using a cell of measured path length. The absorbance of the standard solution at the chosen wavelength is given by:

$$A_s = \varepsilon c_s l$$

and that for the unknown at the same wavelength is:

$$A_u = \varepsilon c_u l$$

then

$$\frac{A_s}{c_s} = \frac{A_u}{c_u} = \varepsilon l$$

Hence, the concentration of the unknown is given by:

$$c_u = \frac{c_s A_u}{A_s}$$

It should be noted that the calculation is based upon an assumption of a linear absorbance/concentration relationship and this may only apply over short concentration ranges.

(**b**) **The use of a calibration graph.** This overcomes any problems created due to non-linear absorbance/concentration features and means that any unknown concentration run under the same conditions as the series of standards can be determined directly from the graph. The procedure requires that all standards and samples are measured in the same fixed-path-length cell, although the dimensions of the cell and the molar absorption coefficient for the chosen absorption band are not needed as these are constant throughout all the measurements.

The results in Table 19.3 are typical for the concentration of an antioxidant, measured at 3655 cm^{-1}, used as an additive in oil. The results are presented in graphical form in Fig. 19.9.

Table 19.3

Antioxidant concentration (per cent w/v)	A
0.129	0.024
0.251	0.044
0.514	0.094
0.755	0.138
1.016	0.186
1.273	0.233
Unknown	0.104

(**c**) **Standard addition methods.** These are not widely applied in quantitative infrared spectrophotometry, being limited to determinations of low concentration components in multicomponent mixtures. This procedure also involves the preparation of a series of solutions in a solvent which does not absorb at the wavelength of the chosen absorption band. The solutions are made from a series of increasing concentrations of the pure analyte (similar to a normal calibration graph set of concentrations) but to each is added a constant, known amount

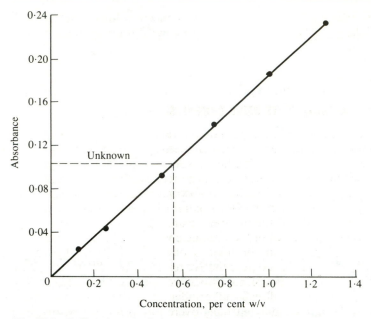

Fig. 19.9 Calibration graph for antioxidant determination.

of the sample containing the unknown concentration. All the solutions are diluted to a fixed volume and their absorbances measured in a fixed-path-length cell by scanning over the chosen absorption band.

A plot of the absorbance against the concentration of the pure analyte does not pass through zero as all the absorbance values are enhanced by an equal amount due to the presence of the unknown concentration in the added sample. Extrapolation of the graph back to the abscissa (the horizontal axis) gives the concentration of the unknown as a negative value. Alternatively it can be determined from the slope of the line by taking any two points on the line, as shown in Fig. 19.10. From this it can be seen that:

$$\frac{A_2}{A_1} = \frac{c_u + c_2}{c_u + c_1}$$

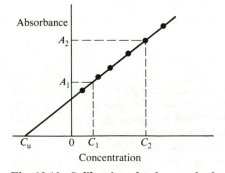

Fig. 19.10 Calibration plot for standard addition procedure.

754

Hence

$$A_2 c_u - A_1 c_u = A_1 c_2 - A_2 c_1$$

$$c_u = \frac{A_1 c_2 - A_2 c_1}{A_2 - A_1}$$

19.8 QUANTITATIVE MEASUREMENTS USING COMPRESSED DISCS

Infrared spectra for solid organic compounds are frequently obtained by mixing and grinding a small sample of the material with specially dry and pure potassium bromide (the carrier), then compressing the powder in a special metal die under a pressure of 15–30 tonnes to produce a transparent potassium bromide disc. As the potassium bromide has virtually no absorption in the middle-infrared region, a very well-resolved spectrum of the organic compound is obtained when the disc is placed in the path of the infrared beam.

In theory, increased quantities of the organic compound finely ground with constant quantities of potassium bromide should give infrared spectra of increasing intensity. However, good quantitative results by this direct procedure are difficult to obtain due to problems associated with the non-quantitative transfer of powder from the small ball-mill grinder (or pestle and mortar) into the compression die. These are only partially overcome by using a micrometer to measure the final disc thickness.

To use KBr discs for quantitative measurements it is best to employ an internal standard procedure in which a substance possessing a prominent isolated infrared absorption band is mixed with the potassium bromide. The substance most commonly used is potassium thiocyanate, KSCN, which is intimately mixed and ground to give a uniform concentration, usually 0.1–0.2 per cent, in the potassium bromide. A KBr/KSCN disc will give a characteristic absorption band at 2125 cm^{-1}. Before quantitative measurements can be carried out it is necessary to prepare a calibration curve from a series of standards made using different amounts of the pure organic compound with the KBr/KSCN. A practical application of this is given in Section 19.9.

19.9 DETERMINATION OF THE PURITY OF COMMERCIAL BENZOIC ACID USING COMPRESSED DISCS

To obtain a calibration curve for benzoic acid, six discs should be prepared using potassium bromide containing 0.1 per cent potassium thiocyanate as described in Section 19.8 and increasing quantities of pure benzoic acid using the following quantities:

| KBr/KSCN | 1.000 | 1.000 | 1.000 | 1.000 | 1.000 | 1.000 g |
| Benzoic acid | 0.000 | 0.050 | 0.075 | 0.100 | 0.150 | 0.200 g |

It should be noted that the weighed amount of KBr/KSCN is constant and that although the problem of non-quantitative transfer of powder from the ball-mill grinder still exists it affects both the carrier and the organic compound equally. When the infrared spectra for the six discs have been obtained the calibration curve is prepared by plotting the **ratio** of the intensity of the selected

755

benzoic acid band (carbonyl, $1695\,\text{cm}^{-1}$) and the KSCN $2125\,\text{cm}^{-1}$ peak against the benzoic acid concentration in the discs. As a result a calibration plot of the type shown in Fig. 19.11 is obtained. This can then be used to assay an impure benzoic acid sample by weighing, say, 0.125 g of the sample and grinding it with 1.000 g of the KBr/KSCN carrier and preparing a compressed disc as before. The peak ratio of the measured absorbances can then be referred to the calibration curve to give a value for the true amount of benzoic acid in the sample.

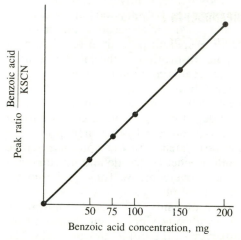

Fig. 19.11   Calibration plot for internal standard procedure for benzoic acid.

19.10   PREPARATION OF A CALIBRATION CURVE FOR CYCLOHEXANE

Run infrared spectra for pure cyclohexane and pure nitromethane. From the spectra select a cyclohexane absorption which is not affected by or overlapping with those of the nitromethane. Prepare a series of solutions of known concentrations of cyclohexane in nitromethane covering the range from 0 per cent to 20 per cent w/v. Using a 0.1 mm fixed path length cell measure the absorbances for the solutions at the chosen peak absorption using the baseline method (Section 19.6) and plot the calibration graph. Use this graph to determine the unknown concentration of cyclohexane in the sample.

19.11   DETERMINATION OF 2-, 3-, AND 4-METHYLPHENOLS (CRESOLS) IN A MIXTURE

Prepare solutions containing weighed amounts of the individual pure cresols (0.5 g) in cyclohexane (20 mL) and use these to obtain infrared spectra for the three cresols in the cyclohexane. Also, prepare a single solution containing the three cresols by mixing together 5 mL of each of the individual solutions. Record the infrared spectra for each of the four solutions using a 0.1 mm or 0.25 mm fixed path length cell. From the spectra select absorption bands suitable for the separate measurement of each isomer. You should find the most appropriate to be: 2-methylphenol $750\,\text{cm}^{-1}$, 3-methylphenol $773\,\text{cm}^{-1}$, 4-methylphenol $815\,\text{cm}^{-1}$. Use the solutions of the individual cresols to prepare a series of calibration standards at appropriate dilutions with cyclohexane for each of the individual cresols and construct the three calibration curves.

Take some crude cresol mixture (1 g) and dissolve it in cyclohexane (20 mL). Obtain the infrared spectrum for the mixture; if necessary, dilute the solution further with cyclohexane to obtain absorbances which will lie on the calibration graphs. From the selected absorption peaks calculate the absorbances for the three individual isomers and use the calibration graphs to calculate the percentage composition of the cresol mixture.

19.12 DETERMINATION OF ACETONE (PROPANONE) IN PROPAN-2-OL

Due to atmospheric oxidation it is common for commercial propan-2-ol to contain a small amount of acetone.

$$CH_3\overset{\overset{\displaystyle OH}{|}}{C}HCH_3 \xrightarrow{\ O_2\ } CH_3\overset{\overset{\displaystyle O}{\|}}{C}CH_3$$

As the greatest accuracy is achieved in any determination by measuring the smaller component, in this case it is the acetone which is determined quantitatively rather than the propan-2-ol.

Run infrared spectra of pure acetone and of pure propan-2-ol. From them select an absorption band for acetone which does not overlap significantly with any of those for the propan-2-ol. The best band is most probably that at $1718\ cm^{-1}$, the carbonyl stretching frequency.

Prepare a 10 per cent v/v bulk solution by dissolving pure acetone (25 mL) in carbon tetrachloride and diluting to 250 mL in a graduated flask. From this prepare a series of dilutions to acetone in carbon tetrachloride covering the concentration range 0.1–2.5 per cent v/v (**carry out this work in a fume cupboard**). Measure the percentage transmittance for each solution at $1718\ cm^{-1}$ using a 0.1 mm fixed path length cell. From the spectra use the baseline method (Section 19.6) to calculate the absorbance for each concentration and plot a calibration curve of absorbance against concentration.

Take 10 mL of commercial propan-2-ol and dilute to 100 mL with carbon tetrachloride in a graduated flask. Record the infrared spectrum and calculate the absorbance for the peak at $1718\ cm^{-1}$. Obtain a value for the acetone concentration from the calibration graph. The true value for the acetone in the propan-2-ol will be 10 times the figure obtained from the graph (this allows for the dilution) and the percentage v/v value can be converted to a molar concentration $(mol\ L^{-1})$ by dividing the percentage v/v by 7.326; e.g. 1.25 per cent v/v $= 1.25/7.326 = 0.171\ mol\ L^{-1}$.

For References and Bibliography see Sections 21.26 and 21.27.

CHAPTER 20
ATOMIC EMISSION SPECTROSCOPY

20.1 INTRODUCTION

This chapter describes the basic principles and practice of emission spectroscopy using non-flame atomisation sources. [Details on flame emission spectroscopy (FES) are to be found in Chapter 21.] The first part of this chapter (Sections 20.2–20.6) is devoted to emission spectroscopy based on electric arc and electric spark sources and is often described as emission spectrography. The final part of the chapter (Sections 20.7–20.11) deals with emission spectroscopy based on plasma sources.

20.2 GENERAL DISCUSSION (EMISSION SPECTROGRAPHY)

When certain metals are introduced as salts into the Bunsen flame characteristic colours are produced; this procedure has long been used for detecting elements qualitatively. If the light from such a flame is passed through a spectroscope several lines may be seen, each of which has a characteristic colour: thus calcium gives red, green, and blue radiations, of which the red are largely responsible for the typical colour that this element imparts to the flame. A definite wavelength can be assigned to each radiation, corresponding with its fixed position in the spectrum. Although the flame colours of, for example, calcium, strontium, and lithium are very similar, it is possible to differentiate with certainty between them by observations on their spectra and to detect each in the presence of the others. By extending and amplifying the principles inherent in the qualitative flame test, analytical applications of emission spectrography have been developed. Thus, more powerful methods of excitation, such as electric spark or electric arc, are used, and the spectra are recorded photographically by means of a spectrograph: also, since the characteristic spectra of many elements occur in the ultraviolet, the optical system used to disperse the radiation is generally made of quartz.

A detailed discussion of the origin of emission spectra is beyond the scope of this book but a simplified treatment is given in Chapter 21, Sections 21.1 and 21.2.

It may be stated, however, that there are three kinds of emission spectra: continuous spectra, band spectra, and line spectra. The continuous spectra are emitted by incandescent solids, and sharply defined lines are absent. The band spectra consist of groups of lines that come closer and closer together as they approach a limit, the head of the band: these are caused by excited molecules.

758

Line spectra consist of definite, usually widely and seemingly irregularly spaced, lines; these are characteristic of atoms or atomic ions which have been excited and emit their energy in the form of light of definite wavelengths. The quantum theory predicts that each atom or ion possesses definite energy states in which the various electrons can exist; in the normal or ground state the electrons have the lowest energy. Upon the application of sufficient energy by electrical, thermal, or other means, one or more electrons may be removed to a higher energy state further from the nucleus; these excited electrons tend to return to the ground state, and in so doing emit the extra energy as a photon of radiant energy. Since there are definite energy states and since only certain changes are possible according to the quantum theory, there are a limited number of wavelengths possible in the emission spectrum. The greater the energy of the exciting source, the higher the energy of the excited electrons, and therefore the more numerous the lines that may appear. The intensity of a spectral line depends largely upon the probability of the required energy transition or 'jump' taking place. The intensity of some of the stronger lines may occasionally be decreased by self-absorption caused by reabsorption of energy by the cool gaseous atoms in the outer regions of the source. With high-energy sources the atoms may be ionised by the loss of one or more electrons; the spectrum of an ionised atom is different from that of a neutral atom and, indeed, the spectrum of a singly ionised atom resembles that of the neutral atom with an atomic number one less than its own.

The lines in the spectrum from any element always occur in the same positions relative to each other. When sufficient amounts of several elements are present in the source of radiation, each emits its characteristic spectrum; this is the basis for qualitative analysis by the spectrochemical method. It is not necessary to examine and identify all the lines in the spectrum, because the strongest lines will be present in definite positions, and they serve to identify unequivocally the presence of the corresponding element. As the quantity of the element in the source is reduced, these lines are the last to disappear from the spectrum: they have therefore been called the 'persistent lines' or the *'raies ultimes'* (R.U. lines), and simplify greatly the qualitative examination of spectra.

Lines in an unknown spectrum may be identified by comparing them with those on a spectrum containing a number of lines of known wavelengths. This may be performed either by comparison with charts of spectra of metallic elements such as iron or copper, or by the use of R.U. powder (see Section 20.4).

The number of lines appearing in the spectrum varies considerably from element to element. The spectra of the transition elements, the lanthanides and such elements as titanium and molybdenum, produce complex spectra; copper, antimony, tin, and lead are intermediate; whilst boron, magnesium, aluminium, zinc, and the alkaline-earth metals give relatively simple spectra. The practical result of these differences is that spectrographs with greater dispersion and resolution are required to separate adequately the lines in complex spectra, e.g. iron, nickel, cobalt, or manganese: such spectrographs are necessarily large and expensive.

For quantitative analysis it is necessary to assess the densities of blackening of lines in a spectrogram due to the constituents being determined; this may be done by comparing the spectra from samples of known and unknown composition. Comparisons may be made either visually (best with the aid of a spectrum projector: see Fig. 20.6) when no great accuracy is desired, or by photoelectric measurement of line densities with a microphotometer (see Section 20.3). Details of the procedure are described in Sections 20.5 and 20.8.

The applications of emission spectrography include:

1. the examination of a single metal or an alloy for impurities;
2. the analysis of an alloy for its general composition, including a search for minor components and traces of impurities;
3. the analysis of ash of organic substances and other materials (e.g. natural waters) amenable to similar treatment; and
4. the detection of contaminants in food.

The chief advantages of the spectrographic method of analysis are:

(a) The procedure is specific for the element being determined, although difficulties occasionally arise when a line of another element overlaps that of the unknown.
(b) The method is time-saving; a quantitative determination of traces of the elements in a sample, especially an alloy or a metal, may be made without any preliminary treatment. Most metals and some non-metals (e.g. phosphorus, silicon, arsenic, and boron) may be determined.
(c) A permanent record may be obtained on a photographic plate.
(d) It may be (and is usually) applied to the determination of small quantities of added constituents or of traces of impurities where conventional methods of analysis are difficult, fail, or give less accurate results. Lengthy and difficult separations by chemical methods, e.g. of zirconium and hafnium and of niobium and tantalum, can be avoided.

The apparent disadvantages are:

(a) Successful use requires wide experience, both in the operation of equipment and in reading and interpreting spectra.
(b) The spectrograph is essentially a comparator; for quantitative analysis, standards (usually of similar composition to the material under analysis) are necessary. Unknown samples, therefore, present a relatively difficult problem when quantitative results are required.
(c) The accuracy and precision are not as high as gravimetric, titrimetric, and some spectrophotometric methods for elements present in quantities greater than 2–5 per cent of the total; indeed, spectrographic methods are not usually applied for elements present to a greater extent than about 3 per cent.

20.3 EQUIPMENT FOR EMISSION SPECTROGRAPHIC ANALYSIS

This section is concerned with describing the equipment which is necessary for an introduction to spectrographic techniques for the analyst. In this instance the practical work will be described for instruments manufactured by Rank Hilger, Margate, Kent, UK, but the comparable products of other manufacturers may also be used.

The essential parts of a spectrograph are a slit, an optical system, and a camera for recording the spectrum. The light from the source of radiation passes through the slit, which is a narrow vertical aperture, then through the optical system, which includes a prism or grating. An image of the slit is produced by means of lenses at the point where the light is recorded. One such image is produced for each radiation having a specific wavelength, and the result is a series of vertical line images which constitute the spectrum of the element being investigated. The optical system may be either glass or quartz; the latter transmits in the ultraviolet region, where many useful lines occur, as well as in the visible. The range of wavelengths employed with quartz extends from about 2000 to 10 000 Å.*

Two sizes of prism spectrographs are widely used for analysis, the medium and the large. The large spectrograph is necessary for the analysis of iron, chromium, cobalt, molybdenum, titanium, tungsten, uranium, and zirconium owing to the numerous lines in the spectra and the need for maximum resolution; it utilises a 25.4 cm × 10.2 cm (10 in × 4 in) plate, but adjustments must be made to bring different regions of the spectrum on the plate. The complete spectrum is 76 cm long, so that it is recorded in three separate sections. The Hilger and Watts medium spectrograph has ample dispersion for most work with non-ferrous metals and light alloys; it has the advantage that the whole of the spectrum range can be photographed in a single exposure on one plate (25.4 cm × 10.2 cm). The essential features are shown in Fig. 20.1. The light produced from the sample by either means of excitation enumerated below is received by a narrow slit and passes through a lens system to a prism which deviates each radiation from a direct path by an amount depending upon its wavelength. A second lens system forms an image of the narrow slit upon a photographic plate in the order determined by the prism. The prism employed is known as a Cornu prism†; the 60° prism is composed of two half-prisms of quartz in optical contact. The two halves are cut so that they compensate each other, with the result that the combination functions as if it were an equilateral quartz prism.

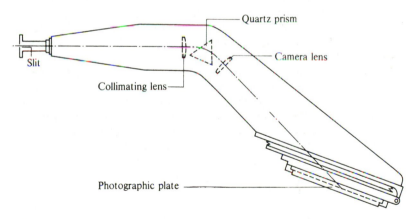

Fig. 20.1

---

* Strict compliance with SI units should require that these values are expressed in nanometres; however, the Ångstrom is still commonly employed and it is felt desirable to retain it for this particular chapter ($10 Å = 1 nm = 10^{-9} m$).
† The Littrow prism is also widely used. It is a 30° prism with a mirror back face: the light passes through the prism to the mirror face and is reflected back through the prism, the total path being equal to a 60° prism. The prism mounting results in a fairly compact instrument (Section 17.7).

The slit (20 mm long) is formed of two parallel metal jaws, which are correctly ground flat; the width is controlled to 0.001 mm by means of a micrometer screw which causes both jaws to move symmetrically. For most determinations with spectrographs a slit width of about 0.02 mm is usually satisfactory. The slit is equipped with a Hartmann diaphragm for placing spectra in juxtaposition on the plate; this involves a number (usually three) of square openings arranged in echelon, the square holes being cut with the bottom of one in line with the top of the next. When these holes are placed successively in front of the slit it is possible to record three spectrograms in juxtaposition without moving the plate holder. The shutter is usually placed between the slit jaws and is operated by a lever.

The spectrograph is provided with a scale graduated in wavelengths, which can be illuminated and printed directly on the spectrogram plate. It is incorporated in the back of the camera and is controlled by a lever. The camera slide for carrying the plate holder is fitted at the end of the instrument casting. Provision is made for turning the plate-holder carrier mounting through a small angle about a vertical axis. The plate-holder slide is operated by a rack-and-pinion motion that raises or lowers the plate carrier over a range of 75 mm in 1 mm steps. A number of spectra (up to 40, but depending on the length of the slit) can be taken on one plate.

The length of the instrument from the slit to the end of the plate holder is about 1.2 metres, and it is supported on a massive base which raises the optical parts about 30 cm above bench level. An optical bar of steel is attached to the base of the instrument, from which it projects about 90 cm; it is parallel with the optical axis. The bar serves to carry lenses, an arc and spark stand (Gramont stand) for holding samples, and other ancillary equipment.

The holders for arc and spark excitation fit directly on the optical bar attached to the spectrograph. In modern instruments the electrode holders are housed in a metal box fitted with a safety shield. The electrodes are carried in strong screw clamps on horizontal arms which are insulated from the supporting base; the movements include vertical adjustment of the upper electrode with rack and pinion, vertical and horizontal movements of the lower electrode through a collar which may be clamped to the supporting rod, and vertical and rotary movements of both electrodes together. The combinations are such that the discharge can be rapidly located on the optical axis. The holders are primarily designed for cylindrical rods not greater than 1.25 cm diameter.

The slit should be uniformly illuminated along its length. For this purpose a quartz lens, of such focal length that it throws an image of the source on the collimating lens (see Fig. 20.1) is placed 2 cm from the entrance slit and located by means of the optical bar. The rapid location of the source of light (i.e. position of the electrodes) on the optical axis of the spectrograph and the adjustment of the length of the discharge gap are conveniently carried out by projecting an image of the source on to a calibrated screen (gauge plate) provided with the instrument. Both the lens and the screen are mounted on the optical bar at the end remote from the slit (Fig. 20.8). After the initial adjustment with the light source, a reading lamp is arranged so that the lamp can be brought quickly into position near the electrodes on the slit side of the stand; it will be found that the electrodes can be seen on the screen. For subsequent work the electrodes can be set in optical alignment with the spectrograph simply by setting the image of the electrodes to agree with the screen gauge without actually passing any current between them. The two outer horizontal lines on the screen gauge are equivalent to an electrode gap of 4 mm and the inner lines are equivalent to a 2 mm gap.

The two most commonly used excitation sources are the low-voltage d.c. arc and the high-voltage a.c. spark. For the **low-voltage d.c.** arc the essential requirements are a source of d.c. at 110–250 volts, a regulating resistance R to control the current (2–12 amperes), an inductance L (this helps to steady the arc and maintain a more constant voltage), a d.c. ammeter A in series with the supply, and an arc gap (see Fig. 20.2). The sample under investigation may consist of electrically conducting rods, which then become the electrodes of the arc. In order to strike the arc, the two electrodes must be brought into contact or can be short-circuited by touching them with an insulated third (carbon) electrode. If the sample is in the form of a powder or small pieces it may be

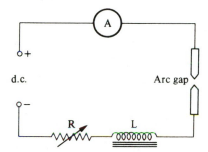

Fig. 20.2

supported on a pure carbon or graphite rod which has been hollowed out to hold the specimen. Another similar rod, generally with a pointed or conical tip, serves as the upper electrode. Sometimes pure metal rods, e.g. copper or silver, are used as the supporting electrodes, and the sample may be made either the cathode or the anode. It is advisable to make exposures with the sample respectively positive or negative in order to determine which gives the better results for reproducibility and sensitivity. With two carbon electrodes it is usual to place the sample in the anode (positive electrode), which serves as the lower electrode. The temperature in the arc stream ranges from 2000 °C to about 5000 °C. When an arc is operated between carbon electrodes in air, some cyanogen molecules are formed, and these may emit molecular band spectra in the region 3200–4200 Å.

The d.c. arc is a sensitive source and of wide application, but its reproducibility is not of the highest order; it is generally used for the identification and determination of elements present in very small concentrations. A comparatively large amount of the substance being analysed passes through the arc, and consequently an average or more representative value of the concentration is shown, provided that the complete sample is burned. This is usually achieved by placing a small amount of the substance on a graphite electrode with a recess drilled to a depth of 4–5 mm,* and continuing the excitation until the sample is completely volatilised.

---

* In practice, it is found desirable to have a small 'pip' in the centre of the graphite electrode and to place the powder in a depression around it. The arc tends to pass between the upper pointed electrode and the small projection.

The **high-voltage a.c. spark** is perhaps the simplest excitation source for many quantitative identifications; its sensitivity is not as high as the d.c. arc source. It is more reproducible and stable than the arc and less material is consumed; this source is well adapted for the analysis of low-melting materials, since the heating effect is less. The basic circuit for spark excitation is shown in Fig. 20.3. Alternating current from the mains is fed into a high-tension transformer T that gives up to 15 kilovolts at its secondary terminals. The condenser C across the secondary of the transformer is charged by the current from the latter until the potential is high enough to cause a discharge across the spark gap: with a suitable value of the condenser (about 0.005 microfarad), a brilliant spark is obtained. By placing an inductance L (0.02–1.5 millihenrys) in series with the spark gap, a high-frequency oscillating discharge is produced.

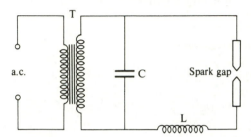

Fig. 20.3

In most spectrographs the intensity of the lines is ultimately registered on a photographic emulsion; the nature of the photographic process is, therefore, of considerable importance. It is assumed that the recommended plate, developer, developing time, temperature, etc., have been followed and that the spectra are obtained on the plate, together with a wavelength scale. Part of such a spectrum (obtained with the aid of a Hartmann diaphragm) is shown in Fig. 20.4; this gives the spectra, from the top downwards, of cadmium, spelter, and zinc. A very satisfactory method of examination of spectra, particularly for qualitative analysis, is to employ a Rank Hilger projection comparator (Fig. 20.5). In this instrument the two plates for comparison are mounted in spring-loaded clips on adjustable stages. The two plates are illuminated from below by means of

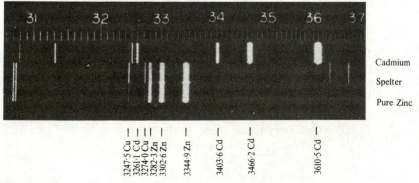

Fig. 20.4

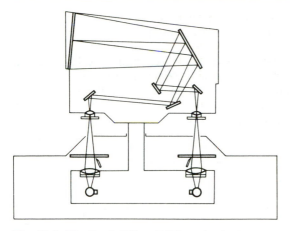

Fig. 20.5 The Rank Hilger L150 projection comparator.

12 V lamps and the transmitted light projected on to a large screen which presents the two images one above the other, each occupying half the screen. Separate focusing lenses enable clear definitions of the individual spectra to be obtained and a ten-fold magnification is possible. Accurate alignment of the spectra is possible by means of a fine adjustment screw.

For quantitative analysis it is necessary to compare the relative blackening of lines with one another and with those produced by standard elements. The density of blackening (or simply blackening) B may be defined as:

$$B = \log \frac{i_0}{i}$$

where i_0 is the intensity of the radiation transmitted by a perfectly clear part of the photographic plate and i is the light transmitted by the line in question.

20.4 QUALITATIVE SPECTROGRAPHIC ANALYSIS

At least 55 elements can be identified under normal conditions of excitation. In a qualitative analysis it is desirable to have a high sensitivity so that the presence of trace elements may be revealed. The d.c. arc usually gives the highest sensitivity. In many cases satisfactory results may be obtained with the a.c. high-voltage spark as the source of excitation; although the sensitivity is lower, the reproducibility is greater. The d.c. arc is preferred for the most difficultly excitable elements and for non-volatile and refractory compounds; a 'complete burn' of the sample may be obtained with comparative ease. A widely used system of qualitative spectrographic analysis is to arc the sample in a depression in a graphite electrode, using a pointed counter-electrode. The graphite electrode in air gives a cyanogen band spectrum (headings at 4216, 3883, and 3590 Å) in the near ultraviolet, which may mask some of the lines of interest. A sample of a few milligrams is sufficient; the exposure conditions must be determined by experiment. Non-conducting samples may be more easily excited by being mixed with an equal volume of graphite powder, and these, as well as all refractory materials, are often treated with a carrier to bring them into the arc

765

stream. This carrier, usually a volatile salt such as ammonium chloride or ammonium sulphate, helps to propel the entire sample smoothly up into the arc gap.

The amount of an element that is detectable varies with its concentration, its relative volatility, the energy of excitation, etc.: approximate sensitivity figures with arc excitation for the common elements are collected in Table 20.1.

Table 20.1 Arc sensitivities of some elements

Percentage detectable	Elements
0.1–1	As, Cs, Nb, P, Ta, W
0.01–0.1	B, Bi, Cd, La, Rb, Sb, Si, Tl, Y, Zn, Zr
0.001–0.01	Al, Au, Ba, Be, Ca, Fe, Ga, Ge, Hg, Ir, K, Mn, Mo, Pb, Sc, Sn, Sr, Ti, V
0.0001–0.001	Ag, Co, Cr, Cu, In, Li, Mg, Na, Ni, Os, Pd, Pt, Rh, Ru

As one dilutes the amount of an element in an arc, the number of lines observable is reduced, and ultimately there remain only a few lines of the element which is diluted. These lines are referred to in Section 20.2 as the '*raies ultimes*' or persistent lines and tables of their wavelengths may be found in chemical handbooks. The identification of these lines will permit detection of elements present in low concentration, and all qualitative methods utilise the persistent lines.

The simplest and most direct procedure for the qualitative analysis of an unknown sample is the **R.U. powder method**. The R.U. powder is a powder marketed by Johnson, Matthey and Co. of London. It consists of small quantities of 50 elements incorporated in a base material composed of calcium oxide, magnesium oxide, and zinc oxide. The quantity of each element present has been carefully adjusted so that only the *raies ultimes* and the most important sensitive lines appear when the spectrum is excited by placing some of the powder (10–20 mg) in the lower (and positive) pole of an arc between graphite electrodes. A current of 5–7 amperes and an arc length of 6 mm is recommended. It is advisable not to expose for a longer time than the powder lasts, or else parts of the spectrum will be unnecessarily masked by the CN bands. A set of seven enlargements of the arc spectra of the R.U. powder is marketed, and these cover a wavelength range of 2284–8000 Å. The spectrum of the iron arc (together with the important wavelengths) is given alongside that of the R.U. powder; this enables the positions of the persistent lines relative to the iron-arc spectra to be seen immediately, and will also permit the position of any sensitive line to be found readily. A qualitative analysis is carried out by producing contiguous arc spectra with the aid of a Hartmann diaphragm of the R.U. powder, the sample, and optionally (if known) of a pure main element present in the sample. The plate is developed, and the lines present in both the R.U. powder and sample spectra are noted; the latter is most simply observed on a Rank Hilger projection comparator (Fig. 20.5). The elements are tabulated with the number of lines appearing: three lines,* which are free from interference, are considered proof of the presence of the element. A portion of the spectrum of R.U. powder

---

* With R.U. powder two lines may be acceptable for some elements; these include B, Cu, Au, P, Ag, and Cs.

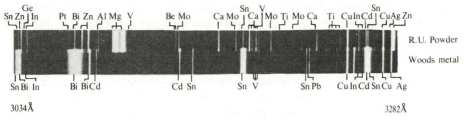

Fig. 20.6

and of a sample of Wood's metal is given in Fig. 20.6; both major and minor constituents are readily identified.

Another problem that frequently arises is to decide whether a substance contains a given element or a small number of specified elements. As an example we may take the presence of cadmium in spelter. Contiguous spectra are taken using the Hartmann diaphragm of: (1) the spectroscopically pure metal known to be present in the specimen under test in considerable quantity (e.g. zinc); (2) the sample under test (e.g. spelter); and (3) the spectroscopically pure metal whose presence or absence is to be determined (e.g. cadmium). Part of such a spectrum is shown in Fig. 20.4. The graduated scale of wavelengths is also photographed on the plots. Examination of the three spectra with a Rank Hilger projection comparator will reveal at once the presence or absence of the specified element (cadmium in the present example) This instrument, which is described in detail above (Section 20.3), also enables comparisons to be made between images on two different spectrographic plates. Thus if a set of reference plates is available, the need for taking comparison spectra of metals looked for against that of the unknown is obviated and a great deal of tedious measurement is likewise rendered unnecessary. Standard spectrographic samples are available *inter alia* from the Bureau of Standards at Washington (DC), USA, and the Bureau of Standards, Middlesborough, UK, also supplies a range of 'Spectroscopic Standard Certified Reference Materials'.

Mention may be made of the great advantage that the spectrograph offers in the following operations.

(*a*) Rapid qualitative analysis of all the metallic constituents of a substance as a basis for planning a chemical analysis.

(*b*) Approximate analysis of minor components by sight (after some experience has been obtained).

(*c*) Examination of precipitates (after weighing) for freedom from constituents which should have been separated.

(*d*) Detection of traces of metallic impurities or constituents in inorganic residues and powders; in organic substances (foodstuffs, textiles, etc.); in vitreous substances (glasses and slags); and in refractories and clays.

(*e*) Testing the purity of analytical reagents.

(*f*) Analysis of substances of which only small quantities are available.

(*g*) Detection of rare or trace metals in minerals.

20.5 QUANTITATIVE SPECTROGRAPHIC ANALYSIS

If the excitation conditions are kept constant and the sample composition is varied over a narrow range, the energy emitted for a given spectral line of an element is proportional to the number of atoms that are excited and thus to the concentration of the element in the sample. The energy emitted (i.e. the intensity of the light) is usually measured by the photographic method: the concentration of the unknown is determined from the blackening of the plate for certain lines in the spectrum. The quantitative determination of the blackening of the individual lines is made with a microphotometer. Measurements are made of the light transmitted by the line in question (i) and the light transmitted by the clear portion of the plate (i_0); the density D (strictly the density of blackening, also represented by B) may be defined by the expression $D = \log_{10}(i_0/i)$. It is assumed that the galvanometer deflection obtained on the microphotometer is directly proportional to the light falling on the photocell.

The density of the image of the spectral line should ideally be proportional to the concentration of the corresponding element in the sample if the exposure time and conditions of excitation, etc., are held constant. This is often not strictly true; hence, wherever possible, it is desirable to photograph spectra of several samples of varying known composition on the same plate with the unknown. The unknown may then be evaluated by interpolation on a graph of density against concentration. The fact that the unknown and all the standards are on the same plate prevents errors due to differences in sensitivity between plates as well as those due to differences in the time or temperature of photographic processing.

Since the intensity of the lines is ultimately registered on a photographic plate, a brief discussion of the nature of the photographic process is desirable. If one plots the density as a function of the logarithm of the exposure,* a curve such as is shown in Fig. 20.7 results. A certain threshold exposure, denoted by A, is necessary before an image is produced. It will be noted that there is a region BC over which the density is proportional to the logarithm of the exposure; this is the useful range of the plate. The slope of this linear portion of the curve is

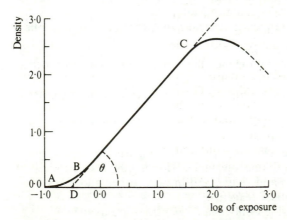

Fig. 20.7

---

* The exposure of the photographic plate is defined as the product of the intensity of the light incident on the undeveloped plate and the time for which it is acting.

known as the gamma (or contrast) of the emulsion: $\gamma = \tan \theta$. Emulsions with a high gamma give images with strong contrast because a small difference in exposure causes a large variation in density: low values of gamma indicate low contrast. The point D (the intercept of BC on the horizontal axis) measures the inertia of the emulsion: its reciprocal is related to the 'speed' of the emulsion. For our purpose the speed is an approximate measure of the minimum amount of light required to produce a useful image. The slope of the characteristic curve of an emulsion varies from emulsion to emulsion, with wavelength, and with the conditions of excitation and of development. In order to determine the curve for any given emulsion, the conditions of excitation (shape of the ends of the electrodes, their distance apart, the electrical circuit, etc.) and also the conditions of development (the type of developer, temperature, and time of development) must be standardised. In selecting a plate on which to photograph the spectrum, one decides first whether a fast emulsion is needed on the basis of the light intensity available and the permissible time of exposure. If very faint spectrum lines are to be detected, high sensitivity at low intensities is needed, which suggests the use of a fast plate. To reproduce both weak and strong spectrum lines on the same spectrogram with correct indication of their relative intensities, medium contrast is needed. For sharp spectrum lines with a clear background, a plate of high contrast is used. Slow contrast plates exhibit high resolving power: such plates are of greatest value for use with spectrographs having low dispersion with high resolving power, as is often the case with certain prism instruments of short focus. The most common type of plate used for emission spectrography has a gamma of about 1.

We may now deal with some of the procedures employed in quantitative spectrographic analysis. In the **comparison sample method**, the spectrum of an unknown sample is compared with the spectra of a range of samples of known composition (e.g. those supplied by the US Bureau of Standards) with respect to a particular component or components. The spectra of the unknown and of the various standards are photographed on the same plate under the same conditions. The concentrations of the desired constituent can then be estimated by comparing the blackening of the lines of the particular constituent with the same lines on the standards; visual or photometric comparison of blackening may be used.

In the **internal standard method** the intensity of the unknown line is measured relative to that of an internal standard line. The internal line may be a weak line of the main constituent. Alternatively, it may be a strong line of an element known not to be present in the sample and furnished by adding a fixed small amount of a compound of the element in question to the sample. The ratios of the intensities of these lines — the unknown line and the internal standard line — will be unaffected by the exposure and development conditions. This method will provide lines of suitable wavelength and intensity by variations of the added element and the amount added, due regard being paid to the relative volatility of the selected internal standard element. It is important to use as internal standard pairs only those lines of which the relative intensities are insensitive to variations in excitation conditions. The line selected as standard should have a wavelength close to that of the unknown and should, if possible, have roughly the same intensity.

20.6   QUALITATIVE SPECTROGRAPHIC ANALYSIS* OF (A) A NON–FERROUS ALLOY AND (B) A COMPLEX INORGANIC MIXTURE

Discussion.   One procedure for the identification of an alloy is to measure the wavelength values of the observed lines and compare these with the recorded data on known elements. A wavelength scale, which has been calibrated with the spectrograph, is photographically reproduced on the plate; this is, of course, only useful as a guide, since the wavelengths cannot be read with sufficient accuracy. A simple method for use with brass, or most other simple non-ferrous alloys, is to determine the spectra with pure samples of the component metals and to compare the spectra by a projection method (see Fig. 20.5). At least three persistent lines must be present for positive identification. The spark technique may be used for metals and alloys, but is not altogether satisfactory for powders, including R.U. powder. The d.c. carbon arc is preferred for the qualitative analysis of powders, and it also gives good results for most alloys: low-melting alloys (e.g. Wood's metal) may be dissolved in nitric acid, evaporated to dryness, then evaporated with concentrated sulphuric acid, and the dry sulphated residue employed in the carbon arc.

To gain experience in qualitative analysis, full details will be given for the analysis of brass and an artificial seven-radical inorganic mixture.

Adjustment of the optical system.   The condensing lens is set between the light source and the slit of the spectrograph so that the beam of light from the d.c. arc source passing through the slit forms a real image at the collimating lens (compare Fig. 20.1) of the spectrograph. The adjustments for a Hilger and Watts medium spectrograph will be evident from Fig. 20.8.

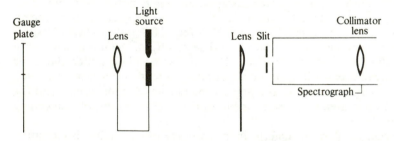

Fig. 20.8

1. Place the Gramont stand so that the electrodes are 38 cm from the jaws of the slit and align them for height, the gap between the electrodes to be 4 mm.
2. Set the condensing lens in position at 2 cm from the slit jaws and at the correct height.
3. Set the gauge plate on its stand at the end of the bar at about 23 cm from the lens on the Gramont stand.

An image of the light source will now be seen on the gauge plate, and the height should be set so that the image agrees with the two outer horizontal lines on the gauge. If a reading lamp is so arranged that it can be brought into position

---

* The experiments in qualitative analysis described utilise a Hilger and Watts medium spectrograph. They can easily be adapted to other similar spectrographs with the aid of the instruction manuals supplied by the manufacturers of the instruments.

near the electrodes on the slit side of the stand it will be found that an image of the electrodes can be seen on the screen of the gauge plate. Thus for any subsequent work the electrodes can be set in optical alignment with the spectrograph without actually passing any current between them. The two outer horizontal lines of the gauge are equivalent to an electrode gap of 4 mm and the inner lines to a 2 mm gap.

Direct-current arc. A 230 volt arc at 4 amperes is suitable for qualitative analysis (see Fig. 20.2). The arc gap may be 2 mm and the slit width 0.02 mm.

Electrodes for d.c. arc. The two electrodes are shown in Fig. 20.9. They are conveniently shaped on a lathe from graphite electrodes (Johnson, Matthey; 30 cm long; JM 3B, 10 mm diameter; JM 4B, 6.5 mm diameter). The maximum depression on the lower electrode is 3 mm: the small projection in the centre helps to ensure that the arc passes between it and the upper electrode and does not 'wander' appreciably to the edges of the electrode. A small quantity (about 20 mg) of the alloy or powder is placed on the lower electrode.

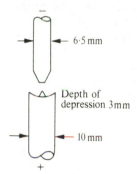

Fig. 20.9

Photographic details. *Darkroom.* A fully equipped darkroom is desirable. Ideally the spectrograph should be set up in the darkroom. The student should become familiar with its facilities — stainless-steel trays (26 cm × 21 cm), running water, safelights, etc.

Plates. Ilford R.40 or Kodak V-F. Charge holder in darkroom.

Photographic developer. Ilford ID-2* or equivalent. Dilute 1 volume with 2 volumes of water.

Photographic fixer. Kodak tropical acid hardening–fixing bath F5† or equivalent.

Procedure.

(*a*) Load the plate holder with plate, making certain that the sensitised side is placed face down in the open holder. Return the holder to the spectrograph.

---

* This has the following composition: Metol, 2 g; sodium sulphite (anhydrous) 75 g; hydroquinone, 8 g; sodium carbonate (anhydrous), 37.5 g; potassium bromide, 2 g; water to make 1 L. Dissolve chemicals in order given.

† This has the following composition: sodium thiosulphate, 240.0 g; sodium sulphite (crystals), 30.0 g; acetic acid (glacial), 17.0 mL; boric acid 7.5 g; potash alum 15.0 g; water to make 1 L. Dissolve chemicals in order given.

Withdraw the safety slide. Expose for several seconds to obtain the wavelength scale in the upper part of the plate; replace slide.

(b) Charge the lower electrode (anode) with about 20 mg of brass or the inorganic mixture.

(c) Withdraw the safety slide which covers the plate. Strike arc between the electrodes. Make an exposure of 6 seconds by opening the shutter of the spectrograph with the Hartman diaphragm in position. (The best exposure time is evaluated by making a number of consecutive exposures, the plate being lowered to the next position after each exposure.)

CAUTION: do not look directly at the arc unless wearing the special dark goggles or dark glasses provided.

(d) Replace the safety slide. Turn off the current to the arc and allow to cool. Insert fresh graphite electrodes. Charge the lower electrode with 20 mg of R.U. powder. Move the Hartmann diaphragm to the second position. Strike the arc again and expose for 5 seconds. (The best exposure is determined as above.)

(e) If brass was used in the first exposure, change the electrodes, charge the lower electrode with about 20 mg of the inorganic seven-radical mixture (say, a mixture of equal weights of magnesium phosphate, zinc borate, copper arsenate, and cadmium oxide), set the Hartmann diaphragm in position, and expose for 5 seconds.

(f) Develop the plate in the usual manner in the darkroom, say for 9 minutes with Ilford ID-2 developer at 18 °C, then rinse it rapidly with water, and place it in the fixer (Kodak F5) for *ca* 20 minutes at 18 °C. Remove the plate from the fixer, wash it with running water for at least 30 minutes, wash with distilled water, and dry in the air.

(g) Identify the lines in the spectra of the brass and the inorganic mixture by comparison with those on the spectrum of the R.U. powder with the aid of the Rank Hilger projection comparator (Fig. 20.5). Use the series of enlarged photographs of the spectra of R.U. powder for exact identification.

The following are some of the lines which should be identified (wavelengths are given to the nearest Ångstrom):

(A) Brass
Cu: 3248, 3274.
Zn: 3282, 3303, 3346, 4680, 4722, 4811.
Pb: 2614, 2833, 4058.
Sn: 2707, 2840, 3175.
Fe: 2483, 2488, 2599, 2973, 2984.
Mn: 2795, 2798, and 'triplet' 4031, 4033, 4035.
Ni: 3002, 3415, 3446.
P: 2536, 2555.

(B) Inorganic mixture containing Cd, Zn, Cu, Mg, phosphate, arsenate, borate
Cd: 2288, 3261, 3404, 3466, 5086.
Zn: 3282, 3303, 3346, 4680, 4722, 4811.
Cu: 3248, 3274.
Mg: 2796, 2803, 2852, and 'quintet' 2777, 2778, 2780, 2781, 2783.
P: 2534, 2536, 2553, 2555.
As: 2350, 2369, 2457.
B: 2497, 2498.

20.7 INTRODUCTION TO PLASMA EMISSION SPECTROSCOPY

The use of a plasma as an atomisation source for emission spectroscopy has been developed largely in the last 20 years. As a result, the scope of atomic emission spectroscopy has been considerably enhanced by the application of plasma techniques.

A plasma may be defined as a cloud of highly ionised gas, composed of ions, electrons and neutral particles. Typically, in a plasma, over 1 per cent of the total atoms in a gas are ionised.

In plasma emission spectroscopy the gas, usually argon, is ionised by the influence of a strong electrical field either by a direct current or by radio frequency. Both types of discharge produce a plasma, the **direct current plasma** (DCP) or the **inductively coupled plasma** (ICP). Plasma sources operate at high temperatures somewhere between 7000 and 15 000 K. Thus, the plasma source produces a greater number of excited emitted atoms, especially in the ultraviolet region, than that obtained by the relative low temperatures used in flame emission spectroscopy (see Section 21.1). Further, the plasma source is able to reproduce atomisation conditions with a far greater degree of precision than that obtained by classical arc and spark spectroscopy. As a result, spectra are produced for a large number of elements, which makes the plasma source amenable for simultaneous elemental determinations. This feature is especially important for the multi-element determinations over a wide concentration range.

20.8 THE DIRECT CURRENT PLASMA (DCP)

The DCP plasma source is shown in Fig. 20.10. Basically, it consists of a high-voltage discharge between two graphite electrodes. The recent design employs a third electrode arranged in an inverted **Y**-shape which improves the stability of the discharge. The sample is nebulised (see Section 20.10) at a flow rate of 1 mL min^{-1} using argon as the carrier gas. The argon ionised by the high-voltage discharge is able to sustain a current of ~ 20 amps indefinitely.

The DCP generally has inferior detection limits to the inductively coupled

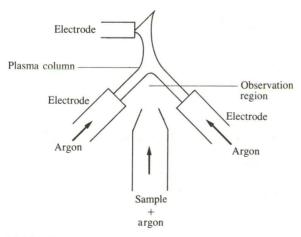

Fig. 20.10

plasma (ICP). Although the DCP is comparatively less expensive than the ICP, the graphite electrodes need replacing after a few hours' use.

20.9 THE INDUCTIVELY COUPLED PLASMA (ICP)

The inductively coupled plasma source (Fig. 20.11) comprises three concentric silica quartz tubes, each of which is open at the top. The argon stream that carries the sample, in the form of an aerosol, passes through the central tube. The excitation is provided by two or three turns of a metal induction tube through which flows a radio-frequency current (frequency ~ 27 MHz). The second gas flow of argon of rate between 10 and 15 L min^{-1} maintains the plasma. It is this gas stream that is excited by the radio-frequency power. The plasma gas flows in a helical pattern which provides stability and helps to isolate thermally the outside quartz tube.

The plasma is initiated by a spark from a Tesla coil probe and is thereafter self-sustaining. The plasma itself has a characteristic torroidal 'doughnut' shape

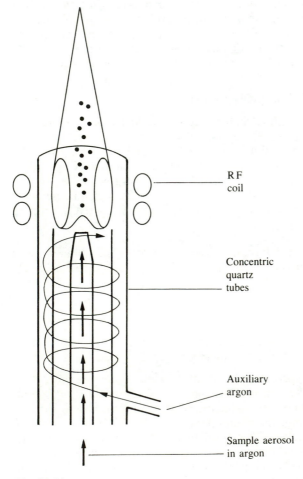

RF
coil

Concentric
quartz
tubes

Auxiliary
argon

Sample aerosol
in argon

Fig. 20.11

in which the sample is introduced into the relatively cool central hole of the 'doughnut'.

20.10 SAMPLE INTRODUCTION

The sample, usually in the form of a solution, is carried into the hot plasma by a nebuliser system similar to that employed for flame methods (see Section 21.5) although for ICP a much slower flow rate of 1 mL min^{-1} is used.

The most widely used nebuliser system in ICP is the crossed-flow nebuliser shown in Fig. 20.12. The sample is forced into the mixing chamber at a flow rate of 1 mL min^{-1} by the peristaltic pump and nebulised by the stream of argon flowing at about 1 L min^{-1}.

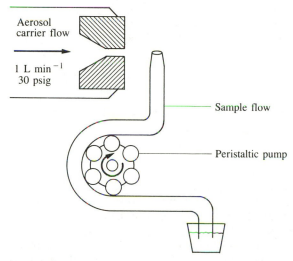

Fig. 20.12

Another kind of nebuliser, the Babington type, is used for handling slurries that can contain up to 10 per cent solids. This design of nebuliser is less likely to suffer from blockage.

20.11 ICP INSTRUMENTATION

There are two basic types of instrument used in plasma emission analysis, namely the simultaneous and the sequential multi-element spectrometer.

Simultaneous multi-element spectrometer. A diagram of the light path of a simultaneous multi-element system is shown in Fig. 20.13. In this instrument the radiation from the plasma enters through a single slit and is then dispersed by a concave reflection grating; the component wavelengths reach a series of exit slits which isolate the selected emission lines for specific elements. The entrance and exit slits and the diffraction grating surface are positioned along the circumference of the 'Rowland circle', the curvature of which is equal to the radius of curvature of the concave grating. The light from each exit slit is

775

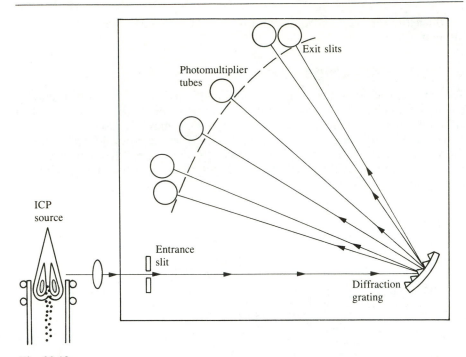

Fig. 20.13

directed to fall on the cathode of a photomultiplier tube, one for each spectral line isolated. The light falling on the photomultiplier gives an output which is integrated on a capacitor; thus the resulting voltage is converted to the concentration of each element present.

Multichannel instruments are capable of measuring the intensities of the emission lines of up to 60 elements simultaneously. To overcome the effects of possible non-specific background radiation, one or more additional wavelengths may be measured and background correction (see Section 21.12) can be achieved.

It is also possible to subtract background at each specific wavelength by using vibrating quartz plates which result in a small shift of the wavelength. The emission of the shifted wavelength is then subtracted from the original value. The main advantage of photoelectric multichannel instruments is that with the modern computational facilities available rapid simultaneous analysis may be achieved with a better precision than that obtained from the spectrograph using a photographic plate. Thus, results for 25 elements may be obtained within a short time of about 1–2 minutes.

Prior to the use of plasma excitation, arc and spark sources were used on multichannel spectrometers, the so-called direct-reading instruments.

Sequential instruments. The diagram of the light path of the Thermo Electron-200 ICP spectrometer is shown in Fig. 20.14. The plasma is located in the upper centre of the instrument just above the nebuliser, which is powered by a computer-controlled peristaltic pump. Communication with the instrument takes place on a video display, which not only guides the operator through the use of the system, but also provides graphics to simplify methods development.

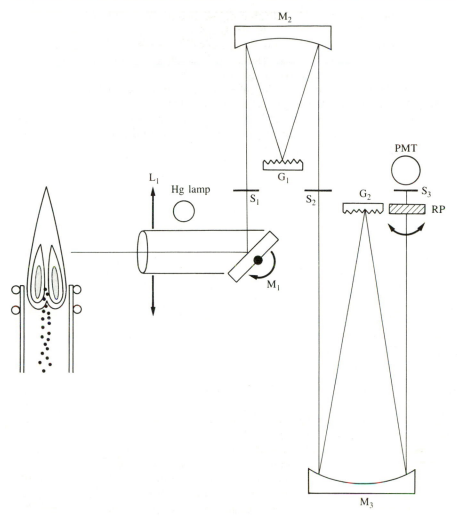

Fig. 20.14

The optical system is as follows. Light from the plasma passes through a lens on to mirror M_1. The angle of M_1 is adjustable under computer control, changing the viewing height of the plasma for each analyte element. Following M_1, the light travels through a monochromator defined by mirror M_2 and grating G_1. The main function of this monochromator is to reduce the level of stray light, which can be a problem in many plasma systems. The light then travels through a second monochromator, grating G_2 and mirror M_3 which improves the resolution.

The system provides a resolution of approximately 0.02 nm, comparable with emission line width.

Before each wavelength scan, the grating locates itself by finding the 365 nm triplet of a built-in mercury light source, and then moves sequentially to the wavelengths for the elements to be determined. As an option the instrument can be equipped with a second double monochromator, making it possible to determine 24 elements per minute. In addition, a high-resolution vacuum monochromator is also available which extends the wavelength range down to 170 nm, enabling elements to be determined which give emission lines in the vacuum ultraviolet region, such as sulphur.

To produce an analytical method, the operator must select the power level of the plasma, the wavelength for each element (preferably free from spectral interferences), and the vewing height at which the plasma is to be seen for each element. Further, it may be necessary to apply background correction; intervals are set using the graphics capability.

For References and Bibliography see Sections 21.26 and 21.27.

CHAPTER 21
ATOMIC ABSORPTION AND FLAME EMISSION SPECTROSCOPY

FLAME SPECTROSCOPY

21.1 INTRODUCTION

If a solution containing a metallic salt (or some other metallic compound) is aspirated into a flame (e.g. of acetylene burning in air), a vapour which contains atoms of the metal may be formed. Some of these gaseous metal atoms may be raised to an energy level which is sufficiently high to permit the emission of radiation characteristic of the metal, e.g. the characteristic yellow colour imparted to flames by compounds of sodium. This is the basis of **flame emission spectroscopy** (FES) which was formerly referred to as **flame photometry**. However, a much larger number of the gaseous metal atoms will normally remain in an unexcited state or, in other words, in the ground state. These ground-state atoms are capable of absorbing radiant energy of their own specific resonance wavelength, which in general is the wavelength of the radiation that the atoms would emit if excited from the ground state. Hence if light of the resonance wavelength is passed through a flame containing the atoms in question, then part of the light will be absorbed, and the extent of absorption will be proportional to the number of ground-state atoms present in the flame. This is the underlying principle of **atomic absorption spectroscopy** (AAS). **Atomic fluorescence spectroscopy** (AFS) is based on the re-emission of absorbed energy by free atoms.

The procedure by which gaseous metal atoms are produced in the flame may be summarised as follows. When a solution containing a suitable compound of the metal to be investigated is aspirated into a flame, the following events occur in rapid succession:

1. evaporation of solvent leaving a solid residue;
2. vaporisation of the solid with dissociation into its constituent atoms, which initially, will be in the ground state; and
3. some atoms may be excited by the thermal energy of the flame to higher energy levels, and attain a condition in which they radiate energy.

The resulting emission spectrum thus consists of lines originating from excited atoms or ions. These processes are conveniently represented diagrammatically as in Fig. 21.1.

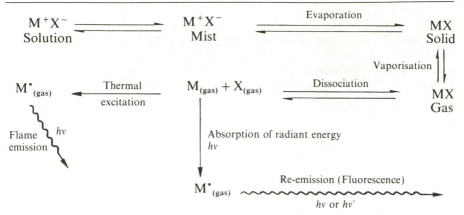

Fig. 21.1

21.1 ELEMENTARY THEORY

Consider the simplified energy-level diagram shown in Fig. 21.2, where E_0 represents the ground state in which the electrons of a given atom are at their lowest energy level and E_1, E_2, E_3, etc., represent higher or excited energy levels.

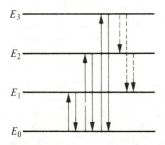

Fig. 21.2

Transitions between two quantised energy levels, say from E_0 to E_1, correspond to the absorption of radiant energy, and the amount of energy absorbed (ΔE) is determined by Bohr's equation

$$\Delta E = E_1 - E_0 = hv = hc/\lambda$$

where c is the velocity of light, h is Planck's constant, and v is the frequency and λ the wavelength of the radiation absorbed. Clearly, the transition from E_1 to E_0 corresponds to the *emission* of radiation of frequency v.

Since an atom of a given element gives rise to a definite, characteristic line spectrum, it follows that there are different excitation states associated with different elements. The consequent emission spectra involve not only transitions from excited states to the ground state, e.g. E_3 to E_0, E_2 to E_0 (indicated by the full lines in Fig. 21.2), but also transitions such as E_3 to E_2, E_3 to E_1, etc. (indicated by the broken lines). Thus it follows that the emission spectrum of a given element may be quite complex. In theory it is also possible for absorption of radiation by already excited states to occur, e.g. E_1 to E_2, E_2 to E_3, etc., but in practice the ratio of excited to ground state atoms is extremely small, and thus the absorption spectrum of a given element is usually only associated with transitions from the ground state to higher energy states and is consequently much simpler in character than the emission spectrum.

The relationship between the ground-state and excited-state populations is given by the Boltzmann equation

$$N_1/N_0 = (g_1/g_0)e^{-\Delta E/kT}$$

where

N_1 = number of atoms in the excited state,
N_0 = number of ground state atoms,
g_1/g_0 = ratio of statistical weights for ground and excited states,
ΔE = energy of excitation = hv,
k = the Boltzmann constant,
T = the temperature in Kelvin.

It can be seen from this equation that the ratio N_1/N_0 is dependent upon both the excitation energy ΔE and the temperature T. An increase in temperature and a decrease in ΔE (i.e. when dealing with transitions which occur at longer wavelengths) will both result in a higher value for the ratio N_1/N_0.

Calculation shows that only a small fraction of the atoms are excited, even under the most favourable conditions, i.e. when the temperature is high and the excitation energy low. This is illustrated by the data in Table 21.1 for some typical resonance lines.

Table 21.1 Variation of atomic excitation with wavelength and with temperature

Element	Wavelength (nm)	N_1/N_0	
		2000K	**4000K**
Na	589.0	9.86×10^{-6}	4.44×10^{-3}
Ca	422.7	1.21×10^{-7}	6.03×10^{-4}
Zn	213.9	7.31×10^{-15}	1.48×10^{-7}

Since, as already explained, the absorption spectra of most elements are simple in character as compared with the emission spectra, it follows that atomic absorption spectroscopy is less prone to inter-element interferences than is flame emission spectroscopy. Further, in view of the high proportion of ground state to excited atoms, it would appear that atomic absorption spectroscopy should also be more sensitive than flame emission spectroscopy. However, in this respect the wavelength of the resonance line is a critical factor, and elements whose resonance lines are associated with relatively low energy values are more sensitive as far as flame emission spectroscopy is concerned than those whose resonance lines are associated with higher energy values. Thus sodium with an emission line of wavelength 589.0 nm shows great sensitivity in flame emission spectroscopy, whereas zinc (emission line wavelength 213.9 nm) is relatively insensitive.

The integrated absorption is given by the expression

$$Kdv = fN_0(\pi e^2/mc)$$

where

K is the absorption coefficient at frequency v,
e is the electronic charge,
m the mass of an electron,
c the velocity of light,
f the oscillator strength of the absorbing line (this is inversely proportional to the lifetime of the excited state),
N_0 is the number of metal atoms per mL capable of absorbing the radiation.

In this expression the only variable is N_0 and it is this which governs the extent of absorption. Thus it follows that the integrated absorption coefficient is directly proportional to the concentration of the absorbing species.

It would appear that measurement of the integrated absorption coefficient should furnish an ideal method of quantitative analysis. In practice, however, the absolute measurement of the absorption coefficients of atomic spectral lines is extremely difficult. The natural line width of an atomic spectral line is about 10^{-5} nm, but owing to the influence of Doppler and pressure effects, the line is broadened to about 0.002 nm at flame temperatures of 2000–3000 K. To measure the absorption coefficient of a line thus broadened would require a spectrometer with a resolving power of 500 000. This difficulty was overcome by Walsh,[41] who used a source of sharp emission lines with a much smaller half width than the absorption line, and the radiation frequency of which is centred on the absorption frequency. In this way, the absorption coefficient at the centre of the line, K_{max}, may be measured. If the profile of the absorption line is assumed to be due only to Doppler broadening, then there is a relationship between K_{max} and N_0. Thus the only requirement of the spectrometer is that it shall be capable of isolating the required resonance line from all other lines emitted by the source.

It should be noted that in atomic absorption spectroscopy, as with molecular absorption, the absorbance A is given by the logarithmic ratio of the intensity of the incident light signal I_0 to that of the transmitted light I_t, i.e.

$$A = \log I_0/I_t = KLN_0$$

where

N_0 is the concentration of atoms in the flame (number of atoms per mL);
L is the path length through the flame (cm),
K is a constant related to the absorption coefficient.

For small values of the absorbance, this is a linear function.

With flame emission spectroscopy, the detector response E is given by the expression

$$E = k\alpha c,$$

where

k is related to a variety of factors including the efficiency of atomisation and of self-absorption,
α is the efficiency of atomic excitation,
c is the concentration of the test solution.

It follows that any electrical method of increasing E, as for example improved amplification, will make the technique more sensitive.

The basic equation for atomic fluorescence is given by

$$F = QI_0 kc$$

where

Q is the quantum efficiency of the atomic fluorescence process,
I_0 is the intensity of the incident radiation,
k is a constant which is governed by the efficiency of the atomisation process,
c is the concentration of the element concerned in the test solution.

It follows that the more powerful the radiation source, the greater will be the sensitivity of the technique.

To summarise, in both atomic absorption spectroscopy and in atomic fluorescence spectroscopy, the factors which favour the production of gaseous atoms in the ground state determine the success of the techniques. In flame emission spectroscopy, there is an additional requirement, namely the production of excited atoms in the vapour state. It should be noted that the conversion of the original solid MX into gaseous metal atoms (M_{gas}) will be governed by a variety of factors including the rate of vaporisation, flame composition and flame temperature, and further, if MX is replaced by a new solid, MY, then the formation of M_{gas} may proceed in a different manner, and with a different efficiency from that observed with MX.

21.3 INSTRUMENTATION

The three flame spectrophotometric procedures require the following essential apparatus.

(a) For flame emission spectroscopy a **nebuliser-burner system** which produces gaseous metal atoms by using a suitable combustion flame involving a fuel gas–oxidant gas mixture is needed. Note however that with so-called non-flame cells, the burner is not required.

(b) A **spectrophotometer system** which includes a suitable optical train, a photosensitive detector and appropriate display device for the output from the detector.

(c) For both atomic absorption spectroscopy and atomic fluorescence spectroscopy, a **resonance line source** is required for each element to be determined, these line sources are usually modulated (see Section 21.9).

A schematic diagram showing the disposition of these essential components for the different techniques is given in Fig. 21.3. The components included within the frame drawn in broken lines represent the apparatus required for flame emission spectroscopy. For atomic absorption spectroscopy and for atomic fluorescence spectroscopy there is the additional requirement of a resonance line source, In atomic absorption spectroscopy this source is placed in line with the detector, but in atomic fluorescence spectroscopy it is placed in a position at right angles to the detector as shown in the diagram. The essential components of the apparatus required for flame spectrophotometric techniques will be considered in detail in the following sections.

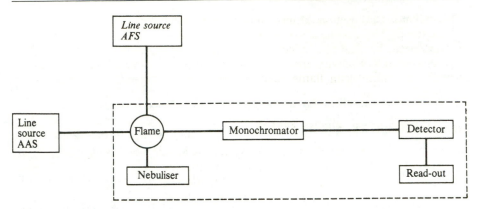

Fig. 21.3

21.4 FLAMES

For flame spectroscopy an essential requirement is that the flame used shall produce temperatures in excess of 2000 K. In most cases this requirement can only be met by burning the fuel gas in an oxidant gas which is usually air, nitrous oxide, or oxygen diluted with either nitrogen or argon. The flame temperatures attained by the common fuel gases burning in (1) air and (2) nitrous oxide are given in Table 21.2. The flow rates of both the fuel gas and the oxidant gas should be measured, for some flames are required to be rich in the fuel gas, whilst other flames should be lean in fuel gas: these requirements are discussed in Section 21.20. The concentration of gaseous atoms within the flame, both in the ground and in the excited states, may be influenced by (*a*) the flame composition, and (*b*) the position considered within the flame.

Table 21.2 Flame temperatures with various fuels

Fuel gas	Temperature (K)	
	Air	Nitrous oxide
Acetylene	2400	3200
Hydrogen	2300	2900
Propane	2200	3000

As far as flame composition is concerned, it may be noted that an acetylene–air mixture is suitable for the determination of some 30 metals, but a propane–air flame is to be preferred for metals which are easily converted into an atomic vapour state. For metals such as aluminium and titanium which form refractory oxides, the higher temperature of the acetylene–nitrous oxide flame is essential, and the sensitivity is found to be enhanced if the flame is fuel-rich.

With regard to position within the flame, it can be shown that in certain cases the concentration of atoms may vary widely if the flame is moved either vertically or laterally relative to the light path from the resonance line source. Rann and Hambly[42] have shown that with certain metals (e.g. calcium and molybdenum), the region of maximum absorption is restricted to specific areas of the flame, whereas the absorption of silver atoms does not alter appreciably within the flame, and is unaffected by the fuel gas/oxidant gas ratio.

For the sake of brevity, the so-called 'cool flame' techniques based upon the use of an oxidant-lean flame such as hydrogen/nitrogen-air, have not been included.

21.5 THE NEBULISER–BURNER SYSTEM

The purpose of the nebuliser–burner system is to convert the test solution to gaseous atoms as indicated in Fig. 21.2, and the success of flame photometric methods is dependent upon the correct functioning of the nebuliser–burner system. It should, however, be noted that some flame photometers have a very simple burner system (see Section 21.13).

The function of the nebuliser is to produce a mist or aerosol of the test solution. The solution to be nebulised is drawn up a capillary tube by the Venturi action of a jet of air blowing across the top of the capillary; a gas flow at high pressure is necessary in order to produce a fine aerosol.

There are two main types of burner system: (a) the **pre-mix or laminar-flow burner**, and (b) the **total consumption or turbulent-flow burner**. In the pre-mix type of burner, the aerosol is produced in a vaporising chamber where the larger droplets of liquid fall out from the gas stream and are discharged to waste. The resulting fine mist is mixed with the fuel gas and the carrier (oxidant) gas, and the mixed gases then flow to the burner head. In atomic absorption spectroscopy the burner is a long horizontal tube with a narrow slit along its length. This produces a thin flame of long path length which can be turned into or away from the beam of radiant energy. The flame path of a burner using air–acetylene, air–propane or air–hydrogen mixtures is about 10–12 cm in length, but with a nitrous oxide–acetylene burner it is usually reduced to about 5 cm because of the higher burning velocity of this gas mixture. In addition to a long light path, this type of burner has the advantages of being quiet in action and with little danger of incrustation around the burner head since large droplets of solution have been eliminated from the stream of gas reaching the burner. Its disadvantages are (1) that with solutions made up in mixed solvents, the more volatile solvents are evaporated preferentially; (2) a potential explosion hazard exists since the burner uses relatively large volumes of gas, but in modern versions of this type of burner this hazard is minimised.

A typical burner of this type is shown in Fig. 21.4. In this particular burner (Perkin–Elmer Corporation), the mixing chamber is a steel casting lined with a plastic ('Penton') which is extremely resistant to corrosion. The burner head is manufactured from titanium, thus avoiding the occasional high readings which are encountered when solutions containing iron and copper in presence of acid are examined with burners having a stainless-steel head. The nebuliser is capable of adjustment so that it can handle sample uptake rates of from $1-5\,\mathrm{mL\,min^{-1}}$. The burner can be adjusted in three directions, and horizontal and vertical scales are provided so that its position can be recorded. The head may be turned through an angle of $90°$ with respect to the light beam, and so the path length of the flame traversed by the resonance line radiation may be varied considerably: by choosing a small path length it becomes possible to analyse solutions of relatively high concentration without the need for prior dilution.

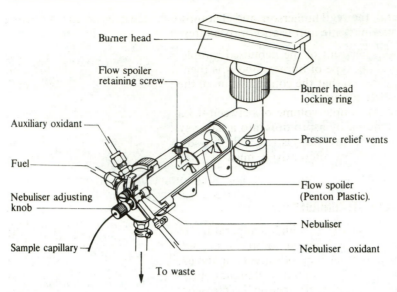

Burner head

Flow spoiler retaining screw

Burner head locking ring

Auxiliary oxidant

Pressure relief vents

Fuel

Nebuliser adjusting knob

Flow spoiler (Penton Plastic).

Nebuliser

Sample capillary

Nebuliser oxidant

To waste

Fig. 21.4

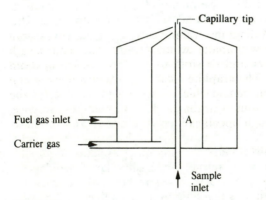

Capillary tip

Fuel gas inlet

A

Carrier gas

Sample inlet

Fig. 21.5

The **total consumption type** of burner consists of three concentric tubes as shown in Fig. 21.5. The sample solution is carried by a fine capillary tube A directly into the flame. The fuel gas and the oxidant gas are carried along separate tubes so that they only mix at the tip of the burner. Since all the liquid sample which is aspirated by the capillary tube reaches the flame, it would appear that this type of burner should be more efficient that the pre-mix type of burner. However, the total consumption burner gives a flame of relatively short path length, and hence such burners are predominantly used for flame emission studies. This type of burner has the advantages that (1) it is simple to manufacture, (2) it allows a totally representative sample to reach the flame, and (3) it is free from explosion hazards arising from unburnt gas mixtures. Its disadvantages are that (1) the aspiration rate varies with different solvents, and (2) there is a tendency for incrustations to form at the tip of the burner which can lead to variations in the signal recorded.

In general terms, Thomerson and Thompson[43] have cited the following disadvantages of flame atomisation procedures:

1. Only 5–15 per cent of the nebulised sample reaches the flame (in the case of the pre-mix type of burner) and it is then further diluted by the fuel and oxidant gases so that the concentration of the test material in the flame may be extremely minute.
2. A minimum sample volume of between 0.5 and 1.0 mL is needed to give a reliable reading by aspiration into a flame system.
3. Samples which are viscous (e.g. oils, blood, blood serum) require dilution with a solvent, or alternatively must be 'wet ashed' before the sample can be nebulised.

21.6 NON–FLAME TECHNIQUES

Instead of employing the high temperature of a flame to bring about the production of atoms from the sample, it is possible in some cases to make use of either (*a*) non-flame methods involving the use of electrically heated graphite tubes or rods, or (*b*) vapour techniques. Procedures (*a*) and (*b*) both find applications in atomic absorption spectroscopy and in atomic fluorescence spectroscopy.

(*a*) Electrothermal atomisers

(1) *The graphite tube furnace.* A diagram of a graphite tube furnace is shown in Fig. 21.6. It consists of a hollow graphite cylinder about 50 mm in length and about 9 mm internal diameter and so situated that the radiation beam passes along the axis of the tube. The graphite tube is surrounded by a metal jacket through which water is circulated and which is separated from the graphite tube by a gas space. An inert gas, usually argon, is circulated in the gas space, and enters the graphite tube through openings in the cylinder wall.

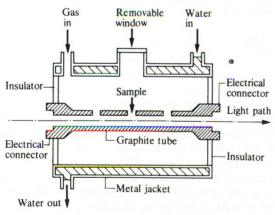

Fig. 21.6

The solution of the sample to be analysed (1–100 μL) is introduced by inserting the tip of a micropipette through a port in the outer (water) jacket, and into the gas inlet orifice in the centre of the graphite tube. The graphite cylinder is then heated by the passage of an electric current to a temperature which is high enough to evaporate the solvent from the solution. The current is then increased so that firstly the sample is ashed, and then ultimately it is vaporised so that metal atoms are produced, typically at a temperature of about 3000 K. For reproducibility, the temperatures and the timing of the drying, ashing and atomisation processes must be carefully selected according to the metal which is to be determined. The absorption signals produced by this method may last for several seconds and can be recorded on a chart recorder. Each graphite tube can be used for 100–200 analyses depending upon the nature of the material to be determined.

(2) *The graphite rod.* A graphite rod of 2 mm diameter was introduced by West and Williams[44] as a means of producing atoms from the sample. The sample is placed upon the rod which is heated, typically by a current of 100 A from a low-voltage (5 V) supply. The rod is placed just below the path of the beam from the radiation source so that vapour from the sample can move upwards into the beam and its absorbance be measured. The whole assembly is contained in a chamber fitted with quartz windows which is purged with argon.

In some circumstances it is found advantageous to coat graphite rods (or tubes) with a layer of pyrolytic graphite: this leads to improved sensitivity with elements such as vanadium and titanium which are prone to carbide formation.

The main advantages of electrothermal atomisers are that (*a*) very small samples (as low as 0.5 μL) can be analysed; (*b*) often very little or no sample preparation is needed, in fact certain solid samples can be analysed without prior dissolution; (*c*) there is enhanced sensitivity, particularly with elements with a short-wavelength resonance line; in practice there is an improvement of between 10^2- and 10^3-fold in the detection limits for furnace AAS compared with flame AAS.

There are, however, disadvantages in the electrothermal atomisation technique compared with flame atomic absorption. Background absorption effects (see Section 21.11) are usually far more serious. There is also a danger of the loss of analyte during the ashing stage. This can be a serious effect when dealing with volatile salts, e.g. the chlorides of cadmium and mercury. On some occasions the sample may not be completely volatilised during the atomisation stage, which can result in 'memory effects' within the graphite furnace. Generally speaking, the precision of furnace AAS (typically 5 per cent when the sample is injected manually) is much poorer than that obtained by flame AAS (often 1 per cent). However, the recent introduction of furnace auto-samplers has considerably enhanced the precision of furnace AAS.

Amongst other devices used to produce the required atoms in the vapour state are the **Delves cup** which enables the determination of lead in blood samples; the sample is placed in a small nickel cup which is inserted directly into an acetylene–air flame. The tantalum boat is a similar device to the Delves cup; in this case the sample is placed into a small tantalum dish which is then inserted into an acetylene–air flame. The use of these devices, especially for small sample volumes, has now been largely superseded by the graphite furnace.

(*b*) **Cold vapour technique.** This procedure is strictly confined to the determination of mercury,[45] which in the elemental state has an appreciable vapour pressure at room temperature so that gaseous atoms exist without the need for any special treatment. As a method for determining mercury compounds the procedure consists in the reduction of a mercury(II) compound with either sodium borohydride or (more usually) tin(II) chloride to form elemental mercury.

$$Hg^{2+} + Sn^{2+} \rightleftharpoons Hg + Sn(IV)$$

A diagram of a suitable apparatus is shown in Fig. 21.7. The mercury vapour is flushed out of the reaction vessel by bubbling argon through the solution, into the absorption tube.

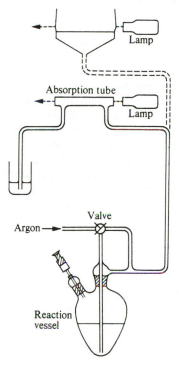

Fig. 21.7

This apparatus may also be adapted for what are termed '**hydride generation methods**' (which are strictly speaking flame-assisted methods). Elements such as arsenic, antimony, and selenium are difficult to analyse by flame AAS because it is difficult to reduce compounds of these elements (especially those in the higher oxidation states) to the gaseous atomic state.

Although electrothermal atomisation methods can be applied to the determination of arsenic, antimony, and selenium, the alternative approach of hydride generation is often preferred. Compounds of the above three elements may be converted to their volatile hydrides by the use of sodium borohydride as reducing agent. The hydride can then be dissociated into an atomic vapour by the relatively moderate temperatures of an argon–hydrogen flame.

The reaction sequence, in the case or arsenic, may be represented as follows:

$$\underset{\text{solution}}{As(V)} \xrightarrow[\text{[H}^+]]{\text{NaBH}_4} AsH_3 \xrightarrow[\text{in flame}]{\text{Heat}} As^0_{(gas)} + H_2$$

789

The requisite additional apparatus is indicated by the broken lines in Fig. 21.7.

It should be noted that the hydride generation method may also be applied to the determination of other elements forming volatile covalent hydrides that are easily thermally dissociated. Thus, the hydride generation method has also been used for the determination of lead, bismuth, tin, and germanium.

21.7 RESONANCE LINE SOURCES

As indicated in Fig. 21.3, for both atomic absorption spectroscopy and atomic fluorescence spectroscopy a resonance line source is required, and the most important of these is the **hollow cathode lamp** which is shown diagrammatically in Fig. 21.8. For any given determination the hollow cathode lamp used has an emitting cathode of the same element as that being studied in the flame. The cathode is in the form of a cylinder, and the electrodes are enclosed in a borosilicate or quartz envelope which contains an inert gas (neon or argon) at a pressure of approximately 5 torr. The application of a high potential across the electrodes causes a discharge which creates ions of the noble gas. These ions are accelerated to the cathode and, on collision, excite the cathode element to emission. Multi-element lamps are available in which the cathodes are made from alloys, but in these lamps the resonance line intensities of individual elements are somewhat reduced.

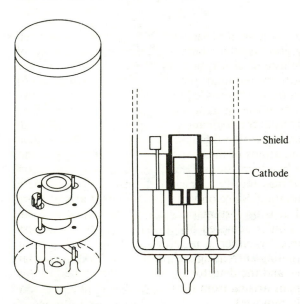

Fig. 21.8

Electrodeless discharge lamps were originally developed as radiation sources for atomic absorption spectroscopy and atomic fluorescence spectroscopy by Dagnall *et al.*;[46] they give radiation intensities which are much greater than those given by hollow cathode lamps. The electrodeless discharge lamp consists of a quartz tube 2–7 cm in length and 8 mm in internal diameter, containing up to 20 mg of the required element or of a volatile salt of the element, commonly the iodide; the tube also contains argon at low pressure (about 2 torr). Under operating conditions the material placed in the tube must have a vapour pressure of about 1 mm at a temperature of 200–400 °C. A microwave frequency of 2000–3000 MHz applied through a wave guide cavity provides the energy of excitation.

21.8 MONOCHROMATOR

The purpose of the monochromator is to select a given emission line and to isolate it from other lines, and occasionally, from molecular band emissions.

In atomic absorption spectroscopy the function of the monochromator is to isolate the resonance line from all non-absorbed lines emitted by the radiation source. In most commercial instruments diffraction gratings (Section 17.7) are used because the dispersion provided by a grating is more uniform than that given by prisms, and consequently grating instruments can maintain a higher resolution over a longer range of wavelengths.

In a simple flame (emission) photometer an interference filter (Section 17.7) can be used. In more sophisticated flame emission spectrophotometers which require better isolation of the emitted frequency, a prism or a grating monochromator is employed.

21.9 DETECTORS

For the simple flame emission photometer (Section 21.13) a barrier-layer cell (Section 17.6) is a sufficiently good detector because an intense wide band of energy reaches the detector. In atomic absorption spectrophotometers, in view of the improved spectral sensitivity required, photomultipliers (Section 17.6) are employed. The output from the detector is fed to a suitable read-out system, and in this connection it must be borne in mind that the radiation received by the detector originates not only from the resonance line which has been selected, but it may also arise from emission within the flame. This emission can be due to atomic emission arising from atoms of the element under investigation, and may also arise from molecular band emissions. Hence instead of an absorption signal intensity I_A, the detector may receive a signal of intensity $(I_A + S)$ where S is the intensity of emitted radiation. Since only the measurement arising from the resonance line is required, it is important that this be distinguished from the effects of flame emission. This is achieved by modulation of the emission from the resonance line source by either a mechanical chopper device, or electronically, by using an alternating-current signal appropriate to the particular frequency of the resonance line, and the detector amplifier is then tuned to this frequency: in this way, the signals arising from the flame, which are essentially d.c. in character, are effectively removed.

The read-out systems available include meters, chart recorders, and digital display; meters have now been virtually superseded by the alternative methods of data presentation.

21.10 INTERFERENCES

Various factors may affect the flame emission of a given element and lead to interference with the determination of the concentration of a given element.

791

These factors may be broadly classified as (a) **spectral interferences** and (b) **chemical interferences**.

Spectral interferences in AAS arise mainly from overlap between the frequencies of a selected resonance line with lines emitted by some other element; this arises because in practice a chosen line has in fact a finite 'bandwidth'. Since in fact the line width of an absorption line is about 0.005 nm, only a few cases of spectral overlap between the emitted lines of a hollow cathode lamp and the absorption lines of metal atoms in flames have been reported. Table 21.3 includes some typical examples of spectral interferences which have been observed.[47-50] However, most of these data relate to relatively minor resonance lines and the only interferences which occur with preferred resonance lines are with copper where europium at a concentration of about 150 mg L^{-1} would interfere, and mercury where concentrations of cobalt higher than 200 mg L^{-1} would cause interference.

Table 21.3 Some typical spectral interferences

Resonance source	Wavelength, λ (nm)	Analyte	Wavelength, λ (nm)
Aluminium	308.216	Vanadium	308.211
Antimony	231.147	Nickel	231.095
Copper	324.754	Europium	324.755
Gallium	403.307	Manganese	403.307
Iron	271.903	Platinum	271.904
Mercury	253.652	Cobalt	253.649

With flame emission spectroscopy, there is greater likelihood of spectral interferences when the line emission of the element to be determined and those due to interfering substances are of similar wavelength, than with atomic absorption spectroscopy. Obviously some of such interferences may be eliminated by improved resolution of the instrument, e.g. by use of a prism rather than a filter, but in certain cases it may be necessary to select other, non-interfering, lines for the determination. In some cases it may even be necessary to separate the element to be determined from interfering elements by a separation process such as ion exchange or solvent extraction (see Chapters 6, 7).

Apart from the interferences which may arise from other elements present in the substance to be analysed, some interference may arise from the emission band spectra produced by molecules or molecular fragments present in the flame gases: in particular, band spectra due to hydroxyl and cyanogen radicals arise in many flames. Although in AAS these flame signals are modulated (Section 21.9), in practice care should be taken to select an absorption line which does not correspond with the wavelengths due to any molecular bands because of the excessive 'noise' produced by the latter: this leads to decreased sensitivity and to poor precision of analysis.

21.11 CHEMICAL INTERFERENCES

The production of ground-state gaseous atoms which is the basis of flame spectroscopy may be inhibited by two main forms of chemical interference: (a) by stable compound formation, or (b) by ionisation.

Stable compound formation. This leads to incomplete dissociation of the substance to be analysed when placed in the flame, or it may arise from the formation within the flame of refractory compounds which fail to dissociate into the constituent atoms. Examples of these types of behaviour are shown by (1) the determination of calcium in the presence of sulphate or phosphate, and (2) the formation of stable refractory oxides of titanium, vanadium, and aluminium. Chemical interferences can usually be overcome in one of the following ways.

1. Increase in flame temperature often leads to the formation of free gaseous atoms, and for example aluminium oxide is more readily dissociated in an acetylene–nitrous oxide flame than it is in an acetylene–air flame. A calcium–aluminium interference arising from the formation of calcium aluminate can also be overcome by working at the higher temperature of an acetylene–nitrous oxide flame.

2. By the use of 'releasing agents'. If we consider the reaction

$$M-X + R \rightleftharpoons R-X + M$$

   then it is clear that an excess of the releasing agent (R) will lead to an enhanced concentration of the required gaseous metal atoms M: this will be especially true if the product R–X is a stable compound. Thus in the determination of calcium in the presence of phosphate, the addition of an excess of lanthanum chloride or of strontium chloride to the test solution will lead to formation of lanthanum (or strontium) phosphate, and the calcium can then be determined in an acetylene–air flame without any interference due to phosphate. The addition of EDTA to a calcium solution before analysis may increase the sensitivity of the subsequent flame spectrophotometric determination: this is possibly due to the formation of an EDTA complex of calcium which is readily dissociated in the flame.

3. Extraction of the analyte or of the interfering element(s) is an obvious method of overcoming the effect of 'interferences'. It is frequently sufficient to perform a simple solvent extraction to remove the major portion of an interfering substance so that, at the concentration at which it then exists in the solution, the interference becomes negligible. If necessary, repeated solvent extraction will reduce the effect of the interference even further and, equally, a quantitative solvent extraction procedure may be carried out so as to isolate the substance to be determined from interfering substances.

Ionisation. Ionisation of the ground-state gaseous atoms within a flame

$$M = M^+ + e$$

will reduce the intensity of the emission of the atomic spectra lines in flame emission spectroscopy, or will reduce the extent of absorption in atomic absorption spectroscopy. It is therefore clearly necessary to reduce to a minimum the possibility of ionisation occurring, and an obvious precaution to take is to use a flame operating at the lowest possible temperature which is satisfactory for the element to be determined. Thus, the high temperature of an acetylene–air or of an acetylene–nitrous oxide flame may result in the appreciable ionisation of elements such as the alkali metals and of calcium, strontium, and barium. The ionisation of the element to be determined may also be reduced by the addition of an excess of an **ionisation suppressant**; this is usually a solution containing a cation having a lower ionisation potential than that of the analyte.

Thus, for example a solution containing potassium ions at a concentration of 2000 mg L^{-1} added to a solution containing calcium, barium, or strontium ions creates an excess of electrons when the resulting solution is nebulised into the flame, and this has the result that the ionisation of the metal to be determined is virtually completely suppressed.

Other effects. In addition to the compound formation and ionisation effects which have been considered, it is also necessary to take account of so-called **matrix effects**. These are predominantly physical factors which will influence the amount of sample reaching the flame, and are related in particular to factors such as the viscosity, the density, the surface tension and the volatility of the solvent used to prepare the test solution. If we wish to compare a series of solutions, e.g. a series of standards to be compared with a test solution, it is clearly essential that the same solvent be used for each, and the solutions should not differ too widely in their bulk composition. This procedure is commonly termed **matrix matching**.

In some circumstances interference may result from **molecular absorption**. Thus, for example, in an acetylene–air flame a high concentration of sodium chloride will absorb radiation at wavelengths in the neighbourhood of 213.9 nm, which is the wavelength of the major zinc resonance line: hence sodium chloride would represent an interference in the determination of zinc under these conditions. Such interferences can usually be avoided by choosing a different resonance line, or alternatively by using a hotter flame resulting in an increase in the operating temperature, thus leading to dissociation of the interfering molecules. The interference referred to as **background absorption** which arises from the presence in the flame of gaseous molecules, molecular fragments and in some instances, where organic solvents are used, of smoke, are dealt with instrumentally by the incorporation of a **background correction** facility. In addition, background effects can be caused by light scatter. The degree of scatter is inversely proportional to the fourth power of the radiation wavelength. Hence, background effects due to scatter are a particularly important interference at high-energy wavelengths in the UV (between 185 and 230 nm).

Background effects especially due to the generation of particulate material to form a smoke are a major problem in furnace AAS.

It should be stressed that background correction methods should always be used in furnace AAS. The background effect in this case may be as high as 85 per cent of the total absorption signal.

To summarise, it may be stated that almost all interferences encountered in atomic absorption spectroscopy can be reduced, if not completely eliminated, by the following procedures.

1. Ensure if possible that standard and sample solutions are of similar bulk composition to eliminate matrix effects (matrix matching).
2. Alteration of flame composition or of flame temperature can be used to reduce the likelihood of stable compound formation within the flame.
3. Selection of an alternative resonance line will overcome spectral interferences from other atoms or molecules and from molecular fragments.
4. Occasionally, separation, e.g. by solvent extraction or by an ion exchange process, may be necessary to remove an interfering element; such separations are most frequently necessary when dealing with flame emission spectroscopy.
5. Use an appropriate background correction facility (see Section 21.12).

21.12 BACKGROUND CORRECTION METHODS

In the previous section it has been shown that the measured sample absorbance may be higher than the true absorbance signal of the analyte to be determined. This elevated absorbance value can occur by molecular absorption or by light scattering. There are three techniques that can be used for background correction: the deuterium arc; the Zeeman effect; and the Smith–Hieftje system.

(*a*) **Deuterium arc background correction.** This system uses two lamps, a high-intensity deuterium arc lamp producing an emission continuum over a wide wavelength range and the hollow cathode lamp of the element to be determined.

The deuterium arc continuum travels the same double-beam path as does the light from the resonance source (see Fig. 21.9). The background absorption affects both the sample and reference beams and so when the ratio of the intensities of the two beams is taken, the background effects are eliminated.

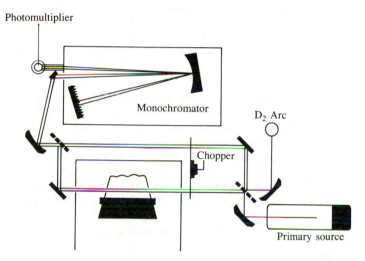

Photomultiplier

Monochromator

D₂ Arc

Chopper

Primary source

Fig. 21.9

The deuterium arc method is widely used in many instruments and is often satisfactory for background correction. It suffers, however, from the difficulty of achieving perfect alignment of the two lamps along an identical light path through the sample cell. In addition, due to the low output of a deuterium lamp in the visible region of the spectrum, its use is limited to wavelengths of less than 340 nm.

Ideally, the background absorption should be measured as near as possible to that of the analyte line. This approach has been achieved in the subsequent methods described below.

(*b*) **Zeeman background correction.** In a strong magnetic field, the electronic energy levels of atoms may be split, resulting in the formation of several absorption lines for each electronic transition (the Zeeman effect). The simplest form of splitting pattern is shown in Fig. 21.10.

In the presence of the magnetic field three components are observed, the π component having the same energy as the transition in the absence of the

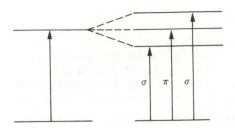

Fig. 21.10

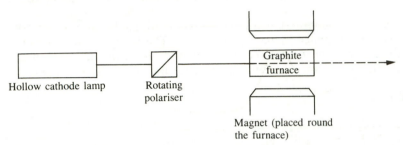

Fig. 21.11

magnetic field, and the σ components observed at lower and higher energies, typically 0.01 nm distant from the π component. The π peak is plane-polarised parallel to the direction of the magnetic field, whilst the σ peaks are polarised perpendicular to the direction of the magnetic field (see Fig. 21.11).

In practice, the emission line is split into three peaks by the magnetic field. The polariser is then used to isolate the central line which measures the absorption A_π, which also includes absorption of radiation by the analyte. The polariser is then rotated and the absorption of the background A_σ is measured. The analyte absorption is given by $A_\pi - A_\sigma$. A detailed discussion of the application of the Zeeman effect in atomic absorption is given in Ref. 51.

(*c*) **The Smith–Hieftje system.** This method[52] is based on the principle of self-absorption. If a hollow cathode lamp is operated at low current the normal emission line is obtained. At high lamp currents the emission band is broadened with a minimum appearing in the emission profile that corresponds exactly to the wavelength of the absorption peak. Hence, at low current the total absorbance due to the analyte and the background is measured, whilst at a high lamp current essentially only the background absorbance is obtained. Thus, the hollow cathode lamp is run alternatively at low and high current (see Fig. 21.12).

Thus, the analyte absorption is given by the difference between the absorbance measured at low lamp current and the absorbance at high lamp current.

Both the Zeeman and the Smith–Hieftje systems have the advantages that only one light source is used and that the background is measured very close to the sample absorption. Neither technique is restricted (as is the deuterium arc) in operating solely in the ultraviolet region of the spectrum. The Zeeman background correction is, however, normally limited to use in furnace AAS and can suffer from a lack of sensitivity. The Smith–Hieftje system, although less expensive than the Zeeman method, suffers the disadvantage that the lifetime

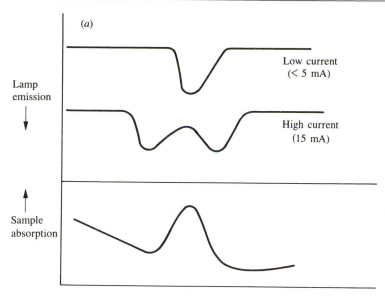

(a)

Lamp emission

Low current (< 5 mA)

High current (15 mA)

Sample absorption

Fig. 21.12

of hollow cathode lamps may be shortened, particularly those incorporating the more volatile elements.

21.13 FLAME EMISSION SPECTROSCOPY

There are now two main methods used for flame emission spectroscopy. The original method, known as flame photometry, is now used mainly for the analysis of alkali metals.

At present, however, the usual flame emission method is obtained by simply operating a flame atomic absorption spectrometer in the emission mode (see Fig. 21.3).

Flame photometers. A flame photometer can be compared with a photoelectric absorptiometer and the intensity of the filtered radiation from the flame is measured with a photoelectric detector. The filter, interposed between the flame and the detector, transmits only a strong line of the element. The simplest and least expensive detector is a barrier-layer cell (Section 17.6): if sufficient energy reaches the cell no amplification or external power supply is necessary, and only a sensitive galvanometer is required. The barrier-layer cell has a high temperature coefficient: it must, therefore, be placed at a cool part of the photometer. In some cases the precision is improved by the use of an internal standard and two filters and, in general, two photocells (one for the standard and one for the unknown) are utilised; the electronic circuit can be devised to give a direct reading of the ratio of line intensities. Flame photometers are intended primarily for analysis of sodium and potassium and also for calcium and lithium, i.e. elements which have an easily excited flame spectrum of sufficient intensity for detection by a photocell. The layout of a simple flame photometer is shown in Fig. 21.13. Air at a given pressure is passed into an atomiser and the suction this produces draws a solution of the sample into the atomiser, where it joins

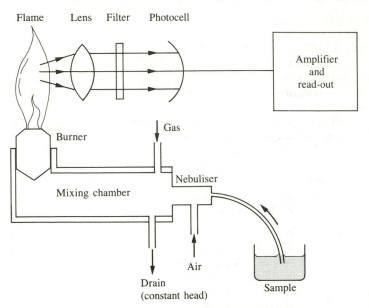

Fig. 21.13

the air stream as a fine mist and passes into the burner. Here, in a small mixing chamber, the air meets the fuel gas supplied to the burner at a given pressure and the mixture is burnt. Radiation from the resulting flame passes through a lens, and finally through an optical filter which permits only the radiation characteristic of the element under investigation to pass through the photocell. The output from the photocell is measured on a suitable digital read-out system.

An example of a modern instrument of this type is the Corning Model 410 flame photometer. This model can incorporate a 'lineariser module' which provides a direct concentration read-out for a range of clinical specimens. Flame photometers are still widely used especially for the determination of alkali metals in body fluids, but are now being replaced in clinical laboratories by ion-selective electrode procedures (see Section 15.7).

21.14 ATOMIC ABSORPTION SPECTROPHOTOMETERS

A large number of commercial instruments are now available and are based either on a single- or double-beam design. Important instrumental features of a modern atomic absorption instrument include the following facilities.

1. It should have a lamp turret capable of holding at least four hollow cathode lamps with an independent current stabilised supply to each lamp.
2. The sample area should be able to incorporate an auto-sampler which can work with both flame and furnace atomisers. Improved analytical precision is obtained when an auto-sampler is used in conjunction with a furnace atomiser.
3. The monochromator should be capable of high resolution, typically 0.04 nm. This feature is most desirable if the AAS is adapted for flame emission work; good resolution is also desirable for many elements in atomic absorption.

4. The photomultiplier should be able to function over a wide wavelength range from 188 to 800 nm.

5. All instruments should be equipped with a background correction facility. Virtually all instruments now have a deuterium arc background correction. The Zeeman system is also available in instruments marketed by the Perkin–Elmer Corporation and the Smith–Hieftje system by Thermo Electron Ltd.

6. The use of an integral video screen in instruments presents very great advantages, both in the ease of operation and in the ability to develop and understand analytical methods. Complete analytical records can be stored in the instrument and a visual display of good calibration curves can be stored in memory and recalled at will. It is most useful to have a graphical display of atomisation peaks when using a furnace where a distinction can be made of the total absorbance peak and that due to the analyte absorbance.

The optical design used in the Thermo Electron Video II, a single-beam instrument, is shown in Fig. 21.14. The optical system is very simple, consisting of just two achromatic lenses A_1 and A_2, located between the lamp L and the monochromator. A more efficient light throughput is obtained when a relatively simple optical system is employed. The monochromator is an Ebert–Fastie design to give high resolution (0.04 nm) suitable for both atomic absorption and flame emission spectroscopy.

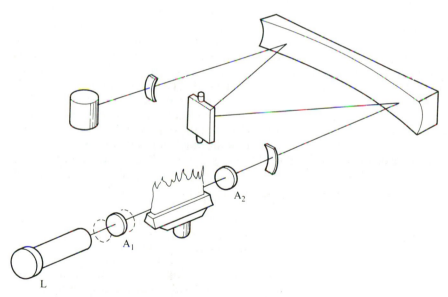

Fig. 21.14 Single-beam system.

Double-beam AA spectrophotometers are still marketed by instrument manufacturers. A double-beam system compensates for changes in lamp intensity and may require less frequent re-zeroing than a single-beam instrument. These considerations had more merit some years ago when hollow cathode lamps suffered from some instability. It should be noted, however, that the optical system of a double-beam instrument is necessarily more complex, resulting in a loss of light efficiency.

21.15 ATOMIC FLUORESCENCE SPECTROSCOPY

Within the confines of the present volume it is not possible to provide a detailed discussion of instrumentation for atomic fluorescence spectroscopy. An instrument for simultaneous multi-element determination described by Mitchell and Johansson[53] has been developed commercially. Many atomic absorption spectrophotometers can be adapted for fluorescence measurements and details are available from the manufacturers. Detailed descriptions of atomic fluorescence spectroscopy are to be found in many of the volumes listed in the Bibliography (Section 21.27).

EXPERIMENTAL

21.16 EVALUATION METHODS

Before dealing with the experimental details of AAS or FES determinations it is necessary to consider the mode of treatment of the experimental data obtained. To convert the measured absorption values into the concentration of the substance being determined it is necessary either to make use of a calibration curve, or to carry out the 'standard addition' procedure.

(*a*) **Calibration curve procedure.** A calibration curve for use in atomic absorption or in flame emission measurements is plotted by aspirating into the flame samples of solutions containing known concentrations of the element to be determined, measuring the absorption (emission) of each solution, and then constructing a graph in which the measured absorption (emission) is plotted against the concentration of the solutions. If we are dealing with a test solution which contains a single component, then the standard solutions are prepared by dissolving a weighed quantity of a salt of the element to be determined in a known volume of distilled (de-ionised) water in a graduated flask. If, however, other substances are present in the test solution, then these should also be incorporated in the standard solutions and at a similar concentration to that existing in the test solution. At least four standard solutions should be used covering the optimum absorbance range of 0.1 to 0.4, and if the calibration curve is found to be non-linear (this often happens at high absorbance values), then measurements with additional standard solutions should be carried out. In common with all absorbance measurements, the readings must be taken after the instrument zero has been adjusted against a 'blank' which may be either distilled water, or a solution of similar composition to the test solution but minus the component to be determined. It is usual to examine the standard solutions in order of increasing concentration, and after making the measurements with one solution, distilled water is aspirated into the flame to remove all traces of solution before proceeding to the next solution. At least two, and preferably three, separate absorption readings should be made with each solution, and an average value taken.

If necessary, the test solution must be suitably diluted using a pipette and a graduated flask, so that it too gives absorbance readings in the range 0.1–0.4.

Using the calibration curve it is a simple matter to interpolate from the measured absorbance of the test solution the concentration of the relevant element in the solution. The working graph should be checked occasionally by making measurements with the standard solutions, and if necessary a new calibration curve must be drawn.

Note. Most instruments now include a microcomputer which stores the calibration curve and allows a direct read-out of concentration.

(b) The standard addition technique. When dealing with a test solution which is complex in character, or one whose exact composition is unknown, it may be very difficult and even impossible to prepare standard solutions having a similar composition to the sample. In such a case the method of standard addition can be employed. As described in Section 16.5, this involves the addition of known amounts of the ion to be determined to a number of aliquots of the sample solution; the solutions thus obtained should all be diluted to the same final volume. Naturally, if the absorbance of the test solution is too high, a quantitative dilution must be carried out, and the measurements made with this diluted solution. The absorbance of the test solution is first measured, and then each of the prepared solutions is examined in turn, leading up to the solution of highest concentration, and remembering to aspirate distilled water into the flame between each solution. The absorbance values are then plotted against the added concentration values; a straight-line plot should result and the straight line can be extrapolated to the concentration axis — the point where the axis is cut gives the concentration of the test solution. If the graph is non-linear, then extrapolation cannot be undertaken with any confidence. It is important to realise that an extrapolation procedure is never as reliable as interpolation, and the latter should, therefore, be chosen if at all possible.

21.17 PREPARATION OF SAMPLE SOLUTIONS

For the application of flame spectroscopic methods the sample must be prepared in the form of a suitable solution unless it is already presented in this form: exceptionally, solid samples can be handled directly in some of the non-flame techniques (Section 21.6).

Aqueous solutions may sometimes be analysed directly without any pre-treatment, but it is a matter of chance that the given solution should contain the correct amount of material to give a satisfactory absorbance reading. If the existing concentration of the element to be determined is too high then the solution must be diluted quantitatively before commencing the absorption measurements. Conversely, if the concentration of the metal in the test solution is too low, then a concentration procedure must be carried out by one of the methods outlined at the end of this section.

Solutions in organic solvents may, with certain reservations, be used directly, provided that the viscosity of the solution is not very different from that of an aqueous solution. The important consideration is that the solvent should not lead to any disturbance of the flame; an extreme example of this is carbon tetrachloride, which may extinguish an air–acetylene flame. In many cases, suitable organic solvents [e.g. 4-methylpentan-2-one (methyl isobutyl ketone) and the hydrocarbon mixture sold as 'white spirit'] give enhanced production of ground-state gaseous atoms and lead to about three times the sensitivity

801

which is achieved with aqueous solutions. Due regard must of course be paid to the question of safety: see Section 21.19.

Inorganic solids such as metallic alloys, minerals, cements, etc., must be brought into solution by the usual standard techniques, the aim being to produce a clear solution with no loss of the element to be determined. Generally speaking, the final solution should not contain acid at a greater concentration than about $1M$ since the aspiration of extremely corrosive solutions into the burner of the apparatus should be avoided as far as possible: the instruction manual supplied with the instrument will normally give guidance in this direction.

Organic solids which contain trace elements can sometimes be dissolved in a suitable organic solvent, or alternatively the organic material may be oxidised and the residue treated to give an aqueous solution of the element to be determined.

Separation techniques may have to be applied if the given sample contains substances which act as interferences (Section 21.10), or, as explained above, if the concentration of the element to be determined in the test solution is too low to give satisfactory absorbance readings. As already indicated (Section 21.10), the separation methods most commonly used in conjunction with flame spectrophotometric methods are solvent extraction (see Chapter 6) and ion exchange (Chapter 7). When a solvent extraction method is used, it may happen that the element to be determined is extracted into an organic solvent, and as discussed above it may be possible to use this solution directly for the flame photometric measurement.

21.18 PREPARATION OF STANDARD SOLUTIONS

In flame spectrophotometric measurements we are concerned with solutions having very small concentrations of the element to be determined. It follows that the standard solutions which will be required for the analyses must also contain very small concentrations of the relevant elements, and it is rarely practicable to prepare the standard solutions by weighing out directly the required reference substance. The usual practice therefore is to prepare stock solutions which contain about $1000\,\mu g\,mL^{-1}$ of the required element, and then the working standard solutions are prepared by suitable dilution of the stock solutions. Solutions which contain less than $10\,\mu g\,mL^{-1}$ are often found to deteriorate on standing owing to adsorption of the solute on to the walls of glass vessels. Consequently, standard solutions in which the solute concentration is of this order should not be stored for more than 1 to 2 days.

The stock solutions are ideally prepared from the pure metal or from the pure metal oxide by dissolution in a suitable acid solution; the solids used must be of the highest purity, e.g. the Johnson Matthey 'Specpure' range, and the acids should ideally be of Aristar quality supplied by BDH Ltd, Poole, Dorset, UK.

A wide variety of SpectrosoL stock solutions ($1000\,\mu g\,mL^{-1}$) suitable for AAS can be obtained from BDH Ltd.

21.19 SAFETY PRACTICES

Before commencing any experimental work with either a flame (emission) photometer or an atomic absorption spectrophotometer, the following guidelines on safety practices should be studied. These recommendations are a summary of the Code of Practice recommended by the Scientific Apparatus Makers' Association (SAMA) of the USA; for full details see Ref. 54.

1. Ensure that the laboratory in which the apparatus is housed is well ventilated and is provided with an adequate exhaust system having air-tight joints on the discharge side; some organic solvents, especially those containing chlorine, give toxic products in a flame.
2. Gas cylinders must be fastened securely in an adequately ventilated room well away from any heat or ignition sources. The cylinders must be clearly marked so that the contents can be immediately identified.
3. When the equipment is turned off, close the fuel gas cylinder valve tightly and bleed the gas line to the atmosphere via the exhaust system.
4. The piping which carries the gases from the cylinders must be securely fixed in such a position that it is unlikely to suffer damage.
5. Make periodic checks for leaks by applying soap solution to joints and seals.
6. The following special precautions should be observed with acetylene.
    (a) Never run acetylene gas at a pressure higher than 15 psi ($103 \, kN \, m^{-2}$); at higher pressures acetylene can explode spontaneously
    (b) Avoid the use of copper tubing. Use tubing made from brass containing less than 65 per cent copper, from galvanised iron or from any other material that does not react with acetylene.
    (c) Avoid contact between gaseous acetylene and silver, mercury or chlorine.
    (d) Never run an acetylene **cylinder** after the pressure has dropped to 50 psi ($3430 \, kN \, m^{-2}$); at lower pressures the gas will be contaminated with acetone.
7. A nitrous oxide cylinder should not be used after the regulator gauge has dropped to a reading of 100 psi ($6860 \, kN \, m^{-2}$).
8. A burner which utilises a mixture of fuel and oxidant gases and which is attached to a waste vessel (liquid trap) should be provided with a **U**-shaped connection between the trap and the burner chamber. The head of liquid in the connecting tube should be greater than the operating pressure of the burner: if this is not achieved, mixtures of fuel and oxidant gas may be vented to the atmosphere and form an explosive mixture. The trap should be made of a material that will not shatter in the event of an explosive flash-back in the burner chamber.
9. Care must be exercised when using volatile flammable organic solvents for aspiration into the flame. A container fitted with a cover which is provided with a small hole for the sample capillary is recommended.
10. Never view the flame or hollow cathode lamps directly; protective eye wear should always be worn. Safety spectacles will usually provide adequate protection from ultraviolet light, and will also provide protection for the eyes in the event of the apparatus being shattered by an explosion.
11. Never leave a flame unattended.

SELECTED DETERMINATIONS

21.20 INTRODUCTION

It is impossible in the present volume for the determination of a wide range of elements by atomic absorption spectroscopy to be discussed in detail. A few detailed examples of the application of atomic absorption are given in Sections 21.21–21.26 these have been chosen to illustrate the general procedures involved, including the manner in which certain interferences may be overcome and how chemical pre-treatment is often necessary in order to perform a successful analysis by this technique.

In Table 21.4 is listed the wavelength of the most widely used resonance line for each of the common elements, together with the normal composition of the flame gases. The optimum working range of concentrations is quoted, and although this can vary with the instrument used, the values cited may be regarded as typical. The term '**sensitivity**' in atomic absorption spectroscopy is defined as the concentration of an aqueous solution of the element which absorbs 1 per cent of the incident resonance radiation; in other words, it is the concentration which gives an absorbance of 0.0044. As a rough guide, the sensitivity may be taken as about one-fiftieth of the lower value of the optimum absorbance range [Section 21.16(a)]. It should be noted that sensitivity is largely dependent upon the reactions occurring in the flame, and is not strictly a characteristic of a given instrument.

The **detection limit** is another value which is often quoted, and this may be defined in a variety of ways. The most widely accepted definition is that the detection limit is the smallest concentration of a solution of an element that can be detected with 95 per cent certainty. This is the quantity of the element that gives a reading equal to twice the standard deviation of a series of at least ten determinations taken with solutions of concentrations which are close to the level of the blank.

A more detailed table of the resonance lines is given in Appendix 9.

The data presented in Table 21.4, in conjunction with the experimental details given in Sections 21.21–21.26, will enable the determination of most elements to be carried out successfully. For detailed accounts of the determination of individual elements by atomic absorption spectroscopy, the Bibliography (Section 21.27) should be consulted. In addition, most instrument manufacturers supply applications handbooks relative to the apparatus in which full experimental details are given.

ATOMIC ABSORPTION EXPERIMENTS

21.21 DETERMINATION OF MAGNESIUM AND CALCIUM IN TAP WATER

The determination of magnesium in potable water is very straightforward; very few interferences are encountered when using an acetylene–air flame. The determination of calcium is however more complicated; many chemical interferences are encountered in the acetylene–air flame and the use of 'releasing agents' such as strontium chloride, lanthanum chloride, or EDTA is necessary. Using the hotter acetylene–nitrous oxide flame the only significant interference arises from the ionisation of calcium, and under these conditions an 'ionisation buffer' such as potassium chloride is added to the test solutions.

(*a*) **Determination of magnesium. Preparation of the standard solutions.** A magnesium stock solution (1000 mg L^{-1}) is prepared by dissolving 1.000 g magnesium metal in 50 mL of 5M hydrochloric acid. After dissolution of the metal the solution is transferred to a 1 L graduated flask and made up to the

Table 21.4 FAAS data for the common elements

Element	Wavelength of main resonance line λ (nm)	Flame*	Working range (μg mL^{-1})
Ag	328.1	AA(L)	1–5
Al	309.3	NA(R)	40–200
As (3)	193.7	AH(R)	50–200
B	249.8	NA(R)	400–600
Ba	553.6	NA(R)	10–40
Be	234.9	NA(R)	1–5
Bi	223.1	AA(L)	10–40
Ca	422.7	NA(R)	1–4
Cd	228.8	AA(L)	0.5–2
Co	240.7	AA(L)	3–12
Cr	357.9	AA(R)	2–8
Cs	852.1	AP(L)	5–20
Cu	324.7	AA(L)	2–8
Fe	248.3	AA(L)	2.5–10
Ga	294.4	AA(L)	50–200
Ge (3)	265.2	NA(R)	70–280
Hg (2)	253.7	AA(L)	100–400
In	303.9	AA(L)	15–60
Ir	208.9	AA(R)	40–160
K	766.5	AP(L)	0.5–2
Li	670.8	AP(L)	1–4
Mg	285.2	AA(L)	0.1–0.4
Mn	279.5	AA(L)	1–4
Mo	313.3	NA(R)	15–60
Na	589.0	AP(L)	0.15–0.60
Ni	232.0	AA(L)	3–12
Os	290.9	NA(R)	50–200
Pb	217.0	AA(L)	5–20
Pd	244.8	AA(L)	4–16
Pt	265.9	AA(L)	50–200
Rb	780.0	AP(L)	2–10
Rh	343.5	AA(L)	5–25
Ru	349.9	AA(L)	30–120
Sb (3)	217.6	AA(L)	10–40
Sc	391.2	NA(R)	15–60
Se (3)	196.0	AH(R)	20–90
Si	251.6	NA(R)	70–280
Sn	224.6	AH(R) (1)	15–60
Sr	460.7	NA(L)	2–10
Te	214.3	AA(L)	10–40
Ti	364.3	NA(R)	60–240
Tl	276.8	AA(L)	10–50
V	318.5	NA(R)	40–120
W	255.1	NA(R)	250–1000
Y	410.2	NA(R)	200–800
Zn	213.9	AA(L)	0.4–1.6

*Key: L = fuel-lean; R = fuel-rich; AA = air/acetylene; AP = air/propane; NA = nitrous oxide/acetylene; AH = air/hydrogen

Notes: (1) If there are many interferences then NA is to be preferred. (2) The use of the non-flame mercury cell (Section 21.6) is far more sensitive for the determination of mercury.

(3) The use of the hydride generation method (Section 21.6) is far more sensitive for the determination of the listed elements.

mark with distilled water. An intermediate stock solution containing $50\,mg\,L^{-1}$ is prepared by pipetting 50 mL of the stock solution into a 1 L graduated flask and diluting to the mark. Dilute accurately four portions of this solution to give four standard solutions of magnesium with known magnesium concentrations lying within the optimum working range of the instrument to be used (typically $0.1-0.4\,\mu g\,Mg^{2+}\,mL^{-1}$).

Procedure. Although the precise mode of operation may vary according to the particular instrument used, the following procedure may be regarded as typical. Place a magnesium hollow cathode lamp in the operating position, adjust the current to the recommended value (usually 2–3 mA), and select the magnesium line at 285.2 nm using the appropriate monochromator slit width.

Connect the appropriate gas supplies to the burner following the instructions detailed for the instrument, and adjust the operating conditions to give a fuel-lean acetylene–air flame.

Starting with the least concentrated solution, aspirate in turn the standard magnesium solutions into the flame, and for each take three readings of the absorbance; between each solution, remember to aspirate de-ionised water into the burner. Finally read the absorbance of the sample of tap water; this will usually require considerable dilution in order to give an absorbance reading lying within the range of values recorded for the standard solutions. Plot the calibration curve and use this to determine the magnesium concentration of the tap water.

If the magnesium content of the water is greater than $5\,\mu g\,mL^{-1}$ it might be considered preferable to work with the less sensitive magnesium line at wavelength 202.5 nm.

(*b*) Determination of calcium. Two procedures are described, (i) involving the use of releasing agents, and (ii) involving the use of an 'ionisation buffer': the latter is the preferred technique provided that an acetylene–nitrous oxide flame is available.

Preparation of the standard solutions. For procedure (i) it is necessary to incorporate a releasing agent in the standard solutions. Three different releasing agents may be used for calcium, (*a*) lanthanum chloride, (*b*) strontium chloride and (*c*) EDTA; of these (*a*) is the preferred reagent, but (*b*) or (*c*) make satisfactory alternatives.

(*a*) Prepare a lanthanum stock solution ($50\,000\,mg\,L^{-1}$) by dissolving 67 g of lanthanum chloride ($LaCl_3,7H_2O$) in 100 mL of $1M$ nitric acid. Warm gently to dissolve the salt, then cool the solution and make up to 500 mL in a graduated flask.

(*b*) A strontium stock solution is prepared by dissolving 76 g of strontium chloride ($SrCl_2,6H_2O$) in 250 mL of de-ionised water and then making up to 500 mL in a graduated flask.

(*c*) An EDTA stock solution is prepared by dissolving 75 g of EDTA disodium salt (analytical grade) in 800 mL of de-ionised water. Warm gently until the salt is dissolved, then cool and make up to 1 L in a graduated flask.

For procedure (ii) an ionisation buffer is required and this involves preparing a potassium stock solution ($10\,000\,mg\,L^{-1}$). Dissolve 9.6 g of potassium chloride in de-ionised water and make up to 500 mL in a graduated flask.

Prepare a calcium stock solution (1000 mg L^{-1}) by dissolving 2.497 g of dried calcium carbonate in a minimum volume of 1 M hydrochloric acid: about 50 mL will be required. When dissolution is complete, transfer the solution to a 1 L graduated flask and make up to the mark with de-ionised water. An intermediate calcium stock solution is prepared by pipetting 50 cm of the stock solution into a 1 L flask and making up to the mark with de-ionised water.

The working standard solutions for procedure (i) contain between 1 μg Ca^{2+} mL^{-1} to 5 μg Ca^{2+} mL^{-1} and are prepared by mixing appropriate volumes of the intermediate stock solution (measured with a grade A pipette), with suitable volumes of the chosen releasing agent solution, and then making up to 50 mL in a graduated flask; the releasing agent solution is measured in a 25 mL measuring cylinder. Five standard solutions are prepared containing respectively 1.0, 2.0, 3.0, 4.0, and 5.0 mL of the intermediate stock solution and 10 mL of releasing agent (a) or 5 mL of either reagent (b) or (c). A blank solution is similarly prepared but without the addition of any of the intermediate calcium stock solution.

For procedure (ii) the working standard solutions are prepared as detailed for procedure (i) except that the releasing agent solution is replaced by 10 mL of the stock potassium solution.

The unknown calcium solution (the tap water) will normally require to be diluted in order that its absorbance reading shall lie on the calibration curve, and the same amount of releasing agent [procedure (i)], or of ionisation buffer [procedure (ii)], must be added as in the standard solutions. So, for example, if the tap water contains about 100 μg mL^{-1} of calcium, 25 mL of it are pipetted into a 100 mL graduated flask and made up to the mark with de-ionised water. Then 5 mL of this solution is pipetted into a 50 mL graduated flask, and if procedure (i) is being followed, 10 mL of reagent (a) is added, or 5 mL of either reagent (b) or (c) and then the solution is made up to the mark. If procedure (ii) is being followed, then 10 mL of the stock potassium solution are used in place of the releasing agent. If any cloudiness should develop during the preparation of the final solution, add 1 mL of 1 M hydrochloric acid before making up to the mark.

Procedure (i). Set up a calcium hollow cathode lamp selecting the resonance line of wavelength 422.7 nm, and a fuel-lean acetylene–air flame following the details given in the instrument manual. The calibration procedure is similar to that described above for magnesium, but the aspiration of de-ionised water into the burner after taking the readings for each solution is even more important in this case owing to the relatively high concentrations of salts present as releasing agent; remember that de-ionised water should be aspirated into the burner for a few minutes at the conclusion of the series of readings.

Procedure (ii). Make certain that the instrument is fitted with the correct burner for an acetylene–nitrous oxide flame, then set the instrument up with the calcium hollow cathode lamp, select the resonance line of wavelength 422.7 nm, and adjust the gas controls as specified in the instrument manual to give a fuel-rich flame. Take measurements with the blank, and the standard solutions, and with the test solution, all of which contain the 'ionisation buffer'; the need, mentioned under procedure (i), for adequate treatment with de-ionised water after each measurement applies with equal force in this case. Plot the calibration graph and ascertain the concentration of the unknown solution.

21.22 DETERMINATION OF VANADIUM IN LUBRICATING OIL

The oil is dissolved in white spirit and the absorption to which this solution gives rise is compared with that produced from standards made up from vanadium naphthenate dissolved in white spirit.

Preparation of the standard solutions. The standard solutions are prepared from a solution of vanadium naphthenate in white spirit which contains about 3 per cent of vanadium. Weigh out accurately about 0.6 g of the vanadium naphthenate into a 100 mL graduated flask and make up to the mark with white spirit: this stock solution contains about $180\,\mu g\,mL^{-1}$ of vanadium. Dilute portions of this stock solution measured with the aid of a Grade A 50 mL burette to obtain a series of working standards containing from $10–40\,\mu g\,mL^{-1}$ of vanadium.

Procedure. Weigh out accurately about 5 g of the oil sample, dissolve in a small volume of white spirit and transfer to a 50 mL graduated flask; use the same solvent to wash out the weighing bottle and finally to make up the solution to the mark.

A double-beam atomic absorption spectrophotometer should be used. Set up a vanadium hollow cathode lamp selecting the resonance line of wavelength 318.5 nm, and adjust the gas controls to give a fuel-rich acetylene–nitrous oxide flame in accordance with the instruction manual. Aspirate successively into the flame the solvent blank, the standard solutions, and finally the test solution, in each case recording the absorbance reading. Plot the calibration curve and ascertain the vanadium content of the oil.

21.23 DETERMINATION OF TRACE LEAD IN A FERROUS ALLOY

The procedure followed entails the removal of gross interferences by solvent extraction, and the selective extraction and concentration of the trace metal by use of a chelating agent. The alloy used should not contain more than 0.1 g of copper in the sample weighed out.

Preparation of solutions. The following solutions are required.

Ammonia solution (concentrated, '0.880', about 35 per cent NH_3). Preferably the special atomic absorption spectroscopy reagent grade should be used.

Hydrochloric acid, concentrated. Also a solution prepared by measuring 50 mL (measuring cylinder) of the concentrated acid into a 1 L graduated flask and making up with de-ionised water.

Nitric acid, concentrated. Analytical reagent grade.

Ammonium citrate. Dissolve 50 g tri-ammonium citrate in 50 mL of concentrated ammonia solution added with care. Cool, and make up to 100 mL with de-ionised water.

Ascorbic acid. Dissolve 20 g of the solid in 100 mL of de-ionised water. This reagent must be freshly prepared.

Potassium cyanide. **(CAUTION!)** Dissolve 25 g of the salt in 35 mL of de-ionised water to which has been added 5 mL of concentrated ammonia solution. Make up to 50 mL with de-ionised water and filter if necessary.

Sodium diethyldithiocarbamate (NaDDC). Dissolve 1 g in 50 mL of de-ionised water and filter if necessary. This reagent must be freshly prepared.

Lead caprate. Prepare a standard stock solution by dissolving 0.1323 g of the solid in 2 mL of naphthenic acid with warming. Add 20 mL of 4-methylpentan-2-one (methyl isobutyl ketone), cool and then make up to the mark in a 100 mL graduated flask with more of the ketone.

Procedure. Weigh accurately 1 g of the alloy and dissolve in 10 mL of concentrated hydrochloric acid; warm gently, and if necessary add concentrated nitric acid dropwise (about 3 mL) to assist the dissolution. When the vigorous reaction is complete, digest the solution with gentle heat for about 15 minutes. Cool, if necessary filter through a Whatman No. 541 filter paper, washing the beaker and filter paper with small portions of concentrated hydrochloric acid so that a final volume of about 20 mL is attained. Transfer the solution to a 250 mL separatory funnel using a further 10 mL of concentrated hydrochloric acid to effect a quantitative transfer. Add 50 mL of butyl acetate, shake for one minute and allow to separate; iron and molybdenum are extracted into the organic layer. Separate the two layers, collecting the acid layer and transferring, with the aid of a further 10 mL of concentrated hydrochloric acid, to a clean 250 mL separatory funnel; extract with a 25 mL portion of butyl acetate. Again separate the two layers, collecting the acid layer in a 250 mL beaker.

Add cautiously (**FUME CUPBOARD**), and with constant stirring, 10 mL of the ammonium citrate solution; this will prevent the precipitation of metals when, at a later stage, the pH value of the solution is increased. Then add 10 mL of the 20 per cent ascorbic acid, and adjust to pH 4 (BDH narrow-range indicator paper), by the cautious addition of concentrated ammonia solution down the side of the beaker while stirring continuously. Then add 10 mL of the 50 per cent potassium cyanide solution (**CAUTION!**) and *immediately* adjust to a pH of 9–10 (BDH indicator paper) by the addition of concentrated ammonia solution.

Transfer the solution to a 250 mL separatory funnel, rinsing out the beaker with a little water. Add 5 mL of the 2 per cent NaDDC reagent and allow to stand for one minute, and then add a 10 mL portion of 4-methylpentan-2-one (methyl isobutyl ketone), shake for one minute and then separate and collect the organic layer. Return the aqueous phase to the funnel, extract with a further 10 mL portion of methyl isobutyl ketone, separate and combine the organic layer with that already collected. Finally, rinse the funnel with a little fresh ketone and add this rinse liquid to the organic extract. In these operations the lead is converted into a chelate which is extracted into the organic solvent.

In order to concentrate the lead extract, remove the lead from the organic solvent by shaking this with three successive 10 mL portions of the dilute hydrochloric acid solution, collecting the aqueous extracts in a 250 mL beaker. To the combined extracts add 5 mL of 20 per cent ascorbic acid solution and adjust to pH 4 by the addition of concentrated ammonia solution. Place the beaker in a fume cupboard, add 3 mL of the 50 per cent potassium cyanide solution and *immediately* adjust the pH to 9–10 with concentrated ammonia solution. Transfer the solution to a 250 mL separatory funnel with the aid of a little de-ionised water, add 5 mL of the 2 per cent NaDDC reagent, allow to stand for one minute and then add 10 mL of methyl isobutyl ketone. Shake for one minute and then separate and collect the organic phase, filtering it through a fluted filter paper. This solution now contains the lead and is ready for the absorption measurement.

Set up a double-beam atomic absorption spectrophotometer with a lead hollow cathode lamp and isolate the resonance line at 283.3 nm; adjust the gas controls to give a fuel-lean acetylene–air flame in accordance with the operating manual supplied with the instrument.

Prepare a blank solution by carrying through all the sequences of the separation procedures using a hydrochloric acid solution to which no alloy has been added, and then measure the absorption given by this blank solution, by a series of standard solutions containing from 1 to 10 μg Pb mL^{-1} prepared by suitable dilution of the lead caprate stock solution (see Note), and finally of the extract prepared from the sample of alloy. Plot the calibration curve and determine the lead content of the alloy.

Note. If lead caprate is not available, standard lead solutions can be prepared from aqueous solutions containing known weights of lead nitrate and following through the extraction procedure as detailed for the final extraction of lead into methyl isobutyl ketone for the alloy. It should also be noted that steps should be taken to avoid excessive inhalation of the vapour of the methyl isobutyl ketone, which can cause a headache.

21.24 DETERMINATION OF TRACE ELEMENTS IN CONTAMINATED SOIL

The procedure followed describes methods for the determination of *total* levels, and in certain cases, *available* amounts of trace elements in soils. The determination of arsenic in soil by hydride generation AAS is included.

(*a*) **Sampling.** Incremental samples (see Section 5.2) of approximately 50 g should be taken from specified sampling points on the site. The sampling points should include surface soil and two further samples taken at depth, typically at 0.5 and 1.0 m. The exact location of these points should be noted, for it may be necessary to take further samples subsequently. The individual samples, carefully labelled, must be stored in separate containers to avoid cross-contamination. When received by the laboratory the samples are air-dried for a period to remove excess moisture. The individual dried samples are passed through a 0.5 mm sieve. The soil passing through the sieve is mixed and used to obtain the analytical sample.

(*b*) **Sample treatment for total element determination.** Weigh out accurately about 1 g of the sieved soil and transfer to a 100 mL tall-form beaker. Add, from a measuring cylinder, about 20 mL of 1:1 nitric acid (SpectrosoL grade) and boil *gently* on a hotplate until the volume of nitric acid is reduced to about 5 mL. Then add about 20 mL of de-ionised water and boil gently again until the volume is approximately 10 mL. Cool the suspension, and filter through a Whatman No. 540 filter paper, washing the beaker and the filter paper with small portions of de-ionised water until a volume of about 25 mL is obtained. Transfer the filtrate to a 50 mL graduated flask and make up to the mark with de-ionised water.

(*c*) **Sample treatment for 'available' metals.** The metals zinc, copper, and nickel are phytotoxic and it is necessary to ascertain, in addition to total levels, the available amounts of these metals that can be taken up by plants. The procedure adopted is as follows. Add from a measuring cylinder about 25 mL of approximately 0.05 M EDTA solution to about 1 g of an accurately weighed soil sample. Shake the suspension mechanically for a period of about four hours, Continue with the filtering procedure described above under (*b*).

(*d*) **Analysis of total metals by flame AAS.** (*i*) **Lead.** Use a fuel-lean acetylene–air flame with either the 217.0 nm resonance line (for samples containing a low lead concentration) or the 283.3 nm resonance line. Standard lead solutions containing from 1 to 10 μg Pb mL^{-1} are suitable for the measurement at 217.0 nm, and from 10 to 30 μg Pb mL^{-1} at 283.3 nm (see Notes 1 and 2).

Notes. (1) If the lead concentration is too high to be measured directly using the 283.3 nm resonance line then further dilution of the sample solution is necessary.

(2) It is advisable to employ background correction (see Section 21.11), especially when the 217.0 nm line is used.

(*ii*) **Cadmium, copper, zinc and nickel.** These can all be determined using an acetylene–air flame with the appropriate resonance lines and working range for the standard solutions given in Table 21.4. Again, further dilution of the sample solutions and background correction may be necessary.

(*iii*) **Zinc, copper and nickel.** These can be determined as available metals using the conditions given in Table 21.4. The standard solutions in this case should contain 0.05 *M* EDTA.

(*iv*) **Arsenic in soil by hydride generation.** The procedure followed entails the use of an atomic absorption spectrometer such as the Thermo Electron 951 AA, equipped with a Thermo Electron AVA 440 (Atomic Vapour Accessory) or an equivalent instrument fitted with a hydride generation accessory. It is important that the atomic absorption spectrometer has a background correction system.

Reagents. All reagents should be AnalaR or SpectrosoL (BDH Ltd, Poole, Dorset, UK).

Sodium tetrahydroborate (III) (sodium borohydride), 1 per cent w/v. Dissolve sodium hydroxide pellets (5.0 g) in 300 mL of de-ionised water and cool. Add sodium tetrahydroborate(III) (5.0 g) directly to the sodium hydroxide solution and make up the total volume to 500 mL with de-ionised water. Shake the solution thoroughly and filter through a Whatman No. 541 filter paper. (The resulting solution is stable for at least one week.)

Hydrochloric acid, 4 M. Dilute SpectrosoL hydrochloric acid (365 mL) to 1 L using de-ionised water.

Arsenic standards. Prepare a 1 mg L^{-1} working standard solution from a 1000 mg L^{-1} SpectrosoL solution of arsenic trichloride in a 4 *M* hydrochloric acid.

Sample digestion procedure (aqua-regia method). Place an accurately weighed sample (about 1 g) of the soil into a Pyrex tube (50 mL) and a small quantity (2–3 mL) of de-ionised water to obtain a slurry. Then add SpectrosoL hydrochloric acid (7.5 mL) followed by SpectrosoL nitric acid (2.5 mL). Cover the tube with 'Clingfilm' overnight and then digest the sample in a Techne block digester for 2 hours under reflux conditions using a cold-finger condenser. Filter the cooled solution through a Whatman No. 540 filter paper into a 50 mL graduated flask, and wash the residue with warm nitric acid (2*M*). Make the filtrate up to the mark with de-ionised water.

Procedure. Follow the conditions recommended by the instrument manufacturers for the determination of arsenic by hydride generation. Typical instrumental parameters recommended by Thermo Electron are as follows:

AVA 440

Reagent setting	'2' (1 mL 1 per cent $NaBH_4$)
Argon cylinder pressure	20 psi
Argon flow rate	2.5 L min^{-1}
Hydrochloric acid concentration	$4M$
Reaction time	0.5 min

951 AA

Resonance line	193.7 nm
Dual beam	On
Background correction	Deuterium arc
Integration time	8 s
Mode	Peak height

To obtain the calibration standards, take aliquots ranging from 50 μL to 300 μL As, from the working standard solution, using an Eppendorf micropipette. Add the appropriate microlitre quantities to the reaction vessel of the vapour generation system, together with 10 mL of hydrochloric acid ($4M$), delivered from a calibrated dispenser.

Start the vapour generator cycle so that the absorption cell is flushed with argon gas and the pre-set volume of $NaBH_4$ (1 mL) is pumped into the sample vessel. After the pre-selected reaction time (0.5 minute), AsH_3 vapour is flushed into the absorption tube. Record the value of each arsenic signal as a peak height measurement. Read off the arsenic concentration of the sample, which is displayed on the instrument video screen.

FLAME EMISSION EXPERIMENTS

21.25 DETERMINATION OF ALKALI METALS BY FLAME PHOTOMETRY

Although flame emission measurements can be made by using an atomic absorption spectrometer in the emission mode, the following account refers to the use of a simple flame photometer (the Corning Model 410 flame photometer). Before attempting to use the instrument read the instruction manual supplied by the manufacturers.

Preparation of standard solutions for calibration curves. The following concentrations are suitable:

(*a*) *Sodium.* Dissolve 2.542 g sodium chloride in 1 L de-ionised water in a graduated flask. This solution contains the equivalent of 1.000 mg Na per mL (i.e. 1000 ppm). Dilute this stock solution to give four solutions containing 10, 5, 2.5, and 1 ppm of sodium ions.

(*b*) *Potassium.* Dissolve 1.909 g potassium chloride in 1 L de-ionised water. This solution contains the equivalent of 1.000 mg K per mL (i.e. 1000 ppm). Dilute this stock solution to give four solutions containing 20, 10, 5, and 2 ppm of potassium ions.

(*c*) *Calcium.* Dissolve 2.497 g calcium carbonate in a little dilute hydrochloric acid, and dilute to 1 L with de-ionised water. This stock solution contains the equivalent of 1.000 mg Ca per mL. Dilute this solution to give solutions containing 100, 50, 25, and 10 ppm of calcium ions.

(d) *Lithium*. Dissolve 5.324 g pure lithium carbonate in a little dilute hydrochloric acid and dilute to 1 L with de-ionised water. This solution contains 1.000 mg Li per mL (i.e. 1000 ppm). Dilute the stock solution to give solutions containing 20, 10, 5, and 2 ppm, of lithium ions.

Prepare calibration curves for each of the above four elements. With the aid of these calibration curves, carry out the following simple determinations.

1. *Potassium in potassium sulphate*. Weigh out accurately about 0.20 g potassium sulphate and dissolve it in 1 L de-ionised water. Dilute 10.0 mL of this solution to 100 mL, and determine the potassium with the flame photometer using the potassium filter.

2. *Potassium and sodium in a mixture*. Mix suitable volumes of the above stock solutions so that the resulting solution contains, say, 4–10 ppm Na and 10–15 ppm K. Determine the Na and K with the aid of the appropriate filters. Compare the results obtained with the true values.

3. *Sodium, potassium, and calcium in a mixture*. Mix appropriate volumes of the above stock solutions so that the test solution contains, say, 5 ppm Na, 10 ppm K, and 40 ppm Ca. Determine the Na, K, and Ca with the aid of the appropriate filters. Compare the results obtained with the true values.

4. *Calcium in calcium carbonate*. Determine the calcium in an analysed sample of dolomite. Dissolve about 0.38 g, accurately weighed in 1:1 hydrochloric acid, warm gently, filter through a quantitative filter paper, wash, dilute the combined filtrate and washings to 1 L. Measure the calcium content of the resulting solution: use a calcium filter. Compare the value for Ca thus obtained with the known Ca content.

Prepare a calibration curve for each element (see Section 21.16) and use this to evaluate the concentration of any unknown sample (see Note).

Note. The unknown solution may require dilution to give a reading on the calibration curve.

21.26 REFERENCES FOR PART F

1. I H Lambert, *Photometria de Mensura et Gradibus Luminus, Colorum et Umbrae*, Augsburg, 1760; reprinted in W Ostwald, *Klassiker der Exakten Wissenschaften*, No. 32, 64, 1892

2. M Bouguer, *Essai d'Optique sur la Gradation de la Lumière*. Paris, 1729; see also W Oswald *Klassiker der Exakten Wissenschaften*, No. 33, **58**, 1891. M Bouguer, *Traite d'Optique sur la Gradation de la Lumière*, ouvrage posthume, Pub. de Lacaille, 1760

3. A Beer, *Ann. Physik. Chem.* (*J. C. Poggendorff*), 1852, **86**, 78. See also H G Pfeiffer and H A Liebhafsky, *J. Chem. Educ.*, 1951, **23**, 123

4. F Bernard. *Ann. Chim. Phys.* 1852, **35**, 385

5. D R Malinin and J H Yoe, *J. Chem. Educ.*, 1961, **38**, 129

6. F H Lohman, *J. Chem. Educ.*, 1955, **32**, 155

7. J Talmi, *Appl. Spectrosc.*, 1982, **36**, 1

8. A T Giese and C S French, *Appl Spectrosc.*, 1955, **9**, 78

9. T C O'Haver, *Anal. Chem.*, 1979, **51**, 91A

10. J E Cahill and F C Padera, *Am. Lab.*, 1980, **12**(4), 101

11. T C O'Haver and G L Green, *Anal. Chem.*, 1976, **48**, 312

12. R A Alberty, *Physical Chemistry*, 7th edn, Wiley, New York, 1987
13. C N Banwell, *Fundamentals of Molecular Spectroscopy*, 3rd edn, McGraw-Hill, London, 1983
14. J M Hollas, *Modern Spectroscopy*, Wiley, Chichester, 1987
15. D L Pavia, G M Lampman and G S Kriz, *Introduction to Spectroscopy*, Holt, Rhinehart and Winston, New York, 1979
16. D H Williams and I Fleming, *Spectroscopic Methods in Organic Chemistry*, 4th edn, McGraw-Hill, London, 1987
17. D A Skoog, *Principles of Instrumental Analysis*, 3rd edn, CBS College Publishing, Philadelphia, 1984
18. J Coetzee (Ed), *Recommended Methods for Purification of Solvents and Tests for Impurities*, Pergamon, Oxford, 1982
19. C Burgess and A Knowles, *Standards in Absorption Spectrometry*, Chapman Hall, London, 1981
20. Analytical Methods Committee, *Determination of Arsenic in Organic Materials*, Society for Analytical Chemistry, London, 1960
21. J H Jones, *J. Assoc. Offic. Anal. Chemists.*, 1945, **28**, 398
22. J M Martin, Jr, C R Orr, C B Kincannon and J L Bishop, *J. Water Pollution Control*, 1967, **39**, 21
23. P B Sweetser and C E Bricker, *Anal. Chem.*, 1952, **24**, 1107
24. L E Hummelstedt and D N Hume, *Anal. Chem.*, 1960, **32**, 1792
25. J B Headridge, *Photometric Titrations*, Pergamon, Oxford 1961
26. *Wilson and Wilson Comprehensive Analytical Chemistry*, G Svehla (Ed), Vol VIII, Elsevier Science, Amsterdam, 1977
27. H Egan, R Sawyer and R S Kirk, *Pearson's Chemical Analysis of Foods*, 8th edn, Longman, Harlow, 1981, p 240
28. N Evcim and L A Reber, *Anal. Chem.*, 1954, **26**, 936
29. D F H Wallach, D M Surgenor, J Soderberg and E Delano, *Anal. Chem.*, 1959, **31**, 456
30. B L Kepner and D M Hercules, *Anal. Chem.*, 1963, **35**, 1238
31. H M von Hattingberg, W Klaus, H Lüllmann and S Zepf, *Experientia*, 1966, **2**, 553
32. B Buděšínsky and T S West, *Talanta*, 1969, **16**, 399
33. R A Chelmers and G A Wadds, *Analyst*, 1970, **95**, 234
34. R C Denney, *A Dictionary of Spectroscopy*, 2nd edn, Macmillan, London, 1982, p 89
35. J M Hollas, *High Resolution Spectroscopy*, Butterworths, London, 1982
36. R C J Osland, *Principles and Practices of Infrared Spectroscopy*, 2nd edn, Philips Ltd, 1985
37. P R Griffiths and J A de Haseth, *Fourier Transform Infrared Spectroscopy — Chemical Analysis Series*, Vol. 83, Wiley, Chichester, 1986
38. R C Denney, *Alcohol and Accidents*, Sigma Press, Wilmslow/Wiley, Chichester, 1986
39. Anon, *Lion Intoximeter 3000 — Operators' Handbook*, Lion Laboratories Ltd, Barry, 1982
40. W O George and P S McIntyre, *Infrared Spectroscopy — Analytical Chemistry by Open Learning*, ACOL, Thames Polytechnic/Wiley, Chichester, 1987, p 238
41. A Walsh, *Spectrochim. Acta*, 1955, **7**, 108
42. C S Rann and A N Hambly, *Anal. Chem.*, 1965, **37**, 879
43. D R Thomerson and K C Thompson, *Chemistry in Britain*, 1975, **11**, 316
44. T S West and X K Williams, *Anal. Chim. Acta*, 1969, **45**, 27
45. W R Hatch and W L Ott, *Anal. Chem.*, 1968, **40**, 2085
46. R M Dagnall, K C Thompson and T S West, *Talanta*, 1967, **14**, 551
47. C W Frank, W G Schrenk and C E McLoan, *Anal. Chem.*, 1966, **38**, 1005
48. V A Fassel, J A Rasmuson and T G Cowley, *Spectrochim. Acta*, 1968, **23B**, 579
49. J E Allen, *Spectrochim. Acta*, 1969, **24B**, 13
50. D C Manning and F Fernandex, *Atom. Absorption Newsletter*, 1968, **7**, 24
51. F J Fernandez, S A Myers and W Slavin, *Anal. Chem.*, 1980, **52**, 741
52. S Smith, R G Schleicher and G M Hieftje, New atomic absorption background

correction technique, Paper no. 442, 33rd Pittsburgh Conference on Analytical Chemistry and Applied Spectroscopy, Atlantic City, USA, 1982

53. D G Mitchell and A Johansson, *Spectrochim. Acta*, 1970, **25B**, 175
54. Anon., safety practices for atomic absorption spectrophotometers. *International Laboratory*, 1974, May/June, 63. International Scientific Communications Inc, Fairfield, Conn., USA

21.27 SELECTED BIBLIOGRAPHY FOR PART F

1. L C Thomas and G J Chamberlin (revised by G Shute), *Colorimetric Chemical Analytical Methods*, 9th edn, Tintometer Ltd, Salisbury, 1970
2. Y Talmi, *Multichannel Image Detectors*, Vol 1 (1979), Vol II (1982), American Chemical Society, Washington DC, USA
3. C T Cottrell, D Irish, V M Masters and J E Steward (Eds), *Introduction to Ultraviolet and Visible Spectrophotometry*, 2nd edn, Pye Unicam Ltd, Cambridge, 1985
4. Z Marczenko, *Separation and Spectrophotometric Determination of Elements*, 2nd edn, Wiley, Chichester, 1986
5. Y K Agarwal and G D Mehd, Solvent extraction and spectrophotometric methods for the determination of some toxic metal ions, *Rev. Anal. Chem.*, 1983, **6**(3), 185
6. E B Sandell and H Onishi, *Colorimetric Determination of Traces of Metals*, 4th edn, Interscience, New York, 1978
7. F D Snell, *Photometric and Fluorometric Methods of Analysis*, Parts 1/2, *Metals*; Part 3, *Non-metals*, Wiley, New York, 1978–1981
8. R M Silverstein, C C Bassler and T C Morrell, *Spectrometric Identification of Organic Compounds*, 4th edn, Wiley, New York, 1981
9. J A Howell and L Harges, Ultraviolet and light absorption spectrometry, *Anal. Chem.*, 1982, **54**, 171R
10. C Burgess and A Knowles, *Techniques in Visible and Ultraviolet Absorption Spectroscopy*, Chapman and Hall, London, 1981
11. A F Fell and G Smith, Higher derivative methods in ultraviolet, visible and infrared spectrophotometry, *Anal. Proc.*, 1982, **19**, 28
12. Wilson and Wilson *Comprehensive Analytical Chemistry*, G Svehla (Ed), Vol XIX (1986); Vol XX (1985); Vol XXII (1986), Elsevier Science Publishers, Amsterdam
13. P J Ewing, E J Meehan and I M Kolthoff (Eds), *Treatise on Analytical Chemistry*, Part 1, Vol 5, 1964, 2nd edn, Vol 7, 1981, Interscience, New York
14. H H Willard, L L Merritt, J R Dean and F A Settle, Instrumental methods of analysis, *Molecular Fluorescence and Phosphorescence Methods*, 6th edn, Van Nostrand Reinhold, New York, 1981, Chapter 5
15. R B Cundall and A Gilbert, *Photochemistry*, Nelson, London, 1970
16. J S Fritz and G H Schenk, *Quantitative Analytical Chemistry*, 3rd edn, Allyn and Bacon, Boston, 1974, Chapter 22, p 438
17. M Pinta, *Detection and Determination of Trace Elements*, Wiley, Chichester, 1974
18. G G Guilbault, *Fluorescence — Theory, Instrumentation and Practice*, London; Edward Arnold, Marcel Dekker, New York, 1967
19. C E White and R J Argauer, *Fluorescence Analysis — A Practical Approach*, Marcel Dekker, New York, 1970
20. C A Parker, *Photoluminescence of Solutions*, Elsevier, Amsterdam, 1968
21. S G Schulman, *Molecular Luminescence Spectroscopy*, Wiley, New York, 1985
22. J R Lakowicz, *Principles of Fluoresence Spectroscopy*, Plenum, New York, 1983
23. E L Wehry (Ed), *Modern Fluorescence Spectroscopy*, Vols 1–4, Plenum, New York, 1976–1981
24. Fluorimetric Analysis, *Anal. Chem.*, Biannual Reviews, alternate years
25. E F H Brittain, W O George and C H J Wells, *Introduction to Molecular*

Spectroscopy — Theory and Experiment, Academic Press, London and New York, 1975

26. P R Griffiths, *Fourier Transform Infrared Spectrometry*, 2nd edn, Wiley, New York, 1986
27. L J Bellamy, *The Infrared Spectra of Complex Molecules*, Vol I (1975); Vol II (1980), Chapman and Hall, London
28. J P Ferraro, and L J Basile, *Fourier Transform Infrared Spectroscopy*, Academic Press, London and New York, 1978
29. D H Williams and I Fleming, *Spectroscopic Methods in Organic Chemistry*, 4th edn, McGraw–Hill, London, 1987
30. L Ebdon, *An Introduction to Atomic Absorption Spectroscopy*, Heyden, London, 1982
31. W J Price, *Spectrochemical Analysis by Atomic Absorption*, 2nd edn, Heyden, London, 1979
32. G F Kirkbright and M Sargent, *Atomic Absorption and Fluorescence Spectroscopy*, 2nd edn, Academic Press London, 1975
33. W Slavin, *Atomic Absorption Spectroscopy*, 2nd edn, Wiley, Chichester, 1978
34. B Welz, *Atomic Absorption Spectroscopy*, Verlag Chemie, Berlin, 1976
35. M Thompson and J N Walsh, *A Handbook of Inductively Coupled Plasma Spectrometry*, Blackie, London, 1983
36. E Metcalfe, *Atomic Absorption and Emission Spectroscopy*, ACOL–Wiley, Chichester, 1987
37. D A Skoog, *Principles of Instrumental Analysis*, 3rd edn, Saunders College, London, 1985
38. G D Christian, *Analytical Chemistry*, 4th edn, Wiley, Chichester, 1986
39. D C Harris, *Quantitative Chemical Analysis*, Freeman, San Fransisco, 1982

APPENDICES

APPENDIX 1   RELATIVE ATOMIC MASSES, 1985

Element	Symbol	Atomic no.	Atomic weight	Element	Symbol	Atomic no.	Atomic weight
Actinium	Ac	89	(227)	Mercury	Hg	80	200.59
Aluminium	Al	13	26.981 539	Molybdenum	Mo	42	95.94
Americium	Am	95	(243)	Neodymium	Nd	60	144.24
Antimony	Sb	51	121.75	Neon	Ne	10	20.179 7
Argon	Ar	18	39.948	Neptunium	Np	93	(237)
Arsenic	As	33	74.921 59	Nickel	Ni	28	58.69
Astatine	At	85	(210)	Niobium	Nb	41	92.906 38
Barium	Ba	56	137.327	Nitrogen	N	7	14.006 74
Berkelium	Bk	97	(247)	Nobelium	No	102	(255)
Beryllium	Be	4	9.012 182	Osmium	Os	76	190.2
Bismuth	Bi	83	208.980 37	Oxygen	O	8	15.999 4
Boron	B	5	10.811	Palladium	Pd	46	106.42
Bromine	Br	35	79.904	Phosphorus	P	15	30.973 762
Cadmium	Cd	48	112.411	Platinum	Pt	78	195.08
Caesium	Cs	55	132.905 43	Plutonium	Pu	94	(244)
Calcium	Ca	20	40.078	Polonium	Po	84	(209)
Californium	Cf	98	(251)	Potassium	K	19	39.098 3
Carbon	C	6	12.011	Praseodymium	Pr	59	140.907 65
Cerium	Ce	58	140.115	Promethium	Pm	61	(145)
Chlorine	Cl	17	35.452 7	Protactinium	Pa	91	231.035
Chromium	Cr	24	51.996 1	Radium	Ra	88	226.025 4
Cobalt	Co	27	58.933 20	Radon	Rn	86	(222)
Copper	Cu	29	63.546	Rhenium	Re	75	186.207
Curium	Cm	96	(247)	Rhodium	Rh	45	102.905 50
Dysprosium	Dy	66	162.50	Rubidium	Rb	37	85.467 8
Einsteinium	Es	99	(254)	Ruthenium	Ru	44	101.07
Erbium	Er	68	167.26	Samarium	Sm	62	150.36
Europium	Eu	63	151.965	Scandium	Sc	21	44.955 910
Fermium	Fm	100	(257)	Selenium	Se	34	78.96
Fluorine	F	9	18.998 403 2	Silicon	Si	14	28.085 5
Francium	Fr	87	(223)	Silver	Ag	47	107.8682
Gadolinium	Gd	64	157.25	Sodium	Na	11	22.989 768
Gallium	Ga	31	69.723	Strontium	Sr	38	87.62
Germanium	Ge	32	72.61	Sulphur	S	16	32.066
Gold	Au	79	196.966 54	Tantalum	Ta	73	180.947 9
Hafnium	Hf	72	178.49	Technetium	Tc	43	(97)
Helium	He	2	4.002 602	Tellurium	Te	52	127.60
Holmium	Ho	67	164.930 32	Terbium	Tb	65	158.925 34
Hydrogen	H	1	1.007 94	Thallium	Tl	81	204.383 3
Iodine	I	53	126.904 47	Thulium	Tm	69	168.934 21
Indium	In	49	114.82	Thorium	Th	90	232.038 1
Iridium	Ir	77	192.22	Tin	Sn	50	118.710
Iron	Fe	26	55.847	Titanium	Ti	22	47.88
Krypton	Kr	36	83.80	Tungsten	W	74	183.85
Lanthanum	La	57	138.905 5	Uranium	U	92	238.028 9
Lawrencium	Lr	103	(260)	Vanadium	V	23	50.941 5
Lead	Pb	82	207.2	Xenon	Xe	54	131.29
Lithium	Li	3	6.941	Ytterbium	Yb	70	173.04
Lutetium	Lu	71	174.967	Yttrium	Y	39	88.905 85
Magnesium	Mg	12	24.305 0	Zinc	Zn	30	65.38
Manganese	Mn	25	54.938 05	Zirconium	Zr	40	91.224
Mendelevium	Md	101	(258)				

Notes:

(1) This table is scaled to the relative atomic mass $A_r(^{12}C) = 12$.

(2) Values in parentheses refer to the isotope of longest known half-life for radioactive elements.

(3) Information provided here is based mainly upon the Report of the Commission on Relative Atomic Masses, *Pure and Applied Chemistry*, 1986, **58**, 1678.

APPENDIX 2 INDEX OF ORGANIC CHEMICAL REAGENTS

The following list has been modified from one formerly published by Hopkin and Williams Ltd (now part of BDH Ltd). Although the authors have not checked all the individual reagents, the table has been included to enable the reader to find what reagents are available for the detection or determination of the more common ions.

The figures in the table refer to the literature items in the list on page 827 after the table. Italic figures indicate that the methods referred to are qualitative only. Some of the reagents and methods are sufficiently well established to appear in standard textbooks and some of these are cited where they seem appropriate.

Only one reference is generally given for each method and this has normally been selected from a number that existed, either because it appeared to be the most important or because it provides a key to earlier papers on the subject.

Reagent	Aluminium	Ammonium	Antimony	Arsenic	Barium	Beryllium	Bismuth	Boron	Cadmium	Caesium	Calcium	Cerium	Chloride, Chlorine	Chromium	Cobalt	Copper	Cyanide	Fluoride	Gallium	Germanium	Gold	Hafnium	Hydrogen sulphide	Indium	Iridium	Iron	Lanthanons	Lead	Lithium	Magnesium	Manganese	Mercury	Molybdenum
Alizarin fluorine blue																		2															
Alizarin fluorine blue lanthanum complex preparation																		184															
Alizarin red S	12					13												14															
Aluminon	1			159										31																			
4-Aminophenazone																																	
Ammonium pyrrolidine-dithiocarbamate			190	190			190		190					190	190	190					190					190						190	190
Arsenazo											161																182						
Arsenazo III																																	
Astrazone pink FG																																	
Barium chloranilate							7																										
Benzoin α-oxime																1																	1
5,6-Benzoquinoline									90										91														
N-Benzoyl-N-phenyl-hydroxylamine	2		2						2					2	2			2						2	2								
Beryllon II			23																														
2,2'-Bipyridyl																											47						43
2,2'-Biquinolyl							2																										
NN'-Bisallylthio-carbamoyl-hydrazine							25								131													24				26	
Biscyclohexanone oxalyldihydrazone																1																	
Bis-(3-methyl-1-phenyl-pyrazol-5-one)	2															2																	
Bithionol																										191							
4-Bromomandelic acid																																	
Bromopyrogallol red			193				193												193					193				193					
Cacotheline																																	
Cadion									30																					29			
Calmagite																																	
Carmine								2																									
Chloranilic acid												1																					
2-Chloro-4-nitro-benzenediazonium naphthalene-2-sulphonate																																	
N-Cinnamoyl-N-phenyl-hydroxylamine																																	
Cleve's acid																																	
Cupferron												36							32							33	36						
Copper(II) ethyl acetoacetate															188																		
Curcumin								2																									
Diacetyl dithiol* (see also Toluene-3,4-dithiol)			2	2			2		2							2															2	2	2
Diaminoethane-NN-di-(2-hydroxyphenylacetic acid)																										166							
1,1'-Dianthrimide							1											167															
Dibenzoyl dithiol (see also Toluene-3,4-dithiol)																																	
NN'-Dibenzyldithio-oxamide																																	
4-Diethylaminobenzyl-idene-rhodanine																			41														
Diethylammonium diethyldithiocarbamate																1																	
NN-Diethyl-p-phenylene-diamine sulphate													189																				
Di-(2-hydroxy-phenyl-imino)ethane	2								2		2																						

* For the use of diacetyl dithiol as a coagulant, catalyst and precipitant for sulphur and Group II sulphides, see Ref. 2.

APPENDIX 2   INDEX OF ORGANIC CHEMICAL REAGENTS

Reagent	Nickel	Niobium	Nitrate, Nitrite	Osmium	Palladium	Perchlorate	Phenols	Phosphate	Platinum	Potassium	Rhenium	Rhodium	Rubidium	Ruthenium	Scandium	Selenium	Silver	Sodium	Strontium	Sulphate	Sulphides, organic	Sulphur dioxide	Tantalum	Tellurium	Thallium	Thorium	Tin	Titanium	Tungsten	Uranium	Vanadium	Yttrium	Zinc	Zirconium
Alizarin fluorine blue																																		
Alizarin fluorine blue lanthanum complex preparation																																		
Alizarin red S																								15										16
Aluminon																																		
4-Aminophenazone							1																											
Ammonium pyrrolidine-dithiocarbamate	190															190							190	190		190								
Arsenazo																								182					182					
Arsenazo III															165											58				165				113
Astrazone pink FG																						17												
Barium chloranilate																				18														
Benzoin α-oxime																										1								
5,6-Benzoquinoline																																		
N-Benzoyl-N-phenyl-hydroxylamine	2	2													2				2				2	2	2			2	2					2
Beryllon II																																		
2,2'-Bipyridyl																																		
2,2'-Biquinolyl																																		
NN'-Bisallylthio-carbamoyl-hydrazine	131				28											27																	131	
Biscyclohexanone oxalyldihydrazone																																		
Bis-(3-methyl-1-phenyl-pyrazol-5-one)																																		
Bithionol																																		
4-Bromomandelic acid																																		2
Bromopyrogallol red		193													193	193										193					193			
Cacotheline																									1									
Cadion																																		
Calmagite																								5										
Carmine																																		
Chloranilic acid																	1																	1
2-Chloro-4-nitro-benzenediazonium naphthalene-2-sulphonate							170																											
N-Cinnamoyl-N-phenyl-hydroxylamine		2															2														2			
Cleve's acid			192																															
Cupferron	57																										34	33		61	35			
Copper(II) ethyl acetoacetate																																		
Curcumin																																		
Diacetyl dithiol* (see also Toluene-3,4-dithiol)					2				2					2									2					2	2					
Diaminoethane-NN-di-(2-hydroxyphenylacetic acid)																																		
1,1'-Dianthrimide																173								172										
Dibenzoyl dithiol (see also Toluene-3,4-dithiol)					2																													
NN'-Dibenzyldithio-oxamide					144				144																									
4-Diethylaminobenzyl-idene-rhodanine																																		
Diethylammonium diethyldithiocarbamate																																	1	
NN-Diethyl-p-phenylene-diamine sulphate																																		
Di-(2-hydroxy-phenyl-imino)ethane																															4			

*

Reagent	Aluminium	Ammonium	Antimony	Arsenic	Barium	Beryllium	Bismuth	Boron	Cadmium	Cæsium	Calcium	Cerium	Chloride, Chlorine	Chromium	Cobalt	Copper	Cyanide	Fluoride	Gallium	Germanium	Gold	Hafnium	Hydrogen sulphide	Indium	Iridium	Iron	Lanthanons	Lead	Lithium	Magnesium	Manganese	Mercury	Molybdenum
Dihydroxytartaric acid																																	
Dimercaptothiadiazole							63																										
4-Dimethylaminoazo-benzene-4'-arsonic acid																		43															
4-Dimethylaminobenzyl-idene-rhodanine																1			*43*														1
9-(4-Dimethylamino-phenyl)-2,3,7-tri-hydroxy-6-fluorone																		179															
4-Dimethylaminostyryl-β-naphthiazole methiodide																																	
2,9-Dimethyl-4,7-diphenyl-1,10-phenanthroline																2																	
Dimethylglyoxime						1																						1					
2,9-Dimethyl-1,10-phenanthroline																2																	
4,4'-Dinitrodiphenyl-carbazide							*43*																										
1,5-Diphenylcarbazide							1							1															1				
Diphenylcarbazone																																	1
NN-Diphenylhydrazine hydrochloride																																	
4,7-Diphenyl-1,10-phenanthroline																										2							
4,7-Diphenyl-1,10-phenanthrolinedisulphonic acid disodium salt																										160							
1,3-Diphenylpropane-1,3-dione																																	
Dithio-oxamide						1									96	1											*1*						*1*
Dithizone									2																				2				2
Eriochrome cyanine R	2																	119	117							117				118			
Formaldoxime hydrochloride																																9	
Furil α-dioxime																											53						
Gallein																																	
Hæmatoxylin	2																																
Hexanitrodiphenylamine										56																							
2-(Hydroxymercury)-benzoic acid																							155										
8-Hydroxyquinaldine						70			65	138					68			62			59						65	59			60	22	
8-Hydroxyquinoline	33								33	65					33			66						67			65			33			
Lanthanum chloranilate																		63															
Magnesium blue																														*185*			
Magnesium uranyl acetate																																	
Magnesons I and II																														1			
Mandelic acid																																	
Mercaptoacetic acid															152												49						
2-Mercaptobenzothiazole							79		78							77				76						73			75			74	
Mercury(II) chloranilate													157																				
3-Methoxynitrosophenol															2											2							
α-Methoxyphenylacetic acid																																	
Methylfluorone			1																														
Mordant red 74				19																													
Morin	84					85		88																									
Naphthalhydroxamic acid											89																						
Nickel uranyl acetate																																	
Nioxime																																	
Nitron																																	
4-p-Nitrophenylazo-orcinol				93																													
1-Nitroso-2-naphthol															33	176											176						

Reagent	Nickel	Niobium	Nitrate	Osmium	Palladium	Perchlorate	Phenols	Phosphate	Platinum	Potassium	Rhenium	Rhodium	Rubidium	Ruthenium	Scandium	Selenium	Silver	Sodium	Strontium	Sulphate	Sulphides, organic	Sulphur dioxide	Tantalum	Tellurium	Thallium	Thorium	Tin	Titanium	Tungsten	Uranium	Vanadium	Yttrium	Zinc	Zirconium
Dihydroxytartaric acid																	42																	
Dimercaptothiadiazole					64																													
4-Dimethylaminoazo-benzene-4'-arsonic acid																																		44
4-Dimethylaminobenzyl-idenerhodanine																	1																	
9-(4-Dimethylamino-phenyl)-2,3,7-tri-hydroxy-6-fluorone																							183											
4-Dimethylaminostyryl-β-naphthiazole methiodide																																	1	
2,9-Dimethyl-4,7-diphenyl-1,10-phenanthroline																																		
Dimethylglyoxime	1				1																													
2,9-Dimethyl-1,10-phenanthroline																																		
4,4'-Dinitrodiphenyl-carbazide																																		
1,5-Diphenylcarbazide																																		
Diphenylcarbazone																																		
NN-Diphenylhydrazine hydrochloride																*43*																		
4,7-Diphenyl-1,10-phenanthroline																																		
4,7-Diphenyl-1,10-phenanthrolinedi-sulphonic acid disodium salt																																		
1,3-Diphenylpropane-1,3-dione																														2				
Dithio-oxamide	96				*1*			*1*			48						*1*																	
Dithizone																	2																2	
Eriochrome cyanine R					\																													
Formaldoxime hydrochloride																																		
Furil α-dioxime	2				52			2	2																									
Gallein																																		
Haematoxylin																																		
Hexanitrodiphenylamine										55	56																							
2-(Hydroxymercury)-benzoic acid																	180																	
8-Hydroxyquinaldine											59														59	69	59				60			
8-Hydroxyquinoline	33	71																									102	178		67		33		
Lanthanum chloranilate							164																											
Magnesium blue																																		
Magnesium uranyl acetate																		33																
Magnesons I and II																																		
Mandelic acid																																		60
Mercaptoacetic acid																																		168
2-Mercaptobenzothiazole					80							74										79												
Mercury(II) chloranilate																																		
3-Methoxynitrosophenol																																		
α-Methoxyphenylacetic acid																	2																	
Methylfluorone																																		140
Mordant red 74																																		
Morin																														87				86
Naphthalhydroxamic acid																																		
Nickel uranyl acetate																	92																	
Nioxime	1				124																													
Nitron			1			1				1																	1							
4-p-Nitrophenylazo-orcinol																																		
1-Nitroso-2-naphthol	176				94																									176			176	

Reagent	Aluminium	Ammonium	Antimony	Arsenic	Barium	Beryllium	Bismuth	Boron	Cadmium	Caesium	Calcium	Cerium	Chloride, Chlorine	Chromium	Cobalt	Copper	Cyanide	Fluoride	Gallium	Germanium	Gold	Hafnium	Hydrogen sulphide	Indium	Iridium	Iron	Lanthanons	Lead	Lithium	Magnesium	Manganese	Mercury	Molybdenum
2-Nitroso-1-naphthol															95																		
Nitroso-R salt															2											2							
1,10-Phenanthroline hydrate							169								1											1							21
Pheolphthalein															98	97																	
Phenylarsonic acid							99																										
p-Phenylenediamine dihydrochloride																	145																
Phenylfluorone																					1												116
Phenyl-2-pyridyl ketoxime																										2							
Phenylthiohydantoic acid															50																		
Picrolonic acid										103																							
Potassium dibenzyldithiocarbamate															104																		
1-(2-Pyridylazo)-2-napththol									40							40								175	11	40					40	40	
4-(2-Pyridylazo)resorcinol															174												174						
2-(2-Pyridyl)-benzimidazole																										105							
2-(2-Pyridyl)-imidazoline																										105							
Quinaldic acid									33						33																		
Quinoxaline-2,3-dithiol															107																		
Rhodamine B			108													110	109																
Rhodamine S			54																														
Salicylaldehyde oxime							111								33													112					
Salicylideneamino-2-thiophenol																																	
Silver diethyldithiocarbamate				142																													
Sodium diethyldithiocarbamate																1																	
Sodium rhodizonate					43						43																	43					
Sodium tetraphenylboron		2								2																						2	
SPADNS																		122															
Stilbazo	123																							154									153
Tannic acid	128						128												129	125													
2,2':6',2''-Terpyridyl															141																		
1-(2-Thenoyl)-3,3,3-trifluoroacetone														51	6	82										72							
Thioacetamide	For the precipitation of analytical group II and IV metals. Ref. 2																																
Thiourea							148		147																							147	
Thiourone																										83							
Thorin				2														2												2			
Tiron																										2							2
Titan yellow																														1			
Toluene-3,4-dithiol (see also Zinc dithiol)			38	2			149		149						38	38			149	2	149			149	149	38		2				2	1
2,4,6-Tri(2-pyridyl)-1,3,5-triazine																										81							
Uranyl zinc acetate																																	
Victoria violet							137																										
2,4-Xylenol																																	
2,6-Xylenol																																	
Xylenol orange							135														20			134									
Zinc dibenzyldithiocarbamate																1																	
Zinc OO-di-iso-propyl phosphorodithioate																162																	
Zinc dithiol* (see also Toluene-3,4-dithiol)			2	2			2		2						2	2			2	2	2			2	2	2		2				2	2	2
Zincon																																2	

	Nickel	Niobium	Nitrate	Osmium	Palladium	Perchlorate	Phenols	Phosphate	Platinum	Potassium	Rhenium	Rhodium	Rubidium	Ruthenium	Scandium	Selenium	Silver	Sodium	Strontium	Sulphate	Sulphides, organic	Sulphur dioxide	Tantalum	Tellurium	Thallium	Thorium	Tin	Titanium	Tungsten	Uranium	Vanadium	Yttrium	Zinc	Zirconium
2-Nitroso-1-naphthol				94										37																				
Nitroso-R salt																																		
1,10-Phenanthroline hydrate				1																										1				
Phenolphthalein																																		
Phenylarsonic acid																											101							100
p-Phenylenediamine dihydrochloride																																		
Phenylfluorone																																		
Phenyl-2-pyridyl ketoxime																																		
Phenylthiohydantoic acid																																		
Picrolonic acid																																		
Potassium dibenzyldithiocarbamate																																		
1-(2-Pyridylazo)-2-naphthol										11																	158		40					139
4-(2-Pyridylazo)resorcinol		143																									174							
2-(2-Pyridyl)benzimidazole																																		
2-(2-Pyridyl)imidazoline																																		
Quinaldic acid				106																													33	
Quinoxaline-2,3-dithiol	107			181																														
Rhodamine B																	3								171									
Rhodamine S																																		
Salicylaldehyde oxime	33																																111	
Salicylideneamino-2-thiophenol																								187										
Silver diethyldithiocarbamate																																		
Sodium diethyldithiocarbamate																																	1	
Sodium rhodizonate																	*43*		114															
Sodium tetraphenylboron							2			2															2									
SPADNS																										120								121
Stilbazo																																		
Tannic acid		126															126								126	130								127
2,2':6',2''-Terpyridyl																																		
1-(2-Thenoyl)-3,3,3-trifluoroacetone	8																													39				
Thioacetamide	For the precipitation of analytical group II and IV metals. Ref. 2.																																	
Thiourea																																		
Thiurone																																		
Thorin																				177							2							2
Tiron																												2						
Titan yellow																																		
Toluene-3,4-dithiol (see also Zinc dithiol)	2		*149*	2				2		2	*149*	2		136										*149*	38		1	1			38		2	
2,4,6-Tri(2-pyridyl)-1,3,5-triazine																																		
Uranyl zinc acetate																		33																
Victoria violet																																		
2,4-Xylenol			2																															
2,6-Xylenol			186																															
Xylenol orange		10																															133	132
Zinc dibenzyldithiocarbamate																																		
Zinc OO-di-iso-propyl phosphorodithioate																																		
Zinc dithiol* (see also Toluene-3,4-dithiol)	2			2	2			2		2		2		2										2	2		2	2			156		2	
Zincon																																		2

* Zinc dithiol is more stable than toluene-3,4-dithiol. It can be used to prepare a solution of toluene-3,4-dithiol and for many purposes it is added as the solid or as a suspension in ethanol (see Ref. 2).

LIST OF LITERATURE REFERENCES. KEY TO APPENDIX 2

1. *Organic Reagents for Metals and other Reagent Monographs*, Vol. 1, 5th edn, 1955. Hopkin & Williams.
2. *Organic Reagents for Metals and for Certain Radicals*, Vol. II, 1964. Hopkin & Williams.
3. *Anal. Chem.*, **33**, 1128 (1961).
4. *Analyst*, **87**, 703 (1962).
5. *Anal. Chim. Acta*, **26**, 487 (1962).
6. *Anal. Chim. Acta*, **27**, 153 (1962).
7. *Anal. Chem.*, **34**, 209 (1962).
8. *Anal. Chim. Acta*, **27**, 591 (1962).
9. *Anal. Chim. Acta*, **27**, 331 (1962).
10. *Talanta*, **9**, 987 (1962).
11. *Anal. Chem.*, **35**, 149 (1963).
12. *Ind. Eng. Chem. (Anal.)*, **15**, 57 (1943); *J.S.C.I.*, **62**, 187 (1943).
13. *Analyst*, **73**, 395 (1948).
14. *Analyst*, **87**, 197 (1962).
15. *Anal. Chim. Acta*, **13**, 142 (1955).
16. *Analyst*, **87**, 880 (1962).
17. *Analyst*, **80**, 901 (1955).
18. *Anal. Chem.*, **29**, 281 (1957).
19. *Analyst*, **94**, 262 (1969).
20. *Mikrochim. Acta*, 29 (1962).
21. *Anal. Chim. Acta*, **26**, 326 (1962).
22. *Anal. Chem.*, **34**, 571 (1962).
23. *Anal. Abstr.*, **4**, 1448 (1957).
24. *Anal. Chim. Acta*, **15**, 21, 102 (1956).
25. *J. Indian Chem. Soc.*, **28**, 89 (1952); *Chem. Abstr.*, **45**, 8394 (1951).
26. *Anal. Abstr.*, **4**, 3870 (1957).
27. *Anal. Abstr.*, **4**, 3858 (1957).
28. *Anal. Chim. Acta*, **19**, 202 (1958).
29. *J. Proc. Austral. Chem. Inst.*, **3**, 184 (1936); *Brit. Abstr.*, A. I. 1082 (1936).
30. *Anal. Chim. Acta*, **19**, 377 (1958).
31. *Anal. Chim. Acta*, **18**, 546 (1958).
32. *Z. anal. Chem.*, **140**, 245 (1953); *Anal. Abstr.*, **1**, 659 (1954).
33. A. I. Vogel (1961). *Text-Book of Quantitative Inorganic Analysis*, 3rd edn, Longman.
34. *J. Res. Nat. Bur. Stand.*, **33**, 307 (1944); *Brit. Abstr.*, C, 91 (1945).
35. *Ann. Chim., Roma*, **43**, 730 (1953); *Anal. Abstr.*, **1**, 471 (1954).
36. *Anal. Chem.*, **26**, 883 (1954).
37. *Anal. Chem.*, **34**, 94 (1962).
38. *Analyst*, **82**, 177 (1957).
39. *Analyst*, **85**, 376 (1960)
40. *Anal. Chim. Acta*, **25**, 348 (1961).
41. *Anal. Chem.*, **23**, 653 (1951).
42. *J. Inst. Petroleum Tech.*, **19**, 845 (1933); *Brit. Abstra.*, B, 1088 (1933).
43. F. Feigl (1958). *Spot Tests in Inorganic Analysis*, 5th edn, Elsevier.
44. (a) *Ind. Eng. Chem. (Anal.)*, **13**, 603 (1941). (b) *Anal. Abstr.*, **4**, 2935 (1957).
45. *Talanta*, **8**, 579 (1961).
46. *Talanta*, **5**, 231 (1960).
47. *Aanalyst*, **85**, 823 (1960).
48. *Anal. Chem.*, **22**, 1281 (1950).
49. *A Handbook of Colorimetric Chemical Analytical Methods*, 6th edn, The Tintometer Ltd.
50. A. I. Vogel (1951). *Text-Book of Quantitative Inorganic Analysis*, 2nd edn, Longman.
51. *Anal. Chem.*, **32**, 1337 (1960).
52. *Anal. Chem.*, **27**, 1932 (1955).
53. 'ANALAR' *Standards for Laboratory Chemicals*, 5th edn, 1957. Hopkin & Williams and The British Drug Houses Ltd, p. 71.
54. *Analyst*, **95**, 131 (1970).
55. (a) *Anal. Chem.*, **25**, 808 (1953). (b) *Analyst*, **80**, 768 (1955).
56. *Brit. Abstr.*, A. I. 315 (1943).
57. *Anal. Chim. Acta*, **19**, 18 (1958).
58. *Anal. Chim. Acta*, **30**, 176 (1964).
59. (a) *Anal. Chem.*, **24**, 1033 (1952). (b) *Anal. Abstr.*, **4**, 2135 (1957).
60. R. Belcher and C. L. Wilson (1956). *New Methods of Analytical Chemistry*, Chapman & Hall.
61. (a) *Anal. Abstr.*, **6**, 147, 916 (1959). (b) *Anal. Chim. Acta*, **19**, 576 (1958).
62. *Anal. Abstr.*, **6**, 869 (1959).
63. *Analyst (abstr)*, **70**, 189 (1945).
64. *Anal. Chim. Acta*, **19**, 372 (1958).
65. *Science*, **125**, 1042 (1957); *Anal. Abstr.*, **5**, 2560 (1958).
66. *Z. anal. Chem.*, **140**, 252 (1953); *Anal. Abstr.*, **1**, 658 (1954).
67. *Anal. Abstr.*, **4**, 2135 (1957).
68. *Anal. Chim. Acta*, **16**, 121 (1957).
69. *Anal. Abstr.*, **4**, 1494 (1957).
70. *Anal. Abstr.*, **4**, 389 (1957).
71. *Analyst*, **82**, 630 (1957).
72. *Anal. Chim. Acta*, **22**, 223 (1960).
73. *Anal. Chem.*, **23**, 514 (1951).
74. *Anal. Abstr.*, **4**, 386 (1957).
75. *Anal. Chim. Acta*, **12**, 218 (1955).
76. *Anal. Abstr.*, **4**, 1444 (1957).
77. *Z. anal. Chem.*, **102**, 24 (1935).
78. *Z. anal. Chem.*, **102**, 108 (1935).
79. *Z. anal. Chem.*, **104**, 88 (1936).
80. (a) *Z. anal. Chem.*, **162**, 96 (1958); *Anal. Abstr.*, **6**, 944 (1959). (b) *Anal. Chim. Acta*, **20**, 379 (1959).
81. *Anal. Chem.*, **32**, 1117 (1960).
82. *Z. anal. Chem.*, **171**, 241 (1959); *Anal. Abstr.*, **7**, 3135 (1960).
83. *Analyst*, **89**, 707 (1964).
84. *Ind. Eng. Chem. (Anal.)*, **12**, 229 (1940).
85. *Anal. Chem.*, **33**, 1671 (1961).
86. *Talanta*, **9**, 749 (1961).
87. *Anal. Abstr.*, **3**, 1344 (1956).
88. *Anal. Chim. Acta*, **3**, 481 (1949).
89. *Analyst*, **85**, 889 (1960).
90. *Analyst*, **58**, 667 (1933).
91. *Ind. Eng. Chem. (anal.)*, **16**, 322 (1944).

92. (*a*) *Analyst*, **56**, 245 (1931). (*b*) *Anal. Abstr.*, **3**, 3271 (1956).
93. *Anal. Chem.*, **28**, 1728 (1956).
94. *Chem. Abstr.*, **45**, 69 (1951).
95. *Anal. Chim. Acta*, **20**, 340 (1959).
96. *Anal. Chim. Acta*, **20**, 332 (1959).
97. *Bull. Soc. chim. Belg.*, **54**, 186 (1945); *Brit. Abstr.*, C, 2 (1947).
98. *J. Chem. Soc. Japan*, **66**, 37 (1945); *Chem. Abstr.*, **43**, 1682 (1949).
99. (*a*) *J. Indian Chem. Soc.*, **21**, 119 (1944); *Analyst Abstr.*, **69**, 383 (1944); (*b*) *J. Indian Chem. Soc.*, **21**, 187, 188 (1944); *Brit. Abstr.*, C, 92 (1945).
100. *Zavod Lab.*, **11**, 254 (1945); *Chem. Abstr.*, **40**, 1418 (1946).
101. *Anal. Chim. Acta*, **21**, 58 (1959).
102. *Analyst*, **76**, 485 (1951).
103. *Analyst*, **76**, 482 (1951).
104. *Analyst*, **79**, 548 (1954).
105. *Anal. Chem.*, **26**, 217 (1954).
106. *Anal. Abstr.*, **6**, 186, 187, 188 (1959).
107. *Anal. Chem.*, **35**, 33 (1963).
108. *Anal. Chem.*, **31**, 1783 (1959).
109. *Anal. Chim. Acta*, **13**, 154 (1955).
110. *Annal. Chim. Acta*, **13**, 159 (1955).
111. *Ind. Eng. Chem. (Anal.)*, **12**, 663 (1940).
112. *Ind. Eng. Chem. (Anal.)*, **14**, 359 (1942).
113. *Talanta*, **8**, 673 (1961).
114. *Brit. Abstr.*, C, 39 (1944).
115. *Talanta*, **8**, 293 (1961).
116. *Anal. Chem.*, **33**, 431 (1961).
117. *Anal. Abstr.*, **6**, 871 (1959).
118. *Metallurgia*, **44**, 207 (1951); *Brit. Abstr.*, C, 134 (1952).
119. *Anal. Chem.*, **22**, 918 (1950).
120. *Anal. Chim. Acta*, **23**, 351 (1960).
121. *Anal. Chim. Acta*, **16**, 62 (1957).
122. *Anal. Chim. Acta*, **13**, 409 (1955).
123. *Anal. Chim. Acta*, **24**, 294 (1961).
124. *Anal. Abstr.*, **5**, 1530 (1958).
125. *Anal. Chim. Acta*, **2**, 254 (1948); see also Ref. 60.
126. *Analyst*, **80**, 380 (1955).
127. *Analyst*, **74**, 505 (1949); **75**, 555 (1950).
128. *Ind. Eng. Chem. (Anal.)*, **16**, 598 (1944).
129. *Anal. Chim. Acta*, **3**, 324 (1949).
130. *Analyst*, **70**, 124 (1945).
131. *Mikrochim. Acta*, 571 (1961).
132. *Talanta*, **2**, 266 (1959).
133. *Talanta*, **8**, 203 (1961).
134. *Talanta*, **3**, 81 (1959).
135. *Talanta*, **8**, 753 (1961).
136. *Anal. Chem.*, **33**, 445 (1961).
137. *Anal. Chem.*, **31**, 1102 (1959).
138. *Anal. Chem.*, **33**, 239 (1961).
139. *Anal. Chem.*, **33**, 125 (1961).
140. *Z. anal. Chem.*, **178**, 352 (1961); *Anal. Abstr.*, **8**, 3225 (1961).
141. *Anal. Chem.*, **26**, 1968 (1954).
142. *Analyst*, **88**, 380 (1963).
143. *Talanta*, **10**, 1013 (1963).
144. *Talanta*, **9**, 761 (1962).
145. *Talanta*, **11**, 621 (1964).
146. *Anal. Chim. Acta*, **9**, 86 (1953).
147. *Anal. Abstr.*, **4**, 1424 (1957).
148. *Angew. Chem.*, **64**, 608 (1952); *Brit. Abstr.*, C, 246 (1953).
149. *Analyst*, **83**, 396 (1958).
150. *Anal. Chem.*, **25**, 1125 (1953).
151. *Analyst*, **58**, 667 (1933); see also ref. 150.
152. *Anal. Chem.*, **33**, 1933 (1961).
153. *Anal. Abstr.*, **9**, 106 (1962).
154. *Anal. Abstr.*, **9**, 58 (1962).
155. *Analyst*, **83**, 314 (1958).
156. *Analyst*, **84**, 16 (1959).
157. *Anal. Chem.*, **29**, 1187 (1957) (for chloride).
158. *Anal. Chim. Acta*, **22**, 479 (1960).
159. *Anal. Chim. Acta*, **22**, 413 (1960).
160. *Talanta*, **7**, 163 (1961).
161. *Zavod. Lab.*, **27**, 803 (1961); *Anal. Abstr.*, **9**, 558 (1962).
162. *Analyst*, **86**, 407 (1961).
163. *Talanta*, **4**, 126 (1960).
164. *Talanta*, **4**, 244 (1960).
165. *Talanta*, **11**, 1 (1964).
166. *Anal. Chem.*, **30**, 44 (1958).
167. *Anal. Chim. Acta*, **21**, 370 (1959).
168. *Talanta*, **3**, 95 (1959).
169. *Anal. Chim. Acta*, **24**, 167 (1961).
170. (*a*) *Helv. Chim. Acta*, **31**, 320 (1948). (*b*) *Analyst*, **74**, 274 (1949).
171. *Analyst*, **83**, 516 (1958).
172. *Anal. Chim. Acta*, **23**, 175 (1960).
173. *Anal. Chim. Acta*, **23**, 565 (1960).
174. *Anal. Chim. Acta*, **20**, 26 (1959).
175. *Anal. Chim. Acta*, **23**, 434 (1960).
176. *Anal. Chem.*, **32**, 1350 (1960).
177. *Anal. Chim. Acta*, **23**, 538 (1960).
178. *Anal. Chem.*, **32**, 1083 (1960).
179. *Talanta*, **8**, 453 (1961).
180. *Analyst*, **86**, 543 (1961).
181. *Anal. Chem.*, **31**, 1985 (1959).
182. *Anal. Chim. Acta*, **26**, 538 (1962).
183. *Zavod. Lab.*, **2B**, 1283 (1957); *Chem. Abstr.*, **53**, 1387c (1959).
184. *Anal. Chim. Acta*, **45**, 341 (1969).
185. *Mikrochim. Acta*, 512 (1961).
186. *Organic Chemical Reagents*, Monograph No. 70, 1967. Hopkin & Williams.
187. *Organic Chemical Reagents*. Monograph No. 71, 1967. Hopkin & Williams.
188. *Analyst*, **91**, 282 (1966).
189. *J. Am. Water Wks. Assocn.*, **49**, 873 (1957); **51**, 15045c (1957) (for chlorine).
190. Monograph No. 74, 1969. Hopkin & Williams.
191. *Anal. Chim. Acta*, **47**, 151 (1969).
192. 'ANALAR' *Standards for Laboratory Chemicals*, 6th edn., 1967. 'ANALAR' Standards Ltd, p. 617.
193. Monograph No. 75, 1969. Hopkin & Williams.

APPENDIX 3   CONCENTRATIONS OF AQUEOUS SOLUTIONS OF THE COMMON ACIDS AND OF AQUEOUS AMMONIA

Reagent	Approximate			
	Per cent by weight	Specific gravity	Molarity	Vol. required to make 1 L of approx. 1 M solution (mL)
Hydrochloric acid	35	1.18	11.3	89
Nitric acid	70	1.42	16.0	63
Sulphuric acid	96	1.84	18.0	56
Perchloric acid	70	1.66	11.6	86
Hydrofluoric acid	46	1.15	26.5	38
Phosphoric(V) acid	85	1.69	13.7	69
Acetic (ethanoic) acid	99.5	1.05	17.4	58
Aqueous ammonia	27(NH_3)	0.90	14.3	71

APPENDIX 4   SATURATED SOLUTIONS OF SOME REAGENTS AT 20°C

Reagent	Formula	Specific gravity	Molarity	Quantities required for 1 L of saturated solution	
				Grams of reagent	mL of water
Ammonium chloride	NH_4Cl	1.075	5.44	291	784
Ammonium nitrate	NH_4NO_3	1.312	10.80	863	449
Ammonium oxalate	$(NH_4)_2C_2O_4,H_2O$	1.030	0.295	48	982
Ammonium sulphate	$(NH_4)_2SO_4$	1.243	4.06	535	708
Barium chloride	$BaCl_2,2H_2O$	1.290	1.63	398	892
Barium hydroxide	$Ba(OH)_2$	1.037	0.228	39	998
Barium hydroxide	$Ba(OH)_2,8H_2O$	1.037	0.228	72	965
Calcium hydroxide	$Ca(OH)_2$	1.000	0.022	1.6	1000
Mercury(II) chloride	$HgCl_2$	1.050	0.236	64	986
Potassium chloride	KCl	1.174	4.00	298	876
Potassium chromate	K_2CrO_4	1.396	3.00	583	858
Potassium dichromate	$K_2Cr_2O_7$	1.077	0.39	115	962
Potassium hydroxide	KOH	1.540	14.50	813	727
Sodium acetate	CH_3COONa	1.205	5.67	465	740
Sodium carbonate	Na_2CO_3	1.178	1.97	209	869
Sodium carbonate	$Na_2CO_3,10H_2O$	1.178	1.97	563	515
Sodium chloride	$NaCl$	1.197	5.40	316	881
Sodium hydroxide	$NaOH$	1.539	20.07	803	736

APPENDIX 5   SOURCES OF ANALYSED SAMPLES

Throughout this book the use of a number of standard analytical samples is recommended in order that practical experience may be gained on substances of known composition. In addition, standard reference materials of environmental samples for trace analysis are used for calibration standards, and pure organic compounds are employed as standard materials for elemental analysis.

(a) The Bureau of Analytical Samples Ltd (BAS), Newham Hall, Newby, Middlesborough, Cleveland, UK, supplies samples suitable for metallurgical, chemical and spectroscopic analysis. A detailed list of British Chemical Standard (BCS) and EURONORM Certified Reference Materials (ECRM) is available. The Bureau of Analytical Samples distributes in the UK from the following overseas sources.

Alcan International, Arvida Laboratories, Canada.
Bundesanstalt für Materialforschung und prüfung (BAM), Federal Republic of Germany.
Canada Centre for Mineral and Energy Technology (CANMET), Canada.
Centre Technique des Industries de la Fonderie (CTIF), France.
Institut de Recherches de la Sidérurgie Francaise (IRSID), France.
National Bureau of Standards (NBS), USA.
Research Institute CKD, Czechoslovakia.
South African Bureau of Standards (SABS), South Africa.
Swedish Institute for Metal Research (Jernkontoret), Sweden.
SKF Steel (SKF), Sweden.
Vasipari Kutato es Fejleszto Vallalat (VASKUT), Hungary.

(b) The Community Bureau of Reference (BCR), 200 rue de la Loi, B-1049 Brussels, Belgium, supples geological, environmental, organic compounds for elemental analysis and artificated samples for trace metal analysis.
(c) In the USA a wide range of standards may be obtained from the US Department of Commerce, National Bureau of Standards, Washington, DC, 20234, USA.
(d) Environmental reference materials are available from the National Institute for Environmental Studies, Yatabe-machi Tsukuba, Ibarasi 305, Japan.
(e) Elements and compounds of high purity and known composition are marketed by Johnson Matthey Chemicals, Ltd, Orchard Road, Royston, Hertfordshire, UK.

APPENDIX 6   BUFFER SOLUTIONS AND SECONDARY pH STANDARDS

The British standard for the pH scale is a $0.05M$ solution of potassium hydrogenphthalate (British Standard 1647: 1984, Parts 1, 2) which has a pH of 4.001 at 20°C. The IUPAC standard is $0.05m$ potassium hydrogenphthalate: see Section 15.15.

Subsidiary pH standards at 25°C include:

	pH
$0.05M$ HCl $+ 0.09M$ KCl	2.07
$0.1M$ Potassium tetroxalate	1.48
$0.1M$ Potassium dihydrogen citrate	3.72
$0.1M$ Acetic acid $+ 0.1M$ sodium acetate	4.64
$0.01M$ Acetic acid $+ 0.01M$ sodium acetate	4.70
$0.01M$ $KH_2PO_4 + 0.01M$ Na_2HPO_4	6.85
$0.05M$ Borax	9.18
$0.025M$ $NaHCO_3 + 0.025M$ Na_2CO_3	10.00
$0.01M$ Na_3PO_4	11.72

The following table covering the pH range 2.6–12.0 (18°C) is included as an example of a **universal buffer mixture**.

A mixture of 6.008 g of citric acid, 3.893 g of potassium dihydrogenphosphate, 1.769 g of boric acid, and 5.266 g of pure diethylbarbituric acid is dissolved in water and made up to 1 L. The pH values at 18°C of mixtures of 100 mL of this solution with various volumes (X) of $0.2M$ sodium hydroxide solution (free from carbonate) are tabulated below.

pH	X(mL)	pH	X(mL)	pH	X(mL)
2.6	2.0	5.9	36.5	9.0	72.7
2.8	4.3	6.0	38.9	9.2	74.0
3.0	6.4	6.2	41.2	9.4	75.9
3.2	8.3	6.4	43.5	9.6	77.6
3.4	10.1	6.6	46.0	9.8	79.3
3.6	11.8	6.8	48.3	10.0	80.8
3.8	13.7	7.0	50.6	10.2	82.0
4.0	15.5	7.2	52.9	10.4	82.9
4.2	17.6	7.4	55.8	10.6	83.9
4.4	19.9	7.6	58.6	10.8	84.9
4.6	22.4	7.8	61.7	11.0	86.0
4.8	24.8	8.0	63.7	11.2	87.7
5.0	27.1	8.2	65.6	11.4	89.7
5.2	29.5	8.4	67.5	11.6	92.0
5.4	31.8	8.6	69.3	11.8	95.0
5.6	34.2	8.8	71.0	12.0	99.6

For many purposes an exact pH value is not required; it suffices if the pH of the solution lies within an appropriate pH range and examples will be found in the text of details of buffer solutions required for particular procedures.

The following are illustrative of buffer solutions covering a range of pH values.

	pH range
1. Hydrochloric acid–sodium citrate	1.0– 5.0
2. Citric acid–sodium citrate	2.5– 5.6
3. Acetic acid–sodium acetate	3.7– 5.6
4. Disodium hydrogenorthophosphate–sodium dihydrogenorthophosphate	6.0– 9.0
5. Aqueous ammonia–hydrochloric acid	8.2–10.2
6. Sodium tetraborate–sodium hydroxide	9.2–11.0

The simplest method of preparing such solutions is to take a $0.1\,M$ solution of the acid component, and then to add $0.1M$ sodium hydroxide solution until the appropriate pH value is attained (use a pH meter); for solutions 1 and 5, $0.1\,M$ hydrochloric acid will be added to the second component.

APPENDIX 7a   DISSOCIATION CONSTANTS OF SOME ACIDS IN WATER AT 25°C

Dissociation constants are expressed as $pK_a (= -\log K_a)$.

Acid	pK_a	Acid		pK_a
Aliphatic acids				
Formic (methanoic)	3.75	Succinic	K_1	4.21
Acetic (ethanoic)	4.76		K_2	5.64
Propanoic	4.88	Glutaric	K_1	4.34
Butanoic	4.82		K_2	5.27
3-Methylpropanoic	4.85	Adipic	K_1	4.43
Pentanoic	4.84		K_2	5.28
Fluoroacetic	2.58	Methylmalonic	K_1	3.07
Chloroacetic	2.86		K_2	5.87
Bromoacetic	2.90	Ethylmalonic	K_1	2.96
Iodoacetic	3.17		K_2	5.90
Cyanoacetic	2.47	Dimethylmalonic	K_1	3.15
Diethylacetic	4.73		K_2	6.20
Lactic	3.86	Diethylmalonic	K_1	2.15
Pyruvic	2.49		K_2	7.47
Acrylic	4.26	Fumaric	K_1	3.02
Vinylacetic	4.34		K_2	4.38
Tetrolic	2.65	Maleic	K_1	1.92
trans-Crotonic	4.69		K_2	6.23
Furoic	3.17	Tartaric	K_1	3.03
Oxalic K_1	1.27		K_2	4.37
K_2	4.27	Citric	K_1	3.13
Malonic K_1	2.85		K_2	4.76
K_2	5.70		K_3	6.40
Aromatic acids				
Benzoic	4.20			
Phenylacetic	4.31	2-Benzoylbenzoic		3.54
Sulphanilic	3.23	Phthalic K_1		2.95
Phenoxyacetic	3.17	K_2		5.41
Mandelic	3.41	*cis*-Cinnamic		3.88
1-Naphthoic	3.70	*trans*-Cinnamic		4.44
2-Naphthoic	4.16	Phenol		10.00
1-Naphthylacetic	4.24	1-Nitroso-2-naphthol		7.77
2-Naphthylacetic	4.26	2-Nitroso-1-naphthol		7.38

	ortho (2-)	meta (3-)	para (4-)
Fluorobenzoic	3.27	3.86	4.14
Chlorobenzoic	2.94	3.83	3.98
Bromobenzoic	2.85	3.81	3.97
Iodobenzoic	2.86	3.85	3.93
Hydroxybenzoic	3.00	4.08	4.53
Methoxybenzoic	4.09	4.09	4.47
Nitrobenzoic	2.17	3.49	3.42
Aminobenzoic	4.98	4.79	4.92
Toluic	3.91	4.24	4.34
Chlorophenol	8.48	9.02	9.38
Nitrophenol	7.23	8.40	7.15
Methylphenol (cresol)	10.29	10.09	10.26
Methoxyphenol	9.98	9.65	10.21

Acid		pK_a	Acid		pK_a
Inorganic acids					
Arsenious		9.22	Nitrous		3.35
Arsenic	K_1	2.30	Phosphoric(V)	K_1	2.12
	K_2	7.08		K_2	7.21
	K_3	9.22		K_3	12.30
Boric		9.24	Phosphorous	K_1	1.8
Carbonic	K_1	6.37	(Phosphonic)	K_2	6.15
	K_2	10.33	Sulphuric	K_1	1.92
Hydrocyanic		9.14	Sulphurous	K_1	1.92
Hydrofluoric		4.77		K_2	7.20
Hydrogen sulphide	K_1	7.00	Thiosulphuric	K_1	1.7
	K_2	14.00		K_2	2.5
Hypochlorous {Chloric(I)}		7.25			

APPENDIX 7b   ACIDIC DISSOCIATION CONSTANTS OF SOME BASES IN WATER AT 25°C

The data for bases are expressed as acidic dissociation constants, e.g. for ammonia, the value $pK_a = 9.24$ is given for the ammonium ion:

$$NH_4^+ + H_2O \rightleftharpoons NH_3 + H_3O^+$$

Otherwise expressed, bases are considered from the standpoint of the ionisation of the conjugated acids. The basic dissociation constant for the reaction

$$NH_3 + H_2O \rightleftharpoons NH_4^+ + OH^-$$

may then be obtained from the relation:

$$pK_a \text{ (acidic)} + pK_b \text{ (basic)} = pK_w \text{ (water)}$$

where pK_w is 14.00 at 25°C.* For simplicity, the name of the base will be expressed in the 'basic' form, e.g. ammonia for ammonium ion, propylamine for propylammonium ion, piperidine for piperidinium ion, aniline for anilinium ion, etc., although it is appreciated that this is not strictly correct: no difficulty should be experienced in writing down the correct name, if required.

Base	pK_a	Base	pK_a
Ammonia	9.24	Hydrazine	7.93
Methylamine	10.64	Hydroxylamine	5.82
Ethylamine	10.63	Benzylamine	9.35
Propylamine	10.57	Aniline	4.58
Butylamine	10.62	o-Toluidine	4.39
Cyclohexylamine	10.64	m-Toluidine	4.68
Dimethylamine	10.77	p-Toluidine	5.09
Diethylamine	10.93	2-Chloroaniline	2.62
Monoethanolamine	9.50	3-Chloroaniline	3.32
Triethanolamine	7.77	4-Chloroaniline	3.81
Trimethylamine	9.80	N-Methylaniline	4.85
Triethylamine	10.72	NN-Dimethylaniline	5.15
Tris(hydroxymethyl)aminomethane	8.08	Pyridine	5.17
Piperidine	11.12	2-Methylpyridine	5.97
Ethylenediamine K_1	7.50	3-Methylpyridine	5.68
K_2	10.09	4-Methylpyridine	6.02
1,3-Propylenediamine K_1	8.64	Benzidine $.K_1$	4.97
K_2	10.62	K_2	3.75
1,4-Butylenediamine K_1	9.35	1,10-Phenanthroline	4.86
K_2	10.80		

* The values at 20°C and 30°C are 14.17 and 13.83 respectively.

APPENDIX 8   POLAROGRAPHIC HALF-WAVE POTENTIALS

Ion	Supporting electrolyte	$E_{1/2}$ (volts vs S.C.E.)
Ba^{2+}	$0.1\,M\ N(CH_3)_4Cl$	-1.94
Bi^{3+}	$1\,M$ HCl	-0.09
	$0.5\,M\ H_2SO_4$	-0.04
	$0.5\,M$ tartrate $+ 0.1\,M$ NaOH	-1.0
	$0.5\,M$ Sodium hydrogentartrate, pH 4.5	-0.23
Cd^{2+}	$0.1\,M$ KCl	-0.64
	$1\,M\ NH_3 + 1\,M\ NH_4^+$	-0.81
	$1\,M\ HNO_3$	-0.59
	$1\,M$ KI	-0.74
	$1\,M$ KCN	-1.18
Co^{2+}	$0.1\,M$ KCl	-1.20
	$0.1\,M$ Pyridine $+ 0.1\,M$ pyridinium ion	-1.07
Cu^{2+}	$0.1\,M$ KCl	$+0.04$
	$1\,M\ NH_3 + 1\,M\ NH_4Cl$	-0.24 (1st wave)
		-0.50 (2nd wave)
	$0.5\,M$ Sodium hydrogentartrate, pH 4.5	-0.09
Fe^{2+}	$0.1\,M$ KCl	-1.3
Fe^{3+}	$0.5\,M$ Tartrate, pH 9.4	-1.20 (1st wave)
		-1.73 (2nd wave)
	$0.1\,M$ EDTA $+ 2\,M\ CH_3COONa$	-0.13 (1st wave)
		-1.3 (2nd wave)
K^+	$0.1\,M\ N(CH_3)_4OH$ in 50% ethanol	-2.10
Li^+	$0.1\,M\ N(CH_3)_3OH$ in 50% ethanol	-2.31
Mn^{2+}	$1\,M$ KCl	-1.51
	$0.2\,M\ H_2P_2O_7^{2-}$, pH 2.2	$+0.1$
Na^+	$0.1\,M\ N(CH_3)_4Cl$	-2.07
Ni^{2+}	$1\,M$ KCl	-1.1
	$1\,M$ KSCN	-0.70
	$1\,M$ KCN	-1.36
	$1\,M$ Pyridine $+$ HCl, pH 7.0	-0.78
	$1\,M\ NH_3 + 0.2\,M\ NH_4^+$	-1.06
O_2	Most buffers, pH 1–10	-0.05 (1st wave)
		-0.9 (2nd wave)
Pb^{2+}	$0.1\,M$ KCl	-0.40
	$1\,M\ HNO_3$	-0.40
	$1\,M$ NaOH	-0.75
	$0.5\,M$ Sodium hydrogentartrate, pH 4.5	-0.48
	$0.5\,M$ Tartrate $+ 0.1\,M$ NaOH	-0.75
Sn^{2+}	$1\,M$ HCl	-0.47
Sn^{4+}	$1\,M$ HCl $+ 4\,M\ NH_4^+$	-0.25 (1st wave)
		-0.52 (2nd wave)
Zn^{2+}	$0.1\,M$ KCl	-1.00
	$1\,M$ NaOH	-1.53
	$1\,M\ NH_3 + 1\,M\ NH_4^+$	-1.33
	$0.5\,M$ Tartrate, pH 9	-1.15

In stripping voltammetry the stripping potential of a given ion is generally close to the polarographic half-wave potential of that ion in solutions with similar supporting electrolytes. Thus, typical stripping potentials in a $0.05\,M$ potassium chloride base solution are as follows: Zn, -1.00 V; Cd, -0.07 V; Pb, -0.45 V; Bi, -0.10 V; Cu(II), -0.05 V.

The half-wave potentials observed with some selected organic compounds are shown below. Where a percentage is quoted against a single organic solvent, the other solvent is water.

Compound	Supporting electrolyte	$E_{1/2}$ (volts vs S.C.E.)
Acids		
Acetic	Et_4NClO_4; CH_3CN	-2.3
Ascorbic	Acetic acid/acetate buffer (pH 3.4)	$+0.17$
Fumaric	$0.1\,M$ NH_4OH; $0.1\,M$ NH_4Cl (pH 8.2)	-1.57
Maleic	$0.1\,M$ NH_4OH; $0.1\,M$ NH_4Cl (pH 8.2)	-1.36
Thioglycollic	Phosphate buffer (pH 6.8)	-0.38
Carbonyl		
Acetaldehyde	$0.6\,M$ LiOH; $0.07\,M$ LiCl	-1.89
Acetone	$0.1\,M$ Bu_4NCl; $0.1\,M$ Bu_4NOH; 80 per cent EtOH	-2.53
Benzaldehyde	$0.1\,M$ Bu_4NCl; $0.1\,M$ Bu_4NOH; 80 per cent EtOH	-1.57
Fructose	$0.1\,M$ LiCl	-1.76
Glucose	$0.1\,M$ KCl	-1.55
Anthraquinone	NH_4OH/NH_4Cl buffer (pH 7.4); 40 per cent dioxan	-0.54
Benzoquinone	Acetic acid/acetate buffer (pH 5.4); 50 per cent MeOH	$+0.15$
Benzoyl peroxide	$0.3\,M$ LiCl; 50 per cent MeOH; 50 per cent C_6H_6	0.0
Nitro		
Nitrobenzene	Acetic acid/acetate buffer (pH 3); 50 per cent EtOH	-0.43
2-nitrophenol	Acetic acid/acetate buffer (pH 3); 50 per cent EtOH	-0.23
3-nitrophenol	Acetic acid/acetate buffer (pH 3); 50 per cent EtOH	-0.37
4-nitrophenol	Acetic acid/acetate buffer (pH 3); 50 per cent EtOH	-0.35
Unsaturated hydrocarbons		
Naphthalene	$0.18\,M$ Bu_4NI; 75 per cent dioxan	-2.50
Anthracene	$0.18\,M$ Bu_4NI; 75 per cent dioxan	-1.94
Phenylacetylene	$0.18\,M$ Bu_4NI; 75 per cent dioxan	-2.37
Stilbene	$0.18\,M$ Bu_4NI; 75 per cent dioxan	-2.26
Styrene	$0.18\,M$ Bu_4NI; 75 per cent dioxan	-2.35
Heterocyclic		
Pyridine	$0.1\,M$ HCl; 50 per cent EtOH	-1.49
Quinoline	$0.2\,M$ Me_4NOH; 50 per cent MeOH	-1.50
Miscellaneous		
Methyl chloride	$0.05\,M$ Et_4NBr; dimethylformamide	-2.23
Diethyl sulphide	$0.025\,M$ Bu_4NOH; MeOH; PrOH; H_2O (2:2:1)	-1.78

836

APPENDIX 9 RESONANCE LINES FOR ATOMIC ABSORPTION

Element	Symbol	Absorbing lines (nm)		Element	Symbol	Absorbing lines (nm)	
		Most sensitive	Alternatives			Most sensitive	Alternatives
Aluminium	Al	396.2	308.2	Mercury	Hg	253.7	—
			309.3	Molybdenum	Mo	313.3	320.9
			394.4	Neodymium	Nd	492.5	463.4
Antimony	Sb	217.6	206.8	Nickel	Ni	232.0	231.1
			217.9				341.5
			231.2				351.5
Arsenic	As	193.7	189.0				352.4
			197.2				362.5
Barium	Ba	553.5	455.4	Niobium	Nb	334.9	405.9
			493.4				408.0
Beryllium	Be	234.9	—				412.4
Bismuth	Bi	223.1	222.8	Osmium	Os	290.9	305.9
			227.7				426.0
			306.8	Palladium	Pd	247.6	244.8
Boron	B	249.8	208.9				340.5
Cadmium	Cd	228.8	326.1	Phosphorus	P	213.6	214.9
Caesium	Cs	852.1	455.6	Platinum	Pt	265.9	264.7
Calcium	Ca	422.7	—				299.8
Cerium	Ce	520.0	569.7				306.5
Chromium	Cr	357.9	425.4	Potassium	K	766.5	404.4
			427.5				769.9
			429.0	Praseodymium	Pr	495.1	513.3
			520.4	Rhenium	Re	346.0	346.5
			520.8	Rhodium	Rh	343.5	328.1
Cobalt	Co	240.7	304.4				369.2
			346.6	Rubidium	Rb	780.0	794.8
			347.4	Ruthenium	Ru	349.9	392.6
			391.0	Samarium	Sm	429.7	476.0
Copper	Cu	324.8	217.9	Scandium	Sc	391.2	390.8
			218.2	Selenium	Se	196.0	204.0
			222.6	Silicon	Si	251.6	250.7
			244.2				251.4
			249.2				252.4
			327.4				288.1
Dysprosium	Dy	419.5	404.6	Silver	Ag	328.1	338.3
Erbium	Er	400.8	389.3	Sodium	Na	589.0	330.2
Europium	Eu	459.4	462.7				589.6
Gadolinium	Gd	368.4	405.8	Strontium	Sr	460.7	407.8
			407.9	Tantalum	Ta	271.5	275.8
Gallium	Ga	287.4	403.3	Tellurium	Te	214.3	225.9
			417.2	Terbium	Tb	432.7	431.9
Germanium	Ge	265.1	271.0				433.8
Gold	Au	242.8	267.6	Thallium	Tl	276.7	258.0
Hafnium	Hf	307.8	268.2	Thorium	Th	371.9	—
Holmium	Ho	410.4	425.4	Thulium	Tm	371.8	436.0
			405.4				410.6
Hydrogen	H	Continuum		Tin	Sn	233.5	224.6
Indium	In	303.9	325.6				266.1
			410.2	Titanium	Ti	364.3	365.4
			451.1				399.0
							399.8

APPENDIX 9 (CONTINUED)

Element	Symbol	Absorbing lines (nm)		Element	Symbol	Absorbing lines (nm)	
		Most sensitive	Alternatives			Most sensitive	Alternatives
Iridium	Ir	208.9	264.0	Tungsten	W	255.1	294.7
			266.5				400.9
Iron	Fe	248.3	248.8				407.4
			372.0	Uranium	U	358.5	356.6
			386.0				351.4
			392.0	Vanadium	V	318.5	306.6
Lanthanum	La	550.1	403.7				318.4
Lead	Pb	217.0	261.4	Ytterbium	Yb	398.8	346.4
			283.3				246.4
Lithium	Li	670.8	323.3	Yttrium	Y	410.2	414.2
Lutetium	Lu	335.9	356.7	Zinc	Zn	213.9	307.6
			337.6	Zirconium	Zr	360.1	468.7
Magnesium	Mg	285.2	202.5				354.8
Manganese	Mn	279.5	279.8				
			280.1				
			403.1				

APPENDIX 10   ELECTRONIC ABSORPTION CHARACTERISTICS OF SOME COMMON CHROMOPHORES

Chromophore	λ_{max} (nm)	ε_{max}
Acetylide ($-C\equiv C-$)	175–180	6000
Aldehyde ($-CHO$)	210	Strong
Amine ($-NH_2$)	195	2800
Bromide ($-Br$)	208	300
Carboxyl ($-COOH$)	200–210	50–70
Ethylene ($-C\equiv C-$)	190	8000
Esters ($-COOR$)	205	50
Ether ($-O-$)	185	1000
Ketone ($\diagdown C=O$)	195	1000
Iodide ($-I$)	260	400
Nitrate ($-ONO_2$)	270	12
Nitrite ($-ONO$)	220–230	1000–2000
Nitro ($-NO_2$)	210	Strong
Nitroso ($-N=O$)	302	100
Oxime ($-NOH$)	190	5000

ABSORPTION CHARACTERISTICS OF AROMATIC AND HETEROCYCLIC SPECIES

	λ_{max} (nm)	ε_{max}	λ_{max} (nm)	ε_{max}
Benzene	184	46 700	202	6900
Naphthalene	220	112 000	275	5600
Pyridine	174	80 000	195	6000
Quinoline	227	37 000	270	3600

APPENDIX 11   CHARACTERISTIC INFRARED ABSORPTION BANDS

	4000	3000	2000	1600	1200	800	400 cm⁻¹
O−H Phenols, alcohols							
Free	□ m						
H-bonded	m□						
Carboxylic acids	m▭						
N−H Amides, primary	m□			□m-s			
and secondary amines							
C−H Aromatic							
(stretch)	s□						
(out-of-plane bend)						▭s	
Alkanes (stretch)	□s						
CH₃− (bend)				m□	□m		
−CH₂− (bend)				m□			
Alkenes							
(stretch)	m▯						
(out-of-plane bend)						▭s	
C≡C Alkyne		□ m-w					
C≡N Nitriles		m□					
C=O* Aldehyde				□s			
Ketone				□s			
Acid				□s			
Ester				▯s			
Amide				□s			
Anhydride			s▯ □s				
C=C Alkene				□m-w			
Aromatic				m-w▯ □m-w			
C−O Esters, ethers,					▭s		
Anhydrides, alcohols					▭s		
Carboxylic acids					▭s		
N=C Nitro (R-NO₂)				□s	□s		
C−Hal. Fluoride					▭s		
Chloride						▭s	
Bromide, iodide							▭s
	4000	3000	2000	1600	1200	800	400 cm⁻¹

*C=O stretching frequencies are typically lowered by about $20-30$ cm⁻¹ from the values given when the carbonyl group is conjugated with an aromatic ring or an alkene group.

w = weak   m = medium   s = strong

Reproduced from R. Davis and C. H. J. Wells, *Spectral Problems In Organic Chemistry*, International Textbook Co., Chapman and Hall, New York, 1984.

APPENDIX 12 PERCENTAGE POINTS OF THE *t*-DISTRIBUTION

The table gives the value of $t_{\alpha;v}$ — the 100α percentage point of the *t*-distribution for v degrees of freedom.

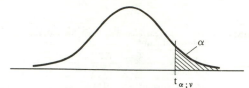

Note. The tabulation is for one tail only, i.e. for positive values of t. For $|t|$ the column headings for α must be doubled.

v	α						
	0.10	0.05	0.025	0.01	0.005	0.001	0.0005
1	3.078	6.314	12.706	31.821	63.657	318.31	636.62
2	1.886	2.920	4.303	6.965	9.925	22.326	31.598
3	1.638	2.353	3.182	4.541	5.841	10.213	12.924
4	1.533	2.132	2.776	3.747	4.604	7.173	8.610
5	1.476	2.015	2.571	3.365	4.032	5.893	6.869
6	1.440	1.943	2.447	3.143	3.707	5.208	5.959
7	1.415	1.895	2.365	2.998	3.499	4.785	5.408
8	1.397	1.860	2.306	2.896	3.355	4.501	5.041
9	1.383	1.833	2.262	2.821	3.250	4.297	4.781
10	1.372	1.812	2.228	2.764	3.169	4.144	4.587
11	1.363	1.796	2.201	2.718	3.106	4.025	4.437
12	1.356	1.782	2.179	2.681	3.055	3.930	4.318
13	1.350	1.771	2.160	2.650	3.012	3.852	4.221
14	1.345	1.761	2.145	2.624	2.977	3.787	4.140
15	1.341	1.753	2.131	2.602	2.947	3.733	4.073
16	1.337	1.746	2.120	2.583	2.921	3.686	4.015
17	1.333	1.740	2.110	2.567	2.898	3.646	3.965
18	1.330	1.734	2.101	2.552	2.878	3.610	3.922
19	1.328	1.729	2.093	2.539	2.861	3.579	3.883
20	1.325	1.725	2.086	2.528	2.845	3.552	3.850
21	1.323	1.721	2.080	2.518	2.831	3.527	3.819
22	1.321	1.717	2.074	2.508	2.819	3.505	3.792
23	1.319	1.714	2.069	2.500	2.807	3.485	3.767
24	1.318	1.711	2.064	2.492	2.797	3.467	3.745
25	1.316	1.708	2.060	2.485	2.787	3.450	3.725
26	1.315	1.706	2.056	2.479	2.779	3.435	3.707
27	1.314	1.703	2.052	2.473	2.771	3.421	3.690
28	1.313	1.701	2.048	2.467	2.763	3.408	3.674
29	1.311	1.699	2.045	2.462	2.756	3.396	3.659
30	1.310	1.697	2.042	2.457	2.750	3.385	3.646
40	1.303	1.684	2.021	2.423	2.704	3.307	3.551
60	1.296	1.671	2.000	2.390	2.660	3.232	3.460
120	1.289	1.658	1.980	2.358	2.617	3.160	3.373
∞	1.282	1.645	1.960	2.326	2.576	3.090	3.291

This table is derived from Table III of R. A. Fisher and F. Yates, *Statistical Tables for Biological, Agricultural and Medical Research*, published by Oliver & Boyd Ltd, Edinburgh, and by permission of the authors and publishers, and also from Table 12 of *Biometrika Tables for Statisticians*, Vol. 1, by permission of the Biometrika Trustees.

APPENDIX 13   *F*–DISTRIBUTION

Probability level	ϕ_2	ϕ_1 (corresponding to greater mean square)											
		1	2	3	4	5	6	7	8	9	10	15	∞
0.10	1	39.9	49.5	53.6	55.8	57.2	58.2	58.9	59.4	59.9	60.2	61.2	63.3
0.05		161.4	199.5	215.7	224.6	230.2	234.0	236.8	238.9	240.5	241.9	246.0	254.3
0.01		4052	4999	5403	5625	5764	5859	5928	5981	6023	6056	6157	6366
0.10	2	8.53	9.00	9.16	9.24	9.29	9.33	9.35	9.37	9.38	9.39	9.42	9.49
0.05		18.5	19.0	19.2	19.2	19.3	19.3	19.4	19.4	19.4	19.4	19.4	19.5
0.01		98.5	99.0	99.2	99.2	99.3	99.3	99.4	99.4	99.4	99.4	99.4	99.5
0.10	3	5.54	5.46	5.39	5.34	5.31	5.28	5.27	5.25	5.24	5.23	5.20	5.13
0.05		10.1	9.55	9.28	9.12	9.01	8.94	8.89	8.85	8.81	8.79	8.70	8.53
0.01		34.1	30.8	29.5	28.7	28.2	27.9	27.7	27.5	27.3	27.2	26.9	26.1
0.10	4	4.54	4.32	4.19	4.11	4.05	4.01	3.98	3.95	3.94	3.92	3.87	3.76
0.05		7.71	6.94	6.59	6.39	6.26	6.16	6.09	6.04	6.00	5.96	5.86	5.62
0.01		21.2	18.0	16.7	16.0	15.5	15.2	15.0	14.8	14.7	14.5	14.2	13.5
0.10	5	4.06	3.78	3.62	3.52	3.45	3.40	3.37	3.34	3.32	3.30	3.24	3.10
0.05		6.61	5.79	5.41	5.19	5.05	4.95	4.88	4.82	4.77	4.74	4.62	4.36
0.01		16.3	13.3	12.1	11.4	11.0	10.7	10.5	10.3	10.2	10.1	9.72	9.02
0.10	6	3.78	3.46	3.29	3.18	3.11	3.05	3.01	2.98	2.96	2.94	2.87	2.72
0.05		5.99	5.14	4.76	4.53	4.39	4.28	4.21	4.15	4.10	4.06	3.94	3.67
0.01		13.7	10.9	9.78	9.15	8.75	8.47	8.26	8.10	7.98	7.87	7.56	6.88
0.10	7	3.59	3.26	3.07	2.96	2.88	2.83	2.78	2.75	2.72	2.70	2.63	2.47
0.05		5.59	4.74	4.35	4.12	3.97	3.87	3.79	3.73	3.68	3.64	3.51	3.23
0.01		12.2	9.55	8.45	7.85	7.46	7.19	6.99	6.84	6.72	6.62	6.31	5.65
0.10	8	3.46	3.11	2.92	2.81	2.73	2.67	2.62	2.59	2.56	2.54	2.46	2.29
0.05		5.32	4.46	4.07	3.84	3.69	3.58	3.50	3.44	3.39	3.35	3.22	2.93
0.01		11.3	8.65	7.59	7.01	6.63	6.37	6.18	6.03	5.91	5.81	5.52	4.86
0.10	9	3.36	3.01	2.81	2.69	2.61	2.55	2.51	2.47	2.44	2.42	2.34	2.16
0.05		5.12	4.26	3.86	3.63	3.48	3.37	3.29	3.23	3.18	3.14	3.01	2.71
0.01		10.6	8.02	6.99	6.42	6.06	5.80	5.61	5.47	5.35	5.26	4.96	4.31
0.10	10	3.29	2.92	2.73	2.61	2.52	2.46	2.41	2.38	2.35	2.32	2.24	2.06
0.05		4.96	4.10	3.71	3.48	3.33	3.22	3.14	3.07	3.02	2.98	2.85	2.54
0.01		10.0	7.56	6.55	5.99	5.64	5.39	5.20	5.06	4.94	4.85	4.56	3.91
0.10	12	3.18	2.81	2.61	2.48	2.39	2.33	2.28	2.24	2.21	2.19	2.10	1.90
0.05		4.75	3.89	3.49	3.26	3.11	3.00	2.91	2.85	2.80	2.75	2.62	2.30
0.01		9.33	6.93	5.95	5.41	5.06	4.82	4.64	4.50	4.39	4.30	4.01	3.36
0.10	15	3.07	2.70	2.49	2.36	2.27	2.21	2.16	2.12	2.09	2.06	1.97	1.76
0.05		4.54	3.68	3.29	3.06	2.90	2.79	2.71	2.64	2.59	2.54	2.40	2.07
0.01		8.68	6.36	5.42	4.89	4.56	4.32	4.14	4.00	3.89	3.80	3.52	2.87
0.10	16	3.05	2.67	2.46	2.33	2.24	2.18	2.13	2.09	2.06	2.03	1.94	1.72
0.05		4.49	3.63	3.24	3.01	2.85	2.74	2.66	2.59	2.54	2.49	2.35	2.01
0.01		8.53	6.23	5.29	4.77	4.44	4.20	4.03	3.89	3.78	3.69	3.41	2.75
0.10	24	2.93	2.54	2.33	2.19	2.10	2.04	1.98	1.94	1.91	1.88	1.78	1.53
0.05		4.26	3.40	3.01	2.78	2.62	2.51	2.42	2.36	2.30	2.25	2.11	1.73
0.01		7.82	5.61	4.72	4.22	3.90	3.67	3.50	3.36	3.26	3.17	2.89	2.21
0.10	60	2.79	2.39	2.18	2.04	1.95	1.87	1.82	1.77	1.74	1.71	1.60	1.29
0.05		4.00	3.15	2.76	2.53	2.37	2.25	2.17	2.10	2.04	1.99	1.84	1.39
0.01		7.08	4.98	4.13	3.65	3.34	3.12	2.95	2.82	2.72	2.63	2.35	1.60
0.10	∞	2.71	2.30	2.08	1.94	1.85	1.77	1.72	1.67	1.63	1.60	1.49	1.00
0.05		3.84	3.00	2.60	2.37	2.21	2.10	2.01	1.94	1.88	1.83	1.67	1.00
0.01		6.63	4.61	3.78	3.32	3.02	2.80	2.64	2.51	2.41	2.32	2.04	1.00

APPENDIX 14   CRITICAL VALUES OF Q $(P = 0.05)$

Sample size	Critical value
4	0.831
5	0.717
6	0.621
7	0.570
8	0.524
9	0.492
10	0.464

Taken from E. P. King, *J. Am. Statist. Assoc.*, 1958, **48**, 531, by permission of the American Statistical Association.

APPENDIX 15   CRITICAL VALUES OF THE CORRELATION COEFFICIENT ρ $(P = 0.05)$

No. of data pairs (x, y)	Critical value
5	0.88
6	0.82
7	0.76
8	0.71
9	0.67
10	0.64
11	0.61
12	0.58

APPENDIX 16   FOUR–FIGURE LOGARITHMS

	0	1	2	3	4	5	6	7	8	9	1	2	3	4	5	6	7	8	9
														Mean differences					
10	0000	0043	0086	0128	0170	0212	0253	0294	9334	0374	4	8	12	17	21	25	29	33	37
11	0414	0453	0492	0531	0569	0607	0645	0682	0719	0755	4	8	11	15	19	23	26	30	34
12	0792	0828	0864	0899	0934	0969	1004	1038	1072	1106	3	7	10	14	17	21	24	28	31
13	1139	1173	1206	1239	1271	1303	1335	1367	1399	1430	3	6	10	13	16	19	23	26	29
14	1461	1492	1523	1553	1584	1614	1644	1673	1703	1732	3	6	9	12	15	18	21	24	27
15	1761	1790	1818	1847	1875	1903	1931	1959	1987	2014	3	6	8	11	14	17	20	22	25
16	2041	2068	2095	2122	2148	2175	2201	2227	2253	2279	3	5	8	11	13	16	18	21	24
17	2304	2330	2355	2380	2405	2430	2455	2480	2504	2529	2	5	7	10	12	15	17	20	22
18	2553	2577	2601	2625	2648	2672	2695	2718	2742	2765	2	5	7	9	12	14	16	19	21
19	2788	2810	2833	2856	2878	2900	2923	2945	2967	2989	2	4	7	9	11	13	16	18	20
20	3010	3032	3054	3075	3096	3118	3239	3160	3181	3201	2	4	6	8	11	13	15	17	19
21	3222	3243	3263	3284	3304	3324	3345	3365	3385	3404	2	4	6	8	10	12	14	16	18
22	3424	3444	3464	3483	3502	3522	3541	3560	3579	3598	2	4	6	8	10	12	14	15	17
23	3617	3636	3655	3674	3692	3711	3729	3747	3766	3784	2	4	6	7	9	11	13	15	17
24	3802	3820	3838	3856	3874	3892	3909	3927	3945	3962	2	4	5	7	9	11	12	14	16
25	3979	3997	4014	4031	4048	4065	4082	4099	4116	4133	2	3	5	7	9	10	12	14	15
26	4150	4166	4183	4200	4216	4232	4249	4265	4281	4298	2	3	5	7	8	10	11	13	15
27	4314	4330	4346	4362	4378	4393	4409	4425	4440	4456	2	3	5	6	8	9	11	13	14
28	4472	4487	4502	4518	4533	4548	4564	4579	4594	4609	2	3	5	6	8	9	11	12	14
29	4624	4639	4654	4669	4683	4698	4713	4728	4742	4757	1	3	4	6	7	9	10	12	13
30	4771	4786	4800	4814	4829	4843	4857	4871	4886	4900	1	3	4	6	7	9	10	11	13
31	4914	4928	4942	4955	4969	4983	4997	5011	5024	5038	1	3	4	6	7	8	10	11	12
32	5051	5065	5079	5092	5105	5119	5132	5145	5159	5172	1	3	4	5	7	8	9	11	12
33	5185	5198	5211	5224	5237	5250	5263	5276	5289	5302	1	3	4	5	6	8	9	10	12
34	5315	5328	5340	5353	5366	5378	5391	5403	5416	5428	1	3	4	5	6	8	9	10	11
35	5441	5453	5465	5478	5490	5502	5514	5527	5539	5551	1	2	4	5	6	7	9	10	11
36	5563	5575	5587	5599	5611	5623	5635	5647	5658	5670	1	2	4	5	6	7	8	10	11
37	5563	5575	5587	5599	5611	5623	5635	5647	5658	5670	1	2	4	5	6	7	8	10	11
37	5682	5694	5705	5717	5729	5740	5752	5763	5775	5786	1	2	3	5	6	7	8	9	10
38	5798	5809	5821	5832	5843	5855	5866	5877	5888	5899	1	2	3	5	6	7	8	9	10
39	5911	5922	5933	5944	5955	5966	5977	5988	5999	6010	1	2	3	4	5	7	8	9	10
40	6021	6031	6042	6053	6064	6075	6085	6096	6107	6117	1	2	3	4	5	6	8	9	10
41	6128	6138	6149	6160	6170	6180	6191	6201	6212	6222	1	2	3	4	5	6	7	8	9
42	6232	6243	6253	6263	6274	6284	6294	6304	6314	6325	1	2	3	4	5	6	7	8	9
43	6335	6345	6355	6365	6375	6385	6395	6405	6415	6425	1	2	3	4	5	6	7	8	9
44	6435	6444	6454	6464	6474	6484	6493	6503	6513	6522	1	2	3	4	5	6	7	8	9
45	6532	6542	6561	6561	6571	6580	6590	6599	6609	6618	1	2	3	4	5	6	7	8	9
46	6628	6637	6646	6656	6665	6675	6684	6693	6702	6712	1	2	3	4	5	6	7	7	8
47	6721	6730	6739	6749	6758	6767	6776	6785	6794	6803	1	2	3	4	5	5	6	7	8
48	6812	6821	6830	6839	6848	6857	6866	6875	6884	6893	1	2	3	4	5	5	6	7	8
49	6902	6911	6920	6928	6937	6946	6955	6964	6972	6981	1	2	3	4	4	5	6	7	8
50	6990	6998	7007	7016	7024	7033	7042	7050	7059	7067	1	2	3	3	4	5	6	7	8
51	7076	7084	7093	7101	7110	7118	7126	7135	7143	7152	1	2	3	3	4	5	6	7	8
52	7160	7168	7177	7185	7193	7202	7210	7218	7226	7235	1	2	2	3	4	5	6	7	7
53	7243	7251	7259	7267	7275	7284	7292	7300	7308	7316	1	2	2	3	4	5	6	6	7
54	7324	7332	7340	7348	7356	7364	7372	7380	7388	7396	1	2	2	3	4	5	6	6	7
	0	1	2	3	4	5	6	7	8	9	1	2	3	4	5	6	7	8	9

	0	1	2	3	4	5	6	7	8	9	Mean differences								
											1	2	3	4	5	6	7	8	9
55	7404	7412	7419	7427	7435	7443	7451	7459	7466	7474	1	2	2	3	4	5	5	6	7
56	7482	7490	7497	7505	7513	7520	7528	7536	7543	7551	1	2	2	3	4	5	5	6	7
57	7559	7566	7574	7582	7589	7597	7604	7612	7619	7627	1	2	2	3	4	5	5	6	7
58	7634	7642	7649	7657	7664	7672	7679	7686	7694	7701	1	1	2	3	4	4	5	6	7
59	7709	7716	7723	7731	7738	7745	7752	7760	7767	7774	1	1	2	3	4	4	5	6	7
60	7782	7789	7796	7803	7810	7818	7825	7832	7839	7846	1	1	2	3	4	4	5	6	6
61	7853	7860	7868	7875	7882	7889	7896	7903	7910	7917	1	1	2	3	4	4	5	6	6
62	7924	7931	7938	7945	7952	7959	7966	7973	7980	7987	1	1	2	3	3	4	5	6	6
63	7993	8000	8007	8014	8021	8028	8035	8041	8048	8055	1	1	2	3	3	4	5	5	6
64	8062	8069	8075	8082	8089	8096	8102	8109	8116	8122	1	1	2	3	3	4	5	5	6
65	8129	8136	8142	8149	8156	8162	8169	8176	8182	8189	1	1	2	3	3	4	5	5	6
66	8195	8202	8209	8215	8222	8228	8235	8241	8248	8254	1	1	2	3	3	4	5	5	6
67	8261	8267	8274	8280	8287	8293	8299	8306	8312	8319	1	1	2	3	3	4	5	5	6
68	8325	8331	8338	8344	8351	8357	8363	8370	8376	8382	1	1	2	3	3	4	4	5	6
69	8388	8395	8401	8407	8414	8420	8426	8432	8439	8445	1	1	2	2	3	4	4	5	6
70	8451	8457	8463	8470	8476	8482	8488	8494	8500	8506	1	1	2	2	3	4	4	5	6
71	8513	8519	8525	8531	8537	8543	8549	8555	8561	8567	1	1	2	2	3	4	4	5	5
72	8573	8579	8585	8591	8597	8603	8609	8615	8621	8627	1	1	2	2	3	4	4	5	5
73	8633	8639	8645	8651	8657	8663	8669	8675	8681	8686	1	1	2	2	3	4	4	5	5
74	8692	8698	8704	8710	8716	8722	8727	8733	8739	8745	1	1	2	2	3	4	4	5	5
75	8751	8756	8762	8768	8774	8779	8785	8791	8797	8802	1	1	2	2	3	3	4	5	5
76	8808	8814	8820	8825	8831	8837	8842	8848	8854	8859	1	1	2	2	3	3	4	5	5
77	8865	8871	8876	8882	8887	8893	8899	8904	8910	8915	1	1	2	2	3	3	4	4	5
78	8921	8927	8932	8938	8943	8949	8954	8960	8965	8971	1	1	2	2	3	3	4	4	5
79	8976	8982	8987	8993	8998	9004	9009	9015	9020	9025	1	1	2	2	3	3	4	4	5
80	9031	9036	9042	9047	9053	9058	9063	9069	9074	9079	1	1	2	2	3	3	4	4	5
81	9085	9090	9096	9101	9106	9112	9117	9122	9128	9133	1	1	2	2	3	3	4	4	5
82	9138	9143	9149	9154	9159	9165	9170	9175	9180	9186	1	1	2	2	3	3	4	4	5
83	9191	9196	9201	9206	9212	9217	9222	9227	9232	9238	1	1	2	2	3	3	4	4	5
84	9243	9248	9253	9258	9263	9269	9274	9279	9284	9289	1	1	2	2	3	3	4	4	5
85	9294	9299	9304	9309	9315	9320	9325	9330	9335	9340	1	1	2	2	3	3	4	4	5
86	9345	9350	9355	9360	9365	9370	9375	9380	9385	9390	1	1	2	2	3	3	4	4	5
87	9395	9400	9405	9410	9415	9420	9425	9430	9435	9440	0	1	1	2	2	3	3	4	4
88	9445	9450	9455	9460	9465	9469	9474	9479	9484	9489	0	1	1	2	2	3	3	4	4
89	9494	9499	9504	9509	9513	9518	9523	9528	9533	9538	0	1	1	2	2	3	3	4	4
90	9542	9547	9552	9557	9562	9566	9571	9576	9581	9586	0	1	1	2	2	3	3	4	4
91	9590	9595	9600	9605	9609	9614	9619	9624	9628	9633	0	1	1	2	2	3	3	4	4
92	9638	9643	9647	9652	9657	9661	9666	9671	9675	9680	0	1	1	2	2	3	3	4	4
93	9685	9689	9694	9699	9703	9708	9713	9717	9722	9727	0	1	1	2	2	3	3	4	4
94	9731	9736	9741	9745	9750	9754	9759	9763	9768	9773	0	1	1	2	2	3	3	4	4
95	9777	9782	9786	9791	9795	9800	9805	9809	9814	9818	0	1	1	2	2	3	3	4	
96	9823	9827	9832	9836	9841	9845	9850	9854	9859	9863	0	1	1	2	2	3	3	4	4
97	9868	9872	9877	9881	9886	9890	9894	9899	9903	9908	0	1	1	2	2	3	3	4	4
98	9912	9917	9921	9926	9930	9934	9939	9943	9948	9952	0	1	1	2	2	3	3	4	4
99	9956	9961	9965	9969	9974	9978	9983	9987	9991	9996	0	1	1	2	2	3	3	3	4
	0	1	2	3	4	5	6	7	8	9	1	2	3	4	5	6	7	8	9

APPENDIX 17 EQUIVALENTS AND NORMALITIES

Throughout the main text of this book standard solutions and quantities have all been expressed in terms of molarities, moles and relative molecular masses. However, there are still many chemists who have traditionally used what are known as normal solutions and equivalents as the basis for calculations, especially in titrimetry. Because of this it has been considered desirable to include this appendix defining the terms used and illustrating how they are employed in the various types of determinations.

The IUPAC* has given the following definition

'The **equivalent** of a substance is that amount of it which, in a specified reaction, combines with, releases or replaces that amount of hydrogen which is combined with 3 grams of carbon-12 in methane $^{12}CH_4$.'

From this it follows that a **normal solution** is a solution containing one equivalent of a defined species per litre according to the specified reaction. In this definition, the amount of hydrogen referred to may be replaced by the equivalent amount of electricity or by one equivalent of any other substance, but the reaction to which the definition is applied must be clearly specified.

The most important advantage of the equivalent system is that the calculations of titrimetric analysis are rendered very simple, since at the end point the number of equivalents of the substance titrated is equal to the number of equivalents of the standard solution employed. We may write:

$$\text{Normality} = \frac{\text{Number of equivalents}}{\text{Number of L}}$$

$$= \frac{\text{Number of milli-equivalents}}{\text{Number of mL}}$$

Hence: number of milli-equivalents = number of mL × normality. If the volumes of solutions of two different substances A and B which exactly react with one another are V_A mL and V_B mL respectively, then these volumes severally contain the same number of equivalents or milli-equivalents of A and B. Thus:

$$V_A \times \text{normality}_A = V_B \times \text{normality}_B \tag{1}$$

In practice V_A, V_B, and normality$_A$ (the standard solution) are known, hence normality$_B$ (the unknown solution) can be readily calculated.

Example 1. How many mL of $0.2N$ hydrochloric acid are required to neutralise 25.0 mL of $0.1N$ sodium hydroxide?

Substituting in equation (1), we obtain:

$x \times 0.2 = 25.0 \times 0.1$, whence $x = 12.5$ mL

Example 2. How many mL of N hydrochloric acid are required to precipitate completely 1 g of silver nitrate?

The equivalent of $AgNO_3$ in a precipitation reaction is 1 mole or 169.89 g. Hence 1 g of $AgNO_3 = 1 \times 1000/169.89 = 5.886$ milli-equivalents.

---

* International Union of Pure and Applied Chemistry, *Information Bulletin* No. 36, Aug. 1974.

Now number of milli-equivalents of HCl = number of milli-equivalents of $AgNO_3$:

$x \times 1 = 5.886$, whence $x = 5.90 \, mL$

Example 3. 25 mL of an iron(II) sulphate solution react complete with 30.0 mL of $0.125 \, N$ potassium permanganate. Calculate the strength of the iron solution in grams of $FeSO_4$ per L.

A normal solution of $FeSO_4$ as a reductant contains 1 mol L^{-1} or 151.90 g (Table 17A.2). Let the normality of the iron solution be n_A. Then:

$25 \times n_A = 30 \times 0.125$

or

$\qquad n_A = 30 \times 0.125/55 = 0.150 \, N$

Hence the solution will contain $0.150 \times 151.90 = 22.78$ g $FeSO_4 \, L^{-1}$.

Example 4. What volume of $0.127 \, N$ reagent is required for the preparation of 1000 mL of $0.1 \, N$ solution?

$V_A \times \text{normality}_A = V_B \times \text{normality}_B$

$\qquad V_A \times 0.127 = 1000 \times 0.1$

or

$$V_A = 1000 \times 0.1/0.127 = 787.4 \, mL$$

In other words, the required solution can be obtained by diluting 787.4 mL of $0.127 \, N$ reagent to 1 L.

The above definition of normal solution utilises the term 'equivalent'. This quantity varies with the type of reaction, and, since it is difficult to give a clear definition of 'equivalent' which will cover all reactions, it is proposed to discuss this subject in some detail below. It often happens that the same compound possesses different equivalents in different chemical reactions. The situation may therefore arise in which a solution has normal concentration when employed for one purpose, and a different normality when used in another chemical reaction.

Neutralisation reactions. The equivalent of an acid is that mass of it which contains 1.008 (more accurately 1.0078) g of replaceable hydrogen. The equivalent of a monoprotic acid, such as hydrochloric, hydrobromic, hydriodic, nitric, perchloric, or acetic acid, is identical with the mole. A normal solution of a monoprotic acid will therefore contain 1 mole per L of solution. The equivalent of a diprotic acid (e.g. sulphuric or oxalic acid), or of a triprotic acid (e.g. phosphoric(V) acid) is likewise one-half or one-third respectively, of the mole.

The equivalent of a base is that mass of it which contains one replaceable hydroxyl group, i.e. 17.008 g of ionisable hydroxyl; 17.008 g of hydroxyl are equivalent to 1.008 g of hydrogen. The equivalents of sodium hydroxide and potassium hydroxide are the mole, of calcium hydroxide, strontium hydroxide, and barium hydroxide half a mole.

Salts of strong bases and weak acids possess alkaline reactions in aqueous solution because of hydrolysis (Section 2.18). A solution containing one mole of sodium carbonate, with methyl orange as indicator, reacts with 2 moles of hydrochloric acid to form 2 moles of sodium chloride; hence its equivalent is half a mole. Sodium tetraborate, under similar conditions, also reacts with 2 moles of hydrochloric acid, and its equivalent is, likewise, half a mole.

Complex formation and precipitation reactions. Here the equivalent is the mass of the substance which contains or reacts with 1 mole of a single charged cation M^+ (which is equivalent to 1.008 g of hydrogen), $\frac{1}{2}$ mole of a doubly charged cation M^{2+}, $\frac{1}{3}$ mole of a triply charged cation M^{3+}, etc. For the cation, the equivalent is the mole divided by the charge number. For a reagent which reacts with this cation, the equivalent is the mass of it which reacts with one equivalent of the cation. The equivalent of a salt in a precipitation reaction is the mole divided by the total charge number of the *reacting* ion. Thus the equivalent of silver nitrate in the titration of chloride ion is the mole.

In a complex-formation reaction the equivalent is most simply deduced by writing down the ionic equation of the reaction. For example, the equivalent of potassium cyanide in the titration with silver ions is 2 moles, since the reaction is:

$$2CN^- + Ag^+ \rightleftharpoons [Ag(CN)_2]^-$$

In the titration of zinc ion with potassium hexacyanoferrate(II) solution:

$$3Zn^{2+} + 2K_4Fe(CN)_6 = 6K^+ + K_2Zn_3[Fe(CN)_6]_2$$

the equivalent of the hexacyanoferrate(II) is one-third of the mole. For other examples of complex-formation reactions, see Sections 10.43–10.73; it is apparent that in many complexation reactions it is preferable to work in moles rather than equivalents.

Oxidation–reduction reactions. The equivalent of an oxidising or reducing agent is most simply defined as that mass of the reagent which reacts with or contains 1.008 g of available hydrogen or 8.000 g of available oxygen. By 'available' is meant capable of being utilised in oxidation or reduction. The amount of available oxygen may be indicated by the equation:

$$MnO_4^- + 8H^+ + 5e \rightarrow Mn^{2+} + 4H_2O$$

Hence the equivalent is $KMnO_4/5$. For potassium dichromate in acid solution, the equation is:

$$Cr_2O_7^{2-} + 14H^+ + 6e \rightarrow 2Cr^{3+} + 7H_2O$$

The equivalent is $K_2Cr_2O_7/6$.

A more general and fundamental view is obtained by a consideration of: (*a*) the number of electrons involved in the partial ionic equation representing the reaction, and (*b*) the change in the 'oxidation number' of a significant element in the oxidant or reductant. Both methods will be considered in some detail.

In quantitative analysis we are chiefly concerned with reactions which take place in solution, i.e. ionic reactions. We shall therefore limit our discussion of oxidation–reduction to such reactions. The oxidation of iron(II) chloride by chlorine in aqueous solution may be written:

$$2FeCl_2 + Cl_2 = 2FeCl_3$$

or it may be expressed ionically:

$$2Fe^{2+} + Cl_2 = 2Fe^{3+} + 2Cl^-$$

847

The ion Fe^{2+} is converted into ion Fe^{3+} (oxidation), and the neutral chlorine molecule into negatively charged chloride ions Cl^- (reduction); the conversion of Fe^{2+} into Fe^{3+} requires the loss of one electron, and the transformation of the neutral chlorine molecule into chloride ions necessitates the gain of two electrons. This leads to the view that, for reactions in solutions, oxidation is a process involving a loss of electrons, as in

$$Fe^{2+} - e = Fe^{3+}$$

and reduction is the process resulting in a gain of electrons, as in

$$Cl_2 + 2e = 2Cl^-$$

In the actual oxidation–reduction process electrons are transferred from the reducing agent to the oxidising agent. This leads to the following definitions.

Oxidation is the process which results in the loss of one or more electrons by atoms or ions.

Reduction is the process which results in the gain of one or more electrons by atoms or ions.

An oxidising agent is one that gains electrons and is reduced; a reducing agent is one that loses electrons and is oxidised.

In all oxidation–reduction processes (or redox processes) there will be a reactant undergoing oxidation and one undergoing reduction, since the two reactions are complementary to one another and occur simultaneously — one cannot take place without the other. The reagent suffering oxidation is termed the reducing agent or reductant, and the reagent undergoing reduction is called the oxidising agent or oxidant. The study of the electron changes in the oxidant and reductant forms the basis of the ion-electron method for balancing ionic equations. The equation is accordingly first divided into two balanced, partial equations representing the oxidation and reduction respectively. It must be remembered that the reactions take place in aqueous solution so that in addition to the ions supplied by the oxidant and reductant, molecules of water H_2O, hydrogen ions H^+, and hydroxide ions OH^- are also present, and may be utilised in balancing the partial ionic equation. The unit change in oxidation or reduction is a charge of one electron, which will be denoted by e. To appreciate the principles involved, let us consider first the reaction between iron(III) chloride and tin(II) chloride in aqueous solution. The partial ionic equation for the reduction is:

$$Fe^{3+} \rightarrow Fe^{2+} \tag{2}$$

and for the oxidation is:

$$Sn^{2+} \rightarrow Sn^{4+} \tag{3}$$

The equations must be balanced not only with regard to the number and kind of atoms, but also electrically, i.e. the net electric charge on each side must be the same. Equation (2) can be balanced by adding one electron to the left-hand side:

$$Fe^{3+} + e \rightleftharpoons Fe^{2+} \tag{4}$$

and equation (3) by adding two electrons to the right-hand side:

$$Sn^{2+} \rightleftharpoons Sn^{4+} + 2e \qquad (5)$$

These partial equations must then be multiplied by coefficients which result in the number of electrons utilised in one reaction being equal to those liberated in the other. Thus equation (4) must be multiplied by two, and we have:

$$2Fe^{3+} + 2e \rightleftharpoons 2Fe^{2+} \qquad (6)$$

Adding (5) and (6), we obtain:

$$2Fe^{3+} + Sn^{2+} + 2e \rightleftharpoons 2Fe^{2+} + Sn^{4+} + 2e$$

and by cancelling the electrons common to both sides, the simple ionic equation is obtained:

$$2Fe^{3+} + Sn^{2+} = 2Fe^{2+} + Sn^{4+}$$

The following facts must be borne in mind. All strong electrolytes are completely dissociated; hence only the ions actually taking part or resulting from the reaction need appear in the equation. Substances which are only slightly ionised, such as water, or which are sparingly soluble and thus yield only a small concentration of ions, e.g. silver chloride and barium sulphate, are, in general, written as molecular formulae because they are present mainly in the undissociated state.

The complete rules for the application of the ion-electron method may be expressed as follows:

(*a*) ascertain the products of the reaction;
(*b*) set up a partial equation for the oxidising agent;
(*c*) set up a partial equation for the reducing agent in the same way;
(*d*) multiply each partial equation by a factor so that when the two are added the electrons just compensate each other;
(*e*) add the partial equations and cancel out substances which appear on both sides of the equation.

A few examples follow.

Reaction I. The reduction of potassium permanganate by iron(II) sulphate in the presence of dilute sulphuric acid.

The first partial equation (reduction) is:

$$MnO_4^- \rightarrow Mn^{2+}$$

To balance atomically, $8H^+$ are required:

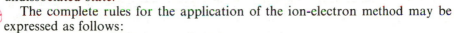

$$MnO_4^- + 8H^+ \rightarrow Mn^{2+} + 4H_2O$$

and to balance it electrically $5e$ are needed on the left-hand side:

$$MnO_4^- + 8H^+ + 5e \rightleftharpoons Mn^{2+} + 4H_2O$$

The second partial equation (oxidation) is:

$$Fe^{2+} \rightarrow Fe^{3+}$$

To balance this electrically one electron must be added to the right-hand side or subtracted from the left-hand side:

$Fe^{2+} \rightleftharpoons Fe^{3+} + e$

Now the gain and loss of electrons must be equal. One permanganate ion utilises 5 electrons, and one iron(II) ion liberates 1 electron; hence the two partial equations must apply in the ratio of 1:5.

$MnO_4^- + 8H^+ + 5e \rightleftharpoons Mn^{2+} + 4H_2O$

$5(Fe^{2+} \rightleftharpoons Fe^{3+} + e)$

or $\overline{MnO_4^- + 8H^+ + 5Fe^{2+} = Mn^{2+} + 5Fe^{3+} + 4H_2O}$

Reaction II. The interaction of potassium dichromate and potassium iodide in the presence of dilute sulphuric acid.

$Cr_2O_7^{2-} \rightarrow Cr^{3+}$

$Cr_2O_7^{2-} + 14H^+ \rightarrow 2Cr^{3+} + 7H_2O$

To balance electrically, add $6e$ to the left-hand side:

$Cr_2O_7^{2-} + 14H^+ + 6e \rightleftharpoons 2Cr^{3+} + 7H_2O$

The various stages in the deduction of the second partial equation are

$I^- \rightarrow I_2$

$2I^- \rightarrow I_2$

$2I^- \rightleftharpoons I_2 + 2e$

One dichromate ion uses $6e$, and two iodide ions liberate $2e$; hence the two partial equations apply in the ratio of 1:3:

$Cr_2O_7^{2-} + 14H^+ + 6e \rightleftharpoons 2Cr^{3+} + 7H_2O$

$3(2I^- \rightleftharpoons I_2 + 2e)$

or $\overline{Cr_2O_7^{2-} + 14H^+ + 6I^- = 2Cr^{3+} + 7H_2O + 3I_2}$

We can now apply our knowledge of partial ionic equations to the subject of equivalents. The standard oxidation–reduction process is $H \rightleftharpoons H^+ + e$, where e represents an electron per atom, or the Avogadro number of electrons per mole. If we know the change in the number of electrons per ion in any oxidation–reduction reaction, the equivalent may be calculated. The **equivalent** of an **oxidant** or a **reductant** is the mole divided by the number of electrons which 1 mole of the substance gains or loses in the reaction, e.g.:

$MnO_4^- + 8H^+ + 5e \rightleftharpoons Mn^{2+} + 4H_2O;$ Eq. $= MnO_4^-/5 = KMnO_4/5$

$Cr_2O_7^{2-} + 14H^+ + 6e \rightleftharpoons 2Cr^{3+} + 7H_2O;$ Eq. $= Cr_2O_7^{2-}/6 = K_2Cr_2O_7/6$

$Fe^{2+} \rightleftharpoons Fe^{3+} + e;$ Eq. $= Fe^{2+}/1 = FeSO_4/1$

$C_2O_4^{2-} \rightleftharpoons 2CO_2 + 2e;$ Eq. $= C_2O_4^{2-}/2 = H_2C_2O_4/2$

$SO_3^{2-} + H_2O \rightleftharpoons SO_4^{2-} + 2H^+ + 2e;$ Eq. $= SO_3^{2-}/2 = Na_2SO_3/2$

Table 17A.1 Ionic equations for use in the calculation of the equivalents of oxidising and reducing agents

Substance	Partial ionic equation
Oxidants	
Potassium permanganate (acid)	$MnO_4^- + 8H^+ + 5e \rightleftharpoons Mn^{2+} + 4H_2O$
Potassium permanganate (neutral)	$MnO_4^- + 2H_2O + 3e \rightleftharpoons MnO_2 + 4OH^-$
Potassium permanganate (strongly alkaline)	$MnO_4^- + e \rightleftharpoons MnO_4^{2-}$
Cerium(IV) sulphate	$Ce^{4+} + e \rightleftharpoons Ce^{3+}$
Potassium dichromate	$Cr_2O_7^{2-} + 14H^+ + 6e \rightleftharpoons 2Cr^{3+} + 7H_2O$
Chlorine	$Cl_2 + 2e \rightleftharpoons 2Cl^-$
Bromine	$Br_2 + 2e \rightleftharpoons 2Br^-$
Iodine	$I_2 + 2e \rightleftharpoons 2I^-$
Iron(III) chloride	$Fe^{3+} + e \rightleftharpoons Fe^{2+}$
Potassium bromate	$BrO_3^- + 6H^+ + 6e \rightleftharpoons Br^- + 3H_2O$
Potassium iodate (**dilute** acid solution)	$IO_3^- + 6H^+ + 6e \rightleftharpoons I^- + 3H_2O$
Sodium hypochlorite	$ClO^- + H_2O + 2e \rightleftharpoons Cl^- + 2OH^-$
Hydrogen peroxide	$H_2O_2 + 2H^+ + 2e \rightleftharpoons 2H_2O$
Manganese dioxide	$MnO_2 + 4H^+ + 2e \rightleftharpoons Mn^{2+} + 2H_2O$
Sodium bismuthate	$BiO_3^- + 6H^+ + 2e \rightleftharpoons Bi^{3+} + 3H_2O$
Nitric acid (conc.)	$NO_3^- + 2H^+ + e \rightleftharpoons NO_2 + H_2O$
Nitric acid (dilute)	$NO_3^- + 4H^+ + 3e \rightleftharpoons NO + 2H_2O$
Reductants	
Hydrogen	$H_2 \rightleftharpoons 2H^+ + 2e$
Zinc	$Zn \rightleftharpoons Zn^{2+} + 2e$
Hydrogen sulphide	$H_2S \rightleftharpoons 2H^+ + S + 2e$
Hydrogen iodide	$2HI \rightleftharpoons I_2 + 2H^+ + 2e$
Oxalic acid	$C_2O_4^{2-} \rightleftharpoons 2CO_2 + 2e$
Iron(II) sulphate	$Fe^{2+} \rightleftharpoons Fe^{3+} + e$
Sulphurous acid	$H_2SO_3 + H_2O \rightleftharpoons SO_4^{2-} + 4H^+ + 2e$
Sodium thiosulphate	$2S_2O_3^{2-} \rightleftharpoons S_4O_6^{2-} + 2e$
Titanium(III) sulphate	$Ti^{3+} \rightleftharpoons Ti^{4+} + e$
Tin(II) chloride	$Sn^{2+} \rightleftharpoons Sn^{4+} + 2e$
Tin(II) chloride (in presence of hydrochloric acid)	$Sn^{2+} + 6Cl^- \rightleftharpoons SnCl_6^{2-} + 2e$
Hydrogen peroxide	$H_2O_2 \rightleftharpoons 2H^+ + O_2 + 2e$

For convenience of reference the partial ionic equations for a number of oxidising and reducing agents are collected in Table 17A.1

The other procedure which is of value in the calculation of the equivalents of substances is the '**oxidation number**' method. This is a development of the view that oxidation and reduction are attended by changes in electronic charge and was originally developed from an examination of the formulae of the initial and final compounds in a reaction. The oxidation number (this will be abbreviated to O.N.) of an element is a number which, applied to that element in a particular compound, indicates the amount of oxidation or reduction which is required to convert one atom of the element from the *free state* to that in the compound. If oxidation is necessary to effect the change, the oxidation number is positive, and if reduction is necessary, the oxidation number is negative.

The following rules apply to the determination of oxidation numbers.

1. The O.N. of the free or uncombined element is zero.
2. The O.N. of hydrogen (except in certain hydrides) has a value of $+1$.
3. The O.N. of oxygen (except in peroxides) is -2.
4. The O.N. of a metal in combination (except in hydrides) is usually positive.

851

5. The O. N. of a radical or ion is that of its electronic charge with the correct sign attached.
6. The O.N. of a compound is always zero, and is determined by the sum of the oxidation numbers of the individual atoms each multiplied by the number of atoms of the element in the molecule.

The equivalent of an oxidising agent is determined by the change in oxidation number which the reduced element experiences. It is the quantity of oxidant which involves a change of one unit in the oxidation number. Thus in the normal reduction of potassium permanganate in the presence of dilute sulphuric acid to an Mn(II) salt:

$$\overset{+1\ +7\ -8}{K\ MnO_4} \rightarrow \overset{+2\ +6\ -8}{Mn\ S\ O_4}$$

the change in the oxidation number of the manganese is from $+7$ to $+2$. The equivalent of potassium permanganate is therefore $\frac{1}{5}$ mole. Similarly for the reduction of potassium dichromate in acid solution:

$$\overset{+2\ +12\ -14}{K_2Cr_2O_7} \rightarrow \overset{+6\quad -6}{Cr_2(SO_4)_3}$$

the change in oxidation number of *two* atoms of chromium is from $+12$ to $+6$, or by 6 units of reduction. The equivalent of potassium dichromate is accordingly $\frac{1}{6}$ mole. In order to find the equivalent of an oxidising agent, we divide the mole by the change in oxidation number *per molecule* which some key element in the substance undergoes.

The equivalent of a reducing agent is similarly determined by the change in oxidation number which the oxidised element suffers. Consider the conversion of iron(II) into iron(III) sulphate:

$$\overset{+2\ -2}{2(FeSO_4)} \rightarrow \overset{+6\quad -6}{Fe_2(SO_4)_3}$$

Here the change in oxidation number *per atom* of iron is from $+2$ to $+3$, or by 1 unit of oxidation, hence the equivalent of iron(II) sulphate is 1 mole. Another important reaction is the oxidation of oxalic acid to carbon dioxide and water:

$$\overset{+2\ +6\ -8}{H_2C_2O_4} \rightarrow \overset{+4\ -4}{2CO_2}$$

The change in oxidation number of two atoms of carbon is from $+6$ to $+8$, or by 2 units of oxidation. The equivalent of oxalic acid is therefore $\frac{1}{2}$ mole.

In general, it may be stated:
(i) The equivalent of an element taking part in an oxidation–reduction (redox) reaction is the relative atomic mass divided by the change in oxidation number.
(ii) When an atom in any complex molecule suffers a change in oxidation number (oxidation or reduction), the equivalent of the substance is the mole divided by the change in oxidation number of the oxidised or reduced element. If more than one atom of the reactive element is present, the mole is divided by the total change in oxidation number.

A useful summary of common oxidising and reducing agents, together with the various transformations which they undergo is given in Table 17A.2.

We are now in a position to understand more clearly why the equivalents of some substances vary with the reaction. We will consider two familiar

Table 17.A.2

Substance	Radical or element involved	O.N. of 'effective' element	Reduction product	New O.N.	Change in O.N.	Gain or loss in electrons
Common oxidising agents						
$KMnO_4$ (acid)	MnO_4^-	$+7$	Mn^{2+}	$+2$	-5	5
$KMnO_4$ (neutral)	MnO_4^-	$+7$	MnO_2 or Mn^{4+}	$+4$	-3	3
$KMnO_4$ (strongly alkaline)	MnO_4^-	$+7$	MnO_4^{2-}	$+6$	-1	1
$K_2Cr_2O_7$	$Cr_2O_7^{2-}$	$+6$	Cr^{3+}	$+3$	-3	3
HNO_3 (dil.)	NO_3^-	$+5$	NO	$+2$	-3	3
HNO_3 (conc.)	NO_3^-	$+5$	NO_2	$+4$	-1	1
Cl_2	Cl	0	Cl^-	-1	-1	1
Br_2	Br	0	Br^-	-1	-1	1
I_2	I	0	I^-	-1	-1	1
$3HCl:1HNO_3$	Cl	0	Cl^-	-1	-1	1
H_2O_2	O_2	-1	O^{2-}	-2	-1	1
Na_2O_2	O_2	-1	O^{2-}	-2	-1	1
$KClO_3$	ClO_3^-	$+5$	Cl^-	-1	-6	6
$KBrO_3$	BrO_3^-	$+5$	Br^-	-1	-6	6
KIO_3	IO_3^-	$+5$	I^-	-1	-6	6
$NaOCl$	OCl^-	$+1$	Cl^-	-1	-2	2
$FeCl_3$	Fe^{3+}	$+3$	Fe^{2+}	$+2$	-1	1
$Ce(SO_4)_2$	Ce^{4+}	$+4$	Ce^{3+}	$+3$	-1	1
Common reducing agents			**Oxidation product**			
H_2SO_3 or Na_2SO_3	SO_3^{2-}	$+4$	SO_4^{2-}	$+6$	$+2$	2
H_2S	S^{2-}	-2	S^0	0	$+2$	2
HI	I^-	-1	I^0	0	$+1$	1
$SnCl_2$	Sn^{2+}	$+2$	Sn^{4+}	$+4$	$+2$	2
Metals, e.g. Zn	Zn	0	Zn^{2+}	$+2$	$+2$	2
Hydrogen	H	0	H^+	$+1$	$+1$	1
$FeSO_4$ [or any iron(II) salt]	Fe^{2+}	$+2$	Fe^{3+}	$+3$	$+1$	1
Na_3AsO_3	AsO_3^{3-}	$+3$	AsO_4^{3-}	$+5$	$+2$	2
$H_2C_2O_4$	$C_2O_4^{2-}$	$+3$	CO_2	$+4$	$+1$	1
$Ti_2(SO_4)_3$	Ti^{3+}	$+3$	Ti^{4+}	$+4$	$+1$	1

examples by way of illustration. A normal solution of iron(II) sulphate $FeSO_4,7H_2O$ will have an equivalent of 1 mole when employed as a reductant, and $\frac{1}{2}$ mole when employed as a precipitant with aqueous ammonia. A solution of iron(II) sulphate which is normal as a precipitant will be half normal as a reductant. Potassium tetroxalate $KHC_2O_4,H_2C_2O_4,2H_2O$ contains three replaceable hydrogen atoms; its equivalent in neutralisation reactions is therefore $\frac{1}{3}$ mole:

$$KHC_2O_4,H_2C_2O_4,2H_2O + 3KOH = 2K_2C_2O_4 + 5H_2O$$

As a reducing agent, a mole contains $2C_2O_4^{2-}$, and the equivalent is accordingly $\frac{1}{4}$ mole:

$$C_2O_4^{2-} - 2e = 2CO_2$$

A solution of the salt which is $3N$ as an acid is $4N$ as a reducing agent.

When a sequence of reactions is involved in a chemical process the reaction which determines the equivalent is the one in which the standard solution is actually used. Thus if sodium nitrate is reduced to ammonia with Devarda's alloy and the ammonia is titrated with standard acid, the equivalent of the sodium nitrate is not determined by the reduction but by the reaction between ammonia and the acid. Since the equivalent of ammonia NH_3 is 1 mole, that of sodium nitrate, $NaNO_3$, is also 1 mole because 1 mole of $NaNO_3$ yields 1 mole of NH_3.

INDEX

The following abbreviations are used:

aa	= atomic absorption	**fu**	= fluorimetric	**sepn.**	= separation
am	= amperometry	**g**	= gravimetric	**soln.**	= solution
ch	= chromatographic	**hf**	= high frequency	**stdn.**	= standardisation
cm	= coulometric	**ir**	= infrared	**temp.**	= temperature
cn	= conductimetric	**p**	= potentiometric	**T**	= table
D	= determination	**prep.**	= preparation	**th**	= thermal
eg	= electrogravimetric	**s**	= spectrophotometric	**ti**	= titrimetric
em	= emission spectrographic	**se**	= solvent extraction	**v**	= voltammetry
fl	= flame emission				

Absorbance: 648, 708, 713
  additivity of, 713
  characteristics of common chromophores,
    (T) 838
  conversion to transmission, 709
Absorbancy: 649
Absorbents: 477
Absorptiometers: 653, 658
Absorption coefficient: 648, 649, 650
Absorption curve, D, of: with a
    spectrophotometer of aromatic
    hydrocarbons, 715
  of methyl red, 718
  of potassium dichromate, 712
  of potassium nitrate, 708
  of potassium permanganate, 712
  of substituted benzoic acids, 710
Absorption spectra:
  auxochromes, 672
  bathochromic shift, 672
  chromophore, 671
  origins of, 671
a.c. spark source: 764
Accelerator: 450
Accuracy: 11, 14, 128
  absolute, 128
  comparative, 129
  in quantitative analysis, 127
Acetic acid: D. of strength of, (ti) 296
  specific gravities of aqueous solns.,
    (T) 829
Acetic anhydride: 306
Acetone, D. of: in propan-2-ol, (ir) 757

Acetylacetone: 169
  chelation complexes with, 169, 248
Acid-base indicators: 262
  prep. of solns. of, 266
  (T) 265
Acid-base titrations: 262, 286
  theory of, 262
Acidimetry and alkalimetry: 236, 258, 262
  theory of, 262
Acids: Brønsted–Lowry theory of, 21
  common, concentration of, (T) 829
  dissociation constants of, (T) 832
  hard, 54
  ionisation of, 20
  Lewis, 22
  polyprotic, 20
  prep. of standard solns., 285, 286
  soft, 54
  specific gravities of selected, (T) 829
  strengths of, 31, 286, (T) 832
  strong, 21
  pK values in aqueous soln., (T) 832
  weak, 21
Acids, standardisation of, 286, 288, 386, 401
Acids, titration of: by hydroxide ion (cm), 544
Activation overpotential: 506
Activity: 23
  coefficient, 23
Adsorption: 422
Adsorption indicators: 345
  applications of, 350, 351, 352
  choice of, 347
  (T) 348